ENCYCLOPEDIC HANDBOOK OF BIOMATERIALS AND BIOENGINEERING

Part B: Applications

Volume 2

ENCYCLOPEDIC HANDBOOK OF BIOMATERIALS AND BIOENGINEERING

Part B: Applications

Volume 2

edited by

Donald L. Wise
Northeastern University
Boston, Massachusetts

Debra J. Trantolo
Cambridge Scientific, Inc.
Belmont, Massachusetts

David E. Altobelli
Harvard School of Dental Medicine
Boston, Massachusetts

Michael J. Yaszemski
United States Air Force
Lackland Air Force Base, Texas

Joseph D. Gresser
Cambridge Scientific, Inc.
Belmont, Massachusetts

Edith R. Schwartz
National Institute of Standards and Technology
Gaithersburg, Maryland

MARCEL DEKKER, INC.   NEW YORK · BASEL · HONG KONG

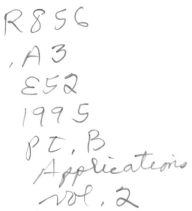

Library of Congress Cataloging-in-Publication Data

Encyclopedic handbook of biomaterials and bioengineering / edited by Donald L. Wise
... [et al.].
    p. cm.
    Contents: Pt. A., v. 1-2. Materials -- Pt. B., v. 1-2. Applications.
    ISBN 0-8247-9593-8 (v. 1 : alk. paper) — ISBN 0-8247-9594-6 (v. 2 : alk. paper)
— ISBN 0-8247-9595-4 (v. 1 : alk. paper) — ISBN 0-8247-9596-2 (v. 2 : alk. paper).
    1. Biomedical engineering -- Encyclopedias. 2. Biomedical materials--
Encyclopedias. I. Wise, Donald L. (Donald Lee)

R856.A3E52   1995
610'.28--dc20   95-21232
                                                                    CIP

The publisher offers discounts on this book when ordered in bulk quantities. For more information, write to Special Sales/Professional Marketing at the address below.

This book is printed on acid-free paper.

Marcel Dekker, Inc.
270 Madison Avenue, New York, New York 10016

Current printing (last digit):
10 9 8 7 6 5 4 3

PRINTED IN THE UNITED STATES OF AMERICA

Preface

The medical device and drug industry is consistently one of the strongest performers. Materials are a key ingredient to this industry. Development of these materials is in a constant state of activity with the burdens of old materials not withstanding the tests of time and new materials needs coming to the forefront of modern applications. This handbook focuses on materials used in or on the human body—materials that define the world of "biomaterials."

The *Encyclopedic Handbook of Biomaterials and Bioengineering* covers the range of biomaterials from polymers to metals to ceramics. The depth of the field necessitated careful integration of basic science, engineering, and practical medical experience in a variety of applied disciplines. As a result, scientists, engineers, and physicians are among the chapter authors, as well as the editors. The handbook provides a detailed accounting of the state of the art in the rapidly growing biomaterials arena. Its organization reflects the diversity of the field.

The encyclopedia is a four-volume reference: Part A, "Materials," in two volumes and Part B, "Applications," in two volumes. In the Applications texts, the focus is on the actual use of the biomaterials in their applied settings. Volume 1 deals first with the general requirements in selecting a proper biomaterial for successful application, then moves to one of the original applications areas for biomaterials—orthopedics. Volume 2 focuses on biomaterials in bone cement, vascular, ophthalmic, and dental applications. Integral to all of these chapters are evaluations of the performance of the biomaterials in the projected clinical setting.

The users of this encyclopedia will represent a broad base of backgrounds ranging from the basic sciences (e.g., polymer chemistry and biochemistry) to the more applied disciplines (e.g., orthopedics and pharmaceutics). To meet varied needs, each chapter provides clear and fully detailed discussions. This in-depth coverage should also assist recent inductees to the biomaterials circle. The editors trust that this handbook conveys the intensity of this fast-moving field in an enthusiastic presentation.

The editors are grateful for the cooperation of many friends and colleagues in their support of this work. Our appreciation extends to each of the contributors for suggestions and comments as the project developed. Their interest and enthusiasm in pulling together a comprehensive reference for all our associates in the biomaterials area have been most gratifying. The editors are especially thankful to Ms. Wanda O'Connell for her patience and competence in dealing with manuscripts from more than 100 authors.

Donald L. Wise, Debra J. Trantolo, David E. Altobelli,
Michael J. Yaszemski, Joseph D. Gresser, and Edith R. Schwartz

Contents of Part B: Applications

Part II. Orthopedic Biomaterials

Part III. Metals in Orthopedics

Part IV. Bone Repair and Joint Replacement

Part V. Tissue Response and Growth

VOLUME 2

Part VI. Bone Cements

Part IX.   Ocular Applications

Part X.   Dental Applications

Contents of Part A: Materials

ENCYCLOPEDIC HANDBOOK OF BIOMATERIALS AND BIOENGINEERING

Part B: Applications

Volume 2

VI
BONE CEMENTS

31
Calcium Phosphate Bone Cements

F. C. M. Driessens, J. A. Planell, and F. J. Gil
Universitat Politècnica de Catalunya
Barcelona, Spain

I. INTRODUCTION

Many materials have been proposed for use in bone surgery, either for repair of bone defects or for bone augmentation, and in joint surgery, which is applied most frequently to hips and knees after arthrosis. In both cases the material in direct contact with the bone is critical for the biocompatibility and performance of the surgical device. It is thought not to be desirable for this material to be inert, because in this case the device is not well fixed to the surrounding living bone. Properties such as biodegradability or bioactive behavior have been proposed as most desirable. Discussions include concepts such as osteoconductivity versus osteoinductivity, and osseointegration versus bone replacement. Many applications of the materials under consideration have been proposed, but only a few have been performed on a large scale, as most of them did not work out satisfactorily in clinical practice. It is questionable whether the replacement of the bone tissue is most desirable or whether one should aim at the development of materials that sooner or later are replaced by new healthy, bone tissue. This latter property may be called *osteotransductivity*. As is shown in this chapter, the only group of materials that might be osteotransductive are thought to be some calcium phosphates. These materials are available in the form of ceramic blocks, granules, or cements. The calcium phosphate cements are the subject of this chapter.

A. Materials Proposed for Bone Repair and Augmentation

The first material proposed for bone repair, as early as the 19th century, was gypsum, or plaster of Paris. By 1965 it was clear that this material resorbed faster than the ongrowth of new bone, and therefore it has been abandoned.

Several ceramics have been proposed since then [1]. The products Bioglass (University of Florida, Gainesville) and Ceravital (Leitz, Wetzlar) belong to a group of surface

855

active ceramics that form a strong bond with bone, mediated by a layer of apatite crystals [2,3]. In this line apatite- and wollastonite-containing glass ceramics have also been developed [41], which form a strong bond to bone [5], and are also mediated by apatite [6]. Further developments in this field are still going on [7]. Ceramics such as alumina do not bond to bone at all [81] under the same conditions. The interface bond strength between bioactive glasses and bone is still increasing after 1 year of implantation in acetabular dog bone [9]. These materials are not biodegradable.

Glass-ionomer cements, which are widely used as dental restorative materials, have also been proposed as bone substitutes [10]. Xenobiotic components such as aluminum are leached from such materials and may accumulate in the soft tissues. Whether this type of material is also bioactive is still questionable. Recently, a really bioactive cement has been developed, which is based on the system $CaO–SiO_2–P_2O_5–CaF_2$ as glass powder in combination with an aqueous solution of ammonium phosphate [11–13]. These materials are not biodegradable either.

More recently, some biodegradable formulations have been proposed for repair of bone defects. A chitosan sol was used as a carrier for a powder containing a mixture of hydroxyapatite, zinc oxide and calcium oxide [14]. Its biological behavior has not yet been reported. A composite of poly (L-lactide) and hydroxyapatite has been studied in a transcortical implantation model in goats [15]. Up to 3 months the interface bonding increased. However, later the bonding diminished due to the dominating effect of poly (L-lactide) resorption without sufficient new bone ongrowth.

Several other materials for bone repair and bone substitution have been proposed. However, up to now none of them has proved to be of major value in surgery. One of the drawbacks is that most of these materials cannot be formed or molded into the desired form during the operation. In this respect, it is too early to judge the practical value of bioglass cements [11–13].

B.   Bone Cements for Fixation of Hip and Knee Endoprostheses

The most successful device for joint replacement is that consisting of an endoprosthesis made of a metal or alloy fixed into bone with a bone cement based on polymethylmethacrylate (PMMA); this is despite the apparent disadvantages of the PMMA bone cement [16]. These are: (1) the high setting temperature, which may lead to bone necrosis, and hence to bone lysis; (2) a setting shrinkage leading to a certain misfit of the prosthesis in the bone cavity—the gap is closed with a layer of fibrous tissue that allows for micromovement of the prosthesis with respect to the bone and that ultimately leads to formation of particulate debris and osteolysis; (3) in some cases fat embolism may lead to cardiac arrest due to monomer toxicity [17,18]; (4) the bone surgery as such may lead to heterotopic ossification and thus to immobilization [19] and the need for surgical revision.

Despite these disadvantages, the annual number of hip and knee replacements is about 1 million and the average service life is about 12 years. At present, about 60% of these prostheses need revision. Several attempts have been made to improve the biocompatibility of PMMA bone cements. From a materials point of view, the incorporation of a hydroxyapatite filler is successful [20,21]. It has also been shown that incorporation of a bioactive glass ceramic powder into the PMMA cement leads to improved bone bonding after implantation [22]. Similar encouraging results were reported for a set of materials derived from a *bis*-phenol-glycidyl methacrylate containing Bioglass filler particles [23].

The main problem with these formulations is that the surrounding bone still sees a foreign body that, at the interface, consists of the polymeric part of the cement.

A somewhat different approach was taken by the incorporation of growth factor into a PMMA bone cement. This elicited an increased tendency to osseointegration and reduced the risk of fibrous tissue formation at the interface with bone [24]. However, the long-term behavior of such cements is still unknown. In summary, the behavior of PMMA bone cement indicates that it is not very biocompatible.

One approach to avoiding the use of these bone cements has been relatively successful in the last 8 years, that is, fixation by press-fit of a metal femoral stem that is covered by a layer of plasma-sprayed hydroxyapatite and that allows for a tight ongrowth of bone without the occurrence of an encapsulation with fibrous tissue.

It is said that a PMMA-cemented hip prosthesis has good initial fixation and bad final fixation, whereas a plasma-sprayed prosthesis gives bad initial fixation and good final fixation. The attempt to solve this problem is still going on.

C.   Why Calcium Phosphate Cements?

First, the question "why calcium phosphate?" is investigated. The biocompatibility of calcium phosphates in the range $1 < Ca/P < 2$ is outstanding, and they have been studied primarily in relation to bone repair and augmentation [25,26]. The main interest has been in ceramics made of hydroxyapatite (HA) and β-tricalcium phosphate (β-TCP): HA implants appear to become integrated but not resorbed into a bone structure, whereas β-TCP is biodegradable and is replaced by new bone tissue after some time [27–32]. Macroporosity plays a role in the rate of bone substitution [33]. It has been pointed out that the rate of bioresorption is related to the relative solubility of calcium phosphates [34–38]. The effect of calcium phosphates on cellular action is not simply resorption, but rather transformation [39–41]. Structural forms such as rods, cones, H-bars, blocks, disks and irregular shapes have been introduced for many different surgical indications in bone surgery [42]. However, for certain applications calcium phosphate granules have been advocated, which for the sake of spacing have been mixed with collagen and bone inductive protein [43]. β-TCP granules have been applied in the repair of periodontal osseous defects and for the filling of periapical endodontic lesions. HA has been recommended for alveolar ridge augmentation, spinal fusion, and bone tumor surgery [44–49]. Yet, although β-TCP is as osteoconductive as autologous bone, the application of calcium phosphate ceramics has never become very popular. One of the drawbacks is that bioceramics must be delivered in certain forms, whereas granules often drift away from the site of implantation. The first step in overcoming the disadvantages was made in 1983 by the introduction of cement formulations consisting of calcium phosphates [50]. These cements can be molded during the operation or simply injected into the bone defect. This answers the question posed above: Why calcium phosphate cements?

II.   CHEMISTRY OF CALCIUM PHOSPHATE CEMENTS: DEFINITION

The setting of calcium phosphate cements bears some resemblance to that of gypsum. One starts with one component having a relatively high solubility in water: calcium sulfate hemihydrate, $CaSO_4 \cdot \frac{1}{2}H_2O$. When this is mixed with water, it starts to dissolve and after a while the aqueous solution becomes supersaturated with calcium sulfate dihydrate,

$CaSO_4 \cdot 2H_2O$, which then precipitates. The setting reaction is finished when all hemihydrate is transformed into dihydrate. Thus the overall chemical reaction for this mechanism is the following:

$$CaSO_4 \cdot \tfrac{1}{2}H_2O \text{ (s)} + 1\tfrac{1}{2}H_2O \text{ (l)} \rightarrow CaSO_4 \cdot 2H_2O \text{ (s)}$$

in which s means solid and l means liquid. Physically, the setting occurs in the following way: (1) the hemihydrate crystals function as nuclei for the precipitating dihydrate; (2) these dihydrate crystals grow in the form of needles so that around each hemihydrate nucleus a cluster of dihydrate crystals is formed; (3) these clusters have some rigidity and they become entangled, and this entanglement gives rigidity to the material as such.

As is shown below, the main difference from calcium phosphate cements is that for this latter group of materials, generally, not one but two active ingredients are necessary, with a relatively high solubility so that both dissolve to make the aqueous phase supersaturated with the calcium phosphate(s) desired for precipitation.

A calcium phosphate cement may be defined as a combination of a powder or a mixture of powders with water or an aqueous solution that sets upon mixing at room or body temperature due to the formation of at least one calcium phosphate, and that retains strength upon soaking in water or Ringer's solution.

A. Ingredients for Calcium Phosphate Cements

As pyrophosphates and metaphosphates are hydrolyzed in the body fluids only orthophosphates are expected to form good calcium phosphate cements. Furthermore, high concentrations of pyrophosphate or metaphosphate ions are not desirable in the body fluids because they can lead to extraosseous calcifications [53]. Hence in the following only calcium orthophosphate cements are considered.

In general, solids occurring in the system H_3PO_4–CaO–H_2O near room and body temperature, come into consideration as ingredients for these cements. They are listed in Table 1, which also gives the abbreviations for materials discussed in the text. Certain general considerations apply as far as relative solubility are concerned. In the range

Table 1   Solids Occurring in the System H_3PO_4–CaO–H_2O Near Room and Body Temperature

Ca/P ratio	Compound	Formula	Abbreviation
0	Phosphoric acid, anhydrous	H_3PO_4	PAA
0	Phosporic acid, hemihydrate	$H_3PO_4 \cdot \tfrac{1}{2}H_2O$	PAH
0.5	Monocalcium phosphate, anhydrous	$Ca(H_2PO_4)_2$	MCPA
0.5	Monocalcium phosphate, monohydrate	$Ca(H_2PO_4)_2 \cdot H_2O$	MCPM
1	Dicalcium phosphate, anhydrous	$CaHPO_4$	DCP
1	Dicalcium phosphate, dihydrate	$CaHPO_4 2H_2O$	DCPD
1.33	Octocalcium phosphate	$Ca_8(HPO_4)_2(PO_4)_4 \cdot 5H_2O$	OCP
1.5	Cation-deficient hydroxyapatite	$Ca_9(HPO_4)(PO_4)_5OH$	CDHA
1.67	Precipitated hydroxyapatite	$Ca_{10}(PO_4)_6(OH)_2$	PHA
∞	Calcium oxide	CaO	—
∞	Calcium hydroxide	$Ca(OH)_2$	—

Source: Ref. 51

Table 2 Orthophosphates that Can Be Prepared by High-Temperature Solid-State Reactions and Contain Only Biocompatible Components

Ca/P ratio	Compound	Formula	Abbreviation
0.5	Calcium zinc phosphate	$CaZn_2(PO_4)_2$	CZP
0.67	Calcium magnesium phosphate	$Ca_4Mg_5(PO_4)_6$	CMP
1.0	Rhenanite	$CaNaPO_4$	RH
1.0	Calcium potassium phosphate	$CaKPO_4$	CPP
1.33	Magnesium-containing TCP	$Ca_8Mg(PO_4)_6$	MTCP
1.42	Sodium whitlockite	$Ca_{10}Na(PO_4)_7$	SWH
1.5	β-Tertiary calcium phosphate	β-$Ca_3(PO_4)_2$	β-TCP
1.5	α-Tertiary calcium phosphate	α-$Ca_3(PO_4)_2$	α-TCP
1.67	Sintered hydroxyapatite	$Ca_{10}(PO_4)_6(OH)_2$	SHA
1.67	Fluoroapatite	$Ca_{10}(PO_4)_6F_2$	FA
1.67	Chloroapatite	$Ca_{10}(PO_4)_6Cl_2$	CA
2.0	Spodiosite	Ca_2PO_4Cl	SP

Source: Ref. 51

2 < pH < 4, DCPD has the lowest solubility and will be the preferred phase of precipitation. In the range 5 < pH < 7, OCP will be the preferred phase of precipitation (for kinetic reasons), although there PHA has a lower solubility. At higher pH values (7 to 8) CDHA, and at still higher pH, PHA, will precipitate [54–60]. The solubility of the other solids given in Table 1 is higher than that of these phases, which eventually precipitate and are, therefore, not likely to be used as active ingredients of the cement powder.

A list of orthophosphates that can be prepared by solid-state reaction and contain only biocompatible components is given in Table 2. They also come into consideration for the preparation of calcium phosphate cements. Not much is known about the relative solubilities of those compounds except for that of the apatites listed in Table 2:

$$\text{solubility FA} < \text{solubility SHA} < \text{solubility CA}$$

One can assume that it is not possible to use FA or SHA as active ingredients in the powder of calcium phosphate cements due to their low solubility [51].

There are many ways to modify the composition of calcium phosphate cements and to improve some of their properties. Table 3 gives an impression of the compounds that might be used for this purpose. Developments in this work are still in their early stages. As is shown below, some of the compounds listed act as accelerators of the setting reaction of calcium phosphate cements, whereas others can be used as retarders.

B. Calcium Orthophosphates Formed by Precipitation

In the previous section DCPD, OCP, CDHA, and PHA have been identified as calcium phosphates that can precipitate directly from aqueous solutions somewhere in the pH range with biological relevance (4 < pH < 9). In synthetic systems containing not only calcium and phosphate ions but also magnesium, sodium, potassium, or carbonate ions there are more possibilities [51], as summarized in Table 4. Indications for the occurrence of magnesium whitlockite, sodium- and carbonate-containing apatite, and heavily carbonated hydroxyapatite in bone and dental tissues have been given elsewhere [51]. In

Table 3   Other Compounds Suitable as Retarders, Accelerators, Bioactivity Promotors, or Simply as Additives Improving the Properties of Calcium Phosphate Cements

Component	Compounds
Sodium	NaF, Na_2CO_3, $NaHCO_3$, Na_2SO_4, orthophosphates of Na, Na_2SiO_3
Potassium	KF, K_2CO_3, K_2SO_4, orthophosphates of K, K_2SiO_3
Magnesium	$MgHPO_4$, $Mg_3(PO_4)_2 \cdot xH_2O$, MgF_2, $MgCO_3$, MgO, $CaMg(CO_3)_2$, $Mg(OH)_2$, $MgSO_4$
Zinc	$Zn_3(PO_4)_2 \cdot 4H_2O$, ZnF_2, $ZnCO_3$, $ZnSO_4$, ZnO, $Zn(OH)_2$
Calcium	$CaSO_4$, $CaSO_4 \cdot \frac{1}{2}H_2O$, $CaSO_4 \cdot 2H_2O$, CaF_2, $CaCO_3$
Biopolymers	Proteins, peptides, proteoglycans, glycosaminoglycans, carbohydrates
Organic acids	Citric acid, malonic acid, pyruvic acid, tartaric acid
Inorganic acids	Phosphoric acid
Synthetic polymers	Polylactic acid, polyglycolic acid
Growth factors	TGF-β, osteocalcine, GLA proteins

Source: Ref. 52

fact, all compounds mentioned in Table 4 come into consideration as calcium phosphates constituting a calcium phosphate cement after setting.

The genesis of bone mineral [51] may be summarized as follows. The precursor mineral is OCP. This mineral is not stable in the bone extracellular fluid and is spontaneously transformed with a half-life of about 1 month into a mixture of magnesium whitlockite, sodium- and carbonate-containing apatite, and heavily carbonated hydroxyapatite. Remodeling occurs in the bone, whereby old bone is resorbed by the osteoclasts (including its mineral), after which osteoblasts regenerate new bone matrix and consecutively induce formation of new OCP. This describes the recycling that occurs in our living bone.

For bone repair or bone augmentation it might thus be preferable to work especially with cements in which OCP is the precipitating phase. A second preference are the CDHA cements, which belong to the same cation-deficient apatite family as the heavily carbonated hydroxyapatite occurring in mature bone.

Table 4   Calcium Phosphates Reported in the Literature as Being Formed by Precipitation from Aqueous Solutions at Room or Body Temperature

Ca/P	Precipitate	Formula
1	Dicalcium phosphate dihydrate	$CaHPO_4 \cdot 2H_2O$
1.28	Magnesium whitlockite	$Ca_9Mg(HPO_4)(PO_4)_6$
1.33	Octocalcium phosphate	$Ca_8(HPO_4)_2(PO_4)_4 \cdot 5H_2O$
1.5	Cation-deficient hydroxyapatite	$Ca_9(HPO_4)(PO_4)_5OH$
1.67	Precipitated hydroxyapatite	$Ca_{10}(PO_4)_6(OH)_2$
1.89	Na- and CO_3-containing apatite	$Ca_{8.5}Na_{1.5}(PO_4)_{4.5}(CO_3)_{2.5}$
1.8	K- and CO_3-containing apatite	$Ca_9K(PO_4)_5(CO_3)_2$
2.0	Heavily carbonated hydroxyapatite	$Ca_9(PO_4)_{4.5}(CO_3)_{1.5}(OH)_{1.5}$

Source: Ref. 51

In the following sections we classify the cements as OCP cements, DCPD cements, CDHA cements, PHA cements, or other, depending on the type of calcium phosphate that has precipitated during the setting reaction.

C. Setting Times, pH, and Reaction Mechanisms

Setting times have been measured by several methods on different calcium phosphate cements (see Table 5). When one wishes to conform to existing methods, the choice is between a Vicat needle and the two Gilmore needles. The rationale for the two Gilmore needles is [52] that with the light-and-wide needle one can measure the initial setting time I, which indicates the end of moldability without serious damage to the cement structure, whereas with the heavy-and-fine needle one can measure the final setting time F beyond which it is possible to touch the cement without causing serious damage.

As far as clinical applications are concerned, the desired ranges may be

$$4 < I < 8 \text{ minutes}$$

for the initial setting time and

$$10 < F < 15 \text{ minutes}$$

for the final setting time. From the few data available in the literature [52, 62–68] it is not clear yet whether these conditions can be provided by the calcium phosphate cements that are being developed at the moment. This matter should be clarified by further developments (see also Section II.D. on common ion effects).

One way to regulate the setting time is by variation of the liquid/powder ratio. The setting times I and F increase with this ratio for an OCP cement [81] and for a magnesium-containing calcium phosphate cement [84], as should be expected theoretically [72]. Other ways to regulate the setting time are dealt with below.

The pH of setting of PHA cement made of a mixture of TTCP and DCP has been mentioned in the literature (see Table 6). It can vary from about 6 to 10.5. This also depends on the additives used, for example, CaF_2, Na_2SiO_3, or H_3PO_4 solution. In general, however, the pH is in agreement with the fact that according to x-ray diffraction the final product is apatitic [62,69,73]. The pH during setting of an OCP cement varies between 7 and 8 [67,82].

Table 5 Different Calcium Phosphate Cements and Different Methods of Measurement for the Setting Time (*s*)

Ca/P ratio	Active ingredients	Type of cement	Setting time (*s*)	Ref.
1.67	DCP + TTCP	PHA	Heavy Gilmore	62
1.67	DCP + TTCP	PHA	Vicat needle	63
1.67	DCP + TTCP	PHA	Light Gilmore needle	64
1.0	MCPM + β-TCP	DCPD	Needle diam.: 0.18 cm, weight: 60 g	65, 66
1.33	DCPD + α-TCP	OCP	Vicat needle	67
1.40–1.498	α-TCP + β-TCP + DCP	?	According to JIS T 6604	68
Several	Several	Several	Light and heavy Gilmore needles	69

Table 6   The pH During Setting of Calcium Phosphate Cement Formulations as Reported in the Literature

Ca/P ratio of the cement powder	Active ingredients	Type of cement	pH during setting	Ref.
1.67	TTCP + DCP	PHA	6.5–9.0	70
1.67	TTCP + DCP	PHA	7–8	62
1.67	TTCP + DCP	PHA	6.5–9.5	71
1.67	TTCP + DCP	PHA	7–10	72
1.67	TTCP + DCP	PHA	7–10.5	73
1.67	TTCP + DCP	PHA	6–8	64

From considerations based on the solubility isotherms for different calcium phosphates it was determined that mixtures of TTCP and DCP or DCPD at Ca/P = 1.67 should set in a one-step mechanism comprising the dissolution of TTCP and DCP and the simultaneous precipitation of HA [62,72,74,75]. This mechanism seems to be corroborated by data from x-ray diffraction [731], but from solution chemistry data [70] it is obvious that OCP, or more probably CDHA, might occur as an intermediate. By variation of the TTCP/DCP ratio in the mixture it could be made clear from solution chemistry data that the first precipitate is nearly pure PHA (pH of formation between 10 and 11) and that further reaction of PHA with excess DCP results in the formation of CDHA (pH of formation around 7.5) [57].

Fulmer et al. [59] formed a CDHA cement from a mixture of MCPM, TTCP, and $Ca(OH)_2$, according to the procedure of Constantz et al. [64]. They determined the following cascade of phase compositions:

$$MCPM + TTCP + Ca(OH)_2$$
$$\downarrow$$
$$DCPD + PHA + TTCP + Ca(OH)_2$$
$$\downarrow$$
$$DCPD + PHA + TTCP$$
$$\downarrow$$
$$CDHA$$

This means that calcium phosphate cement chemistry can be considerably more complex than gypsum chemistry.

For the DCPD cement formed from a combination of MCPM with β-TCP, it is claimed that a one-step setting mechanism is involved [76]. X-ray diffraction data seem to confirm this [65]. Solution chemistry with this cement is hardly possible due to its very short setting time (1 to 2 minutes). For the OCP cement formed from a combination of DCPD and α-TCP a one-step reaction also is claimed, which was confirmed by x-ray diffraction [67].

D.   Effects of Seeds, Common Ions and Particle Size

So far we have mentioned cements of the type DCPD, OCP, CDHA, and PHA. The DCPD cement sets so quickly that it does not need any addition of seed crystals

[65,66,76]. Due to the good epitaxy of OCP on CDHA and HA [77,78], one can use either CDHA or HA, which are readily available as seed crystals for OCP cements [52].

Chow et al. [62,72,74,75] used HA as seeds for their PHA cement formed from combinations of TTCP with DCP. They found that the final setting time F decreased from 22 to 8 minutes on adding up to 43% HA to the cement powder. This was confirmed by Xie and Monroe [71]. Takezawa et al. [63] kept the amount of PHA added to the cement powder constant and varied the particle size of the seed crystals. They found that a large number of small crystal seeds shortened the setting time more than a small number large crystal seeds. Accordingly, the setting times I and F of an OCP cement decreased with the addition of increasing amounts of PHA to the cement powder [83].

Only Driessens et al. [79] carried out an experiment to determine the common ion effect. They found such an effect when using solutions of Na_2HPO_4 or $NaH_2PO_4 \cdot 2H_2O$ for making a CDHA cement out of α-TCP, to which 2% of PHA seed crystals had already been added. Their results for initial setting time I and final setting time F as a function of the concentration of the sodium phosphates in the cement liquid are given in Table 7. Hence, common ions can have an accelerating effect on the setting reaction. Addition of $CaCl_2$ to the aqueous solution of this cement did not have any effect [79].

From a theoretical point of view, smaller particle sizes of the components should result in shorter setting times [72]. This has been confirmed by the particle size of TTCP in HA cements made from TTCP/DCP mixtures [64] and by the effect of milling time on setting time for a number of magnesium-containing calcium phosphate cements [84].

E. Other Additives for Regulating the Setting Mechanism

According to Brown and Chow [62], the use of diluted phosphoric acid instead of water should give a faster approach to the steady-state setting reaction in the preparation of PHA cements from TTCP/DCP mixtures or TTCP/DCPD mixtures. In view of the results of Driessens et al. [79], this effect might be interpreted as a common ion effect.

Additions of CaF_2 to the powder of PHA cements has been applied to accelerate the setting reaction [69,71,75]. It may act by the formation of FA nuclei.

The use of a solution of 2 M Na_2SiO_3 with these cements also has an accelerating effect [71]. A retarding effect is provided by mixtures of water with glycerol or with polyethylene glycol [80]. In the case of DCPD cements, a retarding effect was obtained by addition of calcium pyrophosphate, calcium sulfate dehydrate, and/or calcium sulfate hemihydrate to the powder [65].

Table 7   Initial Setting Time I and Final Setting Time F of a
CDHA Cement Using Aqueous Solutions of Na_2HPO_4 or
$NaH_2PO_4 \cdot 2H_2O$ at a Liquid/Powder Ratio of 0.35

% $NaH_2PO_4 \cdot 2H_2O$	0	2	4	6
I (min)	25	7	4	2·5
F (min)	55	23	14	8
% Na_2HPO_4	0	2	4	—
I (min)	25	7	4	—
F (min)	55	30	19	—

III. PHYSICAL PROPERTIES AND WORKABILITY

For the performance of calcium phosphate cements as biomaterials, some physical properties like compressive strength, diametral tensile strength, and fracture toughness may be important. For their application *in situ* their workability is of great interest to the clinicians. Both aspects are considered in this section.

A. Compressive Strength and Diametral Tensile Strength

Up to now the main criterion used to describe the quality of calcium phosphate cements has been the compressive strength (see Table 8). One must bear in mind that the value found for this property depends, not only on the product, but also on the storage conditions [63,85]. Several of the papers extracted in Table 8 did not mention storage conditions.

Table 8 Compressive Strength C and Diametral Tensile Strength T (MPa) Reported for Several Calcium Phosphate Cements

Cement type	Active ingredients	Modifier added	C	T	Ref.
PHA	TTCP + DCP	—	34 ± 4	—	62
PHA	TTCP + DCP	PHA	31 ± 7	—	62
PHA	TTCP + DCP	PHA/FA	5–28	—	70
PHA	TTCP + DCP	PHA	10–42	—	63
PHA	TTCP + DCP	—	36	—	74
PHA	TTCP + DCP	—	0–51	—	73
PHA	TTCP + DCPD	—	31 ± 5	—	62
PHA	TTCP + DCPD	—	9 ± 3	—	67
OCP	α-TCP + DCPD	—	17 ± 3	—	67
OCP	α-TCP + DCP	PHA	50 ± 5	8 ± 1	83
?	α-TCP	collagen	15 ± 5	—	87
CDHA	α-TCP	PHA	70 ± 5	11 ± 2	80, 84
DCPD	β-TCP + MCPM	—	—	0.1 ± 1.1	77
DCPD	β-TCP + MCPM	plaster	—	0.9 ± 3.2	65
PHA	DCPD + CaCO$_3$	β-TCP	—	0.6 ± 1.7	66
DCPD	α-TCP + citric acid	—	110 ± 20	—	87
DCPD	α-TCP + citric acid	—	90–120	—	88
?	α-TCP + DCP + β-TCP	—	18–35	—	89
CDHA	MCPM + TTCP + Ca(OH)$_2$	—	50–90	—	64
OCP	MCPM + α-TCP	—	12 ± 2	3 ± 1	84
CDHA	MCPM + CaO	PHA	6 ± 1	2 ± 1	84
OCP	MCPM + CaO	PHA	6 ± 1	1.5 ± 0.5	82
CaMg$_2$(PO$_4$)$_2$	DCP + MgO + MgHPO$_4$·3H$_2$O	—	11 ± 1	2.3 ± 0.3	85
OCP	α-TCP + DCPD	PHA	6 ± 1	1.2 ± 0.3	86
DCPD	TTCP + MCPM	—	4 ± 1	—	86
OCP	CaKPO$_4$ + DCP	PHA	2	1	86
OCP	CaKPO$_4$ + DCPD	PHA	2	1	86

In particular, the condition of soaking for at least 1 day in water or Ringer's solution at 37°C is thought to be valuable for the use of these cements as biomaterials. The dimensions of the samples might also be important. Most authors have adopted the dimensions used in International Standards Organization (ISO) standards for dental cements and for acrylic bone cements: cylinders with a diameter of 6 mm and a height of 12 mm. According to our experience [87], it does not make much difference which crosshead speed is used from 1 to 25 mm min^{-1} although we mostly assumed 1 mm min^{-1}.

Among the product-related factors affecting the strength, the factor having the highest influence may be the particle size of the components of the cement powder: Chow et al. [72] obtained variations in strength from 0 up to 51 MPa solely by variations in particle size in a PHA cement made out of DCP and TTCP mixtures. The same holds for a CDHA cement made out of α-TCP with addition of PHA [79].

Another important factor is the size of the seed crystals. Takezawa et al. [63] showed that the compressive strength reached with small -grain PHA added to a TTCP + DCP mixture was much higher (up to 42 MPa) than that obtained when larger seed crystals were added (as low as 10 MPa). The amount of seed crystals can also be important. Brown and Chow [62] did not find much difference in the strength when they added up to 10% of PHA to their TTCP + DCP mixture. However, Bermúdez et al. [85] found a continuous decrease in strength of DCPD cements made of β-TCP + MCPM when up to 40% DCPD seed crystals were added to the powder mixture. Driessens et al. [79] found that the strength of CDHA cements made out of α-TCP to which PHA was added, was maximum when 2% PHA was used [79].

A third product-bound factor is the liquid/powder ratio: in general the higher this ratio, the lower the strength [84,85]. However, one reaches a maximum in the strength somewhere in the direction of lower liquid/powder ratios beyond which the wetting of the powder by the liquid becomes insufficient and the cement paste becomes inhomogeneous. The liquid/powder ratio of the paste is of prime importance for the degree of microporosity obtained in the resulting cement (see Section III.C).

A fourth product-bound factor important to strength is the chemical stability of the cement during storage in near neutral water or Ringer's solution. In general, DCPD cements will disintegrate due to the instability of DCPD in neutral solutions and the natural tendency to hydrolyze [66]. Although OCP tends to be transformed into CDHA in contact with a neutral aqueous solution [51], we did not find any change in the strength of an OCP cement made of α-TCP to which some PHA was added during a storage of more than 90 days [821]. However, Hirano and Takeuchi [88] found a decrease in strength of a mixed OCP + CDHA cement made out of α-TCP + DCP mixtures when stored in an unspecified artificial body fluid during 90 days [88]. As shown by Driessens [87], the strength of DCPD cements obtained by mixing calcium phosphates such as α-TCP or TTCP with acids or acidic solutions [86,89] depends critically on their storage conditions. Due to the low solubility of PHA, and its relative stability in comparison with other calcium phosphates, PHA cements are expected to hold their strength indefinitely during their storage in near neutral aqueous solutions at 37°C [51].

The time necessary to reach maximum strength is less than 1 h with DCPD cements, but Fukase et al. [73] reported a value of 4 hrs for a PHA cement. In an OCP cement this time was reduced from 36 h to 14 h by the addition of 2% seed crystals to the cement powder. For the sake of clinical applications, it is desirable to reach 90% of the final strength within 1 to 4 h. This means that the calcium phosphates under development now must still be optimized in this respect.

From our experience [79,81–85] we can conclude that the ratio C/T for compressive strength over diametral tensile strength for this class of materials is about 7. Due to the form of the stress-strain curve one would classify these materials as brittle. However, as the maximum compression before failure is nearly 1%, the resilience of these materials is considerable. Up to now we have no idea what the fracture toughness might be or what value would be necessary for load-bearing applications.

B. Setting Shrinkage or Expansion and Setting Temperature

Due to the fact that in storage in air the cement mass might shrink simply due to evaporation of water, the dimensional change of these cements during setting should be registered while they are soaked in water or Ringer's solution. Takezawa et al. [63] found a linear expansion of 0.06 ± 0.01 % of a PHA cement made of TTCP + DCP. Mostly a linear shrinkage of 0.1 ± 0.05 % was found for OCP and CDHA cements [87]. This behavior is much better than that of classical PMMA bone cements, which show a linear shrinkage of 1.2 ± 0.2 % with this method of measurement [87].

Brown and Fulmer [90] measured an exothermic effect in the reaction of TTCP and DCP mixtures in a PHA cement. However, the increase in temperature during setting in most of the cements mentioned in Table 8 is less than 1°C [87]. A notable exception is formed by the mixtures of MCPM with CaO, which heat up to about 45°C when mixed with water at 25°C [87]. Also in this respect, the behavior of calcium phosphate cements is much better than that of the classical PMMA bone cements, which might develop temperatures as high as 100°C.

C. Crystal Morphology and Porosity of the Set Cements

Research into the crystal morphology of the set cements may reveal the geometric relation of the different crystals in the microstructure. As explained in Section II. gypsum sets due to the formation and entanglement of crystal clusters. A similar behavior is shown by PHA cements made from TTCP/DCP mixtures [57,59,72,73,90] and CDHA cements made either from MCPM/CaO mixtures or from α-TCP [91]. A somewhat different microstructure is shown by DCPD cements made from β-TCP/MCPM mixtures: Here it seems as if the DCPD crystals form bridges between the excess crystals of β-TCP [76,92]. Similar pictures were found for a PHA-type cement prepared from DCPD/CaCO$_3$/β-TCP mixtures [66]. However, the joint addition of calcium sulfate hemihydrate and calcium pyrophosphate to the DCPD cement of Mirtchi et al. [92] resulted in a microstructure in which the entanglement of crystals was clearly visible again.

Brown and Chow [75] reported a porosity of 47% in total for their PHA cement produced from TTCP/DCP mixtures. Chow et al. [72] pointed out that the main origin of the porosity was the value of the liquid-to-powder ratio. They expected a porosity of 60%, 43%, and 33% for L/P ratios of 0.50, 0.25, and 0.17, respectively. Fukase et al. [73] showed that the lower limit of the L/P ratio was 0.25 for this cement, which means that the minimum porosity is 43%.

For a PHA cement prepared from DCPD/CaCO$_3$/β-TCP mixtures, Mirtchi et al. [66] reported a porosity of 54%, which is quite high. During the investigation of OCP cements from α-TCP/DCP mixtures and from CDHA cements prepared from α-TCP, Bermúdez [93] found a microporosity in the range from 32% to 35% and a macroporosity of 1% to 4%. The microporosity (pores of the order of 0.5 μm) is apparently related to the liquid/powder ratio. However, the macropores having a size from 3 to 15 μm have their origin in the inclusion of air bubbles in the cement paste during mixing. The micropores form a contingent, three-dimensional structure through which diffusion of small molecules can take place [80].

D. Workability, Moldability, and Injectability

As explained in Section I.C., the attractivity of cement formulations depends on their moldability during the surgical or dental operation. As far as workability of the cement is concerned, it is our experience that mixing is done most easily in a mortar. It should take not more than 1 minute to prepare a good paste using the pestle and a cement spatula (the latter as used in dentistry).

As explained in Section II.C., the initial setting time should not be shorter than about 4 minutes. This allows for a period of about 3 minutes after mixing to transfer the cement paste into the field of operation and to bring it to the desired form, while deformation is still possible without breaking up the microstructure because there is still no entanglement. This transfer could be done by hand or by syringe: Most calcium phosphate cements can be made injectable. The syringe technique may be slower because it takes some time to transfer the paste into the syringe. In this case an initial setting time of up to 8 minutes may be desirable.

After insertion and molding of the material, it is desirable for the cement to make its final setting as soon as possible. Our experience is that the final setting time can be made to be about 10 to 15 minutes. After that the wound can be closed without seriously damaging the cement structure when the material is touched.

Another advantage of calcium phosphate cements over classical PMMA bone cement with respect to their handling is that they will not cause any allergy, neither to the dentist or surgeon nor to the patient.

IV. BIOLOGICAL PROPERTIES

As calcium phosphates implanted in bone or dental tissues do not have any systemic effect [94], it is sufficient to establish the tissue and cellular reactions for investigation of biocompatibility. It is also worth studying the effect of the biological environment on the material properties of the calcium phosphate cements.

A. Cellular Reactions to Calcium Phosphate Cements

Gruninger et al. [95] tested a PHA cement obtained from a mixture of TTCP + DCP both *in vitro* and *in vivo. In vitro* testing included human red blood cell hemolysis, acute mouse oral toxicity, mouse fibroblast cloning efficiency, and the Ames test for mutagenicity. *In vivo* testing included subcutaneous guinea pig and rat tibia implantations of compressed, preformed cement implants. Implanted rats were sacrificed weekly through 7 weeks, and the tibial bone was removed for evaluation by radiograph and standard histology. The cement appeared to be neither toxic nor mutagenic. Preformed implants were well tolerated by the animals and showed no adverse tissue reaction. After 7 weeks the tibial implants were not resorbed but were tightly situated within the holes.

The biocompatibility of this same cement was tested by Sugawara et al. [96] by subcutaneous implantation in Donryu rats. They found very slight inflammatory reactions around the cement implants as well as around HA ceramic implants used as controls.

B. Bone Tissue Reactions to Calcium Phosphate Cements

Before we start with the discussion of the literature available on this point, we want to distinguish two types of bioresorbability: passive and active. A material might be bioresorbable because it is not stable in the body fluids, this being without consideration of cellular activity. This type of bioresorbability might be called *passive* in comparison with the active type mediated by cellular activity. A typical example of a material having passive bioresorbability is plaster of Paris, or gypsum. According to its solubility in near-neutral solutions [51], a DCPD cement is also expected to have passive bioresorbability. However, calcium phosphate cements of the types OCP or CDHA are expected to show only active bioresorbability: The activity of the osteoclasts is necessary and sufficient to make them resorb. It is questionable whether PHA cement is bioresorbable, because it is not evident [97,98] that osteoclast activity is sufficient to make it resorb. Compare also the fact that implants of HA ceramic granules are integrated but not resorbed into a bone structure [27-32].

As mentioned in Section I.C., β-TCP implants are as osteoconductive as HA ceramic implants are, but simultaneously they are replaced by new bone tissue after some time [27-32]. This type of biodegradability is certainly not passive [34-38]. Hence, β-TCP combines the property of active biodegradability with that of osteoconductivity, in such a sense that the process taking care of its biodegradation is exactly the process that also induces the new formation of bone. For this reason one might call this unique combination of biological properties *osteotransductivity*. On the basis of their natural involvement in bone mineral turnover [99, 100], we expect the same osteotransductivity for calcium phosphate cements of the type OCP or CDHA.

Sugawara et al. [96] implanted a PHA cement made of TTCP + DCP in surgically formed pockets in the lower jaws of dogs. They found no inflammatory reactions in tissue areas adjacent to the implants. The cement mass was covered with new bone and periosteum whereas some CPC particles were replaced by bone. They did not mention the retrieval time of the implants.

Xie and Monroe [71] placed a PHA cement made of TTCP + DCPD in the jaws of rats. Retrieval times were 4 and 12 weeks. When the cement had been mixed with a $Ca(OH)_2$ solution for setting, there was good osteoconductivity. However, if a H_3PO_4 solution was used connective tissue was found between the implant and the bone.

Constantz et al. [101] implanted a DCPD cement in the femora of rabbits. They confirmed that DCPD cements have a passive bioresorbability. However, some transformation into bone apatite occurs simultaneously, apparently due to the fact that the dissolution of DCPD goes so fast that the bone extracellular fluid becomes supersaturated with bone mineral apatite. After about 8 weeks a Ca/P ratio of 1.61 is found in the remnants of the implant.

Gunasekaran et al. [102] implanted a PHA cement and an aggregate of PHA cement with 50% (vol) $CaCO_3$ in the femora of rabbits. It is known that $CaCO_3$ is passively resorbable. However, surprisingly, its presence in the aggregate seemed to activate the active biodegradation of the PHA part of the aggregate as compared to the PHA cement control. The effect may be similar to that of macroporosity in calcium phosphate ceramics [33].

Hong et al. [103] used a PHA cement made of TTCP + DCP for endodontic treatment in the teeth of monkeys. After 1 month they observed minimal adverse tissue reactions. However, adjacent to the cement new bone formation was also observed.

Constantino et al. [104] evaluated the histologic response of a PHA cement by implanting disks within the heads of 9 cats. Three sets of 12 PHA cement disks were made containing either 0%, 10%, or 20 % macropores by volume. The disks were implanted subcutaneously, intramuscularly, above the periosteum of the skull, or directly onto the surface of the calvarium. Animals were killed up to 9 months postoperatively. There were no toxic reactions, extruded implants, or wound infections. Histologic examination of the implant-soft tissue interfaces revealed a transient inflammatory response without foreign-body reaction. The disks were resorbed during implantation proportionally to their macropore content in all groups except for those disks placed directly onto the surface of the calvarium below the periosteum. In this group, foci of bone formed at the skull-implant interface, with variable replacement of the deep surface of these implants by bone. According to these authors [104], implant replacement by bone might occur through a combination of implant resorption coupled with osteoconduction.

From these data it is clear that even "PHA" cements are osteotransductive: In older papers Chow et al. [72,75] held that their PHA cement would not be resorbable, but recently they explained [74] that it appeared to be "the first calcium-based cement that sets to resorbable apatite and is replaced by bone in an approximate one-to-one relationship." In other words, Chow [74] explains that these cements are osteotransductive. This may also be apprehensible if these cements were considered as CDHA cements rather than PHA cements, as has already been argued by Brown et al. [55,56,57,59].

Further proof for the osteotransductivity of calcium phosphate cements was found by Munting et al. [111]. They filled bone defects made metaphysially in the long bones of adult mongrel dogs with a cement consisting mainly of β-TCP. Microradiography, histology, and scanning electron microscopy were used to evaluate the bone structures obtained after 4- or 7-month retrieval times. These studies demonstrated the slow resorption of the cement and the simultaneous ingrowth of bone into the defect, in such a way that the original structural pattern of the bone tended to be restored 7 months after implantation.

C. Material Properties After Implant Retrieval

Only in one study [88] have the properties of a calcium phosphate cement been determined after implantation. Subcutaneous implantation of a CDHA cement made of α-TCP was carried out in stainless steel gauges. Retrieval times were either 0, 1, 2, 4, or 8 weeks. The compressive strength C and the diametral strength T did not change with time (see Table 9). However, the Na and CO_3 content increased significantly (see Table 10). This behavior is the same as that of natural bone mineral [99,100].

Table 9 Compressive Strength C and Diametral Tensile Strength T of a CDHA Cement After Different Periods of Subcutaneous Implantation in Rats

	Week of implantation				
	0	1	2	4	8
C (MPa)	36 (4)[a]	39 (9)	36 (7)	32 (7)	41 (6)
T (MPa)	9.0 (0.8)	7.8 (0.9)	7.1 (1.2)	7.1 (1.2)	7.7 (0.6)

[a]Standard deviations are given in parentheses.

Table 10   Chemical Composition of a CDHA Cement Before Implantation and 8 Weeks After Subcutaneous Implantation in Rats

Implanta- tion times (weeks)	% Na	% Mg	% Ca	% P	% CO₃	Ca/P
0	0.25	0.10	36.35	18.60	0.06	1.510
8	0.49	0.11	36.17	18.38	0.46	1.521

Source: Ref. 105

V. SURGICAL AND DENTAL APPLICATIONS

Brown and Chow [62,75] envisaged the following applications for this type of material:

1. In the orthopedic field:
    a. replacement of bone that has been removed surgically or lost due to trauma
    b. luting cement in orthopaedic surgery
    c. material which will promote bone mineral growth in its vicinity
2. In the field of dentistry and oral surgery:
    a. cavity base or liner to protect the pulp
    b. material for capping exposed pulp
    c. material for replacing or promoting regeneration of bone mineral lost due to periodontal disease
    d. temporary filling material
    e. cement for building up alveolar ridges in edentulous patients
    f. endodontic filling material for root canals
    g. material for cementing retention pins
    h. material for filling sockets after tooth extraction
    i. cement for implanting or replanting teeth
    j. luting cement in dentistry
    k. root cement for remineralizing and desensitizing exposed root surfaces

Very few of these applications have been developed yet. Application 2-f was successfully tried in beagle dogs by Chohayeb et al. [106]. Application 1-a was also done, namely as frontal sinus obliteration and reconstruction in cats [107].

A new subject was approached by *in vitro* investigation by Coutts et al. [108]. They determined the strength of a combination using an intramedullary rod cemented in a long bone on both sides of a fracture. In this way they tried to avoid the use of screws normally used for interlocking of intramedullary nails. They compared fixation by a calcium phosphate cement with that by a classical PMMA cement. Torsional and compressive stiffness of the combination appeared to be practically equal.

Moore et al. [109] tried to use a calcium phosphate cement to augment the fixation of unstable hip fractures and to improve their fixation by transpedicular screws used to anchor posterior spined instrumentation. It appeared to compare favorably *in vitro* in compression testing and pull-out tests with a classical PMMA cement for this application.

Schwarz et al. [110] used a calcium phosphate cement for cementation of a hip prosthesis in canine femora. The results were quite satisfactory and compared well under *in vitro* dynamic loading with PMMA bone cement.

In conclusion, there is still a long way to go in the development of clinical applications for these materials.

VI. SUMMARY AND CONCLUSIONS

The first synthesis of a calcium phosphate cement was reported in 1983. After reaction it consisted mainly of precipitated hydroxyapatite. Since then other formulations have been reported, resulting in the formation of either dicalcium phosphate dihydrate, octocalcium phosphate, or calcium-deficient hydroxyapatite. Pronounced effects of seeds, common ions, and particle size on setting times and reaction mechanisms have been reported.

Compressive strengths of up to 90 MPa have been reported. The highest value for the tensile strength is only 11 MPa. Factors affecting the strength are particle size of the active ingredients of the cement powder, the size and the amount of seed crystals, the liquid/powder ratio, and the amount and type of accelerator added to the cement liquid. Most calcium phosphate cements do not exhibit a dimensional change or a heat effect during setting, in contrast with the conventional PMMA bone cements. The total porosity varies from 32% to 54%, of which about 2% is due to macropores formed by the incorporation of air bubbles during the mixing of powder and liquid. The workability of the cement is enhanced by mixing in a mortar. The mix can be molded for some time before the cement reaches its initial setting time. Also, injection from a syringe directly in the field of operation is possible.

Calcium phosphate cements are neither toxic nor mutagenic. Soft tissue reactions show good biocompatibility. Bone tissue reactions are excellent: These materials are osteotransductive; that is, they are transformed rapidly by passive reaction with the body fluids into bone-like mineral, which within less than 1 year is replaced by newly formed bone, as several animal experiments have shown. Thus, the rate of replacement by bone equals that of the resorption of the cement due to the fact that they are both caused by the processes involved in bone remodelling.

The development of surgical and dental applications is in progress. Up to now the main fields of application are thought to be bone repair, bone augmentation, and surgery of bone growth defects. Future developments may lead to load-bearing applications in orthopedic surgery.

ACKNOWLEDGMENTS

This study was supported by a grant from the Dirección Científica y Técnica of Spain. The authors would also like to thank the CICYT for funding this work through project MAT90-0755.

REFERENCES

1. R. H. Doremus, Review bioceramics, *J. Mater. Sci.*, *27* (1992) 285–297.
2. L. L. Hench, R. J. Splinter, T. K. Greenlee, and W. C. Allen, Bonding mechanisms at the interface of ceramic prosthetic material, *J. Biomed. Mat. Res. Symp.*, *5* (1971) 117–141.
3. C.P.A.T. Klein, Y. Abe, H. Hosono, and K. de Groot, Different calcium phosphate bioglass ceramics implanted in rabbit cortical bone: an interface study, *Biomaterials*, *5* (1984) 362–364.
4. T. Kokubo, S. Ito, M. Shigematsu, S. Sakka, and T. Yamamuro, Mechanical properties of a new type of apatite-containing glass-ceramics for prosthetic application, *J. Mater. Sci.*, *20* (1985) 2001–2004.

5. T. Nakamura, T. Yamamuro, S. Higashi, T. Kokubo, and S. Ito, A new glass-ceramic for bone replacement: evaluation of its bonding to bone tissue, 1, *Biomed. Mater. Res.*, *19* (1985) 685–698.

6. T. Kitsugi, T. Nakamura, T. Yamamuro, T. Kokubo, T. Shibuya, and M. Takagi, SEM-EPMA observation of three types of apatite-containing glass-ceramics implanted in bone: the variance of a Ca/P-rich layer, *J. Biomed. Mater. Res.*, *21* (1987) 1255–1271.

7. L. A. Wolfe and A. Boyde, Biocompatibility tests on a novel glass-ceramic system, *J. Appl. Biomaterials*, *3* (1992) 217–224.

8. Y. Iwashita, T. Yamamuro, R. Kasai, T. Kitsugi, T. Nakamura, H. Okumura, and T. Kokubo, Osteoconduction of bioceramics in normal and osteopenic rats: comparison between bioactive and bioinert ceramics, *J. Appl. Biomaterials*, *3* (1992) 259–268.

9. S. Yoshii, T. Yamamuro, T. Nakamura, M. Oka, H. Takagi, and S. Kotani, Glass-ceramic implant in acetabular bone defect: an experimental study, *J. Appl. Biomaterials*, *3* (1992) 245–249.

10. I. M. Brook, G. T. Craig, and D. J. Lamb, *In vitro* interaction between primary, bone organ cultures, glass-ionomer cements and hydroxyapatite/tricalcium phosphate ceramics, *Biomaterials*, *12* (1991) 179–186.

11. Y. Taguchi, T. Yamamuro, T. Nakamura, N. Nishimura, T. Kokubo, E. Takahata, and S. Yoshihara, A bioactive glass powder-ammonium hydrogen phosphate composite for repairing bone defects, *J. Appl. Biomaterials*, *1* (1990) 217–223.

12. N. Nishimura, T. Yamamuro, Y. Taguchi, M. Ikenaga, T. Nakamura, T. Kokubo and S. Yoshihara, A new bioactive bone cement: its histological and mechanical characterization, *J. Appl. Biomaterials*, *2* (1991) 219–229.

13. N. Nishimura, T. Yamamuro, T. Nakamura Y. Taguchi, T. Kokubo, and S. Yoshihara, A novel bioactive bone cement based on CaO–SiO_2–P_2O_5–CaF_2 glass, in *Bioceramics* (W. Bonfield, G. W. Hastings, and K. E. Tanner, eds.), Vol. 4, Butterworth-Heinemann, 1991, pp. 295–299.

14. M. Ito, *In vivo* properties of a chitosan-bonded hydroxyapatite bone-filling paste, *Biomaterials*, *12* (1991) 41–45.

15. C. C. P. M. Verheyen, J. R. de Wijn, C. A. van Blitterswyk, K. de Groot, and P.M. Rozing, Hydroxylapatite/poly(L-lactide) composites: an animal study on push-out strengths and interface histology, *J. Biomed. Mat. Res.*, *27* (1993) 433–444.

16. R. S. M. Ling, *Complicaciones de las artoplastias totales de cadera*, Salvat Editores, Barcelona, 1987.

17. A. F. Newens and R. G. Volz, Severe hypotension during prosthethic hip surgery with acrylic bone cement, *Anesthesiology*, *36* (1972) 298–300.

18. I. S. Milne, Hazards of acrylic bone cement, *Anesthesia*, *28* (1973) 538–543.

19. W. J. Maloney, R. J. Krushell, M. Jasty, and W. H. Harris, Incidence of heterotopic ossification after total hip replacement: effect of the type of fixation of the femoral component, *J. Bone Joint Surg.*, *73* (1991) 191–193.

20. J. Dandurand, V. Delpech, A. Lebugle, A. Lamure, and C. Lacabanne, Study of the mineral-organic linkage in an apatitic reinforced bone cement, *J. Biomed. Mater. Res.*, *24* (1990) 1377–1384.

21. K. Ishihara, H. Arai, N. Nakabayashi, S. Morita, and K. Furuya, Adhesive bone cement containing hydroxyapatite particle as bone compatible filler, *J. Biomed. Mater. Res.*, *26* (1992) 937–945.

22. W. Hennig, B. A. Blemke, H. Brômer, K. K. Deutscher, A. Gross and W. Ege, Investigations with bioactivated polymethyl methacrylates, *J. Biomed. Mater. Res.*, *13* (1979) 89–99.

23. J. Raveh, H. Stick, P. Schawalder, C. Ruchti, and H. Cottier, Biocement, a new material, *Acta Oto-Laringologica*, *94* (1982) 371–384.

24. S. Downes, Improved bone integration with biomaterials used in orthopaedic surgery, in *Biosis* (D. Komitowski, ed.), Laub, Elztal-Dachau, 1992, pp. 51–58.

25. S. N. Bhaskar, D. E. Cutright, M. J. Knapp, J. D. Beasley, B. Perez, and T. D. Driskell,

Tissue reaction to intrabony ceramic implants, *Oral Surg. Oral Med. Oral Pathol.*, *31* (1971) 282–289.

26. K. Köster, E. Karbe, H. Kramer, H. Heide. and K. König, Experimenteller Knochenersatz durch resorbierbare Calciumphosphat-Keramik, *Langenbecks Arch. Chir.*, *341* (1976) 77–86.

27. P. Ducheyne and K. de Groot, *In vivo* surface activity of a hydroxyapatite alveolar bone substitute, *J. Biomed. Mater. Res.*, *15* (1981) 441–445.

28. J. F. Osborn and K. Donath, Die enossale Implantation von Hydroxylapatitkeramik und Tricalciumphosphatkeramik: Integration versus Substitution, *Dtsch. Zahmärztl. Z.*, *39* (1984) 970–976.

29. C. P. A. T. Klein, A. A. Driessen, K. de Groot, and A. van den Hooff, Biodegradation behavior of various calcium phosphate materials in bone tissue, *J. Biomed. Mater. Res.*, *17* (1983) 769–784.

30. M. Winter, P. Griss, K. de Groot, H. Tagai, G. Heimke, H. J. A. van Dijk, and K. Sawai, Comparative histocompatibility testing of seven calcium phosphate ceramics, *Biomaterials*, *2* (1981), 159–160.

31. H. Newesely, *Grundprinzipien bioreaktiver Implantatwerkstoffe*, *Zahn-, Mund-, und Kieferheilk.*, *72* (1984) 230–239.

32. C. M. Büsing, C. Zöllner, and G. Heimke, The degradation of calcium phosphate ceramics, *Clinical Materials*, *2* (1987) 303–307.

33. C. P. A. T. Klein, P. Patka, and W. den Hollander, Macroporous calcium phosphate bioceramics in dog femora: a histological study of interface and biodegradation, *Biomaterials*, *10* (1989) 59–62.

34. F. C. M. Driessens and R. M. H. Verbeeck, Relation between physicochemical solubility and biodegradability of calcium phosphates, in *Implant materials in Biofunction*, (C. de Putter, G. L. de Lange, K. de Groot, and A. J. C. Lee, eds.), Elsevier, Amsterdam, 1988, pp. 105–111.

35. F. C. M. Driessens, Physiology of hard tissues in comparison with the solubility of synthetic calcium phosphates, *Ann. NY Acad. Sci.*, *523* (1988) 131–136.

36. F. C. M. Driessens, Is the solubility product of synthetic calcium phosphates a good predictor for their biodegradability? in *Euroceramics*, Vol. 3 (G. de With, R. A. Terpstra, and R. Metselaar, eds.), *Engineering Ceramics*, Elsevier, London, 1989, pp. 3.48–3.52.

37. F. C. M. Driessens, M. M. A. Ramselaar, H. U. Schaeken, A. L. H. Stols, and P. J. van Mullem, Chemical reactions of calcium phosphate implants after implantation *in vivo*, *J. Mater. Sci. Mat. Med.*, *3* (1992) 413–417.

38. M. M. A. Ramselaar, F. C. M. Driessens, W. Kalk, J. R. de Wyn, and P. J. van Mullem, Biodegradation of four calcium phosphate ceramics: *in vivo* rates and tissue interactions, *J. Mater. Sci. Mat. Med.*, *2* (1991) 63–70.

39. G. Daculsi, R. Z. LeGeros, E. Nery, K. Lynch, and B. Kerebel, Transformation of biphasic calcium phosphate ceramics *in vivo*: ultrastructural and physicochemical characterization, *J. Biomed. Mater. Res.*, *2* (1989) 883–894.

40. G. Daculsi, R. Z. LeGeros, M. Heughebaert, and I. Barbieux, Formation of carbonate-apatite crystals after implantation of calcium phosphate ceramics, *Calcif. Tissue Int.*, *46* (1990) 20–27.

41. I. Orly, M. Gregoire, J. Menanteau, M. Heughebaert, and B. Kerebel, Chemical changes in hydroxyapatite biomaterial under *in vivo* and *in vitro* biological conditions, *Calcif. Tissue Int.*, *45* (1984) 20–26.

42. J. Koeneman, J. Lemons, P. Ducheyne, W. Lacefield, F. Magee, T. Calaham, and J. Kay, Workshop on characterization of calcium phosphate materials, *J. Appl. Biomaterials*, *1* (1990) 79–90.

43. M. Watanabe, K. Harada, I. Asahira, and S. Enomoto, Implantation of hydroxyapatite granules mixed with atelocollagen and bone inductive protein in rat skull defects, in *Handbook of Bioactive Ceramics*, Vol. 11, *Calcium Phosphate and Hydroxyapatite Ceramics* (T.

Yamamuro, L. L. Hench, and J. Wilson, eds.), CRC Press, Boca Raton, FL, 1990, pp. 223–228.

44. G. D. Barrett and A. B. Schaffer, Alveolar ridge augmentation: using nonresorbable hydroxyapatite as an implant bone graft substituted, *J. Michigan Dent. Assoc.*, *66* (1984) 193–198.

45. G. Riess, Bioreaktive TCP-Impantate, Suprastruktur und Resultate, *Dtsch. Zahnärztl. Z.*, *38* (1983) 100–103.

46. D. S. Metsger, T. D. Driskell, and J. R. Paulsrud, Tricalciumphosphate ceramic—a resorbable bone implant: review and current status, *J. Am. Dent. Assoc.*, *105* (1982) 1035–1038.

47. T. Han, F. A. Carranza, and E. B. Kenney, Calcium phosphate ceramics in dentistry: a review of the literature, *J. Western Soc. Periodont.*, *32* (1984) 88–108.

48. N. Pasutti, G. Daculsi, S. Martin, and C. Deudon, Macroporous polycrystalline calcium phosphate implant for spinal fusion in man and dogs, in *Handbook of Bioactive Ceramics*, Vol. 11, *Calcium Phosphate and Hydroxyapatite Ceramics* (T. Yamamuro, L. L. Hench, and J. Wilson, eds.), CRC Press, Boca Raton, FL, 1990, pp. 345–354.

49. A. Uchida, E. Kurisaki, and K. Ono, The use of bioactive ceramics for bone tumor surgery, in *Handbook of Bioactive Ceramics*, Vol. 11, *Calcium Phosphate and Hydroxyapatite Ceramics* (T. Yamamuro, L. L. Hench, and J. Wilson, eds.), CRC Press, Boca Raton, FL, 1990, pp. 345–354.

50. W. E. Brown and L.C. Chow, A new calcium phosphate setting cement, *J. Dent. Res.*, *62* (1983) 672.

51. F. C. M. Driessens and R. M. H. Verbeeck, *Biominerals*, CRC Press, Boca Raton, FL, 1990.

52. F. C. M. Driessens, M. G. Boltong, O. Bermúdez, and J. A. Planell, Formulation and setting times of some calcium orthophosphate cements: a pilot study, *J. Mater. Sci. Mat. Med.*, *4* (1993) 503–508.

53. N. S. Mandel, The structural basis of crystal induced membranolysis, *Arthritis Rheum.*, *19* (1976) 439–447.

54. H. Monma and T. Kanazawa, The hydration of α-tricalcium phosphate, *Yogyo Kyokai-Shi*, *84* (1976) 209–213.

55. P. W. Brown, D. Sample, and N. Hocker, The low temperature formation of synthetic bone, *Mat. Res. Soc. Symp. Proc.*, *179* (1991) 41–48.

56. P. W. Brown and M. Fulmer, Kinetics of hydroxyapatite formation at low temperature, *J. Am. Ceram. Soc.*, *74* (1991) 934–940.

57. P. W. Brown, N. Hocker, and S. Hoyle, Variations in solution chemistry during the low-temperature formation of hydroxyapatite, *J. Am. Ceram. Soc.*, *74* (1991) 1848–1854.

58. P. W. Brown, Phase relationships in the ternary system $CaO-P_2O_5-H_2O$, at 25°C, *J. Am. Ceram. Soc.*, *75* (1992) 17–22.

59. M. T. Fulmer, R. I. Martin, and P.W. Brown, Formation of calcium deficient hydroxy apatite at near-physiological temperature, *J. Mater. Sci. Mat. Med.*, *3* (1992) 299–305.

60. L. Xie and E. A. Monroe, The hydrolysis of tetracalcium phosphate and other calcium orthophosphates, in *Handbook of Bioactive Ceramics*, Vol. 11, *Calcium Phosphate and Hydroxyapatite Ceramics*, CRC Press, Boca Raton, FL, 1990, pp. 29–37.

61. H. Monma, M. Goto, and T. Kohmura, Effect of additives on hydration and hardening of tricalcium phosphate, *Gypsum and Lime*, *188* (1984) 11–16.

62. W. E. Brown and L-C. Chow, Dental restorative cement pastes, US Patent 4,518,430, May 21, 1985.

63. Y. Takezawa, Y. Doi, S. Shibata, N. Wakamatsu, H. Kamemica, T. Goto, M. Ijima, Y. Moriwaki, K. Uno, F. Kubo, and Y. Haeuchi, Self-setting apatite cement. II. Hydroxyapatite as setting accelerator, *J. Japan. Soc. Dent. Mater. Devices*, *6* (1987) 426–431.

64. B. R. Constantz, B. Barr, and K. McVicker, Intimate mixture of calcium phosphate sources as precursor to hydroxyapatite, US Patent 5, 053, 212, October 1, 1991.

65. A. A. Mirtchi, J. Lemaitre, and E. Munting, Calcium phosphate cements: action of setting regulators on the properties of the β-tricalcium phosphate—monocalcium phosphate cements, *Biomaterials*, *10* (1989) 634–638.

66. A. A. Mirtchi, J. Lemaitre, and E. Munting, Calcium phosphate cements: study of the β-tricalcium phosphate– dicalcium phosphate–calcite cements, *Biomaterials, 11* (1990) 83–88.

67. H. Monma, A. Makishima, M. Mitomo, and T. Ikegami, Hydraulic properties of the tricalcium phosphate–dicalcium phosphate mixture, *Nippon–Seramikkusu–Kyokai–Gakujutsu–Ronbushi, 96* (1988) 878–880.

68. M. Hirano and H. Takeuchi, Hydraulic calcium phosphate cement composition and cement composition containing hardening liquid, US Patent 5,152,836, October 6, 1992.

69. Y. Doi, Y. Takezawa, S. Shibata, N. Wakamatsu, H. Kamemizu, T. Goto, M. Ijima, Y. Moriwaki, K. Uno, F. Kubo, and Y. Haeuchi, Self-setting apatite cement, I. Physicochemical properties, *J. Japan. Soc. Dent. Mater. Devices, 6* (1987) 53–58.

70. Y. Doi, S. Shibata, Y. Takezawa, N. Wakamatsu, H. Kamemizu, M. Ijima, Y. Moriwaki, K. Uno, F. Kubo, and Y. Haeuchi, Self-setting apatite cement, III. Setting mechanism, *J. Japan. Soc. Dent. Mater. Devices, 7* (1988) 176–183.

71. L. Xie and E.A. Monroe, Calcium phosphate cements, *Mat. Res. Soc. Symp. Proc., 179* (1991) 25–39.

72. L. C. Chow, S. Takagi, P. D. Constantino, and C. D. Friedman, Self-setting calcium phosphate cements, *Mat. Res. Soc. Symp. Proc., 179* (1991) 3–24.

73. Y. Fukase, E. E. Eanes, S. Takagi, L. C. Chow, and W. E. Brown, Setting reactions and compressive strength of calcium phosphate cements, *J. Dent. Res., 69* (1990) 1852–1856.

74. L. C. Chow, Development of self-setting calcium phosphate cements, *J. Ceram. Soc. Japan, Int. Ed., 99* (1992) 927–936.

75. W. E. Brown and L. C. Chow, A new calcium phosphate water-setting cement, in *Cements Research Progress*, (P. W. Brown, ed.), Am. Ceram. Soc., Westerville, OH, 1986, pp. 351–379.

76. J. Lemaitre, A. Mirtchi, and A. Mortier, Calcium phosphate cements for medical use: state of the art and perspectives of development, *Silicates Industriels Ceramique Science et Technology, 52* (1987) 141–146.

77. W. E. Brown, J. P. Smith, J. R. Lehr, and A. W. Frazier, Crystallographic and chemical relations between octocalcium phosphate and hydroxyapatite, *Nature, 196* (1962) 1050–1058.

78. W. E. Brown, L. W. Schroeder, and J. S. Ferris, Interlayering of crystalline octocalcium phosphate and hydroxyapatite, *J. Phys. Chem., 83* (1979) 1385–1388.

79. F. C. M. Driessens, M. G. Boltong, J. A. Planell, O. Bermúdez, M. P. Ginebra, and E. Fernández, A new apatitic calcium phosphate bone cement: preliminary results, *Bioceramics*, Vol. 6 (P. Ducheyne and D. Christiansen, (eds.), Butterworth-Heinemann, Oxford, 1993, pp. 469–473.

80. A. Sugawara, L. C. Chow, S. Takagi, and H. Chohayeb, *In vitro* evaluation of the sealing ability of a calcium phosphate cement when used as a root canal sealer-filler, *J. Endodontics, 16* (1990) 162–165.

81. O. Bermúdez, M. G. Boltong, F. C. M. Driessens, and J. A. Planell, Optimization of a calcium orthophosphate cement formulation occurring in the combination of monocalcium phosphate monohydrate with calcium oxide, *J. Mater. Sci. Mat. Med., 5* (1994) 67–71.

82. O. Bermúdez, M. G. Boltong, F. C. M. Driessens, and J. A. Planell, Development of an octocalcium phosphate cement, *J. Mater. Sci. Mat. Med., 5* (1994) 144–146.

83. O. Bermúdez, M. G. Boltong, F. C. M. Driessens, and J. A. Planell, Development of some calcium phosphate cements from combinations of α-TCP, MCPM, and CaO, *J. Mater. Sci. Mat. Med., 5* (1994) 160–163.

84. M. P. Ginebra, M. G. Boltong, F. C. M. Driessens, O. Bermúdez, E. Fernández, and J. A. Planell, Preparation and properties of some magnesium containing calcium phosphate cements, *J. Mater. Sci. Mat. Med., 5* (1994) 103–107.

85. O. Bermúdez, M. G. Boltong, F. C. M. Driessens, and J. A. Planell, Compressive strength and diametral tensile strength of some calcium orthophosphate cements: a pilot study, *J. Mater. Sci. Mat. Med., 4* (1993) 389–393.

86. H. Oonishi, Orthopaedic applications of hydroxyapatite, *Biomaterials*, *12B* (1991) 171–178.

87. F. C. M. Driessens, unpublished results.

88. M. Hirano and H. Takeuchi, Hydraulic calcium phosphate cement, composition and cement composition containing hardening liquid, US Patent 5,152,836, October 6, 1992.

89. S. T. Liu and H. H. Chung, Resorbable bioactive calcium phosphate cement, US Patent 5,149,368, September 22, 1992.

90. P. W. Brown and M. Fulmer, Kinetics of hydroxyapatite formation at low temperature, *J. Am. Ceram. Soc.*, *74* (1991) 934–940.

91. O. Bermúdez, F. C. M. Driessens, J. A. Planell, and M. G. Boltong, Comparison of calcium phosphate cements with dental plaster, *Fourth World Biomaterials Congress*, Berlin, April, 1992, p. 428.

92. A. A. Mirtchi, J. Lemaitre, and N. Terao, Calcium phosphate cements: study of the β-tricalcium phosphate–monocalcium phosphate system, *Biomaterials*, *10* (1989) 475–480.

93. O. Bermúdez, unpublished results.

94. E. Fischer-Brandies, E. Dielert, G. Bauer, F. H. Feder, G. Reidel, and S. Mollenstedt, Zur organspezifischen Ablagerung von ^{45}Ca-markierter Kalziumphosphat-keramik, *Dtsch. Zahnärtzl., Z.*, *44* (1989) 436–437.

95. S. E. Gruninger, C. Siew, L. Chow, A. O'Young, N. K. Tsao, and W. Brown, Evaluation of the biocompatibility of a new calcium phosphate setting cement, *J. Dent. Res.*, *63* (1984) 200.

96. A. Sugawara, K. Kusama, S. Nishimura, M. Nishimura, M. Ohashi, I. Moro, I. C. Chow, and S. Takagi, Biocompatibility and osteoconductivity of calcium phosphate cement, *J. Dent. Res.*, *69* (1990) 312.

97. F. C. M. Driessens, J. W. E. van Dijk, and R. M. H. Verbeeck, The role of bone mineral in calcium and phosphate homeostasis, *Bull. Soc. Chim. Belg.*, *95* (1986) 337–342.

98. F. C. M. Driessens and R. M. H. Verbeeck, Evidence for intermediate metastable states during equilibrium of bone and dental tissues, *Z. Naturforsch.*, *35 C* (1980) 262–266.

99. F. C. M. Driessens, R. M. H. Verbeeck, J. W. E. van Dijk, and J.M.P.M. Borggreven, Degree of saturation of blood plasma in vertebrates with octocalcium phosphate, *Z. Naturforsch.*, *43 C* (1988) 74–76.

100. F. C. M. Driessens, J. M. P. M. Borggreven, and R. M. H. Verbeeck, The dynamics of biomineral systems, *Bull. Soc. Chim. Belg.*, *96* (1987) 173–179.

101. B. R. Constantz, B. M. Barr, J. Quiaoit, I. C. Ison, J. T. Baker, L. McKinney, S. B. Goodman, D. R. Sumner, and S. Gunasekaran, Conversion of brushite, bone cement to hydroxyapatite in rabbit femoral defects, *Fourth World Biomaterials Congress*, Berlin, April 1992, p. 56.

102. S. Gunasekaran, B. M. Barr, I. C. Ison, J. Quiaoit, J. T. Baker, L. McKinney, S. B. Goodman, D. R. Sumner and B. R. Constantz, *In vivo* response to hydroxyapatite bone cement: effect of calcium carbonate aggregates on resorption: *Fourth World Biomaterials Congress*, Berlin, April 1992, p. 432.

103. Y. C. Hong, J. T. Wang, C. Y. Hong, W. E. Brown, and L. C. Chow, The periapical tissue reactions to a calcium phosphate cement in the teeth of monkeys, *J. Biomed. Mat. Res.*, *25* (1991) 485–498.

104. P. D. Constantino, C. D. Friedman, K. Jones, L. C. Chow, H. J. Polner and G. A. Sisson, Hydroxyapatite cement. I. Basic chemistry and histologic properties, *Arch. Otolaryngol. Head Neck Surg.*, *117* (1991) 379–384.

105. R. M. H. Verbeeck, unpublished results.

106. A. A. Chohayeb, L. C. Chow, and P. J. Tsaknis, Evaluation of calcium phosphate as a root canal sealer-filler material, *J. Endodontics*, *13* (1987) 384–387.

107. C. D. Friedman, P. D. Constantino, K. Jones, L. C. Chow, H. J. Polner, and G. A. Sisson, Hydroxyapatite cement. II. Obliteration and reconstruction of the cat frontal sinus, *Arch. Otolaryngol. Head Neck Surg.*, *117* (1991) 385–389.

108. R. D. Coutts, S. A. Hacker, B. R. Constantz, J. M. Kerine, and V. N. Nguyen, The

evaluation of a bioresorbable cement for the temporary fixation of intramedullary rods, 10th European Conference on Biomaterials, Davos, September 1993, p. 24.

109. D. C. Moore, E. P. Frankenburg, J. A. Goulet, G. P. Graziano, R. S. Maitre, L. A. Farjo, and S. A. Goldstein, Fixation enhancement with a new, *in situ* setting Ca-P cement, 10th European Conference on Biomaterials, Davos, September 1993, p. 94.

110. P. D. Schwarz, S. Tuner, T. W. Bauer, and B. R. Constantz, Evaluation of an *in-situ* setting carbonated calcium phosphate cement in a dynamically loaded canine femoral component, 10th European Conference on Biomaterials, Davos, September 1993, p. 8C.

111. E. Munting, A. A. Mirtchi, and J. Lemaitre, Bone repair of defects filled with a phosphocalcic hydraulic cement: an *in vivo* study, *J. Mater. Sci. Mat. Med.*, *4* (1993) 337–344.

32
Acrylic Bone Cements

J. A. Planell, M. M. Vila, F. J. Gil, and F. C. M. Driessens
Universitat Politècnica de Catalunya
Barcelona, Spain

I. INTRODUCTION

A great revolution in orthopedic surgery was about to take place in the early 1960s, when Sir John Charnley presented the preliminary results of a new method for the fixation of joint prostheses to bone [1–3]. The idea was to distribute the contact stresses between the implant and the bone over a large area by means of a filler material, called bone cement and consisting of self-curing polymethylmethacrylate (PMMA). This idea led to the development of a worldwide successful technique in orthopedics. However, it has been seen with time that the fixation of cemented joint prostheses does not last forever, and revision operations become necessary in order to replace the loosened prostheses. Such a second operation has a high cost in both economic and social terms, and moreover, the chances of success diminish severely if a second revision operation is needed.

The main advantages of the cemented prostheses lay in the excellent primary fixation, in the even load distribution between the implant and the bone, and in the fact that the technique allows a fast recovery of the patient. However, the main disadvantage is that about 10% of the patients may need revision in less than 10 years after implantation. The observed aseptic loosening that appears with time is attributed to causes such as the lack of secondary fixation; the mechanical failure of the cement; the initial formation of a fibrous tissue between the cement and the bone, partly due to the necrosis of bone induced by the heat liberated during the setting stage; and the osteolysis caused by the foreign body reaction induced by wear particles and debris, part of which could come from the bone cement. The toxicity of the liquid monomer of the cement should also be mentioned as another disadvantage. In the balance acrylic bone cements still have a future, mainly when the moderate success of other fixation techniques such as metallic porous coatings and plasma-sprayed hydroxyapatite coatings is taken into account.

This chapter reviews the basic chemistry of self-curing acrylic bone cements and their main chemical and physical properties. (See the tables in later sections for listings of the cements discussed.) Special attention is given to their mechanical properties and the variables that affect them, from a microstructural point of view, since aseptic loosening has been currently explained in terms of the mechanical failure of bone cement. The biological interactions in terms of monomer release and the formation of a fibrous capsule are described, and finally the recent developments and ideas for the modification of bone cements, in order to achieve a longer life in service, are reviewed. Acrylic bone cements are presented from the point of view of biomaterials science and engineering, and although clinical results are reported, it is not the aim of this chapter to describe or discuss surgical techniques. Finally, although the chapter reviews 275 references, many others are available and can be traced from those listed. This gives a clear idea of how alive is research in the bone cement area.

II. CHEMISTRY OF ACRYLIC BONE CEMENTS

A. Chemical Composition: Powder and Liquid

Acrylic bone cements are based in PMMA, which is accepted as a biocompatible polymer when cured. The methylmethacrylate monomer (MMA) can either polymerize spontaneously very slowly, or its polymerization can be activated by means of ultraviolet light, heat, or an initiator. The last method is used in orthopaedic surgery where a reasonably short setting time is required. The bone cement is usually prepared by mixing the two components of the dose: a transparent liquid and a white powder. The main properties of each component are described below (for an excellent review, see Ref. 5).

1. Liquid Component

The liquid is transparent, volatile, and has a characteristic penetrant smell. Its viscosity is low and its boiling temperature is approximately 100°C at 760 mm Hg. Its density is 0.94 g/cm^3. It contains three basic ingredients:

1. MMA monomer: 97% v/v

$$CH_2=\underset{\underset{CH_3}{|}}{C}-\overset{\overset{O}{\|}}{C}-O-CH_3$$

2. *N,N*-Dimethyl-*p*-toluidine: 2.7% v/v

$$\underset{CH_3}{\overset{CH_3}{>}}N-\!\!\left\langle\bigcirc\right\rangle\!\!-CH_3$$

3. Hydroquinone: 750 ppm

$$OH-\!\!\left\langle\bigcirc\right\rangle\!\!-OH$$

The *N,N*-dimethyl-*p*-toluidine acts as accelerant of the polymerization reaction that activates the initiator mixed with the powder. The hydroquinone is an inhibitor that prevents the premature polymerization of the monomer. The volume of liquid is usually 20 ml.

2. Solid Component

The basic ingredients of the powder are:

1. PMMA: 89% w/w

In some instances, instead of PMMA beads, other polymers or copolymers are used.
2. Benzoyl Peroxide (BP): 0.75% w/w

3. Barium sulfate or zirconium dioxide: 10% w/w

The BP acts as initiator, producing free radicals when it reacts with the amine. $BaSO_4$ is added in order to obtain a radiopaque cement. The diameter of most of the PMMA particles in the cement ranges between 30 and 150 μm and their shape depends on the manufacture process used. Therefore the shape and size of the PMMA particles depend on the commercial brand. The weight of solid material in a dose is 40 g.

The PMMA particles or beads constitute approximately 70% by weight of the polymerized cement. Their microstructure, both physical and chemical, plays a relevant role in the final properties of the cement. When the PMMA particles appear as spherical beads, their preparation process is known as suspension polymerization [4]. The size and shape of the beads control the viscosity of the cement [5].

B. Chemical Reactions: Polymerization

The preparation process of the cement consists in the mixing of the two components, solid and liquid. At this moment the polymerization of the liquid monomer starts as a typical reaction of addition polymerization [5]. Three steps take place: initiation, propagation, and termination. In this process the double bonds of the monomer are attacked by the free radicals of the initiator and an active center is produced. It is from this active center that the chain is formed, growing, by the addition of monomer molecules [4,5].

The initiation process consists in the activation of the initiator, the benzoyl peroxide (BP), by the accelerant, the *N,N*-dimethyl-*p*-toluidine (DMPT), producing a benzoyl peroxide free radical R·:

This radical can then react with the MMA monomer and produce the primary free radical RM·:

$$\text{\langle O \rangle} - \overset{\overset{\text{O}}{\|}}{\text{C}} - \text{O} - \overset{\overset{\text{H}}{|}}{\underset{\underset{\text{H}}{|}}{\text{C}}} - \overset{\overset{\text{CH}_3}{|}}{\underset{\underset{\underset{\underset{\underset{\text{CH}_3}{|}}{\text{O}}}{|}}{\underset{\text{C}=\text{O}}{|}}}{\text{C}\cdot}}$$

The propagation process takes place by the continuous addition of MMA monomer units. The chains form in a fraction of a second and for each monomer addition energy is liberated. This energy is responsible for the polymerization temperature rise of the setting bone cement:

$$RM\cdot + M \rightarrow RMM\cdot + \text{heat}$$
$$RMM\cdot + RMMM\cdot + \text{heat}$$

and therefore:

$$RM_i\cdot + M \rightarrow RM_{i+1}\cdot + \text{heat}$$

Finally, the termination process may take place in three different ways:

1. A chain of free radical $RM_i\cdot$ reacts with a free radical $R\cdot$ and a single bond is formed.
2. Two chains of free radicals $RM_i\cdot$ and $RM_j\cdot$ react to form a single bond.
3. Disproportion through a β-hydrogen atom transfer, where a hydrogen of the CH_2 group next to the carbon with the free radical of the $RM_i\cdot$ chain is transferred to the $RM_j\cdot$ chain and a single bond is formed giving a RM_j chain, while the RM_i chain gets a double bond formed by the missing hydrogen and the free radical.

C. The Setting Process

When the mixture of the powder and the liquid components of the bone cement starts, both physical and chemical phenomena take place simultaneously, which will affect the setting process as well as the microstructure and the mechanical properties of the set material, and which depend on variables such as the chemical composition and concentration of the initial powder and liquid components, the physical mixing method, and the chemical environment.

The time elapsed from the moment in which the powder and liquid components are mixed until the cement is set is known as *setting time* [6–9]. If the evolution of the temperature with time is recorded, the setting time is the time when the temperature of the polymerizing mass is: $T_{amb} + (T_{max} - T_{amb})/2$, where T_{amb} is the ambient temperature, taken as $23° \pm 1°C$, and T_{max} is the maximum temperature reached by the polymerizing cement in degrees Celsius [7,9]. The maximum temperature, or peak temperature, is produced by the exothermic propagation reactions that take place during polymerization. The cement sets before the peak temperature is reached. The time at which the mixed cement mass does not adhere to a surgically gloved finger is known as the *dough time*. Finally, the difference between the setting time and the dough time is called the *handling time*, and it corresponds to the period of time during which the cement is workable and can be implanted (or molded). The leaflets provided with commercial brands of bone cements usually give the working time of the cement in terms of environmental temperature.

The chemical composition and the concentration of the products in the powder and the liquid components cannot be controlled in a clinical application, although they will play a very strong role in the kind of material that will be obtained after setting [5]. For the cement to be produced, it is necessary that the liquid monomer wets the powder particles of PMMA, which should then swell and allow the diffusion of the liquid into the organic matrix of the particles. Simultaneously, it is necessary that a component such as BP is completely diluted into the liquid. Monomer evaporation during these early stages of mixing is well known, and the clinical consequences have been considered [3,8,10,11]. Finally, some degree of polymer–polymer interdiffusion should take place from the liquid to the solid, or monomer-swollen, phase. The good adhesion between the polymerized monomer and the PMMA particles in the set cement means that chain entanglement of the polymer powder and the polymerized monomer should exist at the interface. Consequently, it can be expected that the physical processes of swelling of the powder, diffusion of the monomer, dilution of the BP into the liquid and polymer–polymer interdiffusion will affect the setting dynamics.

From the chemical point of view it seems that the Trommsdorff–Norrish effect, or gel effect [12,13], takes place. This effect is caused by an increase in viscosity of the reaction medium, which delays dramatically the termination reactions. The increase in viscosity is due to the formation of polymer molecules. Since viscosity increases, monomer diffusion decreases, although at a lower rate than the diffusion of growing polymer chains and other active species. The decrease of the termination rate, while the initiation and propagation rates remain practically unchanged, leads to an increase in the polymerization rate and consequently to a rapid rise in temperature. This effect explains both the steep temperature rise when the cement sets and the high molecular weight of the polymerized monomer in the cement matrix. The BP–DMPT reactions may produce other free radicals or by-products different from the benzoyl peroxide free radical. The kinetics of the free radicals and their slow decay have been well studied [14–16]. It has been suggested that unreacted radicals and residual components in the set mass may affect the biocompatibility of the cement and its eventual degradation [17–19]. Moreover, the production of CO, from BP decomposition could play a role in the cement microporosity [5].

A very important residual component is the monomer left unpolymerized after the setting of the cement, excluding the unreacted monomer evaporated before the cement sets and cools down to its environmental temperature. The amount of residual monomer left in the cement mass depends on the type of cement, since different commercial brands give different results, but in all cases it decreases with time. The actual values reported depend also on the environment in which the specimens are kept. After 215 days in air, monomer concentration is 2.42%, and after being held for 1.5 months in water at 37°C and then dried for 1.5 months under 10 mm vacuum at 37°C, the monomer concentration is 1.08% [17]. After 21 days in water at 37°C the average monomer concentration was 0.05%, 1.3%, and 1.2% for Boneloc, Palacos, and Simplex bone cements, respectively [19]. Taking into consideration the volume and the geometry of the specimens, the material interfacing the dough cement and the rate of diffusion of the monomer into the surrounding media, it is possible to assess the portion of residual monomer that will dissolve in water or tissue fluids. After 6 days a value of 0.43% has been reported [17].

D. Variables That Affect the Polymerization and the Setting Processes

Chemical composition controls the setting process and even the physical and mechanical properties of bone cement [18], but there are other variables such as concentration, mixing method, environment, and others that play a strong role in relevant properties such as workability, setting time, and mechanical strength of the cement. From a clinical point of view, the control of the dough time and the setting time is very important, and therefore, the understanding of all the events that take place during the setting process is essential.

The polymerization process is strongly dependent upon the environmental temperature at which the reaction takes place [5,6,9,20]. Both dough time and setting time depend, not only on chemical composition and concentration, but also on environmental factors such as relative humidity or kneading frequency of the mix. In order to be able to make comparisons, the ASTM Standard on acrylic bone cements [7] states that the dough time and the setting time evaluation should be conducted at a room temperature of 23° ± 1°C and a relative humidity of 50% ± 10%, and it allows a maximum dough time of 5 min and a setting time ranging from 5 to 15 min. The idea of accelerating the polymerization process has been proposed and the use of preheated implants has been considered [21]. It seems that for a stem heated up to 50°C, the curing time of the cement is reduced in 50% while the temperature at the cement–bone interface does not go over 57°C, and the compressive strength of the resulting cement is only reduced by 10%.

A parameter that has a very strong effect on the working time of the cement and that is under the control of the manufacturer is the powder-to-liquid ratio of the product, which is taken as the ratio between the weight of the powder in grams and the volume of the liquid in milliliters. As the powder-to-liquid ratio increases, the peak temperature decreases [5,6,9,14]. These results can be understood in terms of the relative amounts of initiator and monomer present in the various powder-to-liquid ratios. Free-radical growth and temperature rise take place faster as the powder-to-liquid ratio increases, although the lesser amount of monomer available to react results in a lower peak temperature [14]. The decrease in the peak polymerizing temperature, as the powder-to-liquid ratio increases, has to be also understood in terms of the role played by unreactive particles in absorbing heat. As a consequence, the setting time will also decrease as the powder-to-liquid ratio increases. It has been reported that for a powder-to-liquid ratio of 3:1, the effect on the dough time is practically nil [6].

E. Exotherm

There is extensive discussion in the literature about the role of exotherm in the necrosis of the surrounding bone. The amount of heat released depends on the weight of reacting monomer in the mixture. However, the peak temperature reached depends also on the volume of cement and its surface-to-volume ratio, in other words, its thickness in service. This is the reason that the ASTM Standard [7] defines very clearly the geometry of the mould and the place where the thermocouple is to be introduced in the cement for the evaluation of the maximum temperature.

More important than the temperature reached by the cement is the temperature at the bone–cement interface, since bone necrosis will be possible if such temperature reaches values over 56°C, which corresponds to the onset of coagulation of albumin [22]. However, it seems that this temperature should not be taken as a threshold since cell damage is not only a question of temperature but also of time. Between 48° and 60°C, exposure time is critical [23, 24]. A wide variety of temperatures at the interface of bone and bone cement have been reported: 83°C [25], 70°C [9], 7°C [21], 48°C [26], 39° to 45°C [27], from 37° up to 69°C [28], from 37° to 65°C [29] depending on the thickness and the position of the thermocouple [29], and from 41° to 67°C in the acetabulum, although this range is reduced to 36.5°–47.5°C with continuous water irrigation [30].

In order to reduce the exotherm of bone cements, alternative monomers have been proposed or have been mixed with MMA [18,31,32].

F. Molecular Weight

It is well known that the molecular weight and the molecular weight distribution of linear polymers play a very strong role in their mechanical properties. Such parameters have been reported by many authors working with a wide variety of commercial brands [6,17,32–37], and they are shown in Table 1.

Depending on the authors, the molecular weight is given as the viscosity average molecular weight (M_v) or as the number-average (M_n) and the weight-average (M_w) molecular weight obtained from gel permeation chromatograms. The ratio M_w/M_n provides information about the molecular weight distribution: a large value indicates that the weight-average molecular weight and the number-average molecular weight are very different, and that the molecular weight distribution is broad. It is usually accepted for PMMA that most mechanical properties are practically independent of molecular weight when this is higher than 1 to 1.5×10^5, although even then the molecular weight distribution may play a strong role [17,33,35,38,39]. It is known that fracture surface energy of PMMA is closely related to molecular weight [39,40], although for high molecular weights (M_v over 10^5), it becomes practically independent [41].

It should be noticed that in most cases the molecular weight of the cement is higher than that of the PMMA powder. This means that the polymerized monomer, which binds together the PMMA particles, forms longer molecular chains than those present in the initial PMMA powder [34]. The interpretation of this has to be approached through the Trommsdorff effect already discussed.

It has also been shown that very similar molecular weights are obtained by polymerizing the bone cement *in vitro* or *in vivo*, in different positions of hip prostheses implanted in dogs [36].

III. MECHANICAL PROPERTIES

Bone cement in service performs mainly a mechanical task, by distributing evenly the contact forces and transfering them from the prosthesis to the bone. This is the reason why mechanical properties of bone cements have been measured and reported by many authors. Since acrylic bone cements are based on PMMA polymers, with a glass transition temperature $T_g \approx 110$°C, their mechanical properties will be very sensitive to the environmental temperature, and strain rate will deeply affect their mechanical behavior due to their viscoelastic nature. Finally, fracture and fatigue mechanisms have to be carefully analyzed. Bone cement is a self-cured two-phase material [5,35,37] when compared to industrial PMMA. These microstructural and process differences explain their poorer mechanical properties. A good review of the mechanical properties of acrylic bone cements can be found in Ref. 42.

Table 1 Reported Values of Molecular Weights of Different Commerical Brands of Acrylic Bone Cements

	Powder				Cement				Ref.
	M_v	M_n	M_w	M_w/M_n	M_v	M_n	M_w	M_w/M_n	
CMW	1.01×10^5				1.45×10^5				[35]
CMW		5.46×10^4	1.43×10^5			4.7×10^4	1.43×10^5	3.0	[32]
CMW	1.14×10^5	4.3×10^4	1.06×10^5	2.5					[37]
Palacos R	4.59×10^5				3.88×10^5				[35]
Sulfix-6	1.19×10^5				1.31×10^5				[35]
Simplex P	8.90×10^4				1.19×10^5				[35]
Simplex P	1.95×10^5	4.4×10^4	1.98×10^5	4.49	2.03×10^5	5.1×10^4	2.42×10^5	4.78	[6,17]
Simplex P, 0-*in vitro*					1.38×10^5				[36]
Simplex P, 0-proximal					1.38×10^5				[36]
Simplex P, 0-distal					1.45×10^5				[36]
Simplex P, 2w-*in vitro*					1.41×10^5				[36]
Simplex P, 2w-proximal					1.28×10^5				[36]
Simplex P, 2w-distal					1.33×10^5				[36]
Simplex P, 8w-*in vitro*					1.32×10^5				[36]
Simplex P, 8w-proximal					1.32×10^5				[36]
Simplex P, 8w-distal					1.37×10^5				[36]
Kallodent			5.4×10^5	2.5			7.5×10^5	3.0	[33]
Stellon			7.2×10^5	2.5			9.0×10^5	3.3	[33]
Rostal	4.74×10^5	1.09×10^5	3.24×10^5	3.0	5.77×10^5	1.3×10^5	3.16×10^5	2.5	[37]

A. Strength

The strength of bone cements has been measured using different kinds of tests such as tension, compression, bending, torsion, and shearing, and by controlling different parameters such as temperature, environment, crosshead speed of the testing machine, and specimen geometry and size. Consequently, a wide variety of results are available. Moreover, the results reported refer to different commercial cement brands. Other factors related with the manufacture of the specimens, which will be analyzed in Sec. IV, contribute to the scatter of results. Table 2 provides a summary of static mechanical properties reported by different authors. These results are relevant from the point of view of characterizing the bone cements, but it is doubtful that they provide any information about the failure mechanims of bone cements in service

It should be mentioned that varying the powder-to-liquid ratio from 2:1 to 3:1 and 3:2 does not change the compressive modulus and the compressive strength significantly [6].

B. Viscolelastic Behavior

In the evaluation of some mechanical properties, such as fracture toughness or elastic moduli, acrylic bone cements are taken as linear elastic materials. However, due to their polymeric nature, they exhibit a viscoelastic behaviour and, consequently, their mechanical properties are strongly dependent on temperature and strain rate. From a practical point of view, creep and stress relaxation will be very important parameters when assessing the life in service of acrylic bone cements.

The viscoelastic characterization performed on an autopolymerized acrylic bone cement by means of dynamic: mechanical thermal analysis (DMTA) showed that the glass transition temperature increases from 93° to 130°C as the testing frequency increases from 0.1 to 100.0 rad/s when measured from the dissipation factor tan δ, and it increases from 78° to 97°C for the same testing conditions when measured from the loss modulus G'' [57].

The strain rate sensitivity plays a very strong role in the mechanical properties of viscoelastic materials. When the strain rate increases, an increase in elastic modulus and in strength has to be expected. It plays a role similar to temperature but in the opposite direction, since an increase in temperature produces a decrease in elastic modulus and in strength. It has been shown for AKZ cement that as strain rate increases from quasistatic values to 1.8 s^{-1}, Young's modulus increases by about 45 % and the compressive strength increases by about 50%. On the other hand, when the testing temperature increases from 20° to 37°C, the Young's modulus and the compressive strength decrease, respectively, by about 5% and 10% if the tests are conducted at a constant strain rate, and Simplex RO and AKZ cements are used [45,58]. For Simplex P cement the proportional limit increases by 63% while the compressive strength increases by 24% when the strain rate increases by a factor of 20 from 0.6173 × 10^{-3} s^{-1} [59].

The good creep behavior and the stress relaxation of bone cements are related with the prosthesis fixation and the ability of the cement to evenly distribute the contact loads from the prosthesis to the bone. The stress relaxation depends on the magnitude of the initial deformation, since for a 1% initial deformation, it is shown that about 49% of the load is relieved during the first 2 h, while from then on a steady state is reached where load is relaxed at a rate of 0.024%/h [60]. For initial deformations corresponding to 10

Table 2 Summary of Static Mechanical Properties Reported by Different Authors Working with Different Bone Cements

Ref.	Cement	Elastic modulus[a] (GPa)	Ultimate strength (MPa)				
			Tension	Compression	Bending	Shear	Torsion
[44]	Simplex P	0.85 (Tor)	25	77		41	37
[34]	Simplex P	2.14 (Comp)	13.2	76.5			
[6]	Simplex radiopaque	2.07 (Bend)	28.9	91.7	51.1		
		2.20 (Comp)					
		2.20 (Bend)					
[45]	Simplex radiolucent	2.30 (Comp)	32.6	93.1	56.9		
	Simplex P	2.4 (Comp)		85-90			
		2.6 (Comp)					
	Simplex RO	2.3-2.4 (Bend)		85-90	60-70		
		2.5-2.5 (Comp)					
	AKZ	2.3-2.4 (Bend)		85-90	50-60		
	CMW + $BaSO_4$	2.5-2.6 (Comp)		85-90			
	Sulfix 6	2.6-2.7 (Comp)		85-90			
	Sulfix 6 + Nebacetin	2.6-2.7 (Comp)		85-90	60-70		
		2.2 (Bend)					
	Palacos R	2.4-2.5 (Comp)		85-90			
		2.5-2.6 (Comp)					
	Palacos R + Refobacin	1.6-1.8 (Bend)		85-90	50-60		
[35]	CMW	2.76 (Tens)	42.7				
	Palacos R	2.62 (Tens)	46.2				
	Palacos RG	2.55 (Tens)	46.2				
	Sulfix 6	2.41 (Tens)	48.2				
	Simplex P	2.55 (Tens)	33.8				
[36]	Simplex P in vitro (8 weeks)	3.0 (Bend)			60-65		
	Simplex P proximal (8 weeks)	1.8-2.0 (Bend)			50-55		
	Simplex P distal (8 weeks)	2.2-2.4 (Bend)			62-67		
[20]	CMW		21.8		49		
	Zimmer		22.9		48.3		

Ref	Material				
	Palacos		22.6		41.4
	Simplex P		25.1		49.3
	Zimmer LVC		25.8		48.3
	Sulfix 6		26.6		48.2
	Simplex R		27.1		50.8
[46]	Zimmer			98	
	Palacos R			100	
	Palacos R + Garamycin			89	
	CMW 1			97	
	CMW 2			120	
	Simplex P			100	
	Simplex R			102	
	Sulfix 6			108	
[47]	Sevriton	2.35 (Comp)	22.8	71.8	
	Sevriton III	1.93 (Comp)	31.6	73.1	
[48]	CMW	2.76	24		
[49]	CMW	2.4–2.9 (Bend)			51.3–61.3
	Simplex P	2.4–2.7 (Bend)			52–70.2
	Palacos R	2.2–2.4 (Bend)			48.8–68.4
		2.7 (Bend)			
[50]	Simplex P	3.2 (Comp)		74	65
		3.0 (Bend)			
[51]	CMW	3.7 (Comp)		83	48.5
	Zimmer			84.4	
	Simplex P			89.2	
	Zimmer LVC			84.7	
[52]	Simplex P	2.324 (Tens)	27.7		
[18]	Simplex P			72.6	
[53]	Simplex P	2.53 (Tens)	36.19		
	Zimmer R	2.81 (Tens)	32.26		
	Zimmer LVC	3.07 (Tens)	39.75		
[54]	Palacos R	2.300		100	85.0
	Simplex P	2.088		90	81.1
	CMW	2.218		100	74.3
[55]	Rostal	2.64	43.8	87.9	

Table 2 Continued

Ref.	Cement	Elastic modulus[a] (GPa)	Ultimate strength (MPa)				
			Tension	Compression	Bending	Shear	Torsion
[56]	CMW 1	2.317 (4 pt. bend)		87	61 (4 pt. bend)		
	Palacos G	2.237 (4 pt. bend)		86	61 (4 pt. bend)		
	Palacos R	2.348 (4 pt. bend)		84	66 (4 pt. bend)		
	Simplex P	2.522 (4 pt. bend)		100	74 (4 pt. bend)		
	Simplex RO	2.563 (4 pt. bend)		99	71 (4 pt. bend)		
	Zimmer D	2.214 (4 pt. bend)		77	48 (4 pt. bend)		
	Cerafix	2.528 (4 pt. bend)		98	71 (4 pt. bend)		
	CMW 3	2.771 (4 pt. bend)		100	65		
	Palacos E	2.647 (4 pt. bend)		95	74		

[a]Tor, torsion; Comp, compression; Bend, bending; Tens, tension

MPa and 20 MPa tensile stresses, the relaxed loads corresponded to 27% and 32%, respectively, after 10^4 s [37].

The creep behavior seems to be correlated with the molecular weight distribution and the glass transition temperature [61]. It has been shown that creep resistance increases with density and large PMMA powder particle size, although residual monomer, radio-paque fillers, and a plasticizing environment such as water decrease it [62]. Creep will also depend upon the polymerizing process of the cement and the temperature at which the test is carried out [63].

Different models to represent creep behavior of bone cements have been proposed. A simple potential model relating deformation and time predicts that a shrinkage of the cement, representing a stress of 1 MPa, will be reached in about 1 to 2 years for Simplex and CMW bone cements, while it will take only 40 days for Palacos [64]. Linear visco-elastic models of a five-parameter equation representing two Kelvin elements in series and of a three-parameter power law, respectively, seem to adjust the results obtained perform-ing tensile creep tests with Zimmer R cement [65], and a four-parameter nonlinear double power law seems to adjust the results obtained in torsional creep for the same cement [66].

C. Fracture Toughness

The fracture features of bone cements have been deeply studied, although some fracture mechanisms have not been well explained yet. The bone cements are taken as linear elastic solids, since PMMA behaves as a brittle material, and K_{IC} can be easily evaluated. An excellent review on the fracture toughness of acrylic bone cements is available [67].

The basic ideas of fractography of acrylic bone cements, based on the two-phase microstructure of the material, were understood [68] in terms of transgranular fracture through the PMMA particles. The fracture surface energy has been reported by several authors, some of them giving also the mean inherent flaw size of the material [35,47,69], as shown in Table 3. Fracture energy has also been evaluated by means of impact methods and the reported values are 1.9 kJ/m^2 and 0.4 kJ/m^2 for unnotched and notched Izod specimens of CMW cement, respectively [48], and 7.0, 2.6, and 3.2 kJ/m^2 for Palacos R, Simplex P, and CMW bone cements, respectively [54]. Some authors prefer to evaluate the energy necessary to propagate a crack in a brittle material by means of the work of fracture using a square cross-sectioned bar in which a chevron notch has been machined

Table 3   Fracture Surface Energy and Mean Inherent Flaw Size of Different Acrylic Bone Cements

Fracture surface energy (J/m^2)	Mean inherent flaw size (μm)	Cement	Ref.
140	340	CMW	[35]
290	830	Palacos R	
310	660	Palacos R0	
380	980	Sulfix 6	
490	1600	Simplex P	
23		Sevriton	[47]
140	83	Stellon	[69]

[70,71]. After 1-week storage in air at 21° and 37°C, the work of fracture of Simplex P cement is 407 J/m^2 and 311 J/m^2, respectively.

Fracture toughness has been evaluated using different kinds of specimens: SENB (single edge notch bend), CT (compact tension), DT (double torsion), and short rod (Barker type). The results reported are summarized in Table 4. The brittleness of acrylic bone cements is apparent since the K_{IC} values lie between 1 and 1.5 MPa·m$^{1/2}$, which are even lower than those of ceramic materials. It can be noticed that the highest fracture toughness reported is about double that of the lowest one. This means that the type of cement, the preparation and the moulding conditions, and the type of test conducted will play a major role in the values obtained.

It has been shown that the crack propagation velocity is related to the stress intensity factor that the material sustains, and therefore it is possible to calculate a failure time prediction in terms of the stress applied, environmental conditions, and content of other phases such as BaSO$_4$ [77].

The fracture behavior of acrylic bone cement under triaxial loading has not been deeply studied. It has been shown, however, that CMW and Rostal bone cements behave according to a Coulomb–Mohr failure citerion when tubular specimens are tested in compression under a constant hydrostatic pressure [78]. The mode of failure is very similar to that produced in a bursting test, where a uniform pressure inside a tubular specimen produces longitudinal cracks [79].

A cause of failure of materials is wear, and abrasive wear is very important in dental acrylic cements. It has been shown by means of empirical relationships obtained testing 23 dental acrylic cements that abrasive wear is not dependent on hardness [43].

D. Fatigue

Among the causes of loosening of cemented joint prostheses, the fatigue failure of the cement is currently understood as one of the most important. A good review of the

Table 4 Fracture Toughness Values Given as K_{IC} in MPa·m$^{1/2}$

K_{IC}	Cement	Specimen	Molding pressure	Ref.
0.877	Simplex P	SENB	0.035 MPa	[72]
1.033	Zimmer	SENB	0.035 MPa	
1.46	Simplex P	SENB	0.035 MPa	[73]
1.318	Zimmer	SENB	0.035 MPa	
1.15–1.6	Simplex P	DT	0.07–0.7 MPa	[74]
1.56	Simplex P radiolucent	CT	No pressure	[52]
1.59		CT	Clamp pressure	
1.19		CT	Mold 37°C, no pressure	
1.55		CT	Mold 37°C, clamp pressure	
1.29	Simplex P radiopaque	CT	No pressure	
1.42	Zimmer	SENB	5 kg on mold	[51]
1.51	Simplex P	SENB	5 kg on mold	
1.23	Zimmer LVC	SENB	5 kg on mold	
1.74	CMW	Short rod	No pressure	[75]
1.53	3M-Concise	Short rod	No pressure	
1.26	Simplex P	Short rod	No pressure	[76]
0.98	Zimmer	Short rod	No pressure	
1.39	Rostal	CT	Clamp pressure	[55]

fatigue properties of acrylic bone cements was published in 1988 [80]. A relevant aspect that should be taken into account in fatigue testing is the effect of cyclic loading and frequency on the temperature of the bone cement due to the poor thermal conductivity of PMMA and to its viscoelastic properties. For a range of cyclic loads up to 4.5 kN, which would be about 6.6 times body weight, and for a range of frequencies up to 6 Hz, which is well above physiological gait frequency, there is no rise in temperature that could affect the fatigue strength of bone cement [81].

The most common approach taken when studying the fatigue failure of acrylic bone cements has been to assess the fatigue endurance at a given number of cycles. *S-N* curves have been obtained in rotating–bending fatigue [73,82–84], in tension–tension [48,85–87], in tension–compression [88,89], in three-point bending [83,84], in four-point bending [90], and even in compression [50], usually under load control. Tests have been also carried out in tension–compression under strain control [91]. Most workers carry out their tests in sinusoidal loading, although this information is not always reported.

Taking fatigue endurance at 10^6 cycles, values of 6.89 MPa [73,82], about 10 MPa [48], 10.84 MPa, 11.72 MPa, 13.43 MPa [85], and 17–20 MPa [50]; and for a fatigue endurance at 10^7 cycles, 15–26 MPa [90] have been reported for Simplex P, Zimmer R, Zimmer LVC, CMW, Palacos R, and Palacos E-Flow bone cements. Such a wide scatter of results will depend not only on the type of cement used but also on its preparation process and its final microstructure. Some specimens were prepared at molding pressures of 0.0345, 0.172, and 0.345 MPa [73,82]; the porosity of the materials was different; the testing frequencies used were different; the tests were different; the time elapsed between molding and testing was probably different; and even the environmental conditions of temperature and humidity were different. Since fatigue has a statistical nature and the probability that two specimens from the same batch will exhibit a different behavior is high, a possible approach is to analyze the fatigue results using the Weibull distribution of survival probability, the failure probability or reliability at a given number of cycles, for a constant stress amplitude and a defined frequency [84,85,88]. Here again there is a scatter of results because different authors use different fatigue tests (i.e., tension–tension, tension–compression, and four-point bending), the type and geometry of the specimens is different and the maximum stresses and frequencies used are also different. It has been observed that the data scatter is lower when representing strain against number of cycles than when stress against number of cycles is used [91].

It is well known that the smoothness of a fatigue specimen will play a major role in the number of cycles to failure that it will sustain. The surface quality of the specimens will also contribute to the scatter of data, mainly porosity existing at the surface or very near the surface of the bone cement specimen. All the tests reported have been conducted with unnotched specimens, which means that the fatigue failure process has consisted of a crack nucleation stage and a crack propagation stage. The number of cycles spent in the nucleating stage will depend on the smoothness and the lack of porosity of the surface. In fact when the cement will set in service, it should be expected that its surface will not be smooth, and porosity and voids at the metal–cement and the bone–cement interfaces should be expected. This means that surface flaws exist and will be responsible for a fast crack nucleation. Therefore, the fatigue behavior of bone cement in service may be dominated by the crack propagation behaviour [37,92,93] as in most engineering structural components. The information available related to fatigue crack propagation in bone cement is not extensive, being the first references related to dental acrylic resins [94]. Fatigue crack propagation in bone cements behaves according to the Paris law: $da/dN = A(\Delta K)^m$, where da/dN is the crack propagation rate in one cycle; ΔK is the amplitude of the stress intensity factor, which depends on the load applied, the crack length and the geometry of the crack; and A and m are constants that depend on the material. When this equation is represented in log-log axes, m is the slope of a straight line and its physical meaning would be related to the rate of increase of the crack velocity as the crack increases its length. This means that for materials with a large m the crack accelerates quickly. The m values obtained for orthopedic bone cements [37,92,93] range between 6.5 and 11.8. On the other hand, the threshold stress intensity factor ranges between 0.3 and 0.5 MPa·m$^{1/2}$ for which the initial propagation rates of the fatigue cracks range between 10^{-8} and 10^{-7} m·cycle^{-1}. The tests were conducted under different conditions, since a maximum load of 156 N and a ratio $R = 0.06$ (minimum load/maximum load) were used in Ref. 92, and a maximum load and a ratio $R = 0.5$ were used in Refs. 37 and 93. For the dental resins, m ranges between 8.8 and 15.6 [95], although the lack of information about the range of loads applied makes it difficult to perform a proper interpretation of the results. Compact tension (CT) [92,93] and double cantilever specimens (DC) [95] have been used; the cements tested were Zimmer R and Zimmer LVC [92], Rostal [93], and six commercial dental resins: Occlusin, P30, Adaptic 11, Ful-Fil, Isomolar, and LC33, plus four experimental ones [95].

The fractographic analysis of acrylic bone cement explanted from patients at their revision operation has shown that fatigue crack growth is most likely the leading *in vivo* failure mechanism [83,84].The fracture surfaces of the explanted specimens compare very well with those obtained from *in vitro* fatigue tests. Such results, together with the crack propagation models proposed [37,83,84], give a very good insight about the way in which the bone cement microstructure should be modified in order to improve the fatigue behavior.

IV. FACTORS AFFECTING THE MICROSTRUCTURE AND THE MICROSTRUCTURE–MECHANICAL PROPERTIES RELATIONSHIPS

The microstructure of acrylic bone cements has been deeply investigated and interpreted [96–98] treating the cements as two-phase materials consisting of PMMA beads and polymerized monomer [97]. A thorough classification of the surface aspects of bone cement in terms of the effects produced by the surrounding tissues, mechanical loads and the changes in volume has been proposed [98]. In the microstructure of bone cements other phases such as BaSO$_4$ and pores should be considered. Moreover, there are factors related with the mixing technique; the presence of blood, grease, or body fluids; the laminations produced when introducing the cement into the bone cavity, the interfaces with the old cement, with the metallic prosthesis, or with the cancellous bone; the thickness of the cement; and the water sorption, which will affect the bulk or the interfacial microstructure of the cements and consequently their mechanical behavior.

A. Handling

When considering the handling of the bone cement, probably the mixing stage is the most important since it is directly related with the final porosity of the cement. The porosity is analyzed in a later section, together with the techniques used to reduce it: centrifuging and vacuum mixing.

An extensive study investigates the influence of the mixing technique on some properties of acrylic bone cements and discusses the role of variables such as the mixing vessel, the order in which the components are mixed together, the mixing time, the rate of mixing, the application of pressure to the mixed bone cement, the kneading, the cement thickness, the pouring into the syringe, the contact force during polymerization, and the quantity of cement prepared [99]. After improving the mixing technique, the authors claim even better results than by centrifuging the cement. The aim of the discussion on the mixing and handling techniques for the cement is to find ways to increase density and reduce porosity. This is the reason why various authors have analyzed the chilling of the monomer, in order to reduce its viscosity; the setting of environmental temperature; or the pressure applied when introducing the cement in the bone cavity [29,58,79,99,100–105]. The application of pressure when cement in the dough state is introduced into the bone cavity increases the interdigitation into cancellous bone [101] and the mechanical strength [100, 102], and seems to decrease the thickness of the cement mantle, which reduces the exotherm [29]. The fact that an increasing thickness of the cement reduces the mechanical strength has to be understood not only in terms of the high exotherm reached [79,102] and the increase in molecular weight of the polymerized monomer, but also in terms of fracture mechanics and the population of defects introduced when a high exotherm takes place. A very important aspect related with the thickness of the cement is its shrinkage. It has been shown that shrinkage increases as porosity decreases, density increases, or moulding pressure increases [6,106–108]. It is interesting to see that Charnley [3] recommended aeration of the dough in order to counteract shrinkage and to eliminate, by evaporation, excessive monomer.

B. *In Vivo* Environment

The influence of body temperature in *in vivo* tests and of temperature of 37°C in *in vitro* tests is important, mainly when aging of the cement is considered. Mechanical tests carried out with specimens prepared in the laboratory and aged for different periods of time and with specimens explanted from patients show that the mechanical properties increase during an initial period of time, ranging from a few days up to a few months, and then decrease at a slow rate during years. The flexural strength of Simplex P cement was found to reach a maximum after 2 weeks and then to decrease very slowly during the next 6 weeks that the experiment lasted, for both *in vitro* and *in vivo* specimens [36,109]. A maximum strength in terms of modulus of fracture and compression is reached in 1–2 weeks, while a continuous decrease in strength is observed during 6 months [110]. It is also shown that cement takes longer to cure *in vivo* than *in vitro*, by measuring the number of free radicals in the samples. The compressive strength of Rostal cement increased during the initial 3 months and then decreased slowly during the next 9 months [37]. Cements Simplex RO, CMW, and Palacos reached a maximum compressive strength, after 7 days, of about 90 MPa, which decreased down to about 80 MPa after about 2 years [45,58]. However, for specimens explanted after 7.5 years in service, the decrease in mechanical properties was marginal [45,58]. The interpretation of such result comes from the fact that the implanted cement, in contact with fat where the residual monomer is soluble, loses monomer that would act as a plasticizer. Simplex P cement increased its bending fracture stress from 48.3 to 50.9 MPa after 12 months implanted in rabbits, and this stress decreased down to 46.6 MPa after 26 months implantation [111]. Young's moduli of Simplex P, CMW, Palacos, and Sulfix 6 also decrease initially and reach an approximately constant value ranging from 4 to 4.5 GPa after 3 to 4 weeks in Ringer's solution at 40°C [79]. Specimens machined from explanted cement in service up to 8 years show also a decrease in the compressive strength and the number of fatigue cycles to failure [112].

Fracture mechanics studies show that the crack velocity is slower in water than in air, and that fracture toughness is about 15–20% higher in water than in air [37,74]. Other results show that the work to fracture increases with the time of storage of the cement in liquids such as water, Ringer's solution, and lipids, although in each case the work to fracture after storage at 21°C is higher than after storage at 37°C [70,71]. One of the most enlightening discussions about fracture toughness of bone cements [76], based in a deep understanding of the material microstructure, shows that parameters such as mixing or preparation method, porosity, and storage or aging play simultaneous strong roles (i.e., aging may increase or reduce fracture toughness depending on the mixing method used and the porosity present in the material). In relation with the fatigue behavior, it has been observed that the number of cycles to failure does not change for specimens kept in bovine serum at 37°C up to 2 years [50]. Fatigue tests carried out in air at 24°C and in saline solution at 24°C and 37°C show that a Weibull life increase occurs when testing in saline solution at room temperature in relation to air, while when testing in saline solution at 37°C the Weibull life decreases when compared to testing in air [86]. This result is difficult to compare with experiments of fatigue crack propagation [37,95], where the liquid seems to reduce the fatigue crack propagation rate. The different behavior should be related with the fatigue crack nucleation stage.

Part of the present discussion should be understood in terms of the fluid sorption ability of the bone cement. The water diffusion coefficient and the equilibrium water absorption concentration for different dental acrylics have been evaluated [113]. Water in bone cement acts as a plasticizer [5], and its effects on mechanical properties have just been described. Taking into account the ability of water to reduce the crack propagation rate and to increase the ductility of the bone cement, the improvement of bone cements properties by using hydrophilic 2-hydroxyethylmethacrylate has been suggested [114] and investigated [115].

The presence and the mixture of body fluids, mainly blood, with the bone cement, has been investigated and their effects seem to be detrimental for the mechanical properties of the material. Reductions of 77% and 69% in tensile and shear strength, respectively, have been reported [116]. Cements deliberately prepared to contain blood showed a reduction in the compressive strength between 8% and 16% [58]. However, the same authors show that specimens prepared with 2 ml blood in the mixing bowl did not show a reduction in the compressive strength [44]. Such results are explained in terms of the laminations caused by the blood–cement interface. The addition of up to 3 ml of blood reduces the elastic modulus of Simplex P, Palacos, and CMW bone cements by up to 21% and reduces the flexural strength of the latter [49].

C. Porosity

The presence of porosity in bone cements has always been detected [3,34], with a volume fraction usually ranging between 2% and 8% [35], although values up to 14.5% have also been reported [20]. The causes of porosity may be described as the space filled by air in the initial powder, which has not been completely filled by polymerizing monomer; as the air entrapped between lumps of the mixture of particles and monomer before the mixture becomes homogeneous in the dough stage; as the air introduced during spatulation of the dough mixture; and finally, as bubbles of boiling or evaporated monomer during setting, when maximum temperature is reached. The evaluation of porosity is usually performed by measuring the density of the bone cement and relating it to that of industrial PMMA, although stereological and image analysis techniques have also been used. Porosity is related to the shrinkage on setting of the bone cement, but it is the influence of porosity on the mechanical properties that has been more deeply investigated. The discussion about the real and significant influence of porosity on the mechanical properties goes on, although it is clear for everyone that it is always convenient to reduce the porosity of bone cements. Consequently, the improvement of the mixing techniques has undergone a fast and spectacular evolution.

The techniques used to reduce porosity include the improvement of hand mixing [76,99], mechanical or ultrasound mixing [104], pressurization of the cement [100,117], centrifugation of the mixture [53,76,88,89,117–123], and vacuum mixing [103,117,123–132] . All these techniques result in a reduction of porosity from the approximate 8% that is achieved by conventional hand mixing to values below 1% for vacuum mixing. Good reviews and discussions about the results on the fight against porosity have been published [76,133]. It should be kept in mind that the possibility of reducing the porosity, or increasing the density, of cements by curing them under pressure had already been considered before 1980 [29,34,62,72–74,82,134].

Most of the reported results show that mechanical properties improve as porosity is decreased. However, certain authors find that the new mixing techniques that reduce porosity do not improve the mechanical properties of every bone cement brand [123], while others find that certain mechanical properties do not improve at all, mainly fracture toughness and the fatigue crack propagation rate [37,76,119,129,135]. The main argument used to explain the detrimental effect of porosity is that pores act as stress concentrators. However, it has to be borne in mind that under a tensile state of stress, polymers have a maximum inherent flaw size that will control the fracture load of the material. On the other hand, the main deformation mechanism of PMMA in tension is crazing and, therefore, the crack will propagate with a Dugdale plastic region ahead of it, which will account for the crazed region. This model of crack propagation in bone cements [37,84] would explain the finding that, more important than porosity is the size of the largest pore in the plane of the advancing crack. A pore will blunt or will not blunt the tip of the crack, depending on the Dugdale zone size and the size of the pore.

Pores have been described as fatigue crack initiators [87], but more research is still required in order to clarify the mechanism [136]. In any case, it is probably the control of the fatigue crack propagation rate that is more relevant in order to delay the fatigue failure of bone cements [37,92,93].

D. Interfaces

The mechanical failure of the cement mantle as a main cause of loosening of joint prosthesis is well accepted, and the failure mechanism is fatigue [84,122,137–139]. The problem of aseptic loosening of cemented joint prostheses has been focused either on the failure of bone cements due to their properties, including mechanical, or on the cementing techniques [140–142]. However, a difficult mechanical problem like the failure of the fixation of a cemented joint implant, where different materials with different mechanical properties work together and transmit loads through two interfaces, cannot be explained by just one cause. This is why an extensive research effort has been developed during these last few years in order to understand all the parameters that take part in the process of failure. Leaving aside the need to improve the bulk mechanical properties of bone cements, the understanding and the improvement of the mechanical behavior of the interfaces is paramount. It is now accepted that the fatigue failure of the cement mantle takes place by cracks emanating from both the cement–prosthesis and the cement–bone interfaces [122,139,143,144].

1. The Prosthesis–Cement Interface

The interfacial adhesion of acrylic bone cement to a metallic surface has been measured in different conditions [37,145-151]. The interfacial strength seems to depend slightly on the metal used when all the conditions of surface preparation and interface formation are the same, giving 316 L stainless steel the highest strength and Ti–6Al–4V the lowest, with Co–Cr–Mo in between. For a surface roughness ranging from 1 μm to that corresponding to grit-blasting, maximum torsional interfacial strength is observed for the coarsest and for the finest surface finish [149], while for a surface roughness ranging from polished and passivated to 15 μm, the tensile interfacial strength is maximum for the polished surface [146]. These results seem to indicate that mechanical interlocking predominates for rough surfaces and that atomic interactions dominate for smooth surfaces [149]. When the cement is molded around the metallic rods in a pre-dough stage [146] or when high pressure is employed during setting [149], the interfacial strength increases. Tests carried out on a hip stem show that the most efficient transmission of load through the interface is achieved when a structured pattern of 50 μm roughness has been produced on the anterior and posterior faces [152]. A thorough stress analysis leading to the conclusion that the failure crack at the interface opens in a combination of modes I and II makes it possible to evaluate the fracture toughness of the interface and to state that the failure of the implant–bone cement interface is highly likely [147]. The fatigue endurance of the interface ranges between 1 and 3 MPa at 5×10^6 cycles, which does not compare safely with the 7 MPa of the bulk bone cement [147].

When the bonding strength is evaluated numerically by means of the finite element method [153,154], the results show that the interfacial stresses come to be of the same order or even higher than those obtained experimentally, and therefore the hypothesis of debonding at the prosthesis–cement interface becomes plausible [145,153]. This does not mean that debonding takes place all along the metallic surface but just in some places where peak stresses occur. This suggests that the local failure of the interface may take place shortly after implantation [153]. In fact acoustic emission and ultrasound techniques are able to determine when, where, and to what extent debonding of the cement–metal interface has occurred [155]. Therefore, loads will not be transferred effectively and the distribution of stresses on the adjacent bonded areas will increase as well as that on the cement itself [154]. In these circumstances the possibility of nucleating fatigue cracks at interfacial defects should increase. In fact, cracks emanating from defects in PMMA near the bone-implant interface [156] are being analyzed, and the porosity at the cement femoral prosthesis is being studied [157-159] and efforts are underway to reduce it [160]. It seems that a great number of pores are created as the cement comes into contact with the prosthesis, and that these pores increase their size as the polymerization exotherm increases [157,158]. A porosity reduction at the metal-cement interface could be achieved by modifying the rheology of the bone cement [160]. Vacuum mixing or centrifuging seem to increase the interfacial strength [161,162].

In order to increase the interfacial strength of the prosthesis–cement interface, the idea of precoating the metallic implant with PMMA was developed and the effect of the coating on the mechanical behaviour of the interface was evaluated [37,148,150,163-168]. Two precoating techniques have been proposed: (1) the moulding of a coating of bone cement [163,164] and (2) the dipping of the metal into a PMMA–MMA solution, which will produce a thin coating, followed by drying and curing at high temperature in order to relieve internal stresses [166]. In all cases the interfacial strength increases even after immersion in saline solution at 37°C for a certain period of time. Fracture toughness and fatigue endurance, unidirectional [166] and torsional [168], improve also, from an average of 2 MPa for the uncoated specimens to over 4 MPa for the coated ones, for the former; and from failure of 15 uncoated specimens below 1×10^6 cycles to the failure of only 4 coated specimens at this same number of cycles, when testing at a torsional loading of ± 2.8 N·m, for the latter. The treatment of the metallic surface with a silane coupling agent increases the interfacial strength [37,166]. When the precoating is applied to a very rough surface (indentations), 7 specimens out of 15 survived 1×10^6 cycles in a push-out fatigue test at 2 MPa, while only 3 specimens of those precoated on a smooth metallic surface and none of the uncoated specimens survived [169]. The specimens that survived were then tested at 3 MPa and the 3 precoated on a smooth surface survived an average of 130 cycles while the 7 precoated on a rough surface survived an average of 212,000 cycles. When precoated stems are implanted in dogs for up to 6 months [165], it is noticed that the fibrous capsule thickness increases from 33.8 μm the first month to 71.8 μm in the 6th month for the precoated stem while the fibrous capsule increases from 52.0 μm the first month to 114.5 μm in the 6th month for the uncoated stem.

It has been shown that the presence of fat or liquids at the metal–cement interface [150,170] and at the PMMA coating–cement interface [171] reduces the interfacial strength. New bone cement seems to bond with old, previously implanted cement or with cement from the coating without producing a weak layer [170–172]. Finally, in pre-formed PMMA devices it seems that sterilization with gamma-irradiation provides the highest bond strength when compared to ethylene oxide and to steam [171].

2. The Bone–Cement Interface

The important role that the bone–cement interface plays in the life in service of the cement and the prosthesis is well known and interpretations about the way in which this interface may fail [173] and how cracks may nucleate and propagate, either in the cancellous bone or in the cement, have been put forward [174,175]. The failure of the interface may be explained by the geometry of the bone cement intrusions into the cancellous bone pores, which may lead either to the fracture of the small cement intrusions or to the fracture of the small bone trabecullar ligaments [173]. This model would fit with the results that show paths followed by the crack in fracture toughness evaluation: (1) through the interface, (2) through the penetrated cement, (3) through the penetrated cement–nonpenetrated bone interface, and (4) through the bulk nonpenetrated bone [175]. In all cases fracture toughness did not show statistical differences and ranged between 0.67 and 0.87 MPa·m$^{1/2}$. An erratic path of the crack breaking PMMA pedicles and trabeculae, and pulling out PMMA pedicles has also been observed [174]. Simultaneous with this failure process, the abrasion of the cement surface may be taking place, producing cement debris and triggering a foreign-body biological reaction [173].

In order to achieve a good fixation and a good interfacial strength, the penetration of the cement into the trabecular bone is a major parameter to control. This parameter has been studied in detail [101,176–180]. When the penetration of bone cement into cancellous bone is analyzed, parameters such as bone cement viscosity, intrusion pressure applied, time and rate of application of the pressure, friction with trabecular bone, and back pressure of the compressed intertrabecular contents have to be taken into account [101]. The penetration of the cement, measured as a ratio of areas of cancellous bone, reamed hole, and cement, depends through a nonlinear function of the applied pressure. For safety reasons related with cardiopulmonary functional impairment, a maximum pressure of 0.52 MPa that produces a 65% penetration is recommended. When penetration is evaluated taking into account the variation of viscosity of the cement with working time and a pressure of 0.035 MPa is applied, values ranging from 2.72 mm for CMW cement up to 5.40 mm for Simplex P cement are obtained [176], while Zimmer, Palacos, and Sulfix 6 show values of 5.01, 3.01, and 3.00 mm, respectively. Different penetrations have been observed for the different segments along the femur, depending also on the cement brand [177]. This is why, for a given segment, depending on the brand and the penetration, an interfacial shear strength ranging from 2 to about 10 MPa can be obtained. In order to achieve good penetration, the preparation of the surface of the cancellous bone is very important. When the effects of nine different techniques are compared and penetration and shear strength are measured, it is noticed that pressurized lavage is effective, either on its own or combined with brushing [179]. Variables such as the jet, continuous or pulsed, and the temperature, 21° or 37°C, do not seem to play any role. The combination of cleaning thoroughly strong trabecular bone and using low-viscosity cement has also been recommended [178]. Intramedullary reaming is an important parameter in the preparation technique of the surface. Maximum shear strength is achieved when intramedullary reaming is minimized [180].

The study of the fatigue behavior of the bone–cement interface has been undertaken [88,181-183]. Centrifugation of the bone cement increases the fatigue life of the interface from 237,969 cycles to 641,056 cycles when tested at 7 MPa or 0.0022 strain in tension–compression [88,182]. A fatigue endurance shear stress of about 0.4 MPa has been reported at 10^6 cycles [183].

The reverse problem of extracting acrylic bone cement from the bone–cement interface is posed in revision surgery. The use of extracorporeal shockwaves to loosen the interface has been proposed. Contradictory results have been reported, showing a decrease between 30% and 70% of the shear strength [184] and no disintegration of the bone–PMMA interface [185], but bone marrow particles which could cause fat embolism were liberated.

E. Additives

The main additives usually found in bone cements are radiopacifier particles such as $BaSO_4$ and ZrO_2, and antibiotics. Their effect can be considered as detrimental for the mechanical properties of bone cements, although it can be concluded that the decrease in mechanical properties that they produce is not the cause of the mechanical failure of bone cements.

The effect of radiopacifiers on bone cements was taken into account very early [6,35,49,50,58,72,82,186,187]. It is usually accepted that $BaSO_4$, in amounts of about 10% in weight, reduces the compressive strength about 5% [58], the compressive and flexural moduli and strength up to 20% [50], and the tensile strength about 10% [6,35]. The addition of 2.5 g of filler increases the elastic modulus of CMW cement about 10%, but the addition of 5 g leads to a reduction of about the same amount, while the flexural strength is reduced another 10% [49]. $BaSO_4$ reduces more the tensile strength than ZrO_2

in the same concentration, and this effect is explained by the agglomeration of ZrO_2 particles in cauliflower-like inclusions in the bone cement while $BaSO_4$ forms a disseminated network [187]. The presence of this filler tends to increase the density of the cement, depending on the curing pressure, but its effect on porosity is not significant, and it plays a minor role in shrinkage [6]. The fracture toughness of Simplex P and Zimmer bone cements is not affected by the addition of 10% $BaSO_4$ when tested in air, but it increases between 10% and 20% when tested in bovine serum [72]. Zimmer cement shows a better fatigue life when tested without $BaSO_4$, but it does not have a great effect in Simplex P cement [82]. Other reported results show that fracture toughness decreases with increasing filler content in such a way that the crack propagation rate may increase by an order of magnitude [77,186]. The filler particles behave like voids for crack nucleation and may grow for crack propagation [77]. When the fracture surface of bone cement containing a radiopacifier is observed at high magnification in the scanning electron microscope, it can be noticed that there is no adhesion between the polymerized PMMA and the filler particles [37]. This suggests that the filler particles will behave like pores when a tensional state of stress is applied to the cement. In fact, the chemical coupling of the radiopacifier to the PMMA increases the tensile strength of the cement back to the values of the cement without any radiopacifier filler [188].

The need to tackle the problem of postoperative infections led to the idea to mix antibiotics with the bone cement and allow them to leach out into the surrounding tissues. A major research effort was devoted to the study of the effect of antibiotic addition on the mechanical properties of bone cements, and a very good review was published [189]. It seems clear that the addition of antibiotics to bone cements reduces their mechanical properties, although the reduction depends very strongly on the amount of antibiotic added. If the amount is kept below 1 g in 40 g of powder, only 4% decrease in the compressive strength can be expected [45,58], while 2 g of different antibiotics do not affect the compressive and the diametral tensile strengths [190]. As the amount of antibiotic increases up to 5 g, the compressive strength may decrease from 20% to 25%, and large doses may lead to overall cement failure [191]. The addition of cefuroxine reduces the tensile strength measured in three-point bending at a rate of 5.1–6.5% for each additional 1% of antibiotic in the mix for CMW, Palacos, and Simplex bone cements [193]. The way in which the cement is mixed plays a role when an antibiotic is present. The addition of antibiotics, such as vancomycin and tobramycin in amounts of either 0.5 or 1 g, to vacuum-mixed bone cement reduces its bending strength and modulus when compared to air-mixed cement, while the addition of antibiotic to air-mixed cement does not change its strength and modulus [128]. The fatigue properties of Palacos R cement are not altered by the addition of 0.5 g of gentamicin while the addition of 0.5 g of erythromycin and 0.24 g of colistin do not decrease the fatigue life of Simplex P. Centrifugation does not improve the fatigue lifes of Palacos R and Palacos R with gentamicin, while the fatigue properties of Simplex P and AKZ (Simplex P with antibiotics) cements are improved [89]. The fatigue life of Simplex P, centrifuged or uncentrifuged and tested in tension–compression at ± 15 MPa and 2 Hz, is not significantly reduced by adding 1.2 g of tobramycin, although the centrifuged cements, with and without antibiotic, increased their fatigue life by a factor of 8 [194].

The addition of antibiotics as liquids has a strong deleterious effect on the mechanical properties of bone cements [190,192].

V. BIOLOGICAL PROPERTIES

It has been reported that one of the hazards in the use of acrylic bone cements is the liberation of the residual monomer to the adjacent tissues. Monomer is lost by evaporation during the mixing stage depending on the beating frequency, and a maximum loss of about 14% has been reported [44] at a frequency of 260 beats per minute. During the molding or working stage, monomer is lost at a lower rate, and during setting, evaporation loss of monomer also takes place at a high rate [8,44]. This means that the exposure of the patient to monomer will depend very strongly on the time of placement of the cement and the frequency and number of beats during the mixing stage. In an early placement of the cement, the patient will be exposed to the monomer released during molding and setting, while a late placement will limit the exposure to the setting loss. For a short period of mixing, about 100 stirs, an early placement would mean an exposure to 3.5% of monomer and a late placement would reduce it to 2.2%. For a long mixing period of 400 stirs, the amounts would be 1.8% and 1.2% respectively [8]. Values up to 21% monomer in the dough mass after 4 min mixing have also been reported [195].

The concentration of residual monomer in the polymerized cement varies from 3.25% after 30 min storage down to 2.42% after 215 days [6,17]. The amount of monomer leached into an aqueous medium has been evaluated in terms of the area and thickness of the cement, from the powder-to-liquid ratio, and from the onset of mixing to the time of immersion in aqueous medium [17,196]. A minimum in the monomer loss is obtained when the powder-to-liquid ratio is about 2:1 [5,196]. It is also observed that the monomer is leached out at a fast rate during the initial 15 min and from then on the process slows down significantly [17,196]. In experimental bone cements it has been shown that *n*-butylmethacrylate is much less extractable in saline solution than methylmethacrylate [197], while higher levels of ethylmethacrylate monomer are leached out in saline solution [32]. The exudation of monomer from cyclically loaded low-modulus cement, containing polybutylmethacrylate beads, is not higher than for conventional PMMA bone cement [198]. Tests carried out with human bone marrow showed that the highest content of monomer was always found in the fat and the lowest in the red blood cells, while intermediate amounts were found in bone marrow fibers and cells, for the five different cements studied (i.e., CMW, Simplex, Palacos E, Sulfix 9, and Palacos R), and for different times and mechanical consistencies of the cements [199].

Certain clinical complications have been associated to the use of bone cement and particularly to the monomer release. A very good review on the peroperative complications and during anaesthesia was published in 1987 [200]. A great deal of information regarding hypotension, hazards, local tissue damage, and even the burn of the sciatic nerve caused by bone cements has been reported by different authors, some of which can be found in Refs. 201–204. Recent studies show that intramedullary pressure may be responsible for hypotension, instead of monomer content in blood [206], that the local presence of MMA does not alter blood cholesterol and triglyceride levels [207], and that MMA is not responsible for a direct toxic effect in clinical practice [11].

From infrared spectroscopy studies it has been claimed that some kind of chemical bond is formed between bone cement and bone powder *in vitro* [208], and that a certain reaction also takes place with bone marrow *in vitro* [209]. However, it has been clearly demonstrated from histological studies that an interface membrane is always present in animal models and in the aseptically loosened cemented joint implants. The pseudosynovial membrane contains macrophages, giant cells, and frequent granuloma formation.

Only a few references are reported from a very abundant literature [140,173,210–218]. Some of the interpretations of the findings reported show that the products of wear and fracture or fragmentation of bone cement, as well as wear debris of polyethylene from the articulating surface, may cause a foreign-body response consisting of a macrophage, giant cell foreign-body granulomatous reaction. This tissue can produce a variety of chemical mediators of inflammation and eventually bone resorption. This process of osteolysis induced by bone cement fragmentation may be the biological cause for the loosening of the cemented joint prostheses [139,140,173,212,213,218–222]. The macrophage exposure to PMMA debris and the release of inflammatory mediators such as prostaglandins and collagenase are related with bone resorption [139,213,218–226]. After rabbit implantations, particulate PMMA provoked a greater histiocytic and giant cell response than bulk PMMA [227]. Lymphocyte and immunological response may also be responsible for some aseptic loosening [228]. It has also been reported that the *in vitro* effects of polymethylmethacrylate are to enhance fibroblast proliferative capacity without involving prostaglandin E_2, [229].

In vitro tests have shown that cement particles inhibit bone ingrowth into titanium chambers [230], although the need to perform the tests with retrieved bone cement particles instead of fabricated ones has been pointed out, due to the effect of size and shape on the living tissues [231]. Radiopacifier materials, such as $BaSO_4$ and ZrO_2, have been observed in the tissues around prosthetic joints [232], and a greater release of inflammatory mediators and bone loss has been reported in cement debris containing $BaSO_4$ than in particles of plain PMMA cement [233].

The addition of antibiotics to bone cement is a well-known practice and the characteristics of elution or release *in vitro* and *in vivo* have been studied [234,235]; and methods for improving drug release from the bone cement, such as the increase of the powder-to-liquid ratio and the use a crystalline formulation of the drug [236], have been proposed. Recent studies show that bacteria adhere to PMMA and although there is inhibition of bacterial adhesion to tobramycin-impregnated bone cement, the resistance of bacteria after adherence to PMMA surfaces increases [237–239].

VI. MODIFICATION OF ACRYLIC BONE CEMENTS

The need to assure a reasonable life in service of joint prostheses means that premature loosening should be avoided. Consequently, fixation methods alternative to bone cement were devised. However, while cementless fixation has not been able to replace bone cement, an extensive research effort has been devoted to improve the properties of this material. At present a wide variety of possible modifications of conventional bone cements are being investigated. Some of them have been described already since they are related to the improvement of the preparation techniques such as porosity reduction, surgical techniques such as reaming and cleaning the bone cavity or pressurization, and the improvement of the interfaces, mainly the metal–cement interface.

The rheological properties of the cement are particularly important in improving interdigitation in the bone–cement interface [240]. However, the poor fatigue behavior of these cements [121] and the lack of evidence of a lower rate of loosening when using a low-viscosity cement after a 5-year follow-up [214] seem to show that this may be not a proper improvement. In fact, a pressurized high-viscosity bone cement, prepared with a powder-to-liquid ratio of 2.7:1, provides less penetration but similar interfacial strength, reduction of 1 min in the dough time and 1.5 min in the setting time, and a decrease in the peak temperature down to 72°C in relation to the conventional powder-to-liquid ratio 2:1 [242].

The idea of achieving good tissue ingrowth into the bone cement has led to the development of porous acrylic cements [243,244]. In both cases a filler, sucrose or tricalcium phosphate [243], and an aqueous gel based on carboxymethyl cellulose [244] have been dispersed through the dough of a conventional bone cement. Convenient volume fractions of filler are required in order to have either interconnected particles or an interconnected filament network that will biodegrade *in vivo* and will allow tissue ingrowth into the cement. The peak temperature decreases down to 70°C for a 30% volume fraction of second phase for the former and down to 60°C for a 50% second phase for the latter. Porosities of 40% and 55% are obtained in either case. The diametral tensile strength lies between 7 and 15 MPa [243], and the tensile strength and the elastic modulus are kept down to 3 MPa and 0.3 GPa, respectivley, although it is claimed that the ingrowth of hard tissue results in a composite with mechanical properties comparable to those of spongy bone [244].

The main path followed to improve the mechanical properties of conventional bone cements has been the reinforcement of the matrix either with particles or with fibers. The use of bioactive particles is being considered since they could improve both the mechanical properties of the bulk cement and the interfacial bone–cement strength. Alternative chemical modifications of the bone cement matrix could also improve other physical, chemical, and biological properties.

A. Matrix Modifications

Alternative monomers, polymers, initiators, and accelerators have been proposed, and the results reported show that they can change the properties of conventional bone cements. The substitution of the inhibitor hydroquinone by the more biocompatible di-*tert*-butyl-*p*-cresol lowers the setting time by 3 min but raises the peak temperature from 71° to 75°C, and when instead of the accelerator *N,N*-dimethyl-*p*-toluidine, *p*-(dimethylamino)phenethanol is used, the setting time is further reduced by 1 min down to 6.8 min, the exotherm is kept at 75°C, and the compressive strength is raised from 73.5 to 78 MPa [18]. An alternative, less toxic accelerator, the acryloyl-*N*-phenylpiperazine gives a similar compressive strength to the cement and it also gives a better aging behaviour than *N,N*-dimethyl-*p*-toluidine [245]. Partial substitution of the monomer with high molecular weight methacrylates such as dicyclopentenyloxyethyl methacrylate with a low concentration (up to 2%) of pentaerythritol tetra-(3-mercaptopropionate) as a chain transfer agent, increases the setting time to 9.3 min, reduces the exotherm down to 62°C, and keeps the compressive strength around 78 MPa. The addition of cross-linking agents such as ethylene dimethacrylate or hexamethylene dimethacrylate, does not affect strongly the setting time nor the peak temperature, but the compressive strength is slightly reduced [18]. When different cross-linking agents such as dimethacrylates of ethylene glycol, 1,3-propanediol, 1,4-butanediol, diethylene-glycol, and triethyleneglycol are added to a mixture of 30 wt% methacrylate monomer and 70 wt% PMMA powder, the glass transition temperature increases [57] and the creep behavior is not affected at low stresses, while at high stresses the strain and the time to fracture increase [63]. In an experimental bone cement where the monomer was 2.2-*bis*-[4(2-hydroxy-3-methacryloylpropoxyphenyl)]-propane with triethylene glycol dimethacrylate as cross-linking agent, and instead of *N,N*-dimethyl-*p*-toluidine, a barbituric acid derivative was used, a working time of more than 5 min was obtained, together with mechanical properties similar to those of conventional bone cements, but with a peak temperature of only 33.5°C [54]. Achievement of a reduced modulus bone cement has been proposed by using polybutylmethacrylate beads instead of PMMA, which gives an elastic modulus of only 0.27 GPa [246]. This material acts as a compliant layer between the implant and the bone as proposed by certain authors [247]. Conventional bone cement has been modified by substituting methylmethacrylate monomer by different amounts of hydroxyethylmethacrylate in order to increase water sorbtion [115,248]. The tensile strength decreases from 32 MPa down to 14 MPa, while the strain to fracture increases from 2% to 3% [115]. Using monomers such as ethylmethacrylate and butylmethacrylate, the creep strain increases while the fracture toughness decreases [249]. Cements elaborated with polyethylmethacrylate powder and N-butylmethacrylate monomer implanted in dogs have been shown to be biologically suitable for implantation into bone [250]. The addition of methylene blue to the cement has been proposed in order to aid cement removal during revision arthroplasty. There are no significant differences in the fatigue life of centrifuged Simplex P cement with and without methylene blue [251]. When human growth hormone is loaded into the bone cement, an early response of osteoid cells to growth hormone takes place, providing an increased bone–cement interfacial strength and the stabilization of the implant is improved [252].

B. Particle Reinforced Bone Cements

The improvement of the mechanical properties of bone cement by particle addition has two main aspects: first, the reinforcement with hard particles such as glass beads [186,253], bioactive glass ceramic particles [254], inorganic bone and demineralized bone matrix [255–257] and hydroxyapatite (HA) [258–264], and second, the reinforcement with tough or rubber-toughened particles [32,37,55,93,265,266].

When glass beads of 75 μm diameter are added to conventional bone cement, the Young's modulus and the fracture toughness increase with volume fraction up to approximately 5 GPa and 2.5 MPa·m$^{1/2}$, respectively, for 25% volume fraction of the second phase [186]. Glass beads of 40 μm diameter in a proportion of 60% in weight decrease the peak temperature down to 68°C, and improve the dimensional stability and the compressive modulus by 70%, up to 3.5 GPa, without decreasing the compressive strength [253]. Animal experiments conducted with a mixture of ground Ceravital bioactive glass ceramic with Palacos cement in the range of 50–70% by weight, and with particle sizes ranging between 50 and 125 μm, showed that the glass ceramic particles at the cement surface bonded durably to newly formed osseous tissue [254]. The bone cement impregnated with demineralized bone matrix showed better results than the cement impregnated with inorganic bone. Bone regrowth took place at the surface of plugs of the former, implanted in canine femora. The interfacial shear strengths were 2.02, 6.10 and 8.52 times higher than the control plugs made of conventional bone cement, after 3, 8, and 11 weeks of implantation [257].

The reinforcement with HA particles up to 25% in weight, gives a maximum flexural strength of about 68 MPa and a corresponding Young's modulus of 3.5 GPa for 2.6% of filler, creep being significantly reduced [258]. The peak temperature decreases between 10% and 15%. The presence of the second phase has a strong influence on the pore distribution and size, and in fact, the HA crystals become a pore nucleation factor [259]. The cause could be the lack of chemical bonding between the crystals and the polymerized monomer, since when HA particles are treated with a silane bonding agent, the mechanical properties of an HA-reinforced compliant polyethylmethacrylate cement increase significantly [262]. Compressive strength and modulus tend to increase when mixtures of 10% and 20% HA granules of three different sizes with different conventional bone cements are tested. When bone cement with 2 or 3 layers of HA particles on its surface is tested in tension, the tensile strength and modulus tend to decrease with increasing particle size [263]. The interposition of 1 or 2 layers of HA particles of 100–300 μm diameter between the bone and the bone cement provides a bioactive interface to the bone cement that increases substantially the interfacial strength measured with push-out tests [260,261]. The chemical linkage of apatitic-octacalcic phosphate and hydroxyethylmethacrylate has been achieved and this grafted apatite can then be bonded to PMMA [264]. This material may have a great potential, because both bioactivity and mechanical property improvement should be expected.

An alternative philosophy on the increase of the modulus and the strength of the cement is to increase the fracture toughness and to decrease the fatigue crack propagation rate. When tough particles of ABS (acrylonitrile–butadiene–styrene) copolymer are mixed with the bone cement in volume fractions up to 20%, the tensile modulus decreases fom 2.64 GPa down to 1.26 GPa, the tensile strength from 43.8 MPa down to 28.8 MPa, and the compressive strength from 87.9 MPa down to 49.9 MPa. However, the fracture toughness increases from 1.39 up to 2.24 MPa·m$^{1/2}$, and for a 10% ABS roughened bone cement, the threshold for fatigue crack propagation increases from 0.3 up to 0.5 MPa·m$^{1/2}$, while the fatigue crack propagation rate is 100 times slower than for vacuum-mixed conventional bone cement [37,55,93]. The latter formulation shows a peak temperature of 70°C, a setting time of 10 min, and a compressive strength of 78.5 MPa. When industrial rubbertoughened PMMA particles containing 40% in volume of rubber are mixed in different proportions with a conventional bone cement, a peak temperature of 64.5°C, a setting time up to 12 min, and a maximum fracture toughness of 2.46 MPa·m$^{1/2}$ are reached [265]. Similar results have been obtained using a different type of industrial rubber-toughened PMMA particles [266]. Latex emulsion polymerization of polybutylacrylate and PMMA has been used to prepare rubber-toughened cements and a fracture toughness of 2.15 MPa·m$^{1/2}$ has been obtained [32].

C. Fiber-Reinforced Bone Cements

The fiber reinforcement of bone cements has followed two main paths: the reinforcement with metallic fibers [267–270] and the reinforcement with polymeric fibers, including carbon fibers [48,51,59,92,271–275]. The reinforcement with 316 stainless steel short fibers (0.5–1 mm) shows that the compressive strength increases from 73 to 95 MPa when the percentage in weight of fibers increases up to 30%. The mechanical properties, including flexural and impact strength are dramatically increased when 6 mm long fibers are used, and a decrease in peak temperature is reached in both cases [267]. When long stainless steel wires of 0.5 and 1 mm diameter are used in different numbers to reinforce the bone cement, the strength to failure measured by three-point bending increases by about 15% with the number of wires used, when the cross-sectional area occupied by the wires reaches 1% [268]. After cement fracture, the wires are still able to sustain load.

Probably more important than the strength characteristics of the composite materials is their fracture toughness and their fatigue behavior. Bone cement has been reinforced with short, tough titanium fibers [269,270] of different diameters and different lengths.

As the fiber content increases, the fracture toughness increases up to 2.372 MPa·m$^{1/2}$. The rough and irregular surface of the fibers shows mechanical interlock with the matrix, which is confirmed by the ductile failure of the fibers [269]. The fatigue life of these materials is enhanced, in both notched and unnotched specimens, as the porosity reduction of the PMMA causes a further improvement [270].

One of the main results that emerges from the results of the different authors who have investigated the fracture surface of the carbon fiber reinforced bone cements, is that there is no adhesion or linkage between the PMMA matrix and the carbon fibers [48,51,92,271]. This explains the moderate improvements reported in fracture toughness. When 1% in weight of chopped graphite fibers, 6 mm in length and 8 μm in diameter, are mixed with Zimmer bone cement, and the material is tested in compression at different strain rates ranging from about 10^{-3} s^{-1} up to 10 s^{-1}, the compressive strength and the elastic modulus increase from 80 to 130 MPa and from 1.7 to 2.7 GPa [59]. In relation to a conventional bone cement, a cement reinforced with 1.33% in weight carbon fibers shows increases in tensile strength and Young's modulus of 30% and 35.8%, respectively; improvements in compressive strength and elastic modulus of 10.7%; increases of 29.5% and 18.5% in bending strength and shear strength, respectively; a decrease in peak temperature of about 13.2%; and finally an increase in the stress relaxation from 11.9% to 13.06% after 12 min at an initial strain of 1% [272]. The reinforcement with 2% in volume chopped fibers, 6 mm long and 7 μm in diameter, increases the tensile strength and the Young's modulus from 24 MPa and 2.76 GPa up to 38 MPa and 5.52 GPa, respectively, the impact resistance for notched specimens from 0.013 J to 0.070 J, and the fatigue endurance at 10^6 cycles from about 10 MPa to more than 15 MPa, while other physical properties such as setting time and peak temperature remain practically unchanged [48]. When 2% in volume carbon fibers of 1.5 mm length reinforce Zimmer and Zimmer LVC bone cements, the compressive strength is not significantly affected, but the fracture toughness increases from 1.42 to 1.88 MPa·m$^{1/2}$ and from 1.23 to 1.61 MPa·m$^{1/2}$, respectively [51]. Fractographic analysis shows no fractured fibers, a lack of intimate contact between the matrix and the fibers, and their pull-out. Similar fractographic results are reported for fatigue-fractured surfaces [92,271]. PMMA cement reinforced with 2% in volume carbon fibers of 1.5 mm length exhibits a fatigue crack propagation rate an order of magnitude lower than for conventional bone cements. It is also shown that fatigue crack propagation is controlled by strain and not by stress, in agreement with the conclusion that fatigue life in bone cements is strain controlled [48], and in agreement with the lower scatter of fatigue results [91].

Alternatively, different kind of polymeric fibers such as aramid [273], polyethylene [274], and PMMA [275] have been used for bone cement reinforcement. The addition of 7% in weight aramid Kevlar-29 fibers increases the tensile strength from 30.8 up to 42.8 MPa, and the fracture toughness from 1.53 up to 2.85 MPa·m$^{1/2}$ [273]. When polyethylene Spectra 900 fibers are mixed in different percentages in weight up to 7%, the mechanical properties obtained are always kept below those yielded by the same proportion of aramid fibres [274]. Like for carbon fibers, poor interfacial bonding with the PMMA matrix is observed, but fracture energy is absorbed by fiber failure. PMMA fibers have been elaborated by melt extrusion followed by a drawing process, and diameters ranging from 0.635 mm down to 25 μm have been obtained [275]. Weight percentages between 20% and 30% of these fibers, with a tensile strength of 225 MPa and a Young's modulus of 7.7 GPa, have been incorporated as short-length fibres (3–5 mm) into a conventional cement, but the fracture toughness has only reached 1.56 MPa·m$^{1/2}$.

VII. SUMMARY AND CONCLUSIONS

Acrylic bone cements have been widely used for about three decades and a very rough estimate would indicate that, probably by the end of the century, the total number of patients with cemented joint implants could be about 10 million, worldwide and for the 40-year period. The current statistics on loosening of implants show that percentages of about 10% in 10 years mean millions of clinical problems with far-reaching consequences, both social and economic. It does not seem that significantly better alternatives to acrylic bone cements are yet available, while a great amount of sound work, both in the clinical and in the engineering areas, has made it possible to gain expertise in the preparation and implantation of acrylic bone cements, and knowledge of their behavior in service. Their chemical structure, their physical properties, and the biological reactions they induce are well described. The optimization of the preparation and the implantation methods seems to yield an improved *in vivo* behavior. Finally, a wide variety of interesting modification ideas meant to improve the mechanical properties have been put forward. However, there is still much research to be done, since it is still necessary to understand properly their mechanical role in terms of the biomechanical design of the prosthesis and the mechanical quality of the bone, and moreover, a deep understanding of their mechanisms of failure is still needed. In fact, a better understanding of the microstructure–mechanical properties of bone cements and a good mechanical model of the material would avoid expensive research only meant to collect data generated in very complicated experiments.

The mechanical properties measured and reported are mainly related to unidirectional tests. Since bone cements are triaxially loaded in service, it would be of great interest to gain a better knowledge about their properties under such complex real situations. Although some good stress analysis has been carried out by means of the finite element method, additional experimental work should be carried out. Moreover, cracks may propagate and be loaded in modes II and III while all the fracture and fatigue studies have been carried out under mode I. In this sense, it is doubtful that cements work in service under plane strain conditions, due to the variable thickness of the cement mantle. A good knowledge of the deformation and failure mechanisms of these two-phase PMMA materials—in terms of the one-phase, or industrial, PMMA—would make it possible to design the optimum microstructure, which would fit the requirements of the mechanical loading conditions in service.

ACKNOWLEDGMENT

The authors are grateful to the Spanish Comisión Interministerial de Ciencia y Tecnología (CICYT) for funding the present research line in bone cements.

REFERENCES

1. Charnley, J., Anchorage of the femoral head prosthesis to the shaft of the femur, *J. Bone Joint Surg., 42B*, 28 (1964).
2. Charnley, J., Bonding of prosthesis to bone by cement, *J. Bone Joint Surg., 46B*, 518 (1964).
3. Charnley, J., *Acrylic Cement in Orthopaedic Surgery*, E.&S. Livingstone, London (1970).
4. Billmeyer, F.W., *Textbook of Polymer Science*, 3rd ed., Wiley, New York (1984).
5. Lautenschlager, I.P., Stupp, S.I., and Keller, J.C., Structure and properties of acrylic bone

cement, in *Functional Behaviour of Orthopaedic Biomaterials*, Vol. II, *Applications* (P. Ducheyne and G.W. Hastings, eds.), CRC Press, Boca Raton, FL, 88 (1984).

6. Haas, S.S., Brauer, G.M., and Dickson, G. A characterization of polymethyl methacrylate bone cement, *Bone Joint Surg., 57-A*(3), 380 (1975).

7. ASTM Standard, Medical Devices, Section 13, 13.01, Philadelphia (1986).

8. Bayne, S.C., Lautenschlager, E.P., Greener, E.H., and Meyer, P.R., Clinical influences on bone cement monomer release, *J. Biomed. Mater. Res., 11*, 859 (1977).

9. Meyer, P.R., Lautenschlager, E.P., and Moore, B.K. On the setting properties of acrylic bone cement, *J. Bone Joint Surg. 55-A*(1), 149 (1973).

10. Smith, R.E., and Turner, R.J. Total hip replacement using methylmethacrylate cement, *Clin. Orthop. Rel. Res., 95*, 231 (1973).

11. Gentil, B., Paugam, C., Wolf, C., Lienhart, A., and Augereau, B. Methylmethacrylate plasma levels during total hip arthroplasty, *Clin. Orthop. Rel. Res., 287*, 112 (1993).

12. Young, R.J., *Introduction to Polymers*, Chapman and Hall, New York (1981)

13. Williams, D.J., *Polymer Science and Engineering*, Prentice-Hall, Englewood, NJ (1971).

14. Turner, R.C., Atkins, P.E., Ackley, M.A., and Park J.B. Molecular and macroscopic properties of PMMA bone cement: Free radical generation and temperature change versus mixing ratio, *J. Biomed. Mater. Res., 15*, 425 (1981).

15. Turner, R.C., White, F.B., and Park, J.B., The effect of initial temperature on free radical decay in PMMA bone cement, *J. Biomed. Mater. Res., 16*, 639 (1982).

16. Turner, R.C. Free radical decay kinetics in PMMA bone cement, *J. Biomed. Mater. Res., 18*, 467 (1984).

17. Brauer, G.M., Termini, D.J., and Dickson, G. Analysis of the ingredients and determination of the residual components of acrylic bone cements, *J. Biomed. Mater. Res., 11*, 577 (1977).

18. Brauer, G.M., Steinberger, D.R., and Stansbury J.W. Dependence of curing time, peak temperature, and mechanical properties on the composition of bone cement, *J. Biomed. Mater. Res., 20*, 839 (1986).

19. Trap, B., Wolff, P., and Jensen, J.S. Acrylic bone cements: Residuals and extractability of methacrylate monomers and aromatic amines, *J. Appl. Biomat., 3*, 51 (1992).

20. Noble, P.C. Selection of acrylic bone cements for use in joint replacement, *Biomaterials, 4*, 94, (1983).

21. Dall, D.M., Miles, A.W., and Juby, G. Accelerated polymerization of acrylic bone cement using preheated implants, *Clin. Orthop. Rel. Res., 211*, 148 (1986).

22. Lehnartz, E., *Chemical Physiologie*, S.87, Springer-Verlag, Berlin (1959).

23. Lundskog, J., Heat and bone tissue, *Scand. J. Plastics Reconstr. Surg.*, Suppl. 9.

24. Moritz, A.R., and Henriques, Jr., F.C., Studies of thermal injury. II, *Am. J. Path., 23*, 695.

25. Homsy, C.A., Tullos, H.S., Anderson, M.S., Diferrante, M.N., and King J.W., Some physiological aspects of prosthesis stabilization with acrylic polymer, *Clin. Orthop., 67*, 121, (1976).

26. Reckling, F.W., and Dillon W.L., The bone–cement interface temperature during total joint replacement, *J. Bone Joint Surg., 59A*, 80 (1977).

27. Toksvig-Larsen, S., Franzen, H., and Ryd, L. Cement interface temperature in hip arthroplasty, *Acta Orthop. Scand., 62*(2), 102 (1991).

28. Zygmunt, S., Larsen, S.T., Saveland, H., Rydhohm, U., and Ryd L. Hyperthermia during occipito-cervical fusion with acrylic cement, *Acta Orthop. Scand., 63*(5), 545 (1992).

29. Sih, G.C., Connelly, G.M., and Berman A.T., The effect of thickness and pressure on the curing of PMMA bone cement for the total hip joint replacement, *J. Biomechanics, 13*, 347 (1980)

30. Wykman, A., Acetabular cement temperature in arthroplasty, *Acta Orthop. Scand., 63*(5), 543 (1992).

31. Lee, H.B., and Turner, D.T. Temperature control of a bone cement by addition of a crystalline monomer, *J. Biomed. Mater. Res., 11*, 671 (1977).

32. Khorasani, S.N., Ph.D. Thesis, University of London (1991).
33. Beech, D.R., Molecular weight distribution of denture base acrylic, *J. Dentistry, 3*(1), 19 (1975).
34. Bayne, S.C., Lautenschlager, E.P., Compere, C.L., and Wildes, R. Degree of polymerization of acrylic bone cement, *J. Biomed. Mater. Res., 9,* 27 (1975).
35. Kusy, R.P. Characterization of self-curing acrylic bone cements, *J. Biomed. Mater. Res., 12B,* 271 (1978).
36. Bargar, W.L., Brown, S.A., Paul, H.A., Voegli, T., Hseih, Y., and Sharkey N., *In vivo* versus *in vitro* polymerization of acrylic bone cement: Effect on material properties, *J. Orthop. Res., 4,* 86 (1986).
37. Vila M.M. Ph.D. Thesis, Universitat Politècnica de Catalunya (1992)
38. Martin, R.J., Cooper, A.R., and Johnson, J.F., Mechanical properties of polymers; influence of molecule weight and molecular weight distribution, *Rev. Macromol. Chem., 9,* 57 (1973).
39. Ward, I.M., *Mechanical Properties of Solid Polymers,* 2nd ed., Wiley & Sons, Bristol (1983).
40. Berry, J., Fracture processes in polymeric materials. V. Dependence of the ultimate properties of polymethylmethacrylate on molecular weight, *J. Polymer Sci. A, 2,* 4069 (1964).
41. Kusy, R.P., and Katz, M.J. Effect of molecular weight on the fracture surface energy of poly(methyl methacrylate) in cleavage, *J. Mater. Sci., 11,* 1475 (1976).
42. Saha, S., and Pal, S. Mechanical properties of bone cement: A review, *J. Biomed. Mater. Res., 18,* 435 (1984).
43. Harrison, A., Huggett, R., and Handley, R.W. A correlation between abrasion resistance and other properties of some acrylic resins used in dentistry, *J. Biomed. Mater. Res., 13,* 23 (1979).
44. Lee, A.J.C., Ling, R.S.M., and Wrighton, J.D. Some properties of polymethylmethacrylate with reference to its use in orthopedic surgery, *Clin. Orthop. Rel. Res., 95,* 281 (1973).
45. Lee, A.J.C., Ling, R.S.M., and Vangala, S.S. The mechanical properties of bone cements, *J. Med. Eng. Technol., 1* (1977).
46. Edwards, R.O., and Thomasz, F.G.V. Evaluation of acrylic bone cements and their performance standards, *J. Biomed. Mater. Res., 15,* 543 (1981).
47. Kusy, R.P., Maim, J.R., and Turner, D.T. Influence of application technique on microstructure and strength of acrylic restorations, *J. Biomed. Mater. Res., 10,* 77 (1976).
48. Pilliar, R.M., Blackwell, R., Macnab, I., and Cameron, H.U., Carbon fiber reinforced bone cement in orthopedic surgery, *J. Biomed. Mater. Res., 10,* 893 (1976).
49. Holm, N.J. The modulus of elasticity and flexural strength of some acrylic bone cements, *Acta Orthop. Scand., 48,* 436 (1977).
50. Jaffe, W.L., Rose, R.M., and Radin, E.L. On the stability of the mechanical properties of self-curing acrylic bone cement, *J. Bone Joint Surg., 56A*(8), 1711 (1974).
51. Robinson, R.P., Wright, T.M., and Burstein, A.H., Mechanical properties of poly(methyl methacrylate) bone cements, *J. Biomed. Mater. Res., 15,* 203 (1981).
52. Sih, G.C., and Berman, A.T., Fracture toughness concept applied to methyl methacrylate, *J. Biomed. Mater. Res., 14,* 311 (1980).
53. Davies, J.P., O'Connor, D., Greer, J., and Harris, W., Comparison of the mechanical properties of Simplex P, Zimmer regular and LVC bone cements, *J. Biomed. Mater. Res., 21,* 719 (1987).
54. Ege, W., and Tuchscherer, Ch. Properties of a new bone cement formulation, in *Proceedings of the Third World Biomaterials Congress,* 582 (1988).
55. Vila, M.M., Raya, A., and Planell, J.A., Mechanical behaviour of a rubber modified bone cement, *Adv. Biomater., 9,* 155, (1990).
56. Hansen, D., and Steen Jensen, J., Additional mechanical tests of bone cements *Acta Orthopaedica Belgica, 58,* 268 (1992).

57. Oyasaed, H., Dynamic mechanical properties of multiphase acrylic systems, *Biomed. Mater. Res., 24,* 1037 (1990).
58. Lee, A.J.C., Ling, R.S.M., and Vangala, S.S. Some clinically relevant variables affecting the mechanical behaviour of bone cement, *Arch. Orthop. Traumat. Surg., 92,* 1 (1978).
59. Saha, S., and Pal, S. Strain dependence of the compressive properties of normal and carbon-fiber-reinforced bone cement, *J. Biomed. Mater. Res., 17,* 1041 (1983).
60. Litsky, A.S., and Yetkinler, D.N., Time dependent stress relation behavior of standard and reduced-modulus acrylic bone cements, *Proceedings of the 17th Annual Meeting of the Society for Biomaterials,* 47 (1991).
61. Migliaresi, C., and Obici, R., Long term creep of acrylic bone cement, *Proceedings of the 3rd World Biomaterials Congress,* 580 (1988).
62. Treharne, R.W., and Brown, N. Factors influencing the creep behavior of poly(methyl methacrylate) cements, *J. Biomed. Mater. Res. Symp., 6,* 81 (1975).
63. Oysaed, H., and Ruyter, I.E. Creep studies of multiphase acrylic systems, *J. Biomed. Mater. Res., 23,* 719 (1989).
64. Holm, N.J. The relaxation of some acrylic bone cements, *Acta Orthop. Scand., 51,* 727 (1980).
65. McKellop, H., Narayan, S., Ebramzadeh, E., and Sarmiento, A., Viscoelastic creep properties of PMMA surgical cement, *Proceedings of the 3rd World Biomaterials Congress,* 328 (1988).
66. McKellop, H., Narayan, S., Lu, B., Ebramzadeh, E., and Sarmiento, A., Viscoelastic models for creep of high and low modulus acrylic cements, *Proceedings of the 16th Annual Meeting of the Society for Biomaterials,* 116 (1990).
67. Lewis, G., The fracture toughness of biomaterials: 1. Acrylic bone cements, *J. Mater. Educ., 11,* 429 (1989).
68. Kusy, R.P., and Turner, D.T. Intergranular cracking of a weak two-phase polymethyl methacrylate, *J. Biomed. Mater. Res., 8,* 185 (1974).
69. Causton, B.E., Fracture mechanics of dental poly(methylmethacrylate), *J. Dent. Res., 54*(2), 339 (1975).
70. Hailey, J.L., Turner, I.G., and Miles, A.W., The effect of storage environment on the mechanical properties of acrylic bone cement, in *Biomaterial-Tissue Interfaces* (P.J. Doherty et al., eds.), Advances in Biomaterials 10, Elsevier Science Publishers, 325 (1992).
71. Hailey, J.L., Turner, I.G., and Miles, A.W., *Proceedings of 4th World Biomaterials Congress,* 89 (1992).
72. Freitag, T.A., and Cannon, S.L. Fracture characteristics of acrylic bone cements. 1. Fracture toughness, *J. Biomed. Mater. Res., 10,* 805 (1976).
73. Stark, C.F., Fracture and fatigue characteristics of bone cements, *J. Biomed. Mater. Res., 13,* 339 (1979).
74. Beaumont, P.W.R., and Young, R.J., Slow crack growth in acrylic bone cement, *J. Biomed. Mater. Res., 9,* 423 (1975).
75. Pilliar, R.M., Vowles, R., and Williams, D.F. Note: Fracture toughness testing of biomaterials using a mini-short rod specimen design, *J. Biomed. Mater. Res., 21,* 145 (1987).
76. Wang, C.T., and Pilliar, R.M. Fracture toughness of acrylic bone cements, *J. Mater. Sci., 24,* 3725 (1989).
77. Owen, A.B., and Beaumont, P.W., Fracture characteristics of surgical acrylic bone cements, in *Mechanical Properties of Biomaterials* (Hastings, G.W., and Williams, D.F., eds.), Wiley & Sons, 277 (1980).
78. Silvestre, A., Raya, A., Fernández-Fairén, M., Anglada, M., and Planell, J.A., Failure of acrylic bone cements under triaxial stresses, *J. Mater. Sci., 25,* 1050 (1990).
79. Muller, K., Use and properties of PMMA bone cement, in *Evaluation of Biomaterials* (Winter, G.D., Leray, J.L., and de Groot, K., eds.), Wiley & Sons, 175 (1980).
80. Krause, W., and Matins, R. Fatigue properties of acrylic bone cements: Review of the literature, *J. Biomed. Mater. Res., 22,* 37 (1988).

81. Humphreys, P.E., Orr, J.F., and Bahrani, A.S., An investigation into the effect of cyclic loading and frequency on the temperature of PMMA bone cement in hip prostheses, *Proc. Instn. Mech. Engrs., 203,* 167 (1989).

82. Freitag, T.A., and Cannon, S.L., Fracture characteristics of acrylic bone cements. II. Fatigue, *J. Biomed. Mater. Res., 11,* 609 (1977).

83. Topoleski, L.D.T., Ducheyne, P., and Cuckler, J.M., *16th Annual Meeting of the Society for Biomaterials,* 108 (1990).

84. Topoleski L.D.T, Ducheyne, P., and Cuckler, J.M., A fractographic analysis of *in vivo* poly(methyl methacrylate) bone cement failure mechanisms, *J. Biomed. Mater. Res., 24,* 135 (1990).

85. Krause, W., Mathis, R., and Grimes, L. Fatigue properties of acrylic bone cement: S-N, P-N and P-S-N data, *J. Biomed. Mater. Res., 22(A3),* 221 (1988).

86. Johnson, J.A., Provan, J.W., Krygier, J.J., Chan, K.H., and Miller J., Fatigue of acrylic bone cement. Effect of frequency and environment, *J. Biomed. Mater. Res., 23,* 819 (1989).

87. Gilbert, J.L., Menis, D.W., Smith, S.N., Lautenschlager, E.P., and Wixson, R.W., Effect of pore size and morphology on fatigue crack initiation in acrylic bone cements, *Proceedings of the 16th Annual Meeting of the Society for Biomaterials,* 103 (1990).

88. Davies, J.P., O'Connor, D.O., Burke, D.W., and Harris, W.H., Does centrifugation improve the fatigue life on bone cement when injected into trabecular bone? *Proceedings of the 13th Annual Meeting of the Society for Biomaterials,* 115 (1987).

89. Davies, J.P., O'Connor, D., Burke, D., and Harris, W. Influence of antibiotic impregnation on fatigue life of Simplex P and Palacos R acrylic bone cements with and without centrifugation, *J. Biomed. Mater. Res., 23,* 379 (1989).

90. Soltész, U., and Ege, W., Fatigue behaviour of different acrylic bone cements, *Proceedings of the 4th World Biomaterials Congress,* 90 (1992).

91. Carter, D.R., Gates, E.I., and Harris, W.H., Strain-controlled fatigue of acrylic bone cement, *J. Biomed. Mater. Res., 16,* 647 (1982).

92. Wright, T.M., and Robinson, R.P. Fatigue crack propagation in polymethylmethacrylate bone cements, *J. Mater. Sci., 17,* 2463 (1982).

93. Vila, M.M., Behiri, J.C., and Planell, J.A., Fatigue crack propagation in acrylic bone cements, in *Biomaterials–Tissue Interfaces* (P.J. Doherty et al., eds.), Advances in Biomaterials 10, Elsevier, 187 (1992).

94. Hertzberg, R., and Manson, J., *Fatigue of Engineering Plastics,* Academic Press, New York (1980).

95. Truong, V.T., Cock, D.J., and Padmanathan, N., Fatigue crack propagation in posterior dental composites and prediction of clinical wear, *J. Appl. Biomat., 1,* 21 (1990).

96. Charosky, C.B., and Walker, P.S., The microstructure of polymethylmethacrylate cement, *Clin. Orthop. Rel. Res.,* 222 (1973).

97. Cameron, H.U., Mills, R.H., Jackson, R.W., and Macnab, I., The structure of polymethylmethacrylate cement, *Clin. Orthop. Rel. Res., 100,* 287 (1974).

98. Willert, H.G., Mueller, K., and Semlitsch, M., The morphology of polymethylmethacrylate (PMMA) bone cement *Arch. Orthop. Traumat. Surg., 94,* 265 (1979).

99. Eyerer, P., and Jin, R. Influence of mixing technique on some properties of PMMA bone cement, *J. Biomed. Mater. Res., 20,* 1057 (1986).

100. Cerulli, G., Tranquilli-Leali, P., Moriconi, F., and Paoletti, G.C., Experimental study of THR cement techniques, in *Biomaterials and Biomechanics* (P. Ducheyne, G. Van der Perre, and A.E. Aubert, eds.), Elsevier, Amsterdam, 25 (1984).

101. Panjabi, M.M., Goel, V.K., Drinker, H., Wong, J., Kamire, G., and Walter, S.D., Effect of pressurization on methylmethacrylate–bone interdigitation: An *in vitro* study of canine femora, *J. Biomechanics, 16*(7), 473 (1983).

102. Brown, S.A., and Bargar, W.L., The influence of temperature and specimen size on the flexural properties of PMMA bone cement, *J. Biomed. Mater. Res., 18,* 523 (1984).

103. Lidgren, L., Bodelind, B., and Moller, J., Bone cement improved by vacuum mixing and chilling, *Acta Orthop. Scand., 57,* 27 (1987).

104. Linden, U., Mechanical versus manual mixing of bone cement, *Acta Orthop. Scand., 59*(4), 400 (1988).

105. Song, Y., Goodman, S.B., and Jaffe, R.A., Femoral Intamedullary pressures during hip replacement using modem cement technique, *Proceedings of the 19th Annual Meeting of the Society for Biomaterials,* 110 (1993).

106. Wijn, J.R., Driessens, F.C.M., and Sloof, T.J.J.H., Dimensional behavior of curing bone cement masses, *J. Biomed. Mater. Res. 6,* 99 (1975).

107. Connelly, T.J., Lautenschlager, E.P., and Wixson, R.L., The role of porosity in the shrinkage of acrylic bone cements, *Proceedings of the 13th Annual Meeting of the Society for Biomaterials,* 114 (1987).

108. Jay, J.L., Noble, P.C., Lindahl, L.J., and Tullos, H.S., Porosity and the polymerization shrinkage of acrylic bone cement, *Proceedings of the 13th Annual Meeting of the Society for Biomaterials,* 113 (1987).

109. Brown, SA., Sharkey, N.A., Bargar, W.L., Paul, H.A., Voegli, T.L., and Hsieh, Y.L., Properties of *in vitro* and *in vivo* cured acrylic bone cement, in *Biomaterials and Biomechanics* (P. Ducheyne, G. Van der Pete, and A.E. Aubert, eds.), Elsevier, Amsterdam, 13 (1984).

110. Looney, M.A., and Park, J.B., Molecular and mechanical property changes during aging of bone cement *in vitro* and *in vivo, J. Biomat. Mater. Res., 20,* 555 (1986).

111. Rostoker, W., Lereim, P., and Galante, J.O. Effect of an *in vivo* environment on the strength of bone cement, *J. Biomed. Mater. Res., 13,* 365 (1979).

112. Fernández Fairén, M., and Vazquez, J.J., The aging of polymethyl methacrylate bone cement, *Acta Orthop. Belg., 49,* 512 (1983).

113. Braden, M., The absorption of H_2O by acrylic resin and other materials, *J. Prosthet. Dent., 14,* 307, (1964).

114. Murray, D.E., and Daw, J.S., An ultrastructural study of the biocompatibility of poly(2-hydroxyethylmethacrylate) in bone, *J. Biomed. Mater. Res., 9,* 699, (1975).

115. Migliaresi, C., and Capuana, P., 2-Hydroxyethylmethacrylate modified bone cement, in *Clinical Implant Materials,* Advances in Biomaterials 9, Elsevier, Amsterdam, 141 (1990).

116. Gruen, T.A., Markolf, K.L., and Amstutz, H.C., Effects of lamination and blood entrapment on the strength of acrylic bone cement, *Clin. Orthop., 119,* 250, (1976).

117. Schreurs, B.W., Spierings, P.T.J., Huiskes, R., and Slooff, T.J., Effects of preparation techniques on the porosity of acrylic cements, *Acta Orthop. Scand., 59*(4), 403 (1988).

118. Burke, D.W., Gates, E.I., and Harris, W.H., Centrifugation as a method of improving tensile and fatigue properties of acrylic bone cement, *J. Bone Joint Surg., 66A*(8), 1265 (1984).

119. Rimnac, C., Wright, T., and McGill, D., The effect of centrifugation on the fracture properties of acrylic bone cements, *J. Bone Joint Surg., 68A,* 281 (1986).

120. Chandeler, D., McKellop, H., Narayan, S., and Sarmiento, A., Effect of porosity on fracture toughness of PMMA surgical cement, *Proceedings of the 3rd World Biomaterials Congress,* 329 (1988).

121. Davies, J.P., Jasty, M., O'Connor, D.O., Burke, D.W., Harrigan, T.P., and Harris, W.H. The effect of centrifuging bone cement, *J. Bone Joint Surg., 71-B,* 39 (1989).

122. James, S.P., Jasty, M., Davies, J., Piehler, H., and Harris, W., A fractographic investigation of PMMA bone cement focusing on the relationship between porosity reduction and increased fatigue life, *J. Biomed. Mater. Res., 26,* 651 (1992).

123. Hansen, D., and Steen Jensen, J., Mixing does not improve mechanical properties of all bone cements, *Acta Orthop. Scand., 63*(1), 13 (1992).

124. Lidgren, L., Drar, H., and Möller, J., Strength of polymethylmethacrylate increased by vacuum mixing, *Acta Orthop. Scand., 55,* 536 (1984).

125. Rajaram, A., and C-C. Chu, A vacuum method for preparing acrylic dental and bone cements, *Proceedings of the 13th Annual Meeting of the Society for Biomaterials*, 110 (1987).

126. Wixson, R.L., Lautenschlager, E.P., and Novak, M.A., Vacuum mixing of acrylic bone cement, *J. Arthoplasty, 2*(2), 141 (1987).

127. Lautenschlager, E.P., Menis, D.L., Wixson, R.L., and Vajda, E., Fatigue crack testing of vacuum and regular mixed Simplex-P, *Proceedings of the 3rd World Biomaterials Congress*, 331 (1988).

128. Askew, M.J., Kufel, M.F., Fleissner, P.R., Gradissar, I.A., Salstrom, S.J., and Tan, J.S. Effect of vacuum mixing on the mechanical properties of antibiotic impregnated polymethyl-methacrylate bone cement, *J. Biomed. Mater. Res., 24*, 573 (1990).

129. Vila, M.M., and Planell, J.A., Influencia de la porosidad en la tenacidad a fracture de los cementos óseos, *Anales de Mecánica de la Fracture, 7*, 134 (1990).

130. Wang, J.S., Franzen, H., Jonsson, E., and Lidgren, L., Porosity of bone cement reduced by mixing and collecting under vacuum, *Acta Orthop. Scand., 64*(2), 143 (1993).

131. Friis, E.A., Stromberg, L.J., Cooke, F.W, and McQueen, D.A., Fracture toughness of vacuum mixed PMMA bone cement, *Proceedings of the 19th Annual Meeting of the Society for Biomaterials*, 301 (1993).

132. Soltesz, U., and Ege, W., Influence of mixing conditions on the fatigue behaviour of an acrylic bone cement, *Proceedings of the 10th European Conference on Biomaterials*, 138 (1993).

133. Wixson, R.L., Do we need to vacuum mix or centrifuge cement? *Clin. Orthop. Rel. Res., 285*, 84 (1992).

134. Keller, J.C., Marshall, G.W., and Lautenschlager, E.P., Morphological identity of porosity in chemically activated acrylic cements, *Scanning Electron Microscopy, 1*, 425 (1979).

135. Sharkey, N.A., and Bargar, W.L., Material properties of retrieved polymethacrylate bone cement, *Proceedings of the Symposium on Retrieval and Analysis of Surgical Implants and Biomaterials*, 53 (1988).

136. Topoleski L.D.T., and Lu, X., An investigation of fatigue crack initiation mechanisms in polymethylmethacrylate bone cement, *Proceedings of the 19th Annual Meeting of the Society for Biomaterials*, 300 (1993).

137. Amstutz, H.C., Markolf, K.L., Magneice, G.M., and Gruen, T.A., Loosening of total hip components: Cause and prevention, in *The Hip*, C.V. Mosby, St. Louis, 102 (1976).

138. Pacheco, V., Shelley, P., and Wroblewski, B.M., Mechanical loosening of the stem in Charnley arthroplasties. Identification of the "at risk, factors," *J. Bone Joint Surg., 70-B*, (1988).

139. Jasty, M., Jiranek, W., and Harris, W., Acrylic fragmentation in total hip replacements and its biological consequences, *Clin. Orthop. Rel. Res., 285*, 116 (1992).

140. Jones, L.C., and Hungerford, D.S., Cement disease, *Clin. Orthop. Rel. Res., 225*, 192 (1987).

141. Hungerford, D. and Jones, L.C., The rationale of cementless revision of cemented ar-throplasty failures, *Clin. Orthop. Rel. Res., 235*, 12 (1988).

142. Stromberg, C.N., Herberts, P., and Palmertz, B., Cemented revision hip arthroplasty, *Acta Orthop. Scand., 63*(2), 111 (1992).

143. Harris, W.H., Clinical and surgical overview, in *The Bone–Implant Interface*, American Academy of Orthopaedic Surgeons, Chicago, 1 (1985).

144. Bramlett, K., Hamilton, G., Devrnja, R., and Lemons, J., PMMA porosity reduction tech-niques and their role in implant mantle failure, *Proceedings of the 3rd World Biomaterials Congress*, 330 (1988).

145. Beaumont, P.W.R., and Plumpton, B., The strength of acrylic bone cements and acrylic cement stainless-steel interfaces. Part II: The shear strength of an acrylic cement-stainless steel interface, *J. Mater. Sci., 12*, 1853 (1977)

146. Keller, J.C., Lautenschlager, E.P., Marshall, G.W., and Meyer, P.R. Factors affecting surgical alloy/bone cement interface adhesion, *J. Biomed. Mater. Res., 14*, 539 (1980).

147. Raab, S., Ahmed, A.M., and Provan, J.W., The quasistatic and fatigue performance of the implant/bone cement interface, *J. Biomed. Mater. Res., 15,* 159 (1981).
148. Ahmed, A.M., Raab, S., and Miller, J.E., Metal/cement interface strength in cemented stem fixation, *J. Orthop. Res., 2*(2), 105 (1984).
149. Bundy, K.J., and Penn, R.W., The effect of surface preparation on metal/bone cement interfacial strength, *J. Biomed. Mater. Res., 21,* 773 (1987).
150. Stone, M.H., Wilkinson, R., and Stother, I.G., Some factors affecting the strength of the cement-metal interface, *J. Bone Joint Surg., 71-B,* 217 (1989).
151. Davies, J.P., and Harris, W.H., Tensile bonding strength of bone cement to the metal surface of a prosthesis, *Proceedings of the 17th Annual Meeting of the Society for Biomaterials,* 204 (1991).
152. Niederer, P.G., Chiquet, C., Semlitsch, M., and Panic, B., The influence of stem surface on load transmission of total hip prosthesis at their interface with acrylic bone cement, in *Evaluation of Biomaterials* (Winter, G.D., Leray, J.L., and de Groot, K., eds.), Advances in Biomaterials 1, Wiley & Sons, 205 (1980).
153. Huiskes, R., Properties of the stem–cement interface and artificial hip joint failure, in *The Bone-Implant Interface,* American Academy of Orthopaedic Surgeons, Chicago, 86 (1985).
154. Ebramzadeh, E., Lu, Z., McKellop, H., Zahiri, C., and Sarmiento, A., Influence of interface bonding strength on the cement stresses in total hip arthroplasty, *Proceedings of the 4th World Biomaterials Congress,* 92 (1992).
155. Davies, J.P., Tse, M.K., and Harris, W.H., Monitoring the bonding of the cement-metal interface using acoustic emission and ultrasound *in situ, Proceedings of the 19th Annual Meeting of the Society for Biomaterials,* 244 (1993).
156. Gharpuray, V.M., Keer, L.M., and Lewis, J.L., Cracks emanating from defects in PMMA, *Proceedings of the 17th Annual Meeting of the Society for Biomaterials,* 203 (1991).
157. James, S., Schmalzried, T., McGarry, F.J., and Harris W., Porosity reduction at the femoral prosthesis–cement interface, *Proceedings of the 17th Annual Meeting of the Society for Biomaterials,* 50 (1991).
158. James, S.P., Schmalzried, T.P., McGarry, F.J., and Harris, W.H., Extensive porosity at the cement–femoral prosthesis interface: A preliminary study, *J. Biomed. Mater. Res., 27,* 71 (1993).
159. Noble, P.C., Ward, K.A., Helmke, H.W., Lednicky, C.L., and Tullos, H.S., Porosity of the cement/metal interface following cemented hip replacement, *Proceedings of the 19th Annual Meeting of the Society for Biomaterials,* 243 (1993).
160. James, S.P., Karydas, D., McGarry, F.J., and Harris, W.H., Reduction of the extensive porosity in the cement at the femoral component-bone cement interface, *Proceedings of the 19th Annual Meeting of the Society for Biomaterials,* 242 (1993).
161. Noble, P.C., Dreeben, S.N., Jay, J.L., and Tullos, H.S., The effects of methods of cement enhancement on the strength of acrylic cement interfaces, *Proceedings of the 13th Annual Meeting of the Society for Biomaterials,* 217 (1987).
162. Noble, P.C., Jay, J.L., Lindahl, L.J., Maltry, J., Dreeben, S.N., and Tullos, H.S., Methods of enhancing acrylic bone cement, *Proceedings of the 13th Annual Meeting of the Society for Biomaterials,* 169 (1987).
163. Park, J.B., von Recum, A.F., and Gratzick, G.E., Pre-coated orthopedic implants with bone cement, *Biomat. Med. Dev. Art. Org., 7*(1), 41 (1979).
164. Barb, W., Park, J.B., Kenner, G.H., and von Recum, A.F. Intramedullary fixation of artificial hip joints with bone cement-precoated implants. 1. Interfacial strengths, *J. Biomed. Mater. Res., 16,* 447 (1982).
165. Barb, W., Park, J.B., Kenner, G.H., and von Recum, A.F. Intramedullary fixation of artificial hip joints with bone cement precoated implants. II. Density and histological study, *J. Biomed. Mater. Res., 16,* 459 (1982).
166. Raab, S., Ahmed, A.M., and Provan, J.W., Thin film PMMA precoating for improved implant bone cement fixation, *J. Biomed. Mater. Res., 16,* 679 (1982).

167. Ahmed, A.M., Characterization and improvement of the metal–cement interface performance: An overview, in *The Bone-Implant Interface*, American Academy of Orthopaedic Surgeons, Chicago, 102 (1985).

168. Davies, J.P., Singer, G., and Harris, W.H., The effect of a thin coating of polymethylmethacrylate on the torsional fatigue strength of the cement-metal interface, *J. Appl. Biomat., 3,* 45 (1992).

169. Davies, J.P., and Harris, W.H., Fatigue strength of cement/metal interface: Comparison of metal, metal with precoating and metal with rough surface and precoating, *Proceedings of the 16th Annual Meeting of the Society for Biomaterials,* 34 (1990).

170. Stone, M.H., Wilkinson, R., and Stother, I.G., Factors affecting the surgical alloy–bone cement interface, *Proceedings of the 13th Annual Meeting of the Society for Biomaterials,* 171 (1987).

171. Hammerman, S.M., Noble, P.C., Alexander, J.W., and Tullos, H.S., The bonding of acrylic bone cement to preformed PMMA positioning devices, *Proceedings of the 13th Annual Meeting of the Society for Biomaterials,* 219 (1987).

172. Wang, C.T., and Pillar R.M. Bone cement bonding. Interfacial fracture toughness determination, *Clin. Mater., 4,* 135 (1981).

173. Johanson, N.A, Bullough, P.G., Wilson, P.D., Salvati, E.A., and Ranawat, C.S. The microscopic anatomy of the bone interface in failed total hip arthroplasties, *Clin. Orthop. Rel. Res., 218,* 123 (1987).

174. Lewis, J.L., Mechanical processes in bone–cement interface failure, in *The Bone-Implant Interface*, American Academy of Orthopaedic Surgeons, Chicago, 34 (1985).

175. Menis, D.L., Wixson, R.L., Gilbert, J.L., and Lautenschlager, E.P., Effect of vacuum mixing on interfacial bone cement fracture toughness, *Proceedings of the 16th Annual Meeting of the Society for Biomaterials,* 67 (1990).

176. Noble, P.C., and Swarts, E. Penetration of acrylic bone cements into cancellous bone, *Acta Orthop. Scand., 56,* 566 (1983).

177. MacDonald, W., Swarts, E., and Beaver, R., Penetration and shear strength of cement–bone interfaces *in vivo, Clin. Orthop. Rel. Res., 286,* 283 (1993).

178. Halawa, M., Lee, A.J.C., Ling, R.S.M., and Vangala, S.S. The shear strength of trabecular bone from the femur, and some factors affecting the shear strength of the cement–bone interface, *Acta Orthop. Traumat. Surg., 92,* 19 (1978).

179. Majkowski, R.S., Miles, A.W., Bannister, G.C., Perkins, J., and Taylor, G.J.S. Bone surface preparation in cemented joint replacement, *J. Bone Joint Surg., 75-B,* 459 (1993).

180. Balu, G., Noble, P.C., and Alexander, J.W., The effect of intramedullary reaming on the cement/bone interface, *Proceedings of the 19th Annual Meeting of the Society for Biomaterials,* 111 (1993).

181. Bergmann, G., Kölbel, R., and Rohlmann, A., Static and dynamic testing of the bone-PMMA implant interface, in *Evaluation of Biomaterials* (G.D.Winter, J.L.Leray, and K. de Groot, eds.), Advances in Biomaterials 1, Wiley & Sons, 197 (1980).

182. Harrigan, T.P., Davies, J.P., Burke, D.W., O'Connor, D.O., Jasty, M., and Harris, W.H., On the presence or easy initiation of fracture in bone cement at the bone cement interface in total joint arthroplasty, *Proceedings of the 13th Annual Meeting of the Society for Biomaterials,* 170 (1987).

183. Kasman, R.A., Hollis, J.M., Devine, S., and Reindel, E., A new methodology for investigation of the fatigue performance of the bone/cement interface, *Proceedings of the 13th Annual Meeting of the Society for Biomaterials,* 172 (1987).

184. Park, J.B., Wenstein, J.N., Park, S.H., Oster, D., and Loaning, S., Effect of shock-wave treatment on the bone/cement interface strength, *Proceedings of the 3rd World Biomaterials Congress,* 578 (1988).

185. Brown, W., Claes, L., Rater, A., and Paschke, D., Effects of extracorporeal shockwaves on

the stability of the interface between bone and polymethylmethacrylate: An *in vitro* study on human femoral segments, *Clinical Biomechanics, 7*, 47 (1992).

186. Beaumont, P.W.R., The strength of acrylic bone cements and acrylic cement–stainless steel interfaces. Part 1: The strength of acrylic bone cement containing second phase dispersions, *J. Mater. Sci., 12*, 1845 (1977).

187. Rudigier, J., Kirschner, P., Richter, I.E., and Schweikert, C.H., Influence of different x-ray contrast materials on structure and strength of bone cements, in *Mechanical Properties of Biomaterials* (G.W. Hastings and D.F. Williams, eds.), Wiley & Sons, 289 (1980).

188. Cooke, F.W., Tsai, Y.H., Marrero, T.R., and Yasuda, H.K., Improved bone cement strength by chemically coupling radiopacifier additions to the PMMA, *Proceedings of the 17th Annual Meeting of the Society for Biomaterials*, 46 (1991).

189. Nelson, R.C., Hoffman, R.O., and Burton, J.A., The effect of antibiotic addition on the mechanical properties of acrylic cement, *J. Biomed. Mater. Res., 12*, 473, (1978).

190. Lautenschlager, E.P., Marshall, G.W., Marks, K.E., Schwartz, J., and Nelson, C.L., Mechanical strength of acrylic bone cements impregnated with antibiotics, *J. Biomed. Mater. Res., 10*, 837, (1976).

191. Lautenschlager, E.P., Jacobs, J.J., Marshall, G.W., and Meyer, P.R., Mechanical properties of bone cements containing large doses of antibiotic powders, *J. Biomed. Mater. Res., 10*, 929, (1976).

192. Marks, K.E., Nelson, C.L., and Lautenschlager, E.P., Antibiotic impregnated bone cement, *J. Bone Joint Surg., 58A*, 358, (1976).

193. Law, H.T., Biomechanics of bone cement, *Seminars in Orthopaedics, 1*, 23 (1986).

194. Davies, J.P., and Harris, W.H., Effect of hand-mixing tobramycin on the fatigue strength of Simplex P, *J. Biomed. Mater. Res., 25*, 1409 (1991).

195. Sheinin, E.B., Benson, W.R., and Brannon, W.L., Determination of methylmethacrylate in surgical acrylic cement, *J. Pharm. Sci., 65*(2), 280 (1976).

196. Schoenfeld, C. M., Conard, G.J., and Lautenschlager, E.P., Monomer release from methacrylate bone cements during simulated *in vivo* polymerization, *J. Biomed. Mater. Res., 13*, (1979).

197. Davy, K.W.M., and Braden, M., Residual monomer in acrylic polymers, *Biomaterials, 12*, 540 (1991).

198. Litsky, A.S.,Frazier, N.C. Monomer exudation from standard and reduced-modulus bone cements under cyclic loading, *Proceedings of the 19th Annual Meeting of the Society for Biomaterials*, 303 (1993).

199. Willert, H.G., Frech, H.A., and Bechtel, A., Measurements of the quantity of monomer leaching out of acrylic bone cement into the surrounding tissues during the process of polymerization, in *Biomedical Applications of Polymers* (H.P. Gregor, ed.), Plenum Press, New York, 121 (1976).

200. James, M.L., Complicaciones anestésicas y metabólicas, in *Complicaciones de las artroplastias totales de cadera* (R.S.M. Ling, ed.), Salvat Editores, 1 (1987).

201. Newens, A.F., and Volz, R.G. Severe hypotension during prosthetic hip surgery with acrylic bone cement, *Anesthesiology, 36*, 298 (1972).

202. Milne, I.S., Hazards of acrylic bone cement, *Anaesthesia, 28*, 538 (1973)

203. Linder, L., and Romanus, M. Acute local tissue effects of polymerizing acrylic bone cement, *Clin. Orthop. Rel. Res., 115*, 303 (1976).

204. Birch, R., Wilkinson, M.C.P., Vijayan, K.P., and Gschmeissner, S., Cement burn of the sciatic nerve, *J. Bone Joint Surg., 74-B*, 731 (1992).

205. Wenda, K., Issendorff, W.D., Rudigier, J., and Ahlers, J., Blood pressure decrease after bone cement—Effect of monomer or intramedullary pressure? *Proceedings of the 13th Annual Meeting of the Society for Biomaterials*, 220 (1987).

206. Wenda, K., Degreif, J., Rudigier, J., and Issendorff, W.D. Pathogenesis and effective

prophylaxis of the cement implantation syndrome, *Proceedings of the 3rd World Biomaterials Congress*, 579 (1988).

207. Anderson, G.I., Humeniuk, B., Gordon ,R.G., and Richards, R.R., Femoral vein cholesterol, triglyceride and methylmethacrylate levels after reaming, lavage and cement pressuration of the intramedullary canal: A canine total hip model, *Proceedings of the 4th World Biomaterials Congress*, 15 (1992).

208. Moharram, M.A., and Khalil, S. Infrared study of the interaction of acrylic bone cement with bone structure *in vitro*, *Int. J. Infrared Millimeter Waves*, *13*(8), 1217 (1992).

209. Moharram, M.A., Higazy, H., and Khalil, S., Study of the interaction of acrylic bone cement with bone marrow, *J. Mater. Sci.*, *28*, 4010 (1993).

210. Vernon-Roberts, B., and Freeman, M.A.R., The tissue response to total hip replacement prosthesis, in *The Scientific Basis of Joint Replacement* (S.A.V. Swanson and M.A.R. Freeman, eds.), Wiley, New York, 86 (1977).

211. Radin, E.L., and Rose, R.M., The relationship between mechanical and biological processes in the loosening of total joint replacements, in *The Bone-Implant Interface*, American Academy of Orthopaedic Surgeons, Chicago, 23 (1985).

212. Goldring, S.R., Roelke, M., Rourke, C.M., Jasty, M., Schiller, A.L., and Harris, W.H., Synovial-like membrane at the bone-cement interface: Its role in implant loosening after total hip replacement, in *The Bone-Implant Interface*, American Academy of Orthopaedic Surgeons, Chicago, 56 (1985).

213. Lennox, D.W., Schofield, B.H., McDonald, D.F., and Riley, L.H. A histologic comparison of aseptic loosening of cemented, press-fit and biologic ingrowth prostheses, *Clin. Orthop. Rel. Res.*, *225*, 171 (1987).

214. Coe, M.R., Fechner, R.E., Jefrey, J.J., Balian, G., and Whitehill, R. Characterization of tissue from the bone-polymethylmethacrylate interface in a rat experimental model, *J. Bone Joint Surg.*, *71-A*(6), 863 (1989).

215. Learmonth, I. D., Heywood, A.W.B., Kaye, J., and Dall, D. Radiological loosening after cemented hip replacement for juvenile chronic arthritis, *J. Bone Joint Surg.*, *71-B*, 209 (1989).

216. Rodriguez Vela, J., Serrano Ostariz, J.L., Canales Cortes, V., Herrera Rodríguez, A., and Suñen Sánchez, E., Estudio de las propiedades mecánicas y quimicas del cements acrílico extraído de prótesis de cadera, *Rev. Orthop. Traum.*, *361B(3)*, 354 (1992).

217. Zafra, M., Casado, J.M., López, F., Gala, M., and Fernández, L., Estudio histopatológico de la membrana de interfase cemento-hueso o prótesis de hueso en protesis totales de cadera con aflojamiento, *Rev. Orthop. Traum.*, *361B(3)*, 267 (1992).

218. Kim, K.J., Rubash, H.E., Wilson, S.C., D'Antonio, J.A., and McClain, J. A histologic and biochemical comparison of the interface tissues in cementless and cemented hip prostheses, *Clin. Orthop. Rel. Res.*, *287*, 142 (1993).

219. Howie, D.W., Vernon-Roberts, B., Oakeshott, R., and Manthey, B., A rat model of resorption of bone at the cement-bone interface in the presence of polyethylene wear particles, *J. Bone Joint Surg.*, *70-A*(2) (1988).

220. Willert, H.G., Bertram, H., and Buchhorn, G.H., Osteolysis in alloarthroplasty of the hip. The role of bone cement fragmentation, *Clin. Ortho. Rel. Res.*, *258*, 108 (1990).

221. Horowitz, S.M., Gautsch, T.L., Frondoza, C.G., and Riley, L., Macrophage exposure to polymethyl methacrylate leads to mediator release and injury, *J. Orthop. Res.*, *9*(3) 406 (1991).

222. Horowitz, S.M., Doty, S.B., Lane, J.M., and Burstein, A.H., Studies of the mechanism by which the mechanical failure of polymethylmethacrylate leads to bone resorption, *J. Bone Joint Surg.*, *75-A*(6), 802 (1993).

223. Jee, W.S.S., Ueno, K., Haba, T., and Deng, Y.P., Prostglandins and bone formation, in *The Bone-Implant Interface*, American Academy of Orthopaedic Surgeons, Chicago, 43 (1985).

224. Welgus, H.G., Collagen degradation by human collagenase, in *The Bone-Implant Interface*, American Academy of Orthopaedic Surgeons, Chicago, 69 (1985).

225. Tietelbaum, S.L., and Kahn, A.J., Macrophages and orthopaedic implant loosening, in *The Bone-Implant Interface*, American Academy of Orthopaedic Surgeons, Chicago, 80 (1985).

226. Goodman, S.B., Chin, R.C., Chiou, S.S., and Lee, J., Modulation of the membrane surrounding particulate polymethyl methacrylate in the rabbit tibia, *Proceedings of the 16th Annual Meeting of the Society for Biomaterials*, 289, (1990).

227. Goodman, S.E., Fornasier, V.L., and Kei, J., The effects of bulk versus particulate biomaterial on bone, *Proceedings of the 13th Annual Meeting of the Society for Biomaterials*, 129 (1987).

228. Gil-Albarova, J., Laclériga, A., Barrios, C., and Cañadell, J., Lymphocite response to polymethylmethacrylate in loose total hip prostheses, *J. Bone Joint. Surg., 74-B*, 825 (1992).

229. Tanner, K.T., Frondoza, C.G., Jones, L., and Hungerford, D.S. *In vitro* effects of polymethylmethacrylate on normal human fibroblasts, *Proceedings of the 16th Annual Meeting of the Society for Biomaterials*, 101, (1990).

230. Goodman, S.B., Aspenberg, P., Doshi, A., Wang, J.S., Emmanuel, J., and Lidgren, L., Cement particles inhibit bone ingrowth into titanium chambers implanted in the rabbit tibia, *Proceedings of the 19th Annual Meeting of the Society for Biomaterials*, 105, (1993).

231. Emmanual, J., Emmanual, J.G., Sauer, B., and Hedley, A., Bone cement: Retrieved wear particles vs. fabricated wear particles, *Proceedings of the 16th Annual Meeting of the Society for Biomaterials*, 10, (1990).

232. Keen, C.E., Brady, P.K., Spencer, J.D., and Levison, D.A., Histopathological and microanalytical study of zirconium dioxide and barium sulphate in bone cement, *J. Clin. Pathol., 45*, 984 (1992).

233. Lazarus, M.D, Cuckler, J.M., Baker, D.G., Ducheyne, P., and Schumacher, H.R. The role of $BaSO_4$ in osteolysis associated with PMMA particulate debris in two *in vivo* models, *Proceedings of the 17th Annual Meeting of the Society for Biomaterials*, 33 (1991).

234. Argenson, J.N., Seyral, P., Drancourt, M., Raoult, D., and Aubaniac, J.M., An *in vivo* and *in vitro* study of the release of vancomycin and tobramycin from acrylic bone cement, *Proceedings of the 4th World Biomaterials Congress*, 431 (1992).

235. Nelson, C.L., Griffin, F.M., Harrison, M.D., and Cooper, R.E., *In vitro* elution characteristics of commercially and noncommercially prepared antibiotic PMMA beads, *Clin. Orthop. Rel. Res., 284*, 303 (1992).

236. Downes, S., Methods for improving drug release from polymethylmethacrylate bone cement, *Clin. Mater., 7*, 227 (1991).

237. 0ga, M., Arizono, T., and Sugioka, Y., Inhibition of bacterial adhesion by tobramycin-impregnated PMMA bone cement, *Acta Orthop. Scand., 63*(3), 301 (1992).

238. Arizono, T., Oga, M., and Sugioka, Y., Increased resistance of bacteria after adherence to polymethyl methacrylate, *Acta Orthop. Scand., 63*(6), 661 (1992).

239. Chang, C.C., and Merritt, K., Microbial adherence on polymethylmethacrylate PMMA surfaces, *J. Biomed. Mater. Res., 26*, 197 (1992).

240. Miller, J., Krause, W.R., Krug, W.H., and Kelebay, L.C., Low viscosity cement, *Orthop. Trans., 5*, 352 (1981); abridged version in *Clin. Orthop. Rel. Res., 276*, 4 (1992).

241. Carlsson, A.S., Nilsson, J.A., Blomgren, G., Josefsson, G., Lindberg, L.T., and Önnerfält, R., Low vs. high viscosity cement in high arthroplasty, *Acta Orthop. Scand., 64*(3), 257 (1993).

242. Hadjari, M., Reindel, E.S., Hollis, J.M., and Convery, F.R., Characterization of high viscosity acrylic bone cement, *Proceedings of the 16th Annual Meeting of the Society for Biomaterials*, 105, (1990).

243. Rijke, A.M., and Rieger, M.R., Porous acrylic cements, *J. Biomed. Mater. Res., 11*, 373 (1977).

244. Van Mullem, P. J., de Wijn, J.R., and Vaandrager, J.M., Porous acrylic cement: Evaluation of a novel implant material, *Ann. Plastic Surg., 21* (6), 576 (1988).

245. Tanzi, M.C., Sket, I., Gatti, A.M., and Monari, E., Physical characterization of acrylic bone cement cured with new accelerator system, *Clin. Mater., 8*, 131 (1991).

246. Litsky, A.S., Rose, R.M., Rubin, C.T., and Thrasher, E.L., A reduced-modulus acrylic bone cement: preliminary results, *J. Orthop. Res., 8*(4), 623 (1990).

247. Ling, R.S.M., Observations of the fixation of implants to the bony skeleton, *Clin. Orthop. Rel. Res., 210*, 80 (1986).

248. Binderman, I., Fine, N., Horowitz, I., and Ashman, A., HTR-polymer bone grafting material—Experimental and clinical studies, *Proceedings of the 13th Annual Meeting of the Society for Biomaterials*, 111 (1987).

249. Johnson, J.A., and Jones, D.W., Influence of composition of methacrylate copolymers on mechanical properties, *Proceedings of the 17th Annual Meeting of the Society for Biomaterials*, 49 (1991).

250. Revell, P.A., Freeman, M., Weightman, B., and Braden, M., The intraosseous implantation of a new bone cement, polyethylmethacrylate*N*-butyl methacrylate, in the dog, *Proceedings of the 4th World Biomaterials Congress*, 166 (1992).

251. Davies, J.P., and Harris, W.H., The effect of the addition of methylene blue on the fatigue strength of simplex P bone-cement, *J. Appl. Biomat., 3*, 81 (1992).

252. Downes, S., Wood, D.J., Malcolm, A.J., and Ali, S.Y., Growth hormone in polymethylmethacrylate cement, *Clin. Orthop. Rel. Res., 252*, 294, (1990).

253. Guida, G., Riccio, V., Gatto, S., Migliaresi, C., Nicodemo, L., Nicolais, L., and Palomba, C., A glass bead composite acrylic bone cement, in *Biomaterials and Biomechanics* (P. Ducheyne, G. Van der Perre, and A.E. Aubert, eds.), Elsevier, Amsterdam, 19 (1984).

254. Henning, W., Blencke, B.A., Brömer, H., Deutscher, K.K., Gross, A., and Ege, W., Investigations with bioactivated polymethylmethacrylates, *J. Biomed. Mater. Res., 13*, 89 (1979).

255. Liu, Y.K., Park, J.B., Njus, G.O., and Steinstra, D., Bone particle impregnated bone cement 1. *In vitro* studies, *J. Biomed. Mat. Res., 21*, 247, (1987).

256. Dai, K.R., Lit, Y.K., Park, J.B., Clark, C.R., Nishiyama, K., and Zheng, Z.K., Bone particle impregnated bone cement. II. *In vivo* weight bearing study, *J. Biomed. Mater. Res., 25*, 141, (1991).

257. Henrich, D.E., Cram, A.E., Park, J.B., Liu, Y.K., and Reddi, H., Inorganic bone and demineralized bone matrix impregnated bone cement: A preliminary *in vivo* study, *J. Biomed. Mater. Res., 27*, 277 (1993).

258. Olmi, R., Moroni, A., Castaldini, A., Cavallini, and Romagnoli, R., Hydroxy apatites alloyed with bone cement: Physical and biological characterization, in *Ceramics in Surgery* (P. Vincenzini, ed.), Elsevier, 91 (1983).

259. Castaldini, A., Cavallini, A., Moroni, A., and Olmi, R., Young's modulus of hydroxyapatite mixed bone cement, *Biomaterials, and Biomechanics*, 427 (1984).

260. Oonishi, H., Kushitani, S., Aono, M., Maeda, K., and Tsuji, E., Interface bioactive bone cement by using PMMA and HAp granules, *Proceedings of the 3rd World Biomaterials Congress*, 581 (1988).

261. Oonishi, H., Kushitani, S., Aono, M., Tsuji, T., Mizukoshi, T., Ishimaru, H., and Delecrin, J., Adhesive effect of hydroxyapatite interposed between bone and PMMA cement in the interface bioactive bone cement technique, *Proceedings of the 17th Annual Meeting of the Society for Biomaterials*, 1 (1991).

262. Behiri, J.C., Braden, M., Khorasani, S.N., Wiwattanadate, D., and Bonfield, W., Advanced bone cement for long term orthopaedic implantations, in *Bioceramics*, Vol. 4 (W. Bonfield, G.W. Hastings, and K.E. Tanner, eds.), Butterworth-Heinemann, 301 (1991).

263. Low, R.F., Hulbert, S.F., and Sogal, A., Mechanical properties of hydroxyapatite–polymethylmethacrilate bone cement composite: Hydroxiapatite embedded on surface and throughout cement matrix, in *Bioceramics*, Vol. 6 (P. Ducheyne and D. Christiansen, eds.), Butterworth-Heinemann, 339 (1993).

264. Dandurand, J., Delpech, V., Lebugle, A., Lamure, A., and Lacabanne, C., Study of the

mineral–organic linkage in an apatitic reinforced bone cement, *J. Biomed. Mater. Res., 24*, 1377 (1990).

265. Murakami, A., Behiri, J.C., and Bonfield, W., Rubber-modified bone cement, *J. Mater. Sci., 23*, 2029 (1988).

266. Moseley, J.P., and Lemons, J.E., An investigation of a fracture toughened autopolymerizing, PMMA, *Proceedings of the 19th Annual Meeting of the Society for Biomaterials*, 302 (1993).

267. Fishbane, B.M., and Pond, R.B., Stainless steel fiber reinforcement of polymethylmethacrylate, *Clin. Orthop. Rel. Res., 128*, 194 (1977).

268. Saha, S., and Kraay, J. Bending properties of wire-reinforced bone cement for applications in spinal fixation, *J. Biomed. Mater. Res., 13*, 443 (1979).

269. Topoleski, L.D.T., Ducheyne, P., and Cuckler, J.M., The fracture toughness of titanium-fiber reinforced bone cement, *J. Biomed. Mater. Res., 26*, 1599 (1992).

270. Topoleski, L.D.T., Ducheyne, P., and Cuckler, J.M., Fatigue properties and failure mechanisms of titanium fiber reinforced and pore reduced polymethylmethacrylate bone cement, *Proceedings of the 17th Annual Meeting of the Society for Biomaterials*, 48 (1991).

271. Wright, T.M., Conelly, G.M., Rimnac, C.M., Hertzberg, R.W., and Burstein, A.H., Carbon fibers of polymeric materials for total joint arthroplasty, *Biomaterials and Biomechanics*, 67 (1984).

272. Saha, S., and Pal, S., Mechanical characterization of commercially made carbon-fiber-reinforced polymethylmethacrylate, *J. Biomed. Mater. Res., 20*, 817 (1986).

273. Wright, T.M., and Trent, P.S., Mechanical properties aramid fiber reinforced acrylic bone cement, *J. Mater. Sci., 14*, 503 (1979).

274. Pourdeyhimi, B., and Wagner, H.D., Elastic and ultimate properties of acrylic bone cement reinforced with ultra-high-molecular weight polyethylene fibers, *J. Biomed. Mater. Res., 23*, 63 (1989).

275. Buckley, C.A., Lautenschlager, E.P., and Gilbert, J.L., High strength PMMA fibers for use in a self-reinforced acrylic cement: Fiber tensile properties and composite toughness, *Proceedings of the 17th Annual Meeting of the Society for Biomaterials*, 45 (1991).

33

Preclinical Testing of Polymer Matrix Orthopedic Bone Cements

Ken McDermott
U.S. Food and Drug Administration
Rockville, Maryland

I. INTRODUCTION

The Orthopedic Devices Branch of the Food and Drug Administration (FDA) is responsible for evaluating the safety and effectiveness of bone cements. A manufacturer or sponsor of a bone cement must submit preclinical laboratory test data and clinical data to obtain FDA approval before marketing a new bone cement. Preclinical data are important because this information helps to predict the long-term durability of cemented prostheses, which clinical data, with an average follow-up of only 2 years (current minimum FDA requirement for orthopedic prostheses), may not do. This chapter focuses on the preclinical testing that a device manufacturer or sponsor of a bone cement should consider submitting to FDA. It covers bone cements intended for orthopedic applications and having a polymer matrix (e.g., polymethylmethacrylate—PMMA), with or without additives (e.g, dye, antibiotic) and/or reinforcement (e.g., fibers).

Preclinical testing may be divided into the following categories:

1. Materials characterization (Table 1)
2. Materials integrity testing and failure analysis (i.e., mechanical and chemical degradation) (Table 2)
3. Materials biocompatibility (Table 2)

In this chapter properties that pertain to polymer matrix bone cements are listed for each of these basic test categories in Tables 1 and 2, along with important test parameters and examples of test methods that may be used to evaluate each test parameter. Some general test variables (Table 3) and suggestions for the organization of this information are also presented. This information should provide FDA and sponsors of bone cements with a common list of possible tests required to determine safety and effectiveness and reduce requests for additional information by FDA.

Table 1   Properties, Parameters, and Test Methods for the Material Characterization of Bone Cements

Property/component	Possible parameters	Examples of test methods
1. Mass/volume	Liquids and solids to be mixed	ASTM F 451, 5
2. Chemical composition of the monomer	Main ingredients	NMR, GC, HPLC, IR
	Additives	
	Purity or trace elements	Residual ignition, AA, ICP
3. Chemical composition of the solid components	Main ingredients	NMR
	Additives	
	Purity or trace elements	Residual ignition, AA, ICP
	Low-MW molecules	GC, HPLC, IR, NMR, GPC
4. Molecular structure of the solid components	Polymer blending	NMR, GPC
	Branched, linear, crosslinked	Solubility, swelling, viscosity
	Copolymer branch length	NMR
	Copolymer conversion	NMR
	Polydispersity, M_n, M_w	GPC, specific viscosity
	MWD	GPC
5. Morphology of the solid components	% crystallinity	X-ray diffraction, DSC, density
	Types and amounts of phases	Optical microscopy, birefringence
	Orientation of phases	X-ray diffraction, draw ratio
6. Microstructure of the solid components	Shape and size distribution	SEM
	Surface texture and treatment	Surface analysis
7. Macrostructure of the cement	Volume/weight fractions	
	Contacts and agglomerations	
	Location within the cement	
	3-dimensional orientation	
8. Physical properties in the dough stage	Dimensional changes	
	Loss of low-MW molecules	HPLC, GC
	Shear rate and shear stress	ASTM F 451, 7.8.1.5
	Dough time	ASTM F 451, 7.5
	Setting time	ASTM F 451, 7.7
	Viscosity:	
	Pre-dough stage extrusion	ASTM F 451, 7.8.1
	Dough stage intrusion	ASTM F 451, 7.8.2
9. Physical properties of the cement	Porosity distribution:	Porosimetry, SEM, microscopy
	Size	
	Location	
	Dimensional changes	
	% water absorption (swelling)	Sorption
	Leaching of low-MW molecules	Solubility
10. Thermal properties	Polymerization temperature	ASTM F 451, 7.6
	Crystallization temperature	DSC
	Glass transition temperature	

Table 2   Properties, Parameters, and Test Methods Regarding the Material Integrity and Biocompatibility of Bone Cements

Property	Possible parameters	Examples of test methods
1. Elastic modulus	Bending	ISO 5833
	Compression	
	Tension	ASTM D 638
2. Cyclic fatigue	S-N or fatigue limit data	ISO 5833, ASTM D 638
3. Fracture	Impact toughness	Notched 3-point bend test
		ASTM E 399 (tension disk specimen)
4. Static strength	Compression or Tension	ASTM F 451, 7.9
		ASTM D 638 (tensile dog bone)
		Diametrical tension
	Shear:	ASTM D 732
	Cement–cement	ASTM D 790
	Cement–bone	
	Cement–implant	
	Bending	
5. Residual stress	Fluid absorption	
	Polymerization	
	Composite	
6. Viscoelasticity	Indentation and recovery	ASTM F 451, 7.10
	Creep	ASTM D 2990
	Stress relaxation	
7. Wear	Micromotion and fretting	
8. Degradation	Clinical radiation therapies	
	Shelf life	
	Sterilization	Use sterilized samples for all evaluations or demonstrate no significant effect on all properties
9. Biocompatibility	General toxicity	Tripartite guidance
	Radiopacifier radioactivity	
	Animal implant study	Animal implant study
	Bone necrosis and remodeling	
	Fibrous membrane formation	
	Strength of bone and cement	
	Particulate formation	
	Cardiovascular effects	Animal implant or clinical study
	Blood hemolysis	
	Heart rate	
	Monomer concentrations	
	Blood pressure	
10. Other	Other failure mechanisms	Try to predict and test other failure mechanisms

Table 3   Test Variables Common to Some of the Properties in Tables 1 and 2

Issues	Concerns	Possible test variables
1. Components tested or the state of the cement when tested	Components before mixing	Liquid monomer Polymer powder Additives Reinforcement
	Dough	Dough
	Components within the cement	Polymer matrix Additives Reinforcement
	Polymerized cement as a whole	Polymerized cement as a whole
2. Clinical mixing	Effect of clinical processing variability on appropriate parameters	Centrifugation Hand mixing Vacuum mixing
3. Implantation technique	Effect of clinical processing variability on appropriate parameters	Cooling Pressurized implantation
4. Environment	Temperature Medium Loads	21°C, 37°C, accelerated Humidity, saline, serum, *in vivo* Cyclic, static, magnitude, duration
5. Elapsed times	Changes in mechanical properties and effects of environment	Chronology, duration, rates, cycles

A specific list of tests necessary to determine the safety and effectiveness of all possible types of bone cements is not possible because certain tests that are appropriate for one type of cement may not be appropriate for another. For example: a biodegradable cement would require more emphasis on strength versus time data, a cement with an antibiotic would require more testing on leaching kinetics and analysis of infection, and a cement that bonds to bone would require data on the cement–bone bond strength. Also, alternative tests may be substituted if they provide the same answers, or a better test method may be developed. Therefore, this chapter suggests some important concerns, issues, mechanisms, evaluation criteria, and conflicting conclusions that have been discussed in the literature, and that should be considered when developing test methods and interpreting test results in submissions of data to FDA. The properties, parameters, and test methods tabulated in this paper can be used to evaluate any type of polymer matrix cement, even if it is not specifically discussed in the text (e.g., absorbable cements, reinforced composite cements, etc.). The absence or presence of a particular topic does not imply a relative importance of one subject or parameter compared to another, nor does the mention of particular brands of cement imply that FDA considers one better than another.

Another function of this chapter is to present a list of properties and test methods for bone cements that industry, clinicians, government, and academia can use as a basis for the exchange of ideas and opinions on how to best evaluate these materials prior to clinical use. This chapter can be used as a framework for an FDA guideline and for new standards for bone cements. These may provide a mechanism for any future reclassification of acrylic bone cements from Class III, requiring Premarket Approval (PMA); to Class II, requiring a Premarket Notification [510(k)].

II. GENERAL TEST VARIABLES

Certain test variables, such as those listed in Table 3, are critical to a number of different material test parameters discussed throughout this chapter and should be evaluated. Perhaps no single test variable is as important or has been studied as much in recent literature as cement mixing (centrifugation, vacuum mixing, and hand mixing) and clinical implantation technique (cooling and pressurization), though their clinical significance, as summarized in Appendix 1, may depend on the implant site and other factors. Cement processing variables are important because:

1. They directly affect the material structure and hence, its properties (e.g., Eyerer and Jin, 1986).
2. Different cement brands can be affected differently by the same mixing technique.
3. All types of mixing techniques will be applied by surgeons regardless of the cement's labeling.

At least some material properties (e.g., morphology, macrostructure, physical properties, and thermal properties) and/or mechanical properties (e.g., fatigue) should be measured after processing by standardized methods of cooling, centrifugation, vacuum mixing, hand mixing, and specimen molding.

The effects of test temperature, surrounding medium, and mechanical loading on the dough or polymerized cement are documented throughout this chapter. At least some mechanical properties of bone cement aged under a simulated worst-case clinical condition (e.g., cyclic loading in a 37°C saline solution) must be evaluated to determine the combined effects of load, temperature, and solution. Because bone cement strength increases rapidly the first 1–2 weeks after curing, then increases at a slower rate for up to 6 months when stored in air (Looney and Park, 1986), the time elapsed between mixing and testing must be reported for all test specimens, whether aged or not.

III. MATERIALS CHARACTERIZATION

Materials characterization is a description of a cement that distinguishes it from other cements. This information also provides data that may be useful in understanding the material's mechanical and clinical performance, failure mechanisms, and biocompatibility.

A. Composition

Even a modest change in the composition of a cement may significantly alter certain properties (e.g., Brauer, Steinberger and Stansbury, 1986; Brauer, Termini and Dickson, 1977). The effect of additives on properties vary. For example, Davies and Harris (1992) reported that the fatigue properties of bone cement were not affected by the addition of 1 ml of 1% aqueous methylene blue to 40 g of Simplex P or Zimmer bone cement. Weinstein, Bingham, Sauer, and Lunceford (1976) concluded that admixing antibiotic to the powder–liquid mixture was not detrimental to the cement's compressive and tensile strength. The addition of antibiotic to hand-mixed Palacos R and Simplex P did not

significantly affect their fatigue properties, though the addition of antibiotic to centrifuged Palacos R and Simplex P significantly lowered their fatigue properties (Davies,
Nagai, and Takeshita, 1989).

On the other hand, Lautenschlager and Marshall (1976) reported that 4 ml of aqueous antibiotic solutions added to 40 g of cement particles significantly lowered the static
mechanical properties of bone cement. Wijn, Slooff, and Driessens (1975) reported negative effects due to additives such as antibiotic and radiopacifiers on curing time, water
resorption, solubility, degradation, flexural modulus and strength, compressive strength
and proportional limit, and impact strength on three commonly used bone cements.
Haas, Brauer, and Dickson (1975) reported that 10 wt% $BaSO_4$ radiopacifying agent in
bone cement reduced tensile strength, transverse strength, and modulus of rupture by
about 10%.

B. Molecular Weight

Molecular weight (MW) has been shown to influence fracture surface energy, Young's
modulus, tensile strength, and inherent flaw size in PMMA (Berry, 1964).

C. Microstructure

The shape and size distribution of the polymer powder affects handling and final mechanical properties. For example, irregularly shaped powder particles have a greater surface
area, which enhances monomer absorption and hastens curing (Kusy, 1978). A bimodal
particle size distribution results in tighter bead packing and diminished free volume between particles, and reduces the amount of monomer needed to fill this interparticle space
to as low as 11.5 ml of monomer per 40 g polymer (Mjoberg and Selvik, 1988). A lower
monomer content may, in turn, lower the exothermal temperature rise. However, a certain amount of free volume in the prepolymerized cement powder may be necessary so
the monomer will be able to flow and be incorporated throughout the powder.

Powder particle size after polymerization may affect mechanical properties (see the
discussion in Sec. IV.B—Fracture Mechanics). The shape and size of PMMA powder
particles change only slightly as a result of polymerization (Cameron, Jackson, and
MacNab, 1974). The smallest particles are completely dissolved while the diameters of
the larger beads (those greater than 160 μm in diameter) are reduced to around 142 μm
(Kusy and Turner, 1974).

D. Agglomeration

Radiopacifying or antibiotic powders may agglomerate before or during mixing or settle
during centrifugation (e.g., Skinner and Murray, 1985). Resulting inhomogeneities may
worsen mechanical properties.

E. Physical Properties in the Dough Stage

1. *Dimensional Changes*

If the volume of the cured cement is less than the volume of the dough, gaps can form
between the cement and bone or implant, resulting in eventual implant loosening. The
methylmethacylate monomer (MMM) alone shrinks 21% (Haas et al., 1975) and its
density increases from 0.94 to 1.19 g/cm^3 during polymerization. As the normal powder:
monomer mix ratio contains only about 37% MMM, the maximum that bone cement
dough could shrink during polymerization would be 7.6% (with $BaSO_4$) or 8% (without

$BaSO_4$) Actual shrinkage of implanted cement that was not vigorously mixed was only about 2–5% because:

1. Unpolymerized, residual monomer does not shrink.
2. The presence of porosity increases the bulk volume of the cement.
3. Some shrinkage of the cement dough occurs before it is implanted.

Charnley (1970) and Learmonth, Spirakis, Gryzagoridis, and Sher (1992) reported a 3–6% expansion of the initial volume of the dough. This increase was probably due to increased internal porosity from vigorous beating, which compensated for the shrinkage of the polymerizing monomer (Charnley and Smith, 1968). Vacuum mixing or centrifugation decreases porosity, resulting in a shrinkage of the cement volume (Connelly, Lautenschlager, and Wixson, 1987). Shrinkage due to centrifugation and vacuum mixing do not compromise prosthesis fixation according to Davies and Harris (1990).

2. Loss of Low-MW Molecules

It is preferable that monomer loss from the polymerizing cement be high before and low after insertion of the dough to reduce tissue necrosis and take advantage of the plasticizing effect of the monomer during handling. MMM evaporation occurs mainly during mixing of the dough (Lee and Ling, 1975). Petty (1980) measured no free MMM in tissues adjacent to bone cement 4 h after *in vivo* polymerization. Darre, Vedel, and Jensen (1987) used gas chromotagraphy (GC) to measure MMM evaporation from seven commercially available bone cements during the dough stage. MMM air concentration increased with increased handling (e.g., increasing the mixing period from 40 sec to 4 min doubled the MMM air concentration). Lee, Ling, and Wrighton (1973) and Bayne and Lautenschlager (1977) determined that monomer loss was related primarily to the physical nature of the mixing process rather than to its duration. Monomer loss increased with increased mixing speeds (e.g., up to 14% was lost at 260 beats/min).

3. Setting Time

The study by Bos, Johannisson, Lohrs, Lindner, and Seydel (1990) is one example of many that concern the effects of cooling and mixing on cement handling. They measured setting times for cement mixed manually and under vacuum at room temperature or chilled to 5°C. Viscosity was high after vacuum mixing at room temperature (except for the low-viscosity brands). Prechilling and vacuum mixing decreased the rate of polymerization and so prolonged the setting time, resulting in a lower cement viscosity (due to a lower-MW polymer) during the handling period.

4. Viscosity

The optimum cement penetration into bone for good cement–bone attachment is about 3–4 mm (Hadjari, Reindel, Hollis, and Convey, 1990). The most important parameter affecting cement penetration into cancellous bone is viscosity (lower viscosity, greater penetration). Noble and Swarts (1983) found that cement penetration into cancellous bone varied significantly for various brands of cement. Niwa, Miki, Sawai, and Hattori (1993) determined that bone cement penetration during hip replacement was facilitated by the use of:

1. An intramedullary plug and calcar dam.
2. A socket flange and acetabular dam.
3. An early insertion (less than 5 min after mixing).
4. Vibrations applied to the socket during insertion.

Cement viscosity testing provides a relative measure of the:

1. Ability of the bone cement to penetrate pores in bone for a strong fixation.
2. Time available to maneuver the cement without the introduction of built-in cracks.
3. Time available to position the prosthesis before the cement hardens.

Viscosity of cements implanted in the predough stage is measured by the extrusion viscosity test. Viscosity of cements implanted in the dough stage is measured by the intrusion viscosity test (see Table 1). For most cements, the viscosity steadily increases with time. In only one cement did the viscosity remain relatively constant between 3 and 5 min (the working time period) before the viscosity began to increase rapidly (Krause, Miller, and Ng, 1982).

F. Physical Properties of the Cement

1. Porosity Distribution

Porosity may be due to the following:

1. The introduction of air during mixing, resulting in porosity values of 5% to 16% for conventional hand-mixed cement and 0.1% to 3.4% for vacuum-mixed or centrifuged cement (Burke, Gates, and Harris, 1984; Wixson, 1992).
2. The expansion of gaseous monomer (MMM boils at 100.3 °C) due to high exothermic temperatures during polymerization (Tuckfield, Worner, and Guerin, 1943).
3. Incomplete monomer flow throughout the polymer powder.
4. Leaching of water-soluble, low-MW substances over a period of months or years, resulting in an additional porosity of 2–5% (internal) and 1–2% (surface) — due to leaching of antibiotic and monomer, respectively (Kusy, 1978).
5. CO_2 production from the degradation of benzoyl peroxide during setting (Lautenschlager, Stupp, and Keller, 1984).

The elimination of pores from bone cement may reduce the number of stress concentrations and improve mechanical properties, but it might also eliminate their functioning as crack termination points (Eftekhar, 1992). Ideally, the internal porosity of bone cement should be lowered to the point where crack initiation occurs at the metal–cement or cement–bone interfaces rather than at pores in the cement (Wixson, 1992).

Internal Porosity.   The method of cement preparation has a major effect on porosity. Jasty, Davies, O'Connor, Burke, Harrigan, and Harris (1990) reported that porosity varied from 5% to 16% when prepared by hand. Centrifugation for 30 sec resulted in a substantial reduction in the overall porosity of Simplex P, AKZ, Zimmer Regular, and CMW bone cements by reducing both the mean pore size and the number of pores per unit area. The porosity of LVC, Palacos R, and Palacos R with gentamicin was not significantly decreased by centrifugation. Chilling the monomer before mixing resulted in higher porosity for both the centrifuged and uncentrifuged Simplex P, Zimmer Regular, and CMW bone cements.

A partial vacuum removes air from cement (Davies and Harris, 1990; Hansen and Jensen, 1990). Noble, Jay, Lindahl, Maltry, Dreeben, and Tullos (1987) reported cement porosities after vacuum mixing to be 30% and 56% of the porosity of hand-mixed and centrifuged cement, respectively. Though vacuum mixing reduced porosity, several relatively large (0.5–3.0 mm) pores occurred regardless of the technique used. Too high a vacuum (550 mm Hg) will evaporate monomer and increase porosity. The time of vacuum application is also important as gas bubbles that form may become trapped before they have time to escape from the hardening cement.

Due to the large influence of mixing technique and porosity on mechanical properties (as discussed later in Sec. IV), it is important to report these variables for all test specimens. A worst-case standardized hand-mixing method might be chosen to evaluate the effects of high amounts of porosity on cement properties. For example, Linden (1988) found that manual mixing of 46 samples of acrylic cement by seven nurses resulted in a mean porosity of 26.7% ± 9.21% with large variations among specimens. There was also a large variation in pore size, which exceeded 300 μm in 23% of the specimens.

Bone Cement–Implant Interface.   Recent autopsy studies have suggested that bone cement failure in hip arthroplasties started at the bone cement–metal interface in hips, more often than at the bone cement–bone interface. Helmke, Lednicky, and Tullos (1992) examined porosity at the cement–implant surface at various positions around the stem for seven cemented femoral explants. The prostheses were implanted an average of 9.2 years (range 7–13 years). Percent porosity varied greatly at the interface: from less than 1% to almost 90% (average 18% ± 18% for all seven explants). Pore densities were the same all around the stems except for a greater proximal and distal porosity in two specimens due to motion prior to complete hardening. Pore size was also consistent, averaging 105 ± 56 μm. Pores appeared to accumulate around the corners of the prostheses and adjacent to other surface features with a small radius of curvature. Rough implant surfaces seemed to have lower pore densities compared to smooth stems. The authors concluded that the interfacial porosity was due to air entrained during insertion of the stem into the cement and not due to monomer evaporation because of the uniform void size and elongated pore distribution.

Extensive porosity was observed over about 20% of the cement-metal hip stem interface by James, Schmalzried, McGarry, and Harris (1993). Porosity covered an even greater area (30–40%) at the interface between bone cement applied during surgery and the precoated bone cement applied to stems by the manufacturer. The interfacial porosity resulted from inadequate flow and the inability of the cement to wet the prosthesis and displace air. Prewetting the stem with monomer reduced the interfacial porosity, but resulted in large standard deviations in porosity due to monomer evaporation. Thickening the monomer with either Cabosil (amorphous fumed silica thickening agent) or PMMA powder reduced the variability in porosity. However, the monomer did not polymerize, resulting in a weak interface.

Centrifugation made the interfacial porosity worse by driving pore nucleation sites out of the bulk cement to the surface (James, Schmalzried, McGarry, and Harris, 1993). The authors found that a higher concentration of pores occurred at the proximal and distal areas, as well as along lines of flow at the corners of hip implants. The 100- to 300-μm pores at the bone cement–stem interface were not visible on radiographs. Cracks were observed aligned along the edge of the implant and running through the pores.

Porosity at the cement–implant interface and its effect on interfacial strength are not inherent properties of bone cement, and measurements of these parameters for the purposes of evaluating safety and effectiveness may be superfluous. This may be true when there is little bone cement–stem contact or if mechanical interlocking can be achieved despite the porosity. However, this may be an important failure mechanism and should be kept in mind when interpreting the clinical significance of other test results.

On the other hand, cement–implant interfacial porosity may be worth studying if a new cement has significantly different wettability or rheological property that might affect the formation of pores at this interface. This effect could be studied directly by observing the amount of pore formation at the interface by a standardized method. For example, James, Karydas, McGarry, and Harris (1993) and James, Schmalzried, Mc-Garry, and Harris (1993) measured pore formation in bone cement adjacent to a glass tube by staining the pores and observing them from inside the glass tube through an arthroscope. Because pore formation was unaffected by the substrate surface chemistry and thermal conductivity, glass could be used in place of metal. Porosity was 1.5% and 9% if the glass was inserted into the bone cement 2 and 7 min respectively, after mixing. Inserting the glass tubes through a tight-fitting diaphragm reduced porosity from 9% to 1%.

2. Dimensional Changes

Sorption is the increase in weight of a cement specimen when immersed in a liquid (Haas et al., 1975). PMMA absorbs about 2–4% water. Sorption may be beneficial if it results in expansion and a better seating of the prosthesis. Water absorption also results in plasticization of the cement, which affects mechanical properties as discussed later in Sec. IV.A.1.

3. Leaching of Low-MW Molecules

James and McGarry (1992) concluded that the degree of polymerization of the monomer of bone cements is quite low and that there is a substantial amount of residual monomer, (10% of the original monomer) that does not polymerize and leaches out of the cement over time. Depending on the amounts, leaching of monomer, accelerator, and stabilizer out of cured bone cement may cause problems with biocompatibility.

Bone cement solubility is the change in weight after soluble molecules have been leached out (Kusy, 1978). Haas et al. (1975) measured residual MMM contents during storage in air of 3.3% after an hour, 2.7% after 24 h and 2.4% after 215 days. After 4.5 months in 37°C water, residual MMM content was 1.4%. Litsky and Frazier (1993) measured monomer release from bone cements made of either PMMA or polybutylmeth-acrylate beads in a methyl methacrylate matrix (PBMMA) during cyclic loading to 2000 N (three times body weight) in 37°C saline. Both cements had a peak monomer release early on (1136 ± 660 ppb for PMMA and 355 ± 309 ppb for PBMMA). The lower-modulus PBMMA cement did not release significantly more monomer (0.33 mg for PMMA and 0.42 mg for PBMMA) though PBMMA deformed more readily during loading.

G. Thermal Properties

1. Cement Temperature During Polymerization

Temperatures in bone cement during polymerization range from 37°C to 70°C *in vivo* and up to 122°C *in vitro* (Toksvig-Larsen, Franzen, and Ryd, 1991). Haas et al. (1975) found a difference of 9°C in the peak exothermic temperature among ten different types of bone cements. A high cement temperature is a disadvantage because it can affect the properties of the cured cement (e.g., increase porosity) or cause tissue necrosis. The heat generated by a setting bone cement is less likely to cause thermal necrosis if:

1. The cement layer between the prosthesis and bone is thin enough.
2. The hardening cement is a poor heat conductor.
3. The prosthesis is a good heat conductor.
4. The implants are adequately cooled.

Biehl, Harms, and Hanser (1974) and Huiskes (1980) obtained a 5°C lower bone cement temperature at the bone–cement interface as a result of cooling the femoral prosthesis. Wykman, A.G.M. (1992) lowered the bone–cement interfacial temperature from 49°C (41°–67°C) to 41°C (37°–48°C) by irrigation of an acetabular cup during cement curing. However, Meyer, Lautenschlager, and Moore (1973); Toksvig-Larsen et al. (1992) and Zygmunt, Toksvig-Larson, Saveland, Rydholm, and Ryd (1992) found no effect on the bone cement temperature at the bone–cement interface as a result of cooling.

Hansen and Jensen (1990) reported that in seven of nine commercially available bone cements, prechilling monomer and powder to 5°C and vacuum mixing resulted in a 10° to 12°C increase in *in vitro* exotherms, often exceeding 88°C. The rate of the polymerization reaction was halved for every 7°–10°C decrease in monomer and powder temperature. The slower polymerization reaction increased the availability of residual monomer and free radicals later in the reaction, resulting in a higher temperature. The presence of PMMA–styrene copolymer or a higher content of benzoylperoxide may also lead to a higher degree of autocatalyzation and higher cement temperature.

Swensen, Schurman, and Pisiali (1981) used finite element analysis to evaluate the thermal conduction around hip implants during cementation. Cooling the hip stem to 0°C made the cement–bone interface cooler by only 3°–5°C due to the low thermal heat flow through the PMMA cement. The lower cement temperature adjacent to a cooled hip stem extended the curing time, giving the implant a greater chance for motion and hence loosening prior to solidification of the cement. In general, cement peak temperatures were not problematic because they were below temperatures that reduce cell regeneration. Cement peak temperatures barely reached the lower end of the coagulation range (56°–70°C) of collagen proteins, and then, only in small areas of the cement–bone interface.

2. Glass Transition Temperature

As a polymer hardens, it becomes increasingly more difficult for monomer to diffuse in order to continue to polymerize. Diffusion is limited by the restricted motion of the solidifying polymer molecules. Molecular motion is restricted when the temperature of the reactants is below the glass transition temperature (T_g). T_g is the temperature at which the polymer chains have enough thermal energy to move freely and behave like a viscous liquid (Cowie, 1973). The T_g of a polymer is effectively lowered by monomers which act as plasticizers. This effect permits the polymerization of doughy bone cement to continue until 10% of the MMM is left though the bone cement temperature is below the actual T_g of PMMA (Flory, 1953). Polymerization ceases at about 10°C below T_g. The type and quantity of crosslinking agents also have a major effect on T_g for autopolymerized types of PMMA such as bone cement (Oysaed, 1990).

A lower T_g permits greater molecular mobility so that low-MW molecules may diffuse through the material at a greater rate, permitting more monomer to react to form polymer more quickly. This increases the rate of polymerization so the cement temperature is lower and more of the monomer is polymerized before implantation rather than escaping into the tissues after implantation (Trap, Wolff, and Jensen, 1992).

IV. MATERIALS INTEGRITY

Materials integrity testing determines the conditions under which a cement fails (e.g., mechanical and chemical degradation). Clinical performance may be predicted by making appropriate comparisons of the integrity of the cement under investigation to other commonly used cements. Saha and Pal (1984), Eftekhar (1978), and Crissman and McKenna (1984) reviewed various mechanical test methods for bone cement and summarized the effects of different variables on mechanical properties.

A. Fatigue

1. Test Methods

Fatigue plays a critical role in bone cement failure (O'Connor, Burke, Zalenski, and Harris, 1989). Krause, Mathis, and Grimes (1988) and Krause and Mathis (1988) reviewed bone cement fatigue properties in the literature and made some suggestions regarding test methods. For example, mixing and handling of bone cements result in many flaws that affects cement fatigue strength. It is important that bone cement fatigue test specimens undergo consistent methods of preparation. If the purpose is to understand the inherent fatigue behavior of a bone cement (e.g., to evaluate the effects of filler particles or various phases), nearly pore-free samples must be used because of the pronounced inverse correlation between porosity and the number of cycles to failure.

Cyclic loading may cause an increase in the cement temperature, which could affect test results. Humphreys, Orr, and Bahrani (1989) cemented a titanium alloy hip prosthesis into a tube (modeling the upper part of the femur). Five thermocouples were embedded in the bone cement and the assembly was subjected to cyclic loading with a range of 0.3–4.5 kN at a frequency of 6 Hz. Temperature measurements over a 48-hour period indicated that the temperature rise in the bone cement was less than 4°C. It was concluded that such tests can be carried out at 6 Hz without significantly affecting the mechanical properties of PMMA bone cement.

The temperature or surrounding medium may also influence fatigue life. Freitag and Cannon (1977) loaded rotating beam bone cement specimens with and without $BaSO_4$ in either air or 37°C serum. The presence of $BaSO_4$ had no effect on fatigue life. The fatigue life of specimens tested in air was less than those tested in 37°C serum, possibly due to the plasticizing effects of absorbed fluids (as described later in Sec. IV.C.1). Stark (1979) reported that the fatigue properties of Surgical Simplex P bone cement were superior to those of Zimmer bone cement when tested at 22°C in air, in contradiction to the results of Freitag and Cannon (1976).

Cement may be cyclically loaded under constant stress or constant strain. Krause, Grimes, and Mathis (1988) discussed methods of presenting fatigue data on acrylic bone cement using the probability-stress-number of cycles to failure (P-S-N) relationship. Carter, Gates, and Harris (1982) reported that strain control tension–compression cyclic fatigue loading of wet bone cement at 37°C more accurately described failure than stress control.

2. Mixing

Centrifugation and vacuum mixing can increase the fatigue life of notched and unnotched Simplex P specimens by a factor of 5 and 100, respectively, over the manufacturer's suggested techniques (Davies and Harris, 1989). Gates, Carter, and Harris (1984); Burke, Gates, and Harris (1984); and James, Jasty, Davies, Piehler, and Harris (1992) found that centrifugation increased fatigue life due to the fewer number and smaller sizes of internal pores and better pore distribution.

3. Porosity

James et al. (1992) correlated fatigue strength with porosity at the fracture surface after cyclically loading notched PMMA bone cement specimens. There may be a pore size and pore volume in bone cement at which fatigue strength is optimum (Topoleski and Lu,

1993). Though porosity can lower some mechanical properties such as strength, pores in the discontinuous damage zone in front of the tip of a propagating crack may dissipate the energy of the crack and slow down its propagation. The crack may deflect to pores or microcracks may form between pores. Cracks initiate near notches, but not necessarily at them due to the discontinuous damage zone in front of the notch.

B. Fracture Mechanics

1. Test Methods

Fracture mechanics may provide useful information regarding bone cement failure mechanisms such as how microstructure affects strength. For example, intergranular cracking around the powder particles within the matrix indicates poor particle–matrix interfacial bonding. Poor bonding may be due to retention of a suspension agent in the original polymer powder or due to a lack of monomer diffusion into the particles before curing as a result of too high a polymerization rate (Kusy, 1978). Bone cement strength may be increased by the weak interfaces between PMMA powder particles because of crack deviation along these interfaces (Beaumont and Young, 1975), or bone cement strength may be decreased if these interfaces act as sites of secondary crack initiation (Kies, Sullivan, and Irwin, 1950; Kusy and Turner, 1974).

Work of fracture has been useful in evaluating the effects of aging and environment. For example, Watson, Miles, and Clift (1990) found no change in work of fracture with time of storage in air at 21°C and 37°C. Compared to air curing, curing in water resulted in a significantly higher work of fracture due to plastication (discussed later in Sec. IV.C.1). For specimens cured in water, storage at 21°C resulted in a 40% increase in work of fracture. No change in work of fracture occurred in specimens stored at 37°C. The work of fracture of water-cured cement was double that of air-cured cement after 21 days. Work of fracture was greater for specimens cured in air at lower temperatures because faster cure rates at higher temperatures resulted in less residual monomer and, therefore, a less ductile bone cement.

Freitag and Cannon (1976) reported values of fracture toughness of bone cements for various environments and as a function of pressure during setting. The $BaSO_4$ particles reacted with bovine serum, resulting in an increase in cement toughness. When fabricated at 5 psi, the fracture toughness of Zimmer cement was greater than that of Simplex P cement for all additive concentrations. The superior Zimmer cement fracture toughness was hypothesized to have been due either to polymer chain ordering or to a lowering of the toughness of the Simplex P by the methacrylate–styrene copolymer in that cement. Stark (1979) did not find a significant difference between these cements.

Due to the complexity of *in vivo* loading conditions, *in vitro* fractography studies of a variety of loading conditions must be performed for adequate comparisons to in vivo fracture samples (James et al., 1992). Another problem is the variation in fracture toughness, which can be as much as 200% for a relatively homogeneous PMMA material (Berry, 1972).

2. Mixing

Rapid, uncontrolled hand mixing of Zimmer LVC bone cement resulted in lower K_{IC} and higher porosity values compared to controlled hand mixing and centrifugation (Wang and Pilliar, 1989). Rimnac, Wright, and McGill (1986) reported no significant difference in fracture toughness between centrifuged and hand-mixed cements. Friis, Stromberg, Cooke, and McQueen (1993) measured the fracture toughness (ASTM E 399, 1990) of Simplex P bone cement, notched with a razor and precracked by a 20-Hz cyclic load.

Vacuum-mixed cements had a 19% higher fracture toughness and a 17% higher tensile strength than hand-mixed cement.

C. Static Strength

1. Test Methods

Cement strength decreases in the presence of water and higher temperatures. The strength of cement can drop by as much as 10% when the test temperature is increased from 20°C to 37°C (Lee, Ling, and Vangala, 1977). Decreases of dental cement tensile strength in 37°C water have been reported to be as high as 50% after 10 months (Kusy, 1978). A 50% reduction in diametral strength of Surgical Simplex P cement mixed with up to 12 cc of water was reported by Lautenschlager and Marshall (1976) and a 3.5% decrease in bending strength of explanted bone cement after 26 months in rabbit muscle was reported by Lautenschlager et al. (1984). Jaffee, Rose, and Radin (1974) found no significant change in the compressive, flexural and fatigue properties of CMW and Simplex cements stored in bovine serum at 37°C for up to 2 years.

The plasticizing effect of water that has diffused into the cement may lower strength because more deformation can occur due to increased molecular mobility (Lauterschlager et al., 1984). However, water plastication may also dissipate stresses at defects and so inhibit crack growth (Beaumont and Young, 1975).

The most appropriate mode of loading bone cements is uncertain. Cement is weaker in tension as cracks propagate under tension. Tensile hoop stresses occur in cement surrounding a hip stem. However, because cement acts as a grout, it should be primarily loaded in compression and bending. Indentation tests have been used to make repeated nondestructive measurements on a single sample as it ages (Haas et al., 1975).

2. Mixing

Noble et al. (1987) reported that vacuum-mixed bore cement was 35% stronger than either hand-mixed or centrifuged bone cement. Arroyo (1986) reported an 80% reduction in porosity and 25% increase in ultimate tensile properties for Radiopaque Simplex P cement vacuum mixed at room temperature. There was no change in setting time, setting temperature, or viscosity. Laminations in cement caused by folding the dough upon itself during implantation create lines of weakness in the cured cement. Laminations that were formed in air and in the presence of blood reduced cement strength by 54% and 77%, respectively (Gruen, Markolf, and Amstutz, 1976).

3. Interfacial Strength

Bone Cement-Bone.   Cement fracture may initiate at the bone cement-bone interface (Bannister, Miles, and May, 1989) though failure at the cement-implant interface is more likely as discussed below. Shear strength at the interface between bone cement and bone has been found to be affected by many parameters (Greenwald, 1979), such as the depth of penetration of the cement into the bone up to 3–4 mm (Hadjari et al., 1990; MacDonald, Swarts, and Beaver, 1993), reaming (Oh, Bushelow, and Sander, 1985), and cancellous bone strength (Bartel and Wright, 1991).

Bone Cement-Implant.   Jasty et al. (1990b) found that cracks in bone cement were initiated at the cement-prosthesis interface in all 16 clinically successful hip explants that they examined. Clinical explant studies have shown that debonding occurred at the bone cement-implant interface while, in the same hip, the bone-bone cement interface remained intact (Harris, 1992). Hip stem separation from the bone cement occurs at a rate of about 26%. The bone cement–metal hip stem interface is weaker than bone cement by a factor of 5 or more (Harrigan, Kareh, and Harris, 1990b). The cement–implant interface is more likely to fail in tension (tensile strength of 1.8 MPa) rather than in shear (shear strength of 3.3 MPa) (Clech, Keer, and Lewis, 1984).

Bundy and Penn (1987) found that the most significant increases in strength at the cement–implant interface occurred when high pressures were applied to the cement during curing. Roughened or highly polished surfaces had significantly higher cement–metal interfacial strengths (the former due to mechanical interlocking and the latter due to atomic interaction effects). Arroyo and Stark (1987) measured a 1.4 MPa increase in cement–Vitallium shear strength for every 1 μm of increased metal surface roughness. Keller, Lautenschlager, Marshall, and Meyer (1980) discovered that higher cement–implant bond strengths were achieved when the cement was placed onto treated metal surfaces before or at the beginning of the dough stage due to the better flow of the cement. Rough surfaces weakened the cement–bone interface by causing air bubble formation there. The cement–metal bond strength was greatest for surface-treated stainless steel and worst for titanium.

D. Stress in the Cement

Jaffee, Rose, and Radin (1974) recommended that cement compressive stress be limited to less than 1.4 MPa (based on compressive fatigue data) to ensure long-term viability of any design or implantation technique. Bone cement about a hip implant sees stresses of about ± 7 MPa (Davies, Daniel, O'Connor, Burke, Jasty, and Harris, 1988). The stress in bone cement is affected by:

1. Stem design (Burke, 1985; Crowninshield, Brand, Johnston, and Milroy, 1980a, 1980b; Weightman, Freeman, Revell, Braden, Albrektsson, and Carlson, 1987).
2. The presence of a porous coating (Manley, Stern, and Gurtowski, 1985).
3. Cement modulus (Crowninshield, Brand, Johnston, and Milroy, 1980b; Fagan and Lee, 1986; Prendergast, Monaghan, and Taylor, 1989; Rohlmann, Mossner, Bergmann, Hees, and Kolbert, 1987; Tarr, Clark, Gruen, and Saramento, 1983).
4. Location along the stem (Davidson, Dingman, and Lynch, 1986; Ebramzadeh, Mina-Araghi, Clarke, and Ashford, 1985; Harrigan, Kareh, and Harris, 1990a).

E. Viscoelasticity

PMMA bone cement exhibits creep (Chwirut, 1984), stress relaxation (Pal and Saha, 1982) and strain rate sensitivity (Lee et al., 1977). Bone cement is very sensitive to strain-rate, so it is important to specify the strain rate in mechanical testing (Saha and Pal, 1983). For example, the ultimate compressive strength of Simplex P increased from 37 to 61 MPa with a 10-fold increase in strain rate (0.0012/sec to 0.012/sec).

Creep of PMMA bone cement may allow a redistribution and lowering of stresses (Litsky, Rose, Rubin, and Thrasher, 1990; Litsky, and Yetkinler, 1991). Cement creep radially around the stem may also result in subsidence in smooth tapered hip prostheses (Fowler, Gie, Lee, and Ling, 1988). Chwirut (1984) reported cracking in five commercially available acrylic bone cements resulting from long-term compressive creep loading in 37°C saline. The cracks resulting from the compressive loading ran between the largest pores present in the cement. Creep strain varied greatly between the five types of cements tested. A carbon-reinforced cement was by far the most creep resistant, even at a fiber content of only 2%. However, the measured creep strains exceeded the elastic strain limit of all the cements.

F. Wear

Wear has been observed on explanted fracture surfaces (Topoleski, Ducheyne, and Cuckler, 1990), and the surfaces of the cement mantle in failed implants are usually smooth and polished due to motion of the loose cement mantle (DiCarlo and Bullough, 1992). The effect of cement particulates on wear of joint prosthetic surfaces and on bone resorption is of concern (see Appendix 2).

G. Degradation Due to Radiation

Scullin, Greenwald, Wilde, and Beck (1977) reported that there was no effect on mechanical properties of bone cement after irradiation at levels comparable to 1–6 times a typical therapeutic dose for the treatment of metastatic disease. Eftekhar and Thurston (1975) found no effect of 10,000 rads on the mechanical properties of bone cement.

V. MATERIALS BIOCOMPATIBILITY

To address the huge amount of literature on acrylic cement biocompatibility would require another chapter. The types of biocompatibility tests necessary (Table 2) depend mainly on the composition of the cement and the types and amounts of materials leaching or degrading *in vivo*. The effect of cement particulates is especially important (see Appendix 2).

VI. REPORTING

To help FDA in its evaluation of safety and effectiveness, all preclinical test methods and results should be organized in the order outlined in Tables 1 and 2. Any additional information not specifically mentioned should be inserted into this organization where appropriate. Numerical values should be summarized in tables with descriptions of materials and methods in just enough detail to identify and distinguish different values (examples given in Appendix 3). Detailed test reports from which the summarized data originated should be organized in a similar manner (as much as possible) and included in the submission to FDA. Each detailed report should be organized as outlined in Appendix 4. Detailed descriptions of possible mechanisms that might explain and tie together the results should be provided.

A materials computer data base should be established to further expedite the submission and review of information. A common data base structure can be distributed to interested persons on an electronic computer disk. The structure could contain a field for every possible parameter (e.g., Tables 1 and 2) for which data are collected. Published and unpublished test results can be entered into the data base so that specific information can be accessed and organized by FDA personnel to allow easy comparisons of different data.

VII. CONCLUSIONS

This chapter lists bone cement properties, parameters, and test methods; and discusses some concerns, issues, mechanisms, evaluation criteria, and conflicting conclusions that may be important to consider when developing test methods and interpreting test results for future submission to FDA. This document should be used by FDA personnel and persons submitting data to FDA as a common guide to providing and organizing information on bone cements. A sponsor of a bone cement is also responsible for addressing pertinent issues, mechanisms, and evaluation criterion that are not specifically discussed in this chapter. Interested persons, societies, and standards organizations are also encouraged to suggest to FDA, additional information and opinions on safety and effectiveness testing that will help industry, clinicians, government, and academia evaluate these materials prior to their clinical use. That manuscript would be converted into a FDA guidance document based on a consensus obtained from persons interested in commenting. These documents might stimulate activity in writing new standards for bone cements. A materials computer data base should be established to further expedite the submission and review of information. These sources will facilitate any future efforts to reclassify acrylic bone cement from Class III, requiring Premarket Approval (PMA); to Class II, requiring a Premarket Notification [510(k)].

ACKNOWLEDGMENTS

I wish to acknowledge the following persons who reviewed this manuscript and provided very helpful suggestions: Mark Melkerson, Debbie Lumbardo, Nestor Arroyo, William Harris, Subarta Saha, Stephen Rhodes, Samie Niver, Daniel B. Smith, Daniel Chwirut, Joon Park, Ted Stevens, and Dan McGunagle.

APPENDIX 1: THE SIGNIFICANCE OF MIXING TECHNIQUE ON BONE CEMENT PROPERTIES

Conclusions Based on Preclinical Testing

Rimnac et al. (1986) concluded that centrifugation may not improve the durability of bone cement much because:

1. Fracture toughness and fatigue crack propagation in the presence of surface imperfections was not improved by centrifugation.
2. Centrifuged cement shrinkage was significant during setting.
3. The stress concentration effects caused by trabecular bone was greater than that caused by the pores.

However, James et al. (1992) and Harrigan, Davies, Burke, O'Connor, Jasty, and Harris (1987) found that centrifugation improved fatigue life, even in the presence of notches formed by trabecular bone because the stress concentrations around the porosity were more severe than around the notches. Both porosity and trabecular bone were major factors in causing fatigue failure of bone cement. Removal of one parameter resulted in the predominance of the other. Centrifugation significantly improved cement fatigue strength even in the presence of trabecular bone. Wixson (1992) stated that improvements in methods of bone preparation and administration of bone cement have resulted in a marked improvement in cement properties.

Conclusions Based on Clinical Data

Sharkey and Bargar (1987) observed no notable improvements in cement mechanical properties due to improvements in cementing technique in explanted bone cement samples over the decade of the 1970s. Wroblewski (1986) reported that none of the improvements in bone-cementing techniques have improved the long-term results of acetabular components. In cases where failure occurs at the cement–bone interface, as in cemented acetabular migration, decreased porosity from centrifugation or vacuum mixing may not be expected to improve mechanical properties (Wixson, 1992).

On the other hand, Mulroy and Harris (1990) reported that improved cementing techniques have greatly reduced aseptic loosening of the femoral hip component, though not the acetabular hip component. Harris and Davies (1988) also concluded that improved cementing techniques seem to have resulted in a marked improvement in the longevity of femoral component fixation. The loosening rate at 6 years had been reduced from 20% using old techniques, to less than 2%.

APPENDIX 2: BIOCOMPATIBILITY AND BONE CEMENT PARTICLES

PMMA particles ingested by macrophages cannot be degraded by lysozymal enzymes (Amstutz, Campbell, Kossovsky, and Clarke, 1992). The cells die and the ultimate tissue response is necrosis and chronic inflammation. Bone in the vicinity undergoes osteolysis, which may lead to loosening. This sequence of events has been referred to as *cement disease* in the past, but now it is more appropriately referred to as *particulate disease*.

Emmanual and Hedley (1991) evaluated the cytotoxicity of bone cement wear particles fabricated in a press or retrieved from femoral and acetabular membranes during revision arthroplasty. The fabricated particles were 11–180 μm in size, white, and had angular and jagged edges. The retrieved particles were 5–200 μm in size, yellow, smooth, and rounded. Both particles stimulated monocytes to a great extent. Papatheofanis and Barmada (1991) incubated ground and washed PMMA bone cement particles (50–60 nm in diameter) with polymorphonucleocytes (PMNs). The centrifuged supernatant was assayed for enzyme release and distance of cell migration. There was a dose-dependent increase in degranulation and decrease in PMN migration. The toxic effect of PMMA may have stimulated the otherwise unstimulated, nascent PMNs.

Lazarus, Cuckler, Baker, Ducheyne, and Schumacher (1991) injected bone cement particles ground to a size of less than 50 μm into subcutaneous air pouch cavities and knee joints of rats. Particles originating from bone cement containing $BaSO_4$ resulted in a greater release of inflammatory mediators and a greater bone loss. The results suggested that PMMA particles cause a significantly greater inflammatory response than titanium particles or saline.

PMMA beads (sized either 1–10 μm, mean 5.5 μm; or 10–126 μm, mean 55 μm) were injected into subcutaneous air pouch cavities of rats (Gelb, Baker, Shumacher, and Cuckler, 1992). The dosage of the large particles was four times the dose of small particles. A significantly higher inflammatory response (white blood cell count, protease and prostaglandin E2 [PGE2]) was observed around the larger particles compared to the smaller. Significantly higher levels of tumor necrosis factor was measured around the smaller particles compared to the larger.

Goodman, Fornasier, and Kei (1988) reported that particulate PMMA debris (10–100 μm) at the cement–bone interface stimulated a foreign-body histiocytic and giant-cell response after a few weeks of implantation. PMNs, which are involved with the initial phagocytic attack at the wound site, may persist due to the presence of bone cement particles. The influx of macrophages, fibroblasts, and vascular endothelial cells, the second stage of wound repair, does not occur because of the PMN persistent release of inflammatory mediators. This cyclical acute inflammation results in osteolysis within only a few weeks after prosthetic implantation according to Maloney, Jasty, Bragdon, and Harris (1990).

Goodman, Lee, and Fornasier (1991) implanted particulate PMMA, $BaSO_4$ (both 1–100 μm) and polyethylene (PE) (an average 16 μm) into a 6-mm drill hole in the tibia of rabbits, 1 cm distal to the knee. After 3 days, PGE2 was significantly higher in the fibrous membrane harvested from the side implanted with particulate bone cement compared to PE. A diet that included naproxen dissolved in drinking water significantly reduced PGE2 levels. PGE2 has been shown to cause increased bone resorption *in vivo* and *in vitro*. More PGE2 has been found in membranes surrounding loose prostheses compared with non-loosened implants (Goodman, Chink, Chiou, Schurman, Woolson, and Masada, 1989).

APPENDIX 3: HEADINGS OF COLUMNS IN A TABLE SUMMARIZING NUMERICAL RESULTS

Numerical results should be summarized into tables containing, at a minimum, the following columns of information:

1. Test method.
2. Reference number.
3. Page in the reference.
4. Sample identity (e.g., Brand X, $BaSO_4$, centrifuged).
5. Type of data (e.g., data point, average, standard deviation, minimum, maximum).
6. Actual value in exponential notation (e.g., 1.4×10^8).
7. Units of measure (e.g., MPa, psi).

APPENDIX 4: THE CONTENT OF DETAILED TEST REPORTS

Detailed reports should be organized and subdivided into separate sections (some sections may be combined if clarification is enhanced) having the following headings:

Report Title
Investigators' Names
Facility Performing the Test
    Name
    Address
    Phone number
Dates
    Test initiation
    Test completion
    Completion of final report
Objective/Hypothesis
Test and control samples
    Sample selection criterion
    Composition
    Structure

Lot numbers
Processing methods
Location of test specimens in device
Differences between test samples and marketed device
Methods
Test system
General description
Test setup schematic or photograph
Test parameters (dependent, independent, uncontrolled, errors)
Environmental medium and temperature
Load directions and magnitude
Times (e.g., chronology, durations, rates, cycles)
Materials processing
Other
Measurement instrumentation accuracy
Methods of specimen examination
Statistical justification for the number of samples
Test procedures
Deviations from referenced protocols, standards, or controls
Results
Failure analysis
Statistical evaluation of results
Discussion of the data and possible mechanisms
Conclusions
List of conclusions
Discussion of the hypothesis/objective
Clinical implications of results, simplifications, and assumptions
Appendixes
Experimental data
Calculations
Bibliography of all references pertinent to the report

APPENDIX 5: ABBREVIATIONS AND ACRONYMS

510(k)	Premarket Notification
AA	atomic absorption
DSC	differential scanning calometry
FDA	Food and Drug Administration
GC	gas chromotagraphy
GPC	gel permeation chromotagraphy
HPLC	high-pressure liquid chromotagraphy
ICP	inductively coupled plasma
IR	infrared chromotagraphy
MMM	methylmethacylate monomer
MW	molecular weight
NMR	nuclear magnetic resonance
PBMMA	polybutylmethacrylate beads in a methyl methacrylate matrix
PE	polyethylene

PGE2   prostaglandin E2
PMA   Premarket Approval Application
PMMA   polymethylmethacrylate
PMN   polymorphonucleocyte
S-N   stress or strain versus the number of cycles
SEM   scanning electron microscopy
T_g glass transition temperature

REFERENCES

American Society for Testing and Materials (1986). ASTM F 451, Specification for Acrylic Bone Cement.

American Society for Testing and Materials (1990a). ASTM D 732, Test Method for Shear Strength of Plastics by Punch Tool.

American Society for Testing and Materials (1990b). ASTM E 399, Test Method for Plane–Strain Fracture Toughness of Metallic Materials.

American Society for Testing and Materials (1991). ASTM D 638, Test Method for Tensile Properties of Plastics.

American Society for Testing and Materials (1992a). ASTM D 790, Test Method for Flexural Properties of Unreinforced and Reinforced Plastics and Electrical Insulating Materials.

American Society for Testing and Materials (1992b). ASTM D 2990, Tensile, Compressive, and Flexural Creep and Creep Rupture of Plastics.

Amstutz, H.C.; Campbell, P.; Kossovsky, N.; and Clarke, I.C. (1992). Mechanism and clinical significance of wear debris-induced osteolysis. *Clin. Orthop., 276,* 7–18.

Arroyo, N.A. (1986). Physical and mechanical properties of vacuum mixed bone cement. *Transactions of the 12th Annual Meeting of the Society for Biomaterials*, Society for Biomaterials, p. 187.

Arroyo, N.A.; and Stark, C.F. (1987). The effect of textures, surface finish and precoating on the strength of bone cement/stem interfaces. *Transactions of the 13th Annual Meeting of the Society for Biomaterials*, Society for Biomaterials, p. 218.

Bannister, G.C.; Miles, A.W.; and May, P.C. (1989). Properties of bone cement prepared under operating theatre conditions. *Clin Mater., 4,* 343–347.

Bartel, D.L.; and Wright, T.M. (1991). Design of total knee replacement. In *Total Joint Replacement* (Petty, W., ed.). W.B. Saunders, p. 467.

Bayne, S.C.; and Lautenschlager, E.P. (1977). Clinical influences on bone cement monomer release. *JBMR, 11,* 859–869.

Beaumont, P.W.R.; and Young, R.J. (1975). Slow crack growth in acrylic bone cement. *JBMR, 9,* 423–439.

Berry, J.P. (1964). Fracture processes in polymeric materials. V. Dependence of the ultimate properties of poly(methyl methacrylate) on molecular weight. *J. Polym. Sci., 2,* 4069–4076.

Berry, J.P. (1972). In *Fracture of Nonmetals and Composites* (Liebowitz, H., ed.). Academic Press, New York.

Biehl, G.; Harms, J.; and Hanser, U. (1974). Experimental studies on heat development in bone during polymerization of bone cement. Intraoperative measurement of temperature in normal blood circulation and in bloodlessness. *Arch. Orthop. Unfallchir., 78,* 62–69.

Bos, I.; Johannisson, R.; Lohrs, U.; Lindner, B.; and Seydel, U. (1990). Comparative investigations of regional lymph nodes and pseudocapsules after implantation of joint endoprostheses. *Pathol. Res. Pract., 186,* 707–711.

Brauer, G.M.; Steinberger, D.R.; and Stansbury, J.W. (1986). Dependence of curing time, peak temperature, and mechanical properties on the composition of bone cement. *JBMR, 20,* 839–852.

Brauer, G.M.; Termini, D.J.; and Dickson, G. (1977). Analysis of the ingredients and determinations of the residual components of acrylic bone cements. *JBMR, 11*, 577–607.

Bundy, K.J.; and Penn, R.W. (1987). The effect of surface preparation on metal/bone cement interfacial strength. *JBMR, 21*, 773–805.

Burke, D. (1985). Stress, strain, porosity and fatigue of acrylic bone cement. In *Advanced Concepts in Total Hip Replacement* (Harris, W.H., ed.). SLACK, Inc., p. 21.

Burke, D.W.; Gates, E.I.; and Harris, W.H. (1984). Centrifugation as a method of improving the tensile and fatigue properties of acrylic bone cement. *JBJS, 66A*, 1265–1273.

Cameron, H.; Jackson, R.W.; and MacNab, J. (1974). The structure of polymethylmethacrylate cement. *Clin. Orthop., 100*, 287.

Carter, D.R.; Gates, E.I.; and Harris, W.H. (1982). Strain-controlled fatigue of acrylic bone cement. *JBMR, 16*, 647–657.

Charnley, J. (1970). *Acrylic Cement in Orthopaedic Surgery*. Williams and Wilkins.

Charnley, J.; and Smith, D.C. (1968). The physical and chemical properties of self-curing acrylic cement. In *Internal Publication No. 16*. Center for Hip Surgery.

Chwirut, D.J. (1984). Long-term compressive creep deformation and damage in acrylic bone cements. *JBMR, 18*, 25–37.

Clech, J.P.; Keer, L.M.; and Lewis, J.L. (1984). A crack model of a bone cement interface. *J. Biomech. Enqr., 106*, 235–238.

Connelly, T.J.; Lautenschlager, E.P.; and Wixson, R.L. (1987). The role of porosity in the shrinkage of acrylic bone cement. *Transactions of the 13th Annual Meeting of the Society for Biomaterials*. Society for Biomaterials, p. 114.

Cowie, J.M.G. (1973). *Polymers: Chemistry and Physics of Modern Materials*. International Textbook Co.

Crissman, J.M.; and McKenna, G.B. (1984). *Information on Polymeric Materials Used in Orthopedic Devices*. NBS Report NBSIR 84-2820 National Bureau of Standards.

Crowninshield, R.D.; Brand, R.A.; Johnston, R.C.; and Milroy, J.C. (1980a). The effect of femoral stem cross-sectional geometry on cement stresses in total hip reconstruction. *Clin. Orthop., 146*, 71–77.

Crowninshield, R.D.; Brand, R.A.; Johnston, R.C.; and Milroy, J.C. (1980b). An analysis of femoral component design in total hip arthroplasty. *JBJS, 62A*, 68–78.

Darre, E.; Vedel, P.; and Jensen, J.S. (1987). Air concentrations of methylmethacrylate monomer after mixing different acrylic bone cements. In *Advances in Orthopaedic Surgery*. Williams & Wilkins, pp. 258–259.

Davidson, J.A.; Dingman, C.A.; and Lynch, G. (1986). Cement hoop strains generated along the medial surface of a Charnley-type femoral component. *Transactions of the 12th Annual Meeting of the Society for Biomaterials*. Society for Biomaterials, p. 108.

Davies, J.E.; Nagai, N.; and Takeshita, N. (1989). Osteogenesis and osteoclasis at the interface with a bioactive bone-substitute. *Transactions of the 15th Annual Meeting of the Society for Biomaterials*. Society for Biomaterials, p. 5.

Davies, J.P.; Daniel, M.S.; O'Connor, O.; Burke, D.W.; Jasty, M.; and Harris, W.H. (1988). The effect of centrifugation on the fatigue life of bone cement in the presence of surface irregularities. *Clin. Orthop., 229*, 156–161.

Davies, J.P.; and Harris, W.H. (1989). Severe weakness of bone cement retrieved after *in vivo* service in man. *Transactions of the 15th Annual Meeting of the Society for Biomaterials*. Society for Biomaterials, p. 52.

Davies, J.P.; and Harris, W.H. (1990). Optimization and comparison of three vacuum mixing systems for porosity reduction of Simplex P cement. *Clin. Orthop., 254*, 261.

Davies, J.P.; and Harris, W.H. (1992). The effect of the addition of methylene blue on the fatigue strength of Simplex P bone-cement. *J. Appl. Biomater., 3*, 81.

DiCarlo, E.F.; and Bullough, P.G. (1992). The biologic responses to orthopedic implants and their wear debris. *Clin. Mater., 9*, 235.

Ebramzadeh, E.; Mina-Araghi, M.; Clarke; I.C.; and Ashford, R. (1985). Loosening of well-cemented total-hip femoral prosthesis due to creep of the cement. In *Corrosion and Degradation of Implant Materials: Second Symposium, ASTM STP 859* (Fraker, A.C., and Griffen, C.D., eds.). American Society for Testing and Materials, pp. 373–399.

Eftekhar, N.S. (1978). *Principles of Total Hip Arthroplasty*. C.V. Mosby.

Eftekhar, N.S. (1992). Do we need to vacuum mix or centrifuge cement? No, if the cement is properly used. *The Hip Society 20th Open Scientific Meeting at the AAOS*, p. 15.

Eftekhar, N.S.; and Thurston, C.W. (1975). Effect of irradiation on acrylic cement with special reference to fixation of pathological fractures. *J. Biomech., 8*, 53–56.

Emmanual, J.; and Hedley, A.K. (1991). Fabricated and actual bone cement wear particles in an *in vitro* system. *Transactions of the 17th Annual Meeting of the Society for Biomaterials*. Society for Biomaterials, p. 207.

Eyerer, P.; and Jin, R. (1985). Influence of mixing technique on some properties of PMMA bone cement. *Transactions of the 11th Annual Meeting of the Society for Biomaterials*. Society for Biomaterials. p. 130.

Eyerer, P.; and Jin, R. (1986). Influence of mixing technique on some properties of PMMA bone cement. *JBMR, 20*, 1057–1094.

Fagan, M.J.; and Lee, A.J.C. (1986). Material selection in the design of the femoral component of cemented total hip replacements. *Clin. Mater. 1*, 151–167.

Flory, P.J. (1953). *Principles of Polymer Chemistry*. Cornell University Press.

Fowler, J.L.; Gie, G.A.; Lee, A.J.C.; and Ling, R.S.M. (1988). Experience with the Exeter total hip replacement since 1970. *Ortho. Clin. N. Amer., 19*, 477–489.

Freitag, T.A.; and Cannon, S.L. (1976). Fracture characteristics of acrylic bone cements. I. Fracture toughness. *JBMR, 10*, 805–828.

Freitag, T.A.; and Cannon, S.L. (1977). Fracture characteristics of acrylic bone cements. II. Fatigue. *JBMR, 11*, 609–624.

Friis, E.A.; Stromberg, L.J.; Cooke, F.W.; and McQueen, D.A. (1993). Fracture toughness of vacuum mixed PMMA bone cement. *Transactions of the 19th Annual Meeting of the Society for Biomaterials*. Society for Biomaterials, p. 301.

Gates, E.I.; Carter, D.R.; and Harris, W.H. (1984). Comparative fatigue behavior of different bone cements. *Clin. Orthop., 189*, 194–583.

Gelb, H.; Baker, D.; Shumacher, R.; and Cuckler, J. (1992). The effects of PMMA particle size and surface area on the inflammatory response in a quantitative *in vivo* model. *38th Annual Meeting, Orthopaedic Research Society*, Washington, DC. Rider Dickerson, p. 345.

Goodman, S.B.; Chin, R.C.; Chiou, S.S.; Schurman, D.J.; Woolson, S.T.; and Masada, M.P. (1989). A clinical–pathological–biochemical study of the membrane surrounding loose and nonloosened joint arthroplasty. *Clin. Orthop., 244*, 182–187.

Goodman, S.B.; Fornasier, V.L.; and Kei, J. (1988). The effects of bulk versus particulate polymethylmethacrylate on bone. *Clin. Orthop., 232*, 255–262.

Goodman, S.B.; Lee, J.; and Fornasier, V.L. (1991). Bone accretion around metallic implants in the rabbit tibia. *Transactions of the 17th Annual Meeting of the Society for Biomaterials*. Society for Biomaterials, p. 17.

Greenwald, A.S. (1979). Properties and applications of acrylic bone cement. *Bull. Hosp. Jnt. Dis., 40*, 72–78.

Gruen, T.A.; Markolf, K.L.; and Amstutz, H.C. (1976). Effects of laminations and blood entrapment on the strength of acrylic bone cement. *Clin. Orthop., 119*, 250–255.

Haas, S.S.; Brauer, G.M.; and Dickson, G. (1975). A characterization of polymethylmethacrylate bone cement. *JBJS, 57A*, 380–391.

Hadjari, M.; Reindel, E.S.; and Convey, F.R. (1990). High viscosity acrylic bone cement: An *in vitro* and *in vivo* biomechanical evaluation. *Transactions of the 36th Annual Meeting of the Orthopedic Research Society*. Rider Dickerson, p. 29.

Hadjari, M.; Reindel, E.S.; Hollis, J.M.; and Convey, F.R. (1990). Characterization of high viscosity acrylic bone cement. *Transactions of the 16th Annual Meeting of the Society for Biomaterials*. Society for Biomaterials, p. 105.

Hansen, D.; and Jensen, J.S. (1990). Prechilling and vacuum mixing not suitable for all bone cements. Handling characteristics and exotherms of bone cements. *J. Arthroplasty, 5*, 287–290.

Harrigan, T.P.; Davies, J.P.; Burke, D.W.; O'Connor, D.O.; Jasty, M.; and Harris, W.H. (1987). On the presence or easy initiation or fracture in bone cement at the bone cement interface in total joint arthroplasty. *Transactions of thie 13th Annual Meeting of the Society for Biomaterials.* Society for Biomaterials, p. 170.

Harrigan, T.P.; Kareh, J.A.; and Harris, W.H. (1990a). A failure criterion based on crack initiation at pores in cement applied to the loosening of femoral stems. *Transactions of the 16th Annual Meeting of the Society for Biomaterials.* Society for Biomaterials, p. 52.

Harrigan, T.P.; Kareh, J.A.; and Harris, W.H. (1990b). Initial loosening mechanisms in cemented femoral stems: Debonding at the cement-metal interface. *Transactions of the 16th Annual Meeting of the Society for Biomaterials.* Society for Biomaterials, p. 189.

Harris, W.H. (1992). Will stress shielding limit the longevity of cemented femoral components of total hip replacement? *Clin. Orthop., 274*, 120.

Harris W.H.; and Davies J.P. (1988). Modern use of modern cement for total hip replacement. *Orthop. Clin. N. Amer., 19*, 581–589.

Helmke, H.W.; Lednicky, C.L.; and Tullos, H.S. (1992). Porosity of the cement/metal interface following cemented hip replacement. *38th Annual Meeting, Orthopaedic Research Society*, Washington, DC. Rider Dickerson, p. 364.

Huiskes, R. (1980). Some fundamental aspects of human joint replacement. *Acta Orthop. Scand.*, Suppl. 185.

Humphreys, P.K.; Orr, J.F.; and Bahrani A.S. (1989). An investigation into the effect of cyclic loading and frequency on the temperature of PMMA bone cement in hip prostheses. *Proc. Inst. Mech. Eng. H., 203*, 167–170.

International Standards Organization (1992). ISO 5833, International Standard for Implants for Surgery: Acrylic Resin Cement.

Jaffe, W.L.; Rose, R.M.; and Radin, S.L. (1974). On the stability of the mechanical properties of self-curing acrylic bone cement. *JBJS, 56A*, 1711–1714.

James, S.P.; Jasty, M.; Davies, J.; Piehler, H.; and Harris, W.H. (1992). A fractographic investigation of PMMA bone cement focusing on the relationship between porosity reduction and increased fatigue life. *JBMR, 26*, 651–662.

James, S.P.; Karydas, D.; McGarry, F.J.; and Harris, W.H. (1993). Reduction of the extensive porosity in the cement at the femoral component–bone cement interface. *Transactions of the 19th Annual Meeting of the Society for Biomaterials.* Society for Biomaterials, p. 242.

James, S.P.; and McGarry, F.J. (1992). The contribution of incomplete polymerization and residual monomer to the weakness of PMMA bone cement. *38th Annual Meeting, Orthopaedic Research Society*, Washington, DC. Rider Dickerson, p. 373.

James, S.P.; Schmalzried, T.P.; McGarry, F.J.; and Harris, W.H. (1993). Extensive porosity at the cement–femoral prosthesis interface: A preliminary study. *JBMR, 27*, 71.

Jasty, M.; Davies, J.P.; O'Connor, D.O.; Burke, D.W.; Harrigan, T.P.; and Harris, W.H. (1990). Porosity of various preparations of acrylic bone cements. *Clin. Orthop., 259*, 122–129.

Jasty, M.; Maloney, W.J.; Bragdon, C.R.; Haire, T.; and Harris, W.H. (1990). Histomorphological studies of the long-term skeletal responses to well fixed cemented femoral components. *JBJS, 72A*, 1220–1229.

Keller, J.C.; Lautenschlager, E.P.; Marshall, G.W.; and Meyer, P.R. (1980). Factors affecting surgical alloy/bone cement interface adhesion. *JBMR, 14*, 639–651.

Kies, J.A.; Sullivan, A.M.; and Irwin, G.R. (1950). *J. Appl. Phys., 21*, 716.

Krause, W.; Grimes, L.W.; and Mathis, R.S. (1988). Fatigue testing of acrylic bone cements: Statistical concepts and proposed test methodology. *JBMR, 22*, 179–190.

Krause, W.; and Mathis, R.S. (1988). Fatigue properties of acrylic bone cements: Review of the literature. *JBMR, 22*, 37.

Krause, W.; Mathis, R.S.; and Grimes, L.W. (1988). Fatigue properties of acrylic bone cement: S-N, P-N, and P-S-N data. *JBMR, 22*, 221–244.

Krause, W.R.; Miller, J.; and Ng, P. (1982). The viscosity of acrylic bone cements. *JBMR, 16*, 219–243.

Kummer, F.J. (1985). Improved mixing of bone cements. *31st Annual Meeting, Orthopaedic Research Society*, Las Vegas, NV. Rider Dickerson, p. 238.

Kusy, R.P. (1978). Characterization of self-curing acrylic bone cements. *JBMR, 12*, 271.

Kusy, R.P.; and Turner, D.T. (1974). Characterization of the microstructure of a cold-cured acrylic resin. *J. Dent. Res., 53*, 948.

Lautenschlager, E.P.; and Marshall, G.W. (1976). Mechanical strength of acrylic bone cements impregnated with antibiotics. *JBMR, 10*, p. 837.

Lautenschlager, E.P.; Stupp, S.I.; and Keller, J.C. (1984). Structure and properties of acrylic bone cement. In *Functional Behavior of Orthopedic Biomaterials*, Vol. 2: *Applications* (Ducheyne, P.; and Hastings, G.W., eds.). CRC Press, pp. 87–119.

Lazarus, M.D.; Cuckler, J.M.; Baker, D.G.; Ducheyne, P.; and Schumacher, H.R. (1991). The role of $BaSO_4$ in osteolysis associated with PMMA particulate debris in two *in vivo* models. *Transactions of the 17th Annual Meeting of the Society for Biomaterials*. Society for Biomaterials, p. 33.

Learmonth, I.D.; Spirakis, A.; Gryzagoridis, J.; and Sher, A. (1992). The effect of polymerization of methylmethacrylate on metal-backed and non-metal-backed acetabular components. *J. Arthroplasty, 7*, 165–171.

Lee, A.J.C.; and Ling, R.S.M. (1975). Further studies of monomer loss by evaporation during the preparation of acrylic cement for use in orthopaedic surgery. *Clin. Orthop., 106*, 122–125.

Lee, A.J.C.; Ling, R.S.M.; and Vangala, S.S. (1977). Mechanical properties of bone cement. *J. Med. Enqr. Tech., 1*, 137–140.

Lee, A.J.C.; Ling, R.S.M.; and Wrighton, J.D. (1973). Some properties of polymethylmethacrylate with reference to its use in orthopedic surgery. *Clin. Orthop., 95*, 281–287.

Linden, U. (1988). Porosity in manually mixed bone cement. *Clin. Orthop., 231*, 110–112.

Litsky, A.S.; and Frazier, N.C. (1993). Monomer exudation from standard and reduced-modulus bone cements under cyclic loading. *Transactions of the 19th Annual Meeting of the Society for Biomaterials*. Society for Biomaterials, p. 303.

Litsky, A.S.; Rose, R.M.; Rubin, C.T.; and Thrasher, E.L. (1990). A reduced-modulus acrylic bone cement: Preliminary results. *J. Orthop. Res., 8*, 623–626.

Litsky, A.S.; and Yetkinler, D.N. (1991). Time dependent stress relaxation behavior of standard and reduced-modulus acrylic bone cements. *Transactions of the 17th Annual Meeting of the Society for Biomaterials*. Society for Biomaterials, p. 47.

Looney, M.A.; and Park, J.P. (1986). Molecular and mechanical property changes during aging of bone cement *in vitro* and *in vivo*. *JBMR, 20*, 555–563.

MacDonald, W.; Swarts, E.; and Beaver, R. (1993). Penetration and shear strength of cement-bone interfaces *in vivo*. *Clin. Orthop., 286*, 283–288.

Maloney, W.J.; Jasty, M.; Bragdon, C.R.; and Harris, W.H. (1990). Osteolysis in association with well-fixed cemented femoral components. *Transactions of the 16th Annual Meeting of the Society for Biomaterials*. Society for Biomaterials, p. 176.

Manley, M.T.; Stern, L.S.; and Gurtowski, J. (1985). The load carrying and fatigue properties of the stem–cement interface with smooth and porous coated femoral components. *JBMR, 19*, 563–575.

Meyer, P.R.; Lautenschlager, E.P.; and Moore, B.K. (1973). On the setting properties of acrylic bone cement. *JBJS, 55A*, 149–156.

Mjoberg, B.; and Selvik, G. (1988). Reduced risk of loosening of hip prostheses with a new cold-curing bone cement. *Acta. Orthop. Scand., 59*, 343–345.

Mulroy, R.D.; and Harris, W.H. (1990). Improved cementing techniques: Effect on femoral and socket fixation at 11 years. *Transactions of the 16th Annual Meeting of the Society for Biomaterials*. Society for Biomaterials, p. 147.

Niwa, S.; Miki, S.; Sawai, K.; and Hattori, T. (1993). Experimental evaluation of the anchoring efficiency of bone cement under various conditions, and improved cementing techniques and instrumentations for total hip replacement. *Engr. Med., 13*, 197–202.

Noble, P.C.; Jay, J.L.; Lindahl, L.J.; Maltry, J.; Dreeben, S.M.; and Tullos, H.S. (1987). Methods of enhancing acrylic bone cement. *Transactions of the 13th Annual Meeting of the Society for Biomaterials*. Society for Biomaterials, p. 169.

Noble, P.C.; and Swarts, E. (1983). Penetration of acrylic bone cement into cancellous bone. *Acta Orthop. Scand., 54*, 566.

O'Connor, D.O.; Burke, D.W.; Zalenski, E.B.; and Harris, W.H. (1989). Study of the fatigue behavior of cemented total hip replacements under conditions simulating gait. *Transactions of the 15th Annual Meeting of the Society for Biomaterials*. Society for Biomaterials, p. 50.

Oh, I.; Bushelow, M.B.; and Sander, T.W. (1985). The role of cancellous bone on cement fixation. *Transactions of the 11th Annual Meeting of the Society for Biomaterials*. Society for Biomaterials, p. 129.

Oysaed, H. (1990). Dynamic mechanical properties of multiphase acrylic systems. *JBMR, 24*, 1037–1048.

Pal, S.; and Saha, S. (1982). Stress relaxation and creep behavior of normal and carbon fibre reinforced acrylic bone cement. *Biomaterials, 3*, 93–96.

Papatheofanis, F.J.; and Barmada, R. (1991). Polymorphonuclear leukocyte degranulation with exposure to polymethylmethacrylate nanoparticles. *JBMR, 25*, 761–771.

Petty, W. (1980). Methyl methacrylate concentrations in tissues adjacent to bone cement. *JBMR, 14*, 427–434.

Prendergast, P.J.; Monaghan, J.; and Taylor, D. (1989). Materials selection in the artificial hip joint using finite element stress analysis. *Clin. Mater., 4*, 361–376.

Rimnac, C.M.; Wright, T.M.; and McGill, D.L. (1986). The effect of centrifugation on the fracture properties of acrylic bone cements. *JBJS, 68A*, 281–287.

Rohlmann, A.; Mossner, U.; Bergmann, G.; Hees, G.; and Kolbert, R. (1987). Effects of stem length and material properties on stresses in hip endoprostheses. *J. Biomed. Enqr., 9*, 7783.

Saha, S.; and Pal, S. (1983). Strain-rate dependence of the compressive properties of normal and carbon-fiber-reinforced bone cement. *JBMR, 17*, 1041–1047.

Saha, S.; and Pal, S. (1984). Mechanical properties of bone cement: A review. *JBMR, 18*, 435–462.

Scuderi, G.R.; Insall, J.N.; Windsor, R.E.; and Moran, M.C. (1989). Survivorship of cemented knee replacements. *JBJS, 71B*, 798–780.

Scullin, J.P.; Greenwald, A.S.; Wilde, A.H.; and Beck, R.D. (1977). The effect of radiation on the shear strength of acrylic bone cement. *Clin. Orthop., 129*, 201–204.

Sharkey, N.A.; and Bargar, W.L. (1987). Properties of retrieved acrylic bone cement. *33rd Annual Meeting, Orthopaedic Research Society*, San Francisco, CA. Rider Dickerson, p. 55.

Skinner, H.B.; and Murray, W.R. (1985). Density gradients in bone cement after centrifugation. *31st Annual Meeting, Orthopaedic Research Society*, Las Vegas, NV. Rider Dickerson, p. 243.

Stark, C.F. (1979). Fracture and fatigue characteristics of bone cements. *JBMR, 13*, 339–342.

Swenson, L.W.; Schurman, D.J.; and Piziali, R.L. (1981). Finite element temperature analysis of a total hip replacement and measurement of PMMA curing temperatures. *JBMR, 15*, 83–96.

Tarr, R.R.; Clark, I.C.; Gruen, T.A.; and Saramento, A. (1983). Comparison of loading behavior of femoral stems of Ti-6Al-4V and chromium–cobalt alloys: A 3-dimensional finite element analysis. In *Titanium Alloys in Surgical Implants* (Lucky, H.A.; and Fubli, F., eds.). American Society for Testing and Materials, pp. 38–101.

Toksvig-Larsen, S.; Franzen, H.; and Ryd, L. (1991). Cement interface temperature in hip arthroplasty. *Acta Orthop. Scand., 62*, 102–105.

Topoleski, L.D.T.; Ducheyne, P.; and Cuckler J.M. (1990). A fractographic analysis of *in vivo* poly(methylmethacrylate) bone cement failure mechanisms. *JBMR, 24*, 135–154.

Topoleski, L.D.T.; and Lu, X. (1993). An investigation of fatigue crack initiation mechanisms in poly(methyl methacrylate) bone cement. *Transactions of the 19th Annual Meeting of the Society for Biomaterials*. Society for Biomaterials, p. 300.

Trap, B.; Wolff, P.; and Jensen, J.S. (1992). Acrylic bone cements: Residuals and extractability of methacrylate monomers and aromatic amines. *J. Appl. Biomater., 3*, 51–57.

Tuckfield, W.J.; Worner, H.K.; and Guerin, B.D. (1943). Acrylic resins in dentistry. II. Their use for denture construction. *Aust. J. Dent., 47*, 1.

Wang, C.T.; and Pilliar, R.M. (1989). Fracture toughness of acrylic bone cements: Cements-elastic–plastic analysis of miniature short rod specimens. *Transactions of the 15th Annual Meeting of the Society for Biomaterials*. Society for Biomaterials, p. 182.

Watson, M.B.; Miles, A.W.; and Clift, S.E. (1990). The influence of curing time and environment on the fracture properties of bone cement. *Clin. Mater., 6*, 299–305.

Weightman, B.; Freeman, M.A.R.; Revell, P.A.; Braden, M.; Albrektsson, B.E.J.; and Carlson, L.V. (1987). The mechanical properties of cement and loosening of the femoral component of hip replacements. *JBJS, 69B*, 558–564.

Weinstein, A.M.; Bingham, D.N.; Sauer, B.W.; and Lunceford, E.M. (1976). The effect of high pressure insertion and antibiotic inclusions upon the mechanical properties of olymethylmethacrylate. *Clin. Orthop., 121*, 68–73.

Wijn, J.R. de; Slooff, T.J.J.H.; and Driessens, C.M. (1975). Characterization of bone cements. *Acta Orthop. Scand., 46*, 38–51.

Wixson, R.L. (1992). Do we need to vacuum mix or centrifuge cement? *Clin. Orthop., 285*, 84–90.

Wroblewski, B.M. (1986). 15-21 year results of the Charnley low-friction arthroplasty. *Clin. Orthop., 211*, 30.

Wykman, A.G.M. (1992). Acetabular cement temperature in arthroplasty. *Acta Orthop. Scand., 63*, 543.

Zygmunt, S.; Toksvig-Larsen, S.; Saveland, H.; Rydholm, U.; and Ryd, L. (1992). Hyperthermia during occipito-cervical fusion with acrylic cement. *Acta Orthop. Scand., 63*, 545–548.

VII
VASCULAR APPLICATIONS

34
Tissue–Polymer Composite Vascular Prostheses

John A. M. Ramshaw and Jerome A. Werkmeister
CSIRO Division of Biomolecular Engineering, Parkville, Australia

Glenn A. Edwards
University of Melbourne, Werribee, Australia

I. INTRODUCTION

A variety of materials of synthetic and biological origin have now been used to meet the medical need for replacement blood vessels [1]. For large-diameter applications where there is high blood flow, such as aortic artery replacement, synthetic materials such as Dacron™ have proved very satisfactory. For smaller arterial applications, such as femoral and popliteal artery replacement, various synthetic and biological materials have been used. These include expanded polytetrafluoroethylene (ePTFE) [2], Dacron in various different weave or knit configurations and textures [3], modified human umbilical vein (MHUV) [4] and biosynthetic materials [5,6]. For small-diameter vessels, such as coronary artery, no adequate material is yet available, and autologous saphenous vein or mammary artery remain the materials of choice. However, since these vessels may be unavailable because of prior use, or through unsuitability due to disease, there is an on-going medical demand for an effective small-diameter vessel replacement.

Of the various materials available for artery replacement, only two types of material are used widely, ePTFE and Dacron, with neither being fully satisfactory, particularly for low blood flow locations such as below the knee. For ePTFE devices, intimal hyperplasia of smooth muscle cells at the anastomosis frequently leads to severe stenosis and failure [7]. For Dacron devices, thrombosis is a major problem [8]. Also, due to their porous nature intraoperative preclotting is necessary to prevent blood loss, and this procedure can increase the risk of infection. To overcome these problems in part, Dacron grafts may be coated, for example, with fibrin [9]. Although this improves the function of these devices, they are still not optimal for below-the-knee procedures.

Similarly, the biological devices have also proved unsatisfactory. For example, various umbilical vein devices failed due to aneurysms and dilatations [4], which were caused by loss of structural integrity of the stabilized collagen material [10]. Subsequently, external reinforcement of the devices by mesh was examined, but since this was not an integral component of the device, similar modes of failure were still observed [11,12].

An alternative concept was also proposed, that of combining the tissue compatibility advantages of natural materials with the durability of synthetic materials by forming an integral composite (see below). This concept, the *biosynthetic prosthesis*, was defined by Baier [13] as the combination of tissue and synthetic materials, usually stabilized with some crosslinking reagent, that is meant to function from the instant of implantation as a thromboresistant conduit eliciting minimal reactions in their blood flow surface while fostering integration with surrounding host tissue.

While this definition may be extended to include prostheses such as the MHUV device with its external Dacron mesh or coated synthetic devices, in the present review, discussion is restricted to materials where the two components form a single, fully integrated structure.

II. DEVELOPMENT OF TISSUE-POLYMER (BIOSYNTHETIC) PROSTHESES

The development of tissue–polymer vascular prostheses emerged from early work that recognized that the implantation of synthetic materials into the body led to a variable degree of inflammation and fibrous tissue response depending on the nature of the material. As such, the development of fibrous tubes, without the benefit of the additional polymeric component, was an integral part of the development of the current biosynthetic devices.

Early studies on medical application of fibrous tissue formation include, for example, the work of Yeager and Cowley [14], who studied polyethylene as a fibrous tissue stimulant and recognized its potential in repairing hernias and supporting vascular aneurysms. Later, in 1953, Pierce and Baltimore [15] reported on the first production and aortic implantation of a fibrocollagenous tube in dogs. In this experiment 6- to 10-mm diameter, 50- to 80-mm long polyethylene tubes were implanted in the canine rectus sheath. After 5 weeks, the newly formed fibrous tube and surrounding fascia were removed *en bloc* and anastomosed in the canine infrarenal aorta. In this study, 10 grafts were followed for up to 30 months, with patency remaining in all grafts. Histology at 12 months revealed a compact fibrous layer with circular orientation of collagen bundles and a thin lining of normal endothelium.

Subsequently, in 1961, Eiken and Norden [16] described satisfactory results using connective tissue tubes that formed around subcutaneously implanted polyvinyl rods (mandrels) for the replacement of carotid arteries in dogs. However, when these tubes were placed in the canine aorta, they noted an unacceptable incidence of aneurysm formation and rupture. Also, by 1964, Schilling and colleagues [17,18] had reported extensive histologic and biochemical studies of fibrocollagenous tubes, noting the crossed pattern of successive collagen layers. In an extended study in 12 dogs, no aneurysm nor degenerative changes were evident in autologous or homologous grafts.

The development of a tissue–polymer composite device emerged when Hufnagel [19] suggested that implanted PTFE mandrels could be surrounded with a polyester (Dacron) mesh to reinforce the fibrocollagenous prosthesis. At a similar time, Assefi and Parsonnet [20] independently reported the necessity of reinforcing collagen tubes with polypropylene mesh to avoid aneurysm formation, and reported patencies of up to 8 months in canine aortas. They also suggested that fibrocollagenous tubes could be made *in situ*, freeing up only the ends of the tube for anastomosis, while leaving the body of the graft in its original implantation bed.

A variation of the subcutaneously grown fibrocollagenous tube was reported by Lindstrom and Larmi [21], who implanted a Dacron net into the peritoneal cavity of dogs to form a tubular graft for aortic implantation. The peritoneal epithelium was thromboresistant and appeared to act as normal vascular endothelium. However, the unpredictability of peritoneal versus subcutaneous implantation discouraged further development of this type of device.

In 1969 Sparks [22] experimented with a device called a *tissue die* consisting of an outer tubular shell of stainless steel punctured by many small holes, an inner tubular mandrel of stainless steel with a highly polished surface, and a middle anchoring skeleton tube of Dacron, all over a central tie rod through the length of the mandrel. The assembled die was implanted subcutaneously through a stab incision and removed after 5 weeks with a special cylindrical cutter. Using this strategy, satisfactory results were reported in 48 tubes implanted in dogs for up to 2 years, with no evidence of degenerative changes [23]. At the same time, Sparks [23] also reported the first clinical implantations. The clinical implant for superficial femoral reconstructions consisted of two 180-mm fibrocollagenous tubes sewn together, which remained patent without complication for up to 18 months follow-up. Successful autogenous grafts produced by this method were also reported in humans for femoropopliteal bypasses [23].

Continuing from their earlier work on mesh-reinforced tubes [20], Parsonnet and colleagues [24] produced fibrocollagenous tubes around polyester (Dacron) or polypropylene mesh over silicone mandrels, which did not use rigid metal parts, placed under the panniculus muscle of the thoracoabdominal wall. When implanted between the canine thoracic and abdominal aorta, these devices produced no aneurysm formation in the 4-year follow-up. Histological examination revealed that the polypropylene supported tube was smooth, thin, and transparent in most areas, consisting of collagen with almost no cellular reaction.

In 1973 Sparks [25] modified his metal die technique by placing subcutaneously a flexible silicone mandrel covered by two layers of knitted polyester (Dacron) mesh, without use of rigid metal stents. Six to eight weeks after implantation, the silicone mandrel was removed and the arterial bypass completed by mobilization of the ends of the conduit and construction of anastomoses proximal and distal to the abstracted section of artery. Sparks [25] reported the use of this prosthesis in 6 patients for femoropopliteal reconstruction, 5 of which remained patent for 2–7 months.

Following these favorable animal and early clinical results, the "Sparks mandrel" remained popular until subsequent clinical reports described poor results. Thus, in a detailed review—which included their own patients, as well as those of Sparks and associates—Hallin and Sweetman [26] reported on the follow-up of 60 "Sparks mandrel" devices in 54 patients for up to 7 years. The overall data showed 100% patency in the ilioiliac and iliofemoral sites, 26% (10/38) in the femoropopliteal, and 0% in the femorotibial and axillofemoral grafts. The complications that were reported included early thrombosis (17%), postoperative haemorrhage (5%), aneurysm formation (20%), and late thrombosis due to lumenal narrowing (44%). Subsequently, Parsonnet and colleagues [27] also reported an overall patency of only 18.5% in 81 clinical cases for up to 2 years, and Christenson and Eklof [28] reported an 18% patency for femoropopliteal bypasses at 12 months.

Thus, despite the ingenuity of the "Sparks mandrel," the initial aim of achieving low thrombogenicity and durability had not been achieved. The unacceptable incidence of degenerative and aneurysmal changes suggested that the fibrocollagenous component of the device, even in the presence of a reinforcing mesh, had insufficient strength and durability for use within the arterial system. Based on these reports, autogenous mandrel-grown fibrocollagenous grafts have no longer been used clinically.

In subsequent reports, Guidoin and colleagues [29,30] proposed that this lack of dimensional stability was due to the absence of elastin fibers or smooth muscle cells in the autogenous tissue; the aneurysmal and dilatory changes appeared to result from stretching of both the tissue and mesh components. They reported on large areas of lumenal thrombosis in areas of exposed mesh, and the surface was found to be more thrombogenic than preclotted Dacron prostheses. They also commented that tissue repair varies from species to species, between individuals, and from site to site within an individual. Thus, strict equivalence of mandrel performance in diseased humans could not be expected, and this may explain the good animal results as compared with the poor performance experienced clinically.

Concurrently with the development of the "Sparks mandrel" technology, Dardik and colleagues [4,31] were developing the use of mesh-reinforced human umbilical vein-based devices. In this work, after initial unsuccessful attempts to use the unmodified vein, alternative strategies were developed to enhance the durability of the vessel and to mask unwanted immunogenic proteins. Dardik and colleagues subsequently selected glutaraldehyde as the stabilization agent with the use of an external polyester (Dacron) mesh support [4]. Although the initial short-term patencies with this MHUV were acceptable, the long-term results were not encouraging. Intimal breakdown, infection, and frequent aneurysms and dilations were the main reasons for low patencies [10–12].

Although a range of potential tissue stabilization agents have been proposed, glutaraldehyde remains the agent of choice (see Ref. 32). However, it is not without problems in clinical use, notably its propensity to induce nonspecific calcification of the implanted device. Nevertheless, for various tissue-based heart valve replacements, the reagent has provided tissue that is stable for over a decade, with durability problems arising predominantly from the calcification [33].

Consideration of the key aspects of the different "Sparks mandrel" and glutaraldehyde stabilization technologies led Ketharanathan to suggest the use of polymer mesh reinforced collagen tubes, grown subcutaneously using a silicone mandrel, in combination with glutaraldehyde stabilization. This work, first reported in 1980 [5,6,34], has led to the development of the Omniflow Vascular Prosthesis™. The use of glutaraldehyde stabilization of the collagen also minimized the extent of cross-species immunological reactivity. This enabled the development of a mandrel-grown ovine collagen prosthesis that may be transplanted between species. The result is the production of tissue tubes with fully incorporated reinforced meshes, displaying high mechanical integrity, and lumenal wall properties associated with good hemocompatibility [13]. Following the initial reports of the use of this prosthesis in animal models [6,35], it has undergone an extensive series of *in vitro* and *in vivo* tests (see below, Secs. III.A–III.C) and is now used successfully in clinical applications, giving a performance that is comparable to, or better than, other available materials (see Secs. III.D–III.E).

III. CURRENT TISSUE–POLYMER (BIOSYNTHETIC) PROSTHESES

Although others have subsequently produced devices [36] similar to those described by Ketharanathan and colleagues [5,6,35], the Omniflow Vascular Prosthesis designed by Ketharanathan remains the only integrated tissue–polymer biosynthetic device currently available commercially. Thus the following sections focus on the properties of this device.

A. Manufacture

The Omniflow Vascular Prosthesis[TM] (Bio Nova International, North Melbourne, Australia) is formed when silicone mandrels covered with a polyester (Dacron) mesh become encapsulated with ovine collagen following implantation for 12–13 weeks beneath the cutaneous trunci muscles of adult sheep. This process ensures that the polyester mesh forms an integral part of the prosthesis structure. After this implant period, the tubes are excised and trimmed of the excess fat and connective tissue, then fixed with 2% glutaraldehyde [37]. The silicone mandrel can then be removed, leaving the fiber-reinforced tube (Fig. 1), which, after sterilization, is stored in 50% v/v ethanol [37].

Since the initial concept was described [5,6], the Omniflow Vascular Prothesis has undergone several modifications [38], leading to three distinct forms of the device. (The various forms are referred to in this chapter as Omniflow (Prototype), Omniflow I, and Omniflow II.) Thus, while the initial Omniflow (Prototype) devices used a loosely woven mesh, the next device, Omniflow I, used a new mesh design, with a multi filament Dacron fiber (Fig. 2). Subsequently, Omniflow II was introduced, where additional meticulous attention was paid to all aspects of stabilization. This resulted in a flexible (kink-free), compliant prosthesis with enhanced thromboresistance [38].

A major advantage of the Omniflow technology has been the possibility of producing "off-the-shelf" prostheses of predetermined dimensions. Also, unlike the homologous composite prostheses, these xenogenic, biosynthetic devices are uniform, with little device-to-device variation. The judicious selection of the mesh and mandrel combination, and of the stabilization protocol enables devices with a range of different characteristics to be produced [37]. This flexibility means that there are excellent prospects for further improvements to the device, and that an effective device suitable for small-diameter vessel replacement, for example, coronary artery, may be possible. By contrast, the

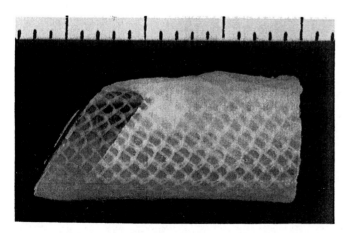

Figure 1 The Omniflow Vascular Prosthesis[TM] with an internal diameter of 6 mm showing total integration of the synthetic and biological components.

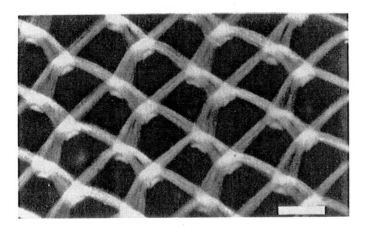

Figure 2   Geometry of the polyester mesh (Dacron™) knitted design used in the manufacture of the Omniflow Vascular Prosthesis I and II. Bar = 1 mm. (From Ref. 39, with permission.)

opportunities for further improvement of ePTFE- and Dacron-based devices may be more limited.

B.   Structural Properties and Biocompatibility

1.   *Morphology*

A variety of histological and electron microscopy approaches have been used to examine the structure of the Omniflow Vascular Prosthesis.

Light microscopy, using, for example, hematoxylin and eosin (Fig. 3), has shown that the mesh is fully encapsulated by the newly deposited ovine collagen, giving an inner surface of collagen with the Dacron fully covered and not able to make blood contact.

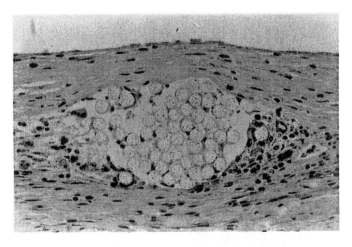

Figure 3   Cross section of the Omniflow II prosthesis, stained with hematoxylin and eosin, showing complete coverage of the polyester filament bundles with ovine collagen. The inner surface of the device (top) shows no exposed breakthrough fibers. Bar = 100 μm.

The thickness of the collagen layer shows some variation, being least at positions where there are cross-over points in the mesh. The collagen has a circular appearance in the regions surrounding the Dacron, but further away from the position of the original silicone mandrel the collagen forms an elongated layered structure. Various cells are associated with the tissue of the device. In particular, fibroblasts aligned with individual synthetic and collagen fibers are found throughout the tissue matrix, and foreign-body giant cells and macrophages may be found adjacent to the mesh fibers [6,35,39]. The inner surface of the material is associated with closely packed elongated fibroblasts. Small vessels are also occasionally observed within the wall of the prosthesis. Specific stains for elastin indicate that the tissue matrix contains little, if any, of this protein [39]. Where elastin is present, it is associated with the small vessels or in the outermost layers, where it is probably derived from the original host tissue.

Polarized light microscopy (Fig. 4) shows that, overall, the collagen is highly aligned, oriented parallel to the longitudinal axis of the device, and that regions of collagen crimp are visible [40]. X-ray fiber diffraction studies also indicate that the bulk of the collagen of the device has a highly oriented structure for the collagen, and indicate that the collagen packing, with a 65.5-nm D-period, is similar to that found in skin and heart valves, and other tissues containing collagen III, and that it is distinct from that found in tendon, cartilage and bone, which have a 67.0-nm D-period [40].

Scanning electron microscopy (SEM) has shown not only that the mesh is fully encapsulated by the collagen, but also that the collagen has infiltrated within the individual filaments of the mesh [39,41] (Fig. 5). SEM shows the different structural zones of the device. These are the inner layer, formerly in contact with the silicone mandrel; the collagen that surrounds and integrates with the polyester mesh; and the layers of collagen that form the majority of the wall of the device. The layered, sheet-like structure of the

Figure 4 Polarized light microscopy of the Omniflow II, indicating that the collagen fibers are oriented parallel to the silicone mandrel. Bar = 100 μm. (From Ref. 40, with permission.)

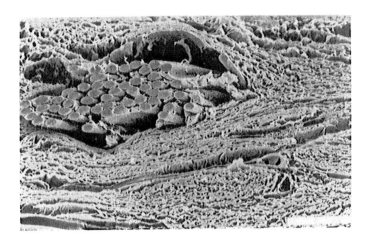

Figure 5 Cross section through a mesh knot of the Omniflow I prosthesis using scanning electron microscopy, showing different zones throughout the device. The polyester fibers are completely integrated with collagen that has infiltrated within the individual filaments. Bar = 0.2 mm. (From Ref. 39, with permission.)

wall collagen is quite distinct from the large fiber bundle structure of collagen found in the skin proximal to the position of the original mandrel implant [39,41].

Transmission electron microscopy (TEM) shows that the collagenous structure of the device is quite distinct for the three zones identified by SEM. The structure of the inner surface of the device—which had been in close contact with the silicone mandrel, and which is to form the new blood contact zone of the device—appears loosely packed, consisting of poorly organized, thin collagen fibrils (Fig. 6a). This thin, open zone, which overlays a discontinuous layer of cells that can be seen by histology (Fig. 3) [41], contained other connective tissue components, including elastic tissue, microfibrillar pro-

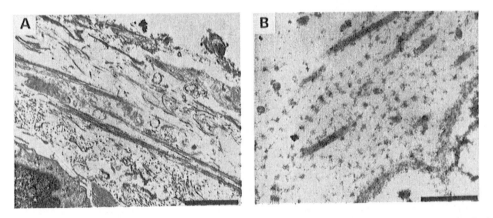

Figure 6 Transmission electron microscopy of the inner surface of an Omniflow II prosthesis that had been in contact with the silicone mandrel (top right), showing (A) loosely packed, poorly organized, thin collagen fibrils, bar = 2 μm,; and (B) collagen VI, bar = 0.5 μm. (From Ref. 41, with permission.)

teins, and nonfibrillar collagens including, for example, collagen VI with its characteristic moniliform structure (Fig. 6b). The open structure of this region suggests that it may be rich in proteoglycan components.

The connective tissue surrounding the polyester fibers was quite distinct from that in the surface layer. The microenvironment surrounding the filaments that make up the fiber was heterogenous. In some areas, mainly at the periphery of the fiber bundle, cells with extended processes surrounded by a diverse connective tissue matrix were found (Fig. 7a). In other areas, frequently within the center of the fiber bundle, a poorly organized collagen network with few cells was found, with the collagen going up to and making contact with the polyester filaments (Fig. 7b). The diameter of these collagen fibrils was fairly uniform and small, being approximately 20–25 nm.

Away from the polyester, a dense collagenous network is found, with bundles of well-organized, interweaving collagen fibers (Fig. 8a). Additional connective tissue components, including proteoglycan, were also evident. Further into the wall of the device, the majority of the collagen bundles were oriented in a single direction (Fig. 8b). Whereas the collagen bundles closely associated with the polyester mesh are aligned in the same direction as the polymer fibers, away from the polyester this alignment is lost and the collagen is mostly aligned in sheets parallel to the axis of the silicone mandrel. This collagen also showed a fairly uniform fibril size distribution, with the fibrils being about 70–80 nm in diameter, about three times those found within the polyester bundles. This region is predominantly composed of interstitial collagens, with little elastic tissue or other non-fibril-forming collagen being evident. A few cells were observed, surrounded by the well-organized collagen network [41].

2. Collagen Content and Properties

The collagen content of Omniflow II has been reported to be about 45% [42], which is comparable to the collagen content of other tissues. To give durability and to minimize or control absorption, the collagen in the device is extensively crosslinked by glutaraldehyde, raising the shrinkage temperature from about 60°C for the untreated tissue to around 81.3° ± 0.9°C [42] in the final device.

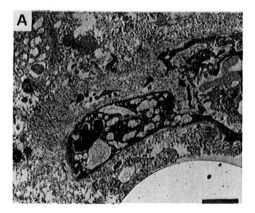 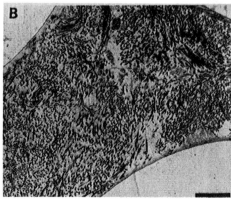

Figure 7 Transmission electron microscopy of the tissue around the polyester filaments of the Omniflow II prosthesis, showing (A) cells and collagens at the periphery of the polyester bundle, bar = 2 μm; and (B) poorly organized collagen network between polyester filaments within the central region of the bundle, bar = 0.5 μm. (From Ref. 41, with permission.)

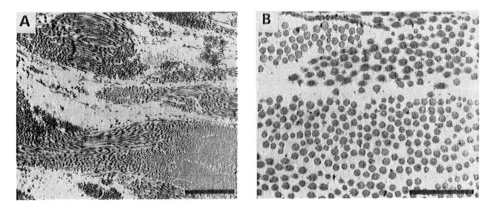

Figure 8   Transmission electron microscopy of the tissue further into the wall of the Omniflow II prosthesis, showing (A) a dense, well-organized collagenous network, bar = 5 μm; and (B) higher magnification of this region with large uniform collagen fibrils, bar = 1 μm. (From Ref. 41, with permission.)

A variety of different collagen types are present in the tissue component of the device. The major ones present are collagen I and collagen III; electrophoretic analysis has shown a high ratio of collagen III relative to collagen I, between 0.7 and 1.2, with the greatest proportion of this collagen being found close to the synthetic materials (Fig. 9). This high proportion of collagen III is much greater than found in other nonpathologic tissues, including dermis. Collagen III is an important structural component of natural blood vessels, and diseases where the amount of this collagen is reduced are characterized by aneurysms, particularly of the aorta [43]. Analysis of tissue surrounding a silicone mandrel without mesh also shows the high proportion collagen III, but with low amounts of total collagen close to the silicone surface, consistent with the TEM findings [44].

TEM had also indicated that a range of minor collagens, such as collagen VI, were present in the device (Fig. 6b). In our laboratory we have developed a wide range of

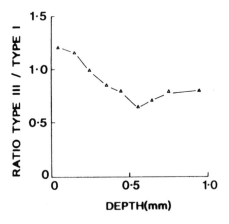

Figure 9   The ratio of collagen I to III at various stratigraphic depths through the Omniflow prosthesis. (From Ref. 39, with permission.)

monoclonal antibodies (MAbs) that are highly specific for different collagen types and other connective tissue components. These MAbs are suitable for immunohistological examination of tissue from a variety of mammalian species, including sheep [45–48], and allow the distributions of the various connective tissue molecules to be assessed.

These MAbs show the biological component of the device to be a complex yet well-organized structure, fully integrated with the synthetic component of the device. Collagen I and collagen III (Fig. 10a) are the predominant components of the prosthesis and are present throughout the entire wall of the vessel [39,41]. Collagen III was particularly prominent within and around the immediate vicinity of the polyester fiber bundles (Fig. 10a), and showed thorough tissue integration with the synthetic component, consistent with the chemical analyses [39]. Collagen V was not readily observed, even when the tissue sections were pretreated with acetic acid in order to expose potential reactive sites [49]. Collagen VI was present at low levels throughout the device, apparently associated with cells. It was most evident in the innermost region, which had been adjacent to the silicone mandrel, and was associated with the cells surrounding the Dacron fibers (Fig. 10b). Of the noncollagenous components, elastin, which is a natural component of most blood vessels, was notably absent by immunohistology, consistent with the standard histological staining data. However, other elastic tissue components were present, including microfibril-associated glycoprotein (MAGP) at moderate levels, and a higher molecular weight, fibrillin-like component that was more abundant and found throughout the entire wall [41]. The aortic wall of the primitive vertebrate lamprey is also deficient in elastin but instead contains these other microfibrillar elastic tissue proteins [50] that provide a mechanically sound elastic architecture similar to that of higher vertebrates.

3. Mechanical Properties

As noted previously, a major concern with biological prostheses has been the long-term structural durability of the device, since structural weakening of the wall can lead to development of dilatations and aneurysms.

Examination of the mechanical durability of the Omniflow II prosthesis *in vitro* has

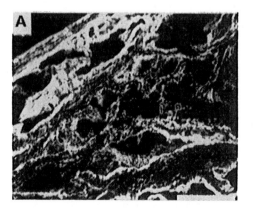

Figure 10   Immunofluorescent detection of the biological components of the Omniflow prosthesis using MAbs to (A) collagen III, and (B) collagen VI. Bars = 100 μm. (From Ref. 41, with permission.)

been described using an accelerated fatigue test, which measured changes in porosity with cyclic pumping [37]. These data showed that the vessel wall was resistant to breakdown under continual, pulsatile stress [37]. The initial porosity of the device is low so preclotting to prevent blood loss through the wall is not needed.

As well as being durable, a vascular replacement should have a compliance that is close to the natural vessel it is intended to replace [51]. The average compliance values (% change in diameter/mm Hg $\times$ 10^2) of Omniflow I and II are 1.9 and 3.9 [37,42], compared to 2.7 for human saphenous vein, 1.9 for Dacron grafts, and 1.6 for ePTFE [51]. It is believed that compliance mismatch can result in anastomotic intimal hyperplasia leading to severe stenosis and loss of function. This is a particularly common problem associated with the synthetic grafts but has not been reported as a problem with Omniflow I or II. The orientation of the collagen along the device provides for larger and better control of compliance of the device compared with a structure where the collagen fibers had formed a circumferential array around the device.

C. Animal Model Studies

The performance of the various forms of the Omniflow prosthesis have been extensively evaluated in dogs and other models, using various diameter devices and a range of surgical bypass locations; in all studies, high patency rates have been observed.

The first animal studies by Ketharanathan and Christie [34], used 8 mm diameter Omniflow (Prototype) prostheses for infrarenal aortic placements in 7 dogs. These implants demonstrated a 100% patency of the prototype device after 3-year implantation. During the period of evaluation, there was no angiographic evidence of dilatation or of progressive stenosis. Excellent host tissue incorporation was reported externally and complete endothelialization occurred by 9 months.

Christie, Perloff, and colleagues [6,35] also reported on the use of Omniflow (Prototype) prostheses of various diameters and lengths placed into the infrarenal aorta ($n = 9$), common iliac artery ($n = 7$), and aorta-iliac ($n = 10$) bypass positions in dogs. These authors also implanted 0.8 mm $\times$ 10 mm grafts into the infrarenal aortas of rats. A 100% patency rate was maintained in the 3 positions in dogs for up to 3 years. Of the prostheses implanted into rats, 72% (23/29) remained patent at 6 months.

Ketharanathan and Christie [52] also reported on a comparative study in which the Omniflow (Prototype) and ePTFE prostheses were implanted as parallel aortoiliac bypass grafts in 10 dogs. All Omniflow prostheses remained patent for periods up to and exceeding 23 months, and no prosthesis showed aneurysmal dilatation. Complete transmural fibrinous tissue ingrowth through the wall of the ePTFE prostheses resulted in progressive and irregular luminal narrowing. This was not seen in the Omniflow prostheses.

Wilson and Klement [53] later reported on a comparative study in which Omniflow I prostheses and MHUV prostheses were individually implanted as aortoiliac arterial bypasses in dogs for up to 3 years without the use of anticoagulation or antiplatelet agents. A total of 47 Omniflow vessels were implanted, 12 of which were examined at over 1-year follow-up and 4 examined after 3 years. Of the 15 MHUV prostheses, 5 were explanted at 1 to 2.5-year follow-up. This study showed that Omniflow patency was 80.8% (38/47), with no fusiform aneurysm formation being evident. On the other hand, the MHUV prostheses showed a lower patency, 73.3% (11/15), and were distinguished by extensive wrinkling and ridging of the flow surface in association with much more extensive coverage with mural thrombi, presumably due to turbulent blood flow induced by the irregular flow surface. These authors speculated that the presence in the Omniflow prosthesis of the reinforcing mesh completely integrated within the inner surface was important in preventing this wrinkling and ridging problem. In contrast, the reinforcing polyester mesh in the MHUV prostheses, applied external to the device, allows considerable deformations to occur at the inner aspect of the wall, which results in the wrinkling and ridging of the flow surface and focal out-pouching that effectively become "minianeurysms." Histopathological examination showed that foci of microcalcification and cartilaginous metaplasia were present in both prosthesis types, as were areas of hemorrhage and fibrin infiltration into the graft wall. These findings were, however, considerably more extensive in the MHUV prostheses than the Omniflow prostheses [53]. With respect to these problems, cartilaginous metaplasia is a very common phenomenon in the dog and some other species, but is a relatively uncommon tissue reaction in man [53], while microcalcification has been reported in numerous glutaraldehyde-fixed connective tissue structures implanted in the body for long periods [32]. There was no evidence in this study that this had any detrimental effect on function.

A major study of the Omniflow I prosthesis by Edwards has been reported [54]. In this study, a group of 63 young adult greyhound dogs weighing 25–40 kg were each implanted with a 100 to 120 mm long, 6 mm internal diameter Omniflow I prostheses as a single aorto-left external iliac bypass with ligation of the terminal aorta. The anastomoses were constructed as 20 mm long end prosthesis to side aorta and iliac artery configuration using a single continuous suture of 6-0 polypropylene. No anticoagulant or anti-platelet agents were administered throughout the study. Patency was assessed by regular palpation of the femoral pulses and by angiographic assessment at 1 month, 3 months, and then at 3-monthly intervals postoperatively, up to 4 years follow-up. Prostheses were explanted at various intervals or when a prosthesis was found to have thrombosed.

In this study, a cumulative patency, determined by Life Table analysis, of 66% at 4 years was achieved [54]. Gross examination of the patent prostheses generally showed a smooth neointimal surface and good host tissue infiltration on the adventitial surface (Fig. 11). No significant dilatation or fusiform aneurysms were found, nor were any gross morphological changes that could threaten the function of the prostheses.

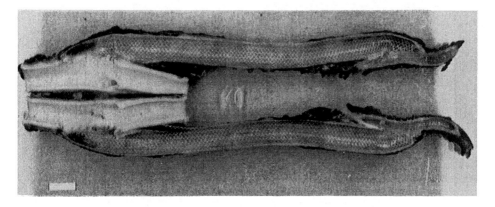

Figure 11 Gross morphology of an explant used as a canine aortoiliac bypass, cut longitudinally to show an essentially thrombus-free surface and lack of aneurysm formation. Polyester mesh is visible throughout the prosthesis. Bar = 6 mm. (From Ref. 54, with permission.)

Examination of the failed prostheses indicated that short-term failures up to 36 days (8 implants) were due either to technical errors at the anastomoses (6 cases) or to infection (2 cases). Of the 3 prostheses that failed after 1 year, 2 appeared to have been caused by buildup of thrombus in the distal, blind end of the aorta. These 2 failures would appear, therefore, to be due to the design of the animal model and would not occur in normal clinical applications. The other failed prosthesis showed evidence of a slowly progressive thrombotic occlusion, possibly associated with an area of exposed polyester fibers. Some thrombus formation was also observed on the patent prosthesis explants with the thrombus-free surface area (TFSA) varying from 80% to 100% in all but 3 prostheses. Of those with a lower TFSA, two showed organized thrombus accumulation secondary to midprosthesis kinks with the thrombus distal to the kink site, while in the other, thrombus was apparently due to exposed polyester mesh fibers on the intimal surface.

After implantation in the host, new connective tissue formed in association with the prosthesis. Thin pannus associated with the intimal surface of the anastomoses and tissue infiltration of the adventitia could be observed by gross morphological examination. Pannus growth increased in length up to 20 mm with implant duration and was much more pronounced at the proximal anastomosis compared with the distal end. Histology (Fig. 12) showed that this anastomotic pannus was composed of fibroblasts and smooth muscle cells extending from the tunica intima and media of the host arteries over the surface of the prosthesis. Cartilaginous metaplasia also appeared within the pannus of a few samples.

Adventitial growth around the prosthesis by host connective tissue appeared more advanced after longer implant periods, with histology showing fibroblasts, vessels,

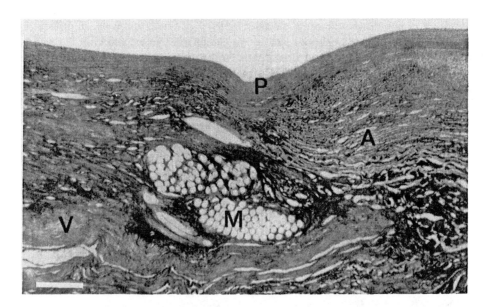

Figure 12   Histology, using Sirius Red staining, of an explant showing a thin layer of pannus (P) at the junction of the prosthesis and the aorta, extending left over the prosthesis from the anastomosis. A = aorta; V = vascular prosthesis; M = mesh bundle of polyester. Intimal surface nearest the blood flow is at top. Bar = 100 μm. (From Ref. 54, with permission.)

nerves, and some fat in the outermost layer of tissue around the prosthesis [54]. Foci of microcalcification, associated with the anastomoses, were also seen in a number of prostheses of greater than 1-year implant duration. The frequency and extent of calcification was low and considered unlikely to lead to device failure. In a more recent study using the same animal model system on Omniflow II (Edwards, unpublished data) 100% patency was achieved for 12 implants over a period of 1 year. In addition, a nonfunctional, subcutaneous biocompatibility study in rats has demonstrated that the Omniflow II prosthesis does not have a tendency to calcify [55].

In all these studies, where the implant is a xenograft, the potential exists for an immunological response to develop. However, in addition to providing long-term stability and durability, the glutaraldehyde stabilization process also minimizes the potential immunogenic effect of the xenogenic tissue. Experiments have shown little evidence of circulating antibodies [35], and histological analyses demonstrated the absence of any marked immunologically mediated lymphocytic response to these implants (unpublished data).

D. Clinical Studies

Although a variety of vascular prostheses have shown good or even excellent results in animal model studies, these models do not always provide a true indication of how the device will perform in clinical situations, and often apparently excellent prostheses have later been found to fail in clinical practice [10–12,26–28]. In the case of the Omniflow prosthesis, it has performed well in clinical use.

The Omniflow I prosthesis has been in clinical use since 1982 and over 3000 units have been sold worldwide. This prosthesis was replaced in 1989 by the improved Omniflow II prosthesis, and to date 1800 units of this prosthesis have been sold. Published results (see below) are available for 511 peripheral implants and 91 arteriovenous (AV) access prostheses. No aneurysm formation has been reported in the published papers. In the case of early thrombosis (17 patients), successful thrombectomy has been reported to have effectively restored the function of the prosthesis (personal communication, Bio Nova International).

The initial clinical reports by Ketharanathan and colleagues [6,35,56] examined the use of the Omniflow (Prototype) in lower limb salvage. These studies revealed a number of problems with the prototype device, including occlusions, such that subsequent low patency of around 30% after 4 years was observed [38]. The prototype showed variations in the proximity of the Dacron mesh to the flow surface, with the possibility in regions of poor coverage that the Dacron mesh could break through into the blood flow and provide a focal point for thrombus buildup. The prototype also showed variations in the overall thickness of the collagen component, and this affected handling characteristics and the response to suturing [56].

Further research, including modifications in mesh design, produced a graft of improved handling characteristics, more uniform collagen thickness, and better collagen coverage of the mesh. This improved device, Omniflow I, showed improved performance. Thus, clinical reports by Ketharanathan and Field and colleagues indicated a 2-year cumulative patency of 61%, 51% above-knee patency at 5 years, 20% for below-knee at 5 years, and 32% for femorocrural bypasses at 3 years for 62 vascular reconstructions (Fig. 13a) [38,57]. In these studies early thrombosis was reported to remain a problem, as was kinking over the knee joint due to poor elasticity; no aneurysms were reported.

In an independent study, Raithel and colleagues [58] reported their 5-year experience

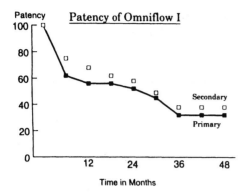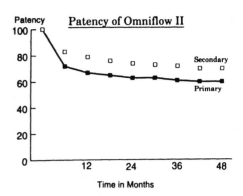

Figure 13   Cumulative clinical patency of the Omniflow I and II Vascular Prostheses up to 4 years implantation. (From Ref. 38, with permission.)

with 191 Omniflow I prostheses in the femoropopliteal and femoro-infrapopliteal positions for limb salvage. They reported no aneurysms. The above-knee cumulative patencies were 68% at 1 year, 53% at 2 years, and 42% at 3 years, while below-knee reconstructions had cumulative patency rates of 58% at 1 year and 44% at 2 years. In this study failures due to kinking across the knee joint were highlighted.

Other clinical studies over 1 to 5 years with the Omniflow I were also encouraging, particularly in comparison to other materials [59–62]. Higher patencies were associated with above-the-knee surgical replacements. In those studies where Omniflow I was compared directly with vein substitutes, the Omniflow I performed equally as well [63]. Since the apparent effectiveness of a device will often depend on the severity of the disease state, comparisons between materials and procedures where comparable cohorts of patients are used are of most value. In single studies, the moderate success of the vascular replacements may be due to the advanced disease state of the patients undergoing limb salvage.

Initial trials with the latest, Omniflow II, prosthesis suggest that the improved physical characteristics have reduced the problems due to kinking that were observed with Omniflow I. Higher patency rates are observed for the Omniflow II prosthesis [38]. Thus, an interim report of ongoing clinical trials in 77 patients, comprising 67% above-the-knee and 33% below-the-knee surgery, shows a primary patency of 68% and a secondary patency of 78% after 4 years (Fig. 13b) [38,57]. In an independent study of femoro-popliteal above-the-knee bypasses, Omniflow II was shown to perform with a cumulative patency of 94.7% after 1 year [64].

To date, the application of Omniflow prostheses has focused on peripheral arterial replacement. However, the capacity to further improve the prosthesis suggests that it has potential for development as a coronary artery replacement. Indeed, even in an early study using the original Omniflow (Prototype), a patient was reported to have had successful left and right main coronary artery bypasses that were patent at 19 months [6].

E.   Explant Analysis

1.   *Animal Studies*

While the host cellular response is well documented for many biomaterials, little is known about the formation and remodeling of the connective tissue components, particularly the collagens, within the vicinity of the device. Detailed analyses of the connective tissue components in explant samples from the dog studies by Edwards using Omniflow I [54] and Omniflow II (unpublished data) have been described [47,48,54,65–67].

All Omniflow prostheses become lined by a cellular neointimal lining, which is in direct contact with the blood, and which can be readily identified by light microscopy (Fig. 14). The neointima is slightly thicker at the two anastomoses, although this does not lead to functional occlusion of the prosthesis. With long-term implants, the neointimal thickness is fixed and generally around 150–200 μm near the anastomotic junction and around 50 μm at the central region of the prosthesis [67]. Analysis of the individual connective tissue components of this new tissue requires MAbs that are specific for the different connective tissue components (Fig. 14). However, within the main vessel wall, the persistence of the original collagen components of the prosthesis and their augmentation with new host tissue are critical to the structural integrity of the prosthesis. To study the individual components of the wall of the device, more selective MAbs are required that are not only specific for particular molecular components, but also specific for the molecules from an individual species (Fig. 14) [54,65].

We have developed, and characterized by a range of methods, a range of MAb probes highly specific to distinct collagen types that allow the interactions between host and implant to be examined [45–48,54,65]. Since the Omniflow prosthesis is manufactured in sheep and thus contains a variety of ovine collagen types, evaluation of the performance of the prosthesis in xenogenic species including dogs and humans is possible with probes

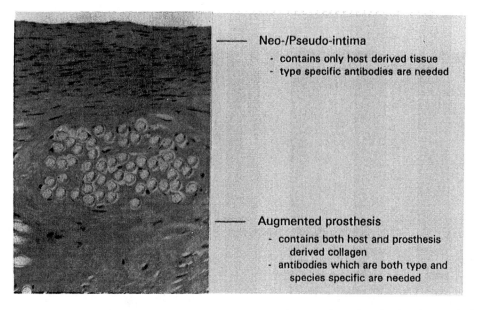

Neo-/Pseudo-intima
- contains only host derived tissue
- type specific antibodies are needed

Augmented prosthesis
- contains both host and prosthesis derived collagen
- antibodies which are both type and species specific are needed

Figure 14   Cross section of an explant stained with hematoxylin and eosin showing newly formed cellular neo/pseudointima and the wall of the device adjacent to the polyester filaments. Immunohistological evaluation of collagen in the neointima requires MAbs that are specific to the type of collagen. The "augmented prosthesis" contains both new host-derived connective tissue and any original ovine implant tissue. In these investigations, MAbs again have to be specific to the type of collagen, but must also be able to distinguish between ovine collagen from the implant and newly deposited dog collagen. (From Ref. 71, with permission.)

that are reactive with either the donor (always ovine) or recipient components but not both [54,65].

Neointima.   After 6 months implantation in the dog, a small but well-defined pseudo-neointima is formed that does not become appreciably thicker with increasing implant time. This region, which is comprised entirely of new host-derived tissue, contains the major fibrillar collagens, types I and III, along with collagen VI, laminin, fibronectin, and microfibril-associated elastic proteins uniformly distributed through the layer [48,65,67]. These components are present in the neointimal region of explants retrieved up to 4 years (Fig. 15).

Collagen V can also be detected at certain times in the newly formed tissue, but, unlike the other components, its distribution is apparently not uniform. In explants up to 6 months, collagen V could be detected in the developing neointima, predominantly in the inner blood contact zone (Fig. 16). Collagen V may form the core around which larger collagen fibrils are built [68], so its presence may indicate new tissue formation [47]. Thus, once new tissue formation stops, it is no longer observed, and when fibrillar collagen accumulates on the existing collagen V, its presence becomes masked. Collagen V is not readily detected in explants examined at longer periods of implantation and appears to be readily detected only with the early tissue growth, consistent with the growth of the neo-intimal layer.

The formation of an endothelial cell layer is considered to be important for the function of vascular prostheses. In the dog model, with the most recent Omniflow II, endothelialization occurs rapidly at the anastomotic junctions, presumably from adjacent

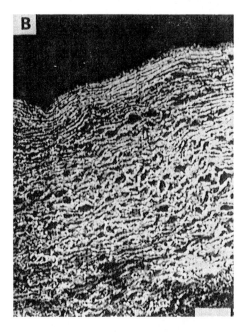

Figure 15   Immunofluorescent detection of newly deposited collagen type III (A) and type VI (B) in the neointima near the anastomosis of a canine explant retrieved after 4 years implantation. Bars = 50 μm.

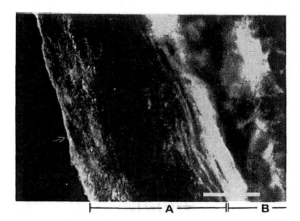

Figure 16   Immunofluorescent detection of newly deposited collagen type V within the neo-intima (A), particularly near the inner blood contact surface (arrow), of a canine explant retrieved after 6 months implantation. The original prosthesis (B) also shows background autofluorescence caused by the processing treatment. Bar = 50 μm. (From Ref. 47, with permission.)

artery, and can be clearly demonstrated by immunohistological staining with an antibody against von Willebrand factor [67]. The Omniflow prostheses also become highly vascularized throughout the vessel wall [67], and by 3–6 months complete endothelialization is evident along the full length of the 120-mm explant [67].

Explants from 3 months to 4 years also demonstrate that smooth muscle cell (SMC) hyperplasia is not a problem with the Omniflow Prostheses. SMCs are detected in all explants, but their proliferation is mild and restricted solely to the anastomotic junctions [67].

Vessel Wall.   With the use of our species-specific anticollagen MAb probes, we have examined explants of Omniflow I and Omniflow II retrieved from dogs after periods up to 4 years. Because a strong autofluorescence can sometimes be present in the explanted prosthesis material [66], a nonfluorescent detection system — for example, horseradish peroxidase — can be used for immunohistological evaluation.

Examination of Omniflow II samples that had been in dogs for 1–6 months showed deposition of new, host-derived collagen VI, infiltrating from around the periphery of the original implant (Fig. 17a) [48], and by 6 months this collagen type, like collagen III, had infiltrated the full thickness of the wall. The function of type VI collagen is unknown, but it may indicate areas of new cell migration and collagen synthesis [69]. The appearance of collagen VI in these explants was associated with the presence of a sizeable number of cells. In the original prosthesis, only relatively few cells are seen in this vessel wall region (Fig. 17b), indicating that the cells observed in the explant were almost entirely due to infiltration from the host [48]. However, it is unlikely that augmentation by collagen VI would confer a substantial increase in structural stability. For this, augmentation by the major fibril-forming collagens would be preferable. Analysis of 4-year explants of Omniflow I has demonstrated synthesis of new dog collagen III, prominent throughout the entire wall of the device (Fig. 18) [54]. Similar augmentation has also been observed in 1-year explants of Omniflow II (unpublished data).

The structural integrity of the Omniflow prosthesis is likely to be dependent upon the

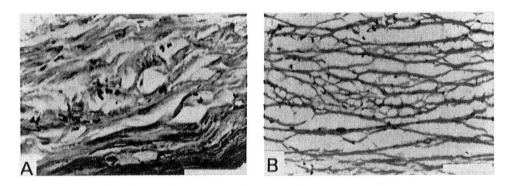

Figure 17   Dog explant retrieved after 3 months implantation, showing (A) immunoperoxidase localization of new host-derived collagen type VI augmentation in the wall, bar = 50 μm. (B) control section of the original Omniflow Prosthesis showing few cells and loosely packed collagen, bar = 25 μm. (From Ref. 48, with permission.)

degree of persistence of the original collagen in the prosthesis [54], as well as the extent of augmentation by new host-derived collagen. While persistence of collagen VI has been shown in 6-month explants of Omniflow II [67], this is only a minor collagen component of the original device and has a limited distribution. However, using an MAb that detects only sheep collagen III, and that does not react with the new dog collagen, the sheep

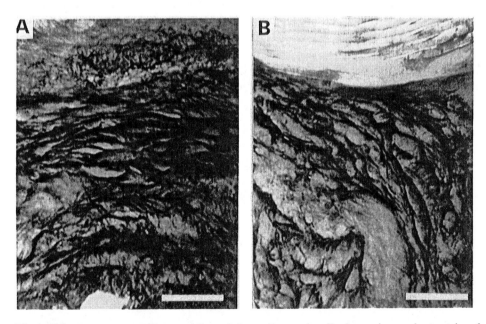

Figure 18   Immunoperoxidase staining of the main vessel wall of a canine explant retrieved after 4 years implantation, showing (A) persistence of the original ovine collagen type III, and (B) augmentation of the device by new host-derived collagen type III. Bars = 100 μm. (From Ref. 54, with permission.)

collagen III from the original prosthesis has been shown to be present after 4 years of implantation (Fig. 18) [54].

2. Clinical Studies

Reports on a limited number of clinical samples, explanted after between 2.5 and 4 years, have been described [48,67]. These samples, which have generally been obtained at revision, focus on the response at the anastomotic region of the Omniflow II prosthesis and do not give data on the response in the midsection of the prosthesis.

Neo-intima. Immunohistological examination of the intimal layer of the human explants showed that a complex connective tissue matrix was formed comprising collagens I, III, IV, and VI; and various other connective tissue components including laminin, fibronectin, MAGP, and a fibrillin-like elastic tissue protein [67], similar to that found in the dog studies [48,65,67]. Extensive intimal hyperplasia was not evident in these samples, smooth muscle cells being confined marginally at the anastomosis site and not extending significantly towards the central region of the graft [67]. Proximal to the anastomosis, von Willebrand factor, indicative of endothelialization, was detected along the intimal lining although the staining pattern was not always continuous. It did not appear as a discrete singular layer, but instead was present as a complex, multilayered structure, which TEM showed to be discontinuous, containing cells, fibrin, and other amorphous protein layers [67].

Vessel Wall. Immunohistology has shown that new, host-derived collagen VI is present in all samples (>2.5 years), uniformly distributed throughout the vessel wall (Fig. 19) [48,67]. Augmentation of the structure by the fibrillar component collagen III has also been reported [67], distributed through the full thickness of the vessel wall, analogous to that observed in dog model studies [54]. TEM examination of the vessel wall also shows that active cellular infiltration has occurred, accompanied by deposition of fibrillar collagens that augment the collagens of the original device [67]. Preliminary immunohistological data also indicate the persistence of the collagen III derived from the original implant [67]. Like the animal explants, the human explants also show excellent signs of neovascu-

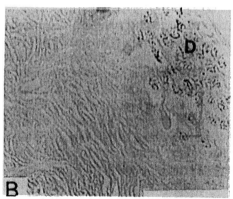

Figure 19   Immunoperoxidase staining of the main vessel wall of a human explant retrieved after 2.5 years implantation, showing (A) augmentation of the device by new host-derived collagen type VI. (B) Negative control of the same explant stained with an irrelevant MAb showing no reaction. Bars = 80 μm. (From Ref. 48, with permission.)

larization that would be beneficial to efficient healing and could potentiate the likelihood of midgraft endothelialization [70].

IV. CONCLUSION

The problems associated with the use of both biological and synthetic vascular prostheses are well documented. While both devices have individual and advantageous attributes, they both fall well short of the performance of the autologous saphenous vein. By contrast, at a diameter of 5 to 6 mm, the Omniflow Vascular Prosthesis has emerged as a sound alternative with a proven track record equivalent to the "gold standard" autograft. Unlike other prostheses, it represents a truly integrated biosynthetic vessel comprising polyester (Dacron) mesh and ovine connective tissue components. As predicted, then, the Omniflow provides the combined advantages of both a biological and a synthetic prosthesis.

Research and development of the Omniflow concept has led to the marketing of the latest, the Omniflow Vascular Prosthesis II, a flexible and compliant device with excellent durability and enhanced thromboresistance. The fine integration of the collagen throughout the polyester filament bundles provides strength to the prosthesis as well as a uniform hemocompatible blood flow surface. The highly ordered collagen structure in the main part of the vessel wall, with the fibers parallel to the longitudinal axis of the device, allows good compliance comparable to the natural vessel. Additionally, the open sheet-like connective tissue network provides a beneficial environment for rapid cellular infiltration and tissue augmentation.

In animal studies, the prosthesis has demonstrated good long-term patency in various surgical replacements. Immunohistological and TEM analysis of explants provide good evidence for the excellent *in vivo* performance. Thrombus formation was rare and consistent with the explant evaluations showing the formation of a continuous thin layer of neointimal matrix of collagens, laminin, microfibrillar elastic proteins, fibronectin, and importantly, von Willebrand factor. The latter is indicative of endothelialization, which was confirmed by SEM and TEM. Smooth muscle cells were absent from midgraft sections but were present at the anastomotic junction, albeit at levels that would hardly be expected to occlude the device. Aneurysms and dilatations were not found; and their absence is likely to reflect the combination of excellent new tissue remodeling and integration, and the persistence of the original stabilized collagen components of the prosthesis. Evaluation of these explants has only been possible with the development of a library of highly specific MAbs that can discriminate between collagens from different species.

Initial studies of retrieved human explants confirm the animal studies, showing good tissue augmentation relating to the structural durability and clinical patency of the device. Consistent with this, proliferation of SMCs was minimal and only seen at the anastomotic regions. Similar to the animal studies, aneurysms and dilatations were also rare or absent. Immunohistological examination of retrieved samples shows excellent tissue augmentation consistent with a sound vessel structure impervious to weakening. At the blood contact zone, von Willebrand factor was found although at present it appears that the inner layer is a mixture of true neointima and pseudointima.

The experimental and clinical studies provide good evidence that the Omniflow II prosthesis is a good choice for below-the-knee revascularization. The data are encouraging. for further development towards the smaller-diameter vessels.

ACKNOWLEDGMENTS

We thank the principal scientist, V. Ketharanathan, and the staff scientists of Bio Nova International who developed the Omnifow II. Support was provided in part under the Generic Technology component of the Industry Research and Development Act 1986.

REFERENCES

1. Greisler, H.P., *Medical Intelligence Unit. New Biologic and Synthetic Vascular Prostheses*, R.G. Landes Co., 1991.
2. Soyer, T., M. Lempinen, P. Cooper, L. Norton, and B. Eiseman, A new venous prosthesis, *Surgery, 72*:864–872 (1972).
3. Pourdeyhimi, B., Vascular grafts: Textile structures and their performance, *Textile Progress, 15*: 1–30 (1987).
4. Dardik, H., I.M. Ibrahim, S. Srayregen, and Dardik, I.I., Clinical experience with modified human umbilical cord vein for arterial bypass, *Surgery, 79*:618–624 (1976).
5. Ketharanathan, V., U.S. Patent 4319363 (1982).
6. Christie, B., V. Ketharanathan, and L.J. Perloff, Patency rates of minute vascular replacements: The glutaraldehyde modified mandrel grown conduits, *J. Surg. Res., 28*:519–532 (1980).
7. O'Donnell, T.F., W. Mackey, J.L. McCullough, S.L. Maxwell, S.P. Farber, R.A. Deterling, and A.D. Callas, Correlation of operative findings with angiographic and non-invasive hemodynamic factors associated with failure of polytetrafluoroethylene grafts, *J. Vasc. Surg., 1*: 136–148 (1984).
8. Stephen M., J. Loewenthal, J.M. Little, J. May, and A.G.R. Shiel, Autogenous veins and velour Dacron in femoro-popliteal arterial bypass, *Surgery, 81*:314–318 (1977).
9. Gundry, S.R., and D.M. Behrendt, A quantitative and qualitative comparison of fibrin glue, albumin and blood as agents to pretreat porous vascular grafts, *J. Surg. Res., 43*:75–77 (1987).
10. Hasson, J.E., D. Newton, A.C. Waltman, J.T. Fallon, D.C. Brewster, C. Darling, and W.M. Abbott, Mural degeneration in glutaraldehyde-tanned umbilical vein graft: Incidence and implications, *J. Vasc. Surg., 4*:243–250, (1986).
11. Dardik, H., N. Miller, A. Dardik, I.M. Ibrahim, B. Sussman, S.M. Berry, F. Wolodiger, M. Kahn, and I. Dardik, A decade of experience with the glutaraldehyde-tanned human umbilical cord vein graft for revascularization of the lower limb, *J. Vasc. Surg., 7*:336–346 (1988).
12. Karkow, W.S., J.J.Cranley, R.D. Cranley, C.D. Hafner, and B.A. Rouff, Extended study of aneurysm formation in HUV grafts, *J. Vasc. Surg., 4*:486–492 (1986).
13. Baier, R.E., Properties and characterization of bioprosthetic grafts. In *Vascular Graft Update: Safety and Performance, ASTM STP 898* (Kambic, H.E., Kantrowitz, A., and Surg, P., eds.). Philadelphia: American Society for Testing and Materials,1986, pp. 95–107.
14. Yeager, G.H., and R.A. Cowley, Studies on the use of polyethylene as a fibrous tissue stimulant, *Ann. Surg., 12*:509–520 (1948).
15. Pierce, E.C., and M.D. Baltimore, Autologous tissue tubes for aortic grafts in dogs, *Surgery, 33*:648–657 (1953).
16. Eiken, O., and G. Norden, Bridging small artery defects in the dog with *in situ* preformed autologous connective tissue tubes, *Acta. Chir. Scand., 121*:90–102 (1961).
17. Schilling, J.A., H.M. Shurley, W. Joel, K.M. Richter, and B.N. White, Fibrocollagenous tubes structured *in vivo, Arch. Pathol., 71*:94–99 (1961).
18. Schilling, J.A., B.A. Shurley, W. Joel, B.N. White, and R.H. Bradford, Abdominal aortic grafts: Use of *in vivo* structured autologous and homologous fibrocollagenous tubes, *Ann. Surg., 159*:819–828 (1964).
19. Hufnagel, C.A., Discussion, *Ann. Surg., 159*:828 (1964).

20. Assefi, I., and V. Parsonnet, An arterial prosthesis composed of an autogenous fibrocollagenous tube with incorporated polypropylene mesh, *J. Newark Beth Israel Hosp., 15*:161–170 (1964).
21. Lindstrom, B.L., and T.K.I. Larmi, The compound arterial graft, *Acta. Chir. Scand., 134*: 195–198 (1968).
22. Sparks, C.H., Autogenous grafts made to order, *Ann. Thor. Surg., 8*:104–113 (1969).
23. Sparks, C.H., Die-grown reinforced arterial grafts: Observations on long-term animal grafts and clinical experience, *Ann. Surg., 172*:787–794 (1970).
24. Parsonnet, V., J. Alpert, and D.K. Brief, Autogenous polypropylene-supported collagen tubes for long-term arterial replacement, *Surgery, 70*:935–939 (1971).
25. Sparks, C.H., Silicone mandrel method for growing reinforced autogenous femoro-popliteal artery grafts *in situ, Ann. Surg., 177*:293–300 (1973).
26. Hallin, R.W., and W.R. Sweetman, The Sparks' mandrel graft. A seven year follow-up of mandrel grafts placed by Charles H. Sparks and his associates, *Am. J. Surg., 132*:221–223 (1976).
27. Parsonnet, V., A.C. Tiro, D.K. Brief et al. The fibrocollagenous tube as a small arterial prosthesis. In *Graft Materials in Vascular Surgery* (H. Dardik, ed.), Symposia Specialists, Miami, 1978
28. Christenson, J.T., and B. Eklof, Sparks mandrel, velour Dacron and autogenous saphenous vein grafts in femoropopliteal bypass, *Br. J. Surg., 66*:514–517 (1979).
29. Guidoin, R., H.P. Noel, M. Marois, L. Martin, F. Laroche, L. Beland, R. Cote, and C. Gosselin, Another look at the Sparks-mandrel arterial graft precursor for vascular repair-pathology by scanning election microscopy, *Biomat. Med. Dev. Art. Org., 8*:145–167 (1980).
30. Guidoin, R., A. Thevenet, H.P. Noel, H. Mary, M. Marois, C. Gosselin, and M. King, Le Sparks-mandrel comme prethese arterielle, *J. Maladies Vasculaires, 9*:277–283 (1984).
31. Dardik, H., R.E. Baier, M. Meenaghan, J. Natiella, S. Weinberg, R. Turner, R. Sussman, M. Kahn, I.M. lbrahim, and I.I. Dradik, Morphologic and biophysical assessment of long term human umbilical cord vein implants used as vascular conduits, *Surg. Gynecol. Obstet., 154*:17–26 (1982).
32. Ramshaw, J.A.M., J.A. Werkmeister, and D.E. Peters, Collagen as a biomaterial. In *Current Perspectives on Implantable Devices* (Williams, D.F., ed.), London: JAI Press, 1990, pp. 151–220.
33. Ferrans, V.J., S.L. Hilbert, Y. Tomita, M. Jones, and W.C. Roberts, Morphology of collagen in bioprosthetic heart valves. In *Collagen*, Vol.3 (Nimni, M.E., ed.), Boca Raton, FL: CRC Press, 1988, pp. 145–190.
34. Ketharanathan, V., and B.A. Christie, Glutaraldehyde-tanned ovine collagen conduits as vascular xenografts in dogs, *Arch. Surg., 115*:967–969 (1980).
35. Perloff, L.J., B.A. Christie, V. Ketharanathan, P.L. Field, P.Y. Milne, D.G. MacLeish, and G. Royal, A new replacement for small vessels, *Surgery, 89*:31–41 (1981).
36. Petruzzo, P., A.M. Asunis, G. Gualtiero, G.Farci, L. Pibiri, M.A. De Giudici, and G. Brotzu, New collagen graft from sardinian sheep. In *Il Congresso Internazionale sulle Protesti Vascolari*, Bologna: Monduzzi Editori, 1989, pp. 147–152.
37. Edwards, G.A., and G. Roberts, Development of an ovine collagen-based composite vascular prosthesis, *Clin. Mater., 9*:211–223 (1992).
38. *Brief History of the Development of Omniflow*, Bio Nova International, North Melbourne, 1993.
39. Ramshaw, J.A.M., D.E. Peters, J.A. Werkmeister, and V. Ketharanathan, Collagen organization in mandrel-grown vascular grafts, *J. Biomed. Mater. Res., 23*:649–660 (1989).
40. Brodsky, B., and J.A.M. Ramshaw, Collagen organisation in fibrous capsule, *Int. J. Biol. Macromol.*, 16:27–30 (1994).
41. White, J.F., J.A. Werkmeister, G.A. Edwards, and J.A.M. Ramshaw, Ultrastructural examination of a collagen–polyester composite vascular prosthesis, *Clin. Mater., 14*:271–276 (1993).

42. Roberts, G., H. McCormack, V. Ketharanathan, D.G. Macleish, P.L. Field, and P.Y. Milne, The role of physical and chemical characteristics in assessing the performance of a new biological vascular graft, *J. Biomed. Mater. Res., 23*:443–450 (1989).

43. Pope, F.M., G.R. Martin, J.R. Lichtenstein, R.P. Penttinen, B. Gerson, D.W. Rowe, and V.A. McKusick, Patients with Ehlers–Danlos syndrome type IV lack type III collagen, *Proc. Natl. Acad. Sci. USA, 72*:1314–1316 (1975).

44. Ramshaw, J.A.M., F. Casagranda, J.F. White, and J.A. Werkmeister, Stratigraphic evaluation of the collagen surrounding a biomaterial implant, *Clin. Mater., 16*:9–13 (1994)

45. Werkmeister, J.A., J.A.M. Ramshaw, and G. Ellender, Characterization of a monoclonal antibody against native human type I collagen, *Eur. J. Biochem., 187*:439–443 (1990).

46. Werkmeister, J.A., and J.A.M. Ramshaw, Multiple antigenic determinants on type III collagen, *Biochem. J., 274*:895–898 (1991).

47. Werkmeister, J.A., and J.A.M. Ramshaw, Monoclonal antibodies to type V collagen as markers for new tissue deposition associated with biomaterial implants, *J. Histochem. Cytochem., 39*:1215–1220 (1991).

48. Werkmeister, J.A., T.A. Tebb, J.F. White, and J.A.M. Ramshaw, Monoclonal antibodies to type VI collagen demonstrate new tissue augmentation of a collagen-based biomaterial implant, *J. Histochem. Cytochem., 41*:1701–1706 (1993).

49. Linsemmayer, T.F., J.M. Fitch, T.M. Schmid, N.B. Zak, E. Gibney, R.D. Sanderson, and R. Mayne, Monoclonal antibodies against chicken type V collagen: Production, specificity, and use for immunocytochemical localization in embryonic cornea and other organs, *J. Cell Biol., 96*:124–132 (1983).

50. DeMont, M.E., and G.M. Wright, Elastic arteries in a primitive vertebrate: Mechanics of the lamprey ventral aorta, *Experientia, 49*:43–46 (1993).

51. Kidson, I.G., The effect of wall mechanical properties on patency of arterial grafts, *Ann. Roy. Coll. Surg. England, 65*:24–29 (1983).

52. Ketharanathan, V., and B.A. Christie, Glutaraldehyde tanned ovine collagen compared with polytetrafluoroethylene (Gore-Tex) as a conduit for small calibre artery substitution; an experimental study in dogs, *Aust. N.Z. J. Surg., 51*:556–561 (1981).

53. Wilson, G.J., and P. Klement, Histopathological comparative long-term evaluation of reinforced biosynthetic ovine collagen vascular prostheses with Dardik human umbilical vein graft. In *Proc, XXVI World Congress of the International College of Surgeons*, Milan, 1988, pp. 464–465.

54. Werkmeister, J.A., V. Glattauer, T.A. Tebb, J.A.M. Ramshaw, G.A. Edwards, and G. Roberts, Structural stability of long-term implants of a collagen-based vascular prosthesis, *J. Long-Term Effects Med. Implants, 1*:107–119 (1991).

55. Casagranda, F., G. Ellender, J.A. Werkmeister, and J.A.M. Ramshaw, Evaluation of alternative glutaraldehyde stabilization strategies for collagenous biomaterials, *J. Mater. Sci. Mater. Med., 5*:332–337 (1994).

56. MacLeish, D.G., P.L. Field, and P.Y. Milne, Clinical evaluation of the tanned ovine biosynthetic vascular prosthesis. In *Proc. XXVI World Congress of the International College of Surgeons*, Milan 1988.

57. Field, P.L., P.Y. Milne, N.R. Atkinson, and D.G. Macleish, Omniflow vascular prosthesis— Medium term results. In *Proc. Royal Australasian College of Surgeons and Faculty of Anaesthetics*, Melbourne, 1989, p. 356.

58. Raithel, D., T. Noppeney, J. Dix, and P. Kasprzak, Long-term results of peripheral reconstructions with an ovine collagen prosthesis. In *Proc. XIX World Congress of the International Society for Cardiovascular Surgery*, Toronto, 1989.

59. Battisti, G., F. Stio, and M. Marigliani, La protesi biosintetica Omniflow: esperienza iniziale, *Ann. Ital. Chir., 60*:431–433 (1989).

60. Koch, G., Limb salvage with polyester collagen arterial prosthesis; a four year experience, In: *Proc. 5th Congress of the European Chapter of the International Union of Angiology*, 1990, p. 324.

61. Matticari, S., G. Covucci, E. Chiti, G. Credi, R. Landini, I. Fusco, G.C. Mascherini, and C. Pratesi, The femoropopliteal revascularisation with a biological graft, In *Proc. Second World Week of Professional Updating in Surgery and in Surgical and Oncological Disciplines of the University of Milan*. Bologna: Monduzzi Editori, 1990, Vol. 8, p. 233.

62. Sarcina, A., P. Bavera, and R. Bellosta, Omniflow ovine collagen grafts in lower limbs revascularisation, *Prog. Angiol.*, 1991, pp. 199–202.

63. Bull, P.G., G.W. Hagmueller, M. Hold, and W. Wandschneider, Clinical comparison of biological prostheses for femoropopliteal and tibioperoneal artery reconstruction: Ovine collagen biograft vs. human umbilical vein. In *Modern Vascular Surgery* (Chang, J.B., ed.), Dana Point, CA: PMA Publishing Corp., 1991 Vol. 4.

64. Yoshida, H., T. Sasajima, M. Inaba, N. Morimoto, N. Otani, and Y. Kubo, An experimental and clinical study of the reinforced biosynthetic ovine collagen vascular prosthesis (Omniflow) for small vessel reconstruction, *Jpn. J. Artif. Org.*, 21:1231–1235 (1992).

65. Werkmeister, J.A., D.E. Peters, and J.A.M. Ramshaw, Development of monoclonal antibodies to collagens for assessing host–implant interactions, *J. Biomed. Mater. Res.*, 23A(3):273–283 (1989).

66. Werkmeister, J.A., T.A. Tebb, D.E. Peters, and J.A.M. Ramshaw, The use of quenching agents to enable immunofluorescent examination of collagen-based biomaterials showing glutaraldehyde-derived autofluorescence, *Clin. Mater.*, 6:13–20 (1990).

67. Werkmeister, J.A., J.F. White, and J.A.M. Ramshaw, Evaluation of the Omniflow collagen-polymer vascular prosthesis, *Med. Progr. Through Technol.*, in press.

68. Linsenmayer, T.F., E. Gibney, F. Igoe, M.K. Gordon, J.M. Fitch, L.I. Fessler, and D.E. Birk, Type V collagen: Molecular structure and fibrillar organization of the chicken alpha 1(V) NH_2-terminal domain, a putative regulator of corneal fibrillogenesis, *J. Cell Biol.*, 121:1181–1189 (1993).

69. Murata, Y., H. Yoshioka, M. Kitaoka, K. Iyama, R. Okamura, and G. Usuku, Type VI collagen in healing rabbit corneal wounds, *Ophthal. Res.*, 22:144–151 (1990).

70. Menger, M.D., P. Walter, F. Hammersen, and K. Messmer, Angiogenesis and neovascualrization in different prosthetic vascular grafts: The significance of graft porosity. In *Clinical Implant Materials* (Heimke, G., Soltesz, U., and Lee, A.J.C., eds.), Advances in Biomaterials Vol. 9, Amsterdam: Elsevier, 1990, pp. 511–516.

71. Werkmeister, J.A., G.A. Edwards, V. Glattauer, T.A. Tebb, F. Casagranda, G. Roberts, and J.A.M. Ramshaw, Evaluation of collagen turnover and structural stability in long term implants of a collagen-based vascular prosthesis, *Adv. Biomater.*, 10:287–291 (1992).

35

Vascular Grafts: Materials, Methods, and Clinical Applications

Rajagopal R. Kowligi, Tarun J. Edwin, Chris Banas,
and Robert W. Calcote
IMPRA, Inc.
Tempe, Arizona

I. INTRODUCTION

Vascular grafts are conduits that are used to replace or bypass diseased arteries and veins in the human body. Historical development of material for vascular grafts dates back to the early 1900s (Table 1). There are several designs of vascular grafts used today, but

Table 1 Historical Development of Materials for Vascular Grafts

Year, author	Description
1906, Carrel	Homologous and heterologous artery and vein transplant in dogs
1906, Goyanes	First autologous vein transplant in man
1915, Tuffier	Paraffin-lined silver tubes
1942, Blakemore	Vitallium tubes
1947, Hufnagel	Polished methyl methacrylate tubes
1948, Gross	Arterial allografts
1949, Donovan	Polyethylene tubes
1952, Voorhees	Vinyon-N, first fabric prosthesis
1955, Egdahl	Siliconized rubber
1955, Edwards and Tapp	Crimped nylon
1957, Edwards	Teflon
1960, Debakey	Dacron
1966, Rosenberg	Bovine heterograft
1968, Sparks	Dacron-supported autogenous fibrous tubes
1972, Soyer	Expanded polytetrafluoroethylene (ePTFE)
1975, Dardik	Human umbilical cord vein

there is no clinically accepted graft that is suitable for smaller vessels such as the coronary artery. There are in excess of 350,000 vascular repair procedures done in the United States every year. Application sites include but are not limited to:

- Abdominal aneurysms
- Aortoiliac
- Femoropopliteal
- Femorofemoral
- Subclavian artery
- Thoracic aneurysms
- Aortobi-femoral
- Femorotibial
- Arteriovenous fistula
- Venous replacements
- Pulmonary artery
- Renal
- Axillofemoral
- Carotid artery

It is believed that the ideal vascular graft should have the following characteristics:

- Variety of sizes and lengths
- Durable, nontoxic, nonallergic
- Conformable
- Ease of suturing
- No kinking
- Nontraumatic to blood elements
- Resistant to infection
- Sterilized repeatedly
- Use throughout the body
- Elastic
- Pliable
- No fraying
- Smooth luminal surface
- Good long-term patencies
- Nonthrombogenic
- Reasonable cost

It is the objective of this chapter to review the state of the art in vascular grafts. A brief description is given with respect to the materials used in graft manufacture, various manufacturing methods, methods of testing these grafts, and their clinical performance. A summary of the recent work to improve vascular graft performance is presented at the end of the chapter.

II. MATERIALS OF CONSTRUCTION

A. Polytetrafluoroethylene

Polytetrafluoroethylene (PTFE) is a fully fluorinated polymer with the chemical formula $-(CF_2-CF_2)n$, discovered in 1938 by R.J. Plunkett [1]. Commercial PTFE has a molecular weight in the range of 20–120 million. It has the following attributes: excellent chemical resistance, high-temperature stability, good nonstick properties, and excellent resistance to degradation of mechanical properties under severe conditions [2]. In 1963 Sumitomo Industries patented a process [3] in which paste-extruded sheets, tubes, and rods of PTFE could be dried and expanded at a temperature below the crystalline melting point to produce uniform and continuous porous structures. After sintering, the structure was preserved and the mechanical strength greatly improved. The coagulated dispersion form of PTFE resins is processed by this method and these resins are currently used in the manufacture of vascular grafts.

B. Polyethylene Terephthalate

Polyethylene terephthalate (PET) was invented by J. R. Whinfield and J. T. Dickson of the Calico Printer's Association (CPA–USA) and was a direct development of the work done by W. H. Carothers. Later E. I. du Pont de Nemours, Inc., purchased the patent from CPA and was issued continuing patent No. 2,465,319 [4]. PET is usually converted into fibers by the conventional extrusion process through dies. The material is available in a variety of linear densities, filament counts, filament diameters, and modified textures. These fibers are put together to form yarn, which is then either woven or knitted into fabric material from which vascular grafts are constructed.

C. Polyurethane

Polyurethanes were first discovered by Otto Bayer of Germany in 1937 in response to the work done by Carothers of the United States on polyamides. It was the reaction of aromatic diisocyanates with the high molecular weight gylcols that led to the formation of polyurethane elastomers that are widely used today [6]. Both aromatic and aliphatic polyurethanes have been investigated as vascular grafts. Polyurethanes are thought to be relatively biocompatible, with excellent physical and mechanical properties. Their material characteristics allow for a design of vascular grafts with greater mechanical similarity to natural vessels than those of expanded polytetrafluoroethylene (ePTFE) or Dacron [6]. Recently, new formulations of polyurethanes have been developed using polycarbonates that address the issue of biostability and degradation [7].

Commercially, polyurethanes are available from various manufacturers such as Dow Corning, Thermedics, Polymedica Industries, Kontron Instruments, and Corvita Corporation (all from the United States), and from Toyobo Corp (Japan).

D. Biodegradable Polymers

Polymers such as polylactic acid (PLLA) and polyglycolic acid (PGA) have been tried as graft materials by themselves or in combination with other polymers, without clinical success to date. Gogolewski and his group in Switzerland have developed a compliant biodegradable microporous polyurethane graft [8], prepared by the commonly known dipping precipitation method. The graft, however, is not clinically used at this time. For more information on resorbable polymers, the reader is referred to other chapters of this encyclopedia.

III. METHODS OF MANUFACTURE

A. Expanded PTFE Grafts

Expanded PTFE vascular grafts are manufactured as follows: The PTFE resin is mixed with an extrusion aid (typically mineral spirits and hydrocarbons) and the mixture is formed into a cylindrical billet. The billet is then paste extruded into tubes, which are then expanded to multiple lengths at temperatures below the melting point. These expanded tubes are finally exposed to temperatures above the melting point (342°C) to sinter the material. The end product is a soft microporous tube (Fig. 1) that is commonly used to repair arteries and veins. These grafts can then be further processed to create a variety of stepped, tapered, and externally reinforced grafts. Details of PTFE material processing are well described in literature [9–11]. Grafts made by this process have been characterized and reported elsewhere [12,13]. For clinical use, ePTFE grafts are most popular in the 6 to 10 mm internal diameter range. Currently, vascular grafts from PTFE are manufactured by IMPRA, Inc., W. L. Gore & Assoc., and Atrium Medical Corp. Grafts manufactured by W. L. Gore have a very thin but dense external PTFE reinforcing layer provided on all product types [13].

B.   Dacron Grafts

Dacron vascular grafts are manufactured as follows: First, the fibers are converted into yarns. Depending on the characteristics required in the grafts, the yarn is made up of 24 or 108 fibers. They may also be texturized. Texturized yarn is used to make the velour form of Dacron grafts. The yarn material is finally converted into fabric for grafts by the weaving or knitting processes.

The weaving process consists of yarns going crosswise (warp), passing alternately over and under another yarn that goes lengthwise (Fig. 2). Some of the advantages include the provision of low porosities (a less expensive method of fabric formation) and the possibility of preventing preclotting in very low porosity grafts. However, this fabric has a major problem of fraying when it is cut during shaping or sizing. The fabric material is rigid when the porosity is low and may cause problems suturing the graft to the host vessel. The low porosity of the material will hinder tissue ingrowth and inhibit

Figure 1   Microstructure of ePTFE vascular graft. Solid areas are nodes, interconnected by fibrils. Graft longitudinal axis is from top to bottom (original magnification: 150×).

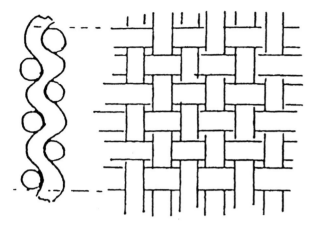

Figure 2   Woven Dacron structure.

anchoring of any tissue growing on either blood contact (lumenal) surface or ablumenal surface. This has the potential of distal embolization if the porosity is too low.

The knitting process consists of passing the yarn material around a needle in such a way that it is knotted to the adjacent row of stitches (Fig. 3). As every stitch is obtained by wrapping the yarn around the needle, there is a gap in the fabric when the needle is removed. The resulting material fabric is rather loose. Porosity is adjusted by changing the needle dimensions. Since the yarn in this fabric is not stretched, the fabric retains the shape much better than the woven fabric. Porosity can also be adjusted by methods such

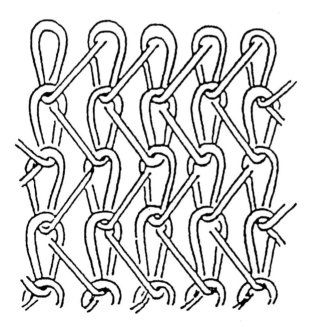

Figure 3   Knitted Dacron structure.

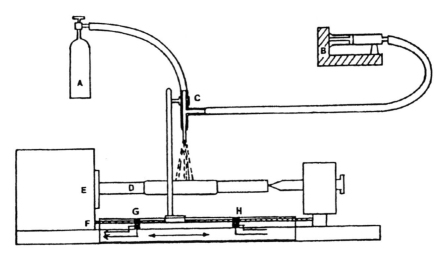

Figure 4 Graft fabrication apparatus. A: nitrogen gas, B: syringe pump to feed solution, C: spray nozzle, D: mandrel, E and F: motors, G and H: microswitches. (From Ref. 24.)

as chemical compaction of the fabric. Major advantages of this fabric are that it is softer and conforms to the shape of the anastomosis, it does not fray when cut, it is easy to suture, and its porosity is adjustable.

Clinically these grafts are most popular in the 8 to 20 mm internal diameter range; and have been characterized and reported elsewhere [14–19]. Dacron vascular grafts are manufactured by several companies:

- Bard–USCI, USA
- B. Braun, Germany
- Meadox Medicals, USA
- Vascutek, Scotland
- Golaski, USA
- Intervascular, USA

C. Spray Coating

A variety of polyurethane grafts have been described using spray coating [20,21]. In one method, a Teflon-coated steel mandrel is mounted on the spindle of a modified lathe and rotated at speeds of 0 to 300 rpm (Fig. 4). A modified three-way stopcock is used as the spray nozzle. This nozzle is clamped to a stand on a lathe that can move along the mandrel. Other types of spray nozzles can also be used in this method. During fabrication, spray from the nozzle is directed onto the rotating mandrel and is moved back and forth along its length. After a predetermined number of passes, the mandrel is removed and dried. The polymer tube is slid off and is ready for use as a vascular graft. Such grafts have been characterized and tested in an animal model [22–24]. The main advantage of this method is the formation of closed-celled pores in the graft wall, which should help in retaining graft compliance after implantation. Grafts made by this procedure are not in clinical use.

D. Flotation–Precipitation Method

Fabrication of polyurethane grafts has been reported using flotation–precipitation [25–27]. In this method, a spray nozzle similar to the one described in the previous section is used to spray polyurethane solution onto flowing water. The polymer precipitates and is collected downstream by a rotating mandrel placed on the surface of the flowing water. The flowing water tank is passed back and forth along the mandrel, and after a predetermined number of passes, the mandrel is removed and dried. The polymer tube is then slid off the mandrel (Fig. 5). Grafts of various water characteristics can be fabricated from this method by changing the fabrication conditions. Kink-resistant grafts can also be made by this method. These grafts have been characterized and reported elsewhere [28,29] but are not available for clinical use.

E. Electrostatically Spun Grafts

Any polymer that can be dissolved in a solvent can be used in electrostatic spinning, but only polyurethanes have been reported [30,31]. The method consists of forcing a solution of polyurethane through a series of hollow metallic needles into a strong electrostatic field. There is a rotating mandrel located in this field. As the polymer comes out through the needles it becomes elongated due to the electrostatic field, separates from the needle, and is collected on the mandrel. During fabrication the mandrel is moved back and forth in front of the needle assembly, and layers of the fiber are gradually built up to form a fibrous polyurethane tube that is porous through the wall. Extensive commercial evaluations have been carried out, but these grafts are not available for clinical use at this time.

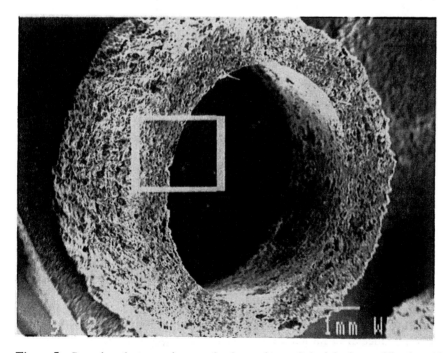

Figure 5 Scanning electron micrograph of vascular graft (original magnification: 10×).

F.   Plasma-Polymerized TFE Grafts

Plasma-polymerized grafts were commercially developed by Atrium Medical Corporation (New Hampshire), based on published work by Garfinkle and Hoffman [32]. The manufacturing process consists of surface treating a woven polyester graft by tetrafluoroethylene (TFE) to modify the blood contact surface. It has been previously shown that this modification reduces the thrombogenic potential of Dacron grafts [33]. The woven Dacron graft is placed in a plasma polymerization chamber, and upon initiation of the plasma reaction, the TFE gas polymerizes in a random fashion into highly crosslinked material and deposits onto the substrate (Dacron). These polymer compositions are highly reactive and easily bond to the Dacron substrate [34]. This polymerization process is very critical because under highly stressed conditions the polymer surface may crack and delaminate from the substrate [24]. Even though this graft was approved by the Food and Drug Administration (FDA), it is not currently available because of unsatisfactory clinical performance.

G.   Modified Dacron Grafts

One of the major drawbacks of Dacron grafts is the need for preclotting the graft with autologous blood before implantation. The preclotting procedure:

- Is time consuming
- Increases the possibility of graft contamination
- Is difficult to perform satisfactorily under conditions of cardiopulmonary bypass
- Increases blood volume used in the operation
- Increases possibility of microemboli formation
- May not be effective if the patient has coagulopathy

To eliminate the need for this preclotting procedure, manufacturers have treated the grafts with a variety of materials to reduce or eliminate the water porosity. Several types of grafts are clinically used:

- GELSEAL and GELSOFT from Vascutek (Scotland): These grafts are impregnated with bovine gelatin and lightly cross-linked using formaldehyde.
- UNIGRAFT from Braun (Germany): This graft is also impregnated with bovine gelatin and crosslinked using isocyanate.
- HEMASHIELD from Meadox Medicals (USA): This graft is impregnated with bovine collagen and crosslinked using formaldehyde.
- HEMAGUARD from Intervascular (USA): This graft is impregnated with bovine collagen and crosslinked using gluteraldehyde.
- USCI ALBUMIN COAT from Bard (USA): This graft is coated with human albumin and crosslinked using glutaraldehyde.

All these grafts have been characterized extensively (Table 2) and have been used clinically.

Other methods of vascular graft fabrication include dip coating, salt casting [35,36], spun polymer grafts [37–39], and processed biologic vascular grafts. Examples of biological grafts include the Dardik Biograft (Meadox Medicals, Inc.), Bioflow graft (BioVascular, Inc), and the Biopolymeric Bovine Heterograft (St. Jude Medical, Inc.).

Table 2 Some Characteristics of Modified Dacron Grafts

Graft	Porosity	Base fabric and structure	Base fabric burst strength (kPa)	Resistance of sealant	Sealant removal mechanism
Vascutek Gelseal	0	Triaxial warp knit, low-internal/high-external velour	2770	*Low:* matching fibrin removal rate healing unimpaired	Hydrolysis
Vascutek	0	Warp knit, low-internal/high-external velour	1790	*Low:* matching fibrin removal rate healing unimpaired	Hydrolysis
Braun Unigraft	0	Warp knit similar to Microvel with trilobal yarn (as abandoned by Meadox)	1050	*Low moderate:* demonstrated adequate healing	Enzymatic + Hydrolysis
Meadox Hemashield	0	Microvel, warp knit, high-internal/high-external velour	770	*Moderate:* some evidence of healing in animal studies	Enzymatic
Intervascular Hemaguard	0	Warp knit and others (woven)	1575	*Very high*	Enzymatic (if at all)
USCI Albumin Coat	Ex-low	Vasculour II warp knit, high-internal/low-external velour	1900	*High:* documented healing impairment in animal studies	Enzymatic

IV. METHODS OF CHARACTERIZATION

A. Internodal Distance

Internodal distance (IND) is a measure of the distance between the nodes throughout the surface of an expanded PTFE graft. This measurement is an indication of the structural properties of the graft, not necessarily a measure of porosity.

A small section of the graft is cut and mounted on a glass slide and viewed under a standard microscope using a known magnification. The average IND is obtained by measuring the distance between two solid PTFE nodes at several locations throughout the surface (Fig. 1). The value is expressed in micrometers.

B. Water Entry Pressure

Water entry pressure (WEP) is a measure of the graft's (PTFE) ability to retain water (or blood) under normal physiological pressures. It is defined as the minimum pressure at which water initially passes through the porous wall of the graft. The WEP must be high enough to prevent seepage of blood or blood plasma through the wall of the graft during and after implantation. Low WEP values (<180 mm Hg or 3.5 psi) can potentially prolong a surgical procedure or force a reoperation.

An approximately 3-in. section of the graft is cut and the ends are clamped with hemostats. A 16 to 20 gauge needle is pierced through one wall of the graft, and water is slowly injected at a constant rate until beads of water are visible on the outside surface of the graft. The pressure recorded at that time is the water entry pressure. The value is expressed in pounds per square inch (psi).

C. Water Permeability

Water permeability is a measure of the rate at which the graft seeps water under a standard physiological pressure (120 mm Hg). It is defined by the ratio of the fluid volume rate of flow divided by the cross-sectional area.

A graft segment with a known total surface area is mounted into an apparatus containing a constant flow syringe pump and a pressure transducer. The segment is closed at one end with a stopcock and attached to the syringe pump on the other end through the pressure transducer. The flow rate is set to obtain a constant pressure of 120 mm Hg. A pan mounted on a scale is placed under the segment to collect the water for a specified period of time. The water permeability is then calculated by:

$$\text{Permeability} = (\text{volume collected}/\text{time collected}) \div \text{total surface area}$$

where the units are expressed in milliliters per minute per square centimeter.

D. Porosity (Void Volume)

Porosity is a measure of the ratio of air (void volume) to solid volume in a graft segment, and is calculated by:

$$\%\ \text{porosity} = 1 - (\text{volume of solid}/\text{volume of segment})$$

where volume of solid = weight of solid/density of solid.

E. Burst Strength

Burst strength (similar to radial tensile strength) is a measure of the graft's ability to withstand internal pressure. It is the actual pressure at which the graft bursts when inflated with water.

The burst strength is also a measure of the graft's ability to withstand the full range of physiological blood pressures and is a key indicator of the graft's ability to maintain the original dimensions after implantation.

A latex balloon is placed inside a 3- to 4-in. segment. Water is pumped into the balloon at a constant rate until the sample ruptures. The peak pressure recorded is defined as the burst strength. The value is expressed in pounds per square inch.

F. Radial Tensile Strength

Radial tensile strength (RTS) is a measure of the strength of the graft wall in the radial direction. It is often used as a substitute for burst strength in determining the graft's ability to withstand the normal physiological pressures. RTS is defined as the ratio between the peak force and the cross-sectional area of the graft.

A short segment of graft with a known cross-sectional area is slipped over a pair of horizontal pins that are mounted in a tensile testing device. One pin is attached to the stationary clamp and the other pin is clamped to the movable section of the Instron. The pins (with graft attached) are pulled apart at a constant speed of 1 in./min until the sample breaks. The peak force is recorded in pounds. The RTS is then calculated by:

$$RTS = peak force/area of cross section$$

and is expressed in pounds per square inch.

G. Longitudinal Tensile Strength

Longitudinal tensile strength (LTS) is a measure of the graft's resistance to the internal longitudinal stresses encountered after implantation. LTS is defined as the ratio between the peak force and the cross-sectional area of the graft.

A segment of graft with a known cross-sectional area is mounted into a tensile testing device. The segment is pulled until the sample breaks. The peak force in recorded in pounds. The LTS is calculated by:

$$LTS = peak force/area of cross section$$

and is expressed in pounds per square inch.

H. Suture Retention Strength

Suture retention strength is a measure of the graft's ability to retain sutures during and after implantation. This is important in preventing failure by aneurysmal dilatation and rupture.

A graft segment is mounted in a tensile testing machine and a 5-0 monofilament suture with a tapered needle is passed through one wall 2 mm from the end. The suture is pulled at a constant rate until the wall ruptures. The resulting peak force is measured in grams. The test is performed in both the longitudinal and transverse directions.

I. Kink Resistance

Resistance to kinking is important when grafts are implanted in a loop or across joints. Resistance to kinking may be expressed in several ways: the bending stress required to produce kinking of samples; the angle formed by the two halves of the tube after buckling occurred; or the radius of curvature of a cylindrical tube, on which the graft kinks when it is wrapped around it. Recently an AAMI standard for vascular prostheses has been published. This standard describes methods to evaluate various types of vascular grafts.

V. VASCULAR GRAFTS: *IN VIVO* PERFORMANCE

A. ePTFE: Lower Limb Bypass and Vascular Reconstruction

Expanded PTFE grafts dominate in access and smaller caliber graft applications. Their use in the infrainguinal position is supported by excellent 2- to 3-year patency rates and they have shown good results in the treatment of lower extremity arterial occlusive disease [40]. Although reversed saphenous vein is still the material of choice for these procedures, between 20% and 40% of patients requiring infrainguinal bypass grafting will not have an adequate ipsilateral saphenous vein.

Experiences with above-knee femoropopliteal bypasses using ePTFE have indicated results that are comparable to reversed saphenous vein [41]. Sterpetti and others found a 75.6% patency rate for ePTFE in the above-knee position after 5 and 7 years [42]. Patterson et al. [41] reported on their experiences with 138 grafts in 124 patients, all above-knee popliteal bypasses. Cumulative primary patency rates are shown in Table 3. These results are comparable to the findings of other investigators (see Table 4). A comparison of vein and ePTFE show similar patency rates at 2 years, but at 5 years vein had superior primary patency rates. Charlesworth and others confirmed [43] these findings in a study with 134 infrainguinal bypass grafts using ePTFE over a 6-year period. They found acceptable results for ePTFE over a 2- to 3-year period, but patency rates of only 24% (versus 70% for vein) after 6 years. Similar findings were obtained by Veith et al. [44] who found divergence in patency rates of vein and ePTFE after only 2 years implant time.

Below-knee and above-knee femoropopliteal bypasses do not have comparable success rates even in the first 2 years. Quinones-Baldrich et al. [45] reported that reconstructions with ePTFE in above-knee and below-knee positions had 5-year patency rates of 63% and 44%, respectively ($p < 0.03$). This is confirmed in Veith's study, [44] which indicates a marked difference in favor of vein (49% primary patency) over ePTFE (21% primary patency) for infrapopliteal (tibial) reconstruction in both early and late results ($p < 0.001$).

Typical causes of failure in ePTFE bypass grafting has been documented. In one study, progression of distal disease was the primary cause of failure in 42% of cases and was associated with intimal hyperplasia in 53% of these. Severity of distal disease, inflow failures, poor distal runoff, infection, and isolated intimal hyperplasia were also touted as typical, but less frequent, modes of failure [40]. It was also reported that 63% of all failures were early and occurred within the first year. Late failures were analyzed by Sterpetti et al. [2] and compared to autogenous saphenous vein failure (see Table 5). Failure modes seem to coincide for the two materials.

Table 3 Life Table Primary Patency of Above-Knee Femoropopliteal Bypass Grafts with ePTFE

Interval (months)	No. at risk at start	Failed	Duration	Lost to follow-up	Died	Interval patency rate	Cumulative patency (%)	SE[a](%)
0–1	138	12	5	2	3	0.09	100	0.00
1–3	116	4	1	0	1	0.97	91	2.6
3–6	110	6	8	3	1	0.94	87	3.0
6–9	92	7	6	1	2	0.92	82	3.6
9–12	76	5	5	0	1	0.93	75	4.3
12–18	65	5	9	1	2	0.92	70	4.8
18–24	48	2	6	0	2	0.96	64	5.5
24–36	38	4	7	0	3	0.88	61	6.2
36–48	24	0	5	0	2	1.00	54	7.5
48–60	17	1	7	0	2	0.92	54	8.9
60–72	7	1	4	0	0	1.00	50	13.4

Source: Ref. 41
[a]Standard error of the mean

Table 4 Primary Long-Term Patency of ePTFE Femoropopliteal Bypass Grafts from Comparable Series in the Literature

Year, author	No. of grafts	Primary patency		
		12 months	36 months	60 months
1980, Gupta	92	NA	83%	NA[b]
1981, Evans	73	69%	57%	NA
1982, Bergan	33	82%	NA	NA
1984, McCauley	90	67%	50%	43%
1985, Christenson	153	84%	80%	76%
1985, Sterpetti	90	82%	64%	58%
1986, Veith	91	84%	60%	NA
1987, Raggerty		79%	52%	29%
Claudication	43			
Limb salvage	63	79%	48%	43%
1988, Rutherford	36[a]	69%	46%	NA
1988, Quiñones-Baldrich	101[a]	82%	74%	63%
Present series	138	75%	61%	54%

Source: Ref. 41
[a]Above-knee grafts only
[b]NA: information not available

B. ePTFE: Angioaccess

Access to the vascular system for the purpose of either injecting nutrients and medication or withdrawing fluids is necessary in both therapeutic and diagnostic medicine. The availability of an efficacious vascular site that enables repeated puncture with minimal pain and discomfort has permitted the application of hemodialysis to end-stage renal disease. With techniques available today, the patient with renal failure that can adjust to a rigorous hemodialysis regimen may anticipate survival for decades [46]. Shunts, such

Table 5 Causes of Late Failure

	Type of graft material[a]	
	ASV	PTFE
Distal progression disease	10	12
Proximal progression disease	2	2
Intimal hyperplasia		
Distal anastomosis	2	6
Proximal anastomosis	—	1
Graft stenosis	2	
Infection	—	2
Pseudoaneurysm	1	—
Unknown	3	2

Source: Ref. 42
[a]ASV: autogenous saphenous vein; PTFE: polytetrafluoroethylene

as the internal Brescia–Cimino shunt, have enabled regular access to the vascular system. To avoid continuously sticking the patient, synthetic vessels such as ePTFE prostheses for angioaccess have gained popularity. As an arteriovenous graft fistula or interposition graft, ePTFE access sites can also be used immediately. In patients prone to infection, autogenous saphenous vein is still recommended, but in all other patients, ePTFE prostheses are recommended [46].

Raju [47] reported on a series of 602 ePTFE hemodialysis grafts, inserted in 532 patients. The placement method on 90% of these grafts was as an interposition between radial artery and cubital vein. Follow-up ranged from 1 to 10 years but involved only 312 of the original grafts. The author found that the most common cause of complication was thrombosis, usually occurring due to stenosis at the venous graft end. Infection and false aneurysms were other complications encountered (see Table 6). Salvage procedures for these grafts included thrombectomy/revision for thrombosis, incision and drainage, and a local bypass for infection [47,50]. In spite of these complications, the modified patency rate (after repair) was 93% at 1 year and 77% at 2 years. When the lifetime of an ePTFE graft is finally expended from repeated/recurrent complications, a site in the forearm of the same or opposite side for secondary insertion should be easily achieved. This can enable adequate access sites even for patients on dialysis for 10 years or more [47]. These results are in agreement with those obtained by other authors. Rapaport et al. summarize 36-month modified patency in a life table, shown in Table 7 [48].

Problems with ePTFE do exist. Seroma formation has been reported in ePTFE grafts used for hemodialysis access, although in 42,000 recorded applications, there were only 33 reports of this side effect [49]. When ePTFE grafts were first popularized, problems with dilatation were reported. These have been remedied either by an external wrap, or by an optimal sintering cycle [12]. Complications with ePTFE grafts occur also as a result of improper usage. Correct puncture technique is important, as is care and maintenance of the graft. In spite of this, ePTFE grafts are still the prostheses of choice when primary fistulas cannot be constructed [50].

C. ePTFE: Other Applications

There are several other situations in which the vascular surgeon may be forced to use a synthetic conduit either for reconstruction or for replacement of sections of the vascular system. The choice of ePTFE is due to its inertness in the physiological environment and its superior performance over other synthetic conduits, especially in the 6 to 10 mm

Table 6   312 ePTFE Grafts for Dialysis Complications in 1- to 10-Year Follow-up

	n	Incidence (%)
Thrombosis	203	64
Infection	110	35
False aneurysm (repeated needle puncture)	50	16
Steal/gangrene	6/3	3
Heart failure	0	0

Source: Ref. 47

Table 7 Modified Life Table Function Analysis

Observation point (months)	Number of grafts	Number of grafts occluding	Deaths	Renal transplants	Cumulative patency (%)
1	103	4	4	0	96
3	95	2	3	1	94
6	85	0	0	2	94
9	78	3	4	1	90
12	65	1	2	0	89
15	57	0	0	0	89
18	51	2	0	2	85
21	42	0	0	1	85
24	36	0	3	1	85
27	26	0	0	0	85
30	14	0	0	0	85
36	5	0	0	0	85
Totals		12 (11.7%)	16 (15.5%)	8 (8.2%)	

Source: Ref. 48

internal diameter applications. In some applications, however, both ePTFE and Dacron can be used.

Symptomatic branch occlusions of the aortic arch often necessitates a carotid-subclavian bypass. Cerebrovascular insufficiency was completely relieved in 88.8% of 18 patients with ePTFE or Dacron grafts in this position [51]. A life table analysis indicated a 91% patency rate for these grafts compared to the 57% patency rate for vein grafts ($p < 0.01$) after 5 years. Similar results for cerebral revascularization are also documented by Yamamoto et al. [52] and Story et al. [53].

For infants with cyanotic congenital heart malformations, the Blalock–Taussig shunt procedure has become the palliative option of choice [55]. A study of survival and event-free rates of 47 ePTFE shunts placed in 42 children resulted in an actuarial estimate of 86% percent patient survival rate at 2 years [54].

Although ePTFE prosthetic grafts have been used for coronary bypass operations, these conduits are actually unsuitable for revascularization of smaller coronary arteries. A continuous decline in patency rate from 86% at 1 week postoperatively to 14% percent at 45 months was reported by Nishida et al. [55]. In spite of this, in cases where coronary revascularization is indicated as a life-saving procedure and neither vein nor internal mammary artery is available or applicable, ePTFE grafts combined with aggressive anti-coagulant and antiplatelet therapies can be effective [56]. However, close observation and regular cardiac catheterization are essential as part of an extensive follow-up program.

Other miscellaneous applications of ePTFE include vena cava reconstruction after resection of malignant tumors [57,58], for repair of any small- to medium-diameter vessels in trauma cases, and in other experimental surgical techniques. Expanded polytetrafluorethylene is also used as a suture material, a surgical membrane, a soft tissue patch for hernia repair, and as a vessel or myocardium repair patch.

D. Dacron

The softness and biocompatibility of Dacron, as well as its ability to be formed into a variety of structures, make it ideal for a vascular prosthesis. Fabrication of Dacron grafts has already been described, and it is indicated that they are the preferred synthetic conduit for thoracic applications, especially for the aorta and other vessels that are larger than 10 mm in internal diameter. The different types of Dacron grafts include woven, knitted, velour, and sealed or coated types.

A clinical evaluation of Dacron double velour grafts done by Lindenauer et al. [59] over a 4-year period studied 318 patients that had undergone elective operations on the aortoiliac vessel for abdominal aortic aneurysm and/or aortoiliac disease. Aortofemoral bypass and aortic-aneurysm resection revealed excellent patency with a low incidence of prosthesis-related complications for up to 5 years. No clinical evidence of overt prosthesis failure, dilation, aneurysm formation, or pseudointimal embolization occurred. The patency rates by life table analyses are shown in Figs. 6 and 7.

Rosenthal et al. [60] reported on a series of 200 above-knee femoropopliteal bypasses done to relieve intermittent claudication. One hundred Dacron and 100 PTFE conduits were used. Analysis by the life table method demonstrated statistically similar primary patency rates at 5 and 10 years ($p > 0.10$). Overall survival rates were 79% at 5 and 42% at 10 years.

In a similar study, Burrell and others [61] used Dacron and PTFE grafts in axillofemoral and axillobifemoral positions. Cumulative patency for axillofemoral procedures was 66% at 32 months and 61% at 58 months. Patency for bifemoral procedures was 97% at 32 months. Dacron and PTFE axillobifemoral graft patency were not significantly different.

Among the other applications of Dacron are its use for resection and reconstruction of the transverse aortic arch [62] and in the portal system as a mesocaval shunt [63].

As mentioned in an earlier section, the main drawback to the use of Dacron in clinical

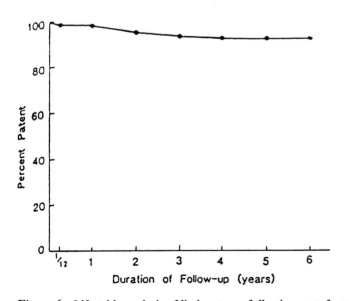

Figure 6   Life table analysis of limb patency following aortofemoral bypass. (From Ref. 60.)

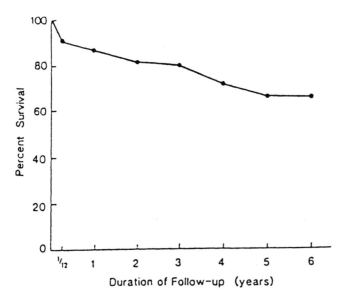

Figure 7 Life table analysis of patient survival following abdominal aortic aneurysm resection. (From Ref. 60.)

applications is the need to preclot the graft in the patient's own blood prior to use. Recent technical advances in biocompatible coatings for these grafts have led to the creation of several coated/impregnated Dacron grafts that allow this step to be eliminated. Some of these are described in a previous section. Of these grafts, the three most popular appear to be the albumin-coated, collagen-coated, and gel-impregnated Dacron grafts.

A Dacron knitted double velour graft impregnated with crosslinked bovine collagen (Hemasheild, Meadox Medical) was implanted in 589 patients over a period of 3 years. Twenty percent were straight grafts, and 69% were bifurcated. Sixty-eight patients had the graft implanted in the aortic position as well. The percent survival after 2 years is shown in Table 8 [64].

A similar graft fabricated by Vascutek of Scotland is the Gelseal Triaxial, which is gelatin coated. Successful animal studies that revealed considerable erosion of the gelatin coating after 9 days to 2 weeks of implantation led to clinical trials [65]. One hundred of these grafts were implanted in the aortic position in 77 men and 23 women for symptoms

Table 8 Late Survival at 1 and 2 Years

	Percent Survival	
	1 year	2 year
Collagen-impregnated graft		
Expected	95.6	91.7
Observed	94.5	91.6
Noncollagenated graft: observed	90.9	84.9

Source: Ref. 65

Table 9 Patency for Aortic Surgery

Year, author	Material[a]	No. of patients	Years follow-up	Patency rate (%)
1981, Chang	Dacron	188	6	84
1978, Brewster	Dacron	261	5	91
1989, Petrovic	PTFE	132	4	98.4
1988, Karner	PTFE/Dacron	60/52	2	97/95
1988, Cintora	PTFE/Dacron	170/142	4	97/90
1980, Nevelsteen	Dacron	352	5	80
			10	62
1975, Vanttinen	Dacron	177	5	91

Source: Ref. 67
[a]PTFE: polytetrafluoroethylene

including aneurysm, intermittent claudication, rest pain, and gangrene. With a follow-up period of 57 months, perioperative mortality was 1% and cumulative primary graft patency was 99% with no measurable blood loss at implementation [66]. Comparative data are reproduced in Table 9.

Dacron grafts are susceptible to failure for a variety of reasons, and like most arterial grafts, the true incidence of failure is believed to be significantly greater than that reported in the literature. A survey of 390 documented cases of graft failure by Pordeyhemi and Wagner indicated that the four most reported complications associated with the use of Dacron grafts are dilation, suture line failure, structural defects, and bleeding or infection [19]. The grafts these apply to are medium to large caliber, so they rarely include intimal hyperplasia or thrombosis as a failure mode. These complications appear to be more prevalent in small-diameter grafts. In conclusion, it should be pointed out that continuous advances in technology and surgical techniques will continue to diminish failure rates and increase performance at all levels.

E. Polyurethanes

Polyurethane elastomers are used extensively throughout the spectrum of medical devices (see Table 10). The flexibility and high compliance that can be designed into these materials make them ideal for use in small-diameter synthetic arteries where artery–graft compliance matching is considered important [67]. Perianastomotic events including increased suture line stress, local flow disturbances, and the development of intimal

Table 10 Applications of Polyurethanes in Medical Devices

- Lead wire coatings for pacemakers
- Acute implantable devices such as hemodialysis catheters
- Temporary and permanent artificial hearts
- In blood dialyzers to bind together bundles of hollow fibers
- Tissue adhesives, wound dressings, and denture materials
- Vascular grafts primarily 6 mm ID or less

Source: Ref. 27

hyperplasia, have been suggested as results of an artery-to-graft compliance mismatch [68].

Many configurations and compositions of polyurethanes have been experimented with in vascular graft applications, with varying results. It is because of this inconsistent performance that no polyurethane vascular grafts have been approved for long-term implantation.

Large ID (10 mm) polyurethane grafts fabricated by an electrostatic spinning technique were implanted into porcine aortas by Annis and others for 1 to 12 months with an almost 100% patency rate [69]. Lyman and others manufactured 4 mm ID polyether polyurethane grafts covered with a knit Dacron tube, and implanted these in canine femoral arteries with a 45% patency rate [70]. They attributed the patent grafts to compliance matching and proved this in a later study [71]. Five millimeter ID Mitrathane microporous grafts were implanted by Marois et al. in the infrarenal location in canines, with close to a 0% patency rate [72]. Fujimoto and others manufactured vascular grafts from a segmented polyurethane using an extrusion technique with NaCl to provide porosity. Three millimeter ID canine implants in the carotid position showed good patency at 6 months but occluded when graft compliancy was decreased by 30% [73]. Wilson et al. constructed medium caliber (6 mm and 10 mm ID) vascular grafts by wrapping extruded polyurethane fibers on a rotating mandrel to create a tube. A canine *in vivo* study in the aortic and aortoiliac position showed all grafts patent at 3- and 6-month follow-up [74]. Kowligi and others manufactured small-diameter (3 mm and 4 mm ID) grafts by a direct spray-coating method and implanted these in the femoral artery position of canines, also with poor results [24]. The causes of failure in this study were similar to those reported by others and can be attributed to: (1) surgical trauma to the host vessel at the time of implantation, exposing thrombus-inducing components to blood; (2) blood-flow disturbances at the host–graft interface caused by inversion and eversion of the graft ends; (3) compliance mismatch [27].

The experience with small-diameter vascular grafts fabricated from polyurethanes confirm the opinion that large-diameter vascular grafts, irrespective of material, have inherently better patency. The real challenge is in the small-diameter (coronary artery size) classification where compliance mismatch is more of an issue. Research reports indicate that a closer match to the host vessel compliance, less trauma to the anastomotic site such as a sutureless anastomosis, and an inert, porous lumen to enhance cellular ingrowth and subsequent endothelialization, would all contribute to an increased patency rate in small-diameter synthetic vascular conduits.

VI. METHODS TO IMPROVE VASCULAR GRAFT PERFORMANCE

A. Endothelial Cell Seeding on Grafts

Among the many ways to improve vascular graft performance are the provision of a nonthrombogenic lining by seeding the graft with endothelial cells [75,76]. Considerable research has been reported with Dacron and ePTFE [77,78] but this technology is not clinically used on a regular basis. Research areas include the selection of cell sources [79] techniques of cell culture and seeding [80], and treatment of the graft substrate to improve seeding efficiency [81,82]. Even though confluency of seeding can be achieved before implantation, no long-term data are available on the viability of the seeded cells after implantation. Results of studies have been mixed, with the treatments showing improvements in patency, resistance to infection, and increased thrombus-free surface area [83].

Seeding techniques have been investigated in humans with varying results. Herring and others showed that there was no difference in patency between seeded and control ePTFE grafts in the femoropopliteal location [84]. Another study, in this location, showed a significant improvement in the 2-year patency of seeded (74%) versus non-seeded (44.5%) control grafts [85]. Clinical implantations in the femorotibial site [86] and femoral artery [87] have been reported, with preliminary indication of benefit with the seeded grafts.

B. Varying Porosity of ePTFE Grafts

Considerable research has been performed to identify the influence of changing internodal distance (IND) on the performance of ePTFE grafts. Kusaba and others [88], showed that pretreatment of ePTFE grafts by multiple large perforations significantly hastened neointima formation in dogs. Clowes and others [89] investigated the role of the IND in early endothelial cell (EC) coverage in a baboon model and found that the 60-micron IND grafts had healed significantly faster (2 weeks) than the regular 30-micron ePTFE grafts (Table 11). They found that the EC source was capillary ingrowth from outside to the lumen of the graft. To determine the optimum IND for healing of ePTFE grafts, Golden and others [90] implanted grafts of various INDs in a baboon model. They found that the EC coverage on the lumen was hastened by increasing IND (Fig. 8). Even though the 90-micron graft had healed completely, there were focal defects in the EC layer at 3 months, and the authors concluded that the 60-micron IND was the optimum IND. Both Clowes and Holden used ePTFE grafts that had no external reinforcing wrap (supplied by W. L. Gore and Associates). To determine if this accelerated EC growth also occurs in humans, Kohler and others [91] implanted 10 femoropopliteal ePTFE grafts composed of equal lengths of 60- and 30-micron portions sutured together, in eight patients. These grafts were reinforced by the thin external ePTFE wrap. Using ^{111}In-labeled platelet imaging techniques, they found that there was no difference in platelet uptake between 60- and 30-micron segments at 1 week or 12 weeks. If there was good EC growth, platelet uptake there should be less than in the areas where there was no EC growth. There was no neointima formation on either segment of the grafts, even though capillary formation was noticed in some portions of the 60-micron segments. It was concluded that the failure to endothelialize is probably due to retarded angiogenesis

Table 11   Early Endothelial Cell (EC) Coverage in ePTFE Grafts

Graft IND[a] (μm)	Time elapsed[b] (weeks)		
	2	4	12
30	NA[a]	0/8	2/13
60	5/5	6/6	6/6

Source: Ref. 91
[a]IND: internodal distance; NA: information not available
[b]Values are the number of grafts completely covered with EC compared to the total number of grafts tested

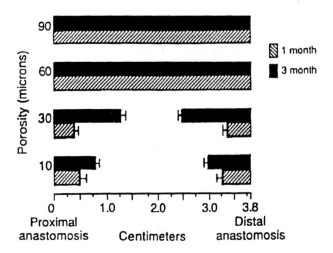

Figure 8 Graph displaying the extent of luminal endothelial coverage of the PTFE grafts at 1 and 3 months after implantation. The coverage is measured in centimeters ± SEM from the heel of the anastomosis to the growing edge of endothelium for the 10- and 30-μ grafts porosity grafts, which derive their endothelium from the anastomoses. ($n = 3$ for all groups except $n = 5$ for 90-μ grafts at 3 months.) (From Ref. 90.)

in adult humans or retardation of ingrowth by the presence of the low-IND reinforcing wrap. Further studies are in progress.

C. Incorporation of Growth Factors

In the previous sections, we have presented evidence that graft performance can be improved by seeding endothelial cells (EC) onto surfaces or by changing the graft micro-structure. Recently Greisler et al. [92] have evaluated the idea that pretreatment of the biomaterial with endothelial cell mitogens may enhance endothelialization. They investigated the influence of heparin-binding growth factor type 1, or fibroblast growth factor type 1 (FGF-1) impregnation on graft healing with reference to EC proliferation and surface coverage. Modified fibrin glue (FG) containing radiolabeled FGF-1, heparin, fibrinogen, and thrombin [96] was impregnated into 60-micron internodal distance ePTFE grafts by a pressure perfusion procedure, and these were implanted in the aorto-iliac positions in dogs (4 mm ID by 5 cm long). Group I contained the experimental FG, Group II had FG without FGF-1 and Group III contained untreated control grafts. Retention studies of the FGF-1 showed that there was only about 4% left at 30 days. However, every one of the experimental grafts showed considerable capillary ingrowth and an EC-covered luminal surface, which was not seen in either of the other groups of grafts. They also reported that the neointima was thicker in the experimental group in comparison to the other two groups, indicating the possibility of replicating activity of other types of cells, possibly smooth muscle cells.

D. Methods to Minimize Suture Hole Bleeding in PTFE

One of the drawbacks to the clinical use of ePTFE is the possibility of suture hole bleeding at the anastomotic site. A comparative evaluation of bleeding rates for PTFE, Dacron, and autogenous vein was performed by Carney and Lilly. They reported blood loss in excess of 300 ml in 43% of the PTFE group in contrast to 22% of the vein group and only 17% of the Dacron group [93].

Tsuchida and others developed an elastomer end-coated PTFE graft designed to provide longitudinal compliance and to reduce suture line bleeding. An external polyurethane coating was applied to 2 cm of the ends of 9 cm long, 6-mm ID PTFE grafts, which were then tested in a canine model. The authors reported that although handling was improved and performance was unaffected, the reduction in suture line bleeding was not statistically different from the uncoated controls. Furthermore, the lack of tissue ingrowth at the coated areas created possible sites for infection or seroma [94].

Although the application of sealants and hemostatic agents such as fibrin adhesive or oxidized regenerated cellulose [95] may be effective in controlling suture line bleeding in PTFE, it is hoped that an efficacious PTFE prosthesis with leak-proof qualities can be designed and marketed.

E. Incorporation of Anticoagulants

It is speculated that small-diameter (<4 mm ID) vascular prostheses fail due to platelet deposition resulting in acute thrombosis and late graft failure through the stimulation of intimal hyperplasia [96]. Pursuing this line of thought, attempts at suppressing platelet deposition and eventual thrombi formation by reducing the inherent thrombogenicity of graft materials are ongoing. Modifications include heparin bonding, albumin binding, endothelial cell seeding, and graft denucleation. Parallel to these studies are attempts to inhibit platelet aggregation onto a biomaterial. The importance of prostacyclin (prostaglandin PGI2) in these applications has recently been realized. Prostacyclin is produced by arterial endothelial cells and is a powerful vasodilator, as well as the most potent inhibitor of platelet aggregation known [97]. A synthetic prostacyclin analog, Iloprost (Schering AG, Berlin, Germany) has allowed experimentation in this area. Eddy and others achieved a 100% patency rate in venous anastomoses using sutures containing Iloprost as compared to a 35% patency rate of prostacyclin-free sutures [98]. Edwin [26] built on this concept and designed porous, water-permeable, small-diameter vascular grafts out of polyurethane, and loaded these with Iloprost-containing polyvinylpyrrolidone (PVP) as a drug delivery device. The grafts were implanted in the femoral and carotid circulations of a canine model. It was hoped that the timed release of prostacyclin analog would delay thrombosis long enough for healing to take place, and an eventual endogenous supply of prostacyclin to evolve. High occlusion rates indicated a more rigorous regimen of prostacyclin may be necessary in order to prevent thrombosis in small-diameter vascular prostheses. However, this study and others have established substantial groundwork for further research in this area.

VIII. FUTURE OF VASCULAR GRAFTS

A. Endovascular Grafts

A recent trend in almost every surgical discipline, including the treatment of vascular disease, is towards minimally invasive surgery (MIS). Leading the conversion of surgical procedures to minimally invasive interventional procedures is the treatment for abdominal aortic aneurysms (AAA). AAAs represent the 14th leading killer in the U.S., claiming approximately 15,000 lives per year. The mortality rate is 75% to 80% when the aneurysm bursts. Approximately 180,000 to 200,000 cases are diagnosed annually at a prevalence

Table 12   Potential U.S. Endovascular Abdominal Aortic Aneurysm (AAA) Market, 1998

Factor	Totals
Number of elective surgical cases	45,000
Number of ruptured surgical cases	10,000
Number of patients diagnosed and untreated	185,000
Total potential domestic patient pool	240,000
Contraidindicated cases[a]	33%
Adjusted patient pool	160,000
Penetration rate	10–20%
Number of cases treated	16,000–32,000
Estimated endovascular AAA selling price[b]	$5,000
1998 market potential	$80–160 million

Source: Ref. 100

[a]Includes anatomic conditions that would prevent passage of the graft to the aneurysm site (e.g., small, occluded, or excessively tortuous arteries) or prevent implant of an adequate prosthesis (e.g., extremely calcified or insufficient aneurysm necks)
[b]Endovascular Technologies is currently the only company in active clinicals with this type of device

estimated at 1.5 million Americans. Growth in minimally invasive surgery will also be driven by the cost-sensitive environment motivated by health care reform. Table 12 shows the potential U.S. endovascular AAA market for 1998 [99].

The current worldwide market for surgically implanted vascular grafts is 460,000 grafts [100]. This number represents approximately 160,000 grafts for abdominal aortic aneurysms and 300,000 grafts for peripheral vascular disease, which includes iliac, femoral, and popliteal aneurysms, as well as total occlusions.

In a study by Szilagyi et al., small, surgically untreated aneurysms were the cause of death in 29.5% of the patients [101]. Darling's study showed that 18.1% of 182 ruptured aneurysms were less than 5 cm in diameter [102]. Cronenwett et al. demonstrated that AAAs as small as 4 cm in diameter can be associated with a rate of rupture as high as 20% per year if hypertension is present [103]. These studies show that the presence rather than the size of an AAA should be the indication for exclusion of the aneurysm from the circulation [105]. The larger the aneurysm, the greater the risk of aneurysm rupture.

Surgical treatment of AAA consists of a 2- to 4-hour procedure, in which a tubular bifurcated synthetic vascular graft is implanted. The surgical procedure requires trauma to the periaortic and periiliac autonomic plexes, aortic cross-clamping, and the use of general inhalation anesthetic [105]. Additionally, the surgical treatment allows for an increase of operative mortality with noncorrectable myocardial ischemia, cardiomyopathy, pulmonary insufficiency, and renal insufficiency. Table 13 shows the comparison between a surgical repair and an endovascular AAA repair. Overall, the data in the table indicate that lower cardiovascular, respiratory, and renal morbidity should decrease mortality [107].

Elective repair of AAA is regularly performed with an operative mortality of under 5% with the expectation that long-term survival is markedly extended [106,107]. However, vascular surgeons are encountering older patients with severe comorbid conditions.

Table 13 Comparison of Surgical and Endovascular Abdominal
Aortic Aneurysm (AAA) Repair

	Surgical AAA repair	Endovascular AAA repair
Procedure time	2–4 h	30 min
Anesthesia	General	Local/epidural
Transfused blood	2-3 units	1 unit
Length of hospital stay	14 days	1–2 days
Home convalescence	1–2 months	1 week
Mortality rate	10.5%	<1%
Complication rate	15–40%	2–3%
Nature of complications	Major	Minor
Cost per case	$25,000–35,000	$10,000–12,000

Source: Endovascular Technologies at EMT

These can increase operative morbidity and may elevate mortality of aortic surgery in excess of 60% [108].

Dotter pioneered the concept of endovascular prosthesis in 1969 when he inserted stainless steel coils into the popliteal arteries of dogs [109]. In 1986 Balko et al. showed that AAAs could be treated *in vivo* using a polyurethane-coated stent [110]. Parodi, Palmaz, and Barone reported a treatment for AAA using an intravascular stent-anchored, Dacron prosthetic graft in animals and discussed the case histories of 5 clinical patients [104]. Figure 9 illustrates the AAA and prosthetic graft–stent prosthesis [104].

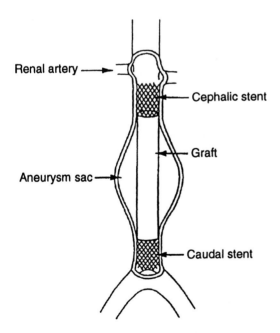

Figure 9   Diagram illustrating an abdominal aortic aneurysm treated with an intraluminal prosthesis anchored to the normal aortic wall by cephalic and caudal stents. (From Ref. 104.)

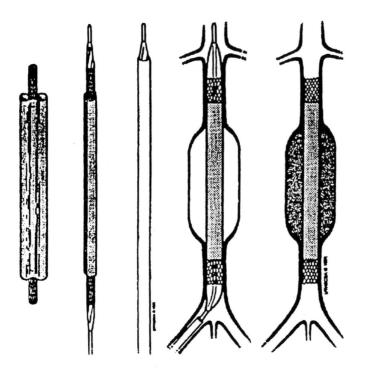

Figure 10   Left to right: endovascular graft with stents attached at both ends, graft and stent folded over angioplasty catheter, sheath placed over assembly, device positioned and proximal/distal stents deployed anchoring graft, aneurysm excluded from circulation. (From Ref. 104.)

Figure 10 shows the noncrimped Dacron graft, with balloon-expandable stents attached to each end. The graft is folded over the balloon angioplasty catheter and introduced into the lumen of a 14-F Teflon sheath. The delivery catheter is moved into place at a position where the proximal stent is distal to the renal arteries. The balloon is inflated, expanding the stent and thus forming a watertight friction seal, which traps the endovascular graft against the aorta using approximately one third of the stent overlapping the graft. After deflating the balloon and withdrawing it, the blood flows through the graft lumen, excluding the aneurysm. In the five clinical AAA procedures reported by Parodi et al., two employed a proximal stent only, while three utilized both a proximal and distal stent to secure the endovascular graft [104].

   Initial clinical results indicate this minimally invasive procedure to exclude the AAA is an effective means in treating patients with comorbid conditions. However, many fundamental questions require further investigation:

- Long-term patency of the endovascular graft
- Effect of occlusion of the lumbar arteries
- Effect of occlusion of the inferior mesenteric arteries
- Potential of distal embolization (thrombus exterior to prosthetic graft may be source of distal embolization by intraluminal manipulation)
- Potential of intimal hyperplasia (proximal/distal to stent placement)

These are just a few areas of investigation that need to be addressed before this procedure can become a viable alternative to surgical repair.

ACKNOWLEDGMENTS

The authors would like to acknowledge the assistance of Melody McLaren in the preparation of this manuscript. The authors would also like to thank the management of IMPRA, Inc. for providing the facilities to complete this work.

REFERENCES

1. Plunkett, R., U.S. Patent 2,230,654, issued February 4, 1941.
2. Blair, J., Flourocarbon polymers, in *Encyclopedia of Industrial Chemical Analysis*, Vol. 13, Wiley, New York, 1971, pp. 73–93.
3. Oshige, S., Japanese Patent No. 42-13560 (67-13560), August 1967.
4. Moncrieff, R. W., Polyesters, in *Man-Made Fibres*, Butterworth, London, 1975, pp. 434–481.
5. Ulrich, H., Urethane polymers, in *Kirk-Othmer: Encyclopedia of Chemical Technology*, 3rd ed., Wiley, New York, 1983, Vol. 23, pp. 576–608.
6. Lelah, M. D., and S. L. Cooper, Vascular prostheses, in *Polyurethanes in Medicine*, CRC Press, Boca Raton, FL, 1987, pp. 151–156.
7. Szycher, M., U.S. Patent 5,254,662, issued October 19, 1993.
8. Gogolewski, S., and A. J. Pennings, Compliant biodegradable vascular prosthesis, in *Polyurethanes in Biomedical Engineering* (H. Planck, G. Egbers, and I. Syre, eds.), Elsevier, Amsterdam, 1984, pp. 279–285.
9. *Processing of PTFE Coagulated Dispersion Powders*, Technical Service Note F3/4/5, 4th ed., ICI Americas, Exton, PA.
10. Gore, R. W., Process for Producing Porous Products, U.S. Patent 3,953,566, April 27, 1976.
11. Okita, K., Process for Producing Porous Polytetrafluoroethylene Tubing, U.S. Patent 4,234,535, November 18, 1980.
12. McClurken, M. E, J. M. McHaney, and W. M. Colone, Physical properties and test methods for expanded polytetrafluoroethylene grafts, in *Vascular Graft Update: Safety and Performance, ASTM STP 898* (H. E. Kambic, A. Kantrowitz, and P. Sung, eds.), Philadelphia, ASTM, 1986, pp.. 86–92.
13. Charara, J., et al., Expanded PTFE prostheses as substitutes in the abdominal aorta of dogs: A comparative study of eleven different grafts, in *High Performance Biomaterials: A Comprehensive Guide to Medical and Pharmaceutical Applications* (M. Szycher, ed.), Technomic Publishing, Lancaster, PA, 1991, pp. 287–312.
14. Hasegawa, M., and T. Azuma, Mechanical properties of synthetic arterial grafts, *J. Biomechanics, 12*, 509–517, 1979.
15. King, M., et al., An evaluation of Czechoslovakian arterial prosthesis, *ASAIO J., 7*(3),114–133, 1984.
16. Guidoin, R., et al., Textile arterial prostheses: Is water permeability equivalent to porosity? *J. Biomed. Mater. Res., 21*, 65–87, 1987.
17. Guidoin, R., et al., Woven polyester arterial grafts with polypropylene wrap: A cosmetic change or improved design? in *High Performance Biomaterials: A Comprehensive Guide to Medical and Pharmaceutical Applications* (M. Szycher, ed.), Technomic Publishing, Lancaster, PA, 1991, pp. 449–474.
18. Guidoin, R., et al., Dacron as arterial prosthetic material: Nature, properties, brands, fate and perspectives, *Biomater. Med. Dev., Art. Org., 5*, 177–203, 1977.
19. Pourdeyhimi, B., and D. Wagner, On the correlation between the failure of vascular grafts and their structural and material properties: A critical analysis, *J. Biomed. Mater. Res., 20*, 375–409, 1986.
20. Kowligi, R. R., W. W. von Maltzahn, and R. C. Eberhart, Manufacture and evaluation of

small diameter vascular grafts, in *Biomedical Engineering. IV. Recent Developments* (B. W. Sauer, ed.), Pergamon Press, New York, 1985, p. 245.

21. Hess, F., et al., Patency rate of small caliber fibrous polyurethane vascular prostheses implanted in dog carotid and femoral artery improved by use of acetylsalicylic acid and dipyridamole, *Thorac. Cardiovasc. Surgeon, 36*, 221–226, 1988.

22. Kowligi, R. R., W. W. von Maltzahn, and R. C. Eberhart, Fabrication and characterization of small diameter vascular prostheses, *J. Biomed. Mater. Res.: Appl. Mater. 22*(A3), 245–256, 1988.

23. Kowligi, R. R., W.W. von Maltzahn, and R. C. Eberhart, Small diameter vascular grafts: Physical property evaluation, in *Biomedical Engineering. V. Recent Developments* (S. Saha, ed.), Pergamon Press, New York, 1986, p. 125.

24. Kowligi, R. R., *Design, Characterization and In-vivo Evaluation of A New Type of Small Diameter Vascular Prosthesis*, Ph.D. Thesis, The University of Texas at Arlington and The University of Texas Health Science Center at Dallas, 1988.

25. Kowligi, R. R., W. W. von Maltzahn, and R. C. Eberhart, Synthetic vascular graft fabrication by a precipitation-flotation method, *ASAIO Trans., 34*, 800–804, 1988.

26. Edwin, T. J., *Fabrication and Evaluation of a Small Diameter Vascular Graft Loaded With Iloprost*, M.S. Thesis, University of Texas at Arlington, and University of Texas Health Science Center at Dallas, 1991.

27. Kowligi, R. R., and R.W. Calcote, Vascular prostheses from polyurethanes: Methods for fabrication and evaluation, in *High Performance Biomaterials: A Comprehensive Guide to Medical and Pharmaceutical Applications* (M. Szycher, ed.), Technomic Publishing, PA, 1991, pp. 425-442.

28. R. W. Calcote, *Identification of Process Control Variables for the Fabrication of Small Diameter Vascular Grafts*, M.S. Thesis, University of Texas at Arlington and University of Texas Health Science Center at Dallas, 1989.

29. Jayaraman, S., and W. W. von Maltzahn, Fabrication process variables of the flotation-precipitation method, in *Proceedings of the 7th Southern Biomedical Engineering Conference*, Clemson, SC, 1988, pp. 87–90.

30. Annis, D., T. V. How, and A. C. Fisher, Recent advances in the development of artificial devices to replace diseased arteries in man: A new elastomeric synthetic artery graft, in *Polyurethanes in Biomedical Engineering* (H. Planck et. al., eds.), Elsevier, Amsterdam, 1984, pp. 287–300.

31. How, T. V., U.S. Patent 4,552,707, issued October 9, 1984.

32. Hoffman, A. S., Ionizing radiation and gas plasma (or glow) discharge treatments for preparation of novel polymeric biomaterials, *Adv. Polym. Sci., 57*, 141–157, 1984.

33. Garfinkle, A. M., et al., Effects of a tetrafluoroethylene glow discharge on patency of small diameter Dacron vascular grafts, *ASAIO Trans., 30*, 432–439, 1984.

34. Guidoin, R., et al., The atrium plasma TFE arterial prosthesis: Physical and chemical characterization, in *High Performance Biomaterials: A Comprehensive Guide to Medical and Pharmaceutical Applications* (M. Szycher, ed.), Technomic Publishing, Lancaster, PA, 1991, pp. 381–399.

35. Murabayashi, S., et al., Biolized polyurethane sponge graft for the study of compliance effect, in *Trans. 13th Annual Meeting of Society for Biomaterials*, New York, June 2-6, 1987, p. 227.

36. Uchida, N., et al., Compliance effect on patency of small diameter vascular grafts, *ASAIO Trans., 35*, 556–558, 1989.

37. Leidner, J., E. W. C. Wong, D. C. MacGregor, and G. J. Wilson, A novel process for manufacture of porous grafts: Process description and product evaluation, *J. Biomed. Mater. Res., 17*, 229–247, 1983.

38. Wong, E.W.C., U.S. Patent 4,475,972, 1984.

39. Pinchuk, L., J. B. Martin, Jr., M. C. Esquivel and D. C. MacGregor, The use of silicone/

polyurethanes graft polymers as a means of eliminating surface cracking of polyurethane prostheses, *J. Biomater. Appl., 3*(2), 260–296, 1988.

40. Quinones-Baldrich, W. J., et al., Failure of PTFE infrainguinal revascularization: Patterns, management, alternatives, and outcome, *Ann. Vasc. Surg., 5*, 163–169, 1991.

41. Patterson, R. B., et al., Preferential use of ePTFE for above-knee femeropopliteal bypass grafts, *Ann. Vasc. Surg., 4*, 338–343, 1990.

42. Sterpetti, A. et al., Seven year experience with polytetrafluoroethylene as above-knee femoropopliteal bypass graft, *J. Vasc. Surg., 2*(6), 907–912, November 1985.

43. Charlesworth, P. M., et al., The fate of polytetrafluoroethylene grafts in lower limb bypass surgery: A six year follow-up, *Br. J. Surg., 72*, 896–899, November 1985.

44. Veith, F. J., et al., Six-year prospective multicenter randomized comparison of autologous saphenous vein and expanded polytetrafluoroethylene grafts in infrainguinal arterial reconstructions, *J. Vasc. Surg., 3*(1), 104–114, January 1986.

45. Quinones-Baldrich, W. J., et al., Is the preferential use of polytetrafluoroethylene grafts for femoropopliteal bypass Justified? *J. Vasc.. Surg., 8*(3), 219–228, September 1988.

46. Butt, K. M., and Friedman, E. A., Evolution of vascular access, *Artif. Organs, 10*(4), 285–297, 1986.

47. Raju, S., PTFE grafts for hemodialysis access, *Ann. Surg., 206*(5), 666–673B, November 1987.

48. Rapaport, A., et al., Polytetrafluoroethylene (PTFE) grafts for hemodialysis in chronic renal failure: Assessment of durability and function at three years, *Austral.NZ J. Surg., 51*(6), 562–567.

49. Bolton, W., and Cannon, J. A., Seroma formation associated with PTFE vascular grafts used as arteriovenous fistulae, *Dialysis and Transplantation, 10*(1), 60–66, January 1981.

50. IMPRA Technical Report TR-l 04, *Access Complications and Revisions*, Impra Inc., 1989.

51. Ziomek, S. et al., The superiority of synthetic arterial grafts over autologous veins in carotid-subclavian bypass, *J. Vasc. Surg., 3*(1), 140–145, January 1986.

52. Yamamoto, S., et al., Femero-internal carotid artery bypass for cerebral ischemia in Takayasu's arteritis, *Arch. Surg., 119*,1426–1429, December 1984.

53. Story, J. L., et al., Cerebral revascularization: Proximal external carotid to distal middle cerebral artery bypass with a synthetic tube graft, *Neurosurgery, 3*(1), 61–65, 1978.

54. Opie, J. C., et al., Experience with polytetrafluoroethylene grafts in children with cyanotic congenital heart disease, *Ann. Thoracic Surg., 41*(2), 164–168, February 1986.

55. Nishida, H., et al., Clinical alternative bypass conduits and methods for surgical coronary revascularization, *Surg. Gynecol.. Obstet., 172*, 161–174, February 1991.

56. Hartman, A. R., et al., Emergency coronary revascularization using polytetrafluoroethylene conduits in a patient in cardiogenic shock, *Clin. Cardiol., 14*, 75–78, 1991.

57. Motta, G., et al., Healing and long term viability of grafts in the venae cavae reconstruction, *Vasc. Surg., 21*(5), 316–330, September–October 1987.

58. Bower, T. C., et al., Vena cava replacement for malignant disease: Is there a role? *Ann. Vasc. Surg., 7*, 51–62, 1993.

59. Lindenauer, S. M., et al., Aorto-iliac reconstruction with Dacron double velour, *J. Cardiovasc. Surg., 25*, 36–42, 1984.

60. Rosenthal, D., et al., Prosthetic above-knee femeropopliteal bypass for intermittent claudication, *J. Cardiovasc. Surg. Torino, 31*(4), 462–468, July–August 1990.

61. Burrell, M. J., et al., Axillofemoral bypass, *Ann. Surg., 195*(6), 796–799, June 1982.

62. Livesay, J. J., et al., Open aortic anastomosis: Improved results in the treatment of aneurysms of the aortic arch, *Circulation, 66*(1), 122–127.

63. Mulcare, R. J., et al., Experience with 49 consecutive Dacron interposition mesocaval shunts, *Am. J. Surg., 147*, 393–399, March 1994.

64. Reigel, M. M., et al., Early experience with a new collagen impregnated aortic graft, *Am. Surgeon, 54*, 134–136, March 1988.

65. Drury, J. K., et al., Experimental and clinical experience with a gelatin impregnated Dacron prosthesis, *Ann. Vasc. Surg., 1*(5), 542–547, 1987.
66. Reid, D. B., and Pollock, J. G., A prospective study of 100 gelatin sealed aortic grafts, *Ann. Vasc. Surg., 5*(4), 320–324, 1991.
67. Szycher, M., et al., Polyurethanes in medical devices, *Med. Des. Mater.*, February 1991, pp. 18–25.
68. Hasson, J. E., et al., Postsurgical changes in arterial compliance, *Arch. Surg., 119*,788–791, July 1984.
69. Annis, D., et al., An elastomeric vascular prothesis, *Trans. Am. Soc. Artif. Intern. Organs, 24*, 209–214, 1978.
70. Lyman, D. J., et al., Development of small diameter vascular prostheses, *Trans. Am. Soc. Artif. Intern. Organs, 23*, 253–260, 1977.
71. Lyman, D. J., et al., Small diameter grafts: comparison of patency between PTFE grafts and copolyurethane grafts (abstract), *Am. Soc. Artif. Intern. Organs, 12*, 30, 1983.
72. Marois, Y., et al., *In vivo* Evaluation of hydrophobic and fibrillar microporous polyetherurethane urea graft, *Biomaterials, 10*, 521–529, October 1989.
73. Fujomoto, K., et al., Porous polyurethane tubes as vascular graft, *J. Appl. Biomater., 4*(4), 347–354, 1993.
74. Wilson, G. J., et al., Anisotropic polyurethane nonwoven conduits: A new approach to the design of a vascular prosthesis, *Am. Soc. Artif. Intern. Organs, 24*, 260–268, 1983.
75. Herring, M., et al., Endothelium-lined small artery prostheses: A preliminary report, *ASAIO J., 6*(2), 93–102, 1983.
76. Rupnick, M. A., et al., Endothelialization of vascular prosthetic surfaces after seeding or sodding with human microvascular endothelial cells, *J. Vasc. Surg., 9*, 788–795, 1989.
77. Schneider, A., et al., An improved method for endothelial cell seeding on polytetrafluoroethylene small caliber vascular grafts, *J. Vasc. Surg., 15*, 649–656, 1992.
78. Wang, Zhong-Gao, et al., Enhanced patency of venous Dacron grafts by endothelial cell sodding, *Ann. Vasc. Surg., 7*, 429–436, 1993.
79. Welch, M., et al., Endothelial cell seeding: A review, *Ann. Vasc. Surg., 6*, 473–484, 1992.
80. Kent, K. C., A. Oshima, and A. D. Whittemore, Optimal seeding conditions for human endothelial cells, *Ann. Vasc. Surg., 6*,, 258–264, 1992.
81. Ito, Y., M. Kajihara, and Y. Imanishi, Materials for enhancing cell adhesion by immobilization of cell-adhesive peptide, *J. Biomed. Mater. Res., 25*, 1325–1337, 1991.
82. Seeger, J. M., and N. Klingman, Improved endothelial cell seeding with cultured cells and fibronectin-coated grafts, *J. Surg. Res., 38*, 641–647, 1985.
83. Carabasi, R. A., S. K. Williams, and B. E. Jarrell, Cultured and immediately procured endothelial cells: Current and future clinical applications, *Ann. Vasc. Surg., 5*, 477–484, 1991.
84. Herring, M. B., A. Gardner, and J. Glover, Seeding human arterial prostheses with mechanically derived endothelium. The detrimental effect of smoking, *J. Vasc. Surg., 1*, 279–289, 1984.
85. Herring, M. B., et al., Endothelial seeding of polytetrafluoroethylene popliteal bypasses. A preliminary report, *J. Vasc. Surg., 6*, 114–118, 1987.
86. Kadletz, M., et al., Implantation of *in vitro* endothelialized polytetrafluoroethylene grafts in human beings, *J. Thorac. Cardiovasc. Surg., 104*, 736–742, 1992.
87. Magometschnigg, H., et al., Prospective clinical study with *in vitro* endothelial cell lining of expanded polytetrafluoroethylene grafts in crural repeat construction, *J. Vasc. Surg., 15*, 527–535, 1992.
88. Kusaba, A., C. R. Fisher III, T. J. Matuiewski, and T. Matsumoto, Experimental study of the influence of porosity on development of neointima in Gore-Tex grafts: A method to increase long term patency rates, *Am. Surgeon, 48*(8), 347–354, 1981.
89. Clowes, A. W., R. K. Zacharias, and T. R. Kirkman, Early endothelial coverage of synthetic arterial grafts: Porosity revisited, *Am. J. Surg., 153*(5), 501–504, 1987.

90. Golden, M. A., et al., Healing of polytetrafluoroethylene arterial grafts is influenced by graft porosity, *J. Vasc. Surg., 11*(6), 838–845, 1990.

91. Kohler, T. et al., Conventional versus high-porosity polytetrafluoroethylene graft: Clinical evaluation, *Surgery, 112*, 901–907, 1992.

92. Greisler, H. P., et al., Enhanced endothelialization of expanded polytetrafluoroethylene grafts by fibroblast growth factor type 1 Pretreatment, *Surgery, 112*, 244–255, 1992.

93. Carney, W. I., and Lilly, M. P., Intraoperative evaluation of PTFE, Dacron, and autogenous vein as carotid patch materials, *Ann. Vasc. Surg., 1*(5), 583–586, July 1987.

94. Tsuchida, H., et al., *In vivo* study of an elastomer end-coated polytetrafluoroethylene vascular prosthesis, *J. Invest. Surg., 4*, 1–10, 1991.

95. Barbalinardo, R. J., et al. A comparison of isobutyl-2 cyanoacrylate glue, fibrin adhesive, and oxidized regenerated cellulose for control of needle hole bleeding from polytetrafluoroethylene vascular prostheses, *J. Vasc. Surg., 4*, 220–223, 1986.

96. Allen, B. T., et al., Platelet deposition on vascular grafts, *Ann. Surgery, 203*(3), 319–328, 1986.

97. Gorog, P., and Kovacs, I., Prostacyclin is a more potent stimulator of thrombolysis than inhibitor of haemostasis, *Haemostasis, 16*, 337–345, 1986.

98. Eddy, C., et al., The use of prostacyclin analog-containing sutures for the prevention of postoperative venous thrombosis in the rat, *Plastic and Reconstructive Surgery*, October 1986, pp. 504–510.

99. *The BBI Newsletter*, Vol 16, No. 3, March 1993, p. 44.

100. *MedPro Month*, November and December 1993, p. 169.

101. Szilagyi, D. E., Smith, R. F., Deruso, F. L., et al., Contribution of abdominal aortic aneurysmectomy to propagation of life, *Am. Surg.*, 1966, pp. 164–169.

102. Darling, R. C., and Brewster, D. C., Elective treatment of abdominal aortic aneurysm. *World J. Surg., 4*(6), 661–666, 1980.

103. Cronenwett, J. L., Murphy, T.F., Zelenock, G. N., et al., Actuarial analysis of variables associated with rupture of small abdominal aortic aneurysms, *Surgery, 98*, 472–483, 1985.

104. Parodi, J. C., Paimaz, J. C., and Barone, H. D. F., Transfemoral intraluminal graft implantation for abdominal aortic aneurysms. *Ann. Vasc. Surg., 5*, 491–499, 1991.

105. Laborde, J. C., Parodi, J. C., and Clem, M. F., Intraluminal bypass of abdominal aortic aneurysm: Feasibility study, *Radiology, 184*(1), 185–190, 1992.

106. Brown, O. W., Hollier, L. H., Pairolero, P. C., et al., Abdominal aortic aneurysm and coronary artery disease: A reassessment, *Arch. Surg., 116*, 148, 1981.

107. Johansson, G., Nydahl, S., Olofsson, P., et al., Survival in patients with abdominal aortic aneurysms: Comparison between operative and nonoperative management, *Eur. J. Vasc. Surg., 4*, 497–502, 1990.

108. McCombs, R. P., and Roberts, B., Acute renal failure after resection of abdominal aortic aneurysms, *Surg. Gynecol. Obstet.*, 1979, pp. 175–179.

109. Dotter, C. T., Transluminally-placed coilspring end arterial tube grafts: Long-term patency in canine popliteal artery, *Invest. Radiol., 4*, 329, 1969.

110. Balko, A., Piasecki, G .H., Shah, D. M., Carney, W. I., Hopkins, R. W., and Jackson, B. T., Transfemoral placement of intraluminal polyurethane prosthesis for abdominal aneurysm, *J. Surg. Res., 40*, 305–309, 1986.

36
Hemocompatible Materials: Surface and Interface Aspects

Agnese Magnani and Rolando Barbucci
University of Siena
Siena, Italy

I. INTRODUCTION

The development and the study of blood-compatible materials has been one of the most investigated areas in the past two decades. Contact of blood with a foreign surface triggers the activation of a number of proteases of the so-called intrinsic pathway of the coagulation cascade, which terminates in the formation of clots of fibrin, originated by thrombin-catalyzed cleavage of fibrinogen. The clot formation is the most important concern preventing even more widespread applications of biomaterials that contribute significantly to the quality and effectiveness of the world's health care system.

All synthetic foreign materials are blood compatible in the sense that they do not have the intrinsic capacity of inducing the specific, evolution-selected mechanisms that lead to the thrombus formation. These mechanisms occur directly only with natural materials such as collagen and tissue factors. Platelet adhesion, activation, and aggregation in the vasculature are in fact normally mediated by the platelet membrane receptors that bind to soluble and insoluble adhesive proteins, including collagen, Von Willebrand factor (vWF), fibrinogen, and fibronectin. Moreover, prothrombin conversion to thrombin and subsequent fibrin formation usually involve natural surfaces such as platelet membranes that initiate and accelerate the clotting enzyme cascade.

In contrast, synthetic foreign materials acquire thrombogenicity only after first interacting with blood, and the event responsible for the transformation of the inert polymer to an active surface is the interaction of plasma proteins with the surface that then initiates foreign-surface thrombosis.

Although these statements pose an oversimplification (e.g., the role of flow disturbances inherent in surface roughness or device design is ignored), they suggest a critical role for protein adsorption in thrombosis on foreign materials. Once a fibrin coating has occurred, it will, if resolution of the fibrin has not taken place, become organized and converted into fibrous-like tissue.

Processes such as protein adsorption, platelet and complement activation, and fibrin polymerization are all surface-induced phenomena and depend to a large extent on the properties and composition of the outer few atomic layers of a material. The successful design of materials for use as blood-contacting devices, thus, demands both advanced techniques for surface analysis and adequate criteria for surface property determination, with the intent of correlating the molecular events occurring at the interface with the chemistry and the hemocompatibility of the biomaterial. Insights into the fundamental interactions of proteins, cells, and surface will come from an enhanced knowledge of surface structure. In developing a rationale for surface characterization of biomaterials, an important question is: How does the material transduce its structural makeup to guide or influence the response of proteins, cells, and the whole organism to it? In the case of materials that are not used as drug delivery systems, this transduction occurs through the surface structure. The body reads the surface structure and responds. Supposing that the relationship between surface makeup and biological response is established, one should then decide which surface properties have to be measured to better understand the biological reactions. Although much effort has been expended in this direction and some useful correlations have been produced [1], up to now well-established, fundamental rules are not available for relating the surface properties of specific materials to their biological performance. Fundamental progress in design of new materials with improved hemocompatibility will, moreover, be determined by a deeper understanding of the nature of interfacial phenomena, mainly protein and platelet interactions with surfaces.

II. SURFACE MODIFICATION FOR IMPROVING HEMOCOMPATIBILITY

Current research is concerned with investigating surface modifications to improve the "performance" of blood-contacting devices. Different strategies have been followed for creating surfaces with enhanced hemocompatibility such as: (1) immobilization and/or release of bioactive agents that prevent the thrombus formation, (2) surface grafting of hydrophilic or hydrophobic polymer chains, and (3) design of polymers containing a microdomain surface structure.

Heparin [2–10], prostacyclins [11,12], prostaglandins and heparin–prostaglandins conjugates [13–18], fibrinolytic enzymes [19–21], and albumin and heparin–albumin conjugates [22–27] have been extensively immobilized on polymeric surfaces. The hydrophilic character of polyethylene oxide (PEO) has been largely used to create surfaces with increased hydrophilicity, by grafting the polymer chains on hydrophobic substrates [28–30], and these grafted surfaces have been shown to adhere and activate fewer platelets than other surfaces. PEO chains have also been used as spacer arms in the immobilization of bioactive molecules onto the surfaces [31]. The advantages of PEO as spacer arms include a higher water content of the surface to inhibit protein and platelet adhesion, and high dynamic motion and extended conformation to allow the drug to interact favorably with the receptor.

Block copolymers with different hydrophilic/hydrophobic natures (amphiphilic polymers) have been used with the intent of creating surfaces showing selective adsorption properties with regard to plasma proteins [32,33]. In general these systems have successfully regulated the extent of protein adsorption and platelet activation *in vitro* and *ex vivo*.

Even though each of these approaches positively contributed to the enhancement of the blood compatibility of polymeric materials, the immobilization of antithrombogenic coatings onto appropriate polymeric substrates has been thought of as the best way to obtain hemocompatible surfaces. The immobilization of biological molecules, known to inhibit the thrombus formation or to lyse it, has been obtained by simple adsorption or by ionic or covalent binding and may be preceded by either surface chemical modification or surface treatment with an energy source. Among the biologically active molecules, heparin was the first immobilized onto synthetic surfaces and, even today, it remains the most widely used agent for achieving polymer antithrombogenicity.

Research on heparinized materials was initiated with the graphite–benzalkonium–heparin coatings [34]. Since the graphite–benzalkonium–chloride detergent with the electrostatically bound heparin desorbed from the surface when exposed to plasma or other protein-containing solutions, many attempts to improve the duration of the immobilized heparin action followed in the 1960s. The anionic groups of heparin have been most frequently bound with quaternary ammonium groups fixed to the surface [35–44]. Polyaminoetherurethaneureas containing tertiary amino groups in the main or side chains have also been synthesized, quaternized with different alkyl halides and heparinized [45,46]. The *in vitro* and *in vivo* biological tests performed on such polyaminoetherurethaneureas showed that the bovine serum albumin was easily adsorbed onto these quaternized and heparinized polymers without serious conformational changes [47].

Since electrostatically bound heparin desorbs easily from the surface when exposed to plasma, the covalent binding of heparin onto synthetic surfaces has also been investigated [48–52]. Coatings of polyurethanes and polyethylene with a heparin–polyvinyl alcohol hydrogel have been studied [53–54]; here the biological activity of heparin was combined with the good hemocompatibility of hydrogel. In this case the immobilized heparin showed a reduced bioactivity compared to a solution of heparin as measured by clotting tests, by *in vitro* platelet interaction, by chromogenic substrate assay, and by its ability to prolong arteriovenous (A-V) shunt patency in animals [55–57]. A new polyetherurethaneurea to which heparin was covalently bound has also been synthesized [58] and the blood compatibility tests showed a deactivation of the blood coagulation system [59]. Heparin has been covalently immobilized onto polyurethane surfaces also by using different spacer arms that enhanced the heparin availability and bioactivity once attached to the surface [31,60–63]. The use of alkyl spacers provided significant surface concentrations of stably attached, active heparin [60]. The use of hydrophilic PEO spacers of different chain lengths reduced protein adsorption and subsequent platelet adhesion on the surface, whereas the biological activity of the bound heparin was enhanced by the incorporation of these spacers [31]. An alternative route to covalent immobilization of heparin onto synthetic surfaces has been the use of heparin-containing triblock copolymers. A hydrophobic block provides the interaction with the surface and insolubility in blood or plasma after application. The combination with a hydrophilic block and the active heparin block is used to create a phase-separated structure at the surface with the heparin moiety partially exposed. These heparin-containing block copolymers have shown an enhanced blood compatibility as measured by the prolongation of the plasma recalcification times (compared to control surfaces), the minimal platelet reaction, and the prolonged occlusion times in *ex vivo* arterioarterius (A-A) shunt experiments performed under low flow and low shear conditions [64]. The most notable disadvantage of covalent immobilization of heparin on polymer surfaces is, however, the reduction of the functionality of the biological moiety.

To overcome this problem, the use of heparin-complexing polymers has been adopted either for grafting on different surfaces or for the synthesis of new heparinizable materials. In these cases, the polyanion heparin is bound to a polycation (heparin-complexing polymer) and the resulting interaction between the two polymers is stronger than the electrostatic interaction among quaternary amines fixed to the surface and the anionic groups of heparin. Surface grafting of a heparin-complexing poly(amidoamine) on a given substrate or its crosslinking with a commercial polymer to obtain a new material resulted in heparinizable surfaces. This poly(amidoamine) has been surface grafted on different materials by using different spacer arms leading to an heparinizable surface without altering the size, shape, and mechanical properties of the substrate [65–72]. Nevertheless, the chemical steps necessary to modify the surface and to graft the heparin-complexing polymer were not easily realized, especially in view of industrial applications. The heparin-complexing poly(amidoamine) has thus been crosslinked to a commercial polyurethane and a partially hydrolized ethylvinylacetate, obtaining new heparinizable materials (PUPA) [10] and EVAPA [73], respectively. As regards the heparin-complexing capacity of the grafted and crosslinked materials, a large quantity of the immobilized heparin is released at physiological pH by saline solution, but even after several washings a high amount of heparin remains bound to the surface and this last portion can be removed only partially by using 0.1 M NaOH solution [74]. PUPA solution has also been used to coat different substrates with the aim of creating heparinizable surfaces without altering the bulk properties of the substrates [75,76]. The *in vitro* results suggested that the heparin-coated devices offer significantly improved blood compatibility. The presence of the ionically bound heparin on the surfaces is responsible for the excellent nonthrombogenic properties as measured by the inactivation of factor C5a, prolongation of clotting time, and absence of thrombus formation, revealed by resonance thrombography [77]. The heparinized surface as well as the native one does not induce any hemolytic effect and exhibits a good cytocompatibility after sterilization with γ-rays. The good blood compatibility of the heparinized PUPA and EVAPA surfaces has been explained also in terms of a weak interaction of the heparinized surface with the plasma proteins. The adsorption of both human serum albumin and fibrinogen on the heparinized surface results, in fact, in a layer of protein that maintains its native conformation, as revealed by the *in situ* attenuated total reflection infrared spectroscopy (ATR/IR) measurements [78]. Both albumin and fibrinogen unfold instead on the bare native surfaces (see Sec. IV.C.).

III. SURFACE ANALYSIS

A. Determination of Surface Properties

In order to achieve a more complete understanding of the molecular-level interactions between foreign materials and the biological systems, subtle surface parameters (surface compliance, functional group distribution and orientation, surface domain distribution, surface dynamics, surface contamination, etc.) should be addressed together with the commonly analyzed surface properties, such as surface structure, surface hydrophobicity/hydrophilicity ratio, and surface charge.

Surfaces have some peculiar characteristics: (1) The surface of a material is uniquely

n = 4

POLY(AMIDO-AMINE) STRUCTURE

reactive. (2) The surface is inevitably different from the bulk; thus, traditional techniques used for bulk structure analysis are not suitable for surface determination. (3) Surfaces are readily contaminated; the contamination, which can be retarded under ultrahigh vacuum conditions, must, however, be accepted due to the fact that biomedical devices work under atmospheric pressure conditions. (4) The surface structure of a material is often mobile; surfaces can restructure in response of the local environment. Enrichment or depletion of surface components may occur as a result of both diffusion processes from/to the inside of the matrix of the material and reactions with the environment adjacent to the material surface. For these reasons, the techniques that can directly probe this interface are of extreme importance. A probe of the solid–liquid interface is, for example, absolutely necessary for direct measurement of the *in situ* interactions of ions, biomolecules, and cells with solid surfaces and for establishing how these interactions are affected by surface structure and properties.

The analysis of polymer surfaces should provide information about their chemical composition before and after exposure to a biological fluid (blood in this case) as a function of the depth from the surface and of the two directions normal to the depth. This analysis should provide data on: elemental composition of the surface region, surface functional groups, extent of the preferential orientation of the functional groups, elemental depth profiles, surface crystalline order, surface domain structure, surface energy, and surface topography.

None of the existing analytical techniques can provide by itself all these data. The combination of many analytical techniques, each having its specificity, is thus required to achieve this goal.

B. Surface Property Measurements

Many methods are available for surface characterization, with each providing a unique piece of information about the nature of the surface. Table 1 summarizes the characteristics of many common surface analysis methods. Figure 1 presents a useful scheme for investigators studying blood–material interactions. The flow sheet (Fig. 1) is organized so that general surface analysis methods are followed by more specific techniques. Advanced surface characterization methods are necessary when subtle surface factors have to be understood to gain molecular-level information of the material interaction with the biological environment.

Table 1   Common Methods for Biomaterial Surfaces Characterization

Method	Depth analyzed	Spatial resolution	Analytical sensitivity
Contact angles	3–20 Å	1 mm [a]	Low or high depending on the chemistry
Scanning force microscopy (SFM)	5 Å	1 Å	Single atoms
Scanning electron microscopy (SEM)	5 Å	40 Å , typically	High; not quantitative
Electron spectroscopy for chemical analysis (ESCA)	10–250 Å	8–150 μm	0.1 atomic%
Secondary ion mass spectrometry (SIMS)	10 Å to 1 μm [b]	500 Å	Very high
Attenuated total reflection infrared (ATR-IR) spectroscopy	1–5 μm	10 μm	1 mol%

[a]The size of a small drop is 1 mm. However, contact angles actually probe the interfacial line at the edge of the drop. The spatial resolution of this zone might be approximately 0.1 μm
[b]Static SIMS ~ 10 Å ; dynamic SIMS to 1 μm

Sample analysis is guided by two general principles:

1. All methods used to analyze surfaces may also alter the surface; the surface analyst must be aware of damage potential of the method.
2. More than one surface analysis method should be used whenever possible to reduce the possibility of artifacts and to obtain the several pieces of information necessary to build a complete picture of the surface.

Information from multiple surface analysis methods is often corroborative and synergistic in developing an understanding of surface structure. A real picture of the surface structure and behavior is often obtained by the combination of the information provided by different surface analysis methods. The data derived from two or more methods should always agree; if they do not, one should try to determine the reason.

Critical steps in surface analysis are represented by the sample preparation and storage. It is, of course, desirable that the sample for surface analysis should resemble, as closely as possible, that which is being subjected to biological testing or implantation. Fingerprints on the surface of the sample can mask structures of interest, and this masking, as well as abrasive contact with the sample surface, must be avoided. Moreover, if the sample is placed in a package for storage or shipping prior to surface analysis, it is important that the materials used for sample storage and shipment are first examined to ascertain their purity because they can induce sample surface contamination.

C.   Specific Methods for Surface Analysis

Of the different methods presented in Fig. 1, only the most commonly used are considered here. Since many review articles have been published on each method, they are briefly discussed only in terms of the information they provide. These methods can be grouped as:

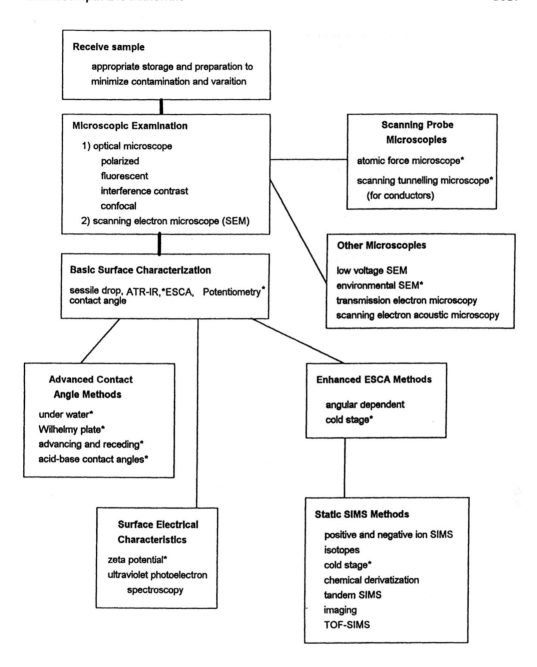

* useful for examining the solid-aqueous interface

Figure 1   Flow sheet for the surface characterization of biomaterials.

1. Imaging specimen methods
2. Spectroscopic/spectrometric methods
3. Thermodynamic methods

1. Specimen Imaging Methods

The simplest picture of a sample surface is provided by the optical microscopic observation, which tells us how homogeneous the surface is. The utility of optical microscopy can be extended by the use of specialized methods including polarized microscopy, interference and differential contrast microscopy, fluorescence microscopy, and confocal microscopy [79–81].

A more detailed picture of the surface is provided by scanning electron microscopy (SEM). The SEM image of surfaces, with a three-dimensional quality, has great resolution and depth of field. The major disadvantage of this microscopy is that nonconductive materials have to be coated with a thin, electrically grounded layer of metal to be observed. Therefore, in SEM analysis of these samples, it is the surface of the metal coating that is imaged and, even if a good representation of the surface geometry is obtained, effects of the specimen surface chemistry on secondary electron emission are lost. The recent development of low-voltage SEM allowed us to overcome this problem [82,83]. Low-voltage SEM has been used to study platelets and phase separation in polymers. In spite of these limitations, SEM observation gives true surface information and it is very useful if used in conjunction with other surface analysis methods; for example, data on surface roughness obtained by SEM are very useful in the interpretation of electron spectroscopy for chemical analysis (ESCA), secondary ion mass spectrometry (SIMS), and contact angle measurements, which are known to be strongly affected by the surface roughness and texture.

A much more advanced image of surfaces at the molecular level is obtained by the so-called scanning probe microscopies such as scanning tunneling (ST) and atomic forces (AF) (or scanning forces – SF) microscopies. STM is actually not very useful for surface characterization of polymeric biomaterials because of its conductivity requirements. In contrast to STM, AFM (or SFM) does not require the studied surface to be electrically conductive. The AFM image is obtained simply from the interaction of a fine tip mounted on a small cantilever spring and the surface atoms, when the tip is scanned across the surface (Fig. 2). Since many molecules and materials (including polymers) are soft, the scanning of the probe tip over the surface may induce deformation or even destruction of the sample. To avoid damage to the sample, the interaction force must be kept below 10^{-8} N. This is difficult to achieve in air. One way to overcome this problem is to operate the AFM with the sample, tip, and cantilever immersed in a liquid such as water [84]. Besides allowing smaller interaction forces, operating the AFM under water also opens the possibility of observing biological samples in a physiological environment, surface restructuring of polymeric materials in aqueous media, and interfacial phenomena such as protein adsorption on solid surfaces [85,86].

2. Spectroscopic/Spectrometric Measurements

Surface IR Spectroscopy. Among the vibrational spectroscopic methods, that mostly used in biomaterials characterization is the infrared (IR) technique, which provides information on the molecular structure of materials by observing the vibrations of atomic and molecular units. These vibrations are excited by the absorption of specific frequencies of energy in the IR range, a generally nondestructive process. Traditionally, IR methods are

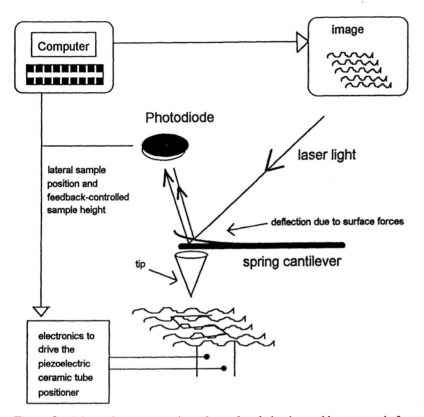

Figure 2 Schematic representation of a surface being imaged by an atomic force microscope.

used to study the bulk of materials. Surface Fourier transform – infrared (FT-IR) studies use methods for coupling the infrared radiation to the sample in a way that increases the intensity of the surface signal and reduces the signal coming from the bulk. Some of these methods are summarized in Table 2, together with their limitations. The attenuated total reflection infrared method (ATR-IR) is the most widely used technique for biomaterial surface characterization. Here the analytical method of infrared spectroscopy is coupled with the physical phenomenon of total internal reflection to enable the study of molecular vibrations within the surface regions of the material [87]. The study of these molecular vibrations absorbing infrared radiation can provide molecular structure, conformation, crystalline, and secondary bonding information about the polymer surface regions. The variation of experimental variables allow the ATR-IR to vary from surface to subsurface sensitivity by altering the sampling depth of the IR beam into the sample [88]. The ATR-IR is now successfully used to study the solid–aqueous interface [89,90] by monitoring interfacial phenomena such as protein adsorption to surfaces with the intent of correlating the molecular events occurring with the surface chemistry and hemocompatibility. This method is designed for "on-line" analysis of the process under physiological conditions, and the information obtained are related to: (1) the type of protein adsorbed onto the surface (qualitative aspect); (2) the rate and the amounts of adsorption of proteins (quantitative aspect); (3) the conformational changes occurring in the adsorbing species (structural issue).

Table 2 Some Surface-Sensitive Infrared Sampling Modes

Mode	Sampling depth	Concerns
Attenuated total reflectance (ATR)	1–5 μm	Sample must make intimate contact with the crystal
ATR (liquid cell)	1–5 μm	Absorption bulk liquid interferes with adsorbate analysis
External reflectance (ER)	1–100 Å	Sample must be on a specular (mirror) surface
Specular reflectance (SR)	1–100 Å	Sample must be on a specular (mirror) surface
Diffuse reflectance (DR)	1–5 μm (not well defined)	Sample should be rough
Transmission	Typically not for surface analysis	Surface information can be obtained for samples with high surface/volume or for ultrathin films

Structural changes in proteins upon adsorption on synthetic surfaces can be observed directly from their IR spectra, without requiring added probes of protein structure. They can be analyzed also as a function of residence time and correlated to surface properties. Figure 3 shows a schematic representation of the ATR flow cell and plumbing setup for protein adsorption experiments [91,92]. The limitations of the method are mainly related to the inherent complexity of the system to be analyzed and the technical problem of the quantitative water subtraction.

Other IR techniques can be used as tools for surface analysis. The external reflection IR method provides extreme surface sensitivity and also offers information on the orientation of molecules at the surface [93,94]. This method, however, requires a highly reflective metal substrate for the absorbing molecules and often suffers from very poor

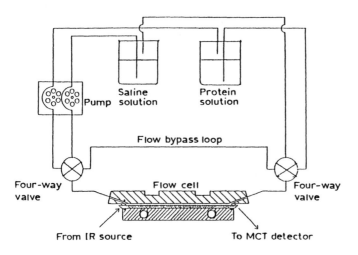

Figure 3 ATR flow-cell and plumbing setup for protein adsorption experiments. (From Ref. 113.)

S/N. Photoacoustic methods use IR heating and sample expansion to probe the surface zone [95]. The diffuse reflectance infrared technique requires little or no sample preparation prior to analysis [96], but the sampling depth is here poorly defined. Finally the microscopy IR methods can provide some surface sensitivity, while also offering spatial resolution [97].

Electron Spectroscopy for Chemical Analysis (ESCA). ESCA, or x-ray photoelectron spectroscopy (XPS—a method based on the photoelectric effect), is generally regarded as an important key technique for the surface characterization and analysis of biomaterials [98,99]. This technique provides a total elemental analysis except for H and He, of the top 10–100 Å (depending on the sample and instrumental conditions) of any solid surface that is vacuum stable or can be made vacuum stable by cooling. Chemical bonding information is also provided. Of all the instrumental techniques for surface analysis, ESCA is generally regarded as being the most quantitative and the most informative with regard to chemical information. Figure 4 is a schematic diagram of an x-ray photoelectron spectroscopy experiment. The measured energies of the emitted photoelectrons provide information about the nature and the environment of the atoms from which they come. The major types of information obtained by ESCA are listed in Table 3. The speed of analysis, the high information content, the low sample damage potential, and the ability to analyze samples with no specimen preparation are the major advantages of this analytical method. The disadvantages related to this technique include the need for vacuum compatibility, the sample damage if long analysis times are used, and the cost associated with the analysis. The vacuum compatibility limitation is of less concern when using an ESCA system with cryogenic sample stage. At liquid nitrogen temperature, samples with volatile components or hydrated samples can be analyzed.

This method is successfully used in the analysis of restructuring phenomena of polymer surfaces in aqueous media and the interaction of proteins and cells with synthetic surfaces [100,101]. Another enhancement to ESCA that increases its utility is the use of

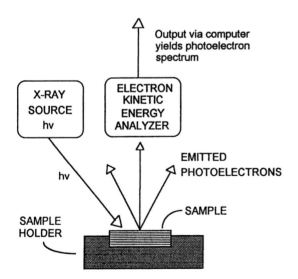

Figure 4   Schematic representation of an XPS experiment.

Table 3   Main Information Provided by ESCA

- Identification of all elements (except H and He) present at concentrations >0.1 atomic%
- Semiquantitative determination of the approximate elemental surface composition (± 10%)
- Information about the molecular environment (oxidation state, bonding atoms, etc.)
- Nondestructive elemental depth profiles 100 Å into the sample and surface heterogeneity assessment by using (1) angular-dependent ESCA studies and (2) photoelectrons with differing escape depths
- Lateral variations in surface composition (8–150 μm spatial resolution, depending upon the instrument)
- "Fingerprinting" of materials by using valence band spectra and identification of bonding orbitals
- Studies on hydrated samples

---

the angle-dependent XPS, which allows a nondestructive depth profiling analysis of the outermost 100 Å of a material [102].

Secondary Ions Mass Spectrometry (SIMS).   SIMS produces a mass spectrum of the outermost 10 Å of a surface. Like ESCA it requires complex instrumentation and an ultrahigh vacuum system. The method provides unique information complementary to ESCA for an understanding of surface composition [103–106]. Some of the analytical capabilities of this technique are summarized in Table 4. SIMS methods can be distinguished in static and dynamic SIMS depending on the ion dose used. Dynamic SIMS is surface erosive and enables the analyst to perform a depth profile of a specimen extending from the outermost atoms to a micron or more into the sample. However, because of the damaging nature of the high-flux ion beam, only atomic fragments can be detected. As the erosion of the specimen occurs, additional artifacts can be introduced into the data because of scrambling of surface atoms. Static SIMS, by comparison, induces minimal surface destruction (less than a monolayer of surface atoms is sputtered during the experiment under controlled conditions) so that extensive degradation and rearrangements of surface chemistry are not observed and intact molecular fragments can be detected. These molecular fragments contain specific chemical information closely related to the surface composition. Spectral interpretation is assisted by specialized methods

Table 4   Information Provided by Static and Dynamic SIMS

Information	Static SIMS	Dynamic SIMS
Identification of elements including hydrogen	+	+
Possible molecular structures	+	
Observation of extremely high mass fragment (proteins, polymers)	+	
Detection of extremely low concentrations	+	+
Depth profiling up to 1 μm into the sample		+
Observation of the outermost 1–2 atomic layers	+	
High spatial resolution (features as small as ~ 500 Å)	+	+
Semiquantitative analysis (for limited sets of specimens)	+	
Useful for polymers	+	
Useful for inorganics (metals, ceramics, etc.)	+	+
Useful for powders, films, fibers, etc.	+	+

[107–110]. Concerns with SIMS include sample damage, the need for careful control of instrument parameters, the complexity associated with spectral interpretation, and the high cost of the instrumentation.

3. Thermodynamic Methods

Contact Angle Measurements. This method provides the surface energy values of specimen surfaces [111]. The surface energy is directly related to its wettability and is a useful parameter that frequently correlates with biological interactions. Experimentally there are a number of ways to measure the contact angle. These are listed in Table 5. The contact angle methods are unique in that the equipment required is relatively simple and inexpensive. Although interpretation of the results obtained is dependent on a number of assumptions, each of which is somewhat controversial, a first-order interpretation is possible and has proven to be useful in many areas of surface science. However, in performing such measurements, a number of concerns such as operator experience, surface roughness and heterogeneity, sample geometry, liquid contamination, and liquid interaction with the sample, must be addressed to obtain meaningful data.

Potentiometry. Potentiometry is a simple and inexpensive technique through which a wide variety of reactions can be followed. The only requirement is that the reactions involve the addition or removal of ions to which the electrode is sensitive [112]. Generally, the electrode is pH sensitive, so that the surfaces investigated contain basic and acid groups. The electromotive force (emf) of the electrode changes during the course of the reaction as dictated by the Nernst equation, and a titration curve is obtained by plotting the voltage against the amount of the added reagent. The information obtained from the analysis of the titration curve of a biomaterial concern: (1) the types and the amounts of titrable groups present on the surface; (2) the basicity/acidity constants, or in a more general case, the equilibrium constants, useful in determining the state of the material surface in different pH environments; (3) the kinetic information of the possible interactions between the grafted polymer and the substrate in the case of grafted materials. The time lag necessary to reach the equilibrium condition at each step of the titrant addition changes according to the interaction of the grafted polymer with the surface of the substrate [113].

D. General Criteria for Surface Analysis

Since not all the above-mentioned techniques for surface analysis may be available in each research laboratory, any researcher involved in such an area needs useful criteria to adopt for doing his or her work correctly. Therefore, some comments on criteria for surface analysis and which methods to choose are in order.

Table 5 Common Methods of Measuring the Contact Angle

- Direct microscopic measurement of the three-phase interface with a goniometer or protractor which directly measures the angle
- Measurements of the dimensions of a drop profile on a surface from which the contact angle can be calculated from spherical trigonometric relationships
- Measurement of the diameter of a drop of known volume on a surface
- Rise in a capillary or on a vertical plate of liquid of known surface tension
- The DuNouy ring technique
- Wilhelmy plate procedure with a liquid of known surface tension

Some methods such as contact angles and ATR-IR can be performed at low cost in almost any laboratory. These techniques provide data that can be compared to those available in the literature to ensure that the sample we are looking at is approximately what it claims to be. In any case, contact angle data provide a limited amount of information and ATR-IR looks too deeply into the sample compared to the surface layer seen from the biological environment (0–10 Å).

A more detailed set of information about the surface chemistry can be obtained by ESCA measurements. The limitation concerns the requirement of operating under vacuum, far from physiological conditions. Moreover, because of the complexity and the cost of instrumentation, the ESCA will generally not be done in any laboratory. A work done by the joint efforts of four research groups of the European Community on the surface characterization of a new class of biomaterials such as hyaluronic acid derivatives [114] can be referenced as an example of the synergistic contribution of different analytical methods in the understanding of surface properties and behavior. In this case the data derived from contact angle, ESCA, and ATR/IR measurements performed on the samples in both their dry and wet states, contributed to an emerging picture of composition, properties, and behavior of these materials (see Fig. 5). SEM can provide a general picture of the surface topographical features. Still more details on surface roughness and order may be obtained from AFM, but analysis must be performed with concern for tip-induced artifacts. Static SIMS provides extremely detailed surface chemistry information, but suffers the same limitations as ESCA.

The fundamental question now is: When should one go beyond the relatively simple in-house characterization? First of all, if two materials that are expected to be the same exhibit significantly different biological reaction, the reason for that should be determined. If contact angles, SEM images, and ATR-IR spectra for the two materials do not distinguish between them, more advanced surface techniques should be applied until the nature of the surface difference leading to different biological reactions is discerned. Moreover, any time the material interaction with a biological environment is analyzed, if useful correlation parameters are not extracted from data coming from the in-house methods, one may seek more detailed data in more advanced surface techniques. The unique important recommendation is that in performing such a sophisticated analysis, one should always be aware and critical of the results obtained.

IV. PROTEIN ADSORPTION

Protein adsorption to foreign materials placed in the bloodstream occurs under complex conditions due to the fact that the plasma phase of blood contains many different proteins and each protein is present at a characteristic concentration that can differ greatly from that of the other proteins (see Table 6).

Several models have been developed to describe the adsorption of proteins onto solid surfaces [115–118]. In these models both the thermodynamic and/or kinetic parameters have been evaluated. The very rapid initial protein adsorption typically observed appears to be diffusion limited. That is, every protein molecule that arrives and contacts the initially empty surface attaches to it; at later times, protein adsorption slows, because it is less likely that the arriving protein molecules will find and fit into an empty spot on the surface. The thermodynamics of protein adsorption is not easily described in terms of the familiar concepts of free energy, enthalpy, and entropy because these properties are measurable in reversible processes, whereas protein adsorption appears to be essentially

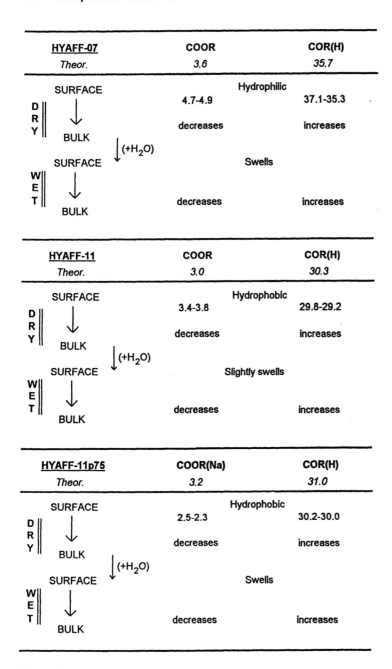

Figure 5 Schematic representation of a picture of composition and properties of three different hyaluronic acid derivative (HYAFF) samples obtained from ESCA, ATR/IR, and contact angle measurements performed in both the dry and the wet state.

Table 6   Some Plasma Proteins

Protein	mg/100 ml serum	MW
Albumin	3500–5500	69,000
IgG	800–1800	150,000
IgA	90–450	150,000–200,000
IgM	60–250	900,000
IgE	~300	190,000
Fibrinogen[a]	200–450	340,000
vWF[a]	1.5–5.0	80,000

[a] Adhesive plasma proteins responsible for platelet interaction.

irreversible. The irreversibility of the process can be explained in terms of protein size. Multiple bonding between the relatively large contact face of the protein molecule and the surface can in fact occur, and it is unlikely that all these bonds will break simultaneously. The irreversible adsorption isotherms cannot, however, be correctly analyzed to calculate a true equilibrium constant and free energy.

A more reliable insight into the thermodynamics can be obtained by the heats of adsorption. The studies in this direction [118] support the importance of entropic factors in the adsorption process, which arise from changes in water binding to the surface and the protein. The irreversibility of protein adsorption does not prevent, however, the exchange of a previously adsorbed protein with other bulk proteins. A generally accepted description of the adsorption of proteins is the multivalent interaction between a protein molecule and a solid surface; this is a qualitative model that explains both the irreversibility of the adsorption process and the exchange effects [119].

In the understanding of the adsorption events occurring at the blood–solid surface interface, four principles should be mainly taken into account.

1. Adsorption is essentially limited to a monolayer; i.e., the protein molecules that directly interact with the surface are within a monolayer and the excess proteins on the surface should be considered as a deposit.
2. The driving forces determining the composition of the adsorbed protein layer are the intrinsic surface activity of each plasma protein and its bulk concentration in the plasma.
3. Surfaces vary in their ability to adsorb selectively and cause changes in the reactivity of proteins.
4. The ability of an adsorbed protein to influence platelet adhesion and activation depends not only on the amount of that protein, but also on its reactivity, as determined by changes in its orientation or conformation.

A.   Monolayer Model

The evidence for the existence of a monolayer of adsorbed proteins comes most clearly from studies with single-protein solutions in which a saturation can often be observed in the adsorption isotherm (Fig. 6). Usually the plateau value falls within the range expected for a close-packed monolayer of protein (0.1–0.5 μg/cm^2 depending on the diameter and orientation assumed for the protein). Moreover, studies on the adsorption from complex protein mixtures or plasma also give values that fall in this range, supporting the mono-

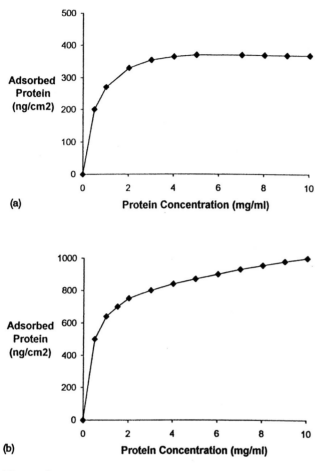

Figure 6   Monolayer concept: Langmuir (a) and Freundlich (b) adsorption isotherms. The amount of the adsorbed protein is plotted against the bulk protein concentration in the solution from which adsorption took place.

layer model [120]. Finally, the competition between the different plasma proteins adsorbing to a surface reflects the availability of a limited number of sites, and the multilayer deposition model would not be expected to lead to the observed protein discrimination by the surface. There are some limitations in the use of the monolayer concept even though it appears to be a good "zero-order" approximation to what really happens in blood. These limitations concern protein–protein interactions, which may occur in whole blood, and the fact that both the orientation of the protein and the true surface area available for adsorption per protein molecule are not known. Anyway, the observed saturation adsorption values, based on the use of the plane equivalent area of the sample, are, within these limitations, fairly close to the expected monolayer value. It should also be noted that well-defined plateaus are not always observed, and if higher concentrations of bulk phase protein are used, the adsorption rises much more slowly than at low concentrations. Both the classical Langmuir isotherm (true plateau) and the Freundlich isotherm (slow rise) (Fig. 6) have been observed for protein adsorption to solid surfaces [121]. The monolayer model does not give any information about protein orientation and fibrin formation. Finally, if the reported values for fibrinogen "adsorption" to surfaces exposed to blood are much higher than the monolayer values, they should be analyzed carefully in the sense that the excess of fibrinogen should not be considered as adsorbing to a surface site, but rather as bound to the surface in the form of macroscopic fibrin clots. It would, thus, be better regarded as fibrinogen "deposit" instead of fibrinogen "adsorbate."

B.   Competitive Adsorption

Adsorption from blood plasma is selective and the surface phase results enriched in certain proteins. Because the number of adsorption sites per unit area is limited, the solid surface can only accommodate a small fraction of the total protein typically present in the bulk phase. Since the plasma proteins vary greatly both in their concentration in the bulk phase and in their intrinsic abilities to adsorb to surfaces, some proteins will be adsorbed preferentially. Examples of surface enrichment are readily available. Despite the fact that fibrinogen is only the third most concentrated protein in plasma, after IgG and albumin, materials exposed to blood are typically enriched in fibrinogen in the adsorbed phase. As an example, adsorption from equal mixtures (by weight) of albumin and fibrinogen on polyethylene results in fibrinogen surface enrichment [122]. An opposite behavior is observed for albumin: Even if its concentration is much higher than fibrinogen, the surface concentration of albumin adsorbed from plasma is typically about the same as fibrinogen. Since the intrinsic surface activity of albumin is low, it is its relatively high concentration in plasma that literally forces it onto the surface according to the law of mass action. The adsorption of fibrinogen from plasma, on the other hand, exhibits some unusual behaviors. On certain surfaces, fibrinogen adsorption is maximum at intermediate dilutions of plasma (Fig. 7). In addition, fibrinogen adsorption from whole or moderately diluted plasma is higher at the very early adsorption times. This is termed the "Vroman effect" [123] and it is a very clear example of the unique effects of competitive adsorption on both the steady-state and the transient composition of the adsorbed layer that forms from plasma. The mechanisms of displacement of plasma

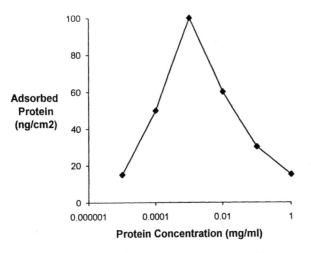

Figure 7   Representation of the "Vroman effect" for fibrinogen adsorption from plasma.

fibrinogen remain somewhat controversial but appear to involve the contact activation system [123-126]. However, studies of surfaces exposed to plasma showed that several different proteins are present in the adsorbate even if enrichment in certain proteins can occur; thus the concept of a complete "domination" of the adsorbed layer by only one protein (such as fibrinogen) is not very realistic [127,128].

Fundamental understanding of the variation in the intrinsic surface activity between various proteins is so far incomplete. Both the overall properties — such as hydrophobicity, charge, and localized sticky patches on the protein surfaces [129] — and the overall structural stability of a protein [118,121] have to be considered as key variables for the adsorption process. The adsorption behavior of four different proteins has been analyzed in terms of their various molecular properties (including their stability), and the authors concluded that the contribution from structural rearrangements is a strong factor in the competition resulting in preferential adsorption of the less stable proteins over the most stable ones [118].

C. The Influence of Surface Chemistry

The outcome of the competitive process of adsorption is an adsorbed film that is richer in some proteins than others. Because each protein has a different tendency to adsorb onto each type of surface, the outcome of the protein competition is surface chemistry dependent, and the composition of the adsorbed film is different on each type of biomaterial. Moreover, the composition of the adsorbed layer may change strongly with time. These are the fundamental facts that are believed to underlie at least some of the differences in blood compatibility among various biomaterials, and many of the strategies for improvements of the biomaterial hemocompatibility have been based upon these facts.

Since the interactions of materials with platelets seem to be mediated by the adsorbed protein layer, the variations in the adsorbed protein film due to changes in the surface composition is the point on which one should focus in order to understand why platelets respond differently to various biomaterials. The influence of changes in biomaterial surface chemistry on the composition of the adsorbed protein layer after surface exposure to blood plasma has been shown by several studies [120,130]. These observations resulted in attempts to produce biomaterials that would preferentially adsorb the "right" protein (typically one thought to be inert, such as albumin), excluding more reactive proteins, such as fibrinogen, that tend to favor thrombus formation [131], or attempts to prevent the formation of the natural forming mixed layer by the preadsorption of an inert protein (e.g., albumin) to the surface. Efforts have also been made to evade the entire question of the adsorbed protein layer composition by perfecting polymers such as acrylic hydrogels [132] and polyethylene oxides [133,134] that seem to bind protein weakly and thus adsorb only small quantities. Also the surface heparinization through an ionic bond yields surfaces that adsorb low amounts of protein without inducing conformational changes in the adsorbed proteins.

As an example, the adsorption of human serum albumin and fibrinogen from single-protein solution on heparinized PUPA and EVAPA surface (see Sec. II) results in a protein layer in which the protein maintains the same conformation as in the bulk phase, even after 4 h of its surface residence (see Fig. 8 for the case of heparinized PUPA). On the contrary, both HSA and fibrinogen unfold on the bare surfaces, as well as on the native polyurethane (see Fig. 9 for the case of PUPA and polyurethane) [77]. The ionically bound heparin seems, thus, to act in a manner that prevents the unfolding and

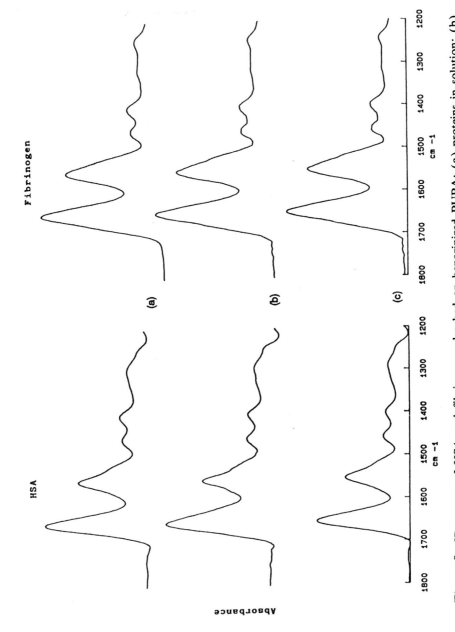

Figure 8   IR spectra of HSA and fibrinogen adsorbed on heparinized PUPA: (a) proteins in solution; (b) proteins adsorbed on heparinized PUPA; RT (residence time) = 1 min; (c) proteins adsorbed on heparinized PUPA; RT = 4 h.

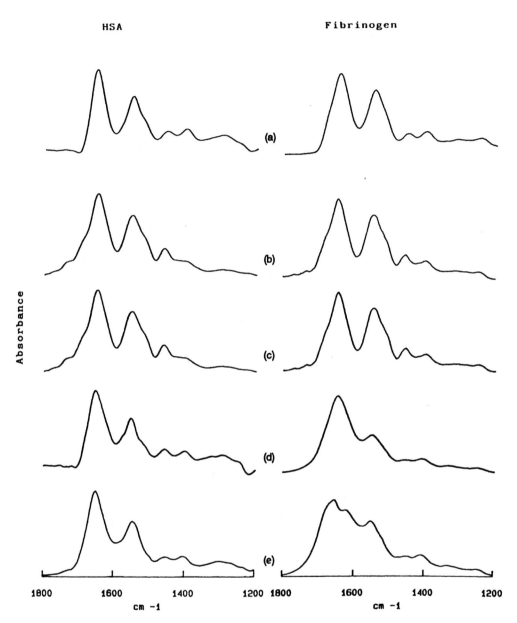

Figure 9 IR spectra of HSA and fibrinogen adsorbed on PUPA and polyurethane: (a) proteins in solution; (b) proteins adsorbed on PUPA; RT = 1 min; (c) proteins adsorbed on PUPA; RT = 4 h; (d) proteins adsorbed on PU, RT = 1 min; (e) proteins adsorbed on PU; RT = 4 h.

denaturation of protein, emphasizing the role of surface heparinization in enhancing the blood compatibility of synthetic materials.

D. Influence of the Conformation and Orientation of the Adsorbed Proteins on Their Bioactivity

Proteins in the adsorbed phase may undergo both noncovalent structural transitions and covalent changes. Although the covalent changes (e.g., fibrin formation and crosslinking, plasmin-catalyzed degradation of adsorbed fibrinogen [135], and cleavage of adsorbed factor XII [136]) are undoubtedly a major determinant in the response of blood of synthetic materials, the noncovalent transitions of the adsorbed proteins also play an important role in the interaction with blood.

When proteins adsorb to solid surfaces, they may change their native conformation because of their relatively low structural stability and their tendency to unfold to allow further bond formation with the surface. In spite of these conformational changes, many proteins retain at least some of their activity in the adsorbed state [137,138]. Thus the noncovalent variations in adsorbed proteins seem to be limited in nature, with only a small fraction of the adsorbed protein population extensively unfolded and the rest only slightly altered, if at all. In addition, proteins may undergo time-dependent transitions upon their adsorption to foreign surfaces. Many indirect evidences suggest the existence of multiple "states" of adsorbed proteins, which implies a transition step in the formation of such mixed populations [121]. The concept of transition from a reversibly adsorbed protein to a more tightly held one, is included in several models of protein adsorption from single-protein solutions at the solid–liquid interface [129,139], and it is also used to explain the time-dependent fibrinogen displacement occurring at the plasma–surface interface [125]. The residence time is the time the protein spends adsorbed on the surface before its "state" is probed. During this period many orientational transitions may occur if the protein continues to interact with the surface provoking the unfolding of the protein and largely reducing the adsorbate elutability [140,141]. A correlation between time-dependent transitions and protein elutability by sodium dodecyl sulfate has been found [142] and the effects of the residence time of fibrinogen adsorbates on the binding of polyclonal antifibrinogen antibodies have been observed as well [143,144]. Generally, proteins adsorbed from plasma or protein mixtures remain more elutable than when they are adsorbed from single-protein solutions. Also, the time-dependent transitions of the adsorbed proteins are reduced if some other proteins are present [143,144], suggesting that the presence and the nature of the coadsorbed proteins prevents or reduces further contacts of the protein with the surface.

As already mentioned, the nature of the surface to which the protein adsorbs also strongly affects the protein biological activity, simply by inducing transitions in the adsorbed protein structure. The surface properties that are most important in determining transitions remain unclear. Thus far no correlation has been found between a measurable surface property and transition rate [145], although it appears that some surfaces induce essentially instantaneous changes in the structure of the adsorbed proteins. In this case the molecule changes from the solution structure to the surface structure immediately upon contact with the surface, while other surfaces take much longer to induce similar changes (as an example, see the previously discussed case of adsorption of HSA and fibrinogen on PUPA and polyurethane polymers [78] – Fig. 9).

Finally, the orientation of the proteins in the adsorbed phase is an important concept because proteins are not uniform across their surface. Since proteins are not very free to rotate once adsorbed, due to multiple bonding, a fixed array of the protein surface is exposed to the bulk phase. The degree of uniformity in such orientation is not known, although calculations of protein–surface interaction energies in a wide array of orientations of the adsorbed protein against a theoretical plane [146] suggest that the orientations favoring the greatest possible degree of contact between the face of the protein and the plane, are preferred.

These findings are important because they appear to suggest a way that differences in hemocompatibility among different biomaterials may be generated (e.g., by altering the availability of epitopes on adhesive proteins for the cell- and platelet-surface receptors). These process are important even though the *in vivo* significance of all these changes is currently not known, and whether the transition changes play a significant role in determining blood compatibility is not clear.

V. PLATELET INTERACTION WITH BIOMATERIAL SURFACES: INFLUENCE OF THE ADSORBED PROTEINS

The adhesion, activation, and aggregation of platelets, followed in some cases by embolization of platelet thrombi, is an accepted description of the sequence of events that occur upon exposure of foreign surfaces to flowing blood [147]. The results of the previously mentioned studies on protein adsorption clearly suggest that platelet reactivity to a polymer is at least partly determined by the "state" of the adsorbed proteins. The platelet-surface interactions with adsorbed proteins are based on the presence of adhesion receptors in the platelet membrane that bind to certain plasma proteins. The platelet–protein interaction appears to be further accentuated when the protein is adsorbed to a surface because of the concentration and localization effects derived from the immobilization of the protein at the interface.

Specific proteins are present in plasma (some are reported in Table 6) that first adsorb on the surface and then mediate platelet adhesion through receptor binding. The vWF is believed to play a dominant role in platelet binding to subendothelial tissues under high shear conditions, while fibrinogen mediates adhesion to foreign surfaces [148]. The "platelet Vroman effect" (a maximum in platelet adhesion on surfaces preadsorbed with diluted plasma) also supports the importance of platelet–fibrinogen interactions in mediating platelet adhesion to foreign surfaces [149]. A recent work suggests that the observable events associated with platelet activation may be caused by the adsorbed proteins [150]. Exactly what properties of the material and/or the adsorbed protein layer are critical in causing such activation, and what role is played by locally generated concentrations of soluble agonist such as thrombin, remain unclear. It is thought, however, that the behavior of fibrinogen—the major protein in plasma known to promote platelet adhesion to foreign surfaces—is especially important. Fibrinogen may in fact lose the ability to bind platelets to different degrees and at different rates once adsorbed on biomaterials. Studies of the time- and surface-dependent transitions of fibrinogen and the subsequent observation that these transitions can influence the ability of platelets to adhere [125,140,141,143,145] are all consistent with a specific mechanism of the adhesion and activation of platelets. The rate and the degree of transformation of adsorbed platelet-adhesive proteins, such as fibrinogen, appear to be controlled by the surface properties of the polymer, so that the reactivity of a surface to the platelets is in turn dictated by the degree to which the adhesive protein has been converted from a reactive form to an unreactive form. That is, the properties of the surface to which the adhesive protein is adsorbed influence both the "amount" and the "bioreactivity" of the protein itself, which, in turn, affect the platelet response.

ACKNOWLEDGMENT

This work was supported by the "Progetto Finalizzato Chimica Fine II" of the Italian National Research Council (CNR).

REFERENCES

1. Ratner, B. D., A. B. Johnston, and T. J. Lenk, Biomaterial surfaces, *J Biomed. Mater. Res. Appl. Biomat., 21*, 59–90 (1987).
2. Kim, S. W., C. D. Ebert, J. Y. Lin, and J. C. McRea, Nonthrombogenic polymers: Pharmaceutical approaches, *Trans. Am. Soc. Artif. Inter. Organs, 6*, 76–80 (1983).
3. Fougnot, C., D. Labarre, J. Jozefowicz, and M. Jozefowicz, Modifications to polymer surfaces to improve blood compatibility, in *Macromolecular Biomaterials* (G. W. Hastings and P. Ducheyne, eds.), CRC Press, Boca Raton, FL, 1984, pp. 217–238.
4. Merril, W. E. W., E. W. Salzanan, P. S. L. Wory, P. T. Ashford, and W. G. Austen, Polyvinyl alcohol–heparin hydrogel 'G,' *J. Appl. Phys., 29*, 723–730 (1970).
5. Labarre, D., M. C. Boffa, and M. Jozefowicz, Preparation and properties of heparin-poly(methyl-methacrylate) copolymers, *J. Polym. Sci., 47*, 131–137 (1974).
6. Miura, Y., S. Aoyugi, Y. Kusada, and K. Miyamoto, The characteristics of anticoagulation by covalently immobilized heparin, *J. Biomed. Mater. Res., 14*, 619–630 (1980).
7. Goose, M. F. A., and M. V. Sefton, Properties of a heparin poly-(vinyl alcohol hydrogel) coating, *J. Biomed. Mater. Res., 17*, 359–373 (1983).
8. Larm, O., R. Larsson, and P. Olsson, A new non-thrombogenic surface prepared by selective covalent binding of heparin via a modified reducing termal residue, *Biomater. Med. Dev. Art. Org., 11*, 161–173 (1983).
9. Ito, Y., Antithrombogenic heparin-bound polyurethanes, *J. Biomater. Appl., 2*, 235–265 (1987).
10. Barbucci, R., M. Benvenuti, G. Dal Maso, M. Nocentini, F. Tempesti, M. Losi, and R. Russo, Synthesis and physico-chemical characterization of a new material (PUPA) based on polyurethane and poly(amido-amine) components, capable of strongly adsorbing quantities of heparin, *Biomaterials, 10*, 299–308 (1989).
11. Ebert, C. D., E. S. Lee, and S. W. Kim, The antiplatelet activity of immobilised prostacyclin, *J. Biomed. Mater. Res., 16*, 629–638 (1982).
12. McRea, J. C., and S. W. Kim. Characterization of controlled release of prostaglandin from polymer matrices for thrombus prevention, *Trans. Am. Artif. Intern. Organs, 24*, 746–750 (1971).
13. Bamford, C. H., I. P. Middleton, Y. Satake, and R. G. Al-Lamee, Studies on the synthesis of non-thrombogenic polymers, in *Blood Compatible Materials and Their Testing* (S. Dawids and A. Bantjes, eds.), Martinus Nijhaff, Boston, 1985, pp. 159–168.
14. Llanos, G., and M. V. Sefton, Immobilization of prostaglandin PGF22 on poly(vinyl alcohol), *Biomaterials, 9*, 429–434 (1988).
15. Chandy, T., and C. P. Sharma, The antithrombotic effect of prostaglandin E_1 immobilized on albuminated polymer matrix, *J. Biomater. Res., 18*, 1115–1124 (1984).
16. Jacobs, H., T. Okano, J. Y. Lin, and S. W. Kim, $PGEI_1$–heparin conjugate releasing polymers, *J. Controlled Rel., 2*, 313–319 (1985).
17. Jacobs, H., D. Grainger, T. Okano, and S. W. Kim, Surface modification for improved blood compatibility, *Artif. Organs, 12*, 500–501 (1988).

18. Jacobs, H., T. Okano, and S. W. Kim, Antithrombogenic surfaces: Characterization and bioactivity of surface immobilized PGE$_1$-heparin conjugate, *J. Biomed. Mater. Res., 23,* 611–630 (1989).
19. Ohshiro, T., Antithrombogenic characteristics of immobilised urokinase on synthetic polymers, in *Biocompatible Polymers, Metals, and Composites* (M. Szycher, ed.), Technical Publishing, Lancaster, PA, 1983, pp. 275–299.
20. Senators, F., F. Bernath, and K. Meisner, Clinical study of urokinase-bound fibrocollagenous tubes, *J. Biomed. Mater. Res., 20,* 177–188 (1986).
21. Ohshiro, T., M. C. Lin, J. Kambayashi, and T. Mori, Clinical applications of urokinase treated material, *Meth. Enzymol., 137,* 529–545 (1988).
22. Guidoin, R. G., R. Snyder, L. Martin, K. Botskoo, M. Marois, J. Award, M. King, D. Domurado, M. Bedros, and C. Gosselin, Albumin coating of a knitted polyester arterial prosthesis: An alternative to preclotting, *Ann. Thorac. Surg., 31,* 457–465 (1984).
23. Guidoin, R. G., M. W. King, J. Award, L. Martin, D. Domurado, M. Marois, M. F. Sigot-Luizard, C. Gosselin, K. Gunasekera, and D. Gagnon, Albumin coated and critical point dried polyester prostheses as substitutes in the thoracic aorta of dogs, *Trans. Am. Soc. Artif. Intern. Organs, 29,* 290–295 (1983).
24. Domurado, D., R. G. Guidoin, M. Morois, L. Martin, C. Gosselin, and J. Award, Albuminated Dacron prostheses as improved blood vessel substitutes, *J. Bioeng., 2,* 79–91 (1978).
25. Benslimane, S., R. G. Guidoin, P. E. Roy, J. Friede, J. Hebert, D. Domurado, and M. F. Sigot-Luizard, Degradability of crosslinked albumin as an artificial polyester prosthesis coating in *in vitro* and *in vivo* rat studies, *Biomaterials, 7,* 268–272 (1986).
26. Kottke-Marchant, K., J. M. Anderson, Y. Umemura, and R. E. Marchant, Effect of albumin coating on the *in vitro* blood compatibility of Dacron arterial prostheses, *Biomaterials, 10,* 147–155 (1989).
27. Hennik, W. E., S. W. Kim, and J. Feijen, Inhibition of surface induced coagulation by preadsorption of albumin–heparin conjugates, *J. Biomed. Mater. Res., 18,* 911–926 (1984).
28. Mori, Y., and S. Nagaoka, A non thrombogenic material with long polyethylene oxide chains, *Trans. Am. Soc. Artif. Intern. Organs, 28,* 459–463 (1982).
29. Andrade, J. D., Surface and blood compatibility, *J. Am. Soc. Artif. Intern. Organs, 10,* 75–84 (1987).
30. Merrill, E. W., and S. W. Salzman, Polyethylene oxide as a biomaterial, *J. Am. Soc. Artif. Intern. Organs, 6,* 80–84 (1983).
31. Park, K. O., T. Okano, C. Nojiri, and S. W. Kim, Heparin immobilization onto segmented polyurethaneurea surfaces: Effect of hydrophilic spacers, *J. Biomed. Mater. Res., 22,* 977–992 (1988).
32. Grainger, D., T. Okano, and S. W. Kim, Surface characteristics of poly(ethylene oxide) polystyrene multiblock copolymers, in *Advances in Biomedical Polymers* (C. G. Gebelein, ed.), Plenum Press, New York, 1987, pp. 229–247.
33. Okano, T., M. Shimada, T. Aoyagi, N. Shinohara, K. Kataoka, and Y. Sakurai, Suppression of platelet activation on microdomain surfaces of 2-hydroxyethyl methacrylate–polystyrene block copolymers, *J. Biomed. Mater. Res., 20,* 1035–1047 (1986).
34. Gott, V. L., J. D. Whiten, and R. C. Dutton, Heparin binding on colloidal graphite surfaces, *Science, 142,* 1297–1298 (1963).
35. Tanzawa, H., Y. Mori, N. Harumiya, H. Miyama, M. Hori, N. Ohshima, and Y. Idezuki, Preparation and evaluation of a new antithrombogenic heparinized hydrophilic polymer for use in cardiovascular system, *Trans. Am. Soc. Artif. Intern. Organs, 19,* 188–194 (1973).
36. Idezuki, Y., H. Watanabe, M. Hagiwara, K. Kanasugi, Y. Mori, S. Nagaoka, M. Hagio, K. Yamamoto, and H. Tanzawa, Mechanism of antithrombogenicity of a new heparinized hydrophilic polymer: Chronic *in vivo* studies and clinical application, *Trans. Am. Soc. Artif. Intern. Organs, 21,* 436–441 (1975).

37. Miyama, H., N. Harumiya, Y. Mori, and H. Tanzawa, A new antithrombogenic heparinized polymer, *J. Biomed. Mater. Res., 11*, 251–265 (1977).
38. Mori, Y., S. Nagaoka, Y. Masubuchi, M. Itoga, H. Tanzawa, T. Kikuchi, Y. Yamada, T. Yonaha, H. Watanabe, and Y. Idezuki, The effect of released heparin from the heparinized hydrophilic polymer (HRDS) on the process of thrombus formation, *Trans. Am. Soc. Artif. Intern. Organs, 24*, 736–744 (1978).
39. Miyama, H., N. Fujii, A. Kuwano, S. Nagaoka, Y. Mori, and Y. Noishiki, Antithrombogenic heparinized polyacrylonitrile copolymer, *J. Biomed. Mater. Res., 20*, 895–901 (1986).
40. Nagaoka, S., and Y. Noishiki, Development of Anthron®, an antithrombogenic coating for angiographic catheters, *J. Biomater. Appl., 4*, 3–21 (1989).
41. Grode, G. A., S. J. Anderson, H. M. Grotta, and R. D. Falb, Nonthrombogenic materials via a simple coating process, *ASAIO Trans., 15*, 1–6 (1969).
42. Gardner, D. L., W. V. Sharp, K. L. Ewing, and A. F. Finelli, Stability of heparin S35 attached to a modified polyurethane vascular prosthetic, *ASAIO Trans., 15*, 7–17 (1969).
43. Noishiki, Y., S. Nagaoka, T. Kikuchi, and Y. Mori, Application of porous heparinized polymer to vascular graft, *ASAIO Trans., 27*, 213–218 (1981).
44. Yen, S. P. S., and A. Rembaum, Complexes of heparin with elastomeric positive polyelectrolytes, *J. Biomed. Mater. Res. Symp., 1*, 83–97 (1971).
45. Shibuta, R., M. Tanaka, M. Sisido, and Y. Imanishi, Synthesis of novel polyaminoetherurethaneureas and development of antithrombogenic material by their chemical modifications, *J. Biomed. Mater. Res., 20*, 971–987 (1986).
46. Yoshihiro, I., M. Sisido, and Y. Imanishi, Synthesis and antithrombogenicity of polyetherurethaneurea containing quaternary ammonium groups in the side chains and of the polymer/heparin complex, *J. Biomed. Mater. Res., 20*, 1017–1033 (1986).
47. Ito, Y., M. Sisido, and Y. Imanishi, Platelet adhesion onto protein-coated and uncoated polyetherurethaneurea having tertiary amino groups in the substituents and its derivatives, *Biomaterials, 23*, 191–206 (1989).
48. Larm, O., R. Larsson, and P. Olsson, A new non-thrombogenic surface prepared by selective covalent binding of heparin via a modified reducing terminal residue, *Biomat. Med. Dev. Art. Org., 11*(2&3), 161–173 (1983).
49. Arnander, C., M. Dryjski, R. Larsson, P. Olsson, and J. Swedenborg, Thrombin uptake and inhibition on endothelium and surfaces with a stable heparin coating: A comparative *in vitro* study, *J. Biomed. Mater. Res., 20*, 235–246 (1986).
50. Arnander, C., P. Olsson, and O. Larm, Influence of blood flow and the effect of protamine on the thromboresistant properties of a covalently bonded heparin surface, *J. Biomed. Mater. Res., 22*, 859–868 (1988).
51. Arnander, C., Enhanced patency of small-diameter tubings after surface immobilization of heparin fragments. A study in the dog, *J. Biomed. Mater. Res., 23*, 285–294 (1989).
52. Tay, S. W., E. W. Merrill, E. W. Salzman, and J. Lindon, Activity toward thrombin-antithrombin of heparin immobilized on two hydrogels, *Biomaterials, 10*, 11–15 (1989).
53. Ip, W. F., W. Zingg, and M. V. Sefton, Parallel flow arteriovenous shunt for the *ex vivo* evaluation of heparinized materials, *J. Biomed. Mater. Res., 19*, 161–178 (1955).
54. Ramon, A. E., and M. V. Sefton, Coating of two polyether–polyurethanes and polyethylene with a heparin–poly(vinyl alcohol) hydrogel, *Biomaterials, 7*, 206–211 (1986).
55. Ip, W. F., and M. V. Sefton, Patency of heparin–PVA coated tubes at low flow rates, *Biomaterials, 10*, 313–317 (1989).
56. Cholakis, C. H., and M. V. Sefton, *In vitro* platelet interactions with a heparin–polyvinyl alcohol hydrogel, *J. Biomed. Mater. Res., 23*, 399–415 (1989).
57. Cholakis, C. H., W. Zingg, and M. V. Setton, Effect of heparin–PVA hydrogel on platelets in a chronic canine arterio-venous shunt, *J. Biomed. Mater. Res., 23*, 417–441 (1989).
58. Ito, Y., M. Sisido, and Y. Imanishi, Synthesis and antithrombogenicity of anionic polyurethanes and heparin-bound polyurethanes, *J. Biomed. Mater. Res., 20*, 1157–1177 (1986).

59. Ito, Y., Y. Imanishi, and M. Sisido, Platelet adhesion onto polyetherurethane urea derivatives: Effect of cytoskeleton proteins of the platelet, *Biomaterials, 8*, 458–463 (1987).

60. Ebert, C. D., and S. W. Kim, Immobilized heparin spacer arm effect on biological interaction, *Thromb. Res., 26*, 43–57 (1979).

61. Heyman, P. W., C. S. Cho, J. C. McRea, D. B. Olsen, and S. W. Kim, Heparinized polyurethanes: *In vitro* and *in vivo* studies, *J. Biomed. Mater. Res., 19*, 419–436 (1985).

62. Han, D. K., K. D. Park, K. D. Ahn, S. Y. Jeong, and Y. H. Kim, Preparation and surface characterization of PEO-grafted and heparin-immobilized polyurethanes, *J. Biomed. Mater. Res.: Appl. Biomater., 23*(A1), 87–104 (1989).

63. Han, D. K., S. Y. Jeong, and Y. H. Kim, Evaluation of blood compatibility of PEO-grafted and heparin-immobilized polyurethanes, *J. Biomed. Mater. Res.: Appl. Biomater., 23*(A2), 211–228 (1989).

64. Vulic, I., T. Okano, F. J. Van der Gaag, S. W. Kim, and J. Feijen, Heparin-containing block copolymers. Part II. *In vitro* and *ex vivo* blood compatibility, *J. Mater. Sci.: Mater. in Med., 4*, 448–459 (1993).

65. Barbucci, R., G. Casini, P. Ferruti, and F. Tempesti, Surface-grafted heparinizable materials, *Polymer, 26*, 1349–1352 (1985).

66. Ferruti, P., I. Domini, R. Barbucci, M. C. Beni, E. Dispensa, S. Sancasciani, M. A. Marchisio, and M. C. Tanzi, Heparin adsorbing capacities at physiological pH of three poly(amido-amine) resins, and of poly(amido-amine)-surface-grafted glass microspheres, *Biomaterials, 4*, 218–221 (1983).

67. Barbucci, R., P. Ferruti, and L. Provenzale, Ital. Pat. Appl. 23610A/80.

68. Ferruti, P., R. Barbucci, N. Danzo, A. Torrisi, O. Puglisi, S. Pigataro, and P. Spartano, Preparation and ESCA characterization of poly(vinyl chloride) surface-grafted with heparin-complexing poly(amido-amine) chains, *Biomaterials, 3*, 33–37 (1982).

69. Ferruti, P., G. Casini, F. Tempesti, R. Barbucci, R. Mastacchi, and M. Sarret, Heparinizable materials (III). Heparin retention power of a poly(amido-amine) either as crosslinked resin, or surface-grafted on PVC, *Biomaterials, 5*, 234–236 (1984).

70. Barbucci, R., M. Benvenuti, G. Casini, P. Ferruti, and F. Tempesti, Heparinizable materials (IV). Surface-grafting on poly(ethylene terephthalate) of heparin-complexing poly(amido-amine) chains, *Biomaterials, 6*, 102–104 (1985).

71. Barbucci, R., M. Benvenuti, G. Casini, P. Ferruti, and M. Nocentini, Heparinized materials V. Preparation and FTIR characterization of polyurethane surface grafted heparin-complexing poly(amido-amine) chains, *Makromol. Chem., 186*, 2291–2300 (1985).

72. Azzuoli, G., R. Barbucci, M. Benvenuti, P. Ferruti, and M. Nocentini, Chemical and biological evaluation of heparinized poly(amido-amine) grafted polyurethane, *Biomaterials, 7*, 61–66 (1987).

73. Barbucci, R., A. Magnani, F. Tempesti, and M. Benvenuti, Synthesis of two novel heparinizable polymeric materials starting from an ethylene vinylalcohol vinylacetate terpoylmer, *Makromol. Chem., 193*, 2979–2988 (1992).

74. Barbucci, R., A. Magnani, and C. Roncolini, Thermodynamic and FTIR spectroscopic studies on heparin–polycation interaction, *Clin. Mater., 8*, 17–23 (1991).

75. Barbucci, R., and A. Magnani, Physiochemical characterization and coating of polyurethane with a new heparin-adsorbing material, *Biomaterials, 10*, 429–431 (1989).

76. Barbucci, R., A. Albanese, M. Magnani, and F. Tempesti, Coating of commercial devices with a new heparinizable material, *J. Biomed. Mater. Res., 25*, 1259–1274 (1991).

77. Albanese, A., R. Barbucci, J. Belleville, S. Bowry, R. Eloy, H. D. Lemke, and L. Sabatini, *In vitro* biocompatibility evaluation of a heparinizable material (PUPA) based on polyurethane and poly(amido-amine) components, *Biomaterials, 15*, 129–136 (1994).

78. Barbucci, R., and A. Magnani, Conformation of human plasma proteins at polymer surfaces: The effectiveness of surface heparinization, *Biomaterials, 15*(12), 955–962 (1994).

79. Shuman, H., J. M. Murray, and C. DiLullo, Confocal microscopy: An overview, *Biotechniques, 7*, 154–160 (1989).

80. Salmon, T., R. A. Walker, and N. K. Pryer. Video-enhanced differential interference contrast light microscopy, *Biotechniques, 7*, 624–633 (1989).

81. Dixon, A. J., and G. S. Benham, Application of the confocal scanning fluorescence microscope in biomedical research, *Am. Biotech. Lab., 5*, 20–25 (1987).

82. Goodman, S. L., C. Li, J. B. Pawley, S. L. Cooper, and R. M. Albrecht, Surface and bulk morphology of polyurethanes by electron microscopies, in *Surface Characterization of Biomaterials* (B. D. Ratner, ed.), Elsevier, Amsterdam, 1988, pp. 281–295.

83. Goodman, S. L., K. Park, and R. M. Albrecht, A correlative approach to colloidal gold labeling with video-enhanced light microscopy, low-voltage scanning electron microscopy, and high-voltage electron microscopy, in *Colloidal Gold — Principles, Methods, and Applications,* Academic Press, San Diego, CA, 1991, pp. 369–409.

84. Drake, B., C. B. Prater, A. L. Weisenhorn, S. A. C. Gould, T. R. Albrecht, C. F. Quate, D. S. Cannell, H. G. Hansma, and P. K. Hansma, Imaging crystals, polymers, and processes in water with the atomic force microscope, *Science, 234*, 1586–1589 (1989).

85. Lea, A. S., A. Pungor, V. Hlady, J. D. Andrade, J. N. Herron, and E. W. Voss, Jr., Manipulation of proteins on mica by atomic force microscopy, *Langmuir, 8*, 68–73 (1992).

86. Hansma, P. K., V. B. Elings, O. Marti, and C. E. Bracker, Scanning tunneling microscopy, and atomic force microscopy: Application to biology and technology, *Science, 242*, 209–216 (1988).

87. Harrick, N. J., *Internal Reflection Spectroscopy*, Interscience, New York, 1967.

88. Knutson, K., and D. J. Lyman, Surface infrared spectroscopy, in *Surface and Interfacial Aspects of Biomedical Polymers*, Vol. 1: *Surface, Chemistry and Physics* (J. D. Andrade, ed.), Plenum Press, New York, 1985, pp. 197–247.

89. Gendreau, R. M., R. I. Leininger, S. Winters, and R. J. Jakobsen, Fourier transform infrared spectroscopy for protein-surface studies, in *Biomaterials: Interfacial Phenomena and Applications*, (S. L. Cooper and N. A. Peppas, eds.), ACS Advances in Chemistry Series 199, American Chemical Society, Washington, DC, 1982, pp. 371–394.

90. Gendreau, R. M., Biomedical Fourier transform infrared spectroscopy: Applications to proteins, in *Spectroscopy in the Biomedical Sciences* (R. M. Gendreau, ed.), CRC Press, Boca Raton, FL, 1986, pp. 21–52.

91. Chittur, K. K., D. J. Fink, R. I. Leininger, and T. B. Hutson, Fourier transform infrared spectroscopy/attenuated total reflection studies of protein adsorption in flowing systems: Approaches for bulk correction and compositional analysis of adsorbed and bulk proteins in mixtures, *J. Colloid Interface Sci., 111*, 419–433 (1986).

92. Magnani, A., and R. Barbucci, Fourier transform attenuated total reflection infrared spectroscopy (ATR/FT-IR): Applications to proteins adsorption studies, in *Test Procedures for the Blood Compatibility of Biomaterials* (S. Dawids, ed.), Kluwer Academic Publishers, The Netherlands, 1993, pp. 171–184.

93. Allara, D. L., Organic monolayer studies using Fourier transform infrared reflection spectroscopy, in *Vibrational Spectroscopy for Adsorbed Species* (A. T. Bell and M. L. Hair, eds.), ACS Symposium Series, Washington, DC, 1980, Vol. 137, pp. 37–49.

94. Porter, M. D., IR external reflection spectroscopy: A probe for chemically modified surfaces, *Anal. Chem., 60*, 1143A–1154A (1988).

95. Graham, J. A., W. M. Grim III, and W. G. Fateley, Fourier transform photoacoustic spectroscopy of condensed-phase samples, in *Fourier Transform Infrared Spectroscopy: Applications to Chemical Systems* (J. R. Ferraro and L. J. Basile, eds.), Academic Press, Orlando, FL, 1985, Vol. 4, pp. 345–392.

96. Chalmers, J. M., and M. W. Mackenzie, Some industrial applications of FT-IR diffuse reflectance spectroscopy, *Appl. Spectrosc., 39*, 634–641 (1985).

97. Katon, J. E., G. E. Pacey, and J. F. O'Keefe, Vibrational molecular microspectroscopy, *Anal. Chem., 58*, 464A–481A (1986).

98. Andrade, J. D., X-ray photoelectron spectroscopy, in *Surface and Interfacial Aspects of*

Biomedical Polymers, Vol. 1: *Surface Chemistry and Physics* (J. D. Andrade, ed.), Plenum Press, New York, 1985, pp. 105–195.

99. Dilks, A., X-ray photoelectron spectroscopy for the investigation of polymeric materials, in *Electron Spectroscopy: Theory, Techniques, and Applications* (A. D. Baker and C.R. Brundle, eds.), Academic Press, London, 1981, Vol. 4, pp. 277–359.

100. Ertel, S. I., A. Chilkoti, T. A. Horbett, and B. D. Ratner, Endothelial cell growth on oxygen-containing films deposited by radio-frequency plasmas: The role of surface carbonyl groups, *J. Biomater. Sci. Polym. Ed., 3*, 163–183 (1991).

101. Lin, H. B., K. B. Lewis, D. Leach-Scampavia, B. D. Ratner, and S. L. Cooper, Surface properties of RGD-peptide grafted polyurethane block copolymers: Variable take-off angle and cold-stage ESCA studies, *J. Biomater. Sci. Polym. Ed., 4*, 183–198 (1993).

102. Tyler, B. J., D. G. Castner, and B. D. Ratner, Regularization: A stable and accurate method for generating depth profiles from angle dependent XPS data, *Surf. Interface Anal., 14*, 443–450 (1989).

103. Briggs, D., SIMS for the study of polymer surfaces: A review, *Surf. Interface Anal., 9*, 391–404 (1986).

104. Davis, M. C., and R. A. P. Lynn, Static secondary ion mass spectrometry of polymeric biomaterials, *CRC Crit. Rev. Biocompat., 5*, 297–341 (1990).

105. Vickerman, J. C., A. Brown, and N. M. Reed, *Secondary Ion Mass Spectrometry: Principles and Applications*, Clarendon Press, Oxford, 1989.

106. Benninghoven, A., Secondary ion mass spectrometry of organic compounds, in *Springer Series of Chemical Physics: Ion Formation from Organic Solids* (A. Benninghoven, ed.), Springer-Verlag, Berlin, 1983, Vol. 25, pp. 64–89.

107. Chilkoti, A., B. D. Ratner, and D. Briggs, A static secondary ion mass spectrometric investigation of the surface structure of organic plasma-deposited films prepared from stable isotope-labeled precursors: Part I. Carbonyl precursors, *Anal. Chem., 63*, 1612–1620 (1991).

108. Chilkoti, A., D. G. Castner, B. D. Ratner, and D. Briggs, Surface characterization of a poly(styrene/*p*-hydroxystyrene) copolymer series using XPS, static SIMS, and chemical derivatization techniques, *J. Vac. Sci. Technol. A, 8*, 2274–2282 (1990).

109. Leggett, G. J., J. C. Vickerman, and D. Briggs, Application of tandem quadrupole mass spectrometry in SIMS, *Surf. Interface Anal., 16*, 3–8 (1990).

110. Price, D., The resurgence in time-of-flight mass spectrometry, *Trends Anal. Chem., 9*, 21–25 (1990).

111. Andrade, J. D., M. L. Smith, and D. E. Gregonis, The contact angle and interface energetics, in *Surface and Interfacial Aspects of Biomedical Polymers*, Vol. 1: *Surface Chemistry and Physics* (J. D. Andrade, ed.), Plenum Press, New York, 1985, pp. 249–292.

112. Metal complexes in solution, in *Proceedings of the International School in Metal Complexes in Solution* (J. A. Everett, E. Rizzarelli, V. Romano, and S. Sammartano, eds.), Piccin, Padova, 1986.

113. Barbucci, R., M. Casolaro, and A. Magnani, Characterization of biomaterial surfaces: ATR-FTIR, potentiometric and calorimetric analysis, *Clin. Mater., 11*, 37–51 (1992).

114. Barbucci, R., A. Magnani, A. Baszkin, M. L. Da Costa, H. Bauser, G. Hellwig, E. Martuscelli, and S. Cimmino, Physico-chemical surface characterization of hyaluronic acid derivatives as a new class of biomaterials, *J. Biomater. Sci. Polym. Ed., 4*, 245–273 (1993).

115. Norde, W., and J. Lyklema, The adsorption of human plasma albumin and bovine pancreas ribonuclease at negatively charged polystyrene surfaces. V. Microcalorimetry, *J. Colloid Interface Sci., 66*, 295 (1978).

116. Norde, W., and J. Lyklema, Thermodynamics of protein adsorption. Theory and special reference to the adsorption of human plasma albumin and bovine pancreas ribonuclease at polystyrene surfaces, *J. Colloid Interface Sci., 71*, 350 (1979).

117. Bissinger, R. L., and E. F. Leonard, Plasma protein adsorption and desorption rates on

quartz: Approach to multi-component systems, *Trans. Am. Soc. Artif. Intern. Organs, 27,* 225 (1981).

118. Norde, W., and J. Lyklema, Why proteins prefer interfaces, *J. Biomater. Sci. Polym. Ed., 2,* 183–202 (1991).

119. van Damme, H. S., *Protein adsorption at solid–liquid interfaces: Influence of surface structure, hydrophobicity and charge,* FEBO-druk, Enschede, 1990.

120. Horbett, T. A., Adsorption of proteins from plasma to a series of hydrophilic–hydrophobic copolymers. 2. Compositional analysis with the prelabeled protein technique, *J. Biomed. Mater. Res., 15,* 673–695 (1981).

121. Horbett, T. A., and J. L. Brash, Proteins at interfaces: Current issues and future prospects, in *Physiochemical and Biochemical Studies,* ACS Symposium Series 343, American Chemical Society, Washington, DC, 1987, pp. 1–33.

122. Horbett, T. A., P. K. Weathersby, and A. S. Hoffman, The preferential adsorption of hemoglobin to polyethylene, *J. Bioeng., 1,* 61–78 (1977).

123. Horbett, T. A., Mass action effects on competitive adsorption of fibrinogen from hemoglobin solutions and from plasma, *Thromb. Haemostas., 51,* 174–181 (1984).

124. Slack, S. M., J. L. Bohnert, and T. A. Horbett, The effects of surface chemistry and coagulation factors on fibrinogen adsorption from plasma, *Ann. NY Acad. Sci., 516,* 223–243 (1987).

125. Slack, S. M., and T. A. Horbett, Changes in the strength of fibrinogen attachment to solid surfaces: An explanation of the influence of surface chemistry on the Vroman effect, *J. Colloid Interface Sci., 133,* 148–165 (1989).

126. Schmaier, A. H., L. Silver, A. L. Adams, et al., The effect of high molecular weight kininogen on surface-adsorbed fibrinogen, *Thromb. Res., 33,* 51–67 (1983).

127. Horbett, T. A., and P. K. Weathersby, Adsorption of proteins from plasma to a series of hydrophilic–hydrophobic copolymers. 1. Analysis with the *in situ* radioiodination technique, *J. Biomed. Mater. Res., 15,* 403–423 (1981).

128. Ziats, N. P., D. A. Pankoowsky, B. P. Tierney, O. D. Rantoff, and J. M. Anderson, Adsorption of Hageman factor (factor XII) and other human plasma proteins to biomedical polymers, *J. Lab. Clin. Med., 116,* 687–696 (1990).

129. Andrade, J. D., Principles of protein adsorption, in *Surface and Interfacial Aspects of Biomedical Polymers* (J. D. Andrade, ed.), Plenum Press, New York, 1985, pp. 1–80.

130. Brash, J. L., and P. ten Hove, Effect of plasma dilution on adsorption of fibrinogen to solid surfaces, *Thromb. Haemostas., 51,* 326–330 (1984).

131. Tsai, C.-C., M. L. Dollar, A. Constantinescu, P. V. Kulkarni, and R. C. Eberhart, Performance evaluation of hydroxylated and acylated silicone rubber coatings, *Trans. Am. Soc. Artif. Intern. Organs, 37,* m192–m193 (1991).

132. Horbett, T. A., Protein adsorption to hydrogels, in *Hydrogel in Medicine and Pharmacy* (N. L. Peppas, ed.), CRC Press, Boca Raton, FL, 1986, pp. 127–171.

133. Golander, C. G., and E. Kiss, Protein adsorption functionalized and ESCA-characterized polymer films studied by ellipsometry, *J. Colloid. Interface Sci., 121,* 240–253 (1988).

134. Lee, J. H., P. Kopeckova, J. Kopecek, and J. D. Andrade, Surface properties of copolymers of alkyl methacrylates with methoxy-(polyethylene oxide) methacrylates and their application as protein-resistant coatings, *Biomaterials, 11,* 455–464 (1990).

135. Brash, J. L., and J. A. Thibodeau, Identification of proteins adsorbed from human plasma to glass bead columns: Plasmin-induced degradation of adsorbed fibrinogen, *J. Biomed. Mater. Res., 20,* 1263–1275 (1986).

136. Kaplan, A. P., Initiation of the intrinsic coagulation and fibrinolitic pathways of man: The role of surfaces, Hageman factor, prekallikrein, high molecular weight kininogen, and Factor XI, *Prog. Hemostasis Thromb., 4,* 127–175 (1978).

137. Sandwick, R. K., and K. J. Schray, Conformational states of enzymes bound to surfaces, *J. Colloid Interface Sci., 121,* 1–12 (1988).

138. Benedek, K., Thermodynamics of alpha-lactalbumin denaturation in hydrophobic-interaction chromatography and stationary phase comparison, *J. Chromatogr., 458*, 93–104 (1988).

139. Lundstrom, I., Models of protein adsorption on solid surfaces, *Progr. Coll. Polym. Sci., 70*, 76–82 (1985).

140. Bohnert, J. L., and T. A. Horbett, Changes in adsorbed fibrinogen and albumin interactions with polymers indicated by decreases in detergent elutability, *J. Colloid. Interface Sci., 111*, 363–377 (1986).

141. Rapoza, R. J., and T. A. Horbett, Changes in the SDS elutability of fibrinogen adsorbed from plasma to polymers, *J. Biomater. Sci. Polym. Ed., 1*, 99–110 (1989).

142. Lenk, T. J., T. A. Horbett, B. D. Ratner, and K. K. Chittur, Infrared spectroscopic studies of time-dependent changes in fibrinogen adsorbed to polyurethanes, *Langmuir, 7*, 1755–1764 (1991).

143. Chinn, J. A., S. E. Posso, T. A. Horbett, and B. D. Ratner, Post-adsorptive transitions in fibrinogen adsorbed to Biomer: Changes in baboon platelet adhesion, antibody binding, and sodium dodecyl sulfate elutability, *J. Biomed. Mater. Res., 25*, 535–555 (1991).

144. Chinn, J. A., S. E. Posso, T. A. Horbett, and B. D. Ratner, Post-adsorptive transitions in fibrinogen adsorbed to polyurethanes: Changes in antibody binding and sodium dodecyl sulfate elutability, *J. Biomed. Mater. Res., 26*, 757–778 (1992).

145. Rapoza, R., and T. A. Horbett, Post-adsorptive transitions in fibrinogen: Influence of polymer properties, *J. Biomed. Mater. Res., 24*, 1263–1287 (1990).

146. Lu, D. R., and K. Park, Protein adsorption on polymer surfaces: Calculation of adsorption energies, *J. Biomater. Sci. Polym. Ed., 1*, 243–260 (1990).

147. Packham, M. A., Minireview. The behavior of platelets at foreign surfaces, *Proc. Soc. Exp. Biol. Med., 189*, 261–274 (1988).

148. Sixma, J. J., G. Hindriks, H. Van Breugel, R. Hantgan, and P. G. de Groot, Vessel wall proteins adhesive for platelets, *J. Biomater. Sci. Polym. Ed., 3*, 17–26 (1991).

149. Chinn, J. A., T. A. Horbett, and B. D. Ratner, Baboon fibrinogen adsorption and platelet adhesion to polymeric materials, *Thromb. Haemostas., 65*, 608–617 (1991).

150. McManama, G., J. N. Lindon, M. Kloczewiak, et al., Platelet aggregation by fibrinogen polymers crosslinked across E domain, *Blood, 68*, 363–371 (1986).

37
Perivascular Polymeric Controlled-Release Therapy

Gershon Golomb, Dorit Moscovitz, Ilia Fishbein, and David Mishaly
The Hebrew University of Jerusalem
Jerusalem, Israel

I. INTRODUCTION

Intravascular thrombosis is a significant complication in any vascular procedure, such as prosthetic heart valve implantation, insertion of vascular prostheses, and various angioplasty procedures. Any form of vascular injury, whether accidental or deliberate, induces a reparative response that includes proliferation of smooth muscle cells (SMCs) [1]. The interactions among cells of the blood vessel, blood cells, and biomaterial often result in processes that limit the long-term success of a number of contemporary cardiovascular therapeutic interventions. For example, these interactions contribute to anastomotic hyperplasia following insertions of small-caliber vascular prostheses, restenosis following balloon or other types of angioplasty or saphenous vein bypass grafting, and arteriosclerosis that complicates solid organ transplantation. Intravascular thrombosis and delayed stenosis secondary to intimal proliferation are potential major complications of over 1.2 million vascular surgical procedures performed annually in the United States.

Over the past decade, mechanical means of achieving revascularization of obstructive atherosclerotic vessels have been greatly improved. Percutaneous transluminal coronary angioplasty (PTCA) procedures include balloon dilation, excisional atheroctomy, endoluminal stenting, and laser ablation [2]. However, revascularization induces thrombosis and neointimal hyperplasia, which in turn cause restenosis in 30% to 40% of coronary arteries within the first 6 months after successful balloon angioplasty and in over 60% of aorta coronary saphenous vein bypass grafts within 5 years post [3,4]. Furthermore, intimal hyperplasia causes restenosis in approximately 30% to 50% of coronary and superficial femoral angioplasties, 20% of carotid end arterectomies, and 10% to 20% of femoro-distal vein bypasses [5]. Despite extensive research on the incidence, timing, mechanisms, and pharmacological interventions in humans and animal models, no therapy consistently prevents this difficult problem [6–9].

II. MECHANISM OF RESTENOSIS

Several researchers claim that treatment failures probably result from a lack of understanding of the cellular mechanisms in restenotic neointimal formation. Accumulation of smooth muscle cells (SMCs) in the intima of human arteries is a characteristic feature of atherosclerosis. These cells, which originate in the intima and media, undergo excessive and abnormally regulated proliferation that determines the size of the obstructive plaque [10]. The carotid artery injured in the rat by a balloon catheter has been widely used as a model of angioplasty.

According to Clowes [11], the endothelial cells are stripped away and the underlying media is stretched following the passage of a balloon embolectomy catheter. This form of injury causes immediate coagulation and thrombosis cascade in which platelets adhere, spread, and degranulate on the denuded surface of the artery. Approximately 24 hours (h) later, SMCs begin to proliferate. After four days, movement of SMCs from the media to the intima can be detected; they continue to proliferate in the intima. The intimal thickening process is further augmented by the deposition of matrix around the intimal cells. After two weeks, when the cellular proliferation is largely complete, and by three months, a steady state is reached in which the intima is approximately 20% cells and 80% extracellular matrix.

The underlying biochemical mechanisms controlling the proliferation and migration are as follows. Injury to the artery causes endothelial and SMC disruption and release of intracellular mitogens such as basic fibroblast growth factor (bFGF), which stimulate SMC proliferation. Factors from platelets (such as platelet-derived growth factor [PDGF] and thrombin) regulate proliferation and migration of SMCs. In addition, the mural thrombus caused by the injury may contribute cytokines, various chemoattractants and growth factors, and angiotensin II, affecting the intimal thickening process.

Histopathological studies of human arteries obtained after PTCA [12], or of restenotic lesions obtained by the use of transluminal atherectomy, revealed that SMC proliferation seems to play a pivotal role in the restenosis process and that, in a small but distinctive group, only intimal hyperplasia was observed without atherosclerotic plaque [8].

Schwartz, Holmes, and Topol proposed an alternative model based on observations in the porcine coronary injury model [7]. In response to arterial injury by an oversize stent, thrombus is formed. In the next stage, termed the *cellular recruitment phase*, the thrombus becomes covered by a layer of endothelium at roughly 3 to 4 days, followed by infiltration, from the lumen side first, of mononuclear cells consisting of macrophages and lymphocytes. In Stage III, the proliferative phase, cellular proliferation occurs in the final stage of arterial healing, approximately 7 to 9 days after injury. Beginning again at the lumen surface, cells that stain for alpha actin (SMC or myofibroblasts) colonize the degenerating thrombus mass as a thin "cap" just beneath the endothelium and progressively thicken. These cells do not arise from the media at the injury site. Extracellular matrix secretion and additional recruitment likely add to the neointimal volume. The healing process is complete when all the residual thrombus is resorbed and replaced by mature neointima.

III. PHARMACOLOGICAL INTERVENTIONS

Each of the steps involved in the formation of a thrombus and neointimal hyperplasia could be sites of pharmacological interventions that may prevent the restenosis process. The alternative hypothesis of Schwartz, Holmes, and Topol mentioned above implies that the thrombus is the important factor in restenosis because it provides the volume into which SMC proliferate [7]. Thus, a recommended therapeutic strategy is to limit mural thrombus size after angioplasty. However, many therapies under consideration involve the inhibition of SMC functions, including migration, proliferation, or matrix synthesis.

A large number of drugs were investigated for restenosis prevention and could be categorized as antithrombotic, antiplatelet, antiproliferative, anti-inflammatory, calcium antagonist, and lipid-lowering drugs. Comprehensive reviews of the various pharmacological approaches to prevent restenosis following angioplasty in humans and animal models were written recently by Herrman et al. [8,9]. However, despite 15 years of clinical experience and research in the field of restenosis prevention, no unequivocal beneficial effects have been observed from any particular drug.

The only agent that has given a hint of promise in the prevention of restenosis is fish oil. Seven trials with fish oil in regard to this indication have been conducted; three have given positive results, but have had methodological problems, and the other four were inconclusive or negative. The results of another large-scale trial, FORT (fish oil restenosis trial), are expected soon [13].

Critical evaluation of the value of drug trials that have been performed in the past is extremely difficult because of differences in selection of patients, methods of analysis, definition of restenosis, and poor statistical design [2,6,14]. The disappointing results obtained with the large number of drugs tested may be due in part to the use of suboptimal dosages in an effort to avoid systemic intolerance or difficulty in providing controlled administration of the drugs for an adequate period of time. Another serious problem is the poor statistical design in significant numbers of clinical pharmacological studies, resulting in lack of power to detect a moderate therapeutic effect of the drug therapy [2,6,14]. Other newer approaches likely to receive more attention in the future include antibodies to growth factors, gene transfer therapy, and antisense oligonucleotides.

The first reported use of antisense oligonucleotides to inhibit synthesis of a normal gene product *in vivo* was by Simons et al. [15]. However, despite intensive efforts of other research groups, the exciting results of this group's demonstration of the inhibitory effect on cell proliferation should be reconfirmed. Previous studies carried out in surgically exposed normal iliofemoral pig arteries demonstrated that genetic material can be successfully transferred to the arterial wall at the site of injury [16–18]. Indirect transfection was also accomplished by Wilson et al. in dogs using an implanted carotid interposition prosthetic graft seeded with genetically engineered endothelial cells [19]. Direct liposome-mediated transfer of the luciferase reported gene to surgically exposed normal dog arteries was also demonstrated [20]. More recently, gene transfer (luciferase) was achieved by using the percutaneous approach in balloon-dilated atherosclerotic arteries [21] and by fluoroscopic-guided direct percutaneous delivery to the myocardium of rabbits and pigs [22]. However, more studies are required to clarify such important questions as the specificity of oligodeoxynucleotide action, the timing and time needed for gene expression in active and rapidly developing lesions, and whether a brief exposure of an injured vessel to gene therapy can permanently inhibit SMC proliferation, which is developed from several days to few weeks [23].

IV. ANIMAL MODELS

Another serious obstacle in identifying effective drugs is the lack of adequate animal models [6]. The quality and quantity of drugs entering clinical studies directly depend on successful animal studies. Animal models are of limited value in restenosis research because it is impossible to create stenoses in animals that resemble human coronary artery disease. At least four animal models of arterial injury restenosis are in current use: (1) balloon injury of a rat carotid artery [24,25], (2) balloon injury of a rabbit iliac artery with or without atherosclerotic diet [26,27], (3) oversize balloon, and (4) stenting of atherosclerotic or normal pig coronaries or carotid [6,28,29].

The rat and rabbit models were found disappointing since many therapies that proved effective in these models failed in human studies. Many possible reasons exist for the discrepancies in these results, including (1) different pathophysiologic responses to arterial injury, (2) differences at the cellular level, (3) different arteries (elastic iliac or carotid as opposed to muscular coronaries), and (4) different restenosis measurements and definition. It was recently demonstrated that the antiproliferative and antithrombotic effects of heparin in rabbits differ markedly, depending on the type of arterial injury (oversized balloon or stenting) [30].

It is important to note that no pharmacological intervention examined for the inhibition of restenosis was found consistently effective in the pig [7,29] or in the crashed rabbit ear [31] models. It is clear, however, that animal models are the best tool to understand the cellular and molecular events of restenosis. The discrepancy between animal and human trials highlights the need for further characterization of restenosis in animal models. Careful animal studies will continue to serve the important task of testing hypotheses and evaluating the safety of new therapies.

In summary, the reasons for the failure of identifying an effective therapy despite the numerous pharmacological trials could be one of the following: (1) an effective agent has not been tested yet, (2) an effective agent has been overlooked due to incorrect dosage or timing, (3) more than a single agent is needed, and/or (4) restenosis cannot effectively be treated by systemic therapy.

V. LOCAL DRUG DELIVERY

The unsuccessful attempts to control restenosis by systemic pharmacological intervention prompt many researchers to look for more promising therapeutic approaches such as local drug delivery. Wolinsky and Thung were the first to use a perforated balloon catheter to deliver heparin into the wall of normal dog arteries [32]. Similarly, heparin was administered by the Wolinsky balloon in femoral hypercholesterolemic rabbits [33], and methotrexate was used by Muller and coworkers in minipigs [34]. Several other studies have been performed using the microporous balloon technique [8,9], but, despite these promising approaches, there remains a high incidence of additional injury and rupture of the lamina elastica interna following the jet stream used to deliver the drug solution to the artery wall.

Another concern is the rapid elimination of the drug or drug solution from the artery wall. For this reason, Wilensky et al. employed microparticles as carriers for drugs [35]. The microparticulate drug delivery system could have more residence time in the vasa vasorum due to outward diffusion. They succeeded in achieving a residence time of 14 days for 5 micron (μ) diameter particles detected in all three layers of the artery. Recently, the effect of the same Wolinsky balloon delivering angiopeptin was examined in rabbits [36]. Unexpectedly, the reduction of myointimal hyperplasia was detected only in the area distal to the local delivery (and to the injury). Angiopeptin, a somatostatin analog, has been shown to inhibit cellular proliferation in a number of *in vitro* and *in vivo* studies when given by subcutaneous (sc) injections [37–40].

Thus, it seems that several limitations in the study, such as the small and variable amount of myointimal hyperplasia produced, the methods of defining the exact area of injury and therapy, and the inherent limitation of the restricted volume of the aorta receiving the drug solution (approximately 0.1 milliliter [ml]) [36], should be resolved, if possible, before concluding whether the microporous balloon delivery is a viable opportunity in restenosis treatment.

More recent preliminary approaches are hydrogel-coated balloon catheters [41]. Antithrombotic and antiproliferative release from polymeric stents is another interesting approach. However, preliminary studies employing pig percutaneous insertion of plain polymeric stents failed because of accelerated and profound intimal proliferation and inflammation reactions to the biomaterial [42–44]. Moreover, it should be noted that only very potent drugs are suitable for such a delivery system due to the restricted size of the stent, which enables very restricted amounts of drug loading.

A. Controlled-Release Local Drug Delivery

As discussed above, there are several important limitations (e.g., timing and duration) in local drug delivery. For instance, for gene therapy, strategies can only address relatively late events in restenosis since the time required for transcription procedures precludes potential therapy of the acute thrombotic events. The results of several studies suggest that the proliferative process of SMC triggered by the vascular injury reaches a plateau after at least several weeks. Therefore, it would be reasonable to assume that an antiproliferative drug needs to be administered to the region of injury over at least the same period of time. Controlled-release drug delivery systems could overcome these important limitations.

B. Adventitial Drug Delivery

Sustained release systems have numerous advantages over oral or intravenous drug therapies. Implantation of controlled-release polymer systems at the site of a cardiovascular disease process offers the advantages of high regional levels of drug and, at the same time, lowered systemic drug exposure, thereby minimizing the possibility of side effects [45]. In addition, local delivery using controlled-release matrices permits the evaluation of novel agents administered in low doses that would otherwise degrade or create severe adverse effects if administered systemically. Periadventitial drug delivery enables the systematic study of potential drugs. Moreover, the site-specific therapy of different pharmacological agents affecting the various steps in restenosis provides an extremely valuable means and a rational approach to studying the pathogenesis of arterial restenosis. The technique of localized perivascular drug delivery may be applicable to a number of surgical settings, including endarterectomy, large-vessel and microvascular anastomosis, cerebral and systemic venous procedures, and arteriovenous shunts.

Controlled release may be defined as formulations of drug polymer composites, either as monolithic matrices or reservoirs with rate-limiting membrane configurations, in which drug administration can be sustained through the use of polymeric materials. A general requirement for investigating the controlled release of biologically active agents has been the ability to evaluate the drug incorporation using assays of both *in vitro* release and *in vitro* biological activity. Following this, experiments may be carried out to assess persisting *in vitro* activity. Cardiovascular controlled-release implants have been shown, by our group and others, to be effective in preventing cardiovascular calcification in bioprosthetic heart valves [46–50], inhibiting life-threatening ventricular arrhythmias in a variety of experimental settings [51], and in preventing heart transplant rejection [52].

A variety of base polymers for use as controlled-release systems has become available, allowing greater flexibility with respect to configuration and desired duration of release. Our group, in particular, has had experience with silicone rubbers [47,48,53], ethylenevinyl acetate (EVA) copolymers [46], and polyurethane (PU) [54,55] in anticalcification and anti-infection studies. These nondegradable and biocompatible polymers offer an array of formulations and release rates since they differ in their hydrophilicity (silicone and ethylenevinyl acetate copolymers are more hydrophobic than polyurethane), in their preparation techniques (EVA and PU are solvent cast and silicone is catalyst polymerized), and in their mechanical properties. In addition, the polymer type, molecular weight fractionation, rate-limiting membrane, drug loading, and formulations with additives make possible an array of controlled-release systems that provides a broad range of release rates [48,50,53,56,57].

Biodegradable controlled-release systems have the advantage of gradual biological elimination without a residual implant structure remaining. A large number of biodegradable polymers have been used, including polyorthoesters, poly (α-hydroxy esters), polyanhydrides, and various polysaccharides. All are potentially compatible with proteins and peptides and might be utilized in the future for fabricating drug-loaded biodegradable stents or periadventitial drug delivery systems. However, until engineering problems are solved, synthetic polymers represent the rational choice for studying drug delivery approaches. Synthetic polymers examined in various biocompatibility systems are readily available, so no specific biocompatibility studies are needed as would be the case if new biodegradable polymers are used. In addition, the perivascular implantation of stable polymers enables quantitative recovery of the polymer for residual drug analyses.

1. Adventitial Delivery of Heparin

Okada, Bark, and Mayberg were the first to evaluate advential heparin release as a means of inhibiting thrombosis and neointimal thickening [58,59]. This approach was also studied by others [60,61]. Heparin has been shown to inhibit intravascular thrombosis effectively after vascular surgical procedures [62,63], either by preventing platelet aggregation [64] or by a directly inhibiting smooth muscle proliferation [25,65,66].

The release kinetics and distribution of heparin after periadventitial delivery were determined by applying tritium-labeled heparin in polyvinyl alcohol (PVA) to the carotid artery in rats [67]. Autoradiographic measurements detected the radioactive-labeled heparin in polymers and vessel walls of specimens at all time periods. Transmission electron microscopy of these sections demonstrated heparin uptake by SMCs, where it was concentrated in the region of nuclear chromatin. There was a rapid logarithmic decrease in labeled heparin concentration for both polymers and vessel walls, with over 80% reductions in concentration one day after application.

Following endothelial damage with balloon denudation in a rat carotid [59], the injured segment was surrounded with a Silastic cuff containing heparin in PVA or PVA alone. There was no significant difference in coagulation parameters for each time period compared with rats treated with PVA alone. Beginning 5 days after endothelial injury, and progressively more prominent 10 and 20 days after the injury, there was a marked reduction in intimal proliferation for arteries treated with heparin/PVA compared with PVA alone. There was a significant reduction in the thymidine index for the vessel wall in treated animals compared with controls 5 days after endothelial injury.

Edelman and colleagues confirmed these observations using adventitial sustained release of heparin from ethylenevinyl acetate (EVA) copolymer matrices and showed that this approach is more effective in the rat carotid injury model than either intravenous or subcutaneous delivery of comparable heparin doses [60]. Moreover, matrix delivery of nonanticoagulant low molecular weight heparin was also found to suppress SMC proliferation effectively. The successful implementation of their technique was demonstrated in the study of antisense oligonucleotides periadvential delivery.

In another investigation utilizing biodegradable polymer, the effect of heparin released from a new polyanhydide polymer, a dimer of oleic and sebacic acid, was studied [68]. The polyanhydride matrices containing heparin were placed around an inverted vein graft in rats and patency rates were compared with those of control animals. Anastomotic patency was significantly greater in the groups treated with transmural heparin, measured both at 24 h and 7 days.

2. Adventitial Delivery of Hirulog

Hirulog, a synthetic peptide with 20 amino acids, is a specific inhibitor of thrombin and is derived from the naturally occurring polypeptide hirudin, which is extracted from leeches [69]. Hirulog has been shown to inhibit thrombin adsorbed to a fibrin clot *in vitro*, whereas clot-bound thrombin is inaccessible to the heparin-antithrombin III complex. Thus, it is a potentially more potent antithrombin agent than heparin, which is an important feature based on the recent hypothesis of Schwartz that thrombin and thrombus formation play a pivotal role in restenosis [7]. Studies on the atherosclerotic rabbit iliac injury model have suggested that hirulog may be more effective than heparin as an inhibitor of intimal proliferation following balloon angioplasty [27].

Hirulog is a stable peptide and its anticoagulant activity is maintained despite sterilization using either heat or gamma irradiation [45]. Tritium-labeled hirulog was used to characterize *in vitro* drug release from a hirulog-silicone rubber system. Silicone rubber matrices with hirulog (1%) and bovine serum albumin (19%) were formulated using Q7-4840. Slab matrices were incubated *in vitro* under perfect sink conditions in buffer solution at pH 7.4. After a burst phase of release, constant release properties (pseudo zero order) were observed after 12 hours *in vitro*. The amount of hirulog predicted to be available during the "zero-order" release phase is 1.54 micrograms/milligram (μg/mg) matrix/day. Thus, a 2-gram matrix, which could be wrapped around a balloon-injured or stented carotid artery, would be predicted to inhibit thrombin-mediated coagulation over a 5-cm arterial segment by providing a continuous regional dose of 3.1 mg hirulog/day.

For the *in vivo* studies, a silicone polymer containing hirulog was implanted around the adventitial surface of a pig carotid artery following oversize stenting, and was compared with a control polymer implanted contralaterally [70]. In 4 pigs sacrificed on Days 3-5, less thrombus was apparent macroscopically on the stents treated with adventitial hirulog, but electron microscopy showed thrombus adherent on all stents. In 10 pigs, carotid segments were resected 32 ± 4 days after stenting. Histologic analysis showed no difference between treated and nontreated sides in the volume of neointima.

In another *in vivo* study [71], rabbits underwent bilateral local femoral injury by air desiccation, followed by an atherogenic diet for 28 days. On Day 28, hirulog-containing and plain polymers as described above were implanted around both femoral arteries. Bilateral femoral balloon angioplasty was performed and the cholesterol diet was discontinued. The data of the intimal and lumen area revealed no significant difference between the treated and nontreated arterial segments.

It may be concluded that, in both of these models, local adventitial hirulog delivery at the dose and delivery rate used does not prevent thrombus formation, and does not reduce the severity of neointimal thickening.

3. Adventitial Dexamethasone Delivery

Dexamethasone is a highly potent, water-insoluble synthetic glucocorticoid that has powerful anti-inflammatory activity, but chronic systemic therapy of the drug causes major adverse side effects. Previous studies have shown that glucocorticoids inhibit human aortic smooth muscle cell proliferation in culture [72] and inhibit atherosclerotic lesion formation in the hypercholesterolemic rabbit [73,74]. When given as a single high dose by intravenous infusion, however, methylprednisolone did not inhibit restenosis after balloon injury in humans [75]. Dexamethasone delivery from polymer coatings of pacemaker leads has been successfully used clinically to prevent adjacent myocardial fibrosis and a rise in the pacing lead threshold [76].

In three groups of rats, dexamethasone 0%, 0.5% and 0.5% in silicone polymer (15 mm length) were implanted around the left common carotid artery following deendothelialization with a balloon catheter [77–79]. Morphometric and histologic analyses of all arterial segments were performed after 21 days. Analysis of the data revealed significant inhibition of neointimal formation in treated arteries. However, dexamethasone serum levels were noted in both 0.5% and 5% drug loads 24 h and three weeks after implantation. Therefore, it was concluded that the activity of the drug-eluting polymer was chiefly, but not exclusively, site specific.

In another study [80], dexamethasone polymers were formulated as described above and were placed around the adventitial surface of stented porcine carotid arteries as described for the hirulog studies. Each animal was sacrificed 30 days after stent implantation. In each of the animals examined, a dramatic reduction in the severity of the fibrotic reaction adjacent to the polymer was apparent. In contrast, the control polymer was again surrounded by a dense fibrous capsule, as had been noted in the hirulog studies. However, histologic analysis revealed no significant reduction in the severity of neointimal thickening.

VI. SUMMARY

The ultimate aim still remains to identify pharmacological agents that are consistently effective in the therapy of restenosis. This conclusion is valid also for local delivery of drugs in animal models. The contradicting results obtained in different animal models highlight the need to evaluate possible therapies in more than one animal model.

The findings on the effectiveness of heparin and dexamethasone in the rat and rabbit models following local delivery warrants further evaluation. The successful inhibition of neointimal proliferation following local delivery of heparin has to be demonstrated in the pig model. The most plausible explanation for the apparent discrepancy between the results obtained in the rat and pig models following local delivery of dexamethasone is the high resistance of the pig model to pharmacological interventions. This conclusion is supported by the lack of any of the numerous pharmacological interventions tested so far to demonstrate consistent therapy in the pig model. Differences in timing, duration, and dose may explain why inconsistent therapeutic results are obtained.

Following the critical evaluation of different forms of arterial injury by Rogers, Karnovský, Karnovsky, and Edelman [30] and the data presented above, it can be suggested that a therapeutic strategy should be examined with the following considerations: (1) rodent (rat or rabbit iliac) and pig models should be used, (2) the effects of timing, duration, and dose should be closely examined. Periadventitial delivery of possible pharmacological agents will remain to serve as the best means for evaluating possible therapy and elucidating the restenosis mechanism.

REFERENCES

1. Schoen, F.J., and J.J. Castellot, Vascular graft intimal fibrous hyperplasia: Prospects for pharmacological inhibition, *J. Vasc. Surg.*, 13:758–761 (1991).
2. Calif, R.M., et al., Restenosis after coronary angioplasty: An overview, *J. Am. Coll. Cardiol.*, 17(6):2B–13B (1991).
3. Willerson, J.T., P. Golino, J. Eidt, W.B. Campbell, and L.M. Buja, Specific platelet mediators and unstable coronary artery lesions: Experimental evidence and clinical implications, *Circulation*, 80:198–205 (1989).
4. Hirsh, P.D., L.D. Hillis, W.B. Campbell, B.G. Firth, and J.T. Willerson, Release of prostaglandins and thromboxane into the coronary circulation in patients with ischemic heart disease, *N. Engl. J. Med.*, 304:685–691 (1981).
5. Clowes, A.W., M.M. Clowes, S.C. Vergel, R.K.M. Muller, J.S. Powell, F. Hefti, and H.R. Baumgartner, Heparin and cilazapril together inhibit injury-induced intimal hyperplasia, *Hypertension*, 18:II-65–II-69 (1991).
6. Muller, D.W.M., S.G. Ellis, and E.J. Topol, Experimental models of coronary artery restenosis, *J. Am. Coll. Cardiol.*, 19(2):418–432 (1992).
7. Schwartz, R.S., D.R. Holmes, and E.J. Topol, The restenosis paradigm revisited: An alternative proposal for cellular mechanisms, *J. Am. Coll. Cardiol.*, 20:1284–1293 (1992).
8. Herrman, J.P.R., W.R.M. Hermans, J. Vos, and P.W. Serruys, Pharmacological approaches to the prevention of restenosis following angioplasty. Part I, *Drugs*, 46:1852 (1993).
9. Herrman, J.P.R., W.R.M. Hermans, J. Vos, and P.W. Serruys, Pharmacological approaches to the prevention of restenosis following angioplasty, *Drugs*, 46:249–262 (1993).
10. Ross, R., The pathogenesis of atherosclerosis: A perspective for the 1990s, *Nature*, 362:801–809 (1993).
11. Clowes, A.W., Intimal hyperplesia and graft failure, *Cardiovasc. Pathol.*, 2:1795–1975 (1993).
12. Waller, B.F., C.A. Pinkerton, C.M. Orr, J.D. Slack, J.W. VanTassel, and T. Peters, Restenosis 1 to 24 months after clinically successful coronary balloon angioplasty: A necroscopy study of 20 patients, *J. Am. Coll. Cardiol.*, 17:58B–70B (1991).
13. More disappointment in restenosis prevention, *SCRIP*, 1754:24 (1992).
14. Beatt, K.J., P.W. Serruys, and P.G. Hugenholtz, Restenosis after coronary angioplasty: New standards for clinical studies, *J. Am. Coll. Cardiol.*, 15:491–498 (1990).
15. Simons, M., E.R. Edelman, J. DeKeyser, R. Langer, and R.D. Rosenberg, Antisense c-myb oligonucleotides inhibit intimal arterial smooth muscle cell accumulation *in vivo*, *Nature*, 359:67–70 (1992).
16. Nabel, E.G., G. Plautz, F.M. Boyce, J.C. Stanley, and G.J. Nabel, Recombinant gene expression *in vivo* within endothelial cells of the arterial wall, *Science*, 244:1342–1344 (1989).
17. Nabel, E.G., G. Plautz, and G.J. Nabel, Site-specific expression *in vivo* by direct gene transfer into the arterial wall, *Science*, 249:1285–1288 (1990).
18. Nabel, E.G., G. Plautz, and G.J. Nabel, Gene transfer into vascular cells, *J. Am. Coll. Cardiol.*, 17:189B–194B (1991).
19. Wilson, J.M., L.K. Birinyi, R.N. Salomon, P. Libby, A.D. Callow, and R.C. Mulligan, Implantation of vascular grafts lined with genetically modified endothelial cells, *Science*, 244:1344–1346 (1989).

20. Lim, C.S., G.D. Chapman, R.S. Gammon, J.B. Mulhestein, R.P. Bauman, R.S. Stack, and J.L. Swain, Direct *in vivo* gene transfer into the coronary and peripheral vasculatures of the intact dog, *Circulation*, 83:2007–2011 (1991).

21. Guy, L., D. Gal, S. Takeshita, S. Nikol, L. Weir, and J.M. Isner, Percutaneous arterial gene transfer in a rabbit model, *J. Clin. Invest.*, 90:936–944 (1992).

22. Gal, D., L. Weir, L. Guy, J.G. Pickering, J. Hogan, and J.M. Isner, Direct myocardial transfection in two animal models, *Lab. Invest.*, 68:18–25 (1993).

23. Epstein, S.E., E. Speir, and T. Finkel, Do antisense approaches to the problem of restenosis make sense? *Circulation*, 88/3:1351–1353 (1993).

24. Fishman, J., G. Ryan, and M. Karnovsky, Endothelial regeneration in the rat carotid artery and the significance of endothelial denudation in the pathogenesis of myointimal proliferation, *Lab. Invest.*, 32:339–351 (1975).

25. Clowes, A.W., M.A. Reidy, and M.M. Clowes, Kinetics of cellular proliferation after arterial injury, *Lab. Invest.*, 49(3):327–333 (1983).

26. Faxon, D.P., T.A. Sanborn, V.J. Weber, C. Haudenschild, S.B. Gottsman, W.A. McGovern, and T.J. Ryan, Restenosis following transluminal angioplasty in experimental atherosclerosis, *Arteriosclerosis*, 4:189–195 (1984).

27. Sarembock, I.J., S.D. Gertz, L.W. Gimpel, R.M. Owen, E.R. Powers, and W.C. Roberts, Effectiveness of recombinant desulphatohirudin in reducing restenosis after balloon angioplasty of atherosclerotic femoral arteries in rabbits, *Circulation*, 84(1):232–243 (1991).

28. Scott, R., D. Kim, J. Schemes, and W. Thomas, Atherosclerotic lesions in coronary arteries of hyperlipidemic swine, *Atherosclerosis*, 62:1–14 (1986).

29. Schwartz, R.S., J.G. Murphy, W.D. Edwards, A.R. Camrud, R.E. Vliestra, and D.R. Holmes, Restenosis after balloon angioplasty. A practical proliferative model in porcine coronary arteries, *Circulation*, 82:2190–2200 (1990).

30. Rogers, C., M.J. Karnovsky, and E.R. Edelman, Inhibition of experimental neointimal hyperplasia and thrombosis depends on the type of vascular injury and the site of drug administration, *Circulation*, 88:1215–1221 (1993).

31. Banai, S., M. Shou, R. Correa, M.T. Jaklitsch, P.C. Douek, R.F. Bonner, S.E. Epstein, and E.F. Unger, Rabbit ear model of injury-induced arterial smooth muscle cell proliferation: Kinetics, reproducibility, and implications, *Circ. Res.*, 69:748–756 (1991).

32. Wolinsky, H., and S.N. Thung, Use of a perforated balloon catheter to deliver concentrated heparin into the wall of the normal canin artery, *J. Am. Coll. Cardiol.*, 15:475–481 (1990).

33. Gimple, L.W., S.D. Gertz, H.L. Haber, M. Ragosta, E.R. Powers, W.C. Roberts, and I.J. Sarembock, Effect of chronic subcutaneous or intramural administration of heparin on femoral artery restenosis after balloon angioplasty in hypercholesterolemic rabbits: A quantitative angiographic and histopathological study, *Circulation*, 86:1536–1546 (1992).

34. Muller, D.W.M., E.J. Topol, G.D. Abrams, K.P. Gallagher, and S.G. Ellis, Intramural methotrexate therapy for the prevention of neointimal thickening after balloon angioplasty, *J. Am. Coll. Cardiol.*, 20:460–466 (1992).

35. Wilensky, R.L., K.L. March, and D.R. Hathaway, Direct intraarterial wall injection of microparticles via a catheter: A potential drug delivery strategy following angioplasty, *Am. Heart J.*, 22:1136–1140 (1991).

36. Hong, M.K., T. Bhatti, B.J. Matthews, K.S. Stark, S.S. Catharpermal, M.L. Foegh, P.W. Ramwell, and K.M. Kent, The effect of porous infusion balloon-delivered angiopeptin on myointimal hyperplasia after balloon injury in the rabbit, *Circulation*, 88:638–648 (1993).

37. Vargas, R., G.W. Bormes, B. Wroblewska, A. Rego, J.L. Foegh, P.A. Kot, and P.W. Ramwell, Angiopeptin inhibits thymidine incorporation in rat carotid artery *in vitro*, *Transplant Proc.*, 21:3702–3704 (1989).

38. Asotra, S., J.L. Foegh, J.V. Conte, B.R. Cai, and P.W. Ramwell, Inhibition of 3H-thymidine incorporation by angiopeptin in the aorta of rabbits after balloon angioplasty, *Transplant Proc.*, 21:3695–3696 (1989).

39. Conte, J.V., J.L. Foegh, D. Calcagno, R.B. Wallace, and P.W. Ramwell, Peptide inhibition of proliferation following angioplasty in rabbits, *Transplant Proc.*, 21:3686–3688 (1989).

40. Lundergan, D., M.L. Foegh, R. Vargas, M. Eufemio, G.W. Bormes, P.A. Kot, and P.W. Ramwell, Inhibition of myointimal proliferation of the rat carotid artery by the peptides angiopeptin and BIM 23034, *Atherosclerosis*, 80:49–55 (1989).

41. Nunes, G.L., S.R. Hanson, and S.B. King, Local heparin delivery with a hydrogel-coated PTCA balloon catheter inhibits platelet-dependent thrombosis, *J. Am. Coll. Cardiol.*, 21: 117A (1993).

42. Lincoff, A.M., R.S. Schwartz, W.J. van der Giessen, H.M.M. van Beusekom, P.W. Serruys, D.R. Holmes, S.G. Ellis, and E.J. Topol, Biodegradable polymers can evoke a unique inflammatory response when implanted in the coronary artery, *65th Scientific Sessions*, 86/4/1 (1992).

43. Lincoff, A.M., W.J. van der Giessen, R.S. Schwartz, H.M.M. van Beusekom, P.W. Serruys, D.R. Holmes, Jr., S.G. Ellis, and E.J. Topol, Biodegradable and biostable polymers may both cause vigorous inflammatory response when implanted in the porcine coronary artery, *J. Am. Coll. Cardiol.*, 21:179A (1993).

44. Murphy, J.G., R.S. Schwartz, W.D. Edwards, A.R. Camrud, R.E. Vliestra, and D.R. Holmes, Percutaneous polymeric stents in porcine coronary arteries, *Circulation*, 86:1596–1604 (1992).

45. Levy, R.J., G. Golomb, J. Trachy, V. Labhasetwar, D. Muller, and E. Topol, Strategies for treating arterial restenosis using polymeric controlled release implants, in *Biotechnology and Bioactive Polymers* (C. Gebelein, ed.), Plenum Press, New York, 1992.

46. Golomb, G., R. Langer, F.J. Schoen, M.S. Smith, Y.M. Choi, and R.J. Levy, Controlled release of diphosphonate to inhibit bioprosthetic heart valve calcification: Dose-response and mechanistic studies, *J. Controlled Rel.*, 4:181–194 (1986).

47. Golomb, G., M. Dixon, M.S. Smith, F.J. Schoen, and R.J. Levy, Controlled release drug delivery of diphosphonate to inhibit bioprosthetic heart valve calcification: Release rate modulation with silicone matrices via drug solubility and membrane coating, *J. Pharm. Sci.*, 76: 271–276 (1987).

48. Golomb, G., Controlled-release of diphosphonate from synthetic polymers to inhibit calcification, *J. Biomat. Applications*, 2:266–288 (1987).

49. Johnston, T.P., J.A. Boyd, B.L. Ciesliga, F.J. Schoen, G. Amidon, and R.J. Levy, Controlled release of ethanehydroxy diphosphonate from polyurethane reservoirs to inhibit calcification of bovine pericardium used in bioprosthetic heart valves, *Inter. J. Pharm.*, 59:95–104 (1990).

50. Levy, R.J., T.P. Johnston, A. Sintov, and G. Golomb, Controlled release implants for cardiovascular disease, *J. Controlled Rel.*, 11:245–254 (1990).

51. Sintov, A., W.A. Scott, R. Siden, and R.J. Levy, Efficacy of epicardial controlled-release lidocaine for ventricular tachycardia induced by rapid ventricular pacing in dogs, *J. Cardiovasc. Pharm.*, 16:812–817 (1990).

52. Bolling, S.F., H. Lin, T.M. Annesley, J.A. Boyd, K.P. Gallagher, and R.J. Levy, Local cyclosporine immunotherapy of cardiac transplants in rats enhances survival, *J. Heart and Lung Transplant.*, 10:577–583 (1990).

53. Golomb, G., P. Fisher, and E. Rahamim, The relationship between drug release rate, particle size and matrix swelling of silicone matrices, *J. Controlled Rel.*, 12:121–132 (1990).

54. Golomb, G., and D. Wagner, Characterization and anticalcification effects of implantable polyurethane matrices containing amorphous dispersion of bisphosphonic acid, *Clin. Mater.*, 8:33–42 (1991).

55. Golomb, G., and A. Shpigelman, Prevention of bacterial colonization on polyurethane *in vitro* by incorporated antibacterial agent, *J. Biomed. Mater. Res.*, 25:937–952 (1991).

56. Langer, R., and J. Folkman, Polymers for sustained release of proteins and other macromolecules, *Nature*, 263:797–800 (1976).

57. Levy, R.J., S.F. Bolling, R. Siden, A. Kadish, Y. Patak, P. Dorostkar, A. Sintov, G. Golomb, and T.P. Johnston. Polymeric controlled release of cardiovascular drugs, in *Cosmetic and Pharmaceutic Polymer Applications* (C. Gebelein, ed.), Plenum Press, New York, 1991, pp. 231–238.

58. Okada, T., D.H. Bark, and M.R. Mayberg, Local anticoagulation without systemic effect using a polymer heparin delivery system, *Stroke*, 19:1470–1476 (1988).

59. Okada, T., D.H. Bark, and M.R. Mayberg, Localized release of perivascular heparin inhibits intimal proliferation after endothelial injury without systemic anticoagulation, *Neurosurgery*, 25:892–898 (1989).

60. Edelman, E.R., D.H. Adams, and M.J. Karnovsky, Effect of controlled adventitial heparin delivery on smooth muscle cell proliferation following endothelial injury, *Proc. Natl. Acad. Sci. USA*, 87:3773–3777 (1990).

61. Edelman, E.R., M.A. Nugent, L.T. Smith, and M.J. Kaarnovsky, Basic fibroblast growth factor enhances the coupling of intimal hyperplasia and proliferation of vasa vasorum in injured rat arteries, *J. Clin. Invest.*, 89:465–473 (1992).

62. Dirrenberger, R.A., and T.M. Sundt, Jr., Carotid endarterectomy. Temporal profile of the healing process and effects of anticoagulation therapy, *J. Neurosurg.*, 48:201–219 (1978).

63. Greenberg, B.M., M. Masem, and J.W. May, Jr., Therapeutic value of intravenous heparin in microvascular surgery: An experimental vascular thrombosis study, *Plast. Reconstr. Surg.*, 82:463–469 (1988).

64. Rosenberg, R.D., Biologic actions of heparin, *Semin. Hematol.*, 14:427–440 (1977).

65. Clowes, G.K., and M.J. Karnovsky, Suppression by heparin of smooth muscle cell proliferation in injured arteries, *Nature*, 265:625–626 (1977).

66. Clowes, A.W., and M.M. Clowes, Kinetics of cellular proliferation after arterial injury. IV. Heparin inhibits rat smooth muscle mitogenesis and migration, *Circ. Res.*, 58:839–845 (1986).

67. Mayberg, M.R., Localized release of perivascular heparin, *Perspectives in Neurological Surgery*, 1:77–95 (1990).

68. Orloff, L.A., M.G. Glenn, and R.A. Esclamado, Anti-thrombotic effects of heparin released transmurally from a polyanhydride polymer, *American Academy of Otolaryngology—Head and Neck Surgery*, San Diego, California (1990).

69. Maraganore, J.M., P. Bourdon, J. Jabonski, K.L. Ramachandran, and J.W. Fenton, Design and characteristics of Hirulogs: A novel class of bivalent peptide inhibitors of thrombin, *Biochemistry*, 29:7095–7101 (1990).

70. Muller, D.W.M., G. Golomb, D. Gordon, J.M. Maraganore, and R.J. Levy, Local adventitial hirulog delivery for the prevention of stent thrombosis and neointimal thickening, *Abstracts of the 65th American Heart Association Meeting*, New Orleans, Louisiana, November 16–19, 1992.

71. Guzman, L.A., A.E. Villa, G. Golomb, R.J. Levy, J. Furst, and E.J. Topol, Effect of *in-situ* polymeric hirulog release on restenosis following balloon angioplasty in the atherosclerotic rabbit, *Abstracts of the 42nd Am. Coll. of Cardiol. Meeting*, March 14–18, 1993.

72. Jarvelainen, H., T. Halme, and Ronnemaa, Effect of cortisol on the proliferation and protein synthesis of human aortic smooth muscle cells in culture, *Acta Med. Scand.*, 560:114–122 (1982).

73. Gordon, D., G.C. Kobernick, G.C. McMillan, and G.L. Duff, The effect of cortisone on the serum lipids and on the development of experimental cholesterol atherosclerosis in the rabbit, *J. Exp. Med.*, 99:371–386 (1954).

74. Hollander, W., D. Kramsch, C. Franzbalu, J. Paddock, and M.A. Colombo, Suppression of atheromatous fibrous plaque formation by antiproliferative and anti-inflammatory drugs, *Circ. Res.*, 34–35(Suppl. 1):I-131–I-140 (1974).

75. Pepine, C.J., et al., A controlled trial of corticosteroids to prevent restenosis after coronary angioplasty, *Circulation*, 81:1753–1761 (1990).

76. Mond, H., K. Stokes, J. Helland, L. Grigg, P. Kertes, B. Pate, and D. Hunt, The porous

titanium steroid eluting electrode: A double blind study assessing the stimulation threshold effects of steroid, *PACE*, 11:214–219 (1988).

77. Villa, A.E., L.A. Guzman, G. Golomb, R.J. Levy, and E.J. Topol, Local delivery of dexamethasone for prevention of neointimal proliferation after balloon injury in the rat carotid model, *J. Am. Coll. Cardiol.*, 21:179A (1993).

78. Villa, A.E., L.A. Guzman, G. Golomb, R.J. Levy, J. Furst, and E.J. Topol, Local delivery of dexamethasone for prevention of neointimal proliferation after balloon injury in the rat carotid model, *Abstracts of the 42nd Am. Coll. of Cardiol. Meeting*, March 14–18, 1993.

79. Villa, V.E., L.A. Guzman, W. Chen, G. Golomb, R.J. Levy, and E.J. Topol, Local delivery of dexamethasone for prevention of neointimal proliferation after balloon arterial injury in the rat carotid model, *J. Clin. Invest.*, 93:1243–1249 (1993).

80. Muller, D.W.M., G. Golomb, D. Gordon, and R.J. Levy, Sustained local dexamethasone delivery for the prevention of neointimal thickening after stent implantation, *Abstracts of the Australian and New Zealand Journal of Medicine* (1993).

38
Albumin-Modified Biomaterial Surfaces for Reduced Thrombogenicity

Mansoor M. Amiji
Northeastern University, Boston, Massachusetts
Kalpana R. Kamath
South Dakota State University, Brookings, South Dakota
Kinam Park
Purdue University, West Lafayette, Indiana

I. INTRODUCTION

Natural and synthetic materials such as polymer, metal, ceramic, glass, carbon, and composites have been utilized for a wide variety of biomedical applications [1]. Of these materials, polymer has been one of the most frequently used biomaterials. Polymers are widely used because of their versatility, ease of fabrication and processing, control over the physical properties necessary for biomedical applications, and ease in surface modification to improve biocompatibility or blood compatibility [2]. Blood-contacting polymeric biomaterials can be divided into three main classes depending on the duration of contact [3]. Biomedical implants such as intravascular catheters may be in contact with blood only once and for a short period of time. Other biomaterials such as membranes for hemodialyzers and blood oxygenators may be exposed for hours at a time. Finally, implants such as heart valves and vascular grafts have to last for years or the lifetime of the patient. Long-term use of blood-contacting biomaterials is significantly limited by the surface-induced thrombosis [4].

Figure 1 illustrates the proposed sequence of events leading to surface-induced thrombosis and embolization on biomaterials [5]. The first event is the adsorption of blood proteins. Protein adsorption is known to produce a "conditioning film" that determines the outcome of other processes (e.g., cell adhesion and activation of the serum complement system) [6,7]. The type and the quantity of the adsorbed proteins in the conditioning layer determine whether or not platelets can adhere to the surface. The presence of platelet-adhesive proteins such as fibrinogen and fibronectin in the conditioning film results in the adhesion of platelets. The adherent platelets may undergo spreading, which is a process of activation, on the biomaterial surface [8]. The spreading platelets release the biochemical contents such as adenosine diphosphate (ADP) from their granules. ADP and other platelet agonists activate circulating platelets to induce thrombus for-

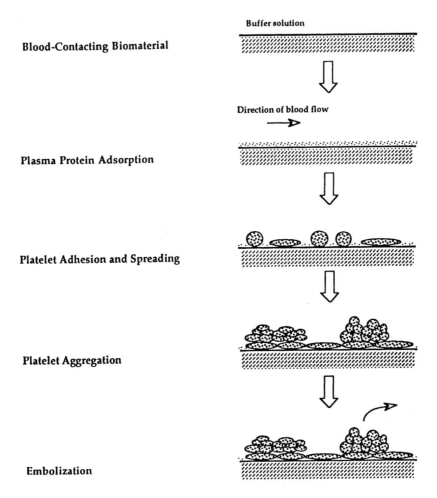

Figure 1 Sequence of events leading to surface-induced thrombosis and embolization on blood-contacting biomaterial surfaces. (From Ref. 5.)

mation on top of the fully spread platelets. Thrombi formed on the biomaterial surface are comprised primarily of platelets and other blood cells trapped in the fibrin network [9].

A number of investigations have indicated that albumin on the surface tends to decrease the surface-induced platelet activation [10–13]. Both the total number of adherent platelets and the extent of platelet activation are reduced on albumin-coated surfaces. For this reason, albumin has been widely used to modify the biomaterial surfaces. Human and bovine serum albumin have been used most often in surface modification of biomaterials [14,15]. The properties of human and bovine serum albumin are summarized in Table 1.

Human serum albumin (HSA) is a 585-amino-acid globular protein with an ellipsoidal shape (38 × 150 angstroms [Å]) and an average molecular weight of 67,000 daltons [16]. Albumin is the most abundant protein in human plasma. The plasma concentration of albumin in normal adults ranges from 35 milligrams per milliliter (mg/ml) to 47 mg/

Table 1 Properties of Human and Bovine Serum Albumin

Property	Human	Bovine
Molecular weight	66,700	69,000
Dimension (Å)	38×150	40×140
Diffusion coefficient (cm^2/s)	6.1×10^{-7}	5.9×10^{-7}
Partial specific volume	0.733	0.733
Intrinsic viscosity	0.042	0.041
Isoelectric point	4.7	4.7
Plasma concentrations (mg/ml)	35.0–47.0	24.5–42.0

Source: Refs. 16 and 17

ml [17]. Albumin has some unique properties that distinguish it from other plasma proteins. Albumin is highly soluble in the aqueous medium due to the polar surface of the molecule and has high thermal stability as a result of 17 disulfide bonds [18]. Albumin has three important functions in the body. It maintains the normal osmotic pressure difference between the intracellular and extracellular environments. Albumin also binds and transports low molecular weight hydrophobic compounds such as bilirubin, fatty acids, or some drugs in plasma. Albumin is synthesized in the liver to serve as a source of amino acids in cases of malnutrition.

II. MODIFICATION OF BIOMATERIAL SURFACES WITH ALBUMIN

Due to the pronounced effect in reducing platelet adhesion and activation, albumin has been frequently immobilized on biomaterial surfaces by various methods. The most common and simple approach for albumin immobilization is physical adsorption of albumin on biomaterials prior to blood contact. Since the physical adsorption does not result in a stable albumin layer on the surface, albumin needs to be covalently grafted to the surface for long-term applications in blood.

A. Physical Adsorption of Albumin

Because of the high surface activity property, albumin can be easily adsorbed to most hydrophobic biomaterial surfaces. The thermodynamics of adsorbed albumin has been treated theoretically and experimentally by various investigators to understand the mechanism of protein adsorption on different surfaces [19]. Various factors, such as surface charge [20–22], surface free energy and hydrophilicity [23–25], chemical structure [26,27], and surface morphology [28], affect the albumin adsorption behavior.

Some of the early *in vitro* studies conducted by Packham et al. showed that platelets had little tendency to adhere to the albumin-coated glass tubes [29]. This was supported by the *in vivo* canine shunt experiments performed by Chang [30]. In the study, platelet-surface interactions were examined using activated charcoal granules coated with cellulose nitrate, followed by albumin. The study demonstrated that albumin-coating completely eliminated the problems associated with platelet adhesion. Studies of Mulvihill et al. indicated that albumin adsorption on plasmapheresis circuits reduced platelet accumulation and thrombus formation [13]. In their studies, plasmapheresis circuits were preadsorbed with a 4% human serum albumin solution for 15–20 min before exposure to the patient's blood. Platelet deposition was significantly reduced by simple adsorption of albumin as compared to the control.

Sigot-Luizard et al. examined the compatibility of albuminated polyester fabrics with newly formed endothelial cells [31]. These fabric samples were incubated in a solution of canine albumin and glutaraldehyde. Because of glutaraldehyde, albumin molecules were adsorbed onto the surfaces in a cross-linked state. After seven days of culture, the number, nature, morphology, and organization of the endothelial cell population were assessed by scanning electron microscopy, histology, and cell counting. Based on this *in vitro* cytocompatibility assay, the authors concluded that cross-linked albumin lost its surface passivating effect since the cell population on untreated samples was higher than on albumin-treated samples. Contradictory results, however, were obtained by Kottke-Marchant et al., who used a similar method to coat albumin on Dacron® arterial prostheses [11]. The material was immersed in a solution of albumin and glutaraldehyde, followed by treatment with glycine to neutralize the free aldehyde groups. The material was then dried in graded ethanol solutions prior to freeze drying. Albumin coating was found to improve the short-term blood compatibility of Dacron in a recirculating *in vitro* perfusion system.

The surface passivation by physically adsorbed albumin is expected to be effective only for short periods of blood contact. The physically adsorbed albumin can be displaced by other blood proteins that have higher binding affinity to the surface. The albumin-surface affinity can be increased by the glow discharge treatment of biomaterials [32–34]. Although the retention of albumin can be increased by such surface modifications, use of biomaterials with physically adsorbed albumin for long-term contact with blood is debatable.

B.   Selective Adsorption of Albumin

An alternative to physical adsorption of albumin was proposed by Munro et al. [35]. Polyurethane tubes were grafted with C16 or C18 hydrocarbon chains for increasing albumin affinity to the surface. The rationale behind using this approach was the high affinity of albumin for the circulating free fatty acids. The hydrocarbon chains grafted to the surface were expected to mimic the nonpolar structure of saturated fatty acids and develop hydrophobic interaction with endogenous albumin. It was also speculated that, even though desorption of the albumin from such surfaces might occur, the grafted hydrocarbon layer would still remain on the surface, thus allowing formation of a renewable albumin layer between the polymer and blood. This hypothesis was supported by the *in vivo* experiments using canine arterial grafts [36]. Fibrin formation and platelet aggregation on C18-alkylated polyetherurethanes was inhibited for a time period of 7.5 hours (h).

Pitt and Cooper conducted a study in order to determine the molecular mechanisms promoting increased adsorption of albumin to such alkylated polymers [37]. According to the study, the initial adsorption rate of delipidized HSA was increased by the addition of C18 alkyl chains to polyurethane; this was attributed to the interaction between the albumin and the alkyl chains through the alkyl binding sites. The alkyl chains, however, did not influence the total amount of HSA adsorbed over a period of 1 h exposure to a 5 mg/ml HSA solution.

A similar approach was used by other investigators for improving blood compatibility

[38,39]. The study by Tsai et al. involved coating of a silicone rubber-based film on such surfaces as methylated polymers, glass, and metals to improve their biocompatibility [38]. The vinyl-methyl siloxane comonomer of the silicone rubber film was modified by either hydroxyl groups or C16 acyl groups. Kinetics, isotherm, and competitive protein adsorption studies suggested that both methods markedly improved the albumin affinity, but not the fibrinogen affinity, to the surface.

Although selective adsorption of albumin from blood to the biomaterial surface appears to be feasible and useful, the adsorbed albumin is still not covalently grafted. Thus, selective adsorption of albumin using hydrophobic chains has the same problem of displacement by other blood proteins as the physically adsorbed albumin. Thus, it is desirable to immobilize albumin to the surface through covalent bonding.

C. Covalent Grafting of Albumin by Chemical Reaction

Covalent grafting of albumin usually requires chemically active groups, such as amine, hydroxyl, or carboxyl groups, on the biomaterial surface. These groups can be introduced to the surface by plasma surface modification [32] or by chemical modifications of the polymer surface [40,41]. In the study by Guidoin et al., silicone substrates were activated with amino silanes to introduce reactive amino groups [40]. Albumin was linked to the amino groups on the surfaces through glutaraldehyde cross-linking. Two methods were used for the albumin grafting: (1) reaction of the surface amino groups with glutaraldehyde followed by cross-linking with albumin, and (2) reaction of the albumin solution containing glutaraldehyde with the surface amine groups. A drawback of the second treatment could be that albumin molecules may cross-link with themselves prior to binding to the desired surface. The silicone rings prepared in this manner were compared with the untreated rings by inserting them in the abdominal aorta and in the vena cava. The preliminary data indicated that coating of the silicone rings with cross-linked polyalbumin reduced kidney infarcts but was inefficient in preventing clotting in the venous flow. No significant qualitative differences were reported between the albumin-grafted silicone rings prepared by the two methods.

Hoffman et al. used ε-amino caproic acid as a spacer to attach albumin chemically to hydrogels grafted on polymer surfaces [41]. Silicone rubber films were immersed in a mixture of hydroxyethyl methacrylate (HEMA) and N-vinyl pyrrolidone (NVP) monomers and then exposed to 0.25 megarads (Mrad) of gamma irradiation. The OH groups of the poly(HEMA) chains in the grafted hydrogel copolymer were activated by reaction with BrCN followed by attachment of albumin. Albumin was grafted to the activated hydrogel surface by either direct chemical bonding to the surface or through spacers. The extent of albumin bonding was found to increase with the hydrogel water content or with the presence of the spacers.

The above-mentioned approaches for covalent grafting of albumin necessitate premodification of the surface to introduce reactive functional groups. It would be more desirable if albumin can be grafted without premodification of the surface. For this reason, methods were developed to graft albumin to the surface without chemically active functional groups.

D. Albumin Grafting by Ultraviolet Irradiation

Albumin can be covalently grafted to the surface without premodification of the surface if albumin is modified with reagents that are ultraviolet (UV) activatable. Matsuda and Inoue reported covalent grafting of phenylazide-modified albumin to surfaces by UV irradiation [42]. Tseng, Kim, and Park also used azide-derivatized albumin for photografting of albumin to polymeric surfaces [43]. 4-Azido-2-nitrophenyl albumin (ANP-albumin) was prepared by reacting albumin with 4-fluoro-3-nitrophenyl azide. Photolysis of the phenyl azide group of ANP-albumin by UV light yielded highly reactive triplet nitrene, which formed covalent bonding with the surface [43]. Thus, ANP-albumin can be covalently grafted to any polymer surface simply by exposure to UV light. Effects of the concentration of ANP-albumin, adsorption time, and UV irradiation time on the grafting efficiency were examined. The concentration of ANP-albumin at 0.05 mg/ml in the adsorption solution was enough to prevent platelet adhesion completely on dimethyl-dichlorosilane-treated glass (DDS-glass) (Fig. 2). The presence of multiazido groups on albumin resulted in an improved grafting efficiency.

E.   Albumin Grafting by Thermal Activation

Although albumin grafting by UV irradiation is simple and highly useful, it has a limitation that the surface has to be exposed to the light. Since UV light does not penetrate most materials, albumin grafting onto the surface of biomaterials inside of the assembled devices is not possible. To overcome such limitation, the ANP-albumin was grafted onto chemically inert materials such as polypropylene by thermal activation of the azide group [44]. The azide was activated to yield nitrene at 100°C. The minimum concentration of ANP-albumin in the adsorption solution required for the surface passivation was 5 mg/ml, which was higher than that required by the photoactivation approach. Unlike photolysis, however, the thermografting could be used for surfaces that are not exposed.

Platelets could not adhere on the albumin-grafted surface when albumin was adsorbed for 1 h and incubated at 100°C for 5 h (Fig. 3). Even though the conformation of albumin is expected to be altered by introduction of phenyl azides and heat treatment, the grafted albumin was effective in the prevention of platelet adhesion. Albumin does not have to maintain its tertiary structure to be effective in the prevention of surface-induced platelet activation.

F.   Albumin Grafting by Gamma Irradiation

Gamma irradiation has been used for the grafting of hydrophilic polymers to various substrates [45–47]. Gamma irradiation was also used to graft albumin covalently to the biomaterial surface. Albumin was first functionalized by reacting with glycidyl acrylate to introduce double bonds [48]. Preliminary studies in the grafting of albumin on DDS-glass provided a set of grafting conditions that resulted in the prevention of platelet activation. In this study, functionalized albumin (FA) was adsorbed to the DDS-glass surface and then exposed to gamma irradiation. For the adsorbed albumin to be grafted to the surface, it has to be in an aqueous solution. If the surface was dried after adsorption of FA and exposed to gamma irradiation in the dried state, the surface was not able to prevent platelet adhesion and activation.

In a study by Sharma and Kurian, albumin was adsorbed on polyetherurethane surfaces and then dried before gamma irradiation [49]. The surfaces prepared in this manner were not able to prevent platelet adhesion and activation. This indicates that either albumin could not be grafted or the amount of the grafted albumin was not high enough to prevent platelet activation.

The study of FA grafting on DDS-glass showed that the concentration of FA in adsorption solution and the gamma irradiation time were the two most important vari-

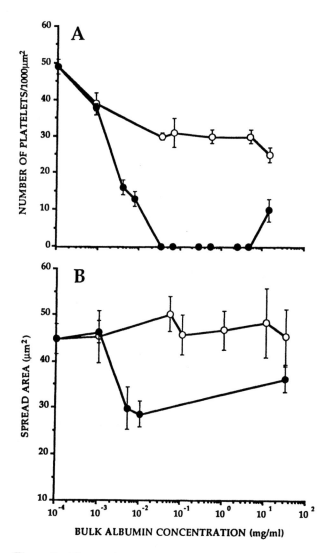

Figure 2   The number (A) and spread area (B) of platelets adherent on albumin-treated dimethyl dichlorosilane-treated glass (DDS-glass) (○) and albumin-grafted DDS-glass (●) as a function of the bulk albumin concentration used for adsorption. Albumin or 4-azido-2-nitrophenyl albumin (ANP-albumin) were adsorbed for 1 h and irradiated for 15 min. Platelets in platelet-rich plasma (PRP) were allowed to adhere for 1 h at room temperature. (From Ref. 43.)

ables that influenced the extent of platelet adhesion and activation. As the concentration of FA increased from 1 mg/ml to 30 mg/ml, platelet activation decreased progressively.

The effect of the gamma irradiation time on platelet activation, however, was not that simple as shown in Fig. 4. While the gamma irradiation time increased up to 4 h (at the dose rate of 0.094 Mrad/h), the platelet activation decreased, as indicated by a decrease in the spread area of adherent platelets. No platelets could adhere to the surface if the adsorbed FA was exposed to gamma irradiation for less than 8–10 h. The extent of platelet activation increased again if the gamma irradiation time exceeded 8–10 h. This

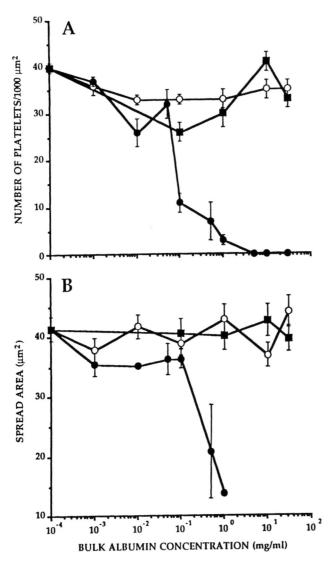

Figure 3   The number (A) and spread area (B) of platelets adherent on the surface treated with 4-azido-2-nitrophenyl albumin (ANP-albumin) at 100°C (●), ANP-albumin at room temperature (■), and albumin at 100°C (○) as a function of the bulk albumin concentration used for adsorption. Albumin or ANP-albumin was adsorbed for 1 h and incubated at 100°C for 5 h. Platelets in platelet-rich plasma (PRP) were allowed to adhere for 1 h at room temperature. (From Ref. 44.)

was attributed to the extensive intermolecular cross-linking induced at higher irradiation times.

Similar results were obtained when FA was grafted on polymers such as polypropylene, polycarbonate, or poly(vinyl chloride) by gamma irradiation [50]. When the FA-grafted polypropylene fibers were exposed to blood, only about 5% of the grafted albu-

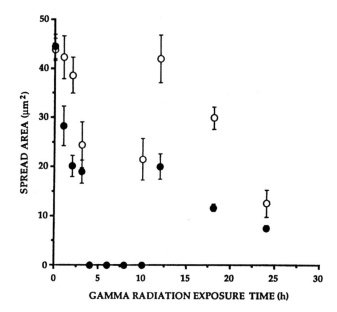

Figure 4 Effect of gamma radiation exposure time on the spread area of platelets adherent on dimethyl dichlorosilane glass (DDS-glass). The gamma irradiation dose was 0.094 Mrads/h. Functionalized albumin (FA) at the bulk concentration of 30 mg/ml was adsorbed on DDS-glass for 30 min. DDS-glass was gamma irradiated in the presence of FA in bulk solution (○) or in the presence of phosphate buffered saline (PBS) after the unadsorbed FA was removed (●). (From Ref. 48.)

min was released from the surface after 6 h of exposure. On the other hand, about 70% of the adsorbed albumin was removed from the surface during the same time period. Albumin grafting on blood oxygenators using this approach brought about almost 70% reduction in platelet adhesion and thrombus formation as compared with the control oxygenators.

III. MECHANISMS OF SURFACE PASSIVATION BY THE GRAFTED ALBUMIN

Although the prevention of surface-induced platelet activation by albumin on the surface is well known, the beneficial effect of albumin is not always realized. The presence of albumin on the surface is not enough to prevent platelet activation. The immobilized albumin should meet the following three requirements to be effective. First, albumin has to be tightly bound to the surface, if not covalently grafted. Second, albumin must cover most of the surface so that no significant portion of the surface is exposed to platelets. Third, the immobilized albumin should maintain its flexibility. These three requirements are critical to render the surface platelet-resistant.

The tight binding of albumin to the surface is important since otherwise albumin can be removed from the surface. Albumin adsorbs weakly to hydrophilic surfaces such as clean glass. Platelets were found to adhere and fully activate on albumin-adsorbed glass even when albumin was present in the bulk solution at a concentration as high as 50 mg/

ml [51]. Platelets can easily remove weakly adsorbed albumin from the surface to interact directly with the surface. In the presence of other proteins, such as in blood, weakly adsorbed albumin can be easily displaced by other proteins and cells to create a surface favorable for platelet activation. On hydrophobic surfaces, however, albumin adsorbs with high affinity, almost irreversibly [14,15]. When a monolayer (0.24 micrograms per square centimeter [μg/cm^2]) of albumin was adsorbed to DDS-glass, platelets could not adhere to the surface. Even if platelets adhered to the surface, they remained contact adherent without any further activation [52]. Although albumin can be tightly adsorbed to the hydrophobic surface, it can be displaced from the surface by other proteins with stronger affinity to the surface if a sufficiently long time is provided. Thus, covalent grafting of albumin to the surface is the most preferred.

The surface coverage by the immobilized albumin has to be such that no significant portion of the surface is exposed to platelets. Other studies showed that platelets could adhere and fully activate on DDS-glass even when only 2–15% of the surface was covered with fibrinogen while the remaining surface was coated with albumin [53,54]. The complete surface coverage by albumin is important. If the surface albumin concentration is far below the monolayer coverage, the presence of albumin on the surface has little influence in the prevention of platelet activation. This is why comparing the surface albumin concentration alone fails to rank order thrombogenicity of different biomaterials.

If the immobilized albumin is cross-linked with either glutaraldehyde or gamma irradiation as described above, albumin can no longer prevent platelet activation [54]. This happens despite the fact that the surface albumin concentration does not change by cross-linking. The effect of the cross-linking on the flexibility of the immobilized albumin needs more systematic study. However, it is expected that the cross-linking makes albumin molecules less flexible so that the albumin layer becomes another surface for platelet adhesion and activation.

Basically, albumin has to have high affinity to the surface, cover the surface completely, and maintain its flexibility. These three requirements are the same as those for the stabilization of colloidal particles, which are stabilized by the steric repulsion mechanism.

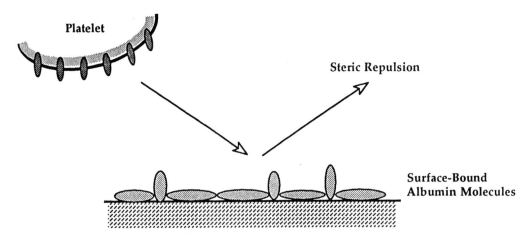

Figure 5   Steric repulsion of platelet adhesion by the surface-adsorbed albumin. (From Ref. 56.)

Steric repulsion occurs due to a loss in configurational entropy resulting from volume restriction and/or the osmotic repulsion between two interdigitating macromolecular layers. Surface-bound water-soluble polymers such as poly(ethylene oxide) have been very effective in steric repulsion of proteins and cells from biomaterial surfaces [55]. Albumin on the surface also prevents the adsorption of other proteins and the adhesion of platelets to the surface by the same steric repulsion mechanism. It is important to note that albumin is not a rigid object, but a flexible molecule that is able to exert steric repulsion.

The similarity between the colloidal stabilization and the prevention of platelet adhesion by the surface-adsorbed albumin was first noted by Morrissey [56]. Recent study using the surface-force apparatus showed that the surface-adsorbed albumin has significant steric repulsion forces [57]. The condition that causes albumin to prevent aggregation of colloidal particles also prevents the adhesion of platelets [58–60]. An artistic description of the steric repulsion by the immobilized albumin is described in Fig. 5. The steric repulsion mechanism is very helpful in optimizing the condition for albumin grafting and in understanding why certain albumin grafting techniques work better than the others.

REFERENCES

1. A. S. Hoffman. Synthetic polymeric biomaterials. In C. G. Gebelein (ed.), *Polymeric Materials and Artificial Organs. ACS Symposium Series*, Vol. 256. Am. Chem. Soc., Washington, DC, 1984, pp. 13–29.
2. M. Szycher and W. J. Robinson (eds.). *Synthetic Biomedical Polymers: Concepts and Applications*. Technomic, Westport, CT, 1980.
3. A. S. Hoffman. Blood-biomaterial interactions: An overview. In S. L. Copper and N. A. Peppas (eds.), *Biomaterials: Interfacial Phenomena and Applications*, Vol. 199. Am. Chem. Soc., Washington, DC, 1982, pp. 3–8.
4. J. D. Andrade, S. Nagaoka, S. L. Cooper, T. Okano, and S. W. Kim. Surfaces and blood compatibility. Current hypothesis. *Trans. Am. Soc. Artif. Intern. Organs* 33:75–84 (1987).
5. K. Park and S. L. Cooper. Importance of composition of the initial protein layer and platelet spreading in acute surface-induced thrombosis. *Trans. Am. Soc. Artif. Intern. Organs* 31: 483–488 (1985).
6. B. R. Young, L. K. Lambrecht, and S. L. Cooper. Plasma proteins: Their role in initiating platelet and fibrin deposition on biomaterials. In S. L. Copper and N. A. Peppas (eds.), *Biomaterials: Interfacial Phenomena and Applications*, Vol. 199. Am. Chem. Soc., Washington, DC, 1982, pp. 317–350.
7. J. D. Andrade and V. Hlady. Protein adsorption and material biocompatibility: A tutorial review and suggested hypothesis. *Adv. Polymer Sci.* 79:1–63 (1986).
8. J. M. Anderson and K. Kottke-Merchant. Platelet interaction with biomaterials and artificial devices. *CRC Crit. Revs. Biocomp.* 1:111–204 (1985).
9. J. S. Shultz, S. M. Lindenauer, and J. A. Penner. Thrombus formation on surfaces in contact with blood. In S. L. Cooper and N. A. Peppas (eds.), *Biomaterials: Interfacial Phenomena and Applications*, Vol. 199. Am. Chem. Soc., Washington, DC, 1982, pp. 43–58.
10. S. W. Kim and J. Feijen. Surface modification of polymers for improved blood compatibility. *CRC Crit. Revs. Biocomp.* 1:229–260 (1985).
11. K. Kottke-Marchant, J. M. Anderson, Y. Umemura, and R. E. Marchant. Effect of albumin coating on the *in vitro* blood compatibility of Dacron arterial prostheses. *Biomaterials* 10: 147–155 (1989).
12. K. Park, D. F. Mosher, and S. L. Cooper. Acute surface-induced thrombosis in the canine *ex vivo* model: Importance of protein composition of the initial monolayer and platelet activation. *J. Biomed. Mater. Res.* 20:589–612 (1986).

13. J. N. Mulvihill, A. Faradji, F. Oberling, and J. P. Cazenave. Surface passivation by human serum albumin of plasmapheresis circuits reduces platelet accumulation and thrombus formation. Experimental and clinical studies. *J. Biomed. Mater. Res.* 24:155-163 (1990).

14. E. Brynda, M. Houska, Z. Pokorna, N. A. Cepalova, Y. V. Moiseev, and J. Kalal. Irreversible adsorption of human serum albumin onto polyethylene film. *J. Bioeng.* 2:411-418 (1978).

15. S. H. Lee and E. Ruckenstein. Adsorption of proteins onto polymeric surfaces with different hydrophilicities: A case study with bovine serum albumin. *J. Colloid Interface Sci.* 125:365-379 (1988).

16. T. Peters, Jr. Serum albumin. In F. W. Putnam (ed.), *The Plasma Proteins: Structure, Function, and Genetic Control*, Vol. 1. Academic Press, New York, 1975, pp. 133-181.

17. J. Belleville and R. Eloy. Guidelines for selecting the proper animal species for a particular test. In S. Dawids (ed.), *Test Procedures for the Blood Compatibility of Biomaterials*. Kluwer Academic Press, Dordrecht, The Netherlands, 1993, pp. 13-34.

18. T. Peters, Jr. Serum albumin. *Adv. Protein Chem.* 37:161-245 (1985).

19. W. Norde. Energy and entropy of protein adsorption. *J. Disp. Sci. and Technology* 13:363-377 (1992).

20. S. D. Bruck. Fundamental phenomena. In *Properties of Biomaterials in the Physiological Environment*. CRC Press, Boca Raton, FL, 1980, pp. 1-21.

21. P. N. Sawyer. Electrode-biologic tissue interactions at interfaces: A review. *Biomat. Med. Dev. Art. Org.* 12:161-196 (1984-1985).

22. H. L. Nossell, G. D. Wilneer, and E. C. LeRoy. Importance of polar groups for initiating blood coagulation and aggregating platelets. *Nature* 221:75-76 (1969).

23. C. P. Sharma. Surface modification: Blood compatibility of small diameter vascular graft. In C. P. Sharma and M. Szycher (eds.), *Blood Compatible Materials and Devices*. Technomic, Lancaster-Basel, 1991, pp. 25-31.

24. J. D. Andrade, H. B. Lee, M. S. John, S. W. Kim, and J. B. Hibbs, Jr. Water as a biomaterial. *Trans. Am. Soc. Artif. Intern. Organs* 19:1-7 (1973).

25. N. Larsson, L.-E. Linder, I. Curelaru, P. Buscemi, R. Sherman, and E. Eriksson. Initial platelet adhesion and platelet shape on polymer surfaces with different carbon bonding characteristics (an *in vitro* study of Teflon, Pellethane and XLON intravenous cannulae). *J. Mater. Sci., Materials in Medicine* 1:157-162 (1990).

26. D. J. Lyman. Bulk and surface effects on blood compatibility. *J. Bioact. Compat. Polymers* 6:283-295 (1991).

27. M. Kumakura, M. Yoshida, and M. Asano. Adsorption of γ-globulin on polymer surfaces having various chemical and physical structures. *J. Appl. Poly. Sci.* 41:177-184 (1990).

28. E. Wilkins and W. Radford. Biomaterials for implanted closed loop insulin delivery system: A Review. *Biosensors and Bioelectronics* 5:167-213 (1990).

29. M. A. Packham, G. Evans, M. F. Glynn, and J. F. Mustard. The effect of plasma proteins on the interaction of platelets with glass surfaces. *J. Lab. Clin. Med.* 73:686-697 (1969).

30. T. M. S. Chang. Platelet-surface interactions: Effect of albumin coating and heparin complexing on thrombogenic surfaces. *Can. J. Physiol. Pharmacol.* 52:275-285 (1974).

31. M. F. Sigot-Luizard, D. Domurado, M. Sigot, R. Guidion, C. Gosselin, M. Marois, J. F. Girard, M. King, and B. Badour. Cytocompatibility of albuminated polyester. *J. Biomed. Mater. Res.* 18:895-909 (1984).

32. R. Sipehia, A. S. Chawla, and T. M. S. Chang. Enhanced albumin binding to polypropylene beads via anhydrous ammonia gaseous plasma. *Biomaterials* 7:471-473 (1986).

33. R. Sipehia and A. S. Chawla. Albuminated polymer surfaces for biomedical application. *Biomat., Med. Dev., Artif. Organs* 10:229-246 (1982).

34. D. Kiaei, T. A. Horbett, and A. S. Hoffman. Albumin retention by glow discharge deposited polymers. *Artif. Organs* 15:302 (1991).

35. M. S. Munro, A. J. Quattrone, S. R. Ellsworth, P. Kulkarni, and R. C. Eberhart. Alkyl

substituted polymers with enhanced albumin affinity. *Trans. Am. Soc. Artif. Intern. Organs* 27:499–503 (1981).

36. R. C. Eberhart, M. S. Munro, J. R. Frautschi, M. Lubin, F. J. Clubb, Jr., C. W. Miller, and V. I. Sevastianov. Influence of endogenous albumin binding on blood material interactions. *Ann. NY. Acad. Sci.* 516:78–95 (1987).
37. W. G. Pitt and S. L. Cooper. Albumin adsorption on alkyl chain derivatized polyurethanes: I. The effect of C-18 alkylation. *J. Biomed. Mater. Res.* 22:359–382 (1988).
38. C.-C. Tsai, H.-H. Huo, P. Kulkarni, and R. C. Eberhart. Biocompatible coatings with high albumin affinity. *Trans. Am. Soc. Artif. Intern. Organs* 36:M307–M310 (1990).
39. I. Strzinar, A. Duncan, J. L. Brash, and M. V. Sefton. Surface alkylation of a polyvinyl alcohol hydrogel. *Trans. of the Society for Biomaterials* 14:43 (1991).
40. R. C. Guidoin, J. Awad, A. Brassard, D. Domurado, F. Lawny, J. Wetzer, J.-N. Barbotin, C. Calvot, and G. Broun. Blood compatibility of silicone rubber chemically coated with cross-linked albumin. *Biomat., Med. Dev., Artif. Organs* 4:205–224 (1976).
41. A. S. Hoffman, G. Schmer, C. Harris, and W. G. Kraft. Covalent bonding of biomolecules to radiation grafted hydrogels on inert polymer surfaces. *Trans. Am. Soc. Artif. Intern. Organs* 18:10–17 (1972).
42. T. Matsuda and K. Inoue. Novel photoreactive surface modification technology for fabricated devices. *Trans. Am. Soc. Artif. Intern. Organs* 36:161–164 (1990).
43. Y.-C. Tseng, J. Kim, and K. Park. Photografting of albumin onto dimethyldichlorosilane-coated glass. *J. Biomat. Appl.* 7:233–247 (1993).
44. Y.-C. Tseng, W. M. Mullins, and K. Park. Albumin grafting on to polypropylene by thermal activation. *Biomaterials* 14:392–400 (1993).
45. J. Singh, A. R. Ray, J. P. Singhal, and H. Singh. Radiation-induced grafting of methacrylic acid on to poly(vinyl chloride) films and their thrombogenicity. *Biomaterials* 11:473–476 (1990).
46. B. D. Ratner and A. S. Hoffman. Surface-grafted polymers for biomedical applications. In M. Szycher and W. J. Robinson (eds.), *Synthetic Biomedical Polymers: Concepts and Applications*. Technomic, Westport, CT, 1980, pp. 133–151.
47. P. R. Hari and C. P. Sharma. Hydrogel grafted surfaces: Protein interaction and platelet adhesion. *J. Biomat. Appl.* 6:170–180 (1991).
48. K. R. Kamath, H. Park, H. S. Shim, and K. Park. Albumin grafting on dimethyldichlorosilane-coated glass by gamma-irradiation. *Colloids and Surfaces B.* 2:471–479 (1994).
49. C. P. Sharma and G. Kurian. Radiation-induced albuminated surfaces: Their modifications towards blood compatibility. *J. Colloid Interface Sci.* 97:38–40 (1984).
50. K. R. Kamath and K. Park. Surface modification of polymeric biomaterials by albumin grafting using gamma-irradiation. *J. Appl. Biomat.* 5:163–173 (1994).
51. K. Park and H. Park. Application of video-enhanced interference reflection microscopy to the study of platelet-surface interactions. *Scanning Microscopy* Supp. 3:137–146 (1989).
52. K. Park, F. W. Mao, and H. Park. Morphological characterization of surface-induced platelet activation. *Biomaterials* 11:24–31 (1990).
53. K. Park, F. W. Mao, and H. Park. The minimum surface fibrinogen concentration necessary for platelet activation on dimethyldichlorosilane-coated glass. *J. Biomed. Mater. Res.* 23:407–420 (1991).
54. M. Amiji, H. Park, and K. Park. Study on the prevention of surface-induced platelet activation by albumin coating. *J. Biomat. Sci., Polymer Edn.* 3:375–388 (1992).
55. M. Amiji and K. Park. Prevention of protein adsorption and platelet adhesion on surfaces with PEO/PPO/PEO triblock copolymers. *Biomaterials* 13:682–692 (1992).
56. B. W. Morrissey. The adsorption and conformation of plasma proteins: A physical approach. *Ann. N.Y. Acad. Sci.* 283:50–64 (1977).
57. E. Blomberg, P. M. Claesson, and C. G. Golander. Adsorbed layers of human serum albumin investigated by the surface force technique. *Disp. Sci. Tech.* 12:179–200 (1991).

58. K. Park. Factors affecting efficiency of colloidal gold staining: pH-dependent stability of protein-gold conjugates. *Scanning Microscopy* Supp. 3:15–25 (1989).

59. H. Tamai, A. Fujii, and T. Suzawa. Colloidal stability of polymer lattices coated with bovine serum albumin. *J. Colloid Interface Sci.* 118:176–181 (1987).

60. A. van der Scheer, M. A. Tanke, and C. A. Smolders. Influence of adsorbed proteins on the stability of polystyrene latex particles. *Chem. Soc. Faraday Discussion* 65:264–287 (1978).

39

Negative Cilia Concept for Thromboresistance

Young Ha Kim, Dong Keun Han, and Ki Dong Park
Korea Institute of Science and Technology
Seoul, Korea

I. INTRODUCTION

A. Hypothesis in Blood Compatibility

Polymeric materials have contributed significantly to the development and improvement of devices and systems in artificial organs. Polyurethane (PU) has been widely used for blood-contacting devices due to its suitable blood compatible and mechanical properties. However, the inherent blood compatibility of PU is not acceptable for more widespread application of such devices. In addition, *in vitro* and *in vivo* degradation and calcification of PU cause a problem for its performance for long-term implantation. Therefore, the need for the creation of highly blood-compatible and biostable materials has been increasing. Although a substantial amount of work in the improvement of blood compatibility of polymeric materials has been carried out, the results are still not conclusive. This is partly because of the fact that the relationship between surface properties of biomaterials and surface-induced thrombosis has not been thoroughly evaluated.

In the past, researchers have described the mechanisms of thrombus formation at foreign interfaces in terms of surface-related factors and hemodynamic effects. Although some bulk properties, such as mechanical compliance and water absorption, are considered to be important, the major material factors influencing blood interactions at polymer interfaces are surface properties that govern the initial events of thrombus formation and subsequent thrombogenicity [1,2].

A variety of approaches has been taken to improve the blood compatibility of polymer surfaces. Approaches involve (1) biological modification using protein or cell seeding [3,4]; (2) utilization of such bioactive agents as potent anticoagulant (heparin and hirudin) [5–7] and fibrolytic enzymes (urokinase and streptokinase) [8,9], through either chemical immobilization or controlled release; and (3) chemical modification by plasma or chemical treatment [2].

The current hypotheses were explained in the literature [2,10]. Many hydrogels or hydrophilized surfaces exhibit good blood compatibility. The grafting of hydrophilic flexible poly(ethylene oxide) (PEO) was reported to be more blood compatible due to an additional chain motion effect. However, very hydrophobic inert surfaces such as fluoropolymers or silicones are also known to be hemocompatible. Hydrophobic/hydrophilic microdomain structures such as PU or polystyrene/poly(hydroxyethylmethacrylate) (PS/PHEMA) block copolymer were also proven to be more blood compatible. All the modified structures above have more or less evidence that are sometimes controversial, especially with respect to hydrophilicity or hydrophobicity. Such phenomena would result partially from the fact that the thrombus formation mechanism was not completely explained.

It was also reported that a negatively charged surface was more blood compatible than a positive one [11]. Especially, many polymers containing sulfonate groups showed very good blood compatibility. Heparin is a clinically useful anticoagulant and the sulfonate group is reported to be responsible for its biological activity. Recently, Ishihara et al. described that a biomembrane-like surface composed of polymer and phospholipids showed excellent blood compatibility [12].

B. Negative Cilia Model for Blood Compatibility

In the authors' laboratory, a new approach to improve the blood compatibility, biostability, and anticalcification of polymers has been developed. This approach was based on the following concepts. First, the hydrophilic environment of the blood-material interface appears to reduce protein adsorption and platelet adhesion and activation and can be achieved by the grafting of hydrophilic polymers such as PEO. The advantages of using PEO include low interfacial free energy [13], its nonadhesive property, highly dynamic motions, and extended chain conformation at the blood-material interface [14,15]. Second, sulfonated polymers have shown anticoagulant activity like heparin [16,17].

We have grafted sulfonated PEO onto PU (PU-PEO-SO$_3$) to induce a synergistic effect of PEO and sulfonate groups (Fig. 1). In this sulfonated PEO-grafted PU system, termed the *negative cilia model* [18], several individual hypotheses described in the section above seem to be combined.

The hypothesis conceived with this research involves a synergistic effect of PEO and sulfonate groups. Hydrophilic PEO chains are expected to reduce protein adsorption and platelet adhesion due to the unique behaviors of PEO described above. The pendant negatively charged sulfonate group expels blood components further by electrical repulsion and also provides anticoagulant activity. The introduction of negative sulfonate groups at the end of PEO would increase the vertical orientation of PEO chains on the surface due to the electrical repulsions of chains for each other. This would maximize the anticoagulant activity of sulfonate groups. In addition, it might be taken account that the water structure at the interface may be changed to a great extent due to the hydrophilic PEO grafting and the negative ions introduced, and that PU-PEO-SO$_3$ showed the most hydrophilic and smoothest surface. Such effects might contribute to the improved antithrombogenicity.

In order to investigate a synergistic effect of PEO, the sulfonate groups were incorporated directly onto PU or at the end of semiflexible hydrophobic hydrocarbon chains and compared. The grafting of sulfonated PEO can be carried out by a direct surface modification of medical devices (see Section II.A) or by a solution reaction for the preparation as a coating material (see Section II.B).

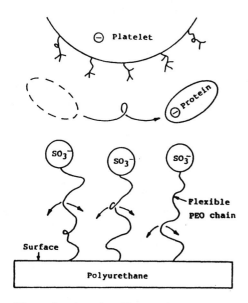

Figure 1 Negative cilia model on PU-PEO-SO₃.

II. CHEMICAL MODIFICATION OF POLYURETHANE

A. *In Situ* Surface Poly(Ethylene Oxide) Grafting and Sulfonation

Surface PEO grafting and sulfonation were performed onto polyurethane (Pellethane®) as shown in Fig. 2 [19]. The synthetic scheme involved the coupling of hexamethylene diisocyanate (HMDI) to PU through an allophanate reaction. The free isocyanate group remaining on the surface was then coupled to terminal hydroxyl end groups on PEO of different molecular weights (200, 1000, and 2000) to obtain PEO-grafted PU (PU-PEO). The free hydroxyl group on PU-PEO was finally sulfonated using propane sultone as the sulfonating agent. Sulfonation was also carried out at the hydroxyl end groups on dodecanediol (DDO) grafted or directly on PU to compare the PEO effect. PU beads and tubings grafted with PEO and PEO-SO₃ were prepared using a similar procedure and

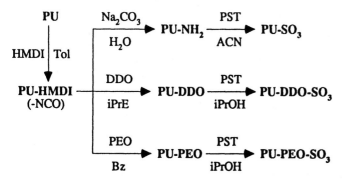

Figure 2 Surface modification scheme of polyurethanes (PUs); HMDI = $OCN(CH_2)_6NCO$, PST = $(CH_2)_3SO_3$, DDO = $HO(CH_2)_{12}OH$, PEO = $HO(CH_2CH_2O)_nH$.

rinsed thoroughly to remove unreacted materials. Surface-modified beads and tubings were used in *in vitro* and *ex vivo* experiments, respectively.

Surface properties of modified PUs were extensively characterized: surface compositions by attenuated total reflectance Fourier transform infrared (ATR-FTIR) and electron spectroscopy for chemical analysis (ESCA), surface morphology by scanning electron microscopy (SEM), and surface energetics by dynamic contact angle measurements [19]. Surface morphology by SEM revealed that PEO and PEO-SO$_3$ grafting improved the surface smoothness to a great extent, as compared with controls. In addition, increased hydrophilicity and contact angle hysteresis (surface reorientation of PEO-SO$_3$ in air/water interface) of surface-modified PUs were confirmed by contact angle measurement [19].

B. Preparation of PU-PEO-SO$_3$ Graft Copolymer

PU-PEO-SO$_3$ graft copolymer was synthesized by a modified procedure of *in situ* surface grafting (see Fig. 3) [20]. The procedure involved the coupling of HMDI to PEO-SO$_3$, which was prepared by sulfonation of diamino-terminated PEO1000 using propane sultone. The free isocyante groups of OCN-PEO-SO$_3$ were then coupled to PU dissolved in dimethylacetamide by an allophanate reaction. Synthetic intermediate and copolymer were confirmed by Fourier transform infrared (FTIR) spectroscopy, nuclear magnetic resonance (NMR), and sulfur analysis. The degree of substitution evaluated by sulfur content in PU-PEO-SO$_3$ was about 7%. Physical characterization methods, such as differential scanning calorimetry (DSC) thermal analysis, mechanical properties, and water uptake, were used to detail the properties of these graft copolymers containing the PEO-SO$_3$ [21].

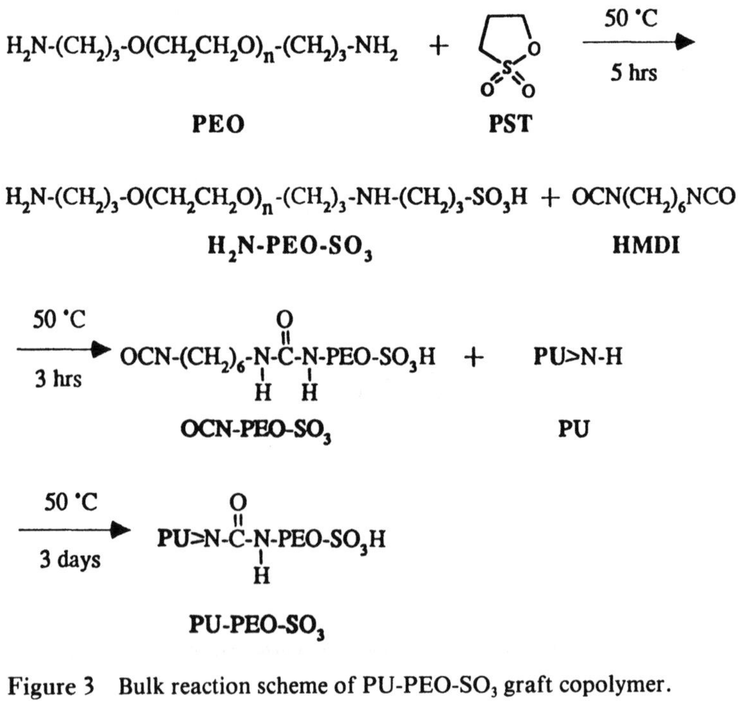

Figure 3 Bulk reaction scheme of PU-PEO-SO$_3$ graft copolymer.

III. RESULTS AND DISCUSSION

A. *In Vitro* Studies

1. *Protein Adsorption*

Protein adsorption occurs instantly on contact with blood. Therefore, the adsorption and subsequent deformation of proteins are regarded as playing an important role in the initial stage of thrombus formation. It is generally understood that fibrinogen is surface active to induce thrombus, but albumin is passive in regard to antithrombogenicity. In fact, in many studies on albuminized surfaces, they were reported to be blood compatible.

Protein adsorption to modified PU surfaces has been investigated [21,22]. Protein adsorption was measured from plasma to which ^{14}C-labeled proteins had been added. Figure 4 shows the adsorption of proteins after 5 minutes (min) incubation with a varied plasma dilution range. On untreated PU, the most adsorbed protein was albumin, and fibrinogen and immunoglobulin G (IgG) were next.

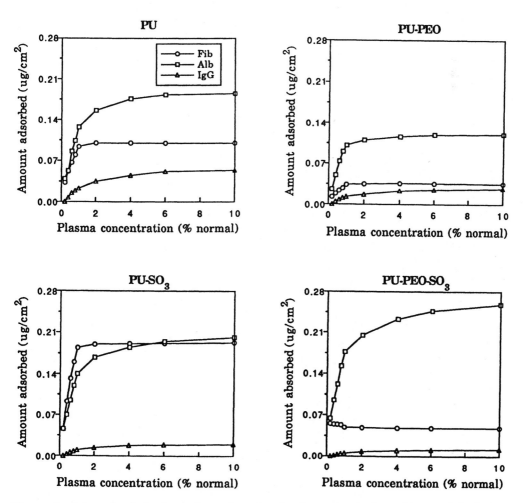

Figure 4 Adsorption behaviors of proteins after 5 minutes incubation with a varied plasma dilution range.

PU-SO$_3$ showed much higher fibrinogen adsorption than the relatively hydrophobic PU control, although PU-SO$_3$ is more hydrophilic than PU. This can be ascribed to the high affinity of the sulfonate group to fibrinogen. These results are consistent with other reports [16,23], in which sulfonated polyurethanes were reported to adsorb much more fibrinogen than the control PU and fibrinogen adsorption increased as the level of sulfonate incorporation increased.

Hydrophilic PU-PEO showed less protein adsorption when fibrinogen adsorption was decreased in a larger amount than albumin adsorption was decreased. In general, it is believed that PEO surfaces reduce protein adsorption and platelet adhesion due to PEO's low interfacial free energy [13], nonadhesive property, and the dynamic motions of PEO chains [14].

PU-PEO-SO$_3$ showed the most unique protein adsorption behavior. Fibrinogen adsorption was decreased substantially, as in the case of PU-PEO; however, the adsorption of albumin was rather increased. Less fibrinogen adsorption of PU-PEO-SO$_3$ than PU-SO$_3$ can be explained by the use of chain-grafted PEO. The protein resistance characteristics of PEO chains overcome the affinity of fibrinogen to a negatively charged SO$_3$ group. PU-PEO-SO$_3$ was the most hydrophilic surface by the contact angle measurements. Nevertheless, the albumin adsorption on PU-PEO-SO$_3$ was the highest, greater than untreated PU. This would indicate a kind of specific affinity of PEO-SO$_3$ to albumin and should be investigated further.

2. Platelet Adhesion

Results from platelet adhesion studies from platelet-rich plasma (PRP) onto modified PU surfaces are shown in Fig. 5. PU-PEO and PU-PEO-SO$_3$ demonstrated lower platelet adhesion than the PU control, suggesting that PEO-modified surfaces, in general, pacified platelet adhesion. These results correlate with the interfacial free energy concept in which a decrease in interfacial free energy decreases protein adsorption, causing lower platelet adhesion and activation. In particular, less platelet adhered on PU-PEO-SO$_3$ compared with PU-PEO. This result is correlated with other reports [24–26] that PU containing sulfonate groups showed less platelet adhesion than a PU control. Such a suppressed interaction of sulfonated polymers with platelets can be partially explained by electrical repulsion between the negatively charged sulfonate group and negatively charged platelets. Moreover, the surface sulfonate group of PU-PEO-SO$_3$ has a high affinity to albumin to contribute to the decrease of platelet adhesion.

Platelet adhesion on the PU-PEO surfaces decreased with increasing PEO chain length (Fig. 6), indicating a specific chain length effect on platelet adhesion [21]. This is consistent with work by Mori et al. [14], who reported a decrease in platelet adhesion with increasing PEO chain length up to 2500. PU-PEO-SO$_3$ surfaces also demonstrated a chain length effect. The decrease in platelets on PU-PEO-SO$_3$ might be ascribed to the synergistic effect of hydrophilic PEO and the negatively charged SO$_3$ group.

A scanning electron micrograph of platelets attached to modified PU surfaces is consistent with a platelet count result as described above. PU-PEO, and especially PU-PEO-SO$_3$, showed less platelet attachment and minimal pseudopodia extension compared with PU.

3. Anticoagulant Activity

Table 1 summarizes the blood clotting times of various modified PUs. In general, the PU-PEO surface demonstrated almost the same activated partial thromboplastin time (APTT) value as PU and pooled plasma. However, an increased APTT was observed in

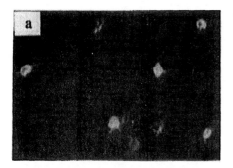

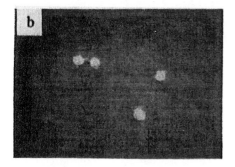

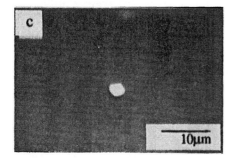

Figure 5 Scanning electron microscopy (SEM) micrographs of *in vitro* platelet adhesion on modified polyurethane (PU) surfaces: (a) PU; (b) PU-poly(ethylene oxide) (PU-PEO); (c) PU-PEO-SO$_3$.

all the sulfonated PUs as compared with the control. This may be explained by a specific anticoagulant activity of SO$_3$ groups, as discussed for other sulfonated polymers [16,17]. It was also reported by other researchers that polymers bearing sulfonate groups prolonged the blood coagulation times [16,27]. Heparin-like anticoagulant activity of SO$_3$ groups is known to be dependent on SO$_3$ concentrations. Although the exact concentrations of sulfonate groups have not been determined, from the sulfur atomic percent from ESCA measurements of sulfonated PUs (1.8% for PU-SO$_3$, 3.1% for PU-PEO200-SO$_3$, 2.2% for PU-PEO1000-SO$_3$, and 1.5% for PU-PEO2000-SO$_3$), it can be seen that these sulfur (SO$_3$) concentrations on the surface corresponded well with the APTT values of sulfonated PUs.

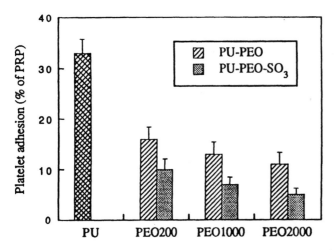

Figure 6 Effect of chain length of poly(ethylene oxide) (PEO) on platelet adhesion after 3 h incubation ($n = 3$).

More detailed studies were performed to measure the anticoagulant activity of PU-PEO-SO$_3$ prepared by a solution reaction [28]. Blocked PEO-SO$_3$ was prepared by reacting the free isocyanate group of OCN-PEO-SO$_3$ with methanol and compared with PU-PEO-SO$_3$. Anticoagulant activities of test materials were measured by APTT, thrombin time (TT), reptilase time measurements, factor Xa assay, and thrombin inhibition assay.

Table 2 summarizes the anticoagulant activity of sulfonated polymers. APTT measurement revealed that the bioactivities of blocked PEO-SO$_3$ and PU-PEO-SO$_3$ show 14% and 2% of free heparin, respectively. In contrast, PEO1000 showed no significant

Table 1 Thromboresistance Data[a] of Modified Polyurethanes

Material	*In vitro* APTT[b] (s)	*Ex vivo* Rabbit A-A shunt occlusion time (min)
PU	35.8 ± 0.2	50 ± 5
PU-DDO	36.2 ± 1.0	70 ± 10
PU-PEO200	33.1 ± 0.5[c]	120 ± 15[c]
PU-PEO1000	34.5 ± 0.6[c]	140 ± 15[c]
PU-PEO2000	35.5 ± 0.8	145 ± 15
PU-SO$_3$	41.9 ± 1.5[c]	90 ± 5[c]
PU-DDO-SO$_3$	40.5 ± 1.2[c]	200 ± 15[c]
PU-PEO200-SO$_3$	49.7 ± 2.5[c]	350 ± 30[c]
PU-PEO1000-SO$_3$	45.5 ± 2.0[c]	360 ± 30[c]
PU-PEO2000-SO$_3$	41.8 ± 1.4[c]	370 ± 30[c]

[a]Mean ± SD ($n = 3$); significance level using an unpaired Student's t test when comparing modified PUs to PU
[b]APTT of pooled plasma was 36.0 s
[c]$p < 0.05$

Table 2 Anticoagulant Activity[a] of Sulfonated Polymers

Polymer	Bioactivity (%)		
	APTT	TT	FXa
Blocked PEO-SO$_3$	14	11	<0.25
PU-PEO-SO$_3$	2.0	1.5	<0.20

[a]Anticoagulant ratio of sulfonated materials to free heparin

bioactivity in the blood clotting system [28]. This result is consistent with other reports that the anticoagulant activity of sulfonated polymers ranged from about 1% to 16% of heparin [29,30]. TT assay showed activity similar to APTT, even though TT measurement demonstrated less value than the APTT test. In addition, factor Xa assay did not show any significant increase in the prolongation of clotting time, suggesting that sulfonated polymers show anticoagulant activity mainly by thrombin inhibition rather than factor Xa inhibition. Similar results have been reported by some investigators that the sulfonated polymers promote the inhibition of either only thrombin [31] or factor Xa [32].

To investigate the mechanism of thrombin inhibition of sulfonated polymers, TT was measured in the presence of antithrombin III (AT III) as well as in its absence. Figure 7 shows thrombin times of blocked PEO-SO$_3$ and PU-PEO-SO$_3$. TT was prolonged to a greater extent in the presence of AT III than in the absence of AT III as the concentration of SO$_3$ groups was increased. This result suggests that AT III contributed to the increased TT in the presence of the sulfonated polymers, which is similar to heparin's action. Heparin catalytically potentiates the action of AT III to inactivate thrombin, a critical process in suppressing the intrinsic blood coagulation cascade. It was also observed that reptilase time was not dependent on the presence of PEO-SO$_3$ or PU-PEO-SO$_3$, which demonstrated a typical heparin-like activity. Our negatively charged sulfonated polymers might act in a similar manner as heparin in thrombin inhibition.

B. *Ex Vivo* Blood Compatibility Evaluation

Nonthrombogenicity of modified polymers in whole blood was evaluated *ex vivo* by the rabbit arterio-arterial (A-A) shunt method [33]. The procedure involved using surface-modified tubing (2.0 mm [millimeters] outside diameter [OD] × 1.5 mm inside diameter [ID], 30 cm in length) as a shunt in the carotid arteries of male rabbits. The experiment measured the time needed for the formation of a stable, nonembolized thrombus large enough to occlude the blood flow in the tube, which was referred to this time as the occlusion time. The shunt flow was maintained at 2.5 ml/min to minimize nonlaminar flow effects through the experiment.

As listed in Table 1 [18], all modified surfaces prolonged the occlusion time longer than the PU control. The introduction of either only PEO or SO$_3$ groups contributed to the enhancement to some extent. In addition, PU-PEO-SO$_3$ surfaces showed increased occlusion time compared with PU-PEO. In general, however, the PEO increase was not so large. The hydrophobic, semiflexible hydrocarbon grafted chain showed less improvement than grafted PEO. The enhanced *ex vivo* blood compatibility of PU-PEO-SO$_3$ compared with PU and PU-PEO is due to the synergistic effect of both hydrophilic PEO

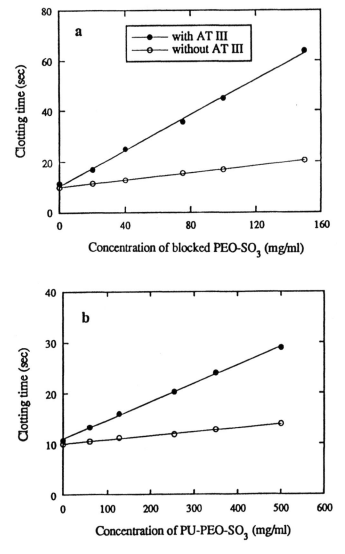

Figure 7 Anticoagulant activity of sulfonated polymers: (a), blocked PEO-SO$_3$; (b), PU-PEO-SO$_3$.

and the negatively charged sulfonate group; consequently, fibrin formation, followed by protein adsorption, platelet adhesion and activation, and subsequent mural thrombus formation, are suppressed.

C. *In Vivo* Studies

1. In Vivo *Biostability and Anticalcification Studies*

The biostability and calcification studies were performed *in vivo* using the rat model [34]. The rats (60–80 grams [g], 4 weeks old) were anesthetized with ketamine and the abdominal area was shaved, cleaned with alcohol, and swabbed with iodine. A midline incision was made in the skin, which was then gently dissected away from the abdominal muscles.

A small intramuscular pouch was made and a 1 × 1 cm specimen sheet inserted in the pouch (four implants, different surfaces in each animal). After times of 2 to 6 months, the specimens were retrieved and processed for electron microscopy and calcium/phosphorous (Ca/P) analysis. The surface morphology and the presence of calcium and phosphorus atoms on retrieved sheets were examined by an energy dispersive analysis of x-ray (EDAX) coupled with scanning electron microscopy (SEM).

Figure 8 demonstrates SEM micrographs of surface-modified PUs implanted for 2, 4, and 6 months. The surface cracking of modified surfaces in general increased with implantation time. As shown in Fig. 8a, the PU surface demonstrated cracks of 10–15 micrometers (μm) in length after 2 months and a few microcracks were also observed on the PU-PEO surface. In contrast, PU-PEO-SO_3 surface showed no significant change and was still intact. After 4 months (Fig. 8b), there were more surface cracks than at 2 months in all modified PUs except for PU-PEO-SO_3, for which no crack was found. More extensive surface cracks were revealed at 6 months (Fig. 8c); these cracks were all over the surfaces. The degree of crack formation after 6 months of implantation decreased in the following order: PU had more cracks than PU-PEO, which had more cracks than PU-PEO-SO_3. PU-PEO-SO_3 showed the least extent of cracks among the test PUs.

The improved biostability of PU-PEO-SO_3 might be attributed to the following [34]. First, the surface has the most smoothness and homogeneity, which can suppress surface cracks that form at the least defect. Second, the nonadhesive property of PEO decreases the adsorption and activation of phagocytic cells [35]. Furthermore, the highly hydrophilic nature of the sulfonated PEO surface provides for rapid hydration, resulting in surrounding the surface with water and a reduction of the reaction of the tissue with foreign materials.

The quantitative analysis of calcium and phosphorus deposited on the sheets was carried out by an inductively coupled plasma (ICP) atomic emission spectrometer. Table 3 summarizes the contents of calcium and phosphorus on modified PUs. All samples demonstrated that the calcium deposition increased as the implantation time increased from 2 to 6 months, while there were no significant differences in phosphorus deposition. The calcium deposition is much higher than phosphorus and the ratio of calcium to phosphorus suggests that this calcium compound is not a hydroxyapatite. The amount of calcium deposition, regardless of implantation time, decreased in the following order: PU deposition was greater than that for PU-PEO, which was greater than that for PU-PEO-SO_3, suggesting that PU-PEO-SO_3 is the most calcification resistant.

It was interesting to find out in this study that PU-PEO-SO_3 exhibited not only better *in vivo* biostability, but also less calcification in addition to the enhanced blood compatibility. The exact mechanism of calcification is not completely explained, although surface defect, calcium ion complexation, adsorbed humoral factors or thrombus, and mechanical stress were discussed as responsible [36]. Van Blitterswijk et al. reported the increase of calcification in the case of PEO/polybutylene terephthalate bioactive copolymer implanted subcutaneously in rats [37]. Therefore, in our study, SO_3 groups introduced might have an important effect on the decrease of calcification, for example, by lowering local pH and dissolving the deposited calcium compounds, in addition to the excellent surface smoothness and improved blood compatibility of PU-PEO-SO_3 [34].

2. In Vivo *Application as Coating*

In vivo biological responses to PU-PEO-SO_3 graft copolymer coating were studied using the dog model [20]. PU-PEO-SO_3 was applied as a coating material over a newly designed

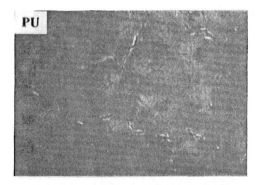

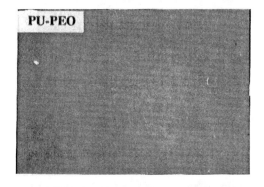

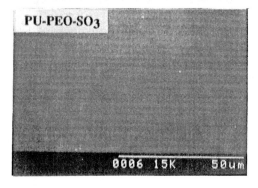

(a)

Figure 8 Scanning electron microscopy (SEM) micrographs of modified polyurethane (PU) surfaces implanted in rats: (a), 2 months; (b), 4 months; (c), 6 months.

Sinkhole bileaflet PU heart valve and a porous PU vascular graft (solution casted, 19 mm in diameter, half-circle, 5 cm in curvature). The fabrication procedures of the PU heart valve and graft were reported elsewhere in detail [21]. *In vivo* canine studies were performed by producing a shunt between the right ventricle (RV) and pulmonary artery (PA) (see Fig. 9). Before implantation, the Sinkhole valve leaflet and vascular graft were treated by half-and-half coating with 2.5% (weight to volume, wt/vol) PU and PU-PEO-SO$_3$ solution in dimethylacetamide. Thrombus and crack formation and surface

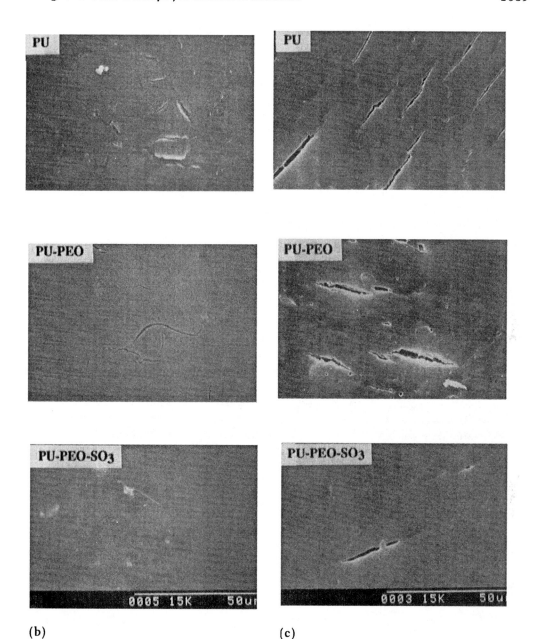

(b)   (c)

morphology of retrieved samples were examined by gross observation and SEM. The quantitative analysis of calcium and phosphorous deposited on test samples was performed by ICP as described above.

Figure 10 shows SEM micrographs of the luminal surface of a vascular graft after 24-days implantation. PU-PEO-SO$_3$ demonstrated much less platelet adhesion and thrombus formation than on the PU vascular graft. The valve leaflet also exhibited a trend similar to the vascular graft. The cracks in the valve leaflet were observed on only

Table 3   The Contents[a] of Calcium and Phosphorus on Modified Polyurethanes

Material	Ca			P		
	2[b]	4	6	2	4	6
PU	79 ± 24	154 ± 43	221 ± 43	0.71 ± 0.46	0.30 ± 0.13	0.37 ± 0.09
PU-PEO	66 ± 7	120 ± 27	162 ± 35	0.58 ± 0.18	0.15 ± 0.07	0.29 ± 0.07
PU-PEO-SO$_3$	57 ± 17[c]	71 ± 24[c]	93 ± 24[c]	0.52 ± 0.04[c]	0.09 ± 0.04[c]	0.27 ± 0.05[c]

[a]$\mu g/cm^2$; mean ± SD (n = 5–7); significance level using an unpaired Student's t test when comparing modified PUs to PU
[b]Months
[c]$p < 0.05$

the PU surface. In contrast, no crack was formed on PU-PEO-SO$_3$ (Fig. 11). These results mean that a close relationship exists between blood compatibility and the biostability of implanted polymers.

Table 4 summarizes the calcium and phosphorus contents of implants at various implantation times. All implants demonstrated that the calcium deposition increased as the implantation time increased from 14 to 39 days, while there were no significant differences in phosphorus deposition. In addition, the degree of calcification on PU-PEO-SO$_3$ was much lower than on PU, suggesting that PU-PEO-SO$_3$ is calcification resistant, which might be due to the synergistic effect of nonadhesive and mobile PEO and negatively charged SO$_3$ groups.

In summary, the above results attest that PU-PEO-SO$_3$ is more blood compatible, biostable, and calcification resistant *in vivo* than untreated PU, and can be applied as a coating material for the polymer valve and vascular graft in canine implantation.

IV. SUMMARY

A new modified polyurethane surface using hydrophilic PEO chains and a negatively charged SO$_3$ group was developed for chemical modification by two different methods.

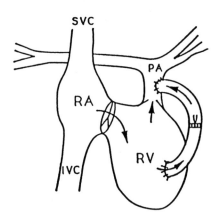

Figure 9   *In vivo* canine right ventricle–pulmonary artery (RV-PA) shunt system.

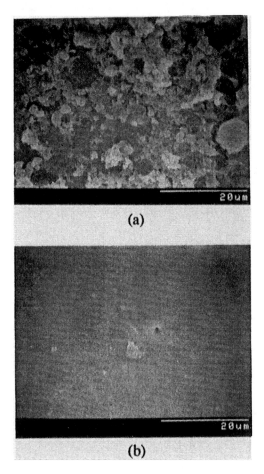

Figure 10   Scanning electron microscopy (SEM) micrographs of the luminal surface of a vascular graft after 24-days implantation in a dog: (a), polyurethane (PU); (b), PU-PEO-SO$_3$.

The first method involved the *in situ* surface grafting of sulfonated PEO (PEO-SO$_3$) onto the PU surface, and the other method utilized the coating of a newly synthesized PU-PEO-SO$_3$ graft copolymer.

In this study, the proposed hypothesis (negative cilia) has been proven both *in vitro* and *ex vivo/in vivo*. The anticoagulant activity of the sulfonate group, in addition to the minimization of protein and platelet interactions with PEO *in vitro*, correlated with the improved blood compatibility seen *ex vivo* and *in vivo*. In addition, PEO-SO$_3$ surfaces demonstrated a reduction in surface crack and calcium deposition. The enhanced blood compatibility correlates with the improved biostability and anticalcification. This is mainly due to the synergistic effect of hydrophilicity and the dynamic motion of PEO chains and negative SO$_3$ groups.

The results obtained (improved blood compatibility as well as biostability and anticalcification) attest to the usefulness of this negative cilia model for the design of blood-contacting medical devices.

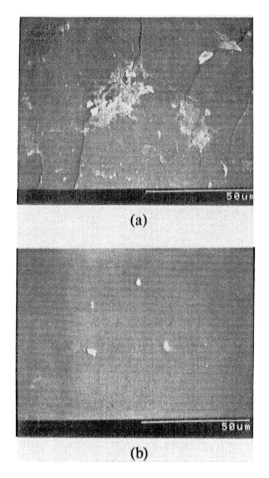

Figure 11   Scanning electron microscopy (SEM) micrographs of Sinkhole valve leaflet after 24-days implantation in a dog: (a), polyurethane (PU); (b), PU-PEO-SO₃.

Table 4   The Contents[a] of Calcium and Phosphorus on Various Implants

	Ca			P		
Material	14[b]	24	39	14	24	39
PU (leaflet)	107 ± 30	183 ± 56	250 ± 72	1.3 ± 0.3	4.1 ± 0.6	7.5 ± 1.0
PU (RV graft)	—	106 ± 30	170 ± 49	—	3.9 ± 0.5	5.1 ± 0.7
PU (PA graft)	53 ± 11	92 ± 25	154 ± 45	0.8 ± 0.2	3.3 ± 0.4	4.7 ± 0.6
PU-PEO-SO₃ (leaflet)	50 ± 10	75 ± 15	98 ± 31	0.4 ± 0.1	1.8 ± 0.3	2.9 ± 0.3
PU-PEO-SO₃ (RV graft)	—	37 ± 6	60 ± 14	—	1.2 ± 0.3	2.1 ± 0.3
PU-PEO-SO₃ (PA graft)	22 ± 5	35 ± 7	52 ± 11	0.3 ± 0.1	0.9 ± 0.2	1.8 ± 0.3

[a]$\mu g/cm^2$, mean ± SD ($n = 3$)
[b]Days

ACKNOWLEDGMENTS

This work was supported by Korea Ministry of Science and Technology grants N0561, N1411, N2760, N6390, N7331, and N8290. The authors thank Drs. S. Y. Jeong, K.-D. Ahn, B. G. Min, and H. M. Kim for their contribution. They also thank Professor S. W. Kim, University of Utah, for valuable advice.

REFERENCES

1. Kim, S. W., and J. Feijen, in *CRC Critical Reviews in Biocompatibility*, Vol. 1, D. Williams (Ed.), CRC Press, Boca Raton, FL, 1985, p. 229.
2. Andrade, J. D., S. Nagaoka, S. L. Cooper, T. Okano, and S. W. Kim, Current hypotheses in blood compatibility, *ASAIO J.*, 10:75–80 (1987).
3. Guidon, R. G., M. W. King, L. Awad, L. Martin, D. Domurado, M. Marois, M. F. Sigot-Luizard, C. Gosselin, K. Gunasekera, and D. Gagnon, Albumin coated and critical point dried polyester prostheses as substitutes in the thoracic aorta of dogs, *Trans. ASAIO*, 29:290–295 (1983).
4. Burkel, W. E., L. M. Graham, and J. C. Stanley, Endothelial linings in prosthetic vascular grafts, *Ann. N.Y. Acad. Sci.*, 516:131–144 (1986).
5. Park, K. D., T. Okano, C. Nojiri, and S. W. Kim, Heparin immobilization onto segmented polyurethaneurea surfaces – Effect of hydrophilic spacers, *J. Biomed. Mater. Res.*, 22:977–992 (1988).
6. Han, D. K., K. D. Park, K.-D. Ahn, S. Y. Jeong, and Y. H. Kim, Preparation and surface characterization of PEO-grafted and heparin immobilized polyurethane, *J. Biomed. Mater. Res.: Appl. Biomater.*, 23(A1):87–104 (1989).
7. Ku, C. S. L., I. Bornstein, J. Breillatt, S. N. Ung-Chhun, R. Johnson, J. Lindon, S. Pokropinski, and D. Rimer, Development of thromboresistant biomaterials: Immobilization of hirudin onto polymers, *Trans. Soc. Biomater.*, 14:44 (1991).
8. Kim, H. P., S. M. Byun, Y. I. Yeom, and S. W. Kim, Immobilization of urokinase on agarose matrices, *J. Pharm. Sci.*, 72:225–228 (1983).
9. Ohshiro, T., and G. Kosaki, *Int. J. Artif. Organs*, 4:58 (1980).
10. Ishihara, K., in *Biomedical Application of Polymeric Materials*, T. Tsuruta (Ed.), CRC Press, Boca Raton, FL, 1993, pp. 89–115.
11. Srinivasan, S., and P. N. Sawyer, Correlation of the surface charge characteristics of polymer with their antithrombogenic characteristics, in *Biomedical Polymers*, A. Rembaum and M. Shen (Eds.), Marcel Dekker, New York, 1971, pp. 51–66.
12. Ishihara, K., N. Nakabayashi, K. Nishida, M. Sakakida, and M. Shichir, Designing biocompatible materials, *Chemtech*, Oct. 19–25 (1993).
13. Andrade, J. D., Interfacial phenomena and biomaterial, *Med. Instrum.*, 7:110–120 (1976).
14. Mori, Y., S. Nagaoka, H. Takiuchi, T. Kikuchi, N. Noguchi, H. Tanzawa, and Y. Noishiki, A new antithrombogenic material with long polyethylene oxide chains, *Trans. ASAIO*, 28:459–463 (1982).
15. Merrill, E. W., and E. W. Salzman, Polyethylene oxide as a biomaterial, *ASAIO*, 6:80–84 (1983).
16. Grasel, T. G., and S. L. Cooper, Properties and biological interaction of polyurethane anionomers: Effect of sulfonate incorporation, *J. Biomed. Mater. Res.*, 23:311–338 (1989).
17. Silver, J. H., A. P. Hart, E. C. Williams, S. L. Cooper, S. Charef, D. Labarre, and M. Jozefowicz, Anticoagulant effects of sulphonated polyurethane, *Biomaterials*, 13:339–344 (1992).
18. Han, D. K., S. Y. Jeong, Y. H. Kim, B. G. Min, and H. I. Cho, Negative cilia concept for thromboresistance: Synergistic effect of PEO and sulfonate groups grafted onto polyurethanes, *J. Biomed. Mater. Res.*, 25:561–575 (1991).

19. Han, D. K., S. Y. Jeong, K.-D. Ahn, Y. H. Kim, and B. G. Min, Preparation and surface properties of PEO-sulfonate grafted polyurethanes for enhanced blood compatibility, *J. Biomater. Sci. Polym. Edn.*, 4:579–589 (1993).

20. Han, D. K., K. B. Lee, K. D. Park, C. S. Kim, S. Y. Jeong, Y. H. Kim, H. M. Kim, and B. G. Min, *In vivo* canine studies of Sinkhole valve and vascular graft coated with blood compatible PU-PEO-SO₃, *ASAIO J.*, 39:537–541 (1993).

21. Han, D. K., Ph.D. dissertation, Seoul National University, Korea (1993).

22. Han, D. K., G. H. Ryu, K. D. Park, S. Y. Jeong, Y. H. Kim, and B. G. Min, Adsorption behavior of fibrinogen to sulfonated polyethyleneoxide grafted polyurethane surfaces, *J. Biomater. Sci. Polym. Edn.*, 4:401–413 (1993).

23. Santerre, J. P., N. H. van der Kamp, and J. L. Brash, Effect of sulfonation of segmented polyurethanes on the transient adsorption of fibrinogen from plasma: Possible correlation with anticoagulant behavior, *J. Biomed. Mater. Res.*, 26:39–57 (1992).

24. Ito, Y., T. Kashiwagi, S. Q. Liu, L. S. Liu, Y. Kawamura, Y. Iguchi, and Y. Imanishi, Design and synthesis of blood-compatible polyurethane derivatives, in *Artificial Heart II*, T. Akusa (Ed.), Springer-Verlag, Tokyo, 1988, pp. 35–42.

25. Chen, W. Y., B. Z. Xu, and X. D. Feng, Synthesis of polysulfohexyl methacrylate with anticoagulant activity, *J. Polym. Sci.: Polym. Chem. Edn.*, 20:547–554 (1982).

26. Kammangne, F. M., H. Serne, D. Labarre, and M. Jozefowicz, Heparin-like activity of insoluble sulphonated polystyrene resins, *J. Colloid Interface Sci.*, 110:21–31 (1986).

27. Mauzac, M., N. Aubert, and J. Jozefonvicz, Antithrombic activity of some polysaccharide resins, *Biomaterials*, 3:221–224 (1982).

28. Han, D. K., N. Y. Lee, K. D. Park, Y. H. Kim, H. I. Cho, and B. G. Min, Heparin-like anticoagulant activity of sulfonated poly(ethylene oxide) and poly(ethylene oxide)-grafted polyurethane, *Biomaterials*, in press (1994).

29. van der Does, L., T. Bevgeling, P. E. Froehling, and A. Bantjes, Synthetic polymers with anticoagulant activity, *J. Polym. Sci. Polym. Symp.*, 66:337–348 (1979).

30. Ito, Y., Y. Iguchi, T. Kashiwagi, and Y. Imanishi, Synthesis and nonthrombogenicity of polyurethaneurea film grafted with poly(sodium vinyl sulfonate), *J. Biomed. Mater. Res.*, 25:1347–1361 (1991).

31. Walker, F. J., and C. T. Esmon, The molecular mechanism of heparin action III. The anticoagulant properties of polyanetholesulfonate, *Biochem. Biophys. Res. Commun.*, 83:1339–1346 (1978).

32. Czapek, E. E., H. C. Kwaan, M. Szczecinski, and R. B. Friedman, The effect of a sulfated polysaccharide on antithrombin III, *Fed. Proc.*, 37:586 (1978).

33. Nojiri, C., T. Okano, D. Grainger, K. D. Park, S. Nakahama, K. Suzuki, and S. W. Kim, Evaluation of nonthrombogenic polymers in a new rabbit A-A shunt model, *Trans. ASAIO*, 33:596–601 (1987).

34. Han, D. K., K. D. Park, S. Y. Jeong, Y. H. Kim, U. Y. Kim, and B. G. Min, *In vivo* biostability and calcification-resistance of surface modified PU-PEO-SO₃, *J. Biomed. Mater. Res.*, 27:1063–1073 (1993).

35. Zhao, Q., N. Topham, J. M. Anderson, A. Hiltner, G. Lonoen, and C. R. Payet, Foreign-body giant cells and polyurethane biostability: *In vivo* correlation of cell adhesion and surface cracking, *J. Biomed. Mater. Res.*, 25:177–183 (1991).

36. Schoen, F. J., H. Harasaki, K. M. Kim, H. C. Anderson, and R. J. Levy, Biomaterial-associated calcification: Pathology, mechanisms, and strategies for prevention, *J. Biomed. Mater. Res.: Appl. Biomater.*, 22(A1):11–36 (1988).

37. van Blitterswijk, C. A., J. van der Brink, H. Leenders, S. C. Hesseling, and D. Bakker, Polyactive: A bone-bonding polymer effect of PEO/PBT proportion, *Trans. Soc. Biomater.*, 14:11 (1991).

40
Endothelial-Lined Vascular Prostheses

Steven P. Schmidt, Sharon O. Meerbaum, and William V. Sharp
Akron City Hospital
Akron, Ohio

I. INTRODUCTION

The use of vascular prostheses in arterial reconstructions has improved life-styles and extended life spans for patients with vascular disease. Using these conduits to replace diseased blood vessels has been the foundation of present-day vascular surgery. Many scientists and surgeons have contributed to the development of currently used vascular protheses during the last 100 years.

The healing processes of the injured arterial wall were first studied by Jones and London in 1805. They identified the intima, renamed it the endothelium, and described its regenerative properties [1]. The majority of research in this field has been focused on vein grafts as arterial replacements. Gluck experimented with autogenous veins as vascular conduits in the arterial circulation in 1894 [2]. Goyanes inserted the first vein graft to replace a segment of artery after excision of an aneurysm of the popliteal artery in 1906, and Lexer, in 1907, used a segment of a patient's own saphenous vein to bridge a defect in the axillary artery [3]. In 1908, Carrel and Guthrie successfully grafted a segment of canine vena cava into the dog's own carotid artery [4]. Bernheim, in 1916, used an autologous saphenous vein to reconstruct a popliteal artery that had been resected because of an aneurysm [5]. In 1949, Kunlin used vein bypass grafts in patients with obliterative atherosclerosis [6], while Julian et al. [7], Lord and Stone [8], Dale, De-Weese, and Merlescott [9], and Linton and Darling [10] reported the use of autogenous veins as bypass conduits in the femoropopliteal region.

Stewart et al. used the saphenous vein in bypass surgery for renal occlusive disease [11], and Garret, Dennis, and De Bakey first used saphenous veins in coronary artery bypass surgery [12]. Today, 10 years following their report, the saphenous vein graft remains the "gold standard" bypass graft for reconstructive surgery of medium- and small-diameter arteries.

Although the saphenous vein graft is the bypass graft standard by which synthetic and other biologic grafts are compared, it does have its deficits. The quality of the vein and its diameter determine, in part, whether or not the graft will succeed as an arterial conduit. Intimal thickening, fibrosis of the valves, development of atherosclerosis, and aneurysmal dilatation are some of the reasons that vein grafts fail as arterial replacements [13]. It is also known that patients bypassed with the internal mammary artery for myocardial revascularization have a higher rate of long-term patency and survival compared with those patients receiving a saphenous vein [14] Presumably, the internal mammary artery is more physiologic than the saphenous vein when implanted in the coronary circulation.

Although progress has been made using autogenous vein grafts in small- and medium-diameter arterial reconstructions, many patients have diseased blood vessels that are inappropriate for use as bypass grafts and all patients have a limited number of "nonessential" veins available for bypass grafting. The size limitation of autogenous vein grafts led to the search to find a treatment for expanding abdominal aortic aneurysms in the 1940s and 1950s and prompted the development of synthetic vascular prostheses. Voorhees, Jaretski, and Blakemore postulated that tubes made from synthetic fabrics could be used as replacements for diseased blood vessels [15]. The origin of this postulate was based on the discovery of a loose silk thread found within the ventricle of a dog that had become coated within a few months with an endothelial-like surface. Blakemore and Voorhees, in 1954, reported the results of a clinical study evaluating the use of synthetic tubes as vascular prostheses [16]. This study was the embarkation of modern-day vascular reconstructive surgery and the quest for the "perfect" vascular prosthesis began.

Szilagyi reviewed the development of arterial prostheses in a 1978 publication and stated that the grafts that had historically been developed were designed to mimic the most simple function of human arteries—that being a mechanical conduit—with virtually no regard for the physiologic and histologic properties of the arterial wall [17]. Despite this lack of scientific understanding of the complex interactions of blood components with either natural or synthetic vascular walls, surgeons have found some success with synthetic grafts formulated from Teflon™, such as PTFE (polytetrafluoroethylene, PTFE), and Dacron™ (polyethyleneterephthalate). Both are extensively used as arterial replacements today by vascular surgeons. When these artificial vascular grafts are placed in high-flow, low-resistance locations such as the aortoiliac position, graft patencies and durabilities remain excellent for as long as 10 years postimplantation [18]. However, when placed in low-flow, small-diameter circulations, the failure rates of synthetic vascular grafts greatly increase. The patency rate at 8 years of Dacron grafts placed below the inguinal ligament is only 10% [19]. The patency rate for PTFE grafts implanted for femorodistal reconstructions is only approximately 20% at 3 years [20]. The debate over the use of synthetic grafts versus autogenous vein grafts in patients with bypassable diseases is still ongoing [21].

The perfect vascular prosthesis has not yet been found and the search for a more suitable graft for use in small and medium arterial reconstruction is in progress. The development of such a graft has been approached from several different perspectives. One approach has been the pharmacological modification of the surface properties of synthetic grafts, such as heparin bonding, or coating the surface with agents that hinder platelet adherence but encourage the endothelialization of the surface. Alternatively, radio-frequency glow discharge (RFGD) has been used to modify the surface of existing graft materials and make them more resistant to thrombotic occlusion [22].

A second approach has been the development of new graft materials that have mechanical properties closer to native blood vessels than the graft materials currently marketed. The vascular surgeon's ideal vascular prosthesis would be a graft that could be an arterial substitute with live intima and an elastic wall according to Szilagyi [17]. However, in order to develop a small-diameter vascular graft properly, a better understanding of the cellular and humoral events occurring at the blood/graft interface in both the natural vessel as well as the synthetic vessel is necessary as is a better understanding of the healing mechanisms involved.

One of the assumptions for the superiority of saphenous vein grafts compared with synthetic grafts in peripheral vascular surgery was the presence of an intact endothelium on the luminal surfaces of the saphenous vein grafts. Under *in vivo* conditions of normal circulatory physiology, endothelial cells (ECs) are the most nonthrombogenic cells known. It has been known from research experiments that synthetic vascular grafts implanted in animals endothelialize, in time, along their entire lengths. This endothelialization occurs only in the anastomotic regions of human vascular grafts, however. The hypothesis of many research groups has thus been to improve graft performance by promoting the endothelialization of the entire length of synthetic grafts implanted in humans by seeding endothelial cells onto their luminal surfaces. How to achieve this endothelialization of synthetic grafts and have them function as natural blood vessels became important research questions.

The purpose of this chapter is to review the research history of endothelial seeding of synthetic vascular grafts from its early beginnings through its current status in clinical trials.

II. *IN VITRO* STUDIES

A. Cell Harvesting

Jaffe et al. described in 1973 the isolation and culture of human umbilical vein endothelial cells by collagenase digestion of the luminal surfaces of umbilical veins [23]. As a result, investigators began an intense study of the endothelial cell and a vast quantity of information concerning endothelial cell growth and behavior *in vitro* became available. Williams published a review of these studies in 1987 [24]. The information gained from these early studies motivated the seeding of endothelial cells onto synthetic vascular graft material in hopes of achieving a live, functioning intima similar to that of a natural blood vessel.

Endothelial cells can be harvested from large vessels by two methods. The intimal lining can be scraped, thus removing cells mechanically [25]. An alternative method of harvest involves enzymatically removing the endothelium with collagenase, which separates the endothelial cells from their basement membrane and extracellular matrix [26]. The mechanical harvesting technique often results in endothelial cell damage and smooth muscle cell contamination. The enzymatic technique has been used more widely and optimized to procure healthy cells in large numbers from large vessel sources [27–32].

In order for endothelial cell seeding to be clinically practical, researchers realized that the efficiency of cell harvest from donor vessels would need to be improved. Rosenman et al. ascertained that only 15% of the available cells were actually harvested from native vessels [33]. Endothelial cell harvesting inefficiencies may also be attributed to inherent viabilities of crude bacterial collagenase preparations derived from *Clostridium histolyticum*. These preparations contain varying amounts of noncollagenolytic substances.

Sharefkin et al. have suggested that pure enzyme preparations with basement membrane lysis activity equivalent to that of crude collagenase preparations are necessary for reliable enzymatic harvest of endothelial cells that must accompany the clinical implementation of the technique [34]. Kirkpatrick, Melzner, and Góler showed the damaging effect of trypsin on cellular membrane phospholipids was higher than that caused by collagenase [35].

Watkins et al. demonstrated that endothelial cells available from adult human saphenous vein segments possessed the growth capacity to cover luminal areas of vascular prostheses commonly used in peripheral vascular surgery [36]. However, in their study, cells harvested from 10 of the 53 vein segments processed did not grow in culture. This supports the theory that factors other than collagenase, both biologic and technical, may be important in endothelial cell derivation. The upper limit of harvest efficiency of endothelial cells from a donor vessel may ultimately depend upon the age and health status of the patient-donor.

B.   Cell Source

Recognizing the limited availability of donor endothelial cells from autogenous veins and the need for relatively high seeding densities to achieve endothelialization of grafts in a clinical setting, researchers began focusing on other sources of cells for graft seeding. Leseche et al. prelined polytetrafluoroethylene grafts with endothelial cells harvested from varicose veins [37]. However, most attention has been directed toward obtaining endothelial cells from fat.

Jarrell et al. reported a technique to derive microvascular endothelial cells from perinephric omental fat [38]. They used an enzymatic method to harvest endothelial cells using collagenase in conjunction with a Percoll gradient to purify the isolated cells. In our laboratory, we found that the use of the Percoll density gradient made the cells lethargic and affected their rate of attachment as well as the rate of growth in culture [39]. Kern and associates used filtration to separate microvascular endothelial cells from contaminating cells obtained from omental and subcutaneous fat [40]. The use of liposuction to obtain autologous fat for the procurement of a purer yield of microvascular endothelial cells was reported by Williams, Jarrell, and Rose [41]. Dynabeads® (Dynal, Oslo, Norway) coated with *Ulex europaeus*-1 (UEA-1) incubated with harvested cell mixture have also been utilized to yield a very pure microvascular endothelial cell isolate [42].

Several workers have doubted that cells isolated from omentum fat are in fact endothelial cells and suggest instead that they are mesothelial cells [43,44]. Monoclonal antibodies against cytokeratins 8 and 18 have been used as markers for human mesothelial cells and studies using these markers to identify cells derived from the omentum have been reported by van Hinsbergh and associates [45]. The results suggest that these cells are in fact mesothelial rather than endothelial.

Comparison of the ability of adult human endothelial and mesothelial cells to attach and spread on PTFE has been investigated by Thompson, Vohra, and Walker [46]. The question of whether the contamination of mesothelial cells in the microvascular cell isolate derived from human autologous omental fat is detrimental or benign has yet to be answered. Baitella-Eberle et al., in their canine studies, found that purification procedures during harvest may not be as crucial as previously thought [47]. In their studies, microvascular cells from omental adipose tissue were seeded onto Dacron grafts and implanted for up to 26 weeks. As early as 5 weeks, the subintimal tissue proliferation reached an equilibrium rather than progressing toward a narrowing of the lumen and, after prolonged implantation, resembled the wall of normal artery in cell differentiation and intercellular matrix.

As an alternative to omental fat, current studies in canines by Williams et al. isolate fat from the falciform ligament rather than omentum, which has resulted in a purer microvessel endothelial cell isolate [48].

C. Surface Treatment of Grafts

In addition to investigating the inefficiencies of cell isolation and cell harvest, researchers have also addressed the issue of retention of endothelial cells on prosthetic grafts after seeding. Endothelial cell attachment and spreading are dependent upon proteins found in the extracellular matrix [49,50]. Because Dacron grafts are porous, whole blood is used to preclot the interstices of the graft. When the preclotting blood is mixed with an inoculum of endothelial cells for seeding, the endothelial cells are presumably harbored within the graft wall as well as on the luminal surface.

Sharefkin et al. labeled endothelial cells with [111]indium-oxine and estimated that 75% of endothelial cells seeded onto Dacron grafts remain attached [51]. In contrast, PTFE grafts are microporous with mean internodal distances of 30 microns (μ), and are thus not preclotted prior to surgical implantation. These grafts also have a hydrophobic luminal surface. Williams et al. evaluated the compatibility of adult human endothelial cells derived from iliac veins with both Dacron and PTFE *in vitro* [52]. Their study concluded that essentially no endothelial cells adhered to untreated grafts. In contrast, endothelial cell adherence increased dramatically on grafts treated with extracellular matrix proteins, plasma, or fibronectin.

The effect of fibronectin upon endothelial cell attachment to prosthetic vascular grafts has received considerable attention. Fibronectin is an adhesive glycoprotein that is a part of the natural basement membrane upon which endothelial cells attach *in vivo*. The significant advances in growing and passaging endothelial cells *in vitro* and the acquired knowledge of endothelial cell growth and behavior on fibronectin-coated polystyrene have justified the interest in fibronectin as a vascular graft coating. Ramalanjaona et al. studied endothelial cell adherence and retention on fibronectin-coated PTFE grafts using [111]indium-oxine-labeled endothelial cells [53,54]. Twice as many cells adhered to fibronectin-coated PTFE grafts compared with uncoated grafts and the losses of cells from the grafts following restoration of blood flow were reduced. These authors also reported the methodology by which fibronectin could be coated onto PTFE grafts to create a stable bond during pulsatile flow.

In addition to the above studies, Kesler et al. also described *in vitro* studies in which fibronectin enhanced the strength of attachment of endothelial cells to both PTFE and polyester elastomer [55]. Seeger and Klingman verified this enhancement of fibronectin coating in the retention of seeded endothelial cells upon PTFE [56]. Other investigators have also found that the modification of the luminal surface of vascular prostheses with fibronectin enhances endothelial cell adhesion [57–59]. Budd et al. studied the effect of various concentrations of fibronectin on endothelial cell attachment to polytetrafluoroethylene vascular grafts and found that a concentration of 20 micrograms/milliliter (μg/ml) was efficient in terms of cell attachment [60]. The time needed to obtain a stable and confluent monolayer of endothelial cells on PTFE pretreated with fibronectin is between 18 hours (h) and 4 days [61].

Other substrates, including collagen, laminin, albumin, gelatin, plasma, fibrin glue, whole blood, serum, and combinations of these substrates, have been studied to surface coat the lumens of a variety of vascular prostheses to improve endothelial cell adherence and growth [62,63,64–81].

D. Cell Retention After Exposure to Blood Flow

Once endothelial cells attach to vascular prostheses, they must withstand the shear stress of blood flow. Haegerstrand, Bengtsson, and Gillis compared serum protein coating to collagen Type I pretreatment of PTFE [82]. Both grafts exhibited confluent areas of endothelial cells that were retained after exposure to pulsatile plasma flow. The methodology employed in this study promoted endothelialization, and also described the use of autologous serum as the main source of protein for adherence of endothelial cells to grafts.

Gourevitch et al. precoated Dacron grafts with cold insoluble globulin or 1% gelatin and inoculated endothelial cells labeled with ^{111}indium-oxine onto the graft walls [83]. They found that, after 102 minutes, 64% of the cells remained attached following exposure to blood flow. Greisler et al. found excellent endothelial cell adherence (90%) on fibronectin-treated polyester elastomer (Hytrel) grafts after 120 minutes of perfusion on a pulse duplicator apparatus under high- and low-shear conditions [84].

Dalsing et al. also reported that Hytrel grafts precoated with fibronectin were superior to collagen-impregnated Dacron for endothelial cell retention during perfusion [71]. In this study, fibronectin-coated PTFE grafts seeded with endothelial cells were either cultured to confluence for 48 h or flow tested immediately in an *in vitro* flow circuit or *ex vivo* carotid-jugular shunt using a sheep model. Greater than 95% cell retention was demonstrated in 1-h seeded and 48-h cultured grafts following *in vitro* flow studies. *Ex vivo* studies of the 48-h cultured grafts showed a cell retention of 81% after 3 hours [85].

Other investigators have speculated on the length of time needed to incubate endothelial cells on precoated grafts for optimum cell retention during blood flow. Vohra et al. found that a 30-minute incubation was sufficient [86], while Prendiville et al. concluded that an incubation period of 72 hours was necessary for retention of endothelial cells on fibronectin-treated PTFE [87]. After a 90-minute incubation period, seeding densities of 1×10^5 and 2×10^5 cells/cm^2 (square centimeter) were found to produce a confluent monolayer with optimal utilization of cells on polytetrafluoroethylene grafts pretreated with plasma or fibronectin. The postperfusion cell attachment increased significantly when compared with a 20-minute incubation time [88]. When human endothelial cells were seeded at confluent densities on fibronectin-coated PTFE and exposed to flow within 90 minutes, a significantly lower cell retention was seen when compared with grafts exposed to flow 24 hours after seeding. The maximal retention was 92%, which occurred 24 hours after seeding [89]. Our own experience with fibronectin-coated grafts is that, although fibronectin enhances cell retention on both plain and carbon-coated PTFE grafts compared with grafts coated with tissue culture media, there is no advantage of fibronectin coating over autologous serum [90].

The method used to seed endothelial cells onto grafts may also affect the spreading and attachment of these cells to surface-treated grafts. Vacuum cell seeding has been used to apply cells from a suspension in culture within 10 minutes in an evenly distributed layer onto the luminal graft surface. The adherent cells immediately began to flatten, covering the luminal surface [91]. Another method used to increase endothelial cell delivery to the luminal surface of grafts is by pressurizing the graft lumen, in contrast to gravity-forced cell deposition [92]. Such a procedure increased endothelial attachment by 2 to 5 times.

Many of the above studies used [111]indium-oxine to label endothelial cells to measure cell retention and investigate the interaction of the cells with prosthetic surfaces during flow. Patterson et al. found that although [111]indium-oxine-labeled endothelial cells' viability was 80% after 24 hours, the loss of the [111]indium marker was 47% at 24 hours, which could lead to an underestimation of cell retention during flow studies [93]. They also found the affinity of graft material for [111]indium increases with the addition of glycoproteins to their surface. This, too, may affect cell retention estimates. The method of using [111]indium-oxine-labeled cells to determine cell retention in flow studies is not without its drawbacks.

Several factors affect the success of establishing an endothelial monolayer on a graft: the origin of the endothelial cells [24,36,94–96], the type of graft material used [62], and the surface treatment of the graft [63,57,97,98]. The responses of endothelial cells on graft material attachment, migration, proliferation, differentiation, and exhibition of normal functions have also been recognized as important factors.

III. ANIMAL STUDIES

Malcolm Herring is recognized as the "father" of endothelial cell seeding as a result of his 1978 publication with Gardner and Glover [27]. This early report described the derivation of canine venous endothelial cells from external jugular veins by mechanical disruption of the cells from the luminal surfaces of the veins using a steel wool pledget. These cells were mixed with blood used to preclot porous 6 mm internal diameter (ID) Dacron prostheses and the grafts were implanted in the infrarenal aortas of dogs. The grafts were evaluated 4 weeks postoperatively. The mean thrombus-free surface area of seeded grafts was 76% compared with 22% for nonseeded grafts; the glistening luminal surface of seeded grafts histologically resembled endothelium.

This report was rapidly followed by two publications involving the same group of investigators that documented the existence of Weibel–Palade bodies, the ultrastructural hallmark of endothelium, on seeded graft linings [99] and confirmed the positive identification of endothelium lining prosthetic grafts by immunofluorescent staining for antigen related to factor VIII [100]. These authors also studied the proliferation of seeded endothelial cells on 14 different graft materials, including Dacrons, Teflon, Orlon®, and polyurethane-backed graft [101]. This last study concluded that weft-knit Dacron grafts were most suitable for endothelial cell seeding and that it was difficult for cellular elements to adhere to Teflon.

Utilizing enzymatic techniques for the harvest of endothelial cells from canine external jugular veins, Graham et al. contributed significantly to the development of the technology of endothelial cell seeding with two reports in 1980 [28,102]. These authors derived endothelial cells by sequential incubations of the vein luminal surfaces in trypsin and collagenase. In one experiment, the enzymatically harvested endothelial cells were seeded immediately following their derivation onto 6-mm double-velour Dacron thoracoabdominal bypass grafts. In the second experiment, the canine venous endothelial cells were cultured *in vitro* following their enzymatic derivation for 14 days prior to their seeding onto 6-mm Dacron thoracoabdominal bypass grafts. This study introduced the option of tissue culture of endothelial cells prior to graft seeding. In both studies, the endothelial cell coverage of the luminal surfaces of the implanted grafts exceeded 80% at

4 weeks postoperatively. These early studies of both Herring and Graham's groups guided the subsequent efforts of researchers in the field of endothelial seeding.

The next phase of the research effort involved evaluating the efficacy of endothelial cell seeding in performances of small-caliber vascular grafts (4 mm ID). The pioneering research of Herring and Graham's groups evaluated 6-mm ID vascular grafts seeded with endothelial cells. These grafts were implanted in high-flow and pressure circulations as infrarenal aortic grafts or thoracoabdominal bypass grafts. In these circulations, under the evaluated rheologic conditions, even nonseeded prosthetic grafts were expected to remain patent and develop a relatively nonthrombogenic pseudointima. In contrast, because of the low flow in circulations in which small-diameter vascular grafts may be of greatest utility, endothelial cell seeding would theoretically offer its greatest advantage in preventing thrombosis.

In 1982, we reported our early data evaluating 4-mm double-velour Dacron (Microvel®) grafts in the canine carotid artery model [103]. Also in that year, Stanley et al. described an experiment in which 10-cm lengths of 4-mm ID endothelial cell seeded and nonseeded externally supported knitted Dacron grafts were evaluated as bilateral iliofemoral bypasses [104]. In our study, the mean patency of successfully seeded grafts was 100% at Week 4 postoperatively and mean thrombus free surface area was 80% by 4 weeks postoperatively. In contrast, the thrombus-free surface area of nonseeded grafts was only 18%. In the Stanley study, 73% of the seeded grafts were patent at the time of harvest; patency in nonseeded grafts was 27%. Both studies concluded that endothelial cells surfaced on the grafts within 2–4 weeks postoperatively and created a confluent monolayer of endothelium.

Further data from our laboratory confirmed the theoretical advantages of endothelial cell seeding in maintaining small-diameter vascular graft patency during conditions of acute reduction in blood flow through the grafts [105,106]. In these studies, we seeded 6-mm lengths of 4-mm ID Dacron double-velour grafts with enzymatically derived endothelial cells, waited for postoperative maturation of the neointima, and then reduced blood flow acutely for 4 hours through each seeded graft and its contralateral nonseeded control. All seeded grafts remained patent during the controlled low flows; in contrast, 50% of the nonseeded grafts thrombosed during low flow when the experiment was performed at three weeks postoperatively and 25% of them thrombosed at 5 weeks postoperatively. Blood flows returned to near initial levels following the experimental period of low flow in seeded grafts, but remained depressed in the nonseeded controls. This experiment generated the objective data that endothelial-cell-seeded grafts did indeed outperform nonseeded grafts under conditions of low flows similar to the circulatory conditions of poor runoff in peripheral bypass grafting.

Graham et al. also reported in 1982 the successful seeding of PTFE prostheses in the canine model [39]. This was an important study in advancing the technology of endothelial cell seeding to the prosthetic graft material preferred by many vascular surgeons because of its handling characteristics and resistance to infection. Subsequent studies have suggested that the endothelium-lined inner capsule that matures in endothelial-cell-seeded PTFE grafts is thinner compared with that in Dacron grafts, which may be of long-term benefit to the performance of the graft [107].

Many investigators have conducted experiments to identify the biologic and physiologic mechanisms by which endothelial cell seeding might improve vascular graft performance. Sharefkin et al. performed platelet survival studies in dogs with seeded and nonseeded thoracoabdominal grafts using [111]indium-oxine-labeled platelets [108]. The research concluded that, when endothelial cell seeding was technically successful, the degree of platelet interaction with Dacron vascular prostheses was reduced and a normal platelet survival time was restored. Clagett et al. reported the parallel between platelet serotonin levels [a monitor of platelet release] and the normalization of platelet survival times in dogs with endothelial-cell-seeded grafts [109]. In addition, there were significant differences in luminal surface production of 6-keto-PGF$_1\alpha$ between seeded and nonseeded grafts. 6-Keto-PGF$_1\alpha$ is the stable hydrolysis product of prostacyclin (PGI$_2$), a prostaglandin synthesized from arachidonic acid in endothelial cells and the most potent biologic antithrombogenic agent known.

Other investigators have also studied the prostaglandin biochemistry of seeded and nonseeded vascular grafts. We reported that endothelial cells seeded onto vascular grafts did indeed synthesize PGI$_2$, but at levels significantly less than the native artery [110]. Sicard et al. did not find a difference in PGI$_2$ levels between seeded and nonseeded vascular grafts [111], but did report that seeding lessened thromboxane A$_2$ (TXA$_2$) production by the walls of seeded grafts compared with nonseeded controls. TXA$_2$ is synthesized from arachidonic acid primarily in platelets and is a potent platelet aggregator. The suggestion was that an alteration in the ratio between PGI$_2$ and TXA$_2$ may then be responsible for the reduction in platelet deposition upon seeded grafts, which ultimately translates into enhanced longer-term graft patencies. The specific effects and consequences of endothelial cell seeding upon the prostaglandin biochemistry of vascular grafts remain controversial, however, and the subject of ongoing research.

The use of the canine model to evaluate vascular graft performances has been debated among researchers. While it has been known for many years that dogs are hypercoagulable compared to humans, Kaplan et al. were the first researchers to categorize dogs according to their thrombotic potentials as low or high responders [112]. In their study, the patency rate of grafts implanted in the carotid circulation of low responders was 100% at 3-weeks postimplantation; patency of grafts in high responders not treated with antiplatelet medications was 10%. High responders treated with antiplatelet medications had graft patency rates of 100%.

We evaluated the efficacy of a variety of medications in combination with endothelial cell seeding upon the performance of both Dacron and PTFE vascular grafts in canines [113,114]. In our study, grafts that were seeded with endothelial cells and implanted in dogs receiving antiplatelet medications performed better than respective controls, even though many of the antiplatelet drugs completely eliminated PGI$_2$ synthesis by the seeded grafts through inhibition of the endothelial cell cyclooxygenase enzyme. These studies suggested that high levels of PGI$_2$ were not necessary to maintain high thrombus-free surface areas on seeded grafts in antiplatelet-medicated dogs.

Based upon the logical desire of researchers to evaluate the efficacy of endothelial-cell-seeded grafts in humans, researchers at Tufts New England Medical Center first evaluated endothelial-cell-seeded grafts in baboons as an alternative to dogs [115]. Although there were no differences in patencies between the seeded and control grafts in this study, in which 5-cm lengths of 4-mm ID Dacron grafts were evaluated, platelet accumulation on seeded grafts was significantly less than on paired controls. In addition, cells on the luminal surfaces of the seeded grafts were identified as endothelial cells based upon morphologic and immunohistochemical characteristics. Although the limitations of these animal models are understood, the dog and baboon remain the choices of animal models for investigators researching endothelial cell seeding.

Other researchers have used animal models to investigate the effects of modification of graft internodal distances or porosities upon the development of the neointima in seeded grafts. We investigated three designs of e-PTFE (enhanced PTFE) with mean internodal distances of 28 μ, 40 μ, and 52 μ in a 1988 report in which these grafts were seeded with venous endothelial cells and implanted in the carotid arteries of dogs [116]. We concluded that grafts with mean internodal distances of 40 μ were most successful in maintaining patencies and thrombus-free surface areas while promoting controlled inner capsule healing. Likewise, Kempczinski et al. have reported excellent results by seeding a highly porous (mean internodal distance, 45 μ), unreinforced PTFE prosthesis [117].

Commercially available PTFE prostheses have mean internodal distances of approximately 30 μ and the internodal spaces remain filled with air when implanted clinically by vascular surgeons. By slightly expanding these internodal spaces and inoculating the cells into the internodal spaces suspended in serum or plasma, it has been assumed that more cells could be retained within the structure of the graft compared with surface seeding alone and that controlled ingrowth through the graft from the external capsule could proceed at a rate to establish a supporting subendothelium for maturation of the seeded cells.

Other investigators have seeded synthetic grafts with microvascular endothelial cells that were evaluated in animal models. Pearce et al. initially reported the successful seeding of 60-μ pore size PTFE grafts with microvascular endothelial cells derived from omentum [118]. Patency of seeded grafts was 58%. We seeded 4-mm Dacron grafts with cells derived from the microvessels of omental fat and implanted the grafts in the canine arteries of dogs for 5 weeks [119]. The mean patencies of both the seeded and nonseeded grafts in our study were 89%. However, the mean thrombus-free surface area of the seeded grafts was 95%, which differed significantly from that of nonseeded grafts (43%), implying a beneficial effect of the seeded microvascular cells. We also seeded omental microvascular cells onto PTFE grafts that were evaluated in canine carotid arteries. In these short-term studies, the patencies and thrombus-free surface areas of seeded grafts were good. However, the inner capsules of the seeded PTFE grafts were thicker than those in grafts from previous experiments in which the PTFE grafts had been seeded with venous endothelium. This thickness presumably resulted from the presence of many contaminating cells in addition to endothelial cells in the seeding inoculum.

In their studies comparing the performances of grafts seeded with endothelial cells from the microvessels of fat and the jugular vein, Sterpetti et al. concluded the technique of derivation of endothelial cells from fat needed refinement in order to reduce the occurrence of contaminating cells [120]. As an alternative technique, Noishiki et al. minced a piece of peripheral vein into small fragments and combined them with saline [121]. This tissue suspension was then sieved through the wall of Dacron protheses and the protheses were implanted in the descending aortas of dogs. After 14 days, the entire luminal surface was lined with endothelial cells and there was no difference in healing between the areas near the anastomotic sites and the midsections of the grafts.

Sterpetti, Schultz, and Bailey investigated the effect of endothelial cell seeding on platelet deposition on endarterectomized arteries [122]. They concluded that endothelial cell seeding on endarterectomized arteries is feasible and reduces platelet uptake.

Because of the reported successes of endothelial cell seeding in lessening thrombogenicity of arterial conduits, several researchers have investigated the effect of endothelial cell seeding upon the performance of venous prostheses. Protheses are rarely used in the venous circulation because the low venous blood flow through these highly thrombogenic grafts results in rapid occlusions. The results of endothelial cell seeding in the venous circulation have been less positive than those in the arterial circulation. Plate et al.

reported that early patency rates were not improved by endothelial cell seeding of iliocaval grafts and that both seeded and nonseeded grafts developed endothelial linings in this model [123]. Herring et al. also reported no benefit of endothelial cell seeding in improving early patencies of e-PTFE inferior vena cava replacements [124]. In contrast, Köveker et al. recently reported that endothelial cell seeding enhanced endothelialization on synthetic vena cava grafts and improved their thromboresistances [125]. Additional studies need to be performed to determine the efficacy of endothelial cell seeding in enhancing venous prosthetic graft performances.

Other researchers have used cells other than endothelial cells to seed grafts. These researchers reported significant coverage of Dacron grafts seeded with peritoneal mesothelial cells at the time of graft implantation in dogs [126]. A follow-up study demonstrated that these grafts seeded with mesothelial cells released more prostacyclin when compared with unseeded grafts [127].

A persistent problem that has plagued researchers in the field of endothelial cell seeding is the confirmation that the cells seeded are truly precursors of endothelial cells that ultimately form the neoendothelium on the graft. The suggestion has often been made that the technique of endothelial cell seeding might provide the stimulus for spontaneous endothelialization. Hollier et al. seeded endothelial cells enzymatically derived from jugular veins of female pigs onto thoracoabdominal bypass grafts implanted in male pig littermates [128]. At 4 weeks postoperatively, the seeded grafts were endothelialized, but chromosome analysis revealed surface endothelium that originated from the male host rather than from the female donor. The pig may not have been a good model for this study, however, as synthetic grafts rapidly endothelialize in the pig by 4 weeks postimplantation even when grafts are not seeded.

In studies comparing homologous and autologous seeding of PTFE arterial-venous prostheses in dogs, grafts seeded with large numbers of homologous endothelial cells generated midgraft endothelial linings [129]. In a similar study, nonautologous endothelial cell seeding in combination with immunosuppression therapy resulted in an endothelial lining comprised primarily of host cells, thus suggesting that nonautologous seeding may indeed enhance host endothelial cells to form a graft neointima [130].

Clowes et al. have suggested that endothelial cells lining graft luminal surfaces may originate from the transmural ingrowth of capillaries through graft pores [131]. In contrast, several authors have reported that endothelial cells grown in culture, genetically altered to carry a morphologic tag, can be transplanted onto a native vessel surface and can be subsequently identified on the luminal surface of the vessel at explant [132,133]. Dailey et al. reported the development of an intravital fluorescent staining technique that permitted isolated, autologous, fat-derived microvascular endothelial cells to be labeled and subsequently detected following their transplantation onto injured rat abdominal aortas [134]. Fluorescence microscopy of the cells in the injured areas established that the cells contained fluorescent dye and were the same cells that were originally transplanted. This technique provided a method to trace the origin and deposition of transplanted cells on vascular surfaces.

IV.　CLINICAL STUDIES

With all the knowledge gained from *in vitro* and animal studies, the true test would be if endothelial-seeded grafts functioned similarly in humans. The first clinical trial of endothelial cell seeding of vascular prostheses in humans was reported by Herring, Gardner, and Glover in 1984 [135]. Grafts were implanted in 161 patients and were seeded with endothelial cells mechanically harvested from adjacent subcutaneous veins in the leg. Their study concluded that endothelial cell seeding improved graft patency rates, but that the results were worse if the patient was a smoker. Herring et al. updated the data from this study in a 1987 report [136]. Seeded Dacron graft patency was superior to nonseeded patency between the first and second postoperative years, but not statistically significant. Seeded graft patency at seven years was 30.8%; patency was 37.1% in nonseeded grafts. Smoking adversely affected seeded graft patency. Also in this publication, the authors reported results of a second clinical trial in which PTFE graft patency averaged 73.9%, which was not significantly different from patency of vein grafts in a concurrent patient population (84.6%). The PTFE grafts were seeded with cells derived from collagenase treatment of external jugular vein segments from each patient. Herring, Baughman, and Glover were able to confirm extensive endothelialization of a seeded PTFE graft by histological observations in a patient at the 90th postoperative day [137].

Zilla et al. also reported results from a series of 18 patients undergoing femoropopliteal bypass [138]. PTFE grafts were seeded with endothelial cells derived from the external jugular vein in 9 of the patients. An average of 3.1×10^3 cells/cm^2 were seeded. Patients were followed by platelet function studies which included scintiscan assessment of ^{111}indium-oxine labeled platelets. On the basis of these follow-ups, the degree of endothelization in the seeded grafts was postulated as only minor. The authors emphasized the concern of all investigators designing clinical trials of endothelial cell seeded grafts: the need for direct biopsy of the graft in order to determine the degree of endothelization.

Risberg et al. seeded one limb of Dacron aortic bifurcation grafts with endothelial cells derived from distal saphenous vein segments [139]. The contralateral limb of each graft was nonseeded. Platelets from each patient were labeled with ^{111}indium-oxine and platelet accumulations on each limb of the grafts were measured at 1 and 4 months postoperatively. The report concluded that there was a significant reduction in the accumulation of labeled platelets on the limbs of the grafts seeded with venous endothelial cells.

In 1990, these investigators reported a follow-up study that showed that platelet deposition on the seeded graft limb remained reduced over a period of 12 months [140]. Preclotted femoropopliteal PTFE grafts in which half the graft was seeded either proximally or distally with endothelial cells were implanted into 23 patients [141]. The seeded graft segments accumulated significantly fewer platelets at 1 and 6 months after implantation.

In our laboratory, 34 patients underwent peripheral vascular reconstructions at Akron City Hospital; 6-mm ID Gore-Tex PTFE grafts were seeded with microvascular endothelial cells derived from autologous fat [142]. In this study, the surgeries were all leg-saving procedures in patients that did not have available autologous veins to use as bypass grafts. These patients were the "worst-case scenarios" of all patients that would potentially receive endothelial-cell-seeded grafts in peripheral vascular reconstructions. All of the procedures required below-the-knee distal graft anastomoses and the runoff conditions ranged between one and three vessels. The mean number of microvascular cells seeded per graft was 8.04×10^6. Based upon these numbers, the seeding density averaged 4.41×10^4 cells/cm^2 for a 50-cm length of 6-mm ID vascular graft. Overall patency of the seeded grafts at 30 months was 42%.

Proving that graft performance is indeed enhanced as a result of endothelial cell seeding is a complicated issue. In order to demonstrate directly the existence of endothelial cells on the graft's luminal lining, a biopsy must be performed at a site removed from

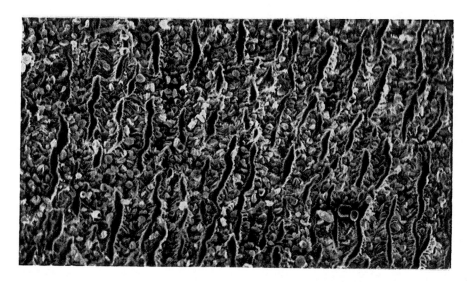

Figure 1   Scanning electron micrograph of unseeded polytetrafluoroethylene (PTFE) hemodialysis access graft one month postimplantation.

either anastomosis. The occurrence of endothelial cells could then be documented by histologic and/or electron microscopic studies. Our laboratory is presently conducting a clinical trial in which patients receive either PTFE grafts seeded with autologous-microvascular-derived cells as arteriovenous (AV) fistulas for hemodialysis access or a nonseeded PTFE graft for hemodialysis access. Biopsy samples are taken at 6 months. The presence or absence of endothelial cells and their morphology and orientation on the luminal surface is evaluated by scanning electron microscopy (Figs. 1–3), while the inti-

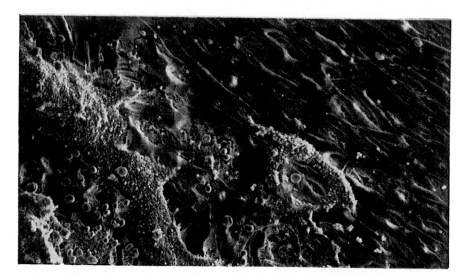

Figure 2   Scanning electron micrograph of microvascular-cell-seeded polytetrafluoroethylene (PTFE) hemodialysis graft one month postimplantation.

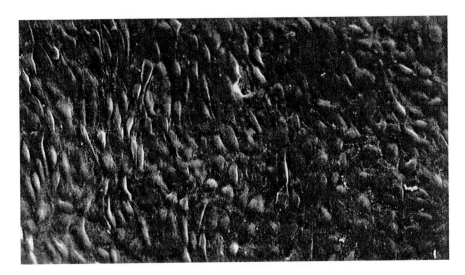

Figure 3   Scanning electron micrograph of microvascular-cell-seeded polytetrafluoroethylene (PTFE) hemodialysis access graft three months postimplantation.

mal cell types, extracellular matrix components, and cytokines are determined with immunohistochemical techniques. Obtaining these pieces of information will allow comparative statements about the relative functional statuses of the seeded versus nonseeded grafts at postimplantation times.

The technology of endothelial cell seeding has evolved along with major advances in the understanding of endothelial cell biology. The endothelial monolayer is a dynamic tissue with a repertoire of functions that have been largely unappreciated. Specific anticoagulant mechanisms exhibited by endothelial cells include the production of heparinlike glycosaminoglycans, prostacyclin production, plasminogen activator production, and expression of thrombomodulin. These normal anticoagulant properties of the vascular endothelium argue in favor of seeding or promoting the growth of an endothelial lining on the lumen of prostheses. In addition to these anticoagulant properties, however, it is now also appreciated that endothelial cells include tissue factor, von Willebrand factor, plasminogen activator inhibitor, thrombospondin, and collagens [143]. These procoagulant functions of endothelial cells are useful physiologically, but may be undesirable in terms of vascular graft performance. It is probable that the techniques for derivation of endothelial cells for vascular graft seeding, the process of seeding the graft, and the evolution of the seeded graft *in vivo* may induce responses in endothelial cells that are different from the functions of unperturbed cells. It can be theorized that the balance between anti- and procoagulant properties of endothelial cells ultimately dictates the thrombotic potential of the graft. It is clear that the focus in the evaluation of vascular graft performance must shift to include critical evaluation of the functional status of the endothelial cells at the blood/material interface.

The technology of endothelial cell seeding has given way to a number of important directions that may affect patient care. Endothelial cells could be manipulated genetically with the potential to deliver many gene products therapeutically directly to the vasculature [144,145]. Genetically engineered endothelial cells could be introduced to the recipient on seeded vascular grafts, promoting and suppressing genes controlling anticoagulant and procoagulant substances, respectively. Such technology could also be used to modify growth factor genes and provide a drug delivery system for treatment of disease. In the future, the use of these techniques in the clinical setting could greatly influence the management and treatment of patients with vascular disease.

REFERENCES

1. Callow, A.D. Historical Overview of Experimental and Clinical Development of Vascular Grafts, in J.C. Stanley, ed., *Biologic and Synthetic Vascular Prostheses*. New York, Grune & Stratton, p. 11, 1982.
2. Gluck, T. Die modene chirugie des circulation apparates. *Berl. Klin.* 70:1, 1898.
3. Dennis, C. Brief History of Development of Vascular Grafts, in P.N. Sawyer, ed., *Modern Vascular Grafts*. New York, McGraw-Hill, 1987.
4. Callow, A.D. Historical Development of Vascular Grafts, in P.N. Sawyer and M.J. Kaplitt, eds., *Vascular Grafts*. New York, Appleton-Century-Crofts, p. 5, 1978.
5. Bernheim, B.M. The ideal operation for aneurysm of the extremity. Report of a case. *Bulletin of Johns Hopkins Hospital* 27:93, 1916.
6. Kunlin, J.L. Le Treatment de l'arterite obliterante à la greffe veineuse. *Arch. Mal Coeur* 42: 371, 1949.
7. Julian, O.C., Dye, W.S., Olwin, J.H., and Jordan, P.H. Direct surgery of arteriosclerosis. *Ann. Surg.* 136:459, 1952.
8. Lord, J.W., and Stone, P.W. The use of autologous venous grafts in the peripheral arterial system. *Arch. Surg.* 74:71, 1957.
9. Dale, W.A., DeWeese, J.A., and Merlescott, W.J. Autogenous venous shunt grafts in femoropopliteal obliterative arterial disease. *Surgery* 51:62, 1962.
10. Linton, R.R., and Darling, R.C. Autogenous saphenous vein bypass grafts in femoropoliteal obliterative arterial disease. *Surgery* 51:62, 1962.
11. Stewart, B.H., Deese, M.S., Conway, J., and Correa, R.J., Jr. Renal hypertension: An appraisal of diagnostic studies and of direct operative treatment. *Arch. Surg.* 85:617, 1962.
12. Garrett, H.E., Dennis, E.W., and De Bakey, M.E. Aortocoronary bypass with saphenous vein graft. Seven year follow-up. *JAMA* 223:792, 1973.
13. Szilagyi, D.E., Hageman, J.H., Smith, R.F., et al. Autogenous vein grafting in femoropopliteal atherosclerosis: The limits of its effectiveness. *Surgery* 86:836–851, 1979.
14. Szilagyi, D.E., Elliott, J.P., and Hageman, J.G. Biologic fate of autogenous vein implants as arterial substitutes: Clinical, angiographic and histopathologic observations in femoropopliteal operations for atherosclerosis. *Annals of Surgery* 178:232, 1973.
15. Voorhees, A.B., Jaretski, A., and Blakemore, A.H. The use of tubes constructed from Vinyon "N" Clorthin bridging arterial defects. *Annals of Surgery* 135:332, 1952.
16. Blakemore, A., and Voorhees, A.B. The use of tube constructed in Vinyon "N" cloth in bridging arterial defects: Experimental and clinical. *Annals of Surgery* 140:324, 1954.
17. Szilagyi, D.E. Perspectives in Vascular Grafting, in P.N. Sawyer and M.J. Kaplitt, eds., *Vascular Grafts*. New York, Appleton-Century-Crofts, p. 23, 1978.
18. Sauvage, L.R., Berger, K., Wood, S.J., et al. *Grafts for the 80's*, The Bob Hope International Heart Research Institute, Seattle, 1981.
19. Darling, R.C., and Linton, R.R. Durability of femoropopliteal reconstruction. *American Journal of Surgery* 123:472, 1972.
20. Bergen, J.J., Veith, F.J., Bernhard, V.M., Yao, J.S.T., Flinn, W.R., Gupta, S.K., Scher, L.A., Samson, R.H., and Towne, J.B. Randomization of autogenous vein and polytetrafluoroethylene grafts in femorodistal reconstructions. *Surgery* 92:921, 1982.
21. Quinoñes-Baldrich, W.J., Busuttil, R.W., Baker, J.D., Vescera, C.L., Ahn, S.S.,

Machleder, H.I., and Moore, W.S. Is the preferential use of polytetrafluoroethylene grafts for femoropoliteal bypass justified? *Journal of Vascular Surgery* 8:291, 1988.

22. Kiaei, D., Hoffman, A.S., and Hanson, S.R. *Ex vivo* and *in vitro* platelet adhesion on RFGD deposited polymers. *Journal of Biomedical Materials Research* 26:357, 1992.

23. Jaffe, E.D., Nachman, R.L., Becker, G., and Minick, C.R. Culture of human endothelial cells derived from umbilical veins. *Journal of Clinical Investigation* 52:2745, 1973.

24. Williams, S.K. Isolation and Culture of Microvessel and Large-Vessel Endothelial Cells: Their Use in Transport and Clinical Studies, in P.F. McDonagh, ed., *Microvascular Perfusion and Transport in Health and Disease*. Basel, Switzerland, Karger, p. 204, 1987.

25. Balconi, G., and Dejana, E. Cultivation of endothelial cells: Limitations and prospectives. *Medical Biology* 64:231, 1986

26. Ryan, U.S., and White, L.A. Varicose veins as a source of adult human endothelial cells in culture. *Science* 181:453, 1973.

27. Herring, M., Gardner, A., and Glover, J. A single-staged technique for seeding vascular grafts with autogenous endothelium. *Surgery* 84:498, 1978.

28. Graham, L.M., Burkel, W.E., Ford, J.W., Vinter, D.W., Kahn, R.H., and Stanley, J.C. Endothelial of enzymatically derived endothelium on Dacron vascular grafts. Early experimental studies with autogenous canine cells. *Arch. Surg.* 115:1289, 1980.

29. Graham, L.M., Burkel, W.E., Ford, J.W., Vinter, D.W., Kahn, R.H., and Stanley, J.C. Expanded polytetrafluoroethylene vascular prostheses seeded with enzymatically derived and cultured canine endothelial cells. *Surgery* 91:550, 1982.

30. Maciag, T.G., Hoover, G.A., Stemerman, M.B., and Weinstein, R. Serial propagation of human endothelial cells *in vitro*. *Journal Cell Biology* 91:420, 1981.

31. Thorton, S.C., Mueller, S.N., and Levine, E.M. Human endothelial cells. Cloning and long-term serial cultivation employing heparin. *Science* 222:623, 1983.

32. Jarrell, B.E., Levine, E., Shapiro, S., Williams, S.K., Carabasi, R.A., Mueller, S., and Thornton, S. Human adult endothelial cell growth in culture. *Journal of Vascular Surgery* 1:757, 1984.

33. Rosenman, J.E., Kempczinski, R.F., Pearce, W.H., and Silberstein, E.B. Kinetics of endothelial cell seeding. *Journal of Vascular Surgery* 2:778, 1985.

34. Sharefkin, J.B., Van Wart, H.E., Cruess, D.F., Albus, R.A., and Levine, E.M. Adult human endothelial cell enzymatic harvesting. *Journal of Vascular Surgery* 4:457, 1986.

35. Kirkpatrick, C.J., Melzner, I., and Göller, T. Comparative effects of trypsin, collagenase, and mechanical harvesting on cell membrane lipids studied in monolayer-cultured endothelial cells and green monkey kidney cell line. *Biochem. Biophys. Acta* 846:120, 1885.

36. Watkins, M.T., Sharefkin, J.B., Zajtchuk, R., Maciag, T.M., D'Amore, P.A., Ryan, U.S., Van Wart, H., and Rich, N.M. Adult human saphenous vein endothelial cells. Assessment of their reproductive capacity for use in endothelial cell seeding of vascular prostheses. *Journal of Surgical Research* 36:588, 1984.

37. Leseche, G., Bikfalvi, A., Dupuy, E., Tobelem, G., Andreassian, B., and Caen, J. Prelining of polytetrafluoroethylene grafts with cultured human endothelial cells isolated from varicose veins. *Surgery* 105:36, 1989.

38. Jarrell, B.E., Williams, S.K., Stokes, G., Hubbard, F.A., Carabasi, R.A., Koope, E., Grenner, D., Pratt, K., Moritz, M.J., Radomski, J., and Speicher, L. Use of an endothelial monolayer on a vascular graft prior to implantation. *Annals of Surgery* 203:671, 1986.

39. Sharp, W.V., Schmidt, S.P., Meerbaum, S.O., and Pippert, T.R. Derivation of human microvascular endothelial cells for prosthetic vascular graft seeding. *Annals of Vascular Surgery* 3:105, 1989.

40. Kern, P.A., Knedler, A., and Eckel, R.H. Isolation and culture of microvascular endothelium from human adipose tissue. *Journal of Clinical Investigation* 71:1822, 1983.

41. Williams, S.K., Jarrell, B.E., and Rose, D.G. Human microvessel endothelial cell isolation and vascular graft sodding in the operating room. *Annals Vascular Surgery* 3:146, 1989.

42. Jackson, C.J., Garbett, P.K., Nissen, B., and Schrieber, L. Binding of human endothelium to *Ulex europeaeus* 1-coated Dynabeads: Application to the isolation of microvascular endothelium. *Journal of Cell Science* 96:257, 1990.

43. Takahashi, K., Goto, T., Mukai, K., Sawasaki, Y., and Hata, J. Cobblestone monolayer cells from human omental adipose tissue are possibly mesothelial, not endothelial. *In Vitro Cellular & Developmental Biology* 25:109, 1989.

44. Clarke, J.M.F., and Pittilo, R.M. Vascular graft seeding. *Surgery* 102:890, 1987.

45. van Hinsbergh, V.W.M., Kooistra, T., Scheffer, Ma., van Bockel, J.H., and van Muijen, G.N.P. Characterization and fibrinolytic properties of human omental tissue mesothelial cells. Comparison with endothelial cells. *Blood* 75:1490, 1990.

46. Thomson, G.J.L., Vohra, R., and Walker, M.G. Cell seeding for small diameter ePTFE grafts: Comparison between adult human endothelial and mesothelial cells. *Annals of Vascular Surgery* 3:140, 1989.

47. Baitella-Eberle, G., Groscurth, P., Zilla, P., Lachat, M., Müller-Glauser, W., Schneider, J., Neudecker, A., von Segesser, L.K., Dardel, E., and Turina, M. Long-term results of tissue development and cell differentiation on Dacron prostheses seeded with microvascular cells in dogs. *Journal of Vascular Surgery* 18:1019, 1993.

48. Williams, S.K., Kleinert, L.B., Rose, D., and McKenney, S. Origin of endothelial cells that line microvessel endothelial cell sodded ePTFE vascular grafts. *Journal of Vascular Surgery* 19:594, 1994.

49. Kleinman, H.K., Klebe, R.J., and Martin, G.R. Role of collagenous matrices in the adhesion and growth of cells. *Journal of Cell Biology* 88:473, 1981.

50. Madri, J.A., and Williams, S.K. Capillary endothelial cell cultures: Phenotypic modulation by matrix components. *Journal of Cell Biology* 97:153, 1983.

51. Sharefkin, J.B., Lather, C., Smith, M., and Rich, N.M. Endothelial cell labeling with ^{111}Indium-oxine as a marker of cell attachment to bioprosthetic surfaces. *Journal of Biomedical Materials Research* 17:345, 1983.

52. Williams, S.K., Jarrell, B.E., Friend, L., Radomski, J.S., Carabase, R.A., Koolpe, E., Mueller, S.N., Thornton, S.C., Marinucci, T., and Levine, E. Adult human endothelial cell compatibility with prosthetic graft material. *Journal of Surgical Research* 38:618, 1985.

53. Ramalanjaona, G., Kempczinski, R.F., Rosenman, J.E., Douville, E.C., and Silberstein, E. The effect of fibronectin coating on endothelial cell kinetics in polytetrafluoroethylene grafts. *Journal of Vascular Surgery* 3:264, 1986.

54. Ramalanjaona, G.R., Kempsczinski, R.F., Ogle, J.D., and Silverstein, E.B. Fibronectin coating of an experimental PTFE vascular prosthesis. *Journal of Surgical Research* 41:479, 1986.

55. Kesler, K.A., Herring, M.B., Arnold, M.P., Glover, J.L., Park, H.-M., Helmus, M.N., and Bendick, P.J. Enhanced strength of endothelial attachment on polyester elastomer and polytetrafluoroethylene graft surfaces with fibronectin substrate. *Journal of Vascular Surgery* 3:58, 1986.

56. Seeger, J.M., and Klingman, N. Improved endothelial cell seeding with cultured cells and fibronectin-coated grafts. *Journal of Surgical Research* 38:641, 1985.

57. Kaehler, J., Zilla, P., Fasol, R., Deutsch, M., and Kadletz, M. Precoating substrate and surface configuration determine adherence and spreading of seeded endothelial cells on polytetrafluoroethylene grafts. *Journal of Vascular Surgery* 9:535, 1989.

58. Hasson, J.E., Wiebe, D.H., Sharefkin, J.B., D'Amore, P.A., and Abbott, W.M. Use of tritiated thymidine as a marker to compare the effects of matrix proteins on adult human vascular endothelial cell attachment: Implications for seeding of vascular prostheses. *Surgery* 100:884, 1986.

59. Sentessi, J.M., Ramberg, K., O'Donnell, T.F., Jr., Connolly, R.J., and Callow, A.D. The effect of flow on vascular endothelial cells grown in tissue culture on polytetrafluoroethylene grafts. *Surgery* 99:337, 1986.

60. Budd, J.S., Allen, K.E., Bell, P.R., and James, R.F. The effect of varying fibronectin concentration on the attachment of endothelial cells to polytetrafluoroethylene vascular grafts. *Journal of Vascular Surgery* 12:126, 1990.

61. Bujan, J., Bellon, J.M., Navlet, J.G., Honduvilla, N., Hernando, A., and Turegano, F. Seeding of expanded polytetrafluoroethylene (ePTFE) vascular grafts. A morphological study of porcine and endothelial and fibroblast cells. *Histology & Histopathology* 7:635, 1992.

62. Padomski, J.S., Jarrell, B.E., Williams, S.K., Koolpe, E.A., Greener, D.A., and Carabase, A. Initial adherence of human capillary endothelial cells to Dacron. *Journal of Surgical Research* 42:133, 1987.

63. Thomson, G.J.L., Vohra, R.K., Carr, M.H., and Walker, M.B. Adult human endothelial cell seeding using expanded polytetrafluoroethylene grafts: A comparison of four substrates. *Surgery* 109:20, 1991.

64. Foxall, T.L., Auger, K.R., Callow, A.D., and Libby, P. Adult human endothelial cell coverage of small-caliber Dacron and polytetrafluoroethylene vascular prostheses *in vitro*. *Journal of Surgical Research* 41:158, 1986.

65. Hasson, J.E., Wiebe, D.H., Sharefkin, J.B., and Abbott, W.M. Migration of adult human vascular endothelial cells: Effect of extracellular matrix proteins. *Surgery* 100:384, 1986.

66. van Wachem, P.B., Breriks, C.M., Beugeling, T., Feijen, J., Bantjes, A., Detmers, J.P., and van Aken, W.G. The influence of protein adsorption on interactions of cultured human endothelial cells with polymers. *Journal of Biomedical Materials Research* 21:701, 1987.

67. Lindblad, B., Wright, S.W., Sell, R.L., Burkel, W.E., Graham, L.M., and Stanley, J.C. Alternative techniques of seeding cultured endothelial cells to ePTFE grafts of different diameters, porosities, and surfaces. *Journal of Biomedical Materials Research* 21:1013, 1987.

68. Absolom, D.R., Hawthorn, L.A., and Chang, G. Endothelialization of polymer surfaces. *Journal of Biomedical Materials Research* 22:271, 1988.

69. Budd, J.S., Bell, P.R., and James, R.F. Attachment of indium[111] labelled endothelial cells to pretreated polytetrafluoroethylene vascular grafts. *British Journal of Surgery* 76:1259, 1989.

70. Radomski, J.S., Jarrell, B.E., Pratt, K.J., and Williams, S.K. Effects of *in vitro* aging on human endothelial cell adherence to Dacron vascular graft material. *Journal of Surgical Research* 47:173, 1989.

71. Dalsing, M.C., Kevorkian, M., Raper, B., Nixon, C., Lalka, S.G., Cikrit, D.F., Unthank, J.L., and Herring, M.B. An experimental collagen-impregnated Dacron graft: potential for endothelial seeding. *Annals of Vascular Surgery* 3:127, 1989.

72. Curti, T., Pasquinelli, G., Preda, P., Freyrie, A., Lasci, R., and D'Addato, M. An ultra-structural and immunocytochemical analysis of human endothelial cell adhesion on coated vascular grafts. *Annals of Vascular Surgery* 3:351, 1989.

73. Scott, W.J., and Mann, P. Substrate effects on endothelial cell adherence rates. *ASAIO Transactions* 36:M739, 1990.

74. Vohra, R., Thomson, G.J., Carr, H.M., Sharma, H., and Walker, M.G. Comparison of different vascular prostheses and matrices in relation to endothelial seeding. *British Journal of Surgery* 78:417, 1991.

75. Stansby, G., Shukla, N., Fuller, B., and Hamilton, G. Seeding of human microvascular endothelial cells onto polytetrafluoroethylene graft material. *British Journal of Surgery* 78: 1189, 1991.

76. Sbarbati, R., Giannessi, D., Cenni, M.C., Lazzerini, G., and Verni, F. Pyrolytic carbon coating enhances Teflon and Dacron fabric compatibility with endothelial growth. *International Journal of Artificial Organs* 14:491, 1991.

77. Newman, K.D., Nguyen, N., and Dichek, D.A. Quantification of vascular graft seeding by use of computer-assisted image analysis and genetically modified endothelial cells. *Journal of Vascular Surgery* 14:140, 1991.

78. Sank, A., Rostami, K., Weaver, F., Ertl, D., Yellin, A., Nimni, M., and Tuan, T.L. New evidence and new hope concerning endothelial seeding of vascular grafts. *American Journal of Surgery* 164:199, 1992.

79. Shrenk, P., Kobinia, G.S., Brucke, P., Syre, G., and Edstadler, A. Fibrin glue coating of e-PTFE prostheses enhances seeding of human endothelial cells. *Thoracic & Cardiovascular Surgeon* 35:6, 1987.

80. Mazzucotelli, J.P., Klein-Soyer, C., Beretz, A., Brisson, C., Archifpoff, G., and Cazenave, J.P. Endothelial cell seeding: Coating Dacron and expanded polytetrafluoroethylene vascular grafts with a biological glue allows adhesion and growth of human saphenous vein endothelial cells. *International Journal of Artificial Organs* 14:482, 1991.

81. Schneider, A., Melmed, R.N., Schwalb, H., Karck, M., Vlodavsky, I., and Uretzky, G. An improved method for endothelial cell seeding on polytetrafluoroethylene small caliber vascular grafts. *Journal of Vascular Surgery* 15:649, 1992.

82. Haegerstrand, A., Bengtsson, L., and Gillis, C. Serum proteins provide a matrix for cultured endothelial cells on expanded polytetrafluoroethylene vascular grafts. *Scandinavian Journal of Thoracic & Cardiovascular Surgery* 27:21, 1993.

83. Gourevitch, D., Jones, C.E., Crocker, J., and Goldman, M. Endothelial cell adhesion to vascular prosthetic surfaces. *Journal of Biomaterials* 9:97, 1988.

84. Greisler, H.P., Endean, E.D., Klosak, J.J., Ellinger, J., Henerson, S.C., Pham, S.M., Durham, S.J., Showalter, D.P., Levine, J., and Borovetz, H.S. Hemodynamic effects on endothelial cell monolayer detachment from vascular prostheses. *Archives of Surgery* 124: 429, 1989.

85. James, N.L., Schindlhelm, K., Slowiaczek, P., Milthorpe, B.K., Dudman, N.P., Johnson, G., and Steele, J.G. Endothelial cell seeding of small diameter vascular grafts. *Artificial Organs* 14:355, 1990.

86. Vohra, R., Thomson, G.J., Carr, H.M., Sharma, H., Welch, M., and Walker, M.G. *In vitro* adherence and kinetics studies of adult human endothelial cell seeded polytetrafluoroethylene and gelatin impregnated Dacron grafts. *European Journal of Vascular Surgery* 5: 93, 1991.

87. Prendiville, E.J., Coleman, J.E., Callow, A.D., Gould, K.E., Laliberte-Verdon, S., Ramberg, K., and Connolly, R.J. Increased *in-vitro* incubation time of endothelial cells on fibronectin-treated ePTFE increases cell retention in blood flow. *European Journal of Vascular Surgery* 5:311, 1991.

88. Kent, K.C., Oshima, A., and Whittemore, A. D. Optimal seeding conditions for human endothelial cells. *Annals of Vascular Surgery* 6:258, 1992.

89. Miyata, T., Conte, M.S., Trudell, L.A., Mason, D., and Whittemore, A.D. Delayed exposure to pulsatile shear stress improves retention of human saphenous vein endothelial cells on seeded eEPTFE grafts. *Journal of Surgical Research* 50:485, 1991.

90. Schmidt, S.P., Boyd, K.L., Pippert, T.R., Hite, S.A., Evancho, M.M., and Sharp, W.V. Endothelial Cell Seeding of Ultralow Temperature Carbon-Coated Polytetrafluoroethylene Grafts, in P. Zilla, R. Fasol, and M. Deutsch, eds., *Endothelialization of Vascular Grafts.* Basel, Karger, pp. 145–159, 1987.

91. van Wachem, P.B., Stronck, J.W., Koers-Zuideveld, R., and Dijk, F. Vacuum cell seeding: An new method for the fast application of an evenly distributed cell layer on porous vascular grafts. *Journal of Biomaterials* 11:602, 1990.

92. Jarrell, B.E., Williams, S.K., Rose, D., Garibaldi, D., Talbot, C., and Kapelan, B. Optimization of human endothelial cell attachment to vascular graft polymers. *Journal of Biomechanical Engineering* 113:120, 1991.

93. Patterson, R.B., Mayfield, G., Silberstein, E.B., and Kempczinski, R.F. The potential unreliability of indium[111]-oxine labeling in studies of endothelial cell kinetics. *Journal of Vascular Surgery* 10:650, 1989.

94. Herring, M.B., Compton, R.S., LeGrand, D.R., Gardner, A.L., Madison, D.L., and

Glover, J.L. Endothelial seeding of polytetrafluoroethylene popliteal bypasses. *Journal of Vascular Surgery* 6:114, 1987.

95. Pearce, W.H., Rutherford, R.B., Whitehill, T.A., Rosales, C., Bell, K.P., Patt, A., and Ramalanjaona, G. Successful endothelial seeding with omentally derived microvascular endothelial cells. *Journal of Vascular Surgery* 5:203, 1987.

96. Jarrell, B.E., Williams, S.K., Stokes, G., Hubbard, F.A., Carabasi, R.A., Koolpe, E., Greener, D., Pratt, K., Moritz, M.J., Radomski, J., and Speicher, L. Use of freshly isolated capillary endothelial cells from the immediate establishment of a monolayer on a vascular graft at surgery. *Surgery* 100:392, 1986.

97. Zilla, P., Fasol, R., Preiss, P., Kadletz, M., Deutsch, M., Schima, H., Tsangaris, S., and Groscurth, P. Use of fibrin glue as a substrate for *in vitro* endothelialization of PTFE vascular grafts. *Surgery* 105:515, 1989.

98. Anderson, J.S., Price, T.M., Hanson, S.R., and Harker, L.A. *In vitro* endothelialization of small-caliber vascular grafts. *Surgery* 101:577, 1987.

99. Herring, M.B., Dilley, R., Jersild, R.A., Jr., Boxer, L., Gardner, A., and Glover, J. Seeding arterial prostheses with vascular endothelium. The nature of the lining. *Annals of Surgery* 190:84, 1979.

100. Dilley, R., Herring, M., Boxer, L., Gardner, A., and Glover, J. Immunofluorescent staining for factor VIII related antigen. *Journal of Surgical Research* 27:149, 1979.

101. Herring, M., Gardner, A., and Glover, J. Seeding endothelium onto canine arterial prostheses. The effects of graft design. *Arch. Surg.* 114:679, 1979.

102. Graham, L.M., Vinter, D.W., Ford, J.W., Kahn, R.H., Burkel, W.E., and Stanley, J.C. Endothelial cell seeding of prosthetic vascular grafts. Early experimental studies with cultured autologous canine endothelium. *Arch. Surg.* 115:929, 1980.

103. Belden, T.A., Schmidt, S.P., Falkow, L.J., and Sharp, W.V. Endothelial cell seeding of small-diameter vascular grafts. *Transactions American Society for Artificial Internal Organs* 28:173, 1982.

104. Stanley, J.C., Burkel, W.E., Ford, J.W., Vinter, D.W., Kahn, R.H., Whitehouse, W.M., Jr., and Graham, L.M. Enhanced patency of small-diameter, externally supported Dacron iliofemoral grafts seeded with endothelial cells. *Surgery* 92:994, 1982.

105. Hunter, T.J., Schmidt, S.P., Sharp, W.V., and Malindzak, G.S. Controlled flow studies in 4 mm endothelialized Dacron grafts. *Transactions American Society for Artificial Internal Organs* 29:177, 1983.

106. Schmidt, S.P., Hunter, T.J., Sharp, W.V., Malindzak, G.S., and Evancho, M.M. Endothelial cell-seeded four-millimeter Dacron vascular grafts: Effects of blood flow manipulation through the grafts. *Journal of Vascular Surgery* 1:434, 1984.

107. Herring, M., Baughman, S., Glover, J., Kisler, K., Joseph, J., Campbell, J., Dilley, R., Evan, A., and Gardner, A. Endothelial seeding of Dacron and polytetrafluoroethylene grafts: The cellular events of healing. *Surgery* 96:745, 1984.

108. Sharefkin, J.B., Latker, C., Smith, M., Cruess, D., Clagett, G.P., and Rich, N.M. Early normalization of platelet survival by endothelial seeding of Dacron arterial prostheses in dogs. *Surgery* 92:385, 1982.

109. Clagett, G.P., Burkel, W.E., Sharefkin, J.B., Ford, J.W., Hufnagel, H., Vinter, D.W., Kahn, R.H., Graham, I.M., and Stanley, J.C. Antithrombotic character of canine endothelial cell-seeded arterial prostheses. *Surgical Forum* 33:471, 1982.

110. Sharp, W.V., Schmidt, S.P., and Donovan, D.L. Prostaglandin biochemistry of seeded endothelial cells on Dacron prostheses. *Journal of Vascular Surgery* 3:256, 1986.

111. Sicard, G.A., Allen, B.T., Long, J.A., Welch, M.J., Griffin, A., Clark, R.E., and Anderson, C.B. Prostaglandin production and platelet reactivity of small-diameter grafts. *Journal of Vascular Surgery* 1:744, 1984.

112. Kaplan, S., Marcoe, K.F., Sauvage, L.R., Zammit, M., Wu, H.-D., Mathisen, S.R., and

Walker, M.W. The effect of predetermined thrombotic potential of the recipient on small-caliber graft performance. *Journal of Vascular Surgery* 3:311, 1986.

113. Schmidt, S.P., Hunter, T.J., Falkow, L.J., Evancho, M.M., and Sharp, W.V. Effects of antiplatelet agents in combination with endothelial cell seeding on small diameter Dacron vascular graft performance in the canine carotid artery model. *Journal of Vascular Surgery* 2:898, 1985.

114. Hirko, M.K., Schmidt, S.P., Evancho, M.M., Sharp, W.V., and Donovan, D.L. Endothelial cell seeding improves 4 mm PTFE vascular graft performance in antiplatelet medicated dogs. *Artery* 14:137, 1987.

115. Shepard, A.D., Eldrup-Jorgenson, J., Keough, E.M., Foxall, T.F., Ramberg, K., Connolly, R.J., Mackey, W.C., Gavris, V., Auger, K.R., Libby, P., O'Donnell, T.F., and Callow, A.D. Endothelial cell seeding of small-caliber synthetic grafts in the baboon. *Surgery* 99:318, 1986.

116. Boyd, K.L., Schmidt, S., Pippert, T.R., Hite, S.A., and Sharp, W.V. The effects of pore size and endothelial cell seeding upon the performance of small-diameter e-PTFE vascular grafts under controlled flow conditions. *Journal of Biomedical Material Research* 22:163, 1988.

117. Kempczinski, R.F., Rosenman, J.E., Pearce, W.-H., Roedersheimer, L.R., Berlatzky, Y., and Ramalanjaona, G. Endothelial cell seeding of a new PTFE vascular prosthesis. *Journal of Vascular Surgery* 2:424, 1985.

118. Pearce, W.H., Rutherford, R.B., Whitehill, T.A., Rosales, C., Bell, K.P., Patt, A., and Ramalanjaona, G. Successful endothelial seeding with omentally derived microvascular endothelial cells. *Journal of Vascular Surgery* 5:203, 1987.

119. Schmidt, S.P., Monajjem, N., Evancho, M.M., Pippert, T.R., and Sharp, W.V. Microvascular endothelial cell seeding of small-diameter Dacron vascular grafts. *Journal of Investigative Surgery* 1:35, 1988.

120. Sterpetti, A.V., Hunter, W.J., Schultz, R.D., Sugimoto, J.T., Blair, E.A., Hacker, K., Chasa, P., and Valentine, J. Seeding with endothelial cells derived from microvessels of the omentum and from the jugular vein. *Journal of Vascular Surgery* 7:677, 1988.

121. Noishiki, Y., Yamane, Y., Satoh, S., Niu, S., Okoshi, T., Tomizawa, Y., and Wildevuur, C.R. Healing process of vascular prostheses seeded with venous tissue fragments. *ASAIO Transactions* 37:M478, 1991.

122. Sterpetti, A.V., Schultz, R.D., and Bailey, R.T. Endothelial cell seeding after carotid endarterectomy in a canine model reduces platelet uptake. *European Journal of Vascular Surgery* 6:390, 1992.

123. Plate, G., Hollier, L.H., Fowl, R.J., Sande, J.R., and Kaye, M.P. Endothelial seeding of venous prostheses. *Surgery* 96:929, 1984.

124. Herring, M., Gardner, A., Peigh, P., Madison, D., Baughman, S., Brown, J., and Glover, J. Patency in canine inferior vena cava grafting: Effects of graft material, size, and endothelial seeding. *Journal of Vascular Surgery* 1:877, 1984.

125. Köveker, G.B., Burkel, W.E., Graham, L.M., Wakefield, T.W., and Stanley, J.C. Endothelial cell seeding of expanded polytetrafluoroethylene vena cava conduits: Effects on luminal production of prostacyclin, platelet adherence, and fibrinogen accumulation. *Journal of Vascular Surgery* 7:600, 1988.

126. Clarke, J.M.F., Pittila, R.M., Nicholson, L.J., Woolf, N., and Marston, A. Seeding Dacron arterial prostheses with peritoneal endothelial cells: A preliminary morphological study. *British Journal of Surgery* 71:492, 1984.

127. Bull, H.A., Pittilo, R.M., Drury, J., Pollock, J.G., Clarke, J.M.F., Woolf, N., Marston, A., and Machin, S.J. Effects of autologous mesothelial cell seeding on prostacyclin production with Dacron arterial prostheses. *British Journal of Surgery* 75:671, 1988.

128. Hollier, L.H., Fowl, R.J., Pennell, R.C., Heck, C.F., Winter, K.A.H., Fass, D.N., and

Kaye, M.P. Are seeded endothelial cells the origin of neointima on prosthetic vascular grafts? *Journal of Vascular Surgery* 3:65, 1986.

129. Zamora, J.L., Navarro, L.T., Ives, C.L., Weilbaecher, D.G., Gao, Z.R., and Noon, G.P. Seeding of arteriovenous prostheses with homologous endothelium. *Journal of Vascular Surgery* 3:860, 1986.

130. Wakefield, T.W., Earley, E.M., Brothers, T.E., Burkel, W.E., Graham, L.M., Fessler, R.D., Saenz, N., Sell, R.M., and Stanley, J.C. Karyotype analysis of cell sex to determine the source of vascular graft luminal linings following autologous and nonautologous endothelial cell seeding. *Transactions American Society of Artificial Internal Organs* 34:864, 1988.

131. Clowes, A.W., Clowes, M.M., Kirkman, T.R., and Reidy, M.A. Capillary endothelium can substitute for arterial endothelium in healing arterial replacements. *Federation Proceedings* 45:473, 1986.

132. Nabel, E.G., Plautz, G., Boyce, F.M., Stanley, J.C., and Nabel, G.J. Recombinant gene expression *in vivo* within endothelial cells of the arterial wall. *Science* 244:1342, 1989.

133. Wilson, J.M., Birinyi, L.K., Salomon, R.N., Libby, P., Callow, A.D., and Mulligan, R.C. Implantation of vascular grafts lined with genetically modified endothelial cells. *Science* 244:1344, 1989.

134. Dailey, S.W., Rose, D.G., Carabasi, R.A., Ahlswede, K., and Williams, S.K. Origin of cells that line damaged native blood vessels following endothelial cell transplantation. *American Journal of Surgery* 162:107, 1991.

135. Herring, M., Gardner, A., and Glover, J. Seeding human arterial prostheses with mechanically derived endothelium. The detrimental effect of smoking. *Journal of Vascular Surgery* 1:279, 1984.

136. Herring, M.B., Compton, R.S., Gardner, A.L., and LeGrand, D.R. Clinical Experiences with Endothelial Seeding in Indianapolis, in P. Zilla, R. Fasol, and M. Deutsch, eds., *Endothelialization of Vascular Grafts*. Basel, Karger, pp. 218–224, 1987.

137. Herring, M., Baughman, S., and Glover, J. Endothelium develops on seeded human arterial prosthesis: A brief clinical note. *Journal of Vascular Surgery* 2:727, 1985.

138. Zilla, P., Fasol, R., Deutsch, M., Fischlien, T., Minar, E., Hammerle, A., Krupicka, O., and Kadletz, M. Endothelial cell seeding of polytetrafluoroethylene vascular grafts in humans: A preliminary report. *Journal of Vascular Surgery* 6:535, 1987.

139. Risberg, B., Ortenwall, P., Wadenvik, H., and Kutti, J. Endothelial Cell Seeding: Experience and First Clinical Results in Göteborg, in P. Zilla, R. Fasol, and M. Deutsch, eds., *Endothelialization of Vascular Grafts*. Basel, Karger, pp. 225–232, 1987.

140. Ortenwall, P., Wadenvik, H., Kutti, J., and Risberg, B. Endothelial cell seeding reduces thrombogenicity of Dacron grafts in humans. *Journal of Vascular Surgery* 11:403, 1990.

141. Ortenwall, P., Wadenvik, H., and Risberg, B. Reduced platelet deposition on seeded versus unseeded segments of expanded polytetrafluoroethylene grafts: Clinical observations after a 6-month follow-up. *Journal of Vascular Surgery* 10:374, 1989.

142. Meerbaum, S.O., Sharp, W.V., and Schmidt, S.P. Lower Extremity Revascularization with Polytetrafluoroethylene Grafts Seeded with Microvascular Endothelial Cells, in P. Zilla, R. Fasol, and A. Callow, eds., *Applied Cardiovascular Biology 1990–91*, Vol. 2. International Society for Applied Cardiovascular Biology, Basel, Karger, pp. 107–119, 1992.

143. Libby, P., Birinyi, L.K., and Callow, A.D. Functions of Endothelial Cells Related to Seeding of Vascular Prostheses: The Unanswered Questions, in M. Herring and J.D. Glover, eds., *Endothelial Cell Seeding in Vascular Surgery*. Orlando, FL, Grune & Stratton, p. 1735, 1987.

144. Zwibel, J.A., Freeman, S.M., Kantoff, P.W., Cornetta, K., Ryan, U.S., and Anderson, W.F. High-level recombinant gene expression in rabbit endothelial cells transduced by retroviral vectors. *Science* 243:220, 1989.

145. Callow, A.D. The vascular endothelial cell as a vehicle for gene therapy. *Journal of Vascular Surgery* 11:793, 1990.

VIII
Coronary Applications

41
Flexible Leaflet Replacement Heart Valves

Stephen L. Hilbert, Michael Jones, and Victor J. Ferrans
Food and Drug Administration
Rockville, Maryland

I. INTRODUCTION

The objective of this chapter is to present a review of preclinical and clinical studies conducted to assess the safety and efficacy of flexible leaflet replacement heart valves. This review focuses primarily on the preclinical evaluation of prototype bioprosthetic heart valves implanted in the mitral position in juvenile sheep; however, pertinent *in vitro* investigations and *in vivo* studies conducted in other animal models also are discussed. Whenever possible, clinical findings are compared with preclinical data on animals. The histologic, biomechanical, and functional alterations that develop in autograft, allograft, xenograft, and polymeric valves as a consequence of tissue harvesting and processing are addressed. Each section concludes with an analysis of the mechanisms responsible for alterations in the durability and the hemodynamic performance of the various replacement heart valves discussed.

The sheep model was developed at the National Heart, Lung, and Blood Institute in 1982 and has become the standard animal model for the preclinical testing of new designs of prosthetic heart valves [1–3]. For several reasons, juvenile sheep are suitable for the study of changes that develop after implantation in prosthetic heart valves. They grow rapidly, increasing in average weight from 29 kilograms (kg) at the time of valve implantation to 45 kg at the time of study. Their blood pressure, heart rate, cardiac output, and intracardiac pressures are essentially the same as those of young human beings. Their valve annular sizes are suitable for implantation of commercially available bioprosthetic valves, and their vessels and body sizes are adequate for standard cardiopulmonary bypass techniques.

The sheep model was originally designed for the study of the accelerated calcification that develops in bioprosthetic xenograft valves, especially in those implanted in children and young adults. This model utilizes lambs that become young adults in four to seven months after valvular implantation. Therefore, they experience within this period of time the growth and the metabolic changes that require nearly a decade in humans. In addition, the morphologic and hemodynamic alterations that develop in the valves implanted in these animals are similar to those recognized in valves implanted in humans. As described in this chapter, the sheep model has demonstrated its usefulness in investigations of a wide variety of types of replacement heart valves.

The first clinical implantation of a replacement heart valve was accomplished in 1950 with the insertion of a ball-in-cage prosthesis in the descending thoracic aorta [4]. This approach to the management of aortic regurgitation did not require cardiopulmonary bypass. The development of cardiopulmonary bypass technology facilitated the first orthotopic heart valve replacement using a ball-in-cage prosthesis. However, the primary disadvantage of ball-in-cage and disk-in-cage prostheses and of mechanical valves in general is their altered patterns of blood flow, which are characterized by restriction of central flow, flow separation, and turbulence (Figs. 1 and 2). The alterations in blood rheology and the presence of synthetic materials greatly increase the potential for thrombosis; for this reason chronic anticoagulation therapy is required when these valves are used clinically.

The thromboembolic events and the sudden, life-threatening modes of failure associated with replacement heart valves of the mechanical type have stimulated interest in the use of tissue valves (allografts and autografts) and the development of bioprosthetic heart valves (xenografts). Tissue and bioprosthetic valves became the valves of choice during the 1970s, having the advantages of not requiring long-term anticoagulation and showing less restriction of central blood flow (Fig. 3).

As long-term clinical experience increased, it became apparent that tissue valves and bioprostheses have limited durability [5] (particularly in the mitral position) due to primary tissue failure (tissue abrasion and wear, calcification). The modes of clinical failure of tissue valves and bioprosthetic valves usually are more gradual and less sudden than those of mechanical valves. The rate of reoperation to replace malfunctioning bioprosthetic heart valves has been reported to be 40% after 8 to 10 years of use [5].

 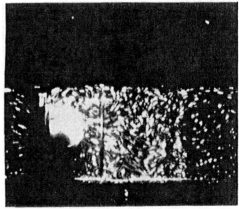

Figure 1   Ball-in-cage valve shown in the open position. Flow visualization (right panel) demonstrates marked obstruction to central flow, as illustrated by the presence of nonstreaming particles immediately downstream of the occluder.

 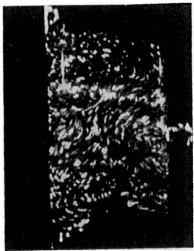

Figure 2   Disk-in-cage valve shown in the open position. Flow visualization (right panel) depicts the obstruction to central flow by the disk-shaped occluder.

While the designs of tissue and bioprosthetic valves were evolving, significant improvements were made in mechanical valves, resulting in less restriction of central blood flow (Figs. 4 and 5) and in more durable and stable occluder materials (e.g., optimal curing of silastic elastomers, pyrolytic carbon); however, lifelong anticoagulation is still required with mechanical valves to prevent catastrophic thrombosis and/or thromboembolism.

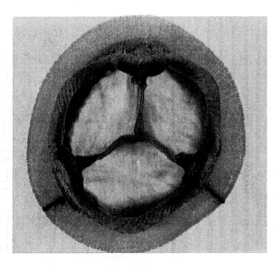 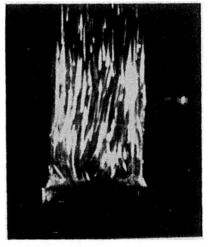

Figure 3   Porcine aortic valve bioprosthesis illustrating the regions of leaflet coaptation. Flow visualization (right panel) demonstrates the presence of streaming particles not impeded by a centrally placed occluder.

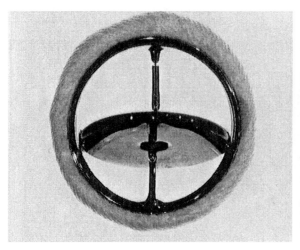

Figure 4 A tilting disk valve shown in the open position. Note the presence of a major (top) and a minor (bottom) outflow orifice. Flow visualization demonstrates (right panel) minimal restriction of central flow as compared with the ball-in-cage or disk-in-cage valve designs.

The materials selected for use as leaflet or occluder components of replacement heart valves include silastic elastomers, polyacetal, pyrolytic carbon, polyurethane, expanded polytetrafluoroethylene, parietal pericardium, dura mater, and aortic, mitral and pulmonic valvular tissues. The development of replacement heart valves made of tissue and of bioprosthetic and flexible leaflet polymeric materials intended to have longer durability has continued to progress. Recent innovations include the use of polyurethane, zero-pressure tissue fixation, nonaldehyde cross-linking agents, anticalcification treatments,

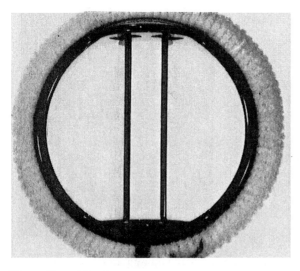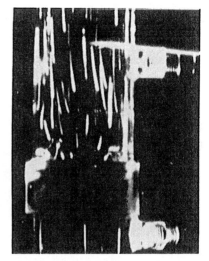

Figure 5 A bileaflet valve, shown in the open position, depicting the minimal restriction of flow by the two centrally placed leaflets.

unstented porcine xenografts, autologous pericardial tissue, and cryopreserved allograft valves. With the exception of cryopreserved allograft aortic valves, which demonstrate a modest increase in long-term durability [6], the clinical efficacy and freedom from primary tissue failure of the next generation of bioprosthetic and autologous tissue valves remains to be demonstrated.

II. AUTOLOGOUS TISSUE VALVES

A variety of other autologous tissues, such as parietal pericardium [7], aortic wall [8], rectus abdominis sheath [9], and superior vena cava [10] have been used in the fabrication of replacement heart valves.

The transplantation of an autologous pulmonary valve to the aortic position, also referred to as the *pulmonary switch*, is a single exception to the general rule that autologous tissues are not suitable for the replacement of diseased valvular tissues [11,12]. Although the performance and durability of the autologous pulmonary valve are notable, the use of this valve has two limitations: (1) a less than ideal prosthetic heart valve must be substituted for the pulmonary valve and (2) it can only be used to replace the aortic valve.

A. Fascia Lata Valves

Fascia lata was the first autologous tissue used for valvuloplasty and for the fabrication of prosthetic heart valves [13]. Autologous fascia lata has been utilized for (1) complete replacement of a diseased aortic valve cusp by a continuous piece of fascia lata, (2) correction of valvular regurgitation by using a strip of fascia lata to lengthen a shortened cusp, and (3) creation of a new free edge of a leaflet for the repair of mitral valve prolapse.

Human fascia lata is composed of two layers: a thin, superficial layer, and a thick, deeper layer. Both layers consist mostly of collagen fibrils (80–250 nanometers [nm] in diameter) and a few sparse fibroblasts, elastic fibers, and capillaries. The collagen fibrils in the superficial layer are oriented perpendicular to the long axis of the fascia lata, while those in the deeper layer are parallel to this axis. The middle third of the length of the human fascia lata is the portion used for the construction of valvular leaflets [14].

The initial experience with autologous fascia lata essentially involved the use of free-hand techniques, which required prolonged intraoperative time. This may have contributed to the significant variability observed in the clinical performance of these valves. Standardized methods were subsequently developed for the intraoperative fabrication of stented autologous fascia lata valves [15,16]. These valves continued to be used clinically until the mid-1970s, when it became evident that death of the connective tissue cells and thickening and contraction of the fascia lata leaflets by fibrous connective tissue of host origin (also referred to as fibrous sheath) continued to be the primary reasons for dysfunction of these valves [17]. Thus, the principal mode of failure observed in autologous tissue valves consists of thickening and contraction of the tissue leaflet, generally after four years of implantation; however, in contrast to xenograft heart valves, leaflet calcification has not been a significant finding in fascia lata valves [18].

B. Dura Mater Valves

Replacement trileaflet heart valves fabricated of glycerol-treated human dura mater were initially reported to give a satisfactory clinical performance (e.g., low thrombogenicity, adequate durability) [19,20]. However, subsequent studies reported a much higher incidence of complications related to primary tissue failure. The use of dura mater valves has decreased sharply during the past few years. Normal human dura mater consists of two fibrous layers, an outer endosteal layer and an inner meningeal layer. The endosteal layer comprises 70% of the full thickness of the dura and is composed of large collagen bundles. In this layer, roughly equal numbers of collagen bundles are oriented either parallel to each other or are multidirectional. The thinner (inner) meningeal layer consists of smaller collagen bundles frequently oriented in an oblique or perpendicular direction with respect to the outer layer. The collagen fibrils have large diameters (average, 110 nm) and display typical collagen periodicity. The two layers are separated by a space of variable width, which typically is occupied by blood vessels and lymphatics. The collagen bundles in both layers demonstrate considerable waviness. Both layers contain few elastic fibers and small amounts of proteoglycans, as assessed by histologic staining (Alcian blue-PAS [Schiff Reagent]). Elongated fibroblasts with extensive profiles of rough endoplasmic reticulum are present throughout both layers [21].

The explant pathological findings on explanted dura mater valves are generally similar to those reported for glutaraldehyde-treated pericardial and porcine aortic valvular bioprostheses. Grossly, calcific nodules and corresponding leaflet perforations are observed as well as linear tears extending from the free edge to the base of the leaflet. These tears were not associated with calcific deposits. Microscopically, collagen bundles had lost their native waviness or crimping and appeared straight. As has been also noted in other long-term bioprosthetic heart valve explants, there were regions within the cusp in which collagen bundles were separated and collagen fibrils had undergone fragmentation and disruption (i.e., electron dense granular material with interspersed collagen fibrils). Ultrastructural studies demonstrated the presence of calcific deposits, either between collagen fibrils and/or associated with collagen fibrils and devitalized cells. Fibrous sheathing was also observed on the outflow surface of the cusps [21]. The fibrous sheath originates from host-derived connective tissue and consists primarily of Type III collagen, proteoglycans, elastic fibers, fibroblasts, and variable numbers of endothelial cells covering the surface.

The effects of glycerol versus glutaraldehyde treatment on the biomechanical properties of bioprosthetic leaflets have not been extensively studied. Glycerol treatment may involve tissue dehydration, protein denaturation, and fixation by aldehyde contaminants present as impurities. The collagen within dura mater shrinks during glycerol treatment; this shrinkage contributes to the prominent extent of collagen waviness observed in dura mater valves; however, it seems likely that glycerol does not increase the degree of crosslinking of collagen fibrils. The influence of these factors on the long-term durability of dura mater bioprostheses remains uncertain.

The transmission of Creutzfeldt–Jakob disease has been associated with the use of cadaveric human dura mater [22]. The potential transmission of virus-related diseases may limit the future use of dura mater in replacement heart valves.

C. Parietal Pericardial Valves

Recently, there has been a renewed interest in autologous tissue heart valves that are fabricated intraoperatively using parietal pericardium as stented trileaflet valves [23]. For the fabrication procedure, a rectangular strip of pericardium is immersed in a solution of buffered 0.625% glutaraldehyde for 5 minutes. A tool is used to define the leaflet geometry and to cut the pericardial tissue accordingly. The cut tissue is precisely oriented and wrapped in an inner stent, and the splayed outer stent is placed over the tissue-wrapped inner stent. The initial preclinical studies suggest that this new method should greatly reduce the variability typically noted with other methods used to fabricate autologous tissue valves [24].

The preclinical evaluation of pericardial autologous tissue valves implanted in the mitral position in juvenile sheep has demonstrated satisfactory hemodynamic performance. Histopathologic studies of valves explanted 20 weeks after implantation showed a marked loss of leaflet cellularity and extensive insudation of plasma proteins (Fig. 6). Variable degrees of fibrous sheath formation had overgrown the stent rail and continued onto the leaflet surface (Fig. 7). Evidence of abrasive wear was not noted; however, tissue thinning and compression secondary to mounting the tissue between the inner and outer stent components was observed. Host tissue ingrowth (i.e., irregular connective tissue) was observed in these regions of tissue compression.

The explant pathological findings in this series of pericardial autologous tissue valves were similar to those frequently observed with glutaraldehyde-treated pericardial bioprosthetic valves; however, there was a remarkable absence of intrinsic calcification in the

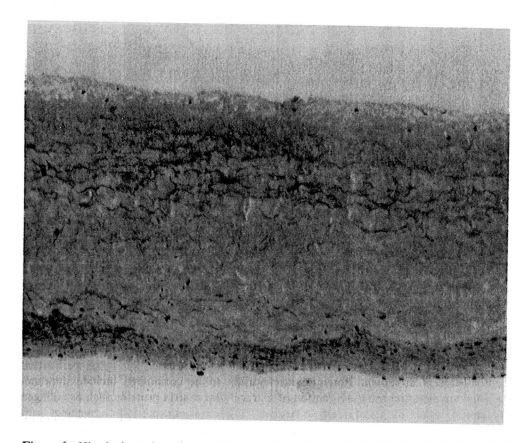

Figure 6 Histologic section of an autologous pericardial valve leaflet implanted for 20 weeks in the mitral position in sheep depicting loss of cellularity, insudated plasma proteins, and fibrous sheath formation on the inflow aspect of the leaflet (bottom) (glycol methacrylate embedding, toluidine blue stain, ×150).

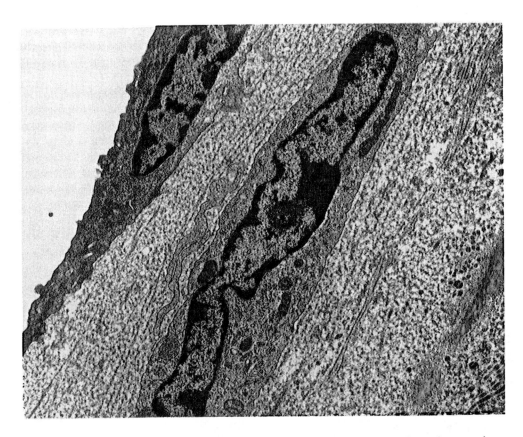

Figure 7   Ultrathin section of the fibrous sheath present on the inflow surface of an autologous
pericardial valve leaflet. The fibrous sheath is composed of fibroblasts, collagen fibrils, elastic
fibers, and proteoglycans. The surface is lined by an endothelial cell (20-week old mitral valve
replacement in sheep, uranyl acetate/lead citrate stain, × 5400).

leaflets made of autologous pericardium. The extent of leaflet calcification ranged from
0.6 milligrams (mg) to 25.7 milligrams per gram (mg/g) of dry tissue. The mean content
of calcium in these valves was 8.36 mg/g, compared with a mean value of 73.1 mg/g in
pericardial xenografts implanted in sheep as mitral valve replacements for a similar period
of time [25].

The mechanism responsible for the marked difference in the extent of calcification
of pericardial autologous tissue valves and glutaraldehyde-fixed xenograft pericardial
bioprostheses is unknown. Potential mechanisms to be considered include immune-
mediated changes and the modification of extracellular matrix proteins such as collagen
and proteoglycans by glutaraldehyde fixation. The latter concept is in agreement with
observations showing that bovine pericardial and porcine aortic valvular tissues treated
with glutaraldehyde undergo extensive calcification when implanted intradermally in rats,
while unfixed tissues of these types do not calcify [26]. Similar findings have been
observed with unfixed and aldehyde-fixed sponges made of purified Type I collagen [27].

III. ALLOGRAFT HEART VALVES

A. Aortic and Pulmonary Valve Allografts

Human cadaver pulmonary and aortic valves (referred to as homografts or allografts) have been used for valvular and ventricular outflow tract reconstruction for approximately 38 years. Before mechanical and bioprosthetic heart valves became readily available, allograft heart valves were implanted in the descending thoracic aorta for the surgical management of aortic regurgitation [28]. With the evolution of allograft technology and the development of cardiopulmonary bypass, the first orthotopic aortic valve replacements were accomplished in the 1960s [29,30]. Currently, allograft heart valves have become the standard of care for reconstruction of the right ventricular outflow tract in children [31].

As the demand for valvular allografts increased, various approaches to process and store allograft valves were developed. However, these early efforts to provide a less restricted supply of allograft valves resulted in a significant reduction in hemodynamic performance and in clinical use [32]. Treatment with and storage in antibiotic-containing solutions (4°C for up to 6 weeks) remained the method of choice for preparing fresh allografts until the recent development of cryopreservation techniques [33,34].

Cryopreservation has minimized the logistical problems associated with allograft supply and storage. Cryopreserved pulmonary and aortic valves have become the replacement heart valves of choice in children and young adults [35,36]. Long-term clinical studies indicate that cryopreserved allograft heart valves demonstrate improved durability and freedom from valve-related complications compared with bioprosthetic valves, particularly in young patients [34,37,38].

1. Morphology

Morphologically, human pulmonary and aortic valves consist of three distinct layers of tissue: the ventricularis, the spongiosa, and the fibrosa (Fig. 8). The ventricularis, an extension of the ventricular endocardium, contains prominent elastic fibers oriented perpendicular to the free edge of the leaflet. The central region, referred to as the *spongiosa*, consists of loosely organized collagen fibers, fibroblasts, and myofibroblasts embedded in a matrix rich in proteoglycans. The fibrosa is subjacent to the outflow surface and is composed of collagen bundles oriented parallel to the free edge, a small number of elastic fibers, proteoglycans, fibroblasts, and myofibroblasts. The elastic fibers typically appear as a distinct layer near the surface of the fibrosa in the basal region of the cusp. This histologic zone is known as the *arterialis*. The fibrosa is increased in thickness along the line of cuspal coaptation, forming a fibroelastic nodule in the center of the free edge. This centrally placed nodule is known as the *nodulus Arantii* in the aortic valve and the *nodulus Morgagni* in the pulmonary valve. The valvular surfaces are covered by a single, continuous layer of endothelial cells [39,40].

Type I and Type III collagen are the most abundant forms of collagen in cardiac valves; however, Type V is also present as a minor component [41]. Recent evidence suggests that collagens may be present as either homogeneous fibrils or as mixed fibrils (e.g., Type I and Type III collagens) [42]. It is currently unknown whether cardiac valves contain mixed fibrils.

Extracellular polymerization of collagen molecules results in the formation of collagen fibrils. The functional effects of proteoglycan-collagen interactions may involve the

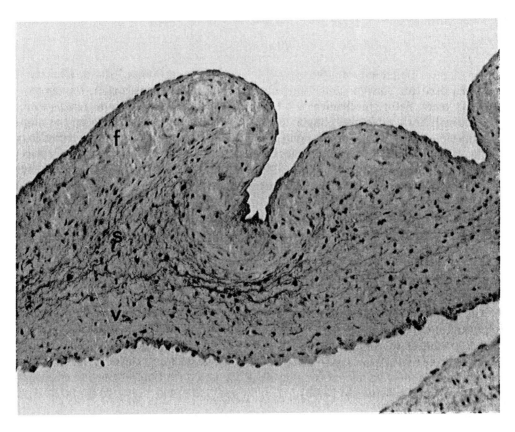

Figure 8 Histologic features of a freshly harvested aortic valve. Note the extent of cellularity and the presence of three histologic layers: the fibrosa (f), the spongiosa (s), and the ventricularis (v) (glycol methacrylate embedding, toluidine blue stain, ×150).

regulation of collagen fibril polymerization since domains of high positive charge density are located within the collagen molecules in the regions of staggered overlap. These sites are critical for the formation of intermolecular cross-links of collagen fibrils. Proteoglycan glucosaminoglycan side chains could also attach to these cationic sites, thus regulating the extent of collagen fibril cross-linking. This concept is supported by the association of dermatan-sulfate-rich proteoglycans with tissues containing collagen fibrils or large diameters and by the presence of chondroitin-sulfate-containing proteoglycans in tissues in which fibrils of small diameter are predominant [43].

A variety of gender- and age-related morphologic changes occur in human aortic valves. These changes are most frequently observed in men and include (1) degeneration of collagen fibers, (2) formation of a fibroelastic "spur" along the coaptation surface, (3) a decrease in the number of fibroblasts and in the content of proteoglycan, (4) accumulation of extracellular lipid, and (5) formation of calcific deposits [44–46]. Because of the prevalence of these changes, current donor criteria for allograft heart valves typically include an age restriction (e.g., less than 50 years of age) and lack of a history of previous cardiac surgery, uncontrolled hypertension, significant cardiac murmurs, rheumatic fever, and malignant, autoimmune, and vascular diseases [47].

2. Tissue Harvesting and Cryopreservation

Allograft heart valves are harvested, disinfected, and cryopreserved in the following manner: First, warm ischemic time (the time elapsed from death to beginning of harvesting) is generally restricted to 24 hours or less. Second, cold temperatures (4°C) are used for dissection to procure the allograft and for subsequent transportation. Third, for disinfection, the valves are exposed for 24 hours at 4°C to antibiotics (e.g., cefoxitin, lincomycin, polymyxin B, vancomycin). Fourth, dimethylsulfoxide (10%) is used as a cryoprotectant. Finally, the valves are then subjected to controlled-rate freezing, 1° per minute to −70°C, and stored in liquid nitrogen vapor (−170°C) until use.

The effects of preharvesting ischemia, disinfection, cryopreservation, and thawing on allograft morphology and intermediate metabolism have been investigated recently [48–50]. These studies indicate that both the duration of harvest-related warm ischemia and the preimplantation processing (disinfection, cryopreservation, and thawing) induce marked alterations in allograft morphology and metabolism, including a decrease in adenine nucleotides and an increase in lactate. As the warm ischemic time is increased, there was a progressive loss of endothelial cells, fibroblasts, and myofibroblasts. Endothelial cell loss was most notable initially, while the morphology of interstitial cells began to deteriorate markedly after approximately 12 hours of warm ischemia. Similarly, an inverse association was also observed between the duration of warm ischemia and the levels of high-energy phosphates.

Fibroblasts and myofibroblasts were the predominant cell types seen within the cusp. These cells developed a spectrum of cellular alterations associated with increasing warm ischemia, including cellular and mitochondrial swelling, dilation of endoplasmic reticulum, degranulation of rough endoplasmic reticulum, mitochondrial flocculent densities, hydropic vacuoles, pyknosis, karyorrhexis and karyolysis, lipid accumulation, and disruption of organelle and plasma membranes. Morphologic changes of irreversible cellular injury, such as mitrochrondrial flocculent densities, karyolysis, and plasma membrane disruption, became progressively more severe in association with increasing warm ischemic time (Fig. 9). All of the morphologic and metabolic changes described above were further aggravated by disinfection, cryopreservation, and thawing.

The morphologic characteristics of elastic fibers and collagen fibrils were retained. However, a progressive loss of proteoglycans was noted as a result of increasing warm ischemic times and preimplantation processing. The extent and magnitude of collagen bundle waviness (crimp) was unaltered by warm ischemia and allograft processing (Fig. 10).

3. Biomechanical Properties

The functional characteristics of the aortic valve are dependent on (1) a dynamic aortic root, (2) the ability of the cusps to close in response to small changes in pressure, and (3) the sharing of mechanical and dynamic stresses between the cusps and the sinuses of Valsalva [51–53]. Stresses are generated as a consequence of the pressure difference across the closed cusps and reversal of cuspal curvature as the valve opens and closes. Studies of aortic valve biomechanics indicate that the height of the commissure and the length of the free edge do not change with increasing pressure; however, the diameter of the sinus and the elastic modulus of the cusp do increase.

With increasing pressure, there is a corresponding decrease in coaptive surface area and cuspal thickness. These geometric changes serve to distribute the pressure load evenly across the coaptive surface. The free edge is also exposed to minimal stress at the time of peak pressure load following valve closure. Bending stresses are also decreased within the

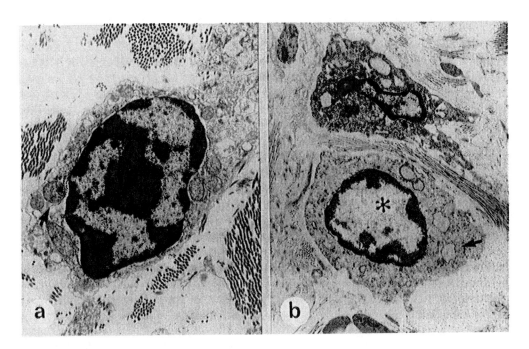

Figure 9   Transmission electron micrograph illustrating various types of cellular injury in aortic valve stromal cells following 24 to 36 hours of warm ischemia (uranyl acetate/lead citrate stain): (a), note the extent of cytoplasmic edema, mitochondrial swelling, and the presence of a mitochondrial flocculent density (arrow) ($\times$4800); (b), profiles of dilated endoplasmic reticulum (arrow) and margination of nuclear heterochromatin (asterisk) ($\times$3500).

coaptive region due to a decrease in cuspal thickness in response to increasing pressure. Bending stresses resulting from the reversal of cuspal curvature during the opening and closing of the valve are less than the static loading stresses present in the tissue when the valve is closed. The static loading stresses are distributed circumferentially.

The organization of the extracellular connective tissue components of the valve (i.e., collagen fibrils, elastic fibers, and proteoglycans) in the sinus of Valsalva minimize the stresses present within the cuspal tissue throughout the cardiac cycle. The basal region of the cuspal attachment to the sinus wall undergoes extensive bending and the spongiosa in this region is expanded into a wedge-shaped configuration rich in proteoglycans. The fibrosa continues into the sinus wall, where the collagen bundles integrate into the aortic wall.

The biomechanical properties of the aortic valve reflect the histologic organization of the extracellular connective tissue components within the cusp [53–56]. Stress-strain studies have demonstrated the nonlinear viscoelastic properties of aortic valve tissue. The elastic modulus is greatest in the circumferential direction and the extensibility is greatest in the radial direction. Thus, aortic valvular tissue is an inhomogenous and anisotropic material. The anisotropic biomechanical properties of the aortic valve reflect the histologic features of the cusp, namely, the circumferential orientation of collagen bundles (fibrosa) and the radial arrangement of elastic fibers (primarily in the ventricularis).

Figure 10   Scanning electron micrograph demonstrating the retention of collagen crimp within the free edge of a cryopreserved pulmonary allograft (×750).

Another morphologic feature, the extent of collagen bundle crimp, is associated with the circumferential compliance. Changes in the circumferential compliance correspond to alterations of collagen crimp rather than to changes in the overall extensibility of collagen bundles. Thus, changes in the collagen crimp (i.e., straightening of the waviness of collagen fibrils) accounts for the majority of the circumferential compliance of the aortic valve [53,55]. Last, the collagen and elastic fibers are dispersed in a gel-like ground substance primarily composed of proteoglycans. As mentioned above, the spongiosa contains the highest concentration of proteoglycans and is positioned between the ventricularis and the fibrosa. The marked concentration of proteoglycans in this region may serve to reduce the shear and bending stresses generated by the relative movements of the ventricularis and the fibrosa during valvular opening and closure [54].

4.   Pathology

Explanted cryopreserved allograft valves demonstrate the following histopathologic findings: (1) marked reduction in the number and integrity of the cellular components (endothelial cells, fibroblasts, and myofibroblasts) of donor origin; (2) increase in cuspal thickness, primarily due to the formation of a layer of fibroelastic connective tissue (fibrous sheath) of host origin, which extends from the basal regions of both sides of the cusps and eventually lines them completely; (3) infiltration of both the donor tissue and the fibrous sheath by inflammatory cells, including lymphocytes, macrophages, and plasma cells; and (4) insudation of plasma proteins since the allograft does not have a continuous endothelial barrier. In contrast to glutaraldehyde-cross-linked xenograft bioprostheses, allograft valves do not undergo extensive cuspal calcification. However, the aortic or pulmonary artery walls can become mineralized [57].

The pathologic changes associated with noncryopreserved allograft valves depend on the type of preimplantation processing to which the tissue is subjected. For example, an increased rate of cuspal rupture was observed in allograft valves sterilized with ethylene oxide or β-propiolactone. In addition to a reduction in tensile strength, β-propiolactone induced tissue shrinkage, cuspal thickening, and a corresponding decrease in surface area [58,59]. The alterations in cuspal geometry caused by sterilization may predispose allograft valves to become regurgitant. Lyophilization also has been reported to decrease the tensile strength of allograft valves and to be associated with an increased incidence of cuspal rupture, with tissue failure occuring either at the cusp-aortic wall junction or at the free edge near the commissure. Both of these sites correspond to regions of high compressive and tensile stress, suggesting that lyophilization significantly reduces allograft mechanical durability [44,60].

The morphologic and metabolic observations discussed above suggest that the improved clinical performance of cryopreserved allograft valves may be related to the retention of critical components of the extracellular matrix rather than to the preservation of viable cuspal cells capable of renewing cuspal connective tissue.

B.   Mitral Valve Allografts

In contrast to aortic valve allografts, mitral valve allografts have had only very limited clinical use. The long-term results of mitral valvular replacement with mitral valve allografts treated with antibiotics have been very poor due to regurgitation caused by rupture of the chordae tendineae. Variable results have been obtained after implantation of fresh mitral valve allografts in dogs [61]. A recent study was made of the postimplantation changes that developed in antibiotic-treated and glutaraldehyde-treated mitral valve allografts after implantation for 20 weeks in the juvenile sheep model [62].

This study showed that the glutaraldehyde-treated allografts underwent rapid, severe calcification, which resulted in valvular failure due to both stenosis (related to the calcific deposits interfering with cuspal mobility) and regurgitation (related to fracture of calcified chordae tendineae). The structural features of the calcific deposits in these valves are similar to those observed in explanted glutaraldehyde-treated xenograft bioprostheses. The antibiotic-treated mitral allografts also failed, albeit to a lesser degree, because of structural deterioration of cuspal and chordal connective tissue leading to cuspal perforation and chordal rupture, respectively. The leaflets of antibiotic-preserved mitral valve allografts showed histologic changes, including structural deterioration of connective tissue components and formation of a fibrous sheath of host origin, that were very similar to those previously found in aortic valve allografts. However, the mitral allografts failed to a much greater extent than would have been expected on the basis of previous experience with aortic valve allografts. The reason for this failure is thought to be related to damage to the mitral chordae tendineae, which are subjected to uniquely high mechanical stresses during left ventricular systole and diastole in a manner that has no counterpart in aortic valve opening and closure.

IV. BIOPROSTHETIC HEART VALVES

The development and clinical evaluation of xenograft valves for the management of valvular heart disease has continued for approximately 30 years [63–66]. Bioprosthetic valves consisting of either glutaraldehyde-treated porcine aortic valves (PAVs) or bovine parietal pericardial valves (BPVs) have been extensively used clinically as replacement heart valves. Bovine and kangaroo aortic valves also have been used in prototype designs. Typically, bioprostheses are configured as trileaflet valves, although bileaflet and mono-leaflet valve designs also have been evaluated [67–69]. Knowledge gained from investigations of the structure and function of the native aortic valve is being applied extensively to the development of the next generation of bioprostheses [69,70]. In addition, efforts are being made to reduce xenograft calcification and to optimize the biomechanical properties of glutaraldehyde-treated tissues [25,26,53].

A. Porcine Aortic Valve Bioprostheses

1. Morphology

The morphology of the PAV is comparable to that described for the human aortic valve (see Section III), with two exceptions. The right coronary cusp of the PAV is larger than the other two cusps and contains a layer of cardiac monocytes. This layer, referred to as the *muscle shelf*, is an extension of the ventricular septal myocytes into the basal region of the cusp. The presence of the muscle shelf results in a delayed opening of the right coronary cusp relative to that of the left and the noncoronary (posterior cusp). This delay is also observed in porcine aortic valve xenografts. Second, the proteoglycan content, as assessed by histologic staining (Alcian blue-PAS), is reduced within the spongiosa in comparison to fresh and cryopreserved aortic valve allografts.

2. Tissue Harvesting

PAVs are harvested in commercial slaughterhouses and shipped (4°C) to valve manufacturers, generally within 24 to 48 hours after collection. During this interval, autolytic damage occurs, resulting in cell lysis, release of intracellular enzymes, loss of proteoglycans, edema, and focal areas of increased eosinophilia. The endothelial cells lining the leaflet surfaces are completely lost as a consequence of tissue harvesting and processing. These morphologic changes resemble those in allograft ischemic injury. The effects of these autolytic changes on PAV durability and the incidence of primary tissue failure, tissue dehiscence, and leaflet calcification are unknown.

3. Fixation

PAV fixation is accomplished using low concentrations (less than 1%) of glutaraldehyde. Fixation reduces xenograft antigenicity, decreases tissue compliance, reduces the rate of degradation of leaflet proteins (particularly of collagen), and increases thermal stability (shrinkage temperature). Glutaraldehyde is cytotoxic even at extremely low concentrations. The conditions of fixation, including aldehyde purity and concentration, temperature, pH, time, and hydrostatic pressure, determine the extent and types of protein cross-links that are formed during this process. Glutaraldehyde reacts with primary amino groups, such as those in lysine, hydroxylysine, and N-terminal amino acids, present in leaflet proteins to form Schiff bases, unsaturated addition reaction products, and pyridinium-type of compounds. These reactions result in stable intra- and intermolecular cross-links, particularly in collagen [71,72].

As a consequence of tissue fixation, bioprosthetic valves exhibit altered biomechanical properties (see "Biomechanical Properties," next) and the cellular components are rendered nonviable. Thus, the ability to regenerate connective tissue components has been lost. The connective tissue in bioprostheses eventually deteriorates because of the wear and tear related to leaflet stresses generated during the cardiac cycle.

Formaldehyde also has been used as a PAV fixative. The cross-links formed by formaldehyde (a monoaldehyde) are less extensive and less stable than those formed by glutaraldehyde, which is a dialdehyde. Formaldehyde-mediated cross-links are initially stable, but progressively dissociate following storage. This may account for the poor clinical durability of formaldehyde-treated PAVs [73–75].

4. Biomechanical Properties

The biomechanical properties of glutaraldehyde-treated PAV tissue are altered by protein cross-linking and the hydrostatic pressure applied to the tissue during fixation. Uniaxial stress-strain studies of glutaraldehyde-treated leaflets indicate that the circumferential and radial compliance are reduced, although the anisotropic properties of the tissue are not affected. The shape of the initial portion of the stress-strain curve (the transitional region) is altered in comparison to that of untreated tissue. Aldehyde fixation results in an alteration of the shape of the transitional region, reflecting the change from a compliant tissue to a tissue that stiffens rapidly at higher stresses; however, the slope of the posttransitional region of the curve is not altered. The characteristic shape of the transitional region is related to the extent of collagen crimp, which is present primarily in the fibrosa. The transition from a compliant tissue to a tissue with increased stiffness is the consequence of the loss of collagen crimp [55].

The ability of the glutaraldehyde-treated PAV tissue to resist structural deterioration during cyclic loading (mechanical durability) is a function of the fatigue characteristics of the tissue. Cyclic loading of glutaraldehyde-treated valvular tissue further reduces the compliance below that observed after aldehyde fixation alone, making the tissue more susceptible to primary tissue failure. Fracture of collagen fibrils will occur from compressive flexure during the unloaded portion of the cyclic fatigue cycle. The retention of collagen crimp is critical to maintaining the desired mechanical properties and resistance to fatigue-induced tissue deterioration [55]. It has been suggested that the durability of PAV xenografts may be enhanced if the native valvular collagen crimp is retained during glutaraldehyde fixation. The magnitude of the hydrostatic pressure applied to the tissue during fixation determines the extent of collagen crimp (Figs. 11 and 12). These considerations have led to the use of low hydrostatic pressures during PAV fixation in the preparation of bioprostheses [76] (Figs. 13 and 14).

Polarized light microscopy and morphometric methods are ideally suited for the quantitative assessment of the effects of pressure fixation on collagen crimp [77,78]. The results of these studies demonstrate that high-pressure fixation (80–100 mmHg) markedly alters the extent of collagen crimp within the commissural cords and the midcuspal region, but not within the free edge of the aortic valve leaflet. In the high-pressure-fixed PAVs, the majority of collagen crimp was eliminated. In contrast to these findings, collagen crimp was retained in low-pressure-fixed PAVs (less than 2 mmHg) in the majority of the commissural cords as well as within the free edge and midcusp regions. The cords in which collagen crimp was not seen were all located close to the commissures. In contrast to high- and low-pressure-fixation conditions, the number of collagen crimps per unit length increased significantly within the cords of PAVs fixed at zero pressure.

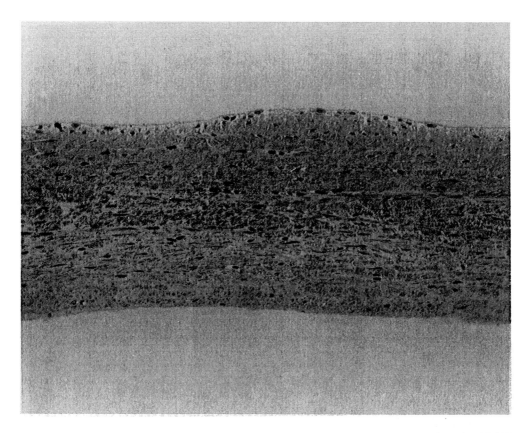

Figure 11 Histologic section of a high-pressure-fixed, unimplanted porcine aortic valve (PAV) xenograft (glycol methacrylate embedding, toluidine blue stain, ×150). The typical convolutions (representing transversely sectioned collagen cords) in the fibrosa have been flattened and the endothelial cells have been lost during preimplantation processing. Compare with Fig. 8.

The results of *in vitro* studies indicate that fixation pressure does significantly affect collagen crimp, primarily in the commissural cords [53,79]. It remains to be demonstrated whether low- or zero-pressure fixation will significantly reduce the incidence of PAV xenograft primary tissue failure observed clinically.

5. Anticalcification Treatments

Calcification is the most frequent cause of xenograft valve dysfunction [80–82]. Calcification may involve intrinsic valve components such as devitalized cells, collagen fibrils, and elastic fibers, or extrinsic components such as thrombi (Figs. 15–17). Calcific deposits are initially associated with devitalized cells, but progress to involve the fibrous components of the valvular connective tissue [83,84]. The phospholipids present in the membranes of devitalized cells may serve as nucleation sites for calcification [85]. The calcification of collagen fibrils is the most significant form of bioprosthetic calcification since collagen is the principal extracellular protein in valvular tissue. The contribution of immune-mediated mechanisms to the process of dystrophic calcification of xenografts is unknown [86]. Tissue processing factors such as glutaraldehyde fixation, the use of

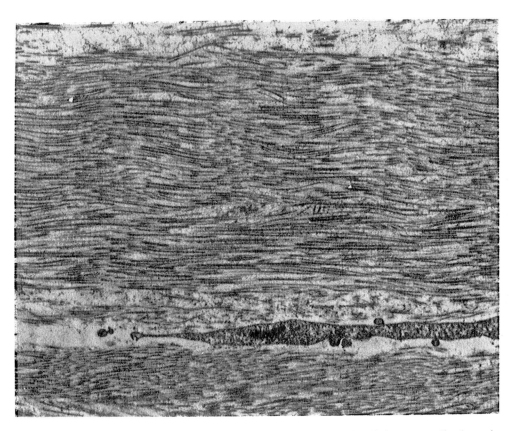

Figure 12 Ultrathin section taken from a region of the fibrosa of a high-pressure-fixed porcine aortic valve (PAV) xenograft (uranyl acetate/lead citrate stain, ×5400). The loss of collagen crimp corresponds to the straightening of collagen bundles.

phosphate buffers, and the extraction of proteoglycans are also thought to contribute to the development of xenograft calcification [26,87]. Eventually, large calcific nodules are formed that cause valvular stenosis and the formation of perforations and tears, with subsequent regurgitation.

Numerous anticalcification treatments have been evaluated in efforts to develop strategies for mitigating xenograft calcification. Agents such as diphosphonates, magnesium chloride, ferric chloride, aluminum chloride, phosphocitrate, protamine, toluidine blue, polyacrylamide, L-glutamic acid, and surfactants effectively reduce the calcification of pieces of valvular tissue implanted intradermally in animals [88–91]. However, only surfactants such as sodium dodecyl sulfate, polysorbate 80, and α-amino-oleic acid have proven to be effective in the mitigation of PAV xenograft calcification in bioprostheses implanted as mitral valve replacements in juvenile sheep [25,92]. Long-term clinical studies will be necessary to determine if these agents will effectively reduce the rate of PAV xenograft calcification in humans.

Studies of the anticalcification effects of surfactants also should include an evaluation of possible detrimental effects of these agents on valvular connective tissue. Some of

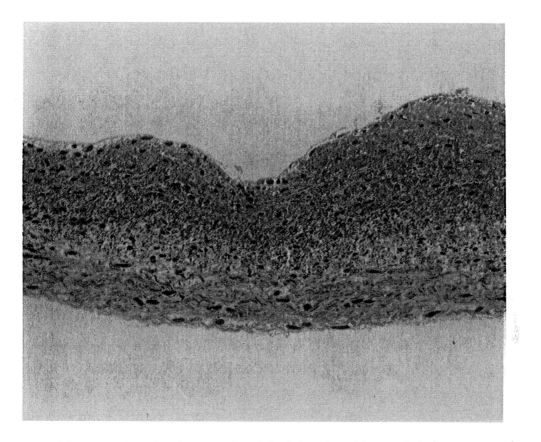

Figure 13   Histologic section demonstrating minimal alteration of the morphologic appearance of the porcine aortic valve (PAV) fibrosa by low-pressure fixation (glycol methacrylate embedding, toluidine blue stain, ×150).

these agents have been shown to reduce the durability of the valves by inducing damage to cuspal connective tissue. PAV bioprostheses treated with Triton X-100 and N-lauryl sarcosine were found to have developed cuspal tears and perforations after implantation in the mitral position for up to 20 weeks in the sheep model [25].

6. Sterilization

The effectiveness of glutaraldehyde and formaldehyde as sterilants as bacteria, fungi, and viruses is dependent upon the conditions of use of these agents. Process validation studies have demonstrated that, at low concentrations, glutaraldehyde is less effective than form-aldehyde against *Mycobacterium* species, bacterial spores, and fungi [71,93]. These find-ings were confirmed by the report of the contamination of a limited number of clinically implanted PAV xenografts by *Mycobacterium chelonii* [94,95]. Due to the documented resistance of *Mycobacterium* species to glutaraldehyde, xenografts are now typically steri-lized with a formaldehyde-containing sterilant solution following glutaraldehyde tissue fixation.

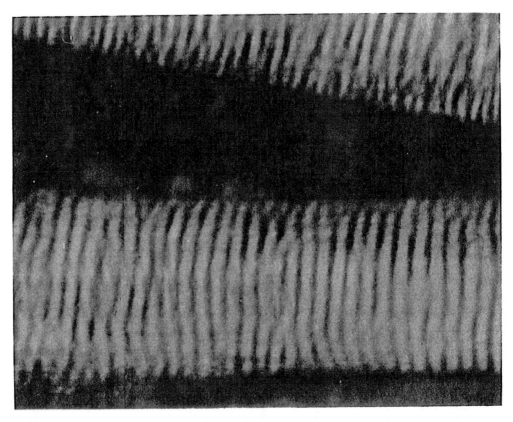

Figure 14   Polarized light micrograph of a collagen cord on the outflow surface of an aortic valve. Polarized light microscopy is ideally suited for the visualization of collagen crimp, as demonstrated in this micrograph by the repeating birefringent pattern (wet specimen, ×75).

7. Mounting

To accommodate their implantation, bioprostheses (BPs) are usually mounted on stents, in contrast to most allografts, which are implanted without the aid of a stent. Stent mounting of tissue valves facilitates their insertion in the mitral and tricuspid position.

Both rigid and flexible stent designs have been used. Rigid stents are composed of metal; flexible stents are made of either polymeric materials or a thinly drawn metal wire (Figs. 18–20). All the currently used clinical xenografts have flexible stents, although the degree of flexibility varies considerably among different stent designs and reflects the configuration and the mechanical properties of the stent materials. Stents typically consist of an outer sewing ring, a circular frame with three stent posts that provide structural support for the valve commissures, and an external cloth covering. Flexible stents result in a slight bending motion of the stent posts during valve opening and closing. The shape of the sewing ring may be modified to allow implantation in either the aortic or the atrioventricular position.

These modifications may be best exemplified by the development of the supraannular aortic valve, which employs a scalloped sewing ring, the shape of which resembles that of the aortic root. All contemporary xenograft designs are configured as trileaflet valves

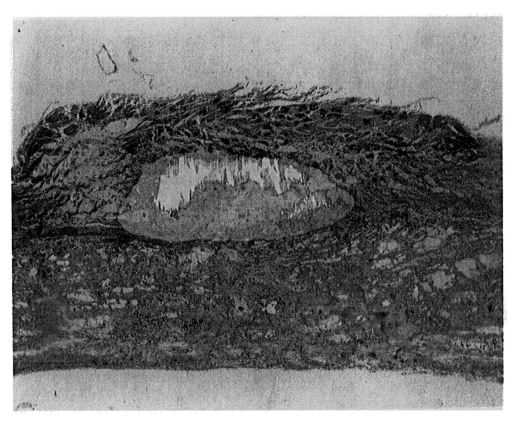

Figure 15 Light micrograph showing a calcific nodule within the cusp of porcine aortic valve (PAV) xenograft implanted for 20 weeks in the mitral position in sheep (glycol methacrylate embedding, toluidine blue stain, ×150).

regardless of their intended site of implantation. Although stent designs duplicate some of the features of the native aortic valve annulus, they are not as expansile as is the native aortic annulus. Therefore, stented xenografts do not undergo the same dimensional and geometric changes as the native cardiac valves (e.g., systolic expansion, diastolic contraction, and the outward deflection of the commissures during systole). Thus, the mechanical forces to which the BP leaflets are subjected during the cardiac cycle are higher and less evenly distributed in BPs than in the native valve [96,97].

Tissue dehiscence (separation of tissue from the stent) and leaflet tears were the principal failure modes observed in PAV xenografts mounted on rigid stents [73,74]. In an effort to reduce the incidence of primary tissue failure, flexible stents were developed to dissipate the stresses associated with leaflet motion, particularly during valvular closure [98]. To reduce the incidence of PAV xenograft dehiscence further, a cloth bias strip (buttress) is also sewn over the PAV-stent interface to reduce the stresses applied to the tissue by the attachment to the stent [99]. The use of flexible stent designs have resulted in a reduction in the incidence of PAV primary tissue failure; however, recently there have been an increasing number of reports of tissue dehiscence associated with the supra-

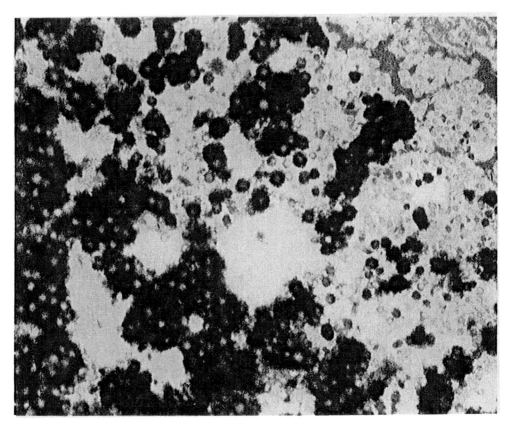

Figure 16 Electron micrograph demonstrating the morphologic appearance of collagen fibril calcification. This alteration may either involve the entire fibril or be limited to the fibril surface (uranyl acetate/lead citrate stain, ×20,000).

annular PAV xenograft. The relationship of these lesions to the stent design of this valve has not been clarified.

Flexible stent designs have evolved from a stent that has a rigid inlet and in which the stent posts undergo some degree of outward motion during valve opening and inward motion during valve closure [100], to a stent consisting of a thin wire configured such that it allows expansion and contraction of the inflow orifice as well as radial deflection of the stent posts [101].

Ideally, the material selected for stent fabrication should have a high fatigue resistance and undergo minimal plastic deformation (creep). Polypropylene, polyacetal, and metal wire have been used as stent materials [98,100,101]. Inward blending of polypropylene stent posts has been observed in explanted bioprostheses. Creep, external compression secondary to the use of an oversize valve (disproportion), and tissue ingrowth onto the stent fabric have been suggested as potential mechanisms culminating in the alteration of polypropylene stent geometry [102–104].

A variety of stent design modifications and tissue mounting techniques have been

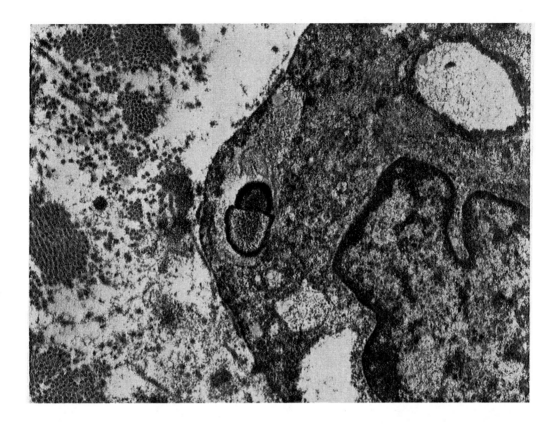

Figure 17 Electron micrograph demonstrating calcification within the cytoplasm of a devitalized cell. Organelle membrane fragments serve as nucleation sites for calcification (uranyl acetate/lead citrate stain, ×38,500).

made based on both empirical observations and computer modeling [105–107]. These modifications include (1) mounting the tissue higher on the stent in order to increase the effective orifice area, (2) incorporating more of the right coronary cusp into the sewing ring to minimize the obstructed effect of the muscle shelf, and (3) configuring a "modified orifice valve" in which the right coronary cusp is removed and replaced by a nonmuscle shelf cusp from another aortic valve [100,108].

Clinical trials are in progress to evaluate the efficacy of unstented PAV xenografts for aortic valvular replacement [109,110]. Hemodynamic studies indicate that the gradient across the valve is less than that observed in stented xenografts of comparable size. The decrease in pressure gradient corresponds to the increase in the effective orifice area of unstented PAV xenografts [110].

The continuing evolution of stent design and the use of unstented PAV xenografts reflect the various approaches being considered to increase the long-term durability of PAV xenografts through efforts to reduce the incidence of primary tissue failure.

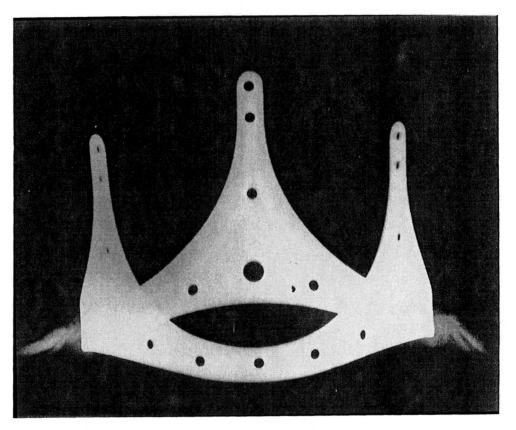

Figure 18   Radiograph of a bovine pericardial valve xenograft mounted on a rigid stent (×3).

B.   Pericardial Bioprostheses

Pericardial bioprostheses are typically constructed from glutaraldehyde-treated bovine parietal pericardium; porcine pericardium has been used for the construction of valved conduits (Polystan valve) [111]. Most bovine pericardial valves in clinical use have been configured as trileaflet valves, although monocusp and bicuspid valves also have been designed [67–69]. Preclinical animal studies suggest that the long-term performance of monocusp pericardial valves may be compromised by the development of leaflet redundancy due to progressive elongation of the pericardial tissue monocusp [112].

The fluid dynamic characteristics of pericardial valves are improved over those of PAV xenografts. The effective orifice areas of BPVs are larger than those of PAV xenografts of comparable sizes [113,114]. The clinical use of pericardial xenografts has decreased considerably because of the increased incidence of early onset primary tissue failure. This failure is manifested by leaflet tears, particularly in valves implanted in the mitral position [115–118]. Recently, new flexible stent designs and tissue mounting techniques have been developed that may significantly reduce the incidence of primary tissue failure and increase the long-term clinical durability of bovine pericardial valves [118]. However, leaflet calcification may still limit the clinical performance of these valves [119,120].

Tissue harvesting, glutaraldehyde cross-linking, and sterilization methods used for pericardial xenografts are similar to those employed for the fabrication of PAV bioprostheses. However, tissue mounting techniques and the efficacy of anticalcification treatments differ significantly for the two types of valves.

1. Morphology

Parietal pericardial tissue consists of three histologic layers, which differ markedly from those of the aortic valve leaflets. These layers are a serosal surface layer of mesothelial cells; the fibrosa, which contains collagen bundles, elastic fibers, nerves, blood vessels, and lymphatics; and the epipericardial layer, formed by loosely arranged collagen bundles and elastic fibers (Fig. 21). Fibroblasts are the principal cell type present in the fibrosa and epipericardial layers. Adipose tissue, mast cells and histiocytes also are present.

Tissue harvesting and preimplantation processing with glutaraldehyde result in loss of the mesothelial cells lining the serosal surface (Fig. 22). Tissue fixation is achieved without the use of hydrostatic pressure. The pericardial tissue is usually mounted on a stent so that the epipericardial surface becomes the inflow surface of the leaflet and the serosal surface becomes the outflow aspect. Macroscopic examination clearly demonstrates that the inflow surface of the leaflet is extremely rough, while the outflow surface

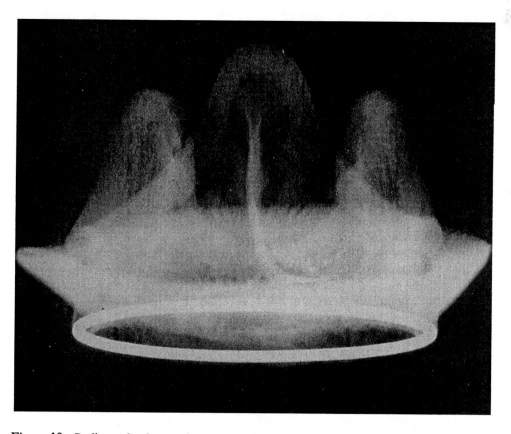

Figure 19 Radiograph of a porcine aortic valve (PAV) xenograft mounted on a flexible stent (×3). Note radiopaque marker incorporated into the sewing ring.

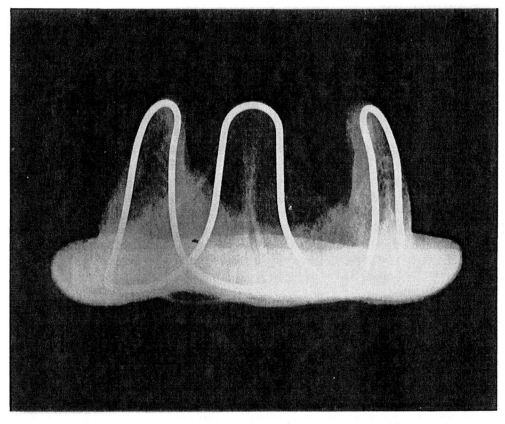

Figure 20   Radiograph of a porcine aortic valve (PAV) xenograft mounted on a flexible stent composed of a thinly drawn wire that also serves as a radiopaque marker (×3).

is quite smooth. Scanning electron microscopy demonstrates these differences in a striking manner. The serosal surface is lined by basement membrane material with collagen bundles visible immediately beneath the surface, while coarse collagen bundles are apparent on the epipericardial surface. The collagen bundles in pericardium are arranged in overlapping, multidirectional layers. The amplitude of the collagen crimp in parietal pericardium is greater than that observed in the aortic valve fibrosa; the collagen bundles in pericardial valves are not always oriented parallel to the free edge as they are in the PAV. The free edge of the pericardial leaflet is formed by cutting the tissue. Thus, it is composed of multiple layers of transected bundles of collagen [121].

2. Biomechanical Properties

The biomechanical properties of pericardium differ significantly from those of the aortic valve leaflets. Pericardial tissue is considered to be an isotropic material. The mechanical characteristics of pericardium reflect the histologic architecture of this tissue, namely, multiple overlapping layers of collagen bundles and elastic fibers. Glutaraldehyde cross-linking increases the tensile strength and percentage strain at fracture and reduces stress relaxation and creep [122]. The biomechanical properties of pericardium vary significantly among different regions of the pericardial sac. Variations in tensile strength, percent strain at fracture, and extensibility have been attributed to the presence of differing amounts of elastic fibers [123]. The collagen fibrils in bovine pericardium are larger in diameter and are more easily fractured by repeated bending during the cardiac cycle.

3. Anticalcification Treatments

Treatment with surfactants, which mitigate the calcification of PAV bioprostheses implanted in the mitral position in the sheep model, does not demonstrate comparable effectiveness when applied to pericardial valves under similar circumstances [124]. Calcification of pericardial xenografts alters valve hemodynamics and has an impact on pericardial valve durability as a consequence of primary tissue failure. Histologic and ultrastructural studies of explanted pericardial valves indicate that devitalized cell components, myelinated nerves, insudated plasma proteins, and collagen fibrils serve as nucleation sites for calcification. In contrast to explanted PAV xenografts, large amounts of insudated plasma proteins are dispersed throughout the extracellular matrix of explanted pericardial valves. Pericardial xenografts may be more prone to the insudation of

Figure 21   Light micrograph of a section of freshly harvested bovine parietal pericardium; note a blood vessel (center) and an intact layer of mesothelial cells (top) (glycol methacrylate embedding, toluidine blue stain, × 150).

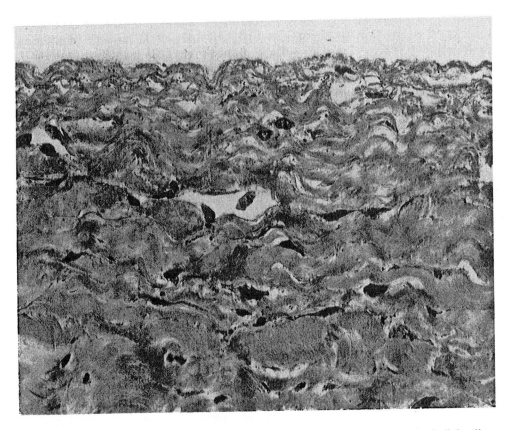

Figure 22 Histologic section illustrating the loss of the surface layer of mesothelial cells as a result of preimplantation processing of bovine pericardial leaflet (glycol methacrylate embedding, toluidine blue stain, × 700).

plasma proteins and lipids than PAV xenografts because the free edge and the rough epipericardial surface do not constitute effective barriers against the penetration of these components of blood [125]. The accumulation of plasma proteins and lipids in pericardial valves may be one factor contributing to the lack of efficacy of anticalcification agents in pericardial xenografts [126].

4. Mounting

The reduced durability of pericardial xenografts, as compared to that of PAV bi prostheses, is the consequence of primary tissue failure occurring principally at the site of tissue mounting to the commissural stent posts [127–129]. The use of an alignment suture as the method of attachment of the pericardial tissue to the stent has been associated with an increased incidence of primary tissue failure at this site (Fig. 23). The mode of primary tissue failure may be best described as an Ishihara Type 1 tear (i.e., in close proximity to the commissure and extending to the free edge) [130] (Fig. 24). These tears are considered to be consequences of localized mechanical stresses concentrated in the tissue by the use of an alignment suture.

Three design approaches have been used to attach the pericardial tissue to the stent:

(1) wrapping the tissue around the stent post and placing an alignment suture a few millimeters below the free edge in the paracommissural region of the leaflet, (2) wrapping the pericardium around the stent post without using an alignment suture, and (3) using a stent design intended to reduce the mechanical stresses within the leaflet tissue and eliminating the use of an alignment suture. Clinical studies indicate that valve designs that reduce leaflet tissue stresses prevent the occurrence of Ishihara Type 1 tears typically seen after 5 or 6 years of use [118–120]. However, the long-term performance of pericardial xenografts may be ultimately limited by altered hemodynamic performance secondary to leaflet calcification [120,124].

V. POLYMERIC HEART VALVES

Synthetic polymeric materials have been widely applied in the development of implantable cardiovascular devices such as pacemaker leads, peripheral vascular grafts, ventricular-assist devices, total artificial hearts, prosthetic heart valves, and valved conduits. The materials used most frequently to fabricate the flexible polymeric leaflet of prosthetic heart valves include polyurethanes, expanded polytetrafluoroethylene (ePTFE), and

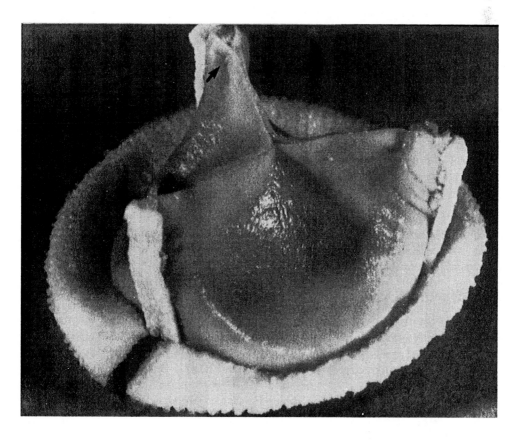

Figure 23   Unimplanted bovine pericardial valve illustrating the use of an alignment suture (arrow) and bias strips applied to the outside of the stent posts (×4).

Figure 24 Scanning electron micrograph depicting an Ishihara Type 1 tear in close proximity to a stent post (bovine pericardial bioprosthesis explanted 75 months after implantation in the aortic portion in a 76-year-old man, ×20).

composite materials (e.g., laminated ePTFE) [131–137]. Of these materials, segmented polyurethanes have been the subject of the most extensive research for their potential application as flexible leaflet material for replacement heart valves and for the valvular components of ventricular-assist devices [137–140].

A. Polyurethane Valves

The continued interest in segmented polyurethanes is based on their physical properties, which are ideally suited for use as leaflet components in replacement heart valves. However, the results of preclinical studies evaluating segmented polyurethane prosthetic valves have been discouraging because of calcification [141,142]. Our discussion of polyurethane valves is limited to a consideration of the explant pathology findings and the pathogenesis of calcification since this mode of valve failure significantly limits the performance of polyurethane and other types of polymeric valves.

The explant pathological findings of a prototype segmented polyurethane trileaflet valve used as a replacement mitral valve in sheep are presented to illustrate the pathologic mechanisms associated with polyurethane heart valves [141]. A similar triad of pathologic features occurs in both xenograft and polyurethane heart valves. This triad consists of leaflet calcification, fibrous sheathing, and thrombus formation (Fig. 25a).

However, the pathogenesis of calcification in the polyurethane leaflet is notably different from that occurring in xenograft valves. Intrinsic calcification, primarily involving collagen fibrils, is responsible for most of the leaflet mineralization observed in xenograft valves. In contrast, calcific plaques develop on the leaflet surface of polyurethane valves (Fig. 25b). The magnitude and distribution of mechanical stresses within the leaflets of both types of valves are believed to contribute to the development of leaflet calcification [143-145].

The membranous components of devitalized cells and the insudated plasma proteins and lipids also serve as potential nucleation sites for calcification. These components are responsible for the calcific deposits typically noted within the fibrous sheath and thrombi on the surfaces of the leaflets of both xenograft and polymeric valves. Thus, the observation of calcific deposits on the surface of polyurethane valves is not an unexpected finding. However, the presence of calcific plaques that are not associated with thrombi or fibrous sheath components may be unique to polyurethane leaflets. This observation suggests that an intrinsic characteristic of the polyurethane surface may be responsible for the initiation of this calcification.

The presence of calcific plaques on the leaflet surface is not associated with surface defects (e.g., microscopic pits) typically produced by the formation of microbubbles

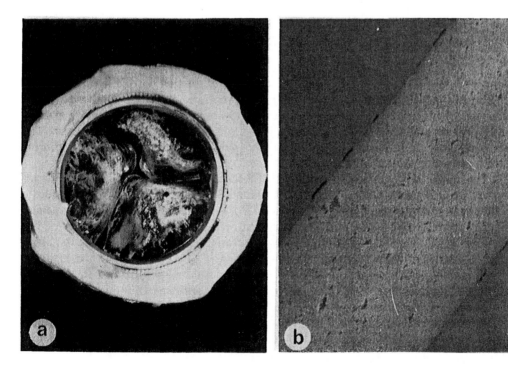

Figure 25 Valve implanted for 20 weeks in the mitral position in sheep (transmitted polarized light microscopy, glycol methacrylate embedding, ×60): (a), the outflow surface of an explanted polyurethane valve covered with fibrous sheath and thrombotic material; (b), histologic section demonstrating adherent plaquelike deposits on both the inflow and the outflow valve surfaces.

during the casting process. Calcific deposits associated with surface defects are usually noted in areas in which thrombus formation has occurred. It is not known whether the polyurethane surface modifications resulting from the casting procedure, the use of chemical adjuncts (e.g., antioxidants incorporated into the polymer), or residual mechanical stresses present within the leaflet induce the formation of calcific plaques.

Segmented polyurethanes and low molecular weight models of polyurethane soft segment regions (e.g., crown ethers) have been reported to form polymer-calcium complexes [145,146]. In addition, adsorbed plasma phospholipids and proteins containing γ-carboxyglutamic acid or other carboxylated amino acids may serve as initiators of calcification in polyurethane heart valves and blood-pumping diaphragms [147,148].

The calcium content of polyurethane valves is within the range of that typically observed in explanted xenograft valves that have not undergone anticalcification treatments (42.7 to 130.1 mg/gm dry weight of leaflet) [25,141]. These findings must be interpreted with caution since there are marked physical and chemical differences between these two classes of prosthetic valves (e.g., high water content of tissue valves versus a polymer material). Developmental research continues in an effort to reduce the extent of polyurethane calcification. These activities include (1) modifications of the soft segment components of the copolymer, (2) changes in leaflet geometry and stent flexibility to minimize the mechanical stresses within the leaflet material, (3) design modifications optimizing the fluid dynamic characteristics to reduce fluid shear stresses and thromboembolic potential further, (4) the incorporation of anticalcification agents into segmented polyurethanes to mitigate calcification, and (5) modification of the polyurethane casting processes to minimize formation of surface defects (e.g., microbubbles, effects of humidity).

B. Polytetrafluoroethylene Valves

Expanded polytetrafluoroethylene (ePTFE) was one of the first polymeric materials used to fabricate aortic valve prostheses [135–137]. However, as in the case of other polymeric materials, the use of ePTFE in prosthetic valves has been unsuccessful. A valve made of an ePTFE has undergone evaluation as a replacement mitral valve in juvenile sheep. This newly designed valve also showed poor hemodynamic performance because of fibrous sheath formation and calcification.

The unique microscopic structure of ePTFE contributes to the development of pathologic changes in valves made with this biomaterial. Scanning electron microscopy ideally demonstrates the morphology of this material (Fig. 26). There are regions of unexpanded PTFE, referred to as nodes, that are connected by expanded regions consisting of filamentous PTFE. The internodal distance may be varied during manufacturing to accommodate tissue ingrowth, if desired. In our experience, ePTFE valves demonstrated altered hemodynamic performance and limited durability after being implanted in the mitral position in juvenile sheep. This dysfunction was a consequence of fibrous sheath formation, calcification, and abrasive wear (Fig. 27). In contrast to the calcification observed in polyurethane valves, that in ePTFE valves was associated with devitalized cells primarily present within microthrombi and the fibrous sheath. This type of calcification is not a function of an intrinsic property of the biomaterial, but a consequence of the host response to the material. The expanded filamentous internodal regions provide an ideal environment for the deposition of connective tissue components that will eventually serve as sites for the initiation of calcification and progress to the formation of calcific nodules and abrasive wear.

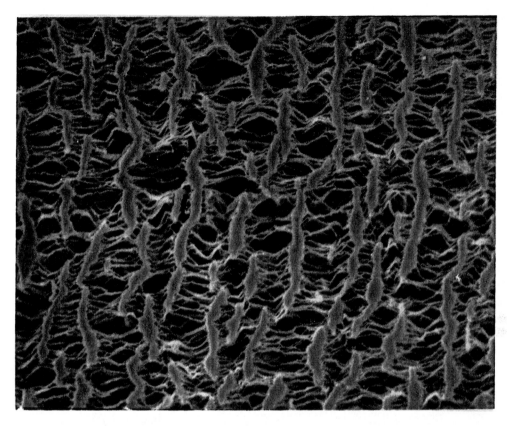

Figure 26 Scanning electron micrograph illustrating the characteristic surface morphology of expanded polytetrafluoroethylene (ePTFE) consisting of nodes and filamentous regions (× 500).

Thus, it is evident that an ideal polymeric biomaterial for use in a replacement heart valve has not yet been developed; however, segmented polyurethanes are emerging as candidate materials for short-term use in ventricular-assist devices.

VI. SUMMARY

This chapter presents an overview of the historical evolution of prosthetic valve designs. It also provides a review of the pathologic findings, including calcification, primary leaflet failure, fibrous sheathing, and thrombus formation, that significantly contribute to prosthetic valve failure. Emphasis is placed on the importance of the selection of biomaterials based on their chemical, physical, and mechanical properties, and on knowledge of the alterations induced in these properties by preimplantation processing. It is hoped that this review will provide guidance for the continuing development of the next generation of prosthetic heart valves utilizing the experience gained from previous preclinical and clinical studies.

Figure 27   Macroscopic photograph of an explanted expanded polytetrafluoroethylene (ePTFE) prosthetic valve (20-week-old mitral valve replacement in sheep, ×6). Calcific nodules are visible within the fibrous sheath near the free edge of the leaflet. Cuspal perforations and tears have resulted from abrasive wear secondary to the presence of calcific nodules.

ACKNOWLEDGMENTS

We wish to thank Dr. Ajit Yoganathan, Georgia Institute of Technology, Atlanta, Georgia, for providing the flow visualization information presented in Figs. 1 through 5. We would also like to express our gratitude to Autogenics, Baxter Healthcare/Edwards Cardiovascular Division, Medtronic Heart Valves, and St. Jude Medical, Inc., for providing the prosthetic heart valves evaluated in our investigational studies.

REFERENCES

1. Barnhart, G.R., Jones, M., Ishihara, T., Rose, D.M., Chavez, A.M., and Ferrans, V.J. Animal model of human disease. Degeneration and calcification of bioprosthetic cardiac valves. Bioprosthetic tricuspid valve implantation in sheep. *Am. J. Pathol.* 106:136–139, 1982.
2. Barnhart, G.R., Jones, M., Ishihara, T., Chavez, A.M., Rose, D.M., and Ferrans, V.J. Bioprosthetic valvular failure: Clinical and pathological observations in an experimental animal model. *J. Thorac. Cardiovasc. Surg.* 83:618–631, 1982.

3. Barnhart, G.R., Jones, M., Ishihara, T., Chavez, A.M., Rose, D.M., and Ferrans, V.J. Failure of porcine aortic and bovine pericardial prosthetic valves: An experimental investigation in young sheep. *Circulation* 66(Supp. 1):I-150–I-153, 1982.
4. Hufnagel, C.A., and Gomes, M.N. Late follow-up of ball-valve prostheses in the descending aorta. *J. Thorac. Cardiovasc. Surg.* 72:900–909, 1976.
5. Foster, A.H., Greenberg, G.J., Underhill, D.J., McIntosh, C.L., and Clark, R.E. Intrinsic failure of Hancock mitral bioprostheses: 10 to 15 year experience. *Ann. Thorac. Surg.* 44: 568–577, 1987.
6. O'Brien, M.F., Stafford, E.G., Gardner, M.A.H., Pahler, P.G., McGiffin, D.C., and Kirklin, J.W. A comparison of aortic valve replacement with viable cryopreserved and fresh allograft valves with a note on chromosomal studies. *J. Thorac. Cardiovasc. Surg.* 94:812–823, 1987.
7. Bjork, V.O., and Hultquist, G. Teflon and pericardial aortic valve prostheses. *J. Thorac. Cardiovasc. Surg.* 47:693–701, 1964.
8. Bailey, C.P., Carstens, H.P., Zimmerman, J., and Hirose, T. Aortic valve replacement with autogenous aortic wall. *Am. J. Cardiol.* 15:367–379, 1965.
9. Athanasuleas, C.L., Anagnostopoulos, C.E., and Kittle, C.F. The autologous rectus sheath cardiac valve. III. Design and physical properties. *J. Thorac. Cardiovasc. Surg.* 65:118–123, 1973.
10. Williams, B.T., Bellhouse, B.J., and Ashton, T. Autologous superior vena cava as a material for heart valve replacement. *J. Thorac. Cardiovasc. Surg.* 66:952–958, 1973.
11. Pillsbury, R.C., and Shumway, N.E. Replacement of the aortic valve with the autologous pulmonic valve. *Surg. Forum* 17:176–177, 1966.
12. Ross, D.N. Replacement of the aortic and mitral valve with a pulmonary autograft. *Lancet* 2:956–958, 1967.
13. Senning, A. Fascia lata replacement of aortic valves. *J. Thorac. Cardiovasc. Surg.* 54:465–470, 1967.
14. Gersbach, P.H., and Wegmann, W. Aorta valve replacement using autologous fascia lata transplants. *Virchows Arch. A Path. Anat. and Histol.* 364:235–247, 1974.
15. Senning, A. Alterations in valvular surgery: Biologic value. In L.H. Cohn and V. Gallucci, eds., *Proceedings of the Second International Symposium on Cardiac Bioprostheses.* New York: Yorke, 1984; pp. 140–153.
16. Edwards, W.S., Karp, R.B., Robbillard, D., and Derr, A.R. Mitral and aortic valve replacement with fascia lata on a frame. *J. Thorac. Cardiovasc. Surg.* 58:854–858, 1969.
17. Inoescu, M.I., Ross, D.N., Dec, R.C., Grimshaw, V.A., Taylor, S.H., Whitaker, W., and Wooler, G.H. Heart valve replacement with autologous fascia lata. *Lancet* 2:335–338, 1970.
18. Silver, M.D., Hudson, R.E.B., and Trimble, A.S. Morphologic observations on heart valve prostheses made of fascia lata. *J. Thorac. Cardiovasc. Surg.* 70:360–366, 1975.
19. Puig, L.B., Verginelli, G., Bellotti, G., Kawabe, L., Frack, C.C.R., Pilleggi, F., Decourt, L.V., and Zerbini, E.J. Homologous dura mater: Preliminary study of 30 cases. *J. Thorac. Cardiovasc. Surg.* 64:154–160, 1972.
20. Puig, L.B., Verginelli, G., Iryia, K., Kawabe, L., Belloti, G., Sosa, E., Pilleggi, F., and Zerbini, E.J. Homologous dura mater cardiac valves. A study of 533 surgical cases. *J. Thorac. Cardiovasc. Surg.* 69:722–728, 1975.
21. Ferrans, V.J., Milei, J., Ishihara, T., and Storino, R. Structural changes in implanted cardiac valvular bioprostheses constructed of glycerol-treated human dura mater. *Eur. J. Cardiothorac. Surg.* 5:144–154, 1991.
22. Center for Disease Control. Rapidly progressive dementia in a patient who received a cadaveric dura mater graft. *Morbid. Mortal. Weekly Report* 36:49–50, 1987.
23. Love, J.W. *Autologous Tissue Heart Valves.* Austin: Landes, 1993.
24. Love, J.W., Schoen, F.J., Breznock, E.M., Shermer, S.P., and Love C.S. Experimental evaluation of an autologous tissue heart valve. *J. Heart Valve Dis.* 1:232–241, 1992.

25. Jones, M., Eidbo, E.E., Hilbert, S.L., Ferrans, V.J., and Clark, R.E. Anticalcification treatments of bioprosthetic heart valves: *In vivo, in situ* studies in the sheep model. *J. Cardiac. Surg.* 4:69–73, 1989.

26. Golomb, G., Schoen, F.J., Smith, M.S., Linden, J., Dixon, M., and Levy, R.J. The role of glutaraldehyde-induced crosslinks in calcification of bovine pericarium used in cardiac valve bioprostheses. *Am. J. Pathol.* 127:122–130, 1987.

27. Levy, R.J., Schoen, F.J., Sherman, F.S., Nichols, J., Hawley, M.A., and Lund, S.A. Calcification of subcutaneously implanted Type I collagen sponges. *Am. J. Pathol.* 122:71–82, 1986.

28. Murray, G. Homologous aortic valve segment transplants as surgical treatment for aortic and mitral insufficiency. *Angiology* 7:466–471, 1956.

29. Ross, D.N. Homograft replacement of the aortic valve. *Lancet* 2:487, 1962.

30. Barratt-Boyes, B.G. Homograft aortic valve replacement and aortic incompetence and stenosis. *Thorax* 19:131–150, 1964.

31. Di Carlo, D., Stark, J., Revignas, A., and de Leval, M.R. Conduits containing antibiotic preserved homografts in the treatment of complex congenital heart defects. In L.H. Cohn and V. Galluci, eds., *Cardiac Bioprostheses.* New York: Yorke, 1982; pp. 259–265.

32. Merin, F., and McGoon, D.C. Reoperation after insertion of aortic homograft-right ventricular outflow tract. *Ann. Thorac. Surg.* 16:122–126, 1973.

33. Angell, W.W., Angell, J.D., Oury, J.H., Lamberti, J.J., and Grehl, T.M. Long-term follow-up of viable frozen aortic homografts: A viable homograft valve bank. *J. Thorac. Cardiovasc. Surg.* 93:815–822, 1987.

34. O'Brien, M.F., Stafford, G., Gardner, M., Pohlner, P., McGiffin, D., Johnston, N., Brosnan, A., and Duffy, P. The viable cryopreserved allograft aortic valve. *J. Cardiac. Surg.* 2(Suppl 1):153–167, 1987.

35. Fontan, F., Choussat, A., Deville, C., Doutremepucih, C., Coupilland, J., and Vosa, C. Aortic valve homografts in surgical treatment of complex cardiac malformations. *J. Thorac. Cardiovasc. Surg.* 87:649–657, 1984.

36. Ray, P.H., and Ross, D.N. Fifteen years experience with the aortic homograft: The conduit of choice for right ventricular outflow tract reconstructions. *Ann. Thorac. Surg.* 40:360–364, 1985.

37. Sanders, S.P., Levy, R.J., Freed, M.D., Norwood, W.I., and Cinstañeda, A.R. Use of Hancock porcine xenografts in children and adolescents. *Am. J. Cardiol.* 46:429–438, 1980.

38. Silver, M.S., Pollock, J., Silver, M.D., Williams, W.G., and Trusler, G.A. Calcification in porcine xenograft valves in children. *Am. J. Cardiol.* 45:685–689, 1980.

39. Gross, L., and Kugel, M.A. Topographic anatomy and histology of the valves in the human heart. *Am. J. Pathol.* 7:445–473, 1931.

40. Ferrans, V.J., and Rodriguez, E.R. Ultrastructure of the normal heart. In M.D. Silver, ed., *Cardiovascular Pathology.* New York: Churchill Livingstone, 1991; pp. 43–101.

41. Bashey, R.I., and Jimenez, S.A. Collagens in heart valves. In M.E. Nimni, ed., *Collagen: Biochemistry,* Vol. 1. Boca Raton, FL: CRC Press, 1988; pp. 257–274.

42. Mendler, M., Eich-Bender, S.G., Vaughan, L., Winterhalter, K.H., and Buckner, P. Cartilage contains mixed fibrils of collagens Types II, IX, X. *J. Cell. Biol.* 108:191–197, 1989.

43. Ruggeri, A., and Benazza, F. Collagen-proteoglycan interactions. In A. Ruggerri and P.M. Motta, eds., *Ultrastructure of the Connective Tissue Matrix.* Boston: Nijhoff, 1984; pp. 113–125.

44. Ross, D., and Yacoub, M. Homograft replacement of the aortic valve: A critical review. *Prog. Cardiovasc. Dis.* 11:926–929, 1965.

45. Smith, J.C. The pathology of human aortic valve homografts. *Thorax* 22:114–138, 1967.

46. Sell, S., and Scully, R.E. Aging changes in the aortic and mitral valves: Histologic and histochemical studies, with observations on the pathogenesis of calcific aortic stenosis and calcification of the mitral annulus. *Am. J. Pathol.* 46:345–365, 1965.

47. Lange, P.L., and Hopkins, R.A. Allograft valve banking techniques and technology. In R.A. Hopkins, ed., *Cardiac Reconstruction with Allograft Valves*. New York: Springer, 1989; pp. 37–63.

48. Hilbert, S.L., Ferrans, V.J., and Jones, M. Effects of preimplantation processing on bioprosthetic and biologic cardiac valve morphology. In R.A. Hopkins, ed., *Cardiac Reconstructions with Allograft Valves*. New York: Springer, 1989; pp. 64–94.

49. Domkowski, P.W., Messier, Jr. R.H., Crescenzo, D.G., Aly, H.M., Abd-Elfattah, A.S., Hilbert, S.L., Wallace, R.B., and Hopkins, R.A. Preimplantation alteration of adenine nucleotides in cryopreserved heart valves. *Ann. Thorac. Surg.* 55:413–419, 1993.

50. Crescenzo, D.G., Hilbert, S.L., Messier, Jr. R.H., Domkowski, P.W., Barrick, M.K., Lange, P.L., Ferrans, V.J., Wallace, R.B., and Hopkins, R.A. Human cryopreserved allografts: electron microscopic analysis of cellular injury. *Ann. Thorac. Surg.* 55:25–31, 1993.

51. Thubrikar, M.J., Nolan, S.P., Aouad, J., and Deck, J.D. Stress sharing between the sinus and leaflets of canine aortic valve. *Ann. Thorac. Surg.* 42:434–440, 1986.

52. Swanson, W.M., and Clark, R.E. Dimensions and geometric relationships of the human aortic valve as a function of pressure. *Circ. Res.* 35:871–882, 1974.

53. Broom, N.D., and Christie, G.W. The structure/function relationship of fresh and glutaraldehyde-fixed aortic valve leaflets. In L.H. Cohn and V. Galluchi, eds., *Cardiac Bioprostheses*. New York: Yorke, 1982; pp. 476–491.

54. Deck, J.D., Thubrikar, M.J., Schneider, P.J., and Nolan, S.P. Structure, stress and tissue repair in aortic valve leaflets. *Cardiovasc. Res.* 22:7–16, 1988.

55. Broom, N.D. The stress/strain and fatigue behavior of glutaraldehyde processed heart valve tissue. *J. Bioeng.* 10:707–724, 1977.

56. Missirlis, M.F., and Chong, M. Aortic valve mechanics. I. Material properties of native porcine aortic valves. *J. Bioeng.* 2:287–300, 1978.

57. Livi, U., Abdulla, A.K., Parker, R., Olsen, E.J., and Ross, D.N. Viability and morphology of aortic and pulmonary homografts. A comparative study. *J. Thorac. Cardiovasc. Surg.* 93:755–760, 1987.

58. Aparicio, S.R., Donnelly, R.J., Dexter, F., and Watson, D.A. Light and electron microscopic studies on homograft and heterograft heart valves. *J. Pathol.* 115:147–162, 1975.

59. Harris, P.D., Kovalik, A.J.W., Marks, J.A., and Malm, J.P. Factors modifying aortic homograft structure and function. *Surgery* 63:45–59, 1968.

60. Reichenbach, D.D., Mohri, H., and Merendino, K.A. Pathological changes in human aortic valve homografts. *Circulation* 39(Suppl.):I-47–I-56, 1969.

61. Van Vliet, P.D., Titus, J.K., Berghis, J., and Ellis, F.H., Jr. Morphologic features of homotransplanted canine mitral valves. *J. Thorac. Cardiovasc. Surg.* 49:504–510, 1965.

62. Tamura, K., Jones, M., Yamada, I., and Ferrans, V.J. A comparison of failure modes of glutaraldehyde-treated versus antibiotic-preserved mitral valve allografts implanted in sheep. *J. Am. Coll. Cardiol.* 248A, 1994.

63. Inoescu, M.I., ed. *Tissue Heart Valves*. London: Butterworths, 1979.

64. DeBakey, M.E. *Advances in Cardiac Valves: Clinical Perspective*. New York: Yorke, 1983.

65. Cohn, L.H., and Gallucci, V., eds. *Cardiac Bioprostheses*. New York: Yorke, 1982.

66. Bodnar, E., and Yacoub, M., eds. *Biologic and Bioprosthetic Valves*. New York: Yorke, 1986.

67. Gabbay, S., and Frater, R.W.M. The unileaflet heart valve biprosthesis: New concept. In L.H. Cohn and V. Galluci, eds., *Cardiac Bioprostheses*. New York: Yorke, 1982; pp. 411–424.

68. Bodnar, E., Bowden, N.L., Drury, P.J., Olsen, E.G.J., Durmaz, I., and Ross, D.N. Bicuspid mitral bioprosthesis. *Thorax* 36:45–51, 1981.

69. Black, M.M., Drury, P.J., Tindale, W.B., and Lawford, P.V. The Sheffield bicuspid valve: concept, design and *in vitro* and *in vivo* assessment. In E. Bodnar and M. Yacoub, eds., *Biologic and Bioprosthetic Valves*. New York: Yorke, 1986; pp. 709–717.

70. Thubrikar, M. *The Aortic Valve.* Boca Raton, FL: CRC Press, 1990.

71. Woodroof, E.A. The chemistry and biology of aldehyde treated tissue heart valve xenografts. In M.I. Ionescu, ed. *Tissue Heart Valves.* London: Butterworths, 1979; pp. 347–362.

72. Cheung, D.T., Perelman, N., Ko, E.C., and Nimni, M.E. Mechanism of cross linking of proteins by glutaraldehyde. III. Reaction with collagen in tissues. *Connect. Tissue Res.* 13: 109–115, 1985.

73. Buch, W.S., Kosek, J.C., Angell, W.W., and Shumway, S.E. Deterioration of formalin treated porcine aortic heterografts. *J. Thorac. Cardiovasc. Surg.* 60:673–682, 1970.

74. Dubiel, W.T., Johansson, L., and Willen, R. Late changes in formalin-treated porcine aortic heterografts replacing human mitral valves. *Scand. J. Thorac. Cardiovasc. Surg.* 9:16–26, 1975.

75. Bortolotti, V., Milano, A., Mazzucco, A., Valfre, C., Fasoli, G., Valente, M., Thiene, G., and Gallucci, V. Longevity of the formaldehyde-preserved Hancock porcine heterograft. *J. Thorac. Cardiovasc. Surg.* 84:451–453, 1982.

76. Broom, N., and Thomson, F.J. Influence of fixation conditions on the performance of glutaraldehyde-treated porcine aortic valves: Towards a more scientific basis. *Thorax* 34: 166–176, 1979.

77. Broom, N.D. Simultaneous morphological and stress-strain studies of the fibrous components in wet heart-valve leaflet tissue. *Connect. Tissue Res.* 6:37–50, 1978.

78. Hilbert, S.L., Ferrans, V.J., and Swanson, W.M. Optical methods for the nondestructive evaluation of collagen morphology in bioprosthetic heart valves. *J. Biomed. Mater. Res.* 20: 1411–1421, 1986.

79. Hilbert, S.L., Barrick, M.K., and Ferrans, V.J. Porcine aortic valve bioprostheses: A morphologic comparison of the effects of fixation pressure. *J. Biomed. Mater. Res.* 24:773–787, 1990.

80. Milano, A., Bortolotti, U., Talenti, E., Valfre, C., Arbustini, E., Valente, M., Mazzucco, A., Gallucci, V., and Thiene, G. Calcific degeneration as the main cause of porcine bioprosthetic valve failure. *Am. J. Cardiol.* 53:1066–1070, 1984.

81. Schoen, F.J., and Hobson, C.E. Anatomic analysis of removed prosthetic heart valves: Causes of failure in 33 mechanical valves and 58 bioprostheses, 1985. *Hum. Pathol.* 16:549–559, 1985.

82. Schoen, F.J., Kujovic, J.L., Levy, R.J., and Sutton, M.S. Bioprosthetic valve failure. *Cardiovasc. Clin.* 18:289–317, 1987.

83. Ferrans, V.J., Boyce, S.W., Billingham, M.E., Jones, M., Ishihara, T., and Roberts, W.C. Calcific deposits in porcine bioprostheses: Structure and pathogenesis. *Am. J. Cardiol.* 46: 721–734, 1980.

84. Schoen, F.J., Levy, R.J., Nelson, A.C., Bernhard, W.F., Nashef, A., and Hawley, M. Onset and progression of experimental bioprosthetic heart valve calcification. *Lab. Invest.* 52:523–532, 1985.

85. Boskey, A.L. The role of calcium phospholipid-phosphate complexes in tissue mineralization. *Metab. Bone Dis. Rel.* 1:137, 1978.

86. Bajpai, P.K. Immunological aspects of treated natural tissue prostheses. In D.F. Williams, ed., *Biocompatibility of Tissue Analogs*, Vol. 1. Boca Raton, FL: CRC Press, 1985; pp. 5–25.

87. Carpentier, A., Nashef, A., Carpentier, S., Ahmed, A., and Goussef, N. Techniques for prevention of calcification of spontaneously degenerated porcine bioprosthetic valves. *J. Thorac. Cardiovasc. Surg.* 90:119–125, 1985.

88. Lentz, D.J., Pollock, E.M., Olsen, D.B., Andrews, E.J., Murashita, J., and Hastings, W.L. Inhibition of mineralization of glutaraldehyde-fixed Hancock bioprosthetic heart valves. In L.H. Cohn and V. Galluci, eds., *Cardiac Bioprostheses.* New York: Yorke, 1982; pp. 306–319.

89. Levy, R.J., Hawley, F.J., Schoen, F.J., Lund, S.A., and Liu, P.Y. Inhibition by diphospho-

nate compounds of calcification of porcine bioprosthetic heart valve cusps implanted subcutaneously in rats. *Circulation* 71:349–356, 1985.

90. Pathak, Y.V., Boyd, J., Levy, R.J., and Schoen, F.J. Prevention of calcification of glutaraldehyde pretreated bovine pericardium through controlled release polymeric implants: Studies of Fe, Al, protoamine sulphate and levamisol. *Biomaterials* 11:718–723, 1990.

91. Grabenwöger, M., Grimm, M., Eybl, E., Leukauf, C., and Müller, M.M. Decreased tissue reaction to bioprosthetic heart valve material after L-glutamic acid treatment. A morphologic study. *J. Biomed. Mat. Res.* 26:1231–1240, 1992.

92. Gott, J.P., Pan-Chih, L.M., Dorsay, M.A., Jay, J.L., Jett, G.K., Schoen, F.J., Girardot, J.M., and Guyton, R.A. Calcification of porcine valves: A successful new method of antimineralization. *Ann. Thorac. Surg.* 53:207–215, 1992.

93. Koorajian, S., Frugard, G., and Stegwell, M.J. Sterilization of tissue valves. In F. Sebening, W.P. Klovekorn, H. Meisner, and E. Struck, eds., *Bioprosthetic Cardiac Valves*. Munich: Deutsches Harzzentrum Munchnen, 1979; pp. 373–378.

94. Center for Disease Control. Isolation of *Mycobacterium* species from porcine heart valve prostheses. *Morbid. Mortal. Weekly Report* 26:42–43, 1977.

95. Laskowski, L.F., Marr, J.J., Spernoga, J.F., Frank, N.J., Barner, H.B., Kaiser, G., and Tyras, D.H. Fastidious *Mycobacterium* grown from porcine prosthetic heart-valve cultures. *N. Engl. J. Med.* 297:101–102, 1977.

96. Thubrikar, M., Piepgrass, W.C., Deck, J.D., and Nolan, S.P. Stresses of natural versus prosthetic aortic valve leaflets *in vivo*. *Ann. Thorac. Surg.* 30:230–239, 1980.

97. Van Steehoven, A.A., Veenstra, P.C., and Reneman, R.S. The effect of some hemodynamic factors on the behavior of the aortic valve. *J. Biomech.* 15:941–950, 1982.

98. Reis, R.L., Hancock, W.D., Yarbrough, J.W., Glancy, D.L., and Morrow, A.G. The flexible stent: A new concept in the fabrication of tissue heart valve prostheses. *J. Thorac. Cardiovasc. Surg.* 62:683–689, 1971.

99. Thomson, F.J., and Barrat-Boyes, B.G. The glutaraldehyde-treated heterograft valve. *J. Thorac. Cardiovasc. Surg.* 74:317–321, 1977.

100. Wright, J.T.M., Eberhart, C.E., and Gibbs, M. Hancock II — An improved bioprosthesis. In L.H. Cohn and V. Gallucci, eds., *Cardiac Bioprostheses*. New York: Yorke, 1982; pp. 425–444.

101. Carpentier, A.F., and Lane, E. Supported bioprosthetic heart valve with compliant orifice ring. U.S. Patent 4,106,129, 1978.

102. Borkon, A.M., McIntosh, C.L., Jones, M., Roberts, W.C., and Morrow, A.G. Inward stent-post bending of a porcine bioprosthesis in the mitral position: Cause of bioprosthetic dysfunction. *J. Thorac. Cardiovasc. Surg.* 83:105–107, 1982.

103. Salomon, N.W., Copeland, J.G., Goldman, S., and Larson, D.R. Unusual complication of the Hancock porcine heterograft: Strut compression in the aortic root. *J. Thorac. Cardiovasc. Surg.* 77:294–296, 1979.

104. Schoen, F.J., Schulman, L.J., and Cohn, L.H. Quantitative anatomic analysis of "stent creep" of explanted Hancock standard porcine bioprostheses used for cardiac valve replacement. *Am. J. Cardiol.* 56:110–114, 1985.

105. Christie, G.W. Computer modeling of bioprosthetic heart valves. *Eur. J. Cardiovasc. Surg.* 6(Suppl. 1):S95–100, 1992.

106. Christie, G.W., and Barratt-Boyes, B.G. On stress reduction in bioprosthetic heart valve leaflets by the use of a flexible stent. *J. Card. Surg.* 6:476–481, 1991.

107. Vesely, I., Krucinski, S., and Campbell, G. Micromechanics and mathematical modeling: An inside look at bioprosthetic heart valve function. *J. Card. Surg.* 7:85–95, 1992.

108. Levine, F.H., Buckley, M.J., and Austen, W.G. Hemodynamic evaluation of the Hancock modified orifice bioprosthesis in the aortic position. *Circulation* 58(Suppl. 1):33–35, 1978.

109. Angell. W.W., Pupello, D.F., Bessone, L.N., Hiro, S.P., Lopez-Cuenca, E., Glatterer,

M.S., Jr., and Brock, J.C. Implantation of the unstented bioprosthetic aortic root: An improved method. *J. Card. Surg.* 8:466–471, 1993.

110. David, T.E., Pollic, C., and Bos, J. Aortic valve replacement with stentless porcine aortic bioprosthesis. *J. Thorac. Cardiovasc. Surg.* 99:113–118, 1990.

111. Polystan A/S. Polystan Biprostheses Information Bulletin. Polystan A/S, Copenhagen, Denmark, 1980.

112. Shemin, R.J., Schoen, F.J., Hein, R., Austin, J., and Cohn, L.H. Hemodynamic and pathologic evaluation of a unileaflet pericardial bioprosthetic valve. *J. Thorac. Cardiovasc. Surg.* 95:912–929, 1988.

113. Yoganathan, A.P., Woo, Y.R., Sung, H.W., Williams, F.P., Franch, R.H., and Jones, M. *In vitro* hemodynamic characteristics of tissue bioprostheses in the aortic position. *J. Thorac. Cardiovasc. Surg.* 92:198–209, 1986.

114. Scotten, L.N., Walker, D.K., and Brownlee, R.T. The *in vitro* function of 19 mm bioprosthetic heart valves in the aortic position. *Life Support Syst.* 5:145–153, 1986.

115. Gallo, I., Nistal, F., Arbe, E., and Artiñano, E. Comparative study of primary tissue failure between porcine (Hancock and Carpentier-Edwards) and bovine pericardial (Ionescu-Shiley) bioprostheses in the aortic position at five- to nine-year follow-up. *Am. J. Cardiol.* 61: 812–816, 1988.

116. Bortolotti, U., Milan, A., Thiene, G., Guerra, F., Mazzucco, A., Valente, M., Talenti, E., and Gallucci, V. Early mechanical failure of the Hancock pericardial xenograft. *J. Thorac. Cardiovasc. Surg.* 94:200–207, 1987.

117. Cooley, D.A., Ott, D.A., Reul, G.J., Duncan, J.M., Frazer, O.H., and Livesay, J.J. Ionescu-Shiley bovine pericardial bioprostheses: Clinical results in 2701 patients. In E. Bodnar and M. Yacoub, eds., *Biologic and Bioprosthetic Valves.* New York: Yorke, 1986; pp. 177–198.

118. Perier, P., Hihaileanu, S., Fabiani, J.N., Deloche, A., Chauvaud, S., Jindani, A., and Carpentier, A. Long-term evaluation of the Carpentier-Edwards pericardial valve in the aortic position. *J. Card. Surg.* 6:589–594, 1991.

119. Otaki, M., and Kitamura, N. Spontaneous and time-related degeneration of Carpentier-Edwards bioprosthesis in mitral position. *Int. Surg.* 78:148–151, 1993.

120. Frater, R.W., Salomon, N.W., Rainer, W.G., Cosgrove, D.M., and Wickham, E. The Carpentier-Edwards pericardial aortic valve: Intermediate results. *Ann. Thorac. Surg.* 53: 764–771, 1992.

121. Ishihara, T., Ferrans, V.J., Jones, M., Boyce, S.W., and Roberts, W.C. Structure of bovine parietal pericardium and of unimplanted Ionescu-Shiley pericardial valvular bioprostheses. *J. Thorac. Cardiovasc. Surg.* 81:747–757, 1981.

122. Crofts, C.E., and Throwbridge, E.A. The tensile strength of natural and chemically modified bovine pericardium. *J. Biomed. Mater. Res.* 22:889–898, 1988.

123. Throwbridge, E.A., Robert, K.M., Crofts, C.E., and Lawford, P.V. Pericardial heterografts. Towards quality control of the mechanical properties of glutaraldehyde fixed leaflets. *J. Thorac. Cardiovasc. Surg.* 92:21–28, 1986.

124. Arbustini, E., Jones, Moses, R.D., Eidbo, E.E., Carroll, J.R., and Ferrans, V.J. Modification by the Hancock T6 process of calcification of bioprosthetic cardiac valves implanted in sheep. *Am. J. Cardiol.* 53:1388–1396, 1984.

125. Goffin, Y.A., Hilbert, S.L., and Bartik, M.A. Morphologic evaluation of 2 new bioprostheses: The Mitroflow bovine pericardial valve and the Xenomedica porcine aortic valve. In E. Bodnar and M. Yacoub, eds., *Biologic and Bioprosthetic Valves.* New York: Yorke, 1986; pp. 365–382.

126. Hilbert, S.L., Ferrans, V.J., McAllister, H.A., and Cooley, D.A. Ionescu-Shiley bovine pericardial bioprostheses: Histologic and ultrastructural studies. *Am. J. Pathol.* 140:1195–1204, 1992.

127. Walley, V.M., and Keon, W.J. Patterns of failure in Ionescu-Shiley bovine pericardial bioprosthetic valves. *J. Thorac. Cardiovasc. Surg.* 93:925–933, 1987.
128. Wheatley, D.J., Fisher, J., Reece, I.J., and Spyt, T. Primary tissue failure in pericardial valves. *J. Thorac. Cardiovasc. Surg.* 94:367–374, 1987.
129. Schoen, F.J., Fernandez, J., Gonzalez-Lavin, and Cemaianu, A. Causes of failure and pathologic findings in surgically-removed Ionescu–Shiley standard bovine pericardial heart valve bioprostheses: Emphasis on progressive structural deterioration. *Circulation* 76:618–627, 1987.
130. Ishihara, T., Ferrans, V.J., Boyce, S.W., and Roberts, W.C. Structure and classification of cuspal tears and perforation in porcine bioprosthetic cardiac valves implanted in patients. *Am. J. Cardiol.* 48:665–678, 1981.
131. Lyman, D.J., Fazzio, F.J., Voorhes, H., Robinson, G., and Albo, D., Jr. Compliance as a factor affecting the patency of a copolyurethane vascular grafts. *J. Biomed. Mater. Res.* 12:337–345, 1978.
132. Coleman, D.L., Meuzelaar, H.L.C., Kesseler, T.R., McLennen, W.H., Richards, J.M., and Gregonis, D.E. Retrival and analysis of a clinical total artificial heart. *J. Biomed. Mater. Res.* 20:417–431, 1986.
133. Akutsu, T., Dreyer, B., and Kolff, W.J. Polyurethane artificial heart valves in animals. *J. Appl. Physiol.* 14:1045–1048, 1959.
134. Bjork, V.O., Cullhed, I., and Lodin, H. Aortic valve prosthesis (Teflon): Two-year follow-up. *J. Thorac. Cardiovasc. Surg.* 45:635–644, 1963.
135. Braunwald, N.S., and Morrow, A.G. A late evaluation of flexible Teflon prostheses utilized for total aortic valve replacement: Postoperative clinical, hemodynamic, and pathologic assessments. *J. Thorac. Cardiovasc. Surg.* 49:485–496, 1965.
136. Fishbein, M.C., Roberts, W.C., Golden, A., and Hufnagel, C.A. Cardiac pathology after aortic valve replacement using Hufnagel trileaflet prostheses: A study of 20 necropsy patients. *Am. Heart J.* 89:443–448, 1975.
137. Imamura, E., and Kaye, M.P. Function of explanded-polytetrafluoroethylene laminated trileaflet valves in animals. *Mayo Clin. Proc.* 52:770–775, 1977.
138. Jansen, J., Willeke, S., Reiners, B., Habott, P., Reul, H., and Rau, G. New J-3 flexible-leaflet polyurethane heart valve prosthesis with improved hydrodynamic performance. *J. Artif. Organs* 14:655–660, 1991.
139. Stewart, S.F., Burte, F., Eidbo, E., Kolff, W.J., Yu, L.S., and Clark, R.E. *In vitro* ultrasound characterization of a polyurethane trileaflet valve. *ASAIO Trans. Am. Soc. Artif. Intern. Organs* 36:M532–535, 1990.
140. Schoephoerster, R., and Chandran, K.B. Velocity and turbulence measurements past mitral valve prostheses in a model left ventricle. *J. Biomech.* 24:549–562, 1991.
141. Hilbert, S.L., Ferrans, V.J., Tomita, Y., Eidbo, E.E., and Jones, M. Evaluation of explaned polyurethane trileaflet cardiac valve prostheses. *J. Thorac. Cardiovasc. Surg.* 94:419–429, 1987.
142. Hoffman, D., Sisto, D., Yu, L.S., Dahm, M., and Kolff, W.J. Evaluation of a stented polyurethane mitral valve prosthesis. *ASAIO Trans.* 37:M354–355, 1991.
143. Deck, J.D., Thubrikar, M.J., Nolan, S.P., and Aouad, J. Role of mechanical stress in calcification of bioprostheses. In L.H. Cohn and V. Galluci, eds., *Cardiac Bioprostheses.* New York: Yorke, 1982; pp. 293–305.
144. Chandran, K.B., Kim, S.H., and Han, G. Stress distribution on the cusps of a polyurethane trileaflet heart valve prosthesis in the closed position. *J. Biomech.* 24:385–395, 1991.
145. Hamon, R.F., Kahn, A.S., and Chow, A. The cation chelation mechanism of metal-ion sorption by polyurethanes. *Talanta* 29:313–326, 1982.
146. Thoma, R.J., Hung, T.Q., Nyilas, E., and Phillips, R.E. Metal ion enchanced enviornmental stress cracking of poly(ether)urethanes. *Trans. Soc. Biomater.* 191, 1986.

147. Lian, J.B., Levy, J.R., Bernhard, W., and Szycher, M. LVAD mineralization and gamma-carboxyglutamic acid containing proteins in normal and pathologically mineralized tissue. *Trans. Am. Soc. Artif. Intern. Organs* 27:683–689, 1981.

148. Dunn, J.M., and Marmon, L. Mechanisms of calcification of tissue valves. *Cardiol. Clin.* 3: 385–396, 1985.

42

New Frontiers of Biomaterials for Cardiovascular Surgery

Takafumi Okoshi
Teikyo University, Tokyo, Japan
Yasuharu Noishiki
Yokohama City University, Yokohama, Japan

In the last three decades, there have been strong demands to develop cardiovascular implants for advanced surgeries. Many types of artificial organs and biomaterials are contributing to the surgical treatments of congenital and acquired heart and blood vessel diseases. In this chapter, some new materials and concepts in the field of cardiovascular implants are reviewed.

I.   ARTIFICIAL HEART VALVES

A.   Introduction

Two types of artificial heart valves have been used for the treatment of cardiac valve diseases. One is a mechanical valve made of metal and pyrolytic carbon [1] and the other is a tissue valve made of porcine aortic valves, bovine pericardium, or other biological tissues [2].

The mechanical valves have better durability than the tissue valves. However, their major drawback is that, after the valve replacement, patients must be on anticoagulant therapy for the remainder of their lives. This means that they are always at risk; they may have a thrombosed valve or thromboembolism if anticoagulant therapy is not sufficient and, conversely, they are at a risk of hemorrhage if it is excessive. Mechanical valves are, therefore, disadvantageous for recipients living far away from medical facilities, young females who want to have children, and anyone who wants to have an occupation or enjoy sports with a risk of injury. Thus, implantation of a mechanical valve significantly restricts a patient's life-style. It appears, however, that patients with a mechanical valve need anticoagulation as long as pyrolytic carbon coating is used for the mechanical valve.

On the other hand, no antithrombotic therapy is usually required with tissue valve recipients and they can enjoy a wider range of daily activities than those with mechanical valves. Because tissue valves are foreign substances to the human body, they need treatment with a reagent in order to inhibit rejection, prevent absorption, and maintain mechanical strength in the human body. Glutaraldehyde has commonly served as a cross-linking reagent for tissue valves and other biologic materials implanted in humans. Thus, the glutaraldehyde (GA) cross-linked bioprosthetic heart valve (GA valve) has been used for those who need to avoid antithrombotic therapy for medical, occupational, or other private reasons. However, a major problem of the GA valve is "calcification," which often leads to valve failure [3,4].

Therefore, tissue pretreatment strategies with agents such as detergents, disphosphonates, and other metallic cations have been tried to prevent calcification in GA valves. For example, the efficacy of aluminum chloride pretreatment was reported, but there was a question about possible systemic toxicity, which may cause Alzheimer's disease or dialysis dementia [5]. Thus, an ideal method to prevent tissue valve calcification has not yet been developed. Moreover, glutaraldehyde cross-linking has other problems that should not be ignored. It makes tissue valves become harder and less pliable so that GA cross-linking of bioprosthetic cardiac valves has a hemodynamic disadvantage [6,7]. Besides, nonthrombogenicity of GA valves seems to be tolerable, but they still need further improvement [8,9].

To overcome the problems encountered with GA valves and improve the quality of life of patients after heart valve replacement, we have developed another strategy by replacing glutaraldehyde, a conventional cross-linking reagent, with a promising new one, a polyepoxy compound [6,7]. In animal experiments, a polyepoxy compound (PC) cross-linked bioprosthetic heart valve (PC valve) displayed a good valve function and reinforced nonthrombogenicity in early results and is expected to show reduced calcium deposition [10–13].

B. Characteristics of Glutaraldehyde and Polyepoxy Compound

Differences in properties between GA and PC valves are readily understood by making a comparison between glutaraldehyde and polyepoxide compound as cross-linking reagents [6,7].

In general, glutaraldehyde has been used as a cross-linking reagent for biologic materials. Glutaraldehyde binds with amino groups of collagen fibers and places itself as a cross-link between them so that collagen fibers are fixed with each other with a distance of a glutaraldehyde molecular length. Glutaraldehyde has a rigid structure and does not have any hydrophilic group. Therefore, glutaraldehyde cross-linking reduces the hydrophilicity, hydration, and pliability of original tissue valves [6]. GA valves with reduced pliability have a hemodynamic disadvantage [6] and often show calcification, although the mechanism has not yet been clarified. In terms of nonthrombogenicity, GA valves, with their structural similarity to human valves and chemically treated biological surfaces, are superior to mechanical valves, with their structural difference from human valves and pyrolytic carbon surfaces. However, both are inferior to the natural heart valve, which has an endothelial lining.

On the other hand, polyepoxy compound cross-linking is done in a similar fashion to glutaraldehyde, but polyepoxy compounds located between collagen fibers have hydroxyl groups, which show hydrophilicity, and ether bonds, which are flexible. As a result, PC valves maintain their pliability and become more hydrophilic and hydrated than GA valves. Pliability provides PC valves with a suitable hemodynamic condition. In addition, biologic materials cross-linked with polyepoxy compound have excellent mechanical strength [6]. Hydrophilicity and hydration give PC valves a reinforced nonthrombogenicity [10,11]. With respect to calcification, we have a hypothesis that hydrophilicity and hydration increase the diffusion rates in the tissue fluid, which contains calcium ions, decreasing the chance of calcium deposition. On the contrary, glutaraldehyde-cross-linked biologic materials show hydrophobicity, which inhibits infiltration and diffusion functions of the tissue fluid. This stagnation of tissue fluids in the valvular tissue is thought to be a major factor of material degeneration [3,4]. PC valves are, therefore, expected to show better durability with less calcification when compared with GA valves.

PC cross-linking maintains the original whitish color of these materials, while GA cross-linking causes biologic materials to lose their original color and become yellowish [6,7]. It has been reported that biologic materials cross-linked with a polyepoxy compound have no serious immunologic problems because polyepoxy compound cross-linking decreases antigenicity by almost the same degree as does glutaraldehyde cross-linking [14]. In order to verify these properties of the PC valve, animal experiments were executed as described below.

C.   Accelerated Calcification Tests

Biologic materials cross-linked with polyepoxy compound are expected to show less calcification than glutaraldehyde-cross-linked ones. Two polyepoxy compounds, three glutaraldehydes, and three non-cross-linked collagen gel disks were prepared and implanted in the subcutaneous layer of the back in growing male Sprague-Dawley rats (4 weeks old). It is well known that calcification of implanted biological materials is accelerated in the body of young animals. The specimens were retrieved 106 days after implantation and quantitative analysis of the calcium content was performed by atomic absorption analysis.

Mean calcium contents of polyepoxy compound, glutaraldehyde, and non-cross-linked collagen gel disks were 0.04, 13.023, and 0.183 milligram/gram (mg/g), respectively. Thus, polyepoxy-compound-cross-linked collagen gel disks implanted in the subcutaneous layer of growing rats showed approximately 300 times less calcium deposition than did glutaraldehyde-cross-linked ones, an encouraging evidence that indicates the superiority of polyepoxy compound cross-linking to glutaraldehyde cross-linking [12].

D.   Fabrication of the Polyepoxy Compound Cross-Linked Bioprosthetic Heart Valve

Considering the suitable properties of polyepoxy compound cross-linking for a bioprosthetic heart valve, PC valves were prepared according to the following procedures. Aortic valves with a small amount of left ventricular outflow tract and aortic root were resected from mongrel dogs and were cross-linked with a hydrophilic polyepoxy compound (Denacol EX-313 [glycerol polyglycidyl ether], Nagase Chemical, Osaka, Japan) [4,5]. The valves were then dipped in 2% Denacol solution (pH 10, 5% ethanol, and 0.07% DMP-30 as catalysts, 0.1% salicylic acid as an accelerator) at room temperature for 48 hours to form PC valves. Then, the unreacted cross-linking reagent, catalysts, and accelerator in the PC valve were washed out by running tap water for more than 24 hours, and the valves were stored in 70% ethanol. When used, they were dipped and rinsed in physiologic saline to replace the ethanol. The PC valves maintained their original white color and pliability [12].

1. Hemodynamics

The PC valve is expected to have good valve function due to its pliability. For preparation of PC valve implantation, the PC valve was properly trimmed and fabric vascular prostheses were anastomosed to both the inflow and the outflow sides of the valve to form a slightly curved conduit (PC-valved conduit). During preparation of a PC-valved conduit, it was easy to handle and suture with little resistance. PC-valved conduits were implanted as right ventricle–pulmonary artery (RV-PA) bypasses in mongrel puppies and the main PA was ligated. The PC valve fitted well to the vascular prostheses, so that there was almost no bleeding from the needle holes or the suture lines on the PC valve after blood started passing through the PC-valved conduit as an RV-PA bypass. Antithrombotic agents were not administered either during or after the operation.

Pressure measurement after bypassing of PC-valved conduit revealed that the pressure gradient across the PC valve was 5 mm Hg and the diastolic pressure of the PA was 19 mm Hg, which indicated almost no stenosis or regurgitation of the PC valve.

X-ray right ventriculography after implantation revealed an excellent open-close performance of the PC valve and there was no visible thrombus in the valve [12].

2. Nonthrombogenicity

The hydrophilicity and hydration are thought to provide reinforced nonthrombogenicity. Nonthrombogenicity of the PC valve was evaluated in the right and left heart systems. Aortic valves harvested from mongrel dogs were cross-linked with PC, and fabric vascular prostheses were anastomosed proximal and distal to the valve (PC-valved conduit). Another type of conduit with a GA valve was fabricated as a control (GA-valved conduit). Both types of conduits were implanted as an RV-PA bypass graft and the main pulmonary artery was ligated (right heart system model). Both types of conduits were also placed as a left ventricle–descending aorta bypass graft (apico-aortic bypass graft) and the descending aorta was ligated at its proximal portion (left heart system model).

The valves implanted in the right and left heart systems were retrieved 14 to 37 days and 7 to 20 days after implantation, respectively. Macroscopical evaluation of each cusp of the valves revealed that PC valves exhibited significantly less thrombus formation than GA valves in the right heart system ($p < 0.05$) and that PC valves showed a tendency to be less thrombogenic than GA valves in the left heart system [15].

3. Conclusions

The goal of this study was to fabricate an artificial heart valve that provided recipients with satisfactory quality of life after valve replacement. It is very important that the patients with an artificial valve can lead their lives in no need of antithrombotic therapy, which significantly restricts their life styles. To achieve this goal, the artificial valve must be installed with excellent durability, good hemodynamics, and sufficient nonthrombogenicity. This study indicated one approach to this issue and suggested that the PC valves are expected to have effective fluid dynamics, reinforced nonthrombogenicity, and to show reduced calcification, as compared with conventional GA valves.

II. CARDIAC WALL SUBSTITUTES

A. Introduction

Materials originally fabricated as vascular prostheses have been used for cardiac wall repair. More recently, various kinds of cardiac wall substitutes, in other words, patch materials, have been prepared and applied according to the purposes and portions of their use. Intracardiac application of those materials includes closure of the atrial septal defect, endocardial cushion defect, ventricular septal defect, and ventricular septal perforation, and buffle repair of transposition of great arteries and more. Free cardiac wall reconstruction contains right ventricular outflow tract enlargement of tetralogy of Fallot and right atrial wall reconstruction in the Senning procedure for transposition of great arteries and the like.

The materials available at present can be used, but are not always satisfactory. Desirable materials for each use are different in detail. For example, materials for free cardiac wall reconstruction may need to be more blood leakproof from needle holes and suture lines than those for intracardiac repair. The materials exposed to high blood pressure must be more durable than those exposed to low blood pressure. However, the requirements for the cardiac wall substitute, regardless of intracardiac or free cardiac wall reconstruction, are essentially almost the same. Properties of the ideal cardiac wall substitute are as follows.

1. Easily stored and immediately supplied when needed
2. Installed with proper thickness, pliability, durability, and easy handling properties
3. Good suturability
4. Almost no bleeding from graft wall, needle holes, and suture lines
5. Sufficient nonthrombogenicity
6. Good neointima formation with endothelialization
7. Good graft wall healing
8. No hypertrophy, shrinkage, aneurysm formation, or degerenation including calcification

In this section, materials used are reviewed, then a new material is introduced that overcomes the problems with conventional materials.

B. Polymer Materials

In the use of polymer materials, Ivalon (polyvinyl alcohol) often developed a false aneurysm [16]. In animal study, it showed severe fibrosis, development of cartilage, calcification, and marked hardening. Woven Teflon had difficulty in suturing due to its considerable stiffness [17], tendency to fray [18], and its frequent association with postoperative hemorrhage [19].

Woven Dacron also has poor suturability and handling properties because of its stiffness [20]. Postoperative bleeding occurred occasionally. It exhibits poor nonthrombogenicity and healing delays due to the low porosity of its graft wall. Expanded polytetrafluoroethylene (ePTFE) is softer and less thrombogenic than woven Dacron [21,22]; however, ePTFE bleeds occasionally from needle holes and suture lines because suture hole elongation and material expansion occur due to its poor resilience when exposed to continuous pressure [20].

C. Autologous Tissues

Autologous membranous tissues such as pericardium [19,23] or fascia lata [24] seem to resolve some of the problems associated with synthetic polymer materials. In general, autologous tissues maintain their viability after implantation, have resistance against infection, show natural pliability with good suturability, and fit well to the implanted position with little bleeding from needle holes or suture lines.

There are, however, some problems, such as hypertrophy [25] and/or shrinkage

[24,25] aneurysm formation [19,26], and degeneration, including calcification [25]. Besides, there are quantitative, qualitative, and morphological restrictions. When harvesting the pericardium, for example, sufficient size in the case of infants or good quality in the case of adhesion may not be obtained.

D. Glutaraldehyde-Cross-linked Pericardia

Heterologous pericardium cross-linked with glutaraldehyde [4,27] was developed as a material that, when needed, is immediately supplied with a certain quality and a sufficient quantity. Although the material becomes harder by glutaraldehyde cross-linking than the natural pericardium, it still maintains a certain degree of pliability and is superior to polymer materials with regard to handling, suturability, and hemorrhage after surgical procedures. Glutaraldehyde-cross-linked heterologous porcine [28], equine [27], and bovine [4] pericardia are commercially available. The major problem of these materials is, however, calcification after implantation [4,19,25,29]. Besides, insufficient nonthrombogenicity and poor healing appear to be other problems [11,30,31].

Harvested autologous pericardium, after being soaked in glutaraldehyde for a few minutes, seems to be used for repair of the host heart or vessel walls [32]. Because natural pericardium is usually thin and too pliable to become wrinkled, pericardium somewhat stiffened by glutaraldehyde cross-linking may be easier to handle during operation than nontreated natural pericardium. Even a few minutes of reaction time, however, is enough for glutaraldehyde to securely cross-link the surface of the pericardium. Autologous pericardium cross-linked with glutaraldehyde is, therefore, thought to be substantially the same as heterologous pericardium cross-linked with glutaraldehyde and may have the same problems.

E. Collagen-Coated Ultrafine Polyester Fabric

Each conventional material has its own merits and demerits. To develop an ideal cardiac wall substitute, we fabricated a new material, collagen-coated ultrafine polyester fabric (CUFP). The characteristics of the CUFP are discussed below.

First, since collagen has an affinity for cells, it provides a good environment for migration and proliferation of host cells and induces better healing [33].

Second, the biologic materials cross-linked with a hydrophilic polyepoxy compound preserve their whitish color and original pliability and become more hydrophilic and hydrated compared with those cross-linked with glutaraldehyde [7]. The hydrophilicity is thought to produce a good affinity for host cells and thus good healing. The polyepoxy compound bonds with the amino group (positive charge) of the collagen, thereby relatively increasing the carboxyl group (negative charge) and resulting in a negative charge. The cross-linked collagen, therefore, repels platelets, which have a negative charge [34,35]. The hydrophilicity [36] and hydration also provide the CUFP with nonthrombogenicity.

Third, the thickness of the ultrafine polyester fiber, which is extremely soft, is less than 3 microns, while the thickness of conventional Dacron fiber is about 20 microns. It has been reported that the finer the polyester fiber is, the better is its affinity for cells [37].

1. Fabrication of Collagen-Coated Ultrafine Polyester Fabric

The CUFP was prepared according to the following procedures [30]. Collagen slurry was compressed into the lumen of a vascular prosthesis (10 mm internal diameter, hydraulic permeability 3650 milliters per minute per square centimeter [ml/min/cm^2] at 120 mmHg) composed of ultrafine polyester fiber (less than 3 microns thick, Ecsaine™, Toray Industries, Inc., Tokyo, Japan), and a Teflon rod was inserted into the lumen of the vessel. The collagen was cross-linked with a hydrophilic polyepoxy compound (Denacol EX-313) [7]. The preparation was then dipped in 2% Denacol solution (pH 11, 50% ethanol) at room temperature for 24 hours. After removal of the Teflon rod, the collagen-coated ultrafine polyester tube with a smooth inner surface was completely washed out with distilled water and stored in 70% ethanol. When used as a patch, it was dipped and rinsed in physiologic saline to replace the ethanol and incised longitudinally to form a membrane.

2. Animal Experiment

CUFPs and glutaraldehyde-cross-linked equine pericardia (GA pericardia) [27] as controls were implanted as patches in the right ventricular outflow tract in dogs. Since GA pericardium is one of the best cardiac wall substitutes commercially available at present, it was used as a control in this study.

The CUFPs were white and easy to handle because they were of suitable thickness and pliability. The GA pericardia were comapratively pliable, but yellowish in color. The CUFPs were easier to suture than the GA pericardia, and fitted well to the native cardiac wall. There was very little bleeding from the needle holes or the suture lines in either the CUFP or GA pericardium.

In the CUFP, almost no thrombi were seen on the inner surface of CUFP up to 21 days. At 28 days, a thin neointima, which was almost endothelialized, had been formed. Fibroblasts and vasa vasorum were seen within both the neointima and the graft wall. At 168 days, smooth musclelike cells were observed in the neointima that was endothelialized. The CUFP showed a white, shiny, smooth, thin, and uniform neointima with endothelialization at 486 days and the neointima was firmly anchored by Day 699. Neither ulcer nor thrombus formation was seen on the surface, and the original pliability of the CUFP was preserved. Fibroblasts with collagen fibers were seen everywhere within the graft wall at 168, 486, and 699 days.

In contrast, GA pericardia showed 2- to 3-mm thick thrombi, mainly in the peripheral area of the patch at 4, 5, and 7 days. At 28 days, there was a large, fresh, red thrombus in the center of the patch. A neointima, with an ingrowth of fibrous tissue and endothelialization, had localized in the peripheral area of the patch. Almost no host cell infiltration into the graft wall was observed. In the specimens up to 28 days after implantation, there seemed to be a tendency toward detachment of the formed thrombus or neointima on the inner surface of the graft. At 147 days, a thick and rough-surfaced neointima, with endothelialization over almost the entire surface, had been formed. The neointima was, however, detached from the GA pericardium except around the suture line, and red thrombi were observed in the slitlike space surrounded by the neointima and the GA pericardium. A cluster of chrondrocytelike cells was seen in the neointima, and the GA pericardium had increased in hardness and showed a ruffled appearance. At 353 days, the neointima was thick around the suture line and thinner in the central area, as if there was pannus formation. Few cell layers and no endothelialization were seen in the central area. At 484 days, the neointima had been formed unevenly and was partly detached from the GA pericardium. In all specimens, minimal infiltration of fibroblasts into the graft wall was observed.

In summary, problems with the GA pericardium were identified in a long-term animal study, as insufficient nonthrombogenicity, incomplete neointima formation, and poor graft wall healing. The CUFP was superior to GA pericardia with regard to suturability, nonthrombogenicity, neointima formation, and graft wall healing [11,30,31].

3. Further Refinement of Collagen-Coated Ultrafine Polyester Fabric

The above-mentioned collagen-coated ultrafine polyester fabric, which is renamed as CUFP-1 in this section, showed satisfactory nonthrombogenicity and excellent neointima formation in an animal study. Although the CUFP-1 with air-dried collagen was an adequate material, CUFP-2 with freeze-dried collagen was developed as a further refinement.

In the same animal model as that in the CUFP-1 study, neointima formation and reconstruction of the material wall were promoted more in the CUFP-2 than in the CUFP-1. The CUFP-2 at 781 days showed a glossy, thin, and uniform neointima with endothelialization and a reconstructed material wall without degeneration.

The collagen of the CUFP-2, which has a spongelike construction with many tiny holes, provided a better environment for cell migration and proliferation. While the CUFP-1 required disinfection and storage in 70% ethanol, the CUFP-2 could be completely sterilized with ethylene oxide and stored in a dry package more easily than the CUFP-1. Therefore, the CUFP-2 is the superior and more practical cardiac wall substitute [38].

4. Expanded Application of Collagen-Coated Ultrafine Polyester Fabric

There is a possibility that the application of the CUFP may be expanded because the concept of the CUFP was originally created for an omnipotent substitute for part or whole of the chamber and conduit with a blood-exposed surface. In fact, it was indicated in the animal experiments that the CUFP was an excellent cardiac wall substitute and that the properties of the CUFP basically coincided with those required by vascular prostheses [11,30,31].

It appears that thrombosis and neointimal hypertrophy tend to develop more on intracardiac surfaces and pulmonary arterial lumens than on systemic arterial inner surfaces. Besides, thick neointima so-called peel was often formed on the internal surface of RV-PA bypass conduit [39], causing conduit stenosis, leading to replacement of the graft [40]. Thus, substitutes for the cardiac walls and pulmonary arteries require much stricter properties. Conversely materials that function and show good healing in the right heart system are thought to work satisfactorily at least as substitutes for large- and medium-size vessels.

Therefore, the CUFP, which seems to be a successful material in the right ventricular outflow tract, may be used for other applications of the cardiovascular system with minor modifications in thickness, strength, and the like. For example, the CUFP may function and heal well as a RV-PA bypass conduit without peel phenomenon because the CUFP does not seem to show neointima hypertrophy in the right heart system. The CUFP also may be implanted as vascular prostheses for superior and inferior vena cavas, and large- and medium-size arteries. Further animal experiments are needed to make a precise and final evaluation.

5. Conclusions

The neointima on the CUFP is rapidly formed with endothelialization, obtains natural antithrombogenicity, and is maintained for a long-term period in the canine model of right ventricular outflow tract patch enlargement. The wall of the CUFP is reconstructed by the host cells after implantation, becomes a kind of host organ, and provides an optimal environment for formation and maintenance of the neointima. As the neointima and the CUFP metabolize, it is thought that there is little possibility of their subsequent degeneration, including calcification.

The CUFP should be immediately supplied as much as needed because the CUFP is a ready-made totally artificial material and may be used not only as patch materials, but also as conduits or other blood-exposed surfaces in various shapes and sizes.

III. SMALL-DIAMETER VASCULAR PROSTHESES

A. Introduction

Autologous vessels such as the great saphenous vein, the interal mammary artery, and the right epigastric artery are well-accepted substitutes for small-diameter arteries. Because of limited supply of the autologous organs and the increasing demand for repeating coronary artery bypass grafting for ischemic heart diseases, limb salvage for atherosclerosis obliterance, aorto-pulmonary artery shunt for congenital heart diseases, and the like, the development of a small-diameter vascular graft with satisfactory handling properties, excellent patency, and durability is more and more critical.

For the past 15 years, some types of vascular prostheses have been investigated. One of them is a synthetic vascular prosthesis made of antithrombogenic polymers such as segmented polyurethane, hydrous polymers, and heparinized hydrophilic polyurethane. This prosthesis had been expected to prevent thrombus adhesion, and showed excellent results in a short period of time. However, almost no long-term results of its patency were reported. It seems, therefore, that most of these prostheses could not achieve long-term patency. The causes of the failures have not been known, but one of the reasons may be a pannus formation problem at the anastomotic sites of their grafts [41]. The pannus disturbs the bloodstream, especially at the distal anastomotic sites, resulting in graft occlusion due to thrombus formation. Only one graft that showed more than one year patency was, however, reported by Nojiri and colleagues [42]. They fabricated a micro-domain polymer graft that could prevent the platelet adhesion on the luminal surface [43].

Herring and coworkers adopted an endothelial cell seeding technique in 1978 [44]. Thereafter, a great number of works were reported using endothelial cell culture technologies *in vitro*. However, the survival rate of these cells after implantation *in vivo* was extremely low and very few clinical applications had been performed. Recently, Miwa and coworkers improved the cell culture technology *in vitro* [45]. They fabricated a vascular prosthesis with a hierarchical structure to mimic a natural blood vessel. A mixture of smooth muscle cells and a gel composed of Type I collagen and delmatan sulfate were coated on the luminal surface of a polyurethane vascular prosthesis. Endothelial cells were seeded on this coating. The grafts showed a good patency and maintained endothelialization [46].

Another element of progress in this field was modification of graft matrices. Dacron fabrics and ePTFE tube have been used; however, they are not available in small-diameter grafts. To modify the structures of these matrices, ultrafine polyester fibers for fabric vascular prostheses [47] and long fibril length ePTFE grafts were developed [48]. They showed adequate improvements. Biodegradable substances such as collagen, gelatin, and albumin also have been used to coat these grafts for prevention of blood loss and improvement of cell adhesion and neointima formation [49,50]. Recent remarkable works were modifications of coating substances of these matrices for cell anchoring. Some growth factors and substances that can accelerate the cell growth are considered for this purpose. A special sequence of amino acids (arginine, glycine, and aspartic acid), which is the major component of cell adhesive protein such as fibronectin, was also used to modify the graft coating substances, resulting in excellent cell anchoring in the graft wall [51,52]. Although these attempts have just started in this field, we expect to have a remarkable progress within a few years.

For the last few years, one successful modification of graft structures was indicated in a microporous polyurethane graft [53]. Okoshi et al. reported small-diameter polyurethane vascular prostheses with observations of satisfactory patency and rapid endothelialization in a rat abdominal aortic replacement model. The graft had a special microporous structure in the luminal surface and graft wall. The hypothesis is that the microporous structure induces the formation of a smooth surface layer of fibrin network. This layer contains a certain amount of albumin that has no relationship with blood coagulation. The naturally biolized surface with host albumin maintains the patency of the graft with minimal thrombus formation and also accelerates the endothelialization.

Biological grafts have been studied for the last 50 years [54]. Recently, some of them showed good results. To reduce antigenicity and to give stability *in vivo*, glutaraldehyde (GA) has been used, but GA is very toxic and GA treatment completely changes the native properties of the original materials. Dialdehyde starch (DAS) has been also used for this purpose [55]; however, the cross-linking rate with DAS is too low to give sufficient mechanical stability, resulting in tissue degeneration after implantation. To overcome these problems, a new cross-linking method using polyepoxy compound (PC) was developed [13]. In this chapter, this new technology is explained in detail.

Autologous tissue transplantation technology was introduced with successful results. Noishiki et al. applied this technique of autologous vein fragment transplantation to a fabric vascular prosthesis [56]. They obtained excellent results in large-diameter grafts, but they encountered a major problem when they applied it to a small-diameter graft since the tissue fragments were thrombogenic. To overcome this problem, they bound heparin ionically to the luminal surface of the graft and achieved a good patency and very rapid endothelialization by, as it were, "*in vitro* tissue culture" on the whole graft luminal surface within 2 to 4 weeks [57].

In the next sections, a polyurethane vascular graft having a microporous luminal structure with a biolized surface, a biological vascular graft cross-linked with polyepoxy compound, and an autologous venous tissue fragmented vascular graft with heparin impregnation are described.

B. Microporous Polyurethane Vascular Grafts

From a biomaterial standpoint, polyurethane-polydimethylsiloxane lends itself to a wide range of microporous structures with communicating voids. In this study, it is indicated that wall porosity of the microporous small-diameter vascular grafts fabricated with that material have significant effects on early graft patency as well as graft healing in the late stage.

1. Fabrication of the Grafts

Three types of microporous polyurethane-polydimethylsiloxane (PU-PDMS, Cardiothane 51™) vascular grafts (1.5-mm internal diameter [ID], 450-μm wall thickness) were fabricated by a spray phase-inversion technique [58,59]. Those grafts had different surface morphology and wall porosity as reflected by hydraulic permeability.

Some grafts had a smooth, dense skin on the luminal surface, a compact porous wall, and a filamentous outer surface with a hydraulic permeability of 0 ml/min/cm^2 (PUG-S-0). Some had a filamentous luminal surface with pores measuring 30 to 70 μm, with a mean hydraulic permeability of 2.7 ml/min/cm^2. The graft wall and outer surface features were the same as PUG-S-0 (PUG-2.7). Some had a filamentous luminal surface, with interfiber spaces ranging from 90 to 130 μm. Wall and outer surface features were widely open with a mean hydraulic permeability of 39 ml/min/cm^2 (PUG-39).

2. Animal Experiment

All of the grafts displayed good handling properties and suturability. Straight grafts of PUG-S-0, PUG-2.7, and PUG-39 (1.5–2.0 cm in length) and one loop PUG-39 (10 cm in length) were implanted by the same surgeon end to end in the infrarenal aorta of male Sprague-Dawley rats weighing 250–350 g. Anticoagulation was not used at any time.

At three months postimplantation, patency of the straight grafts was 0% for PUG-S-0, 8% for PUG-2.7, and 76% for the PUG-39, and the loop PUG-39 was also patent. The patency of PUG-S-0 1 to 2 weeks after implantation was also 0% and there were only two patent PUG-S-0 grafts, one graft retrieved at the first postoperative day and one at the third postoperative day. The sole patent PUG-2.7 showed neointimal hyperplasia and incomplete endothelialization. All but one of the patent PUG-39 grafts showed a glistening and transparent neotinima with complete endothelialization. Numerous host cells had migrated and proliferated within the graft wall. The loop PUG-39 displayed endothelialization from each anastomosis and in many islands in the middle portion of the graft, totaling 47% of the luminal surface by morphometric analysis. Thick mural thrombus, anastomotic hyperplasia, or aneurysm formation were not observed in any patent PUG-39. The three-month graft patency of straight PUG-39 was significantly higher than that of PUG-2.7.

A longer duration of observation will be needed to assess the chemical and mechanical durability of the polyurethane vascular prostheses, although problems of aneurysmal dilatation or calcification were not noticed up to three months.

3. Conclusions

The major factors affecting the patency of very small vascular grafts are thought to be graft material, luminal surface geometry, wall structure and resilience, hydraulic permeability, compliance, preparation techniques, storage conditions, and surgical skills. In these experiments, graft surface morphology and wall porosity as reflected by hydraulic permeability were the main determinants of patency and completeness of the healing process, including endothelialization.

There is a great possibility that the loop graft of PUG-39 may show complete endothelialization with more time because there were many islands of endothelial cells, which may expand with time. The concept of this graft may lead to development of substitutes for small-diameter vascular prostheses for coronary artery bypass grafting or aortopulmonary artery shunt operation.

C. Heparinized Biological Grafts Cross-Linked with Polyepoxy Compound

Biological materials have unique, fine structural and mechanical properties that cannot be simulated by any current technologies. For example, arteries have ideal hemodynamically shaped ramifications. Within the arterial walls, they have unique and fine structures especially suitable for cell inhabitation. They also have unique mechanical properties to accept and pass the pulsatile blood pressure and flow. If we could use these special properties and structures for biomedical materials, we could make excellent artificial organs. One of the problems is, however, antigenicity of the materials (except for those of autologous origin) and another problem is biodegradability of the materials. To reduce the antigenicity and biodegradability of the materials, chemical modifications by glutaraldehyde, dialdehyde starch, formaldehyde, and hexamethylene diisocyanate have been used. However, treatments with these reagents make the materials hydrophobic and stiff. Glutaraldehyde is the most frequently used reagent, but it has cytotoxicity and prevents cell infiltration into the graft wall [60]. To overcome these problems, a new cross-linking reagent, hydrophilic polyepoxy compound (PC), which was already explained in the heart valve section, was introduced [61]. Another technology we have developed is the heparinization of biological materials.

Collagen is one of the major components of the biological materials. It uniquely induces host cell migration and proliferation, but, detrimentally, platelet adhesion and accumulation occur as well [62]. Therefore, these biological materials such as collagen show thrombogenicity when used as cardiovascular artificial organs. To reduce the thrombogenicity, we developed a heparinization method [63]. With the combined use of the new cross-linking and the heparinization methods, a unique small-diameter biological graft was developed.

1. Preparation of the Grafts

For graft preparation, a fresh carotid artery with an inner diameter of 2.5 to 3.0 mm was obtained from dogs. The artery was soaked in distilled water for one hour and submitted to ultrasonic waves to cause cell destruction. Cell debris was then removed by rinsing with distilled water. Thus, a natural tissue tube composed of collagen and elastic laminae was created. A 2% protamine sulfate solution at pH 5.9 was injected into the lumen of the natural tissue tube. Then, the graft was treated with a 5% PC solution for five hours at room temperature to cross-link the tissue and to make the protamine impregnated into the graft wall and covalently immobilized. The graft was rinsed with distilled water, then soaked in a 1% heparin solution and again rinsed with distilled water repeatedly. The graft was then preserved and sterilized in a 70% ethanol solution.

2. Animal Experiments

Mongrel dogs weighing 8 to 12 kg were used for the experiments. About 6-cm segments were resected from bilateral carotid arteries and heparinized grafts, 6 cm long, 2.5 to 3.0 mm internal diameter, were implanted end to end. No anticoagulants were administered at any time. The animals were euthanized electively at selected time intervals. The implantation periods ranged from one hour to 429 days. Heparin concentration in the whole graft wall was measured according to the method by Lagnoff and Warren [64].

All the grafts were patent at the time of the angiographic examination. The inner surface of the grafts was smooth throughout their length, and no stenosis or aneurysmal dilatation was observed in any of the grafts. The overall patency rate of the heparinized graft was 96%. All of the patent grafts were still as soft and pliable as the native artery. Within 100 days after implantation, their inner surfaces were completely free from thrombus deposition. The surfaces were as shiny, white, smooth, and glistening as those of the host arterial intima. The grafts implanted for more than 100 days displayed slightly yellowish and semitransparent small spots sporadically on the luminal surface.

Microscopic observations confirmed that there was neither thrombus nor fibrin deposition. The luminal surface showed the original internal elastic membrane of the graft and no endothelial cells. In the early stage, there was no foreign-body reaction, such as giant cell infiltration on the surface of the graft. A small number of plasma cells was observed at a short period of time after implantation. After six months, smooth-muscle-like cells infiltrated into the graft wall from the adventitial sides. These fusiform cells had occupied the graft wall completely after a long period of time. They were circumferentially arranged in the luminal side three-quarters of the graft wall and were longitudinally oriented in the outer quarter of the graft wall. After 389 days, the central part of the inner surface was covered with endothelial cells, which impinged directly on the surface of the elastic lamina. The structure of the graft following long-term implantation closely resembled that of the native arterial wall. At 389 days, a thin layer of pannus with an endothelial cell lining covered the surface near the anastomotic lines.

Before and after implantation, the total amount of heparin in the graft was measured. The value before implantation was about 76.0 units/cm^2, but no heparin was detected in specimens in place for more than 80 days.

3. Discussion

The unique aspect of this study is that stable and permanent antithrombogenicity of the vascular grafts can be obtained by combination of temporary slow heparin release and permanent natural antithrombogenicity by endothelialization. The slow heparin release completely prevents thrombotic graft obstruction in the early stage after implantation. More than 95% of the immobilized heparin was released in two months after implantation. Endothelial cells appeared 45 days after implantation and a continuously endothelialized layer was observed at 153 days. Temporal antithrombogenicity was thus smoothly transformed into permanent and natural antithrombogenicity during the experimental period and mural thrombi were not observed in any grafts.

The infiltration of smooth-muscle-like cells mainly contributed to reconstruction of the graft with a great similarity to the natural arterial wall. PC-cross-linked grafts obtained a reconstructed graft wall but still maintained their original graft compliance.

Furthermore, the PC-cross-linked grafts are hydrophilic because PC has hydroxyls in its molecular structure, while the GA cross-linked grafts are hydrophobic. In this report, the PC-cross-linked graft displayed superior nonthrombogenic characteristics because the high hydrophilicity might provide the material with nonthrombogenicity. Besides, this graft has another merit with regard to nonthrombogenicity since it becomes slightly negatively charged after the cross-linking. This was because e-NH$_2$ groups (positive charge) in the collagen molecules were consumed during the cross-linking, resulting in relative dominance of the carboxyl groups (negative charge) in number. The weakly negatively-charged surface contributes to the nonthrombogenicity of the materials by prevention of platelet (negative charge) aggregation.

4. Conclusions

It is concluded that short-term antithrombogenicity of slow-release heparin followed by the permanent antithrombogenicity of endothelial cells in combination with the natural tissue compliance of these grafts was the main cause of their success as a small-caliber vascular graft.

D. Autologous Tissue Fragment Transplanted Grafts

In general, endothelialization of vascular grafts is extremely delayed [65]. Most of the grafts implanted are not endothelialized, and are covered with fresh thrombi for a long time after implantation except for areas near anastomotic sites. This delayed healing is due to a kind of protracted ulcer in the vascular wall. Therefore, autologous tissue fragment transplantation might be effective to enhance the healing. To verify this hypothesis, we made a fabric vascular prosthesis transplanted with autologous venous tissue fragments onto the wall.

1. *Preparation of the Grafts*

For preparation of the grafts, a canine jugular vein was resected, minced into tissue fragments as small as possible, and then was stirred into 20 ml of physiological saline containing 1000 international units (IU) heparin. A highly porous fabric Dacron vascular prosthesis (MICROKNIT, 4-mm ID, hydraulic permeability of 4000 ml/min/cm^2) was used as the framework of the graft. The fabric prosthesis was invaginated inside out, the tissue suspension was injected under pressure into the prosthesis, causing tissue fragments to be trapped in the graft wall, and finally the sealed prosthesis was reinvaginated.

The tissue fragmented and heparinized prosthesis (TFH-graft) was implanted end to end into bilateral carotid arteries in dogs. Preclotted grafts without tissue fragments were also implanted into the carotid arteries as controls. Grafts were retrieved 1 hour to 400 days after implantation.

In an *in vivo* study, an average of 1806 IU heparin/g (graft weight) was detected. While rinsing the graft in saline, approximately 92.5% of the heparin was released during the first 5 hours and 94.5% was released after 25 hours.

2. *Animal Experiments*

In animal experiments, in the TFH-grafts, host cells migrated and proliferated actively into the fibrin inner capsule. A single layer of endothelial cells formed on the luminal surface, covering multiple layers of smooth muscle cells underneath. New arterial wall was complete throughout the TFH-grafts within 2 weeks in the patent grafts, but all control grafts were occluded within one week. There was no thrombus formation due to the tissue fragments. After the initial heparin release, only a very small amount of heparin remained and contributed to maintaining the nonthrombogenicity of the fragments. Endothelial cells had started covering the graft luminal surface before the graft completed its heparin release.

3. *Discussion*

The tissue-fragment-seeded graft showed extremely rapid healing along its entire length due to cell migration and proliferation from the venous fragments transplanted. Neointima formation started by the fourth day, and was completed at two weeks. This is the fastest healing of any grafts reported. The healing proceeded equally at all locations, with no difference in healing between the central area and the anastomotic sites.

Transplantation of autologous tissue fragments has already been used in orthopedic and plastic surgery as osteoblasts migrate very rapidly from the cut edges of transplanted bone fragments, and epidermal cells migrate very rapidly from skin fragments [66,67]. The cut edges of multiple tiny fragments have sufficient surface from which cells can migrate and proliferate very rapidly under *in vivo* physiological conditions.

In this experiment, three kinds of cells (fibroblasts, smooth muscle cells, and endothelial cells) migrated and proliferated at the same time from the fragments. This is a very unique phenomenon that has not been observed previously in cell culture. When fibroblasts and endothelial cells are cultured in a petri dish, endothelial cells are suppressed and fibroblasts proliferate to form a confluent layer. In this *in vivo* experiment, however, these three cells migrated and proliferated together. Endothelial cells produced capillaries, and rose to the inner surface of the graft to face the bloodstream. On the other hand, smooth muscle cells formed multilayers beneath the endothelial cells, and fibroblasts moved down under the smooth muscle cell layers in the proximity of the polyester fibers. This phenomenon suggests that the behavior of these cells is determined by the physiological environment in which they are placed.

4. Conclusions

In the tissue fragmented and heparinized vascular grafts, heparin slow release can prevent thrombus formation in the early stage of graft implantation, and neointima formation is enhanced by the tissue fragment transplantation followed by the natural and permanent antithrombogenic property of the endothelialized vascular graft.

REFERENCES

1. Fernandez, J., Laub, G.W., Adkins, M.S., Anderson, W.A., Chen, C., Bailey, B.M., Nealon, L.M., and McGrath, L.B. Early and late-phase events after valve replacement with the St. Jude Medical prosthesis in 1200 patients. *J. Thorac. Cardiovasc. Surg.* 107:394–407, 1994.
2. Glower, D.D., White, W.D., Hatton, A.C., Smith, L.R., Young, W.G., Wolfe, W.G., and Lowe, J.E. Determinants of reoperation after 960 valve replacements with Carpontier-Edwards prostheses. *J. Thorac. Cardiovasc. Surg.* 107:381–393, 1994.
3. Cipriano, P.R., Billingam, M.E., Oyer, P.E., Kutsche, L.M., and Stinson, E.B. Calcification of porcine prosthetic heart valves: A radiographic and light microscopic study. *Circulation* 66:1100–1104, 1982.
4. Gabbay, S., Bortolotti, U., Factor, S., Shore, D.F., and Fraster, R.W.M. Calcification of implanted xenograft pericardium. Influence of site and function. *J. Thorac. Cardiovasc. Surg.* 87:782–787, 1984.
5. Webb, C.L., Flowers, W.E., Horton, C., Schoen, F.J., and Levy, R.J. Long-term efficacy of Al^{3+} for prevention of bioprosthetic heart valve calcification. *Trans. Am. Soc. Artif. Intern. Organs* 36:408–410, 1990.
6. Kodaira, K., Noishiki, Y., Miyata, T., Furuse, M., and Yamane, Y. Characterization of collagenous materials by new crosslinking method. *Jpn. J. Artif. Organs* 16:1346–1349, 1987.
7. Kodaira, K., Miyata, T., Furuse, M., and Noishiki, Y. Characterization of collagenous materials crosslinked by polyepoxy compounds. *Jpn. J. Artif. Organs* 15:239–242, 1986.
8. Edmunds, L.H. Thrombotic and bleeding complications of prosthetic heart valves (collective review). *Ann. Thorac. Surg.* 44:430–445, 1987.
9. Williams, J.B., Karp, R.B., Kirklin, J.W., Kouchoukos, N.T., Pacifico, A.D., Zorn, G.L., Jr., Blackstone, E.H., Brown, R.N., Piantadosi, S., and Bradley, E.L. Consideration in selection and management of patients undergoing valve replacement with glutaraldehyde-fixed porcine bioprostheses. *Ann. Thorac. Surg.* 30:247–258, 1980.
10. Tomizawa, Y., Noishiki, Y., Okoshi, T., Miyata, T., and Koyanagi, H. Aortocoronary bypass grafting with hydrophilic small caliber vascular grafts. *Trans. Am. Soc. Artif. Intern. Organs* 35:199–202, 1989.
11. Okoshi, T., Noishiki, Y., Tomizawa, Y., Morishima, M., and Koyanagi, H. Long-term results of a new antithrombogenic cardiac wall substitute. *Trans. Am. Soc. Artif. Intern. Organs* 35:391–395, 1989.
12. Okoshi, T., Noishiki, Y., Tomizawa, Y., Morishima, M., Taira, T., Kawai, T., Itoh, H., Miyata, T., and Koyanagi, H. A new bioprosthetic cardiac valve with reduced calcification. *Trans. Am. Soc. Artif. Intern. Organs* 36:411–414, 1990.
13. Noishiki, Y., Miyata, T., and Kodaira, K. Development of a small caliber vascular graft by a new crosslinking method incorporating slow heparin release collagen and natural tissue compliance. *Trans. Am. Soc. Artif. Intern. Organs* 32:91–96, 1986.

14. Murayama, Y., Satoh, S., Oka, T., Imanishi, J., and Noishiki, Y. Reduction of the antigenicity and immunogenicity of xenografts by a new crosslinking reagent. *Trans. Am. Soc. Artif. Intern. Organs* 34:546–549, 1988.

15. Okoshi, T., Noishiki, Y., Tomizawa, Y., Egoh, Y., and Koyanagi, H. Improved nonthrombogenicity of a new bioprosthetic heart valve crosslinked with a polyepoxy compound. *9th Congress of International Society for Artificial Organs (ISAO)/20th Congress of European Society for Artificial Organs (ESAO), International Artif. Organs* 16:466, 1993.

16. Payne, W.S., and Kirklin, J.W. Late complications after plastic reconstruction of outflow tract in tetralogy of Fallot. *Ann. Surg.* 154:53–57, 1961.

17. Edwards, W.S., Forrest, C., Martin, P.D., and Lovoy, S. Reconstruction of the cardiac septa and right ventricular outflow tract with Teflon patches. *J. Thorac. Cardiovasc. Surg.* 41:631–634, 1961.

18. Boyd, D.P., and Midell, A.I. Woven Teflon aortic grafts. An unsatisfactory prosthesis. *Vasc. Surg.* 5:148–153, 1971.

19. Seybold-Epting, W., Chiariello, L., Hallman, G.L., and Cooley, D.A. Aneurysm of pericardial right ventricular outflow tract patches. *Ann. Thorac. Surg.* 24:237–240, 1977.

20. Guidoin, R.G., Snyder, R.W., Awad, J.A., and King, M.W. Biostability of vascular prostheses, in *Cardiovascular Biomaterials*, ed. by G.W. Hastings, Springer-Verlag, London, 1992, pp. 143–172.

21. Sakamoto, T., Imai, Y., Koyanagi, H., Hayashi, H., and Hashimoto, A. Clinical use of expanded polytetrafluoroethylene in cardiac surgery. *Jpn. J. Thorac. Surg.* (*Kyobu-geka*) 31:23–29, 1978.

22. Coto, E.O., Norwood, W.I., Lang, P., and Castaneda, A.R. Modified Senning operation for treatment of transposition of the great arteries. *J. Thorac. Cardiovasc. Surg.* 78:721–729, 1979.

23. Sauvage, L.R., Gross, R.E., Rudolph, A.M., Pontitus, R.G., and Watkins, E., Jr. Experimental study of tissue and prosthetic grafts with selected application to clinical intracardiac surgery. *Ann. Surg.* 153:321–343, 1961.

24. Ross, D., and Sommerville, J. Fascia-lata reconstruction of the right ventricular outflow tract. *Lancet* 1:941–943, 1971.

25. Hjelms, E., Pohlner, P., Barratt-Boyes, B.G., and Gavin, J.B. Study of autologous pericardial patch-grafts in the right ventricular outflow tracts in growing and adult dogs. *J. Thorac. Cardiovasc. Surg.* 81:120–123, 1981.

26. Kaplan, S., Helmsworth, J.A., McKinivan, C.E., Benzing, IIIG, Schwartz, D.C., and Sreiber, J.T. The fate of reconstruction of the right ventricular outflow tract. *J. Thorac. Cardiovasc. Surg.* 66:361–374, 1973.

27. Shin-oka, T., Hoshino, S., Kurosawa, H., Takanashi, Y., and Imai, Y. Clinical use of glutaraldehyde preserved pericardial xenograft (Xenomedica). *Jpn. J. Artif. Organs* 14:733–736, 1985.

28. Kaizuka, H., Nishiyama, H., Nagara, H., Tahara, S., Kei, J., and Wada, J. Glutaraldehyde preserved porcine pericardium as a heterogeneous bioprosthesis in cardiovascular surgery. A study of 53 cases. *Jpn. J. Artif. Organs* 11:1183–1186, 1982.

29. Ferrans, V.J., Boyce, S.W., Billingham, M.E., Jones, M., Ishihara, T., and Roberts, W. Calcific deposits in porcine bioprostheses: Structure and pathogenesis. *Am. J. Cardiol.* 46:721–734, 1980.

30. Okoshi, T., Noishiki, Y., Tomizawa, Y., Morishima, M., and Koyanagi, H. Development of an antithrombogenic cardiac wall substitute which can be reconstructed by infiltration of host cells. *Trans. Am. Soc. Artif. Intern. Organs* 34:532–537, 1988.

31. Okoshi, T., Noishiki, Y., Tomizawa, Y., Morishima, M., Terada, R., and Koyanagi, H. Neointima formation on an antithrombogenic cardiac wall substitute that can be reconstructed by host cells. *Trans. Am. Soc. Artif. Intern. Organs* 36:303–306, 1990.

32. Cooly, D.A. *Techniques in Cardiac Surgery*, 2nd edition. W.B. Saunders, Philadelphia, 1984, pp. 134–137.
33. Noishiki, Y., and Chvapil, M. Healing pattern of collagen-impregnated and preclotted vascular grafts in dogs. *J. Vasc. Surg.* 21:401–411, 1987.
34. Miyata, T., Schwartz, A., Wang, C.L., Rubin, A.L., and Stenzel, K.H. Deposition of platelets and fibrin on chemically modified collagen hollow fibers. *Trans. Am. Soc. Artif. Intern. Organs* 22:261–267, 1976.
35. Stenzel, K.H., Miyata, T., and Rubin, A.L. Collagen as a biomaterial. *Ann. Rev. Biophs. Bioeng.* 3:231–253, 1974.
36. Mori, Y., Nagaoka, S., Takiuchi, H., Terada, R., Nishiumi, S., Tanzawa, H., Kuwano, A., and Miyama, H. Interactions between hydrogels containing polyethyleneoxide chains and platelets, in *Progress in Artificial Organs 1983*, ISAO Press, Cleveland, 1984, pp. 825–830.
37. Satoh, S., Naito, K., Shirakata, S., Oka, T., and Noishiki, Y. Effect of polyester fibers on cell proliferation in culture. *Jpn. J. Artif. Organs* 17:643–646, 1988.
38. Okoshi, T., Noishiki, Y., Tomizawa, Y., Morishima, M., Terada, R., and Koyanagi, H. A further refinement of a nonthrombogenic cardiac wall substitute which can be reconstructed by host cells. *Abstracts of Trans. Am. Soc. Artif. Intern. Organs* 37:25, 1991.
39. Agarwal, K.C., Edwards, W.D., Feldt, R.H., Danielson, G.K., Puga, F.J., and Mcgoon, D.C. Clinicopathological correlates of obstructed right-sided porcine-valved extracardiac conduits. *J. Thorac. Cardiovasc. Surg.* 81:591–601, 1981.
40. Mcgoon, D.C., Danielson, G.K., Puga, F.J., Ritter, D.G., Mair, D.D., and Ilstrup, D.M. Late results after extracardiac conduit repair for congenital cardiac defects. *Am. J. Cardiol.* 49:1741–1749, 1982.
41. Noishiki, Y. New concepts and development of vascular graft prostheses, in *Advances in Cardiovascular Engineering*, ed. by N.H.C. Hwang, V.T. Turitto, and M.R.T. Yen, Plenum Press, New York, 1992, pp. 419–434.
42. Nojiri, C., Nakahama, S., Senshu, K., Okano, T., Kawagoshi, N., Kido, T., Sakai, K., Koyanagi, H., and Akutsu, T. A new amphiphilic block co-polymer with improved elastomeric properties for application in various medical devices. *Trans. Am. Soc. Artif. Intern. Organs* 39:322–326, 1993.
43. Okano, T., Aoyagi, T., Kataoka, K., and Sakurai, Y. Hydrophilic-hydrophobic microdomain surfaces having an ability to suppress platelet aggregation and their *in vitro* antithrombogenicity. *J. Biomed. Mater. Res.* 20:919–927, 1986.
44. Herring, M.B., Garner, A.L., and Glover, J. A single stage technique for seeding vascular graft with autologous endothelium. *Surg.* 84:489–504, 1978.
45. Miwa, H., Matsuda, T., and Iida, F. Development of a hierarchically structured hybrid vascular graft biomimicking natural arteries. *Trans. Am. Soc. Artif. Intern. Organs* 39:273–277, 1993.
46. Kanda, K., Matsuda, T., Miwa, H., and Oka, T. Phenotype modulation of smooth muscle cells in intima-medic incorporated hybrid vascular prostheses. *Trans. Am. Soc. Artif. Intern. Organs* 39:278–282, 1993.
47. Noishiki, Y., Watanabe, K., Okamoto, M., Kikuchi, Y., and Mori, Y. Evaluation of a new vascular prosthesis fabricated from ultrafine polyester fiber. *Trans. Am. Soc. Artif. Intern. Organs* 32:309–314, 1986.
48. Hunter, G.C. The relationship between cellular ingrowth, internodular distance, and the healing of small diameter PTFE grafts. *Cardiovascular Science and Technology Conference*, Washington, DC, Dec. 1993.
49. Jonas, R.A., Ziemer, G., Schoen, F.J., Britton, L., and Castaneda, A.R. A new sealant for knitted Dacron prostheses: Minimally cross-linked gelatin. *J. Vasc. Surg.* 7:414–419, 1988.
50. De Mol Van Otterloo J.C.A., Van Bockel, J.H., Ponfoort, E.D., Brommer, E.J.P., Her-

mans, J., and Daha, M.R. The effects of aortic reconstruction and collagen impregnation of dacron prostheses on the complement system. *J. Vasc. Surg.* 16:774–783, 1992.

51. Tweden, K.S., Blevitt, J., Harasaki, H., Helmus, M., Glass, J., Dickenson, K., Craig, W.S., and Piersohbacher, M. RGD modification of cardiovascular prosthetic materials. *Cardiovascular Science and Technology Conference*, Washington, DC, Dec. 1993.

52. Dec, K.C., Anderson, T.T., and Bizic, R. New trends in the design of substrates which promote endothelialization. *Cardiovascular Science and Technology Conference*, Washington, DC, Dec. 1993.

53. Okoshi, T., Soldani, G., Goddard, M., and Galletti, P.M. Very small-diameter polyurethane vascular prostheses with rapid endothelialization for coronary artery bypass grafting. *J. Thorac. Cardiovasc. Surg.* 105:791–795, 1993.

54. Kimoto, S., Sugie, S., and Tsunoda, M. Experimental and clinical studies on arterial homo- and hetero-grafts preserved in alcohol. *AMA Arch. Surg.* 69:549–558, 1954.

55. Mitchell, I.M., Essop, A.R., and Scott, P.J. Bovine internal mammary artery as a conduit for coronary revascularization: Long-term results. *Ann. Thorac. Surg.* 55:120–122, 1993.

56. Noishiki, Y., Yamane, Y., Tomizawa, Y., Okoshi, T., Satoh, S., Widevuur, C.H.R., and Suzuki, K. Rapid endothelialization of vascular prostheses by seeding autologous venous tissue fragments. *J. Thorac. Cardiovasc. Surg.* 104:770–778, 1992.

57. Noishiki, Y., Tomizawa, Y., Yamane, Y., Okoshi, T., Satoh, S., and Matsumoto, A. Acceleration of neointima formation in vascular prostheses by transplantation of autologous venous tissue fragments. Application to small-diameter grafts. *J. Thorac. Cardiovasc. Surg.* 105:796–804, 1993.

58. Strathman, H. Production of microporous media by phase inversion process. *Material Science of Synthetic Membranes* 8:165–195, 1985.

59. Soldani, G., Panol, G., Sasken, F., Goddard, M.B., and Galletti, P.M. Small-diameter polyurethane-polydimethylsiloxane vascular prostheses made by a spraying, phase-inversion process. *Journal of Materials Science: Materials in Medicine* 3:106–113, 1992.

60. Speer, D., Chvapil, M., Eskelson, C.D., and Ulreich, J. Biological effects of residual glutaraldehyde in glutaraldehyde-tanned collagen biomaterials. *J. Biomed. Mater. Res.* 14:753–764, 1980.

61. Tomizawa, Y., Noishiki, Y., Okoshi, T., Miyata, T., and Koyanagi, H. Development of a small caliber biological vascular grafts: Evaluation of its antithrombogenicity and the early healing process. *Trans. Am. Soc. Artif. Intern. Organs* 36:734–737, 1990.

62. Wang, C.L., Miyata, T., Schlear, S., Weksler, B., Rubin, A.L., and Stenzel, K.H. Collagen and glomerular basement membrane effects on platelets. *Trans. Am. Soc. Artif. Intern. Organs* 21:422–425, 1975.

63. Noishiki, Y., and Miyata, T. A simple method to heparanize biological materials. *J. Biomed. Mater. Res.* 20:337–346, 1986.

64. Lagnoff, D., and Warren, G. Determination of 2-deoxy-2-sulfoaminohexose content of mucopolysaccarides. *Arch. Biochem. Biophys.* 99:396–400, 1962.

65. Burger, K., Sauvage, L.R., Rao, A.M., and Wood, S.J. Healing of arterial prostheses in man: Its incompleteness. *Ann. Surg.* 175:118–127, 1992.

66. Barsky, A.J., Kahn, S., and Simon, B.E. *Principles and Practice of Plastic Surgery.* 2nd ed. McGraw-Hill, New York, 1964, pp. 419–420.

67. Heppenstall, R.B., ed. *Fracture Treatment and Healing. Bone Grafting.* Philadelphia, W.B. Saunders, 1980, pp. 97–112.

43

Fabrication Techniques and Polymer Considerations for the Blood Contacting Components of the Penn State Circulatory-Assist Devices

George Felder III and James H. Donachy, Sr.
Hershey Medical Center
Hershey, Pennsylvania

I. INTRODUCTION

At the Pennsylvania State University's School of Medicine, the Milton S. Hershey Medical Center, four types of heart replacement/circulatory assistance have been developed. Heart replacement with a total artificial heart (TAH) has been performed with the Penn State Heart, a pneumatically powered device (P-TAH). This heart is one of only two approved for heart replacement by the Food and Drug Administration (FDA).

Concurrently, an electrically powered TAH (E-TAH) has been developed at Hershey, and is in the second phase of a National Institutes of Health (NIH) contract between Penn State and Sarns/3M Healthcare, one of only three contracts awarded for E-TAH development. It is hoped that E-TAH clinical trials will begin shortly after the year 2000. Animal testing at Hershey has resulted in many long-term survivors, including the world's record for longevity for an animal with an E-TAH. These world-record animals have had complete E-TAH systems implanted; there are no breaks in the skin for potential infection. All system power is transmitted through the skin (painlessly) via induction coupling, a manner of energy transmission known as transcutaneous energy transmission system (TETS).

The other type of device, circulatory assistance, is handled by a different type of pump, a ventricle-assist device (VAD), so named because its purpose is to assist a failing/failed ventricle with the natural heart remaining in place. Most of the time the left ventricle, the chamber that pumps blood throughout the body, is failing, and the assist device is called an LVAD (left ventricle assist device). Of course, the right ventricle may be supported with an RVAD (right ventricle assist device), and both ventricles may be supported by a BiVAD (biventricle assist device). Dr. Rosenberg et al. are presently working on an E-LVAD (electrically powered LVAD), similar to the E-TAH, and clinical trials are expected to begin by 1997.

Meanwhile, the device that has been used the most worldwide, the pneumatic VAD

1171

(P-VAD), was developed by William S. Pierce, chief of the Division of Cardiothoracic Surgery at Hershey, and James H. Donachy, Fabrication Director, Section of Artificial Organs. This device is manufactured by Thoratec Laboratory Corporation and is known as the Pierce/Donachy assist device (Fig. 1).

The derivation of all the device permutations may be noted. It should also be mentioned that, within a device category (e.g., P-TAH), there may be several different sizes of pump (ventricle) to accommodate different-size people, including neo-natal. Hershey (Penn State) is the only institution in the world that is working on all four types of devices.

II. DEVICE DESIGN PHILOSOPHY

Regardless of the type of circulatory-assist device, we have several standards of device design that are applicable to all of the devices. The first is a seamless, ultrasmooth, one-piece blood bag fixed inside a rigid pump case. Pierce and Donachy were the first investigators to use Biomer®, a proprietary segmented polyether urethane urea (SPEUU) licensed by DuPont Corporation to Ethicon Corporation. In the pneumatic devices, air pulses from a driving console are delivered to the pump via a flexible polyvinyl chloride (PVC) driveline of 1/4 inch inside diameter (ID). The pulse then moves an SPEUU air diaphragm that is in direct contact with the blood bag and, via compression, the blood is ejected from the pump. Correct blood flow inside the pump is maintained by one-way mechanical tilting disk valves, inlet and outlet (Fig. 2).

The electric devices have a brushless direct current (DC) motor driving a rollerscrew mechanism with pusher plate(s) affixed. Circular motor motion is converted to translatory motion and the pusher plate pushes directly on the blood bag, accomplishing ejection via the one-way mechanical valves.

In both device, ejection volume is controlled by the air diaphragm (P-TAHs, P-VADs) or pusher plate travel (E-TAHs, E-VADs).

A patient with a permanent TAH of either type will have four mechanical valves in the two ventricles, and will need to be on anticoagulation drugs for the remainder of his or her life. However, the benefits in mechanical strength and ease of design incorporation into the device outweigh the use of tissue valves. In any event, any patient that undergoes heart valve replacement with any type of mechanical valve must also remain on anticoagulation therapy for life; there are thousands of people worldwide with mechanical heart valves.

Another design philosophy is the connection of the device into the body, the mechanical/human interface. At Hershey, Pierce and Rosenberg utilize screw-on connections (union nuts). While they are more difficult to design and machine fabricate, harder for the surgeon to manipulate in the operating room (OR), and may require special tools to align and tighten, the benefits of positive mechanical fixation (with correct design fitting of all components) outweigh any initial inconveniences.

One research group that for years advocated a snap-on, wired connection has switched to screw-on connections after enough experiences with thrombus formation at the interface that can flex (in a snap-on connection) and allow blood to seep behind the flexing connection. Screw connectors provide a positive, nonflexing connection and, if the design requires it, can compress a mechanical valve into a polymer blood bag "seat" in the rigid case, which will allow a smooth interface between the vascular connections and the blood pump.

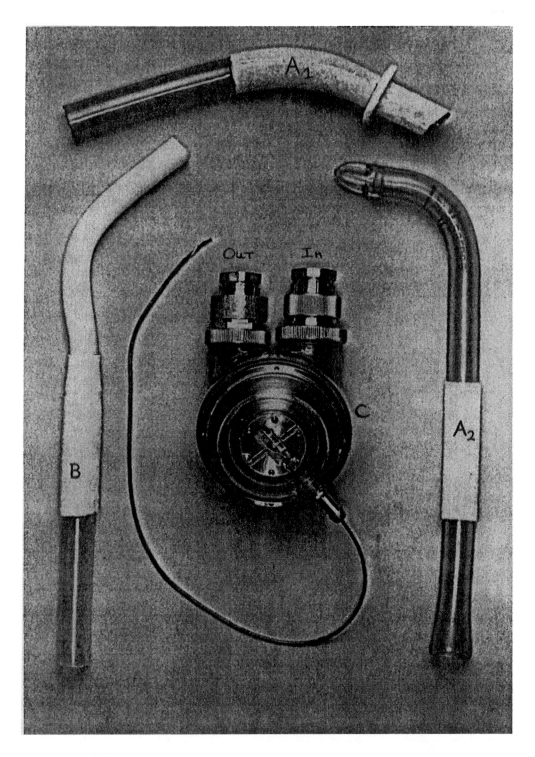

Figure 1   Pierce/Donachy assist device. A_1 = Ventricle cannula, when VAD is used as a bridge-to-transplantation device (the natural heart *will* be removed). A_2 = Atrial cannula, when VAD is used to support a patient with myocardial infarction (the natural heart will *not* be removed). B = Aortic outlet cannula/graft — the same for all VAD applications. C = VAD, cap side up showing fill-to-empty sensing wire.

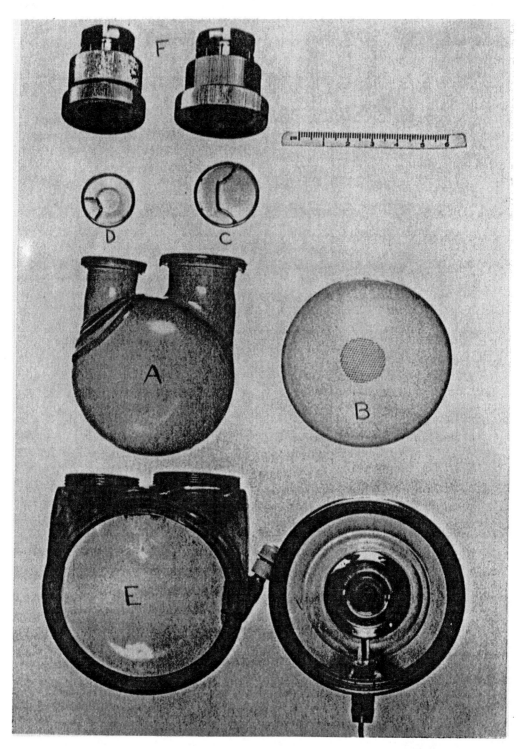

Figure 2 Correct blood flow inside the pump is maintained by one-way mechanical tilting disk valves at the inlet and outlet. A = SPEUU blood sac; B = SPEUU air diaphragm; C = 29 mm Björk–Shiley Delrin® inlet valve; D = 25 mm Björk–Shiley Delrin® outlet valve; E = Radel A® polysulfone VAD case and cap; F = stainless steel (316) 4-pc convectors (cannula to VAD).

The actual connections of a TAH to the body are two (right and left) atrial connections (inlet) and the left outlet–aorta (Ao), right outlet–pulmonary artery (PA) connectors.

VAD connections consist of an inlet and outlet cannula that connect the pump to the patient's natural heart, either the left atrium (LA), left ventricle inflow, and Ao outflow pneumatic/electric LVAD (P/E LVAD), or right atrium (RA) inflow PA outflow pneumatic/electric RVAD (P/E RVAD). These SPEUU cannulas then attach to special titanium (Ti) pump connectors and result in a smooth, positive locking interface.

In this chapter, we are concerned with the fabrication methods and polymers utilized in the blood-contacting components of these Penn State/Hershey devices. These components include the blood bags, the atrial cuffs, the Ao and PA grafts and the VAD cannulas. The only other components the blood "sees" are the mechanical valves, which are obtained commercially with one change made for use in our artificial device; the disks are of Delrin® polyacetal rather than Pyrolytic® carbon, which is the standard valve disk material. The Delrin® is more impact resistant in an artificial device, in which valve-seat impacts are more severe than in the natural heart. As a final note on design philosophies, nonfabrication theory dealing with such things as full-to-empty pump operation, fluid (blood) flow, and the like is outside the purpose of this chapter.

III. CHOICE OF POLYMERS FOR BLOOD-CONTACTING COMPONENTS

Since this chapter concerns fabrication methods of the blood-contacting components in the Penn State heart-assist devices, the polymers discussed are those that comprise the components (i.e., SPEUU, Dacron®-polyethyleneterephthalate [PET]) and the polymer coating, which does not comprise any part of the components, but aids in providing their surface smoothness and is made of polydimethylsiloxane (PDMS) one-component room temperature vulcanizing (RTV) silicone rubber.

Before discussing these component polymers, it should be noted that the rigid blood cases are made of Radel R® (polysulfone), which is an improvement over Radel A®, also polysulfone. Amoco Oil Company makes these polysulfone (PSF) resins, while Westlake Company (Malvern, PA) fabricates the resins into sheets we machine and stress anneal.

Motor housings (for E-TAHs and E-VADs) are made of commercially pure titanium (CP-Ti) and machined to final form, as are most other metal components. The remaining metal components are 316 stainless steel.

Below, we describe SPEUU (several versions) and PET.

IV. POLYMER CHEMISTRY

First, to delve deeply into basic polymer chemistry in a chapter such as this is not possible. An excellent book for polymer chemistry basics is by Saunders, *Organic Polymer Chemistry*, 2nd edition, Chapman and Hall [1].

PET may be prepared from terephthalic acid or dimethyl terephthalate. With either starting material, a two-step polymerization occurs. Either an acid-based process, with initial esterification to yield bis(2-hydroxyethyl) terephthalate, or an ester-based process, with an initial ester interchange also to yield bis(2-hydroxyethyl) terephthalate, yields oligomers that are treated with an excess of ethylene glycol (1:1.5–2.2 molar at 100–

250°C and 0–4 atmospheres) with manganese acetate (catalyst) and methanol (MeOH) removed as the reaction proceeds.

The second step is to form the polyester via an ester interchange reaction that yields ethylene glycol, which is continuously removed as it forms, thereby increasing the molecular weight (MW) to about 20,000 for fibers, which are spun directly.

PET is colorless, rigid, and readily crystallizes. To produce Dacron fibers, the molten PET is forced through spinnerets into room temperature air, yielding a rapidly cooled, amorphous, weak material. These weak filaments are drawn above their glass transition temperature T_g, (80°C) inducing molecular orientation and crystallinity (W_c). Finally, the filaments are heated at 200°C under tension to yield a dimensionally stable material of maximum W_c.

PET is insoluble under any conditions in the human body since those compounds that affect it would be incompatible with life (i.e., concentrated nitric acid, strong ionic bases, chlorinated and fluorinated solvents, etc.).

With the filaments formed, the graft material is woven via several different techniques by different companies in this field, such as Meadox, Cooley-Graft, DeBakey, and the like. In the fabrication section, it is discussed that heat-forming is possible and an asset to the specific forming of the PET graft.

SPEUU synthesis is also described in Saunders' book, but two additional books, Cooper and Lelah's *Polyurethanes in Medicine* [2] and Szycher's *High Performance Biomaterials* [3] will answer any question the reader of this chapter could possibly have on SPEUUs, including how and why the various chemical components are selected to synthesize the SPEUU or segmented polyether urethane (SPEU) for a specific purpose.

Basically, a urethane results from the reaction between an isocyanate and an alcohol, which yields a urethane link, very similar to a peptide bond seen in biochemistry. By using multifunctional forms (e.g., diisocyanates and dialcohols), long chains result. If a diamine is then used to extend the chain, a polyurethane urea results that is mechanically tougher than a polyurethane. The urea and the isocyanate in the chain represent the hard part or "segment," while the polyalcohol or polyether represent the soft segment, hence the name *segmented polyether polyurethane urea.*

The "standard" diisocyanate used in medical-grade SPEUUs is methylene(bis)diisocyanate (MDI), which, with ethylene diamine (ED), comprises the hard segment. The soft segment is polytetramethylene oxide (PTMO), a polyether.

SPEUUs have, over the years, proven to yield a good mixture of mechanical properties with good biocompatibility. One of the reasons for the good mechanical properties of SPEUUs is due to their having the structure of a thermoplastic elastomer (TPE). A true elastomer is usually a thermoset (unable to be solvated or chemically broken down once it has polymerized) that is cross-linked to yield a network of polymer chains that are "tied together" for superior mechanical properties (e.g., modulus, tensile strength, compressibility, creep resistance, etc.). A thermoplastic polymer can be solvated or heated above its melt temperature (T_m) and be reformed. This ability would *not* make a good elastomer as the backbone chains could slip past one another under stress or load conditions.

A thermoplastic elastomer, however, results when segmented polymers exhibit multiphase behavior, and the hard segments coalesce into microdomains (microcrystalline dicontinuous phase) inside the soft segment (continuous phase) matrix. These microcrystallite hard segments then act as "tie points" to hold the chains together in a manner very similar to (but not the same as) actual cross-links. This TPE property allows SPEUUs to be processed from a solution in the fabrication of medical devices.

Polymerization of SPEUUs is reasonably straightforward, but extreme care and technique is needed to ensure that all components are very dry, all solvents triply distilled to dryness, and an inert atmosphere is used. Water vapor will effectively "kill" the isocyanate reaction, lowering molecular weight.

A difunctional isocyanate is "end capped" with a polyether, the reaction is allowed to proceed with mechanical stirring, and, finally, the "prepolymer" chains formed are chain extended with the diamine or, possibly, diols, again with vigorous stirring as the molecular weight rise equals viscosity increase.

The polymer is washed in iced water and MeOH to kill free isocyanate reactions, dried, and resolvated in aprotic solvents such as dimethylacetamide (DMAc) dimethyl formamide (DMF) n-methylpyrollidinone (NMP) or, occasionally, tetrahydrofuran (THF). DMAc seems to be the solvent of choice for most medical-grade SPEUUs, but a higher concentration of urethanes results in a SPEU that also dissolves in THF and DMF. Our medical-grade SPEUUs are processed from DMAc. At this point, it should be restated that Pierce and Donachy were the first investigators to use Biomer® (SPEUU) as a blood bag material.

With the fear of litigation threatening to strangle progress, some companies are withdrawing from the medical device market, leaving researchers to obtain other FDA-approved biomaterials or, in a long, expensive and exhaustive process, synthesize their own biopolymers and obtain FDA approval for clinical trials/implantation. Accordingly, Biomer will no longer be available for medical use.

Given the costs and time involved, most clinicians would be unable to develop their own particular "brand" of Biomer. Robert Ward, chief executive officer of the Polymer Technology Group, Incorporated (Emeryville, CA), has developed an entire line of biopolymers known as Biospan®. As of this writing, Ward has obtained FDA approval of Biospan as a Biomer replacement, and this material is being integrated into our program.

V. SOLUTION CASTING

To fabricate the blood bags, atrial cuffs (for E/P TAHs), and outlet cannulas (for E/P VADs) from SPEUU, solution casting is the only method available to obtain the desired complex shape and thickness of the blood-contacting component. The SPEUU is in solution with DMAc, typically 20% SPEUU to 80% DMAc weight/weight (w/w). Most of the time, we further dilute the solution to 16% solids for less solution viscosity and better dipping control. However, utilizing the new Biospan SPEUU, we have successfully cast up to 23% (supplied at 24% solids).

The SPEUU chain usually has a 2000-MWt soft segment. This represents a reasonably high molecular weight for PTMO, but still gives ease of processability. Both the PTMO, containing an even number of carbon atoms between the ether oxygen, as well as the MDI hard segment are able to crystallize. This resulting microcrystallinity or microdomain formation (mentioned above) would result in difficult processability from a melt. Accordingly, solution casting is employed.

We use three different methods of SPEUU solution casting to fabricate our blood bags, atrial cuffs, and cannulas. They are described in the fabrication section, but basically consist of applying a "wet" layer of SPEUU/DMAc solution to a form or mold and allowing the solvent to "dry off" or evaporate, thereby "depositing" the 20% solid SPEUU. When the drying time is deemed sufficient, another layer may be added to further build up the thickness to design specifications.

An important consideration is the drying time. If excessive, the SPEUU surface may oxidize slightly, especially at higher drying temperatures (80°C and up). If surface oxidation occurs, the next wet layer may not sufficiently "solvate" the dried layer, and delamination of the component may occur, especially if used in a flexing mode, which all of our blood bags are. If too little time or too low a temperature is allowed (e.g., 30°C), it will result in insufficient drying time of the previous coat, and the resulting heavy build up will cause surface wrinkling or even, in extreme cases, sloughing of the layers, ruining the product.

Humidity also plays a very important role. We try to keep the relative humidity (RH) under 15% in our hoods. The drying section of the hood manages to "control" the coating/dipping section, seemingly through stratification of the laminar airflow. In any event, DMAc is hygroscopic enough so that water vapor will turn the wet polymer milky, and then cause the wet layer to drip off the mold due to plasticization (lowering the apparent MW) of the SPEUU by the water.

In principle, solution casting should be easily accomplished. In actuality, the process can be difficult, must be closely monitored, and may even be dependent upon the time of the year fabrication occurs (in the summer, the RH may be excessive for proper solution casting).

To summarize, solution casting is a fabrication method used to buildup a component to desired design specifications by coating with multiple layers of polymer in solution. Chemically, the layers must be solvated sufficiently between coats to allow interpenetration of the polymer chains and the equivalent of a single layer of "very heavy" polymer deposited. In this manner, the bulk properties of the SPEUU are expressed in the component as required by the design.

VI. COMPONENT FABRICATION

A. Blood Bags

We solution cast our blood bags by machine dipping a form into a container of SPEUU at 16–20% solids inside a laminar flow (LF) hood with an RH less than 15% and a curing temperature of 60°C. The temperature may be increased slightly, up to 70°C, to help "lower" RH in summer or during higher humidity conditions. At higher temperatures, the curing time (between coats) of 45–60 minutes is lowered to 40–45 minutes.

Our blood bags each have a shape dictated by the type of cardiac-assist device in which it is to be used, and in the case of TAHs, whether the bag is for a right or left pump. (They are mirror images in the E-TAHs, but have a different outlet angle in P-TAHs) (Fig. 3).

Blood bag fabrication begins with a hollow male low molecular weight polyethylene (LMWPE) form, obtained by filling a hollow aluminum female mold with melted LMWPE, rotating, cooling, and pouring out the excess. The form is then examined for exactness of mold reproduction, correct sizing (LMWPE crystallizes and "shrinks"), and flaws such as holes, pits, or bubbles from incomplete flow and the like. Many times, the form can be "repaired" by heating with an ethanol (EtOH) torch flame to "blend" irregularities and remove the mold flash lines. Holes or pits may be filled with crystals of LMWPE melted in place with the torch and smoothed over.

When the form is graded as satisfactory, it is mounted on a stainless steel stem to allow for dipping (coating) and rotation (drying) in the LF hood. This rod or stem is

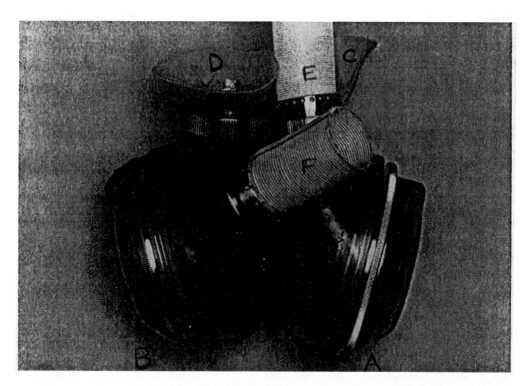

Figure 3 The blood bags each have a shape dictated by the type of cardiac device in which it is to be used and, in the case of total artificial hearts (TAHs), whether the bag is for a right or left pump. A = left pump; B = right pump; C = left atrial cuff; D = right atrial cuff; E = aortic graft; F = pulmonary artery graft.

1179

usually fused into one of the "legs" of the form, usually the larger (in diameter) inlet port leg. LMWPE crystals are melted onto the stem with the torch and, when the stem is in place inside the leg, more crystals are poured on top of the melted "blob" to fill the hollow leg and melt against the stem and the form. By fusing the stem in place, its removal is much easier upon completion of the finished blood bag.

Our LF hood is set up to rotate four blood bags simultaneously. To minimize error, only one type of form (e.g., left 70-cc E-TAH forms) is dipped on a given day. With the need for "interwoven" polymer chains in the solution casting process, all blood bags are completely finished in one day.

At this point, we have four mounted, acceptable LMWPE forms ready for solution casting in SPEUU. However, the PDMS (the third polymer) soon comes into play. The forms are lightly sprayed with fluorocarbon mold release and buffed with gauze to remove the excess. Deionized air is used on the forms to blow all impurities and dust that are attracted there by static electricity charges. As the forms are blown clean, they are examined and carefully placed into the LF hood onto holders. The hood is at both room temperature and RH.

Ultrapure PDMS is allowed to warm to room temperature (all of our polymers are stored under nitrogen gas at 4°C). The PDMS is then decanted into a polyethylene container, usually a 2000-ml (milliliter) beaker.

The mold-released, clean forms are then dipped by hand into the PDMS, immersed totally, and agitated to assure complete coverage. The form is then removed, allowed to drain about 30 seconds, and placed onto a rotisserie that rotates and revolves the form in a horizontal position. The rotation allows even coating of polymer plus prevents sags and runs while the wet polymer is drying. Our LF hood has the rotisserie in the heated side, while the dipping machine is on the unheated side. Thus, we are able to dip by hand or machine on the left (unheated) side and then move the coated form into the right (heated) side and onto the rotisserie without removing the form from the clean LF hood.

The PDMS cures completely in about four hours regardless of the RH in the LF hood. We allow the forms to dry overnight to assure complete cure, then bring the LF hood temperature up to 60°C on the cure/rotisserie side. Within one hour, the temperature "lowers" the RH to less than 20% in summer or less than 10% in winter. We are now ready to begin coating the form with SPEUU.

To summarize, the PDMS coating totally covers and smooths any imperfection in the LMWPE form, giving the form a shiny, glassy appearance on the surface. This coat may be easily visually checked for imperfections due to its high gloss. Obviously, any imperfection would result in a flaw in the SPEUU blood bag, so that form is then rejected. The cured PDMS may be removed (remember the mold release step) from the form and the form later recoated.

A possible benefit from the PDMS layer is that thrombogenicity *may* be reduced due to lowered surface tension on the SPEUU coat; this results when silicone moieties migrate from the PDMS into the SPEUU. Surface chemists have confirmed the presence of silicone and fluorocarbon (mold release) on the actual surface of the SPEUU blood bags. Since we have not had major thrombus formation problems either clinically or in animals, we feel that our surface is quite acceptable.

The SPEUU coating process occurs next. The forms are attached to a dipping device, a Unislide® with attached Bodine® motor and gear drive with external motor control (outside the LF hood). The forms are dipped into the SPEUU container slowly so that air is not drawn down into the polymer, which would create bubbles upon withdrawal.

The manner in which the forms are coated (i.e., leg first, body first, angle of entry, etc.) is dependent upon what form shape we are working with, as is also the total number of coats.

The coating withdrawal speeds from the polymer range from 0.017 inches per second (in/s) to 0.082 in/s. Obviously, the faster speeds will result in more polymer being "dragged" along the form, so that upon completion of the 45-min cure cycle, a "faster" coat will be heavier than a "slower" coat. Again, the percent solid of the SPEUU solution greatly affects solution viscosity. A rise in percent solid from only 16% to 17% will greatly increase the solution viscosity, and create a dried coat 0.002-in heavier.

We derived our coating schedule (% solids, dipping speeds, form entry/removal angles, number of coats, and rotisserie speeds) through years of trial and error, good guesswork, and the like. This area of polymer fabrication does not lend itself well to a didactic approach. The actual polymer must be coated onto a form, dried, and cut off to measure thicknesses at all points over the form to see if design thickness specifications were met. There is no substitute for experience in this area. With the final coats being placed onto the form at 45-min intervals, the SPEUU-coated forms are allowed to dry overnight on the rotisserie.

The next morning, the forms are brought from the heated side of the LF hood to the cooler (40°C) side, and allowed to reach 40°C. At this point, they are cool enough to permit thickness measurements with an ultrasonic measuring gauge (Panametrics Model 5224) to see if the design thicknesses have been attained. The ultrasonic gauge will measure the SPEUU layer down to the PDMS layer while still on the form. It is nearly impossible to get an accurate reading when the form is hot (60–70°C) or has been recently coated (within 45 min) so that when the correct number of coats are completed in one day, the fabricator is really relying on past experience and keeping all of the variables possible in solution casting within standard conditions so that, with the 16-hour overnight cure, the bags are of correct thickness.

If the bags measure very slightly thick (0.001–0.002 in), it is not a major concern as a 24-hour elevated temperature (80°C) heat cure will follow outside the LF hood in a clean oven. If the bags are slightly thin, another coat may be added by figuring out what type of coat would best bring the bag to design specifications and, after lowering it into the SPEUU solution, allowing it to sit in the polymer for about 3 minutes. This time in the SPEUU solution allows the DMAc solvent to solvate the dried surface of the bag to attach the last coat as solidly as the previous coats dipped at the standard 45-min intervals the day before. We prefer not to add a coat a day later; however, if the bag is thin, it is sometimes necessary. We never have seen blood bag delaminations that could be directly traced to overdrying the bag between coats, but it is best not to take chances with an oxidized outer layer that might cause a poor attachment of the last coat to the other coats.

As stated, if the finished bag meets the initial thickness specifications, it undergoes a 24-hour 80°C cure. When the high-temperature cure is finished, the bag is measured over its surface at about 60 locations and thicknesses are recorded on the charts that become the permanent record of each blood bag.

Upon completion of the measurements, the bag is stored by wrapping it in cotton gauze and placing it into a polyethylene (PE) bag in a drawer. The stainless steel stem was removed prior to measuring, after it was determined that no additional coats were needed. We try to leave the completed bags stored on the form as long as possible, as the longer the time that the SPEUU is in contact with the PDMS, the (theoretically) more silicone moieties that can migrate into the SPEUU soft segment.

When the blood bag is needed for a pump, it is removed from the PE outer bag and placed into an oven at 150°C. The legs of the blood bag have been cut open at the tips and the bag is suspended in a gauze "sling" legs down, over a pan. The LMWPE form melts out in about 18–22 min, leaving an SPEUU blood bag with a very thin layer of LMWPE adhering to PDMS, which is still inside the bag. The hot bag is brought outside the oven and allowed to cool to the touch. Surgical hemostats are then used to grip the LMWPE/PDMS layer and peel it from the outer SPEUU layer, producing the finished blood bag. Because there is no actual chemical bonding of SPEUU to PDMS (they are immiscible in one another), the two layers separate readily, leaving the SPEUU blood bag with a very shiny inside surface.

At this point, quality control will examine the bag to see if there are any imperfections that were not visible while the bag was still on the form. If the bag passes quality control, it is now measured to fit the outer case to which it will be joined for a pump assembly. Fitting the blood bag into the case involves a process unique to SPEUU-type polymers: heat settability/memory retention.

The bag/case design of our heart pumps utilize a "step" in the case where the valves sit, in the legs of the bag, and the bag rolls back on itself and is cemented with RTV into a slot in the screw connector attachment. This step and reverse bend formation is done by drawing a line on the outside of the bag leg where the bag fits against the top of the case. Then, the bag leg is stretched over a stainless steel forming jig and the inside plug of the jig is inserted. The result of the jig is to sandwich the bag leg between the male and female jig halves and create the valve seat and the rollback lip. The assembled jig/bag combination is placed inside a steam autoclave and cycled at 260°F/30 lb for 55 min. When the jig and bag are removed, the SPEUU now has the valve seat and rolled lip formed into it.

This phenomenon of heat settability is caused by the segmented block copolymer having two T_g's (one each for the hard and soft segments) and a T_m of the hard segment. The 260°F autoclave temperature increases the SPEUU temperature above the soft segment T_g's, and allows long-range disruption of the hard segment bonds, which allows the component to retain the new shape when it returns to room temperature. With heat setting, any SPEUU component may be formed into almost any shape (the considerations being normal design constraints as to acuteness of angle on a bend, etc.).

After heat setting of both legs, the blood bag is now able to fit the case perfectly, and the mechanical valves will "snap" into the seats in the top of the case. Before the SPEUU blood bag is cemented into the thread groove, it is inspected and thoroughly cleaned in Sparkleen® and sterile distilled water in an ultrasonic cleaner, followed by three rinsings of sterile water.

The pump is assembled in the clean room and is packaged and sterilized with ethylene oxide (EtO) at about 40°C.

In the OR, the pump is removed from the sterile pack, reexamined, and filled with a solution of albumin (bovine or human, depending on animal or clinical use). The albumin adsorbs to the SPEUU, giving excellent resistance to thrombus formation — in fact, a primary reason purported for the SPEUU excellence as a biomaterial is due to its ability to adsorb albumin selectively, thereby decreasing the ability of platelets or red blood cells to adhere.

B. Atrial Cuffs

The P-TAHs and E-TAHs at Hershey use a polymer/Lycra® cloth cone-shaped cuff to attach to the inlets of the pumps and interface to the body, the right and left atria of the natural heart that remain when the ventricles are removed for TAH implantation. By leaving the natural (host) atria, only four connections need to be made to interface the TAH; these connections are both atria, the Ao, and the PA.

As the right and left pumps are different, so are the atrial cuffs. The LA cuff is a medium cone shape, about 8 cm wide at the base, that attaches to the existing atrial tissue; the tip is the correct size (depending on whether the 70- or 100-cc pump is used) to match the inlet valve at the pump/union nut connector. The RA cuff is a flatter cone, 25 cm wide where it can be shaped to fit both the atrial tissue and its junction with the inferior vena cava (IVC); the tip also matches the pump inlet valve (Fig. 4).

To fabricate a cuff of any shape, the correct mold is selected. The mold is made of Delrin to keep dipping and rotation weight down, and resembles a large cone with a smaller cylinder at the cone tip. The mold also disassembles at that juncture.

As in blood bag fabrication, the cuffs are made in the LF hood with SPEUU over PDMS. The mold is cleaned and layered with PDMS by dipping via hand (tip down), rotating, and simply flipping it back upright on a stand inside the LF hood to allow the polymer to flow off. This process is at room temperature with no RH control.

The next day, the hood temperature is increased to 60°C and SPEUU coating commences by hand dipping the cured PDMS-coated mold into the SPEUU polymer solution, rotating and flipping it back upright on its stand in the heated side of the LF hood, and allowing it to drain. After 45 minutes, another coat is applied, followed by a third coat 45 minutes later. We now have a PDMS-coated Delrin mold that has three coats of SPEUU overcoating.

Concurrently, between coats, a pattern is used to cut a piece of Lycra® cloth that will be the atrial cuff backing material. Lycra Spandex® is a DuPont fiber that actually is SPEUU.

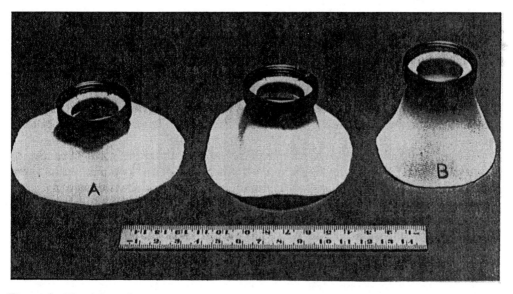

Figure 4 The right atrium (RA) cuff can be shaped to fit both the atrial tissue and its junction with the inferior vena casa (IVC) and the tip also matches the pump inlet valve. A = right atrial cuff; B = "special" sharp left cuff—not often used; C = left atrial cuff.

The pattern for the RA cuff is one piece, round, with a central hole cut in it for the cylinder at the cone tip. The LA pattern is a thick, c-shaped piece because it needs to be wrapped around the LA cone at the cylinder junction at the tip of the cone.

Whichever cuff is being fabricated, the correct pattern Lycra cloth is laid shiny side down onto the SPEUU layered cone inside the LF hood. A cotton-tipped wooden swab is coated with SPEUU and rolled to dryness so that cotton fibers do not come off. When the swab is ready, it is used to dip into the SPEUU solution container and transfer about 1 cc of wet SPEUU solution to the cloth-covered form. The SPEUU is then dripped onto the dry cloth and rolled in with the swab, much like hand laying fiberglass onto a canoe or the like. The SPEUU is transferred, bit by bit, to the cloth and rolled in until the cloth is saturated all the way around the mold surface.

The cloth actually covers the base of the cylinder at the cone tip down to the base of the mold. The RA cloth will lay flat when wetted with the SPEUU because the Lycra cloth is stretchable in both directions, but the LA cloth piece must be "worked" by first sealing one edge to the form tip to base, then rolling wet SPEUU around the form, tip to base, and finishing by going back to the start. The RA cloth, being one piece, is worked from the tip circumferentially down to the base. By wetting, working, and rolling small amounts of SPEUU into the Lycra cloth and saturating it, the SPEUU bonds through the cloth into the SPEUU layers already semidried on the cone.

The cone/cuff is now allowed to dry the usual 45 minutes, and then the small cylinder end at the cone tip is hand dipped into the SPEUU solution only to where the cone starts to flare out right below the tip. This dip will aid in reinforcing the connector end of the cuff when it is removed from the cone. Usually, two of these "tip dips" are performed, with the cuff being rotated on the rotisserie so that the SPEUU solution will not run all the way down the cone to the base like the first three coatings, but will "stay" on the cylinder tip.

When the cuff is finished in one day, it is allowed to dry overnight, high-temperature cured (80°C) the next day for 8 hours, and autoclaved at 260°F for 55 minutes while still on the form as a heat-setting step, similar to the blood bags. This heat-setting step does not actually shape the cuff so much as it helps the cuff to retain its shape and prevents it from rolling up when it is finally removed from the form.

The next day, after autoclaving, the tip cylinder is removed (disassembled) from the main cone part of the mold, and the cuff slips right off; SPEUU and PDMS are immiscible, and act as release agents for each other. The resultant atrial cuff is a cloth cone shape with a very shiny inside surface that leads to a flared SPEUU cylinder that is not cloth backed.

The cuff is then bonded to the stainless steel or CP-Ti atrial cuff screw connector by a Delrin jig that bonds the cylinder inside the flat part of the two-piece atrial screw connector. SPEUU solution is used as the bonding agent to bond the two materials.

When the metal-SPEUU cuff has dried (one day) in heat (60°C), 4-0 polypropylene monofilament suture is used to sew the edges of the metal to the SPEUU/cloth cuff. Finally, SPEUU solution is used to overcoat the suture knots and comprise a sealing bead around the cloth/metal junction.

The surgeon may shape the finished, sterilized cuff with scissors to fit the anatomy of the patient being operated on. When the atrial cuff is sutured in place, the free end with the metal union nut is used to mate with the pump at the inlet valve screw connector.

C. Cannulae

Our P-VADs and E-VADs interface to the body via inlet and outlet cannulae (tubes) that lead from the RA or LA or left ventricle to the pump inlet and the pump outlet to the Ao or PA. Special metal screw connectors fasten one end of the cannula to the pump while the other end is located in an atrium by a purse-string suture, sewn into the ventricle by cloth tabs, or sewn directly to the Ao or PA.

The cannulae are made by solution casting with SPEUU, and many research groups have dip coated them just like the blood bags. The problem with dip coating a cylindrical mandrel, even when alternate-end dipping, is that of uneven coating thickness.

Donachy solved this problem while employed by the NIH in the late 1960s by using a steel or aluminum mandrel in a metal-turning lathe, rotating the mandrel and applying the SPEUU from a pot attached to the feed-screw mechanism of the lathe. The pot had a spring-loaded tip that deposited SPEUU solution onto the mandrel as the pot tip traveled down the length of the lathe bed. A heat lamp over the lathe bed dried the polymer solution, and 45 minutes later, the pot tip traveled back up the lathe bed to apply another layer. This process was repeated until the cannula was the desired thickness.

One end of the mandrel was held by a chuck in the lathe head, but the other end could not be supported by a standard rotating tailstock. Accordingly, Donachy invented a "draw-chuck" [4], which is similar to a drill motor chuck that can be tightened on the free end of the mandrel, then pulled away from the head chuck end of the lathe. This pull or tension on even the smallest-diameter mandrel allowed the pot and tip to evenly distribute SPEUU to the mandrel, rather than bending the mandrel in the middle if it were supported the usual way in a lathe. Without the tension, the cannulae cannot be evenly coated. All of our cannulae are fabricated in this manner, via lathe solution casting.

We have modernized the equipment now, and have a special LF hood over the lathe with a full-length quartz-tube heater to cure the polymer between coats. These adaptations have made cleaner cannulae, and eliminate dangerous DMAc fumes.

Fabrication of the cannula is performed as follows. The correct mandrel is selected and polished on a buffing wheel to a high gloss. It is then wiped with solvent to remove any buffing rouge, and placed onto the lathe as described above. A layer of PDMS is applied (similar to that used on the blood bag form) to form the smooth inner surface of the cannula. A day later, SPEUU coating beings under heat. Coats are applied every 30–45 minutes, until design thickness is reached.

A novel design of some cannulae is wire-winding reinforcement, a process by which stainless steel (304) wire, 0.004–0.006-in OD (outside diameter), is wound along the SPEUU layer when that layer is one-half the design thickness for the cannulae. The wire is applied from a carrier on the back side of the travel mechanism (the pot and tip sit on the front) and the wire is wound onto whatever design length is required. An internal SPEUU air line for a P-TAH would be 1/4-in ID, 3/8-in OD, and wire wound the full length of 36 in. A ventricle cannula for a VAD may be 12-in long with 6-in wire winding in the center. When the wire winding is complete, additional SPEUU coats are made to bring the cannula to the full design thickness.

Upon completion of SPEUU coating, the cannula is dried overnight at 60°C and removed from the mandrel the next day. Again, the SPEUU does not bond to the PDMS, and the cannula is slipped right off one end of the mandrel.

In the area of wire reinforcement, the cannula will be stiff, while it will be flexible where there is only SPEUU, depending upon wall thickness. A heavy (e.g., 0.050 in) wall results in a heavy, almost rigid cannula.

Heat setting plays a very important role in cannula fabrication. When the cannula is off of the mandrel, a smaller (in OD than the cannula ID) mandrel, which usually consists of a piece of malleable copper tubing with silicone tubing forced over it, may be precurved or prebent and placed inside the SPEUU cannula, causing it to take the shape of the precurved inner mandrel. After an autoclave cycle, the cannula takes the same shape as the inner mandrel, which is then removed. We fabricate all sizes, diameters, and shapes of cannulae and make multiple bends in cannulas, we can even make straight ones.

The "final touches" usually are Dacron velour, which may cover the cannula, or felt rings, usually placed near a ventricle cannula tip.

In the OR, the surgeon may clamp across a non-wire-reinforced portion of the cannula for temporary occlusion, while the reinforced section remains stiff and patent.

Again, the VADs only "hookup" to the patient is via cannulae, and any shape cannula may be fabricated by machine (lathe) solution casting.

D. Pulmonary Artery and Aorta Grafts

The final blood-contacting components to have fabrication methods described are the P-TAH and E-TAH outlet connectors from the pumps, the PA and Ao graft-screw connectors. These components are the only non-SPEUU component to "see" the blood (besides the mechanical valves) and are made of PET saturated with PDMS.

The actual screw connectors are very similar to the atrial cuff screw connectors, but proportionally smaller, that is, the outlet valves of all types and sizes of our pumps are fixed at 25 mm (as opposed to varying inlet valve sizes). The connectors are machined from CP-Ti or 316 stainless steel (SS), and are of the same two-piece construction as the atrial cuff design. We use 22- or 24-mm Meadox® woven PET graft, which is supplied nonsterile in 9–10-in lengths that may be stretched to twice the length due to the "corrugations" of the graft. One 9-in graft can be used to fabricate three graft connectors of about 2 in each. A single 3-in piece of graft is stretched over a special mandrel both to "remove" the corrugations and create a 90° flat bend that may be bonded into the metal screw connector. When the graft is correctly located on the mandrel (a process relatively simpler than orienting the blood bags onto the heat-setting mandrel), the assembly is placed into the autoclave for one cycle (260°F for 55 min).

The autoclave cycle heat sets the PET into the desired shape, but not via the same polymer mechanisms as SPEUU heat setting. In SPEUU heat forming, the autoclave temperature is higher than the T_g's of the soft segments, but below the actual T_m of the hard segment, and long-range disruption of the hard segments allows the microcrystalline domains to reform into their "new" shape. However, the autoclave temperature is higher than the T_g of the PET; this resulting chain mobility (flexibility) allows for the permanent set of the PET in the jig.

The graft is then trimmed to fit the metal connector and bonded in by first coating the flat connector flange and heat-formed area of the graft with SPEUU and allowing that layer to dry in the LF hood. Next, wet SPEUU is used to bond the coated flange and graft with a special jig that holds the parts aligned while applying sufficient pressure; they are again placed into the LF hood at 60°C to speed the drying process. When bonding either the atrial cuffs or grafts into the metal connectors, it is very important *not* to use a large amount of pressure to mate the pieces as the wet SPEUU will be forced from the bonding area and will produce an inferior bond.

Finally, when the SPEUU bond has dried, the graft is sewn to the connector rim (reinforcement) with 4-0 polypropylene monofilament suture. The knots and boundary edge of the graft are sealed with SPEUU.

The final step in Ao/PA graft fabrication is to coat the graft portion with PDMS, which is accomplished by lowering the connector graft first into the PDMS solution *only* up to the metal-PET boundary. This will assure a PDMS coating only on the PET graft. PDMS would "flake" off the inside of the metal when dried, with potentially disastrous results for the patient.

When the saturated graft is removed from the PDMS, a hemostat-type clamp is used to grip the graft and stretch it downward from the held metal connector; it is allowed to retract then restretch for about a minute (like a small concertina) to allow excess PDMS to drain out of the graft and work into the interstices of the woven PET. The graft is then hung graft down in the LF hood at 40°C and allowed to dry, with sufficient weight (usually two small hemostats) suspended evenly from the bottom of the graft to stretch the graft enough to spread the corrugations slightly.

After 24 hours, the graft is completely dry, and the lower end is trimmed about 1/2 in to remove the heavier PDMS coat. (The grafts will be trimmed to correct length in the OR by the surgeon.) Upon inspection and quality assurance approval, the grafts are packaged in peel-packs and gas (EtO) sterilized, usually with the pumps and other components.

VII. SUMMARY

In the development of the blood-contacting components for the circulatory-assist devices, the polymer of choice was always SPEUU. The only acceptable method of fabrication with SPEUU is solution casting, which allows total control of all design parameters for each component.

SPEUU and PET are polymers with proven biocompatibility, and they represent the best starting point for the development of the long-term "ultrapolymer" for the E-TAHs and E-LVADs that are to have a rated use of five years or more.

In the fabrication process with SPEUU, even with all of the physical variables controlled as much as possible, experience is the main factor to obtain a usable component. No amount of classroom knowledge will teach how a wet polymer layer will "sag" while a blood bag is being dipped.

Another concern with SPEUU fabrication is to have a very clean area (clean room) or laminar flow hood that will help in controlling humidity, curing the layers, and keeping DMAc vapors away from the fabricator.

Finally, the heat memory of the two polymers allows for great variability in design parameters, to control the ultimate function of the component.

ACKNOWLEDGMENTS

In any long-term development project of this magnitude, many people will have contributed invaluable help over many years. In addition to William S. Pierce, M.D., Chief, Division of Cardiothoracic Surgery, and James H. Donachy, Sr., Fabrication Director, now retired (December 31, 1994) Section of Artificial Organs, who originally designed the pneumatic devices, there is Gerson Rosenberg, Ph.D., Chief, Section of Artificial Organs, who is the principal developer of the electric devices. Assisting him are bioengineers Alan J. Snyder, Ph.D., William Weiss, M.S., Thomas Cleary, M.S., John Reibson, B.S., and David Katz, B.S.

Assisting Donachy are machinists Clifford Weber and James Donachy, Jr., and polymer specialists Felder, the new fabrication director, and Lynford Reichert. Quality control, record maintenance/documentation and OR scrub nursing is by Anne Zarlenga, A.S., C.O.R.T., E.M.T.

Our world-record animals are kept hale and healthy by John Sapirstein, M.D., Allen Prophet, B.S., and Chief Animal Technician, Mark Schwartz, B.S. Our full-time animal technicians are Barbara Bush, Suzanne Swartz, and Jeannette Mohl.

The authors also wish to thank Drs. Pierce and Rosenberg for their assistance with this chapter, and a very large thank you to Susan Dysinger, secretary, Section of Artificial Organs, for all of the manuscript preparations.

There have been many faculty and graduate students at Penn State University's Departments of Engineering and Polymer Science who have been very instrumental, since 1970, in varied aspects of this project.

Our superb animal facility at Hershey, under C. Max Lang, D.V.M., and O.R. nurse Carole Mancuso, B.S., R.N., allows us to get the best possible care for our important "patients."

REFERENCES

1. K. J. Saunders. *Organic Polymer Chemistry*, 2nd ed. Chapman and Hall, New York (1988).
2. S. L. Cooper and M. D. Lelah. *Polyurethanes in Medicine*, CRC Press, Boca Raton, FL (1986).
3. M. Szycher. *High Performance Biomaterials*, Technomic, Lancaster, PA (1991).
4. J. H. Donachy, Sr. U.S. patent 3625529 (December 7, 1971).

44
Preclinical Testing for Antimineralization Treatments of Heart Valve Bioprostheses

Jean-Marie Girardot and Marie-Nadia Girardot
Biomedical Design, Inc., Atlanta, Georgia
John P. Gott
Crawford Long Hospital of Emory University, Atlanta, Georgia
Carol Eberhardt, David Myers, and Mark Torrianni
Medtronic Heart Valve Division, Irvine, California

I. INTRODUCTION

Approximately 40,000 glutaraldehyde-fixed heart valve bioprostheses made of either bovine pericardium or porcine heart valves will be implanted worldwide in 1994, and, although cuspal calcification has been for years identified as the most frequent clinical problem associated with these valves, it still accounts for the failure of more than 50% of implanted heart valves within 15 years after implantation. The age of the patient significantly affects the rate and severity of cuspal calcification; in young children, 50% of the valves fail within 3 years of operation [1–3]. The most frequent mode of calcification-induced failure is regurgitation resulting from tears produced by the calcific nodules; stenosis occurs less frequently, and emboli rarely [4–7].

The mechanism of cuspal mineralization has not been clearly defined yet, mostly because of its complexity. Factors such as stress, the dysfunctional calcium pump of the devitalized cells, the membrane phospholipids, various enzymes, and the glutaraldehyde cross-links have been suggested as contributors to calcification [3,8,9,10]. Histological studies have shown that, in the subdermal rat model of mineralization, calcification occurs in the devitalized cells within 48 hours, then spreads to collagen and elastin fibers, forming calcified nodules [11,12]. Although alkaline phosphatase, an enzyme important in skeletal mineralization, has been found to retain a moderate degree of activity in glutaraldehyde-treated heart valves [13] and its role in the early events of bioprosthetic tissue calcification has been postulated [14], recent studies have deemphasized the role of this enzyme in the early events of cuspal mineralization [15,16].

Several antimineralization treatments have been investigated experimentally or clinically. The most common mode of treatment consists of incubating the glutaraldehyde-fixed valve in a buffered solution containing an anticalcification agent. The list of agents that have been used for this type of treatment is comprised of sodium dodecyl sulfate, polysorbate-80, Triton X-100 and N-lauryl sarcosine, toluidine blue, ferric or aluminum chloride, protamine sulfate, glutamic acid, and aminodiphosphonate [17–22]. Less frequent modes of treatment with anticalcification agents include daily injections; slow release from polymers; site-specific delivery by osmotic pumps such as for diphosphonate compounds [23–26]; and oral administration in the diet of the animal, as for diphenylhydantoin, a vitamin D antagonist [27]. The synergistic effect of several antimineralization agents has been investigated (i.e., ferric chloride and ethanehydroxydiphosphonate) [26]. Other combinations of chemicals have been applied to glutaraldehyde fixation that concomitantly increase the cross-linking within the tissue and prevent calcification [28,29]. These treatments are reviewed in more detail in this chapter.

Before a heart valve treated with an antimineralization agent is implanted in humans, it is essential that the treatment be assessed not only for its anticalcification efficacy, but also for its safety in terms of biocompatibility, durability, and preservation of structure and function. Efficacy is usually assessed initially by implanting the tissue subdermally in small animals for a few weeks, yet preclinical efficacy should not be claimed until treated valves have been implanted for several months in the circulation of larger animals. Safety has to be demonstrated during the preclinical phases of development by performing an extensive battery of preclinical tests that have been developed over the years by investigators and manufacturers following general concerns about the health of patients and rigid regulations from the Food and Drug Administration (FDA).

The tests of efficacy and safety for antimineralization treatments of heart valves are described in this chapter using, as an example, the experiments performed on the Medtronic Freestyle™ heart valve, which is currently under clinical investigation. This valve is a stentless porcine aortic root fixed with glutaraldehyde and treated with the antimineralization agent 2-amino oleic acid (AOA). This treatment consists of incubating the glutaraldehyde-fixed heart valve in a solution containing alpha-aminooleic acid, a C18-chain fatty amino acid derived from oleic acid by adding an amine to the molecule, which allows its binding to the aldehydes of the tissue. The structure of AOA is illustrated in Fig. 1.

The results of the preclinical testing of the AOA-treated Freestyle valve indicated that the AOA process is biocompatible, that it is neither toxic nor mutagenic, and that it alters neither the structure of the tissue nor the function of the valve. It prevents cuspal calcification in standard experimental animal models [30–33]. However, it is not effective in preventing aortic wall calcification [33], which may be of clinical significance for some types of bioprostheses such as the stentless valves.

II. EFFICACY AND SAFETY OF ANTIMINERALIZATION TREATMENTS

To be efficacious and safe, the antimineralization treatment under consideration should provide long-term prevention of mineralization in standard animal models and not mere retardation. It should not negatively affect the strength, physical properties, and function of the valve. Also, it should not elicit adverse reactions from the body after implantation. Shelf-life studies should also be performed to demonstrate that the antimineralization treatment remains efficacious when the treated valve is kept for months or years before implantation, as is often the case.

Once an anticalcification treatment has been identified and its safety demonstrated, an attempt should be made to understand its mode of action. Information gained from

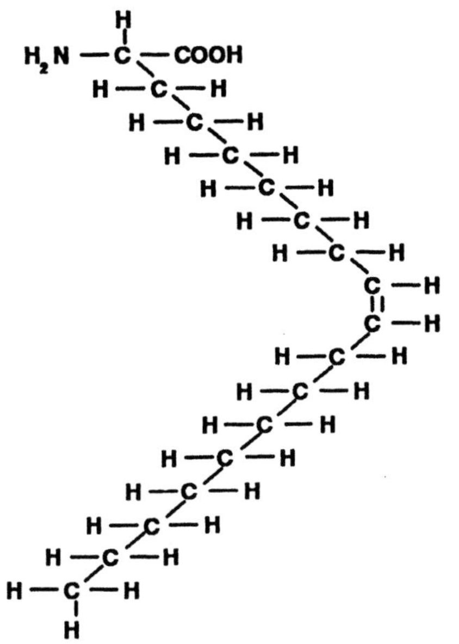

Figure 1 Chemical structure of 2-amino oleic acid (AOA). The alpha proton of oleic acid has been replaced by an amine group through a five-step chemical reaction.

basic research on the mechanism of calcification and on the inhibition of tissue mineralization could lead to new approaches for permanent inhibition of calcification of bioprostheses with increased durability in patients.

A. Anticalcification Efficacy

1. Assessment of Efficacy

Currently, the accepted *in vivo* models for testing anticalcification efficacy of antimineralization treatments on heart valves are (1) the subdermal model, for which treated and untreated cusps or walls are implanted in small animals for several weeks, and (2) the circulatory model, for which treated and untreated valves are implanted in the hearts of larger juvenile animals for several months.

The Subdermal Model of Mineralization. In the subdermal model of mineralization, the implants, in most cases cusps and wall coupons from treated and untreated porcine aortic valves or bovine pericardium, are implanted in subdermal pouches of mice [34], rats [11,32,36], and rabbits [37]. The animal should be of growing age to optimize the rate of calcification. The rat model is the most commonly used because of its relative low cost, convenient size, resistance to infection, and ease of care. In addition, the rate of calcification is faster in the young rat than in the young rabbit since half-maximum calcification generally occurs after 21 days in the rat and after 42 days in the rabbit [12]. The most decisive reason for choosing the rat stems from the reported demonstration that the pathology of mineralization in this animal has been shown to be similar to the dystrophic calcification found in clinically retrieved valves [11,12,35].

It must be emphasized that, because the subdermal mode of implantation does not simulate all aspects of the circulatory environment, it is important that it be considered as a mere screening model for calcification, and that the results be critically analyzed accordingly. When placed in the circulation, a heart valve is submitted to additional factors such as blood wash-out, infiltration of blood components, and dynamic stress, which not only affect calcification but may also reduce the effect of antimineralization agents, particularly when they are not tightly bound to the tissue. Valves treated with polyacrylamide and aminodiphosphonate are typical examples of differential calcification in the subdermal and the circulatory model. The tissues from these valves did not calcify in the subdermal model, yet they calcified extensively when the valves were implanted in juvenile sheep [17].

The method for subdermal implantation in the rat, which has been developed by Levy et al. [35], has become the standard, with slight variations among investigators. Because calcification is affected by many known and as yet unknown factors, it is important that untreated samples always be randomly implanted at the same time as the treated samples. Both the treated and the control samples must be rinsed several times with saline before implantation to eliminate some of the glutaraldehyde from the tissue, which may induce toxic effects. This process, which simulates clinical conditions, generally consists of 3 rinses for 2 minutes each.

At implantation, the animal is anesthetized, the abdomen is shaved, and incisions are made in the skin, which is then separated from the underlying tissue by forming pouches approximately 2 centimeters (cm) apart and 4 square centimeters (cm^2) in size. The rinsed treated and control samples are then individually placed in the pouches, and the wounds are closed using surgical staples. The samples are retrieved after a variable, predetermined number of days, depending on the investigator and the particular intention of the study.

For instance, qualitative results of calcification have been obtained as early as 48 to 72 hours after implantation by light microscopy evaluation of retrieved samples stained with the von Kossa method [11,12,18]. For quantitative evaluation of calcification, the samples are generally retrieved after 3 weeks. However, we recommend that more than one duration point be tested, with the last point at least 8 weeks after implantation in order to determine whether calcification has been merely retarded by the treatment, or whether it has been effectively prevented.

After retrieval, the surrounding tissue is carefully removed from the sample, which is then hydrolyzed. The level of calcium of each sample is then determined using standard methods such as atomic absorption or inductively coupled plasma analysis. We propose that two statistical analyses be performed on the results of subdermal evaluation of anticalcification treatments. The first analysis, a comparison of means between treated and control samples, provides an estimation of the potency of the anticalcification agent in the rat model. The second analysis, a regression analysis of the calcium levels in the treated sample as a function of time of implantation, indicates whether calcification has been effectively prevented or only retarded by the treatment.

A study performed in our laboratory and aimed at evaluating the effect of AOA on cusps of porcine aortic valves and porcine pericardium was selected to illustrate a typical experiment in the rat model. AOA-treated and untreated control cusps from porcine aortic valves and 1-cm^2 pieces of porcine pericardium were implanted subdermally for 8 weeks in 3-week-old male rats according to the method described above. After retrieval and cleaning, the samples were scored visually for calcification. This scoring method, based on the ratio of surface area calcified to the entire surface of the sample (− = 0%, + = <20%, ++ = <40%, +++ = <60%, ++++ = <80%, and +++++

= 80–100%), provides a rapid and very reliable method of evaluation, although less precise than the quantitative analysis.

The samples were then washed 5 times for 5 minutes each time in distilled water, dried under vacuum, weighed, and acid hydrolyzed in vacuum in 1 milliliter (ml) of 6 N ultrapure redistilled HCl for 24 hours at 85°C. After dilution, the samples were submitted to quantitative calcium analysis using the inductively coupled plasma analysis method [32]; this method was selected over atomic absorption spectrometry because of the lack of phosphorous interference.

The results of this experiment, illustrated in Fig. 2, indicated that AOA is a potent inhibitor of calcification of both porcine aortic valve leaflets and porcine pericardium in the rat model. Based on other experiments not illustrated here, we have observed that the level of calcium in AOA-treated cusps after 8 weeks of implantation was not significantly higher than the level found after 3 weeks of implantation, which indicates that AOA prevents calcification.

The Circulatory Model of Mineralization. Small mammal subdermal implants are a relatively rapid, inexpensive, relevant means of screening bioprosthetic antimineralization techniques. The major limitation is lack of exposure to the ultimate working environ-

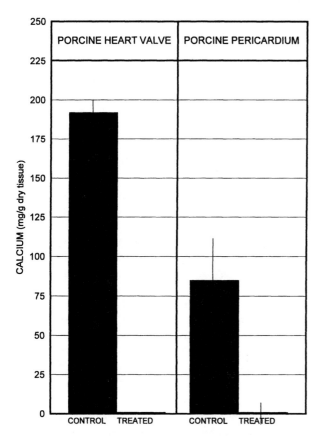

Figure 2 Calcium levels in control leaflets and porcine aortic valve leaflets treated with 2-amino oleic acid and porcine pericardium implanted subdermally for 8 weeks in 3-week-old rats.

ment—the circulatory system. Mechanical wear testing with its repeated *in vitro* hydraulic cycling can mimic some of the loading stresses and uncover the potential for premature loss of structural integrity. However, currently there is no more effective preclinical proving ground than intracardiac implantation of the final bioprosthetic version into a large mammal.

Exposure to the circulation in this manner allows close approximation of the biochemical, immunologic, inflammatory, rheologic, mechanical, and other less obvious host/prosthetic interactions to unfold. Potential deleterious processes that would not have been anticipated in the subdermal model or mechanical hydraulic testing may be uncovered at this stage. For example, calcification is now known to progress more rapidly at areas of bioprosthetic tissue stress such as encountered under the loading conditions imposed by circulation. This circulatory stress and biochemical milieu mimics the ultimate clinical implantation condition. An antimineralization treatment effect may not become obvious until implantation and exposure to circulating blood, which allows interaction of the tissue with circulating blood components. Infiltration of the leaflets with inflammatory cells may provide new sites for calcification. In addition, the phenomenon of leaching of the antimineralization compound from the tissue may also be manifested in long-term *in vivo* implantation.

Many large mammals have been used for prosthetic evaluation, including the porcine, bovine, and ovine models. The sheep model holds a distinct advantage because it has been well characterized under the leadership of Jones and others, formerly of the surgery branch of the National Institutes of Health. Sheep have the advantage of rapid mineralization of implanted bioprostheses, thus compressing into weeks what would take years in a human implantation. The juvenile sheep model has the advantage of only moderate overall body growth over the experimental period compared with the pig or cow. The sheep does very well with general anesthetic and cardiopulmonary bypass, with the need for minimal postoperative intensive care. Animal husbandry issues make the animal very favorable to work with on a day-to-day basis. All of these factors make the sheep an excellent choice for experimental validity as well as economy [17,32,33,38–46].

Technical aspects of implantation in juvenile sheep. The juvenile sheep model has been used to study calcification of heart valve bioprostheses implanted in the mitral [17,32], the tricuspid [17], and the apical-aortic shunt positions [33,43–46]. The general preparation of the animals is similar for the various implantation procedures. The animals are fasted the evening prior to the day of operation. Although the sheep have a generally docile demeanor, they are extremely uncooperative with invasive procedures and can be quite rambunctious. Therefore, it is necessary, prior to preparation for invasive procedures, to give them a premedicating sedative. Atropine to help control the copious gastrointestinal and respiratory secretions (0.2 milligrams per kilogram [mg/kg]) and acetylpromazine are excellent for sedation (0.55 mg/kg). These drugs can be injected intramuscularly and, after a few minutes have elapsed, the animals remain conscious and have satisfactory airway control and may then be prepared for the surgical procedure.

An antimicrobial for prosthetic valve endocarditis prophylaxis is given parenterally 30–45 minutes prior to incision. An endotracheal tube is easily placed using a standard human laryngoscope and endotracheal tube. Gentle retraction of the tongue forward allows visualization for intubation, and the tube is secured with a tie to the lower jaw. Once the airway is secure, the animal is sheared over the jugular vein, which is used for venous access, and the operative site is sheared with standard electric clippers. Actual shaving of the skin is not necessary.

The animal is then placed on the operating table in the right lateral decubitus position and given ketamine (22 mg/kg). Halothane is administered using inhalation techniques. Neuromuscular blockade is provided by pancuronium (0.1 mg/kg). The electrocardiogram is monitored, and arterial pressure is monitored with a small in-dwelling femoral artery catheter placed percutaneously. For this ruminate animal, gastrointestinal decompression is mandatory, and a large-bore rumen tube (3/4-inch PVC tubing may be used) is passed until the return of up to 500 ml of a thin, brownish feculent fluid, which may be simply collected by gravity.

(a) Mitral valve replacement. For the mitral valve replacement procedure, the arterial inflow from the cardiopulmonary bypass circuit is via the right femoral artery. Descending aortic cannulation is a good alternative. The left femoral artery is used for arterial pressure and blood gas monitoring. A left thoracotomy incision exposes the heart. Venous return to the heart/lung machine is accomplished by retrograde cannulation of the right ventricle through the main pulmonary artery. Two purse-string sutures are placed in the main pulmonary artery, a partial exclusion clamp is placed, a longitudinal arteriotomy is made, and a cannula is passed retrograde across the pulmonary valve into the right ventricle. A satisfactory cannula for this purpose can be made by oblique transection of the distal 2 or 3 centimeters of a standard human right atrial single venous cannula.

The sheep is placed on cardiopulmonary bypass, and the temperature is allowed to drift to moderate hypothermia, although any of a variety of myocardial protection techniques would be suitable. We have had excellent results with cold oxygenated crystalloid cardioplegia. An antegrade delivery for this short cross-clamp period (typically 30 minutes or less) remains our technique of choice. If other techniques are chosen, one must remember that the ruminant animal has systemic communication between the coronary sinus and the systemic veins, making a retrograde delivery of cardioplegic solution less practical due to the necessity of isolating the communication between the two venous circulations.

Another technical point with regard to the myocardial protection has to be mentioned. After cross-clamping of the ascending aorta and delivery of the antegrade cardioplegia solution and opening of the left atrium for valve placement, the ascending aorta fills with air due to the position of the animal. This air then fills the left main coronary, circumflex, and left anterior descending arteries. To prevent the coronary air embolus and resultant deterioration of left ventricular function, we have placed a tourniquet around the left main coronary artery, which is occluded immediately after delivery of the cardioplegia solution and subsequently opened after removal of the cross-clamp and deairing of the ascending aorta. This has yielded excellent functional results. The first two animals we operated upon had problems with coronary air embolus and decreased function and arrhythmias. Use of the tourniquet has eliminated this problem.

During preparation for implantation, the bioprosthetic valve is prepared just as one would prepare for human implantation, using multiple rinse cycles in a sterile saline solution. For mitral implantation, we have chosen a supraannular position through a left atriotomy through the left atrial appendage, and use interrupted atrially based pledgeted sutures. The anterior leaflet is excised, and the mural leaflet remains in place for ventriculoannular continuity. After the cardiac deairing maneuvers, the cross-clamp is removed, the left main tourniquet is removed, and the heart usually returns with a coarse ventricular fibrillation, which is electrically defibrillated. The sheep is then usually easily tapered from cardiopulmonary bypass and decannulated. The right ventricular cannula is removed promptly after it is clear that hemodynamics are satisfactory.

For the postimplant hemodynamic studies, high-fidelity micromanometers are placed simultaneously in the left atrium and in the left ventricle after careful zeroing and calibration. The pressure gradient is determined. For cardiac output determination, a balloon-tipped, flow-directed pulmonary artery catheter is placed through the ruminant vein, which communicates with the coronary sinus and thus the right atrium. The flow directs the balloon into the right ventricle and into the pulmonary artery. Alternatively, the catheter may also be placed through the inferior vena cava, through a femoral vein, or through a jugular vein. It may also be passed via the internal thoracic vein, but this is tedious.

The valve is typically retrieved after a 20-week implantation period. The sheep is anesthetized as before. The approach to the heart at this point is through a median sternotomy, which allows easy access for the high-fidelity micromanometers for the final hemodynamic studies. Sheep implanted with glutaraldehyde-fixed valves that have not been protected with antimineralization treatment may not survive the 20-week period due to valve prosthetic stenosis and regurgitation induced by the rapid calcification of the valve leaflets. This may be associated with sudden death, which precludes organized terminal hemodynamic studies. However, close monitoring of the sheep during the study period may detect the onset of valvular dysfunction and allow collection of hemodynamic data. Certainly, the humane care of the animals is utmost in the mind of the researcher, and termination of the experiment for any particular animal prior to the 20-week period would be carried out should the animal suffer from valvular dysfunction. After the hemodynamic data are collected, the animal is killed with commercial euthanasia solution.

The valve is then excised and grossly examined for signs of perivalvular leak, blood clots, mineral deposits, vegetation, tears, and perforations. Cultures of the valve leaflets are taken to evaluate the presence of infection or colonization with microorganisms. The gross examinations are documented photographically. X-ray exposure of the valve using a mammographic technique further documents the extent of calcification. The valve is then placed in 4% formaldehyde and submitted for detailed morphologic and histologic studies. The results of one such study have been reported elsewhere [32].

(b) Aortic valve replacement. For aortic valve replacement, due to the small left ventricular outflow tract of the juvenile sheep, the implantation of clinically relevant valve sizes is a problem. The mismatch in size requires extensive modification of the left ventricular outflow tract to accommodate the larger size. Alternatively, a smaller, less clinically relevant bioprosthesis could be obtained. This defeats the purpose of the experiment, as one would like the sheep implantation to be the last prior to consideration for human implantation. Clinically relevant valve sizes for hemodynamic study and for calcification study should be used. Even 19-mm, nonstented, porcine aortic roots cannot be placed in the very small juvenile sheep left ventricular outflow tract without significant residual obstruction.

To circumvent this problem, we have chosen to make some assumptions and some compromises to separate the study of the orthotopic hydraulics and hemodynamics of aortic bioprostheses from the anticalcification studies by placing a clinically relevant valve size in the circulation. The exposure to the circulation is achieved by fashioning a conduit from the sheep left ventricular apex to the descending aorta. This is a composite conduit composed of a manufactured apical connector and the valve to be studied.

We most recently have used the porcine aortic root with and without an antimineralization treatment in this model. This does allow some hemodynamic study, but its real value is in providing an environment for anticalcification treatment. The formal hydraulic hemodynamic studies can be carried out in a large mammal in the orthotopic position. Although the adult mammals are less suited to antimineralization studies, possibly because of their altered calcium and phosphorus metabolism, they provide excellent models for hemodynamic studies.

(c) Apicoaortic shunt.   For the apicoaortic shunt, previously described techniques [41–46] were modified and used [33]. Similar anesthetic techniques were used, and a left thoracotomy exposure through the fourth intercostal space allowed access to the apex, the left ventricle, and descending aorta. In this procedure, the arterial pressure is monitored with a small catheter passed into the left internal mammary artery, which is easily accessible through the incision, and the pulmonary artery catheter is inserted through the left azygous vein for cardiac output and pressure measurements. The intrapericardial inferior vena cava is encircled with umbilical tape for tourniquet control of venous return. A purse-string suture bolstered by felt pledgets is passed through the myocardium of the left ventricular apex. Also, four interrupted felt pledgeted sutures are placed more superiorly through the myocardium and then through the sewing ring of the 18-mm right-angle apical connector (Fig. 3).

After assurance of large-bore venous access for resuscitation, the availability of donor blood, and premedication with antiarrhythmics, the inferior vena cava is occluded with a tourniquet and a special coring device is used expeditiously to remove a disk of myocardium at the left ventricular apex, at which time the conduit is quickly passed into the apex and secured with a purse string in interrupted sutures. Flow is then allowed back into the heart by releasing the vena caval tourniquet and the animal is resuscitated as needed. Once the animal is stable, a partial exclusion clamp is placed on the ascending aorta, a longitudinal aortotomy is made, and the porcine aortic root or other study valve (in a conduit fashion) is anastomosed to the descending thoracic aorta using a running suture technique. The apical conduit to porcine aortic root anastomosis is then made. The clamp is removed, and the animal is usually quite stable at this point.

In an effort to redirect blood flow through the apical conduit, flow through the ascending aorta must be minimized. A felt-buttressed umbilical tape is therefore passed around the ascending aorta, which is then occluded by tying this ligature, using special care not to be so low as to impinge directly on the coronary ostia or so high as to impinge on the brachiocephalic flow.

This particular mode of implantation is very difficult technically because of the obvious arrhythmogenic and hemorrhagic nature of the left ventricular coring procedure, and the risk of ischemia associated with either the cerebral or the coronary circulation that may result from the ligation of the very short ascending aorta in the sheep.

Techniques similar to those used for the mitral valve study animals are used to study the hemodynamics at explantation for the apical conduit animals. The high-fidelity micromanometers are placed at the level of the valve of the ascending aorta retrograde via the left femoral artery and through the conduit to the area just proximal to the portion of the aortic valve. Thermodilution cardiac outputs are measured as for the mitral replacement.

Evaluation of the Freestyle valve in the sheep model.   AOA-treated and nontreated

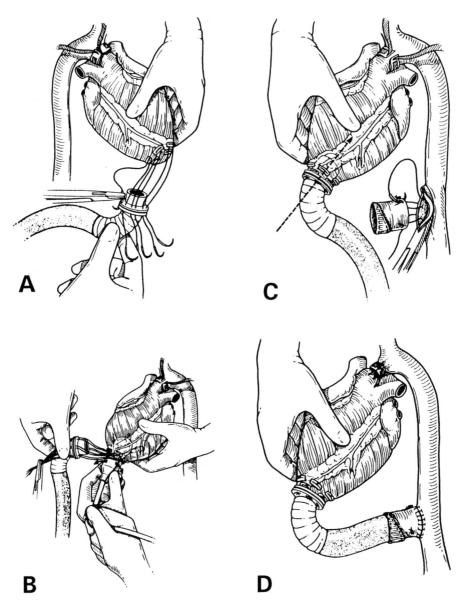

Figure 3 Illustration of the surgical procedure used for the implantation of a Freestyle valve as a terminus of apicoaortic conduit. (A), The apex of the heart is prepared for cannulation by applying pledgeted sutures. The sutures are then passed through the sewing flange of the left ventricle connector. (B), A Foley catheter is passed through the center of an appropriately sized trocar blade and through an incision into the left ventricle. Traction is applied to the saline-filled balloon while the trocar blade is rotated to remove a core of the apical myocardium. (C), Completed left ventricle to apical connector anastomosis. Arrow indicates ideal alignment of connector with respect to left ventricular apex. (D), The cannula is inserted into the ventricle and the sutures are secured. The bioprosthesis is then anastomosed to graft of the left ventricle (LV) connector and to the thoracic aorta. The native aorta has been ligated to redirect LV outflow through the Freestyle aortic root. (Courtesy of Carlyle Frazer Heart Center, Emory/Crawford Long Hospital, Atlanta, Georgia.)

Freestyle aortic roots (19-mm in diameter) were implanted at the Carlyle Fraser Heart Center, Emory/Crawford Long Hospital, Atlanta, Georgia, as apicoaortic shunts according to the surgical method described above. After 20 weeks and completion of the hemodynamic studies, the animals were heparinized and euthanized. The heart of each animal was then excised in continuity with the porcine aortic root.

The conduit and the root were carefully excised and examined for the presence of vegetations, hematoma, thrombotic material, and signs of biological or mechanical degradation such as tears and perforation, fibrous ingrowth, and calcific deposits. The porcine aortic roots were then immersed in 10% neutral buffered formalin and shipped to Dr. Frederick J. Schoen (Pathology, Brigham and Women's Hospital, Boston, MA) for histological studies.

(a) Histological examination. For histological examination, the valves were cut longitudinally to expose the three leaflets, which were examined for signs of calcification, thrombi, and tears. A photograph of AOA-treated and nontreated retrieved valves is shown in Fig. 4. In these photographs, the cusps of the control valves are grossly calcified, while the cusps of the treated valves are free of calcium deposits. For both valves, the aortic wall was cracked because of the pressure induced to the valve during the opening; the wall of both valves was extensively calcified. These findings were confirmed on radiography (Fig. 5).

The three leaflets and three pieces of aortic wall were then dissected from the aortic root and embedded in JB-4 glycol methacrylate medium (Polysciences, Inc., Warrington, PA). Sections 3–5 microns thick were made and stained with the hematoxylin and eosin (H&E) stain for cells and the von Kossa stain for calcium phosphates.

The light microscopy examination of the leaflet sections stained with the von Kossa reagent confirmed that the leaflets from the AOA-treated valves were free of mineralization, while the untreated leaflets had focal calcification. In addition, the aortic walls of both the controls and the AOA-treated valves had moderate-to-severe calcification, particularly in two bands, approximately 200-microns thick, along the intima and the adventitia. Examples of cusps and walls of retrieved AOA-treated and control Freestyle roots are shown in Figs. 5 and 6.

(b) Quantitative calcium analysis. The remaining portions of the leaflets and the walls were sent to Dr. Robert Levy (University of Michigan, Ann Arbor, MI) for quantitative calcium analysis. The samples were hydrolyzed in 6 N ultrapure redistilled HCl and the calcium levels were determined using atomic absorption [11,35,47]. Quantitative calcium analysis confirmed the histological findings and indicated that AOA is a potent inhibitor of heart valve leaflet mineralization, but does not protect the aortic wall [33] (Fig. 7). The lack of AOA effect on the wall is not yet understood, but research aimed at preventing the calcification of the aortic wall is in progress. Recent results from our laboratory indicate that, by doubling the volume (from 100 to 200 ml per valve) of the AOA solution during treatment, a significant increase in AOA incorporation is seen in both leaflets and aortic wall (Fig. 8), resulting in 50% inhibition of aortic wall calcification compared with control in the rat subdermal model (Fig. 9) [48]. More research is needed to enhance further the anticalcification efficacy of AOA in the wall.

Various anticalcification treatments of glutaraldehyde-fixed valves have been tested in the sheep model. The results are reported in Table 1.

2. Shelf-Life Studies

Because the treatment of heart valves with anticalcification agents involves chemical reactions within the tissue, the nature and stability of which are in many cases uncertain,

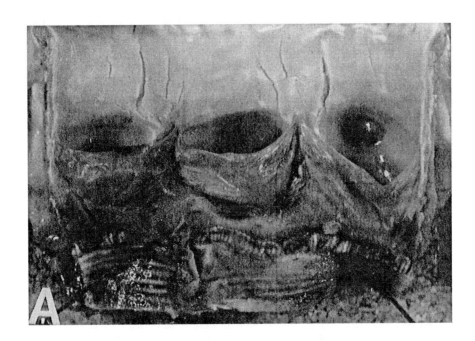

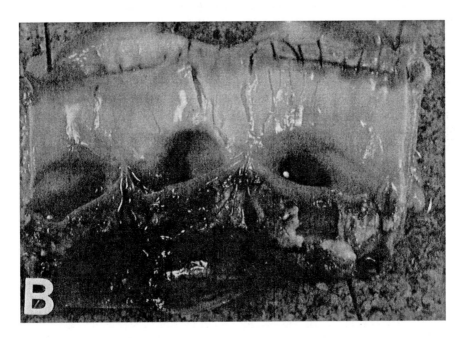

Figure 4   Photographic appearance of retrieved Freestyle valves implanted as apicoaortic shunt. (A), Photograph of valve treated with 2-amino oleic acid (AOA). The cusps are free of calcific deposits. (B), Photograph of nontreated valve. Calcium deposits are associated with the three cusps. The intima side of the aortic walls in (A) and (B) are cracked during opening of the valves due to calcification. (Courtesy of Medtronic Heart Valve Division, Irvine, CA.)

it is important that shelf-life studies be performed on valves that can be treated months or years after fixation, and/or implanted months or years after treatment. Because of undesirable chemical reactions within the tissue, or because anticalcification agents that are not tightly bound to the tissue may be washed out during storage, efficacious anticalcification treatments may lose their potency on the manufacturer's or the surgeon's shelves. The finding, which has since been confirmed in our laboratory, that glutaraldehyde-fixed valves stored for over one year before implantation no longer calcify [49] is a clear illustration that the chemistry of glutaraldehyde-fixed valves is a dynamic phenomenon that must be considered when developing anticalcification treatments.

Shelf-life studies were performed on AOA-treated valves. The two time variables that were considered in this experiment are the storage before treatment and the storage after treatment. The experiment was initially designed as a two-variable matrix with five time levels for each variable: 0, 3, 6, 12, and 24 months for the first variable, and 0, 6, 12, 18, and 36 months for the second variable. The samples were cusps from AOA-treated and control valves, fixed with glutaraldehyde at the same time, and randomly selected for treatment with AOA and for implantation according to the time schedule. The methods for fixation and treatment with AOA were similar to those applied to the Medtronic Freestyle valve. Six cusps dissected from treated and from control valves were implanted at each of the predetermined time conditions. This experiment is still in progress, but the results obtained so far, which are illustrated in Fig. 10, indicate that the level of calcium remains negligible in cusps of valves treated with AOA and stored either before treatment for up to 24 months or after treatment for up to 18 months.

3. Mechanism of Action

For convenience and lack of better categorization, we classified the anticalcification agents that have been effective in the rat and/or sheep models according to their most probable mode of action. This type of categorization may create confusion because there is as yet rarely enough supportive data that demonstrate that the anticalcification effect for any treatment is due to any particular action, and because it tends to narrow the mechanism by which each treatment acts into a single category, disregarding other possible actions as well as interactions. The problems associated with this type of classification, which has been used by other authors [36], are particularly highlighted for the AOA antimineralization process.

AOA has been labeled as a detergent [36], although there is as yet no evidence that it acts as a detergent, that it prevents cuspal calcification because of this detergent effect, or that other mechanisms are not involved, such as ionic effect, charge modification, calcium chelation, or capping of the free aldehydes. The relative contribution of each of these possible mechanisms and their interactive effects are as yet unknown. The results of active ongoing fundamental research may, in the future, provide a stronger and more valid basis for classification of antimineralization treatments.

Surfactants. Sodium dodecyl sulfate, generally referred to as SDS or T6, polysorbate-80, and Triton X-100 in combination with L-lauryl sarcosine, all known surfactants, prevent calcification of porcine aortic valve leaflets in both the subdermal and the circulatory models. However, they do not prevent calcification of bovine pericardium in the sheep circulatory model [17]. Porcine valves treated with sodium dodecyl sulfate have been in clinical testing for several years, and heart valves treated with polysorbate-80 are commercially available in Europe but not in the United States. Triton X-100 and L-lauryl sarcosine seem to have deleterious effects on the tissue; the leaflets of 8 out of 17 valves

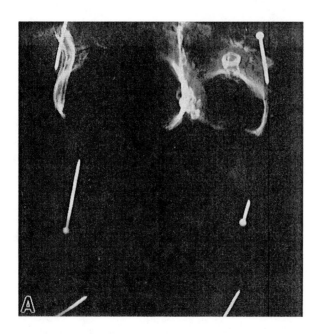

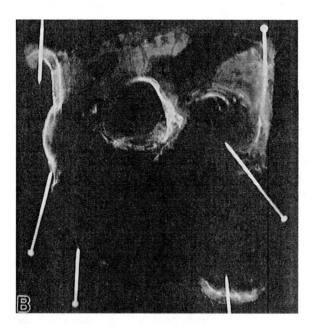

Figure 5 Radiographic appearance and microscopic morphology of calcification of removed Freestyle valves treated with 2-amino oleic acid (AOA) and control Freestyle valves implanted as apicoaortic shunt. (A), Radiograph of AOA-treated valve. No calcification of cusps is present. (B), Radiograph of control valve. Cuspal calcification is noted. Radiographic calcification is equivalent in the aortic walls of the AOA-treated and control specimens (A and B). In (A) and (B), pins in lower portion of photo point toward noncoronary cusp (on left) and right cusp (on right) previously amputated from remainder of specimen. Left coronary cusp remains attached above (center). (C), Photomicrograph of cusp from AOA-treated valve. No calcification is noted. (D), Photomicrograph of cusp from control valve. Focal calcification is designated by arrow. (C) and (D) stained with von Kossa's reagent (calcium phosphates black); approximately ×100. (Courtesy of Dr. Frederick J. Schoen, Brigham and Women's Hospital, Boston, MA.)

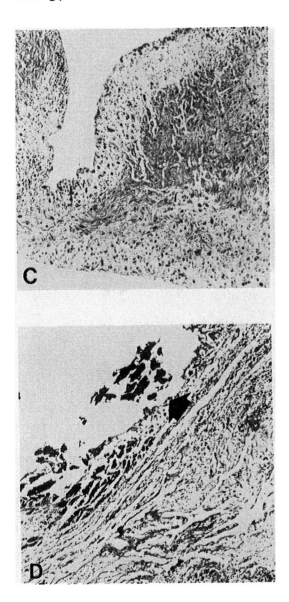

implanted in sheep were damaged after retrieval [17]. SDS was shown to remove phospholipids from the tissue [3,36], which indicates that SDS acts as a detergent. Polysorbate-80, Triton X-100, and L-lauryl sarcosine probably have the same effect on the tissue. In a recent study, it was demonstrated that SDS is effective if porcine aortic valve leaflets are incubated in the solution but not if it is delivered locally [36]. These results strongly suggest that these agents act on the tissue as surfactants, but the causal and unique relationships between detergent and anticalcification effects remain to be established.

Apatitic Crystal Inhibitors. Apatitic crystal growth inhibitors such as diphosphonate and its derivatives were specifically tested because of their known physiological properties in bone metabolism. The major concern with some of these compounds is the toxic effect

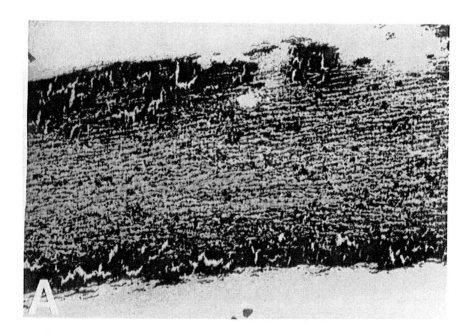

Figure 6 Photomicrographs of aortic walls from Freestyle valves treated with 2-amino oleic acid (AOA) and nontreated Freestyle valves retrieved after implantation as apicoaortic shunt (stained with the von Kossa method, calcium phosphates in black): (A), from AOA-treated valve; (B), from nontreated valve. Predominant calcification of the inner 25% and the outer 25% of the aortic wall, with diffuse calcification of the media of both AOA-treated and nontreated valves. (Courtesy of Medtronic Heart Valve Division, Irvine, CA.)

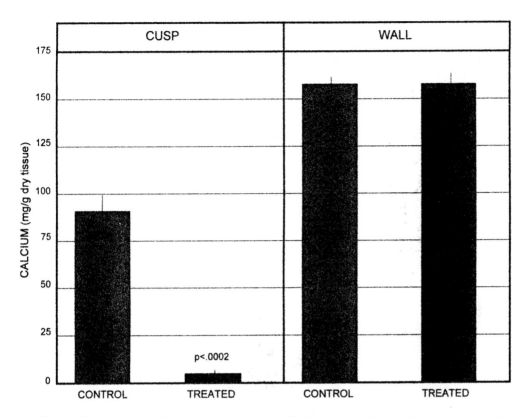

Figure 7 Calcium levels of cusps and aortic walls from control Freestyle aortic roots and those treated with 2-amino oleic acid (AOA) implanted as apicoaortic shunts in sheep and retrieved after 20 weeks.

on bone growth. Several modes of treatment, including binding of aminodiphosphonate to glutaraldehyde [28,50,51], site-specific delivery of diphosphonate using osmotic pumps [22–24], and slow release from polymer matrices [25] have been used successfully to avoid the toxic effects of these agents while effectively preventing calcification in the rat model. To our knowledge, these agents have not been tested in the circulatory model.

Adipoyl- and suberoyl-biphosphonates, analogs of biphosphonic acid, also prevented calcification of heart valve bioprosthetic tissue in the rat subdermal implant model [52], but aminohydroxypropane diphosphonic acid covalently bound to the valve tissue was not effective when tested in the circulation [17]. Although ineffective when administered systemically, phosphocitrate and its synthetic analog N-sulfo-2-amino tricarballylate were effective in preventing subdermal mineralization when delivered locally; phosphocitrate was more potent than N-sulfo-2-amino tricarballylate [53].

Ion Antagonists. Ferric and aluminum chlorides have been successfully used as anticalcification agents of heart valve bioprostheses in the rat subdermal implant model [14,18,19,54], but not in the circulatory sheep model. Transmission electron microscopy coupled with electron energy loss spectroscopy of aluminum-chloride-treated bovine pericardium cross-linked with glutaraldehyde demonstrated that Al^{+++} was localized in the sarcolemma and cytoplasmic and nuclear membranes of devitalized pericardial connective

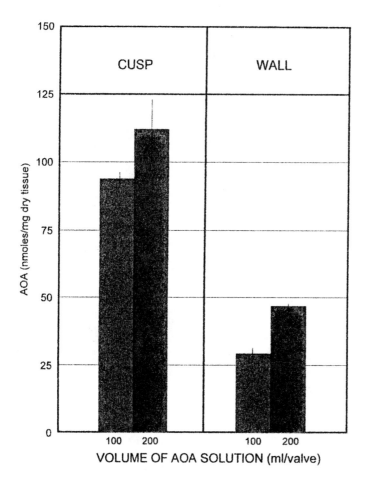

Figure 8   Incorporation of 2-amino oleic acid (AOA) into cusps and aortic wall of Freestyle aortic roots as a function of volume of the treatment solution. (From Ref. 45 with permission.)

tissue cells. This discrete localization at intracellular sites coincided with phosphorus loci. Studies performed using ferric chloride demonstrated that Fe^{+++} was localized within the devitalized connective tissue cells.

These studies suggest that the anticalcification effect of these agents is related to the high affinity of Al^{+++} and Fe^{+++} for membrane-associated and other intracellular phosphorus loci [18]. However, the trivalent cations Ga^{+++} and La^{+++}, also known to interact with the formation of calcium phosphates, were not effective in the prevention of calcification in the rat subdermal implant model [18]. The different anticalcification effects of these cations may be related to their size, electrostatic charge, and reactivity with base [54].

It may be of interest for understanding the mechanisms of calcification to note that both SDS and the trivalent cations (Fe^{+++} or Al^{+++}) seem to prevent calcification of heart valve bioprosthetic tissue by interacting with endogenous phosphates, either by physically removing them from the tissue or by reacting with them.

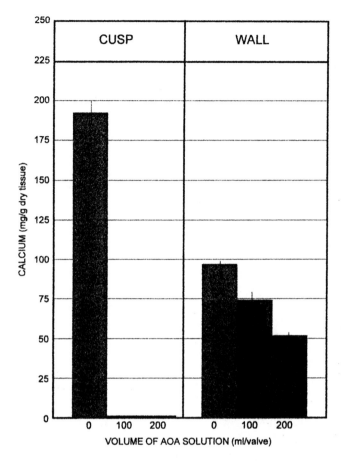

Figure 9 Effect of increased 2-amino oleic acid (AOA) incorporation on calcium levels of cusps and aortic wall from nontreated and AOA-treated Freestyle aortic roots implanted subdermally in 3-week-old rats. (From Ref. 45 with permission.)

Charge Modification.   In the sections below, we discuss agents that have the potential of modifying the charge of the tissue, whether or not they react through Shiff base formation with the residual aldehyde moieties [55–57].

Protamine sulfate.   Protamine sulfate has been covalently bound using formaldehyde before glutaraldehyde cross-linking of the bioprosthetic tissue [20]. Significant inhibition of calcification was observed after implantation in the rat subdermal model. The investigators also found that the rate of transport of calcium across protamine-sulfate-treated tissue was decreased as compared with control tissue. It was hypothesized that protamine sulfate, rich in amines, would restore the positive charge lost during reaction of glutaraldehyde with tissue amines, thereby repulsing calcium anions. However, the rate of calcium transport could also be affected by water transport without being related to the charge of the tissue. More research needs to be done to confirm this hypothesis. To our knowledge, the efficacy of the protamine sulfate treatment has not been tested in the circulatory model.

Table 1   Anticalcification Studies in Sheep

Treatment	Implant position	Ca^{++} (mg/g tissue) mean ± SEM
Control (PAV)	Mitral	99.8 ± 11.1
Polysorbate-80 (PAV)	Mitral	7.6 ± 2.6
Triton X-100 and L-lauryl sarcosine (PAV)	Mitral	24.4 ± 1.8
Control (PAV)	Mitral	64.7 ± 9.6
Sodium dodecyl sulfate (SDS or T6), (PAV)	Mitral	17.7 ± 4.2
Control (BPV)	Mitral	136.2 ± 3.6
Sodium dodecyl sulfate (SDS or T6), (BPV)	Mitral	117.7 ± 5.3
Control (PAV)	Mitral	139.3 ± 14.7
Toluidine blue (PAV)	Mitral	81.6 ± 12.0
Control (BPV)	Tricuspid	66.1 ± 9.6
Polyacrylamide (BPV)	Tricuspid	112.9 ± 15.3
Control (PAV)	Apico-aortic	91.2 ± 19.5
2-Amino oleic acid (PAV)	Apico-aortic	5.5 ± 3.0

Source: Modified from Ref. 17
PAV, porcine aortic valve; BPV, bovine pericardium valve

Glutamic acid.   Glutamic acid has been bound to glutaraldehyde-cross-linked tissue through the residual aldehyde moieties and the primary amine of glutamic acid, thereby reducing the level and toxicity of free aldehydes in the tissue. This method leads to tissue resistant to calcification in the rat subdermal model and provides a nontoxic substrate for the attachment of endothelial cells [21]. The anticalcification results do not support the charge modification hypothesis derived from the protamine experiment since an additional negative charge is provided by the binding of glutamic acid. It is possible that the calcification inhibition obtained is due to the capping of the residual aldehyde groups of the tissue and reaction of calcium ions with the carboxyl moieties of glutamic acid.

However, it has been found by other investigators [58] and in our laboratories that treatment of glutaraldehyde-fixed porcine aortic valve leaflets with lysine under conditions leading to binding to residual aldehyde moieties, which modifies the charge of the tissue and caps the free aldehydes, did not prevent calcification of the leaflets in the subdermal rat model. Moreover, we have found that oleyl-amine-treated glutaraldehyde-fixed porcine aortic valve leaflets calcified severely in the rat subdermal implant model. These results suggest that mere scavenging of free aldehyde moieties by amine groups does not provide adequate protection against calcification. The mechanism by which glutamic acid prevents calcification remains to be elucidated.

Toluidine blue.   The metachromatic nuclear stain toluidine blue binds to residual aldehydes through its primary amine and increases the positive charge of the tissue. Although toluidine blue significantly inhibited calcification of cusps from valves implanted in sheep, the level of calcium remained high, and it has been suggested that this treatment is not clinically applicable [17]. To our knowledge, the mechanism by which toluidine blue reduces calcification has not been investigated.

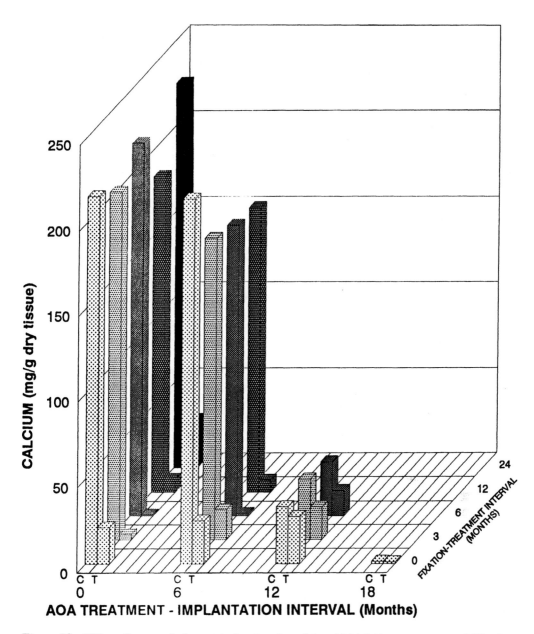

Figure 10   Effect of storage before and after 2-amino oleic acid (AOA) treatment on calcification of cusps from AOA-treated and nontreated Freestyle valves implanted in rats for 8 weeks (C = nontreated; T = AOA treated).

Alpha-Aminooleic Acid. AOA prevents calcification of glutaraldehyde-cross-linked porcine aortic valve leaflets in both the rat subdermal model [31] (Fig. 2), and the sheep circulatory model [30,32,33]. However, AOA did not prevent calcification of the wall of the Freestyle valve implanted in sheep [33]. Although AOA has been classified as a detergent [36] and may have some surfactant activity due to its chemical properties, its mode of action has not been elucidated yet despite intensive research in Dr. Robert Levy's laboratory. It may prevent calcification by removing phospholipids from the tissue during treatment, by reducing the rate of calcium transport, by interacting with calcium ions, and/or by capping residual aldehyde groups in the tissue. Initial results have provided evidence that AOA, like protamine sulfate [20], reduces the rate of calcium transport in a concentration-dependent manner across the treated tissue [58]. Whether this reduced rate of calcium transport is a sufficient and/or necessary cause for the potent anticalcification effect of AOA, and whether it is due to the hydrophobic nature of AOA, to a calcium-carboxyl interaction within the tissue, or to the blocking of channels by AOA remains to be investigated.

It has also been shown that AOA was bound through an amino-aldehyde linkage, and that a significant level of AOA remained in the tissue after incubation in calf serum for eight months [58]. This finding, which indicates that AOA is strongly bound in the tissue, may be significant for the long-term anticalcification properties of AOA.

B. Morphology

The bioprosthetic tissue used in the fabrication of heart valves and treated with antimineralization agents should maintain its structural integrity in order to conserve the durability characteristics needed in valves that will be implanted for several years in patients. The methods used for morphological examination most commonly are (1) light microscopy examination of stained thin sections of tissue and (2) scanning electron microscopy.

1. Light Microscopy

Important information concerning the structure of the tissue can be collected by light microscopy before implantation and after explantation from either the subdermal or the circulatory implant models. The methodology used for staining is similar to the one described in the circulatory model section. However, it is not necessary to stain the nonimplanted samples with the von Kossa stain. Other stains for collagen, elastin, and other structural components of the tissue can also be used. In addition, the biocompatibility of the treated tissue can be assessed by examining the host immunological cell invasion of the retrieved tissue and the surrounding fibrous capsule.

2. Scanning Electron Microscopy

The scanning electron microscopy method permits visualization of a large area of the tissue under investigation as well as selected structures, such as fibrosa, spongiosa, ventricularis, and inflow and outflow surfaces. Possible deleterious effects of anticalcification treatments could be visualized as a roughening of the surface and delamination or loosening of the internal structures. The examination of the inflow and outflow surfaces after implantation in the circulation may provide information on spontaneous endothelization of the tissue, which may be the ultimate biocompatibility test. The cell coverage may also prevent intrinsic calcification by maintaining a physiological calcium level within the tissue.

The method, as applied in our laboratory, consists of cutting the tissue in half to allow exposure of the internal structures as well as of the surfaces. The samples are prepared according to standard techniques, which consist of fixation in aqueous osmium tetroxide, followed by dehydration in ethanol, critical point drying, and coating with AuPd.

As an example, glutaraldehyde-fixed AOA-treated and untreated stentless porcine aortic roots were processed and observed in a Hitachi S-800 field emission scanning electron microscope (SEM) at 15 kilovolts (kV). The photographs presented in Fig. 11 show that the AOA-treated tissue is compact, that there is no delamination of the fibrosa-spongiosa-ventricularis structure, and that there is no evidence of roughening of the surfaces. The lack of endothelial cell coverage onto the inflow surfaces of both AOA-treated and untreated control valves is mainly due to the mechanical manipulation of the valves during processing. This study clearly indicates that the AOA treatment does not alter the structure of the tissue. Due to the limited number of explanted valves, SEM was not performed after the circulatory studies.

C. Function

1. Accelerated-Cycle Testing

Accelerated-cycle testing predicts failure modes for prosthetic valves by exposing them to pulse frequencies equivalent to at least five years of implantation in humans. The validity of the test has been questioned because the valves (1) are not subjected to typical *in vivo* cardiac cycle pressures of fluid flow conditions, (2) are not subjected to a blood environment, and (3) are submitted to excessively high pulse frequencies, resulting in abnormal mechanical stresses. Understanding these limitations helped in the design of the testing equipment to offset some of the testing deficiencies.

The accelerated-cycle testing performed on the AOA-treated Freestyle aortic roots serves as an example to describe the method and the type of results that can be obtained from this test.

Testing Conditions. The accelerated-cycle testing was completed on three clinical-quality AOA-treated Freestyle aortic roots of standard sizes, including six of the largest, 27-mm size. Three 31-mm Hancock standard valves were tested as control valves and were sutured within simulated silicone aorta test fixtures. Three nontreated Freestyle aortic roots were also tested to evaluate any possible effect of the AOA treatment on durability. The test was performed in normal saline at room temperature. Due to the amount of handling required during tests for hydrodynamic and compliance evaluations, sterile filtered saline was used to limit bacterial growth. A peak closed-valve pressure drop of 110 ± 5 mmHg was maintained during the test. Tester speed (1450 cycles per minute) and stroke volume were adjusted to allow typical leaflet excursions (full opening and closing) during the cardiac cycle.

Methods. After insertion of the Freestyle aortic root into the test chambers, the test system was adjusted using strobe observation and pressure drop measurement. All valves were tested for a minimum of 200×10^6 cycles. Every 20×10^6 cycles, the valves were removed from the accelerator for visual examination, photographs, and evaluation.

Temperature, speed, and elapsed hours (total cycles) were monitored and recorded on a regular basis. Peak closed-valve pressure drop was monitored at the start of the test, before and after every visual examination, within 48 hours of each setup, a minimum of once weekly, and at the end of the test. Each valve was photographed prior to initial test onset, during the visual examinations, and at the conclusion of the test. In addition, all

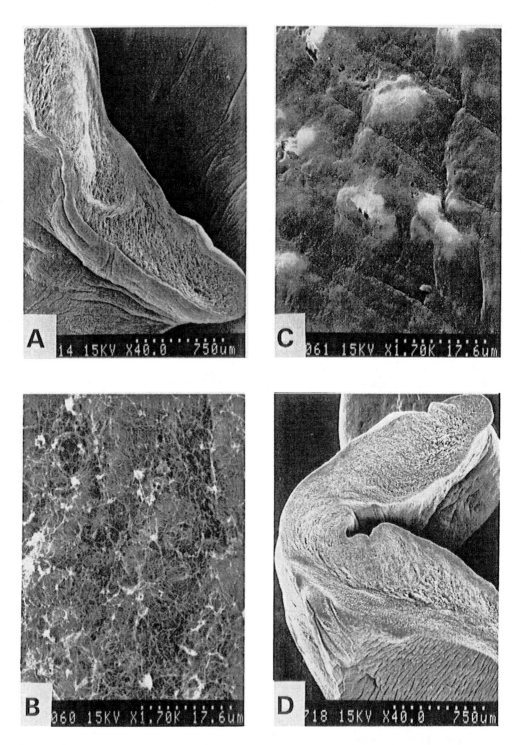

Figure 11 Scanning electron microscopy of nontreated Freestyle aortic root leaflets and those treated with 2-amino oleic acid (AOA): (A), radial cut of control leaflet showing compact structure; (B) inflow surface of nontreated leaflets; (C), outflow surface of nontreated leaflet; (D) radial cut of AOA-treated leaflet showing compact fibrosa-spongiosa-ventricularis structure; (E), inflow surface of AOA-treated leaflet; (F) outflow surface of AOA-treated leaflet. The endothelial cell covering has been removed from the inflow surfaces in (B) and (E) during the manipulation of the valves during manufacturing.

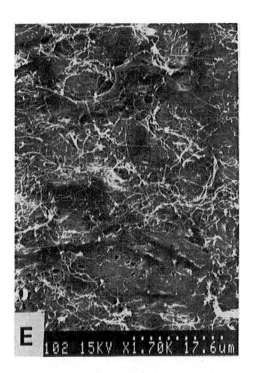

valves were hydrodynamically tested at pretest and at 60×10^6, 120×10^6, and 200×10^6 cycles.

The objective of the hydrodynamic test is to evaluate the performance of valves by providing data on pulsatile flow pressure drop and valvular regurgitation. For the testing of the AOA-treated Freestyle, the pulsatile flow pressure drop test was performed at a cardiac output range of 2.5 to 7.5 liters per minute (l/min) and at a pulse rate of 70 beats per minute (bpm), with systole accounting for approximately 35% of the simulated cardiac cycle. Pulsatile flow regurgitant volumes versus back pressure (closing pressures applied to the valve) were determined at a pulse rate of 70 bpm with a cardiac output of 5.0 l/min, and regurgitant volume tests were performed at back pressures of 80, 100, 120, 140, and 160 mmHg. All tests were conducted at ambient temperature in normal saline.

Results and Discussion.

Visual examination. All Freestyle aortic roots opened and closed normally, and maintained closed-valve pressure drop at test termination ($\geq 200 \times 10^6$ cycles). There was no significant difference between the AOA-treated and the nontreated Freestyle aortic roots, and there was no evidence of change in cusp coaptation or shape. Furthermore, none of the valves exhibited any evidence of delamination or alteration of the leaflet structure. All wear and tear was of technical origin. A tear of one cusp of an AOA-treated root was caused by contact with the attached tube fixturing that holds the valve in the test chamber; the opening of this cusp was consequently reduced.

In most cases, the observed wear resulted from the longitudinal motion of the valve/aorta fixturing assembly held within the noncompliant test chamber. During the closing phase of the cycle, the root is forced forward in the chamber while the end section of the aorta tubing is held stationary and is therefore not completely free to move forward to the full extent of the pressure applied. This stress damaged the valve/aorta attachment sutures and the inflow leaflet attachment. Another, less frequent type of wear was due to broken sutures that abraded the inflow surface of the leaflet. No valve failure (i.e., inability to maintain closed-valve pressure drop) or detachment from the aorta resulted from this type of damage. In rare cases, because of the anatomical shape of the root, wear resulted from contact of the right cusp with the trimmed edge of the silicone tubing. Some wear was also due to the slight forward motion of the root during the closing cycle, which causes the root to contact the trimmed edge of the silicone tubing, producing tears originating from the outside of the valve.

Hydrodynamic performance. Because of the amplitude of the data generated by the hydrodynamic performance test, only the results for the 19-mm AOA-treated and nontreated Freestyle and the 31-mm Hancock valves are presented. This represents the worst-case situation since the smallest sizes have the highest transvalvular gradients.

(a) Pulsatile flow pressure gradient. The pulsatile flow pressure gradient, which represents the transvalvular pressure gradient as a function of pulsatile flow rate, was obtained on all test valves. It was not different for the 19-mm AOA-treated, the untreated Freestyle, and the 31-mm Hancock control valves at a flow rate of 167 milliliters per second (ml/s) (Table 2). When comparisons were made for all sizes tested, the gradients were not significantly different for the AOA-treated and the nontreated Freestyle roots, but they were significantly lower for these valves than for the Hancock valves. The results also indicated that the gradient was slightly reduced as a result of accelerated wear testing, particularly for the 19-mm AOA-treated and nontreated Freestyle and the 31-mm Hancock valves, for which the gradient reduction was higher than 1 mmHg.

Table 2 Freestyle Durability Study: Pulsatile Flow

RMS flow rate (ml/s)	Mean pressure drop (mmHg)			
	Pretest	60×10^6	120×10^6	200×10^6
AOA treated				
167	2.6 ± 0.5	2.1 ± 0.4	2.2 ± 0.5	1.7 ± 0.1
250	6.2 ± 1.1	4.9 ± 1.0	4.9 ± 1.2	4.9 ± 0.1
333	10.9 ± 1.6	7.9 ± 3.5	8.5 ± 2.7	8.1 ± 0.9
417	16.5 ± 1.6	11.5 ± 3.5	13.1 ± 4.8	12.7 ± 0.6
Nontreated				
167	3.0 ± 0.3	2.7 ± 0.5	2.5 ± 0.3	2.3 ± 0.1
250	6.9 ± 0.2	5.5 ± 0.1	4.4 ± 0.7	5.3 ± 0.2
333	12.0 ± 0.6	8.0 ± 0.9	7.8 ± 1.1	8.5 ± 1.1
417	18.0 ± 1.3	12.1 ± 1.1	11.0 ± 1.5	13.3 ± 0.4
Hancock control				
167	2.4 ± 0.4	1.7 ± 0.3	1.7 ± 0.3	1.3 ± 0.4
250	4.3 ± 1.4	3.8 ± 0.4	3.5 ± 0.3	2.9 ± 0.9
333	6.9 ± 1.7	6.4 ± 0.7	5.7 ± 0.6	5.7 ± 1.1
417	9.6 ± 2.9	9.5 ± 1.0	8.1 ± 1.1	8.0 ± 0.9

The results are expressed as mean ± SD of 3 valves.

(b) Valvular regurgitation. The data on regurgitation volume as a function of back pressure (closing pressure applied to the valve) were obtained for all test valves. Total regurgitation is equivalent to the sum of closing and leakage volumes. Closing and leakage volumes must therefore be determined in order to determine valvular regurgitation. The results of this test for the 19-mm AOA-treated and nontreated Freestyle aortic roots are presented in Fig. 12.

The closing volumes were, in most cases, low and remained constant throughout the accelerated-cycle testing for the AOA-treated and for the nontreated 19-mm Freestyle aortic roots. They decreased at the 120×10^6 cycles test point, and increased again with increases as a function of back pressures for both AOA-treated and nontreated valves.

The leakage volume was also reduced at the 120×10^6 cycles test point. This reduction may be attributed to the softening of the leaflets during the accelerated-cycle testing. The mean leakage volume for the AOA-treated root increased considerably after 200×10^6 cycles of accelerated-cycle testing because of the technical wear of the wall of one valve (described above). The leakage volumes for the other 19-mm AOA-treated roots remained low, as reflected by the standard deviation.

The total regurgitant volumes generally remained low from pretest to 60 million cycles, decreased at 120 million cycles, and increased again at 200 million cycles. These variations reflect the variations found for the closing and the leakage volumes, which have been described above.

It was concluded from the accelerated-wear testing of the Freestyle valve that the AOA treatment does not negatively affect the durability of the valve when submitted to the dynamic stress of the accelerated-cycle test.

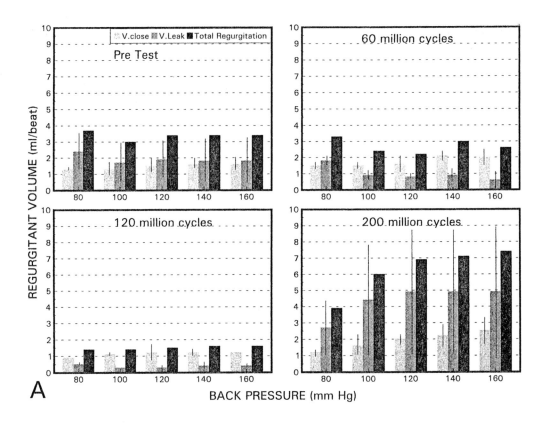

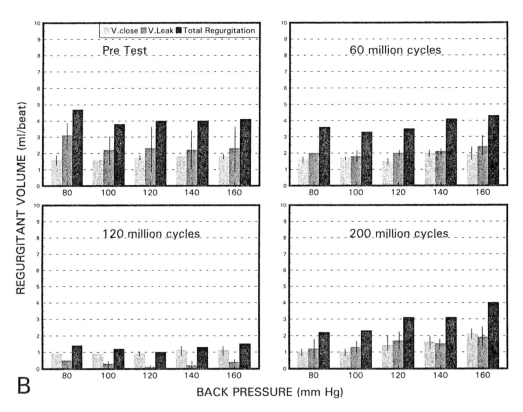

2. Effect of Storage

It may be necessary to store glutaraldehyde-fixed valves for months before they are treated with AOA. As discussed above, it is important that the effect of storage be assessed not only on the efficacy of anticalcification treatments applied months or even years after the tissue has been fixed, but also on the performance of the valve. We have already presented evidence that storage before AOA treatment for up to 18 months (refer to shelf-life studies section) did not reduce the anticalcification potency of AOA.

The effect of pretreatment storage on valve performance was evaluated by hydrodynamic testing of Freestyle roots treated with AOA either within 3 months of fixation (control) or after 24 months of fixation (test). The results indicated that neither the pulsatile flow pressure gradient nor the pulsatile flow regurgitant volume was different for the treated and the nontreated groups. It was concluded that Freestyle aortic roots can be stored for up to 24 months without affecting the performance of these valves.

D. Biocompatibility

This section concerns preclinical biological evaluation of antimineralization-treated materials and devices to determine if they are safe for human use. Specifically, pretreatment designed to prevent mineralization of the bioprosthesis must be shown to yield a tissue that is nontoxic/nonpyrogenic, nonsensitizing, nonirritating, and hemocompatible (nonthrombogenic, nonhemolyzing). Supplementary evaluations may also include tests for genotoxicity, carcinogenicity, and teratogenicity (reproductive/developmental toxicity).

It should be recognized that biological testing, which often relies upon animal models, cannot be conclusively shown to elicit the same response in humans. Also, an adverse reaction either in animals or in an *ex vivo* or *in vitro* test for one device application may not be a problem with the same pretreatment in another application. Finally, actual testing may be supplemented with relevant experience that, with demonstrable evidence of safe use, can obviate the need for additional evaluations. Each device and pretreatment must therefore be considered on its own merits; additional tests may be indicated, as well as certain assays may be waived.

1. General Principles

The following are general principles that should be applied to the biological testing of antimineralization pretreated devices.

1. Testing should be performed on the final product or representative samples of the final product. This means that the device or material should be tested in its implantable or "as used" state (i.e., it must have been sterilized and appropriately rinsed, if required, before preparation of extracts or implantation).

Figure 12 *In vitro* hydrodynamic studies of control Freestyle valves and Freestyle valves treated with 2-amino oleic acid (AOA) during accelerated wear testing (light shade, closing volume; medium shade, leakage volume; black, total regurgitant volume): (A) closing volume, leakage volume, and total regurgitant volume from AOA-treated valves; (B) closing volume, leakage volume, and total regurgitant volume from nontreated control valves. The regurgitant volume is the sum of closing and leakage volumes. The high leakage volume seen for the AOA-treated valves at 200 million cycles is due to abrasion of the aortic root, not poor coaptation of the leaflets.

2. Test procedure selection shall take into account
   a. The nature and conditions of exposure of the device to the patient in the normal intended use.
   b. The chemical and physical nature of the final product.
   c. The toxicological activity of the chemicals making up the final product.
   d. Certain tests (e.g., those designed to assess systemic effects) may not be applicable when the presence of leachable materials has been excluded or when leachables have a known and acceptable degree of toxicity.
   e. The relationship of device wet weight/surface area to recipient body size, blood volume/flow rate.
   f. Information available from the literature, in-house experience, and nonclinical testing.
   g. The primary goal is human safety; a secondary goal is to ensure animal welfare by minimizing the number and exposure of animals.
3. Solvents and conditions of extraction should be appropriate to the nature and conditions of use of the final product.
4. Positive and negative controls should be used when appropriate.
5. Since test results cannot ensure freedom from potential biological hazard, careful observation for unexpected adverse reactions should be made during initial clinical use.

The above general considerations were modified from the Association for the Advancement of Medical Instrumentation and the International Organization for Standardization (AAMI/ISO) 10993-1 proposed new *American National Standard on Biological Evaluation of Medical Devices, Part 1: Guidance on Selection of Tests* [59].

2. Manufacturer-Specific Antimineralization Treatments

Antimineralization pretreatment using sodium dodecyl sulfate (T6) and Polysorbate-80 are currently being used, respectively, by Medtronic and Baxter-Edwards. Medtronic has also utilized a cationic dye, toluidine blue, as an antimineralization treatment on the INTACT bioprosthesis; all three pretreatments are applied to assembled valves just prior to final inspection and packaging. Most recently, Medtronic has employed 2-amino oleic acid (AOA) as an antimineralization pretreatment for a stentless aortic bioprosthesis, the Freestyle valve. As with the other pretreatments, AOA is applied to the glutaraldehyde-fixed valves shortly before final packaging.

Each proprietary antimineralization process was tested for safety and biocompatibility (as well as efficacy in animal trials versus nontreated controls) prior to clinical use. The following sections detail the biocompatibility tests performed on these valves and currently required by the United States Food and Drug Administration (FDA) [60], by the Association for the Advancement of Medical Instrumentation and International Organization for Standardization [61], and by the European Committee for Standardization [62]. Although there are minor differences in the requirements for biological evaluation among these documents, there is general agreement that the following list of tests [62] should be performed upon the final product or representative samples from the final product. Details concerning sample preparation and guidance on actual test methodology can be found in Parts 1 through 10 of the AAMI/ISO document [61].

3. Tests

Cytotoxicity.   With the use of cell culture techniques, cytotoxicity tests determine the lysis of cells (cell death), the inhibition of cell growth, and other effects on cells caused by the device, materials, and/or extracts thereof.

Sensitization. Sensitization tests estimate the potential for contact sensitization of devices, materials, and/or their extracts, using an appropriate model.

Irritation. Irritation tests estimate the irritation potential of devices and/or their extracts, using appropriate sites or implant tissues such as the skin, eyes, and mucous membrane in a suitable model. The tests performed should be appropriate for the route (skin, eye, mucosa) and duration of exposure.

Intracutaneous Reactivity. Intracutaneous reactivity tests assess the localized reaction of tissue to extracts of the device and are applicable where determination of irritation by dermal or mucosal implantation is inappropriate (e.g., devices with access to the blood path). They may also be useful when extractables are hydrophobic.

Systemic Toxicity (Acute). Acute systemic toxicity tests estimate the potential harm of either single or multiple exposures, during a period of less than 24 hours, to devices, materials, and/or their extracts in an animal model. They are appropriate when contact allows potential absorption of toxic leachables and degradation products.

Pyrogenicity tests are included under this category to detect material-mediated pyrogenic reactions of extracts of devices or materials. Note that no single test can differentiate material-mediated pyrogenic reactions from those due to endotoxin contamination. The U.S. FDA also recommends reference to their guidance document [63].

Subchronic Toxicity (Subacute). Subchronic toxicity tests determine the effects of either single or multiple exposures to devices, materials, and/or their extracts during a period of not less than 24 hours to a period not greater than 10% of the total life span of the test animal (i.e., up to 90 days for rats). These tests should be appropriate for the route and duration of administration, and may be waived for materials already having chronic toxicity data.

Genotoxicity. Genotoxicity tests utilize mammalian or nonmammalian cell culture or other techniques to determine gene mutations, changes in chromosome structure and number, or other deoxyribonucleic acid (DNA) and gene toxicities caused by devices, materials, and/or their extracts.

Implantation. Implantation tests assess the local pathological effects on living tissue, at both the gross and microscopic level, of a sample of the material/final product surgically placed into an appropriate implant site. They are equivalent to subchronic toxicity tests if systemic effects are also investigated.

Hemocompatibility. Hemocompatibility tests evaluate effects on blood or blood components by blood-contacting devices and materials using an appropriate model or system. Specific hemocompatibility tests may be designed to simulate the geometry, contact conditions, and flow dynamics of the device or materials during clinical application.

Hemolysis. Hemolysis tests determine the degree of red blood cell lysis and hemoglobin release caused by devices, materials, and/or their extracts *in vitro*.

Thrombogenicity. The thrombogenicity tests determine the potential for the production of blood clots (thrombi) or thrombosis of the device or material when placed in contact with whole blood *in vitro*.

Supplementary Evaluation (As Required).

Chronic toxicity. Chronic toxicity tests determine the effects of either single or multiple exposures to devices, materials, and or their extracts during a period of at least 10% of the life span of the test animal (e.g., over 90 days for rats). Also, these tests should be appropriate for the route and duration of exposure or contact.

Carcinogenicity. Carcinogenicity tests determine the tumorigenic potential of devices, materials, and/or their extracts from either a single or multiple exposures over a period of the major portion of the life span of the test animal. These tests may be designed such that chronic toxicity and tumorigenicity may be assessed in a single experimental study. Carcinogenicity tests should be conducted only if there are suggestive data from other sources. These tests should be appropriate for the route and duration of exposure.

Reproductive and developmental toxicity. Reproductive and developmental toxicity tests evaluate the potential effects of devices, materials, and/or their extracts on reproductive function, embryonic development (teratogenicity), and prenatal and early postnatal development. Reproductive/developmental toxicity tests or bioassays should only be conducted when the device has a potential impact on the reproductive potential of the subject. The application site of the device should be considered.

Biodegradation tests. When applicable, biodegration tests determine the processes of absorption, distribution, biotransformation, and elimination of leachables/degradation products from devices, materials, and/or their extracts.

4. Selection of Biological Evaluation Tests

Tables 3 and 4 identify the initial and supplementary evaluation tests that are to be considered for determining biocompatibility of a permanently implanted blood-contact device (C). Bioprosthetic heart valves, regardless of whether they have been pretreated with an antimineralization (AMT) agent, fall into this category. Due to the different materials and processing protocols employed in the various AMT treatments, not all of the biological tests in each category are to be considered necessary or practical.

The following section details the biological tests performed on the AOA-treated Freestyle aortic root, and discusses the results obtained. As recommended, each device is considered on its own merits; certain tests not indicated in Tables 3 and 4 were carried out, while others were waived with a rationale for the evaluations actually selected.

5. Biocompatibility Tests Performed on the Freestyle Root

The AOA-treated Freestyle Bioprosthetic Heart Valve falls into the category designated as a permanently implanted (over 30 days), blood-contact device. As such, these devices

Table 3   Guide for Initial Evaluation Tests

	Device categories					
	Body contact duration[a]			Blood contact duration[a]		
	A	B	C	A	B	C
Biological effect						
Cytotoxicity	*	*	*	*	*	*
Sensitization	*	*	*	*	*	*
Irritation or intracutaneous reactivity	*	*	*	*	*	*
Systemic toxicity	*	*	*	*	*	*
Subchronic toxicity		*	*		*	*
Genotoxicity		*	*		*	*
Implantation	*	*	*	*	*	*
Hemocompatibility	*	*	*	*	*	*

[a]A = limited (<24 h); B = prolonged (24 h to 30 days); C = permanent (>30 days)

Table 4 Guide for Supplementary Evaluation Tests

	Device categories					
	Body contact duration[a]			Blood contact duration[a]		
	A	B	C	A	B	C
Biological effect						
Chronic toxicity			*			*
Carcinogenicity			*			*
Reproductive/developmental						
Biodegradation						

[a]A = limited (< 24 h); B = prolonged (24 h to 30 days); C = permanent (> 30 days)

are recommended in the American National Standards Institute (ANSI)/AAMI guidance for biological testing under all initial and supplementary evaluations except reproductive/developmental and biodegradation effects (Tables 3 and 4).

Biocompatibility Testing of 2-Amino Oleic Acid.   The antimineralization agent itself was biologically tested first, as a neat compound, prior to testing AOA-treated, glutaraldehyde-fixed tissue. Table 5 lists the tests employed, the laboratories performing the tests, and the test results.

The neat compound was found to be completely biocompatible as a result of these

Table 5 Biocompatibility Evaluation of 2-Amino Oleic Acid (Neat)

Test method	Laboratory	Results
Cytotoxicity: Agar diffusion Mouse L929 fibroblasts	PRL	Noncytotoxic
Static hemolysis: Rabbit RBC	PRL	Nonhemolytic
USP Class V	PRL	
Intracutaneous injection: Rabbit		
Cottonseed oil		Nonirritating
0.9% saline		Nonirritating
Systemic injection: Mouse	PRL	
Cottonseed oil		Nonirritating
0.9% saline		Nonirritating
Mutagenicity, Ames test *Salmonella typhimurium*	Nelson (Utah)	Nonmutagenic
Sensitization: Guinea pig	U.S. Testing (New Jersey)	
Cottonseed oil		Nonsensitizing
0.9% saline		Nonsensitizing
Pyrogenicity: Rabbit	U.S. Testing	Nonpyrogenic
Cage Implant: Rabbit AOA-treated cusps	Case Western Reserve University	Biocompatible

PRL, Medtronic Physiological Research Laboratories, Minneapolis, Minnesota

initial tests. Specimens of AOA-treated cusps were also found to be compatible using the cage implant system in rabbits developed at Case Western Reserve University by Dr. James M. Anderson [64]. This system can quantitatively monitor an animal's inflammatory response to an implanted material by enclosing the specimen in a stainless steel wire cage and implanting it subcutaneously in the back of the test animal. Implanted empty cages act as controls. Cell-implant interactions such as cellular migration, adhesion, and activation that occur at the host tissue-implant interface (believed critical in determining the long-term biocompatibility and biostability of an implanted device) are evaluated by periodically measuring the concentration of various types of leukocytes in cage exudates postimplant.

Biocompatibility Testing of Clinical-Quality Valves. Biocompatibility evaluations of AOA-treated Freestyle valves were conducted next. The valves were completely configured for biological testing as sterile, clinical-quality devices.

Table 6 indicates the results of this initial battery of biological tests. It can be seen that the material failed the test for cytotoxicity (agar diffusion using mouse fibroblasts) and both intracutaneous and systemic injection of saline extracts in rabbits and rats (US Pharmacopoeia [USP] Class V), but passed all of the other assays.

Using valves not treated with AOA as controls, it was shown that even this material caused a cytotoxic cell response in tissue culture, and yielded saline extracts that were irritating upon injection into rabbits and rats (see Table 7). It was concluded from these tests that residual glutaraldehyde was the most likely cause of the toxic/irritant response, not the new antimineralization treatment process.

Table 6   Biocompatibility Evaluation of Freestyle Valves: Treated with 2-Amino Oleic Acid

Test method	Laboratory	Results
Cytotoxicity: Agar diffusion	PRL	Cytotoxic
Mouse L929 fibroblasts		
Static hemolysis: Rabbit RBC	PRL	Nonhemolytic
USP Class V	PRL	
Intracutaneous injection: Rabbit		
Cottonseed oil		Nonirritating
0.9% saline		Irritating
Systemic injection: Mouse	PRL	
Cottonseed oil		Nonirritating
0.9% saline		Irritating
Mutagenicity, Ames Test	Nelson (Utah)	Nonmutagenic
Salmonella typhimurium		
Sensitization: Guinea pig	U.S. Testing (New Jersey)	
Cottonseed oil		Nonsensitizing
0.9% saline		Nonsensitizing
Pyrogenicity: Rabbit	U.S. Testing	Nonpyrogenic
(AOA-treated cusps)		
Cage implant: Rabbit	Case Western Reserve University	Biocompatible (Post Week 2)

PRL, Medtronic Physiological Research Laboratories, Minneapolis, Minnesota

Table 7 Biocompatibility Evaluation: Comparison of Treated and Nontreated Freestyle Valves

	Results	
Test method	AOA treated	Nontreated
Cytotoxicity: Agar diffusion	Cytotoxic	Cytotoxic
Mouse L929 fibroblasts		
USP Class V		
Intracutaneous injection: Rabbit		
Cottonseed oil	Nonirritating	Nonirritating
0.9% saline	Irritating	Irritating
Systemic injection: Mouse		
Cottonseed oil	Nonirritating	Nonirritating
0.9% saline	Irritating	Irritating

Glutaraldehyde Elution Studies. Glutaraldehyde elution studies were performed on AOA-treated valves to determine the extraction profile of this chemical from the tissue. For the studies, 3 clinical-quality 27-mm valves (the largest size to be marketed) were rinsed according to the standard preimplantation protocol (0.9% saline, 3 × 500 ml). The valves were then individually further extracted in a 50-ml aliquot of saline for 2, 4, 6, 8, 24, 50, 80, 100, and 120 hours. Samples of each rinse solution and rinsed valve extract were analyzed by gas chromatography for glutaraldehyde content (0.5 parts per million [ppm] limit of detection).

The results of the rinse and extraction study showed that 3.4 mg of glutaraldehyde per rinsed valve could have been released into the saline extracts prepared for cytotoxicity and USP V testing. For each test, under the extraction conditions used, the calculated glutaraldehyde concentration in the extracts prepared for the standard cytotoxicity and irritant tests was 85 ppm, which is considerably higher than the threshold (a few ppm) known to yield a positive response in these assays.

Health Risk Assessment. A health risk assessment was conducted to determine the potential risk to patients due to the residual glutaraldehyde in the rinsed valve; Table 8 summarizes the findings of this study. Assuming the entire stentless bioprosthesis was implanted (worst case as the device can be tailored by the surgeon resulting in a smaller

Table 8 Results of Health Risk Assessment and Biocompatibility Tests

A glutaraldehyde leach rate constant was calculated to be 0.73 h^{-1}, with a leaching half-life of 1 hour.

Assuming an entire 27-mm valve was implanted (worst case), the peak concentration of glutaraldehyde in the blood was calculated to be 0.069 mg/l and would occur about 4 hours after implantation. This level was considered well below (100 times) that reported to be acutely toxic.

All acute cytotoxicity and USP V testing performed with both 10- and 100-fold dilutions (compared with the original extracts) gave negative responses. These tests utilized solutions of glutaraldehyde up to 300 times greater than the maximum levels expected in the body fluids of patients implanted with the largest valve.

quantity of implanted material), the peak concentration of glutaraldehyde in the blood was calculated to be 0.069 ppm. This level of glutaraldehyde is considered 100 times lower than that reported as toxic.

Reevaluation of Cytotoxicity and USP V Testing.   Cytotoxicity and USP V irritant testing were repeated using both 10-fold and 100-fold dilutions of the extracts compared with those that were prepared originally.

All tests gave negative responses for toxicity and irritancy. Since the extracts containing up to 300 times the maximum level of glutaraldehyde expected in a patient were thus shown nontoxic, the rinsed valves were considered safe for implantation.

Other Biocompatibility Tests on the Freestyle Root.   The thrombotic potential of AOA-treated valves was compared with that of nontreated controls by way of an acute, *in vitro* assay utilizing heparinized whole human blood. The performance of the tissue in this system was evaluated by comparing cellular deposition/adhesion, protein deposition, and the presence or absence of surface thrombus using scanning electron microscopy.

In addition, an *in vitro* fibrinopeptide A (FPA) immunoassay was performed on the blood plasma immediately following exposure of the heparinized whole blood to the heart valve tissues. The level of FPA in the plasma gives an indication of the propensity of the test material to elicit fibrin clot formation, thus allowing comparison of antimineralization-treated tissue versus nontreated control tissue. These tests were developed and performed by Medtronic Physiological Research Laboratories, Minneapolis, Minnesota. Other thromboresistance tests, such as the *ex vivo* Mantovani method, may be used [65].

Examination of the AOA-treated tissue by scanning electron microscopy following exposure to fresh human blood showed no significant differences in thrombus formation or cell adhesion compared with control specimens. Likewise, there was no significant difference in fibrinopeptide A levels in the plasma from the blood samples exposed to the treated and control material. It was concluded from these studies that AOA treatment did not significantly change the thromboresistance of the native (nontreated) tissue.

6.   *Rationale for Biocompatibility Tests Not Performed*

Carcinogenicity.   Biological testing for carcinogenicity was not performed on the AOA-treated valves since (1) these tests were developed for soft-tissue environments and are thus not clinically relevant to replacement heart valves and (2) glutaraldehyde, used to preserve heterografts for more than 20 years, has never been implicated in any known case of carcinogenicity.

Teratogenicity.   Tests for teratogenicity were not performed since such assays would be based upon a finding that extracts of either the neat antimineralization agent or of tissue treated with it were cytotoxic or mutagenic. Since neither the agent itself nor the treated tissue as shown to be cytotoxic or mutagenic, reproductive/developmental testing was waived.

Subchronic/Chronic Toxicity.   Subchronic/chronic toxicity testing was not performed since very little to no adverse toxic effects have been attributed to the long-term (over 20 years) implantation of glutaraldehyde-fixed heart valves. In addition, extracts prepared for biological reactivity tests are prepared under conditions that enhance/accelerate leaching and diffusion. As such, the level of extracted materials may not be clinically relevant. Since no acute responses were observed with proper specimen-to-fluid-extraction ratios, subchronic and chronic toxicity testing was not warranted.

7. Conclusion of Biocompatibility Testing

A framework in which to plan the biological evaluation of a new antimineralization-treated bioprosthesis has been provided. The appropriate selection and interpretation of these tests requires an understanding of the rationale behind such testing as well as an informed decision weighing the advantages/disadvantages of various material and test procedures.

It was demonstrated that a change in bioprosthesis design (e.g., stented to stentless) involved not only implanting a device treated with a new antimineralization agent, but one with a relatively larger mass of glutaraldehyde-fixed tissue. It was determined that residual fixative in aqueous extracts of clinically rinsed stentless valves can elicit an acute toxic response if the extraction/dilution parameters are not relevant to the clinical conditions. Stented valves, tested using "classic" (more concentrated) extraction conditions, did not elicit toxic or irritant responses due to the lower ratio of the tissue to the frame, thus allowing more rapid washout of the potential toxic load at implant.

III. SUMMARY AND DISCUSSION

Noncalcifying heart valve bioprostheses would be of great benefit to cardiac surgery because they would increase the quality of life of many patients because of the high risk, and fast rate, of calcification of the glutaraldehyde-fixed valve, particularly in young children, and because of the anticoagulation therapy associated with the mechanical valve. Over the years, extensive research has been conducted toward providing antimineralization treatments. As a result, many treatments have been proposed that focus on one or more of the causative factors suspected to initiate the calcification process. Many of these processes, although initially promising in certain *in vitro* and *in vivo* models of calcification, have fallen short after further testing, underscoring the frustrations and difficulty of producing a treatment that not only remains efficacious when going from *in vitro* and subdermal models of calcification to long-term circulatory models in larger animals, but is also proven to be biocompatible, not toxic, not mutagenic, and to preserve the structure, function, and durability of the original glutaraldehyde-fixed valve.

These points were illustrated in the evaluation of 2-amino oleic acid (AOA) as an antimineralization treatment in Freestyle porcine aortic root prostheses. All *in vitro* and *in vivo* testing performed indicated that AOA-treated valves would be safe and efficacious for implantation, yet the *in vitro* biocompatibility studies initially showed the product to be cytotoxic. Understanding the product configuration, a whole porcine aortic root as compared to a stented valve, and its potential clinical use allowed for the appropriate selection and interpretation of testing. As shown in the reevaluation of the cytotoxicity testing and the health risk assessment performed, more clinically relevant glutaraldehyde levels, 100 times lower, were no longer found to be toxic. This information corresponds well with the clinical and experimental knowledge gained over the past 20 years in the use of glutaraldehyde-fixed bioprostheses, showing them to be safe and efficacious as heart valve replacements.

AOA is an effective agent in inhibiting cuspal mineralization. However, based on data from 20-week sheep implants, the potent effect of AOA was restricted to the cusp, and tissue of the supporting aortic wall had mineralized as much as nontreated tissue. Based on our current theories of the mechanism of action for AOA and our knowledge of the amino-aldehyde linkages between the tissue and AOA, our research has led to efforts for increasing the level of bound AOA in the aortic wall of Freestyle valves. This attempt has been partially successful since a significant decrease in the level of calcification was obtained in those valves treated with increased amounts of AOA incorporated into aortic wall tissue [48], which lends strength to the possibility that an increasing aortic wall AOA concentration to levels similar to cuspal tissue may possibly lead to a totally noncalcifiable prosthesis.

Strategic in the development of effective antimineralization treatment is the mechanical performance properties of the tissue. The role that stress plays in the mineralization process has not been fully elucidated, but the experimental evidence to date suggests that it is important to maintain treated tissue with structural and functional properties as close to native tissue as possible. The effect of AOA treatment on the structural integrity and physical properties of the resultant tissue has been monitored throughout the development process. The current treatment process has yielded protheses that have fully maintained this structural integrity, as evidenced by the results of the *in vitro* performance testing. All AOA-treated valves tested in this study functioned normally when compared with nontreated controls, a fact further demonstrated by the excellent performance of the AOA-treated valves in the 20-week sheep implant model.

Antimineralization treatments must also be proven to be biocompatible. Failure in any one of the areas of the testing would have adverse results on the *in vivo* performance of the bioprosthetic device. Therefore, testing should be performed on samples that are of clinical quality and in their "as used" state, emphasizing that, although the agent used in the antimineralization treatment may be safe, it is the device submitted to the treatment process that must be evaluated. Product design is also a vital consideration when evaluating the results of biocompatibility testing. This has been clearly demonstrated in our toxicity studies of AOA, for which a modification in prosthetic design (stentless versus stented) could affect the biocompatibility testing outcome. The results of our studies demonstrate that the AOA treatment process is safe and efficacious.

Development of a viable antimineralization treatment process does not merely rely on the selection of a procedure that fulfills the requirements of a single safety test. It is a multifactorial task, and many aspects of biochemistry, physiology, material interactions, tissue structure and performance, host response (biocompatibility), and durability must be considered. The development of AOA as an antimineralization process has shown that it is possible to treat glutaraldehyde-fixed porcine aortic valves effectively without compromising the structural integrity, performance, and durability of the valve. As the development of the AOA-treated Freestyle bioprosthesis continues, it has entered into its next phase of development, which is the clinical evaluation. It is in this final evaluation that any antimineralization treatment will ultimately be tested and proven.

ACKNOWLEDGMENTS

The authors wish to thank Dr. Frederick J. Schoen (Brigham and Women's Hospital, Boston, MA) for providing radiographs and histological photographs with interpretation from Freestyle valves from the sheep implant study; Dr. Robert J. Levy and Dr. Weiliam Chen (University of Michigan, Ann Arbor, MI) for the calcium analysis and kindly providing to us unpublished results on the AOA mechanism; and Materials Analytical Services (Norcross, GA) for the scanning electron microscopy studies. We are also grateful to Medtronic Heart Valve Division, Irvine, California, for providing photographs of retrieved valves. The research on AOA was funded in part by Medtronic, Incorporated, and a grant from the National Heart, Lung, and Blood Institute, National Institutes of Health, R43 HL42748-01.

REFERENCES

1. Foster, A.H., G.J. Greenberg, D.J. Underhill, C.L. McIntosch, and R.R. Clark. Intrinsic failure of Hancock mitral bioprostheses: 10- to 15-year experience, *Ann. Thorac. Surg.*, 44: 568–577 (1987).
2. Jamieson, W.R.E., V. Gallucci, G. Thiene, et al. Porcine valves, in *Replacement cardiac valves* (Eds. E. Bodnar and R. Frater), New York, Pergamon Press, pp. 229–275 (1991).
3. Schoen, F.J., R.J. Levy, and H.R. Piehler. Pathological considerations in replacement heart valves, *J. Soc. Cardiovasc. Pathol.*, 1:29–52 (1992).
4. Schoen, F.J., and R.J. Levy. Bioprosthetic heart valve failure: Pathology and pathogenesis, *Cardiol. Clinics*, 2:717–739 (1984).
5. Schoen, F.J. Cardiac valve prostheses: Pathological and bioengineering considerations, *J. Cardiac Surg.*, 2:65–108 (1987).
6. Schoen, F.J., and C.E. Hobson. Anatomical analysis of removed prosthetic heart valves: Causes of failure of 33 mechanical valves and 58 bioprostheses, 1980 to 1983, *Hum. Pathol.*, 16:549–559 (1985).
7. Walley, V.M., P. Giannoccaro, D.S. Beanlands, and W.J. Keon. Death at cardiac catheterization: Coronary artery embolization of calcium debris from Ionescu–Shiley bioprosthesis, *Catheter. Cardiovasc. Diagn.*, 21:92–94 (1990).
8. Golomb, G., F.J. Schoen, M.S. Smith, J. Linden, M. Dixon, and R.J. Levy. The role of glutaraldehyde-induced cross-links in calcification of bovine pericardium used in cardiac valve bioprostheses, *Am. J. Pathol.*, 127:122–130 (1987).
9. Levy, R.J., F.J. Schoen, H.C. Anderson, H. Harasaki, T.H. Koch, W. Brown, J.B. Lian, R. Cumming, and F.B. Gavin. Cardiovascular implant calcification: a survey and update, *Biomater.*, 12:707–714 (1991).
10. Thubrikar, M.J., J.D. Deck, J. Aouad, and S.P. Nolan. Role of mechanical stress in calcification of aortic bioprosthetic valves, *J. Thorac. Cardiovasc. Surg.*, 86:115–125 (1983).
11. Schoen, F.J., R.J. Levy, A.C. Nelson, W.F. Bernhard, A. Nashef, and M. Hawley. Onset and progression of experimental bioprosthetic heart valve calcification, *Lab. Inves.*, 52:523–532 (1985).
12. Levy, R.J., F.J. Schoen, and G. Golomb. Bioprosthetic heart valve calcification: Clinical features, pathobiology and prospects for prevention, *CRC Clinical Reviews in Biocompatibility*, 2:147–187 (1986).
13. Maranto, A.R., and F.J. Schoen. Alkaline phosphatase activity of glutaraldehyde-treated bovine pericardium used in bioprosthetic cardiac valves, *Circ. Res.*, 63:844 (1988).
14. Levy, R.J., F.J. Schoen, W.B. Flowers, and S.T. Staelin. Initiation of mineralization in bioprosthetic heart valves: Studies of alkaline phosphatase activity and its inhibition by $AlCl_3$ or $FeCl_3$ preincubations, *J. Biomed. Mater. Res.*, 25:905–935 (1991).
15. Hirsh, D., F.J. Schoen, and R.J. Levy. Effect of metallic ions and diphosphonates in inhibition of pericardial bioprosthetic tissue calcification and associated alkaline phosphatase activity, *Biomaterials*, 14:371–377 (1993).
16. Pathak, Y.V., J. Boyd, R.J. Levy, and F.J. Schoen. Prevention of calcification of glutaraldehyde pretreated bovine pericardium through controlled release polymeric implants: Studies of Fe^{+++}, Al^{+++}, protamine sulphate and levamisole, *Biomater.*, 11:718–725 (1990).
17. Jones, M., E.E. Eidbo, S.L. Hilbert, V.J. Ferrans, and R.E. Clark. Anticalcification treatments of bioprosthetic heart valves: *In vivo* studies in sheep, *J. Card. Surg.*, 4:69–73 (1989).
18. Webb, C.L., F.J. Schoen, W.E. Flowers, A.C. Alfrey, C. Horton, and R.J. Levy. Inhibition of mineralization of glutaraldehyde-pretreated bovine pericardium by $AlCl_3$: Mechanisms and

comparisons with FeCl₃, LaCl₃, and Ga(NO₃)₃ in rat subdermal model studies, *Am. J. Pathol.*, 138:971–981 (1991).

19. Baldwin, M.T., B.L. Ciesliga, L.D. Barkasi, and C.L. Webb. Long-term anticalcification effect of Fe3+ in rat subdermal implants of glutaraldehyde preserved bovine pericardium, *ASAIO Trans.*, 37:170–172 (1991).

20. Golomb, G., and V. Ezra. Prevention of bioprosthetic heart valve tissue calcification by charge modification: Effects of protamine binding by formaldehyde, *J. Biomed. Mater. Res.*, 25:85–98 (1991).

21. Grimm, M., M. Grabenwoger, E. Eybl, P. Block, M.M. Muller, and E. Wolner. Improved biocompatibility of bioprosthetic heart valves by L-glutamic acid treatment, *Fifth Inter. Symposium Cardiac Bioprostheses*, Avignon, France, p. 72 (May 24–27, 1991).

22. Levy, R.J., M.A. Hawley, F.J. Schoen, S.A. Lund, and P.Y. Liu. Inhibition by diphosphonate compounds of calcification of porcine bioprosthetic heart valve cusps implanted subcutaneously in rats, *Circulation*, 71:349–356 (1985).

23. Levy, R.J., J. Wolfrum, F.J. Schoen, M.A. Hawley, S.A. Lund, and R. Langer. Inhibition of calcification of bioprosthetic heart valves by local controlled-release diphosphonate, *Science*, 228:190–192 (1985).

24. Golomb, G., R. Langer, F.J. Schoen, M.S. Smith, Y.M. Choi, and R.J. Levy. Controlled release of diphosphonate to inhibit bioprosthetic heart valve calcification: Dose-response and mechanistic studies, *J. Controlled Release*, 4:181–194 (1986).

25. Johnston, T.P., C.L. Webb, F.J. Schoen, and R.J. Levy. Site-specific delivery of ethanehydroxy diphosphonate from refillable polyurethane reservoirs to inhibit bioprosthetic tissue calcification, *J. Controlled Release*, 25:227–240 (1993).

26. Hirsch, D., J. Drader, Y.V. Pathak, R. Yee, F.J. Schoen, and R.J. Levy. Synergistic inhibition of the calcification of glutaraldehyde pretreated bovine pericardium in a rat subdermal model by FeCl₃ and ethanehydroxydiphosphonate — Preincubation and polymeric controlled-release studies, *Biomater.*, 14:705–711 (1993).

27. Lio, K., E. Seifter, S.M. Factor, E.L. Yellin, and R.W.M. Frater. Diphenylhydantoin inhibits calcification of bovine pericardial implants and myocardium: A preliminary study, *J. Surg. Res.*, 53:349–356 (1992).

28. Nimmi, M.E., D. Ertl, J. Villanueva, and B.S. Nimmi. Inhibition of ectopic calcification of glutaraldehyde crosslinked collagen and collagenous tissues by a covalently bound diphosphonate (ADP), *Am. J. Cardiovasc. Pathol.*, 3:237–245 (1990).

29. Nimmi, M.E., D. Cheung, B. Strates, M. Kodama, and K. Sheikh. Chemically modified collagen: A natural biomaterial for tissue replacement, *J. Biomed. Mater. Res.*, 21:741–771 (1987).

30. Girardot, J.-M., J.P. Gott, R.A. Guyton, and F.J. Schoen. A novel technology inhibits calcification of bioprosthetic heart valves: A sheep pilot study, *Trans. Society for Biomaterials*, 14:60 (1991).

31. Girardot, M.N., J.-M. Girardot, and F.J. Shoen. Alpha-aminooleic acid, a new compound, prevents calcification of bioprosthetic heart valves, *Trans. Society for Biomaterials*, 14:114 (1991).

32. Gott, J.P., Pan-Chih, L. Dorsey, J.L. Jay, G.K. Jett, F.J. Schoen, J.-M. Girardot, and R.A. Guyton. Calcification of porcine valves: A successful new method of antimineralization, *Ann. Thorac. Surg.*, 53:207–216 (1992).

33. Hall, J.D., J. Whitlark, S. Horsley, L. Dorsey, Pan-Chih, J.-M. Girardot, M.N. Girardot, F.J. Schoen, J.P. Gott, and R.A. Guyton. Antimineralization of bioprostheses: An improved amino oleic acid technique, *ASAIO Meeting*, New Orleans, p. 93 (1993).

34. Levy, R.J., F.J. Schoen, and S.L. Howard. Mechanism of calcification of porcine bioprosthetic aortic valve cusps: Role of T-lymphocytes, *Amer. J. Cardio.*, 52:629–631 (1983).

35. Levy, R.J., F.J. Schoen, J.T. Levy, A.C. Nelson, S.L. Howard, and L.F. Oshry. Biologic determinants of dystrophic calcification and osteocalcin deposition in glutaraldehyde-preserved porcine aortic valve leaflets implanted subcutaneously in rats, *Amer. J. Pathol.*, 113:143–155 (1983).

36. Hirsh, D., J. Drader, T.J. Thomas, F.J. Schoen, J.T. Levy, and R.J. Levy. Inhibition of calcification of glutaraldehyde pretreated porcine aortic valve cusps with sodium dodecyl sulfate: Preincubation studies and controlled release studies, *J. Biomed. Mater. Res.*, 27: 1477–1484 (1993).

37. Fishbein, F., R.J. Levy, A. Nashef, V.J. Ferrans, L.C. Dearden, A.P. Goodman, and A. Carpentier. Calcification of cardiac valve bioprosthesis: Histologic, ultrastructural, and biochemical studies in a subcutaneous implantation model system, *J. Thorac. Cardiovasc. Surg.*, 83:602–609 (1982).

38. Arbustini, E., M. Jones, and R.D. Moses. Modifications by the Hancock T6 process of calcification of bioprosthetic cardiac valves implanted in sheep, *Am. J. Cardiol.*, 53:1388–1396 (1984).

39. Barnhart, G.R., M. Jones, T. Ishihara, A.M. Chavez, D.M. Rose, and V.J. Ferrans. Failure of porcine aortic and bovine pericardial prosthetic valves: An experimental investigation in young sheep, *Circulation*, 66(1):150–153 (1982).

40. Gallo, I., F. Nistal, E. Artinano, D. Fernandez, R. Cayon, M. Carrion, and V. Garcia-Martinez. The behavior of pericardial versus porcine valve xenografts in the growing sheep model, *J. Thorac. Cardiovasc. Surg.*, 93:281–290 (1987).

41. Brown, J.W., C.A. Salles, and M.M. Kirch. Extraanatomical bypass of the aortic root: An experimental technique, *Ann. Thorac. Surg.*, 24:433–438 (1977).

42. Brown, J.W., and M.M. Kirch. Technique for insertion of apicoaortic conduit, *J. Thorac. Cardiovasc. Surg.*, 76:90–92 (1978).

43. Cooley, D.A., and J.C. Norman. Apical left ventricular-abdominal aortic composite conduits for left ventricular outflow obstructions, *Cardiovascular Diseases, Bulletin of the Texas Heart Institute*, 5:112–127 (1979).

44. Brown, J.W., D.A. Girod, R.A. Hurwitz, R.L. Caldwell, A.P. Rocchini, D.M. Behrendt, and M.M. Kirch. Apicoaortic valved conduits for complex left ventricular outflow obstruction, *Ann. Thorac. Surg.*, 38:162–168 (1984).

45. Murray, J.F. Valve placement in the ventricular apex for complicated left ventricular outflow obstruction, *Ann. Thorac. Surg.*, 25:368–371 (1978).

46. Norman, J.C., M.R. Nihill, and D.A. Cooley. Valved apico-aortic composite conduits for left ventricular outflow tract obstructions. A 4 year experience with 27 patients, *Am. J. Cardiol.*, 45:1265–1271 (1980).

47. Schoen, F.J., J.W. Tsao, and R.J. Levy. Calcification of bovine pericardium used in cardiac valve bioprostheses. Implication for the mechanisms of bioprosthetic tissue mineralization, *Am. J. Pathol.*, 123:134–145 (1986).

48. Girardot, M.N., M. Torrianni, and J.-M. Girardot. Effect of AOA on glutaraldehyde-fixed bioprosthetic heart valve cusps and walls: Binding and calcification studies, *Int. J. Artif. Organs*, 17:76–82 (1994).

49. Schryer, P.J., E.R. Tomasek, J.A. Starr, and J.T.M. Wright. Anticalcification effect of glutaraldehyde-preserved valve tissue stored for increasing time in glutaraldehyde, in *Biologic and Bioprosthetic Valves*, Proceedings of the Third International Symposium, A. Bodnar and M. Yacoub, eds., New York, Yorke Medical Books, pp. 471–472 (1986).

50. Dewanjee, M.K., E. Solis, S.T. Lanker, G.M. Lombardo, C. Tidwell, R.D. Ellefsen, and M.P. Kaye. Effect of diphosphonate binding to collagen upon inhibition of calcification and promotion of spontaneous endothelial cell coverage on tissue valve prostheses, *ASAIO Trans.*, 32:24–29 (1986).

51. Webb, C.L., F.J. Schoen, and R.J. Levy. Covalent binding of aminodiphosphonate to glutaraldehyde residues in pericardial bioprosthetic tissue: Stability and calcification inhibition studies, *Exp. Mol. Pathol.*, 50:291–302 (1989).

52. Golomb, G., M. Levi, and J.M. Vangelder. Controlled release of bisphosphonate from a biodegradable implant — Evaluation of release kinetics and anticalcification effect, *J. Applied Biomaterials*, 3:23–28 (1992).

53. Tsao, J.W., F.J. Schoen, R. Shankar, J.D. Sallis, and R.J. Levy. Retardation of calcification

of bovine pericardium used in bioprosthetic heart valves by phosphocitrate and a synthetic analogue, *Biomater.*, 9:393–397 (1988).

54. Tan, W.M., W.K. Loke, B.L. Tan, A. Wee, E. Khor, and K.S. Goh. Trivalent metal ions in the prevention of calcification in glutaraldehyde treated biological tissues. Is there a chemical correlation? *Biomater.*, 14:1003–1007 (1993).

55. Korn, A.H., S.H. Feairheller, and E.M. Filachione. Glutaraldehyde: Nature of the reagent, *J. Mol. Biol.*, 65:525–529 (1972).

56. Woodroof, E.A. Use of glutaraldehyde and formaldehyde to process tissue heart valve, *J. Bioeng.*, 2:1–9 (1978).

57. Cheung, D.T., and M.E. Nimmi. Mechanism of cross-linking of proteins by glutaraldehyde: I reaction of model compounds, *Connect. Tissue Res.*, 10:187–199 (1982).

58. Chen, W., J.D. Kim, F.J. Schoen, and R.J. Levy. Effect of 2-amino oleic acid exposure conditions on the inhibition of glutaraldehyde cross-linked porcine aortic valves, *J. Biomed. Mater. Res.*, 28:1485–1495 (1994).

59. Association for the Advancement of Medical Instrumentation/International Organization for Standardization (AAMI/ISO), AAMI/ISO Document 10993-1, *American National Standard on Biological Evaluation of Medical Devices, Part 1: Guidance on Selection of Tests*, Arlington, Virginia.

60. US Department of Human Health Service, Public Health Service (HHS/PHS), Food and Drug Administration, *Draft Replacement Heart Valve Guidance*, Version 4, (October 1994) FDA Center for Devices and Radiological Health, Rockville, Maryland.

61. Association for the Advancement of Medical Instrumentation/International Organization for Standardization, AAMI/ISO Document 5840, *Cardiovascular Implants—Cardiac Valve Prostheses*, Arlington, Virginia.

62. European Committee for Standardization, Document CEN/TC 285 WG3 TF 1, *Cardiac Valves*, Cen. P.O. Box 5059, 2600 GB Delft, The Netherlands.

63. US Department of Human Health Service, Public Health Service (HHS/PHS), Food and Drug Administration, *Guideline on Validation of the Limulus Amebocyte Lysate Test as an End-Product Endotoxin Test for Human and Animal Parenteral Drugs, Biological Products, and Medical Devices* (December 1987) FDA Center for Devices and Radiological Health, Rockville, Maryland.

64. Spilezewski, K.L., J.M. Anderson, R.N. Schaap, and D.D. Salomon. *In vivo* biocompatibility of catheter materials, *Biomater.*, 9:253–256 (1988).

65. Mantovani, F., W. Marooni, L. Coglia, and G. Togna. A simple method for *ex vivo* evaluation of biomaterial interaction with blood platelets, *Int. J. Artific. Org.*, 3(5):305–310 (1980).

IX
OCULAR APPLICATIONS

45
Collagen as an Ophthalmic Biomaterial

Dale P. DeVore
Autogenesis Technologies Inc.
Acton, Massachusetts

I. INTRODUCTION

The collagens are a family of proteins, ubiquitous in nature and primary structural components of all connective tissues. Thus far, 13 different collagen gene products have been identified and characterized. These are listed in Table 1. The most abundant collagen is Type I, found in high concentrations in skin, bone, tendon, ligament, vascular tissue, and cornea. Type I collagen is easily extracted from skin and tendon sources, and has been used as a biomaterial for a variety of medical devices including surgical hemostats, injections for correction of dermal imperfections, artificial skin, sutures, dental implants,

Table 1 Family of Collagen Proteins

Type	Form	Location
I	Fibrillar	Skin, bone, tendon, ligament, cornea
II	Fibrillar	Hyaline cartilage, vitreous
III	Fibrillar	Skin, blood vessels, lymph nodes
IV	Nonfibrillar	Basement membrane, placenta
V	Small fibers	Skin, tendon, smooth muscle, cornea
VI	Microfibrils	Aorta, skin, muscle, tendon
VII	Dimer	Anchoring filaments
VIII	?	Endothelial cells, descemet's membrane
IX	Short chains	Cartilage, cornea, vitreous
X	Short chains	Hypertrophic cartilage
XI	Small fibers	Cartilage
XII	Short chains	Skin, tendon, ligaments, periosteum
XIII	?	Epidermis, bone

and powder to treat ulcerations. Ophthalmic implants fabricated from collagen are currently limited commercially to eye shields and punctal plugs for treatment of dry eye [1–5]. One other type of collagen, Type IV, has been extracted from placenta and formulated for use as a viscoelastic solution for ophthalmic surgery [6].

The ready availability of type I collagen from tendon and skin of animal sources and the ability to chemically derivatize soluble collagen make this unique protein a good candidate for use in fabricating new ophthalmic devices and implants [7,8].

II. COLLAGEN EXTRACTION, PURIFICATION, AND DERIVATIZATION

Type I collagen is easily extracted from bovine tendon or skin by treatment with proteolytic enzymes in organic acids. The most common method for extraction of soluble, Type I collagen is by treating skin or tendon with pepsin in 0.5 N acetic acid. Pepsin digests portions of the nonhelical peptides of collagen molecules, allowing the helical chains to separate and disperse in solution. Type I collagen can be purified from other constituents by differential salt precipitation or column chromatography. The resultant pure collagen solution may then be filter sterilized prior to chemical derivatization.

Acid-soluble collagen molecules have the unique capability to reconstitute as the pH of the solutions is brought to neutrality. Thus, clear solutions become an opaque mass of reconstituted collagen fibrils. Most implant materials useful for ophthalmology must be clear and transparent. Collagen fibrillogenesis at neutral pH can be prevented by chemically modifying the molecule to change its pK_a to either side of neutrality. Such modification, particularly of deprotonated free amines, has been used to prepare clear viscous solutions of collagen at neutral pH [9,10]. Films formed from such solutions have been evaluated as corneal replacements and lenticules for modifying the cornea [11,12].

III. POTENTIAL USES OF COLLAGEN IMPLANTS IN THE EYE

The human eye is schematically shown in Fig. 1. Collagen implants have been developed for the following purposes:

1. Eye shields placed on the surface of the cornea to accelerate epithelial healing and to delivery ocular pharmaceuticals [1,2,13–16]
2. Collagen-based onlay lens to correct refractive errors by changing the curvature of the anterior corneal surface [12]
3. Corneal grafts to replace damaged or diseased cornea [11,17]
4. Intracorneal implants of collagen to replace corneal tissue damaged from trauma or disease [18]
5. Collagen punctal plugs/intracanalicular plugs to treat dry eye conditions by slowing the excretion of tears [4,5]
6. Viscoelastic solutions for use during ophthalmic surgery to protect sensitive ocular tissues and cells from damage resulting from surgical manipulation [6,19]
7. Collagen-based viscous solutions for use as vitreous replacements [20,21]
8. Scleral implants to retard or prevent progressive myopia [22]

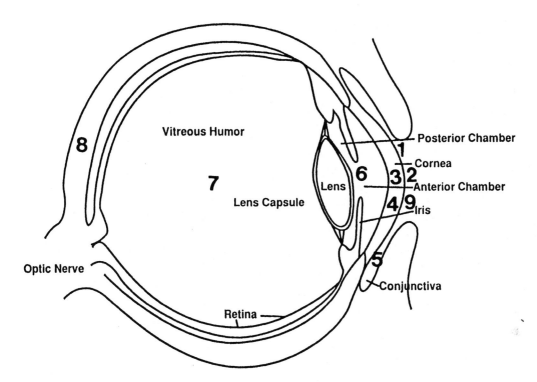

Figure 1   Schematic of the human eye. Locations for collagen implants.

9.   Corneal surface treatment for use during excimer laser keratectomy to provide a smooth corneal surface [23,24]
10. Collagen adhesives to treat epithelial defects and for wound closure [25,25]

Of these possible applications, only eye shields and punctal plugs are currently available commercially. All other applications are in some phase of development or testing. In this chapter, the development and early evaluation of three specific collagen-based materials is presented: (1) collagen smoothing agent for use in excimer laser photoablation, (2) collagen-based corneal grafts to replace damaged cornea, and (3) collagen viscoelastic solution for application in cataract surgery.

A.   Collagen Smoothing Agent for Excimer Laser Photoablation

Excimer laser photoablation procedures employ a 193-nm excimer laser as a surgical tool to ablate or remove a precise amount of tissue from the anterior corneal surface. In addition to its usual application in correcting refractive errors (e.g., myopia, hyperopia, and astigmatism) by altering the curvature of the cornea, this technique is being applied for removal of opacities and irregularities from the corneal surface.

The application of excimer lasers for photoablation has been thoroughly described in the medical literature [26–30]. A far-ultraviolet argon fluoride laser, emitting 193 nm, is successful for corneal photoablation due to its minimal tissue interaction, ablation efficiency, and ease in controlling ablation depth. Moreover, irradiation at 193 nm shows less mutagenic potential compared to longer ultraviolet wavelengths. The extent and depth of ablation depends on a number of parameters including shape and diameter of the laser beam, laser energy fluence (megajoules/square centimeter: MJ/cm^2), duration of irradiation bursts (nanoseconds: nsec), pulse rate, and the number of pulses. For refractive keratectomy, excimer laser ablation should produce a smooth corneal surface allowing reepithelialization, a clear cornea, and an appropriate change in surface curvature to correct refractive errors. For therapeutic keratectomy, excimer laser ablation should only remove the corneal defect without producing photodecomposition of surrounding corneal tissue and without affecting the refractive characteristics of the cornea.

In certain situations, modulators or smoothing agents would be useful in excimer laser photoablation procedures. Surface irregularities are not removed by excimer laser photoablation and become reproduced deeper in the corneal surface. It has been suggested that a fluid absorbing at 193 nm and exhibiting a moderate viscosity to fill in corneal depressions could produce a smooth corneal surface following photoablation [31–33]. The solution or gel would be applied to smooth the corneal surface before laser ablation. In refractive procedures, the agent could act as a mask in facilitating alteration of anterior corneal curvature. In therapeutic procedures, the smoothing agent would protect corneal tissue surrounding defects from unwanted photoablation. In both cases the smoothness of the corneal surface would be improved. The modulator or smoothing agent should be capable of absorbing 193 nm ultraviolet irradiation, flowing evenly across the corneal surface and filling in depressions with little to no meniscus, and resisting rippling effects due to suction or gaseous flow sometimes used with excimer laser photoablation techniques. An additional benefit from a smoothing agent might be its ability to maintain the hydration of the corneal surface. This is of particular importance since it is known that the pulse depth of excimer laser ablation depends on the level of tissue hydration [34,35].

A unique collagen preparation has been formulated that exhibits the functional properties mentioned above. In particular, this collagen preparation is added as a dilute solution and instantly forms a rigid gel upon addition of a cationic buffer solution. Thus, the solution will flow evenly over the corneal surface, filling in depressions, and is then instantly gelled in place to provide resistance to surface imperfections caused by vacuuming or positive air flows.

1.   Methods: Excimer Laser Photoablation

Collagen Smoothing Solutions.   The proprietary collagen solution was prepared from pure atelopeptide collagen extracted from bovine hide. Purified soluble collagen was prepared using standard procedures [36]. Briefly, split calf hides were minced, washed in reagent alcohol, minced into squares of about 1 cm^2, and placed in 20 volumes of 0.5 M acetic acid containing 3% pepsin. Collagen extraction proceeded in an Amicon DC-10 diafiltration system for 72 h, after which the pH was increased to 8.5 to denature the pepsin. Denatured pepsin was removed by centrifugation and the collagen solution reduced in pH to 7.0. Collagen was purified by differential salt precipitation at neutral pH and under acidic conditions. Diafiltration was conducted to remove low molecular weight molecules. The final collagen precipitate was dissolved in 0.1 M acetic acid and sterile filtered through a 0.22-micron cartridge filter. Sterile collagen solutions were processed to remain soluble at physiological pH and to allow spontaneous gel formation when exposed to cationic buffer solutions at room temperature. Solutions were sterilized by filtration through 0.22-micron syringe filters.

The Laser System.   The Taunton Technologies (now VISX, Inc., Sunnyvale, CA) Model LV 2000 laser was used in most experiments. This laser produces 193-nm wavelength output at 10 Hz and was adjusted to deliver a fluence of 100 to 120 mJ/cm^2. The number of pulses was varied depending on the particular experiment. Enucleated porcine or human donor eyes were placed in a holding device and positioned by three-axis alignment. The parameters for ablation were entered into the computer control module and pulses activated using a foot pedal.

Ablation Experiments.

Control (no collagen) vs. collagen smoothing agent. Enucleated porcine eyes were subjected to 600 pulses of 193-nm laser photoablation producing a 5.6-mm ablation zone. Treated eyes were coated with two different concentrations of collagen smoothing agent, 3 mg/ml and 5 mg/ml. A drop of collagen solution was placed on the surface of the de-epithelialized cornea and allowed to flow over the corneal surface. Immediately, a second drop of cationic buffer was applied using the same technique. A rigid gel formed within 30 sec and the eye was subjected to 600 pulses of 193-nm laser photoablation forming a 5.6-mm ablation zone. Specimens were then submitted for scanning electron microscopy (SEM). Micrographs were examined for depth of laser penetration and smoothness of the ablation zone.

Formation of grid patterns on cornea. Stainless steel or plastic grids were placed on the surface of human cornea and exposed to 600 pulses forming a 5.6-mm ablation zone. Corneas were excised and placed in fixative solution for scanning electron microscopy.

Control vs. 5 mg/ml collagen smoothing agent on grid irregularities. Untreated and collagen-coated human cornea and porcine cornea were exposed to 600 pulses in an ablation zone of 5.6-mm diameter. Corneas were then excised and placed in fixative solution for scanning electron microscopy.

Scanning Electron Microscopy. The ablated cornea were removed and placed in 3% glutaraldehyde solution. Specimens were transferred to a mixture of 2% glutaraldehyde and 2% paraformaldehyde in Cacodylate buffer. The specimens were subsequently placed in 1% osmium in Cacodylate buffer, dehydrated in alcohol, subjected to critical point drying in CO$_2$, mounted, coated with gold, and examined using the JEOL 6400 digital scanning electron microscope. Photographs were taken of all specimens.

Effect of Collagen Smoothing Agent on Sandpaper-Roughened Cornea. Sandpaper was used to roughen the surface of cornea of enucleated rabbit eyes. Control cornea were untreated and treated with a 300 pulse ablation. Treated cornea were coated with the collagen smoothing agent and then treated with a 300 pulse ablation. Eyes after excimer laser ablation were examined by slit-lamp microscopy and photographed.

2. Results: Excimer Laser Photoablation

Ablation Experiments.

Control vs. 3 mg/ml and 5 mg/ml collagen smoothing agent. Scanning electron micrographs of control ablation are shown in Fig. 2. The 5.6-mm ablation disk is clear with an ablation depth of about 60–100 μm. Higher-magnification micrographs show a smooth surface in the center of the ablation zone and a relatively smooth transition from ablated to unablated areas [Fig. 2b]. The ablation zone after application of the collagen smoothing agent (3 mg/ml) is shown in Fig. 3. The 5.6-mm ablation disk is visible. However, the depth of ablation is reduced and the ablated zone smoother than that for untreated specimens, as shown in Fig. 3a. Ablation following application of 5 mg/ml collagen is shown in Fig. 4. The ablation depth is even more reduced. It is estimated that the collagen coating reduced ablation penetration by 75–80% on the cornea treated with collagen at 5 mg/ml. Even more dramatic was the smoothness of the corneal

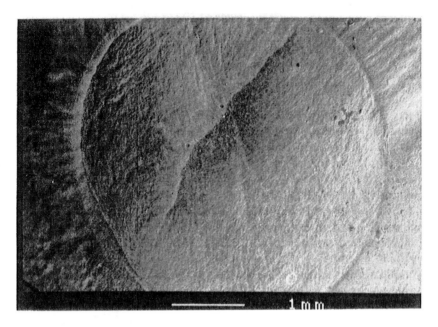

(a)

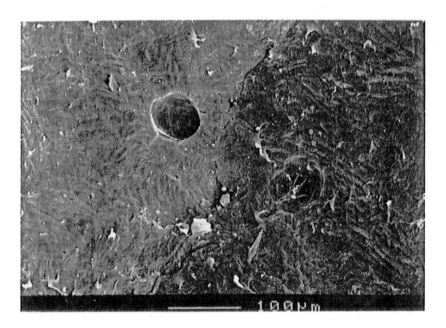

(b)

Figure 2 (a) Untreated porcine cornea ablated using 600 pulses in a 5.6-mm zone. (b) Surface topography of center of ablated zone in (a).

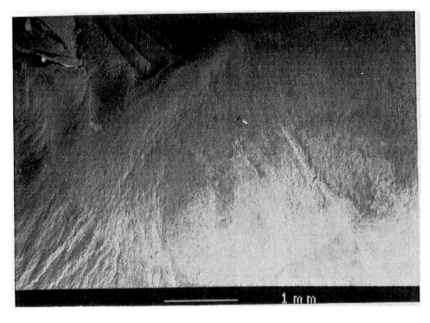

(a)

(b)

Figure 3 (a) Porcine cornea treated with 3 mg/ml collagen gel and ablated using 600 pulses in a 5.6-mm zone. (b) Surface topography of center of ablated zone in 3 mg/ml collagen gel porcine cornea.

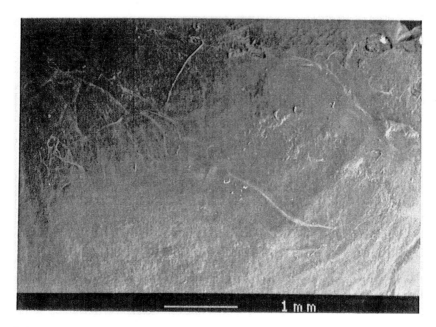

(a)

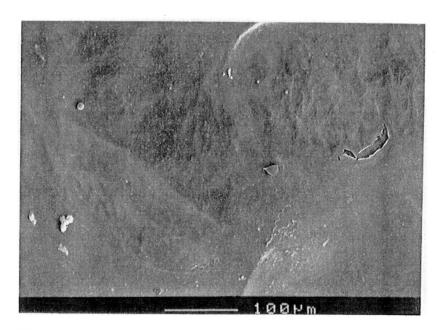

(b)

Figure 4   (a) Porcine cornea treated with 5 mg/ml collagen gel and ablated using 600 pulses in a 5.6-mm zone. (b) Surface topography of center of ablated zone in (a).

surface on the eyes treated with collagen, particularly using the 5-mg/ml formulation (Fig. 4a).

Formation of a grid pattern on cornea. The formation of a plastic screen grid pattern on cornea is shown in Fig. 5. Note the deep crevices produced by ablation of the plastic grid.

Ablation of grid irregularities: control vs. 5 mg/ml collagen smoothing agent. Human cornea and porcine cornea with 300 pulse grid pattern irregularities were again photoablated. Ablation of control (no collagen) porcine eyes with a second 600 pulse burst (Fig. 6) did not remove the surface irregularities, which appeared similar to cornea before ablation. Ablation of collagen-treated human cornea resulted in removal of most of the grid pattern, as shown in Fig. 7.

Effect of collagen smoothing agent on rabbit cornea roughened with sandpaper. Slit-lamp micrographs of untreated and collagen treated rabbit cornea are shown in Figs. 8a and 8b. The smoothing effects of collagen treatment are readily observed.

3. Summary: Excimer Laser Photoablation

Numerous materials have been evaluated as so-called modulators or masking fluids including hydroxymethylpropylcelulose, dextran, carboxymethylcellulose, saline solution, Tears Naturale II, methyl cellulose, sodium hyaluronate, and collagen [8,31–37,38]. Most provide some degree of corneal tissue hydration and exhibit absorbance characteristics at 193 nm although the ablation rate may be dramatically different than that of corneal tissue. Saline and Tears Naturale II exhibit low viscosity, which could prevent effective coating of surface irregularities. The natural polymeric materials can be formulated at nearly any viscosity to optimize flow characteristics. All solutions are subject to ripple and wave irregularities in the presence of air flow or vacuum conditions. Resistance to ripples and waves could be improved by forming a gel on the corneal surface. This

Figure 5 Grid pattern formed by a 300 pulse photoablation of a plastic template on a porcine cornea.

Figure 6   Untreated porcine cornea with grid pattern ablated using 600 pulses.

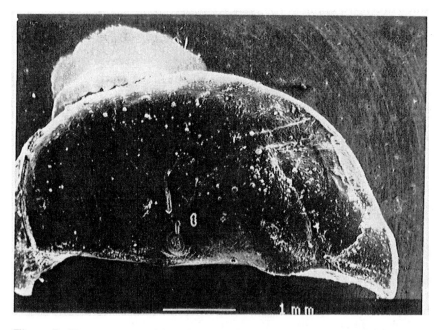

Figure 7   Human cornea with preformed grid irregularities treated with 5 mg/ml collagen gel and ablated using 600 pulses.

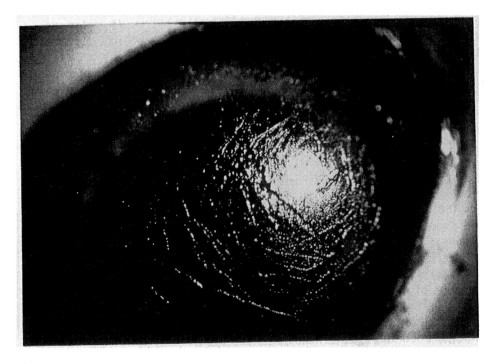

(a)

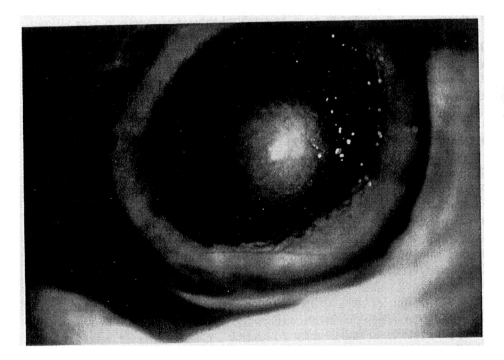

(b)

Figure 8 (a) Slit-lamp photograph of rabbit cornea roughened with sandpaper. (b) Slit-lamp photograph of cornea in (a) following photoablation.

has been accomplished using collagen as a smoothing agent [32,33,37]. Unfortunately, standard neutralized, acid-soluble collagen solutions require up to 45 min to polymerize [7]. In the studies presented herein and previously presented in abstract form, a neutralized collagen solution has been developed that gels instantly upon addition of a cationic buffer solution. Furthermore, this collagen smoothing agent can be formulated to exhibit an ablation rate nearly identical to that of corneal tissue. This collagen formulation appears to exhibit the characteristics required for the optimum smoothing agent. *In vitro* studies demonstrate that the collagen smoothing agent is effective in providing a smooth corneal surface following photoablation of enucleated porcine eyes and human donor cornea with irregular surface properties. It is expected that the collagen material will be equally as effective in improving corneal smoothness following phototherapeutic keratectomy procedures to remove corneal imperfections and opacities.

Thus, the instantly polymerized collagen formulation appears to meet the functional properties desired in a smoothing agent for use in excimer laser photoablation procedures. The formulation exhibits low viscosity when applied to the corneal surface but forms a rigid gel when exposed to cationic solutions, and the formulation can be prepared so that the collagen gel ablates at the same rate as corneal tissue. The gel offers greater resistance to the creation of rippling and waves caused by air currents from gaseous air flow or suction, and due to its high water content would likely maintain hydration of the corneal surface. The clinical significance of these properties will only be realized after evaluation of the collagen smoothing agent in human clinical trials.

B. Collagen Corneal Graft

Pellier de Quengsy first suggested using an artificial substance to restore clear vision to those with opaque corneas in 1771. Materials used for optic cores in keratoprostheses have included glass, polymethylmethacrylate, and polyvinyl alcohol. Materials used as support structures have included metal, ceramic, bone, Teflon, Dacron, and nylon [38–45]. There has been only limited success with these materials and recent work has demonstrated the advantages of using biological materials [46,47]. In this preliminary report, initial experience using collagen-based allografts in the rabbit model is presented.

1. Methods: Corneal Graft

Graft Preparation.   All grafts were composed of collagens extracted from the dermis of New Zealand white rabbits. The clear optic zone contained soluble collagen chemically modified with glutaric anhydride. The fibrous periphery contained intact rabbit collagen fibers immersed in glutaric-derivatized, soluble collagen. Techniques for fabricating the two-part corneal grafts are described in U.S. patent 5,067,961 (1991).

Collagen fibers.   Intact collagen fibers were mechanically dispersed from intact rabbit dermis. Dispersed fibers were subsequently washed with phosphate buffer (0.05 M, pH 7.2), filtered through a 100-micron filter, and concentrated by centrifugation.

Soluble collagen.   Rabbit dermis was harvested and treated in 0.5 N acetic acid containing 3% pepsin. Soluble, telopeptide-poor, predominantly Type I collagen was obtained by differential salt precipitation following pepsin removal. The monomolecular collagen was then sterilized by 0.22-micron filtration.

Modified collagen.   Sterile, soluble collagen was chemically modified at pH 9.0 by well-controlled acylation of a limited number of free amines. This reaction reduces the pK_a to approximately 4.5, providing a physiologically soluble collagen preparation. Modified collagen was precipitated at pH 4.5, washed extensively with sterile water, and reconstituted in phosphate buffer containing a nonionic osmolality enhancer. The viscous preparation was filtered through a 5-micron filter and deaerated by centrifugation.

Graft fabrication. Isolated collagen fibers were dispersed in an equal volume of chemically modified collagen solution. The dispersion was placed in a specially fabricated, polished concave molds, 7 mm in diameter and with base curves from 5.0 to 6.5 mm. The preparation was permitted to partially dehydrate in a laminar flow hood and, at a critical point, a center core of 3 mm was removed and the space filled with modified, soluble collagen. The material was then returned to the laminar flow hood to continue dehydration.

Following complete dehydration, the molds were placed in a controlled atmosphere chamber, devoid of oxygen, and exposed to an 8-W, short-wavelength ultraviolet lamp for 18–24 min. Finally, polymerized, collagen-based grafts were washed with reagent alcohol and placed in sterile phosphate buffer until ready for implantation. The final grafts contained a clear and transparent optic zone of 3 mm diameter and a peripheral, fibrous ring of approximately 2 mm in width. The thickness of the graft ranged from 0.2 to 0.4 mm.

Implantation Techniques. All investigations conformed to the Association for the Research in Vision and Ophthalmology (ARVO) Resolution on the Use of Animals in Research. Collagen grafts were implanted in 12 New Zealand white rabbits weighing 2–3 kg each. The animals were given anesthesia and a 5.5-mm corneal button was trephined from the recipient eye. The collagen graft was trephined to 6 mm. Viscoelastic was placed over the iris and anterior capsule, and the graft secured with 8–16 interrupted 10-0 monofilament nylon sutures. The knots were buried if possible. After surgery, viscoelastic was instilled to retain the anterior chamber. A tarsorrhaphy was performed and the eye treated topically with atropine, dexamethasone, and Polymyxin B.

Graft Evaluation. The grafts were examined daily by slit-lamp microscopy for clarity and extent of reepithelialization. Fluorescein staining was used to estimate epithelialization. The anterior chamber was evaluated for cells and flare, anterior and posterior synechia, and inflammatory membranes. Animals were sacrificed at various time periods using intracardial sodium nembutal. The operated eye was enucleated and immediately placed in Karnovsky's solution. Corneas were removed and processed for light microscopy and for transmission microscopy.

2. Results: Corneal Graft

Slit-Lamp Evaluation.

Graft clarity. Graft clarity varied between animals but generally remained clear up to 24 days postop (Fig. 9). Corneal neovascularization of the peripheral fibrous zone developed in all animals by day 14. Partial reepithelialization of the fibrous zone preceded neovascularization in all of the eyes. Aqueous leaks were observed at the suture sites in 5 of the 12 animals immediately postop and these sealed by day 1 postop. No aqueous leaks developed at the graft–host junction.

Reepithelialization. The extent of reepithelialization of the fibrous periphery increased with implant time. For example, animals sacrificed between 0 and 7 days postop developed 0–70% epithelialization while animals sacrificed at day 17–24 had 100% epithelialization of the fibrous zone. No animals in the study reepithelialized the clear optic core.

Anterior segment inflammation. Trace cell and flare were observed during the first week only. No retrocorneal inflammatory membranes or anterior or posterior synechia developed.

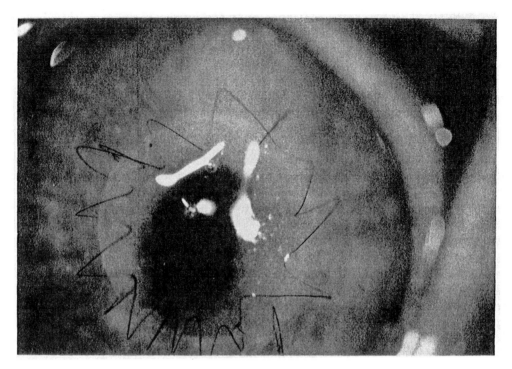

Figure 9   Collagen cornea graft implanted in the rabbit model.

Pathology.

Light microscopy.   Light microscopy confirmed the presence of a continuous epithelial cell layer covering the peripheral graft and host. Migration of fibroblasts from the host to the graft was noted by day 3 (Fig. 10) and was present throughout the peripheral portion of the graft by day 24 (Fig. 11). Fibroblasts were not noted in the clear, optic zone.

Transmission electron microscopy.   TEM demonstrated an amorphous, fine fibrillar appearance in the clear optic core, randomly arranged collagen fibers in the peripheral region, and a regular array of banded collagen fibers at the host–graft junction (Figs. 12–14). The well-organized fiber arrangement in Fig. 14 is characteristic of the collagen lamellae in the corneal stroma.

3.   Summary: Corneal Graft

A number of artificial materials have been used in keratoprostheses. Complications include leakage of aqueous humor around the prosthesis, adjacent tissue necrosis, epithelial downgrowth, prosthesis loosening, inflammatory membranes, and endophthalmitis. Some of these problems are likely secondary to an implant not being compatible with the host tissue. More biocompatible materials have demonstrated fewer complications [46,47].

The ideal donor material would be biocompatible, disease-free, immunologically compatible, biologically stable, and optically clear. The collagen-based allograft appears to meet many of these criteria. The allograft is fully hydrated, containing 90–93% water,

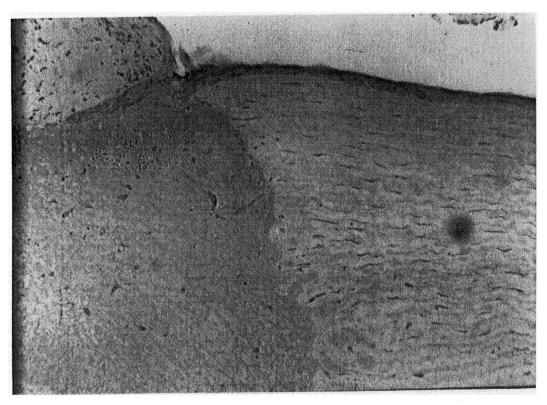

Figure 10   Graft–host junction at day 3. Fibroblasts migrating from host to graft (×42.5).

and has a clear optic zone demonstrating better than 95% transparency over the visible wavelength range. The central core is fully hydrated, acellular, and nonfibrillar, and does not promote nor support fibroblast cell infiltration.

The long-term success of the collagen graft was complicated by the failure to reepithelialize the central optic zone and by apparent thinning and occasional stromal ulceration. The failure to reepithelialize is apparently related to the chemical nature of the collagen used in the optic zone. More recent chemical modifications have been conducted to incorporate reactive groups in the modified collagen. Such preparations formed into lenses of 0.2–0.4 mm thickness have been shown to support epithelial cell spreading and adhesion in *in vitro* studies, with synthesis of hemidesmosomes. Additional formulations have been prepared to contain cell-binding substances including laminin and fibronectin. Studies are in progress to evaluate lenses prepared from these formulations.

Results from this initial pilot study were encouraging, particularly with respect to the rapid reepithelialization and fibroblast ingrowth of the fibrous peripheral region. There was no thinning or ulceration at the graft–host junction and no severe, or even moderate, intraocular inflammation. New generations of the collagen-based allograft are being fabricated for a second series of studies in the rabbit model.

Figure 11   Graft–host junction at day 24. Large fibroblast infiltrate in the graft with indications of attempts of collagen fiber organization.

C.   Viscoelastic Collagen Solution

Viscoelastic solutions are routinely used in cataract surgery procedures to protect sensitive ocular tissues from traumatic damage during surgical manipulation [48–50]. These solutions are used in practically all cataract procedures, which now total about 1.8 million each year in the United States alone. Commercially available solutions are composed of hyaluronan, hyaluronan plus chondroitin sulfate, and hydroxypropylmethyl cellulose. Solutions composed of acrylamide polymers are no longer available on a commercial basis while formulations containing Type IV collagen have only been evaluated in limited clinical investigations [6].

Acid-soluble, Type I collagen can be chemically derivatized, rendering it soluble at neutral pH and physiological osmolality. Solutions of chemically modified collagen at concentrations of about 1% are clear, transparent, and viscous, and might be candidates for so-called viscoelastic solutions.

1.   Methods: Viscoelastic Solution

Preparation of Collagen.   Type I collagen was extracted from bovine hide using standard procedures. Hide was cleaned and washed extensively with sterile, pyrogen-free water and

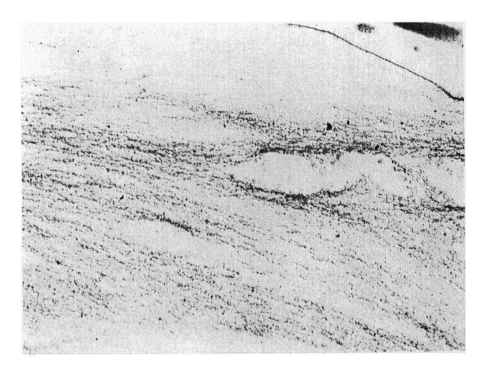

Figure 12   Transmission electron micrograph of the optic core of the collagen graft containing a fine granular material.

then with reagent alcohol. The washed hide was cut into squares of about 1 cm^2 and then placed in 20 volumes of 0.5 N acetic acid. After swelling the hide pieces for 6 h, pepsin (Sigma) was added to 2% of the wet hide weight. The swollen hide pieces were stirred at room temperature for 48 hours to digest collagen into predominantly monomolecular units. Type I collagen was purified using differential salt precipitation, and the final acetic acid solution was passed through a 0.22-micron filter to sterilize the collagen solution.

   Collagen derivatization.   Filtered collagen solutions at approximately 3 mg/ml in 0.1 N acetic acid were chilled to 4°C and brought to pH 9.0 using 10 N NaOH and 1 N NaOH. Excess glutaric anhydride was added (10–20% by weight of collagen solids) while maintaining the pH at 9.0. The reaction was allowed to continue for 30 min, after which the solution was again passed through a 0.2-micron filter. The filtered modified collagen was then adjusted to pH 4.3 by addition of 6 N and 1 N HCl to precipitate the derivatized collagen. The precipitate was recovered by centrifugation at 20,000 × g and washed three times with deionized water at pH 4.3. The final precipitate was dissolved in 0.004 M phosphate buffer containing a nonionic osmolality enhancer. The final, viscous solution exhibited a neutral pH and physiological osmolality.

Viscosity Evaluation.   Samples were submitted to the University of Minnesota for rheometric analysis using the Rheometries System IV rheometer. Dynamic viscosity was conducted at 20% strain using a 50-mm cone at 0.04 radius angle. Viscosity sweeps were

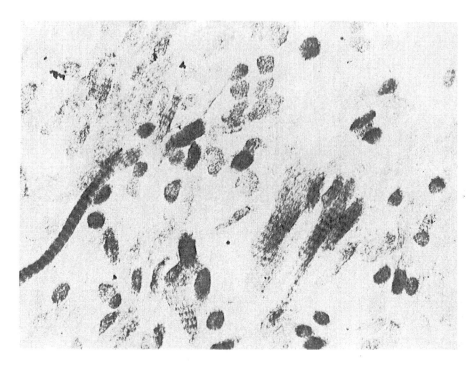

Figure 13   Transmission electron micrograph of the paracentral region of the collagen graft showing numerous banded fibers 133 to 140 nm in diameter.

plotted as dynamic viscosity versus frequency and show both storage (G') and loss (G") modulus curves. Separate sweeps were performed to relate viscosity to shear frequency. Viscosity was also measured at 24° C using the Brookfield digital viscometer with cone spindle, CP-52, at 0.5 rpm.

Animal Testing.   Viscoelastic collagen lots were evaluated as an anterior chamber injection in rabbit models and compared to commercial products. Test parameters included corneal thickness, intraocular pressure, and ocular irritation and inflammation as graded by the McDonald–Shadduck scoring system. Animals were examined prior to and at 0, 1, 3, 5, 24, and 48 h after injection of test material or commercial viscoelastic into the rabbit eyes. Animal studies were conducted in accordance with Good Laboratory Practices and with the ARVO Resolution on the Use of Animals in Research.

2.   Results: Viscoeleastic Solution

Three lots of viscoelastic collagen (VEC) were prepared for evaluation. The characteristics of these lots are presented in Table 2. All lots were sterile, with pyrogen levels less than 0.15 endotoxin unit (EU)/ml. Material was clear and colorless.

Dynamic Viscosity.   Dynamic viscosity sweeps to measure storage and loss modulus characteristics, and to provide viscosity curves are shown in Figs. 15 and 16. Figure 15 demonstrates the dynamic viscoelastic properties of the collagen solutions. The transition from loss modulus (viscous dissipation) to the storage or elastic modulus occurs sharply as the oscillatory frequency increases. Above this transitional point, the solution's mechanical energy predominates (elastic properties). The viscosity curve shown in Fig. 16

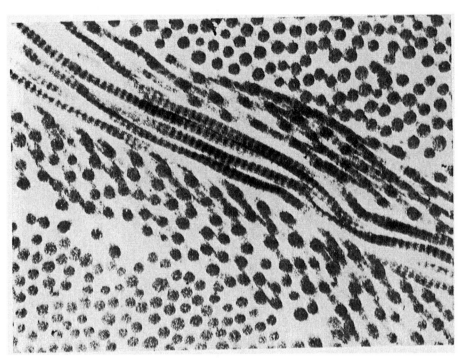

Figure 14 Peripheral fibrous zone of the graft showing a regular array of banded fibers typical of collagen in corneal stroma.

shows decreasing viscosity with increasing shear, demonstrating the thixotropic properties of the collagen solutions. The rheological characteristics of the collagen solutions are very similar to those of high molecular weight hyaluronan.

Animal Evaluations. Viscoelastic collagen samples were evaluated in rabbits and compared to commercial viscoelastic substances, Viscoat® and Healon®. Results for 012290A are shown in Figs. 17–19. Ocular irritation and inflammation, scored using the McDonald–Shadduck system, are shown in Fig. 17. There were no significant differences between scores of VEC and Viscoat up to 48 h. Corneal thickness was well maintained in the VEC sample. Viscoat-treated eyes exhibited some thinning at 1 and 5 h. There were no significant differences in elevation of intraocular pressure (IOP) between the VEC- and Viscoat-treated eyes (Fig. 19). However, pressure elevations were lower in the VEC-treated eyes than in the Viscoat-treated eyes, as shown in Fig. 20.

Table 2 Characteristics of Viscoelastic Collagen Preparations

Lot no.	Viscosity (cps $\times 10^3$)	Concentration (mg/ml)	pH	Osmolality (mOsmol)
101889A	69.2	11.3	7.48	317
12290A	62.8	9.23	7.36	292
12290B	63.5	7.32	7.38	317
12290D	64.6	4.93	7.57	308

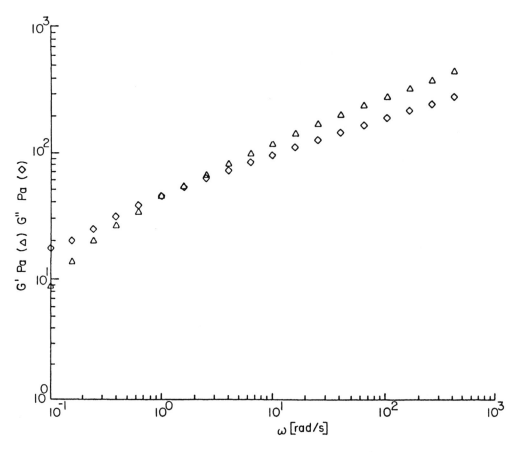

Figure 15 Dynamic viscosity plots of viscoelastic collagen showing transition from viscous (G'') to elastic (G') behavior with increase in strain frequency.

Viscoelastic collagen lot 6989B was compared to Healon in identical evaluations. There were no significant differences in McDonald–Shadduck scores and in corneal thickness between lot 6968B and Healon (Figs. 21 and 22). While intraocular pressures were not significantly different, there was an obvious trend toward lower pressure elevations in the viscoelastic collagen treated eyes. This is particularly evident in Fig. 23, which shows the elevation in IOP above pretreatment levels.

Overall, the viscoelastic collagen lots were equal to and possibly superior to commercial viscoelastics in terms of intraocular pressure elevations and in maintaining corneal thickness.

3. Summary: Viscoelastic Solution

Viscoelastic substances have become an important adjunct to most cataract procedures and to many other ophthalmic procedures: corneal grafting, vitreoretinal surgery, glaucoma surgery, etc. Viscoelastics function to maintain the anterior chamber, separate tissue surfaces, and protect sensitive ocular tissues from surgical trauma. Most viscoelastics perform these functions well. However, most viscoelastics also appear to be associ-

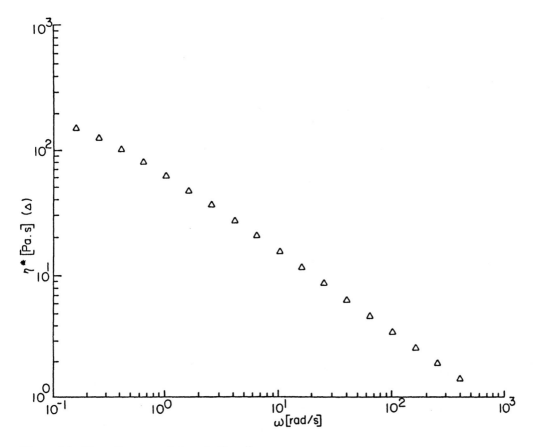

Figure 16 Viscosity curve of viscoelastic collagen.

ated with undesirable elevations in intraocular pressure following surgery, especially in the 2- to 6-h postsurgery time period [51–53]. Such elevations could contribute to development of glaucoma. For this reason, it is always recommended that viscoelastic materials be irrigated at the end of surgery. Even after irrigation, elevated intraocular pressure is common.

Preliminary results presented in this study indicate that specific viscoelastic collagen formulations might provide the functionality of commercial hyaluronan preparations without significant elevations in intraocular pressure. An effective viscoelastic not contributing to pressure elevations would be a significant improvement over current products and will be of great interest to ophthalmic surgeons. Future studies will examine viscoelastic collagen in primates; however, the real clinical benefits of viscoelastic collagen will only be determined in human clinical trials.

V. FINAL COMMENTS

Three different collagen-based formulations proposed for use as ophthalmic implants have been discussed above. The collagen materials exhibit quite different physical and functional characteristics. For example, the collagen smoothing agent is a dilute solution

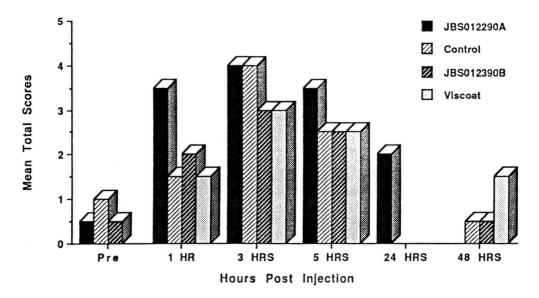

Figure 17   McDonald–Shadduck scores comparing viscoelastic collagen with Viscoat®.

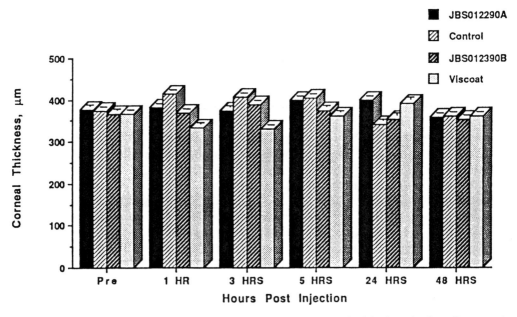

Figure 18   Corneal thickness measurements of rabbit eyes treated with viscoelastic collagen and Viscoat®.

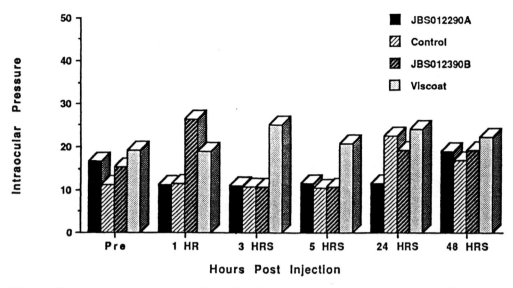

Figure 19 Intraocular pressure readings in rabbit eyes treated with viscoelastic collagen and Viscoat®.

formulated to gel when contacted by a cationic buffer solution. The resultant gel fills in depressions on the corneal surface prior to excimer laser photoablation. Since the collagen gel ablates at rates similar to corneal stroma, a smooth corneal surface is achieved.

The collagen corneal graft is a complex implant composed of chemically derivatized soluble collagen and intact collagen fibers. Chemical derivatization by acylation of deprotonated amine groups produces a new collagen moiety that remains in solution even at physiological pH. This collagen, when molded, dehydrated, and crosslinked, forms a biologically stable and optically clear lens. Biological stability is attained by carefully controlling the degree of acylation. The fibrous "skirt" of the corneal graft is composed

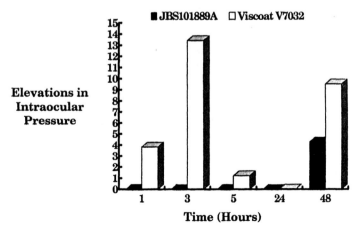

Figure 20 Comparison of IOP elevations in eyes treated with viscoelastic collagen and Viscoat®.

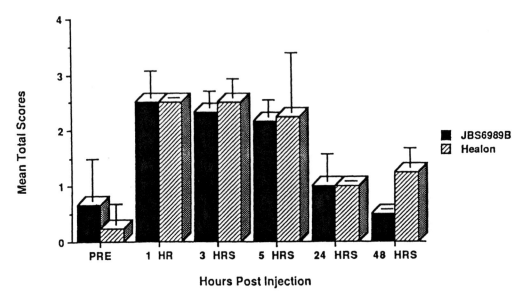

Figure 21   McDonald–Shadduck scores comparing viscoelastic collagen with Healon®.

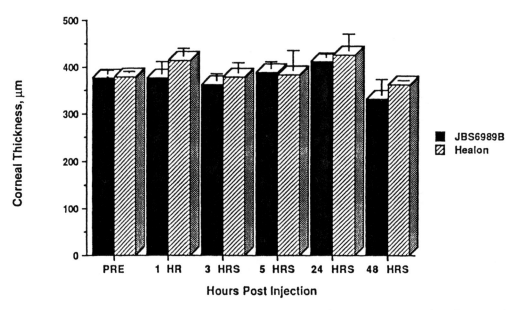

Figure 22   Corneal thickness measurements of rabbit eyes treated with viscoelastic collagen and Healon®.

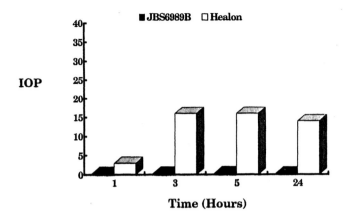

Figure 23   Comparison of IOP elevations in eyes treated with viscoelastic collagen and Healon®.

of intact collagen fibers embedded in the chemically derivatized collagen formulation. The resultant molded graft appears to exhibit many of the physical, biological, and functional properties required for a full-thickness corneal graft.

Finally, chemically derivatized collagens dissolved in a specific, noncationic buffer system exhibit rheological properties similar to high molecular weight hyaluronan solutions. These solutions are functionally equal to or even superior to hyaluronan solutions. In addition, preliminary animal studies show that elevations in intraocular pressure, commonly associated with use of viscoelastic substances, may be significantly reduced with viscoelastic collagen.

Collagen is indeed a versatile biomaterial for preparing a variety of ophthalmic implants. And while only eye shields and punctal/intracanalicular plugs are currently commercialized, new and innovative ways of preparing collagen and of fabricating implant forms will lead to additional collagen-based ophthalmic products in the future, perhaps to replace diseased cornea, to treat myopia, or to provide a vitreous substitute during retinal surgery.

ACKNOWLEDGMENTS

The author wishes to acknowledge the following for their contributions to the development and evaluation of collagen smoothing agents, corneal grafts, and viscoelastic collagen solutions: R. A. Eiferman, M.D. and R. S. Hoffman, M.D., VA Medical Center, Louisville, KY; R. E. Nordquist, Ph.D. and H. Nguyen, University of Oklahoma; E. W. Kornmehl, M.D., M. B. Raizman, M.D., and B. K. Bredvik, M.D., New England Eye Center, Tufts University, Boston, MA; Debra Skelnik, Insight Biomedical, Minneapolis, MN; and J. B. Scott, Autogenesis Technologies Inc., Acton, MA.

REFERENCES

1.   Aquavella, J. V., del Cerro, M., Musco, P. S., Ueda, S., and DePaolis, M. D., The effect of a collagen bandage lens on corneal wound healing: A preliminary report, *Ophthalmology, 18*, 570 (1987).

2. Update I. Clinical experience with the Bio-Cor™ corneal shield, Bausch & Lomb Pharmaceuticals, 1987.

3. Aquavella, J. V., Musco, P. S., Ueda, S., and LoCascio, J. A., Therapeutic applications of a collagen bandage lens: A preliminary report, *CLAO J., 14*, 47 (1988).

4. Punctal occlusion reversible and safe for dry eye management, *Ophthalmology Times*, 1992, p. 17.

5. Temporary Collagen Implants, promotional literature from Lacrimedics, 1985.

6. Bleckmann., H., Vogt, R., and Garus, H.-J., Collagen—a new viscoelastic substance for ophthalmic surgery, *J. Cataract Refract. Surg., 18*, 20 (1992).

7. Rubin, A. L., Miyata, T., and Stenzel, K. H., Collagen: Medical and surgical applications, *J. Macromol. Sci. – Chem., A3*, 113 (1969).

8. Simpson, R. L., Collagen as a biomaterial, *Biomaterials in Reconstructive Surgery* (L. R. Rubin, ed.), C.V. Mosby, St. Louis, MO, 1983, Chap. 11.

9. Miyata, T., Succinylated atelocollagen solution for use in viscosurgery and as a vitreous substitute, U.S. Patent 4,748,152, 1988.

10. DeVore, D. P., Scherrer, R .A., and Scholz, M. T., Viscoelastic collagen solution for ophthalmic use and method of preparation, U.S. Patent 4,851,513, 1989.

11. DeVore, D. P., Kelman, C. D., Stark, W., Green, W. R., Bredvik, B., and Kornmehl, E. W., Evaluation of collagen-based corneal grafts in the rabbit model, *Refract. Corneal Surg., 9*, 208 (1993).

12. Liu, Y. S., Thompson, K. P., Gallitis, R. P., Banks, S. R., and Taylor, G. C., Methods and lenticules for modifying the cornea, U.S. Patent 5,163,956, 1992.

13. Gussler, J. R., Ashton, P., VanMeter, W. S., and Smith, T. J., Collagen shield delivery of trifluorothymidine, *J. Cataract Refract. Surg., 16*, 719 (1990).

14. Murray, T. G., Jaffe, G. L., McKay, B. S., Han, D. P., Burke, J. M., and Abrams, G. W., Collagen shield delivery of tissue plasminogen activator: Functional and pharmacokinetic studies of anterior segment delivery, *Refract. Corneal Surg., 8*, 44 (1992).

15. Hobden, J. A., Reidy, J. J., O'Callaghan, R. J., Insler, M. S., and Hill, J. M., Quinolones in collagen shields to treat aminoglycoside-resistant *Pseudomonal* keratitis, *Invest. Ophthalmol. Vis. Sci., 31*, 2241 (1990).

16. Liang, F.-Q., Viola, R. S., del Cerro, M., and Aquavella, J. V., Non-cross-linked collagen discs and cross-linked collagen shields in the delivery of gentamicin to rabbit eyes, *Invest. Ophthalmol. Vis. Sci., 33*, 2194 (1992).

17. Kornmehl, E. W., Bredvik, B. K., Kelman, C. D., Raizman, M. B., and DeVore, D. P., In vivo *evaluation of a collagen corneal allograft derived from rabbit dermis*, accepted for publication in *Refract. Corneal Surg*.

18. DuPont, D., Gravagna, P., Albinet, P., Tayot, J.-L., Romanet, J.-P., Mouillon, M., and Eloy, R., Biocompatibility of human collagen type IV intracorneal implants, *Cornea, 8*, 251 (1989).

19. DeVore, D. P., Scholz, M. T., Scherrer, R. A., and Mendenhall, H. V., Viscoelastic collagen for use in ophthalmic surgery, in *Proceedings of the First Atlantic Congress on the Future of Collagen*, 1985, p. 253.

20. Stenzel, K. H., Dunn, M. W., Rubin, A. L., and Miyata, T., Collagen gels: Design for a vitreous replacement, *Science, 164*, 1282 (1969).

21. Pruett, R. C., Calabria, G. A., and Schepens, C. L., Collagen vitreous substitute, *Arch. Ophthalmol., 88*, 540 (1972).

22. Safoain, A., Guida, E., Mazzoncini, V., and Menicagli, C., *Sclerotomy with collagenoplasty*, presented at the 7th International Congress on Cataract and Refractive Surgery, Frienze, Italy, June 19, 1993.

23. DeVore, D. P., Scott, J. B., Nordquist, R. E., Hoffman, R. S., Nguyen, M. S., and Eiferman, R. A., *Rapidly polymerized collagen gel as a smoothing agent in excimer laser photoablation*, accepted for publication in *Refract. Corneal Surg.*, 1995.

24. Englanoff, J. S., Kolahdouz-Isfani, A. H., Moreira, H., Cheung, D. T., Nimni, M. E.,

Trokel, S. L., and McDonnell, P. J., *In situ* collagen gel mold as an aid in excimer laser superficial keratectomy, *Ophthalmology, 99*, 1201 (1992).

25. DeToledo, A. R., Witlock, D. R., Kaminski, L. A., and Robin, J. B., Preliminary, evaluation of a new collagen-derived bioadhesive, *Invest. Ophthalmol. Vis. Sci.*, Suppl. 31, 317 (1990). (Abstract)

26. Kelman, C. D., and DeVore, D. P., Collagen-based adhesives and sealants and methods of preparation and use thereof, U.S. Patent 5,219,895, 1993.

27. Sher, N. A., Bowers, R. A., Zabel, R. W., Frantz, J. M., Eiferman, R. A., Brown, D. C., Rowsey, J. J., Parker, P., Chen, V., and Lindstrom, R. L., Clinical use of the 193 nm excimer laser in the treatment of corneal scar, *Arch. Ophthalmol., 109*, 491 (1991).

28. Zabel, R. W., Sher, N. A., Ostrov, C. S., Parker, P., and Lindstrom, R. L., Myopic excimer laser keratectomy: A preliminary report, *Refract. Corneal Surg., 6*, 329 (1990).

29. Marshall, J., Trokel, S., Rothery, S., and Kreuger, R. R., Photoablative reprofiling of the cornea using an excimer laser: Photorefractive keratectomy, *Lasers Ophthalmol., 1*, 21 (1985).

30. Trokel, S. L., Srinivasan, R., and Braren, B., Excimer laser surgery of the cornea, *Am. J. Ophthalmol., 96*, 710 (1983).

31. Kornmehl, E. W., Steinert, R. F., and Puliafito, C. A., A comparative study of masking fluids for excimer laser phototherapeutic keratectomy, *Arch. Ophthalmol., 109*, 860 (1991).

32. Kreuger, R .R., and McDonnell, P. J., New directions in excimer laser surgery, in *Excimer Laser Surgery: The Cornea* (F. B. Thompson and P. J. McDonnell, eds.), Igaku-Shoin, New York, 1993, Chap. 11.

33. Scott, J. B., DeVore, D. P., Eiferman, R. A., and Nordquist, R. E., Collagen modulator fluids for use during photoablation keratectomy, *Invest. Ophthalmol. Vis. Sci., 33*, 765 (1992). (Abstract)

34. Dougherty, P. J., Wellish, K. L. and Maloney, R. K., Excimer laser ablation rate and corneal hydration, *Invest. Ophthalmol. Vis. Sci., 34*, 801 (1993). (Abstract)

35. Russell, R. S., McDonald, M. B., Klyce, S. D., and Varnell, R. J., Excimer laser ablation rate in corneal stroma, *Invest. Ophthalmol. Vis. Sci., 34*, 801 (1993). (Abstract)

36. Miller, E. J., and Rhodes, R. K., Preparation and characterization of different types of collagen, in *Methods of Enzymology* (L. W. Cunningham and D. W. Frederiksen, eds.), Academic Press, New York, 1982, Vol. 82, p. 33.

37. Spadaro, A., Battaglia, F., Mangiafico, S., and Vinciguerra, P., The potential use of sodium hyaluronate as a masking agent in excimer laser ablation and its comparison with hydroxypropylmethylcellulose, *Invest. Ophthalmol. Vis. Sci., 34*, 1245 (1993). (Abstract)

38. Cardona, H., Keratoprosthesis: Acrylic optical cylinder with supporting intralamellar plate., *Am. J. Ophthalmol., 54*, 284 (1962).

39. Dohlman, C. H., Schneider, H. A., and Doane, M. G., Prosthokeratoplasty, *Am. J. Ophthalmol., 77*, 694 (1974).

40. Polack, F. M., and Heimke, G., Ceramic keratoprosthesis, *Ophthalmology, 87*, 693 (1980).

41. Aquavella, J. V., Rao, G. N., Brown, A. C., and Harris, J. K., Keratoprosthesis: Results, complications and management, *Ophthalmology, 89*, 655 (1982).

42. Trinkhaus-Randall, V., Capecchi, J., Newton, A., Vadasz, A., Leibowitz, H., and Franzblau, C., Development of a biopolymeric keratoprosthetic material, *Invest. Ophthalmol. Vis. Sci., 29*, 393 (1988).

43. Sletteberg, O., Hovding, G., and Bertelsen, T., Keratoprosthesis I. Results obtained after implantation of 12 one-piece prostheses, *Acta Ophthalmol., 68*, 369 (1990).

44. Sletteberg, O., Hovding, G., and Bertelsen, T., Keratoprosthesis II. Results obtained after implantation of 27 dismountable two-piece prostheses, *Acta Ophthalmol., 69*, 375 (1990).

45. Caiazza, S., Falcinelli, G., and Pintucci, S., Exceptional case of bone resorption in an osteodonto-keratoprosthesis, *Cornea, 9*, 23 (1990).

46. Kirkham, S. M., and Dangle, M. E., The keratoprosthesis: Improved biocompatibility through design and surface modification, *Ophthalmol. Surg., 22*, 455 (1991).

47. Thompson, K. P., Hanna, K. D., Gipson, I. K., Gravagna, P., Waring, G. O., and Wint-

Johnson, B., Synthetic epikeratoplasty in rhesus monkeys with human type IV collagen, *Cornea, 12,* 35 (1993).

48. Liesegang, T. J., Viscoelastic substances in ophthalmology, *Surv. Ophthalmol., 34,* 268 (1990).

49. Arshinoff, S. A., Viscoelastic substances: Their properties and use when placing an IOL in the capsular bag., *Curr. Can. Ophthalmol. Prac., 4,* 2 (1986).

50. Balazs, E. A., Sodium hyaluronate and viscosurgery, in *Healon: A Guide to Its Use in Ophthalmic Surgery* (D. Miller and R. Stegman, eds.), Wiley, New York, 1993, p. 5.

51. Passo, M. S., Ernest, J. T., and Goldstick, T. K., Hyaluronate increases intraocular pressure when used in cataract extraction, *Br. J. Ophthalmol., 69,* 572 (1985).

52. Barron, B. A., Busin, M., Page, C., Bergsma, D. R., and Kaufman, H. E., Comparison of the effects of Viscoat® and Healon® on postoperative intraocular pressure, *Am. J. Ophthalmol., 100,* 377 (1985).

53. Berson, F. G., Patterson, M. M., and Epstein, D. L., Obstruction of aqueous outflow by sodium hyaluronate in enucleated human eyes, *Am. J. Ophthalmol., 95,* 668 (1983).

46
Biomaterials Used for Intraocular Lenses

F. Richard Christ, Shelley Y. Buchen, Jim Deacon,
Crystal M. Cunanan, Jane Ellen Giamporcaro, Patricia M. Knight,
Joseph I. Weinschenk III, and Stan Yang
Allergan, Inc.
Irvine, California

I. INTRODUCTION

This chapter reviews the materials used in intraocular lenses (IOLs) and the methods used to characterize these materials. The chapter is divided into sections that cover IOL and material performance requirements; traditional and newer materials used in IOLs; evaluation methods for biological, chemical, mechanical, optical, and clinical performance; and a brief discussion of the future of IOL materials and designs. The evaluation sections include test methodologies along with the results of these tests on various IOL materials. This introduction section briefly reviews the basis, history, and development of cataract surgery and IOL designs.

The IOL is a prosthesis for the natural crystalline lens. To fully understand the function of the IOL, it is first necessary to understand the function of the natural lens. The natural lens is located just posterior to the iris (Fig. 1). The lens consists of an outer elastic capsule, a single layer of subcapsular cuboidal epithelial cells underlying the anterior capsule of the lens, and a central predominance of fibers derived from the epithelium. These fibers are composed of a family of proteins known as *crystallin*.

Normal, or *phakic*, vision places three requirements on the crystalline lens: *transparency*, *refractive capability*, and *accommodation*. The transparency of the lens generally deteriorates with age, until the lens becomes opaque and significantly impairs vision. The condition of the lens at this point is referred to as a cataract. The refractive capability of the lens enables a person to see objects at a specific distance. The lens provides approximately 33% of the refractive capability of the eye (the cornea provides the remaining 67%). The ideal refractive condition, called *emmetropia*, enables an in-focus image at 10 meters (m). Refractive deficiencies, or ametropic conditions, include *myopia*, *hyperopia*, and *astigmatism*.

Accommodation is the ability of the crystalline lens to change its shape and achieve

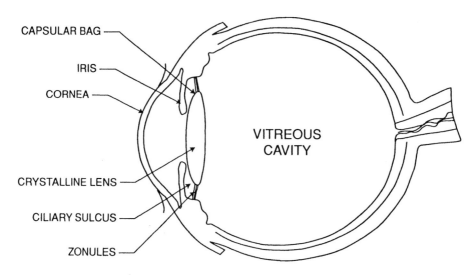

CAPSULAR BAG

IRIS

CORNEA

VITREOUS
CAVITY

CRYSTALLINE LENS

CILIARY SULCUS

ZONULES

Figure 1 Sagittal view of the human eye showing the location of the natural crystalline lens. Cataract is the opacification of the natural lens. In modern surgery, the cataract is removed through an incision peripheral to the cornea.

focus on objects located at different distances from far to near. The condition in which accommodation is significantly diminished is known as *presbyopia*.

The most common type of cataract is age related and occurs most commonly in individuals 60 years or older. Less frequent causes for cataract formation, which can occur at any age, include (1) trauma to the head, (2) exposure to electromagnetic radiation outside the visual range (e.g., microwave), (3) diabetes, and (4) congenital disease. It is estimated that over 15 million people in the world suffer from debilitating cataracts (authors' estimate).

Medical treatment for cataracts existed as far back in time as ancient India when a technique known as *couching* was practiced [1]. Couching involves entering the anterior portion of the eye with a sharp instrument and displacing the entire cataractous crystalline lens into the vitreous cavity and out of the visual axis. Since the mid-1700s, treatment for cataracts has involved removal of the cataract from the eye, rather than couching, in order to restore transparency of the ocular system. Once the cataract is removed, the condition is known as *aphakia*.

Prior to the 1970s, the intracapsular cataract extraction (ICCE) technique was used to remove the entire lens, lens capsule, and associated zonular structures through a large corneal incision. During the 1970s, the technique called extracapsular cataract extraction (ECCE) was popularized. ECCE involves expressing the lens contents through a surgical incision of perhaps 6-10 mm. ECCE leaves the posterior portion of the capsular bag intact as a barrier to the movement of vitreous humor forward into the anterior segment of the eye. While this posterior capsular portion provides structural integrity in many patients, it tends to opacify within a few years after cataract surgery. This condition, known as *posterior capsular opacification* (PCO), is treated by using a Neodymium: YAG (Nd:YAG) laser to perforate this opaque membrane. This perforation, known as a *posterior capsulotomy*, is typically 2 mm to 4 mm in diameter.

A recent ECCE technique, known as *phacoemuisification*, allows the cataractous lens contents to be ultrasonically fragmented and aspirated from the eye. This technique can be accomplished through a much smaller incision than other techniques; approximately 3 mm. The smaller incision allows for much less invasive surgery with the benefits of a potentially safer surgical procedure, more rapid wound healing, faster visual rehabilitation, and better refractive outcome.

Prior to the mid-20th century, surgeons typically prescribed aphakic spectacles to restore refractive capabilities to the aphakic eye. These spectacles were extremely thick, similar in appearance to the bottom of a soda bottle. Aphakic spectacles are still used in developing nations. In most developed nations, they are considered to be cosmetically and functionally unacceptable. Other correction techniques used in the 1970s and early 1980s were contact lenses and refractive keratoplasty, both of which sought to compensate for the loss of natural lens refraction by adjusting the refraction at the cornea. Both of these techniques have been used only to a limited extent, largely due to the widespread acceptance of IOLs.

With the development of the IOL, surgeons are able to surgically restore refractive capabilities to the eye. The first inspiration for IOLs came in 1765 when Tadini, an Italian oculist practicing in Warsaw, reported the concept of implanting a glass ball into the eye to restore refractive power [2]. In 1949 a British ophthalmologist, Harold Ridley, examined an RAF pilot whose cornea had been penetrated by shards from a shattered poly (methyl methacrylate) (PMMA) cockpit canopy [3]. Ridley noted that the PMMA fragments appeared to be well tolerated by the tissue. It occurred to him that a lens made from PMMA might be also be well tolerated in the intraocular environment and could be a suitable prosthesis for an extracted cataractous natural lens. After Ridley developed an IOL design fabricated from PMMA, approximately 1100 of these lenses were implanted. The results were generally unsatisfactory, due to weight, poor design, and poor manufacturing technology [1]. Since the late 1970s, when the implantation of IOLs became widespread due to improved technology, approximately 10 million PMMA IOLs have been implanted worldwide (authors' estimate).

In conjunction with the evolution of surgical techniques and technology (e.g., phacoemulsification) towards less invasive surgery, the IOL designs and materials have also evolved. The early ICCE technique required anterior chamber IOL designs that were implanted anterior to the iris, either with or without attachment to the iris itself (Fig. 2a). Later, the ECCE technique, which preserved the posterior portion of the capsule of the removed natural lens, allowed for posterior chamber IOL designs that could be implanted posterior to the iris (Fig. 2b). The benefit of these lenses was to provide less contact with delicate tissues such as the cornea and iris. Posterior chamber IOLs are now widely preferred over anterior chamber IOLs and account for over 90% of all lenses implanted (authors' estimate). With the transition to small incision surgery, there is currently a trend toward IOLs fabricated from flexible elastomers that can be folded and inserted through smaller surgical incisions of 3–4 mm in length.

All IOL designs consist of an optical portion and a portion used for stable positioning within the eye. These designs are either *plate* designs, usually constructed from one material, or *optic/haptic* designs, in which the optic is traditionally constructed from one material and the haptics are constructed separately from a second material. More recent optic/haptic designs are fabricated from a single material. Figure 3 illustrates various IOL designs. A listing of the major IOL manufacturers is given in Table 1.

IOLs are also categorized as foldable or nonfoldable. Nonfoldable, rigid plastic IOLs

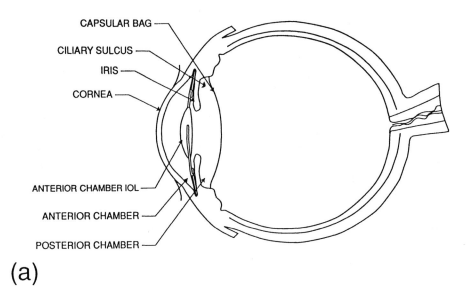

CAPSULAR BAG

CILIARY SULCUS

IRIS

CORNEA

ANTERIOR CHAMBER IOL

ANTERIOR CHAMBER

POSTERIOR CHAMBER

(a)

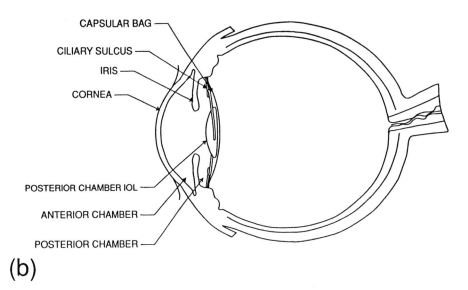

CAPSULAR BAG

CILIARY SULCUS

IRIS

CORNEA

POSTERIOR CHAMBER IOL

ANTERIOR CHAMBER

POSTERIOR CHAMBER

(b)

Figure 2   Sagittal view of the human eye showing the placement of either an anterior chamber or posterior chamber IOL. (a) The anterior chamber IOL lies anterior to the iris. (b) The posterior chamber IOL can sit either in the ciliary sulcus or in the remnants of the capsular bag. Capsular bag placement, which is preferred, is shown here.

are typically made of PMMA although other rigid plastics such as polycarbonate and polysulfone have been considered. Foldable IOLs are made from silicone, acrylic, or hydrogel elastomers and are designed to be folded or otherwise deformed for insertion through a smaller incision than is possible with a rigid, nonfoldable lens.

Before reviewing the details of IOL materials, their properties, and evaluation methodologies, it is important to understand the performance requirements expected for the materials that are used in IOLs. This is the focus of the next section of this chapter.

II. PERFORMANCE REQUIREMENTS FOR IOL MATERIALS

IOL materials must satisfy certain clinical performance requirements for safety and efficacy. Safety-related requirements include biocompatibility, biostability, and various chemical and mechanical criteria. The primary efficacy-related requirements are optical in nature. The various requirements are summarized in Table 2. Laboratory and clinical evaluations are used to measure material performance relative to these requirements.

From a biological perspective, the materials used in the construction of an IOL must be biocompatible — that is, nontoxic, noninflammatory, nonmutagenic, nonsensitizing, and nonirritating — in the intraocular environment. Biocompatibility can be demonstrated through the use of *in vitro* and *in vivo* test methods.

Another requirement is that the material must remain stable to the degradative agents that exist in the eye such as enzymes, aqueous media (hydrolysis), and ultraviolet light. The methods used to screen materials relative to degradation include *in vitro* accelerated aging studies and animal implantation.

Chemically, the materials of construction must be as pure as possible; that is, they must not contain any migratable components that are potentially leachable into the intraocular environment. The method used to verify chemical purity begins with an organic extraction of the material followed by the identification and quantitation of any extractable species. Surface chemistry is evaluated by analytical and biological techniques in order to predict the interaction of the chemical species present in the ocular environment with an implanted material's surface. Finally, IOL biomaterials must be stable towards standard sterilization processes such as ethylene oxide or autoclaving.

Mechanically, the optic and haptics must be designed appropriately and made of suitable materials such that the IOL can be surgically handled, inserted, and manipulated without damage to the IOL or to ocular structures. The material must be resistant to potential damage from procedures such as Nd:YAG posterior capsulotomies. Furthermore, the IOL must remain positionally stable once in the eye. These requirements relate to the basic properties of haptic material flexibility and shape memory, and haptic and optic material strength and surface hardness. These are evaluated in the laboratory and during human clinical investigations of the particular lens designs. Additional mechanical requirements are placed on foldable IOL materials. They must be capable of easy folding with complete return to original prefolded shape, dimensions, and optical performance without damage to the material surfaces.

The efficacy of an IOL relates almost entirely to its optical performance. This distinguishes IOLs from other implantable devices, which typically rely on mechanical performance for efficacy. As noted in the Introduction, the optical objective is to return the ocular system to normal refracting and resolving capabilities. The optic material must be transparent in the visible light range (400–700 nm). At the same time, it should block UV radiation at wavelengths less than 400 nm in order to replace the UV blocking characteristic of the natural lens and thus protect the retina and other structures of the posterior ocular environment. The material must be capable of resolving images to a minimum level when it is fabricated into an appropriate lens body. The refractive index needs to be such that the resultant lens design will have sufficient refracting power while being able to

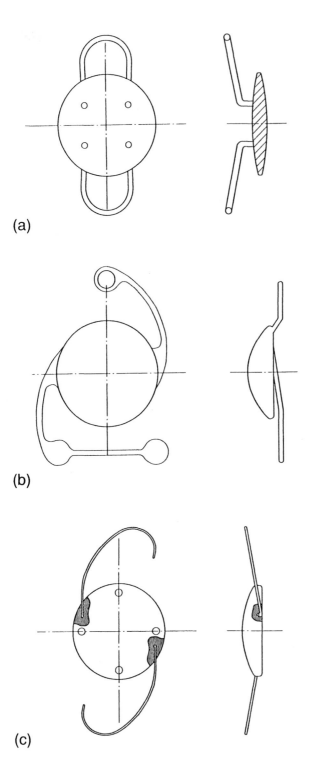

(a)

(b)

(c)

Figure 3 An approximate chronology of IOL styles: (a) iris-fixated lens (mid-1970s, discontinued), (b) one-piece anterior chamber lens (late 1970s), (c) three-piece posterior chamber lens with PMMA optic and poly(propylene) haptics (early 1980s), (d) one-piece foldable posterior chamber lens made from silicone (1984, STAAR), (e) one-piece posterior chamber lens made from PMMA (mid-1980s), (f) three-piece posterior lens with second-generation silicone optic and poly(propylene) haptics (1989, AMO Phacoflex II®).

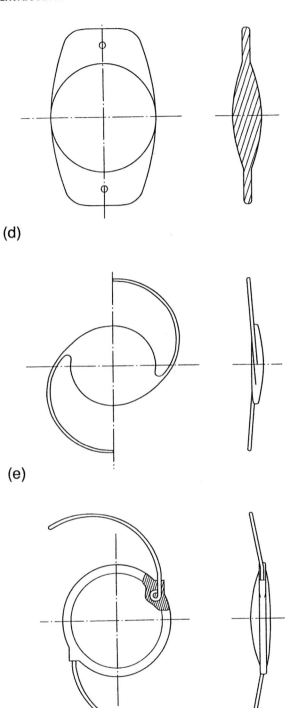

(d)

(e)

(f)

Table 1   A List of Major IOL Manufacturers

Manufacturer	Location	Optic materials
Allergan (AMO)	Irvine, CA	PMMA, silicone, foldable acrylic
Alcon Laboratories, Inc.	Fort Worth, TX	PMMA, foldable acrylic
Chiron IntraOptics	Irvine, CA	PMMA, silicone
Domilens	Lyon, France	PMMA
Hoya	Tokyo, Japan	PMMA
Iolab Corporation	Claremont, CA	PMMA, silicone
Menicon Corporation	Nagoya, Japan	PMMA
Optical Radiation Corp. (ORC)	Azusa, CA	PMMA, acrylic hydrogel
Kabi Pharmacia Ophthalmics	Uppsala, Sweden	PMMA, heparin-coated PMMA
Seed	Tokyo, Japan	PMMA
STAAR Surgical Company	Monrovia, CA	Silicone
Storz Instrument Company	St. Louis, MO	PMMA, acrylic hydrogel

Table 2   Performance Requirements for IOL Biomaterials

Category	Performance requirements
	Safety
Biocompatibility	• Nontoxic
	• Nonmutagenic
	• Nonsensitizing
	• Nonirritating
	• Noninflammatory
Biostability	• Stable to hydrolysis
	• Stable to UV light
	• No enzymatic degradation
Chemistry	• Unbound constituents less than 1%
	• Identify unbound constituents
	• Undetected aqueous extractables
	• Determine polymer molecular weight
	• Identify polymer
	• Sterilization compatibility
	• Evaluate surface chemistry
Mechanical Properties	
All IOLs	• Resistance to surface damage
	• Resistance to haptic breakage (one-piece PMMA)
Additional for foldable IOLs	• Folding/unfolding behavior
	• Resistance to optic deformation
	• Resistance to folding damage
	• Optic recovery after folding
	Efficacy
Optical properties	• Refractive capability
	• Resolution efficiency (>70% in water)
	• Ultraviolet transmission (<1% below 390 nm)

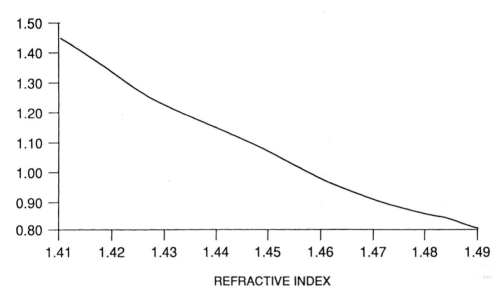

Figure 4 The relationship of IOL center thickness to the refractive index of the optic material for a 20 diopter lens. For foldable IOLs, a thinner optic may provide an opportunity for a smaller surgical incision.

fit within the length of the incision desired and the dimensions of the intraocular spaces. Figure 4 illustrates that materials with higher refractive indexes lead to thinner optics.

III. IOL MATERIALS

Both optic and haptic materials are discussed in this section. Materials that are currently in commercial use as well as those that are currently in human clinical trials are included. Also, materials that were considered for intraocular use but were not pursued are discussed at the end of this section.

The search for suitable materials for intraocular implantation involves either the screening of already available, "off-the-shelf" materials or the synthesis of completely novel materials. If the material is off-the-shelf, prior history as a successful implant material in another location in the body is an advantage.

A. Currently Used Optic Materials

1. PMMA

Rigid PMMA sheet, known commercially as Plexiglas® or Perspex®,* has historically been the material of choice for IOLs. PMMA is a member of the family of addition-polymerized materials known as *acrylics/methacrylics*. The chemical structure of the

---

*Perspex® is a trademark of Imperial Chemical Industries, Ltd.; Plexiglas® is a trademark of Rohm & Haas.

PMMA polymer is shown in Fig. 5. PMMA is typically produced by a free radical initiated polymerization scheme. A generalized scheme is shown in Fig. 6.

The final molecular weight distribution, and consequently, the final mechanical properties, of the PMMA material is dependent upon the conditions of the polymerization process. In general, the greater the average molecular weight, the tougher the material. While there is opportunity to adjust PMMA mechanical performance by changes in polymerization conditions, there is no opportunity to significantly adjust optical properties such as refractive index.

IOLs account for only a small fraction of the total commercial consumption of PMMA materials. PMMA is more generally recognized as the material of which glass-like sheet and other industrial objects are manufactured. The glass-like clarity of PMMA stems from its completely amorphous molecular character. There is virtually no tendency to form microcrystalline domains that would lead to light scatter and haze. In addition to IOLs, PMMA has been used in many medical applications including contact lenses, bone cements, and dental impression materials.

Even though PMMA has been successfully used in IOLs on a regular basis over the past two decades, certain qualities are not completely optimal. First, PMMA cannot be heat sterilized because of its low glass transition temperature (T_g). Ethylene oxide (EtO) gas is the common sterilization technique for PMMA IOLs. Other rigid materials such as polycarbonate and polysulfone have been considered because of their autoclavability. Also, PMMA sheet is brittle and is not inherently compatible with newer one-piece

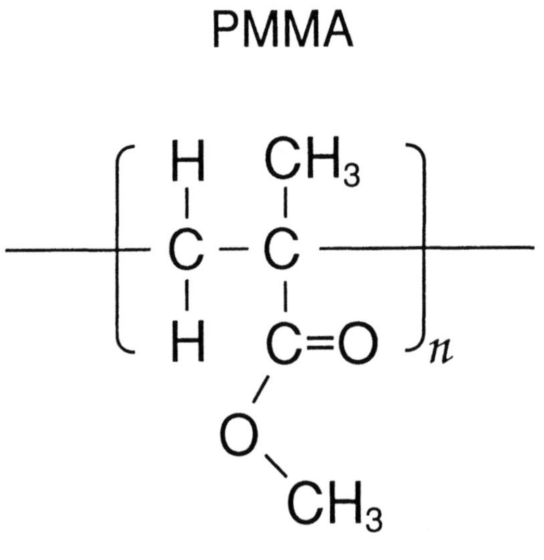

I

Figure 5 The chemical structure of poly(methyl methacrylate) (PMMA) (I). PMMA is a member of the acrylic family of polymers. PMMA, a rigid amorphous plastic, was first used as an IOL biomaterial in 1949. Currently, a high molecular weight grade of PMMA, Perspex CQ® (Imperial Chemical Industries, Ltd., London, England), is most widely used in IOLs.

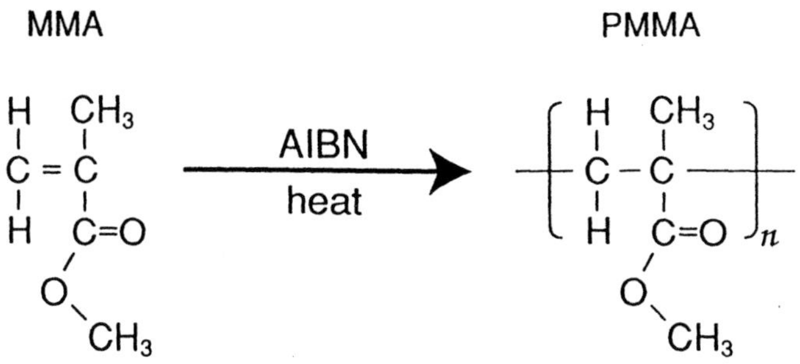

MMA   PMMA

II

Figure 6   A generalized polymerization scheme for PMMA. Methyl methacrylate (MMA) monomer (II) is polymerized using an appropriate initiator such as azo-bis-isobutyryl nitrile (AIBN) via a free-radical addition mechanism.

IOL designs that require flexible, thin-profile haptics (see Fig. 3e). For this reason, polycarbonates and crosslinked PMMAs, with their greater flexibility and impact resistance, have been considered as alternative materials. Postprocessing of the PMMA material to achieve greater molecular alignment, thus enabling its fabrication into a one-piece design, has also been pursued. Overall, however, the balance of properties offered by the PMMA materials has been well suited for IOLs. Thus, PMMA continues to be a popular optic material.

Several IOL fabrication techniques are used with PMMA. Higher molecular weight PMMA materials such as Perspex CQ® are not readily melt processible and therefore are formed into IOLs by mechanical processes such as lathing, milling, and tumble polishing. Most IOL manufacturers use these processes. In the past, certain companies used an injection-molding process to fabricate IOLs using lower molecular weight PMMA resins such as VS-100 [a copolymer of methyl methacrylate (MMA) and ethyl acrylate (EA) from Rohm & Haas, Philadelphia]. Another technique used in the industry involved compression molding of PMMA optic blanks followed by milling of the haptics to form a one-piece IOL. Direct cast molding of MMA monomer was also attempted but was not pursued because of the difficulty of containing the volatile MMA monomer within the molds during polymerization.

2. Silicone Elastomers

As cataract extraction techniques have evolved to smaller and smaller incisions, the need has increased for foldable IOLs that can be inserted through these incisions. Elastomeric materials that return to their original configurations after folding were needed. Silicone elastomers were chosen because they are optically clear materials. Furthermore, solid crosslinked silicone elastomer has been widely used in a variety of implants used in other parts of the body such as finger joints, scleral buckles, heart valves, and hydrocephalic shunts, with excellent biocompatibility and clinical results [4]. Several styles of silicone IOLs have been subjected to extensive human clinical evaluations and are available to the general ophthalmic surgeon. These include one-piece, all-silicone plate lenses (Fig. 3d) and three-piece silicone lenses with either polypropylene, extruded PMMA, or polyamide haptics (Fig. 3f).

Silicones are known chemically as *polysiloxanes* based on their silicon-oxygen molecular backbone (Fig. 7). It is this backbone that confers mechanical flexibility (the T_g for polysiloxane materials can be $-100°C$ or lower) and high vapor permeability to the materials. The backbone is also remarkably stable to oxidation, hydrolysis, and high-energy radiation. This combination of properties has allowed the polysiloxanes to be used as advanced materials in aerospace applications.

Pendant to the siloxane backbone are organic groups such as vinyl, methyl, and phenyl (refer to Fig. 7). These groups determine properties such as refractive index, mechanical strength, and clarity. For example, a higher phenyl content results in a silicone with a higher refractive index. The most commonly used polysiloxane is poly(dimethyl siloxane), or PDMS, in which each silicon backbone atom is appended with two methyl groups. PDMS has applications in industry, residential materials (sealants and protective coatings), and medical implants.

Polysiloxanes are formed by ring-opening condensation polymerization reactions. The *monomers* in these syntheses are siloxane *cyclics*, an example of which is sketched in Fig. 8. A small amount of unreacted cyclics will typically remain in the final silicone product after molding. This can be removed by solvent extraction.

The physical states of silicones range from viscous oils to relatively hard resins depending upon the formulation of the silicone material. The presence or absence of crosslinking agents and reinforcing materials determines the physical characteristics of the finished materials. Free-flowing oils are typically high molecular weight PDMS with no crosslinking. These have been used ophthalmically as tamponades in vitreoretinal surgery [5,6]. Loose crosslinking creates low-modulus gels that are extremely deformable and may not return to their original shape after deformation. Higher levels of crosslinking

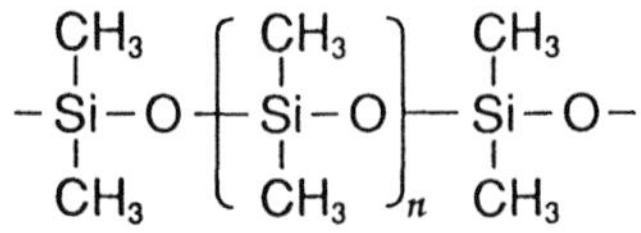

III

SLM-1/UV
POLY(DIMETHYLSILOXANE)

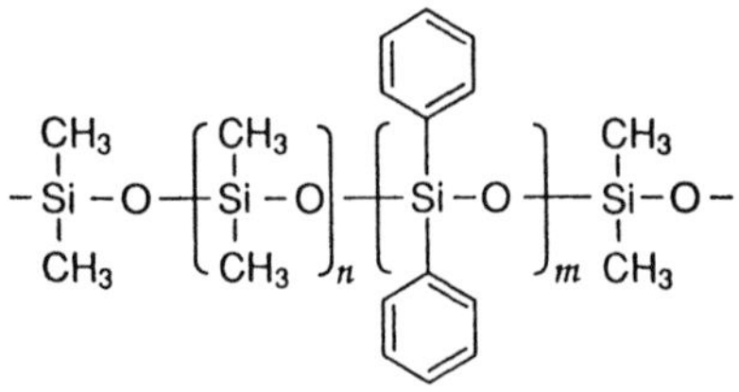

IV

SLM-2/UV
POLY(DIMETHYLDIPHENYLSILOXANE)

Figure 7 The chemical structures of two silicone elastomers used in the fabrication of foldable IOL optics; poly(dimethyl siloxane) (PDMS) (III) and poly(dimethyl diphenyl siloxane) (PDMDPS) (IV). The siloxane backbone is extremely flexible, giving silicone elastomers glass transition temperatures as low as $-100°C$. PDMDPS is a second-generation silicone IOL material that has a higher refractive index than PDMS.

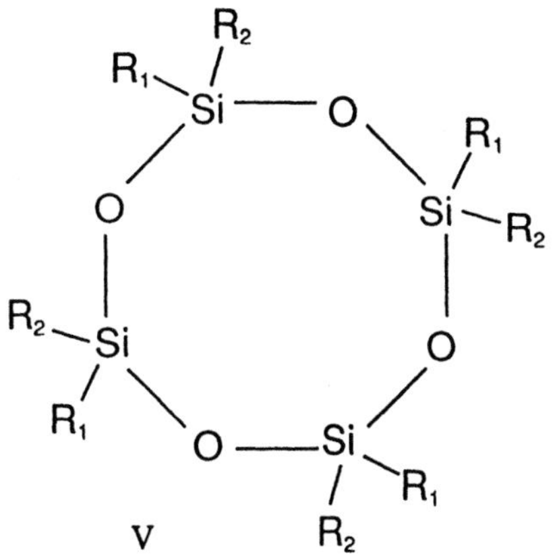

Figure 8 A sketch of a tetrasiloxane cyclic (V). Siloxane cyclics are the building blocks of silicone polymers. They are typically liquid in state. R_1 and R_2 are hydrocarbon groups chosen from methyl, phenyl, hydride, vinyl, or others. The cyclics are polymerized into polysiloxanes via a ring-opening mechanism.

coupled with reinforcing agents such as inorganic silica or siloxane resins leads to more rigid grades of elastomers. This is the class of silicone materials that includes IOL materials. These silicones are characterized by surface hardness, moderate resistance to deformation, complete recovery after deformation, and absence of migratable species.

Several silicone materials are used commercially in IOLs (Table 3). The SLM-1/UV silicone (Allergan, Inc., Irvine, CA) and the Staar and Chiron silicones are simple PDMS elastomers having a refractive index of about 1.41. The IOLab silicone is reported to have a refractive index of 1.43, suggesting that the material has phenyl groups pendant to the polysiloxane backbone. Allergan's SLM-2/UV has the highest refractive index of all known IOL silicones, 1.46. This results from the greater incorporation of diphenyl functionality compared to the other silicones. Since the addition of diphenyl to a silicone material tends to weaken the material, SLM-2/UV is reinforced with fumed inorganic silica, a more effective reinforcing agent than siloxane resins.

Table 3 Commercially Available Silicone IOL Materials

	SLM-1/UV	SLM-2/UV	RMX-3 [7]	Soflex [8]
Manufacturer	Allergan	Allergan	Staar	IOLab
Structure	PDMS[a]	PDMDPS[b]	PDMS	PDMDPS
Refractive index	1.41	1.46	1.41	1.43
Reinforcement	Siloxane resin	Fumed silica	Siloxane resin	Siloxane resin
IOL fabrication	LIM[c]	LIM	Cast molding	—
IOL models	SI-18NGB, SI-26NB	SI-20NB, SI-30NB	AA-4203, others	LI-30U

[a]PDMS, poly(dimethyl siloxane)
[b]PDMDPS, poly(dimethyl diphenyl siloxane)
[c]LIM, liquid injection molding

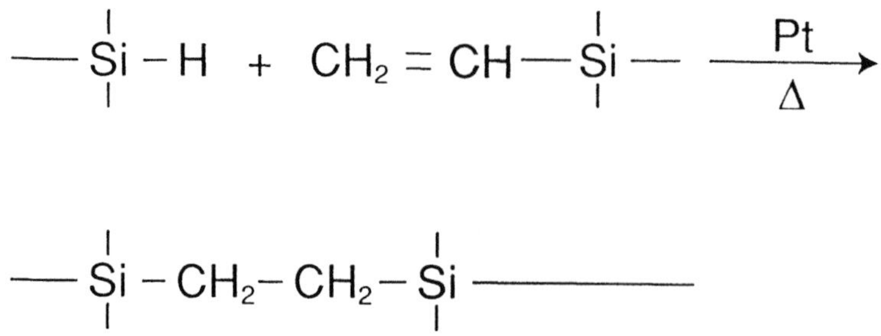

$$\overset{|}{\underset{|}{Si}} - H \;+\; CH_2 = CH - \overset{|}{\underset{|}{Si}} - \quad \xrightarrow[\Delta]{Pt}$$

$$\overset{|}{\underset{|}{Si}} - CH_2 - CH_2 - \overset{|}{\underset{|}{Si}} \rule{4cm}{0.4pt}$$

Figure 9 The hydrosilation reaction for addition-cured silicone elastomers. The reaction is typically catalyzed using a platinum complex and can be accelerated by heat: Elastomers cured by this mechanism are known as *room-temperature vulcanizing* (RTV) silicones.

Silicone IOLs are fabricated by molding. The specific silicones used in IOLs are known as *addition-cure, room temperature vulcanizable* (RTV) materials. The curing mechanism is known as *hydrosilation*, in which a vinyl group on a base polymeric molecule reacts with the hydride group on a cross-linking molecule. This is illustrated in Fig. 9. The hydrosilation reaction is catalyzed by a platinum–organosiloxane complex and can be accelerated by heat.

Typically, the IOL manufacturer will receive the silicone raw materials as a two-part system of viscous *preelastomeric* liquids. One part characteristically will contain catalyst and the other, crosslinking agent. Base polymer can be a component of one or both of the two parts. The curing reaction initiates when the two parts are mixed together, introduced into an IOL mold, and heated. Typically, the curing reaction is complete within minutes when thermally accelerated. Cast molding or liquid injection molding techniques are used. Often an organic solvent extraction process is used on the IOL to remove unreacted polymer and oligomer chains as well as residual cyclic starting materials left over from the base polymer synthesis. The SLM-1/UV and SLM-2/UV silicone materials are processed using such an extraction procedure. Figure 10 is a schematic diagram of a typical silicone IOL manufacturing process.

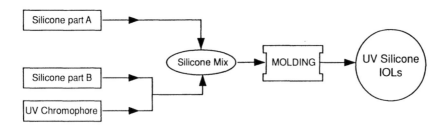

Figure 10 The fabrication process for silicone foldable IOLs. The preelastomer intermediates are typically packaged as a two-part liquid system. An ultraviolet-absorbing chromophore is dispersed into either preelastomer Part A or Part B. Dispersion into Part B is shown in this figure. The two parts are mixed and introduced into an IOL mold. The mixture is heated to accelerate the hydrosilation curing reaction. The resulting solid silicone lens is demolded, solvent extracted, cleaned, and sterilized prior to shipping.

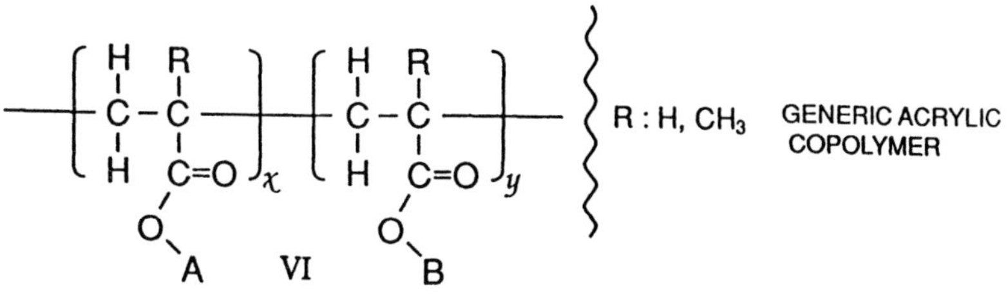

Figure 11 The chemical structure of a generic flexible acrylic copolymer (VI). Side groups; A and B, are chosen to impart properties such as high refractive index, flexibility, and surface lubricity to the cured material. The polymerization scheme for these materials is similar to PMMA (see Fig. 6). Flexible acrylics have been developed specifically for use in IOLs.

3. Foldable Acrylics

Foldable acrylic IOLs have recently been developed and are in human clinical evaluation at this writing. They have not been approved for wide distribution. Foldable acrylics are members of the same family of acrylic/methacrylic polymers as the rigid PMMA materials and the poly(hydroxyethyl methacrylate) (PHEMA) hydrogels used in soft contact lenses. A sketch of the structure of a generic foldable acrylic copolymer is provided in Fig. 11.

The synthetic strategy for foldable acrylics for IOLs is to combine a monomer that will provide a higher refractive index (*refractive index monomer*) than is possible with silicones and a monomer that has a relatively long hydrocarbon side chain to provide flexibility (*flexibility monomer*). Additional monomers can also be included to create ter- or higher-order polymers to gain unique surface properties such as enhanced lubricity. Several useful monomers are highlighted in Table 4.

Foldable acrylics are being developed specifically for IOLs. They are not currently used in other implants. They are distinguishable from silicones primarily in their higher refractive indexes, 1.47 or greater, and their different folding and unfolding behavior. Typically, the silicones will demonstrate undamped recovery after folding whereas foldable acrylics can be significantly damped. Another differentiating property between the silicones and foldable acrylics is glass transition temperature, T_g. As noted earlier, the T_g for a silicone can be significantly below room temperature. Foldable acrylics typically display T_gs well above 0°C and often above room temperature. Certain acrylic materials are therefore difficult to fold at the temperatures characteristic of a typical operating room.

Another challenge in acrylic formulation research is to address the tendency of these

Table 4 Potentially Useful Acrylate/Methacrylate Monomers for Foldable Acrylic IOL Materials

Monomer designation	Typical examples
Softening (or T_g lowering)	• *n*-Alkyl acrylates, e.g., *n*-butyl acrylate
Refractive index increasing	• Aromatic acrylates, e.g., phenylethyl acrylate

materials to display surface tackiness. This can result in the lens sticking to surgical instruments or to itself upon folding. A hydrophilic coating may be one way to overcome this difficulty.

Foldable acrylics can be fabricated into IOLs by either direct molding or low-temperature machining (cryolathing). Direct lens molding can proceed either directly from the monomer mixture or from a partially polymerized *pregel*. Physically, the pregel material is similar to the preelastomeric liquid silicones discussed earlier. The choice between direct-monomer and pregel molding is critically dependent on the volatility of the monomers. Highly volatile monomers are difficult to contain in an IOL mold, often resulting in incomplete mold fills and bubbles or gaps in the final product. A pregel process would be used in this case. When the monomers are of lower volatility, the direct-monomer molding technique can be simpler and faster.

The lathing approach to foldable acrylic IOL fabrication is similar to that used for rigid PMMA IOLs in that, first, an acrylic sheet is cast and thermally cured. Blanks are then cored out of the sheet. These blanks are machined into finished IOLs. A cryolathing technique must be employed because the T_gs typical of foldable acrylics are relatively close to ambient room temperature.

Several foldable acrylic IOLs currently under clinical investigation are presented in Table 5 along with the names of the manufacturers and other details. Note that the ORC and Storz acrylic materials are actually low-water hydrogels.

B. Currently Used Haptic Materials

Currently, four materials are used for the haptics of three-piece IOLs: poly(propylene) (PP), extruded PMMA, poly(vinylidene fluoride) (PVDF), and polyimide. PP has been in use since the 1970s while PMMA, polyimide, and PVDF have more recently been developed. The chemical structures of these materials are provided in Fig. 12. PP, PMMA, and PVDF are used in the form of extruded monofilament. After extrusion, the monofilament materials are heat formed into the proper shapes for use as IOL haptics. Polyimide is not processible by heat-forming methods. Therefore, polyimide haptics are formed directly by etching techniques similar to those used in the microelectronics industry [13].

Table 5   Foldable Acrylic/Acrylic Low-Water Hydrogel IOL Materials

	Alcon [9]	Allergan[a] [10]	ORC [11]	Storz [12]
Tradename	Acrysof®	Acryflex®[a]	Memorylens®	Hydroview
Chemistry	PEA/PEMA[b]	Acrylate/ methacrylate esters	MMA/HEMA[c]	—
Refractive Index (37°C)	1.55	1.47	1.48	1.47
Water content	<2%	<2%	20%	18%
Estimated T_g (°C)	15.5–21.5	11	25	—
Clinical status	U.S. clinical trials completed	U.S. clinical trials started	U.S. clinical trials completed	U.S. clinical trials started

[a]Formerly Ioptex; Acrylens®
[b]PEA, phenylethyl acrylate; PEMA, phenylethyl methacrylate
[c]MMA, methyl methacrylate; HEMA, hydroxyethyl methacrylate

POLY(PROPYLENE) POLY(VINYLIDENE FLUORIDE)

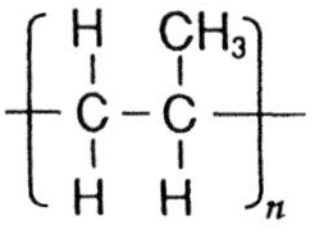

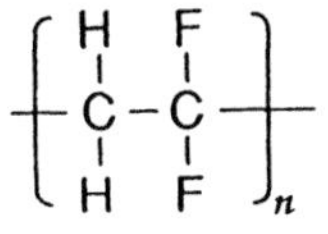

VII IX

POLYIMIDE

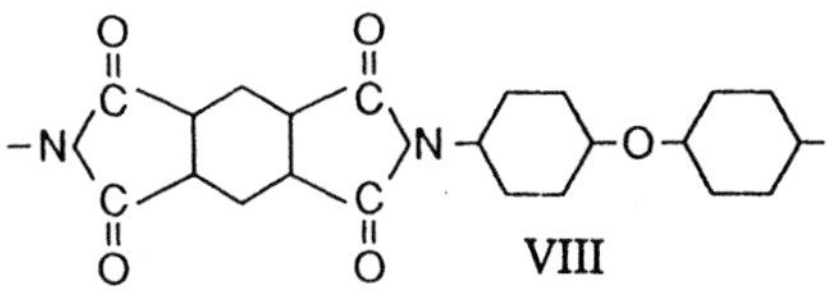

VIII

Figure 12 The chemical structures of three haptic materials; poly(propylene) (PP, VII), polyimide (VIII), and poly(vinylidene fluoride) (PVDF, IX). Extruded PMMA, also used as a haptic material, has the same chemical structure as shown in Fig. 5. PP, PVDF, and PMMA are used in monofilament form and are heat shaped into haptics. Polyimide haptics are photoetched from sheet material.

Each material possesses different mechanical resistance and memory profiles. Studies by Kimura et al., indicate that the memory of PVDF is greater than either PP or extruded PMMA [14,15]. This means that PVDF may be expected to maintain a greater residual pressure against the capsular membrane with time. Etched polyimide haptics tend to be rigid and to have little memory loss. All currently used haptic materials display excellent biocompatibility, good handling characteristics, and satisfactory mechanical performance.

C. Currently Used Ultraviolet-Absorbing Compounds

The first ultraviolet (UV)-absorbing IOL materials were introduced in the early 1980s. This was in response to a desire to protect the retina from harmful ultraviolet radiation in the 300–400 nm range. UV light of wavelengths less than 300 nm are screened out by the cornea. Normally, protection from 300- to 400-nm UV is provided by the crystalline lens; however, cataract extraction surgery removes this natural UV barrier. Several studies have emphasized the need for the addition of an ultraviolet blocker to IOL optic materials [16–18]. The UV–visible transmission spectrum for the natural lens of a 53-year-old human is compared to the spectra of a PMMA without chromophore and PMMA and silicone materials with an incorporated chromophore in Fig. 13. It can be seen that the natural lens is effective at screening UV radiation as are the UV-absorbing IOLS.

The earliest UV-absorbing PMMA materials were formulated with non-bondable benzophenone chromophores in the 5–10 wt% range. This resulted in a cutoff of ultraviolet light with wavelengths less than 400 nm. The chemical structure of such a benzophenone is shown in Fig. 14. A second class of chromophores, the benzotriazoles, was also identified as being useful in IOL materials (also Fig. 14). Benzotriazoles are particularly attractive because of their greater UV-absorbing efficiency relative to the benzophenones.

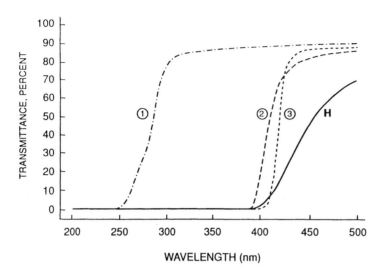

Figure 13 Ultraviolet (UV)–visible light transmission curves for UV-absorbing (2, silicone; 3, PMMA) and non-UV-absorbing IOLs (1, non-uv PMMA). The transmission curve for the natural lens of a 53-year-old human is included for comparison (H). The cutoff for UV-absorbing IOLs in the vicinity of 400 nm may be important for the protection of the retina from UV radiation.

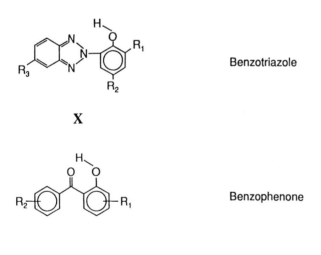

Benzotriazole

Benzophenone

Figure 14 The generic chemical structures of two classes of UV-absorbing chromophores; a benzotriazole (X) and a benzophenone (XI). The R groups are typically hydrocarbon or halogen. These chromophores are dispersed into PMMA materials in order to fabricate UV-absorbing IOLs. The benzotriazoles are more efficient in the absorption of UV radiation. Therefore, a lower concentration is required relative to the benzophenone for the desired UV-blocking capacity.

With a benzotriazole, only 1% or less of the additive is necessary to obtain the desired UV cutoff at 400 nm.

Both benzophenones and benzotriazoles are sufficiently soluble in PMMA to be used as additives without "blooming" to the IOL surface or leaching into the aqueous humor. However, it was not initially obvious that this would be the case. Consequently, research was initiated in the early 1980s to develop so-called bondable chromophores. These were functionalized with a pendant vinyl group to allow the chromophore to participate in the addition polymerization reaction. The structures of three bondable UV-absorbing chromophores are sketched in Fig. 15.

Benzotriazoles are not generally soluble in IOL-grade silicones. Additive chromophores will tend to bloom. Therefore, a bondable chromophore is required. Fortunately, the vinyl-functional chromophores shown in Fig. 15 are completely compatible with the hydrosilation curing reaction used with silicones. The extraction results of Christ et al. [19] indicate that a vinyl-functional benzotriazole can be completely incorporated into either PDMS (SLM-1/UV) or poly(dimethyl diphenyl siloxane) (SLM-2/UV).

A common difficulty in incorporating the simple vinyl-functional chromophores into silicones is maintaining chromophore dispersion in the silicone preelastomer liquids while they are stored on the shelf. This has led to research into chromophores that are more soluble in silicone liquids. One such chromophore has been developed at Allergan and is sketched in Fig. 16. This chromophore has been named *SUVAM* (for silicone-compatible ultraviolet-absorbing monomer). SUVAM has been shown to provide a stable dispersion

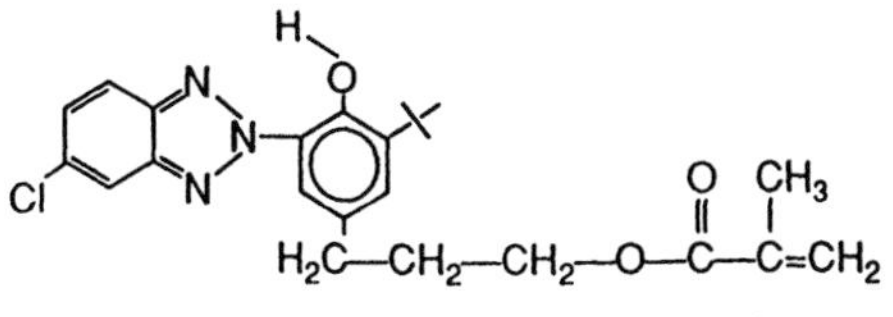

XII — C. Beard et. al. U. S. Patent 4,528,311

XIII — C. Reich et. al. U. S. Patent 4,868,251

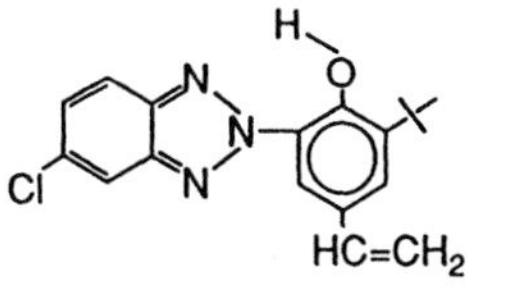

XIV — T. Posin et. al. U. S. Patent 4,636,212

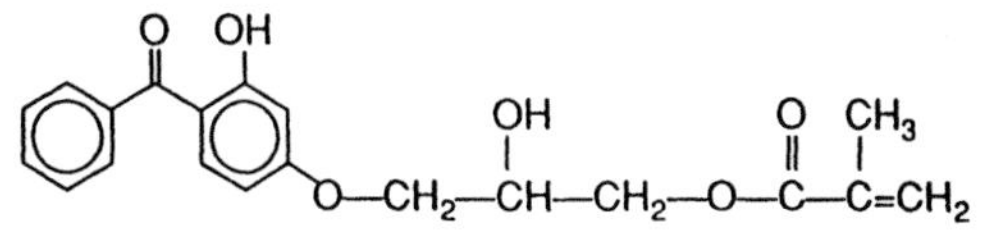

Figure 15 The chemical structures of three bondable UV-absorbing chromophores. Structure XIII has been termed *UVAM*, for ultraviolet absorbing monomer, at Allergan. The vinyl functionality allows these chromophores to participate in either the free-radical addition polymerization scheme of PMMA and the flexible acrylics, or in the hydrosilation curing scheme of the silicone elastomers. The chromophore is thus completely bonded into the material and is not free to be solvent extracted from the material.

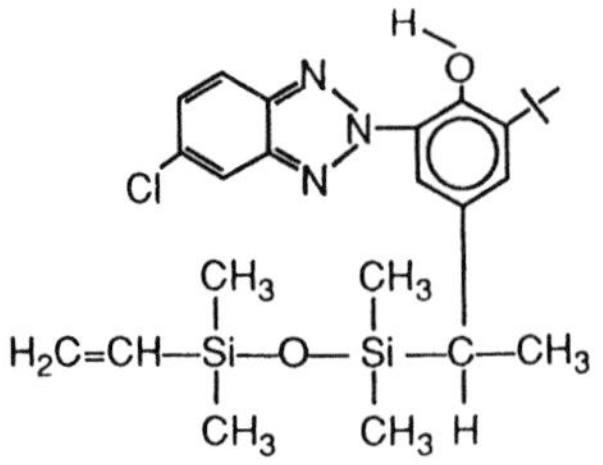

Figure 16 The chemical structures of two isomers of a bondable disiloxane benzothiazole (XV, XVI). This compound has been dubbed *SUVAM*, for silicone-compatible ultraviolet-absorbing monomer, at Allergan. The disiloxane group increases the solubility of the chromophore in silicone preelastomer liquids.

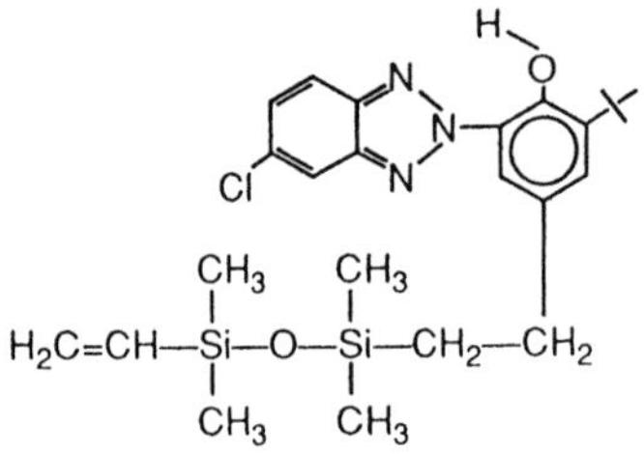

in both SLM-1/UV and SLM-2/UV, to provide the expected UV cutoff, and to be fully incorporated by hydrosilation into the silicone material matrix [20].

D. Other Materials Evaluated for Use in IOLs

Several additional optic and haptic materials were proposed at one time or another but were, for various reasons, not widely used. One of the earlier competitors to PMMA was optical glass as commercialized by Lynell Medical Technology, Inc. Glass was considered an excellent material for IOLs primarily because of its clarity and high refractive index resulting in a thin lens. Disadvantages of glass were its high specific gravity, potentially leading to IOL fixation difficulties, and fabrication and polishing costs. The most significant drawback of the glass IOL, however, was its tendency to crack in the eye when impacted by the Nd:YAG laser beam used in posterior capsulotomy procedures [21].

Foldable hydrogel lenses based on poly(hydroxyethyl methacrylate) (PHEMA) lenses have been made and evaluated in human clinical investigations. There was speculation that this material was more biocompatible than other materials because of the hydrophilic nature of its surface. The PHEMA lenses were found to be well tolerated in human clinical trials [22]. The PHEMA IOL was not commercially pursued, however, because of a design-related tendency to dislocate posteriorly into the vitreous cavity following Nd: YAG laser posterior capsulotomy [23]. There remains a potential for PHEMA to be commercially reintroduced in alternate designs.

Poly(ether urethane) (PEU) was also evaluated for use as an optic material. This was because of the higher refractive index of the PEU material (1.48) relative to PDMS

(1.41) and, thus, the potential to create a thinner, more easily folded lens. In addition, the PEU demonstrated superior mechanical strength relative to silicone. *In vitro* aging studies showed the PEU material to be resistant to hydrolysis and UV degradation. Animal implantation, however, resulted in a deterioration of the material surfaces, which in turn led to reduced light transmittance and poor optical resolution [24]. PEU was thus not pursued into human studies.

Other materials were used for haptics but were discontinued. Various metals including platinum, platinum–iridium, titanium, and stainless steel were incorporated into anterior chamber IOL designs [25]. These metal haptics tended to display a "cheese-slicer" behavior against sensitive intraocular tissues. Polyamide (nylon) monofilament was also tried as a haptic material but was found to degrade by an enzymatic mechanism in the eye [26]. This is not surprising considering the similarity of the molecular backbone of the polyamide to that of polypeptides and proteins.

IV. EVALUATION METHODS RELATED TO IOL SAFETY

A. Biocompatibility/Toxicity Requirements and Evaluation

1. Laboratory Testing

IOL optic and haptic materials must be biocompatible, that is, nontoxic, nonmutagenic, nonsensitizing, nonirritating, and noninflammatory. The biocompatibility requirements/ evaluation methods for IOL materials are summarized in Table 6. Standardized *in vitro* and *in vivo* test methods are used to assess materials relative to these requirements. Typically, the material is first evaluated for toxicity using cell culture methods. Next, aqueous and cottonseed oil extracts of the material are evaluated in the mouse, rabbit, and guinea pig for systemic toxicity, irritation, and sensitization potential, respectively. Additionally, the material is evaluated for mutagenicity using one or more *in vitro* methods such as the Ames test [27]. Failure to pass any of these tests would raise a concern about the biocompatibility, and hence suitability, of the material for use in an IOL. These tests represent relatively low-cost screening methods that are run prior to committing to the larger costs of implant studies.

2. Animal Implantation

The *in vitro* simulation of the implant environment for the evaluation of material biostability as well as the impact of the material on the biological environment (e.g., cell or tissue damage, inflammation, etc.) can be a relatively quick and cost-effective approach

Table 6 Biocompatibility Requirements and Methods

Performance requirement	Method
Nontoxic	• Cell culture
	• Injection of extract into mouse
	• Intramuscular implantation
Nonmutagenic	• Ames test
Nonsensitizing	• Injection of extract into guinea pig
Nonirritating	• Injection of extract into rabbit
Noninflammatory	• Intraocular implantation in cat, primate, or rabbit

to screening potential implantable biomaterials. Unfortunately, the *in vitro* techniques seldom are able to take into account the complex interactions of biological environmental factors. Thus, it is essential that the performance of a biomaterial be additionally assessed in one or more suitably chosen animal implantation models.

Two types of implantation models have proven useful for IOLs. The first is a screening model to provide a preliminary *in vivo* assessment of material biostability and tissue/cellular reaction. Normally, this model would include a cost-effective animal species such as the rabbit or rat. The implantation site is either subcutaneous or intramuscular. Implants are retrieved at specified intervals and evaluated via tissue histology of the implant site and scanning electron microscopy (SEM), UV and visible light transmittance, and mechanical properties testing on the retrieved material specimens [24]. The subcutaneous and intramuscular environments are more vascularized than is the intraocular environment. Implantation at those sites may provide an accelerated *in vivo* material evaluation model relative to intraocular implantation.

The second model is intraocular implantation in a species whose eye reasonably mimics that of the human. Rabbit, feline, and primate models have been used with success. Typically, these models are more costly than are the subcutaneous or intramuscular models described above. Three levels of postsurgical follow-up are used. First, the IOL and the intraocular tissues are examined *in situ* periodically by slit-lamp biomicroscopy for signs of inflammation or other abnormalities. Second, enucleated eyes from animals sacrificed at predetermined intervals are subjected to a complete histological evaluation in which the cornea, iris, retina, and other structures are examined for abnormality. Third, recovered IOLs are examined microscopically for changes in color and clarity, by SEM for signs of surface degradation, and by collimator for retention of optical refractive power and resolution.

Buchen et al. [28] describe a 12-month intraocular implantation study of a silicone IOL in the feline model. The silicone material (SLM-2/UV) was found to be well tolerated by the eye. Furthermore, the material displayed no signs of functional or physical degradation as a result of the 12-month exposure to the intraocular environment. The retention of resolution of the silicone test lenses as well as PMMA control lenses is summarized in Table 7. A visual comparison of explanted PMMA and silicone lenses indicated that cellular debris was less adherent to the silicone lens [28].

Christ et al. [24] applied subcutaneous, intramuscular, and intraocular implantation methods to an experimental PEU as discussed earlier. This material was found to biodegrade. This can be easily seen in Fig. 17, which shows a PEU specimen recovered from the rabbit subcutaneous environment at 3 months. These studies graphically illustrate the importance of animal implantation in qualifying new IOL materials.

Table 7   Retention of Optical Resolution of SLM-2/UV and Perspex CQ® IOLs Following 12 Months in the Feline Eye [28]

Intraocular lens category	*n*	Preoperative (lp/mm)[a]	Postoperative (lp/mm)[a]
SLM-2/UV (implanted)	3	219	235
SLM-2/UV (nonimplanted)	2	213	213
Perspex CQ (implanted control)	3	197	227

[a]lp/mm, line pairs/millimeter [55]

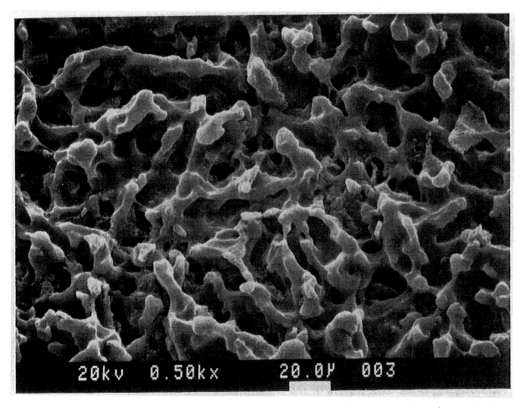

Figure 17 A scanning electron micrograph of a poly(ether urethane) specimen recovered from the subcutaneous region of the rabbit after 3 months implantation (500×). Note the severe surface degradation. This degradation is thought to be mediated by inflammatory cells. It resulted in the opacification of the surface of the sample.

B. Biostability Requirements and Evaluation

The resistance of IOL materials to the degradative agents characteristic of the intraocular environment is an important consideration. The degradative agents include (1) aqueous medium (hydrolysis), (2) ultraviolet radiation from the external environment (sunlight and fluorescent lighting), (3) oxygen (aqueous humor oxygen tension varies from approximately 60 torr at the interface with the cornea to less than 10 torr at a point 2.25 mm posterior from the cornea [29]), (4) stress (compression of haptics by contracting/fibrosing capsular sac), (5) enzymes or other biologically active agents (produced by inflammatory cells), and (6) free radicals or other chemical species such as superoxide (O_2^-). The two parameters most amenable to *in vitro* screening methods are hydrolysis and ultraviolet aging. In addition, these parameters can be modeled to simulate long-term exposure. For the other parameters, animal implantation, as discussed earlier, is a more rational means for assessing biologically oriented degradative mechanisms. Biostability requirements are summarized in Table 8.

Table 8   Biostability Requirements and Methods

Requirement	Method
Stable to UV radiation	Accelerated *in vitro* UV aging (xenon source)
Stable to hydrolytic degradation	Accelerated *in vitro* hydrolytic aging (60° or 80°C)
Stable to enzyme degradation	Intramuscular or intraocular implantation

1. Hydrolytic Aging Studies

Prospective IOL materials are commonly tested at a variety of acceleration temperatures as part of the product development process. These pilot studies indicate which acceleration temperature will provide the highest acceleration and, therefore, the shortest duration for a given desired simulation without introducing unrealistic degradation mechanisms. The accelerated hydrolytic aging model is based on an Arrhenius-type expression in which the rate of hydrolysis doubles for each 10°C above the "use" temperature of the device, that is, 35°C [30] The hydrolytic acceleration model can be written as follows:

$$t_{sim} = A_{hyd} \cdot t_{lab} \tag{1}$$

where:

$$A_{hyd} = 2\exp\{(T_{exp} - T_{use})/10\} \tag{2}$$

and t_{sim} is the simulated *in vivo* duration of aging; t_{lab} is the duration of the laboratory exposure; T_{use} is the temperature of the implant environment, typically 35°C; and T_{exp} is the laboratory exposure temperature.

For silicone materials, 80°C is typically used as the acceleration temperature, T_{exp}, while 60°C is suitable for PMMA. In a study on a potential PEU IOL material, it was found that 80°C melted the material [24]. Therefore, the 60°C conditions were used. The values of t_{sim} for $t_{lab} = 1$ year are presented for several experimental exposure temperatures in Table 9.

Degradation in these studies is monitored by mechanical properties and visible light transmittance for slab samples and by optical resolution, microscope examination, and scanning electron microscopy (SEM) for lens samples. The mechanical and light trans-

Table 9   Equivalent *In Vivo* Hydrolysis Exposures for a 1-Year Exposure at Different Experimental Temperatures

Study temperature (°C)	Equivalent *in vivo* exposure (years)[a]
50	2.8
60	5.6
70	11.3
80	22.6
90	45.3

[a]Using Eqs. (1) and (2) with $T_{use} = 35°C$

Table 10 Properties of SLM-1/UV Slabs and IOLs Before and After Accelerated Hydrolytic Aging Simulating 20 Years *In Vivo* Exposure

	n	Before aging	After aging
SLM-1UV slabs [19]			
Tensile strength (psi)	8	734	857
Transmittance (%, 600 nm)	8	93.7	93.7
SLM-1/UV IOLs [31]			
Optical Resolution (lp/mm)	2	250–290	350

mittance results for two silicones are reported in Christ et al. [19] and are reproduced in Table 10. Silicone lens studies are covered in Francese et al. [31]. Optical resolution data from this study for aged silicone lenses are also given in Table 10. There was no evidence of silicone degradation for immersions simulating 30 years of *in vivo* exposure.

2. Ultraviolet Aging Studies

Accelerated UV aging models are designed along the following pattern:

$$t_{sim} = A_{uv} \cdot t_{lab} \tag{3}$$

where t_{sim} is simulated *in vivo* exposure, t_{lab} is the duration of the aging study, and

$$A_{uv} = (I_s/I_r)^y (D_s/D_r) \tag{4}$$

where I_s is the UV intensity used in the aging study, I_r is the intraocular UV exposure intensity, y is an acceleration exponent, D_s is the daily UV exposure duration for the aging study, and D_r is the daily UV exposure duration of a human subject.

Various ultraviolet aging models have been used to assess the UV stability of polypropylene haptic material [32,33] and silicone optic materials [19,31,34]. In the latter studies, degradation was monitored by mechanical properties testing, visible and UV light transmittance, and retention of UV absorber efficiency for slab samples; and by optical resolution and microscope examination for lens samples. Most recently, Yang et al. reviewed the various accelerated-aging models and proposed a refined model based upon new values for the key parameters, I_r and D_r [35]. UV aging results for two silicone IOLs and a PMMA IOL simulating *in vivo* exposures of 17 and 5 years, respectively, are included in Table 11.

Table 11 Optical Resolution of Silicone IOLs and Perspex CQ® IOLs Before and After Ultraviolet Aging

IOL type	n	Before aging (lp/mm)[a]	After aging (lp/mm)[a]	Equiv. exposure
Silicone [31]				
SLM-1/UV	3	280	270–310	17 years
SLM-2/UV	3	260–330	250–260	17 years
PMMA				
Perspex CQ	6	279–285	279–285	5 years

[a]lp/mm, line pairs/millimeter

C. Chemical Evaluation Methods

Chemical methods of analysis are important for implantable biomedical materials for the following reasons:

1. To assure identity and reproducibility of different lots of raw materials
2. To detect and identify impurities in the materials and to thereby make a determination as to the potential impact of those impurities
3. To link performance characteristics to chemical properties, for example, bulk chemistry to mechanical performance, surface chemistry to implant–bioenvironment interactions (covered in a later section)

The chemical analyses that are covered in this section are: (1) molecular weight, (2) the chromatographic detection and characterization of unbound constituents of the materials, (3) the spectroscopic characterization of IOL polymers and UV-absorbing chromophores, and (4) thermal methods of analysis. Material chemistry requirements are summarized in Table 12.

1. Molecular Weight

The standard approach to determining the molecular weight of the polymers used in intraocular lenses is gel permeation chromatography (GPC). Numerous texts and monographs describe this technique [36,37]. GPC is directly useful in the molecular weight determination of PMMA. A sample of the polymer is dissolved in a solvent such as tetrahydrofuran (THF) and evaluated. Appropriate molecular weight standards (e.g., polystyrene) are used to calibrate the experiment. A GPC scan for Perspex CQ® is shown in Fig. 18. Perspex CQ® has a weight average molecular weight of approximately 2 million.

It is difficult to define the "molecular weight" of elastomeric materials such as the silicones and flexible acrylics. The reason for this is that each material, in finished form, is a single three-dimensional molecule. In the case of silicone, however, it is possible to characterize the base polymer by molecular weight. GPC scans of the preelastomer Parts A and B for SLM-1/UV are shown in Fig. 19. The molecular weights of the base polymer are about 100,000. The additional peaks in the vicinity of 1000 represent crosslinking and reinforcing agents.

Foldable acrylics, since they are typically polymerized and crosslinked in the mold in a single-step process, are even less amenable to the characterization of molecular weight.

Table 12 Material Chemistry Requirements and Methods

Requirement	Method
Unbound constituents less than 1%	• Organic extraction/gravimetric analysis
Identify unbound constituents	• HPLC (for UV chromophores)
	• GC/MS for volatiles
Undetected aqueous extractables	• Aqueous extraction/gravimetric analysis
Determine polymer molecular weight	• GPC
Identify polymer ("fingerprint")	• IR spectroscopy
Sterilization compatibility	• TGA for heat sterilization
	• EtO residuals by GC for gas sterilization
Evaluate surface chemistry	• Surface morphology/chemistry techniques

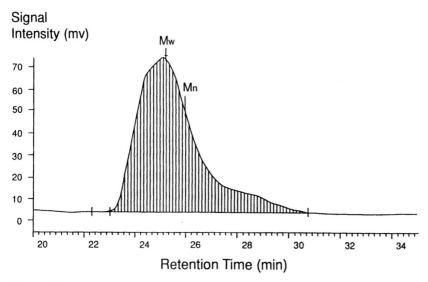

Figure 18 A gel permeation chromatogram for Perspex CQ®. The weight average molecular weight of approximately 2 million corresponds to a retention time of about 25 min in this experiment. Polystyrene molecular weight standards were used to calibrate the experiment.

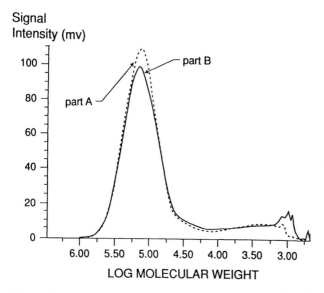

Figure 19 Gel permeation chromatograms for Parts A and B of the SLM-1/UV formulation. The large peak represents the base polymeric molecule with a weight average molecular weight of about 100,000. The smaller peaks in the 1000–3000 region indicate the presence of crosslinking and siloxane resin reinforcing agents.

Under certain conditions, where a polymeric pregel is synthesized followed by crosslinking, the acrylic pregel may be characterized by molecular weight using the techniques described above.

In both silicone and acrylic pregels, molecular weight can also be estimated by measuring viscosity. Predictive plots of molecular weight as a function of viscosity can be constructed and used to track and control pregel synthesis. A simple rotational viscometer is best for this measurement.

2. Detection and Characterization of Unbound Constituents

It is virtually impossible when preparing a finished IOL material to entirely bind up all chemical constituents either into entangled molecular chains (in the case of PMMA, which is thermoplastic) or into the three-dimensional elastomeric network (in the case of foldable acrylics). Typically, there will be unreacted monomers (PMMA and foldable acrylics), cyclics and oligomers (silicones), catalysts, and other components. It is the responsibility of the IOL manufacturer to identify and quantify any unbound constituents as well as to take steps to minimize their levels.

It is typical to use organic solvents to either dissolve (PMMA) or swell (silicones and foldable acrylics) the materials to release all components that are not a part of the materials themselves. THF/dichloromethane has been determined to be a good solvent system for PMMA and is used in the detection of residual methylmethacrylate (MMA) monomer by gas chromatography (GC). IOL-grade PMMA should have a residual MMA monomer content of less than 1.0% by weight. GC is also useful in characterizing the purity of all raw materials (i.e., monomers) that are used in IOL materials and in tracking the progress of distillations designed to increase the purity of these monomeric starting materials.

THF, toluene, and chlorinated hydrocarbons are useful solvents for swelling silicone elastomers and sweeping out the unbound constituents for analysis. Analyses include reverse-phase high-performance liquid chromatography (HPLC) for unbound UV absorbers (see Fig. 20 for a typical chromatogram), gas chromatography–mass spectrometry (GC–MS) for the separation and identification of cyclics and oligomers, and GPC for estimating the molecular weight distribution of unbound oligomeric and polymeric species. The typical level of unbound constituents in SLM-1/UV and SLM-2/UV silicones was found to be less than 0.75% by weight [19].

Solvents that are useful for swelling foldable acrylic elastomers are acetonitrile and acetone. Typical extractable constituents are residual monomers and unincorporated oligomeric and polymeric acrylic species. GC is used to determine the identity and levels of residual monomers, and GPC is used to determine the molecular weight distribution of un-crosslinked polymer chains.

3. Spectroscopic Evaluation of IOL Polymers and UV-Absorbing Chromophores

IOL polymer materials are normally subjected to spectroscopic analysis in order to determine and document the exact identity of the material, that is, to establish a "fingerprint" for future reference and control. Techniques that have proven suitable are infrared (IR) and nuclear magnetic resonance (NMR). IR can be used for thermoplastics such as PMMA by preparing a thin film using a heated press. An IR scan for Perspex CQ® is shown in Fig. 21. IR can be used for silicone preelastomer liquids. A spectrum for Part B of SLM-1/UV is shown in Fig. 22. IR is more difficult for the cured silicone (or foldable

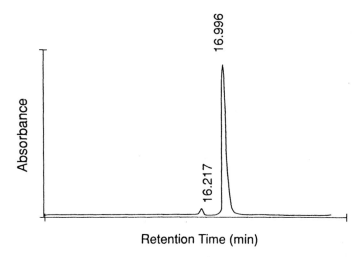

Figure 20 A reverse-phase liquid chromatogram showing the retention time (16.996 min) for UVAM (XIII). The small peak upstream is the nonbondable benzotriazole starting material. This technique is used to check the purity of incoming lots of chromophore as well as to test for the presence of the chromophore in the organic extracts of UV-absorbing silicone lens materials.

acrylic) elastomer and requires the use of an attenuated total reflectance (ATR) accessory or other surface IR technique.

Simple NMR techniques are most suitable for the characterization of the preelastomer liquids for silicones and pregels for flexible acrylic elastomers. The characterization of the finished polymeric materials would require more sophisticated solid-state NMR techniques. ¹H-NMR scans of silicone preelastomer liquids can highlight the presence and abundance of methyl, vinyl, phenyl, and hydride protons.

UV-absorbing chromophores can be characterized by IR, NMR, and MS. ¹H-NMR and MS spectra for SUVAM (structure XV–XVI, Fig. 16) are shown in Figs. 23 and 24, respectively [20]. Furthermore, spectroscopy can be used to track and control both lab-

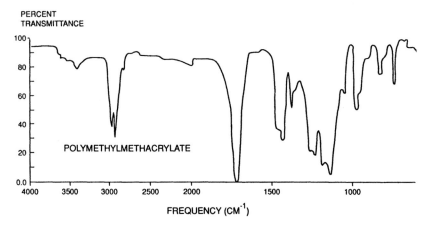

Figure 21 An infrared (IR) spectrum of PMMA. The "fingerprint" of PMMA is seen in the 1000–1500 cm⁻¹ region.

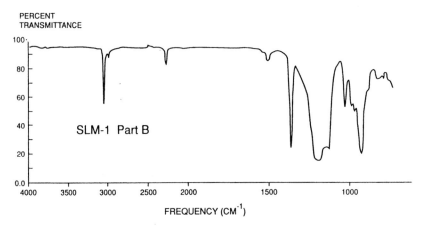

Figure 22 An infrared (IR) spectrum for the silicone preelastomer liquid, SLM-1/UV Part B. The "fingerprint" of PDMS is found in the 800–1500 cm^{-1} region. The band at approximately 2140 cm^{-1} represents the Si—H function of the crosslinking agent.

scale and production synthetic processes. This is illustrated for SUVAM in Fig. 25, where IR shows the addition of hydride in the intermediate and the subsequent loss of this hydride upon addition of a disiloxane tail by the hydrosilation reaction [20].

4. Thermal Methods of Analysis

Thermal methods include differential scanning calorimetry (DSC) and thermal gravimetry (TGA), and are important in assessing the temperature ranges over which material properties are constant. The DSC method is typically used to determine the glass transi-

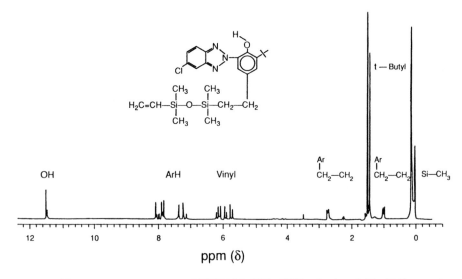

Figure 23 An NMR spectrum of SUVAM (XV, XVI), the bondable disiloxane benzotriazole. This spectrum was used to confirm that the desired product was obtained from the synthesis procedure.

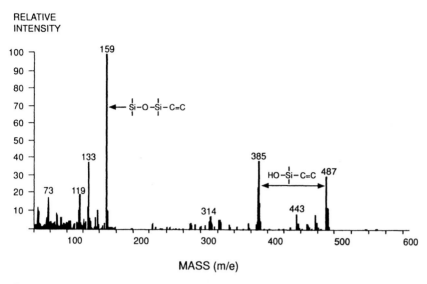

Figure 24 A mass spectrum of SUVAM (XV, XVI) confirming, along with NMR (Fig. 23), that the desired compound was synthesized.

tion temperature, T_g, of polymeric materials as well as the melt temperature, T_m, of polymers such as PP that have some level of crystalline content. A knowledge of T_g (and T_m as appropriate) is important when defining the process temperature for thermoforming monofilament strands into IOL haptics. Typically, the monofilament must be warmed to a temperature above its T_g or T_m, formed into its haptic shape, and then cooled back to room temperature in order for the shape to become permanent. Additionally, the use temperature of the monofilament, 35°C for intraocular use, must not be too close to the material's T_g or the haptic will return to its unformed shape. A DSC scan of PP monofilament is shown in Fig. 26, showing a melt temperature of 166.26°C.

T_g determinations by DSC can also be instrumental in developing foldable acrylic materials because their T_gs tend to be in the room-temperature range. Thus, the researcher can predict whether an experimental acrylic formulation will be flexible at the ambient temperature in the operating room or not. T_g values for several foldable acrylics were provided earlier in Table 5.

TGA provides a means to understand the thermal stability of IOL materials. For instance, there is no loss in mass for the silicone SLM-1/UV, as measured by the TGA experiment, up to 400°C. Thus, one can safely predict that this material will be thermally stable at physiological temperatures and under heat sterilization (i.e., autoclave) conditions.

D. Surface Evaluation Methods

The properties of the surfaces of IOL materials are of critical importance because the surfaces form the interfaces of the IOL with the biological environment. This is true for all biomedical implants. The biological environment of the implanted IOL is the aqueous humor of the eye. The aqueous humor, which is a filtered form of blood plasma, enters the anterior segment of the eye through the so-called ciliary processes. The bimolecular components of the aqueous humor are reviewed in Table 13.

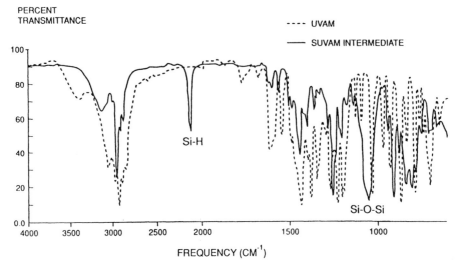

(a)

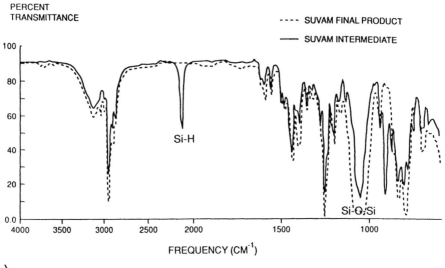

(b)

Figure 25 The use of infra-red spectroscopy to monitor the synthesis of SUVAM (XV, XVI): (a) an overlay of spectra of the starting material, UVAM (XIII), and the hydride-functional intermediate; (b) an overlay of spectra of the intermediate and the final product, SUVAM, that was produced by the hydrosilation of a disiloxane with the intermediate. Note the creation of hydride (2140 cm^{-1}) in the intermediate and the loss of hydride during the hydrosilation reaction producing SUVAM.

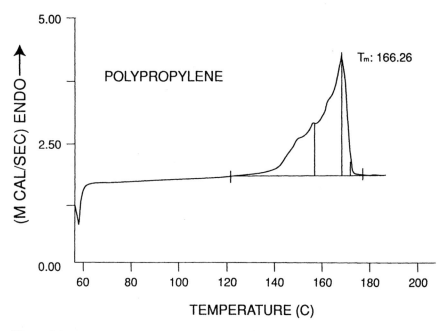

Figure 26   A differential scanning calorimetry (DSC) scan for poly(propylene) (VII) showing the melt temperature to be 166.26°C. DSC can help in identifying the appropriate temperature for heat-shaping process for converting monofilament materials into IOL haptics.

Table 13   Biomolecular Constituents of the Aqueous Humor of the Rabbit [29]

Constituent	Concentration (mM/kg water)
Sodium (Na)	143.5
Potassium (K)	5.25
Calcium (Ca)	1.7
Magnesium (Mg)	0.78
Chlorine (Cl)	109.5
Bicarbonate (HCO_3)	33.6
Lactate	7.4
Pyruvate	0.66
Ascorbate	0.96
Urea	7.0
Reducing value (as glucose)	6.9
Amino acids	0.17

The IOL–aqueous interface undergoes its greatest biologically mediated modification immediately following implantation. At this point, the protein and cellular contents of the aqueous humor are at their highest concentration due to the eye's physiological response to surgery. The initial protein layers become adsorbed onto the surface of the IOL materials, and leukocytes and macrophages begin to adhere in an attempt to engulf and destroy the invading foreign body. The inability of the inflammatory cells to engulf the IOL surface leads to what has been called *frustrated phagocytosis* [38]. At this point, the cells burst and release superoxide as well as cytokines. The cytokines disrupt the blood-aqueous barrier (BAB), resulting in the further release of inflammatory cells into the aqueous humor. This is the time of greatest biological stress on the implanted materials. The response is immediate and decays over a period of several weeks, slowly returning to pre-operative levels. The chemistry and physical morphology of the surfaces of the materials used in the IOL optic and haptic elements are important in determining to what extent inflammatory cells will adhere to the IOL.

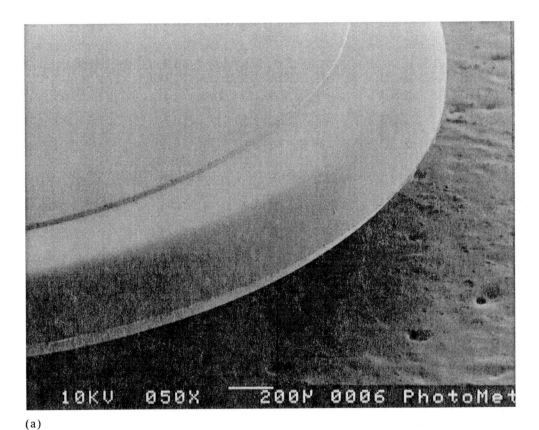

(a)

Figure 27 Scanning electron micrographs (50×) of a three-piece silicone IOL highlighting: (a) surface and edge finish, and (b) haptic insertion zones. SEM is used to check the performance of IOL molds and thus to indicate when molds must be reworked or replaced.

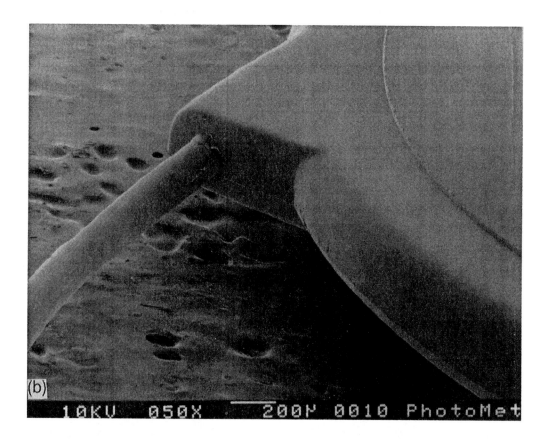

(b) 10KU 050X 200M 0010 PhotoMet

1. Surface Morphology

The most widely used technique for examination of the morphology of IOL material surfaces is scanning electron microscopy (SEM). The technique is described in detail in Goldstein et al. [39] and is used to monitor IOL fabrication processes as well as the presence of degradation in biostability studies. Figure 27 shows SEM photos of the optic edge and haptic attachment region of a silicone IOL. Particular attention is paid to molding flash for silicones and foldable acrylics, and to machining lines for PMMA.

Surface roughness is a factor in biological response. It has been shown that rougher surfaces can cause a greater amount of cellular debris to adhere to a surface [40]. In addition, sharp, rigid edges such as on a poorly finished PMMA IOL can cut into delicate ocular tissues, leading to increased inflammation and cellular response.

Recent surface-imaging techniques that have been used for IOL surface analysis at a molecular level are scanning tunnelling microscopy (STM) and atomic force microscopy (AFM) [41]. These techniques are limited in their view of the overall surface morphology of an IOL, however, since their scanning areas are on the order of only 100 μm^2.

2. Surface Chemistry

Several techniques exist for probing the chemistry of a polymeric implant surface. These include contact angle measurements, x-ray photoelectron spectroscopy (XPS), energy dispersive x-ray spectroscopy (EDS), and static secondary ion mass spectrometry

(SIMS). Contact angle provides a macroscopic probe of the surface and can distinguish between levels of hydrophilicity and hydrophobicity. XPS and EDS provide an atomic profile of the material surface, with XPS additionally providing bonding environment information. SIMS probes molecular structures.

Contact Angle. An important advantage of contact angle analysis over spectroscopic techniques is that the analysis can be carried out in an aqueous environment, whereas the spectroscopic methods generally require vacuum. There are two general categories of contact angle techniques: static and dynamic. Static contact angle measurements characterizing a material–aqueous interface can be taken using either sessile drop or captive bubble methods. Sketches of the sessile drop and captive bubble experiments are provided in Fig. 28. Contact angle measurements by the sessile drop method using fluids of different surface tensions are used to calculate the critical surface tension of the surface. A more physiologically accurate approach, however, is the captive bubble method, which uses different probes (e.g., octane, air) to measure contact angle in an aqueous environment. These data can be used to calculate the interfacial free energy of a material.

The sessile drop and captive bubble contact angles are reported for several materials in Table 14, along with calculated interfacial free energies [42]. It is interesting to note that elastomeric materials display a considerable molecular mobility in their surfaces, as one might expect from their low T_gs. This phenomenon is demonstrated by comparing the sessile drop contact angle for SLM-1/UV, 107°, with the captive bubble contact angle for the same material, 61°. In air, the surface behaves hydrophobically whereas in water, the behavior is more hydrophilic.

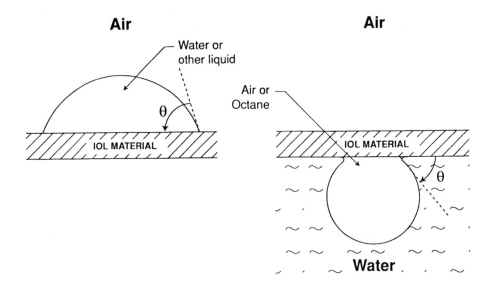

SESSILE DROP   **CAPTIVE BUBBLE**

Figure 28 Sketches of two contact angle techniques: sessile drop, in which a droplet of water or other fluid is placed on the material surface in air; and captive bubble, in which a air bubble or octane drop is applied to the underside of a material immersed in water. In both cases, the contact angle is read using a goniometer.

Table 14   Contact Angle/Surface Free Energy of Several IOL Optic Materials [42]

Material	Sessile drop CA (water)	Captive bubble CA (air)	Interfacial free energy (ergs/cm^2)
Perspex CQ®	71	67	14
SLM-1/UV	107	61	52
SLM-2/UV	106	50	45
PHEMA	66	16	21

Dynamic contact angle analysis in which advancing and receding angles are measured can also highlight the changes in surface hydrophilicity/hydrophobicity as a material is transferred from an air environment to an aqueous environment. This technique allows a probe of the dynamic nature of surface hydration and reorientation through repeated wet/dry cycles. For SLM-1/UV, the advancing and receding contact angles are 125° and 65°, respectively. A comparison of these data with the static data in Table 14 shows that the advancing contact angle, obtained when the dry material is immersed in an aqueous medium, approaches that of the sessile drop water contact angle. The receding contact angle, obtained when the wet material is exposed to air, approaches that of the captive bubble air contact angle.

X-ray Photoelectron Spectroscopy.   XPS provides information about the atomic composition of a polymeric surface as well as the bonding state of each of the elements identified. For example, a hydroxyl oxygen is distinguishable from an acid or backbone oxygen. XPS theory and techniques are thoroughly reviewed in Brundle [43]. The XPS data for several materials are provided in Table 15 along with theoretical atomic abundances [44]. An important drawback for using XPS for the surface analysis of biomedical implants is that the analysis must be done under vacuum, which does not approximate the implant environment. However, XPS has found use in the IOL production environment in assessing the presence and identification of IOL surface contaminants.

Energy Dispersive X-ray Spectroscopy and Secondary Ion Mass Spectrometry.   Two other surface chemistry analysis methods are energy dispersive x-ray spectroscopy (EDS) [45] and secondary ion mass spectrometry (SIMS) [46,47]. EDS has been used to identify particulate contaminants on material surfaces or within materials (by preparing a material cross section), with some success. This technique is most often combined with SEM to probe the elemental composition of surface contaminants such as residual polishing compounds (used with PMMA IOLs). SIMS has been used to detect the presence of

Table 15   Surface Elemental Composition by XPS for Several IOL Optic Materials [44]

Material	Measured composition %C	%O	%Si	Theoretical composition %C	%O	%Si
Perspex CQ®	73	27	0	72	28	0
SLM-1/UV	62	24	12	50	25	25
SLM-2/UV	51	29	20	55	22	22
PHEMA	68	23	4	67	33	0

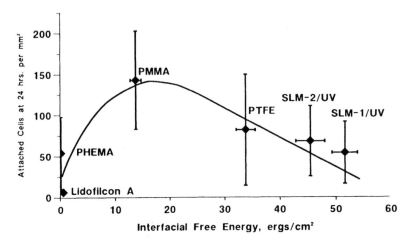

Figure 29 A plot showing the relationship of cell adherence to the interfacial free energy of various IOL materials. Note that the greatest level of cell adherence occurs with materials of intermediate interfacial free energy (e.g., PMMA).

surface coatings on IOL materials. It provides considerable molecular information about the surface of the material by ablating the surface to create molecular ion species that are analyzed by mass spectrometry.

3. Correlation of Cellular Response to Surface Chemistry

The *in vitro* contact of cultured lens epithelial cells constitutes a biological probe of an IOL material's surface. Studies using bovine lens epithelial cells on several materials were described in Cunanan et al. [42]. The authors made an attempt to correlate cell adhesion with interfacial free energy (calculated from contact angle data) and elemental composition by XPS. The most consistent correlation seemed to occur with interfacial free energy. A plot showing the number of cells adhered vs. interfacial free energy is reproduced in Fig. 29. Note that surfaces of particularly high or low interfacial free energy were found to be cell resistant, while surfaces of intermediate interfacial free energy were associated with more cell adhesion. In addition, more spreading of each cell was noted on the intermediate surfaces. A comparison of cell adhesion on silicone vs. PMMA indicated that, as predicted by Fig. 29, the epithelial cells were more numerous and more completely spread on PMMA [42].

Various animal models have been used to observe cell deposits on IOLs. Buchen et al. [28] reported reduced deposits of protein and cells in the feline intraocular model for IOLs made from SLM-2/UV compared to Perspex CQ® PMMA. Okada et al. [48] implanted PMMA lenses that had been half-coated with PDMS into the rabbit anterior chamber. This allowed the elimination of subject-to-subject and procedure-to-procedure variability in the comparison of silicone and PMMA. The authors found that fewer deposits developed on the PDMS coating than on the PMMA substrate. In human clinical studies, Hirokawa et al. [49] and Amon et al. [50], reported fewer cell deposits on silicone and hydrogel IOLs relative to PMMA. These results from both animal and human studies appear to confirm the *in vitro* findings of Cunanan et al. [42].

E. Mechanical Evaluation Methods

1. *The Determination of Fundamental Mechanical Properties*

The fundamental mechanical properties of biomaterials used in IOLs are (1) tensile strength, elongation, and modulus; (2) surface hardness; and (3) tear resistance (for foldable materials). Tensile properties are measured by mechanical testers such as those manufactured by Instron, Inc., using ASTM Method D412 as a guide [51]. The tensile strength, elongation, and modulus for Perspex CQ®, SLM-1/UV, SLM-2/UV, and an investigational foldable acrylic are given in Table 16. A clear differentiation can be made between the rigid (high-modulus) and foldable (lower-modulus) materials based on these data. Folding deformation of foldable lenses involves greater than 100% elongation. Tensile strength and elongation measurements are useful in monitoring the state of cure of foldable elastomeric materials and in designing IOLs with the desired performance characteristics. Incomplete cures will result in materials with lower strength, greater elongation, and compromised clinical performance.

Tensile testing is the primary mechanical characterization method for extruded monofilament haptic materials. The tensile strength and ultimate elongation of PP and extruded PMMA are provided in Table 16.

Surface hardness is measured per ASTM D2240 using standard durometers (Shore Mfg.) [52]. Several hardness scales exist, the most common being the Shore A and D scales for elastomeric materials such as silicone, foldable acrylics, and polyurethanes. The hardness values for two silicones and a foldable acrylic are listed in Table 16. Hardness may be correlated with foldability; that is, harder materials tend to be more rigid and softer materials tend to be more foldable. Hardness also correlates with the resistance of

Table 16   Mechanical Properties for Several IOL Materials

Property	Optic materials			
	Perspex CQ®	SLM-1/UV	SLM-2/UV	Foldable acrylic [9]
Tensile strength (psi)	8–10,000	780	720	—
Ultimate elongation (%)	2–5	300	240	—
Tensile modulus (psi)	14,500	105	180	230 (shear)
Hardness	—	36 (Shore A)	38 (Shore A)	45 (Shore A)
Tear strength (psi)	NA	60	35	—

Property	Haptic materials	
	Polypropylene	Extruded PMMA
Tensile strength (psi)	56,000	50,000
Ultimate elongation (%)	24	60

a material surface to indentation and scratching from forces that would occur during normal surgical handling.

Tear resistance is measured on a mechanical tester using the ASTM D624 test method [53]. Tear resistance has meaning only relative to foldable materials and can be an indicator of the likelihood that a material will be torn during folding or surgical manipulation. The tear resistances of two silicones are given in Table 16.

The critical clinically important mechanical behaviors of the IOL materials are dependent both on the fundamental properties of the materials and on the configurations and dimensions of the lenses. This makes it difficult to directly relate laboratory-determined fundamental mechanical properties (i.e., those measured on a mechanical tester) to clinical performance. Thus, functional testing of the IOL itself is critical to predict clinical performance capability. These functional tests are described briefly in the following paragraphs.

2. Functional Mechanical Requirements and Testing

IOL functional testing is targeted at the specific clinical requirements for IOL mechanical performance. These are summarized in Table 17.

The surface-related mechanical requirements for IOLs revolve around resistance to cracking and marring during handling, and cracking during Nd:YAG laser procedures. A study involving the Nd:YAG damage of two different types of PMMA IOLs and a silicone IOL was reported [54]. The materials were: (1) lathe-cut Perspex CQ®, a high molecular weight PMMA, (2) injection molded PMMA, and (3) SLM-1/UV silicone. Figure 30 shows an SEM analysis of 1mJ Nd:YAG pulses directed at the surfaces of these materials. One can see that cracking occurred at the margin of the impact crater for both the injection-molded and the lathe-cut PMMA materials. The foldable silicone material demonstrated less tendency to crack on impact from the Nd:YAG laser beam.

An additional mechanical requirement for one-piece all-PMMA IOLs, where a milled PMMA haptic is present, is that the haptic elements are flexible enough to resist breaking during in-plane and out-of-plane deformation. These tests are schematically described in

Table 17   Functional Mechanical Requirements and Methods

Performance requirement	Possible fundamental mechanical correlates	Method
Resistance to surface damage/ cracking	• Hardness • Impact resistance	• Scratch testing • Direct YAG laser impact
Resistance to haptic breakage (one-piece PMMA)	• Toughness	• Force and angle of deformation: in and out of plane
Folding/unfolding behavior	• Flexural modulus • Viscoelastic rheometry	• Folding force • Unfold time
Resistance to optic deformation during haptic compression	• Compressive modulus	• Haptic compression with photokeratoscope and resolution
Resistance to folding damage	• Hardness • Tear strength • Tensile strength	• SEM • Resolution • Photokeratoscope
Optic recovery after folding	• Stress–strain cycling	• Folding with resolution and photokeratoscope

Fig. 31. Perspex CQ® PMMA is inherently susceptible to breakage when haptics are subjected to large-angle deformations. Several companies have addressed this by applying compression or stretching treatments to orient the PMMA molecules. The resulting molecular orientation is thought to more closely mimic the orientation that exists in the extruded monofilament materials typically used in three-piece IOLs.

With three-piece IOLs, the resistance of the haptics to being pulled out during surgical manipulation is an important consideration. This is measured as the force necessary to pull the haptic out using a tensile tester. The current standard for haptic pull-out force is 50 g [55].

Foldable IOL materials must possess elastic moduli that allow relatively easy folding and yet, at the same time, will resist the optic deformation from forces exerted by the haptics during postsurgical capsular bag contraction. The *folding modulus* is estimated by taking measurements of the force required to fold IOLs made from various foldable materials. This has been done for two silicone materials where it was found that optic curvature and thickness were stronger determinants of resistance to folding than was a small difference in the modulus of the material [19].

The resistance to foldable optic distortion by haptic compression is also measured. Test IOLs are placed in fixtures designed to compress the haptics. While the IOL is in this test apparatus, its optical resolution is determined. This will indicate whether an optically significant deformation of the surface has occurred. Alternately, a photokeratoscope* can be used to determine if any spherical distortion results from the haptic compression. Both forms of testing demonstrated no changes in resolution or surface sphericity for SLM-1/UV or SLM-2/UV even when the haptics were compressed from 14 mm overall diameter to 9 mm [19].

Two additional requirements related to folding of the optic are that the material is capable of complete recovery of dimensions and optical performance, and that no damage is done to the optic surface, as a result of the folding procedure. Susceptibility to damage is related to the material's hardness and the design of folding instruments. The first step is to subject the optic to sustained periods of folding (i.e., 5–10 min) or to repeated folding (i.e., 25 times). The folded optics are then released, and resolution is measured. Recovery is determined by measuring optical resolution as a function of time after release. Earlier work by Christ et al. showed that lenses made from SLM-1/UV and SLM-2/UV returned completely to their original optical performance within 24 h of being released, without damage to the surface as assessed by SEM as well as by photokeratoscope examination [19].

V. EVALUATION METHODS RELATED TO IOL EFFICACY

A. Optical Evaluation Methods

The optical design and performance of an IOL depends on the material used for the optic portion of the IOL. The three most important properties of these materials are their refractive index, visible light transmittance, and homogeneity. The refractive index of the

---

*A photokeratoscope projects concentric rings of light onto a spherical surface. The regularity of the spacings between the reflected light rings indicates the extent to which the sphericity of the surface is retained during haptic compression or after folding. A loss of sphericity can be detrimental to the optical performance of the IOL.

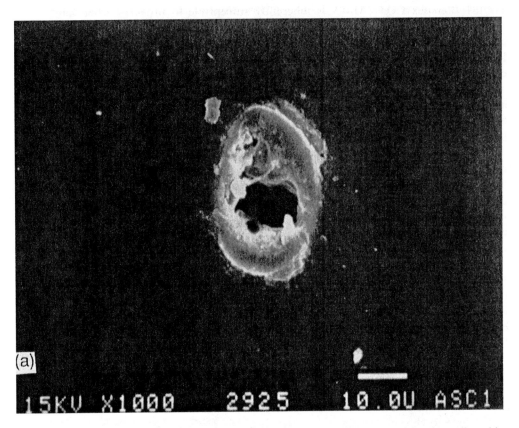

15KV X1000   2925   10.0U ASC1

Figure 30 Scanning electron micrographs of Nd:YAG damage (1000×) resulting from direct hits on three IOL optic materials: (a) injection-molded PMMA, (b) lathed Perspex CQ, and (c) SLM-1/UV silicone. The Nd:YAG beam energy was 1 mJ for all cases. Note that the injection-molded PMMA material and the lathed Perspex CQ material display cracking at the crater margins whereas SLM-1 /UV displays a more rounded crater with no evidence of cracking.

optic material will determine the shape (i.e., surface curvature and overall thickness) of the optic portion of the IOL for a given IOL refractive power. The IOL refractive power, visible light transmittance, and material homogeneity (i.e., freedom from microcrystalline domains, causing scattered light, and refractive index gradients) will together determine the resolution efficiency of the IOL. These relationships are summarized, along with optical performance requirements, in Table 18. A brief discussion of optical test methods follows.

1. Refractive Power

The contributory material property to refractive power, namely refractive index, is measured using a refractometer. ASTM D542 provides a guideline to the procedure [56]. The refractive index of an optical material is determined by the chemical identity of the material (aromatic, brominated, and iodinated groups raise refractive index; fluorinated groups lower refractive index). Refractive index also has temperature and wavelength

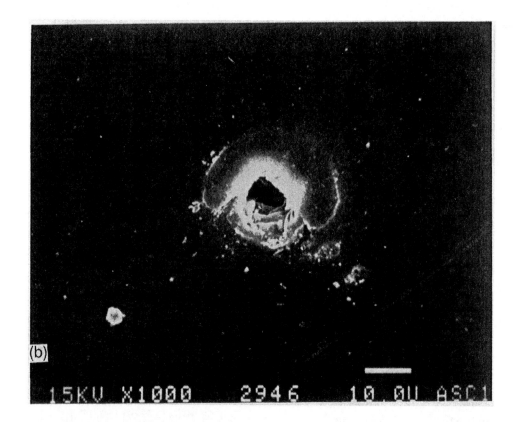

(b)

15KV X1000   2946   10.0U ASC1

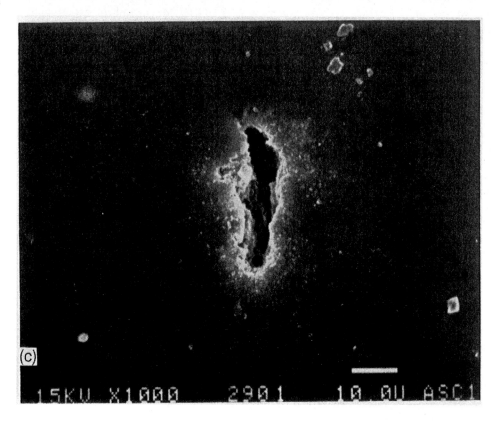

(c)

15KV X1000   2901   10.0U ASC1

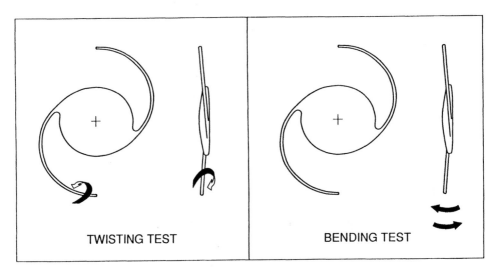

Figure 31 A schematic of the test methodologies used to evaluate PMMA materials for haptic strength in one-piece IOL designs; twisting of the haptic around its axis; bending the haptic out of the optic plane. The tests include measures of deformation force and angle.

dependencies. For example, Holladay et al., have shown that the temperature coefficient of the refractive index of silicone is higher than that of PMMA [57]. The refractive indexes of silicones, PMMA, a 38% water hydrogel (PHEMA), and an experimental foldable acrylate are shown in Table 19.

The actual refractive power of an IOL depends on the shapes of the two optic surfaces and on the refractive index of the optic material. The refractive power is calculated in diopters (D) from measurements of focal length previously determined with a

Table 18 Optical Performance Requirements and Methods

Performance requirement	Contributing material/design properties	Method
Refractive capability	• Refractive index • Surface curvature	• Refractometer • Radiuscope • Interferometer
Resolution efficiency (>70% in water)	• Optical transfer • Refractive power • Refractive index distribution • Surface curvature • Surface smoothness • Light transmittance (>90%) • Material homogeneity	• MTF[a] • Collimator • Profilometry • Spectrophotometry • Haze analysis
Ultraviolent transmission (<1% below 390 nm)	• Added or bonded UV chromophore	• Spectrophotometry

[a]MTF, modulation transfer function (see text)

Table 19 Refractive Index Comparison for
Several IOL Materials

Material	Refractive index (25°C)
PMMA	1.49
PDMS	1.41
PDMDPS	1.43–1.46
PHEMA (38% water)	1.43
Foldable acrylics	1.47–1.55

collimator. The typical range of IOL powers lies between + 10 D and + 30 D depending on the refraction needs of the patient. Thus, the surgeon can correct for preexisting myopia or hyperopia with the IOL. However, since currently available IOLs do not restore accommodative capability to the patient, the patient will likely require reading glasses for near vision.

2. Resolution

As noted earlier, resolution is dependent on the refractive index of an IOL as well as on the material's homogeneity and on the smoothness and sphericity of the optic's surfaces. Homogeneity is measurable as scattered visible light. Both transmitted and scattered light are evaluated on a visible light spectrophotometer using a test method such as ASTM D1003 [58]. The percent visible light transmittances of Perspex CQ® and the SLM-1/UV and SLM-2/UV silicones are greater than 90%. The level of haze (i.e., scattered light) is less than 2% for these materials [19].

Surface topography is determined by SEM, interferometry, and profilometry. Rough or nonspherical surfaces can adversely impact the quality of image formation by the lens, that is, optical resolution. Sphericity is determined on a radiuscope or by interferometer.

Resolution is subjectively measured using the collimator and an Air Force 1951 Resolution Target [55] (Fig. 32). A more objective measurement is the modulation transfer function (MTF). MTF assesses how well a system can transfer the contrast of an object to an image (see Fig. 33) [59,60]. MTF measures the efficiency of this transfer as a percentage. The data are plotted as modulation (percent efficiency) vs. spatial frequency (target size). The MTF curve of a typical PMMA lens is compared to that of an ideal, diffraction-limited lens in Fig. 34. Surface inconsistencies, material inhomogeneities (light scatter), and severe opacity will move the MTF curve significantly downward and to the left.

VI. A BRIEF SUMMARY OF THE CLINICAL PERFORMANCE OF CURRENTLY USED IOL MATERIALS

Clinical performance is considered the ultimate confirmation of the suitability of a new implantable device. It is incumbent upon manufacturers to generate the clinical data and to provide summary reports to regulatory agencies (e.g., U.S. Food and Drug Administration) and to the surgeons who will eventually use the device. The regulatory agencies require that reports of clinical data be submitted prior to widespread commercial distribution in order to demonstrate the safety and efficacy of the device. With a new IOL, these

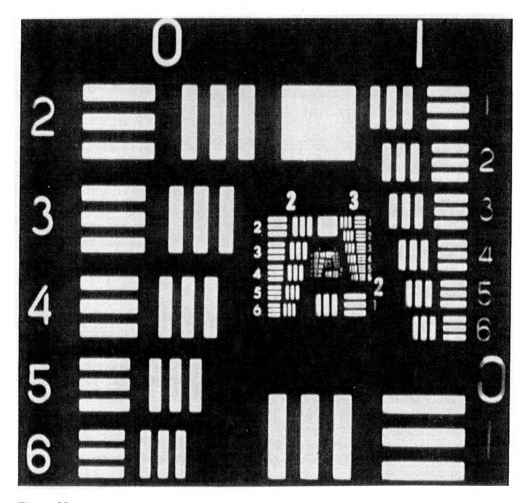

Figure 32   A photograph of the 1951 Air Force target used with a collimator for lens resolution determinations. Optical resolution relates to the ability of the lens to create an image that clearly distinguishes between adjacent bars in the target.

reports often involve extensive clinical investigations of 100 to 700 patients. In these clinical studies, postoperative complications are assessed to measure the IOL safety. Postoperative vision is the primary measure of efficacy. Secondary measures are ease of insertion and incision size, which demonstrate the effectiveness of the IOL design and material in terms of use with required surgical and insertion techniques. The relationship of these performance requirements, preclinical material tests, and clinical measurements are summarized in Table 20.

Long-term data on PMMA IOLs of various designs have been investigated and summarized in the literature [61–65]. These involved measures of postoperative complications and visual acuity, as these were the most important concerns with early IOL materials and designs. The PMMA IOLs were shown to demonstrate long-term safety

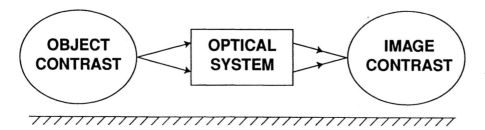

Figure 33 A schematic demonstrating the concept of *optical transfer*, or the efficiency by which an optical system transfers an object into an image. This is the basis of the modulation transfer function (MTF) approach to evaluating optical systems such as IOLs.

and efficacy as defined in Table 20. Clinical studies on more recent, foldable IOL designs comprised of silicone have evaluated these parameters, as well as incision size [66–69]. The AMO® PhacoFlex®, AMO® PhacoFlex® II, STAAR, and Chiron silicone IOLs, which have been used since the mid-1980s in over 500,000 patients, have also been shown to demonstrate successful long-term clinical performance. In addition, the intended benefits of these foldable IOLs, such as smaller incision size leading to faster visual rehabilitation, have been reported. Foldable acrylic IOLs, which have been evaluated since 1990, are currently in clinical investigation and are not commercially available. The clinical investigation and measurements are similar to those parameters evaluated with the silicone IOLs [9,10].

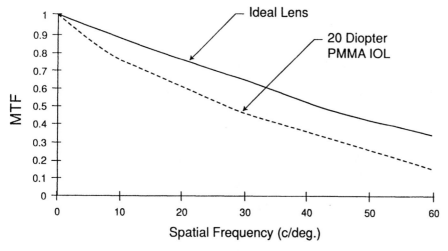

Figure 34 An MTF curve for a PMMA lens relative to an ideal *diffraction-limited* lens. Lenses closer to the *ideal* curve have higher resolution efficiency, that is, more efficient optical transfer. Surface inconsistencies and material inhomogeneity can cause a lens's curve to move away from the ideal curve.

Table 20 Clinical Performance Requirements and Methods

Performance criteria	Preclinical methods	Clinical methods
Safety		
Biocompatibility	*In vitro* toxicity Animal implant	Postoperative inflammation
Biostability	Accelerated aging of material/ lenses Animal implant	Postoperative inflammation; IOL decentration/disloca- tion
Chemical purity	Unbound constituents	Postoperative inflammation
Mechanical stability	Functional tests	IOL decentration/dislocation
Efficacy		
Optical resolution	MTF: resolution efficiency	Visual acuity (best corrected)
Refractive capability	Collimator: refractive capabi- lity	Refractive correction
Biostability	Resolution of aged lenses	Long-term visual acuity
Insertion through required wound size	Functional mechanical tests	Ease of insertion; incision size required for insertion
Resistance to damage	Functional mechanical tests	IOL decentration/dislocation

VII. FUTURE IOL MATERIALS AND DESIGNS

Continued research in IOL materials and designs will likely focus on products that address the following desired outcomes:

1. IOL insertion through incisions smaller than 3.0 mm
2. The reduction of complications such as early postoperative inflammation and posterior capsular opacification
3. Simultaneous correction of presbyopia and cataract in patients
4. The ability to monitor intraocular function (e.g., intraocular pressure) and to deliver targeted medications

The first outcome has been targeted by several manufacturers through the continuing development of advanced IOL materials, designs, and insertion techniques. The objectives are to minimize trauma to the eye, to keep pace with expected advances in cataract extraction technology, and to further enhance safety of the procedure. The availability of an IOL that can be inserted through a 2.5- to 2.8-mm incision is expected within the next 5 years.

Second of the desired outcomes is the reduction of complications such as early postoperative inflammation and posterior capsular opacification. In general, with current IOL materials and designs, inflammation following cataract surgery is considered well within acceptable safety limits for most patients. However, the continued interest of surgeons and manufacturers in further reducing the possibility of inflammation in difficult cataract surgeries has been the impetus for ongoing work in this area. There is also interest in minimizing the occurrence of posterior capsular opacification, reducing the need for Nd:YAG capsulotomy procedures.

To address the challenge of reducing post-operative inflammation and posterior capsular opacification, IOL materials research has involved two approaches: (1) the development of new bulk materials and (2) refinements, such as surface treatments, to existing materials. As already discussed, there has been speculation that the PHEMA hydrogel IOL would lead to reduced inflammation because of its hydrophilic surface and less adherence of inflammatory cells. Unfortunately, the mechanical weakness of the PHEMA material makes it more difficult to manipulate and insert.

More recently, several manufacturers have looked at surface coatings on conventional bulk materials such as PMMA and silicone. Two diverse paths have been taken. Some manufacturers have attempted to coat the IOL with a hydrophobic coating based on poly(tetrafluoroethylene) (PTFE) using gas plasma techniques, while others have tried to mimic the hydrophilic surface of PHEMA. Kabi Pharmacia has developed a PMMA IOL that is coated with a hydrophilic layer of heparin, a sulfated glycosaminoglycan. The heparin is covalently attached to a polymeric interlayer that is in turn attached by van der Waals attraction to the underlying PMMA material. Several clinical investigations have suggested that cellular adhesion has been reduced in the early postoperative period by the presence of the heparin coating [70,71]. Correlations of the heparin coating to reduced inflammation or posterior capsular opacification have not been established, however.

Outcome 3 arises from the limitation of currently available IOLs with regard to monofocality. Current IOLs do not vary their focus in a way similar to that of the prepresbyopic crystalline lens. Multifocal IOL technology is the newest concept to surgically treat presbyopia following cataract surgery. The general concept is to incorporate unique optical surfaces that distribute light to more than one focusing point along the optical axis of the patient's eye. This provides multiple diopter powers to enable focusing of objects at various distances (i.e., far and near). This *pseudoaccommodative* ability in principle would reduce a patient's usage or dependence on visual aids such as spectacles, which are needed for optimum focus during common daily visual tasks. Optic materials considered for multifocal designs must be capable of being shaped into the appropriate multifocal surface patterns. Initial multifocal IOL designs, based upon varied optical theories, have been introduced and studied over the past 5 years [72–74].

An approach to full restoration of accommodative function is the injectable IOL concept. Two different forms have been proposed: (1) the injection of an *in situ* gelling polymer into the natural lens capsule (which had been previously evacuated through small-incision cataract removal) and (2) the combination of an injectable gelling polymer and a polymeric bag fabricated into the shape of a natural capsule. In both cases, the desire is for a replacement of the natural lens with an equally pliable lens to accommodate for near, intermediate, and far vision. Several organizations, including manufacturers and university groups, have been conducting research in this area [75,76]. Injectable silicones and acrylic gels have been investigated. The material and surgical challenges are enormous, however, and a clinical product is unlikely within the next 5 years.

Farther out on the horizon are IOLs that interact chemically, mechanically, or electrically with the intraocular environment. The injectable IOL, described above, is actually a mechanically interactive concept. One can imagine IOLs that monitor physiologic conditions in the eye and subsequently release therapeutic agents in response to those conditions. One can only guess as to the unique IOL material requirements and testing methodologies that might be necessary for these "21st-century" IOLs.

ACKNOWLEDGMENT

The authors would like to thank Alan Lang, Nick Tarantino, and Valdemar Portney for reviewing the text and offering suggestions.

REFERENCES

1. Terry, A. C., W. J. Stark, M. C. Leske, and A. E. Maumenee, History and trends in cataract surgery and intraocular lenses, in *Anterior Segment Surgery* (W. J. Stark, A. C. Terry, and A. E. Maumenee, eds.), Williams & Wilkins, Baltimore, 1987, p. 1.

2. Deutman, A. F., History of lens implantation and current trends in Europe, in *Anterior Segment Surgery* (W. J. Stark, A. C. Terry, and A. E. Maumenee, eds.), Williams & Wilkins, Baltimore, 1987, p. 5.

3. Ridley, H., Further observations on intraocular acrylic lenses, *Trans. Ophthalmol. Soc. UK, 72*, 511–514 (1952).

4. Frisch, E. E., Silicones in artificial organs, in *Polymeric Materials and Artificial Organs* (C. G. Gebelein,ed.), American Chemical Society, Washington, DC, 1984, pp. 63–97.

5. Lean, J. S., Use of silicone oil as an additional technique in vitreoretinal surgery, in *Retina* (S. J. Ryan, B. M. Glaser, and R. G. Michels, eds.), C.V. Mosby, St. Louis, 1989, pp. 279–292.

6. Scott, J. D., Silicone oil as an instrument, in *Retina* (S. J. Ryan, B. M. Glaser, and R. G. Michels, eds.), C. V. Mosby, St. Louis, 1989, pp. 307–315.

7. Grabow, H. B., and R. G. Martin, STAAR AA-4203 one-piece plate-haptic silicone IOL, in *Foldable Intraocular Lenses* (R. G. Martin, J. P. Gills, and D. R. Sanders, eds.), SLACK Incorporated, Thorofare, NJ, 1993, pp. 73–114.

8. Kershner, R. M., IOLAB LI30U three-piece silicone IOL, in *Foldable Intraocular Lenses* (R. G. Martin, J. P. Gills, and D. R. Sanders, eds.), SLACK Incorporated, Thorofare, NJ, 1993, pp. 179–189.

9. Anderson, C., D. D. Koch, G. Green, A. Patel, and S. Van Noy, Alcon AcrySof™ acrylic intraocular lens, in *Foldable Intraocular Lenses* (R.G. Martin, J. P. Gills, and D. R. Sanders, eds.), SLACK Incorporated, Thorofare, NJ, 1993, pp. 161–177.

10. Mackool, R. J., IOPTEX Acrylens™ acrylic IOL, in *Foldable Intraocular Lenses* (R. G. Martin, J. P. Gills, and D. R. Sanders, eds.), SLACK Incorporated, Thorofare, NJ, 1993, pp. 191–196.

11. Fishkind, W. J., ORC MemoryLens™: A thermoplastic IOL, in *Foldable Intraocular Lenses* (R. G. Martin, J. P. Gills, and D. R. Sanders, eds.), SLACK Incorporated, Thorofare, NJ, 1993, pp. 197–212.

12. Green, G. F., Storz Hydroview composite Hydrogel IOL, in *Foldable Intraocular Lenses* (R. G. Martin, J. P. Gills, and D. R. Sanders, eds.), SLACK Incorporated, Thorofare, NJ, 1993, pp. 213–219.

13. Utrata, P. J., and D. C. Brown, STAAR Elastimide three-piece silicone IOL, in *Foldable Intraocular Lenses* (R. G. Martin, J. P. Gills, and D. R. Sanders, eds.), SLACK Incorporated, Thorofare, NJ, 1993, p. 117.

14. Kimura, W., T. Kimura, T. Sawada, T. Kikuchi, H. Toda, Y. Yamada, and H. Nagai, Comparison of shape recovery ratios in various intraocular lens haptics, *J. Cataract Refract. Surg., 18*, 547 (1992).

15. Kimura, W., T. Kimura, T. Sawada, T. Kikuchi, H. Nagai, and Y. Yamada, Comparison of shape recovery ratios of single-piece poly(methyl methacrylate) intraocular lens haptics, *J. Cataract Refract. Surg., 19*, 635 (1993).

16. Peyman, G. A., R. Zak, and H. Sloane, Ultraviolet-absorbing pseudophakos: An efficacy study, *Am. Intra-Ocular Implant Soc. J., 9*, 161–170 (1983).

17. Ham, W. T., H. A. Mueller, J. J. Ruffolo, D. Guerry, and R. K. Guerry, Action spectrum for retinal injury from near-ultraviolet radiation in the aphakic monkey, *Am. J. Ophthalmol., 93*, 299–306 (1982).

18. Mainster, M. A., The spectra, classification, and rationale of ultraviolet-protective intraocular lenses, *Am. J. Ophthalmol., 102*, 727–732 (1986).

19. Christ, F. R., D. A. Fencil, S. Van Gent, and P. M. Knight, Evaluation of the chemical,

optical, and mechanical properties of elastomeric intraocular lens materials and their clinical significance, *J. Cataract Refract. Surg., 15*, 176–184 (1989).

20. Yang, S., H. Makker, J. Kerkmeyer, and R. Christ, UV absorbing silicone compositions for IOL, *Proceedings of the American Chemical Society, Division of Polymeric Materials: Science and Engineering, 69*, 417–118 (1993).

21. Fritch, C. D., Neodymium:YAG laser damage to glass intraocular lens, *J. Am. Intra-Ocular Implant Soc. J., 10*, 225 (1984).

22. Barrett, G., and I. J. Constable, Corneal endothelial cell loss with new intraocular lenses, *Am. J. Ophthalmol., 98*, 157 (1984).

23. Lowe, K. J., and D. L. Easty, A comparison of 141 polymacon (logel) and 140 poly(methyl methacrylate) intraocular lens implants, *Br. J. Ophthalmol., 76*(2), 88–90 (1992).

24. Christ, F. R., S. Y. Buchen, D. A. Fencil, P. M. Knight, K. D. Solomon, and D. J. Apple, A comparative evaluation of the biostability of a poly(ether urethane) in the intra-ocular, intramuscular, and subcutaneous environments, *J. Biomed. Mater. Res., 26*, 607–629 (1992).

25. Clayman, H. M., Materials and manufacturing techniques of intraocular lenses, in *Anterior Segment Surgery* (W. J. Stark, A. C. Terry, and A. E. Maumenee, eds.), Williams & Wilkins, Baltimore, 1987, pp. 19–20.

26. Kronenthal, R. L., Nylon in the anterior chamber, *Ophthalmology, 88*, 965 (1981).

27. Maron, D. M., and B. N. Ames, Revised methods for the Salmonella mutagenicity test, *Mutation Res., 113*, 173–215 (1983).

28. Buchen, S. Y., S. C. Richards, K. D. Solomon, D. J. Apple, P. M. Knight, R. Christ, L. T. Pham, D. L. Nelson, H. M. Clayman, and L. G. Karpinski, Evaluation of the biocompatibility and fixation of a new silicone intraocular lens in the feline model, *J. Cataract Refract. Surg., 15*, 545–553 (1989).

29. Davson, H., *Physiology of the Eye*, 5th ed., Pergamon Press, New York, 1990, p. 19.

30. Szycher, M., and W. J. Robinson (eds.), *Synthetic Biomedical Polymers. Concepts and Applications*, Technomic Publishing, Westport, CT, 1980, pp. 4–5.

31. Francese, J. E., L. Pham, and F. R. Christ, Accelerated hydrolytic and ultraviolet aging studies on SI-18NB and SI-20NB silicone lenses, *J. Cataract Refract. Surg., 18*, 402–405 (1992).

32. Mowbray, S. L., S.-H. Chang, and J. F. Casella, Estimation of the useful lifetime of polypropylene fiber in the anterior chamber, *Am. Intra-Ocular Implant Soc. J., 9*, 143–147 (1983).

33. Lerman, S., Effect of ultraviolet radiation (300–400 nm) on polypropylene, *Am. Intra-Ocular Implant Soc. J., 9*, 25–28 (1983).

34. Yang, S., H. Makker, and F. R. Christ, Accelerated ultraviolet aging of intraocular lenses. Part II: 50-year simulated aging of SI-18NB and SI-20NB silicone intraocular lenses, *J. Cataract Refract. Surg.* (in press).

35. Yang, S., H. Makker, and F. R. Christ, Accelerated ultraviolet aging of intraocular lenses. Part I: A new photo-aging model, *J. Cataract Refract. Surg.* (in press).

36. Provder, T. (ed.), *Size Exclusion Chromatography: Methodology and Characterization of Polymers and Related Materials*, American Chemical Society, Washington, DC, 1984.

37. Cazes, J. (ed.), *Liquid Chromatography of Polymers and Related Materials*, Marcel Dekker, New York, 1977.

38. Marchant, R. E., K. M. Mitler, and J. M. Anderson, *In vivo* biocompatibility studies. V. *In vivo* leukocyte interactions with Biomer, *J. Biomed. Mater. Res., 18*, 1169–1190 (1984).

39. Goldstein, J. I., D. E. Newbury, P. Echlin, D. C. Joy, C. Fish, and E. Lifshin, *Scanning Microscopy and X-Ray Microanalysis*, Plenum Press, New York, 1981.

40. Gomi, K., J. D. de Bruijn, M. Ogura, and J. E. Davies, The effect of substratum roughness on osteoclast-like cells *in vitro*, *Cells Mater., 3*(2), 151 (1993).

41. Howland, R. S., and M. D. Kirk, Scanning tunneling microscopy and scanning force micros-

copy, in *Encyclopedia of Materials Characterization* (C. R. Brundle, C. A. Evans, and S. Wilson, eds.), Butterworth–Heinemann, Boston, 1993, pp. 85–98.

42. Cunanan, C. M., N. M. Tarbaux, and P. M. Knight, Surface properties of intraocular lens materials and their influence on *in vitro* cell adhesion, *J. Cataract Refract. Surg.*, *17*, 767 (1991).

43. Brundle, C. R., X-ray photoelectron spectroscopy, in *Encyclopedia of Materials Characterization* (C. R. Brundle, C. A. Evans, and S. Wilson, eds.). Plenum Press, New York, 1993, pp. 282–299.

44. Cunanan, C. M., M. Ghazizadeh, and P. M. Knight, Surface characterization of intraocular lens materials and their influence on *in vitro* cell adhesion, *Proceedings of the American Chemical Society, Division of Polymeric Materials: Science and Engineering, 69*, 463 (1993).

45. Joy, D. C., A. D. Romig, and J. I. Goldstein (eds.), *Principles of Analytical Electron Microscopy*, Plenum Press, New York, 1986.

46. Chu, P. K., Dynamic secondary ion mass spectrometry, in *Encyclopedia of Materials Characterization* (C. R. Brundle, C. A. Evans, and S. Wilson, eds.) Plenum Press, New York, 1993, pp. 532–548.

47. Katz, B., Static secondary ion mass spectrometry, in *Encyclopedia of Materials Characterization* (C. R. Brundle, C. A. Evans, and S. Wilson, eds.), Plenum Press, New York, 1993, pp. 549–558.

48. Okada, K., M. Funahashi, K. Iseki, and Y. Ishii, Comparing the cell population on different intraocular lens materials in the eye, *J. Cataract Refract. Surg., 19*, 431 (1993).

49. Hirokawa, K., and Y. Majima, *Deposits on PMMA and silicone PCL in 3-pcs model*, presented at Symposium on Cataract, IOL, and Refractive Surgery, Los Angeles, March 1990.

50. Amon, M., and R. Menapace, Cellular invasion on hydrogel and poly(methyl methacrylate) implants: An *in vitro* study, *J. Cataract Refract. Surg., 17*, 774 (1991).

51. Standard test method for rubber properties in tension (D412), in *Annual Book of ASTM Standards*, Volume 08.01. Plastics (I). ASTM Publications, Philadelphia, 1985, pp. 154–167.

52. Standard test method for rubber property – Durometer hardness (D2240), in *Annual Book of ASTM Standards*, Vol. 08.02: *Plastics* (II), ASTM Publications, Philadelphia, 1985, pp. 324–327.

53. Standard test method for rubber property – Tear resistance (D624), in *Annual Book of ASTM Standards*, Vol. 09.01: *Rubber, Natural and Synthetic – General Test Methods: Carbon Black*, ASTM Publications, Philadelphia, 1985, pp. 152–155.

54. Steinert, R. F., *New advances in Nd:YAG laser therapeutics*, presented at American Society for Cataract and Refractive Surgery Symposium, Orlando, 1987.

55. *American National Standard for Ophthalmics – Intraocular Lenses – Optical and Physical Requirements*, ANSI Z80.7-1984, American National Standards Institute, New York, 1984, pp. 10–11.

56. Standard test methods for index of refraction of transparent organic plastics (D542), in *Annual Book of ASTM Standards*, Vol. 08.01. *Plastics* (I), ASTM Publications, Philadelphia, 1985, pp. 189–191.

57. Holladay, J. T., S. Van Gent, A. C. Ting, V. Portney, and T. R. Willis, Silicone intraocular lens power vs. temperature, *Am. J. Ophthalmol., 107*(4), 428–429 (1989).

58. Standard test method for haze and luminous transmittance of transparent plastics (D1003), in *Annual Book of ASTM Standards*, Vol. 08.01: *Plastics* (I), ASTM Publications, Philadelphia, 1985, pp. 513–520.

59. Williams, C. S., and O. A. Becklund, *Introduction to the Optical Transfer Function*, Wiley, New York, 1989.

60. Lang, A., and V. Portney, Interpretation of multifocal IOL modulation transfer functions, *J. Cataract Refract. Surg., 19*, 505–512 (1993).

61. Southwick, P. C., and R. J. Olson, Shearing posterior chamber intraocular lenses: Five-year postoperative results, *Am. Intra-Ocular Implant Soc. J., 10*, 318–323 (1984).

62. Stark, W. J., D. M. Worthen, J. T. Holladay, and P. E. Bath, The FDA report on intraocular lenses, *Ophthalmology, 90*, 311–317 (1983).
63. Jaffe, N. S., H. M. Clayman, M. S. Jaffe, and D. S. Light, The results of extracapsular cataract extraction with a Shearing posterior chamber lens implant 34–40 months after surgery, *Ophthal. Surg., 13* 47–49 (1982).
64. Stark, W. J., L. W. Hirst, R. C. Snip, and A. E. Maumenee, A two-year trial of intraocular lenses at the Wilmer Institute, *Am. J. Ophthalmol., 84*, 769–774 (1977).
65. Jaffe, N. S., D. M. Eichenbaum, H. M. Clayman, and D. S. Light, A comparison of 500 Binkhorst implants with 500 routine intracapsular cataract extractions, *Am. J. Ophthalmol., 85*, 24–27 (1978).
66. Martin, R. G., D. R. Sanders, M. A. Van der Karr, and M. DeLuca, Effect of small incision intraocular lens surgery on postoperative inflammation and astigmatism: A study of the AMO SI-18NB small incision lens, *J. Cataract Refract. Surg., 18*, 51–57 (1992).
67. Steinert, R. F., S. F. Brint, S. M. White, and I. H. Fine, Astigmatism after small incision cataract surgery: A prospective, randomized, multicenter comparison of 4- and 6.5-mm incisions, *Ophthalmology, 98*, 417–424 (1991).
68. Brint, S. F., M. Ostrick, and J. E. Bryan, Keratometric cylinder and visual performance following phacoemulsification and implantation with silicone small incision or poly(methyl methacrylate) intraocular lenses, *J. Cataract Refract. Surg., 17*, 32–36 (1991).
69. Shepherd, J. R., Induced astigmatism in small incision cataract surgery, *J Cataract Refract. Surg., 15*, 85–88 (1989).
70. Steinkogler, F. J., E. Huber, M. Aichmair, V. Huber-Spitzy, and E. Arocker-Mettinger, Heparin surface modified PMMA lenses in a prospective double blind study, *Eur. J. Implant Ref. Surg., 4*, 79 (1992).
71. Amon, M., and R. Menapace, Long-term results and biocompatibility of heparin-surface-modified intraocular lenses, *J. Cataract Refract. Surg., 19*, 258 (1993).
72. Holladay, J. T., H. van Dijk, A. Lang, V. Portney, T. R. Willis, R. Sun, and H. Oksman, Optical performance of multifocal intraocular lenses, *J. Cataract Refract. Surg., 16*, 413–422 (1990).
73. Percival, S. P. B., and S. S. Setty, Prospectively randomized trial comparing the pseudoaccommodation of the AMO ARRAY multifocal lens and a monofocal lens, *J. Cataract Refract. Surg., 19*, 26–31 (1993).
74. Percival, P., An update on multifocal lens implants, *Doc. Ophthalmol., 81*, 285–292 (1992).
75. Haefliger, E., J. M. Parel, F. Fantes, E. W. Norton, D. R. Anderson, R. K. Forster, E. Hernandez, and W. J. Feuer, Accommodation of an endocapsular silicone lens (Phaco-Ersatz) in the nonhuman primate, *Ophthalmology, 94*(5), 471–477 (1987).
76. Nishi, O., Refilling the lens with an inflatable endocapsular balloon: Surgical procedure in animal eyes, *Graefe's Arch. Clin. Exp. Ophthalmol., 230*(1), 47–55 (1992).

47
Prosthetic Materials in Ocular Surgery

P. Robert Ainpour and Jim Christensen
Partners in Biomaterials
Glendora, California

I. INTRODUCTION

The eye is a complex physiological device that has many anatomical details. The abbreviated schematic in Fig. 1 illustrates the salient features.

The human adult eye is approximately 2.5 cm in diameter and forms an image of the environment on its photoreceptive membrane, the retina. After some processing the information is sent to the brain via the optic nerve. In essence the eye is an extension of the brain. In its simplest form the eye may be compared to a camera, the sclera being the box and the retina the film. The lens of the eye, even though it focuses by accommodation (i.e., changes convexity), is similar to the camera lens.

Current technology provides remedies for failure of some of the components of the eye. In terms of biomaterial implants, the plastic replacement of the lens that has opacified has been the greatest achievements in prosthetic surgery. Approximately 1 million cataract surgeries are done in the United States alone, with a success rate of over 97%. One can say that for cataract surgery the evolutionary course has been completed successfully in the span of 40 years. In terms of currently doable things, our success has been with the lens and temporary aqueous replacements, while research is ongoing to find suitable replacements for vitreous and cornea. The following discussion is meant to provide an overview of the current ophthalmic biomaterials that are used routinely.

II. LENS PROSTHESES

A. Hard Lenses

Intraocular lens (IOL) prostheses have evolved since the the pioneering work of Dr. Harold Ridley in the 1950s [1]. These lenses have been and still are made of polymethylmethacrylate (PMMA). Most of the innovations that have occurred since their introduc-

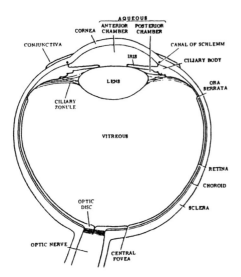

Figure 1   Abbreviated structure of the human eye.

tion have been in design and to a lesser extent in material processing. Certainly a badly designed lens will lead to complications in spite of the superior biocompatibility of PMMA.

$$-(-CH2-CHMe-)_n$$
$$|$$
$$C(O)OMe$$

Structure of PMMA, Me = CH3

The vast majority of the 1 million IOLs currently implanted annually in the United States are lenses that are placed in the capsular bag, that is, in the acellular basement membrane that contains the natural lens. Placing the lens in the bag seems to insulate it well from the surrounding tissue, therefore minimizing the risk of inflammation and other complications. In terms of percentages, more than 90% of the lenses currently implanted are made of PMMA. The PMMA stock is usually cast in the forms of rods or sheets of high molecular weight (MW) (i.e., 2 million). Suppliers of PMMA include ICI (Willmington, DE), Rohm and Haas (Philadelphia, PA), and Polycast (Stamford, CT).

Tissue biocompatibility of PMMA lenses is well documented [1] and usually there are no complications unless the IOL is positioned incorrectly, so that it causes chafing of pigmented uveal tissue. Other problems include: inadequate amount of polishing to remove sharp edges; residual polishing compound left behind on the lens, causing inflammation; or surface contamination.

B.   Soft Lenses

Soft lenses were introduced in the 1980s and the commercial ones so far fall into two categories: silicone elastomers and hydrogels. Some manufacturers are in the process of developing "soft acrylic" lenses, which are acrylic copolymers with a low enough glass transition to be folded at room temperature. The rationale for development of this type of lens is the ability to fold the IOL and then place it inside the eye through a smaller incision. Of course, this advantage accompanies a disadvantage: the need for special holders to hold the lens and additional skills during the surgery. Brief descriptions of the types of soft IOL are provided in the foowing sections.

1. Silicone IOLs

Silicone IOLs are usually manufactured by mixing two liquid resins and curing them inside a suitable mold. The chemistry is as follows:

$$CH2{=}CH-\underset{\underset{Me}{|}}{\overset{\overset{Me}{|}}{Si}}-O{-}(\underset{\underset{Me}{|}}{\overset{\overset{Me}{|}}{Si}}-O{)_n}\underset{\underset{Me}{|}}{\overset{\overset{Me}{|}}{Si}}-CH{=}CH2$$

Part A: Viny terminating polydimethylsiloxane

$$Me-\underset{\underset{Me}{|}}{\overset{\overset{Me}{|}}{Si}}-O{-}(\underset{\underset{Me}{|}}{\overset{\overset{H}{|}}{Si}}-O{)_n}\underset{\underset{Me}{|}}{\overset{\overset{Me}{|}}{Si}}-Me$$

Part B: Polyhydrosiloxane

Usually the manufacturer mixes the platinum (Pt) catalyst with component A. Component B can either be mixed with the A part without any catalyst or used in pure form. The former method has the advantage of mixing the two components in equal amounts whereas in the latter case a much smaller amount of the B part is needed. In order to impart tear strength to the elastomer, a filler is needed. In industrial applications fumed silica is used. This causes a haze because of refractive index mismatch. For IOL application a resin filler is used. Such an elastomer, even though not as strong as a silica-filled elastomer, has adequate tear strength and is optically transparent. Currently Nusil silicone technology (Carpenteria, CA) and Huls of America (Bristol, PA) supply silicone for medical applications. Dow Corning (Midland, MI) has abandoned the implantable silicone market, due to the problem with the silicone breast implants. Silicone elastomer chemistry is highly experimental and many of its aspects are "art." The *Encyclopedia of Polymer Chemistry* (Volume 15, 2nd. ed., John Wiley, New York) has an excellent chapter about silicones.

In order to take full advantage of the folding ability of soft lenses, it is advantageous to make them from a material with a refractive index equal to or higher than PMMA (1.49). The silicone elastomers made of dimethyl polysiloxane have a refractive index (RI) of 1.40. A recent development has been the introduction of silicone IOLs with RI of 1.46 by Allergan Medical Optics (Irvine, CA). This material has about 20% of the methyl groups substituted with phenyl. The advantage of these lenses is that for the same power, a thinner optics can be made, therefore making it easier to fold and insert.

Early silicone lenses had opacification and discoloration problems [2] that became apparent after implantation. These problems were traced to inadequate extraction and filtering of the resin components. Better extraction and filtration seems to have reduced these problems dramatically.

2. *Hydrogel IOLs*

Another category of soft lenses includes IOLSs made of hydrogels. So far the commercial hydrogel lenses have been produced from poly(hydroxyethylmethacrylate) (poly-HEMA).

$$\begin{array}{c} \text{Me} \\ | \\ +\text{CH2}-\text{C}+_n \\ | \\ \text{C(O)}-\text{O}-\text{CH2}-\text{CH2}-\text{OH} \end{array}$$

The water content of this material is about 40% and since its introduction by Alcon Ophthalmics (Forth Worth, TX), it has had problems due to design [3–7] and therefore has not had the popularity of silicone. Alcon had previously withdrawn these lenses from the market. Currently other manufacturers are attempting to introduce this material, not in a one-piece design (i.e., the entire lens is made of the same material), but with conventional haptics to give the lens more support in the eye [8].

Another concept, introduced by Optical Radiation Corporation (Azusa, CA), is the use of a low water content hydrogel (20%). The idea here is to combine the advantage of a foldable lens and the stability of a rigid lens. This lens, called a *memory lens*, is a copolymer of methylmethacrylate and HEMA plus other monomers. When this lens is heated it softens and can be folded. When cooled in the folded state (below 10°C) it retains its folded shape. The lens then can be inserted into the eye and placed in its intended position (usually inside the capsular bag or the posterior chamber). The heat from body will warm up the lens, which causes it to unfold. The advantage here is that no special folding holder is needed to fold and constrain the lens while inserting it into the eye.

In summary, prosthetic lenses are made of hard PMMA or soft silicone or hydrogels. All three seem to have comparable biocompatibility in this application as long as attention is paid to details such as removing impurities, residual surface contaminations, and sharp edges that come into contact with delicate tissues of the eye.

III. AQUEOUS HUMOR SUBSTITUTE

Aqueous humor is a fluid that is continuously secreted by the ciliary body in the posterior chamber. Its composition is mainly salts and water, without much protein (less than 0.02%). The secretion rate is about 2 μl per min, and the total volume it occupies is about 200 μl. Under normal conditions the fluid does not need to be replaced artificially. During cataract surgery it drains from the eye at the site of the incision. This allows the cornea to collapse, which can damage the endothelial cells on the inside of the cornea. The surgeon usually tries to bring the cornea back to its original shape in order to have space to remove the cataractous lens and replace it with a prosthetic IOL, without touching the endothelial cells.

Early attempts at fluid substitution included injection of balanced salt solution or air into the anterior chamber. During the 1980s polymeric solutions of hyaluronic acid were introduced for this purpose. The high viscosity that could be obtained with low concentrations (1% hyaluronic acid) provided a solution that could maintain the anterior chamber, if the polymer had a sufficient molecular weight.

The first hyaluronic acid solution was Healon®, which is a 1% solution of sodium hyaluronic acid of about 1–3 million daltons (Da) [9].

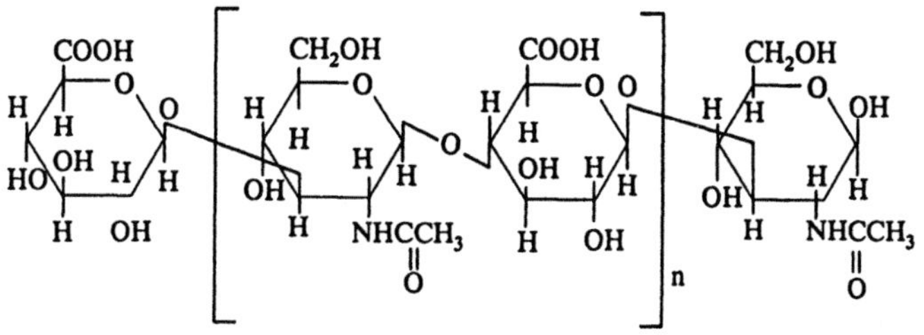

Hyaluronic Acid

This polysaccharide is naturally occurring in diverse organisms and even bacteria. Healon® is purified from rooster combs and is marketed by Pharmacia (Upssala, Sweden). However, recent fermentation technology has made it possible to obtain this material from bacterial sources as well. Other polymeric solutions that have been formulated for this purpose include: chondroitin sulfate [10] (another polysaccharide found in animal tissue), hydroxypropyl methylcellulose [11], and carboxymethylcellulose [12].

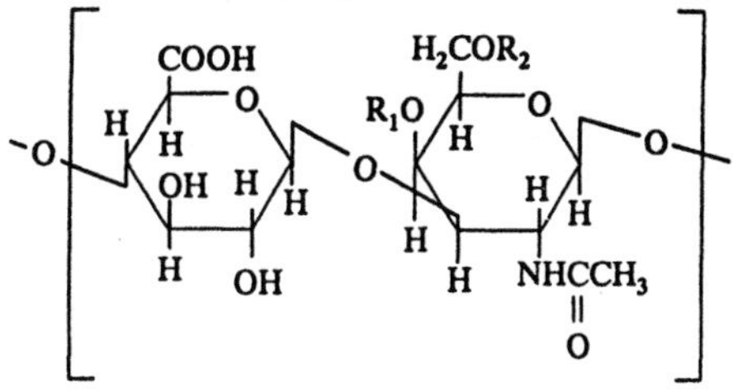

Chondroitin Sulfate

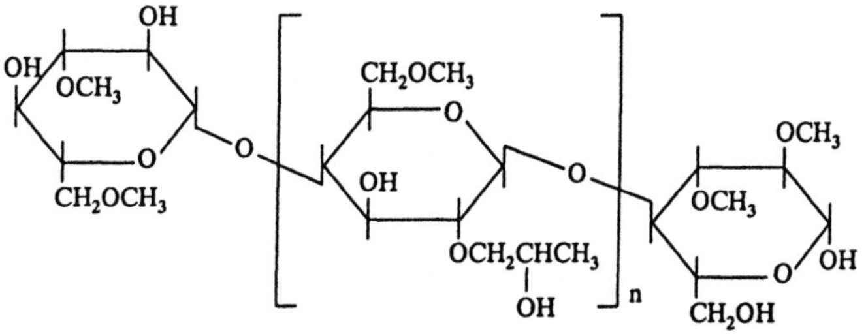

Hydroxypropyl Methylcellulose

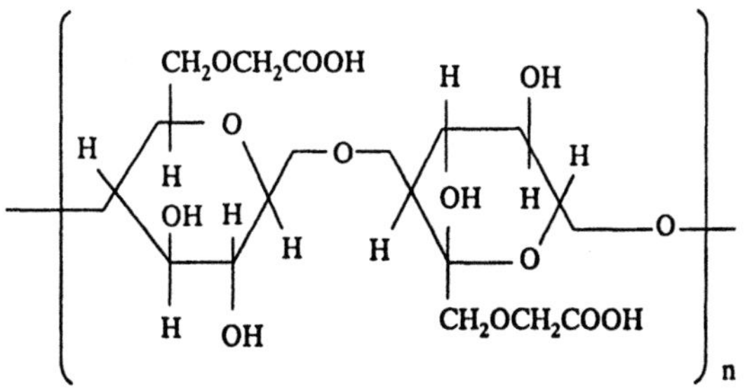

Carboxymethylcellulose

All of these substances need to be removed after the surgery so that clogging of the trabecular meshwork does not occur, which could result in an increased intraocular pressure that could lead to loss of the eye. In addition they must be free of particulates, pyrogenic agents, and inflammatory agents. Finally they must be sterile and osmolar.

IV. COLLAGEN SHIELDS

Another tool that has been added, in the past few years, as an aid in cataract surgery is the collagen coneal shields. These shields resemble contact lenses and are packaged in a dehydrated form. They are rehydrated before application, using balanced salt solutions or medications such as antibiotics and steroids. Since the lenses conform to the shape of the cornea, multiple base curves are not needed. The collagen lens after placement on the eye starts to dissolve gradually by the action of the proteoletic enzymes in the tear. Depending on the degree of crosslinking, the shields can last from 1 day to several days.

Several commercial manufacturers are currently in the market (Alcon, Chiron Ophthalmics, Oasis Medical, and Bausch and Lomb), and they use either bovine tendon or pig sclera (Bausch and Lomb) as the source of collagen. To prepare the lenses, usually the collagen is purified and made into a solution, cast into a suitable mold, and allowed to dry [13]; this results in a brittle collagen film, having the shape of a contact lens.

Studies that have been performed so far indicate that use of the shield results in less postsurgical trauma. Most doctors hydrate the shield in a solution of antibiotics and steroid to deliver medication [14–16].

V. OTHER PROSTHESES – THE FUTURE

Besides the devices mentioned that are in commercial production, research is being conducted to find materials that can permanently replace cornea [17], correct vision by implanting a lenticule inside the stroma of the cornea [18], and replace vitreous [19]. Currently none of these goals has been met due to host–tissue interaction problems. However, excellent long-term biocompatibility has been reported for a 71% water containing hydrogel that has been used as an intrastromal implant [18].

REFERENCES

1. Apple, D., Kincaid, M., Mamalis, N., and Olson, R., *Intraocular Lenses. Evolution, Design, Complications, and Pathology*, Williams and Wilkins, Baltimore, 1980.
2. Newman, D. A., McIntyre, D. J., Apple, D. J., Deacon, J., Popham, J. K., and Isenberg, R. A., Pathological findings of an explanted silicone intraocular lens, *J. Cataract Refractive Surg., 12*, 292–297 (1986).
3. Barrett, G. D., The evolution of hydrogel implants, *Dev. Ophthalmol., 22*, 70–71 (1991).
4. Noble, B. A., Hayward, J. M., and Huber, C., Secondary evaluation of hydrogel implants, *Eye, 4*, 450–455 (1990).
5. Neumann, A. C., and Cobb, B., Advantages and limitations of current intraocular lenses, *J. Cataract Refractive Surg., 15*, 257–263 (1989).
6. Menapace, R., Skorpik, C., Juchem, M., and Scheidel, W., Evaluation of the first 60 cases of polyHEMA posterior chamber lenses implanted in the sulcus, *J. Cataract Refractive Surg., 15*, 264–271 (1989).
7. Menapace, R., Amon, M. R., and Adax, U., Evaluation of 200 consecutive IOGEL 1103 capsular bag lenses implanted through a small incision, *J. Cataract Refractive Surg., 18*, 252–264 (1992).
8. *Ocular Surgery News*, Vol. 11, No. 24, 1993.

9. Balazs, E. A., Ultrapure hyaluronic acid and the use of thereof, U.S. Patent 4,141,973, 1979.
10. Hammer, M. E., and Burch, T. G., Viscous corneal protection by sodium hyaluronate, chondroitin sulfate, and methyl cellulose, *Invest. Ophthalmol. Vis. Sci., 25*, 1329–1332 (1984).
11. Fechner, P. U., and Fechner, M. U., Methylcellulose and lens implantation, *Br. J. Ophthalmol., 67*, 259–263 (1983).
12. Goldberg, E. et al., U.S. Patent 4,819,617, April 11, 1989.
13. Fedorov, S. N., U.K. Patent GB 2196008 A, April 20, 1988.
14. Poland, D. E., and Kaufman, E., Clinical use of collagen shields, *J. Cataract Refract. Surg., 14*, 409 (1988).
15. Pleyer, U., Legmann, A., Mondigo, B. J., and Lee, D. A., Use of collagen shields containing amphotericin B in the treatment of experimental *Candida albicans*-induced keratomycosis in rabbits, *Am. J. Ophthalmology, 113*, 303 (March 1992).
16. Marmer, R., Therapeutic and protective properties of the corneal collagen shields, *J. Cataract Refract. Surg., 14*, 496 (1988).
17. *Ophthalmology Times, 18*(23), 1 (1993).
18. Parks, R., and McCarey, B., Hydrogel keratophakia: Long term morphology in the monkey model, *CLAO J., 17*(3), 216 (1991).
19. Chan, I. M., Tolentino, F. I., Refojo, M. F., Fournier, G., and Albert, D., Vitreous substitute, *Retina, 4*(1), 51 (1984).

X
DENTAL APPLICATIONS

48
Materials Science and Technology in Dentistry

B. K. Moore and Y. Oshida
Indiana University School of Dentistry
Indianapolis, Indiana

I. INTRODUCTION

Research and development in dental materials and manufacturing technologies, and the characterization and testing methods involved in dental materials are some of the best examples of interdisciplinary science and technology. For example, the disciplines might include physics, chemistry, metallurgy, mechanics, and surface science, as well as biological science. Although empiricism and trial and error played a major role in early dental materials research, in the last 20 years progress has been made toward a multidisciplinary scientific approach to the study of dental materials and the development of better products [1].

Presently more than 50 different materials are used for prosthodontic devices. These materials fall into categories of metals, ceramics, and polymers [2]. Recently, more nondental manufacturers (i.e., engineering industries) have begun to produce dental products, using dental materials as an outlet for expertise developed for other industrial applications. For example, certain new ceramics and castable glasses originate primarily from nondental companies [3].

The advances in research and development in dental materials are rapid. Only 4 years ago it was predicted that research on fluoride-releasing agents would result in polymeric materials that could provide a small, steady rate of intraoral fluoride release to reduce the incidence of recurrent caries and enhance the usefulness of certain therapies, such as pit-and-fissure sealants [4]. The development of fluoride-releasing polymers is a major research activity being currently conducted jointly by the Materials Department of Indiana University School of Dentistry and the Chemistry Department of Purdue University [5].

Not only materials but also technologies developed and successfully used in industry can be transferred to dentistry.

The computer-aided design and machining system (CAD/CAM) is an excellent example. Dental researchers are developing computer-aided design (CAD) systems that may be used in the near future to design dental restorations. Investigators are also combining computer-aided design and artificial intelligence (AI) to design complex prosthetic devices such as partial dentures. With the CAD/CAM systems, unconventional materials can be used to fabricate restorations. The dentist and dental laboratory can use any machinable material. Computers have also been used to analyze stress on dental restorations and natural dentition. They can simulate internal stress distributions under different conditions and loading situations, and the results can be used to optimize the computer-aided design process (for example, two-dimensional, 2-D; and three-dimensional, 3-D, finite element modeling, stress analyses). CAD/CAM, as it now is being applied to dentistry, could eliminate the need for impressions and temporary restorations by designing and fabricating a finished appliance within a single patient appointment [6].

The application of lasers to dentistry is another good example of technology transfer. The laser has been used for enamel/dentin preparation prior to restorative material placement and for caries prevention [7–9]. Various ways of laser application for surface modifications of metals have also been introduced; they include cutting, welding, surface hardening, and laser surface alloying [10].

In the following sections, noble metals and their alloys, dental amalgam, metal-ceramic veneers, titanium and its alloys, resins, plastics, composites, and cements are discussed as they are currently being used or evaluated for use in dentistry.

II. REGULATION, SPECIFICATIONS, AND STANDARDS

Until the passage in 1976 of the Medical Device Amendments to the Food and Drug Act, medical and dental materials and devices for use in the human body were not regulated by any agency of the United States government. The only exception was materials for which therapeutic claims were made, which allowed the Food and Drug Administration (FDA) to consider them as a drug.

Long before 1976, the dental profession had realized a need for the development of criteria to help ensure the safety and efficacy of dental material. The first efforts were initiated by the U.S. Army, which commissioned the National Bureau of Standards now the National Institute for Standards and Technology (NIST) to develop specifications for dental amalgam alloys used in federal service. In 1928 the American Dental Association (ADA) assumed sponsorship of this program at NBS to develop standards for dental materials. The ADA adopted these standards as specifications in its Certification Program for Dental Materials. Each specification contained requirements for physical, mechanical, and chemical properties of a material that were felt to be relevant to the clinical application of the material and would ensure safety and efficacy. Materials that met the relevant specification were granted certification by the ADA. Lists of certified materials were published in *The Journal of the American Dental Association* and the manufacturer was permitted to display the ADA Seal of Certification on the product. Compliance with this program was voluntary. The ADA Council on Dental Materials, Instruments, and Equipment has the administrative responsibility for the certification program. This council also acts as administrative sponsor of the American National Standards Institute's (ANSI) Accredited Standards Committee MD156. Committee MD156 appoints subcommittees to formulate and revise specifications. These are submitted to ANSI and, if accepted, become American National Standards (ANS). The ADA Council then has the option of adopting the ANS as a specification under the ADA certification program. There are now 51 specifications for various dental materials.

A similar situation exists at the international level. The Fédération Dentaire Internationale (FDI) represents organized dentistry and acts in cooperation with the International Standards Organization (ISO) through the ISO committee TC106—Dentistry.

As of January 1992 there were 64 ISO standards under TC106, many of which have also been accepted as FDI specifications. The dental materials industry has rapidly developed into a worldwide industry with a world market. As a result, the ISO standards are becoming increasingly important.

The Medical Device Amendments of 1976 gave the FDA jurisdiction over all materials, devices, and instruments used in the diagnosis, cure, mitigation, treatment, or prevention of disease in man. This includes materials used professionally and the over-the-counter products sold directly to the public. Nineteen panels were established representing different areas of medicine and dentistry for the purpose of classifying all medical and dental materials. The Dental Panel places an item into one of three classes:

Class I, materials posing minimum risk: These are subject only to good manufacturing and recordkeeping procedures.

Class II, materials for which safety and efficacy needs to be demonstrated and for which performance standards are available (established by the FDA or other authoritative body such as the ADA in its certification specifications): Materials must be shown to meet the performance standard.

Class III, materials that pose significant risk and materials for which performance standards have not been formulated: This class is subject to premarket approval by the FDA for safety and efficacy, in much the same manner as a new drug.

The regulatory functions of the FDA have made the activities of the ADA and ANSI even more important. The time required to develop and approve a new federal regulation is such that the FDA would have been faced with an almost impossible task to develop new specifications for the large number of dental products that have been placed in Class II.

III. NOBLE METALS

A. Gold Alloys

1. General

Until 1968 gold-based, casting alloys with a high noble metal content were normally used in dentistry for fixed partial prostheses such as inlays, crowns, and bridges. Deregulation of the price of gold resulted in steadily rising Au prices, which were reflected in the cost of these alloys. Lower gold content alloys, alloys based on other noble metals, and base metal alloys were developed as alternatives to the traditional dental gold casting alloys [11]. This section discusses the noble metal casting alloys used in dentistry to fabricate all metal prostheses. Alloys intended for porcelain veneering are discussed in another section.

The lost wax casting process used in dentistry today dates to 1907 and the work of Taggart. The mechanical properties of pure gold were quickly found to be inadequate for dental castings, and the profession turned to gold-based jewelry alloys, which were basically ternary alloys of Au, Cu, and Ag. Trial and error resulted in a number of formulations that were shown to be clinically acceptable. The primary requirements used in alloy evaluation were resistance to tarnish and corrosion in the oral environment, and adequate strength to withstand oral stresses. As the number of available alloys grew they came to be classified into four groups based on their properties, clinical applications, and composition. This classification was later formalized in the American Dental Association Specification No. 5 for Dental Gold Casting Alloys. The current requirements of Specification 5 are shown in Table 1. Note that as one moves from Type I gold alloys to Type IV, the mechanical properties increase and the total noble metal content decreases. Although corrosion and tarnish resistance are main requirements for a dental casting alloy, the specification contains no corrosion criteria. It was determined by clinical experience that a total noble metal content of at least 75% would make a dental gold alloy corrosion resistant. The noble metals included are Au, Pt, and Pt group metals. Silver is not considered a noble metal in the oral environment since it is subject to attack by Cl and S. These traditional gold alloys contain Au as the major constituent. Small amounts of Pt and/or Pd are added to increase the mechanical properties of the alloy. Copper and Ag are added to strengthen the alloy; Ag also helps offset the reddening effect of Cu. Historically the gold color of these dental alloys has been considered desirable. Early development clearly indicated that if too much Cu or Ag were added to strengthen the alloy, the tarnish and corrosion resistance decreased. For example, a 16- or even 18-karat jewelry alloy may not be corrosion resistant in the mouth. The addition of small amounts of Pt or Pd helps to maintain corrosion resistance as the total noble metal content decreases.

Low-Au content alloys may be suitable alternatives for conventional Au casting alloys if two conditions are met. First, to prevent the occurrence of tarnish and corrosion, the absolute Au content must be at least 42 wt%. Such low-Au content can be accommodated only if the Pd content is at least $3 \sim 4\%$. Second, to ensure that an alloy is readily burnishable, the (Au + Ag) to Cu ratio should be at least 10:1. Increasing the Cu content at the expense of Au and Ag reduces ductility and elevates hardness, which decreases the burnishability index [12]. Composition of nine modern Au–Cu–Pd alloys, representing the full range of alloy requirements, is given together with a table of their physical properties [13].

For many years the specification for minimum Au and Pt contents in alloy composition was considered sufficient to guarantee the stability of cast gold dental prostheses in the oral environment. Research has revealed a considerably more complex situation and stresses the need for extensive laboratory and clinical testing of new alloys formulation before these may be unreservedly recommended to practitioners [14]. There is still not a universally accepted laboratory test criterion for tarnish and corrosion resistance of noble metal dental casting alloys.

Developers of alternative casting alloys for the conventional ADA types I–IV gold alloys have tended to use an ADA type to designate the clinical application intended for alternative alloys. Hence, there are marketed type III and type IV alloys that contain no gold.

One possible classification scheme for the numerous alloys available is shown in Table 2. Physical properties of typical alternative alloys are presented in Table 3 [15]. The qualities of alloys required for restorative dental devices are described in [16]. The properties of a range of Au–Ag–Cu, Au–Ag–Cu–(Pt group metal) and Ag–Cu–Pd alloys have been described. It is concluded that there are very few alternatives to Au alloys that can match their outstanding combinations of strength, castability, and corrosion resistance in the oral environment [16].

Table 1   ADA Specification No. 5: Requirements for Gold Casting Alloys—Composition, and Mechanical and Physical Properties

ADA-type gold casting alloy	Composition: Au, Pt, Pd, minimum %	Vickers hardness (VHN)			Yield strength (0.1% offset), MPa		Percent elongation		Melting range, °C	
		Quenched:		Hardened: MIN	Quenched: MIN	Hardened: MIN	Quenched: MIN	Hardened: MIN	Solidus: MIN	Liquidus: MAX
		MIN	MAX							
I	83	50	90	–	–	–	18	–	800	1050
II	78	90	120	–	140	–	12	–	800	1050
III	78	120	150	–	200	–	12	–	800	1050
IV	75	120	–	220	340	500	10	2	800	1050

Table 2 Typical Compositions of Modern Noble Metal Dental Casting Alloys

Type		Percent Au	Percent Cu	Percent Ag	Percent Pt and Pd	Zn, In, Sn, Fe, Ga, other metals
I	Gold	83	6	10	0.5	Balance
II	Gold	77	7	14	1	Balance
III	Gold	75	9	11	3.5	Balance
	Low gold	46	8	39	6	Balance
	Silver–palladium			70	25	Balance
IV	Gold	69	10	12.5	6	Balance
	Low gold	56	14	25	4	Balance
	Silver–palladium	15	14	45	25	Balance

Table 3 Typical Physical Properties of Modern Noble Metal Dental Casting Alloys

Type		Melting range, °C	Density, g/cm³	Yield strength,[a] MPa	Hardness[a] (VHN)	Elongation,[a] %
I	Gold	943–960	16.6	103	80	36
II	Gold	924–960	15.9	186	101	38
III	Gold	932–960	15.5	207	121	39
				H275	H182	H19
	Low-gold	843–916	12.8	241	138	30
				H586	H231	H13
	Silver–palladium	1021–1099	10.6	262	143	10
				H323	H154	H8
IV	Gold	921–943	15.2	275	149	35
				H493	H264	H7
	Low-gold	871–932	13.6	372	186	38
				H720	H254	H2
	Silver–palladium	930–1021	11.3	434	180	10
				H586	H270	H6

[a]H = full age hardened condition. All other values are for quenched (softened) condition.

2. Mechanical Properties

Dental prostheses are usually made by the lost wax process because of its dimensional precision. The four classifications of alloys, as shown in Table 1, are generally employed for the following restorations:

Type I alloys are used for small inlays not subject to high stresses; they have been mostly replaced by direct filling materials such as composite resins, amalgam, and glass ionomer cements.

Type II alloys are used for multiple surface inlays subject to modest occlusal stresses; they also have been largely replaced by direct filling materials.

Type III alloys are used for full coverage crowns and short span bridges.

Type IV alloys are used for large prostheses supporting high stresses and for frameworks for removable partial dentures.

X-ray fluorescence analysis of more than 100 commercial alloys showed that Au content decreased from 86% to 69% on going from type I to IV. Hardening by thermal treatment of binary Au-Cu alloys is discussed as well as ternary Au-Ag-Cu modified by minor additions and complex commercial alloys. Hardness changes in the binary alloys are compared with lattice changes due to superlattice formation during annealing. In the ternary alloys, dilatometric, hardness, and x-ray results show hardening results from decomposition of the solid solution into Ag- and Cu-rich phases and superstructure formation in the Cu-rich phase. Three Ag-Au-Cu alloys with Pt, Pd, and Zn additions were studied to simulate commercial alloys. In general, the same two hardening mechanisms operate in the complete alloys. Zn is a particularly effective hardener due to precipitation of Au_3Zn and formation of the AuCu superlattice [17]. Table 1 shows the requirement that Type IV alloys be hardenable by an appropriate heat treatment. This hardening process results in very significant increases in hardness and yield strength along with decreases in ductility.

Dilatometric, lattice parameter, and microhardness measurements have been reported on three alloys in the Au-Ag-Cu system consisting 58, 60, and 64 at.% Au, respectively. Similar tests were made on three complex Au-Ag-Cu hardenable dental alloys containing 73, 74 and 75 at.% Au with additions of Pt, Pd, and Zn, and the report includes a discussion of the relative contributions to hardening of ordering and precipitation. The effect of additions of up to 1.6% Zn was in particular studied and it was shown that in the hardening of Pt-modified alloys, a martensitic type of transformation may be involved [18].

3. Fatigue

To determine the effects of natural notches corresponding to interdental spaces, a simulated lower maxillary bridge with abutments on a canine and first molar (4 unit) was cast in a Ni-Cr (65Ni, 24Cr, 8Fe, 2.3Si) and a traditional Au alloy (80.5Au + Pt + Pd + Ir, 13Cu, 6.5Ag). A load of 600 N was applied by an Amsler machine at a frequency of 1 Hz. Tests were conducted with and without fixing the end supports. The Au alloy failed in 1000 cycles with unsupported ends and in 50,000 ~ 300,000 cycles supported. The Ni-Cr failed at 10,000 ~ 22,000 cycles unsupported and at more than 70,000 supported. Fatigue cracks appeared in the intertooth spaces. Design of the bridge was highlighted as essential in achieving fatigue life [19].

It is difficult to produce soldered joints without defects such as porosities, voids, and introduction of extraneous particles. Fractography by scanning electron microscopy (SEM) of soldered joints in 18 bridges of dental gold alloys showed failure initiation at a pore plus multicrack initiation sites leading to a fatigue crack with semicircular striations spreading from the pore [20].

Most research on solder joints has been based on static tensile tests. A study generated fatigue data by establishing the S-N diagrams (applied stress vs. number of cycles to failure) of solder joints made using a gold-palladium alloy and a 51.1Au-38.5Pd-8.5In-1.5Ga soldered with 585 solder (58.5Au-21Ag-14Cu-4.7Zn-1.3Sn) [21]. Alloy specimens were cast and soldered, and the solder joint milled and polished. The cantilevered specimen was loaded at 70 strokes per minute with an applied tensile stress between 294.2 and 686.5 MPa. Three specimens were cycled for each stress level until fracture occurred. On a logarithmic scale the relationship between applied stress and cycles to failure was linear. No definite endurance limit could be detected within the stress range examined. The observed mode of failure was identical with that recorded by Wictorin and Fredrikson [20] on joints in partial dentures that had failed clinically. This indicates that fatigue behavior of solder-alloy combinations should be evaluated as well as maximum tensile strength [21].

4. Age Hardening

Gold alloys that can be hardened by thermal aging play an important role in dentistry. But the ordering and precipitation phenomena responsible for hardening are complex and not fully understood.

The age-hardening characteristics of a commercial dental Au alloy—with 11 wt% Pt; 6% Pd; 6% Ag; 9% Cu—were extensively studied by resistometric measurements, hardness tests, x-ray diffraction, and electron microscopy. The resistometric and hardness studies showed that age hardening occurred in two stages. The first stage arose from the nucleation of the AuCu I type ordered structure and was characterized by a slow growth rate of the ordered platelets. In the second stage, simultaneous ordering and precipitation took place. An alternating lamellar structure composed of the AuCu I type ordered platelets and Pt-rich face-centered cubic (FCC) precipitates was observed in electron micrographs. The age hardening in this stage was considered to arise primarily from ordering and to be brought about with the aid of precipitation. The grain boundary precipitates did not contribute to the age hardening in the alloy studied [22-24].

The effect of heat treatment on microstructure, microhardness, tensile properties, and corrosion behavior of two low-Au dental alloys has been studied. After quenching, specimens were heat treated in the temperature range from 300° ~ 500°C. Microhardness, yield strength, and tensile strength showed maximum values between 300° ~ 400°C, and then decreased at higher temperatures. Potentiodynamic curves revealed that nobility affects Ag corrosion activity [25].

Transmission electron microscopy (TEM) and selected area electron diffraction (SAED) studies were conducted to clarify a difference in age-hardening mechanism with changes of Au content (75 ~ 58.3%) in three Au-Cu-Ag ternary alloys for which the ratio of Cu to Ag remained at 65:35. Age hardening in the 75% Au alloy was attributed to the formation of the AuCu I type ordered platelets on the matrix {100}. In the lower Au alloys, a periodic antiphase domain structure of the AuCu II type ordered phase made a major contribution to age hardening [26].

Aging behavior of a low-Au, white dental alloy was investigated by electrical resistivity measurements, x-ray diffraction, and electron microscopy. It was characterized by both grain interior and grain boundary reactions. The coexistence of AuCu I and CuPd in dental alloy was confirmed [27].

Three Cu-free type III yellow low-Au alloys (Ag–Au–In–Pd–Zn) were investigated. Microstructural analysis was performed with x-ray diffraction, light microscopy, and electron x-ray microanalyses. The electrochemical behavior of the alloy was tested with standard potentiodynamic techniques in 0.9% saline solution and an artificial saliva and compared with that of a traditional high-Au alloy. The three alloys turned out to be two-phase mixtures of a FCC (matrix) and a body-centered cubic (BCC—island) phase. The yellow color, present in spite of the low Au content, was caused by the disordered Pd–In, island-type structure. An additional third structure was seen after etching, However, no distinct phase could be correlated with it. The corrosion behavior for the experimental alloys, based on cyclic polarization, can be considered acceptable in both electrolytes used but is inferior to that of a conventional gold control alloy [28].

The structure and concentration gradients in both as-cast and heat-treated low-Au and conventional-type III Au alloys were studied. A much more lamellar eutectic phase was found at the grain boundaries in the low-Au alloy. TEM investigations showed that the interior of the grains consisted of the fine lamellae, which probably were alternating Au–Ag- and Au–Cu-rich bands due to the miscibility gap in the solid state. Microprobe analyses, where the beam overlapped several of the observed lamellae, displayed both the interdendritic and grain boundary segregations to be much larger for the low-Au alloy than for the type III alloy. The lamellae observed in the as-cast state are quickly dissolved at 700°C into one phase, but the relaxation by diffusion of the concentration differences associated with grain boundary segregations required several hours because of the much larger distances involved. Aging at 350°C caused precipitation of ordered face-centered tetragonal (FCT) particles. On the basis of structure and alloy composition, they are most likely AuCu I and may contain some Pd [29].

Aging behavior in a dental low-Au alloy Au–20Ag–17Cu–3Pd–1Pt–4Zn was studied by hardness tests, resistivity measurements, x-ray diffraction, SEM, and TEM. Aging at 500°C produced two stages that were accompanied by hardening and softening. They corresponded to a reaction at the grain interior and at the grain boundary, respectively. Aging at 400°C was completed in three stages. In the intermediate aging period, an additional hardening due to the formation of a CuAu II-type structure was observed. The final microstructure depended on the aging temperature associated with the homologous temperature [30]. In low-gold alloys, 40 ~ 59.5 wt% Au, with elevations in Cu and Pd, heat-hardening curves indicated that the maximum aging effects were attained after 30 min at 400°C [31].

The traditional explanation for the heat treatment hardening mechanism of Au–Ag–Cu dental and jewelry alloys is the conversion of the α-disordered solid solution to the ordered state. Evidence indicates that there is a steep hardness dip at precisely 50 at.% for the Au–Cu binary, indicating that it is the ordered state that is the very hard state upon aging this composition after solution anneal. A series of Au–Ag–Cu alloys were prepared in the vicinity of the 50% Au content and it was found that striations occurred in the microstructures, which increased in fineness further away from stoichiometry. The striations appeared metallographically on aging, suggesting that this was the microstructural form of the ordered state, and that the finer the striations, the harder was the resulting alloy. However, the maximum hardness occurred when the striations had not yet appeared, suggesting that it was the atomic clustering, leading to formation of the ordered state, that has the high hardness, much like the age hardening in certain Al–Cu alloys due to Guinier–Preston zones [32]. The composition and temperature range of the AuCu-type superlattice formation in the Au–Cu–Ag ternary system was determined by x-ray powder diffraction. Four distinguishable phase transformations were detected: disordered FCC phase $\rightarrow$ AuCu I α $\rightarrow$ AuCu II α $\rightarrow$ α_1 (Cu-rich FCC phase) + α_2 (Ag-rich FCC phase) and α_2 $\rightarrow$ AuCu$_3$. The stable temperature range of AuCu I-type superlattice was restricted to a lower temperature range by adding Ag to stoichiometric and off-stoichiometric Au–Cu alloys. Depending on the increase in Ag content, the coexisting region of the AuCu II-type superlattice and the Ag-rich α_2 phase extended to lower temperatures and over a wide composition range [33].

5. Tarnishing and Corrosion

As previously noted, tarnishing of Au alloys, both *in vivo* and *in vitro*, increases when the content of Au and Pt group metals is reduced below 75 wt%. Aqueous solutions of sulfides have in particular been shown to promote tarnishing. Annealing to a single-phase material has been found to significantly improve the tarnish resistance of low-gold alloys. This effect is thought to be mainly due to the removal of local galvanic cells originating from thin lamellae (20 nm) consisting of alternating phases containing predominantly AuAg and AuCu. Even single-phase low Au containing alloys tarnish *in vitro* if the Ag content becomes as high as 61 wt%. Furthermore, it has been shown by tarnish testing of the coarse-grained mixtures of such phases produced by long-time annealing that the AuAg phase tarnished, while the AuCu phase remained bright. The distinct difference in the observed tarnish resistance between these two phases could partly be explained by electron spectroscopy for chemical analysis (ESCA) measurements of the surface composition of binary, single-phase AuAg and AuCu alloys (both with 30 at.% Au) after immersion in chloride- or sulfide-containing solutions. It was found that Cu, in contrast to Ag, was selectively dissolved from the 2 or 3 uppermost atomic layers, leaving behind a surface significantly enriched in Au. The increase in Au content in the surface layer of the AuCu alloy was found to be accompanied by an increase in rest potential. The galvanic cells that are created by the alternating lamellae in low-Au castings probably enhance cathodic reactions on the AuCu phase. In the oral environment, however, the alloy surface is covered with a thin protein film that may increase or decrease the corrosion rate. *In vitro* tests of the same alloys in aqueous solutions of chlorides or sulfides containing proteins (mucin) showed no noticeable effect of the mucin addition. Testing of the same alloys *in vivo* for 8 days revealed the same tendencies to selective dissolution of Cu and tarnish of the AuAg [34].

Single-phase alloys with 30 at.% Au and with different proportions of Ag, Cu, and Zn have been investigated with regard to selective dissolution. Measurement of rest potentials and ESCA analyses show that Cu and Zn are preferentially dissolved in aqueous solutions of chloride and small concentrations of sulfide, leaving behind a surface significantly enriched in Au with a thickness of approximately 1 nm. No significant selective dissolution of Ag was observed in the solutions investigated. Only alloys containing substantial amounts of Ag tended to become tarnished in sulfur containing solutions. It is suggested that Ag$_2$S is more easily formed than sulfides of Cu and Zn because other oxidative reactions to oxides or hydroxides of these elements take place, which is not the case for Ag [35].

Laboratory test methods which have been used to evaluate tarnish and corrosion of dental Au alloys include immersion in various sulfide, chloride, and artificial saliva solutions and examination by Auger electron spectroscopy (AES) and optical instruments. Corrosion tests consist mainly of electrochemical studies. Results of the laboratory tests are related to available clinical data. Independent test methods are shown to give similar conclusions concerning tarnish and corrosion resistance of a particular alloy. The ability of immersion and electrochemical tests to predict clinical behavior was investigated for five low Au content alloys ranging from 44 to 62 wt% (Au + Pd), and a good correlation between *in vivo* and *in vitro* behavior was observed [14].

Studies on tarnishing of a low-Au alloy, both as cast and solid solution annealed, have been carried out *in vivo* and compared with that of a type III Au alloy. Small polished samples were partly embedded in complete dentures for more than 9 months. Generally, the low-Au alloy in the as-cast state was found to be the most prone to tarnish, followed by the same alloy, solid solution annealed. These results are in agreement with previous *in vitro* studies using 2% solutions of Na_2S. The tarnish attacks followed the pattern of Cu segregations *in vitro*, and frequently also *in vivo*. Previous findings *in vitro*, on the other hand, show that tarnishing occurs predominantly on Ag-rich phases. In the Cu-rich regions along grain boundaries and in interdendritic positions, eutectic particles with thin lamellae (~ 0.05 μm) of alternating Ag and Cu phases occurred. In these areas is found a more unstable matrix with increased tendency to splitting into similar, but even thinner, lamellae. Most likely, these lamellae of alternating compositions act as galvanic cells. The cathodic reduction of oxygen probably takes place on the Cu-rich lamellae, which are Au enriched on the surface due to initial selective dissolution of Cu. With increasing time, attacks up to several micrometers in depth and comprising both kinds of lamellae tend to occur at the grain boundary regions in an *in vivo* test, probably by a fretting corrosion mechanism [36].

A wide composition range of commercially available dental casting alloys containing Au, Ag, Pd, In, and Cu were evaluated for corrosion resistance. Based on polarization measurements, high-Au alloys are highly corrosion resistant and exhibit the lowest corrosion rates; intermediate Au, Ag, and Pd alloys with Cu are passive but exhibit small corrosion rates. In-Ag alloys (20%) exhibit active corrosion at potentials only 100 mV above the corrosion potential [37].

Tarnish tests were carried out on Au-Cu-Ag and Au-Au-Pd alloys. This study focused on the individual and combined effects of nobility, Pd content, and microstructure. Tarnish resistance was almost perfect for the alloys with more than 50 at.% noble metals but seems to relate to the Pd:Au atomic ratio for the alloys with low nobility. Palladium inclusion reduced the tarnish susceptibility up to ~ 10 at.%. Tarnishing of alloys with low nobility was very sensitive to microstructure. The tarnishing susceptibility of a dual-phase Au-Cu-Ag alloy was twice as high as that of the single-phase alloy. However, a Pd-bearing alloy showed no increase in the degree of tarnish as a result of phase separation [38].

Electron spectroscopic studies of Au-Ag-Cu alloys of the type used for dental castings show that small additions (< 3 wt.%) of Pd reduce the thickness of the sulfide layer formed on surfaces of samples treated in Na_2S solutions. Relative to Ag, Pd does not enrich in the sulfide layer, but significant Pd enrichment is found immediately below the sulfide layer. The mechanism of the impeding effect of Pd on sulfide formation is assumed due to a decrease in diffusion from the bulk alloy to the surface due to the enriched layer. The effect cannot be explained by change in the electronic structure of the alloy due to Pd alloying [39].

The corrosion resistance of homogeneous gold–palladium–silver experimental alloys was investigated in a chloride solution. Analogous measurements were made with alloys containing intermetallic phases that may be present in dental alloys of more complex composition. The results indicate a high corrosion resistance for the homogeneous Au-Ag alloys, since their free corrosion potentials are within a region of their anodic polarization curves where the current density is extremely low. Within the composition limits of these alloys the substitution of gold by palladium did not reduce significantly the extent of this region of immunity. Intermetallic phases, however, as present in the systems silver-tin, silver-palladium-tin and silver-palladium-indium show a clearly restricted region of immunity and, at the free corrosion potential, are dissolved at a relatively high rate [40].

6. Biocompatibility

Dental alloys are required to be biocompatible, corrosion and tarnish resistant in the mouth, strong, stiff and ductile, and easy to handle in the dental laboratory. These requirements are met by the conventional age-hardening Au alloys with sufficient noble metal content. However, each of them must be examined for noble metal alloys that have compositions outside the range of ADA Specification No. 5.

In the past decade a variety of electrochemical and cytotoxicity tests have been used to assess the biocompatibility of dental alloys. Little evidence has been presented that would substantiate a relationship between alloy corrosion products and alloy toxicity. A study was designed to correlate data from *in vitro* corrosion and cytotoxicity tests on 17 Cu-rich alloys in the Au-Cu-Ag ternary system. Corrosion behavior of the 17 alloys was monitored using potentiodynamic polarization techniques. *In vitro* evaluation of cytotoxicity was conducted as described in ANSI/ADA Specification No.41 for Biological Evaluation of Dental Material (Sec. 4.4.1). The release of Cu into the culture medium was quantified by atomic absorption spectrophotometry. Linear numerical analysis of the data revealed the following: The tendency of an alloy to corrode was related chiefly to alloy Cu content; current density recorded at an overpotential of $+400$ mV vs. standard calomel electrode (SCE) was related to the amount of Cu released into the culture medium; positive correlation between alloy potentiodynamic corrosion behavior and elicited cytotoxic response was displayed. The resultant mathematical model enables accurate prediction of cytotoxic response from electrochemical data [41].

Three ternary systems provide the basis for most of the commercially successful noble metal casting alloys, Au-Ag-Cu, Pd-Ag-Cu, and Au-Pt-Pd. Basic features of these three systems are discussed with respect to alloy formation and performance. Segregation, phase stability, hardenability, melting behavior, thermal expansion, color, castability, tarnish, corrosion, and biocompatibility are compared in these ternary systems. The review focuses on specific applications and details properties of representative chemistries. Data are presented to demonstrate composition and microstructural influences on mechanical, physical, thermal, and chemical properties. An overall emphasis is given to the use of precious metals in dentistry based on their accepted biocompatibility; current trends in the application and formulation of these alloys are included to enhance this point [42].

7. Casting

Dental noble metal alloys are usually cast into a refractory mold using a centrifugal casting machine in air. Casting that reproduce very fine detail (<10 μm) with dimensional accuracies of at least 50 μm are commonly obtained with conventional gold alloys. A desirable property of an alternative alloy is castability, which equals that of the conventional gold alloys with little modification in casting technique and materials.

A series of casting hulls was produced from jewelry and dental alloys in which the liquid fraction was centrifuged off a partially solidified casting 1 ~ 12 sec after filling the mold. Marked differences were found between coarse- and fine-grained types of alloys. Vertices, edges, etc., in the pattern configuration were source of solidification nucleation. These nuclei appeared to burst into existence very rapidly and, in turn, were the sources of new nucleation, producing nucleation chains penetrating into the liquid region. These nucleation chains were in effect fingers of solid enmeshing with other fingers, occasionally entrapping pockets of liquid metal. Entrapped liquid pockets also occurred where hot spots were produced in regions of hot metal impingement on some local point on the mold wall. Such trapped pockets result in localized porosity as a result of solidification shrinkage and cooling to room temperature [43].

Conventional gold alloy dental castings are usually made into gypsum-bonded investment molds containing forms of crystalline SiO_2 to provide high-temperature strength and thermal expansion to compensate for casting shrinkage. These investment materials commonly contain carbon, which acts as a reducing agent to leave a clean surface on the gold casting. Prior to casting, dental investment molds are heated to a temperature that produces sufficient thermal expansion of the mold space to compensate for metal shrinkage. Higher-melting alloys require higher casting temperatures, which in turn require higher mold temperatures. Gypsum-bonded investments should not be used above 700°C due to decomposition of the gypsum binder. Alternative alloys with high casting temperatures (see Table 3) require the use of phosphate or other high-temperature investments. Pd-based and Ag–Pd alloys should not be cast into investments containing carbon since a surface reaction occurs between the alloy and the investment. Since dental castings are usually made by a centrifugal process, alloys with lower densities than the conventional gold alloys (Table 3) may require higher centrifuge speeds to compensate for their lower densities [15]. With appropriate adjustments in casting technique and materials, the alternative alloys as shown in Table 3 result in casting comparable to those obtained with conventional gold alloys.

Time-dependent temperature profiles in dental casting molds were analyzed by an unsteady-state heat conduction model. The thermal conductivity and initial temperature of the mold greatly affected the heat transfer in the mold. The thermal conductivities of gypsum- and phosphate-bonded investments at high temperature were accurately measured by the hot wire method. From the data obtained, the solidification times of several alloys were calculated and compared with the experimental results [44]. Such numerical modeling may ultimately allow for a better understanding of the difficulties encountered in producing acceptable dental castings on a routine basis.

8. Adhesion

Cast dental fixed prostheses are secured in the mouth by cementing them to remaining tooth structure with a liquid adhesive that hardens rapidly. The function of this adhesive is twofold. It retains the casting in place and seals the space remaining between the casting surface and the tooth to prevent penetration of oral fluids and other material. This function is referred to as *luting*. Adhesion of a polycarboxylate cement and a zinc phosphate cement to dental Au casting alloy was examined; pickling the casting surface in acid reduces adhesion. Surface preparation such as sand blasting or electrolytic etching increases adhesion by chemically cleaning and by increasing surface area and roughness. The bond strength of polycarboxylate to Au alloy was one order of magnitude greater than that of the zinc phosphate cement. Bond failure was generally cohesive in nature, occurring at the metal–cement interface for zinc phosphate and within the cement film for polycarboxylate [45]. The bond between dental cements and noble metal casting alloys is generally thought to be mechanical. Polycarboxylate cement is one of the few dental materials shown to produce a chemical bond-to-tooth structure.

The sealing ability of zinc phosphate and glass ionomer cements were studied for cast gold restorations cemented to prepared teeth. Results of testing using an artificial caries model *in vitro* showed apparent leakage around all restorations although tooth demineralization was more advanced when restorations were cemented with zinc phosphate cement. Glass ionomer cements have also been shown to bond chemically to tooth structure [46].

B. Silver Alloys

1. Mechanical Properties and Metallurgy

The mechanical properties and hardenability of a Ag–Pd alloy have been determined, together with an evaluation of the effect of mold and alloy casting temperature on the marginal fit, surface roughness, and castability. The alloy was found to be amenable to heat treatment, and optimum conditions were determined. Mold and, to a lesser extent, alloy casting temperature influenced both the fit and the castability of the alloy but had no effect on surface roughness. The handling of the phosphate-bonded investment material was adjusted to obtain proper casting fit. Within the limitations of this study, the Ag–Pd alloy was found to be an acceptable alternative to high-gold alloys for dental restorations [47].

The structure of a commercial dental Ag–Pd–Cu–Au casting alloy has been studied. After being annealed at 400°, 500° and 600°C for 7 weeks, the alloy consisted of three phases: a Cu- and Pd-rich FCC phase (α_1), a Ag-rich matrix (α_2), and an ordered CsCl-type BCC PdCu phase. The PdCu phase was not observed above 600°C and the proportion of α_1 increased sharply above 700°C. After being annealed at 900°C, the alloy matrix was partly decomposed at the Cu-enriched grain boundaries. The decomposed areas grew into the grain interior during subsequent precipitation hardening. No segregation of Au was detected after casting and the element was evenly distributed throughout the alloy structure after all heat treatments [48].

2. Corrosion

The corrosion behavior of the Ag–Pd binary alloy system was studied using vacuum induction melted alloys 80Ag–20Pd, 60Ag–40Pd, 40Ag–60Pd, and 20Ag–80Pd, and pure Pd and Ag. Anodic potentiodynamic polarization curves, linear polarization curves, and corrosion behavior diagrams were constructed in aqueous solutions of NaCl (0.1 ~ 10%). The results indicated that with increasing Pd concentration there is a shift in the polarization profile in the noble direction. With increasing NaCl concentration, the potential range at which a sharp current density increase occurs shifts in the more active direction [49]. The results of anodic potentiodynamic polarization in 1% sodium chloride solution and alternate immersion and wiping tests in 0.5% Na_2S solution on Ag–In and Ag–In–Pd alloys was reported. Alloying with In does not increase chloride corrosion resistance. With In addition, improvement in tarnish resistance was detectable, but significant tarnish was still observed in Ag–In alloys containing up to 20% In. Palladium enhances the corrosion resistance and tarnish resistance of Ag and Ag–In alloys [49,50].

The correlation between microstructure, tarnish, and corrosion behavior of a commercial dental Ag–Pd–Cu–Au casting alloy after various heat treatments was studied. The results were compared with those from two experimental alloys with compositions similar to the analyzed components in the commercial alloy annealed at 600°C for 7 weeks. The matrix of the commercial alloy was decomposed, according to the heat treatments, into varying proportions of a tarnish- and corrosion-resistant Cu- and Pd-rich component and a nonresistant Ag-rich component. The presence of Cu-Pd component considerably increased the tarnish rate of Ag component due to galvanic coupling. Age hardening of the alloy increased the proportion of the tarnish- and corrosion-prone component [48,51].

Polished specimens of four Ag–Pd–Cu–Zn alloys were embedded in full acrylic dentures for periods ranging from 6 to 8 months *in vivo*. The results have been compared with those from immersion tests using 2% Na$_2$S solutions. In both types of tests, a multiphase alloy consisting of a Pd–Cu–Zn-rich component in a Ag-rich matrix displayed less corrosion and tarnish resistance than did individual alloys with compositions similar to the Pd–Cu–Zn compound or the Ag-rich matrix. Sulfur was detected by a microprobe in the tarnish layer developed in the oral cavity. The microstructural location of the corrosion products was somewhat different *in vivo* from the many dark spots created predominantly on Ag-rich areas during *in vitro* testing. This differences is most likely due to dissimilar conditions with regard to sulfide ion concentrations, the presence of protein films, and the length of exposure time [52].

Cast prosthetic devices of the alloy 65Ag-22Pd-6Au-5Sn-2.2Zn were examined in optical and electron microscopes, and by electron microprobe. Microstructural and compositional inhomogeneities were revealed. Examination of prosthetic devices removed after 3 years use showed intergranular corrosion pitting [53].

IV. DENTAL AMALGAM

A. Introduction

Dental silver amalgam is one of the most important materials at the dentist's disposal. It is a metallurgical triumph (see Fig. 1) [54]; it can be made plastic, and within a short time it will develop almost the strength of a casting. It can take on a high polish, is highly resistant to abrasion, and can withstand the rigors of the oral environment. Finally it is among the most biologically compatible of current restorative materials.

Dental amalgams are the most widely used of all dental restorative materials. After decades of very gradual evolution, amalgam alloys have recently undergone a period of rapid development with the advent of the high copper content amalgam alloys. Along with these compositional developments have come a greater understanding of the reaction mechanism and microstructure of the set amalgam, and of the effects of clinical variables on properties [55].

Generally, amalgam restorations fail because of secondary caries, fracture, dimensional change, or irreversible involvement of the pulp or periodontal tissues. Marginal breakdown and gross fracture of amalgam have been associated with various mechanical properties.

Historically, silver dental amalgam alloys (prior to mixing with mercury) contained at least 65 wt% silver, 29 wt% tin, and less than 6 wt% copper, a composition close to that recommended by Black in 1896 [56]. During the 1970s, many amalgam alloys containing between 6 and 30 wt% copper were developed. Many of these high-copper alloys produce amalgams (high-copper amalgams) that are superior to the traditional low-copper amalgams [57].

A recent classification of amalgam alloys by particle microstructure of the powdered

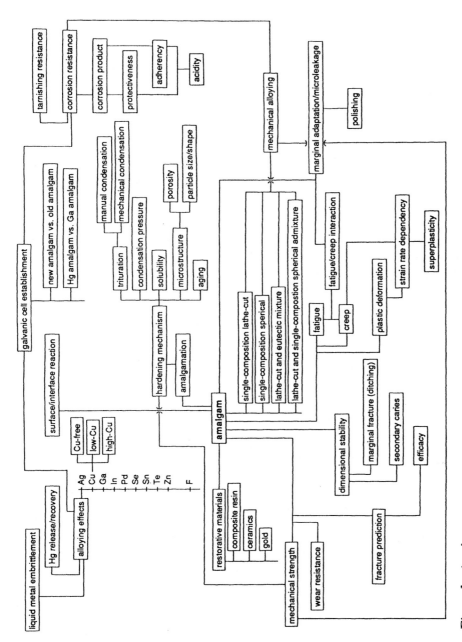

Figure 1 Amalgam map.

alloy lists four categories: single-composition lathe-cut; single-composition spherical; lathe-cut and eutectic (Ag–Cu) admixture; and lathe-cut and single-composition spherical admixture. A representative from each classification was selected and physical properties were tested utilizing varied times for trituration. Results showed little compromise in laboratory values for the properties tested. Additional laboratory and clinical trials are warranted, but this study supports the premise that clinicians may vary trituration time to improve handling characteristics without significantly affecting mechanical properties [58].

Today's dental amalgams have a long service life and seldom fail when properly placed in a cavity. Early amalgams disintegrated or expanded in the mouth. Studies were done to eliminate these deficiencies. The U.S. National Bureau of Standards (NBS) helped to formulate a federal specification for dental amalgam alloys in 1925. Work done in the 1920s and 1930s led to an understanding of the metallurgy of amalgam alloys. Cooperation between the NBS and the ADA (American Dental Association), along with new research tools and methods, produced a basic understanding of these alloys in the 1950s and 1960s. Tests proved that amalgams have high compressive and low tensile strengths. Mechanical properties and phases were also studied in this period. Finally, more durable and corrosion-resistant higher Cu-content alloys were patented based on microstructural studies.

Today, amalgam competes for its role in dentistry with several alternate materials, the most significant of which are resin-based composites (which generally have inferior wear resistance in back teeth) [59]. Table 4 compares four competitive restorative materials in terms of life span, tooth color matching, wear resistance, and risk of fracture [60].

B. Amalgamation

1. Reaction

A silver amalgam is prepared by reacting approximately equal weights of liquid Hg and a size-controlled powder consisting principally of the phase γ (Ag_3Sn) and having the nominal composition 69.4 Ag, 26.2 Sn, 3.6 Cu, 0.8 Zn. After the Hg and powder are ground together, dissolution and interdiffusion occur leading to the formation of γ_1 (Ag_2Hg_3), with ~8% Sn in solid solution, and followed by the precipitation of a second phase, γ_2 ($Sn_{7-8}Hg$), containing traces of Ag. Finally with time, Hg diffuses into the residual Ag_3Sn particles forming a gradient from outside to the center. The final structure is a metal matrix composite in which the γ_1 and γ_2 are the matrix phases and γ is the filler. Powders are prepared either by cutting on a lathe (chip shaped) or atomization (spherical). Mixtures of each may be used. A modern modification consists of chips of Ag_3Sn mixed with a spherical eutectic Ag–Cu powder (28% Cu). Upon amalgamation of the latter, the γ_2 phase disappears and a layer of η' (Cu_6Sn_5) appears around each Ag–Cu

Table 4

	Life span, years	Tooth color match	Wear resistant	Risk of fracture
Amalgam	10–20	No	Yes	Medium
Composite	3–10	Yes	No	Medium
Ceramic	10	Yes	Yes	High
Gold	>20	No	Yes	Low

particle. There are also single-composition, high Cu content amalgam alloys containing 12 ~ 30% Cu which form η' rather than γ_2. Creep properties change with time after amalgamation and are affected by partial transformation of γ_1 to a new Ag–Hg phase, β_1.

Model experiments to explore the amalgamation mechanism in conventional Ag_3Sn and high-Cu single-composition dental amalgam alloys have been conducted by rotating alloy rods in liquid Hg at different speeds and for different durations. After removing the rod from the Hg, the vacuum volatilization technique was used to accelerate the supersaturation of dissolved elements in the liquid Hg. Surface and sectional scanning electron microscope views give direct evidence for new amalgamation mechanisms in high-Cu dental amalgams. The amalgamation reaction begins with the selective dissolution of the Ag_3Sn phase in Hg. Sn getters to the crumbled-off ϵ (Cu_3Sn) phase and heterogeneous nucleation of the Cu_6Sn_5 phase on the seeds of the crumbled-off Cu_3Sn phase occurs followed by nucleation of the Ag_2Hg_3 phase. The tendency for the formation of the γ_2 phase in amalgams with higher content of Hg and the absence of the γ_2 phase in amalgams with higher content of Cu can be understood by applying this new model [61].

The mechanism of the attack of Hg on γ-phase Ag–Sn amalgam alloy has been investigated by applying small drops of Hg to plastically deformed single crystals and polycrystalline specimens of the alloy, and observing the progress of the reaction in a scanning electron microscope. Preferential Hg attack leading to the formation of deep, planar intrusions was observed to occur along well-defined crystal planes. Using electron channeling techniques and transmission electron microscopy, these were identified as being {010} slip planes and {011} and {211} deformation twins in the ordered orthorhombic γ-phase structure. In polycrystalline specimens, preferential grain boundary attack was also observed. Particularly aggressive reaction was found to occur at the intersection of twins and slip planes, and at the points where deformation twins terminated on grain boundaries [62].

The investigations conducted indicated that the amalgamation starts by reaction between Hg and the surface of the Ag_3Sn particles. Silver and tin are dissolved in liquid mercury and precipitated as γ_1 phase. The liquid Hg saturates with Sn and finally γ_2 precipitates in the voids left in the already solid γ_1 phase. The amalgamation reaction is different for high-Cu amalgams. Mercury dissolves Ag and Sn from the Ag_3Sn phase simultaneously with Ag from the Cu–Ag eutectic. Tin is concentrated in the Hg and reacts with dissolved Cu, forming η' phase (Cu_6Sn_5) [63].

In high Cu content amalgams the precipitation of the η' (Cu_6Sn_5) phase instead of the mechanically weak and easily corroded γ_2 ($Sn_{7,8}Hg$) phase is beneficial. The suppression of the latter by the preferential formation of η' largely takes place during the second sintering stage. The overall efficiency with which a particular type of Cu additive particle suppresses γ_2 phase precipitation is dependent on the gettering rate of Sn during the growth of the reaction zones [64].

The condensation pressure of amalgam achieved under simulated *in vivo* conditions was investigated by 42 general practitioners who were asked to fill a precut, standard Class 2 cavity in a mannequin head. A measuring device was designed to allow the maximum and average condensation pressures, and working and resting periods to be obtained. The results showed a maximum condensation pressure of 9.17 ± 3.04 MPa and 4.09 ± 1.41 MPa with a small (diameter 1.15 mm) and a large (diameter 1.8 mm) amalgam plugger, respectively. Only 1 of the 42 dentists reached the recommended condensation pressure of 15 MPa. A significant ($p \leq 0.001$) correlation between the duration of the working periods and the maximum condensation pressure was found ($r = 0.61$). Further investigations are required to determine the influence of these different condensation pressure on the physical properties of various amalgam [65].

2. Hardening Mechanisms

Mössbauer spectroscopy has been used to follow the Sn compounds that form during the amalgamation of Ag–Sn–Hg dental alloys. To measure γ_2 formation kinetics, a powder containing 70Ag, 28Sn, 2Cu mixed with Hg at a ratio of 5/4 was used. The amount of γ_2 formed, $m(t)$, can be represented by an Avrami-type expression, $m(t) = 1 - \exp(-kt^n)$. A plot of $\log_{10} \ln\{1/1 - m(t)\}$ vs. $\log t$ shows two successive mechanisms with n values of 0.7 and 0.95; these mechanisms are also demonstrated by dilatometric measurements. The first stage is attributed to increase in γ_2 nucleation sites and the second stage to γ_2 growth controlled by the γ_1–γ_2 interface after saturation of γ_1 with Sn [66].

The microstructural changes of dental amalgam during aging were studied in an SEM equipped with an energy dispersive x-ray analyzer. Experiments were carried out using two different types of specimens: (1) amalgam/ϵ (Cu_3Sn) diffusion couple and (2) several amalgams with variations in Cu content. Structural changes that occurred on the surface and in the interior were examined. It was revealed that some intermetallic compounds in amalgams decompose, followed by a redistribution of atoms to form more stable compounds during aging. The rate of the redistribution depends upon the phases present in the amalgam and on the types of diffusion processes that may take place [67].

3. Solubility

The metals dissolved in the excess Hg expressed from Cu-rich amalgams ranged from 0.06 to 0.63%. The equilibrium solubility in Hg of the individual metals in alloys for dental amalgam (Ag, Cu, Sn, Zn, and In) cannot be used to predict solubilities when dental alloys are amalgamated since the amalgam is not in equilibrium. The observed solubilities obtained while the amalgam was in the plastic stage may assist in explaining the phase formation in Cu-rich amalgams and the kinetics of the hardening of dental amalgam. They may also be helpful in computing the compositions of test specimens, if comparatively large amounts of Hg are used in making the amalgam. The findings were viewed in regard to the theory that the γ-phases are formed by the precipitation of Ag and Sn in solution in the Hg; the precipitation of γ_1 (Ag–Hg) solid that occurs first should be enhanced by the supersaturation of Ag in Hg. The γ_2 (Sn–Hg) phase is thought to form later. This is not difficult to reconcile with the observation that Ag is supersaturated in the Hg, whereas Sn is not. It may be that the γ_2-phase has not yet formed when excess Hg is expressed from amalgam. The low-Sn and high-Cu content of the expressed Hg may also enhance the formation of the phase (Cu–Sn), which is present in high-Cu amalgam [68].

The solubility of Cu in Hg was measured in the range 10° to 25°C by monitoring the potential of a Cu amalgam whose Cu content was varied from zero to past saturation by coulometric electrolysis. Measurements were performed in a vacuum cell with a saturated Cu amalgam as reference electrode. Saturation concentrations in the range examined are $(1.61 + 0.209t/°C) \times 10^{-4}$ mol (kg Hg)$^{-1}$, giving 6.84×10^{-4} mol kg^{-1} at 25°C [69].

4. Microstructures and Crystallography

Studies of two types of dental amalgam of differing composition (one approximately 3% Cu and the other approximately 13% Cu) have been undertaken using electron optical equipment. Backscattered electron imaging was used to examine the microstructure of the amalgams and showed that there was a complex structure of phases and precipitates formed on amalgamation. This complexity had not been observed at lower resolutions and fine precipitates (approximately 0.05 μm) have been observed throughout the matrix. Energy dispersive x-ray analysis has shown that these precipitates are a copper–tin phase, previously unreported in low-copper amalgam. It has also been possible to prepare thin foils for TEM using ion beam thinning methods. The grain size has been found to be small (0.2 μm across), which necessitated the use of convergent beam techniques to obtain crystallographic information. The precipitates observed in the microstructure of the low-copper amalgam have been found to lie within the grains of the silver–mercury matrix phase. By studying the microstructure of the two amalgams at higher resolution it has been possible to propose a modified setting reaction for amalgam [70–73].

Microstructures of high-Cu commercial amalgams containing varying amounts of Hg, ranging from 20% above to 25% below recommended values, were investigated by x-ray diffraction. Mechanisms relating to the absence or presence of γ_2 (Sn–Hg) phase in these amalgams were discussed in relation to the presence of Cu and Sn in their original alloys. The optimum Hg concentration in some amalgams required to keep them free of the γ_2 (Sn–Hg) phase was determined [74].

Three single-composition, high-Cu amalgams were analyzed for the composition before and after reaction. The method used was electron probe microanalysis. All alloys were essentially of Ag_3Sn (γ), Cu_3Sn (ϵ), and, in some cases, Ag_4Sn (β); when the original alloy contains higher amounts of Cu, more Cu is found in solution in the Ag_3Sn (γ) and Ag_4Sn (β) phases. The reaction phases in amalgams made from these alloys consist essentially of $Ag_{22}SnHg_{27}$ (γ_1) and Cu_6Sn_5 (η') with no detectable Sn_8Hg (γ_2); when indium is present, it is found principally in the Ag_3Sn (γ) and $Ag_{22}SnHg_{27}$ (γ_1) phases, and both the γ and γ_1 phases are of different composition than that for the other alloys. When Pd is present, it is found principally in the Cu_3Sn (ϵ), Cu_6Sn_5 (η') and $Ag_{22}SnHg_{27}$ (γ_1) phases [75].

The phase boundaries of copper-rich dental amalgams for phase equilibria that contain neither the $Sn_{7.8}Hg$ phase (γ_2) nor liquid Hg, were estimated by studying the isothermal section. Ag_3Sn with at most 45 wt% Hg contains neither the $Sn_{7.8}Hg$ phase nor liquid Hg, if the composition of the prealloyed powder is within the two-phase equilibrium Ag_3Sn (γ_0) and Cu_3Sn (ϵ) with 8 to 38 wt% Cu in the system Ag–Cu–Sn [76].

The temperature dependence of microstructure in amalgam also has been studied by the use of scanning electron microscopy. The β_1 phase becomes remarkably visible after 80°C heating. However, after keeping the specimen at room temperature for 150 days, the structure of the specimen is found to be similar to that before heating [77].

C. Effects of Alloying Elements

1. Ag

The reactions of Ag–Sn alloys with compositions spanning the phase fields β, ($\beta + \gamma$), γ, and ($\gamma + Sn$) were examined. The nature and morphology of the reaction products formed on bulk alloy were examined. The reactions with powdered alloys were monitored using a high-sensitivity dilatometer. These results were correlated with direct observations on the development of the microstructures. The reaction of Hg with the β-phase alloy occurred rapidly and resulted in a very marked and rapid expansion during the initial stages of hardening. The γ-phase alloys reacted more slowly and contracted markedly during hardening. The behavior of amalgams made from alloys with compositions lying between these two extremes appeared to have characteristics of the separate phases from which they were constituted [78].

Various methods of preparation have been employed in the quest for producing a new dental amalgam alloy that would exhibit the acceptable properties but contain a minimal amount of Ag. The effects of various base materials, ranging from porous glass particles to Cu and Cu alloys, along with preparation techniques (including lathe cutting, splat cooling, and atomization coupled with various electrodeless and immersion deposition processes, and a varied particle size and shape) were observed. The alloy particles and the amalgams produced from these particles were investigated using the techniques of SEM, EDXS, and x-ray diffraction technique. Amalgams were also subjected to 1 wt% NaCl solutions and tested for compressive strength and expansion. The results showed the feasibility of dramatically decreasing the Ag content in dental alloy while producing an amalgam that exhibits a 24-h compressive strength of $\sim 54,000$ psi with an expansion of $2 \sim 3$ μm/cm [79].

2. Cu

If Cu is present up to 6 wt% then Cu_3Sn (ϵ) or Cu_6Sn_5 (η') forms a solid solution with Ag_3Sn (γ), Ag_5Sn (β), or Sn. When the alloy with Cu is triturated with Hg, copper-containing Ag_3Sn, Ag_5Sn, Sn, and Cu_3Sn are released. When setting starts, part of the Cu available in Hg is incorporated in Ag_2Hg_3, and a part in Sn_7Hg; the rest precipitates as Cu_6Sn_5. The formation of Cu–Hg may occur at isolated areas away from tin. Substitution of Cu for silver may reduce the amount of tin–mercury (γ_2) by the formation of Cu_6Sn_5. An uncontrollable expansion and increase in the amalgamation rate of amalgam with Cu may be related to the presence of more Ag_5Sn (β) phase [80].

Generally, high-Cu amalgams possess higher strength and lower creep, and are usually almost free of the γ_2 phase [81]. Cu is commonly incorporated as a replacement for some of the Ag in conventional single-composition-type amalgams. Commercial amalgam alloys exist that contain of much as 30% Cu, most of which has replaced Ag.

3. Ga

Because of the high surface tension of Hg, which reduces wetting, and its high vapor pressure and toxicity, attention has been directed to its replacement by Ga, which has good wetting properties and low vapor pressure, and is less toxic. To lower the Ga melting point of 30°C, eutectics of 95% Ga–5% Sn (melting point: 20°C) and 83% Ga–17% In (melting point: 15°C) were examined. Alloys made by mixing Ga with Ag were too soft; however, alloys with Pd had adequate hardness ($60 \sim 80$ VHN) [82].

By mixing a liquid gallium alloy and an amalgam alloy powder, a plastic mix can be made that solidifies in as short a time as dental amalgam. The properties of gallium mixes and dental amalgams prepared using the same high-Cu powders for each mix were compared. The reaction between the powders with the liquid Ga alloy was also examined. Two high-Cu powders, 49.5Ag–20Cu–30Sn–0.5Pd and 56–Ag–20Sn–15Cu were triturated at the powder:liquid ratio of 1:0.45 and 1:0.47, respectively, with the ternary eutectic liquid Ga–Sn–In alloy. Compressive strengths (1 h $\sim$ 7 days), creep and dimensional change were found to be similar to those of the corresponding amalgams. The structure of alloys made with Ga–In–Sn consists of unconsumed alloy particles embedded in the mixed matrix of In_4Ag_9 and β-Sn grains. By combining with gallium, copper in the particles formed a reaction layer ($CuGa_2$) on the surface, and $CuGa_2$ platelets dispersed in the matrix. No η'-Cu_6Sn_5 was formed in either of the gallium mixes [83].

4. Pd

Palladium additions to a dispersed phase high-copper amalgam have been shown to suppress markedly the η' (Cu_6Sn_5) concentration and to decrease creep. A detailed study of the creep for Pd-containing high-copper amalgam and six commercial controls as a function of applied temperature and stress was performed. Experimental amalgam containing up to 10 wt% Pd substituted for Ag demonstrated essentially constant creep over the temperature and stress range applied [84].

5. Sn

A low-Cu dental amalgam was examined by electron probe microanalysis. This amalgam exhibited unusually large γ_1 grains (10 μm), which allowed the electron beam to be completely contained within these large grains. The analyses showed significant Sn content of 2 ~ 3 wt%. Analyses of areas that contained grain boundaries showed slight or no differences in Sn content compared with that for the grains. The conclusion supports the hypothesis that the Ag–Hg (γ_1) phase in dental amalgam does contain a significant amount of Sn [85].

6. Zn

Amalgams from 25 Ag–Sn alloys containing various additions of Zn were tested for physical properties pertinent to dental restoration use. The Hg retained, dimensional change, strength, and creep were all functions of the Zn content [86].

To investigate the effect of Zn on Hg ratio and residual Hg in Ag-based dental amalgam, an experiment was performed by setting the range of Zn content from 0% to 8% in the Ag–Sn alloy. It was found that the Hg ratio and residual Hg in Ag-based dental amalgam increased with increased Zn content. The former seems to result from the decrease of wettability with increasing Zn content, the latter from the increase of Hg ratio and the formation of Zn–Hg compound with increasing Zn content [87].

The effect of Zn on the structure and mechanical properties of the Ag–Sn dental amalgam was investigated for 0 ~ 8% Zn. The γ phase in the alloy structure decreased as Zn content increased, whereas free Sn and Ag–Zn phase increased as Zn content increased. The γ_2 phase increased as Zn content increased. Hardness of the γ phase in the alloy increased as Zn content increased. Hardness and compressive strength of amalgam alloy containing Zn up to 3.7% increased as Zn content increased, then decreased with higher Zn content [88].

D. Dimensional Stability

The dimensional changes of the four types of amalgams are explained on the basis of their particle sizes and shapes and their constitutions [89,90]. For example, if amalgam alloy contains more than 30% Sn, it will shrink, if <25% Sn, it will expand; therefore the Sn concentration should be within these limits for proper amalgamation and approximately zero dimensional change on setting [91].

The delayed excessive expansion of amalgam is defined as an excessive expansion beginning 2 ~ 3 days to a week after mixing; the amount of expansion can reach to 400 μm/cm and is due to hydrogen gas generation from the reaction of Zn with water contamination in the mix. Moisture contamination during mixing, handling, or condensation caused delayed excessive expansion and deterioration of the physical properties of conventional low-Cu alloy amalgams containing Zn, but not of high-Cu amalgams. It affected the compressive strength and creep but not the hardness [58].

E. Mechanical Strength and Plastic Deformation

The compressive strength of dental amalgam decreases as the strain rate is reduced. During the compressive strength testing: (1) Plastic deformation increased as strain rate decreased. (2) Grain boundary sliding was probably responsible for plastic deformation. (3) brittle amalgams exhibited increased strength as strain rate increased. (4) In most amalgams, the interface between γ_1 and the alloy particles was the site for crack initiation [92].

The compressive strength of amalgam depends on the loading rate. The tensile strength is also somewhat rate dependent. Amalgam, like concrete, has a high compressive strength but a relatively low tensile strength ($\sim 10\%$). The stress–strain behavior is nonlinear over its entire range. The viscoelastic behavior of the amalgam involves dislocation climb and grain boundary sliding—classic mechanisms for high-temperature deformation [93]. The oral cavity at 37°C is high service temperature considering the solidus temperatures of the γ_1 and γ_2 phases.

At low strain rates, the mechanical properties of Ag–Sn and high-Cu dental amalgams are controlled by the γ_1-based matrix. The plasticity of γ_1 and γ_2 increases at these strain rates. This plasticity provides the accommodating deformation that must occur before grains can slide. When η Cu–Sn crystals interlock γ_1 grains, grain boundary sliding is restricted. For sliding to occur, γ_1 grains must move plastically around η crystals. The effect of η crystals, therefore, is to increase both low strain rate compressive strength and resistance to creep deformation [94].

Dental amalgam is a particulate composite, meaning that ternary Ag–Sn–Cu alloy particles ($5 \sim 40\ \mu m$) together with some intermetallic compounds, such as fine particles of Cu_6Sn_5, are embedded in an equiaxed Ag_2Hg_3 grain ($1 \sim 3\ \mu m$) matrix. The effects of various strain rates on fracture characteristics of dental amalgams were studied. The amalgam specimens were fractured using loading speeds of 0.42 mm/s to 8.47×10^{-4} mm/sec. In addition, the specimens were strained at constant load ($29.4 \sim 88.2$ N) with a slow strain rate ranging from 1.0×10^{-7} to 4.0×10^{-8}/sec. When the specimens were fractured at a high strain rate, the amalgams showed brittle failure without appreciable plastic deformation, while at a lower strain rate, the Ag_2Hg_3 grains exhibited a high degree of plastic deformation, forming many fine needles (1 μm or less in diameter and as long as 5 μm). In addition, significant grain boundary sliding of the Ag_2Hg_3 grains was evident. Dramatic differences in the characteristics of fractures resulted from fast and slow strain rate deformation [95].

Compressive strength of dental amalgam prepared from high Cu and conventional alloys was evaluated as a function of deformation rate. Results showed that the compressive strength rose to a maximum as deformation rate increased and then began to fall at higher deformation rates. Cracks propagated through the γ_1 phase and around the copper-containing particles at lower strain rates. Cracks occurred through copper-containing original particles when the stress at fracture was sufficiently high [96]. The η' (Cu_6Sn_5) is more resistant to the propagation of cracks at the particle–matrix interfaces as well as transgranular cracking. It is also reported that the strength of amalgam phases decreases in the following order: residual Ag–Sn particle $> \gamma_1$ (Ag_2Hg_3) $> \gamma_2$ (Sn_7Hg) [96].

The dynamic mechanical properties of two high-copper amalgams and two traditional amalgams were measured over a temperature range of $0° \sim 70°C$ and at frequencies of 0.1, 1, and 10 Hz. Values of storage modulus (E') for the amalgams were equivalent to the Young's modulus (E) measured from static mechanical test methods. Values of E' decreased with increased temperature. E' for traditional amalgams decreased more rapidly than for of high-Cu amalgams. The value of the loss modulus (E'') for the single-composition high-copper amalgam was smaller than those of the admixed and two traditional amalgams. High values of E'' for the traditional amalgams correspond to a greater viscous behavior. Marked differences between the magnitude of tan δ and its temperature coefficients for traditional and high-Cu amalgams were observed, which is indicative of differences in viscoelastic behavior between these two amalgam systems [97].

Hardness, microstructure, deformation behavior, and fracture behavior of four dental amalgams have been investigated. The deformation tests were performed in compression at 37°C under constant load. Stress dependence of the minimum deformation rate and the activation energy for creep are not constant. At low stresses the stress exponent of the deformation rate approaches a value of 2, which seems to be due to superplastic deformation of the γ_1 matrix, which exhibits a very fine grain size. During creep the microstructure of the dental amalgams changes, leading to softening and to fractures at high stresses. The strain to fracture strongly decreases with increasing stress. The softening of the amalgams is attributed to the formation of pores at the phase boundaries [98]. Low-Cu amalgam (O), admixed high-Cu amalgam (D), and high-Cu single-composition amalgam (ST) were tested, An overall trend for the strength of wetted amalgams was observed: ST > D > O. In D amalgams, intraparticle fracture of Ag–Cu eutectic particles was more frequently observed in the Hg-plated specimens. Hg appears to embrittle the eutectic particle, reducing the amalgam strength [99].

The strength of amalgam is governed by the fracture resistance of the particles and the bond between the particles and the matrix. The embrittlement of solid metal by liquid metal is a well-known metallurgical phenomenon. Liquid metal embrittlement is the decrease in strength or ductility of a solid metal or alloy when subjected to stress during or after contact with liquid metals. Mercury embrittlement is seen as a distinct type of brittle fracture, in which mercury weakens atomic bonds at crack tips, thereby decreasing the initial stress for crack propagation. It is suggested that Hg released during aging or corrosion of amalgam *in vivo* might embrittle unconsumed alloy particles, weakening the amalgam restoration [100].

F. Wear

The wear of dental amalgam by a smear mechanism and amalgam transfer onto the opposing cusp was confirmed by simulated studies in an artificial mouth. The coefficient of wear for dental amalgam was 4.89×10^{-5}. Dental amalgam is distinguished from many other direct restorations by a resistance to occlusal wear [101].

G. Fatigue

During compressive fatigue testing, amalgam cylinders showed fatigue cracks, appearing to initiate at cylinder surfaces in Hg reaction phases and propagating longitudinally. Crack propagation during the fatigue fracture of amalgam occurred predominantly through γ_1 grains in much the same manner as has been reported for slow compressive tests [102]. Surface crack propagation in fatigue followed classical fatigue fracture patterns. Fatigue fracture creates many microcracks that could lead to significant corrosion and precipitate stress-corrosion cracking [103].

A conventional (γ_2-containing) dental amalgam was fatigue tested at 1800 and 80 cycles/min, employing uniaxial, sinusoidal loading, with R (stress ratio) $= -8$. Compressive, tensile, and creep tests were conducted to characterize the static mechanical behavior of the alloy. Tests were performed at 37°C on specimens that were aged for 7 days at 37°C. Fatigue-tested specimens were microscopically examined for fracture surface appearance and crack path. The amalgam demonstrated a frequency dependence and a significant reduction in fracture strength due to fatigue loading. The fatigue crack path was primarily intergranular in the γ_1 phase and inclined at $\sim 45°$ to the principal stress axis [104]. These observations are characteristic of some metals when subjected to low-frequency, elevated-temperature testing where significant grain boundary sliding occurs, and therefore suggest a creep-fatigue interaction. A study was designed to determine whether the mechanical cyclic stressing that occurs during normal mastication contributes to margin breakdown of dental amalgam restorations. The method used appears to duplicate the mechanical stresses developed *in vivo* during mastication as the result of tooth flexing. One low-Cu alloy and three high-Cu alloys were evaluated. Simple amalgam restorations were prepared in a cavity centrally located in an Al beam. Each specimen received 5 periods of 3-point cyclical loading (1.7 Hz, 4200 cycles at 37°C). The margin area was subjected to SEM examination prior to and at the completion of each period of cycling. At the beginning of each period of cycling, beam deflection was set to establish a maximum theoretical stress cycle of 1, 2, 4, 6, and 8 MPa. For all amalgams, cycling resulted in margin gap formation and surface wrinkling. Wrinkling in the high-Cu admixed amalgam occurred as a wide band of shallow wrinkles, whereas that in the low-Cu occurred as a narrow band of deep wrinkles. At 21,000 cycles, very little void formation and fracturing had occurred in either. In contrast, the single-composition high-Cu restorations developed extensive fracturing even after 4200 cycles. Fracture surface analyses indicated that creep fatigue rupture was the fracture mechanism responsible for margin breakdown in these amalgam restorations [105].

H. Marginal Fracture (Ditching)

Amalgam restorations of alloys of traditional composition (less than 6% Cu) show a deterioration or breakdown at their margin (the amalgam tooth interface at the surface of the tooth) after some length of clinical service. This breakdown is progressive and is one of the main reasons for replacement of amalgam restorations. Both creep and corrosion have been related to marginal breakdown. Both creep and corrosion increase with an increase in the final Hg content of a restoration. For one particular alloy a rapid increase in creep starts at 46% Hg and stops at 48.5% Hg, an increase in anodic current begins at 46% Hg, and the γ_2 phase also increases. Results confirm a relationship between creep and corrosion current in that both show high values when γ_2 is present and low values when γ_2 is absent. However, behavior patterns suggest that creep is influenced by γ_1 grain boundary effects and not by the amount of γ_2. Anodic corrosion current is primarily related to the amount of γ_2 present [106,107].

Marginal degradation of two Ag–Sn–Cu amalgams with different creep properties was studied as a function of clinical variables. The amalgam with the highest creep values showed the most marked marginal degradation, irrespective of operator, doubling of the trituration time, and the use of contralateral, opposing or unrelated teeth for comparison [108].

I. Creep

Both corrosion and creep have been identified as possible contributors to marginal fracture of amalgam. The stresses that induce creep may arise from the continued setting expansion of the amalgam [109], the formation of corrosion products, mastication, or from the thermal expansion of the amalgam during ingestion of hot foods. The latter two are low-frequency cyclic stresses. Amalgams made from seven different alloys were condensed into stainless steel dies. After being allowed to set for 7 days, the specimens were thermally cycled between 4° ~ 50°C for 500 and 1000 cycles. Amalgam marginal integrity was evaluated by scanning electron microscopy both before and after each cycling period. The amount of margin fracture was calculated after 1000 cycles. Thermal cycling of amalgam restorations resulted in predominantly intergranular fracture of the amalgam margins, indicating that creep-fatigue failure may be a significant contributor to *in vivo* margin fracture [110].

Creep mechanisms have been investigated in both conventional and non-γ_2 amalgams. The back-stress during steady-state creep was measured by the "stress-dip" technique, and prepolished vertical surfaces of nearly cylindrical specimens were studied after creep. Indications of both superplastic flow and dislocation creep were observed [111].

Between 23° and 58°C, under a compressive stress of 10 ~ 120 MPa up to 2200 sec, creep obeyed the power law $\epsilon = C + At^B$, where $1 > B > 0.5$, thus encompassing the transition from the Andrade creep to a steady-state creep. The activation energy for the steady state creep was 5 ~ 16 kcal/mol [112].

The microstructure of a Ag–Sn amalgam and an admixed high-Cu amalgam have been observed during compressive creep tests. Sliding of γ_1 grains has been observed. The sliding rate is much lower in the γ_2-free high-Cu amalgam than in the γ_2-containing Ag–Sn amalgam [113].

Low-Cu amalgams aged at 37°, 50°, 60°, 70° and 80°C for periods of 7 and 30 days were examined using the SEM and x-ray diffraction methods. Grain size of the γ_1 Ag–Hg phase and γ_2 Sn–Hg volume fraction, and surface area were determined by quantitative metallography. At 60°C and below, γ_1 was the predominant matrix phase. Little β_1 Ag–Hg was found. In this temperature range, both γ_1 grain size and creep were found to obey an Arrhenius law. A linear relationship was found between the log of the mean creep and the mean γ_1 grain size. Above 60°C, the γ_1 phase and the majority of the γ Ag–Sn particles disappeared. In their place, β_1 appeared. The γ_2 volume fraction increased during aging at temperatures higher than 60°C, but a coarsening of γ_2 was reflected in a decrease in γ_2 surface area/unit volume. Aging for 30 days resulted in a decrease in creep. At 60°C and below, the decrease was caused by an increase in γ_1 grain size. Above 60°C, the decrease in creep was caused by several factors, one of which was the appearance of large β_1 grains [114].

When aged at 37°C for 6 months, dental amalgam exhibits a marked decrease in the property of creep. The objective of the study was to investigate the relationships between this decrease in creep and selected metallurgical characteristics. The formation of β_1 (Ag–Hg), the grain size of γ_1 (Ag–Hg), and the composition of γ_1 were chosen for investigation. Creep was determined according to ADA Specification Test No. 1, β_1 was measured by the x-ray diffraction method, and γ_1 grain size and γ_1 composition were determined by electron probe microanalysis. The results showed that decrease in creep was related to β_1 formation and not to changes in either γ_1 grain size or γ_1 composition [112].

The creep strain rate and fracture of several dental amalgams have been shown to be markedly dependent on the type of loading applied. Cyclic loading typical of the *in vivo* situation increases creep rates and reduces times to fracture when analysis is based on a mechanical equation of state. The cyclic strain rate enhancement over static loading is highest for conventional amalgam while the modern ternary alloys show the smallest increases [115].

Rather than the usual compressive dental creep test, various types of 1-week-old amalgams were continuously monitored in tensile creep. Testing was done at 37°, 45°, and 50°C at a constant true tensile stress of 17 MPa. For the first time, the classical four stages of creep were observed at elevated temperatures in the low-Cu amalgams, including creep rupture. The high-Cu systems displayed only transient creep up to 50°C and no rupture. The tensile test required approximately one half the stress needed in compression to provide the equivalent creep [116].

In spite of differences in strength of 35 ~ 56 MPa for 23 proprietary amalgams, no correlation with behavior in service is shown. In case of creep behavior, while a difference is noted in frequency of marginal deterioration of fillings made with high-creep γ_2-containing amalgams compared with high-copper amalgams free of γ_2, no correlation exists between this frequency and creep rates measured for different high-Cu amalgams [117].

Dynamic creep of amalgam is continuing permanent deformation of a material under fatigue loading, and increasing Hg content increases the dynamic creep. Dynamic creep of a 53% Hg containing amalgam is of the order of 1.5 times the dynamic creep of a 48% Hg amalgam. Trituration has only minor influence on the dynamic creep; static creep may be used to predict dynamic creep accurately because it correlates well with the dynamic creep [118].

Another theory for creep-related failure of amalgam is that stress-induced creep produces extrusion of the restoration, resulting in tensile fracture of the margins from occlusal stresses. The creep stress may be due to phase changes within the amalgam or corrosion buildup at the tooth–amalgam interface with the attendant diffusion of Hg into the amalgam [119].

In summary, the γ_1 (Ag–Hg) phase is primarily responsible for creep, while the γ_2 phase is responsible for the corrosion. To the extent that marginal breakdown is an effect of corrosion, creep is not related to marginal fracture. An amalgam with no γ_2 phase and high creep might have excellent marginal integrity. Hence it is extremely important to reduce the γ_2 phase as much as possible [120].

J. Tarnishing and Corrosion

1. *Tarnishing*

Various media are utilized for evaluation of *in vitro* corrosion resistance of dental amalgams. They include Ringer's solution, Hank's solution, Fusayama's solution, arterial Tyrode's solution, artificial physiological solution, 0.9 ~ 1.0% NaCl solution, concentrated artificial saliva, etc. A detailed discussion on chemistry of each media is beyond the scope of this chapter but it is sufficient to say that each corrosive medium has a different pH value and chlorine ion concentration, suggesting that, in electrochemical corrosion tests, the corrosion potential and current will not be the same for different types of media.

General dulling or tarnishing of dental amalgam may result from the formation of tin oxides on γ_1, γ_2, η' whereas γ and ϵ remain unaffected. In a sulfide-containing environment, γ_1, γ_2, and Cu–Sn ($\epsilon + \eta'$) phases probably remain unaffected, while mild attack may take place on the γ phase. Furthermore, the β phase is more prone to sulfide attack than the γ phase. Since the high-Cu amalgam may contain more β than a traditional amalgam, it will be more susceptible to tarnishing or dulling [80,121].

Zn and Sn from amalgam fillings penetrate and may discolor the dentin adjacent to amalgam fillings, and radiopaque areas caused by the presence of these metals can be seen on clinical intraoral radiographs [122]. Discoloration can take place by a penetration process of any amalgam constitutive element (Ag, Zn, Sn, Cu, Hg) through the fine dentinal tubules, but Zn and Sn have higher ionization tendency and are the only elements to penetrate. Zn can penetrate deeper with higher concentrations [123].

2. Chemical Corrosion

Corrosion of dental amalgam has generated a great deal of research interest, focused on the physical and chemical effects of corrosion on the dental amalgam restoration. The corrosion process seems to be complex and involves reactions of the restoration and the surrounding oral environment, including tooth and oral fluids, in which interactions of Sn, Zn, Hg, Ca, P, O, and Cl occur. It has been suggested that the corrosion can lead to demineralization and recurrent caries because the tooth–amalgam interface is acidic and an ideal environment for the deterioration of tooth structure and the amalgam restoration [124].

Dental amalgam shows passive behavior, but the protective films can be destroyed by abrasion [125]. To evaluate the influence of tooth brushing on the *in vivo* corrosion resistance of dental amalgams, immersion tests were carried out *in vitro* in artificial saliva at 37°C under static and dynamic conditions. Three types of dental amalgam were employed: a conventional type, an admixed high-Cu amalgam, and a single-composition amalgam. Brushing of the surfaces prevents the accumulation of corrosion products and causes an accelerated corrosion of the amalgam. This phenomenon is particularly evident in the conventional amalgam. Observation of the surface by SEM after 100,000 brushing strokes shows deep corrosion of the γ_2 phase [126].

After 2- to 5-year immersion of high-Cu amalgam in artificial saliva, in concentrated artificial saliva (10×), and in 1% Na_2S + Ringer's solution in a 1:1 ratio, preferential corrosion attack occurred in the η' Cu_6Sn_5 phase. Cu appears to be preferentially released, resulting in insoluble Sn-containing corrosion products. The γ_1-to-β_1 transformation was greatly enhanced by corrosion [127].

Commercial Cu-rich amalgams were immersed in various solutions to study the effects of Cl concentration, sulfide contamination, and pH effects. The major corrosion products found on the surface of high-Cu amalgams after immersion in Ringer's solution were red-colored prismatic Cu_2O crystals covered by loosely packed green $CuCl_2 \cdot 3Cu(OH)_2$ crystals. The formation sequence of corrosion products in Ringer's solution and diluted Ringer's solution was SnO_2 and/or $ZnSn(OH)_6$ (in Zn-containing amalgams only), CuO_2 and $CuCl_2 \cdot 3Cu(OH)_2$. Increased pH values of the solution around the sample surfaces enhanced the formation of corrosion products. The η' (Cu–Sn) reaction product was found to be preferentially attacked and formed Cu- and Sn-containing corrosion products. After the corrosion attack of the η' phase, the Ag–Cu eutectic particles and Cu_3Sn were also attacked. The corrosion of Cu-rich amalgams also enhanced the transformation of the Ag–Hg matrix from γ_1 to β_1 [128,129].

3. Effect of Sn

The anodic polarization behavior of the Ag–Hg matrix phase of dental amalgam (γ_1) and the rate of Hg dissolution were examined as a function of the Sn content. Anodic polarization curves were recorded for specimens of the γ_1 phase containing $0 \sim 1.5\%$ Sn and for the γ_2 phase and a γ_2-containing dental amalgam, in synthetic saliva. Mercury dissolved in synthetic saliva in 24 h was determined for the γ_1 phase containing 0 ~ 1.0% Sn. Tin induced passive behavior of the γ_1 phase and suppressed Hg dissolution. The passive γ_1 phase did not exhibit the breakdown of passivity that occurs in the γ_2 phase. The results were consistent with the Sn present in γ_1 in a solid solution rather than in Sn-rich precipitates, and indicated that at least 1.5% Sn can be dissolved in γ_1 at 37°C [130].

4. Effect of Zn

In a corrosion test of amalgam in oxygen-saturated, 1% sodium chloride solution, zinc-free amalgams corroded 2 to 3 times more than Zn-containing amalgams. Amalgam prepared from Zn-containing, preamalgamated alloy (alloys containing small amounts of Hg) showed varying degrees of corrosion, probably because the preamalgamation process may deprive the alloy powder of most of its original zinc content [131].

5. Effect of Surface Roughness

The influence of mechanical treatment of the surface and of the chemical composition of amalgams on the corrosion behavior was examined in saliva by using potentiodynamic polarization measurements. The corrosion resistance increased with decreased roughness of the surface and lower concentrations of γ_2 phase. The concentration of γ_2 phase is influenced by the chemical composition and to a certain extent by the mechanical treatment of the surface. Therefore limiting the amount of γ_2 phase and polishing the amalgam will improve the corrosion resistance of the restoration [132].

6. Electrochemical Corrosion

A study was made by electrochemical techniques of 12 amalgams representative of four classes: conventional lathe-cut powder Ag_3Sn, conventional alloy with a small addition of eutectic Ag–Cu microspheres, a mixture of lathe-cut alloy and Ag–Cu microspheres, and spherical or spheroidal Ag–Sn–Cu particles of a single composition. The polarization curves show clearly the peak characteristic of the γ_2 phase of conventional amalgams, the oxidation of the Cu_6Sn_5 compound formed by reaction of Ag–Cu microspheres with γ_2 and the very flat traces obtained with single-composition powders. Mixed amalgams and single-composition amalgams are less affected by aging. The mixed amalgams and those of single composition (Ag–Au–Cu) are most corrosion resistant. Among the latter, an alloy containing indium was least resistant [133].

In an effort to determine the mechanism of improved corrosion resistance of Zn-containing dental amalgams, two Zn-containing conventional amalgams, their Zn-free counterparts, and three experimental amalgams (SnHg, ZnHg, and SnZnHg) were evaluated by potentiodynamic polarization technique in 1% NaCl solution. The main difference between the two types of amalgams was found in their respective breakdown potential at which passivity was destroyed. The breakdown potential of Zn-containing amalgams was ~200 mV more positive than that of the Zn-free amalgams. The improved stability of the Zn-containing amalgams has been attributed to the formation of a Zn stannate passive film. The formation of Zn stannate was not found to affect the oxygen reduction reaction, the major cathodic reaction involved in the corrosion of dental amalgams [134].

Although the superiority of high copper content amalgams has been generally accepted, the comparisons of properties of the different types of high-Cu alloys is of interest. A comparison was made of the electrochemical behavior of high-copper amalgams obtained from a single-composition amalgam alloy with that of those made from an admixed Ag–Sn/Ag–Cu eutectic. The results confirm that: (1) amalgams from additive systems and single-composition amalgams have a different tarnish attack mechanism; and (2) in the single-composition amalgam tested, the tarnish attack mechanism is influenced by a diffusion process in the solid state [135].

In vitro corrosion of two low-Cu and three high-Cu dental amalgams was examined using stressed and unstressed specimens. Stress was applied in tension, and anodic polarization curves and polarization resistance were determined. In the stressed condition, the passive current density was higher and polarization resistance was lower. The changes are attributed mainly to the fracture of passive films by creep strain [136].

7. Electrogalvanism

Dissimilar metal alloys in contact with each other directly or through the saliva establish a galvanic cell. The ensuing electrical current is thought to enhance corrosion of the metals and may result in sensitivity, pain, or injury to the oral mucosa [137]. Electrogalvanism has been recognized for a long time as a potential source of discomfort for patients. In a case where a Ag amalgam was placed adjacent to a tooth restored with a Au alloy, pain began several days after placement of the amalgam restoration. The pain disappeared when a plastic orthodontic separator was placed between the teeth to remove them from contact. When the patient removed the separator several days later, the discomfort returned [137,138].

The galvanic corrosion of new and aged amalgams placed in contact was studied *in vitro*. The amalgams included conventional, admixed high-Cu, single-composition spherical high-Cu, and single-composition lathe-cut high-Cu. Freshly condensed amalgams were placed in contact with aged amalgams in artificial saliva. The corrosion current and electric charge transfer between fresh and old amalgam specimens was measured for 20 h. The fresh amalgam was always anodic. Maximum anodic current densities were reached within the first hour and ranged from $12 \sim 35 \ \mu A/cm^2$ for combinations excluding the fresh admixed alloy and from $165 \sim 238 \ \mu A/cm^2$ in combinations including the fresh admixed alloy. After 20 h all combination were below $8 \ \mu A/cm^2$. The integrated charge transfer showed the same pattern; combinations without the admixed alloy ranged from $0.27 \sim 0.79 \ C/cm^2$, and combinations with fresh admixed amalgam ranged from $2.00 \sim 2.84 \ C/cm^2$. The results indicate that galvanic corrosion occurs when fresh amalgam restorations are placed in contact with old ones, and that various amalgam combinations may give rise to considerably different corrosion rates [139].

8. Corrosion Products

Selective interfacial amalgamation using a cavity liner for dental amalgam that can adhere to the tooth surface and amalgamate with residual Hg in the amalgam has been developed with the intent of increasing the amalgam–tooth adhesion and reducing the effects of corrosion and marginal leakage. It is believed that corrosion products formed from dental amalgams can shield the marginal leakage if they are adherent. Anodic polarization and differential aeration tests were used to examine the effectiveness of this technique. The tests indicate that the selective interfacial amalgamation does enhance the corrosion resistance of a dental amalgam [140]. Eight cavity liner systems using selective interfacial amalgamation were evaluated for their ability to bond dental amalgam to tooth structure. Both punch shear and tensile adhesion were evaluated, along with the fracture path. Results showed maximum mean adhesive tensile and shear strengths to be 3.5 MPa and 15 MPa, respectively [141].

Conventional [143] and high-Cu amalgams [142] were immersed in oxygen-saturated NaCl solution adjusted to pH 4 with HCl; in the conventional amalgam, no corrosion of γ_1 occurred until all of the γ_2 had corroded; in high-Cu amalgam, corrosion of γ_1 occurred from the beginning, concurrent with corrosion of ϵ and η'. Corrosion products observed were AgCl, Hg_2Cl_2, $CuCl_2 \cdot 3Cu(OH)_2$, and SnO_2 [143].

A high-Cu amalgam placed in 37°C Ringer's solution for 10 months formed a double layer of corrosion products. The inner layer was Cu_2O (a red product) while the outer layer was $CuCl_2 \cdot 3Cu(OH)_2$ (green product) [144].

For amalgams placed in normal saliva, or distilled water, Zn and In seemed to dissolve preferentially during the early period of immersion, and Cu and Hg after 3 or 4 months. The high-Cu amalgam released much larger amounts of Cu and Hg than did the conventional ones. No Ag was lost from any amalgams. Sn was selectively attacked to form insoluble corrosion products rather than dissolving in the solution [145].

Analysis has been reported [146] of the soluble solution species, insoluble solution precipitate, adherent corrosion products, and microstructural changes of the substrate amalgam after selected polarization to -0.2 V and $+0.5$ V in a chloride solution. Results indicate only small concentrations of soluble species, high concentrations of an insoluble Sn precipitate at -0.2 V and high concentrations of a Cu precipitate at $+0.5$ V, related to $CuCl_2 \cdot 3Cu(OH)_2$. The completely oxidized amalgam microstructure indicates a thin outermost layer of predominantly Sn–Cl, a thick corroded layer of Ag–Sn–Hg–Cl, and amalgam substrate. The compound $(SnO)_{16}O$ is also associated with the thick corroded layer. The microstructure of the substrate amalgam exhibits four new phases due to the redistribution of Sn, Cl, and oxygen from the γ_2 corrosion products. These phases were examined gravimetrically [146].

The γ_2 phase (Sn_8Hg) and a dental amalgam in a phosphate buffer have been studied by potentiostatic and galvanostatic techniques using a rotating ring-disk electrode. The analysis of results has shown that phosphate ions play an important role in the corrosion of amalgam. The anodic reaction leads to the formation of soluble species and a passivation film that is probably composed of tin hydroxide and tin phosphate [147].

9. In Vitro *Versus* In Vivo *Tests*

Corrosion behavior of not only dental materials but also industrial/structural materials has been investigated in either laboratory (*in vitro*) tests or plant (*in vivo*) tests. Intercorrelation between data from *in vitro* and *in vivo* testing must be demonstrated before relying upon laboratory data.

The relationship between *in vitro* corrosion and *in vivo* marginal fracture of dental amalgams has been studied. The potentiodynamic cyclic polarization technique has been used to determine an *in vitro* corrosion index. The marginal fracture data were obtained from previous clinical studies reported by three independent research groups. A linear regression analysis was employed to determine the degree of correlation between *in vitro* and *in vivo* data. Depending on the method of corrosion index measurements and the source of clinical data, correlations ranging from 0.82 to 0.94 were obtained [148].

Potentiostatic and potentiodynamic (hysteresis) polarization have been employed to evaluate the corrosion resistance of dental amalgams and dental casting alloys in 1.0% sodium chloride solution. Unlike conventional dental amalgams, where corrosion is associated with the presence of the Sn–Hg (γ_2) phase, the corrosion of γ_2-free high-Cu dental amalgams results from the presence of the Ag–Cu eutectic or Cu–Sn phases, or both [149]. Results agree with those obtained from long-term laboratory and clinical studies and demonstrate the usefulness of relatively quick, convenient electrochemical techniques in the evaluation of the corrosion resistance of dental alloys [150].

The corrosion potentials of freshly placed admixed high-Cu and conventional amalgams were measured *in vivo*. Impedance and current measurement methods at controlled potential were used to examine the corrosion of electrodes made from dental amalgams and the binary alloys Hg–Sn and Cu–Sn in a Ringer's solution in the absence of dissolved oxygen. These *in vitro* measurements were used to make some deductions about the nature of the corrosion reaction on the amalgams and to estimate the *in vivo* corrosion rates. It was deduced that the amalgams *in vivo* are protected from electrochemical Hg dissolution by the dissolution of the other metals, such as Zn, Sn, and Cu. It was also suggested that the mechanisms of intraoral Hg release from dental amalgams are by evaporation; by coupling with the electrochemical dissolution of Zn, Sn, and Cu; and by mechanical wear [151].

Although the corrosion of dental amalgam has been studied extensively it has been difficult to identify the products that form *in vivo*. A report demonstrates that $Sn_4(OH)_6Cl_2$ and SnO are formed *in vivo* and *in vitro*. The Sn–Cl product is the predominant phase formed by corrosion of the Sn_8Hg phase of amalgam [152].

K. Fit and Microleakage

Close adaptation of dental amalgam to the cavity wall is important in reducing the risk of secondary caries. The flow stress of triturated amalgam increases with the deformation time [153]. Amalgam must be condensed immediately after trituration and condensation completed while the material is still plastic.

Dental amalgam is one of the few restorative materials for which microleakage between the restoration and tooth actually decreases with time. This phenomenon is thought to be due to the buildup of relatively insoluble corrosion products in the gap between the amalgam and the tooth. In effect these act as a "hydraulic cement" to seal this interface.

Conventional and high-Cu Class V amalgam restorations showed leakage after 7-months storage in artificial saliva and thermal stressing. The rate of microleakage was not significantly affected by the application of a Copal varnish to the cavity walls prior to placement. After 14-month storage and thermal stressing, all varnished and unvarnished high-Cu restorations and the varnished conventional amalgam restorations showed significantly improved sealing properties compared with the 7-month period. The unvarnished conventional amalgam restorations appeared to have reached their peak sealing level after 7 months under the conditions of this experiment. Copal varnish did not significantly improve the sealing properties of either the conventional or high-Cu amalgam restorations after the 14-month period. [154].

A liner of $Ca(OH)_2$ placed beneath an amalgam restoration can be used to monitor microleakage in a reverse leakage model. A pH paper can be used to measure the release rate of OH^- into the solution surrounding the tooth. This test indicated extensive microleakage of restoration immediately after insertion. In all cases, leakage decreased over a period of time. The rate of decrease was substantially dependent upon the form of calcium hydroxide used as well as its pH. Reagent grade $Ca(OH)_2$ generated the greatest percentage of positive results for the longest duration [155].

L. Hg Release

The question of the safety of the dental amalgam restoration has continued to be raised periodically since dental amalgam was first used. Reliable data for the amount of Hg released per unit time from sound amalgam restorations are required to place this source of Hg exposure into context with other environmental sources such as air, water, and dietary intake.

The difficulties associated with estimations of daily doses of inhaled Hg vapor released from dental amalgam are considerable. Existing data are often unreliable, especially if they are based on a single or a small series of samples of intraoral concentrations of Hg vapor before, during, and after chewing stimulation. The aim was to obtain a more representative estimation of the daily dose of Hg vapor inhaled from amalgam fillings by measurement of amounts of Hg vapor released in the oral cavity during 24 h, under conditions that were as normal as possible. A series of measurements was carried out on 15 subjects, with at least 9 occlusal surfaces restored with dental amalgam, and on 5 subjects without amalgam restorations. The subjects followed a standardized schedule for the 24-h period. The amounts of Hg vapor released per unit time were measured at intervals of 30 ~ 45 min by means of atomic absorption spectrophotometry. None of the subjects was professionally exposed to Hg, and all amalgam fillings were more than 1 year old. Study casts were made for each subject, and the area of the amalgam surfaces was measured. Samples of urine and saliva were analyzed so that values for the Hg concentrations and the rate of release of Hg into saliva could be obtained. The average dietary exposure to Hg was estimated. The release of elemental Hg from dental amalgam was corrected for retention of inspired Hg vapor and for oral-to-nasal breathing ratio. The estimated average daily dose of Hg vapor inhaled from the amalgam restorations was 1.7 μg, ~ 1% of the occupational threshold limit value (TLV) for exposure to airborne Hg for a 40-h work week [156,157].

In another study, five restorations made from an amalgam labeled with ^{203}Hg isotope were prepared in plastic molds that were placed in saliva. During the first 24 h, 68% of the total Hg release was observed from the polished fillings and 61% from the unpolished ones; in 2 days, the percentages were 78 and 69%, respectively, and over a period of 1 week 90 and 79%. The average release of Hg was 313 mμg/filling in the first day. During the last month, the average daily release of Hg was 0.06 mμg from the polished fillings and 0.09 mμg from the unpolished ones. Except during the first day, the release of Hg was much lower than the maximum daily exposure limit [158].

M. Clinical Evaluation

The first commercial high-Cu amalgam was developed in the mid-60s. In 1975 a study to evaluate *in vivo/in vitro* performance of 9 of these new alloys was initiated. The mechanical properties and γ_2 content of these high-Cu amalgams plus 4 traditional amalgams were measured and 60 restorations of each material were placed in patients. Marginal fracture was assessed at intervals. After 13 years, approximately 1/3 of the original patients were still available for clinical examination and the mean loss of all restorations at 13 years was 11.3%. The evaluation of the fracture at the margins indicated that the 4 γ_2-containing alloys showed the greatest rate of fracture at the margins. A correlation matrix of the margin fracture data at different time periods, the loss of restorations after 13 years, and the mechanical properties indicated that early (1- and 2- year) fracture at the margins predicts the long-term result. Mechanical properties, however, do not predict the clinical performance of the materials at 13 years [159].

The survival rates and modes of failure of amalgam restorations were investigated retrospectively: 2660 Class I or II lesions were evaluated yearly or half-yearly for failures during the 30- to ~84-month follow-up period. Restorations with unacceptable margins were not counted as failures if no traces of secondary caries could be seen. Eight percent of the restorations were lost because of patient drop-out. Of the remaining restorations, 1% were replaced due to primary caries. Of the remaining number (2431), 9% failed because of all other reasons. The leading mode of failure was bulk fracture (4.6%), followed by tooth fracture (1.9%), and marginal ridge fracture (1.3%). All other failures totaled 0.8%. Only two restorations were replaced because of secondary caries. The alloy selection in both conventional and high-Cu categories significantly influenced the survival of the restorations [160].

A clinical survey of amalgam restorations made with four proprietary alloys with low-Ag and high-Cu content was followed during a 2-year period, according to USPHS criteria. With the exception of one alloy, the marginal integrity was found to be very satisfactory. All the restorations lost their superficial luster in a short period of time [161].

Twenty-five silver amalgam restorations ranging in age from 2 to 25 years were obtained from 5 subjects. An electron microprobe was used to analyze the specimens for bulk elemental composition and phase composition, and the volume fractions of phases were determined by point counting on back-scattered electron micrographs. Twenty-one of the specimens were conventional, low-Cu amalgams, and the remaining 4 were high-Cu amalgams. The bulk elemental composition showed little variation from newly prepared amalgams except for the presence of a small amount of chloride and other contaminants. The compositions of the phases were essentially the same as found in new amalgams, except there was considerable internal amalgamation of the γ particles. The distribution of phases in the clinically aged amalgams was quite different from that of new amalgams. The low-Cu amalgams had decreased amounts of γ, γ_1 and γ_2 phases, and increased β_1; and the high-Cu had enlarged reaction rings (γ_1 and η') [162].

V. METAL CERAMICS

A. Introduction

Porcelain has long been a popular material for the restoration of teeth in parts of the mouth that are normally visible; it was one of the earliest tooth-colored restorative materials. A porcelain jacket crown covers the tooth with an artificial cap made of a dental ceramic material. Unfortunately, these restorations are time consuming to fabricate, do not fit over the remaining tooth structure very well, and are subject to fracture. The metal ceramic system involving the fusion of glasses, porcelain enamels, and ceramics to metal copings combines the corrosion resistance, aesthetics, and brittleness of ceramics with the toughness, strength, and relative ease of fabrication of metals [163]. The popularity of metal ceramic restorations has steadily increased since their invention more than 40 years ago. This is due to their excellent aesthetic appearance in combination with high mechanical stability [164]. To achieve a reliable metal ceramic system, a proper understanding of many factors involved during fabrication is a necessity. A metal ceramic map is shown in Fig. 2 [165]. From this map several important factors governing the success of metal ceramic systems can be pointed out. It is obvious that the adhesive bonding between the fired porcelain and metal substrate seems to be controlled by oxide formation between them and the residual stress developed there. Creep resistance (referred to as *sag resistance* in dentistry) and the strength of the metal substrate are related. Each depends on

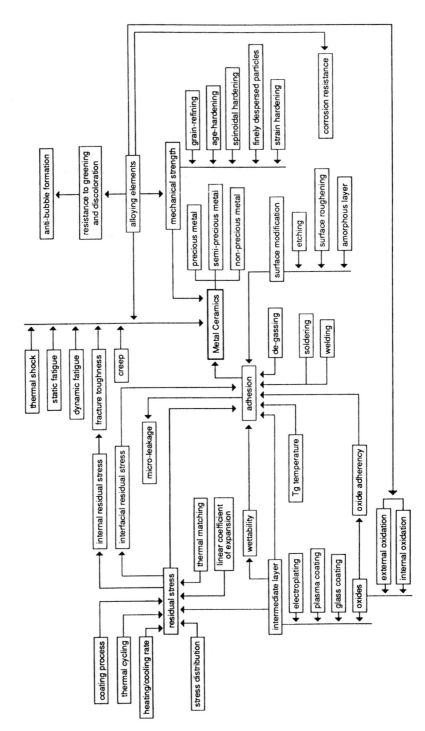

Figure 2 Metallic ceramic system map.

relevant basic principles from physics, chemistry, and metallurgy. For example, in order to achieve excellent bonding, an intermediate ceramic layer or preoxidation of the metal has been proposed. Residual interfacial (sometimes internal as well) stresses can either increase or decrease the bonding strength. Metallurgical manipulation can improve the mechanical properties of the metal substrate and can control the thickness of intermediate metal oxides. Different modifications in metal ceramic systems have been proposed to improve one or more of these factors, as is discussed later.

There are four important criteria for successful bonding of fired ceramics to metal: (1) small differences in linear coefficients of thermal expansion that place the interface in compression, (2) strong bonding between the metal substrate and its oxides, (3) chemical bonding between the metal oxides and ceramic surface, and (4) metal oxides that do not alter properties of the porcelain. In this section, the following are discussed: metal alloys for metal ceramic restorations; mechanical, chemical, and thermal properties of metal ceramics; interfacial characteristics governing the adhesion between metals and ceramics; and surface modification to enhance the bonding.

B. Metal Ceramics

Metal ceramic restorations can be classified into the following six groups [166]:

1. High-Au Alloys

The high-Au alloys are composed of Au 80 ~ 88 wt%, Pt, and Pd, plus small amounts of Sn, In, and reactive metals to promote a good porcelain bond by forming metal oxides. These are usually light yellow in color but also can be white, depending upon the ratios of Au to Pd and Pt and other additions such as Rh, Ta, and W [167]. Fe is usually added to increase the yield strength and promotes age hardening if alloy additions are made to stabilize the hardening reaction [168]. Addition of elements such as Si, B, Ge, and mixtures of these will prevent discoloration of the porcelain during subsequent firing processes [169]. High Au content alloys exhibit superior corrosion resistance but typically have relatively low strength and sag resistance.

2. Medium-Au Alloys

Medium-Au alloy composition is Ag (10 ~ 15 wt%)-Pd (23 ~ wt%)-Au. These alloys show increases in the modulus of elasticity, good corrosion resistance, hardness, strength, and castability, but have a tendency to induce color changes in lighter shades of porcelain. Some alloys contain at least 3.5 wt% of Ga, In, and Sn to promote oxide–ceramic adhesion [170].

3. Low-Au Alloys

The low-Au alloys, Ag-free-50 wt% Au-40 wt% Pd, possess a good modulus of elasticity and hardness, but may have poor thermal expansion compatibility with some higher expansion porcelains. Basically, replacing Au with Pd increases hardness but decreases density [170]. Some alloys contain Sn (1 ~ 12 wt%), Re (0.05 wt%), and Ru (0.05 ~ 1 wt%) [171].

4. Pd–Ag Alloy

The increasing cost of Au has forced the dental industry to look for alternatives to Pd-Au alloys. One result is Pd-based alloys having most of the qualities expected for high Au content alloys. These will discolor porcelain due to the fact that Ag is included.

Discoloration-resistant porcelain are now available for use with such alloys. These alloys have received FDA approval for marketing [172].

With high Pd (70 ~ 80 wt%)-Cu (15 wt%) alloys, porcelain discoloring is not a problem, and casting accuracy, finishing, and soldering properties are acceptable in casting. Cleaning (pickling) of castings in acids should not be done since it depletes Pd at the surface, causing decreased tarnish resistance. Pd-Ag is a single-phase alloy if Cu is added. An amount of Cu up to ~30 wt% was found to lower the melting point (1100° ~ 1400°C) and to raise the thermal expansion coefficient [173].

It was found that the alloying elements Sn, In, Ga, and Al are effective for hardening and oxide formation. Si and Zn eliminate discoloration and bubble formation in the porcelain during the firing process, and Re, Ru, and Ir are grain refiners [174-177]. Grain refining of metal ceramic alloys may impair their sag (creep) resistance.

A metal ceramic Pd based alloy (melting range of 1302 ~ 1399°C) for use with a porcelain having a firing temperature of 954° ~ 1010°C consisting of Ga (7 ~ 12 wt%), Au (2 ~ 5 wt%), and B (0.125 ~ 0.50 wt%) was developed [174].

5. Pd-Ag-In Alloys

This group of alloys (usually containing Ru, Cu, Co, In, Sn, Ga, Ta, W, and Zn) shows strong bonding to porcelain, and increased oxidation and tarnish resistance, but will result in color change [178].

6. Base-Metal Alloys

The base-metal alloys include Ni-Cr-based alloys, Co-Cr-based alloys [166,179], and Ti and Ti-6Al-4V alloy [180]. These alloys exhibit good corrosion resistance due to passivation, biocompatibility, excellent mechanical properties, and high bond strength. However, it is reported that Ni-containing alloys have the potential for causing allergic reactions, and the carcinogenic potential of some of the base metals (e.g., Ni, Cr) is controversial [166]. Ni-Cr metal ceramic alloys often contain 2 ~ 4 wt% Be. Be, Ni, and Cr are all occupational safety concerns.

C. Mechanical Strength and Hardening Mechanism

The majority of noble metal ceramic alloys, except for the Ag-Pd system, are hardened during the firing processes by precipitation of a secondary phase containing Sn and/or In as alloying elements. This precipitation hardening can be more effective if the alloy is quenched in advance. The principal mechanism for hardening of noble metal alloys depends upon the alloy system. For example, an alloy (85Au-5Pt-5Pd) with additions of Ag, Sn, Fe, and Co was subjected to air cooling from 1000°C and hardened by forming a intermetallic (second) phase identified as $PtFe_{0.6}Co_{0.3}$. Hardening is attributed to an ordering reaction of the compound, which begins at approximately 850°C and reaches completion at 450 ~ 200°C [181,182].

An age-hardening phenomenon of Au-based alloys containing Cu during isothermal annealing occurs in two stages; the first stage corresponds to the formation of a metastable AuCu I phase within grains, and the second stage to a cellular reaction at grain boundaries. It was reported that the former contributed to the hardening and the latter to the softening [30,183]. It was found that the Au_3Cu ordered phase and disordered FCC phase were formed by aging at 400°C, and hardening was brought about by the mechanism of the antiphase domain size effect. Below 350°C, the Au_3Cu ordered phase decomposed to the AuCu I ordered phase, and disordered FCC phase was decomposed to the

AuCu I ordered phase and disordered FCC phase by prolonged aging. An appreciable further hardening was generated by the formation of the AuCu I ordered phase in addition to hardening by the Au_3Cu ordering in this temperature range [23]. On the other hand, it was reported that no evidence was found for the presence of an FCC phase at the interface between AuCu I type ordered platelets and the surrounding Au_3Cu ordered phase [184]. Although no evidence was found for second-phase formation in a 52Au–38Pd alloy, a possible superlattice has been identified that is indicative of spinoidal hardening. Examinations at different cold-working levels indicated a tendency for planar slip. Accordingly, large strain hardening exponents are observed at low plastic strains [185].

Phase transformation in Au–Cu–Pd single-phase alloys was also studied. Age hardening of the alloys examined was attributed to the formation of fine domains of long-range-ordered AuCu I type lattice in the interior of the grain. Prolonged aging caused formation of large long-range-ordered domains of a single type at grain boundaries or microtwinning of long-range order in the interior of the grain, depending on the ordering rate of the alloy. The effectiveness of Rh addition on the grain refining was proved experimentally [186].

Age hardening in a Pd-based alloy (e.g., 55Pd–36Ag–5Sn–4In) exhibits a peak aging phenomenon. In early stages of aging, a metastable ordered phase with a face centered tetragonal (FCT) structure based on Pd_3In was formed within grains. Hardening was due to coherency strain between the metastable structure and the matrix with an FCC structure. In the latter stages, a lamellar structure, consisting of stable ordered phase with an FCT structure, based on Pd_3In and a stable phase with an FCC structure, grew from the grain boundaries. Finally the lamellae covered the whole grain and resulted in softening [187].

D. Fracture and Cracking in Metal Ceramics

The apparent fracture toughness and elastic modulus-to-hardness ratio of metal ceramic restorations were found to be significantly greater than those of individual porcelain specimens, presumably due to residual compressive stresses in their surface [188].

The incompatibility stresses that result from thermal contraction differences (for instance between a Ni–Cr alloy and porcelain) are one reason for failure of metal ceramic systems. Delayed failure of metal ceramic restorations due to static fatigue can occur when residual tensile stress is present in porcelain, even in the absence of intraoral forces [189]. A thermal shock test was employed to evaluate the crazing resistance of 55 metal ceramic systems. Metals tested include Ni–Cr, Au–Pd, Pd–Ag, Au–Pt–Pd, and Au–Pd–Ag dental alloys. The effect of porcelain brand used was the most significant determinant of porcelain crazing. The effects of the alloys and alloy ceramic combinations, although also significant, were second-order effects. Crazing of dental porcelain that has been veneered to metal substructures may result from excessive stresses generated during the porcelain firing cycles. Although such fracture patterns may result from differential contraction within the porcelain itself, thermal expansion or incompatibility between the metal and porcelain is believed to be the primary cause of these failures [190].

Heating rate effect on deformation temperature of porcelain during firing was studied. Viscoelastic stress is affected by elastic modulus, viscosity, coefficient of thermal expansion, the effects of cooling and heating rates on the glass transition temperature (T_g), and temperature distribution in the porcelain slab. It was suggested that the temperature where the incompatibility stress develops in metal ceramic strips during cooling can be estimated closely from the deformation point of the heating curve of the porcelain with an applied stress of about 1.2 ~ 3.1 MPa [191].

Conventional metal ceramic crowns turn out to have significantly higher failure loads (2863 N) compared to all ceramic restorations. An injection-molded all-ceramic crown system has corrected failures loads (1353 N) that are approximately 60% of the metal ceramic fracture load. Fracture loads of a castable ceramic system turn out to be about 30% of the metal ceramics [192].

E. Creep

Although two Pd–Cu alloys demonstrated excellent resistance to creep at low stress and high temperature, they were especially susceptible to deformation at high stresses and at temperatures near the T_g of dental porcelains. In comparison, a Ni–Cr alloy and a Pd–Co alloy demonstrated superior creep resistance at high stress and temperatures just below T_g. This indicates a relatively low potential for deformation due to stresses that result from a thermal contraction differential between these two alloys and poorly matched dental porcelains [193].

Near the solidus temperature, creep rates for the alloys investigated (including high-Au-, Au–Pd-, and Pd–Ag-based alloys) differ by an order of magnitude and become increasingly different at lower temperature. Depending upon the presently unknown mechanisms involved in *sag* or *margin opening* during porcelain firing, control of activation energy has varying potential for modifying creep-induced distortion of metal ceramic restorations [194].

F. Stress, Residual Stress, and Stress Distribution

As discussed previously, fracture in metal ceramics is governed by internal stresses as well as externally applied stresses. Calculations of the stress distribution in simple planar metal ceramic systems indicates that metal ceramic restorations are subject to stresses due not only to mastication forces but also to the differential thermal contraction of the two components [195].

Magnitude and sign of residual stress in both metal and ceramic portions as well as at the interfacial layer between metal and ceramic serve as a dominant factor governing marginal fit and bond strength in metal ceramic systems. Stresses in dental castings are widely held to be the cause of change of fit of metal ceramic dental restorations. Residual stresses are thought to result from the casting process and from mismatch of thermal expansion coefficients [196].

One study showed that residual stress occurs in the metal portion of metal ceramic restorations as a result of the casting process and during preparation of the metal prior to porcelain application. Residual stress was relieved during the oxidation and porcelain firings, and was not increased measurably by the application of porcelain. It was also mentioned that composition gradients near the surface of a material might affect residual stress measurements and indicate stresses that are not actually present [197]. Residual stress generated during the fabrication of a metal ceramic crown on Au alloys was studied for various conditions: as cast, sandpaper polished, and oxidized (at 960°C for 10 min), and for each firing process (800° ~ 900°C at a heating rate of 50°C/min). It was reported that the residual stress as measured by x-ray diffraction technique was −36.68

(compressive) to 36.87 (tensile) MPa for the as-cast condition, −190.3 MPa for sandpaper-polished surface, and −98.1 MPa for oxidized/fired surface [198].

A change in the marginal fit of metal ceramic crowns with repeated firings has been thought to be due to the shrinkage of the porcelain. In one theory, interfacial stress generated by a differential in the rate of concentration between the metal substructure and the adhering porcelain in the cooling portion of the firing cycle acts to deform the restoration, producing marginal discrepancy. It usually is implied that stresses are precluded by a match in expansion and contraction coefficients. However, expansion coefficients normally match only at one temperature ranging from 371° to 427°C. Interfacial stresses that result from all other temperatures may be responsible for weakness and adaptation failures that sometimes appear clinically [199]. Apparently, a stronger, more rigid restoration deforms less than does a weaker, less rigid one. It was found that segments of the margin with porcelain adjacent to them did not change more than those without porcelain, and that the marginal opening increased more in the specimens made from the higher-strength metal than it did in those made from the lower-strength metal [200].

Finite element modeling (FEM) has been extensively utilized for stress analysis in metal ceramic systems [201,202]. The stress distribution induced in anterior metal ceramic crowns fabricated with either Au- or Ni-alloy copings of reduced thickness was calculated using plane stress analyses. Two-dimensional finite element models of three crown designs were subjected to a simulated biting force of 200 N that was distributed over porcelain near the lingual metal ceramic junction. Based on plane stress analysis, the maximum tensile and compressive stresses in porcelain were 29.5 and 123.1 MPa, respectively. The highest tensile strains in porcelain for the veneered Ni–Cr and Au–Pd copings with conventional dimensions were 0.016 and 0.014%, respectively. The maximum stresses and strains in porcelain for the crowns with a conventional coping thickness (0.3 mm) and a reduced coping thickness (0.1 mm) were not significantly different. It was reported that all values were below the critical failure values of porcelain [202].

Large amounts of stress relief in the multilayered systems comprising dental restorations apparently take place at temperatures well below the porcelain glass transition temperature. A split ring model, with the use of FEM, holds promise for revealing information on the stress compatibility of metal ceramic systems, and the split ring alone may prove to be a useful, simplified tool for quality control and evaluation of compatibility [203].

G. Thermophysics

The thermal compatibility between metal and ceramic is the most important factor controlling the mechanical stability as well as bonding strength of metal ceramic systems [204]. Thermal expansion is directly proportional to thermal contraction. When metal and ceramic are heated, they expand; when allowed to cool, they contract. For metal, the contraction is equal in magnitude to the expansion. But for the ceramics, the total amount of contraction may vary slightly according to cooling rate, number of previous firings, and the chemistry of the ceramic. Ideally, the contraction of the ceramic should be slightly less than that of metal, generating a small compressive force on the ceramic at the interface, which aids in bonding and improves overall strength. If the thermal expansion-contraction of the metal is too great for the ceramic, a shearing type of fracture may occur. The fracture usually appears as a crescent shape in the incisal area. Conversely if the metal expansion–contraction rate is too small for the ceramic, the ceramic will place the metal in compression. The resulting fracture, which intersects the metal coping, generally appears as a straight line in a horizontal or vertical direction [205].

Several researchers have introduced a compatibility index (C_i) to evaluate the thermal compatibility of metal to ceramics in metal ceramic systems [206,207]. Determination of the thermal expansion compatibility index, C_i, of a metal ceramic system is critically dependent on the glass transition temperature (T_g) of the ceramic. The Au–Pd and Ag-free alloys demonstrate consistently negative C_i values with opaque and body porcelains. Values of C_i for most opaque and body porcelain systems changed significantly between one and five firings. In some systems the C_i value changed sign. Opaque body porcelain systems exhibited C_i values of the same order of magnitude as alloy opaque systems. Large positive C_i values are indicative not only of tangential compressive stresses in porcelain, but also of radial tensile stresses that may contribute to system failure. It may be inferred from large C_i values for opaque body systems that significant residual stress levels can develop between opaque and body porcelains. Such stresses must be considered in conjunction with the stress levels developed between alloy and opaque porcelains in evaluating the alloy–ceramic compatibility [207].

Thermal expansion data from room temperature to above the T_g range are important for matching the thermal expansion of ceramics to alloys. By use of the Moynihan equation (where T_g systematically increases in temperature with an increase in both cooling and heating rates), the T_g temperatures derived from these data were shown to be related to the heating rate [208]. The thermal expansion coefficient (TEC) is reported for about 60 alloys available for metal ceramic restorations. The results within a temperature range between 100°C and 600°C vary [$(14 \sim 16) \times 10^{-6}/°C$]. Both the TEC and the T_g temperatures for dental ceramics depend on the number of firing processes and cooling conditions. The data obtained show that there is a large increase for the TEC of the body porcelain with the number of firing cycles but a slight decrease for the opaque porcelains. Applying slower cooling rates results in an increase of the TEC in both cases [209]. Results on the determination of thermal expansion of several brands of dental porcelain and metal ceramic alloys indicated that there exist significant differences in the thermal expansion of dental ceramics as a function of temperature. An approximately 20% difference in extremes of ceramic expansion up to 500°C reveals the potential for much less compatibility between some combinations of ceramic and alloy if a range of only about 7% is accepted for any given alloy [206].

Cracking or rupturing, due to thermal stress, of porcelain bonded to a Pd–Ag alloy indicated that metal–ceramic thermal compatibility was dependent on the difference in thermal expansion coefficients of the ceramics and metal, the geometry and dimensions of the samples, and the metal-to-ceramic thickness ratio. A higher incidence of cracking resulted from an increase in both specimen size and metal-to-ceramic thickness ratio [210].

H.   Oxides and Oxidation

The majority of metal ceramic restorations now placed use base metal alloys. Controlled oxidation of these metal substrates is essential for bonding to ceramics. The bonding of porcelain to dental alloys (Ni–Cr or Co–Cr based) was found to be closely related to the respective oxide adherence strengths of the alloys. One class of alloys (Pd–Ag based) tended to form an internal oxide. For this latter class of alloys, porcelain adherence seemed to result from mechanical interlocking between the porcelain and Pd–Ag nodules formed on the alloy surface by a Nabarro–Herring creep mechanism. In order to ensure thermal compatibility in a new metal ceramic system, the role of thermal properties of porcelain cooled at high rates in the dental fabrication process must be understood. The glass transition temperature and thermal expansion properties at high cooling rates are needed in order to utilize the Timonshenko equation in evaluating interface bending stress in metal ceramic restorations. Glass transition temperature measurements have been made for a limited number of porcelains using the modified bending beam viscometer for rapid heating and cooling rate measurements [211].

Formation of an oxide layer at the alloy surface is important to metal–ceramic bonding [212]. In general, oxidation of metals can be classified into three groups: (1) general external oxidation with or without sublayers of inner oxide layer, (2) internal oxidation without any external oxide films, and (3) selective oxidation of a certain type of alloying element that shows higher affinity to environmental oxygen under elevated temperatures.

Oxides formed on the surface of Ni–Cr and Co–Cr dental alloys during degassing heat treatments were investigated. The loss of oxide–metal contact appeared to be limited to the areas of localized void formation, predominantly in the Cr-rich layer with minute protrusions of oxide. Large pegs of oxide were found to have extended into the NiBe intermetallic phase of these alloys. Beryllium appears to be the oxygen-active element responsible for peg formation [212]. With Pd–Cu–Ga–In–Sn–Au alloy and Pd–In–Sn–Cu–Ga–Co alloy, degassed for 1 ~ 10 min in vacuum/air at 940 ~ 1020°C, In and Ga were preferentially oxidized [213].

Changes in the morphology of the oxidation zone and the quantities of oxide for Au alloys containing two base metals (In and Sn) were studied by heating alloys at 1000°C for 1 h in air. The content of In and Sn was varied in the range from 0 ~ 1.5%. In the range of 0.3 ~ 1.5% In and 0 ~ 1.15% Sn, In_2O_3 was formed predominantly at the alloy surface. The electron probe x-ray microanalyzer detected no uniform external oxidation zone. A mixture composed of In_2O_3 and SnO_2 precipitated internally in the alloy matrix. An external oxidation zone composed of SnO_2 was formed on the alloys containing >1.2% Sn, and no internal oxidation zone was observed [214].

With high-Pd alloys for metal ceramics, In and Ga components oxidize preferentially, and the effect of Cu on oxide formation is to suppresses oxide formation of Cu and Ga as well as formation of tin oxide. The driving force for oxidation decreases in the order Ga/In, Sn, Cu [215].

Oxidation of Ni–Cr-based alloys (with Al, Be, Si, Nb, and Mo) were studied. Oxidation of these alloys at 1000°C for 10 min produced an oxide layer consisting principally of Cr oxide (more likely Cr_2O_3), but the oxide morphology varied with each alloy depending on the alloy microstructure. Controlling alloy microstructure while keeping the overall composition unchanged may be a means of preventing wrinkled, poorly adherent scales from forming [216].

Additives and/or impurities in NiCr alloys (Ni–20Cr) would be expected to affect the oxidation and adherence of dental glasses. Oxidation characteristics and bonding of several glasses to a commercial impure 80Ni–20Cr alloy and laboratory-prepared higher-purity alloys were studied at 1000°C. The commercial alloy formed a porous and buckled single-layer oxide scale with lenticular voids at the interface. The scales was penetrated by glass that formed a chemical bond at the alloy–glass interface, resulting in excellent adherence with fracture in the glassy phase. The purer alloy formed a complex multilayer oxide scale to which the glass bonded but did not penetrate; the assembly was not satisfactory, since fracture occurred in the oxide scale [217].

The oxides formed at the surface of an alloy during preheating at about 1000°C prior to the firing of ceramic are vital for the formation of strong chemical bonds between alloy and ceramic. Replacing Au as the basic constituent in the alloy with Pd seems to have a rather small effect on the behavior of the elements Sn and In, which were found to oxidize internally. However, no external oxidation of the transition metal Co was found, in contrast to Ni in Au-based alloys. After ceramic firing a zone enriched with Co and In is found in the interaction area between alloy and ceramic. In another alloy, Ga was observed to oxidize internally and much more readily than Cu due to its much lower free energy of formation. A solid solution was produced in both the alloys by ceramic firing [218].

Internal oxidation was observed in Au-rich alloys that contain small amounts of Fe and Sn. The internal oxidation proceeded with oxygen ions diffusing to the inner part of the alloy through Fe_2O_3 formed at the grain boundaries of the alloy matrix. SnO_2 was formed internally together with the Fe_2O_3 [219].

For Au alloys containing only In as a base metal, an external In_2O_3 layer forms uniformly on the alloy surface. However, when the alloy contains Sn and In, no external oxide layer can be detected by electron probe microanalysis, and oxide particles composed of In_2O_3 and SnO_2 precipitate within the alloy. For alloys containing Ni, in addition to In and Sn, the external oxide is composed of NiO; there is little development of internal oxide. For alloys containing Fe and Sn, an oxide layer of only Fe_2O_3 forms on the alloy surface and the internal oxidation zone shows a band-like structure containing SnO_2 and a small amount of Fe_2O_3 [220].

The mechanism of formation of nodular material on the surface of a Pd–Ag-based alloy for porcelain during heat treatment was investigated using scanning electron microscopy, x-ray diffraction, quantitative metallography, and Auger electron spectroscopy. The nodules were found to form by a Nabarro–Herring creep mechanism driven by the internal oxidation of Sn and In [221].

Wettability, which is believed to be related to the adherency of the metal surface oxide layer to the porcelain, of platinum by dental porcelain compositions is improved by an oxidized layer of tin plating that prevents the platinum surface acting as a contact catalyst for forming porcelain compounds [222].

I.   Interface and/or Surface Modification

Several methods have been proposed to enhance the chemical/mechanical reaction and activities at interfacial layers between metal and ceramic. The manner in which this can be achieved by the interposition of a thin gold layer between two surfaces coupled with firing at about 1000°C is described. The presence of the gold reduces the tendency to form an oxidized layer on the metal surface; recent developments have resulted in a gold composition that can be fired over a wider temperature range (950 ~ 1070°C) without softening or blistering [223]. The adhesion at metal–ceramic interfaces was studied in two Au–Pd–Pt–Ag alloys by electron diffraction. The first alloy contained 0.4% Sn while the second contained 0.1% Si without any Sn. The materials were heated in air at about 1000°C. The adhesion of the 0.4% Sn alloy were superior to 0.1% Si alloy because the SnO_2 surface layer can adhere more firmly to the alloy substrate than the SiO_2 [224].

To achieve maximum strength in Pt ceramic systems, an intermediate layer of tin oxide between the porcelain and the Pt foil is used. The tin oxide layer produces an interface with a distinctive structure and bonding mechanism. The effective use of Pt foil in the manufacture of dental jacket crowns is dependent on the creation of either good mechanical interlocking or a smooth chemically bonded dental porcelain surface [225].

Ni-Cr-base alloys used with porcelain were investigated. After oxidizing heat treatments and porcelain firing, phase changes occurred in some alloys near the porcelain interface. These changes were due to the depletion of the matrix in alloying elements as a result of the oxidizing treatments [226].

J. Metal-Ceramic Interface and Oxide Formation

Interfaces between metal alloys and ceramics can be categorized, from the viewpoint of structure of the contact surface, into Group I (relatively sharp contact), Group II (contact zone of a width of 2 ~ 3 μm), and Group III (30 ~ 50 μm) [227]. With Ni-Cr alloys, surface roughening improves oxide adherence. The formation of Cr_2O_3 also assists the wetting of the opaque to the dental alloy and forms a transition zone. Preoxidation of the dental alloy has no beneficial effect on the bonding strength. Firing at 70 torr leads to higher bonding strength [228].

There is considerable controversy concerning the relative importance of strengthening of the oxide film versus surface roughness. Al_2O_3, CuO, Mn_2O_3, and SnO_2 coatings were applied by the rf sputtering to a 85Pd-10Cu-5Ga alloy. The Al_2O_3 coated specimens gave 45% increase in bond strength over uncoated control samples. A "natural" oxide film developed by oxidation in air gave an improvement of 150%. However, surfaces that were roughened mechanically showed improvements as high as 487%. To explain these improvements the interface was examined by SEM and TEM to investigate the fracture path. These examinations disclosed that the principal effect of oxidation was the development of oxide cavities in the metal. These cavities were penetrated by the porcelain and provided mechanical interlocking. In no case was any vestige of the sputtered oxide film found at the interface. However, oxidation during firing is important because samples fired in a reducing atmosphere exhibited only 10% of the typical strength. Because the shear test galled the structure at the interfaces, a new test, the interface separation test, was developed to spall off the porcelain layer from the interface for microexamination. These specimens showed clearly that the strength was proportional to the degree of mechanical interlocking. The results indicate that mechanical interlocking, whether produced by oxide cavities or mechanical roughening, is a major factor in bond strength, and that the role of oxide films is probably just an aid to provide wetting [229].

An interfacial void could seriously weaken a metal ceramic restoration if the void was in an area of high interfacial stress from loading. The predominant mode of catastrophic failure for metal ceramics is cohesive. The SEM fractographs indicated the adhesive bond was adequate [230].

For Ni-Cr-Mo-Si alloys, specimens with poor adherence between metal and ceramic show a high Cr content at the interface. High metal-ceramic reactivity is favored by a controlled release of metallic cations, which substitute for the alkaline and alkaline earth elements in the ceramic oxides [231].

K. Bonding Strength

The literature provides evidence for a strong correlation between the presence of an oxide layer at the metal-ceramic interface and the bond strength of a metal ceramic system [232]. Bonding is thought to result from a combination of mechanical interlocking (due to surface roughness, or residual stresses in porcelain), van der Walls forces, and chemical bonding (between oxides and porcelain surface). All three mechanisms require wetting of the metal with porcelain during sintering [233,234]. Chemical bonding between dental porcelains and alloys results from interdiffusion of porcelain and metal ions. An interfacial diffusion zone is created that, most likely, has properties different from those of bulk materials. This changed interface might affect experimental measurements of thermomechanical strain [235].

A number of clinical reports of failure of metal ceramic restorations indicates that the strength of the bond between dental porcelain and base metal alloys containing Cr is adversely influenced by the formation of chromium oxide. This reduces the expansion coefficient of the porcelain, thus increasing the residual stress at the bond [236]. When a poor bond is formed, a nonadherent oxide scale consisting mainly of Cr and Si oxides forms on the surface of the alloy during preoxidation, before application of the porcelain. This nonprotective barrier allows Cr to diffuse into the specimen during the firing cycle, where it reduces the dispersed tin oxide particles to form molten, metallic Sn. The molten Sn reacts with Ni and Si in the alloy, forming a Ni-Sn-Si intermetallic at the interface between the porcelain and metal. The bonding agent, which consists of Al powder and opaque porcelain frit, promotes the formation of an adherent stratified oxide scale that consists of a predominantly NiO outer layer, a Cr_2O_3- and Al_2O_3-containing subscale and a deeply convoluted oxide–alloy interface [237].

Ni-Cr-Si-Be dental alloys have good metallurgical characteristic. However the metal–ceramic bond is poor, because of the formation of Cr_2O_3 in the metal/ceramic interface during the heating. By a small chemical modification of the opaque (addition of 2% Zn), it is possible to prevent Cr_2O_3 formation, replacing it by a $ZrCr_2O_4$ spinel compatible with the other oxides present in the interface — NiO and $NiCr_2O_4$ — that adhere to the alloy. The mechanical strength of the metal–ceramic bond is then doubled without any modifications of the dental laboratory fabrication techniques [238]. Two 80Ni-20Cr-type alloys, one 99.99% pure and one with 98% purity, formed a multilayer oxide scale with an outer layer of NiO and a single-layer scale with the outer portion richer in Cr_2O_3, respectively. An applied potassium–aluminosilicate glass at 1020°C penetrated the latter oxide scale. Adherence was present at all interfaces due to the existence of chemical equilibrium. Fracture occurred in the multilayer oxide scale and in the glass close to the oxide–glass interface [239].

The bonding of an experimental low-fusing porcelain to Ti and Ti-6Al-4V was evaluated by an x-ray spectrometric technique. Oxide adherence strength values of Ti and Ti-6Al-4V oxidized at 750° and 1000°C were measured in tension with use of high-temperature adhesives. The porcelain was found to delaminate completely from the metal substrate, leaving 1% of the surface covered with porcelain. The oxide adherence of specimens oxidized at 750°C was good, but those at 1000°C exhibited significantly lower oxide adherence. The simulated porcelain-firing treatments also produced significant adherence. A 750°C oxidation treatment produced oxide films too thin to be visualized in the SEM, whereas the 1000°C oxidation produced oxide films ~ 1 μm thick. The lower oxide adherence of the 1 μm thick oxide films is consistent with reports in the Ti literature of oxide delamination when the oxide film reaches 1 μm in thickness [180].

A higher degree of vacuum during firing yields a greater bonding strength of porcelain to Au alloys. The longer the period of hydrofluoric acid treatment, the higher is the strength of the bond. Treatment of the metal surface, especially with a bonding agent, can influence the bonding strength to some extent. A degassing procedure is indispensable for a strong bond and degassing treatment for 30 min is sufficient [240].

Interactive effects of etching and preoxidation on porcelain adherence to base metal alloys were studied by a guided planar shear bond test under noncantilever conditions. The results indicated that electrolyte etching decreases bond strength, but that oxidation of the etched surface restores the bond strength. The results are explained by the hypothesis that depletion of critical elemental species occurs during etching [241].

Au, Pt, Pd, and Ag alloys with minor additions of In, Sn, Zn, or Ga were studied. To determine the effect of surface treatments of these alloys on changes in surface concentration and hence on thermal expansion and metal–ceramic bond strength, cast alloys were oxidized in air, poor vacuum, and high vacuum, followed by no treatment, sand blasting, and chemical attack by sulfuric acid. Highest bond strengths were achieved by specimens preoxidized, followed by chemical attack [242].

A four-point flexural test for metal–ceramic bond strength has been proposed. Specimen geometry dictates whether failure occurs at the porcelain surface or at the interface under a line of force intensification. Finite element stress analysis indicates that bond separation, if it occurs, is probably due to normal tensile stresses [164].

The incidence of mechanical failure of cast metal ceramic restorations (Au–5Pt–4Pd–1.5In–1.5Ag, Pd–38Ag–8Sn, Ni–13Cr–7.5Ga–7Mo–5.5Fe and Pd–20Ag–2Al) has prompted a study to develop reliable methods of characterizing and improving bond strength of devices produced under standard dental laboratory conditions. Single-edge notch beam specimens were prepared by firing porcelain from a single blended frit to either side of a central metal coupon and subjecting these to four-point bending at 20°C and 0.5 mm/min crosshead speed. Specimens of three representative dental alloys and an experimental alloy were prepared in two specimen sizes and three notch widths. The fracture toughness (K_{IC}) values were independent of notch width and specimen size over the range of these variables examined, and good discrimination was obtained. The fracture toughness of all bonds was improved by a commercial hot isostatic pressure cycle. SEM examination of the fractured surfaces suggested that this improvement was due to the reduction in microporosity at the interface [232].

Although most dentists prefer that porcelain surfaces be glazed, numerous circumstances require the removal of glaze either before or after cementation of the restoration. Certain polishing procedures can restore unglazed surfaces to an acceptably smooth condition. Metal ceramic samples were made from high-fusing Au alloys and porcelain. The rank order for roughness of the surfaces produced by the initial reduction did not necessarily coincide with the smoothness of the final finished unglazed surfaces. It is suggested that the optimum matching of initial reduction method to finishing methods, as well as the clinical significance of small differences in surface texture be given future study [233].

L. Coating

In order to achieve optimum thermal matching between metal substrate and ceramics, several coatings have been proposed. When Co–Cr–Ni and Ni–Cr alloys were plasma coated with Ni–Al powder as the first layer and Al_2O_3 or ZrO_2 powder as the second layer, resulting in differences in coefficients of thermal expansion between the alloy and the porcelain larger than $5.4 \times 10^{-6}/°C$, the fused porcelain starts to spall or crack off [234].

Al_2O_3, CuO, Mn_2O_3, and SnO_2 coatings were applied to 85Pd–10Cu–5Ga alloy by radio frequency (RF) sputtering. Of these, the Al_2O_3-coated specimens gave 45% increase in bond strength over uncoated control samples. A "natural" oxide film developed by oxidation in air gave an improvement of 150% [229].

The use of a Ni–Cr–Al alloy as an undercoating improves the adherence of a SiO_2–Al_2O_3 dental ceramic to Ni–Cr alloys. It is also possible to enamel Co–Cr alloys in this way, thus giving mechanical properties appreciably superior to those obtained on Au alloys. The diffusion of the Cr plays a leading part in this adherence in that it ensures a chemical continuity at the interface by developing an oxide that is partly soluble in the ceramic. Si improves the corrosion resistance of the metal and so prevents the formation of too thick a layer of Cr oxide, which can increase the interface brittleness [243].

A titanium surface was coated with tin through the argon ion PVD process, and oxidation kinetics that simulated porcelain firing were investigated. It was found that nitriding the titanium substrate was effective in control of the thickness of surface oxide film of TiO_2 [165].

M. Leakage

Radioactive isotopes (^{45}Ca) were used to investigate the integrity of the metal–ceramic interface for six commercially available casting alloys. Specimens were examined at intervals 1, 6, and 12 months. All exhibited isotope penetration at the metal–ceramic junction. High Au content alloys experienced the least penetration, followed by the Pd and the Ni–Cr metal alloys. Alloys of Pd without Au showed the most leakage. Isotope penetration and leakage increase with decreasing bonding strength. When a true chemical bond exists between metal and ceramic, the interface should be hermetically sealed [244].

N. Corrosion

Five commercial Pd–Ag alloys were investigated. Their electrochemical behavior was tested with standard potentiodynamic techniques. Corrosion characteristics of the Pd–Ag alloys were quite similar regardless of microstructural and compositional differences. For the potential range of oral interest, all alloys showed sufficient corrosion resistances. These alloys, unlike the high-Pd and Ag-free alloys, were adversely affected by a high Cl^- content electrolyte, and a visible corrosion film formed [245].

Quantitative measurements of tarnish and corrosion resistance in low Au content dental alloys in a variety of metallurgical states have shown that the desirable features include a large grain size, a minimum of segregation, and the absence of precipitates and ordered phases [246].

O. Technology Transfer

In the early 1980s a new method was introduced into dentistry to manufacture commercially pure titanium restorations. This technology is known as *electrodischarge machining* (EDM) or *spark erosion* [247]. The advantages of EDM are: (1) absolutely no pressure operation, (2) no heat effect on work piece, (3) accurate parallel walls can be obtained, and (4) operation is automatic [247]. EDM can be further interfaced to a CAD/CAM system to automate the whole fabrication process.

VI. TITANIUM AND Ti-BASED ALLOYS

A. General

Although only certain types of Ti-based alloys are used currently as dental materials, the microstructures and related mechanical properties of all Ti alloys are discussed. Basically, Ti and its alloys fall into one of four major groups: (1) HCP α type, (2) near α type, (3) ($\alpha + \beta$) type, and (4) BCC β type; see Table 5.

Sn and Zr do not significantly alter the β-phase transition temperature. Elements including Al, O, and N appear to stabilize the α phase (which is stable at lower temperature) and increase the β-transition temperature, thus expanding the α-phase region. Alloying elements, such as V, Mo, Fe, Cr, and Mn, stabilize and expand the β-phase region (which is stable at higher temperature).

In pure Ti, the $\alpha \rightarrow \beta$ transition point is 882°C; the α phase is stable below 882°C while the β phase is stable above this temperature. Above the transition, Ti becomes prone to oxidation. Within the β zone, grain growth causes a reduction in the mechanical strength.

Mechanical properties depend on type of alloy (see Table 6). The specific gravity, heat treatability, plastic formability, and strain rate sensitivity increase in order of α type < near α type < ($\alpha + \beta$) type < β type. Weldability and creep strength decrease in the same order [248].

Although cast Ti and Ti-based alloys (mainly Ti–6Al–4V) are used for denture base materials [249] as well as for metal ceramic copings [180], the most common use of titanium is for dental and orthopedic implants. These implants are usually wrought material with various surface conditions, such as smooth surface, porous surface, or spray coated with other materials, including hydroxyapatite. In addition to these applications, some Ti-based alloys are drawn into wire for use as orthodontic archwires.

The excellent tissue–bone compatibility of Ti is due mainly to the properties of its stable surface oxide layer. Biocompatibility of implant materials relies on chemical and electrochemical stability of this surface oxide layer, which interfaces with the soft and hard tissue and bone structure. Even bond strength with porcelain (as discussed in Sec. V) is controlled by the crystalline structure and thermal properties of surface oxides and their wettability by porcelain. Oxide layers formed on titanium change from lower to higher oxides as oxidation progresses and temperature increases. The following phases form in air: $Ti + O \rightarrow Ti(O) \rightarrow Ti_6O \rightarrow Ti_3O \rightarrow Ti_2O \rightarrow TiO \rightarrow Ti_2O_3 \rightarrow Ti_3O_5 \rightarrow TiO_2$. In aqueous solution, TiO oxidizes to TiO_2. Above 100°, the surface is a lower oxide

Table 5

HCP, α type	Near α type	($\alpha + \beta$) type	BCC β type
Pure Ti	Ti–5Al-6Sn–2Zr–1Mo–0.1Si	Ti–6Al–4V	Ti–13V–11Cr–3Al
Ti–5Al–2.5Sn	Ti–8Al–1Mo–1V	Ti–6Al–6V–2Sn	Ti–8Mo–8V–2Fe–3Al
	Ti–6Al–2Sn–4Zr–2Mo	Ti–6Al–2Sn–4Zr–6Mo	Ti–3Al–8V–6Cr–4Mo–4Zr
		Ti–8Mn	Ti–4Mo–4Al–2Sn
			Ti–11.5Mo–6Zr–4.5Sn

Table 6

Alloys	Heat treatment	Tensile strength, MPa	Yield strength, MPa	Elongation, %
Near α				
CP Ti	As annealed	549	481–657	15
8Al–1Mo–1V	Aging	995	947	15
11Sn–2.25Al–50.Zr–1Mo–0.2Si		1098	989	15
6Al–2Sn–4Zr–2Mo	As annealed	975	892	15
5Al–5Sn–2Zr–2Mo–0.25Si	970C (0.5h), AC + 593C (2h), AC	1043	961	13
6Al–2Nb–1Ta–1Mo	As rolled	851	755	13
6Al–2Sn–1.5Zr–1Mo–0.35Bi–0.1Si		1009	940	11
α + β				
8Mn	Annealed	940	858	15
3Al–2.5V	Annealed	686	584	20
6Al–4V	Annealed	989	920	14
	Aging			10
6Al–4V(ELI)	Annealed	892	824	15
6Al–6V–2Sn	Annealed	1064	995	14
	Aging			10
7Al–1Mo	Aging	1098	1030	16
6Al–2Sn–4Zr–6Mo	Aging	1263	1176	10
6Al–2Sn–2Zr–2Mo–2Cr–0.25Si	Aging	1270	1133	11
5Al–2Cr–1Fe	Aging	1153	1087	10
β alloys				
Ti–13V–11Cr–3Al		1236	1196	8
Ti–8Mo–8V–2Fe–3Al		1304	1236	8
Ti–3Al–8V–6Zr–4Mo–4Zr		1442	1373	7
Ti–11.5Mo–6Zr–4.5Sn		1383	1314	11
Ti–15Mo–5Zr		1373	1324	10
Ti–15Mo–5Zr–3Al		1471	1451	14

type, such as TiO. At elevated temperatures (above 200°C), complex oxides are formed, such as Al_2TiO_5 on Ti–6Al–4V and $NiTiO_3$ on TiNi in addition to TiO_2 [250].

B. Commercially Pure Titanium

1. Oxide Formation

XPS and AES complemented with information from SIMS, EDX, and NMA (nuclear microanalysis) were used for surface characterization of pure Ti implant materials prepared by procedures commonly used in clinical practice. The surface was found to consist of a thin surface oxide covered by a carbon-dominated contamination layer. In comparison with reference spectra from single-crystal TiO_2 (rutile), the composition of the surface oxide is mainly TiO_2, with minor amounts of suboxides and TiN_x. The thickness of the surface oxides is 2 ~ 6 nm, depending on the sterilization method. The surface contamination layer varies considerably from sample to sample and consists mainly of hydrocarbons with trace amounts of Ca, N, S, P, and Cl [251].

The residual adhesion of Ti oxide films to the metal substrate has been measured by tensile tests after cooling unalloyed samples from oxidation temperatures between 550° and 700°C. Adhesion measurements were then correlated with the oxidation behavior. These results reveal the important roles of oxidation temperature and thickness of the oxide layer, and the significant influence of nitrogen on oxidation in air and on the adhesion of the oxide layers to the metallic substrate [252].

2. Corrosion

Corrosion reactions around Ti may be provoked by coupling it galvanically with active dental alloys. A study simulating a Ti dental implant or root canal post coupled with an amalgam filling showed corrosion current densities that could lead to local tissue damage. Clinically, a Ti implant used as a single-tooth replacement might contact a proximal amalgam filling. Ti in aerated 0.1% NaCl was coupled with samples of gold alloy, Co–Cr alloy, 304 stainless steel, conventional amalgam, and high-copper amalgam. It was found that Au, Co–Cr alloy, and stainless steel showed no corrosion current and no pH change. The current and pH curves have a comparable shape for both types of amalgam [253].

Commercial pure titanium (CPT) and Ti alloys (TiNi, Ti–13Cu–4Ni, and Ti–4.5Al–5Mo–1.5Cr) were electrochemically tested in Hank's solution. Open-circuit electrode potential measurements indicated that Ti alloys exhibit the same rest-potential after a period of 2 to 3 weeks, regardless of prior treatments, including mechanical polishing, steam sterilization, washing in boiling 3.5% NaCl solution for 30 min, or washing in 30% HNO_3 at 60° for 15 min. The measurements also indicated that the surface oxide film on Ti is stable to 2.4 V, which is 1.98 V above the stable oxide formed on Co–Cr alloy [254].

3. Bone Ingrowth and Biocompatibility

Under certain circumstances, pure Ti implants inserted into human bone will establish and maintain direct contact with the bone tissue; this is called osseointegration [255,256]. One factor believed to be of crucial importance for the success or failure of a particular implant is the outermost surface layer of the implant material [257,258], particularly micromorphological surface properties [259]. For the Ti implants described here, the surface consisted of a 2 ~ 5 nm oxide [260]. Characterization of the surface topography of implant materials also is important for understanding tissue response. The topography of Ti surfaces used in osseointegrated dental implants was measured. STM (scanning tunneling microscopy), which provides three-dimensional real space images, was used. In addition to clinical samples, electropolished and anodically oxidized surfaces were measured. Clinical samples were rather inhomogeneous in character, showing grooves and steps with a maximum depth of 0.11 μm. Micropores with an average diameter of 30 nm also were present. Electropolished samples were homogeneous and very smooth, showing steps of 1 ~ 5 nm in height [261].

Auger electron spectroscopy has been used to study the interface between human tissue and implants of Ti and stainless steel. Both the thickness and the nature of the oxide layers on the implant change during the time of the implantation. Stainless steel implants have a surface oxide about 5 nm in thickness prior to implantation. It was observed that the metal in the oxide was mainly Cr [262]. The change in the oxide thickness depends on the location of the implant in the body. For implants located in cortical bone, the thickness of the interfacial oxide layer remains unaffected, while it increases by a factor of 3 ~ 4 on samples located in bone marrow. In both cases, calcium and phosphorous are incorporated in the oxides. Implants located in soft tissue have an interfacial oxide with a thickness of about 1.5 times that of an unimplanted sample. On these samples, Ca and P are not incorporated in the oxide layer. For Ti, an increase in oxide thickness and an incorporation of Ca and P are found. With Ti implants, the oxidation process occurs over several years. The interfacial oxide layers on both stainless steel and Ti implants contained P strongly bound to oxygen, suggesting the presence of PO_4 groups in the oxides [262].

Loaded, prestressed implants of dense hydroxyapatite (HAP) and nonloaded HAP-coated Ti implants were placed in edentulous regions of the lower jaw of dogs. After 6 months thin nondecalcified ground sections were made for histology. Although the HAP showed histological differences between coated implants and the prestressed solid ones, both had an extensive apposition of normal lamellar bone on the whole surface of the bone-buried part of the implant. The bone contact was very intimate, without any visible intermediate tissue layer. The tissue response observed supports the clinical application of HAP-coated Ti implants [263].

4. Bonding

The bonding between silane-treated Ti and PMMA was studied. Ti was mechanically polished, cleaned, and washed in water followed by cleaning in a plasma unit *in vacuo*. After the Ti surface was silane treated, a heat-polymerizing-type PMMA was applied. It was found that the average bonding strength in four-point bending was 25 MPa after dry storage at 37°; while average strengths were 9 ~ 11 MPa after water storage for 20, 60, and 90 days at 37. Infrared spectroscopy indicated that the adhesion between the silanized Ti interface and the polymer was due to the chemical bonds [264].

C. Ti–6Al–4V Alloy

1. Bone Ingrowth and Biocompatibility

Ti–6Al–4V offers an excellent combination of mechanical properties for load-bearing applications such as hip prostheses, but there is concern about the slow accumulation of potentially harmful metal ions, Al and V, in the soft tissue surrounding the prosthesis [265,266]. Very recently, the uptake of salts or fretting corrosion products from Ti–6Al–4V implants was examined. It was found that vanadium was toxic at levels greater than 10 μg/ml, and the percentage of cellular association of titanium was shown to be about 10 times that of vanadium [267]. Furthermore, evidence [268] indicates that tissue reactions adjacent to this alloy are less natural than those in the vicinity of CPT. Metal ion release is considered to occur via chemical dissolution of the surface oxide film. The dissolution rate might be influenced by passivation treatments and surface coating such as TiN. These effects may be associated with structural changes of the surface oxide [269].

The influence of the surface oxide on the dissolution of the substrate material in saline solution was investigated by using a combination of atomic absorption spectroscopy, ellipsometry, and transmission electron microscopy. It was demonstrated that a substantial reduction in the release of metal ions may be achieved by aging the surface oxide in boiling distilled water or by thermal oxidation. The change in the dissolution behavior is thought to be associated with transformation of the TiO_2 from the anatase form to the more compact rutile structure [269].

The bone mineral-like characteristics of calcium phosphate ceramics (CPC) have led to interest in their clinical use. In contrast to most other materials, the CPC surface makes direct contact with bone tissue [270]. Furthermore, CPC promotes the formation of normal bone tissue at the surface [271]. As a result, they are generally described as *osteoconductive*. They are commonly applied as coatings on metallic substrates by techniques such as plasma spraying, electrophoretic deposition, sputter coating, hot isostatic pressing (HIP), and ion-assisted sputtering [272–274]. Scanning Auger electron spectroscopy was used to determine the compositions of the surface and the interface of calcium phosphate ceramic coatings electrophoretically deposited and sintered on Ti and its alloys before and after 4 weeks of immersion in a simulated physiological solution. At the CPS coating–metal interface, the phosphorous diffused beyond the Ti oxide layer. The phosphorous concentration in the interface followed a Gaussian distribution for both unalloyed and alloyed Ti. This diffusion depleted the P in the ceramic adjacent to the metal. The surface of the ceramic, however, was substantially unchanged. A major change in the compositional depth profiles was induced by immersion. Thick and uniform Ti phosphide layers of constant composition were observed on the Ti-based metal substrates [275].

2. Porous Implants

Porous Ti–6Al–4V implants are fabricated from spherical metal powders that are compacted and sintered. The volume fraction of voids was 0.3 ~ 0.4, regardless of particle sizes, which ranged from fine (75 ~ 420 μm), to medium (225 ~ 500 μm), to coarse (325 ~ 707 μm). It was found that the percent bone growth is proportional to the square root of particle size [276].

A void metal composite is a porous metal developed to fix a prosthesis to bone by tissue ingrowth. Ti–6Al–4V is the metal of choice for void metal composites. It was selected for its corrosion resistance, good mechanical properties, low density, and good tolerance by body tissue. Structures with spherical pore size ranging from 275 to 650 μm have been fabricated with up to 80% theoretical densities. Cylindrical pores with controlled direction and arrangement also were produced. The optimum structure for attachment strength seems to be a pore size of 450 μm and 50% theoretical density [277].

Porous Ti–6Al–4V single-tooth implants were placed in the mandibles of monkeys [278]. The implants, used to support bridges and crowns, were comprised of spherical Ti alloy powders sintered on a solid Ti alloy core to encourage ingrowth of bone. No measurable mobility was observed in relatively long-term studies. Shorter-term results with human patients indicated the possibility of successful application of the implants [279].

3. Corrosion

Static, *in vitro* electrochemical, and gravimetric corrosion studies were conducted to evaluate the effectiveness of 0.5 μm ultra-low-temperature isotropic carbon coatings in reducing the metal ion release rates of porous Ti and solid Ti–6Al–4V alloy. The results indicated significant reductions for both materials, with the corrosion rate of carbon-coated porous Ti being reduced to that of solid uncoated Ti–6Al–4V. Analyses of the results indicate a minimum expected carbon film lifetime of 38 years [280].

4. Superplastic Forming

Superplastic forming of Ti–Al–V alloy for denture bases is in the research and development stage [281]. Metallic denture bases are now mainly manufactured by casting Co–Cr or Ni–Cr alloys. Although Ti–6Al–4V is an attractive material for use as a denture base because of its biocompatibility, high strength, and low density, difficulties with fabrication have discouraged this use. Ti–6Al–4V has superplastic properties at high temperatures [281,282]. Ti–6Al–4V sheet has been formed superplastically using Ar gas pressure at 800° ~ 900°C. Superplastically formed denture bases of Ti–6Al–4V showed some merits compared with conventionally cast or cold-pressed denture bases; they include good fit, light weight, and low production cost, using a smaller number of manufacturing processes [282].

D. β Ti

The β Ti alloy, Beta III, has high strength capability and excellent hot and cold formability, allowing it to be produced in essentially all mill product forms. Applications for Beta III to date include orthodontic appliances and aircraft fasteners and springs. AMS 4977 covers BETA III bar and wire products typically used in aircraft applications. Recent biocompatibility studies have shown encouraging results for future dental and medical uses [283]. Beta-stabilized Ti alloys have great potential in orthodontic appliance fabrication. Proper thermomechanical treatments can produce a ratio of yield strength to the elastic modulus almost twice that of 18-8 metastable austenitic stainless steel [284].

The total cold reduction and reduction per pass used during the drawing of fine wires influences the mechanical properties of β Ti alloys. Both Ti–11.5Mo–6Zr–4.5Sn and Ti–13V–11Cr–3Al showed increased yield strengths and low elastic moduli, resulting in wires appropriate for orthodontic applications [285].

Ti–11.5Mo–6Zr–4.5Sn (metastable β-phase Ti alloy) was introduced as a replacement for stainless steel and Ni–Cr–Co wires in orthodontics. Ti–3Al–8V–6Cr–4Mo–4Zr, Ti–15V–3Cr–3Al–3Sn, and Ti–10V–2Fe–3Al were also tested. The evaluation indicated that Ti–15V–3Cr–3Al–3Sn was the best in formability and springback in clinical usage [286].

The dynamic elastic modulus of Ti–11.5Mo–6Zr–4.5Sn is 10.6×10^6 psi, while that of Ni–45Ti–3Co is 8.32×10^6 psi. The dynamic mechanical properties differ with regard to whether the alloy is a conventional, stabilized-martensitic alloy, conventional shape-memory alloy, a stress-induced martensitic alloy, or a biphasic alloy [287].

E. TiNi Alloys

1. Application

The intermetallic TiNi possesses two unique characteristics: superelasticity (SE) and shape memory effect (SME). Stainless steel and Co–Cr–Ni wires have a long history of use in orthodontics but TiNi has rather recently been generally adopted. Orthodontic applications require low forces to be delivered over long working ranges. TiNi wires can be used in Class I, II, or III malocclusions in both extraction and nonextraction cases. The most important benefits from TiNi wire are realized when a rectangular wire is used in treatment to achieve simultaneous rotation, leveling, tipping, and torquing [288].

The β Ti alloys as described previously have also seen recent popularity for their high springback, low stiffness, and high formability. In addition, components made of β Ti alloys can be assembled by spot welding without appreciably reducing the resilience of the material [289].

Recently the SE properties of equiatomic TiNi alloys have been utilized in clinical orthodontics. The relatively constant forces on unloading produced by orthodontic appliances fabricated from this alloy have the effect of increasing the rate of tooth movement. In addition, the load deflection characteristics of this material may be controlled clinically by controlling the amount the orthodontic appliance is activated. It is also possible to apply the concepts of constant cross section, variable load, and constant load deflection orthodontics as treatment strategies using this alloy system. This is because the force generated by a SE TiNi wire may be altered by varying its Ni content and/or annealing temperature rather than the archwire cross section [290].

2. Mechanical Properties

When TiNi wire, TMA (β-phase Ti-Mo) wire, and austenitic stainless steel wire were compared: (1) TMA showed the greatest plastic strain and springback, followed by TiNi and stainless steel; (2) TiNi wire showed the highest stored energy value in bending–torsion, followed by TMA and stainless steel; and (3) the highest spring ratio (stiffness) in bending–torsion was found in stainless steel, followed by TMA and TiNi [291].

A method for quantifying the torsional elastic properties of orthodontic archwires was investigated. Three wire sizes (0.018 × 0.020 in., 0.017 × 0.025 in., and 0.019 × 0.025 in.) were tested in three different alloys: stainless steel, β Ti, and TiNi. The shear modulus (G) was determined dynamically using a torsion pendulum. The torsional yield strength (T_{ys}) was determined using a static torsion test that generated torque (Γ) versus angular deflection (θ) tracings. The three basic torsional elastic properties (strength, stiffness, and range) were calculated for each of the wires from G, T_{ys}, and precise dimensional data. It was reported that the TiNi wire had the greatest range and the lowest stiffness, and stainless steel archwires showed the greatest stiffness and the least range [292]. A three-point bending test shows the SE TiNi wire provides a lower delivered force and lower permanent deformation than stainless steel and Cr–Co–Ni wires [293].

Eight straight wire materials were studied: an orthodontic β Ti-Mo (TMA), three orthodontic TiNi products (Nitinol, Titanal, and Orthonol), three prototype alloys (a martensitic, austenitic, and biphasic alloy), and a hybrid shape memory effect product (Biometal). Each wire was prepared with a length-to-cross-sectional area of at least 3600 cm^{-1}. Using an Autovibron Model DDV-II-C in tensile mode, each sample was scanned from $-120°$ to $+200°$C at 2°C/min. From the database, plots of the log (storage modulus), log(tan δ), and percent change in length versus temperature were generated. Results showed that the dynamic mechanical properties of these Ti-based alloy systems are quite different. The Ti-Mo alloy was invariant with temperature, having a modulus of 7.30 × 10^{11} dyne/cm^2. The three cold-worked alloys, Nitinol, Titanal, and Othonol, behaved similarly and had a modulus of 5.74 × 10^{11} dyne/cm^2. The biphasic SME alloy had a phase transformation near ambient temperature. The hybrid shape-memory product, Biometal, underwent 35% change in length during its transformation between 95° and 125°C [287].

3. Corrosion Behavior and Biocompatibility

The application of TiNi for medical and dental implants has been attempted, but questions about the biocompatibility of alloys with high Ni content remain. Shape recovery of the alloy Ti–V–Fe–Al was studied. Since this alloy contains more than 80% Ti without Ni, it should be biocompatible. Shape recovery in this alloy system took place for compositions Ti(10.0–12.0)–V(1.5–3.0)–Fe(2.0–4.5)Al with the orthorhombic martensite structure retransforming to the original disordered BCC β phase. With a solution heat treated alloy, recovery was insufficient at lower temperatures. However by aging at 400°C for 3 ~ 5 sec followed by deformation at −20°C, the recovery ratio in Ti–11.5V–1.7Fe–3.3Al was increased to 85% at 80°C. This improvement was due to a fine precipitate of α phase formed by aging, which may suppress slip deformation. A recovery of 92% at 60°C was obtained in Ti–11.5V–1.7Fe–4.0Al alloy. A blade-vent-type implant was fabricated with this alloy and a wing opening angle of 60° in the blade edge part was memorized. After treatment, an opening angle of 50° was recovered at 60°C. The corrosion resistance of Ti–11V–2Fe–3Al was evaluated by anodic polarization measurement in 1% NaCl solution at 37°C. Up to 2.0 V overpotential a constant current density similar to that of pure Ti was obtained. Hence, corrosion resistance comparable to that of pure Ti is expected. In contrast, Ti–Ni showed an abrupt increase in current density above 1.2 V. These results suggested that this Ti–V–Fe–Al alloy has applicability for dental implants [294].

Anodic polarization measurements made in Hank's physiological solution at 37°C and a pH of 7.4 show Ti to be the most passive of the following metals: Ti, Ti–6Al–4V, TiNi, MP35N (Co–Ni–Cr–Mo), Co–Cr–Mo, 316L stainless steel, and pure Ni. The influence of the amino acids cysteine and tryptophan on the corrosion behavior of TiNi and Ti–6Al–4V was studied. Cysteine caused a lower breakdown potential for TiNi, but it did not affect the breakdown of Ti–6Al–4V although an increase in current density for Ti–6Al–4V was observed [295].

4. Implants

In a study of biocompatibility, implants made of SME TiNi alloy were embedded in mandibles of Japanese monkeys. When the SME implant was inserted, the trabecular bone surrounded the implant and anchored tightly to it without an interface of fibrous connective tissue. The oxide film on the surface of the implant acts to prevent Ni ion migration and the TiNi alloy implant would appear a viable dental implant system [296].

5. New Production Techniques

Bundles of fine wires of Ni and Ti are diffusion treated at a controlled temperature and atmosphere to fabricate orthodontic archwires [297]. Some ceramics and intermetallic compounds can be made by the self-propagating high-temperature synthesis (SHS) method in vacuum, under high pressure. Resulting materials synthesized in vacuum have excellent properties with very low oxygen and nitrogen content. The mechanical properties of TiAl synthesized in high vacuum and a hot isostatic press (HIP) are equivalent to those of centrifugally atomized TiAl. TiNi with O content of 700 ppm can be synthesized in high vacuum by using powders with very low O content. Wire and thin plates of TiNi are manufactured by HIPing, hot rolling, cold rolling, and drawing. The shape-memory effect and the SE are equal to those of commercial TiNi alloy [297].

F. Other Ti-Based Alloys

Plasma vapor deposition (PVD) of TiN coatings and N^+ implantation of Ti-6Al-7Nb alloy resulted in surface hardening to a depth of less than 3 μm. The new oxygen diffusion hardening (ODH) treatment increased the hardening depth to 50 μm. PVD TiN coating resulted in improvement in tribological behavior while N^+ implantation increased the wear rate of polyethylene (PE) against the alloy. The wear rate of ultrahigh molecular weight PE and the friction coefficient against the ODH-treated surface were reduced to one-half of the values achieved with PE paired with a Co-Cr-Mo alloy [298].

A new dental casting alloy (Ti-13% Cu) was implanted in the skeletal muscle in rabbits. Routine histopathological and chemical analysis were utilized to study *in vivo* tissue reactions to this alloy. A moderately thick, somewhat cellular fibrous connective tissue capsule surrounded the implants after 2 weeks. Remodeling of the fibrous tissue into a thin acellular tissue capsule occurred at 52 weeks after implantation. Chemical analyses failed to detect deposition of either Ti or Cu corrosion products at the implant sites or within major organs [299].

G. Casting of Titanium Alloys

Although Au alloys are considered the ideal dental casting alloys, the price of Au and its fluctuations have prompted development of alternatives based on other less noble metals such as Ag and various base metals such as Ni-Cr and Co-Cr alloys. In general, these alternative alloys do not possess the corrosion resistance or castability of Au alloys. The use of Ni-based alloys has been questioned due to its potential for allergic reactions. Ti alloys have excellent corrosion resistance and biocompatibility, low cost, and adequate mechanical properties, and should be good candidates for dental castings metal if new casting methods are developed for dental applications [300]. The melting points and reactivity with oxygen of Ti alloys make conventional dental casting techniques and investment mold materials impractical for use with Ti. A review of the Ti casting alloys, their properties, and commercial techniques for precision casting is given in Refs. 301-306.

At the present time, most of the castings made for dental applications are made from pure Ti. Metallurgical structure, mechanical properties, and *in vitro* corrosion resistance were studied for cast Ti and for four selected cast Ti alloys (Ti-6AL-4V, Ti-20Cu, Ti-30Pd, and Ti-15V). A "castmatic" dental casting machine was utilized, which involved Ar arc melting and subsequent Ar vacuum-pressurized casting. Photomicrographs revealed quite large grains in the cast structures. X-ray diffraction analyses on cast Ti alloys showed quasi-equilibrium phases present. The strength of cast Ti can be significantly increased by alloying. Cross-sectional microhardness measurements of cast Ti alloys showed U-type hardness distributions due to surface mold reactions. All the cast Ti alloys examined showed a strong passivity in corrosion tests. The results suggest that cast Ti alloys have promise for future dental uses [307].

Investment casting of Ti is adversely affected by reaction of the melt with the atmosphere in the melting or casting chamber, by reaction of the melt with the melting crucible, and by reaction of the melt with the mold material. The first can be avoided by providing a closed melting and casting chamber using an Ar protective atmosphere; reaction within the melt during melting can be reduced by skull melting in a water-cooled copper crucible. Reaction with the mold materials (SiO_2, Al_2O_3, MgO and ZrO_2) was evaluated. It was found that the last yielded the least reaction with molten Ti [308].

H.   Soldering and Welding

The flat–flat welding configuration is favorable for spot welding TMA (77.8Ti–11.3Mo–Zr6.6–4.3Sn). Ductility was significantly decreased in transformer-welded wire segments compared with specimens jointed with a capacitor welder [309].

Since Ti is very reactive with oxygen, soldering should be performed in vacuum or in argon atmosphere. Soldering materials for Ti and its alloys are Ag–Li system alloys including Ag–3Li (soldering temperature 800°C), Ag–7.5Cu–0.2Li (920°C), Ag–28Cu–0.2Li (830°C), and Ag–20Cu–2Ni–0.2Li (920°C). Ag–Al, Ag–Zn–Cd, Ni-based, Ti-based, and Pd-based alloys are also used as soldering materials. Very active flux materials such as chlorides (AgCl, $CuCl_2$, KCl, NaCl, and $ZnCl_2$) or fluorides (LiF, and HKF_2) are required to dissolve the Ti surface oxide film and permit wetting by the solder alloy [248].

Laser (Nd:YAG) welding was recently introduced to weld titanium and other dental materials [310].

VII.   DENTAL POLYMERS AND RESTORATIVE RESINS

Polymeric materials, both natural and man-made, have a long history of dental applications. Soon after the development of poly(methylmethacrylate) in the 1930s, this polymer was used to make base plates for complete dentures. It was the first really successful denture base material and is still the most commonly employed material for this application. The use of acrylic resins as direct filling restorative materials was introduced in the 1950s. These materials were quickly found to have a number of serious clinical limitations.

Their polymerization shrinkage is quite high, which results in gap formation between the restoration and the tooth. This gap in turn leads to marginal staining and often to recurrent decay (caries). The coefficient of thermal expansion of acrylic resin is more than 8 times greater than that of tooth structure. The oral cavity can experience temperature cycles from $0 \sim 60°C$. The differential expansion between the tooth and resin causes the marginal gap to open and close as the temperature changes. The unfilled acrylic resins have rather poor mechanical properties and exhibit high intraoral wear rates. The chemical activation system that causes these resins to harden (polymerize) in the mouth has poor color stability in the oral environment. There was no mechanism for bonding the acrylic resins to tooth structure, and their retention relied on mechanical undercuts in the cavity preparation. All of these factors resulted in limited use of the acrylic resins as direct restorative materials. Two developments in the 1960s overcame most of these limitations and resulted in successful, direct-filling, resin restorative materials. These developments were the composite dental resin and the acid-etch technique for bonding resin to tooth enamel.

Today plastic materials are used extensively in restorative dentistry, both for the construction of prosthodontic devices such as dentures and as direct-filling restorative materials. The later application is growing rapidly and resin-filling materials are now being used even on chewing surfaces where they are considered alternatives to dental silver amalgams. Since the materials employed, the requirements placed on the materials, and the fabrication techniques are different for direct restorations and for prosthodontic appliances constructed outside the mouth, the two applications are discussed separately. The term *resin* is most commonly employed for the composite materials used for direct restorations. *Plastic* is often used to refer to the polymers used in prosthetic dentistry although the term *resin* is also common.

A. Plastics in Prosthetic Dentistry

Poly(methylmethacrylate), PMMA, and its copolymers are the most commonly employed plastic materials in current use for dental prosthetic devices. PMMA is a thermoplastic polymer with a glass transition (T_g) below 100°C. It is reasonably strong, tough, and stable in the oral environment, and exhibits excellent biocompatibility. The polymer is formed from its monomer MMA by an addition polymerization reaction that produces 12.9 kcal/mol of exothermal heat.

The conventional method of fabrication of dental plastic prostheses is somewhat unusual compared to those employed for industrial molding of thermoplastics. The dental laboratory starts with a fine powder consisting of PMMA spheres mixed with a small amount of benzoyl peroxide (BP). To this powder, liquid MMA is added at a liquid-to-powder ratio of ~1:3 by volume. The monomer wets the surface of the PMMA spheres and dissolves some of the surface material. After a period of time, the rheological properties of the mixture resemble those of molding clay. In this condition it can be packed into the mold that has been constructed for the dental appliance to be formed. The mold halves are closed in a hydraulic press and a clamp is placed around them to keep them in compression during processing. This assembly is then transferred to a hot water bath. When the temperature of the dough reaches ~60°C the BP decomposes to form free radicals. These react with the C–C double bond on the MMA monomer backbone, which initiates the addition polymerization reaction. Processing requires between 1 and 9 h depending on the water bath temperature cycle employed. An increase in temperature to near 100°C at the end of the cycle ensures maximum conversion of the MMA to PMMA. This technique is called *heat processing* (curing) and the resin system used is referred to as a *heat-cured/heat-processed/heat-activated* denture base resin. The end result is a plastic–plastic composite structure. The original polymer beads can be readily seen surrounded by the PMMA matrix using light or scanning electron microscopy.

Numerous variations on the basic PMMA denture resin exist. Copolymers and plasticizers are often added to the prepolymerized PMMA in the beads. Other monomers are mixed with the MMA liquid to serve as crosslinking agents to improve mechanical properties and reduce water sorption of the denture. Monomers such as vinyl chloride, vinyl acetate, and styrene also have been used for denture base resins. Mechanisms for activation of polymerization other than heating can be employed. If a suitable tertiary amine is added to the monomer, it will react with the BP when the liquid and powder are mixed. The result is a chemically activated–self-curing–cold curing–autopolymerizing denture base resin. These resins are usually employed to repair or reline existing heat-processed dentures. A photoactivated initiation system can also be used. An example is camphoroquinone with a suitable amine. Exposure to blue light between 450 ~ 500 nm in wavelength results in free-radical formation, which will initiate the polymerization reaction. These denture resins are called *light-cured* or *light-activated*. Polymerization of a heat-activated denture resin can also be activated by exposure to microwave energy. The resin dough absorbs energy from the microwaves, and its temperature increases until polymerization is initiated. Although conventional heat-activated denture resins can be polymerized in a microwave oven, the resulting properties are usually inferior to those of hot water bath processed resins. Special denture resins have been formulated for use in microwave ovens and result in denture resin properties comparable to water bath processing.

In general, the mechanical properties of a high molecule weight polymer improve as the average molecule weight of the polymer increases. Hence, the properties of a denture base resin are highly dependent on processing conditions as well as the composition of the resin itself. The quality of the finished prosthesis depends upon the properties of the resin and other factors related to processing such as dimensional accuracy and porosity. Reference 311 contains a detailed discussion of some of these interactions for the various types of denture base resins.

Crosslinking of the polymer of the interstitial matrix of a heat-activated resin denture base is said to provide *craze resistance* for the denture base. The mechanism of crazing can be related to both water sorption and solvent attack. The present study shows that the addition of ethylene glycol dimethacrylate in concentrations of 0 ~ 100% of monomer volume has little effect on water sorption but is an efficient method of providing solvent resistance [312].

A visible light activated poly(urethane dimethacrylate) resin, which uses a diketone initiator and an amine accelerator, has been developed for use as a dental impression material. The compositions are similar to those employed in one commercial, visible light activated denture base resin. Exposure to blue light results in polymerization of a single-component impression material to form an elastomeric solid. This elastomeric impression material has unlimited working time, short setting time after placement, good wettability of oral tissue, excellent elastic quality, high accuracy, and excellent biocompatibility [313].

The fracture toughness of a dimethacrylate resin has been found to increase with the degree of cure of the network, which was previously attributed to increased crosslinking and the concomitant reduction in unreacted monomer (sol species). Studies of the dependence of the degree of polymerization on the photoinitiator concentration and cure time illustrate the interrelation of the weight percent monomer and the degree of cure, and reveal that a considerable amount of residual monomer remains in undercured networks, as predicted by simple gelation theory. In an attempt at separating the effects of residual monomer and crosslinking on the fracture behavior, series of dimethacrylate and epoxy resins were studied in which the crosslink density and sol levels were independently varied. Although the fracture energy and toughness were raised by increasing the crosslink density in the epoxy resins, no significant variation was found for the dimethacrylates. Addition of saturated analogs of dimethacrylates (used to represent residual monomer) significantly impaired the fracture resistance, suggesting that the reduction in residual monomer is responsible for the improved fracture toughness observed with postcured dimethacrylate networks [314].

Chemically activated acrylic is widely used in clinical practice. Its porosity when cured under pressure at 23°, 40°, and 60°C is less than 0.5% with 4–6 pores/mm^2 of 20 ~ 30 μm diameter. Without pressure, one acrylic showed porosity of 0.5% (at 23°C), 2% (40°C), and 3% (60°C), and another acrylic showed 6 ~ 10% [315].

The universal method of manufacturing PMMA as a two-phase dough-molded system has some inherent weakness including low fracture strength. It is demonstrated that polymerization using different PMMA powders and curing cycles generally produced higher molecular weight values in the two-phase products than in the original powder phase. Denture base systems with an average molecular weight $> 10^5$ displayed optimum fracture strength properties [316].

Resilient semiplastic polymer materials often are used as "soft" linings for dentures in the short-term management of traumatized denture-bearing mucosa. The effects of various plasticizers on gel strength and gelation rate were studied. PMMA powders were mixed with combinations of ethyl alcohol and various phthalate esters. Gelation rate, gel strength, and level of plasticity of short-term prosthodontic denture-lining materials can be controlled by varying the amounts of ethyl alcohol, phthalate ester, or polymer prior to mixing [317].

Elastomeric polyurethane resins for adhesive liners for restorative materials and resilient denture liners were little affected by thermal shock, with negligible loss of adhesion after 13 months in water under a constant load of 100 psi [318]. A number of useful room temperature polymerizing resins were formulated, based on poly(ethyl methacrylate) powder and a range of low-shrinkage heterocyclic methacrylate monomers. Isobornyl methacrylate is a useful diluent monomer for reducing the curing exotherm [319].

The water absorption and desorption behavior of poly(isobornyl methacrylate) and poly(tetrahydrofuran-2-dimethyl methacrylate) obeyed diffusion laws on repeated absorption–desorption cycles. However, the polymers 2,3-epoxypropyl, tetrahydrofurfuryl, and the tetrahydropyranyl methacrylates did not obey diffusion laws, did not equilibrate after 2 years immersion in water, and exhibited very high water uptake (30 ~ 90%). A clearly detailed structure of the heterocyclic ring is critical. The use of these monomers in room temperature polymerizing poly(ethyl methacrylate)–monomer systems generally reflected the behavior of the related homopolymers [320].

The values for the elastic modulus of heat-activated and chemically activated PMMA denture resins were not found to be statistically different. However, the values for flexural strength from the 3-point bend test were always greater for the heat-activated resin [321].

The fracture toughness of homogeneous PMMA and a two-phase acrylic were measured using compact tension (CT) and double torsion (DT). Materials were tested in air as processed and after immersion for 1 month in water. Most materials demonstrated stable crack propagation with these tests, allowing measurement of crack velocity. Modulus and unnotched fracture strength were determined at a strain rate in bending identical to that used in the fracture toughness tests, allowing accurate calculation of the flaw size for a material and an assessment of how applicable linear elastic fracture mechanics are to these materials [322].

The mechanical properties of a number of heterocyclic and monocyclic methacrylate have been studied for their potential application in low polymerization shrinkage systems. This study included both homopolymers and room temperature polymerizing systems using poly(ethyl methacrylate) powder with a heterocyclic methacrylate monomer. The cyclic methacrylate studied, isobornyl methacrylate, gave an extremely brittle polymer; furthermore, it would not form a dough with poly(ethyl methacrylate). The homopolymers had Young's moduli in the range 1.38 ~ 2.19 GPa. These materials were generally ductile and the mechanical properties indicated a useful class of materials for clinical use [323].

A new commercial denture base material, nylon 12 with 50 wt% glass spheres, was compared to commercial denture base materials for flexural and impact strength, surface hardness, glass transition temperature, and water absorption; and dimensional nylon 12 was considerably stronger than the other polymers tested. It did not fracture in a traverse bend test jig, even at loads of 15 N. Glass beads increased the stiffness of the polymer 2.5 times. The impact strength of the nylon was considerably higher than for microdispersed rubber-phase polymers. Glass spheres harden the nylon 12. The creep properties were all similar [324].

An acrylic denture resin was filled with Kevlar fibers (15 μm diam, 31 mm long) at mixing ratios of 0.5, 1.0, and 2.0 wt%. The use of reinforcing Kevlar fibers appears to enhance the fracture resistance of acrylic resin denture base materials [325]. However, another experiment showed that a Kevlar-reinforced acrylic had lower transverse bending strength due to delamination of the fiber–matrix interface [326].

Short carbon fiber reinforced composites could potentially replace some of the metal alloys used in implants. In particular, polysulfone and, more recently, PEEK (polyether ether ketone) have been considered as the matrix material for carbon fiber reinforced composite implant materials. ASTM standards F813 and F619 for direct contact cell culture evaluation and extraction were employed to determine the *in vitro* biocompatibility of a carbon fiber composite of PEEK, in comparison to a carbon fiber reinforced polysulfone composite. Overall, the cellular responses to the PEEK and polysulfone composites were negligible, indicating that further *in vitro* studies with these materials are appropriate [327].

Composites fabricated from PEEK and graphite fiber have been evaluated. Successful processing has been achieved using commercially available prepreg tape. This composite system offers exceptional properties over state-of-the-art graphite–epoxy systems, especially where toughness is concerned. Composite characterization included tensile, flexure, interlaminar, and in-plane shear strengths as well as first-ply failure, edge delamination, and Mode I fracture using the double cantilever beam test [328].

Heat-cured and cold-cured resin denture bases placed into an ultrasonic bath at 63°C for 10 min showed that ultrasonic cavitation had no permanent or significant effect on the dimensional stability of PMMA denture bases [329].

Acrylic resin has a relatively poor resistance to stresses of impact, bending, and fatigue. To reinforce the fracture-prone acrylic prostheses, three treatments have been used: (1) addition of a mesh of gold wire, (2) a rigid stainless steel bar within the acrylic resin, or (3) cast Co–Cr denture base. Recently, carbon and aramide fiber reinforced acrylics have been shown significantly weaker in transverse loading than conventional denture base acrylic resin. The typical fracture pattern was a vertical crack initiated at the tensile surface of the beam and propagated through the beam until a reinforcing-fiber layer was reached. At that point, horizontal progression of the crack occurred for a variable distance (1 to 5 mm) along the interface. The fracture then progressed toward the compression surface [326].

When a complete or removable partial denture opposes natural teeth, the use of gold occlusal surfaces on acrylic resin denture teeth will minimize occlusal wear, prevent premature loss of occluding vertical dimension, and minimize soft tissue abrasion and resorption associated with malocclusion [330]. These data indicate that the wear properties of existing resin denture teeth need to be improved.

Denture stomatitis caused by acrylic dentures has been ascribed too poorly fitting dentures, dirty dentures, unbalanced occlusion, and fungal infection with *Candida* strains [331]. In addition to these etiological factors, another possible factor is the chemotoxicity of substances leached from denture base acrylic polymers [332]. Leachability of a residual monomer, methyl methacrylate (MMA), has been investigated on resins immersed in distilled water [333] and in aqueous solutions of organic solvents [334]. Tests at pH 4.0

~ 6.8 and 37°C suggest that chemotoxic actions of autopolymerized resins are potentially ascribable to methyl methacrylate at low pH and to methacrylic acid at higher pH [335].

Radiolucency is one disadvantage in the use of acrylic resin for restorative prostheses. Ingestion or aspiration of a radiolucent dental material may result in serious medical problems. Attempts to solve this include the addition of stainless steel fragments and barium compounds to the acrylic polymer. Radiopaque denture acrylic resin containing barium fluoride developed dark staining around the necks of the porcelain teeth. Dentures made from resin with bismuth subnitrate developed diffuse dark stains. Dentures made from resin with barium sulfate were radiopaque [336].

Removable partial dentures are composite structures that include a metal structural framework embedded in an acrylic denture resin base. The absence of chemical bonding between the resin base and the metal framework results in a potential for microleakage, straining, and bond failure at the junction. Recent development of a denture base resin that incorporates 4-methyloxyethyltrimellitic anhydride (4-META) has been reported. The 4-META resin system allows resin bonding to the base metal alloys used in removable partial prosthodontics [337].

4-META and phosphate–methacrylate resins can adhere strongly to dental alloys. However, to ensure durability of the adhesive interface during long-term water exposure, the oxidation of the alloy surface is indispensable. A new oxidation method using ion sputtering was developed, and the effectiveness of this surface treatment on two dental alloys – a Type IV Au alloy and Ni–Cr–Be alloy – was studied. Thermocycling for 100,000 cycles was followed by tensile adhesive bond strength testing. Ion coating of the surface of the alloys resulted in strong bonds with adhesive resins, and after 100,000 thermocycles, a bond strength of >20 MPa was maintained [338].

Some of the techniques presently available that improve the bonding of filled and unfilled resins to base metal alloys have focused on surface treatments of alloys for use in removable partial denture frameworks. PMMA denture base resin was processed against treated surfaces following standard laboratory procedures. Thermocycling was conducted to incorporate the effects of intraoral temperature changes and water sorption on bond strengths. After tensile bond testing, SEM analysis of the fracture patterns revealed a correlation between bond strength and amounts of residual resin on the etched surfaces [339].

The etched metal, resin-bonded (EMRB) fixed partial denture offers a conservative and reversible method for the replacement of missing teeth. The proximal and lingual enamel of the adjacent teeth is used to retain the restoration. Etching of the metal framework creates a micropitted surface similar to that of etched enamel, to provide retention for the bonding adhesive [340].

Composite resin-retained, fixed partial dentures are subjected to a wide range of temperatures in the oral cavity. Shear bond strengths of two retainer designs were adversely affected by thermocycling while tensile bond strengths remained unaltered. Bond strengths of electrolytically etched castings were consistently stronger than those of the perforated design. Most bond failures occurred near the retainer–composite resin interface [341].

Several physical property tests compared microwave energy and conventional hot water bath polymerization techniques. The two methods of polymerization produced similar dimensional accuracy in complete denture bases. No differences were found in traverse strength, Knoop hardness, density, and residual monomer content of resin test strips. Comparable strength was found between microwave-polymerized and autopoly-merized repairs of resin test strips. No porosity was observed in complete or removable partial denture bases polymerized by either technique. The Knoop hardness of micro-wave-polymerized removable partial denture bases was found to be slightly lower near the metal framework [342].

Heat-activated acrylic denture resins polymerized by microwave energy and in a conventional hot water bath base were compared with respect to molecular weight, con-version of monomer, and porosity. Monomer conversion using microwave energy was substantial, but the minimal residual monomer levels attainable with the water bath system were not achieved. Microwave curing at 70 W for 25 min minimized porosity but porosity-free material could only be guaranteed in sections no thicker than 3 mm [343]. This study demonstrates the importance of using denture resins formulated for micro-wave curing if optimal properties are desired.

During visible light activated, exothermic polymerization, temperature peaks occur-ring for curing times of less than 31 sec were higher than those peaks occurring for curing times between 31 and 60 sec. The speed of the exothermic reaction of visible light acti-vated composite resin materials increased with increasing intensity of the light source. A higher-intensity light source is more effective in curing the resins than higher temperatures generated by the light source [344]. DTA (differential thermal analysis) is a convenient method of measuring heats of reaction and obtaining data on degrees of conversion to provide insight into curing of light-activated composite resins. As conversion proceeds, the glass transition temperature of the resin rises until it reaches the temperature of the polymerizing material [345].

Autopolymerizing acrylic resins are used in many areas of dentistry, including pros-thodontic and orthodontic appliances, and occlusal splints. Polymerizing resins under pressure reduces porosity, increases transverse strength, and results in a smooth surface. Both vacuum mixing and pressure polymerization significantly increased transverse strength compared to a bench-cured resin. One disadvantage of vacuum mixing is that it requires an additional $30 \sim 60$ sec of working time, but the monomer can be chilled to increase working time [346].

One promising clinical application of VLC (visible light cured) denture resin is for relining complete and partial dentures. Typically, a VLC resin would be used to reline an existing prosthesis fabricated from heat-cured resin. When a VLC resin was thermocycled between 6° and 60°C at 1 min intervals for 4 h [347], an increase in surface degradation of the VLC composite and the formation of microcracks or failure of interfacial bonds were seen [348]. The water sorption for a VLC resin is higher than for a heat-activated acrylic resin, which could potentially influence bond strength [349]. The bond of autopo-lymerizing resin to a heat-processed denture base using a heat-activated resin bonding agent was significantly stronger than the bond of VLC to the same material [347].

A visible light activated denture relining material was found to have lower values for bending strength than conventional resins, but laminating the denture base and the relin-ing material increased the plasticity and thus the toughness. The fit of dentures relined with the new material was similar to that of an autopolymerizing resin. A strong adhesion was obtained between the relining material and the denture base when a light-activated bonding agent was used. The material would be expected to have better long-term behav-ior than the conventional relining material, an autopolymerizing acrylic resin [350].

Physical properties, clinical applications, and tissue compatibility of a VLC resin material were tested. The matrix was a urethane dimethacrylate with small amounts of colloidal silica to control viscosity; the filler was acrylic beads of various sizes that become part of an interpenetrating polymer network (IPN) structure. Toxicity of the uncured resin is low and the cured resin is nontoxic. VLC tensile strengths were 54% greater than chemically activated acrylic resin and 21% greater than heat-cured resin. Horizontal and vertical dimensional changes were also less. The elastic modulus was greater than either of the acrylic resins. Clinical applications of VLC include: complete dentures, interim complete and partial dentures, permanent removable partial denture bases, dental re-liners, denture repairs, orthodontic appliances including retainers and positioning de-vices, occlusal splints, maxillofacial prostheses, and resin denture teeth [351].

The use of visible light activated resin materials in orthodontics is very promising. The physical and chemical properties of the resin show advantages over autopolymerizing methacrylates. The materials are nearly free of monomer and have good patient accep-tance. Bacteria adherence and the soft tissue response are good. The polymerization shrinkage was equal to that of heat-cured and autopolymerized acrylic resins, and trans-verse deflection was less. Thermal data indicated that heat-cured resin was the most stable, followed by the light-activated resin and then autopolymerized resin [352].

Friction can develop with the use of soft denture liners due to changes within the liner material with time and also as a result of the restricted volume of saliva existing between the liner and supporting mucosa. Any factor that decreases the volume of water at the interface— such as absorption by the liner, dimensional change in the linear, or fluid displacement during mastication or through denture movement—will increase friction. Such frictional effects might be expected to cause patient discomfort [353].

The rheological behavior of a denture base polymer was studied from mixing to setting. In addition, monomer evaporation and the exothermic behavior of the mix were evaluated. The results show that the material behaves as a pseudoplastic fluid. It is shown that the viscosity increases at different rates with time and increases with higher temperatures. Also, it is shown that monomer evaporation and polymerization both play roles in dough formation [354].

Microwave sterilization techniques for dental prostheses have previously been dis-cussed in the literature, but there is need for information regarding the stability of acrylic resins subjected to these techniques. The purpose of one investigation was to measure the influence of microwave sterilization on the dimensional stability of polymerized acrylic resin denture base materials. Samples of heat-polymerized, autopolymerized, and visible light polymerized acrylic resins were measured for changes in weight and length before and after microwave exposure. All three materials maintained excellent stability and had shrinkage values in the range of 0.02 ~ 0.03%. This is clinically insignificant compared to polymerization shrinkage, which averages 0.2% [355].

B. Direct Restorative Resins

Composite resins are being used in all types of dentistry, especially operative and pros-thetic dentistry. Composite resins have eliminated many of the early problems associated with the use of direct restorative acrylic resins such as water absorption, polymerization shrinkage, and dimensional changes due to thermal cycling and rapid wear [356].

The advent of composite resins and the introduction of acid etching of enamel has greatly facilitated the clinical repair of fractured incisors. Findings show that the mechan-ical removal of as little as 0.1 mm of surface enamel (by 600 grit carborundum) will significantly increase the shear strength of the bond between composite resin and etched enamel [357].

The composite resins used in direct restorative dentistry are resin matrix composite materials highly filled with inorganic materials, which include SiO_2, other minerals, various glasses, and ceramics. Polymeric filler materials and colloidal silica are also used. The inorganic fillers are chemically bonded to the resin matrix with a silane coupling system such as 3-methoxy-propyl-trimethoxy-silane. The resin matrices employed are commonly BIS-GMA (an adjunct of bisphenyl-A and glycidal dimethacrylate), urethane dimethacrylate, and similar acrylic resins. The base monomers are heavily crosslinked with acrylics such as triethylene glycol dimethacrylate. Initiation of polymerization can be chemically activated by a peroxide–amine system, but visible light activated systems are now the most common. These composite resins are placed directly into a prepared tooth cavity and exposed to a visible light curing unit, which results in immediate hardening.

Composite resins have a number of distinct advantages over the unfilled acrylics that were used in the 1950s. The addition of 60 ~ 65 vol% inert filler significantly reduces polymerization shrinkage and the coefficient of thermal expansion. All of the mechanical properties are improved, as is the wear resistance to some types of oral abrasion processes. The BIS-GMA and other resins used for matrix materials have much higher molecular weights than does MMA and hence have lower polymerization shrinkage and somewhat improved properties compared to PMMA. The remaining problem was to find a means to adhesively bond the resin restoration to the tooth. This was solved by the discovery that the tooth enamel could be etched with 35 ~ 50% H_3PO_4 to create microporosities in the hydroxyapatite surface. When composite resin is placed against the etched enamel, the resin component flows into these microporosities and hardens. The resultant mechanical interlocking produces a bond whose strength exceeds the cohesive strength of the enamel (22 MPa). The resultant tooth–resin interface also is resistant to microleakage of oral fluids.

Composite resin filling materials have seen extensive development since their introduction into dentistry in the late 1960s. The first resins were chemically activated by a peroxide–amine system. This resulted in very short working times after the two paste components were mixed. Mixing also tended to incorporate air, which resulted in porosities in the restoration. Today most of the direct-filling composites are light activated by exposure to 450 ~ 500 nm light from a visible-light curing unit. The developments that have most greatly improved properties, however, have been in the nature and size of the filler materials used. Early composites were filled with ground quartz with particle sizes as large as 50 ~ 70 μm. The result was a material that could not be finished to a smooth surface by mechanical reduction with any of the common dental abrasives or cutting tools. Even a composite surface cured against a smooth matrix material became rough with time as the softer resin matrix was worn away from the filler material. The wear resistance of these early composites was poor and they were not suitable for use on chewing surfaces due to their high wear rates. These composites with relatively large inorganic fillers are referred to as *conventional*, *traditional*, or *macrofilled* composites. A reduction in both the mean and maximum particle size and the use of somewhat softer fillers such as glasses improved the finishing characteristics but not the wear behavior. A submicron-size SiO_2 filler called *colloidal silica* (0.02 ~ 0.04 μm) was used in further attempts to improve finishing characteristics. This material is commonly employed as a thickening agent and its addition to dental resin increases the viscosity very rapidly. At concentrations of only 20 wt% the resulting composite is too thick to flow into the prepared cavity. In the development of a composite material, an increase in filler content normally improves the physical properties. Higher amounts of colloidal silica can be added to a dental resin by use of solvents or heating to lower the viscosity, or by high shear rate blending apparatus. Resin monomer, which is filled to 60 ~ 70 wt%, is polymerized by the manufacturer and then ground into a powder with particle sizes similar to a conventional composite. The prepolymerized "organic" filler is added to additional resin monomer that contains small amounts of colloidal silica to control handling properties, and the result is called a *microfilled composite resin*. The total inorganic filler content is 35 ~ 60 wt%. The microfilled composites have very smooth finished surfaces and tend to wear smoothly. Their mechanical properties and shrinkage are inferior to a conventional composite (see Table 7). They are still considered materials of choice for use in non-stress-bearing, highly visible locations. If these materials are subjected to cyclic loading, as in a chewing surface restoration, the bond between the organic filler and the resin matrix tends to fail, resulting in the falling out of the relatively large pieces of organic filler. In an effort to obtain the mechanical properties of a conventional composite and the surface smoothness of a microfill, the small-particle composites were developed. These have an average filler size of 1 ~ 5 μm with a very broad distribution of filler sizes that allows for the highest total filler loading (80 ~ 90 wt%) of any dental composite resin. The fillers are various glasses. The small-particle composites can be made radiopaque by the addition of radiopaque glasses; conventional and microfilled composites are usually radiolucent. The small-particle composites have the best overall mechanical properties. These are materials of choice for high-stress locations such as chewing surfaces of posterior teeth or incisal edges of anterior teeth. The surface finish obtainable is inferior to that of a microfilled resin. The hybrid composite resin is an attempt to obtain a material with a smoother finished surface. These resins contain both small particle glass and colloidal silica (10 ~ 20 wt%) fillers. The glass particles are smaller than those used in a small-particle composite (0.06 ~ 1.0 μm), with over 75% of the particle distribution smaller than 1.0 μm. The resulting mechanical properties are slightly inferior to a small-particle composite, but the surface smoothness approaches that of a microfilled resin. Radiopaque glasses are generally employed as fillers [311].

Although composite resins have much lower shrinkage than the unfilled acrylics, they still show significant shrinkage during hardening (1.5 ~ 2%). Some of this contraction is compensated for by flow of the partially hardened resin. From the stress–strain ratio (modulus) and the polymerization contraction measurements, the "theoretical" contraction stress can be calculated. The discrepancy between this value and the actual stress was attributed to flow. The rate of shrinkage is highest during the earliest stage of setting when the material exhibits the lowest flow stress. As the material gains strength it is less able to yield but the rate of contraction decreases. If the adhesive composite–enamel junction can resist the polymerization contraction forces and the material is not damaged internally by these tensile stresses, expansion due to long-term water sorption will partially relieve the residual contraction stress [358].

The mechanical and physical properties of aromatic, thermosetting, composite restorative resins are superior to PMMA unfilled resin and silicate materials [359].

The extent of polymerization of light-cured resins depends on several factors including the concentration of activating light reaching a certain depth, catalyst concentration, and composition of the material [360].

Viscosity is not a limiting factor for the penetration of restorative resin monomer into the pores (50 μm depth, 0.4 μm diam) of etched enamel surfaces. An intermediate layer of

Table 7 Properties of Dental Composite Resins

	Unfilled acrylic	Conventional	Microfilled	Small particle	Hybrid
Inorganic filler					
(vol%)		60–65	20–55	65–77	60–65
(wt%)		70–80	35–60	80–90	75–80
Compressive strength (MPa)	69	250–300	250–350	350–400	300–350
Tensile strength (MPa)	24	50–65	30–50	75–90	70–90
Elastic modulus (GPa)	2.4	8–15	3–6	15–20	7–12
Thermal exp coefficient (10^{-6}/°C)	92.8	25–35	50–60	19–26	30–40
Water sorption (mg/cm^2)	1.7	0.5–0.7	1.4–1.7	0.5–0.6	0.5–0.7
Knoop hardness (KHN)	15	55	25–30	50–60	50–60

low-viscosity, unfilled resin between a composite restorative material and etched enamel is not likely to improve the quality of composite restorations over applying the composite material directly to the etched enamel surface [361,362].

Surface microhardness (Knoop—KHN) of VLC composite materials for anterior restorations was measured as a function of time. The KHN steadily increases with time reaching a maximum after 1 week. The rate of increase is highest in the first hour and was greater for samples stored at 37 °C as compared with 32 °C. These data reflect progressive crosslinking in the resin phase, which continues after photoactivation [363].

For VLC composite resins, residual carbon–carbon double bonds can be further polymerized by heating to elevated temperatures. Postcuring efficiency was studied by DSC and by Knoop hardness. Microfilled materials have a smaller depth of cure and higher amount of unconverted ($C=C$) groups as a function of distance and curing time. The depth of cure and the hardness pattern are extended as inorganic loading increases. The cure profiles (obtained from micro-ATR) change more rapidly than the KHN. The thermal properties of the filler fraction affect the degree of unsaturation [364]. Although this technique cannot be used for direct restorations, it can be used on composite resin inlays that are fabricated outside the mouth.

Generally, the microfilled resins had lower yield strengths and higher creep strains than larger-particle composites. Average yield stress for the microfilled resins (90 MPa) was slightly higher than for unfilled dimethacrylate resins (80 MPa). It was suggested that compressive deformation of dental composite resins is dependent on filler concentration to a greater extent than on degree of conversion in the resin. Homogeneous distribution of the finely ground filler particles is an important factor [365].

Dental composites consist of a polymerizable monomer and a suitable filler. Four materials—(1) tribasic calcium phosphate (50% TCP) and triethylene glycol dimethacrylate (TEGDMA), (2) silanated lithium aluminum silicate (75%) and TEGDMA, (3) barium sulfate (70%) and PMMA, and (4) silane-coated barium glass (75%) and PMMA—were used in water sorption studies at 37 °C. The values of D (diffusion coefficient) were found to be significantly smaller with (2), suggesting effective matrix–filler coupling. The filled materials sorbed nearly twice as much water as an unfilled PMMA. It appears that the filler–matrix interface provides paths of facile diffusion similar to grain boundary diffusion in metals [366].

A light-cured microfilled composite resin containing 37% by weight of agglomerated colloidal silica (15 μm average agglomerate particle size) was used to make intracoronal veneers of laboratory processed composite resin to produce aesthetic restorations. The retention of these materials depends on the bonds formed at the veneer–bonding agent interface and the bonding agent–enamel interface. The bond strength of the veneer–bonding agent interface was investigated using resin bonding agents differing in means of polymerization, filler content, and apparent viscosity. The bond strengths were better with bond agents that were light cured and less highly filled [367].

An empirical curve relating mechanical properties to volumetric filler fractions of composite resin expresses the dramatic influence of the inorganic component on the resin composite properties and is of considerable interest for new composite development. The present work offers a way to predict this influence for several mechanical properties, thus permitting optimization of the filler content with respect to the combination of the considered properties [368].

Degradation of composites with various fillers (quartz, strontium glass, colloidal silica) and the same resin matrix (amine-cured mixture of BIS-GMA 70% + TEGDMA

30%) was studied by toothbrush abrasion and was related to measurements of dissolution and surface roughness and to observations of the subsurface. Effects of water and thermocycling were minimal in a microfilled composite and in a material containing relatively large quartz fillers. Increase in surface degradation of the glass-filled composite was attributed to the formation of microcracks or failures of interfacial bonds [348].

It is desirable for a dental restorative resin to convert all of its monomer to polymer during the polymerization process, but in BIS–GMA-based resins, a significant concentration of unreacted carbon double bonds remains after curing. The degree of curing influences mechanical properties, solubility, dimensional stability, color change, and biocompatibility.

The degree of conversion of carbon double bonds was examined by Knoop hardness measurements. Results indicate that the increase in hardness during the setting of unfilled dental restorative resins correlated well with the increase in degree of conversion for a specific resin. However, an absolute hardness value cannot be used to predict an absolute value for degree of conversion in all resins [369].

Future composites may become stronger and more durable with improved bonding between the organic resin and the inorganic filler particles. The use of three-dimensional glass fiber networks and of semiporous (superficial porosity) glass particles is being studied to improve the bond at the resin–filler interface [370]. The microporous glasses are prepared by the gel route. The gels are prepared from sols containing aluminum and silicon oxides, and, to improve x-ray opacity, zirconium and tin oxides. The resin system used was BIS–GMA, diluted with TEGDMA [371].

After bond strength tests of composite resin bonded to enamel, evidence of localized plastic deformation was observed from the morphology of surfaces formed by predominantly brittle fracture of the resin. Evidence of plastic deformation was also obtained after demineralization of the enamel–resin interface, by noting departures of resin tag morphology from that expected on an entirely brittle replication of the dentin interface. Copolymers formed by photopolymerization of BIS–GMA and TEGDMA could be made more brittle by heating to 160°C, in vacuum, but still gave fractographic indications of some localized plastic deformation [372].

Abrasive wear as a result of masticatory function shows no consistent relationship to mechanical properties of composite resins such as tensile strength and microhardness; it is apparent that abrasive wear is a complex phenomenon and that materials with high values for strength or hardness, or both, do not necessarily have high resistance to abrasive wear [373].

Toothbrush abrasion was measured on three composite resin surfaces: (1) polymerized against glass (smooth surface), (2) finished with alumina oxide disks under dry conditions (polished surface), and (3) obtained after removing ~200 μm from the top of the smooth surface (bulk surface). Surface (1) abraded faster during the first hours of brushing than later and will erode and roughen faster than (3) [374].

For light-activated composites, porosity has a limited effect on the abrasive wear characteristics but has a considerable effect on the resistance to fatigue. Porosity would, therefore, be expected to influence durability [375].

If composite resins are used as alternatives to dental silver amalgam to restore chewing surfaces of posterior teeth, their masticatory wear behavior is of prime importance. Considerable effort has been expended in attempts to simulate *in vivo* wear processes in a laboratory test. One particular wear testing machine was used to evaluate 24 commercially available restorative resins. The rank order of the results appeared to be in agreement with clinical experience. The linear relationship between the *in vivo* and *in vitro* wear data indicates that the wear test used may offer a reasonable prediction of clinical abrasion and attrition due to chewing [376].

Despite significant improvements in the wear resistance of posterior composite restorations, they undergo occlusal wear, color change, and surface staining with time. The ability to repair the surface of resin restorations *in vivo* could considerably extend their clinical lifetimes. The efficacy of repair of aged and resurfaced posterior composite restorations was studied by screening the effects of various mechanical and chemical techniques, and primer agents. The optimal rebond strengths of new composite to old composite was achieved with a combination of mechanical roughening with a diamond burr, chemical conditioning with water and/or phosphoric acid (and rehydration of the surface with water), and use of a dental bonding agent prior to the addition of new composite material [377].

In tests in load control mode with a programmed triangular load function, materials having high compressive strengths have higher fatigue limits. The correlation coefficient between compressive strength and fatigue limit is 0.765, which is significant at the 95% confidence level [378].

Compression, tension, and hardness tests were conducted on some dental composite resins with a BIS–GMA resin matrix. The effects of temperature and aging on these properties were studied. There was a marked increase in the mechanical properties (compressive strength, diametral tensile strength, compressive elastic modulus, and hardness) for all the tested composites with increase of both temperature and time. This was explained in terms of the influence of temperature on the polymerization rate of the materials. The improvement in the mechanical properties of the samples, aged at 37°C, was attributed to continued polymerization of the resin system. Such mechanical improvement was verified by linear regression equations against both temperature and time [379].

When light-cured and heat-cured microfilled resins were compared, the lowest creep rates were obtained for the light-cured specimens. Resin facings of single crowns and fixed partial dentures may be exposed to prolonged stress in individuals with habits of bruxism or clenching [380].

During chewing, the average load in local contact areas on the occlusal surface has been reported to be 66 N, and biting force contact loads can reach 90 N [381]. The use of composite materials in posterior teeth as a substitute for dental amalgam in load-bearing Class I and Class II restorations is increasing, partially due to the tooth-like appearance of these filled polymers.

In a study of five light-cured composites and three chemically cured composites, it was shown that during cyclic loading the creep deformation at the end of the test period was of the same magnitude as that with a comparable static load. The static creep and recovery measurements indicated inelastic deformation in all the test materials. The dynamic measurements demonstrated dissipation of inelastic energy during the cycles (hysteresis), which decreased with increasing the number of cycles. Absorbed water increased creep and decreased creep recovery in the composite materials [382].

The relative degree of cure of light-activated resin was determined by using Knoop hardness measurements. In 100% N_2 gas atmosphere the degree of cure improved. Coloring agents decreased the degree of cure while increasing the irradiation time improved the degree of cure [383].

Newly developed matrix resins, fillers, and filler surface treatments have increased the mechanical properties of composite resins, but low resistance to marginal fracture and wear still remain problems. Composite resins are inherently brittle, and grinding and polishing introduces microscopic surface flaws that can shorten their service life. Rate of surface wear may control the service life of a dental composite, and occlusal wear processes have been related to fracture toughness.

To evaluate fracture toughness of a material, the stress intensity factor, K_I, is used. This parameter reflects stress distribution around a crack tip when load is applied. The critical stress intensity factor (or fracture toughness K_{IC}) is the limiting value of resistance of material to fracture by external force, and is a material constant. Fracture resistance evaluations have been done of dental materials such as acrylic resin, amalgam, porcelain, and glass ionomer cement, and there are many reports on the fracture toughness of dental composite resin. Various methods such as the three-point bending of a single edge notch (SEN) specimen, tensile testing of a short rod chevron notch specimen, double torsion test, and indentation tests are used to calculate the fracture toughness of dental composite resins.

Acoustic emission (AE) patterns generated during fracture toughness testing have been used to understand the microfracture process, and fracture surface findings are correlated to explain fracture behavior. Fracture mechanisms of each type of dental composite resin are proposed [384].

The time to failure by static fatigue was predicted from the K_I (stress intensity factor)–V (crack velocity) diagrams. The assumption that microcracking occurs in the subsurface layer due to cyclic and impact stresses gives four criteria for good wear resistance: (1) high fracture toughness (high critical K_{IC}), (2) large threshold crack length (a_t), (3) small inherent flaw size (a_0), and (4) high crazing stress (σ_c) [385].

Short-rod chevron-notch specimens were thermal cycled between 0° and 60°C. K_{IC} was tested after $10 \sim 10^4$ cycles. The K_{IC} drops for all materials, but the decrease after 10^4 cycles was not significantly different from that after storage in water for an equivalent period of time (42 days), indicating that, in the determination of the K_{IC}, the filler–matrix bond is of secondary important to the presence of fillers. The effects of temperature cycles likely to be encountered in the mouth are not a significant factor in reduction of the fracture strength of composite filling material [386].

Bonds between three metals (Ni- and Au-based alloys) and two resins have been tested in both tensile and shear modes. The excellent bond strengths achieved between a mechanically roughened surface and phosphorus-containing resins are a significant advance for this type of tooth replacement. As reported, all the fractures in both tension and shear were cohesive. This suggests that further research is required into developing composite resins with greater cohesive strength. The metal–resin bond may have additional dental and nondental applications worthy of investigation [387].

The strength of bonding of the composites to enamel was most highly correlated with the proportional limit and the elastic modulus of the materials and, to a lesser extent, with the tensile strength and filler content. It was not correlated with the compressive strength. The use of bonding agents did not change the relative influence of the mechanical properties of the composites on their bond strengths but resulted in higher correlation coefficients [388].

Some investigators state that no interfacial fracture between an adhesive and etched enamel should be classified as such but rather should be seen as a failure occurring within both materials, whereas others assume the occurrence of an adhesive (true interfacial) type of failure. The results of a study indicate that fracture always occurs within the restorative resin. The fracture propagates either through the bulk of the material or through the material very close to the interface. The finding that apparent interfacial fractures were in reality failures within the resin material offers an explanation of the absence of a correlation between bond strength and type of failure [389].

Bonding techniques to enamel and even to dentin have improved to the point where failure sometimes occurs in the adhesive resin instead of at the interface [372].

Thermal diffusivity is important in the protection of the pulp from thermal shock and for characterizing the transient thermal strain response of a restoration to changes in the temperature of the oral cavity. Thermal diffusivity increases with increasing density or increasing inorganic filler content [390].

The relatively large values for polymerization shrinkage (1.5 ~ 2.0%) of composite dental resins are one of their most serious remaining problems. This shrinkage places the bond between resin and tooth under a tensile stress that can produce bond failure and a gap between resin and tooth, which in turn leads to microleakage. If the surrounding tooth structure has been weakened or cracks have formed, the shrinkage stress can result in fracture of tooth structure. In the case of a large intracoronal restoration the stress can produce bending stresses on the adjacent walls of the tooth and can result in pain and sensitivity. For light-activated resins, the polymerization shrinkage is not isotropic but tends to be directed toward the light source. All of these considerations can and do result in clinical difficulties [311].

The reduction of the polymerization shrinkage stress by flow of chemically initiated composites was investigated in relation to the cavity configuration. It was found that the flow was strongly dependent on the type of composite and on the configuration of the cavity [391].

The use of composites for stress-bearing restorations in the posterior region revealed essential shortcomings of these materials with respect to wear resistance and sealing capacity at the margins. Short-term exposure to heat (125°C) effects wear resistance. For both chemical- and light-activated composites, 20 ~ 60% improvement in wear resistance was observed. The heat-induced improvement could not be explained by a continuation of polymerization but rather by stress relief, which is common for annealing process. Polymerization shrinkage stresses, initially concentrated mainly around the filler particles, became more homogeneously distributed by the heat treatment. Long-term improvement of non-heat-treated composites occurs by the same mechanism but proceeds more gradually [392].

Composite resins were alternately thermocycled in silver nitrate and in water from 6°(2.25 min) → 27°(0.75 min) → 60°C(2.25 min) for 1000, 5000, 10,000, and 50,000 cycles. Rapid temperature changes resulted in the formation of layers within the surface that may have resulted from microcracking. Slow rates of change increased the depth and rate of diffusion of silver nitrate. Measurements of the temperature changes at the surface of a restoration in the mouth indicated that *in vivo* rates of temperature change are more likely to increase the depth of diffusion of oral fluids than to cause microcracking of the surface [393].

A study showed that heat-treated composite inlays (formed outside the mouth and cemented into place) allowed significantly less microleakage than light-cured, directly placed composite resin restorations. The cusp fracture resistance of heat-treated inlays was not significantly different from traditionally placed composite resin restorations [394].

Excellent marginal adaptation extends the longevity of restorations. Unfortunately, polymerization shrinkage of composite restorations adversely affects this property. The residual stress within the cured resin compromises the materials properties, causes marginal openings, and flexes cavity walls. Factors that enhanced adaptation optimized marginal quality and reduced the amounts of residual stresses. The latter was measured by intercuspal narrowing after the restoration was completed. The most effective factors in optimizing marginal quality included guidance of the shrinkage vectors; reducing the ratio of bonded to free, unbonded restoration surfaces; and minimizing the mass of *in situ*-cured composite [395].

Temperature variation in the clinical range may adversely affect bonded composite resin retainers. A significant decrease was found in fracture strength of teeth thermocycled between 5° and 55°C. Extensive microleakage was found in both control (uncycled) and thermocycled teeth. Shrinkage of the resins during polymerization may play a greater role in the initial development of microleakage than does variation in temperatures [396].

Tests examined the color stability of composite resins on exposure to irradiation by (1) a 150-klux xenon lamp and (2) a standard RS light source, and (3) on exposure to elevated temperature (60°C), in air or water. Discoloration increased with exposure time. Irradiation of the materials by the RS sunlamp in air produces similar discoloration to that produced by exposure to the xenon in water. Visible light cured materials are more color stable than commonly used chemically cured restoratives containing tertiary aromatic amine accelerators. Light shades are more susceptible to color change than dark shades of the same brand. Exposure of composites to 60°C in the dark leads to more rapid and severe discoloration for materials stored in water than in air [397].

VIII. DENTAL CEMENTS

A. General

Dental cements are used for a number of quite different applications. The properties desired in a cement depend closely upon the application(s). Hence, it is useful to describe briefly the major applications and consider the required properties. Cements can be broadly classed as *luting* or *restorative* materials.

The term *luting* means to use a cement to retain another dental material in place against tooth structure. Ordinarily, a dental luting cement accomplishes this by forming mechanical interlocks between the two solid surfaces resulting in mechanical adhesion or bonding. Of the list in Table 8, only the cements that form polycarboxylate reaction products — zinc polycarboxylate and glass ionomer — have been shown to form true chemical bonds to enamel or dentin.

There is a second equally important function of a dental luting cement. The restorations or appliances to be cemented never exactly fit against the tooth structure; there is always a gap. If this gap were not present, there would be no place for the cement. The luting cement fills this gap and seals the interface to prevent leakage of oral fluids into the tooth. Obviously, the cement physical properties of primary interest are cohesive strength, solubility in oral fluids, and the ability to flow over and wet tooth structure and the materials from which the dental appliance is made. Luting cements can be subdivided into permanent and temporary bonding agents. A permanent cement should have adequate properties to last the lifetime of the restoration with which it is used. A temporary cement is used to retain a restoration and seal its margins for some well-defined, short period of time — usually a few weeks at most. At the end of this time the restoration will be removed. The cement must be weak enough to be fractured so that the appliance can be removed without damaging the oral structures.

Table 8 Classification of Dental Cements

Matrix type	Name	Liquid, or part A	Powder, or part B	Other components
Phosphate	Zinc phosphate	Phosphoric acid	Zinc oxide	
	Zinc silicophosphate	Phosphoric acid	Zinc oxide + aluminofluorosilicate glass	
Phenolate	Zinc oxide–eugenol	Eugenol	Zinc oxide	
	ZOE-EBA	Eugenol + o-ethoxybenzoic acid	Zinc oxide	o-Ethoxybenzoic acid
	ZOE-alumina	Eugenol	Zinc oxide	Alumina
	ZOE-polymer, reinforced	Eugenol	Zinc oxide	Alumina
	Hard-setting calcium hydroxide	Disalicylate ester	Calcium hydroxide	Ethylene toluene sulfonamide
Polycarboxylate	Zinc polycarboxylate	Polyacrylic acid	Zinc oxide	
	Glass ionomer (polyalkenoate)	Polyacrylic acid	Calcium aluminum fluorosilicate glass	Tartaric acid, Other polyalkenoic acids
	Metal-modified GIC	Polyacrylic acid	Calcium aluminum fluorosilicate glass	Silver or silver–tin alloy
Acrylic resin	Acrylic resin	Methyl methacrylate	PMMA	
	Composite resin	BIS-GMA or urethane dimethacrylate + initiator	BIS-GMA or urethane dimethacrylate + activator	Silica, silicate glass, or borosilicate glass fillers
Polycarboxylate + acrylic resin	Light-cured glass ionomer or hybrid GIC	Polyacrylic acid + HEMA + photo initiator	Calcium aluminum fluorosilicate glass + initiator	

Dental cements used as restorative materials also serve several applications. They may be used as direct restorative materials to form a permanent, interim, or temporary restoration. They may be used beneath another direct restorative material to isolate the tooth from undesirable effects of the restorative material. They may be used as a dentin replacement in deep cavity preparations to insulate the dental pulp from thermal changes in the oral cavity. Dentin is a good thermal insulator, whereas most metallic restorative materials are good thermal conductors. Cements may be applied in a relatively thin layer on the cavity floor to protect the pulp against a low-pH luting cement. These later two uses are referred to as *cavity lining agents* and *bases*.

Lists of all of the types of dental cement in current use, along with their applications, are given in Tables 8 and 9. Dental cements can be considered composite materials. The matrix material of the cement composite is formed by a chemical reaction that also converts the liquid adhesive into a rigid solid. Table 8 categorizes the dental cements according to the chemical nature of their matrix phase. This table also describes the

Table 9   Applications of Dental Cements

Cement	Principal application	Other applications
Zinc phosphate	Luting agent for restorations and orthodontics	Thermal insulating bases, intermediate restorations
Zinc silicophosphate	Luting agent for porcelain jacket crowns	Luting orthodontic appliances, intermediate restorations
Zinc oxide–eugenol	Temporary luting agent for restorations, intermediate and temporary restoration, thermal insulating base, cavity liner, pulp cap	Root canal sealer, periodontal surgical dressing
ZOE–EBA	Luting agent for restorations	Temporary restorations, bases
ZOE–alumina	Luting agent for restorations	Temporary restorations, bases
ZOE-improved (resin reinforced)	Intermediate restorations (IRM)	
Hard-setting calcium hydroxide	Pulp capping, liners, insulating bases	
Zinc polycarboxylate	Luting agent for restorations, insulating bases	Luting agent for orthodontics, intermediate restorations
Glass ionomer	Luting agent for restorations and orthodontics, anterior restorative material, liners	Bases, endodontic retrograde sealant
Metal-modified glass ionomer	Core buildups and limited posterior occlusal restorations	
Acrylic resin	Luting agent for restorations, veneers, orthodontic bonding	
Composite resin	Luting agent for orthodontic bonding and etched metal restorations	
LC glass ionomer–resin hybrids	Same as glass ionomers	Pit and fissure sealants, endodontic retrograde sealant

principal components responsible for the chemical reaction. Tables 10 and 11 list typical mechanical and physical properties for dental cements whose principal application is as a luting agent or a restorative material, respectively [398-400].

In addition to the properties required to meet the physical demands upon a dental cement, the material must also possess adequate biocompatibility. The clinical applications of dental cements often place these materials in very close proximity to a vital tooth pulp. If the unset or set cement leaches toxic components, these may readily penetrate the remaining dentin and cause pulp damage or death. Materials that are either very acid or very alkaline can also produce pulp irritation or damage. Both hydrogen and hydroxyl ions can rapidly migrate through the dentin into the pulp. Some dental materials can actually stimulate a favorable pulp response. Calcium hydroxide, for example, will stimulate the formation of secondary dentin and can even bridge across an area where the pulp is completely exposed.

Most of the original dental cements were formed in a reaction between a liquid acid and a powdered metal oxide, which hydrolyzes to form a base. The resulting salt forms the cement matrix material, which bonds together excess, unreacted cement powder. None of the salts formed is totally insoluble. Tables 9 and 10 list solubilities of the matrix materials in water. Unfortunately, clinical evidence shows that solubility measured in water does not accurately reflect solubility in the oral environment. At the best, water solubility can be used only to compare behaviors of cements that have the same setting reaction. Solubility in the mouth leads to washing out of the cement at the restoration-tooth marginal opening. This can result in the development of recurrent dental caries due to oral bacteria accumulating in this gap. Silicate cement, one of the very first dental restorative cements, forms as a reaction between phosphoric acid and an aluminum silicate glass that contains 15 ~ 20% fluorine. The fluorine is not chemically bound into the matrix phase and will leach out at a slow, steady rate over a long period of time. The properties of silicate cement are very poor, so it is no longer used. Silicate cement tends to dissolve and abrades readily. It also stains easily. However, after many years of use, the observation was made that recurrent caries were almost never seen adjacent to a silicate restoration—even one that had been almost completely worn away. Current knowledge about the inhibiting influence of fluoride ion on oral bacteria and the dental caries process makes the action of silicate cement easily understood. One of the goals of dental materials research is to incorporate slow, long-term fluoride release into a broad spectrum of dental materials. Of the dental cements in current use, only the glass ionomers and the silicophosphates have been demonstrated to release fluoride ion in a manner similar to that of silicate cement.

Ideally, a dental luting cement should have a low viscosity and film thickness, long working time and a rapid set at mouth temperature, good resistance to aqueous or acid attack, high compressive and tensile strengths, resistance to plastic deformation, and good adhesion to tooth structure. Dental cements used for restorations should be cariostatic, biocompatible with the pulp, optically translucent, and radiopaque. Some of the current glass ionomer luting cements possess most of these properties [401].

B. Powder Size, Powder/Liquid Ratio, and Other Handling Considerations

The mechanical properties of a cement are important for both luting and restorative applications. Cement powder morphology influences physical and chemical properties. A cement produced from a powder with a fine particle size is fast setting and strong, and

Table 10 Properties of Dental Luting Cements

Material	Film thickness, μm	Setting time,[a] min	Solubility, wt%	Strength, MPa		Modulus of elasticity, GPa	Pulp response
				Compressive	Tensile		
Zinc phosphate	25–35	6	0.06–0.2	80–100	5–7	13.0	Moderate
Zinc silicophosphate	25–35	3.5–4.0	1.0	140–170	8–13	–	Severe
Zinc oxide–eugenol	25–35	4–10	1.5	2–25	1–2	0.22	Mild
ZOE–EBA–alumina	35–45	9.5	1.0	55–70	3–6	5.4	Mild
ZOE, resin reinforced	35–45	6–10	1.0	35–55	5–8	2.7	Mild
Zinc polycarboxylate	20–25	6	0.06	55–85	8–12	5.0	Mild
Glass ionomer	25–35	7	1.0	90–140	6–7	7.0	Mild–moderate
LC glass ionomer–resin[a]	25–35	40 sec[a]	0.06	140–170	25–40	5.0	Mild–moderate
Resin[b]	20–50	2–4[b]	<0.01	70–200	25–40	3.5	Moderate

[a]Setting time for light-activated materials is equal to time required for light exposure, typically 40 s.
[b]Resin cements are available as chemically activated, light-activated, and dual-activated systems.

Table 11 Properties of Dental Cements Used for Restorations, Core Buildups, and Bases

Material	Strength, MPa		Modulus of elasticity, GPa	Solubility, wt%	Anticariogenic properties
	Compressive	Tensile			
Zinc phosphate	100–170	5–14	20–22	0.1	No
Zinc polycarboxylate	70–90	9–14	4–5	0.2	No
Glass ionomer	70–150	6–9	7–8	0.2	Yes
Metal-modified glass ionomer	150	6.0	7.0	–	Yes
LC glass ionomer–resin hybrid	150–200	25–40	5.0	<0.01	Yes
ZOE–EBA–alumina	50–60	3–5	–	1.5	No
ZOE, resin reinforced	35–40	3–5	2	1.5	No
Composite resin	130	25	5.6	<0.01	No
Hard setting Ca(OH)$_2$	10–30	1.0	0.4	0.5–6.0	No

contains little water-soluble material. The effect of particle size distribution on tensile strength, setting time, and resistance to chemical attack is minimal; but the rheology (flow) of the paste and resultant film thickness are influenced [402].

A study of the effects of temperature, rate of incorporation of powder into liquid, and moisture contamination on compressive strength, tensile strength, working time, setting time, solubility, and orthodontic band retention of zinc phosphate and silicophosphate cements showed that by decreasing the mixing temperature, the powder/liquid (P/L) ratio necessary to achieve a standard consistency increased; decreasing mixing time increased working time; decreasing the temperature increased the working time; decreasing the mixing time increased both compressive strength and tensile strength with zinc phosphate but not silicophosphate; decreasing the temperature while maintaining moisture-free mixes increased both compressive and tensile strengths; and mixing zinc phosphate cement on a frozen slab significantly increased band retention [403].

Improvement in the flow characteristics and the compressive strengths of zinc polycarboxylate cement can be achieved by adding 60 wt% aluminum oxide to the powder used in a mix with a 2:1 P/L ratio. The compressive strength (120 MPa) increased by 80% and the tensile strength (15.5 MPa) increased by 100% over the unfilled cements due to the crosslinking of part of the polyacrylic acid molecules by aluminum ions [404].

C. Comparative Properties

A comparison has been reported of (1) zinc phosphate-, (2) hydrophosphate-, (3) polycarboxylate-, and (4) alumina-reinforced ZOE cements. Results indicate that (1), (2), and (4) are equal in tensile strength; and (3) is at least one third stronger. (1), (2), and (3) are equal in retentive ability, and (4) is 50% lower [405].

A comparison of bond strengths of cements (zinc phosphate, zinc polycarboxylate, EBA alumina modified ZOE) to natural teeth with mechanical properties resulted in no correlation between bond strength and compressive strength, tensile strength, or film thickness, therefore implying an interface controlled phenomenon [406].

D. Interaction with Other Substances

Because the dental cements based on polyacrylic acid are the ones known to bond to tooth structure, there is considerable interest in understanding this process. When polyacrylate (PA) ions and hydroxyapatite (HAP) interact in solution, PA ions become irreversibly attached to the surface of HAP by displacing existing phosphate ions. The mechanism is not one of simple ion exchange, since Ca ions are displaced by P ions, maintaining electrical neutrality. A postulated mechanism for adhesion involves polyelectrolyte chains embedded in the enamel surface with the displacement of P and Ca ions. It may be presumed that the surface layer of the adhering cement becomes enriched in P and Ca ions as these diffuse from the enamel surface. There may be other types of ionic exchange across the interface, mainly of cations. It has been suggested that there is a layer of cement attached to the tooth, differing in chemical composition from the bulk cement. A picture of the adhesive bond emerges with an intermediate layer between the bulk enamel and the bulk cement [407].

Zinc phosphate (80 ~ 110 MPa) cement showed no change in compressive strength with aging time, while GIC (90 ~ 140 MPa) and polycarboxylate cements (55 ~ 85 MPa) increased in compressive strength [408,409].

Adhesion tests of (1) zinc polycarboxylate cement and (2) an unfilled dimethacrylate resin cement to porous stainless steel were done. Strength depends on the physical and chemical properties of the substrates; the adhesion of (1) to stainless steel is high [410].

Rheological behavior during the setting of zinc polycarboxylate and glass ionomer dental cements has been studied. The P/L ratio was found to alter the rate of reaction without altering the basic form of the kinetics. Two models were advanced to explain the rheological and chemical differences between the two types of polyelectrolyte cements. The setting of the glass ionomer cements was consistent with the development of a homogeneous polymer network whereas the zinc polycarboxylate cements were viewed as setting by an inhomogeneous core growth reaction [411].

Cementation of stainless steel orthodontic brackets to dental enamel with zinc poly-carboxylate cement gave 50% retention after 4 months. This is not clinically acceptable. An increase in the surface area of the bond by a factor of 2 to 3 and improved cleaning of the enamel increased retention to 88% [412].

In laboratory and clinical evaluations, zinc silicophosphate cement proved compara-ble to zinc phosphate cement in bond strength to enamel when used for the cementation of orthodontic bands. With silicophosphate, no decalcification of enamel was found under bands that had become loose during orthodontic treatment, possibly because of the presence of fluoride in the cement [413].

Zinc polycarboxylate and zinc phosphate cement have tensile strengths in same range, but the compressive strength of the carboxylate cement is lower. The carboxylate cement has a stronger bond to enamel or dentin than zinc phosphate when tested under tensile loading and subjected to thermal fluctuations [414].

As part of an investigation of the setting of dental polyelectrolyte cements, the chemistry of a selection of glass ionomer and zinc polycarboxylate cements was studied by pH, conductivity, and IR measurements. The zinc polycarboxylate cements were found to react at a greater rate than the glass ionomer cements. The effect of reducing the P/L ratio is to decrease the surface area available to attack and hence the reaction rate. The basic form of the kinetics appears to be unaffected except for low P/L ratios where there is a deficiency of available metal cations [415].

The coefficients of thermal expansion of fixed prosthodontic cements and cast resto-rations do not match those of enamel and dentin; as a result, thermally induced stress may cause the cement to crack and subsequently fail. In a test of the effects of thermally induced stress on zinc phosphate (ZP) cement and polycarboxylate (PC) cement, ZP was 13.2% more retentive than PC 96 h after cementation and before thermally induced stress. ZP was more retentive after 9120 thermal cycles; failures with both cements were due to expansion coefficient mismatch [416].

Zinc phosphate, silicophosphate, glass ionomer, zinc oxide eugenol (reinforced), and zinc oxide–eugenol/*o*-ethoxy benzoic acid cements were studied. Slow compressive strength and creep testing were used to examine the flow properties. The extent of flow is the least for phosphate-bonded cements, characterized by a three-dimensional ionic net-work, and greater for cement based on organic units that are linked only by secondary bonds. The phosphate-bonded and glass ionomer cements showed brittle fracture. The creep of phosphate-bonded cements was very small, whereas that of the eugenol cements was considerable [417].

E. Polycarboxylate

Bond strengths of an unfluoridated polycarboxylate cement were inferior to those of the SnF_2-containing cement. Significant increase in bond strength was obtained with the unfluoridated polycarboxylate cement following pretreatment of dentine with $Ca(OH)_2$, or a mineralizing agent, or *in situ* precipitation of $Ca_3(PO_4)_2$. Similar increases were obtained with the polycarboxylate cement containing SnF_2, which showed additional improvement after *in situ* precipitation of $Sn_3(PO_4)_2$ [418].

Polycarboxylate cement has good mechanical strength, resistance to dissolution, and bond strength. The compatibility of this cement on teeth with vital pulps appears to be considerably superior to that of zinc phosphate cement, thus recommending its use for cementation of gold inlays, full crowns, and metal ceramic bridges [419].

It was reported that surface treatment by electrodeposition of nickel followed by activation with 10% nitric acid greatly enhances the bond of carboxylate cement to Au alloys [420].

Implant and pulp studies of the biological reactions to zinc polycarboxylate cement showed this material to be less irritating than ZOE (zinc oxide eugenol) cement. After 32 days there was little evidence of residual inflammation. ZOE is the standard, negative control employed in these studies [421]. A clinicopathologic study investigated pulpal reaction to polycarboxylate cement and concluded that PC cement possesses properties comparatively favorable to dental pulp [422].

F. Zinc Oxide Eugenol

The physical character of zinc oxide powders affects the setting reactions of ZOE cements. Zinc oxide is manufactured either by oxidation of the metal in oxygen or by the direct decomposition of zinc ores in air. It may also be prepared by the thermal decomposition of zinc compounds, for example, acetate, hydroxide, oxalate, and nitrate. These powders react at different rates in the ZOE system. Powders prepared by the thermal decomposition of zinc salts are more reactive with eugenol than those prepared by oxidation of zinc metal, because of the higher water content. Zinc oxides prepared by the oxidation of zinc show little reactivity toward eugenol [423].

The setting reaction of ZOE is very sensitive to moisture and temperature. When the effects of humidity and temperature on rheological and setting properties of ZOE cements are studied, it is recommended that the temperature and humidity should be controlled to ± 1°C and ± 2% RH [424].

ZOE cement disintegration in aqueous media is a consequence of the continual loss of eugenol by leaching from the cement matrix. The zinc eugenolate chelate is of low stability, and the equilibrium between it and the eugenol and zinc oxide contained in the cement is upset when eugenol is removed by aqueous leaching [425].

An extensive review has been made of the impinging acid jet method for evaluating the durability of dental cements. A comprehensive range of 28 dental cements was examined. The initial rate of erosion was found to be essentially constant. Zinc oxide cements were found to be markedly less durable than aluminosilicate cements. Results broadly correlate with those found in clinical studies [426]. This test has been included in the ISO Specification for Dental Water-Based Cements.

Poly(methyl methacrylate)-reinforced zinc oxide–eugenol cement is an effective intermediate type of restorative material [427]. Reinforced ZOE cements containing (1) zinc oxide–eugenol with polycarbonate resin or (2) ZOE with 24% Al_2O_3 and 10% PMMA added to the powder and 50% EBA added to the liquid showed equal strengths but (1) had higher toughness [428]. It was also reported that, when ZOE cement is reinforced with PMMA, the degree of reinforcement depends on particle size and uniformity of distribution [429].

G. Glass Ionomer Cement (GIC)

There are three major uses of GIC: as luting agents (Type I GIC), as restorative materials (Type II GIC), and as liners and bases (Type III GIC). Glass ionomers exhibit a coefficient of thermal expansion that is the closest of any dental material to that of tooth structure, particularly dentin. However, the light-cured (LC) glass ionomer–resin hybrids have a somewhat higher coefficient of thermal expansion due to the resin component.

As luting agents, glass ionomers have been used for more than a decade. In terms of crown retention, they are equivalent to zinc phosphate cement or polycarboxylate cements. *In vivo* tests have shown that GIC exhibit the lowest intraoral solubility of the three cements. This property, along with their adhesion to tooth structure and fluoride release, makes them a material of choice for permanent cementation in most situations. One of few problems associated with glass ionomers as luting cements is postoperative sensitivity, which is occasionally reported. It is of interest that only the Type I GIC has been associated with sensitivity. The incidence of sensitivity with GIC is greater than with polycarboxylate but no greater than that with zinc phosphate.

As liners and bases, glass ionomers (Type III) also bond to the underlying dentin. Using a glass ionomer before placing a posterior composite resin can minimize the effect of polymerization shrinkage of the composite resin. Under such conditions, marginal integrity can be enhanced and the potential for cracks in the adjacent enamel can be decreased.

Due to recent modifications in the structure of glass ionomers, they are now being recommended as aesthetic restorative materials. Although they are capable of being subjected to limited masticatory stresses, they are best used as Class V restorative agents [430]. Significant long-term clinical data have been developed to demonstrate the success of GIC in Class V restorations that have a portion of their margin in dentin or cementum.

GIC is a powder/liquid system. The powder is mainly a calcium aluminosilicate glass, and the liquid is an aqueous solution of polyacrylic and/or other alkenoic acids with a small amount of tartaric acid added. Unset GIC showed a very mild cytotoxicity compared with polycarboxylate (PC) and zinc oxide eugenol (ZOE) cements. This cytotoxicity decreased during the setting process and disappeared after setting, whereas the cytotoxicity of PC and ZOE remained after setting [431].

Microleakage was compared in an *in vitro* study using three restoration methods: (1) a composite resin with its dentin bonding agent, (2) a composite resin based with glass ionomer lining, and (3) GIC restorative cement. In order to remove the smear layer, 10% polyacrylic acid was used. The findings suggest, first, that the smear layer should be removed only when a GIC is used under a composite resin restoration and, second, that the microleakage of the GIC restorations was greater than either the composite resin or composite resin based with GIC liner [432].

There is considerable controversy about whether the surface of a GIC liner/base should be etched before a composite resin is placed over it. One study showed that etching the glass ionomer too soon after the mix was made and failure to use an unfilled bonding resin before placing the composite significantly decreased the reliability of the bond [433].

A small percentage of the liquid resin used in commercial dental composites was added to the liquid used in a commercial glass ionomer restorative in order to produce a fluoride-containing hybrid restorative material that would adhere to dentin and be stronger, less brittle, and less sensitive to desiccation in the oral cavity than glass ionomer.

The addition of 13 wt% of a light-cured resin, such as that used in dental composites, to a GIC liquid produced a hybrid material with significantly improved early mechanical properties, lower solubility in water, lower moisture sensitivity, and reduced brittleness. In addition, the adhesion of this material to dentin was unchanged from that of the unmodified glass ionomer [434].

The mechanism of dissolution of two dental cements of the acid–base type (silicate and glass ionomer) is considered. Dissolution is incongruent, probably because most of the leached species can derive from both the matrix (polysalt gel) and the partly dissolved glass particles. The release occurs by means of three discrete mechanisms: surface wash-off, diffusion through pores and cracks, and diffusion through the bulk. Such behavior can be modeled with extremely high goodness of fit using equations such as $y = $ const $+ at^{1/2} + bt$. Analogies from geochemistry and nuclear fuel storage are made since these systems obey similar relationships. The dental cement systems differ, however, in that dissolution is to some extent reversible. This is explained in terms of formation of insoluble complexes, either by reaction of the constituent ions or by replacement of OH^-, for example, with F^- [435].

Resin-modified GIC have the advantages of a long working time combined with a rapid set and higher early strength, are easily bonded to resins, and have strength properties comparable to conventional GIC rather than composite resins. They share with composite resins the disadvantage of containing free monomers and therefore may not be as biocompatible as conventional GIC [436].

Torsional creep studies of Type I and II GIC at 21°, 37°, and 50°C showed that these materials all exhibited linear viscoelastic behavior at low deformation. With increasing temperature there was an increase in creep and residual strain but a decrease in shear modulus [437].

The strength of the GIC bond was approximately doubled when the dentin was treated with an acidic cleansing agent followed by an application of aqueous ferric chloride solutions (containing 5, 15, 25, and 30% by weight of $FeCl_3 \cdot 6H_2O$) [438]. The glass ionomer cement forms effective adhesive bonds to citric acid treated Pt and Au dental alloys but not to the inert surfaces of dental porcelain or pure Au and Pt [439].

Orthodontic bands cemented with GIC are significantly less likely to fail during orthodontic treatment than those cemented with a polycarboxylate (PC) cement. Both PC and GIC cements have the ability to adhere to dental enamel and stainless steel bands. The irregular space between the crown and band is filled with cement that is exposed to total oral environment at cervical and occlusal margins. The adhesiveness of PC cement seems to have been counteracted by the cement's solubility, and likelihood of enamel demineralization under the band is increased [440].

Four GICs were tested for abrasion resistance, surface hardness, and the effects of hydration/dehydration on wear resistance. The wear of dentin is greater than GIC, which is greater than composite and enamel. The effect of hydration/dehydration is not significant on GIC [441].

A GIC reinforced with sintered Ag powder had significantly greater compressive strength, as well as compressive fatigue limit; was less susceptible to erosion in pH 4.0 buffer; and set more rapidly than conventional material. The Ag-reinforced material was similarly susceptible to two-body abrasion with a fatigue element fatigue, but was more susceptible to three-body abrasion and exhibited a lower modulus of elasticity and flexural strength when compared to the conventional material [442].

A mixture of GIC with silver–tin dental amalgam alloy has been suggested for use as:

(1) a base under amalgam, composite resin, and cast restorations; (2) for cementation and recementation of ill-fitting or poorly castings; (3) as temporary restorations and as an emergency repair of fractured cusps without the need for anesthesia, tooth preparation, or temporary crown construction; (4) in restorations under rest seats and clasps on removable partial dentures; and (5) in crown buildups without the use of pins for retention in restoration of severely fractured, worn, or carious teeth. Most of these suggestions are based on the assumption that addition of the silver–tin alloy improves the mechanical properties of a glass ionomer as well as providing radiopacity. Reference to Table 10 shows that little if any improvement is seen from the addition of metals to a glass ionomer material.

A derivative of an enamel remineralization system was specially formulated to facilitate the bonding of polycarboxylate and glass ionomer cements. Using (1) potassium dihydrogen phosphate saturated with calcium hydrogen phosphate dihydrate and (2) disodium hydrogen phosphate containing fluoride ion, test results show that three applications of this solution to dentin increases the tensile bond strength of both cements by more than 100% [443].

Glass ionomers release fluoride ions, which reduces microbial activity and is beneficial in inhibiting secondary caries. *In vitro*, it was previously demonstrated that there was considerable uptake of F and Al from silicate cement by the cavity walls. The powder of GIC is similar to that of silicate cement; tests indicate that, like silicate cement, GIC has anticariogenic properties, due to the uptake of F and Al by the cavity walls [444].

The fluoride leaching out of a GIC is believed to affect caries resistance in the tooth structure adjacent to the restoration. Artificial caries were created by immersion of restored teeth in a lactic acid gel at pH 4.2 for 10 weeks. This experiment indicates that secondary caries initiation and progression may be reduced significantly by the use of glass ionomer restorations [445].

Calcium hydroxide cement, calcium aluminate cement, and glass ionomer cement were compared for use as cavity liners. GIC were generally stronger than the calcium hydroxide liners in tensile and flexural strengths. GIC were initially acidic, reaching a final pH between 5.4 and 7.3, whereas the other cements were strongly alkaline from 30 sec to 24 h after mixing [446].

The microleakage of adhesive GIC and PC cements was compared with nonadhesive zinc phosphate cements. GIC provided perfect seals, the sealing ability of PC cement was excellent, and all restorations of zinc phosphate cement leaked [447]. In a comparison of zinc phosphate, polycarboxylate, and GIC, the mean value for retention was highest for the GIC [448].

H. Resin Cement

With the exception of polyacrylate cements, traditional cementation, or luting, depends on micromechanical bonding between the restoration and tooth structure. Adhesive resin cements rely on mechanical bonding but may also incorporate adhesive agents. Adhesive bonding of resin-bonded bridges has been achieved by self-cured resin cements containing organic phosphonate groups and 4-META. These cements have demonstrated adhesive bonding to enamel and dentin as well as to metal oxides. Recently, a dual-cured resin cement containing an organic phosphonate was developed for bonding of indirect composite restorations. With the adhesive primers and surface treatments now available for dental porcelain, adhesive resin cements for ceramic restorations are also possible [449]. These materials may also make possible long-term, intraoral repair of fractured ceramic and metal ceramic restorations.

The tensile bond strength of sand-blasted Ni–Cr alloy bonded to a composite resin using a urethane dimethacrylate resin cement with a phosphoric acid derivative metal primer was slightly greater than that formed using a lightly filled composite resin cement and electrolytically etching the surface of the same alloy. Both strengths were nearly double that between the composite and a sand-blasted metal surface alone. The shear bond strength of the electrolytically etched metal surface bonded with the lightly filled composite resin cement was well over double that obtained with the primer–dimethacrylate cement system and a sand-blasted alloy surface. When the primer–dimethacrylate cement was applied to an electrolytically etched alloy surface it consistently failed to set and the resin at the interface was full of voids [450].

The shear strength of acid-etched bridge cements ranges from 44 to 70 MPa. The bulk shear strength of the resin cement exceeds these values; hence, the weak link in the system is the interface at or adjacent to the resin and metal or the resin and enamel [451].

Regression analysis showed no correlation between *in vivo* cement solubility and the buffer capacity or pH of saliva. Therefore, it can be presumed that cement dissolution is caused primarily by acids from dental plaque and/or food, and not by the saliva itself [452].

I. Biocompatibility

ZOE possesses little antimicrobial activity, has toxic side effects, and may actually delay healing [453,454]. Allergic reactions have also been reported to eugenol [455]. Zinc polycarboxylate cement compared with ZOE and zinc phosphate is well tolerated by the dental pulp in humans [456].

The protective role of different dentin fractions and of dentin slices in moderating the cytotoxicity of zinc oxide eugenol (ZOE) was investigated. The collagen fraction of dentin powder provided increased protection compared to complete powdered dentin. Dentin slices offered greater protection, probably by providing a physical barrier to the diffusion of eugenol, which may also bind to the contents of the dentinal tubules. This protection increased with increasing thickness of the dentin slices. ZOE stimulated calcium release from dentin but the low levels released are unlikely to have a significant effect on the protective role of dentin [457].

Cell culture studies indicate that strong cytotoxic components of GIC can be eluted even after hardening for 48 h. In contrast, calcium hydroxide proved to be more biocompatible after the same hardening period [458].

The hypothesis that calcium hydroxide used as a root canal sealer in primary teeth will produce clinical, radiographic, and histologic responses comparable to those produced by ZOE was tested. The teeth treated with calcium hydroxide demonstrated no abscesses after 4 weeks, whereas the group treated with ZOE exhibited four abscesses [459].

A clinicopathological study was designed to clarify the effects of glass ionomer cement on the pulp and the ability of this material to protect the pulp from undesirable reactions caused by composite resin. It was concluded that GIC was not histopathologically irritating to the pulp, and that it protected the pulp from irritation by composite resin [460].

The concentration of zinc ions in dentin *in vitro* beneath cavities filled with zinc oxide–eugenol and zinc phosphate cements was measured by atomic absorption spectrophotometry. No significant difference was recorded between the two materials [461].

The bactericidal qualities of silicate, zinc phosphate, and polycarboxylate cements were compared. The major cause of pulpal inflammation is the presence of bacteria in the cavity at the time of placement of the cement. If the cement is not bactericidal and provides a large gap at the dentin–cement interface, these bacteria can grow and cause subsequent pulpal damage. Zinc phosphate cement prevents pulpal damage by adapting closely to the dentin cavity walls. The fact that zinc phosphate cement can chelate to the calcium in the tooth gives it excellent adaptability to dentin; in addition, it has demonstrable bactericidal qualities that make it favorable as a cement lining material. Unfortunately, the same qualities may result in chemical insult to the pulp [462].

Bacterial penetration of a GIC base was compared with the conventional cement bases, ZOE, and zinc phosphate placed over a Copalite varnish. GIC bases maintain their capacity to inhibit bacterial penetration as postoperative time period increases. Apparently a GIC base minimizes microleakage at the dentin surface [463].

A test of the antimicrobial activity of commercial glass ionomers against *Streptococcus mutans* (*S. mutans*), the known cause of caries, found that GI cements were antimicrobial [464].

GIC, polycarboxylate, and zinc phosphate showed a relatively rapid initial increase in pH followed by a slow but steady rise for the remainder of the test period. The water-mixed GIC had pH values significantly lower than those of the non-water-mixed GIC or the zinc phosphate cement for the first 5 min after mixing. The final pH values at 24 h were grouped together for the PC and GIC between 5.5 and 6.5 while the pH of the zinc phosphate was under 5 [465].

Microblasting the internal surfaces of cast metal crowns with 50 μm aluminum oxide or 50 μm glass beads has become a common alternative practice to cleaning the crowns by pickling in acid. Pickling and glass bead microblasting are less effective treatments than aluminum oxide microblasting in preparing cast gold crowns for cementation [466].

The success of dentin bonding agents (DBA) in bonding composite resins to dentin has led to interest in bonding dental amalgam. Bonded amalgam specimens were prepared, then aged in water at 37°C for 7 days. It was concluded that a strong, reliable bond can be achieved between a LC glass ionomer–resin hybrid and high-copper dental amalgam, but that the use of a resin DBA as an intermediary is contraindicated [467].

ACKNOWLEDGMENT

We would like to dedicate this chapter to the memory of Dr. Ralph Phillips, Research Professor Emeritus of Dental Materials at Indiana University School of Dentistry, who was originally asked to author this chapter on dental materials science. Unfortunately, Dr. Phillips died as plans for this Encyclopedia were being made. We would like to formally acknowledge the contributions Dr. Phillips made to the development of dental materials science. He was one of the pioneers of this field and authored *The Science of Dental Materials*, now in its ninth edition, which is used as a text in dental schools throughout the United States and the world. Ralph Phillips' unique contribution was his ability to bridge the gap between science and its practical application to dentistry. Although he was not a dentist, he was able to relate to the practicing dentist in a manner that was unexcelled and was in demand to make presentations to dental organizations throughout the United States and many other countries.

As long as man-made materials are used by dentists to restore and maintain oral health, the profession will owe Ralph Phillips a large debt of gratitude. He is sorely missed by his fellow researchers in dental materials.

We accepted the privilege of writing this chapter with humility and hope that Dr. Phillips would have approved of our efforts.

REFERENCES

1. Ryge, G., Adhesive restorative dental materials, in *Biomaterials: Bioengineering Applied to Materials for Hard and Soft Tissue Replacement* (A.L. Bement, ed.), University of Washington Press, Seattle, 1971, pp. 105–127.
2. Hench, L.L., and E.C. Ethridge, *Biomaterials: An Interfacial Approach*, Academic Press, 1972, p. 385.
3. Phillips, R.W., Changing trends in dental biomaterials science, *Int. J. Oral Maxillofac. Implants, 3*, 79–80 (1988).
4. Phillips, R.W., *Dental Materials*, internal report (1989).
5. Moore, B.K., unpublished data (1994).
6. Sakaguchi, R.L., and R. DeLong, Computer-aided design in dentistry, *Dentistry, 7*, 7–9 (1987).
7. Melendez, E.J., C.J. Arcoria, J.P. Dewald, and M.J. Wagner, Effect of laser-etch on bond strengths of glass ionomers, *J. Prosth. Dent., 67*, 307–312 (1992).
8. Loiacono, C., D. Shuman, M. Darby, and J.G. Luton, Lasers in dentistry, *Gen. Dent.*, Sept.–Oct. 1993, pp. 378–381.
9. Ferreira, J.M., J. Palamara, P.P. Phakey, W.A. Rachinger, and H.J. Orams, Effects of continuous-wave CO_2 laser on the ultrastructure of human dental enamel, *Arch. Oral Biol., 34*, 551–562 (1989).
10. Singh, J., The constitution and microstructure of laser surface-modified metals, *J. Metals, 34*, 8–14 (1992).
11. Sperner, F., Precious metal dental alloys. *Proc. Conf. Noble Metals Fabrications Technology*, 141–152 (1985).
12. Leinfelder, K.F., R.P. Kusy, and W.G. Price, Physical and clinical characterization of low-gold dental alloys, *Proc. Conf. Precious Metals*, 259–271 (1981).
13. Nielsen, J.P., The modern precious metal dental casting alloys, *Proc. Conf. Precious Metals*, 443–448 (1981).
14. Laub, L.W., and J.W. Stanford, Tarnish and corrosion behavior of dental gold alloys, *Gold Bull., 14*, 13–18 (1981).
15. Phillips, R.W., *The Science of Dental Materials*, 9th ed., W.B. Saunders, Philadelphia, 1991, pp. 359–384.
16. Knosp, H., M. Nawaz, and M. Stumke, Dental gold alloys: Composition, properties and applications, *Gold Bull., 14*, 57–64 (1981).
17. Labarge, J.J., and D. Treheux, Hardening of cast gold-based dental alloys: Effect of minor elements on hardening, *Metallurgic Dentaire*, 1980, pp. 23–38.
18. Labarge, J.J., D..Treheux, and P. Guiraldenq, Hardening of gold-based dental casting alloys, *Gold Bull., 12*, 46–52 (1979).
19. Flourens, J.P., B. Joniot, M. Victor, and J. Verge, Study of the fatigue phenomenon in dental prosthesis, *Metallurgic Dentaire*, 1981, pp. 367–376.
20. Wictorin, L., and H. Fredriksson, Microstructure of the solder-casting zone in bridges of dental gold alloys, *Odontol. Rev., 21*, 187–196 (1976).
21. Wiskott, H.W.A., J.I. Nicholis, and R. Taggart, Fatigue strength of Au–Pd alloy/585 solder combination, *J. Dent. Res., 70*, 140–145 (1991).
22. Yasuda, K., and M. Ohta, Age-hardening characteristics of a commercial dental gold alloy, *J. Less-Common Met., 70*, 75–87 (1980).
23. Udoh, K.I., K. Yasuda, and M. Ohta, Age-hardening characteristics in an 18 carat gold commercial dental alloy containing palladium, *J. Less-Common Met., 118*, 249–259 (1986).

24. Skjerpe, P., J. Gjonnes, E. Sorbroden, and H. Hero, Analytical electron microscopy of cast Au–Ag–Cu alloys, *J. Mater. Sci., 21*, 3986–3992 (1986).

25. Jovanovic, M., G. Lukic, M. Gligic, and B. Lukic, Effect of heat treatment on some properties of two low-gold dental alloys, *Metallography, 19*, 447–459 (1986).

26. Yasuda, K., and M. Ohta, Difference in age-hardening mechanism in dental gold alloys, *J. Dent. Res., 61*, 473–479 (1982).

27. Hisatsune, K., M. Ohta, T. Shiraishi, and M. Yamane, Aging reactions in a low-gold, white dental alloy, *J. Dent. Res., 61*, 805–807 (1982).

28. Mezger, P.R., A.L.H. Stols, M.M.A. Vrijhoef, and E.H. Greener, Metallurgical aspects and corrosion behavior of yellow low-gold alloys, *Dent. Mater., 5*, 350–354 (1989).

29. Hero, H., R. Jorgensen, and E. Sorbroden, A low-gold dental alloy: Structure and segregations, *J. Dent. Res., 61*, 1292–1298 (1982).

30. Hisatsune, K., K. Udoh, M. Nakagawa, and M. Hasaka, Aging behavior in a dental low carat gold alloy and its relation to CuAu II, *J. Less-Common Met., 160*, 247–258 (1990).

31. Kusy, R.P., and K.F. Leinfelder, Age-hardening and tensile properties of low gold (10 to 14 kt.) alloys, *J. Biomed. Mater. Res., 15*, 117–135 (1981).

32. Nielsen, J.P., H. Lipsius, and J. Tucillo, Hardening mechanism in gold–silver–copper alloys, *Proc. Conf. Noble Metals Fabrications Technology*, 1985, pp. 153–158.

33. Uzuka, T., Y. Kanazawa, and K. Yasuda, Determination of the AuCu superlattice formation region in gold–copper–silver ternary systems, *J. Dent. Res., 60*, 883–889 (1981).

34. Hero, H., G. Hultquist, and A. Oden, Selective dissolution of AuCu and AuAg alloys *in vitro* and *in vivo*, *Biomaterials, 6*, 393–395 (1985).

35. Hultquist, G., and H. Hero, Surface ennoblement by dissolution of Cu, Ag and Zn from single phase gold alloys, *Corr. Sci., 24*, 789–805 (1984).

36. Hero, H., and J. Valderhaug, Tarnishing *in vivo* and *in vitro* of a low-gold alloy related to its surface, *J. Dent. Res., 64*, 139–143 (1985).

37. Johnson, D.L., V.W. Rinne, and L.L. Bleich, Polarization-corrosion behavior of commercial gold- and silver-base casting alloys in Fusayama solution, *J. Dent. Res., 62*, 1221–1225 (1983).

38. Ohta, M., M. Nakagawa, and S. Matsuya, Effect of palladium addition on the tarnishing of dental gold alloys, *J. Mater. Sci.: Mater. Med., 1*, 140–145 (1990).

39. Suoninen, E., H. Hero, and E. Minni, Effect of palladium on sulfide tarnishing of noble metal alloys, *J. Biomed. Mater. Res., 19*, 917–934 (1985).

40. Hiller, K., H. Kaiser, W. Braemer, and F. Sperner, Untersuchungen zur Resistenz von Edelmetall-Dentallegierungen, *Wertst. Korros., 33*, 83–88 (1982).

41. Wright, D., R. Gallant, and L. Spangberg, Correlation of corrosion behavior and cytotoxicity in Au–Cu–Ag ternary alloys, *Proc. Conf. Precious Metals*, 1981, pp. 433–441.

42. German, R.M., Precious-metal dental casting alloys, *Int. Metal Rev., 27*, 260–288 (1982).

43. Nielsen, J.P., Solidification fundamentals of jewelry and dental casting, *Proc. Conf. Precious Metals*, 1983, pp. 3–12.

44. Okazaki, M., J. Takahashi, H. Kimura, and K. Ida, Estimation of solidification time during casting by use of a heat transfer model, *J. Dent. Res., 61*, 1188–1191 (1982).

45. Ady, A.B., and C.W. Fairhurst, Bond strength of two types of cement to gold casting alloy, *J. Prosth. Dent., 29*, 217–220 (1973).

46. Kidd, E.A.M., and J.W. McLean, The cavity sealing ability of cemented cast gold restorations, *Br. Dent. J., 147*, 39–41 (1979).

47. Myers, G.W., and D.W. Cruickshanks-Boyd, Mechanical properties and casting characteristics of a silver-palladium bonding alloy, *Br. Dent. J., 163*, 323–326 (1982).

48. Niemi, L., and H. Hero, The structure of a commercial dental Ag–Pd–Cu–Au casting alloy, *J. Dent. Res., 63*, 149–154 (1984).

49. Sastri, W.L., Potentiostatic and potentiodynamic polarization of dental silver-palladium alloys, Polytech Inst. New York, *Diss. Abst.*, 1981, p. 142.

50. Vaidyanathan, T.K., and A. Prasad, *In vitro* corrosion and tarnish analysis of the Ag–Pd binary system, *J. Dent. Res., 60*, 707–715 (1981).

51. Niemi, L., and R.I. Holland, Tarnish and corrosion of a commercial dental Ag–Pd–Cu–Au casting alloy, *J. Dent. Res., 63*, 1014–1018 (1984).

52. Hero, H., and L. Niemi, Tarnishing *in vivo* of Ag–Pd–Cu–Zn alloys, *J. Dent. Res., 65*, 1303–1307 (1986).

53. Biolaru, T., L. Ene, and M.V. Constantinescu, Study of contact corrosion of Ag–Pd dental alloys in the mouth, *Metallurgie Dentaire*, 1981, pp. 327–333.

54. Oshida, Y., unpublished information (1994).

55. Norling, B.K., Dental amalgams: Composition, fabrication and trituration, in *Encyclopedia of Materials Science and Engineering*, Pergamon Press, Oxford, 1986, Vol. 2, pp. 1047–1051.

56. Yamada, T., and T. Fusayama, Effect of moisture contamination on high-copper amalgam, *J. Dent. Res., 60*, 716–723 (1981).

57. Phillips, R.W., *Skinner's Science of Dental Materials*, 9th ed., W.B. Saunders, 1991, p. 303.

58. Murchison, D.F., E.S. Duke, B.K. Norling, and T. Okabe, The effect of trituration time on the mechanical properties of four copper amalgam alloys, *Dent. Mater., 5*, 74–76 (1989).

59. Waterstrat, R.M., Brushing up on the history of intermetallics in dentistry, *J. Metals, 42*, 8–14 (1990).

60. *Consumer Reports*, May 1991, pp. 316–319.

61. Lee, K.H., M.C. Shin, and J.Y. Lee, Amalgamation mechanism in dental amalgam alloys, *J. Mater. Sci., 22*, 3949–3955 (1987).

62. Abbott, J.R., D.R. Miller, and D.J. Netherway, *The Hardening Reaction in Dental Amalgam, Strength of Metals and Alloys*, Pergamon Press, Oxford, 1982, pp. 347–352.

63. Calvo, F.A., M. Concepcion Merino, E. Otero, and A. Pardo, On the mechanism of amalgamation in dental amalgams, *Prakt. Metallogr., 24*, 268–279 (1987).

64. Boswell, P.G., The suppression of gamma-2 precipitation and the kinetics of the gettering of tin in a dispersed phase amalgam, *J. Mater. Sci., 15*, 1311–1314 (1980).

65. Lussi, A.S., and W.B. Buergin, A new method to measure the condensation pressure of amalgam under *in vivo* conditions, *J. Dent. Res., 66*, 737–739 (1987).

66. Dubois, J.M., G. Caer, J.G. Dumagny, and F. Dupont, A Mössbauer spectroscopy study of the amalgamation of a Ag–Sn–Hg dental alloy, *Acta Metall., 29*, 1159–1169 (1981).

67. Okabe, T., R.J. Mitchell, M.B. Butts, and J.L. Ferracane, Microstructural changes on the surfaces of dental amalgam during aging, *Corrosion, Failure Analysis, and Metallography*, 1986, pp. 177–195.

68. Paffenbarger, G.C., N.W. Rupp, and R.M. Waterstrat, Metals in solution in mercury expressed from copper-rich dental amalgams, *J. Dent. Res., 61*, 30–32 (1982).

69. Gronlund, F., and B. Kristensen, Electrochemical determination of the solubility of copper in mercury, *Acta Chem. Scand., A38*, 229–232.

70. Smith, F.E., and R.W. Devenish, The structure of dental amalgam, in *Modern Metallography in Metallurgy Conference and Exhibition*, 1982, p. 31.

71. Smith, F.E., R.W. Devenish, and R.C. Pond, The structure of dental amalgam, *Electron Microscopy, 1*, 729 (1982).

72. Fusayama, T., and K. Hayash, Microstructure of amalgam surfaces, *J. Dent. Res., 49*, 733–741 (1970).

73. Smith, F.E., D.F. Williams, and R. Pond, Backscattered electron imaging of dental amalgam, *J. Mater. Sci., 22*, 2382–2386 (1987).

74. Malhorta, M.L., and K. Asgar, X-ray diffraction analysis of gamma 2 (Sn–Hg) phase in high-copper amalgams of varying mercury content, *J. Dent. Res., 60*, 149–153 (1981).

75. Mahler, D.B., and J.D. Adey, Microprobe analysis of three high-copper amalgams, *J. Dent. Res., 63*, 921–925 (1984).

76. Kraft, W., G. Petzow, and F. Aldinger, Constitution of copper-rich dental amalgams, *Z. Metallkd., 71*, 699–703 (1980).

77. Kishiyanagi, M., J. Onagawa, and T. Abe, Temperature dependence of microstructure of Ag–Sn amalgam after heating, *Tohoku Gakuin Uni, 22*, 43–46 (1987).

78. Abbott, J.R., D.R. Miller, and D.J. Netherway, Influence of alloy composition on the hardening of silver–tin dental amalgam, *J. Biomed. Mater. Res., 20*, 1391–1400 (1986).

79. Boyer, C.E., Lowering the silver content in dental alloy, Dissertation, Univ. of Virginia, *Diss. Abstr. Int., 47*, 172 (1986).

80. Srakar, N.K., Copper in dental amalgams, *J. Oral Rehabil., 6*, 1–8 (1979).

81. Eames, W.B., J.F. MacNamara, and C.O. Palmertree, High-copper amalgam alloys – An update, *J. Ind. Dent. Assoc., 55*, 27–31 (1976).

82. Desportes, J., and F. Caitucoli, Structure and hardness of gallium alloys obtained by diffusion at 37°C, *Metallurgie Dentaire*, 1980, pp. 219–225.

83. Okabe, T., M. Woldu, H. Nakajima, B.H. Miller, and L.K. Mash, Gallium alloys made from alloy powders for dental amalgam, *J. Dent. Res., 71*, 252, Abstract 1175 (1992).

84. Greener, E.H., K.H. Chung, K. Chern, and J.H. Lin, Creep in a palladium-enriched high-copper amalgam, *Biomaterials, 19*, 213–217 (1988).

85. Mahler, D.B., and J.D. Adey, Tin in the Ag–Hg phase of dental amalgam, *J. Dent. Res., 67*, 1275–1277 (1988).

86. Johnson, L.B., and G.C. Paffenbarger, The role of zinc in dental amalgams, *J. Dent. Res., 59*, 1412–1419 (1980).

87. Han, S.W., and J.N. Lee, Study on Ag–Sn system dental amalgam, I. The effect of zinc on mercury ratio and residual mercury in the silver-base dental amalgam, *J. Korean Inst. Met., 21*, 19–24 (1983).

88. Han, S.W., Y.H. Cha, and J.N. Lee, Study on the Ag–Sn system for dental amalgam, II. The effect of zinc on the structure and mechanical properties of the silver-base dental amalgam, *J. Korean Inst. Met., 21*, 635–648 (1983).

89. Brown, I.H., and B.H. Ide, A dilatometer for measuring dimensional change in dental amalgam, *J. Phys. E., Sci. Instrum., 19*, 123–124 (1986).

90. El-Hadary, M.E., A. Kamar, A. El-Kady, S.H. Kandil, M.A. El-Gamal, and S.E. Morsi, Isothermal dimensional changes of some dental materials, *J. Thermal Anal., 35*, 1351–1357 (1989).

91. Black, G.V., *Operative Dentistry*, Medio-Dental Publ. Co., Chicago, 1988, Vol. 2, pp. 299–329.

92. Greener, E.H., and K. Szurgot, Properties of Ag–Cu–Pd dispersed phase amalgam: Compressive strength, creep and corrosion, *J. Dent. Res., 161*, 1192–1194 (1982).

93. Wolfenden, A., The piezoelectric ultrasonic composite oscillator technique (PUCOT) for monitoring metallurgical changes, *Advanced Techniques for Characterizing Microstructures*, 1982, pp. 185–210.

94. Mitchell, R.J., M.B. Butts, and C.W. Fairhurst, Deformation of amalgam at low strain rates, *Metallurgie Dentaire*, 1981, pp. 395–412.

95. Mitchell, R.J., H. Ogura, and K. Nakamura, Characterization of fractured surfaces of dental amalgams, *ISTFA '87, Advanced Materials*, 1987, pp. 179–187.

96. Zardiackas, L.D., and L. Anderson, Crack propagation in conventional and high copper dental amalgam as a function of strain rate, *Biomaterials, 7*, 259–262 (1986).

97. Abbott, J.R., D.R. Miller, and D.J. Netherway, Plastic deformation of beta and gamma phase silver–tin alloys, *J. Mater. Sci., 19*, 3255–3265 (1984).

98. Schumann, G., W. Blum, and H.H. Hasselt, Creep of dental amalgams, *Z. Metallkd., 72*, 251–260 (1981).

99. Okabe, T., J. Staman, J. Ferracane, and R. Mitchell, Effect of free mercury on the strengths of dental amalgams, *Dent. Mater., 1*, 180–184 (1985).

100. Okabe, T., W.W. Tseng, S. Galloway, B. Timmerman, and R. Mitchell, Resistance to mercury embrittlement of some alloys for amalgam, *Dent. Mater., 1*, 19–25 (1985).

101. DeLong, R., R.L. Sakaguchi, W.H. Douglas, and M.R. Pintado, The wear of dental amalgam in an artificial mouth: A clinical correlation, *Dent. Mater., 1*, 238-242 (1985).

102. Lahiri, S.K., and D. Gupta, A kinetic study of platinum–mercury contact reaction, *J. Appl. Phys., 51*, 5555-5560 (1981).

103. Bayne, S.C., L.D. Zardiackas, and M.D. Carlson, Fatigue fracture analysis of amalgam by SEM-EDS, *Dent. Mater., 2*, 53-57 (1986).

104. Sutow, E.J., D.W. Jones, G.C. Hall, and E.L. Milne, The response of dental amalgam to dynamic loading, *J. Dent. Res., 64*, 62-66 (1985).

105. Williams, P.T., and J.R. Cahoon, Amalgam margin breakdown caused by creep fatigue rupture, *J. Dent. Res., 68*, 1188-1193 (1989).

106. Mahler, D.B., J.D. Adey, and M. Marek, Creep and corrosion of amalgam, *J. Dent. Res., 61*, 33-35 (1982).

107. Gale, E.N., J.W. Osborne, and P.G. Winchell, Fracture at the margins of amalgam as predicted by creep, zinc content and gamma-2 content, *J. Dent. Res., 61*, 678-680 (1982).

108. Mjor, I.A., and S. Espevik, Assessment of variables in clinical studies of amalgam restorations, *J. Dent. Res., 59*, 1511-1515 (1980).

109. Holland, R.I., R.B. Jorgensen, and J. Ekstrand, Strength and creep of dental amalgam: The effects of deviations from recommended preparation procedures, *J. Prosthet. Dent., 54*, 189-194 (1985).

110. Williams, P.T., and G.L. Hedge, Creep-fatigue as a possible cause of dental amalgam margin failure, *J. Dent. Res., 64*, 470-475 (1985).

111. Hero, H., On creep mechanisms in amalgams, *J. Dent. Res., 62*, 44-50 (1983).

112. Mahler, D.B., J.D. Adey, and S.J. Marshall, Effect of time at 37°C on the creep and metallurgical characteristics of amalgam, *J. Dent. Res., 66*, 1146-1148 (1987).

113. Okabe, T., M.B. Butts, and R.J. Mitchell, Changes in the microstructures of silver–tin and admixed high-copper amalgams during creep, *J. Dent. Res., 62*, 37-43 (1983).

114. Okabe, T., R.J. Mitchell, M.B. Butts, S.S. Galloway, and W.S. Twiggs, Change in creep rate and microstructure in an aged, low-copper amalgam, *J. Biomed. Mater. Res., 19*, 727-746 (1985).

115. Williams, K.R., Cyclic creep and fracture of dental amalgam, *Biomaterials, 4*, 255-261 (1983).

116. Greener, E.H., K. Szurgot, and E.P. Lautenschlager, Tensile creep of dental amalgam, *Biomaterials, 3*, 101-104 (1982).

117. Vermeersch, A., H. Letzel, and M. Vrighoef, Laboratory physical tests and behavior in the mouth of six high-copper content amalgams after two years, *Metallurgie Dentaire*, 1981, pp. 412-425.

118. Mahler, D.B., and J. Van Eysden, Dynamic creep of dental amalgam, *J. Dent. Res., 48*, 501-508 (1969).

119. Osborne, J.W., P.G. Winchell, and R.W. Phillips, A hypothetical mechanism by which creep causes marginal failure of amalgam restorations, *J. Ind. Dent. Assoc., 57*, 16-17 (1979).

120. Sarkar, N.K., Creep, corrosion and marginal fracture of dental amalgams, *J. Oral Rehabil., 5*, 413-423 (1978).

121. Lemaitre, L., M. Moors, A.P. Van Peteghem, and M. Thys, The corrosion behavior of dental amalgam as related to the silver content of the dental amalgam alloy, *Int. Conf. Precious Metals*, 1987, pp. 293-301.

122. Halse, A., Metals in dentinal tubules beneath amalgam fillings in human teeth, *Arch. Oral Biol., 20*, 87-88 (1975).

123. Ray, N.J., Surface tarnishing *in vitro* of some commercially produced high-copper silver amalgams, *J. Dent., 16*, 145-146 (1988).

124. Sarkar, N.K., R.A. Fuys, C.M. Schoenfeld, L.M. Armstrong, and J.W. Stanford, The tooth-amalgam interaction, *J. Oral Rehabil., 8*, 401-411 (1981).

125. Marek, M., Acceleration of corrosion of dental amalgam by abrasion, *J. Dent. Res., 63*, 1010–1013 (1984).

126. Angelini, E., P. Bianco, and F. Zucchi, Influence of brushing and abrasion on the *in vitro* corrosion resistance of dental amalgams, *Surf. Coat Technol., 34*, 523–535 (1984).

127. Lin, J.H.C., S.J. Marshall, and G.W. Marshall, Microstructures of high copper amalgams after corrosion in various solutions, *Dent. Mater., 3*, 176–181 (1987).

128. Lin, J.H.C., Corrosion product formation and microstructural changes in copper-rich amalgams. Northwestern Univ., *Diss. Abstr. Int., 44*, 287 (1984).

129. Lin, J.H.C., G.W. Marshall, and S.J. Marshall, Microstructures of copper-rich amalgams after corrosion, *J. Dent. Res., 62*, 112–115 (1983).

130. Marek, M., The effect of tin on the corrosion behavior of the Ag–Hg phase of dental amalgam and dissolution of mercury, *J. Dent. Res., 69*, 842–845 (1983).

131. Jorgensen, K.D., and A.L. Esbensen, Gravimetrisk bestemmelse auf dentale amalgamers korrosion, *Tandlaegebladet, 77*, 162–170 (1973).

132. Werner, H., and M. Weiland, Influence of mechanical treatment of the surface and of chemical composition on the corrosion of dental amalgams, *Werkst. Korros., 41*, 210–214 (1990).

133. Meyer, J.M., and A. Wiskott, Electrochemical behavior of high-copper amalgams, *Metallurgie Dentaire*, 1981, pp. 304–314.

134. Sarkar, N.K., and J.R. Part, Mechanism of improved corrosion resistance of zinc-containing dental amalgams, *J. Dent. Res., 67*, 1312–1315 (1988).

135. Lemaitre, L., M. Moors, and A.P. van Peteghaem, AC impedance measurements on high copper dental amalgams, *Biomaterials, 6*, 425–429 (1985).

136. Averette, D.F., and M. Marek, The effect of tensile strain on corrosion of dental amalgam, *J. Dent. Res., 62*, 842–845 (1983).

137. Parvinen, T., Corrosion of a dental alloy in the mouth, *Proc. Finn. Dent. Soc., 72*, 210–212 (1976).

138. Senia, E.S., and D.J. Bales, Dental pain of galvanic origin: Report of case, *J. Endod., 3*, 280–281 (1977).

139. Ravnholt, G., and R.I. Holland, Corrosion current between fresh and old amalgam, *Dent. Mater., 4*, 251–254 (1988).

140. Zardiackas, L.D., and G.E. Stoner, Electrochemical characteristics of the S.I.A. dental amalgam cavity liner, *Biomaterials, 1*, 13–16 (1980).

141. Zardiackas, L.D., and G.E. Stoner, Tensile and shear adhesion of amalgam to tooth structure using selective interfacial amalgamation, *Biomaterials, 4*:9–13 (1983).

142. Angelini, E., and F. Zucchi, High copper dental amalgams: Comparative study of the structure and the *in vitro* corrosion resistance, *Surf. Technol., 25*, 385–396 (1985).

143. Ravnholt, G., Accelerated corrosion analysis of dental amalgams, *Scand. J. Dent. Res., 94*, 553–561 (1986).

144. Marshall, S.J., J.C. Lin, and G.W. Marshall, Cu_2O and $CuCl_2 \cdot 3Cu(OH)_2$ corrosion products on copper rich dental amalgams, *J. Biomed. Mater. Res., 16*, 81–85 (1982).

145. Kozono, Y., B.K. Moore, R.W. Phillips, and M.L. Swartz, Dissolution of amalgam in saliva solution, *J. Biomed. Mater. Res., 16*, 767–774 (1982).

146. Mueller, H.J., Spectral and gravimetric analysis of completely oxidized amalgam systems, *Biomaterials, 8*, 221–256 (1980).

147. Do Duc, H., J.M. Meyer, and P. Tissot, Electrochemical behavior of Sn_8Hg (gamma-2) and dental amalgam in a phosphate buffer solution, *Electrochem. Acta, 25*, 851–856 (1980).

148. Srakar, N.K., J.W. Osborne, and K.F. Leinfelder, *In vitro* corrosion and *in vivo* marginal fracture of dental amalgams, *J. Dent. Res., 61*, 1262–1268 (1982).

149. Silva, M., B.F. Zimmerman, R. Weinberg, and N.K. Sarkar, Corrosion and artificial caries-like lesions around amalgam restorations, *Aust. Dent. J., 32*, 116–119 (1987).

150. Sarkar, N.K., R.A. Fuya, and J.W. Stanford, Application of electrochemical techniques to

characterize the corrosion of dental alloys. Corrosion and degradation of implant materials, *ASTM STP, 684*, 277–294 (1979).

151. Gross, M.J., and J.A. Harrison, Some electrochemical features of the *in vivo* corrosion of dental amalgams, *J. Appl. Electrochem., 19*, 301–310 (1989).
152. Marshall, S.J., and G.W. Marshall, $Sn_4(OH)_6Cl_2$ and SnO corrosion products of amalgams, *J. Dent. Res., 59*, 820–823 (1980).
153. Hero, H., and R.B. Jorgensen, Flow stress and deformation hardening of triturated amalgam, *Dent. Mater., 1*, 145–149 (1985).
154. Liberman, R., A. Ben-Amar, D. Nordenberg, and A. Jodaikin, Long-term sealing properties of amalgam restorations: An *in vitro* study, *Dent. Mater., 5*, 168–170 (1989).
155. Rehfeld, R.L., R.B. Mazer, K.F. Leinfelder, and C.M. Russell, Evaluation of various forms of calcium hydroxide in the monitoring of microleakage, *Dent. Mater., 7*, 202–205 (1991).
156. Berglund, A., Estimation by a 24 h study of the daily dose of intra-oral mercury vapor inhaled after release from dental amalgam, *J. Dent. Res., 69*, 1646–1651 (1990).
157. Mackert, J.R., Factors affecting estimation of dental amalgam mercury exposure from measurements of mercury vapor levels in intra-oral and expired air, *J. Dent. Res., 66*, 1775–1780 (1987).
158. Uusheimo, K., and I. Rytomaa, Mercury release from dental amalgam into saliva, *Tech. Res. Cent. Finland Res. Rep.*, 1988, p. 18.
159. Osborne, J.W., and R.D. Norman, 13-year clinical assessment of ten amalgam alloys, *Dent. Mater., 6*, 189–194 (1990).
160. Letzel, H., M.A. van't Hof, M.M.A. Vrijhoef, G.W. Marshall, and S.J. Marshall, A controlled clinical study of amalgam restorations: Survival, failures, and causes of failure, *Dent. Mater., 5*, 115–121 (1989).
161. Capel Cardoso, P.E., W. Gomes Miranda, and J.F. Ferreira Santos, Low-silver amalgam restorations: A two-year clinical evaluation, *Dent. Mater., 5*, 277–280 (1989).
162. Boyer, D.B., Composition of clinically aged amalgam restoration, *Dent. Mater., 6*, 146–150 (1990).
163. Tuccillo, J.J., and P.J. Cascone, Composition and properties of noble and precious metal alloys, *Biocompatibility Dent. Mater., 4*, 19–36 (1982).
164. DeHoff, P.H., K.J. Anusavice, and P.W. Hathcock, An evaluation of the four-point test for metal/ceramic bond strength, *J. Dent. Res., 61*, 1066–1069 (1982).
165. Oshida, Y., and A. Hashem, Titanium–porcelain system, Part I. Oxidation kinetics of nitrided pure titanium, simulated to porcelain firing process, *J. Bio-Med. Mater. Engr.*, in press (1994).
166. Bertolotti, R.L., Alternative casting alloys for today's crown and bridge restorations, Part II. Restorations, *J. Calif. Dent. Assoc., 11*, 63–69 (1983).
167. Knosp, H., Gold alloy for firing on porcelain for dental purposes, U.S. Patent No. 4218244 (1980).
168. Smith, D.L., A.P. Burnett, M.S. Brooks, and D.H. Anthony, Iron–platinum hardening in casting golds for use with porcelain, *J. Dent. Res., 49*, 283–288 (1970).
169. Tsai, M.H., Gold alloys for fusion to porcelain, U.S. Patent No. 4194907 (1980).
170. Ingersoll, C.E., and D.P. Agarwall, Ceramic substrate alloy, U.S. Patent No. 4201577 (1980).
171. Knosp, H., Silver-free, low-gold-noble metal alloys for firing of dental porcelain. U.S. Patent No. 4179286 (1979).
172. Boyajian, B.K., The development of a new palladium-base alloy system for fusion to dental porcelain, University of Virginia, *Diss. Abstr., 50*, 179 (1990).
173. Prasad, A., Gallium and silver free, palladium based dental alloys for porcelain-fused-to-metal restorations, U.S. Patent No. 4518564 (1985).
174. Prosen, E.M., Palladium-base dental alloys, U.S. Patent No. 4179288 (1979).

175. Schaffer, S.P., and J.M. Ney, Palladium-Silver alloy for use with dental porcelains, U.S. Patent No. 4350526 (1982).
176. Prasad, A., Dental alloys for porcelain-fused-to-metal restorations, U.S. Patent No. 4419325 (1983).
177. Prasad, A., Grain-refined gold-free dental alloys for porcelain-fused-to-metal restorations, U.S. Patent No. 4576789 (1986).
178. Schock, G., J. Hausselt, J. Rothaut, and D. Hathaway, Application of silver–palladium alloys for bonding dental ceramics by firing, Ausz. Eur. Patentanmeld EP No. 0178506 (1986).
179. Jendresen, M.D., Non-precious metals and the ceramometal restoration, *J. Ind. Dent. Assoc., 54,* 6–10 (1975).
180. Adachi, M., J.R. Mackert, E.E. Parry, and C.W. Fairhurst, Oxide adherence and porcelain bonding to titanium and Ti–6Al–4V alloy, *J. Dent. Res., 69,* 1230–1235 (1990).
181. German, R.M., Hardening reactions in a high-gold content ceramo-metal alloy, *J. Dent. Res., 59,* 1960–1965 (1980).
182. German, R.M., Gold alloys for porcelain-fused-to-metal dental restorations: Their hardening by ordering of an FePt-type compound, *Gold Bull., 13,* 57–62 (1980).
183. Hisatsune, K., M. Nakagawa, K. Udoh, B.I. Sorosoedirdjo, and M. Hasaka, Age-hardening reactions and microstructures of a dental gold alloy with palladium and platinum, *J. Mater. Sci.: Mater. Med., 1,* 49–54 (1990).
184. Yasuda, K., G. Van Tendeloo, J. Van Landuyt, and S. Amelinckx, High-resolution electron microscopic study of age-hardening in a commercial dental gold alloy, *J. Dent. Res., 65,* 1179–1185 (1986).
185. Biederman, R.R., R.M. German, and J.R. Toran, The physical metallurgy of a Pd–Au dental alloy, *Conf. Proc. Precious Metals,* 1982, pp. 423–431.
186. Ohta, M., T. Shiraishi, and M. Yamane, Phase transformation and age-hardening of Au–Cu–Pd ternary alloys, *J. Mater. Sci., 21,* 529–535 (1986).
187. Hisatsune, K., M. Hasaka, B.I. Sosrosoedirdjo, and K. Udoh, Age-hardening behavior in a palladium-base dental porcelain-fised alloy, *Mater. Charact., 25,* 177–184 (1990).
188. Rosenstiel, S.F., and S.S. Porter, Apparent fracture toughness of dental porcelain with a metal substrate, *Dent. Mater., 4,* 187–190 (1988).
189. Anusavice, K.J., P.H. Dehoff, A. Gray, and R.B. Lee, Delayed crack development in porcelain due to incompatibility stress, *J. Dent. Res., 67,* 1086–1091 (1988).
190. Anusavice, K.J., R.D. Ringle, P.K. Morse, C.W. Fairhurst, and G.E. King, A thermal shock test for porcelain-metal systems, *J. Dent. Res., 60,* 1686–1691 (1981).
191. Asaoka, K., and J.A. Tesk, Visco-elastic deformation of dental porcelain and porcelain-metal compatibility, *Dent. Mater., 7,* 30–35 (1991).
192. Vrijhoef, M.M.A., A.J. Spanauf, and H.H. Renggli, Axial strengths of foil, all-ceramic and PFM molar crowns, *Dent. Mater., 4,* 15–19 (1988).
193. Anusavice, K.J., C. Shen, D. Hashinger, and S.W. Twiggs, Interactive effect of stress and temperature on creep of PFM alloys, *J. Dent. Res., 64,* 1094–1099 (1985).
194. Bertolotti, R.L., and J.P. Moffe, Creep rate of porcelain-bonding alloys as a function of temperature, *J. Dent. Res., 59,* 2062–2065 (1990).
195. Hausselt, J., and M. Clasing, Contribution to the adhesion and stress relationships for metal–ceramic solid solutions in dental technology, *Metall; Internationale Zeitschrift für Technik und Wirtschaft, 36,* 765–771 (1982).
196. Bagby, M., S.J. Marshall, and G.W. Marshall, Residual stress in two dental alloys during porcelain application, *Conf. Proc. Advances in X-ray Analysis,* 1988, p. 31.
197. Bagby, M., S.J. Marshall, and G.W. Marshall, Residual stress in two alloys during porcelain application, *Adv. X-ray Anal., 31,* 255–260 (1988).
198. Terada, Y., and R. Hirayasu, Radiographic stress measurement of two metal ceramic gold alloys, *Int. J. Pros., 2,* 123–127 (1989).

199. Mintz, V.W., A.A. Caputo, and C.M. Belting, Inherent structural defects of porcelain-fused-to-gold restorations: A preliminary report, *J. Prosthet. Dent., 32*, 544–550 (1974).

200. Buchanan, W.T., C.W. Svare, and K.A. Turner, The effect of repeated firings and strength on marginal distortion in two ceramometal system, *J. Prosthet. Dent., 45*, 502–506 (1981).

201. DeHoff, P.H., K.J. Anusavice, and R.J. Boyce, Analysis of thermally-induced stresses in porcelain-metal systems, *J. Dent. Res., 62*, 593–597 (1983).

202. Anusavice, K.J., B. Hojjatie, and P.H. Dehoff, Influence of metal thickness on stress distribution in metal-ceramic crowns, *J. Dent. Res., 65*, 1173–1178 (1986).

203. Tesk, J.A., R.P. Whitlock, G.E.O. Widera, A. Holmes, and E.E. Parry, Consideration of some actors influencing compatibility of dental porcelains and alloys, Part II. — Porcelain/alloy stress, *Conf. Proc. Precious Metals*, 1981, pp. 283–291.

204. Dorsch, P., Stresses in metal-ceramic systems as a function of thermal history, *Ceram. Forum Int., 58*, 157–163 (1981).

205. Tobey, R., Consideration of thermal compatibility in selection of ceramic alloys, *Trends Tech. Contem. Dent. Lab., 3*, 32–34 (1986).

206. Whitlock, R.P., J.A. Tesk, G.E.O. Widera, A. Holmes, and E.E. Parry, Consideration of some factors influencing compatibility of dental porcelains and alloys, I. Thermo-physical properties, *Conf. Proc. Precious Metals*, 1981, pp. 273–282.

207. Fairhurst, C.W., K.J. Anusavice, R.D. Ringle, and S.W. Twiggs, Porcelain-metal thermal compatibility, *J. Dent. Res., 60*, 815–819 (1981).

208. Fairhurst, C.W., D.T. Hashinger, and S.W. Twiggs, The effect of thermal history on porcelain expansion behavior, *J. Dent. Res., 68*, 1313–1315 (1989).

209. Dorsch, P., Thermal compatibility of materials for porcelain-fused-to-metal (PFM)-restorations, *Ceram. Forum Int., 3*, 159–163 (1982).

210. Walton, T.R., and W.J. O'Brien, Thermal stress failure of porcelain bonded to a palladium-silver alloy, *J. Dent. Res., 64*, 476–480 (1985).

211. Fairhurst, C.W., J.R. Mackert, and S.W. Twiggs, Bonding of ceramics to alloys (retroactive coverage), *Ceram. Eng. Sci. Proc., 6*, 66–83 (1985).

212. Mackert, J.R., E.E. Parry, and C.W. Fairhurst, Oxide morphology and adherence on dental alloys designed for porcelain bonding, *Oxid. Met., 25*, 319–333 (1986).

213. Vrijhoef, M.M.A., and J.M. vander Zel, Oxidation of two high-palladium PFM alloys, *Dent. Mater., 1*, 214–218 (1985).

214. Ohno, H., Y. Kanazawa, I. Kawashima, and N. Shiokawa, Structure of high-temperature oxidation zones of gold alloys for metal/porcelain bonding containing small amounts of indium and tin, *J. Dent. Res., 62*, 774–779 (1983).

215. Vrijhoef, M.M.A., and J.M. vander Zel, Oxidation of two high-palladium alloys, *Dent. Mater., 1*, 214–218 (1986).

216. Baran, G., M. Merancer, and P. Farrell, Transient oxidation of multiphase Ni–Cr base alloys, *Oxid. Met., 29*, 409–418 (1988).

217. Pask, J.A., and A.P. Tomsia, Oxidation and ceramic coatings on 80Ni20Cr alloys, *J. Dent. Res., 67*, 1164–1171 (1988).

218. Suoninen, E., and H. Hero, The structure and oxidation of two palladium ceramic fusing alloys, *Biomaterials, 6*, 133–137 (1985).

219. Ohno, H., and Y. Kanazawa, Internal oxidation in gold alloys containing small amounts of iron and tin, *J. Mater. Sci., 18*, 919–929 (1983).

220. Ohno, H., S. Miyakawa, K. Watanabe, and N. Shiokawa, The structure of oxide formed by high-temperature oxidation of commercially gold alloys for porcelain/metal bonding, *J. Dent. Res., 61*, 1255–1261 (1982).

221. Mackert, J.R., R.D. Ringle, and C.W. Fairhurst, High-temperature behavior of a Pd–Ag alloy for porcelain, *J. Dent. Res., 62*, 1229–1235 (1983).

222. Jones, D.W., Fusion of ceramics to metal in dentistry, *Trans. J. Br. Ceram. Soc., 84*, 40–48 (1985).

223. Sperner, F., Noble metal alloys and dental ceramics today, *Metall; International Zeitschrift für Technik und Wirtschaft, 33*, 875–877 (1979).

224. Yamaguchi, S., and T. Tsuchiya, Mineralized surfaces of the dental alloys, *J. Electrochem. Soc., 128*, 483–484 (1981).

225. Ji, H., and P.M. Marquis, Solid-state reaction between aluminous dental porcelain and platinum, *J. Mater. Sci. Lett., 8*, 670–672 (1989).

226. Baran, G.R., Phase changes in base metal alloys along metal/porcelain interfaces, *J. Dent. Res., 58*, 2095–2104 (1979).

227. Von Radnoth, M.S., and E.P. Lautenschlager, Untersuchungen über die Morphologie der Grenzflache zwischen Edelmetallegierungen und aufgebranuten keramischen Massen an Kronen, *Deutsch Zahnaerzt, 24*, 1029–1036 (1969).

228. Duh, J.G., B.S. Chien, and B.S. Chiou, Interfacial morphology and pre-oxidation effect in dental alloy/opaque bonding, *J. Mater. Sci. Lett., 8*, 355–357 (1989).

229. Wagner, W.C., The effect of interfacial variables on metal-porcelain bonding, University of Michigan, *Diss. Abstr., 51*, 84 (1990).

230. Rasmussen, S.T., and A.A. Doukoudakis, The effects of using recast metal on the bond between porcelain and a gold–palladium alloy, *J. Prosthet. Dent., 55*, 447–453 (1986).

231. Fioravanti, S., Combined use of SEM, EPMA, and XPS for analysis of dental metal/ceramic interfaces, *Proc. Conf., Electron Microscopy and Analysis*, IOP Publishing, 1988, pp. 209–212.

232. Wiley, D.R., and N.E. Waters, A fracture mechanics approach to ceramo-metal bond evaluation, *J. Mater. Sci., 24*, 2337–2342 (1989).

233. NeWitter, D.A., E.R. Schlissel, and M.S. Wolff, An evaluation of adjustment and postadjustment finishing techniques on the surface of porcelain-bonded-to-metal crowns, *J. Prosthet. Dent., 48*, 388–395 (1982).

234. Itoh, M., Studies on the porcelain fusing to Cr–Co–Ni system alloys used ceramics coating method, *Shika Riko, 20*, 167–187 (1979).

235. Tesk, J.A., R.W. Hinman, G.E.O. Widera, A.D. Holmes, and J.M. Cassel, Effects of porcelain/alloy interfacial diffusion zones on thermomechanical strain, *J. Dent. Res., 62*, 585–589 (1983).

236. Sced, I.R., and J.W. McLean, The strength of metal/ceramic bonds with base metals containing chromium, *Br. Dent. J., 132*, 232–234 (1972).

237. Mackert, J.R., Porcelain bonding of a nickel–chromium dental casting alloy, University of Virginia, *Diss. Abstr., 40*, 164 (1980).

238. Ogolnik, R., B. Picard, R. Collongues, and A.M. Lejus, Improvement of the ceramic/metal bonding on non-precious alloy used in odontology, *Rev. Int. Hautes Temp. Refract., 26*, 131–141 (1990).

239. Tomsia, A.P., and J.A. Pask, Bonding of dental glass to nickel-chromium alloys, *J. Am. Ceram., 69*, C239–C240 (1986).

240. Nishiyama, Y., Various factors affecting the bonding strength of porcelain fused to gold alloys, *Shikawa Gaku, 70*, 35–53 (1970).

241. Murakami, I., J. Vaidyanathan, T.K. Vaidyanathan, and A. Schulman, Interactive effects of etching and pre-oxidation on porcelain adherence to non-precious alloys: A guided planar shear test study, *Dent. Mater., 6*, 217–222 (1990).

242. Susz, C.P., J.M. Meyer, and P.F. Orosz, Thermal expansion of precious metal alloys for ceramic-metal use, *Metallurgie Dentaire*, 1981, pp. 124–134.

243. Klimonda, P., O. Lingstuyl, B. Lavelle, and F. Dabosi, The use of a flame-sprayed undercoat to improve the adherence of SiO_2–Al_2O_3 dental ceramics on Ni–Cr and Co–Cr alloys, in *Proc. Conf. Surfaces and Interfaces in Ceramic and Ceramic-Metal Systems*, Plenum Press, 1981, pp. 477–486.

244. Chambless, L.A., J.L. Long, and J.H. Hembree, Leakage of various types of alloys at the porcelain–metal interface, *Int. J. Pros., 1*, 47–50 (1988).

245. Mezger, P.R., M.M.A. Vrijhoef, and E.H. Greener, The corrosion behavior of palladium–silver-ceramic alloys, *Dent. Mater., 5*, 97–100 (1989).

246. Fioravanti, K.J., and R.M. German, Corrosion and tarnishing characteristics of low gold content dental casting alloys, *Gold Bull., 21*, 99–110 (1988).

247. Van Roekel, N., Electrical discharge machining in dentistry, *Int. J. Prosthodont., 5*, 114–121 (1992).

248. Kusamichi, H., ed., *Metallic Titanium and Its Application*, Nikkan-Kogyo, 1983, p. 51.

249. Mitsuya, H., M. Okada, and I. Kato, Superplastic forming of titanium alloy for dental use, *Titanium/Zirconium, 37*, 19–23 (1989).

250. Oshida, Y., R. Sachdeva, and S. Miyazaki, Changes in contact angles as a function of time on some pre-oxidized biomaterials, *J. Mater. Sci.: Mater. Med., 3*, 306–312 (1992).

251. Lausmaa, J., B. Kasemo, and H. Mattsson, Surface spectroscopic characterization of titanium implant materials, *Appl. Surf. Sci., 44*, 133–146 (1990).

252. Coddet, C., Measurements of the adhesion of thermal oxide films: Application to the oxidation of titanium, *J. Mater. Sci., 22*, 2969–2974 (1987).

253. Ravnholt, G., Corrosion current and pH rise around titanium coupled to dental alloys, *Scand. J. Dent. Res., 96*, 466–472 (1988).

254. Fraker, A.C., A.W. Ruff, P. Sung, A.C. Van Orden, and K.M. Speck, Surface preparation and corrosion behavior of titanium alloys for surgical implants, *ASTM STP, 796*, 206–219 (1983).

255. Branemark, P.I., B.O. Hansson, R. Adell, U. Breine, J. Lindstrom, O. Hallen, and A. Ohman, Osseointegrated implants in the treatment of the edentulous jaw. Experience from a 10-year period, *Scand. J. Plast. Reconstr. Surg., 16*, 100–104 (1977).

256. Branemark, P.I., R. Adell, T. Albrektsson, U. Lekholm, S. Lundkvist, and B. Rockler, Osseointegrated titanium fixtures in the treatment of edentulousness, *Biomaterials, 4*, 25–28 (1983).

257. Kasemo, B., Biocompatibility of titanium implants: Surface science aspects, *J. Prosth. Dent., 49*, 832–837 (1983).

258. Albreksson, T., P.I. Branemark, H.A. Hnasson, B. Krasemo, K. Larsson, and I.I. Lunstrom, The interface zone of inorganic implants *in vivo*: Titanium implants in bone, *Ann. Biomed. Eng., 11*, 1–27 (1983).

259. Screoder, A., van der Zypen, H. Stitch, and F. Sutter, The reactions of bone, connective tissue, and epithelium to endosteal implants with titanium-sprayed surfaces, *J. Maxillofac. Surg., 9*, 15–25 (1981).

260. Lausmaa, J., P. Uvdal, and E. Ordell, *Gothenburg Institute of Physics Report No. GIPR-245*, 1983.

261. Baro, A.M., N. Garcia, R. Miranda, L. Vazquez, C. Aparicio, J. Olive, and J. Lausmaa, Characterization of surface roughness in titanium dental implants measured with scanning tunneling microscopy at atmospheric pressure, *Biomaterials, 7*, 463–466 (1986).

262. Sundgren, J.E., P. Bodo, and I. Lunstrom, Auger electron spectroscopic studies of the interface between human tissue and implants of titanium and stainless steel, *J. Colloid. Interface Sci., 110*, 9–20 (1986).

263. de Lange, G.L., and K. Donath, Interface between bone tissue and implants of solid hydroxyapatite or hydroxyapatite-coated titanium implants, *Biomaterials, 10*, 121–125 (1989).

264. Ekstrand, K., I.E. Ruyter, and H. Oysaed, Adhesion to titanium of methacrylate-based polymer materials, *Dent. Mater., 4*, 111–115 (1988).

265. Black, J., Systemic effects of biomaterials, *Biomaterials, 5*, 11–18 (1984).

266. Meachim, G., and D.F. Williams, Changes in nonosseous tissue adjacent to titanium implants, *J. Biomed. Mater. Res., 7*, 555–571 (1973).

267. Maurer, A.M., K. Merritt, and S.A. Brown, Cellular uptake of titanium and vanadium from addition of salts or fretting corrosion *in vitro*, *J. Biomed. Mater. Res., 28*, 241–246 (1994).

268. Johnsson, C., J. Lausmaa, M. Ask, H.A. Hansson, and T. Alberktsson, Ultrastructural differences of the interface zone between bone and Ti-6Al-4V or commercially pure titanium, *J. Biomed. Eng., 11*, 3-8 (1989).

269. Wisbey, A., P.J. Gregson, L.M. Peter, and M. Tuke, Effect of surface treatment on the dissolution of titanium-based implant materials, *Biomaterials, 12*, 470-473 (1991).

270. Jarcho, M., J.F. Kay, K.I. Gunmaer, R.J. Doremus, and H.P. Drobeck, Tissue, cellular and subcellular events at a bone-ceramic hydroxyapatite interface, *J. Bioeng., 1*, 79-92 (1977).

271. Denissen, H.W., K. deGroot, O. Ch. Makkes, A. van den Hooff and P.J. Klopper, Tissue response to dense apatite implants in rats, *J. Biomed. Mater. Res., 14*, 713-721 (1980).

272. Ducheyne, P., L.L. Hench, A. Kagan, M. Martens, A. Burssens, and J.C. Mulier, The effect of hydroxyapatite impregnation on skeletal bonding of porous coated implants, *J. Biomed. Mater. Res., 14*, 225-237 (1980).

273. Eschenroeder, H.C., R.E. McLaughlin, and S.I. Reger, Enhanced stabilization of porous coated metal implants with tricalcium phosphate granules, *Clin. Orthop. Rel. Res., 216*, 234 (1987).

274. Cook, S.D., K.A. Thomas, and J.F. Jarcho, Hydroxyapatite-coated porous titanium for use as an orthopaedic biologic attachment system, *Clin. Orthop. Rel. Res., 230*, 303 (1988).

275. Kim, C.S., and P. Ducheyne, Compositional variations in the surface and interface of calcium phosphate ceramic coatings on Ti and Ti-6Al-4V due to sintering and immersion, *Biomaterials, 12*, 461-469 (1991).

276. Clemon, A.J.T., A.M. Weinstein, J.J. Klawitter, and J. Anderson, Interface mechanics of porous implants, *J. Biomed. Mater. Res., 15*, 78-82 (1981).

277. Wheeler, K.R., M.T. Karagianes, and K.R. Sump, Porous titanium alloy for prosthesis attachment, *ASTM Conf Titanium Alloys in Surgical Implants*, 1983, pp. 241-254.

278. Keller, J.C., F.A. Young, and B. Hansel, Systematic effects of porous Ti dental implants, *Dent. Mater., 1*, 41-42 (1985).

279. Young, F.A., and J.C. Keller, Porous titanium dental implants in primates and humans, *Eng. Med., 13*, 203-206 (1984).

280. Buchanan, R.A., E.D. Rigney, and C.D. Griffin, Biocorrosion studies of ultralow-temperature-isotropic carbon-coated porous titanium, *ASTM STP, 953*, 105-114 (1987).

281. Miura, I., Titanium alloys for biomaterials, especially dental materials, *Titanium Zirconium, 38*, 17-18 (1990).

282. Mitsuya, H., M. Okada, and I. Kato, Superplastic forming titanium alloy for dental use, *Titanium Zirconium, 37*, 19-23 (1989).

283. Backman, J.P., and C.F. Yolton, Beta III (Ti-11.5Mo-6Zr-4.5Sn), *Proc. Conf. on Beta Titanium in the 1980s*, 1984, pp. 401-408.

284. Goldberg, J., and C.J. Burstone, An evaluation of beta titanium alloys for use in orthodontic appliances, *J. Dent. Res., 58*, 593-599 (1979).

285. Shastry, C.V., and A.J. Goldberg, The influence of drawing parameters on the mechanical properties of two beta-titanium alloys, *J. Dent. Res., 62*, 1092-1097 (1983).

286. Wilson, D.F., and A.J. Goldberg, Alternative beta-titanium alloys for orthodontic wires, *Dent. Mater., 3*, 337-341 (1987).

287. Kusy, R.P., and T.W. Wilson, Dynamic mechanical properties of straight titanium alloy arch wires, *Dent. Mater., 6*, 228-236 (1990).

288. Andreasen, G.F., and R.E. Morrow, Laboratory and clinical analyses on Nitinol wires, *Am. J. Orthod., 73*, 142-151 (1978).

289. Burstone, C.J., and A.J. Goldberg, Beta titanium: A new orthodontic alloy, *Am. J. Orthod., 77*, 121-132 (1988).

290. Sachdeva, R., and S. Miyazaki, Application of shape memory nickel-titanium alloys to orthodontics, *Proc. Int. Meeting Advanced Materials*, 1988, pp. 125-129.

291. Drake, S.R., D.M. Wayne, J.M. Powers, and K. Asgar, Mechanical properties of orthodontic wires in tension, bending, and torsion, *Am. J. Orthod., 82*, 206-210 (1982).

292. Larson, B.E., Torsional elastic property measurements of selected orthodontic archwires, *U.S. Air Force Inst. Technol.*, 1987, p. 93.

293. Miura, F., The super-elastic property of the Japanese NiTi alloy wire for use in orthodontics, *Am. J. Orthod., 90*, 1-10 (1986).

294. Shomura, T., and H. Kimura, Shape recovery in Ti-V-Fe-Al alloy and its application to dental implant, *Proc. Int. Conf. Martensitic Transformations ICOMAT-86*, 1987, pp. 1065-1070.

295. Speck, K.M., and A.C. Fraker, Anodic polarization behavior of Ti-Ni and Ti-6Al-4V in simulated physiological solutions, *J. Dent. Res., 59*, 1590-1595 (1980).

296. Choi, J., H. Yoshizawa, K. Suzuki, T. Takamata, and S. Fukuyo, The histological observation of the biocompatibility of titanium and nickel-titanium alloy, *Matsumoto Shigaku, 17*, 189-200 (1991).

297. Ishibe, H., NiTi system shape memory alloy manufactured by a diffusion porcess, *Special Steel, 38*, 46-48 (1989).

298. Streicher, R.M., H. Weber, R. Schon, and M. Semlitsch, New surface modification for Ti-6Al-7Nb alloy: Oxygen diffusion hardening (ODH), *Biomaterials, 12*, 125-129 (1991).

299. Keller, J.C., F.A. Young, C.F. Marcinak, and B. Hansel, Preliminary studies of the histopathological responses to Ti-13% Cu casting alloys, *Biomaterials, 6*, 252-256 (1985).

300. Okuno, K., Can titanium replace gold as dental metal? *Met. Technol., 52*, 6-7 (1982).

301. Ida, K., Titanium and its alloys as biomedical and especially dental materials, *Titanium Zirconium, 34*, 145-150 (1986).

302. Ita, I., Dental casting technique of titanium, *Met. Technol., 55*, 4-9 (1985).

303. Toloui, B., C. Gagg, and J.C. Wood, Some environmental and mechanical properties of various semi-solid cast titanium alloys, *Sci. Technol., 4*, 2527-2532 (1985).

304. Nakamura, S., Ohara's titanium castings for dental use, *Titanium Zirconium, 34*, 154-156 (1986).

305. Ida, K., Casting of titanium for biomedical use, especially on dental castings, *Met. Technol., 58*, 14-15 (1988).

306. Nakamura, S., Casting of titanium for dental use, *Met. Technol., 58*, 16-19 (1988).

307. Taira, M., J.B. Moser, and E.H. Greener, Studies of titanium alloys for dental castings, *Dent. Mater., 5*, 45-50 (1989).

308. Ott, D., Investment casting of titanium in dental laboratory, *Metall; International Zeitschrift für Technik und Wirtschaft, 44*, 366-369 (1990).

309. Hruska, A.R., A novel method for vacuum casting titanium, *Int. J. Pros., 3*, 142-145 (1990).

310. Yamagishi, T., M. Ito, and E. Masuhara, Laser-welding of titanium and other dental alloys, *Shikazairyo-Kiki, 6*, 763-772 (1991).

311. Phillips, R.W., *The Science of Dental Materials*, 9th ed., W.B. Saunders, Philadelphia, 1991, pp. 177-212.

312. Jagger, R.G., and R. Huggett, The effects of cross-linking on sorption properties of a denture-base material, *Dent. Mater., 6*, 276-278 (1990).

313. Craig, R.G., Properties of a new polyether urethane dimethacrylate photoinitiated elastomeric impression material, *J. Prosthet. Dent., 63*, 16-20 (1990).

314. Cook, W.D., Fracture and structure of highly crosslinked polymer composites, *J. Appl. Polym. Sci., 42*, 1259-1269 (1991).

315. Compton, F.H., G.S. Beagrie, and R. Chernecky, Strength determination of periodental splints fabricated from acid-etched retained materials, *J. Periodontal., 48*, 418-420 (1977).

316. Huggett, R., J.F. Bates, and D.E. Packham, The effect of the curing cycle upon the molecular weight and properties of denture base materials, *Dent. Mater., 3*, 107-112 (1987).

317. Jones, D.W., E.J. Sutow, B.S. Graham, E.L. Milne, and D.E. Johnston, Influence of plasticizer on soft polymer gelation, *J. Dent. Res., 20*, 209-216 (1986).

318. Lee, H.L., A.L. Cupples, R.J. Schubert, and M.L. Swartz, An adhesive dental restorative material, *J. Dent. Res., 50*, 125-132 (1971).

319. Patel, M.P., and M. Braden, Heterocyclic methacrylates for clinical applications. II. Room temperature polymerizing systems for potential clinical use, *Biomaterials, 12,* 649-652 (1991).
320. Patel, M.P., and M. Braden, Heterocyclic methacrylates for clinical applications. III. Water absorption characteristics, *Biomaterials, 12,* 653-657 (1991).
321. Chitchumnong, P., S.C. Brooks, and G.D. Stafford, Comparison of three- and four-point flexural strength testing of denture-base polymers, *Dent. Mater., 5,* 2-5 (1989).
322. Hill, R.G., J.F. Bates, T.T. Lewis, and N. Rees, Fracture toughness of acrylic denture base, *Biomaterials, 4,* 112-120 (1983).
323. Patel, M.P., and M. Braden, Heterocyclic methacrylates for clinical applications. I. Mechanical properties, *Biomaterials, 12,* 645-648.(1991).
324. Stafford, G.D., The use of nylon as a denture-base material, *J. Dent., 14,* 18-22 (1986).
325. Berrong, J.M., R.M. Weed, and J.M. Young, Fracture resistance of Kevlar-reinforced poly (methyl methacrylate) resin: A preliminary study, *Int. J. Prosthod., 3,* 391-395 (1990).
326. Grave, A.M.H., H.D. Chandler, and J.F. Wolfaardt, Denture base acrylic reinforced with high modulus fibre, *Dent. Mater., 1,* 185-187 (1985).
327. Wenz, L.M., K. Merritt, S.A. Brown, A. Moet, and K. Steffee, *In vitro* biocompatibility of polyetheretherketone and polysulfone composites, *J. Biomed. Mater. Res., 24,* 207-215 (1990).
328. Hartness, J.T., An evaluation of polyetheretherketone matrix composites fabricated from unidirectional prepreg tape, *SAMPE J.,* 1984, pp. 26-35.
329. Morris, D.R., and R.W. Elliott, Effect of ultrasonic cleaning upon stability of resin denture bases, *J. Prosth. Dent., 27,* 16-20 (1972).
330. McCartney, J.W., Gold occlusal surfaces for acrylic resin denture teeth, *J. Prosthet. Dent., 41,* 582-585 (1979).
331. Budtz-Jorgensen, E., The significance of *Candida albicans* in denture stomatitis, *Scand. J. Dent. Res., 82,* 151-190 (1974).
332. Weaver, R.E., and W.M. Goebel, Reactions to acrylic resin dental prostheses, *J. Prosthet. Dent., 43,* 138-142 (1980).
333. Stafford, G.D., and S.C. Brooks, The loss of residual monomer from acrylic orthodontic resins, *Dent. Mater., 1,* 135-138 (1985).
334. Koda, T., H. Tsuchida, M. Yamauchi, Y. Hoshino, N. Takagi, and J. Kawano, High-performance liquid chromatographic analysis of residual monomer eluted from dental acrylic resin, *J. Gifu. Dent. Soc., 14,* 356-363 (1987).
335. Koda, T., H. Tsuchiya, M. Yamauchi, S. Ohtani, N. Takagi, and J. Kawano, Leachability of denture-base acrylic resins in artificial saliva, *Dent. Mater., 6,* 13-16 (1990).
336. Rezende, J., A radiopaque internal prosthetic acrylic resin, *Oral Surg., 36,* 599-602 (1973).
337. Jacobson, T.H.E., The significance of adhesive denture base resin, *Int. J. Prosthod., 2,* 163-172 (1989).
338. Tanaka, T., M. Hirano, M. Kawahara, H. Matsumura, and M. Atsuta, A new ion-coating surface treatment of alloys for dental adhesive resins, *J. Dent. Res., 67,* 1376-1380 (1988).
339. Tiffany, R.L., Effects of different surface treatments on the tensile bond strength of polymethyl metacrylate processed against chemically etched ticonium 100, *Air Force Inst. Technol.,* AD-A196 323/XAB, 1987, p. 74.
340. LaBarre, E.E., and H.E. Ward, An alternative resin-bonded restorations, *J. Prosthet. Dent., 52,* 247-249 (1984).
341. Brantley, C.F., B.E. Kanoy, and J.R. Sturdevant, Thermal effects on retention of resin-bonded retainers, *Dent. Mater., 2,* 67-71 (1986).
342. Shlosberg, S.R., C.J. Goodacre, C.A. Munoz, B.K. Moore, and R.J. Schnell, Microwave energy polymerization of poly(methyl methacrylate) denture base resin, *Int. J. Prosthod., 2,* 453-458 (1989).

343. Al Doori, D., R. Huggett, J.F. Bates, and S.C. Brooks, A comparison of denture base acrylic resins polymerized by microwave irradiation and by conventional water bath curing system, *Dent. Mater., 6*, 25–32 (1991).

344. Masutani, S., J.C. Setcos, R.J. Schnell, and R.W. Phillips, Temperature rise during polymerization of visible light-activated composite resins, *Dent. Mater., 4*, 174–178 (1988).

345. McCabe, J.F., Cure performance of light-activated composites by differential thermal analysis, *Dent. Mater., 1*, 3136 (1986).

346. Chee, W.W.L., T.H.E. Donovan, F. Daftary, and T.M. Siu, The effect of vacuum-mixed autopolymerizing acrylic resins on porosity and transverse strength, *J. Prosthetic Dent., 60*, 517–519 (1988).

347. Curtis, D.A., T.L. Eggleston, S.J. Marshall, and L.C. Watanabe, Shear bond strength of visible-light-cured resin relative to heat-cured resin, *Dent. Mater., 5*, 314–318 (1989).

348. Montes, G.M., and R.A. Daughan, *In vitro* surface degradation of composites by water and thermal cycling, *Dent. Mater., 2*, 193–197 (1986).

349. Al-Mulla, M.A.S., Some physical and mechanical properties of a visible light-activated material, *Dent. Mater., 4*, 197–200 (1988).

350. Hayakawa, I., M. Nagao, T. Matsumoto, and E. Masuhara, Properties of a new light-polymerized relining material, *Int. J. Prosthod., 3*, 278–284 (1990).

351. Lewis, E.A., R.E. Ogle, and S.E. Sorensen, Orthodontic applications of new visible light curing (VLC) resin system, *NY State Dent. J., 52*, 32–34 (1986).

352. Lewis, E.A., R.E. Ogle, S.E. Sorensen, and D.A. Zysik, Clinical and laboratory evaluation of visible light-cured denture base resins and their application to orthodontics, *Am. J. Orthod. Dentofac. Orthop., 94*, 207–215 (1988).

353. Suchatlampong, C., E. Davies, and J.A. von Fraunhofer, Frictional characteristics of resilient lining materials, *Dent. Mater., 2*, 135–138 (1986).

354. Mutlu, G., R. Huggett, and A. Harrison, Rheology of acrylic denture-base polymers, *Dent. Mater., 6*, 288–293 (1990).

355. Burns, D.R., A. Kazanoglu, R.C. Moon, and J.C. Gunsolley, Dimensional stability of acrylic resin materials after microwave sterilization, *Int. J. Prosthod., 3*, 489–493 (1990).

356. Salvo, C.A., Composite resin used with an etched chrome–cobalt bar, *Compend. Contin. Educ. Dent., 9*, 696–700 (1988).

357. Schneider, P.M., L.B. Messer, and W.H. Douglas, The effect of enamel surface reduction *in vitro* on the bonding of composite resin to permanent human enamel, *J. Dent. Res., 60*, 895–900 (1981).

358. Davidson, C.L., and A.J. de Gee, Relaxation of polymerization contraction stresses by flow in dental composites, *J. Dent. Res., 63*, 146–148 (1984).

359. Lee, H.L., M.L. Swartz, and F.F. Smith, Physical properties of four thermosetting dental restorative resins, *J. Dent. Res., 48*, 526–535 (1969).

360. Ferracane, J.L., Relationship between shade and depth of cure for light-activated dental composite resins, *Dent. Mater., 2*, 80–84 (1986).

361. Asmussen, E., Penetration of restorative resins into acid etched enamel. I. Viscosity, surface tension and contact angle of restorative resin monomers, *Acta Odontol. Scand., 35*, 175–182 (1977).

362. Asmussen, E., Penetration of restorative resins into acid etched enamel. II. Dissolution of entrapped air in restorative resin monomers, *Acta Odontol. Scand., 35*, 183–191 (1977).

363. Watts, D.C., O.M. Amer, and E.C. Combe, Surface hardness development in light-cured composites, *Dent. Mater., 3*, 265–269 (1987).

364. Eliades, G.C., G.J. Vougiouklakis, and A.A. Caputo, Degree of double bond conversion in light-cured composites, *Dent. Mater., 3*, 19–25 (1987).

365. Ferracane, J.L., H. Matsumoto, and T. Okabe, Time-dependent deformation of composite resins—compositional considerations, *J. Dent. Res., 64*, 1332–1336 (1985).

366. Kalachandra, S., Influence of fillers on the water sorption of composites, *Dent. Mater., 5,* 283–288 (1989).

367. Heymann, H.O., V.B. Haywood, S.B. Hndreaus, and S.C. Bayne, Bonding agent strengths with processed composite resin veneers, *Dent. Mater., 3,* 121–124 (1987).

368. Braem, M., W. Finger, V.E. Van Doren, P. Lambrechts, and G. Vanherle, Mechanical properties and filler fraction of dental composites, *Dent. Mater., 5,* 346–349 (1989).

369. Ruyter, I.E., K. Nilner, and B. Moeller, Color stability of dental composite resin materials for crown and bridge veneers, *Dent. Mater., 3,* 246–251 (1987).

370. Bowen, R.L., Composite and sealant resins – Past, present, and future, *Pediatr. Dent., 4,* 10–15 (1982).

371. Mabie, C.P., and D.L. Menis, Microporous glassy fillers for dental composites, *J. Biomed. Mater. Res., 12,* 435–472 (1978).

372. Konishi, R.N., and D.T. Turner, Deformation and fracture of dentinal adhesive resin, *Dent. Mater., 1,* 43–47 (1985).

373. Harrison, A., and R.A. Draughn, Abrasive wear, tensile strength, and hardness of dental composite resins – Is there a relationship? *J. Prosth. Dent., 36,* 395–398 (1976).

374. de Gee, A.J., H.C. Harkel-Hagenaar, and C.L. Davidson, Structural and physical factors affecting the brush-wear of composites, *J. Dent., 13,* 60–70 (1985).

375. McCabe, J.F., and A.R. Ogden, The relationship between porosity, compressive fatigue limit and wear in composite resin restorative materials, *Dent. Mater., 3,* 9–12 (1987).

376. Finger, W., and J. Thiemann, Correlation between *in vitro* and *in vivo* wear of posterior restorative materials, *Dent. Mater., 3,* 280–286 (1987).

377. Crumpler, D.C., S.C. Bayne, S. Sockwell, D. Brunson, and T.M. Roberson, Bonding to resurfaced posterior composites, *Dent. Mater., 5,* 417–424 (1989).

378. Draughn, R.A., Compressive fatigue limits of composite restorative materials, *J. Dent. Res., 58,* 1093–1096 (1979).

379. Kandil, S.H., A.A. Kamar, S.A. Shaaban, N.M. Taymour, and S.E. Morsi, Effect of temperature and aging on the mechanical properties of dental polymeric composite materials, *Biomaterials, 10,* 540–544 (1989).

380. Berge, M., Creep of resin veneer materials, *Dent. Mater., 3,* 158–162 (1987).

381. Laurell, L., and D. Lundgren, A standardized programme for studying the occlusal force pattern during chewing and biting in prosthetically restored dentitions, *J. Oral Rehabil., 11,* 39–44 (1984).

382. Oden, A., I.E. Ruyter, and H. Oysaed, Creep and recovery of composites for use in posterior teeth during static and dynamic compression, *Dent. Mater., 4,* 147–150 (1988).

383. Onose, H., H. Sano, H. Kanto, S. Ando, and T. Hasuike, Selected curing characteristics of light-activated composite resins, *Dent. Mater., 1,* 48–54 (1985).

384. Kim, K.H., J.H. Park, Y. Imai, and T. Kishi, Fracture behavior of dental composite resins, *Bio-Med. Mater. Eng., 1,* 45–57 (1991).

385. Truong, V.T., and M.J. Tyas, Prediction of *in vivo* wear in posterior composite resins: A fracture mechanics approach, *Dent. Mater., 4,* 318–327 (1988).

386. Mair, L.H., and R. Vowles, The effect of thermal cycling on the fracture toughness of seven composite restorative materials, *Dent. Mater., 5,* 23–26 (1989).

387. McConnel, R.J., C. Moriarty, and D. Taylor, Fracture strength of resin/metal bonds, *Key Eng. Mater., 32,* 1–7 (1989).

388. Boyer, D.B., Y. Chalkley, and K.C. Chan, Correlation between strength of bonding to enamel and mechanical properties of dental composites, *J. Biomed. Mat. Res., 16,* 775–783 (1982).

389. Zidan, O., E. Asmussen, and K.D. Jorgensen, Microscopical analysis of fractured restorative resin/etched enamel bonds, *Scand. J. Dent. Res., 90,* 286–291 (1982).

390. Watts, D.C., C.M. Haywood, and R.T. Smith, Thermal diffusion through composite restorative materials, *Br. Dent. J., 154,* 101–103 (1983).

391. Feilzer, A.J., A.J. DeGee, and C.L. Davidson, Quantitative determination of stress reduction by flow in composite restorations, *Dent. Mater., 6,* 167–171 (1990).

392. DeGee, A.J., P. Pallav, A. Werner, and C.L. Davidson, Annealing as a mechanism of increasing wear resistance of composites, *Dent. Mater., 6,* 266–270 (1990).

393. Mair, L.H., Surface permeability and degradation of dental composites resulting from oral temperature changes, *Dent. Mater., 5,* 247–259 (1989).

394. Wendt, S.L., Microleakage and cusp fracture resistance of heat-treated composite resin inlays, *Am. J. Dent., 4,* 10–14 (1991).

395. Lutz, F., I. Krejci, and F. Barbakow, Quality and durability of marginal adaptation in bonded composite restorations, *Dent. Mater., 7,* 107–113 (1991).

396. Eakle, W.S., Effects of thermal cycling on fracture strength and microleakage in teeth restored with a bonded composite resin, *Dent. Mater., 2,* 114–117 (1986).

397. Brauer, G.M., Color changes of composites on exposure to various energy sources, *Dent. Mater., 4,* 55–59 (1988).

398. O'Brien, W.J., *Dental Materials: Properties and Selection,* Quintessence Publ., Chicago, 1989.

399. Craig, R.G., *Restorative Dental Materials,* 9th ed., Mosby, St. Louis, 1993.

400. Phillips, R.W., *Science of Dental Materials,* 9th ed., W.B. Saunders, Philadelphia, 1991.

401. McLean, J.W., A.D. Wilson, and H.J. Prosser, Development and use of water-hardening glass-ionomer luting cements, *J. Prosthet. Dent., 52,* 175–181 (1984).

402. Kent, B.E., and A.D. Wilson, Dental silicate cements: XV. Effects of particle size of the powder, *J. Dent. Res., 50,* 1616–1620 (1971).

403. Shepherd, W.B., B. Walter, K.F. Leinfelder, and H.G. Hershey, The effect of mixing method, slab temperature, and humidity on the properties of zinc phosphate and zinc silicophosphate cement, *Angle Orthod., 48,* 219–226 (1978).

404. Lawrence, L.G., and D.C. Smith, Strength modification of polycarboxylate cements with fillers, *J. Can. Dent. Assn., 39,* 405–409 (1973).

405. Richter, W.A., J.C. Mitchum, and J.D. Brown, Predictability of retentive values of dental cements, *J. Prosth. Dent., 24,* 298–303 (1970).

406. Arfaei, A.H., and K. Asgar, Bond strength of three cements determined by centrifugal testing, *J. Prosthet. Dent., 40,* 294–298 (1978).

407. Simmons, J.J., The miracle mixture: glass ionomer and alloy powder, *Tex. Dent. J., 100,* 6–12 (1983).

408. Mortimer, K.V., and T.C. Tranter, A preliminary laboratory evaluation of polycarboxylate cements, *Br. Dent. J., 127,* 365–370 (1969).

409. Drummond, J.L., J.W. Lenke, and R.G. Randolph, Compressive strength comparison and crystal morphology of dental cements, *Dent. Mater., 4,* 38–40 (1988).

410. Low, T., E.H. Davies, and J.A. van Fraunhofer, On the adhesion of cementing media to stainless steel, *Br. Dent. Orthod., 2,* 179–181 (1975).

411. Cook, W.D., Dental polyelectrolyte cements: II. Effect of powder/liquid ratio on their rheology, *Biomaterials, 4,* 21–24 (1983).

412. Mizrahi, E., and D.C. Smith, Direct attachment of orthodontic brackets to dental enamel: A preliminary clinical report, *Br. Dent. J., 130,* 392–396 (1971).

413. Clark, R.J., R.W. Phillips, and R.D. Norman, An evaluation of silicophosphate as an orthodontic cement, *Am. J. Orthod., 71,* 190–196 (1977).

414. Phillips, R.W., M.L. Swartz, and B. Rhodes, An evaluation of a carboxylate adhesive cement, *JADA, 81,* 1353–1359 (1970).

415. Cook, W.D., Dental polyelectrolyte cements. I. Chemistry of the early stages of the setting reaction, *Biomaterials, 3,* 232–236 (1982).

416. Barghi, N., and E.W. Simmons, The effect of thermocycling on tensile strength of two types of fixed prosthodontic cements, *Tex. Dent. J., 97,* 12–14 (1979).

417. Wilson, A.D., and B.G. Lewis, The flow properties of dental cements, *J. Biomed. Mater. Res., 14,* 383–391 (1980).

418. Beech, D.R., A. Solomon, and R. Bernier, Bond strength of polycarboxylic acid cements to treated dentine, *Dent. Mater., 1*, 154–157 (1985).

419. McLean, J.W., Polycarboxylate cements: Five years' experience in general practice, *Br. Dent. J., 132*, 9–15 (1972).

420. Rogers, O.W., and J.R. Griffth, The bonding of polycarboxylate cement to gold, *Aust. Dent. J., 22*, 371–372 (1977).

421. Truelove, E.L., D.F. Mitchell, and R.W. Phillips, Biological evaluation of a carboxylate cement, *J. Dent. Res., 50*, 166–171 (1971).

422. Takenoshita, S., and T. Ishikawa, A clinico-pathological study on pulp reaction to polycarboxylate cement application, *Bull. Tokyo Dent. Coll., 10*, 119–131 (1978).

423. Prosser, H.J., and A.D. Wilson, Zinc oxide eugenol cements: VI. Effects of zinc oxide type on the setting reactions, *J. Biomed. Mater. Res., 16*, 585–598 (1982).

424. Batchelor, R.F., and A.D. Wilson, Zinc-oxide-eugenol cements: I. The effect of atmospheric condition on rheological properties, *J. Dent. Res., 48*, 883–887 (1969).

425. Wilson, A.D., and R.F. Batchelor, Zinc oxide-eugenol cements: II. Study of erosion and disintegration, *J. Dent. Res., 49*, 593–598 (1970).

426. Wilson, A.D., D.M. Groffman, D.R. Powis, and R.P. Scott, An evaluation of the significance of the impinging jet method for measuring the acid erosion of dental cements, *Biomaterials, 7*, 55–60 (1986).

427. Parker, W.A., and D.W. Hutchins, Improved zinc oxide-eugenol temporary filling material: Two year clinical evaluation, *J. S. Calif. Dent. Assn., 41*, 351–354 (1973).

428. Civjan, S., E.F. Huget, I.B. de Simon, and T.V. Rapheld, Further studies of resin-modified temporary restorative materials, *J. Dent. Res., 52*, 59–64 (1973).

429. Civjan, S., E.F. Huget, G. Wolfhard, and L.S. Waddell, Characterization of zinc oxide-eugenol cements reinforced with acrylic resin, *J. Dent. Res., 51*, 107–114 (1972).

430. Benefits of glass ionomers, *AGD Impact*, December 10, 1993.

431. Kawahara, H., Y. Imanishi, and H. Oshima, Biological evaluation on glass ionomer cement, *J. Dent. Res., 58*, 1080–1086 (1979).

432. Srinsawasdi, S., D.B. Boyer, and J.W. Reinhardt, The effect of removal of the smear layer on microleakage of class V restorations *in vitro, Dent. Mater., 4*, 384–389 (1988).

433. Welbury, R.R., Factors affecting the bond strength of composite resin to etched glass ionomer cement, *J. Dent., 16*, 188–193 (1988).

434. Mathis, R.S., and J.L. Ferracane, Properties of a glass-ionomer/resin-composite hybrid material, *Dent. Mater., 5*, 353–358 (1989).

435. Kuhn, A.T., and A.D. Wilson, The dissolution mechanisms of silicate and glass-ionomer dental cements, *Biomaterials, 6*, 378–382 (1985).

436. Wilson, A.D., Resin-modified glass-ionomer cements, *Int. J. Prosthod., 3*, 425–429 (1990).

437. Paragogiannis, Y., M. Helvatjoglou-Antoniadi, R.C. Lakes, and M. Sapoutjis, The creep behavior of glass-ionomer restorative materials, *Dent. Mater., 7*, 40–43 (1991).

438. Shalabi, H.S., E. Asmussen, and K.D. Jorgensen, Increased bonding of a glass onomer cement to dentin by means of $FeCl_3$, *Scand. J. Dent. Res., 89*, 348–353 (1981).

439. Hotz, P., J.W. McLean, I. Sced, and A.D. Wilson, The bonding of glass ionomer cements to metal and tooth substrates, *Br. Dent. J., 142*, 41–47 (1977).

440. Mizrahi, E., Glass ionomer cements in orthodontics — an update, *Am. J. Orthod. Dentofacial Orthop., 93*, 505–507 (1988).

441. Forss, H., L. Seppa, and R. Lappalainen, *In vitro* abrasion resistance and hardness of glass-ionomer cements, *Dent. Mater., 7*, 36–39 (1991).

442. Pearson, G.J., and A.S. Atkinson, Long-term flexural strength of glass ionomer cements, *Biomaterials, 12*, 658–660 (1991).

443. Levine, R.S., D.R. Beech, and B. Garton, Improving the bond strength of polyacrylate cements to dentine: A rapid technique, *Br. Dent. J., 143*, 275–277 (1977).

444. Wesenberg, G., and E. Hals, The *in vitro* effect of glass inomer cement on dentine and enamel walls: An electron and microradiographic study, *J. Oral Rehabil., 7*, 35–42 (1980).

445. Hicks, M.J., C.M. Flaitz, and L.M. Silverstone, Secondary caries formation *in vitro* around glass ionomer restorations, *Quintessence Int., 17*, 527–532 (1986).

446. Tam, L.E., E. Pulver, D. McComb, and D.C. Smith, Physical properties of calcium hydroxide and glass-ionomer base and lining materials, *Dent. Mater., 5*, 145–149 (1989).

447. Powis, D.R., H.J. Prossier, and A.D. Wilson, Long-term monitoring of microleakage of dental cements by radiochemical diffusion, *J. Prosthet. Dent., 59*, 651–656 (1988).

448. Dahl, B.L., and G. Oilo, Retentive properties of luting cements: An *in vitro* investigation, *Dent. Mater., 2*, 17–20 (1986).

449. Powers, J.M., Adhesive resin cements, *Shigaku, 79*, 1140–1145 (1991).

450. Atta, M.O., D. Brown, and B.G.N. Smith, Bond strength of contemporary bridge cements to a sand-blasted or electrically etched nickel–chrominum alloy, *Dent. Mater., 4*, 201–207 (1988).

451. Drummond, J.L., and M.A. Khalaf, Shear strength and filler particle characterization of Maryland (acid etch) bridge resin cements, *Dent. Mater., 5*, 209–212 (1989).

452. Pluim, L.J., and J. Arends, The relation between salivary properties and *in vivo* solubility of dental cements, *Dent. Mater., 3*, 13–18 (1987).

453. Kozam, G., and G.M. Mantell, The effect of eugenol on oral mucous membranes, *J. Dent. Res., 57*, 954–957 (1957).

454. Haugen, E., and A.J. Mjor, Bone tissue reactions to periodontal dressings, *J. Periodont. Res., 14*, 76–85 (1979).

455. Koch, G., B. Magnusson, N. Nobreus, G. Nyquist, and M. Soderholm, Contact allergy to medicaments and materials used in dentistry (IV), *Odontologisk. Revy, 24*, 109–114 (1973).

456. Jendresen, M.D., and H.O. Trobridge, Biologic and physical properties of zinc polycarboxylate cement, *J. Prosth. Dent., 28*, 264–271 (1972).

457. Meryon, S.D., and K.J. Jakeman, An *in vitro* study of the role of dentine in moderating the cytotoxicity of zinc oxide eugenol cement, *Biomaterials, 7*, 459–462 (1986).

458. Mueller, J., W. Hoerz, D. Bruckner, and E. Kraft, An experimental study on the biocompatibility of lining cements based on glass ionomer as compared with calcium hydroxide, *Dent. Mater., 6*, 35–40 (1990).

459. Hendry, J.A., B.G. Jeansonne, C.O. Dummett, and W. Burrell, Comparison of calcium hydroxide and zinc oxide and eugenol pulsectomies in primary teeth of dogs, *Oral Surg., 54*, 445–451 (1982).

460. Yakushiji, M., T. Kinumatsu, T. Fuchino, and Y. Machida, Effects of glass ionomer cement on the dental pulp and its efficacy as a base material, *Bull. Tokyo Dent. Coll., 20*, 47–59 (1979).

461. Browne, R.M., R.S. Tibias, S.J. Wilson, and M.J. Tyas, An *in vitro* study of zinc in dentine beneath cavities filled with two dental cements, *Biomaterials, 6*, 41–44 (1985).

462. Beagrie, G.S., Pulp irritation and silicate cement, *Dent. J., 45*, 67–70 (1979).

463. Heys, R.J., and M. Fitzgerald, Microleakage of three cement bases, *J. Dent. Res., 70*, 55–58 (1991).

464. DeSchopper, E.J., R.R. White, and W. von der Lehr, Antibacterial effects of glass ionomers, *Am. J. Dent., 2*, 51–56 (1989).

465. Charlton, D.G., B.K. Moore, and M.L. Swartz, Direct surface pH determination of setting cements, *Operat. Dent., 16*, 231–238 (1991).

466. O'Connon, R.P., A. Nayyar, and R.E. Kovarik, Effect of internal microblasting on retention of cemented cast crowns, *J. Prosth. Dent., 64*, 557–562 (1990).

467. Aboush, Y.E.Y., and R.J. Elderton, Bonding of a light-curing glass-ionomer cement to dental amalgam, *Dent. Mater., 7*, 130–132 (1991).

49
The *In Vitro* Testing of Cytotoxicity and Cell Adhesion of Biomaterials in the Dental Field

Masaaki Nakamura, Shoji Takeda, Koichi Imai, Hiroshi Oshima, Dai Kawahara, Yoshiya Hashimoto, and Shungo Miki
Osaka Dental University
Osaka, Japan

I. INTRODUCTION

Dental treatment contributes substantially to the quality of life. If the integrity of the oromaxillofacial region is damaged by disease, injury, congenital malformations, developmental anomalies, or aging, one's whole body will be affected. Moreover, the tissue restoration ability of the region is unfortunately low. Mastication and pronunciation skills, as well as aesthetics, must be preserved if an individual is to enjoy an active and healthy lifestyle throughout his or her life. As a result, the use of quality dental materials and devices is indispensable. This means that current materials and devices must be further improved.

II. NATURE OF BIOMATERIALS IN THE DENTAL FIELD

When their applicability is assessed, dental materials are treated in much the same way as medicine. However, they are actually different from medicine and most medical devices in the following three aspects: (1) the larger number and more varied types of constituent substances; (2) the various modes of their use; and (3) the use of a number of fabrication techniques such as mixing, melting, cooling, chemical activation, and polymerization. Medicine is largely dependent on the application of prefabricated end products. Moreover, dental materials are subject to a fair amount of occlusal force, which reaches a maximum of over 800 N [1], and they have to withstand such stress without breaking or becoming distorted.

Dental materials can be classified into five types on the basis of contact with body tissue: (1) those imbedded within hard tissue for a long time; (2) those contacting soft tissue for a long time; (3) those imbedded for a long time within tissue that maintains the blood system; (4) those contacting hard and/or soft tissue for a short time, followed by removal through washing or replacement; and (5) those not originally used in tissue but rather for fabrication or treatment, although accidentally contacting the skin or being inhaled or ingested. The materials in each group affect body tissue in different ways. Good understanding of these effects and subsequent selection of a proper testing method allows material behavior to be more properly evaluated.

Furthermore, there is yet another unique aspect of biocompatibility in dentistry. Most dental restorations, except for some types of dental implants, are custom-made. Following a dentist's diagnosis and subsequent design of restoration in a particular case, dental technicians fabricate metal crowns or complete dentures, or the dentist directly fills a prepared cavity with amalgam or composite resin. Various biological hazards may arise during fabrication, and these may create problems of occupational hygiene and safety. This contrasts sharply with the use of artificial organs such as artificial intraocular lenses, or artificial blood vessels, or artificial hip joints in the medical field, which are generally produced by manufacturers and distributed ready to use. Doctors and other medical personnel do not take part in the fabrication process in most cases. Their participation generally amounts to simply choosing the best fit. This means that a broader testing method is required to assess every possible effect of dental materials, both as end products and during the fabrication process.

III.   *IN VITRO* EVALUATION OF CYTOTOXICITY AND CELL ADHESION

The phenomena at the interface between the cells and the material must be properly expressed in an *in vitro* environment in order to evaluate a material's behavior. *In vitro* studies are preferable to animal experiments, in terms of reproducibility, quantification, speed, simplicity, and cost. Although we are still a long way from establishing even a partial alternative, the cell culture method represents a promising approach. First, the detrimental effects of materials during contact at the interface may be assessed by cell multiplication and function. Second, cytotoxicity *in vitro* can be evaluated by several methods. However, materials that are not cytotoxic are not necessarily biocompatible. Another aspect of quality material is its interfacial compatibility, that is, adhesion. Dental restoration requires good adhesion, as this is the key to its longevity.

There is a fundamental need to define the aim of *in vitro* testing, that is, to determine whether the study is being carried out to examine the cytotoxic mechanism of a material in depth or for a regulatory purpose. The former approach requires a complex, long-term method, and the cost of testing is limited only by the resources of the institution. On the other hand, the testing method applied for a regulatory purpose should be simple, so that it can efficiently characterize and quantify any possible detrimental elements, at a low cost. Long, complicated procedures result in increased costs and eventually become a burden to the manufacturer and eventually to the user.

IV.   *IN VITRO* EVALUATION OF CYTOTOXICITY

A.   Constituent Factors in Cytotoxicity Testing

The three major factors in cytotoxicity testing *in vitro* are the cells, the medium, and the materials. Each of these factors is discussed below.

1. Types of Cells

The types of cells used in a test are important, namely, whether they are of human or nonhuman origin; cells of matured tissues or those of fetal tissues; diploids or heteroploids; primary cells or established cell lines; epithelial cells or fibroblast cells; and so forth. Some of the most commonly used cells in dentistry are normal pulp and gingival cells, Gin-1, WI-38, NCTC 2544, HEp-2, HeLa S3, 3T3, L-929, Vero, BHK-21, CHO, and V-79. Using cells like those to which the material will be applied is considered an appropriate selection. However, higher reproduction can be obtained by using established cell lines [2]. Our findings also revealed that established cells were more sensitive in expressing the cytotoxicity levels of various dental materials (Fig. 1) [3–5]. Specific toxicity based on the possible effects on hepatocytes or kidney cells can be used. A broad selection of cells for cytotoxicity testing is the result of the recent development of cell culture techniques. The basic principle in the selection of cells is whether one is testing for acute toxicity or specific toxicity.

2. Medium and Serum

The medium has been the basic target of investigation with the aim of gaining a higher yield of cell growth. Studies on cytotoxicity and cell adhesion to biomaterials might not be successful until an effective cell culture is established. Nowadays, one can select media on the basis of higher cell growth or function.

Serum is an indispensable biological substance for successfully cultivating cells *in vitro*. However, it is also true that lot-to-lot variation and type of serum can lead to fluctuations in the *in vitro* data. Efforts have been made to obtain more stable cell culture results by using better-quality serum. At the same time, there have been occasions of interaction between known or unknown factors in the serum and materials. Such a phenomenon may either enhance or weaken the effects of materials *in vitro* and thus lead to different results. When extracting nickel powder from minimum essential medium (MEM) or MEM containing 0.4% albumin, the amount of nickel dissolved in MEM containing 0.4% albumin slightly increased, although the cytotoxicity decreased (Fig. 2) [6]. A serum-free culture could be valuable for analyzing the effects of the materials. The effect of resin monomers, which are widely used for dental restorations on cultured cells in serum-free environments, was also examined. The 50% inhibitory concentration of cell viability (IC_{50}) was 2 to 35 times lower in a serum-free condition (Fig. 3) [7]. Increasing the quantity of serum by 1%, 5%, and 10% resulted in decreases in the cytotoxicity level of the substance. Similar results were obtained when metal ions obtained from various metal salts were used (Fig. 4) [8]. These findings suggest that cytotoxicity tests that avoid the influence of serum are necessary for evaluating the cytotoxicity of the material itself and for determining the interaction between a substance and protein.

3. Materials

The third factor is the material itself. Materials range in form from solid to rubbery, pasty, and fluid. In many cases, they are used in an unset state and biological problems may arise in such a stage. Meanwhile, other materials are fabricated in a dental laboratory and brought to the clinic ready to use. When carrying out a cytotoxicity test *in vitro*, the mode of contact between the cells and the materials is the key to a better understanding of the behavior of the materials, as is discussed later. The volume and surface area of the materials are critical factors that are directly related to the results [9–12]. The results obtained in our laboratory show well the relative importance of each of these factors [9].

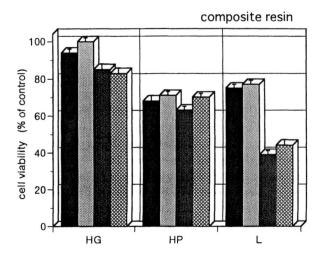

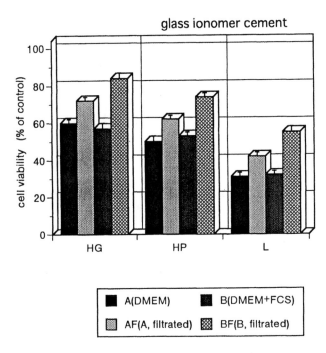

Figure 1   Effects of extracts of composite resin and glass ionomer cement on neutral red uptake of human gingival (HG), human dental pulp (HP), and L 929 cells (L) [3–5]. L 929 cells were more sensitive to the cytotoxic factor than the primary human cells.

One known cytotoxic material was tested under two experimental conditions. In experiment 1, the amount of material was fixed, but the surface area was increased. In experiment 2, on the other hand, the surface area of the material was fixed, but the amount was decreased (Fig. 5). In experiment 1, an increase in the surface area of the material resulted in a marked decrease of cell multiplication. However, in experiment 2 there was

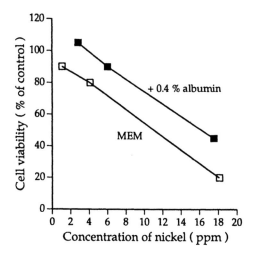

Figure 2   Cell viability against nickel concentration extracted in MEM and MEM containing 0.4% albumin [6]. Nickel extracted in MEM containing 0.4% albumin tended to decrease cytotoxicity.

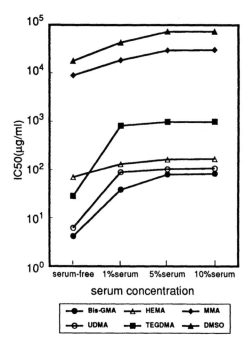

Figure 3   The 50% inhibitory concentration (IC_{50}) of five monomers [7]. Serum-free culture induced reduction in IC_{50}, but the magnitude of reduction was related to the kind of monomer. The IC_{50} of TEGDMA, Bis-GMA, and UDMA was greatly shifted to a low concentration under serum-free culture.

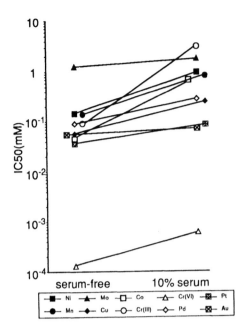

Figure 4   Comparison of the IC_{50} between serum-free and 10% serum containing medium [8]. The IC_{50} of metal ions was reduced in serum-free medium. The shift of IC_{50} to a low concentration was the largest for Cr(III) and the smallest for Au.

little change in cell multiplication (Fig. 6). In other words, a change in the surface area of the material rather than a change in the amount affected cytotoxicity [9–12]. This is a good example of the importance of the surface area of a material in cytotoxicity testing.

4.   Reference Materials

The latest testing methods require the use of reference or control materials (Biological reactivity tests, *in vitro, The United States Pharmacopeia* [13]; F813-83, The American Society for Testing and Materials [14]; ISO/TR 7405-1984 (E), International Organization for Standardization [15]). The rationale behind using a reference substance is to maintain the validity of a test. The selection of an appropriate substance for reference is crucial. Availability, handling, consistency, and safety are some of the important items that must be considered. Table 1 shows eight commercially available rubbers as well as two chemicals that were tested as reference materials. Three extraction methods—dynamic, static, and heat extraction—were used. The results of experiments using commercial rubbers showed that some of them could possibly be used as negative or positive control (Table 1, Figs. 7–9) [16,17]. However, the quality of any particular product may vary with the manufacturing lot. Moreover, the consistency of reference or control materials becomes a matter of concern, even after other requirements are fulfilled.

B.   Cell–Material Contact

The contact between cells and materials is important in all cytotoxicity tests. The types of contact can be roughly divided into three groups: extraction, direct, and indirect cell–

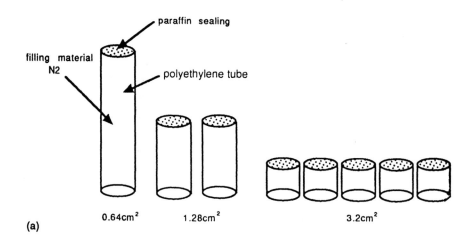

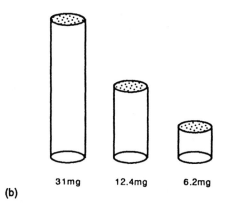

Figure 5 Experimental method for examination of the effects of (a) surface area and (b) volume on cytotoxicity [9].

material contact. Extracts can be obtained by either static or dynamic extraction. Direct cell–material contact can be achieved by placing a material directly on the cells. In the case of indirect cell–material contact, materials are separated from cells by permeable intermediates such as agars or Millipore filters. The three groups are:

1. Extraction
   i. Static extraction methods
   ii. Dynamic extraction methods
2. Direct cell–material contact
   i. Direct contact method

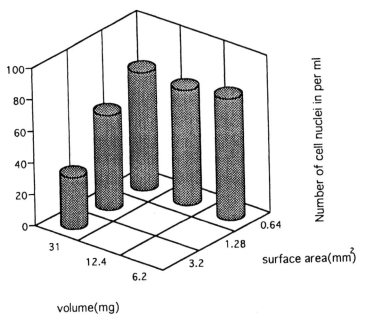

Figure 6 The effect of surface area and volume of the material on the cell multiplication [9]. Cell multiplication was little affected by the volume of filling material N2 but decreased as the surface area of filling material N2 increased.

3. Indirect cell–material contact
   i. Agar overlay method
   ii. Model cavity method

The agar overlay method has been widely used as a method of evaluating cytotoxicity. Using the model cavity method makes it possible to predict the cytotoxicity of chemicals from restorative materials introduced into the cavity being diffused through the dentin.

Table 1 Rubbers and Chemicals Tested from the Viewpoint of Reference Materials

Materials	Code	Manufacturer
Rubbers		
Butyl rubber	IIR	Kureha Co., Ltd.
Chloroprene rubber	CR	Kureha Co., Ltd.
Ethylene–propylene rubber	EPR	Osaka Rubber Co.
Fluorine-containing rubber	FPM	Tiger Polymer Co., Ltd.
India rubber	NR	Osaka Rubber Co.
Isoprene rubber	IR	Osaka Rubber Co.
Nitrile rubber	NBR	Kureha Co., Ltd.
Silicone rubber	SI	Tiger Polymer Co., Ltd.
Chemicals		
Phenol	PH	Ishizu Seiyaku Ltd.
Zinc chloride	ZC	Sigma Chemical Co.

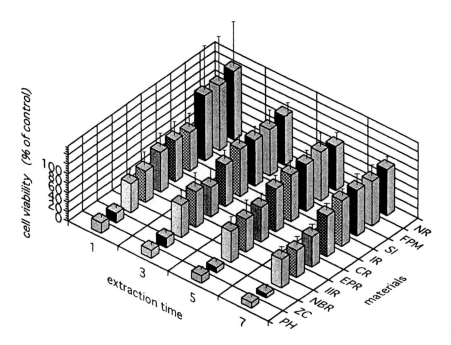

Figure 7 Effects of extracts of test materials obtained by static extraction on cell viability of human gingival tissue [16,17]. Test materials except for PH and ZC gave a cell viability more than 50% of that of the control, but three rubbers—NBR, IIR, and EPR—reduced the cell viability more than the other rubbers.

1. Extraction

The effects of cell–material contact, which are difficult to determine by the direct contact method, can be determined by the extraction method described later. Several extraction factors—for example, type and volume of extraction vehicle, surface area of test specimen, extraction period, temperature, and extraction mode—may exert a direct influence upon the results.

Extraction Vehicles. Selecting the most appropriate extraction vehicle is important in evaluating cytotoxicity because the dissolution of biomaterials is influenced by the vehicle. Distilled water, saline, a balanced salt solution, artificial saliva, a simulated body fluid culture medium, and a culture medium with serum have been used. Sodium chloride solution or water for injection, or even vegetable, sesame, or cotton seed oils have also been used (ASTM F619-79) [18]. However, solvents, chemicals, water, or oils cannot be used in a cell culture environment because they do not sustain cell growth. Even extraction vehicles that support cells are not necessarily appropriate for this purpose. Metal dissolution proved to be particularly affected by the presence of amino acids and proteins (Figs. 10 and 11) [19]. Copper dissolution in an L-glutamine solution became higher in a neutral range than it did in a solution free of L-glutamine (Fig. 10). The negative Cotton effect near 600 nm in a circular dichroism curve, as shown in Fig. 11, indicates that a copper chelate of L-glutamine was produced. Similarly, a negative Cotton effect was recognized near 600 nm when a copper plate was dissolved in a minimum essential medium (Fig. 11).

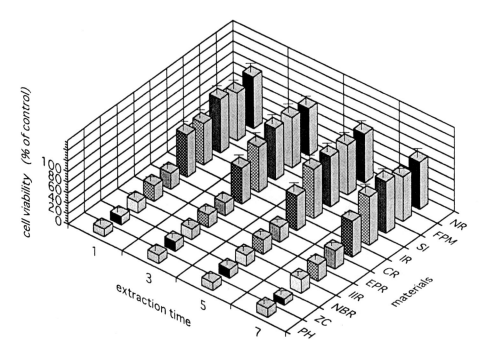

Figure 8 Effects of extracts of test materials obtained by heating on cell viability of human gingival tissue [16,17]. The test materials IIR, EPR, NBR, PH, and ZC reduced cell viability considerably. The cell viability was slightly lowered by CR compared to the control. Extraction by heating was more effective than the static extraction.

Nickel dissolution in artificial saliva and in the minimum essential medium was also higher than in a saline solution such as Ringer's or Hanks' solution (Fig. 12) [20]. The presence of serum and albumin slightly increased nickel dissolution. The possible reasons for these results include the binding of vehicle components with those of the specimen. Therefore, selection of extraction vehicles should be based on the objectives. Solutions resembling bodily fluids, for example, a simulated body fluid, culture medium, or culture medium with serum, can generally be used.

Relationship Between the Surface Area of a Specimen and the Volume of the Extraction Vehicle Used. The relationship between these two factors is crucial for determining cytotoxicity. In an extreme case, one could manipulate results by changing this relationship, as pointed out by Mjör et al. [21]. The relationship which can be found in the dental literature ranges from 1 cm^2/1.93 ml or ca. 0.5 cm^2/ml to 1 cm^2/15.9 ml or ca. 0.06 cm^2/ml for the ratio between the surface area of the specimen and the volume of the extraction vehicles [22,23]. Our ratio was determined to be 1 cm^2/10 ml or 0.1 cm^2/ml through the experimental results obtained using various dental materials [24]. On the other hand, ISO adopted a ratio between 6 cm^2/ml and 0.5 cm^2/ml for conducting cytotoxicity tests on medical devices [25]. Moreover, the latest Japanese guidelines proposed a ratio of 5 cm^2/ml [26]. However, there is no adequate rationale for the adoption of any of these recommended ratios. This means that there is no single universal relationship, but rather a

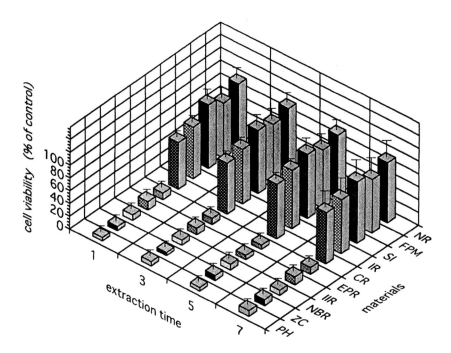

Figure 9 Effects of extracts of test materials obtained by dynamic extraction on cell viability of human gingival tissue [16,17]. Cell viability was not affected by CR, FPM, NR, IR, and SI as compared to the control; but IIR, EPR, NBR, PH, and ZC lowered it to less than 20% of the control. The efficiency of dynamic extraction was approximately the same as extraction by heating.

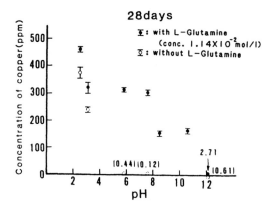

Figure 10 Effect of pH and L-glutamine on copper dissolution [19]. Copper dissolution was rarely found in neutral and alkaline solutions but was accelerated in acidic solution. Copper dissolution was also increased in neutral red and weak alkaline solutions by addition of L-glutamine.

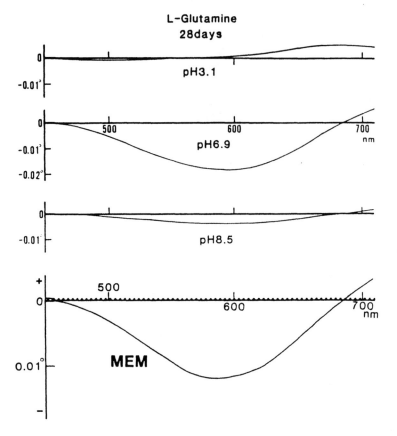

Figure 11   Circular dichroism of each solution when a copper plate was dissolved in L-glutamine solution and MEM [19]. A negative Cotton effect was not recognized in pH 3.1, but a negative Cotton effect was recognized at 600 nm due to a complex between copper and L-glutamine at pH 6.9 and 8.5; in particular, that at 6.9 was larger than that at pH 8.5. A negative Cotton effect was also recognized at 600 nm in MEM. This suggested that the interaction between copper and amino acids in MEM was similar to that between copper and L-glutamine.

different setup is required for each device, because the difference in surface area between artificial heart, blood vessel or hip joint, and an intraocular lens or dental restoration is large. Setting up an inappropriate ratio will lead to vague or erroneous results.

Mode of Extraction.   Generally, extractions have been done under a static condition, probably because most of the tests were conducted on plastics. However, a single mode of extraction does not reflect the true aspects of the material in the tissues. This is quite reasonable, if we consider the long list of materials available and their intended use. Dynamic extraction or agitation during extraction has to be added as another possible mode of extraction. This is true in tests conducted on metallic materials that will create a corrosion problem in many cases and those conducted on stress-bearing usage. Some examples of the latter are artificial hip joints and dental restorations or prostheses. Dynamic extraction has the following features:

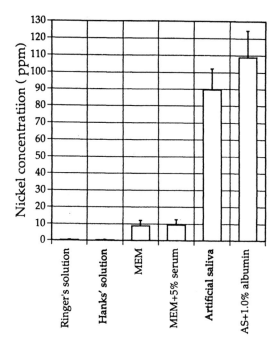

Figure 12 The effects of extraction solutions on nickel dissolution [20]. Extraction was done by gyrating at 200 r/min. Extracts were changed every week and this figure indicated the nickel concentration of extracts after the third extraction.

1. A dynamic load can be applied on a material which is used under a stress-bearing condition such as an occlusion or attrition.
2. Agitation of the extraction vehicle resulting from dynamic extraction prevents a local equilibrium of dissolved component from a specimen. The phenomena of stagnation and a nearly complete prevention of corrosion occur in the metallic materials because of the buildup of corrosive products on the surface. In other words, polarization occurs. Continuing dynamic extraction makes it possible to prevent a decrease in the extracting speed, thus leading to an even distribution within the extraction vehicles.
3. The extraction process can be accelerated. By creating the proper conditions, the behavior of the material during a long period of time can be predicted within a shorter period of time.
4. Extension of dynamic extraction allows further study of the detailed aspects of materials within a specified time.

Duration of Extraction.   The extraction period may range from 1 h to 2 weeks [27,28]. A long extraction period seems to be appropriate, when considering the long use of materials in tissue. On the other hand, denaturation of the extraction vehicle should also be considered, because it may occur in a tissue culture medium or in the presence of serum at 37°C. Studies have shown that longer storage at 37°C resulted in a decrease in cell growth [29]. Agitation of the medium further reduced the growth potency of that medium.

Extraction Temperature.   Extraction temperature is another factor affecting extraction efficiency. Raising the temperature during extraction has been recommended in the testing of plastics [13,18,30]. Recently, raising the temperature has also been adopted as a standard method used in testing the biological properties of medical devices [25]. In such a case, the temperature can be raised to 50°C, 70°C, or 121°C to accelerate dissolution. However, this practice is not necessarily useful for all types of biomaterials. Using various biomaterials, we examined the extraction efficiency of four different extraction methods, that is, static extraction at 37°C for 24 h, extraction by heating in an autoclave for 30 min, dynamic extraction at 37°C for 24 h, and extraction by vibration. The biomaterials we used included metals, plastics, and ceramics for medical and dental use. Figure 13 shows the different degrees of extraction of each of the four types of extraction methods [31,32]. Extraction by heating appeared to be effective for plastics, but not for metals or ceramics (Fig. 13).

Extraction Method.

Static extraction.   The leachable components of materials can be extracted under static conditions. Leachables are distributed evenly within an extract and therefore are easily accessible to cells in a culture system. This is one of the main differences from the direct contact method. Extraction may continue for a maximum of 72 h [25]. However, the efficiency of extraction may decrease as the extraction period is increased, as pointed out earlier. The extraction temperature may be kept at either a physiological or nonphysiological level for the reasons given.

Dynamic extraction.   The two major objectives of dynamic extraction are: (1) to simulate the testing environment of materials that will be used under a stress-bearing condition, and (2) to accelerate the leaching process. However, it is difficult to character-

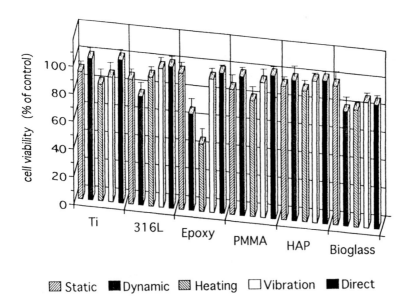

☒ Static   ■ Dynamic   ▨ Heating   ☐ Vibration   ■ Direct

Figure 13   Comparison of cytotoxicity of metals, plastics, and ceramics by four types of extraction methods [31,32]. Extraction by heating decreased the cell viability in plastics, while dynamic extraction decreased the cell viability in metals and plastics. However, static and vibration extraction did not reveal any differences in the cell viability in the three types of materials.

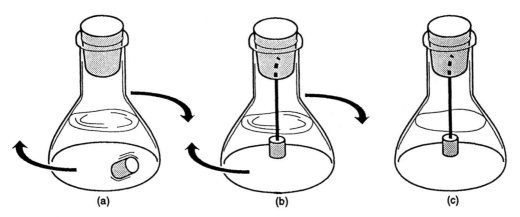

Figure 14   Three extraction methods are shown [33]. (a) Dynamic extraction with a freely moving specimen, cylindrically shaped, 6.0 mm in diameter and 5.0 mm in height at 200 r/min and 37°C. (b) Dynamic extraction with a suspended specimen that was placed in the middle of the extract and gyrated at 200 r/min. The effect of the collision of the specimen was different than that in (a). (c) Static extraction with a suspended specimen at 37°C.

ize the best extraction conditions because sufficient information is lacking. In order to start dynamic extraction, several steps were taken, as described below for two types of the procedure.

Dynamic extraction I.   First, three extraction methods were examined (Fig. 14) [33]:

a. Dynamic extraction with a freely moving specimen that was cylindrical shaped and measured 6.0 mm in diameter and 5.0 mm in height at 200 r/min and 37°C.
b. Dynamic extraction with a suspended specimen placed in the middle of the extract and gyrated at 200 r/min, in which the effect of the collision of the specimen was subtracted from that in (a).
c. Static extraction with a suspended specimen at 37°C.

The test results obtained from common dental restorative materials and dental amalgams are shown in Fig. 15 [34]. Of the five amalgams tested, two showed marked wear and cytotoxicity after 52 weeks when method (a) was used. The others also showed wear, and cytotoxicity fluctuated. However, the marked wear and cytotoxicity shown by method (a) did not clearly appear where the two other methods, (b) and (c), were used. Second, the effect of a specimen's shape on dissolution and cytotoxicity was also examined [35]. Three specimen shapes — cylindrical, spherical, and square — were used to prepare specimens made of nickel–chromium alloy, cobalt–chromium alloy, and titanium. They were extracted one by one in a glass vessel by gyration at 200 r/min at 37°C for 3, 5, and 7 days. The extracts and filtrates, separated using a 0.22 μm filter, were measured for cell viability by cell culture, while the amount of dissolution was measured by atomic absorption spectroscopy. As Fig. 16 shows, no difference in cell viability was detected in the extracts or filtrates obtained from the square and spherical specimens of the three types of metals. On the other hand, the cylindrical specimens of nickel–chromium alloy yielded a markedly lower cell viability. The most efficient extraction condition for *in vitro* cytotoxicity testing was found to be extraction using a cylindrical specimen [35].

In an oral environment, a certain metal often contacts another component of the

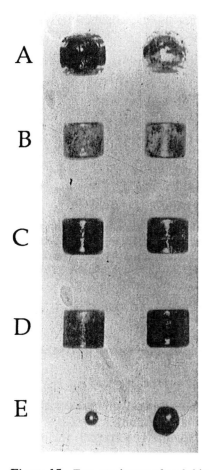

Figure 15 Two specimens of each kind of amalgam are shown [34]. Wear occurred markedly in specimens (E) and (A) during a 52-week period. Total weight loss amounted to an average of 86.4% of the original weight of 1.5 g in specimen (E) and an average of 11.2% in specimen (A). On the other hand, weight losses of other specimens were lower, the averages ranging between 1.7% and 4.4%. (From M. Nakamura, S. Takeda, K. Imai, H. Oshima, D. Kawahara, H. Kosugi, and Y. Hashimoto, *Biomaterials; Mechanical Properties*, ASTM STP 1173, 1994.)

same metal or a different metal. Thus two specimens were placed in a glass vessel and extracted with 200 r/min gyration at 37°C (Fig. 17) [36].

Dynamic extraction II (accelerated dynamic extraction). It is necessary to establish a dynamic extraction method allowing for wear. One such method is the pin-on-disk method of abrasion testing recommended by ASTM (Fig. 18) [37]. According to this method, the pin-type metal, ceramic, or polymer specimen is rotated and reciprocated on a disk made of similar or other materials by using a lubricant such as saline, a simulated body fluid, or serum. Extracts containing wear debris are sterilized and then exposed to cells. Sasada et al. [38] found that noncytotoxic metal in a bulk state became cytotoxic when metal ions were dissolved during wear caused by a combination of 316L stainless steel and 316 stainless steel or titanium alloy and titanium alloy. Ito and Tateishi [39]

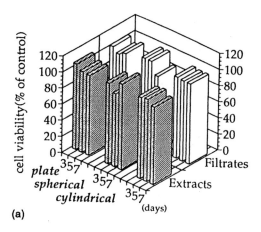

(a)

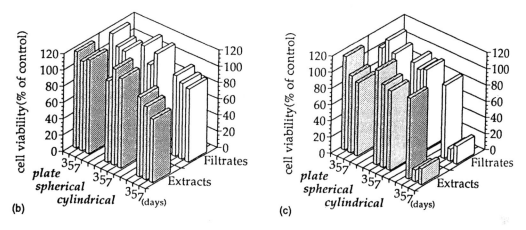

(b)   (c)

Figure 16 The relationship between specimen shape and cytotoxicity of three types of metals [35]. The three specimens, (a) titanium, (b) cobalt–chromium alloy, and (c) nickel–chromium alloy, have the same surface area. Surface area/volume of extract = 1 cm²/10 ml. Each specimen moved freely in glass vessel by gyration.

studied the cytotoxicity of extracts containing wear debris generated by a combination of ultrahigh molecular weight polyethylene (UHMWPE) and various ceramics, and concluded that such cytotoxicity was caused mainly by fine ceramic particles measuring less than 0.22 μm. However, the pin-on-disk method has some disadvantages. First, extracts must be sterilized before being exposed to cells, and the effect of the sterilization procedure on the extract and wear debris might not be negligible. Second, it is difficult to obtain extracts under various conditions simultaneously because machines must be meticulously set.

To use our accelerated dynamic extraction method, a square specimen was placed on pure alumina and zirconia balls, and was subjected to 200–240 r/min gyration in an extract at 37°C (Fig. 19) [40]. Metal ion release was accelerated and wear debris was created by the friction between the alumina or zirconia balls and the specimen. The

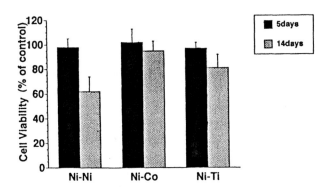

Figure 17 Cell viability of extracts obtained by gyrating the combinations of two metals simultaneously in a glass container [36]. After 5 days of extraction, there was no difference in cell viability of three kinds of combinations: nickel–chromium alloy/nickel–chromium alloy, nickel–chromium alloy/cobalt–chromium alloy, and nickel–chromium alloy/titanium. However, the combination of nickel–chromium alloy and nickel–chromium alloy caused a decrease of the cell viability after 14 days of extraction. The combination of nickel–chromium alloy and titanium also caused a decrease.

filtrate was separated from the extract using a 0.22 μm filter to examine the effect of wear debris. The extract and filtrate obtained were then used to culture cells. Figure 20 is a photograph showing the cells exposed to the extracts of 316L stainless steel and titanium. This experiment showed that the cytotoxicity of the extracts obtained by dynamic extraction was different from that of those obtained by static extraction. Under static extraction for 5 days, the cell viability of each specimen was similar to that of the control. Under

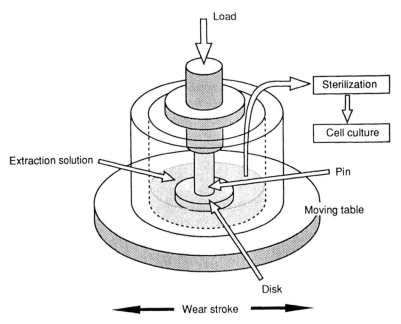

Figure 18 Cytotoxicity evaluation of wear debris according to the abrasion test of ASTM [37].

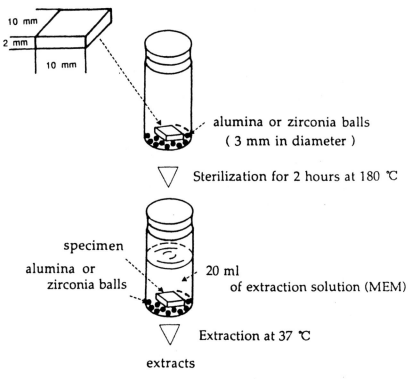

10 mm
2 mm
10 mm

alumina or zirconia balls
(3 mm in diameter)

▽ Sterilization for 2 hours at 180 °C

specimen
alumina or
zirconia balls

20 ml
of extraction solution (MEM)

▽ Extraction at 37 °C

extracts

Figure 19   Accelerated dynamic extraction [40]. Extracts were obtained by gyrating specimens on alumina and zirconia balls.

dynamic extraction, cell viability depended on the type of specimen, gyrating speed, and extraction period (Figs. 21 and 22). The cell viability of each specimen, particularly silver–indium, silver–tin, and nickel–chromium alloys, decreased as the extraction period increased. The cell viability of the extracts obtained by dynamic extraction for 5 days at 200 r/min was slightly but not significantly lower than that obtained by extraction for 5 days at 240 r/min. The cell viability of the extracts obtained by gyrating the specimen on zirconia balls was higher than that obtained on alumina balls. The most efficient dynamic extraction condition for evaluating *in vitro* cytotoxicity was found to be extraction by using gyrating metallic biomaterials on alumina balls at 200 or 240 r/min for 5 days. The cell viability of the filtrates increased in gold and cobalt–chromium alloys, and in titanium. Therefore, wear debris in extracts appears to play an important role in cytotoxicity. Figure 23 shows the cytotoxicity of the wear debris created under accelerated dynamic extraction [41,42]. The wear debris of pure titanium and cobalt–chromium alloy was noncytotoxic, while that of the gold alloy was mildly cytotoxic and that of the alloy containing silver or nickel as its main component was cytotoxic. However, our methods have some disadvantages. First, the uniformity of the surface finish may vary from one specimen to another depending on whether its shape is cylindrical, spherical, or plate. This problem, however, has recently been rectified by the use of a square specimen during accelerated dynamic extraction. Second, the amount of mechanical stress is unknown, although the gyratory movement of the revolutions is controlled. This makes this method

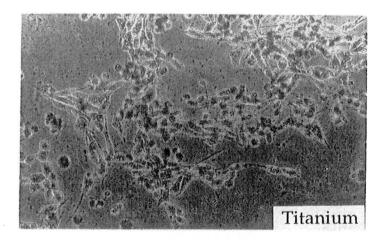

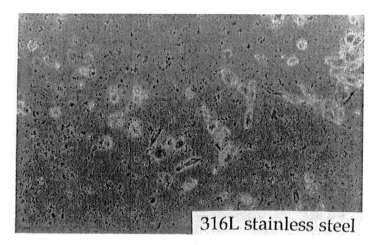

Figure 20   Cells exposed to extracts obtained by accelerated dynamic extraction.

substantially different from the pin-on-disk method, where a predetermined load can be placed on the disk specimen, allowing the force per unit area to be measured. Third, differences arise in the specific gravities of specimens composed of metals, plastics, ceramics, or other substances. Such differences may lead to a difference in the amount of mechanical stress exerted on specimens during extraction. One attempt to solve this problem entails fixing a glass tube to the upper surface of the specimens using tissue-embedding wax that is not cytotoxic in order to maintain an adequate balance between each specimen (Fig. 24).

2.   Direct Cell–Material Contact

Direct Contact Method.   The direct contact method is a simple, but sensitive, way to examine the effect of a material. Cell–material contact is established by either placing the test specimen adjacent to a tissue explant grown *in vitro* or on top of a cell monolayer, or by pouring the cell suspension onto a test specimen that has been placed at the bottom of

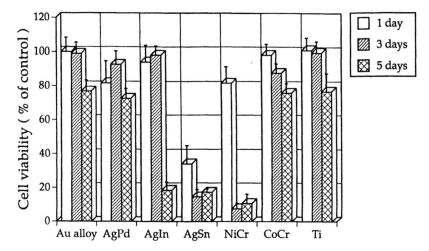

Figure 21   Effects of extraction periods on cell viability of extracts after gyrating specimens on zirconia balls at 240 r/min [40]. Cell viability of silver–indium, silver–tin, and nickel–chromium alloys decreased with extraction periods and was less than 20% after 5 days of extraction.

a culture vessel and has been cultivated for a certain period of time. This method has been adopted in testing standards and has also been used successfully as a means of controlling quality in the health industry [13,14,43].

The components of a specimen are eluted into a medium and are diffused according to a concentration gradient. Thus, the cells affected are dependent on the intensity of material cytotoxicity. One can then make an estimate of the material cytotoxicity by looking from the edge of the test specimen to the far side of the affected cells or by

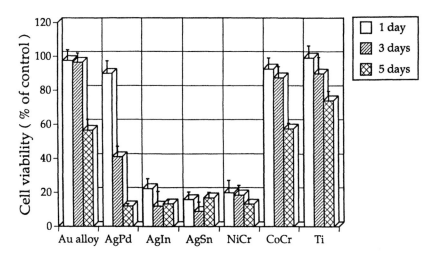

Figure 22   Effects of extraction periods on cell viability of extracts after gyrating specimens on alumina balls at 240 r/min [40]. Cell viability was reduced more markedly by gyration of specimens on alumina balls, compared with gyration of specimens on zirconia balls.

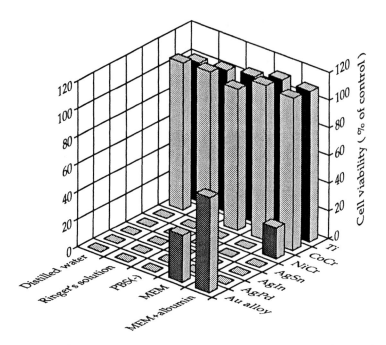

Figure 23 The cytotoxicity of the wear debris borne under accelerated dynamic extraction [41,42].

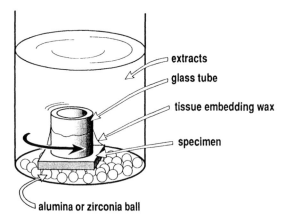

Figure 24 The modification of accelerated dynamic extraction. The specimen was adjusted to a given weight by fixing the glass tube to the upper surface of specimen with tissue-embedding wax. Then it was put into a glass container and gyrated on alumina or zirconia balls.

examining a zone area. Material with intense cytotoxicity may affect remote cells; in other words, a large area or a radius of affected zone will appear. On the other hand, weak cytotoxicity will show only a limited effect within the close vicinity of the cells. For example, cells around titanium appeared normal (Fig. 25), but cells adjacent to pure copper showed severe damage. At the same time, cells far from the test specimen appeared to be damaged by the diffused copper ions. This method has some disadvantages. First, water-soluble components tend to be extensively diffused; thus stronger results tend to be obtained. Second, a rapid decrease of elusion will occur when equilibrium is established following the building up of a higher concentration of components eluted in the medium around the test specimen. Such a phenomenon is possible because the experiment takes place in a static environment. Third, the effect of the culture medium on setting materials does not necessarily reflect what will happen in a clinical situation. Given the above-mentioned points, the direct contact method is best applied to testing implant materials or materials that will directly contact tissue. The method characterized by direct

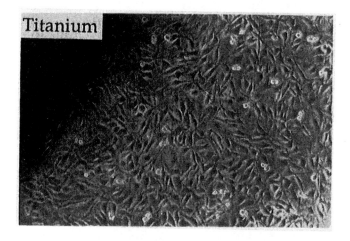

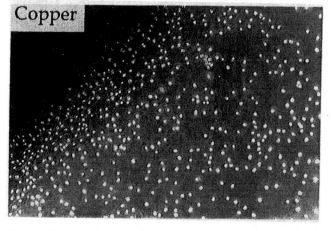

Figure 25 Cell reaction to material due to cell direct contact method. Cells around titanium appeared normal but cells adjacent to copper underwent severe damage.

incubation on the surface of the material offered good cell–material contact, but cell response is dependent on the cytotoxic effect and the surface characteristics of the material. For instance, cells do not adhere to the surface of Teflon due to its low surface energy. Therefore, Teflon would lead to erroneous results if this method was used [44].

Indirect Cell–Material Contact.

Agar overlay method. This method was developed by Guess et al. [45]. In this method, a cell monolayer is cultured on the bottom of a dish for 24 h. This layer is then covered with a layer of fluid agar medium, followed by staining with a neutral red solution. The test specimen and accompanying negative and positive controls are then placed on this agar layer. The results are evaluated 24 h later with a response index measured by the discolored zone and the percentage of the cells lysed (Fig. 26). This method has also been adopted in testing standards and has been successfully used as a means of controlling quality in the health industry [13,14,46]. This method can be applied to materials in a variety of states, solid, powder, film, paste, or fluid. However, heavy materials (such as metallic materials) are not appropriate for this method. The methodology itself is simple, shows sensitivity to acute toxic substances, and has high reproducibility [47]. The cytotoxic components of materials diffuse through the agar layer and react with the cells. Water-soluble substances tend to diffuse through the agar layer easily. On the other hand, certain components absorbed by agar or agarose hardly react with the cells. Therefore, a molecular filter method was designed for application to a wider range of substances irrespective of their water solubility [48]. The method modified by Schmalz is advantageous to present a clearer view of the affected cells by utilizing fluorescein diacetate staining [49]. The collagen gel culture method replaces agar with collagen gel [50].

Model cavity method. Chemical, electrochemical, or physical stimuli from the dental restorative material used and external stimuli causing a failure of adhesion at a tissue-material interface may penetrate dentin and influence dental pulp. Thus, in examining the biological safety of dental restorative materials, test conditions that reproduce the tissue

Figure 26 Agar overlay method.

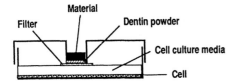

Figure 27   Method using human dentin powder as diffusion layer [51].

environment, as in the so-called model cavity system or dentin barrier test system, are indispensable. Such an approach is unique to dental materials, as unprotected cavities might face a risk deep inside the hard tissues.

In the method using a diffusion layer of human dentin powder, cellulose acetate filter with a pore size of 0.45 μm is attached to one side of a glass tube, which is filled with dentin powder made up of grains measuring less than 250 μm (Fig. 27) [51]. This method has already been included in British Standards Document BS 5828-1989 [52].

Other methods use a diffusion layer of human dentin slice. Tyas [53] suggested the use of a dentin slice in place of dentin powder (Fig. 28). Hanks et al. [54] developed a new system consisting of an *in vitro* pulp chamber by interposing a 0.4–0.5 mm thick dentin slice between carbonate plates, as shown in Fig. 29. In addition, 1 cm H_2O pressure is applied to the chamber on the basis of the fluid dynamics of dentin.

Hume's method used freshly extracted human, third-molar crowns [55]. The root and pulp tissue of the tooth were discarded and the crown was attached to a wax chamber containing the tissue culture medium. Chemicals from the restorative materials introduced into the cavity that diffused through the dentin were collected in the tissue culture medium (Fig. 30). The cytotoxicity of this medium was then examined using standard culture systems.

Such methods may be useful to some extent in testing dental restorative materials; however, they suffer from some disadvantages caused by the use of human dentin. It is well known that quality varies by individual difference, type of teeth, age, and preservation conditions. Also, it is hard to obtain many teeth of uniform quality. As a result, in order to deal with these disadvantages, bovine dentin, collagen, or hydroxyapatite has been used in place of human dentin [56].

C.   End Point Parameters

Different end points have been used to quantify the cytotoxicity caused by biomaterials, for example, permeability, functional, morphological, or reproductive changes. Abnormal membrane permeability has been assessed by dye exclusion [57] by measuring the release of ^{51}Cr from labeled cells [58], or measuring the release of vital stains [45] or LDH (lactate dehydrogenase) [59]. The neutral red uptake assay is a quick, reliable, and

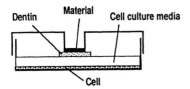

Figure 28   Method using human dentin slice as diffusion layer developed by Tyas [53].

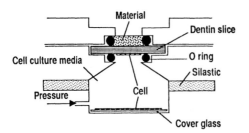

Figure 29 Method using human dentin slice as diffusion layer developed by Hanks et al. [54].

inexpensive type of *in vitro* assay [60]. Under bright-field microscopy, vital cells are stained red by accumulating neutral red dye in the lysosomes. At 540 nm the intensity of the absorption spectrum of the supernatants containing neutral red extracted with 1% acetic acid–50% ethanol from cells is finally measured by a microplate reader.

Functional assays typically assess viability by examining changes in the metabolic functions sustaining cell growth. This method includes DNA [61] or protein synthesis [62], as well as changes in enzyme activity [53,63]. The widely used enzyme assay measures succinate dehydrogenase with a tetrazolium dye. Tetrazolium salt, 3(4,5-dimethylthiazoyl-2-yl) 2,5-diphenyltetrazolium bromide (MTT) is cleaved by the mitochondrial enzyme succinate dehydrogenase and changed into a blue formazan product [64].

Morphological assays examine morphological changes at the cell surface [65], in change of volume [66], or in the cytoskeleton [67].

Reproductive assays measure the ability of cells to proliferate [68]. One of the most generally accepted reproductive assay methods involves measuring the ability of a single cell to form colonies in isolation [23]. This assay assesses survival by simply diluting a single-cell suspension and then counting the colonies that are formed (Fig. 31). This is a simple and highly sensitive method.

The end points of these *in vitro* tests are obtained immediately after the various factors cease to have any effect. However, establishing a follow-up culture, which is the recovery phase, after treatment with cytotoxic factors may express a wider aspect of cytotoxicity. We propose a new *in vitro* cytotoxicity test, the so-called cell recovery test, which allows the affected cells to be replaced by a normal culture environment and thus to recover for a certain period (Figs. 32 and 33) [69–71]. Following the cell-recovery period, the cells are examined to assess cell growth. This method is better for expressing

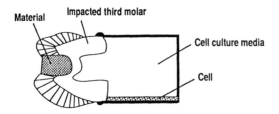

Figure 30 Method using freshly extracted human third-molar tooth crowns [55].

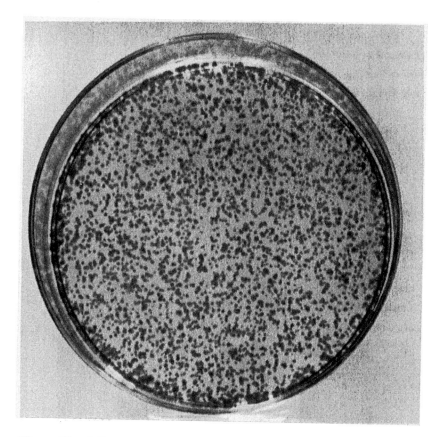

Figure 31 Cell colonies that developed in a petri dish. Number of colonies per unit area is compared with that of a control culture. Data collection of colony area by image analyzer is more efficient.

sublethal damage, as well as delayed cytotoxicity, and thereby may lead to discovery of a wider range of the effects that a material has on cells *in vitro*.

D. Long-Term Cytotoxicity Tests

Because materials are implanted in living systems for a long period of time, a biocompatibility test is required to evaluate the interaction between the living system and biomaterials. It should be noted that biocompatibility is not permanent. Rather it begins to deteriorate immediately after biomaterials are applied to the body. At the same time, our body tissues and organs undergo gradual gerontic changes. The data obtained by the previously mentioned long-term dynamic extraction method showed the necessity for establishing testing and evaluation methods to assess material deterioration.

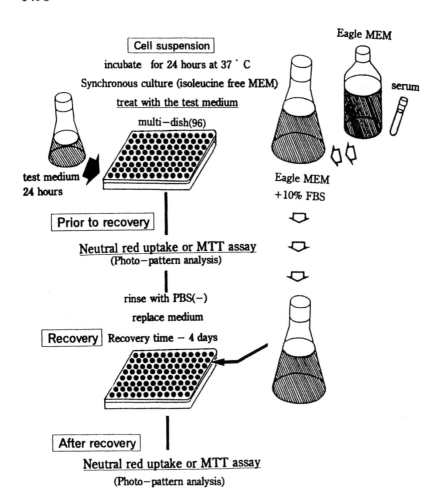

Figure 32 Cell recovery test [69–71]. In a 96-well dish, containing MEM, were plated 5×10^3 cells/ml, and kept in a CO_2 incubator (5% CO_2, 95% air, 37°C) for 24 h, followed by a synchronous culture with a medium free of isoleucine. Following treatment with the test medium containing a test substance for 24 h, the preparations were rinsed with a phosphate buffer solution [PBS(−)] free of calcium and magnesium ions several times, and the values prior to recovery were obtained by neutral red uptake (NR) or 3-[4,5-dimethylthiazol-2-yl]-2,5-diphenyl tetrazolium bromide (MTT) assays or photo pattern analysis. Finally, the treated cells were returned to a normal culture environment, i.e., MEM supplemented with 10% FBS, and cultivated for 4 days. Then the values after recovery were taken with the same assay techniques. For evaluation of cell recovery, three values were used: (1) value prior to recovery, which is represented by the value at 50% inhibitory concentration (IC_{50}); (2) values after recovery, which is represented by a value at 50% inhibitory concentration (IC_{50}); and (3) the area confined by the curve of these values.

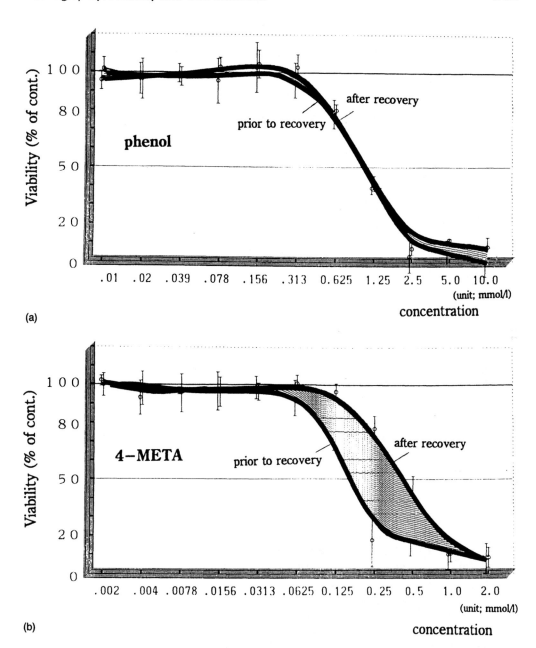

Figure 33 A sample of data from a cell recovery test [69–71]. Cell viability was expressed as the percentage of control against various concentrations of substance: (a) phenol, (b) a monomer for dental use. Different recovery patterns were noticed. In (a) the two IC_{50} values were similar, with a small area, which means almost no cell recovery, while the two IC_{50} values were different, with a large area in (b), in which cell recovery was found.

V.   EVALUATION OF CELL ADHESION

A.   Cell Morphology on Substrate

Interfacial compatibility (i.e., adhesion) is essential for the success of restorations. Adhesion between dental implants and adjacent tissue, and adhesion between restorative materials and hard tissue, are critical factors. Scanning electron microscopy (SEM) has been routinely used to study three-dimensional cell morphology on biomaterials. L929 cells were cultivated for 4 days using titanium and zirconium beads in a gyration shaker at 70 r/min. Then the attachment of cells was examined by SEM. L929 cells appeared to be firmly attached to the titanium beads (Fig. 34) [72].

B.   Measurement of Forces Required to Detach Cells

Many investigators have taken different approaches to evaluation of the force necessary to detach biological cells from biomaterial surfaces. These approaches have been classified, for the sake of convenience, into three categories depending on the nature of the force used during the detachment process [73,74].

1.   *Micromanipulation*

This technique involves detaching adherent cells from biomaterial surfaces through the use of a micromanipulator [75].

2.   *Centrifugation*

In this approach, the biomaterial and substrate are placed in a centrifuge rotor together with the adherent cells and medium. Centrifugal forces are applied in a direction perpendicular to the biomaterial surface [76–78].

3.   *Removal by Hydrodynamic Shear*

In this approach, adherent cells maintained in a shear flow are subject to a viscous force that detaches them from the biomaterial surface. In his experiment, Weiss used a viscometer-type apparatus made of parallel disks [79]. In a more recent work, a viscometer was used, in which the upper disk was machined into the shape of a cone. Weiss et al. [80–82] used a cone with an angle of 1°, while Kawahara et al. [83] used one at an angle of 1° 34′. We employed a cone with an angle of 48′ or 3° (Fig. 35) [84,85]. At first, we measured the dependence of the shear rate on the apparent viscosity and the shear rate of MEM. Table 2 shows the effect of the shear rates on the shear stress and apparent viscosity of MEM at 37 °C. The culture medium exhibited a non-Newton flow. A shear stress of 0.74 Pa was applied to the cells on biomaterial surfaces by using a viscometer with an angle of 48′. Figure 36 shows the effect of cultivation time on the cell adhesion rate. The number of adhered cells was about 60% of those seeded to the glass initially. After cultivation for 24 h, the number of cells rose to 80%. The percentage of adhered cells after shear stress was applied decreased as the applied shear time increased (Fig. 36).

C.   Cell Adhesion and the Surface Structure of a Substance

Implant surface properties such as the nature of a charge, roughness, and the presence of contaminants may affect longevity. The surface characterization of biomaterials has been carried out using methods such as contact angle, electron spectroscopy for chemical analysis (ESCA), Auger electron spectroscopy (AES), secondary ion mass spectroscopy (SIMS), Fourier transform infrared spectroscopy (FTIR-ATR), scanning tunneling mi-

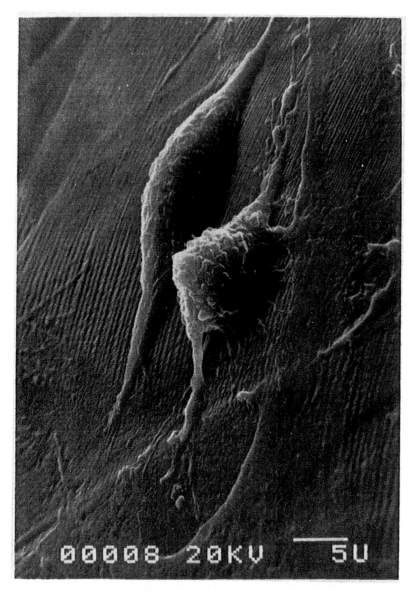

Figure 34   Cell morphology of L929 cells on titanium beads [72].

croscopy (STM), atomic force microscopy (AFM), and scanning electron microscopy (SEM).

The effects of various surface treatments such as cleaning and sterilization when pure titanium was used were examined on the basis of critical surface tensions and cell attachment to treated surfaces. Cleaning using acetone, or acetone with 4% hydrofluoric acid treatment, yielded a high critical surface tension. On the other hand, detergent cleaning yielded a low critical surface tension (Fig. 37) [86]. However, the degree of initial cell attachment did not differ as a result of these surface treatments or sterilization procedures (Fig. 38).

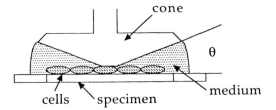

Figure 35 Schematic diagram of the cone–plate viscometer to detach adherent cells from biomaterials [84,85]. A cone with an angle of 48′ or 3° was rotated and the shear stress was applied to the adherent cells through the medium.

Table 2 Apparent Viscosity and Shear Stress
of Eagle's Minimum Essential Medium
Against Shear Rate at 37°C

Shear rate (s^{-1})	Apparent viscosity $(mPa \cdot s)$	Shear stress (Pa)
37.5	0.81	0.03
70.5	0.83	0.06
150.0	0.85	0.13
375.0	0.90	0.34
750.0	0.99	0.74

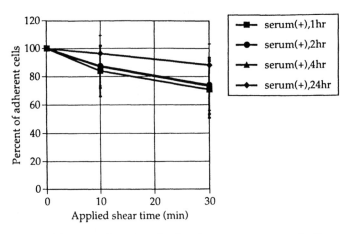

Figure 36 The effect of cultivation time on percentage of adherent cells after applying shear stress [84].

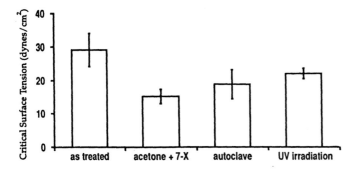

Figure 37   The effects of surface treatment of cleaning and sterilization on critical surface tension of titanium [86].

VI.   NECESSITY FOR COPING WITH CHANGES IN THE TESTING METHODOLOGY

A.   Development of Laboratory Technology

Every test has some disadvantages. Animal experimentation and cell culture are no exception to this rule. The development of new technology is progressive and will help create more effective testing methods in the future. Therefore, testing methods must be continuously improved so that they will remain relevant and valid.

B.   Changes in Social Requirements

Apart from the necessity of upgrading the laboratory technology, social requirements must also be considered. Animal rights and environmental protection are now general public concerns. Such a movement would not have been possible before the 1970s. In

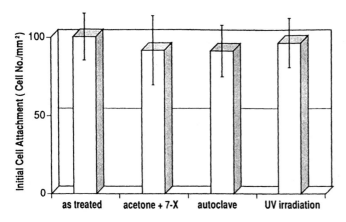

Figure 38   Initial cell attachment for titanium surfaces after various cleaning and sterilization procedures [86].

recent times, medical testing by scientifically based methods has, in the actual sense, been limited by these concerns. This is particularly true in the case of regulatory measures. Mere maintenance of a scientific level may no longer be sufficient, as such requirements must also be met. Furthermore, the rules for such requirements may change over time in an unpredictable manner. Therefore, such matters also require constant attention.

VII. NECESSITY FOR THE FEEDBACK OF CLINICAL EVALUATION DATA

The concept of the testing methods presently applied to medical devices seems only to deal with preclinical biological safety. In other words, the main preoccupation is how safely a device can be introduced into a human body. However, this is only the first step in the whole process of usage of medical devices. Biological responses during a long period of time must also be evaluated. In this respect, three aspects must be considered.

A. Investigation of Side Effects

The biological evaluation of medical devices at the preclinical stage is well developed. However, it is still not perfect. Unexpected side effects may arise after long-term use. Therefore, an extensive and uninterrupted investigation could be important for avoiding unwanted side effects in the future. Both an academic and a regulatory, framework for this purpose should be established.

B. Retrieval Study

Retrieved restorations or prostheses must also be investigated. Such an investigation would be similar to the above-mentioned investigation into side effects, although the material and physical factors involved are not solely concerned with the longevity of dental prostheses such as crowns, dentures, and implants. Factors such as techniques, maintenance, dental care, and posttreatment course are also very important.

C. Development of *In Vitro* Evaluation Techniques

The absence of side effects and the removal of prostheses would be most favorable, though unexpected occurrences are not avoidable. It is, therefore, important to develop an *in vitro* risk evaluation system at the preclinical stage. Such a development would be in accordance with the recent animal rights movement. This system would range from the development of tissue models representing tissue conformation of the oral region and oromaxillofacial functional simulator, to the prediction of the longevity of prostheses.

VIII. SUMMARY

The *in vitro* testing of cytotoxicity and the cell adhesion of biomaterials in the oromaxillofacial region has been discussed. The oral cavity, which is located in the center of this region, interfaces with the external environment and is subject to high-magnitude occlusal force. However, its tissue-restoration capacity is extremely low. Testing conditions to assess the biocompatibility of materials used in this region should, therefore, precisely reflect conditions in the oral cavity. This chapter has dealt solely with *in vitro* testing methods for biomaterials used in the dental field. Considering use conditions, the importance of dynamic extraction methods has been mentioned. Moreover, a new evaluation method for testing cell recovery was also introduced. At the same time, the other aspect of biocompatibility, interface compatibility (i.e., cell adhesion), was also discussed. The development of new testing technology that will meet social requirements is expected. Continued experimentation will help devise more appropriate testing methods, which will lead to more effective and safe products.

ACKNOWLEDGMENT

This work was supported in part by grants from Special Coordination Funds for Promoting Science and Technology from the Science and Technology Agency, Japan (1992–1994), from Scientific Research (B) of the Ministry of Education, Science, and Culture, Japan (No. 03454453 and No. 03454454) (1992), and from the Ministry of Health and Welfare, Japan (1989–1990).

REFERENCES

1. Craig, R. G., Ed., *Restorative Dental Materials*, 7th ed., C.V. Mosby, St. Louis, 1985.
2. Hanks, C. T., M. Anderson, and R. G. Craig, Cytotoxic effects of dental cements on two cell culture systems, *J. Oral. Rehabil., 10*, 101–112 (1981).
3. Taoka, Y., Cytotoxicity of dental filling materials in primary cultured cells derived from human dental pulp, *J. Jpn. Soc. Dent. Mater. Devices, 8*, 324–336 (1989).
4. Yamagata, N., and H. Oshima, Cytotoxic effects of restorative materials on early passage cultured cells derived from human gingiva (*in vitro*), *J. Jpn. Soc. Dent. Mater. Devices, 9*, 541–554 (1990).
5. Tsutsumi, N., and H. Oshima, Cytotoxicity evaluation of crown and bridge materials by extraction (*in vitro*), *J. Jpn. Soc. Dent. Mater. Devices, 10*, 555–565 (1991).
6. Takeda, S., Cytotoxicity of constituent metals in base metal alloys, *J. Osaka Odontol. Soc., 52*, 719 (1989).
7. Takeda, S., Y. Hashimoto, Y. Miura, Y. Kimura, and M. Nakamura, Cytotoxicity test of dental monomers using serum-free cell culture (*in vitro*), *J. Jpn. Soc. Dent. Mater. Devices, 12*, 613–619 (1993).
8. Takeda, S., Y. Hashimoto, Y. Miura, Y. Kimura, and M. Nakamura, Cytotoxicity test of metal ion using serum-free cell culture (*in vitro*), *J. Jpn. Soc. Dent. Mater. Devices, 13*, 134–139 (1994).
9. Kawahara, H., A. Yamagami, T. Kataoka, M. Nishimura, K. Takahashi, and J. Katou, Studies on the tissue irritable action of various canal filling materials by means of tissue culture, *J. Jpn. Soc. Dent. Appar. Mater., 13*, 26–28 (1972).
10. Meryon, S. D., The importance of surface area in the cytotoxicity of zinc phosphate and silicate cements *in vitro, Biomaterials, 4*, 39–43 (1983).
11. Meryon, S. D., The influence of volume on the cytotoxicity of dental materials—An *in vitro* study, *Clin. Mater., 1*, 269–273 (1986).
12. Meryon, S. D., The influence of surface area on the *in vitro* cytotoxicity of a range of dental materials, *J. Biomed. Mater. Res., 21*, 1179–1186 (1987).
13. Biological reactivity tests, *in vitro, United States Pharmacopeia XXII*, 1984.
14. ASTM Designation: F 813-83, Standard practice for direct contact cell culture evaluation of materials for medical devices, *Annual Book of ASTM Standards*, Vol. 13.01, American Society for Testing and Materials, 1993, pp. 239–242.
15. *ISO/TR 7405, Biological evaluation of dental materials*, International Organization for Standardization, 1984.
16. Oshima, H., Y. Kimura, M. Takeda, and M. Nakamura, A search for a reference standard for the cytotoxicity assay of dental materials, *J. Jpn. Soc. Dent. Mater. Devices, 12* (Special Issue 21), 38–39 (1993).

17. Oshima, H., and M. Nakamura, A study on reference standard for cytotoxicity assay of biomaterials, *Bio-Med. Mater. Eng.*, *4*, 327–332 (1994).

18. ASTM Designation: F 619-79, Standard practice for extraction of medical plastics, *Annual Book of ASTM Standards*, Vol. 13.01, American Society for Testing and Materials, 1993, pp. 122-125.

19. Takeda, S., H. Kawahara, S. Yokota, A. Yata, A. Matsumoto, and Y. Taoka, Copper dissolution in basic amino acid solutions, *Proc. 5th Meeting Jpn. Soc. Biomater.*, 53–54 (1983).

20. Isami, T., Dissolution of nickel in various solutions, *J. Jpn. Soc. Dent. Mater. Devices, 7*, 513-524 (1988).

21. Mjör, I. A., A. Hensten-Pettersen, and O. Skogedal, Biological evaluation of filling materials. A comparison of results using cell culture techniques, implantation tests and pulps studies, *Int. Dent. J., 27*, 124-129 (1977).

22. Meryon, S. D., and K. J. Jakeman, An *in vitro* study of the role of dentine in moderating the cytotoxicity of zinc oxide eugenol cement, *Biomaterials, 7*, 459-462 (1986).

23. Nakamura, M., H. Kawahara, Y. Kataoka, S. Maehara, M. Izutani, and H. Taguchi, Biocompatibility of dental amalgam *in vitro* during 52 weeks period, *J. Jpn. Soc. Dent. Appar. Mater., 21*, 228-244 (1980).

24. Valle, G. F., J. F. Taintor, and C. L. Marsh, The effect of varying liquid-to-powder ratio to zinc oxide and eugenol of rat pulpal respiration, *J. Endo., 6*, 400-404 (1980).

25. *ISO 10993-5, Biological evaluation of medical devices Part 5: Tests for cytotoxicity: In vitro methods*, International Organization for Standardization, 1992.

26. Nakamura, M., S. Takeda, K. Imai, H. Oshima, D. Kawahara, H. Kosugi, and Y. Hashimoto, Cell-to-materials interaction—An approach to elucidate biocompatibility of biomaterials *in vitro*, in *Biomaterials' Mechanical Properties* (H. E. Kambic and A. T. Yokobori, eds.), ASTM, 1994, pp. 167-179.

27. Schmalz, G., Ein Vergleich zweier Eluationsverfahren zur biologishen Materialprufung, *Dtsch. Zahnaerztl Z., 33*, 850-855 (1987).

28. Nakamura, M., H. Kawahara, K. Imai, S. Tomoda, Y. Kawata, and S. Hikari, Long-term biocompatibility test of composite resins and glass ionomer cement (*in vitro*), *Dent. Mater. J., 2*, 100-112 (1983).

29. Sanchez, A., K. Imai, M. Nakamura, and H. Kawahara, Alteration of tissue culture mediums following long stock in rotation culture chamber, *Shika J., 12*, 646-647 (1980).

30. The approval standards for intraocular lenses, *The Pharmaceutical Affairs Law of Japan*, 1985.

31. Oshima, H., and M. Nakamura, Cytotoxicity evaluation of biomaterials by extraction, *In Vitro Cellular & Developmental Biology, 27*, 72A (1991).

32. Oshima, H., J. E. Lemons, and M. Nakamura, Investigation of cell contact methods for evaluating cytotoxicity profiles of biomaterials, *Trans. 2nd Int. Congr. Dental Mater.*, 1993, p. 322.

33. Nakamura, M., H. Koda, and H. Kawahara, A proposition for long-term biocompatibility test of dental materials *in vitro*, *Dent. Mater. J., 2*, 113-123 (1983).

34. Nakamura, M., H. Kawahara, Y. Kataoka, S. Maehara, M. Izutani, and H. Taguchi, Biocompatibility of dental amalgams *in vitro* during 52 week period, *J. Jpn. Soc. Dent. Appar. Mater., 21*, 228-244 (1980).

35. Takeda, S., and M. Nakamura, The effects of dynamic extractions on cytotoxicity of dental alloys, in *Abstracts of BIOMAT 91*, 1991.

36. Hashimoto, Y., S. Takeda, T. Kawade, T. Inoue, and M. Nakamura, The effects of dynamic extraction with a combination of two alloys on cytotoxicity, *J. Jpn. Soc. Dent. Mater. Devices, 13* (Special Issue 23), 278-279 (1994).

37. ASTM Designation: F 732-82, Standard practice for reciprocating pin-on-flat evaluation of friction and wear properties of polymeric materials for use in total joint prostheses, *Annual Book of ASTM Standards*, Vol. 13.01, American Society for Testing and Materials, 1993, pp. 189-194.

38. Sasada, T., T. Imaizumi, M. Morita, and K. Mabuchi, Evaluation of wear toxicity for biomedical materials through cell culture method, *Junkatsu, 33,* 288–293 (1987).
39. Ito, A., and T. Tateishi, Cytotoxicity of wear particles, *Proc. 2nd Bioeng. Symp., 920-7,* 167–168 (1992).
40. Takeda, S.,and M. Nakamura, Selection of cell-material contact in cytotoxicity evaluation of metallic biomaterials, *Trans. Soc. Biomater., XVII* (1994).
41. Takeda, S., and M. Nakamura, Cytotoxicity evaluation of wear debris produced by dynamic extraction (*in vitro*), *Proc. 2nd Bioeng. Symp., 920-7,* 165–166 (1992).
42. Sano, Y., and S. Takeda, Study of cytotoxicity and dissolution of metallic biomaterials using dynamic extraction (*in vitro*), *Osaka Odontol. Soc., 55,* 125–140 (1992).
43. Smith, L. M., Direct-contact cell-culture method, in *Cell-Culture Test Methods, ASTM STP 810* (S. A. Brown, ed.), American Society for Testing and Materials, 1983, pp. 5–11.
44. Grinnell, F., M. Milam, and P. A. Srere, Adhesion of cells to surfaces of diverse chemical composition and inhibition of adhesion by sulfhydryl binding reagents, *Arch. Biochem. Biophys., 153,* 193–198 (1972).
45. Guess, W. L., S. A. Rosenbluth, B. Schmidt, and J. Autian, Agar diffusion method for toxicity screening of plastics on cultured cell monolayers, *J. Pharm. Sci., 54,* 1545–1547 (1965).
46. Fehn, J., and D. Schottler, Tissue-culture methods for determining biocompatibility, in *Cell-Culture Test Methods, ASTM STP 810* (S. A. Brown, ed.), American Society for Testing and Materials, 1983, pp. 19–24.
47. Northup, S. J., *USP Committee of Revision Pharmacopeial Forum,* 1987, pp. 2939–2942.
48. Wennberg, A., I. A. Mjör, and A. Hensten-Pettersen, Biological evaluation of dental restorative materials—A comparison of different test methods, *J. Biomed. Mater. Res., 17,* 23–36 (1983).
49. Schmalz, G., A modification of the cell culture agar diffusion test using fluorescein diacetate staining, *J. Biomed. Mater. Res., 19,* 653–661 (1985).
50. Kawahara, D., Usefulness of collagen gel matrix culture for biological evaluation of dental materials (*in vitro*), *J. Jpn. Soc. Dent. Mater. Devices, 8,* 499–516 (1989).
51. Meryon, S. D., and R. M. Browne, Evaluation of the cytotoxicity of four dental materials *in vitro* assessed by cell viability and enzyme cytochemistry, *J. Oral Rehabil., 10,* 363–372 (1983).
52. *BS 5828:1989, British standard methods for biological assessment of dental materials,* British Standard Institution, 1989.
53. Tyas, M. J., A method for the in vitro toxicity testing of dental restorative materials, *J. Dent. Res., 56,* 1285–1290 (1977).
54. Hanks, C. T., R. G. Craig, M. L. Diehl, and D. H. Pashley, Cytotoxicity of dental composites and other materials in a new *in vitro* device, *J. Oral Pathol., 17,* 396–403 (1986).
55. Hume, W. R., A new technique for screening chemical toxicity to the pulp of dental restorative materials and procedures, *J. Dent. Res., 64,* 1322–1325 (1985).
56. Meryon, S. D., and K. J. Jakeman, An *in vitro* study of the role of dentine in moderating the cytotoxicity of zinc oxide eugenol cement, *Biomaterials, 7,* 459–452 (1986).
57. Neupert, G., and D. Welker, Toxicity evaluation of water-soluble substances of dental materials by means of cell populations *in vitro, Arch. Toxicol.,* Suppl. 4, 410–412 (1980).
58. Spangberg, L., Kinetic and quantitative evaluation of material cytotoxicity *in vitro, Oral Surg. Oral Med. Oral Pathol., 35,* 389–401 (1973).
59. Meryon, S. D., Quantitative enzyme spectroscopy in the assessment of cell damage *in vitro, Inter. Endo. J., 21,* 113–119 (1988).
60. Borenfreund, E., and J. A. Puerner, Toxicity determined *in vitro* by morphological alterations and neutral red absorption, *Toxicol. Lett., 24,* 119–124 (1985).
61. Wennberg, A., An *in vitro* method for toxicity evaluation of water-soluble substances, *Acta Odont. Scand., 34,* 33–41 (1976).
62. Kato, T., and R. Nemoto, Rapid assay system for cytotoxity tests using ^{14}C-leucine incorporation into tumor cells, *Tohoku J. Exp. Med., 131,* 261 (1980).

63. Wennberg, A., G. Hasselgren, and L. Tronstand, A method for toxicity screening of biomaterials using cells cultured on Millipore filters, *J. Biomed. Mater. Res., 13*, 109-120 (1979).

64. Mosmann, T., Rapid colorimetric assay for cellular growth and survival: Application to proliferation and cytotoxicity assays, *J. Immunol. Methods, 65*, 55-63 (1983).

65. Borrelli, M. J., R. S. L. Wong, and W. C. Dewey, A direct correlation between hyperthermia-induced membrane blebbing and survival in synchronous G1 CHO cells, *J. Cell. Physiol., 126*, 181-190 (1986).

66. Laiho, K. U., J. D. Shelburne, and B. F. Trump, Observation on cell volume, ultrastructure, mitochondrial conformation and vital-dye uptake in Ehrlich ascites tumor cells, *Am. J. Pathol., 65*, 203-230 (1971).

67. Coss, R A., W. C. Dewey, and J. R. Bamburg, Effects of hyperthermia on dividing Chinese hamster ovary cells and on microtubules *in vitro, Cancer Res., 42*, 1059-1071 (1982).

68. Wilson, A. P., Cytotoxicity and viability assay, in *Animal Cell Culture* (F. I. Freshney, ed.), IRL Press, 1986.

69. Imai, K., M. Nakamura, K. Kokita, and R. Matsumoto, An attempt on cytotoxicity test from the viewpoint of cellular recovery, *J. Jpn. Soc. Dent. Mater. Devices, 6* (Special Issue 10), 167-168 (1987).

70. Imai, K., M. Nakamura, R. Matsumoto, and K. Kokita, Cytotoxicity test using cell recovery test method—Comparison with the conventional cell growth method, *Med. Biol., 125*, 109-112 (1992).

71. Imai, K., and M. Nakamura, Cytotoxicity evaluation from the viewpoint of cell recovery, *In Vitro, 28*, 134A (1992).

72. Imai, K., H. Kawahara, and M. Nakamura, Cell contact with titanium alloys and zirconium *in vitro, Trans. Tissue Culture Soc. Dental Res., 16*, 9-10 (1979).

73. Hubbe, M. A., Adhesion and detachment of biological cells *in vitro, Progr. Surf. Sci., 11*, 65-138 (1981).

74. Bongrand, P., C. Capo, and R. Depieds, Physics of cell adhesion, *Progr. Surf. Sci., 12*, 217-286 (1982).

75. McKeever, P. E., Methods of study pulmonary alveolar macrophage adherence micromanipulation and quantitation, *J. Reticuloendothelial Soc., 16*, 313-317 (1974).

76. Milam, M., F. Grinnell, and P. A. Srere, Effect of centrifugation on cell adhesion, *Nature, 244*, 83-84 (1973).

77. Corry, W. D., and V. Defendi, Centrifugal assessment of cell adhesion, *J. Biochem. Biophys. Methods, 4*, 29-38 (1981).

78. Hertl, W., W. S. Ramsey, and E. D. Nowlan, Assessment of cell-substrate adhesion by a centrifugal method, *In Vitro, 20*, 796-801 (1984).

79. Weiss, L., The measurement of cell adhesion, *Exp. Cell Res. Suppl., 8*, 141-153 (1961).

80. Weiss, L., and D. L. Kapes, Observations on cell adhesion and separation following enzyme treatment, *Exp. Cell Res., 41*, 601-608 (1966).

81. Weiss, L., Studies on cell adhesion in tissue culture. XV. Some effects of cycloheximide on cell detachment, *Exp. Cell Res., 86*, 223-232 (1974).

82. Weiss, L., and D. Huber, Some effects of antimetabolites on cell detachment, *J. Cell Sci., 15*, 217-220 (1974).

83. Kawahara, H., T. Maeda, T. Iseki, and A. J. Sanchez, Studies on cell adhesion to biomaterials by viscometric method, *in vitro, J. Jpn. Soc. Biomater., 2*, 187-192 (1984).

84. Takeda, S., and M. Nakamura, Comparative study of adhesive strength of cells to biomaterials *in vitro, Abstracts of BIOMAT 92*, 1992, pp. 85-86.

85. Kokita, K., and S. Takeda, Cell adhesion to dental materials, *J. Jpn. Soc. Dent. Mater. Devices, 12*, 242-249 (1993).

86. Kawahara, D., Y. Kimura, M. Nakamura, and H. Kawahara, Studies on the tissue adhesive capability to titanium by dynamic wettability test and cell attachment, *in vitro, Clin. Mater., 14*, 229-233 (1993).

50
Biomaterials for Dental Implants

Haruyuki Kawahara
Institute of Clinical Materials
Higashi-osaka-shi, Osaka, Japan

I. INTRODUCTION

One of the main aims of dentistry is to restore the function and form of defective tooth structure by using artificial materials. Dental implants are similar in principle to inlays, crowns, and bridges, for all are placed in contact with oral tissues. The term *dental implant* is used only when the materials are in contact with bone and soft tissue. Adhesion of these tissues is vital to retention of the implant and may prevent infection. This is equivalent to the adhesion of filling materials to enamel and dentin being a factor in the prevention of microleakage and secondary caries.

Nonbiological materials have become indispensable in the management of hard tissue defects. As a result, a unique scientific field has been established in dentistry concerned with issues such as the choice of materials, the design of appliances, the molding of artificial materials, and their harmonization with living tissues. These studies have made great contributions to the development of biomaterials for use in the medical and dental field.

Systematic biological evaluation of dental and surgical materials has been carried out by the author since 1965 [1]. At present, it is impossible to make artificial materials that equate wholly with living tissues. Attempts to raise the quality of implanted biomaterials to the level of the recipient's tissues are likely to be in vain, but better substitutes must continue to be sought [2,3].

II. REQUIREMENTS OF IMPLANTABLE MATERIALS

Two main factors are involved in making successful dental implants. The first is how to achieve close cell adhesion (attachment) of gingival epithelium to the cervical portion of the implant post, that is, the *biological seal* [4]. The second is how to make direct contact between bone and implant surface in alveolar bone tissue, that is, *osseointegration* [5]. To establish the biological seal and osseointegration, the following requirements must be met (Fig. 1).

1. Bioinertness: noncytotoxic, nonirritating, nonallergenic, and noncarcinogenic
2. Biomechanical balance: biomechanically matched with the mechanical properties of the tissue
3. Bioadhesiveness: implant material that is not bioadhesive does not allow close tissue adhesion to the implant and may easily loosen
4. Easy fabrication
5. High esthetical acceptability

III. POLYMERIC MATERIALS

A. Stability of Polymers

The covalent bonds of polymeric materials are attacked and disrupted by changing temperature, pH, osmotic pressure, ionic balance, and so forth, which may interact and accelerate the deterioration process with their synergistic effect. The deterioration affects the main chain, side groups, crosslinks and their original molecular arrangement. Knowledge of the covalent structure of polymers and their degradation *in vitro* can be used to predict the mechanism of deterioration of polymeric implants *in vivo*.

Unlike *in vitro*, the chemical and mechanical stabilities *in vivo* are influenced by biological effects of enzymes and unknown factors, relating to the composition, molecu-

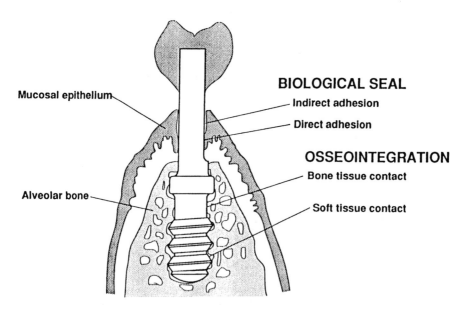

Figure 1 Biological seal and osseointegration. Biological seal is a structure preventing ingress of oral debris, bacteria, and other contaminants into internal milieu. Osseointegration is a structure of implant–bone interface without soft tissue intervention that maintains reliable fixation.

lar weight, and structure of polymer. Of course, although the chemical and mechanical stabilities *in vivo* depend upon the molecular composition and structure of polymer, in general the lower molecular weight polymers are less stable than those of higher molecular weight.

The average molecular weight is a very significant indication in relation to the chemical and mechanical stabilities. And also, the molecular weight distribution plays an important role in determining the chemical and mechanical stabilities of polymeric implants. A narrow molecular weight distribution may contribute greater chemical and mechanical stability in *in vivo* circumstance. However, all synthetic polymers include a wide range of molecular weights [6–8].

Polymeric materials start to deteriorate when they are implanted into living body environment. The most important cause of deterioration is ionic attack of OH^- and dissolved oxygen. Aqueous hydrolysis thus readily results in molecular fragmentation, and also, enzymic attack may introduce its particular degradation with unknown biological reactions. For example, a recent series of experiments with ^{14}C-labeled polymers demonstrated wide-ranging enzymatic activity in relation to polyester, polyamide, and polyurethane substrates [9,10].

Most hydrophilic polymers such as polyamides, nylon, polyvinyl alcohol, and cellulose react easily with body fluids that accelerate their deterioration. On the contrary, hydrophobic polymers of polyethylene, polytetrafluoroethylene (PTFE, Teflon®), and polypropylene are less prone to deteriorate in body environment. Polyesters, epoxies, and polymethacrylates have an intermediate degree of deterioration. They are slowly biodegradable, characteristically over implantation periods of 10 years of more. The rate of deterioration increases with the number of hydrophilic side chains [6].

B. Tissue Response and Molecular Structure

The more the implant material resembles host tissues in molecular structure, the more severe is the reaction in the living body. This is because the host experiences difficulty in discriminating immunologically between implant material and host tissue. Conversely, removal of the antigenic stimulus from transplants or implanted material results in a more inert nonbiological agent, but this helps the host to recognize it easily as foreign. Either way, the host attempts to digest and absorb it, to isolate it within the body by the deposition of a fibrous capsule or to reject it [11].

Materials closely similar to living tissue will be easily dissolved and digested in the host tissue, and therefore implant materials with molecular structures similar to proteins or polysaccharides cannot be expected to be satisfactory. For example, nylon, a polyamide, is likely to become an allergen and is therefore easily digested and absorbed within the body. On the other hand, silicone polymer, polyethylene, and Teflon (polytetrafluoroethylene), which have molecular structures completely different from biological substances, are generally more stable in the body (Fig. 2) [12]. However, these polymers are hydrophobic and offer little potential for adhesion of living cells and tissues in spite of their high stability. It is difficult to fix them firmly in the alveolar bone, to protect the gingival margin from bacterial invasion, and to prevent epithelial down-growth at the junction between implant and mucosa. They are not, therefore, suitable materials for use in the replacement of dental roots. Indeed, there is no polymer having good adhesion to living cells as well as high stability in tissue fluids [13].

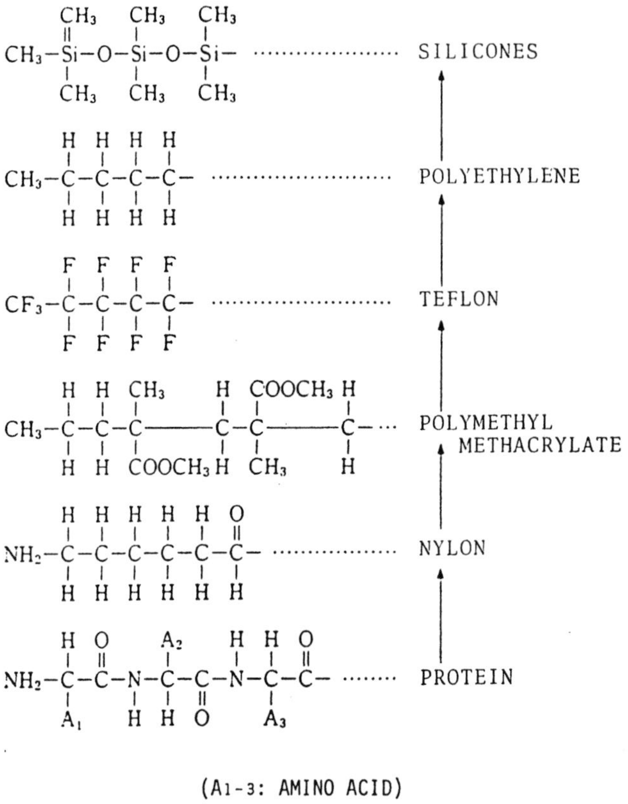

```
      CH₃   CH₃   CH₃
       ‖     |     |
CH₃−Si−O−Si−O−Si−   ···················· SILICONES
       |     |     |
      CH₃   CH₃   CH₃


      H  H  H  H
      |  |  |  |
CH₃−C−C−C−C−   ···················· POLYETHYLENE
      |  |  |  |
      H  H  H  H


      F  F  F  F
      |  |  |  |
CF₃−C−C−C−C−   ···················· TEFLON
      |  |  |  |
      F  F  F  F


      H  H  CH₃      H  COOCH₃ H
      |  |  |        |  |      |
CH₃−C−C−C————C−C————C− ···   POLYMETHYL
      |  |  |        |  |      |        METHACRYLATE
      H  H  COOCH₃ H  CH₃    H


      H  H  H  H  H  O
      |  |  |  |  |  ‖
NH₂−C−C−C−C−C−C−   ···················· NYLON
      |  |  |  |  |  |
      H  H  H  H  H  H


      H  O     A₂     H  H  O
      |  ‖     |      |  |  ‖
NH₂−C−C−N−C−C−N−C−C−   ········ PROTEIN
      |     |  |  ‖     |
      A₁    H  H  O     A₃
```

(A₁₋₃: AMINO ACID)

Figure 2 Molecular structure and biostability. Materials that are closer in molecular structure to proteins and polysaccharides are more unstable in living tissues.

C. Cell Adhesion to Polymers

Attempts have been made to determine bioadhesiveness in tissue culture experiments by observing the profiles of cells in contact with various materials. In tissue culture systems, isolated cells sink in the medium to the bed, which can be made of various polymeric materials. The cells adhere with various profiles to the surfaces of the beds; by observing their contact angles, the wettability of cells on the materials can be determined indirectly (Fig. 3) [14]. Interpretation of the measurements obtained presumes that the contact angle relates closely to the molecular structure of the bed material in that the contact angle decreases in proportion to the bonding strength of hydrophilic radicals, such as carboxyl, hydroxyl, aldehyde, and carbonyl; and also that it decreases as the number of such polar functional groups increases [15]. These findings were obtained in serum-free media. However, cell adhesion mechanisms are more complicated in the living body because the surface condition of the material is easily changed by adsorbed lipids. These have both hydrophilic and hydrophobic functionalities, which the cells contact before they make contact with the surface of the polymer. After 2-h cultivation, the cells showed lower wettability on a hydrophilic surface than on a hydrophobic surface in a medium containing serum, whereas the reverse results were observed in a medium without serum.

Silicone
Polyethylene
Polystyrene

Methylmethacrylate
Polycarbonate
Epoxy

Polyamine
Cellulose acetate

Polyvinyl alcohol
Cellulose
Polysaccharide

Figure 3 Cell adhesion to polymeric materials. The cell at the top shows loose adhesiveness with a large contact angle to the surface of repellent plastics. The middle two cells show medium adhesiveness to the surface of plastic with a small number of hydrophilic radicals. The bottom cell shows strong adhesiveness with a small contact angle of cells. These are hydrophilic plastics.

However, the cells degenerated completely in the serum-free medium after 24-h cultivation. In serum-containing cultures of 24 h or more, the cells came to have higher wettability on hydrophilic surfaces rather than on hydrophobic ones. This is because in the media with serum the cells have a hydrogel shell of glycosaminoglycan and proteoglycan outside their cytoplasmic membrane. These do not bond with residual hydrophobic radicals of the lipids when the hydrophilic radicals are adsorbed onto the hydrophilic material surface, but the hydrogel shell may act as an adhesive material with incorporation of a conformation layer modified by lipids and proteins from the medium (Fig. 4, Table 1) [11,12].

The strong adhesion of cells may be produced by conformational changes in the sandwich layer consisting of the extracellular matrix and conformation layer [12,15]. These results were confirmed by measuring the strength of adhesion of cells, in culture, to the surface of materials using viscometric methods in which the adhering cells were detached by a shear stress of 3.7 dyn/cm^2 created by the flow of the culture medium. The adhesive strength of the cell to the material surface was estimated from the number of cells detached [16]. With the polymeric implant *in vivo*, tissue adhesion to its hydrophilic

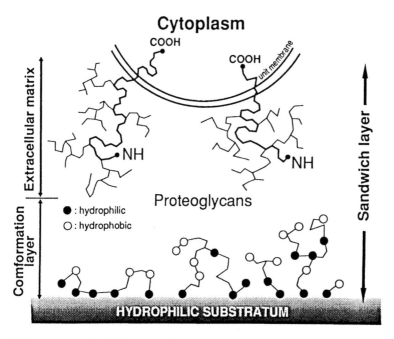

Figure 4   Sandwich layer at cell–material interface consists of extracellular matrix and conformation layer.

Table 1   Cell Shapes After Contact with Polymers in Media with (+) and without (−) Serum After 2 and 24 h

Materials serum	Glass		PVAC		CEL		PST	
	−	+	−	+	−	+	−	+
	2-h Cultivation							
⬤	51–67	72–96	57–72	77–98	65–87	73–92	61–72	65–81
◖	18–28	9–17	8–16	6–17	3–14	11–25	13–31	12–22
◀▬▶	0–5	0	7–21	0	2–7	0	0	8–15
✴	8–19	0	7–21	0	0	0	0	0–3
	24-h Cultivation							
⬤	100	16–22	100	5–11	100	23–39	100	21–35
◖	0	15–21	0	14–20	0	9–15	0	32–46
◀▬▶	0	18–26	0	18–28	0	18–25	0	23–36
✴	0	36–46	0	43–63	0	29–43	0	0–8

Note: PVAC, polyvinyl acetate; CEL, cellulose; PST, polystyrene

surface becomes stronger by gelation of the sandwich layer over time. However, polymers with hydrophilic functionalities are apt to swell and be absorbed when in the body. This problem might be overcome as shown by recent studies on antithrombotic polymer with hydrophilic and hydrophobic heteromicrodomain structures [17–19] and by studies on the grafting of hydrophilic polymers onto the surface of the hydrophobic plastics [20] to make implantable plastics that have high stability and adhere strongly to the surrounding tissues.

D. Clinical Use

1. *Polymethylmethacrylate*

Polymethylmethacrylate (PMMA) is a transparent material that is also referred to as *plastic glass*; it is used widely in medical and dental applications. PMMA has excellent physical properties and is easy to fabricate; it can be cast, molded, or machined, and also has excellent chemical resistance and high biocompatibility. Animal experiments and clinical investigations on PMMA dental implants have been carried out by Hodosh et al. [21–23], Gettlemen et al. [24], and Ehrlich and Azaz [25]. Their animal tests and clinical investigations revealed that PMMA implants assumed reasonably normal function even when unsupported by external fixation to adjacent teeth and could become functionally and cosmetically excellent devices in oral rehabilitation over the fairly short term (Table 2) [22]. In spite of these successful results, PMMA dental implants are not used clinically at present, because PMMA implants were associated with cervical infection and loosening in long-term implantations. We have experienced many failures in long-term implantation of 15 years of more in the human body: for example, around 1945, PMMA ping-pong balls were inserted into the chest cavity to treat pulmonary tuberculosis by pressure, but these are now being removed because they have deteriorated severely and caused adverse tissue responses. These phenomena are observable as soft tissue capsulation, including many macrophages and giant cells around the PMMA implant, 2 years or more postoperation.

Any shape can be easily produced, by cold curing PMMA, which is widely used as a dental filling material. It has been developed as Palacos "R" and "K" by Kulzer, as implant materials. These are being used even today, under the designation *bone cement*, for supporting artificial hip joints, and the same system was used for supporting dental implants in alveolar bone. Also, dental implants made of cold-curing PMMA composite, which mimicks root forms, have been investigated by histological and radiographical

Table 2   Successful Rate of PMMA Implant in Animal Tests

Animal	Number of implants	Time postimplantation (months)	Result[a]			
			E	G	N	D
Baboon	11	1–5	6	1	4	
	15	11–37	9	4	2	
Macaca	15	1–5	4	7	3	1
	25	6–40	7	9	9	
Dog	15	3–36	5	5	5	

[a]E, excellent; G, good; N, no good; D, death

observations, and it has been reported that cold-curing PMMA did not result in recognizable pathologic changes and destruction of hard or soft tissue [26,27]. However, cold-curing PMMA does not reach complete polymerization and retains a great quantity of monomers. The residual monomers give rise to an injurious effect on surrounding tissue through adverse tissue reaction, and they have a destructive and resorptive effect even on the bone tissue.

2. Polyethylene

Unlike the low-density polyethylene, high-density polyethylene (HDP) and ultrahigh molecular weight polyethylene (UHMWPE) do not contain branches, are highly crystalline, and have high glass transition temperatures. HDP and UHMWPE developed by Phillips Petroleum and DuPont and DuPuy have been used for artificial hip joint cups. This requires that they neither deform nor deteriorate when sterilized by heating and implanting into body. Pennisi utilized HDP as artificial mandible and reported successful data [28,29]. Dental implants made of HDP composite with SiO_2, K_2TiO_3 whiskers, which have a Young's modulus of 4.9 GPa, were tested by implantation into Japanese monkey's mandibles. As many as 43% failure cases were experienced 1 year postimplantation, and 68% failure was found 3 years after the implantation, due to severe peri-implantitis caused by poor biological seal and/or bone tissue adhesion [30].

3. Polysulfone

Polysulfone (PS) is a cross linking thermoplastic polymer and has high mechanical strength, compared with those of PMMA. Any form can be fabricated by injection molding at a temperature of 370°C. PS composites containing fillers of glass fiber and potassium titanate whiskers were developed for dental implant by Kawahara; these mimicked the mechanical behavior of natural tooth dentine or hard bone (Table 3) [30].

Ten PS composite dental implants of the screw type have been tested by histopathological observation in dogs (six implants) and in clinical trials (four). In spite of showing no adverse tissue response and a successful rate of 83% in animal tests, a 50% (2/4) failure rate was observed in the clinical test, due to poor biological seal and peri-

Table 3   Mechanical Properties of Polysulfone Composite and Raw Material of PMMA and PS

Tests	PMMA	PS	PS composite
Impact (kgf·cm·cm)	2.5	12.1	12.3
Tensile (MPa)	58.8	63.7	85.7
Bending (MPa)	93.1	107.8	135.2
Young's modulus (MPa)	3038.1	2550.5	6390.2
Fracture energy			
Tensile (kgf·mm)	27.0	<75	280.0
Vending (kgf·mm)	5–6.0	<20	75.0
Hardness (Brinell)	13.9	12.7	15.6
Water absorption (%)	1.28	0.17	0.06
Wear resistance (%)	4.28	0.37	0.02

implantitis probably caused by poor tissue adhesion. The implanted prostheses are apt to be separated from adjacent soft tissue and hard tissue (Figs. 5 and 6) [31,32].

E. Summary

Cold-curing composite resins of MMA + SiO_2, Bis-GMA + SiO_2 [33] and TMM + Si_3N_4 + SiO_2 [34] may be usable as quick, tailor-made dental implants, if the adverse effects of residual monomers upon living tissue are completely confined or controlled. However, this is very difficult to carry out.

Polyester, polymethane, epoxy, and polyamide are not suitable for dental implants because long-term implantation is accompanied by deterioration in relation to number of hydrophilic functional groups.

Polysulfone, polyethylene (high-density polyethylene), and Teflon are stable in long-term implantation and are not toxic or irritating, but they are not usable because their surface is water repellent or water resistant, and they consequently cannot make reliable biological seal and close bone bonding. In short, *no polymeric material is usable for dental implants at present.*

However, two shock-absorbing composite of Teflon + carbon fiber (Proplast) [35–37], with low modulus of elasticity, have been developed to prevent bone resorption caused by impact stress during mastication. However, fatigue life of these materials is unknown in the *in vivo* environment.

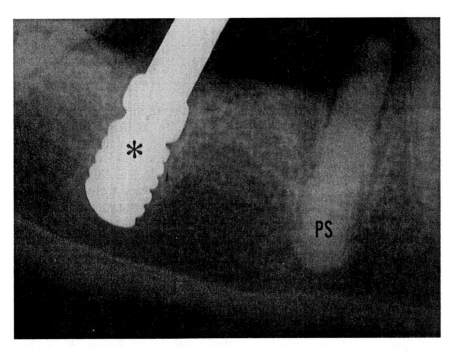

Figure 5 Polysulfone composite implant in dog mandible, 6 months postimplantation. *Bionium (Co–Cr–Mo); PS, polysulfone composite implant.

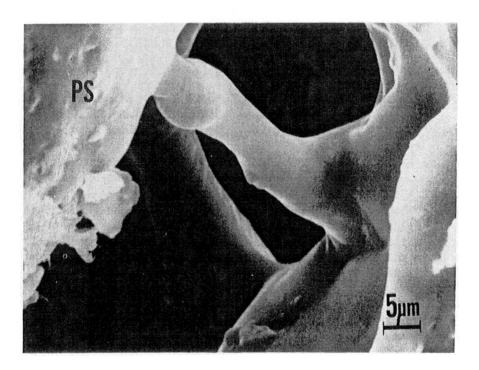

Figure 6   Large space at implant–bone interface, 6 months postimplantation. PS, polysulfone composite implantation.

IV.   METALS AND ALLOYS

A.   Metals

Examination of cytotoxicity and tissue irritability in relation to the position of elements in the periodic table reveals that Be, Mg, Ca, Cd, Sr, Ba, Zn, and Hg (Group II) have strong cytotoxicity; whereas Al, In, and Ga (Group III); Si, Ti, Zr, and Sn (Group IV); and Cr, Mo and W (Group VI) do not cause cytotoxic reactions or tissue irritation in living tissue. Metals of lower atomic weight in some groups—such as Cu (Group I); V, As, and Sb (Group V); and Fe, Co, and Ni (Group VIII)—possess cytotoxicity, tissue irritability and carcinogenicity, whereas metals within the same group with higher atomic weights—for example, Au (Group I), Ta (Group V), and Pd and Pt (Group VIII)—show no cytotoxicity and no irritability. *Thus, the cytotoxicity or irritability of a metal to cells is closely related to its location in the periodic table* (Table 4) [38].

In order to exert effects on cells, the metal must be ionized. This ionization process begins with a reaction between the metal and its surrounding tissue fluid. The ions directly affect the cells, which then degenerate in the altered tissue fluid. That is, when a metal is placed in living tissue, the degree of the injury to tissue depends upon the degree and type of chemical reaction between the metal and the tissue fluid. Kawahara [39] and his coworkers [40] have investigated metallic corrosion in tissue culture conditions that allow the cells to multiply. Corrosion mechanisms have been studied by placing metals in tissue culture medium containing various macromolecular substances such as amino acids, proteins, polysaccharides, and glucoproteins; the investigations used paper chromatography

Table 4 Relationship Between the Periodic Table of Elements and Cellular Responses to Pure Metals [1]

Periodic number	Metal	Atomic weight	Electromotive force	Inhibition index of cell outgrowth	Inhibition index of cell multiplication	Cell contact	Cytotoxicity
Group I	Cu	63.54	+0.34	0.01	0.00	−	+ + +
	Ag	107.88	+0.80	0.76	0.84	+	+
	Au	197.00	+1.50	1.14	0.89	+ +	±
Group II	Mg	24.32	−1.87	0.06	0.00	−	+ + +
	Zn	65.58	−0.76	0.00	0.00	−	+ + +
	Sr	87.63	−2.89	0.08	0.00	−	+ + +
	Cd	112.41	−0.40	0.00	0.00	−	+ + +
	Hg	200.61	+0.80	0.00	0.00	−	+ + +
	Ba	173.36	−2.90	0.00	0.00		+ + +
Group III	Al	26.98	−1.24	0.82	1.14	+ + +	±
	In	114.82		1.42	0.97	+ + +	−
	Ga	69.72		1.23	1.02		−
Group IV	Si	28.09		1.10	1.08		−
	Ti	47.90		1.56	1.04	+ + +	−
	Zr	91.22		1.23	1.08	+ + +	−
	Sn	118.70	−0.13	1.33	1.26	+ + +	−
	Pb	207.21	−0.13	0.78	0.07	−	+ +
Group V	V	50.95		0.00	0.00	−	+ + +
	As	74.91	+0.3	0.00	0.00	−	+ + +
	Sb	121.76	+0.1	0.00	0.00	−	+ + +
	Bi	209.00	+0.2	0.63	0.65	+	+ +
	Ta	180.95		1.19	0.91	+ + +	±
Group VI	Cr	52.01	−0.55	1.16	0.95	+ + +	±
	Mo	95.95		1.21		+ +	±
	W	183.36		1.25			−
	Te	127.61		1.28			
Group VII	Mn	59.94	−1.04	0.52	0.00		+
Group VIII	Fe	55.85	−0.44	0.55	0.00	−	+ +
	Co	58.94	−0.28	0.38		−	+
	Ni	58.71	−0.25	1.30		±	−
	Pd	106.40	+0.82	1.32	0.87	+ + +	−
	Pt	195.09	+0.86	1.23		+ + +	−
Control (to the glass)				1.00	1.00	+ + +	±

[41], atomic absorption spectrometry [42,43], electronprobe microanalysis [44], and circular dichroism [45,46]. These studies have clarified the ionization mechanisms of pure metals and alloys in the body.

Even though their ionizing tendencies are higher than hydrogen, Al and In (Group III); Si, Sn, Ti, and Zr (Group IV), Ta (Group V); and Cr (Group VI) are not so easily ionized because a passive oxidized film is produced on the surface of the metal in tissue fluid at physiological pH. Such metals are therefore comparatively stable in living tissue and exhibit almost no tissue irritability. However, Al, In, Sn, and Si are corroded easily and ionized if the pH at the site of the implant changes markedly to a lower level, perhaps following infection or some other cause [12,14]. Precious metals are not suitable for use in pure form due to their softness, and hard precious alloys usually contain cytotoxic metals, Cu, Cd, and Zn. However, work-hardened Ti, Zr, and Ta are fairly stable in living tissue and are already being acclaimed for use as dental implants in clinical practice. Tissue irritability and cytotoxic properties of metals do not necessarily coincide with their ionization tendency. This may be because the pattern of cellular response to metal differs depending upon the type of ion and the passive film on the metal surface [1,14].

B.   Stability of Alloys

Alloys also ionize in living tissue. Those that have more than two phases form local cells in each phase and, in some cases, ionization may be promoted. On the other hand, alloys that are intermetallic compounds (e.g., Ag–Sn and Cu–Sn) form oxidized films on their surfaces, protecting them from corrosion in living tissue [12]. This also applies to alloys of Co–Cr–Mo and Fe–Cr–Ni. When 10–20% of Cr is mixed with Co or Ni, the cytotoxicity of these elements is completely masked and the resulting alloy is not cytotoxic (Fig. 7) [47]. In this case, the alloy surface becomes passive because a homogeneous oxidized chromium film is formed.

This is also the reason that low-fusible Ag–Sn alloy and dental amalgam are comparatively stable in living tissue [48,49]. Co–Cr–Mo alloys and Fe–Ni–Cr–Mo alloys are widely used in the construction of appliances and dental implants. In this respect, Mg–Al [50], Ti–Al–V [51], Ti–Mo [52], Ti–Pd [52], Ti–Fe [53], Zr alloys [54], and Ti–Ni [55] all deserve to be evaluated as they are light, stable, and corrosion resistant, and have high shape memory qualities.

C.   Cell Adhesion to Metallic Materials *In Vitro*

In assessing cell wettability and adhesion to metallic surfaces, profiles of cells in contact with evaporated metal films and cell–metal interfaces were observed by transmission electron microscopy. Indirect adhesion to evaporated films of Au, Pd, and Pt was demonstrated by clear evidence of an interspace between the cell membranes and the film. There was no degeneration in the cytoplasm and cell organelles remained normal. All cells were able to develop normal mitosis on metal films of Au, Pd, and Pt. Cytoplasmic fusion with the metallic film was occasionally observed over minute areas but elsewhere there were comparatively large spaces without interaction between the cell membranes and the metal film after 24-h adhesion (Fig. 8) [12].

Cells adhering directly to Ti, Zr, and Cr films showed normal cytoplasmic structure as well as cell mitosis, in spite of their strong adhesion at the cell–metal interfaces. The normal structure of the cytoplasm and its organelles was confirmed by electron microscopy after 24-h adhesion, except in the area immediately adjacent to the film. Direct contact between cytoplasm and the Ti film was observed. However, clear membrane was observed elsewhere. In the contact area, electron-dense metallic particles were dispersed into the cytoplasm, but there was no cytoplasmic degeneration beyond the contact (Fig. 9) [12]. The findings with a Cr and a Mo film were similar (Fig. 10). Oxidation film of base metals of Ti, Zr, Cr, and Mo has high biochemical activity and may produce chemical and physical bonding to the adjacent tissues of bone and gingival epithelium,

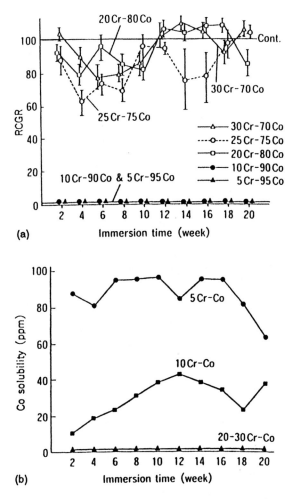

Figure 7 (a) Cytotoxicity of Co–Cr alloys. Relative growth rate: to 100% in control cells without any test piece. Cell, L; medium, MEM + 5% calf serum. (b) Solubility of cobalt from Co–Cr alloys in tissue culture medium. MEM + 5% calf serum.

unlike stable surface of Pt, Au, and Pd. This may be a reason for direct bonding of the base metals to bone tissue.

The oxidation layer of the base metals and alloys adsorbs water, giving a hydrated layer that cannot be removed at normal atmospheric pressure. The surface is not only stable but also nonirritating, noncytotoxic, noncarcinogenic, and superior in cell adhesion. In the tissue culture medium with serum, the metal–cell interface is complex. The hydrated surface layer (MOOH) adsorbs proteins, lipids, polysaccharides, proteoglycans, and other macromolecules, which are associated with metallic and nonmetallic ions, and it constitutes the conformation layer [31]. A recent review [56] on titanium implants reported that Ti-superoxide (TiO_2^-) and Ti-peroxide (TiO_2^{2-} are incorporated into the uppermost part of the oxide (TiOOH) and degrade to finally form TiO_2, molecular oxygen, and water. Furthermore, Ti(III) macromolecular complexes may form in the

Figure 8   Cell–metal interface (Pt). A cell–metal interspace of 1000 Å or more between L cells and a Pt film; the intact unit membrane adhered indirectly to the precious metal of Pt film through a sandwich layer after 24-h cultivation (see Fig. 11). Close adhesion to the metal film with cytoplasmic fusion is indicated by the arrow (↑), N, nucleus.

partially anaerobic milieu when the oxide is reduced. Similar phenomena may be caused at all metallic implant–tissue interfaces (Fig. 11).

Ti alloys with Pd, Pt, Zr, Mo, or Fe were developed in 1979 and studies on cell adhesion at the alloy–cell interface compared them with pure Ti. HeLa S3 cells and Hp cells were cultured with beads of alloy in a gyratory shaking incubator. Cells adhered to the surface of alloy beads after 24-h culture. After 7 days the beads made of Ti alloys were closely bound with cell-to-bead and cell-to-cell adhesion. The cells showed cytoplasmic stretch, close adhesion, and cytoplasmic bridges among the beads (Fig. 12) [12]. The adhesive strength between the beads and cells was tested by attempting to separate them with supersonic vibration. There was strong adhesion of cell to beads and cell to cell with the Ti alloy beads [57].

Figure 9 Cell–metal interface (Ti). Metallic elements from a titanium film dispersed into the cytoplasm and cytoplasmic fusion to Ti film at the location (↑); normal unit membrane at the intact area (↑↑).

D. Tissue Adhesion to Metallic Implant *In Vivo*

From the results of *in vivo* investigations, it appears that the following three steps represent the interfacial reaction between the stable metallic implant and surrounding tissue (Fig. 13) [12].

Step 1. Immediately after the insertion of the metallic implant into bone tissue, the implant surface is coated with a blood clot. Both epithelial and connective tissues are bound to the hydrated layer of the implant surface through the sandwich layer, consisting of the conformation layer and the extracellular matrix (Fig. 11) [58]. The sandwich layer is probably involved in establishing the strong adhesion between the tissue and the implant that develops over time. Gingival epithelial cells, if they retain their vitality, may absorb the water from the sandwich layer, and syneresis of that layer may create high adhesive strength and a good biological seal between the implant surface and the epithelial cells.

Step 2. One month after insertion of implant in alveolar bone, the blood clot layer is reorganized as the phagocytes and also bone tissue injured by drilling and/or overstressed by implant insertion may introduce bone resorption. This is a cause of the initial loosen-

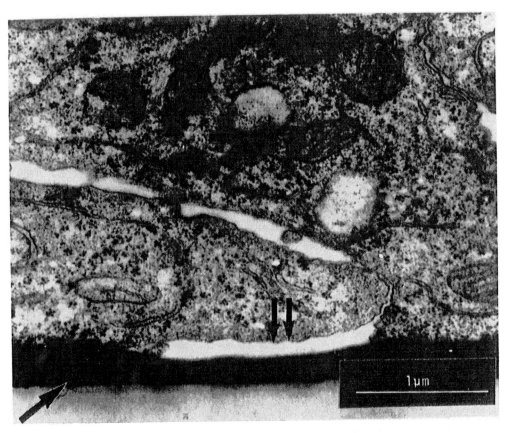

Figure 10   Cell–metal interface (Cr). Metallic elements from chromium film dispersed into cytoplasm and cytoplasmic fusion to Cr film at the locations of (↑); normal unit membrane at intact area (↑↑).

ing frequently observed in clinical practice. After that, fibroblasts and osteoblasts reach the implant surface and come into direct contact with the surface via cytoplasmic extensions.

Step 3. Three months later, 20–50% (in length) direct contact of bone tissue to titanium surface is observed, and in the other part, the collagen fibers bind closely to the implant, lying in parallel with its surface and diagonally between bone tissue and the parallel layer. Gingival epithelial cells adhere closely to the implant surface and may produce hemidesmosomes in their unit membrane.

In observations of experiments on the titanium implant–tissue interface in animals, many reports have noticed an intervention of thin soft tissue layer at the titanium–bone interface, in contrast to the lack of soft tissue intervention in bioactive ceramic implants and coated implants. On the other hand, direct bone contact to titanium has been observed by many investigators since the establishment of low-speed drilling system with suitable irrigation to control thermal necrosis of bone caused by frictional heat of drilling. Recent work has confirmed that osteogenesis started from the Ti implant's surface [59–61] and the bone tissue intimately apposed to the titanium surface through a 50–100 nm

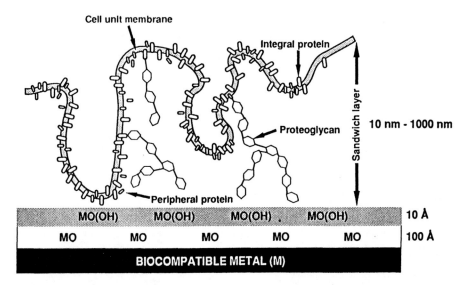

Figure 11 Schema of cell–metal interface. The hydrated surface on a MO(OH) layer produces a conformation layer (see Fig. 4) with adsorption of proteins, polysacchrides, glycoproteins, lipids, and other macromolecules in body fluid containing ions: Ca^{2+}, Mg^{2+}, K^+, N^+, H^+, Cl^-, PO_4^{3-}, O^-, O_2, H_2O, H_2O_2. The conformation layer forms a sandwich layer with fibrous proteins and proteoglycans of extracellular matrix.

thick electron-dense layer [62] consisting of Ti, Ca, P, and macromolecular substances [63], but without any interposed unmineralized connective tissue (Fig. 14) [59–61].

E. Allergenicity and Carcinogenicity of Metal

The possibility of an allergic or carcinogenic effect of a metallic implant is a vital consideration. The transfer of metal from the implant into the surrounding tissues is relevant [64,65]. To determine this, Ti–Al–V and Ti blade implants were inserted into a monkey mandible for 4 months. The transfer of metal from the implant was detected in the soft tissue and bone around the implant by electron-probe plane and linear analysis (Fig. 15). However, there has been no well-substantiated report of a carcinogenic effect of Co–Cr–Mo, Ti, or Ta in dental implants, although a few authors have suggested this.

F. Clinical Use of Metallic Materials

1. Stainless Steel

Medical-use stainless steels, such as SUS-316, SUS-316L, and SUS-317, may be considered as feasible implant materials. However, these metals are ionized and dissolved into the surrounding tissue at a fairly high density [64]. Nickel is not bound tightly in the alloy and is released in living tissue so as to induce cytotoxicity [65]. There is no proof that stainless steel is perfectly safe. Giorgio [66] reported data in support of this fact when he found that stainless steel induced an allergic reaction. McDougall [67] also reported cases of carcinogenesis with 18.8 stainless steel screw implants. Bergman et al. [68], reported high sensitivity to nickel, and the Swedish National Board of Health and Welfare issued a

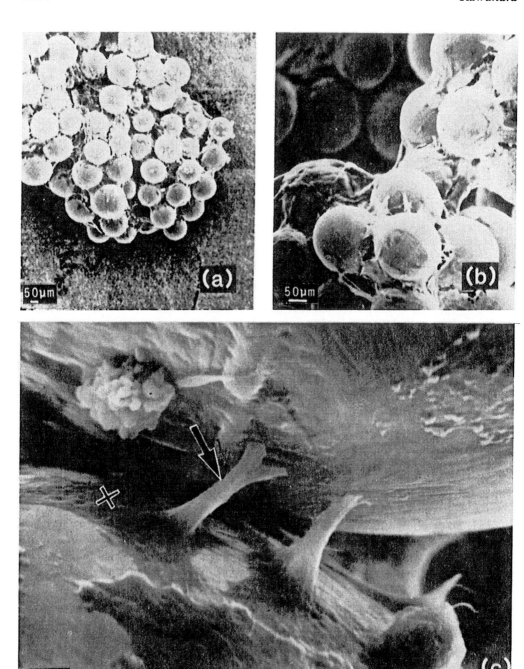

Figure 12   Ti–Mo alloy beads bound together by adhering cells. (a) Bead-cell aggregation after 7-day cultivation in a gyratory incubator. (b) Cytoplasmic bridges among the beads. (c) Cytoplasmic bridges (↑) and cytoplasmic stretch on the bead (X) at higher magnification.

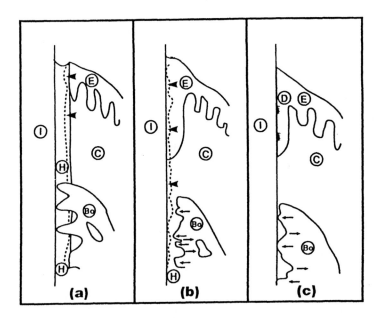

Figure 13   The adhesion process attaching the tissues to the implant surface. (a) Immediately after implantation, (b) one month later, (c) three months later; I, implant; H, bloodclot; arrowhead, sandwich layer; E, epithelial tissue; C, connective tissue; Bo, bone; D, half-desmosome.

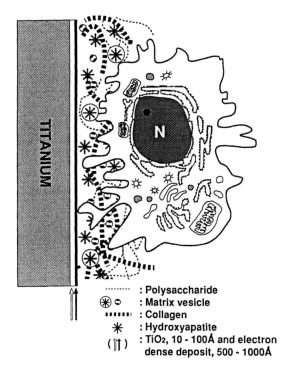

--------   : Polysaccharide

⊛ ◦   : Matrix vesicle

••••••   : Collagen

✳   : Hydroxyapatite

(⇑)   : TiO₂, 10 - 100Å and electron dense deposit, 500 - 1000Å

Figure 14   Bone deposition to titanium.

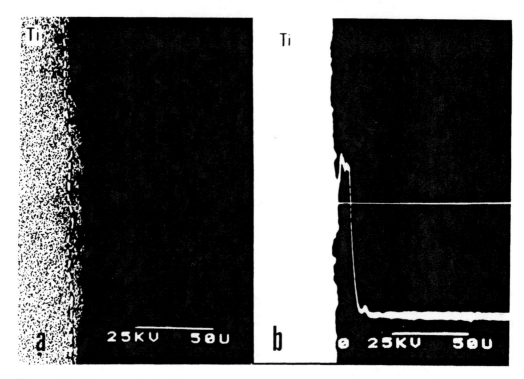

Figure 15 Electron probe analysis on titanium dispersed into the surrounding tissue. Ti implant 3 months postimplantation in monkey mandible: (a) plane analysis; broken line, titanium surface; (b) linear analysis.

statement containing a warning against the use of dental casting alloys containing more than 1% by weight of nickel.

2. Co-Cr-Mo Alloy

Tailor-made implants can be fabricated by casting Co-Cr-Mo alloys that have been made more reliable in mechanical strength and bioinertness by alloying with the purged mother metals; these are the well-known surgical alloys Vitallium and Bionium (Table 5) [70]. The casting technique has been improved in dimensional accuracy and porosity control. At present, the alloy is in wider use as subperiosteal and tailor-made endosseous implants that have a biocompatible surface with favorable oxidation layer of passive state [71,72]. It is important, however, to control the casting microporosity (Figs. 17 and 18).

3. Titanium and Its Alloys

Pure Titanium.   Titanium has suitable oxidation layers of TiO_2 and TiOOH, which have biochemical activities favorable to making a biological seal and osseointegration. Thus it may be recommended as an excellent implantable metal. At present, many types of endosseous implants such as pin, screw, spiral, cylinder, hollow cylinder, blade, plate, anchor, porous, three-dimensional, and so forth, have been made from pure titanium and are widely used. Each implant has achieved a high success rate. Pure titanium is too soft and so has to be hardened in the fabricating process by cold working, which, however, may cause fracture failure due to working strain. And also in the process of fabrica-

Table 5 A. Composition of Co–Cr–Mo Alloy for Dental Implants [70]

Trade name	Co	Cr	Ni	Mo	W	C	Si	Mn	Fe	Other
Vitallium (USA)	62.3	30.8	—	5.0	—	0.4	0.3	0.5	0.7	Cu 3.5
Ticonium (USA)	33.25	31.6	20.1	7.0	—	0.05	0.35	3.0	0.25	Be 0.9
Haynes HS-21 (USA)	60.55	27.5	2.5	5.5	—	0.35	0.6	1.0	2.0	—
Haynes HS-25	48.9	20.0	10.0		15.0	0.1	1.0 (max)	2.0 (max)	3.0 (max)	
Wisil (Germany)	65.45	27.0	—	4.5	—	0.35	0.4	1.1	1.0	0.2
Bionium-S (Japan)	62.0	29.0	—	6.4	—		0.5			2.5
MP 35 N[a]	35.0	20.0	35.0	10.0	—					—
ASTM specification	62.55 … 57.20	27 … 30	2.5 (max)	5 … 7	—	0.20 … 0.35	1.0 (max	1.0 (max)	0.75 (max)	

B. Mechanical Properties of Co–Cr–Mo Alloy for Dental Implants [70]

Alloy	Property				
	Tensile strength	Yield strength	Elongation	Hardness	Modulus of elasticity
Vitallium	72.0	57.2	8	Hv 228	25.2
Haynes HS-25	103.0	49.0	35	Hv 257	24.4
Bionium	81.2	52.5	8.6	Hv 510	25.0
MP 35 N[a]	183.0	161.0	10.0	H_RC 28–34	23.3
ASTM specification	66.0 (min)	45.4 (min)	8 (min)		21.7 (min)
Human Femur	10.5				2.03

[a]I. Miura and H. Hamanaka

1489

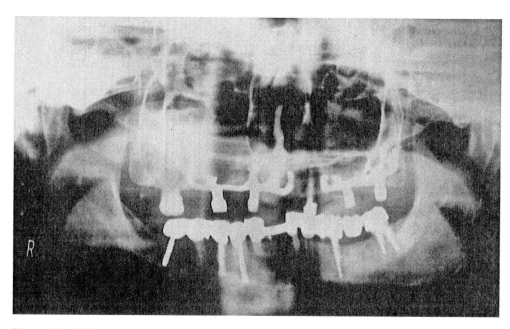

Figure 16 Tailor-made endosseous implant made by Bionium casting system (T. Yamane, 1973).

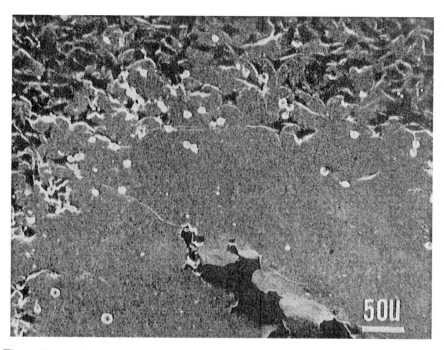

Figure 17 Cell damage around the casting porosity. Test piece: Bionium (Co–Cr–Mo alloy). Cell, L strain; cultivation, 48 h; medium, MEM + 5% calf serum.

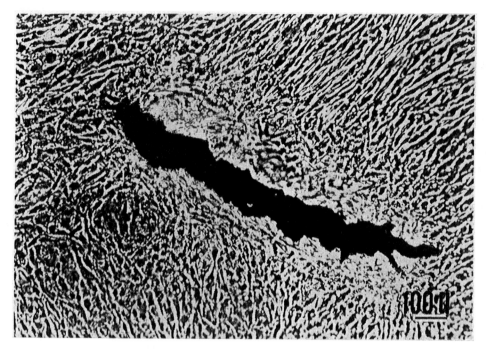

Figure 18 Normal cell growth around the casting porosity. After the test piece was degassed and treated with 4% hydrofluoric acid solution, it did not show any cellular damage around the casting porosity.

tion by lost-wax casting, it is very difficult to maintain uniform surface conditions, mechanical properties, porosity control, and dimensional accuracy [73]. It is not suitable for use as subperiosteal implants at present.

Alloys.   In the past 5 years, Ti–6Al–4V has been utilized for dental implants because of its high mechanical strength and suitable biocompatibility even after casting. However, the possibility of cytotoxicity of vanadium and chronic systemic effects of aluminum and vanadium are of concern in long-term implantation in clinical use [74].

To improve these shortcomings, a new type of titanium alloy of TiX® with low concentrations of iron (0.5%), oxygen (0.5%), and nitrogen (0.1%), was developed for hard tissue replacements. TiX is a useful biometal as wire, plate, screw, and mesh for dental and medical use because of the favorable biocompatibility, high mechanical strength, and easy fabrication by lost-wax casting [30,53]. It is, however, difficult to make TiX with constant composition. We have developed Ti alloys of Ti–Mo, Ti–Pd, Ti–Pt, and Ti–Zr that have excellent biocompatibility and high mechanical strength [52,54]. Porous implants made from these Ti alloys allowed bone ingrowth into pore and created a macroscopic bone anchor for fixing the implant in alveolar bone [30].

The Ti50–Ni50 shape memory alloy Nitinol® has been used in dental and orthopedic surgery. Shape memory implants of Ti–Ni are effective for anchoring bone. However, allergic reaction and tumor formation caused by Ni release should be of concern in long-term implantation in alveolar bone [75–79].

4. Tantalum

Casting of this metal is difficult as its melting point is 2880°C, but it has been used for long time as ribbon, wire, and sheet because it is flexible, forges well, and can be made in any form. Tantalum gauze is especially well known. Kawahara et al. [38] confirmed by *in vitro* testing that tantalum is an inert metal with satisfactory biocompatibility. It is, however, too soft to be used as dental implants. However, tantalum reaches suitable mechanical strength after work hardening and is used, mainly by European implantologists, for Scialom's tripod pin implants [80] and the Tantal–Doppelkliningenimplanta system [81].

V. CERAMIC MATERIALS

A. Stability of Bioceramics

No materials placed within a living tissue can be considered to be completely inert. However, ceramics, by their very nature, do not suffer from corrosion, as do metals or plastics. Extensive progress in ceramic technology has led to the discovery of a variety of materials whose chemical, physical, and mechanical properties confer a high degree of stability, bioactivity, and bioadhesiveness [82,83]. This makes them suitable candidates for long-term implants within living tissue.

As mentioned above, the more the implant material resembles the host tissue in molecular structure, the more severe is the reaction in the living tissue (Figs. 2 and 19) [84]. The reason for this is the difficultly the host experiences in discriminating immunologically between implant material and host tissue. Conversely, removal of the antigenic stimulus from implanted material or transplants results in a more inert nonbiological agent that the host easily recognized as foreign. Ceramics that have molecular structures completely different from those of living substances are generally more stable inside the living body and have a high adhesion to the surrounding tissue as shown by *in vitro* and *in vivo* experiments.

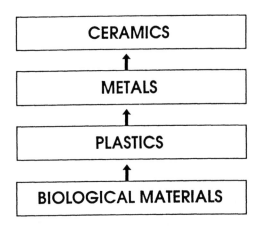

Figure 19   The greater the difference between a material and the living substance, the more inert is the reaction within living tissue.

B. Fabricability Versus Bioinertness

It is interesting to note that those materials that are more easily fabricated are often the most unstable, degradable, irritable, and toxic in living tissue (Table 6). This is a function of material bonding, atomic structure, and defect concentrations. *In vivo* stability increases from self-curing plastics to injection-moldable polymers. Metallic materials with higher melting points are more stable and inert in living tissue but they are more difficult to form. Ceramics with high sintering temperatures (e.g., alumina, zirconia [85,86], silicon nitride, and vitreous carbon) have high stability in living tissue although vitreous carbon implants have not been used in clinical cases due to inappropriate designs and the esthetic problem of its black coloration.

Some bioinert ceramics possessing relatively low crystallization and sintering temperatures demonstrate decreased stability and mechanical strength after the implantation, compared to those bioactive ceramics possessing higher processing temperatures, which are more stable in living tissue. From these findings, one may conclude that single-crystal alumina of sapphire should have the highest stability and bioinertness as a biomaterial for hard tissue replacement [87–89].

C. Mechanical Strength of Bioinert Ceramics and Bioactive Ceramics

1. Bionert Ceramics

The stable ceramics such as alumina, zirconia, silicon nitride, vitreous carbon, and LTI carbon [90–92] exhibit high biocompatibility and adhesion to tissues, and have a higher mechanical strength than bioactive ceramics. Complex and slender designs for hard tissue implants are possible, because no change of mechanical properties or chemical components is observed even after long implantation times. In recent years, endosteal dental implants of sapphire, Bioceram® (Fig. 20) [87] have been reconsidered with buffer action of soft tissue intervention for 50–300 μ thickness, controlled by the suitable biting stress of 2–8 MPa to the surrounding bone tissue.

2. Bioactive Ceramics

Hench and Paschall [93] have developed a glass with soluble additives called *bioglass* (SiO_2 30–45%, Na_2O 24.5%, CaO 12.25–24.5%, P_2O_5 6%, CaF_2 6–12.25%). According to their report, insertion of this material into bone resulted in silica-rich layer formation on the bio-glass surface and promoted apatite crystallization due to the dissolution of calcium and phosphorus from the bioglass surface into the adjacent tissue. The resultant chemical bond between the silica and collagen protein then enabled the crystallization of hydroxyapatite. Brömer et al. [94] also have developed a new bioactive glass ceramics, Ceravitar® (SiO_2 40–45%, CaO 20–30%, Na_2O 15–30%, P_2O_5 5–10%), which contains K_2O and MgO (5–10%) in order to match the ionic constitution of blood and body fluid. Normal bone formation and a tight contact between the new bone tissue and the glass ceramics without soft tissue intervention were observed.

Mouroe et al. [95] proposed the use of sintered, fine-grained, synthetic hydroxyapatite for dental implants, since this material resembles tooth and bone substance. Fisher-Brandies and Dierlert [96] implanted a dense hydroxyapatite [$Ca_{10}(Po_4)_6(OH)_2$] on unreamed canine cortical bone and concluded that a direct contact formed between the hydroxyapatite and bone without any connective tissue intervention. However, these bioactive ceramics have shortcomings in their bending strength and impact strength (Table 6, B). Ceravitar has a strength of 98 MPa and bioglass probably less than 100

Table 6   Fabricability and Bioinertness of Implant Materials
A.   Polymeric and Metallic Materials [83]

Materials	Molding temperature (°C)	Fabrication	Bioinertness	Bending strength (MPa)	Modulus of elasticity (GPa)
Plastics					
Polymethylmethacrylate (composite)	10–15	Self-curing	*	25–188	2.4–5.8
Tetramethylolmethane (composite)	10–15	Self-curing	*	88–90	6.0–6.5
Silicone	10–15	Self-curing	*		
Polymethylmethacrylate (dental)	50–70	Heat-curing	**	104.65	2.3
Polyurethane	80–100	Heat-curing	**		
Segmented polyurethane	80–120	Heat-curing	**	2.45–31.36	0.03–0.17
Polymethylmethacrylate	220–280	Injection	**	88.2–102	0.3–1.9
Polysulfone (composite)	320–370	Injection	***	114.36–135.24	2.1–5.6
Polysulfone			***	89.2–127.4	4.96–4.80
Polyethylene (HDPE)		Injection	***	29.4	0.49
Metals and alloys (bioreactive)					
Co–Cr–Mo[a]	1300–1400	Casting	***	196–294 (deform)	235
Fe–Ni–Cr	1250–1350	Casting	***	98–196 (deform)	205
Ti[b]	1800	Welding Casting	***	30–230 (deform)	
Zr	1700	Welding	***		
Ti–Al–V[c]	1650	Casting	***	293–340	117
Ti–Mo[d]	1850	Sintering	***		
Ti–Zr[e]	1600	Sintering	***		
Ta	2996	Welding	***		

MPa. Fisher-Brandies and Dierlert [96] reported that sintered dense hydroxyapatite has bending strengths of between 113 and 196 MPa. The latter value was, however, obtained from a small, highly polished sample. Therefore, its practical value would, in fact, be about 115 MPa. Kokubo et al. [97] reported a new type of glass ceramic containing apatite and wollastonite (CaO 44.9%, SiO_2 34.2%, P_2O_5 16.3%, MgO 4.6%, CaF 0.5%). It has a consistently high bending strength of 157 ± 8 MPa. Gross and Strunz [98] presented their histological observations of the suitable interfacial reaction between glass ceramics and new bone, but the reported bending strength was still only 113–157 MPa. Strength is clearly a problem with the bioactive ceramics. Furthermore, the solubility and leaching of the bioactive elements may reduce the mechanical properties in long-term implantation [83,98,99]. These bioactive ceramics cannot be used singly due to poor mechanical strength but may be effective as coating materials for metallic and nonmetallic substratum or as a component in composites with polymeric and metallic materials.

It is believed that bioactive ceramics might form an organic bond with new bone

Table 6 (Cont'd)   B.   Ceramics [83]

Materials	Fabrication temperature (°C)	Fabrication method	Bioinertness	Bending strength (MPa)	Modulus of elasticity (GPa)
Bioactive ceramics					
Bioglass[f]			***	98	
Glass ceramics[g]	1050–1450	Melting Sintering	***	137–150	98
Hydroxyapatite[h]	900–1200	Sintering	***	105–215	41.2–121.0
Bioinert ceramics					
Vitreous carbon[i]		Sintering	****	68.6–205	16.8–27.4
LTl carbon[j]	1500	Sintering	****	508	50–70
Zirconia[k]	1500–2000	Sintering	****	780	225
Silicon nitrate[l]	1800–2000	Sintering	****	490	294
Silicon carbide[m]	2000	Sintering	****	490	392
Polycrystal alumina[n]	1700–1800	Sintering	****	372	372
Single-crystal alumina[o]	2050	Melting	*****	1274	392
Hard tissues					
Cortical bone				29.4–186	10.8–17.6
Cancellous bone				0.4	0.09–0.19
Enamel				3.2–12.7	41–56
Dentine				6–16	12–18

[a]Venable (1937); Kawahara H (1973)
[b]Linkow (1968); Kawahara (1965)
[c]Aragon and Hulbert SF (1972)
[d]Miura and Kawahara (1980)
[e]Kawahara and Miura (1984)
[f]Hench (1973)
[g]Blencke and Brömer (1975); Pernot and Zarzycki (1979); Yamamuro et al. (1980)
[h]Mouroe et al. (1970); Aoki et al. (1976)
[i]Dumas M et al. (1974)
[j]Bokros (1972)
[k]Hulbert (1973); Nagai et al. (1982)
[l]Hulbert (1970); Griss (1978)
[m]Hulbert (1970)
[n]Sandhous (1965)
[o]Kawahara (1975)

following osteogenesis and induce close adhesion or osteoankylosis between the new bone and the bioactive ceramics [100]. These expected phenomena do not always occur, because osteogenesis is an extremely complex interfacial phenomenon, especially under a functional load [101]. Instead of conducting new bone growth, some implants may cause bone resorption because of the elements released into the surrounding tissue. Klawitter and Hulbert [102] reported that even a slight hydration of calcium aluminate depressed its osteogenic effect. Thus, it may be that osteogenesis is depressed by using a soluble ceramic material with an unstable surface.

In summary, bioactive ceramics are not suitable materials for single use in dental implant due to the poor mechanical strength and the continued reduction of strength after long implantation times.

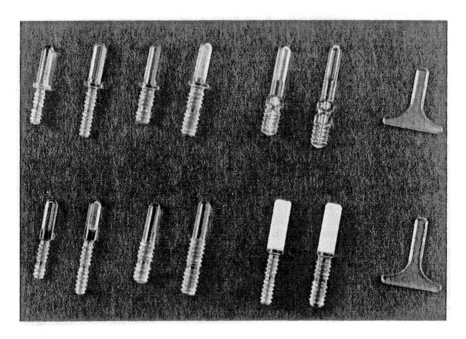

Figure 20   Sapphire dental implants.

D.   Tissue Adhesion to Bioactive Ceramics and Bioinert Ceramics

1.   *Bioactive Glass and Glass Ceramics*

Bioactive reactions to bone tissue have been reported by Hench and Paschall [93] and were summarized as follows [100]:

1. Exchange of Na^+ or K^+ in bioactive glass with H^+ or H_3O^+ in body fluid.
2. Loss of soluble silica from the glass to the body fluids.
3. Concentration and repolymerization of the soluble silica in body fluid on the glass surface results in a SiO_2-rich layer on the glass surface.
4. Migration of Ca^{2+} and PO_4^{3-} groups through the SiO_2-rich layer.
5. Formation of $CaO-P_2O_5$-rich film on top of the SiO_2-rich layer.
6. Growth of a SiO_2-rich layer and a $CaO-P_2O_5$-rich film by incorporation of soluble calcium phosphate from the body fluid.
7. Crystallization and polycrystallization of an amorphous $CaO-P_2O_5$-rich layer.
8. Incorporation of organic components involves agglomeration and chemical bonding of the apatite crystals.

Bioactive glass ceramics, Apoceram, Ceravital®, and HA–AW glass ceramics [97] are almost the same as the bioactive glasses in their constituents as well as in apatite crystallization and bone bonding mechanism to the materials.

2.   *Bioinert Ceramics*

In the light of all these findings, the only base materials suitable for the construction of artificial tooth roots and the replacement of bone are those that are chemically inert in living tissues and have sufficient mechanical strength. The stable ceramics such as sapphire, polycrystalline alumina, silicon nitride, zirconia, and diamond-like carbon exhibit a high degree of biocompatibility and excellent adhesion to tissue, and are useful materials from which to form artificial roots and bone and joint prostheses [103,104].

The surfaces of these ceramics adsorb water, giving an aqueous layer that cannot be removed even in a vacuum at a temperature of 400°C (Fig. 21). The surfaces are not only stable but also nonirritating, noncytotoxic, noncarcinogenic, and superior in adhesion to the surrounding tissue. For example, when such ceramic beads were placed in an agitated cell suspension of Hp mesenchymal cells or HeLa S3 epithelial cells, a monolayer of cells closely adhered to the surface after 2 days *in vitro*. After 3 days, the tissue layer built up on the surface of ceramic beads, with cell-to-bead and cell-to-cell adhesion. After 4 days, the ceramic beads were bonded together with strong cell-to-cell adhesion (Fig. 22) [11]. If the ceramic beads were coated with oily materials, no cell aggregation was observed. Another interesting factor was investigated using culture techniques. When a sapphire plate was placed in an MEM (minimum essential medium) suspension of the cells, the cells showed cytoplasmic adhesion and stretching due to their high wettability with respect to the plate, and they adhered closely to the plate with a strong adhesion of 3.7 dyn/cm^2 for 10 min. This was measured using viscometric methods [105,106]. If the plate was contaminated with greasy materials, the cell adhesion was disturbed and a decrease of adhesive strength resulted; therefore, it is important that all implant surfaces should be kept clean during operations (Figs. 23 and 24).

E. Need of Chemical Bond?

From *in vitro* and *in vivo* observations, the interfacial reaction between bioinert ceramic implant and surrounding tissue is different from that of bioactive ceramics, as are metallic implants or Ti, Zr, and Ta, which have virtually no soluble components or released elements. These released elements may conduct new bone formation and support chemical bond formation between implant and bone. Unlike bioactive ceramics, the chemical bond of bone tissue to bioinert ceramics cannot be anticipated, but the physical bond is required. Is the chemical bond needed to make reliable fixation of dental implant in alveolar bone? Mechanical interlocking at the micron to millimeter level at the bone–implant interface may play a dominant role in implant fixation, rather than chemical bonding, for long-term implantation, as evidenced by 10 years of clinical radiographic investigations on the implant–bone interface [107]. And also, gingival epithelium adhered to the bioinert ceramic implant and produced a suitable biological seal with hemidesmosomes in their unit membrane [4].

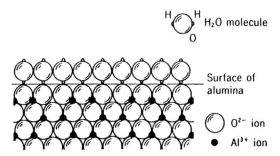

Figure 21   A model for water adsorption on the surface of crystal alumina.

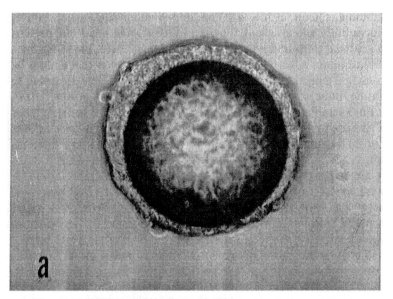

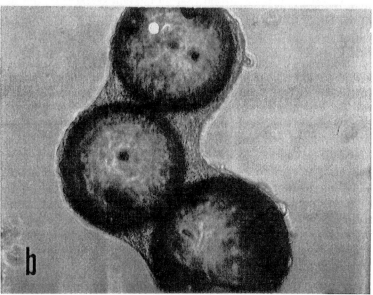

Figure 22   Cell adhesion to glass beads in gyratory cultivation. Cell, HP; medium, MEM + 5%
calf serum. (a) Tissue reconstruction on glass bead, 3-day cultivation. (b) Cell–bead aggregation,
4-day cultivation.

VI.   SURFACE MODIFICATION

The aim of surface modification is to ensure reliable fixation of implant in alveolar
bone and to establish a biological seal of epithelial tissue adhesion (attachment), by
establishment of chemical and physical bonding (angstrom to nanometer level) and/or
mechanical interlocking (micrometer to millimeter level) with bone ingrowth into rough

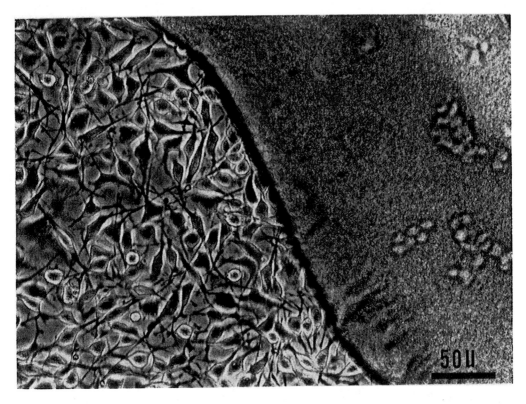

Figure 23 Cell adhesion to a single crystal plate. No cell adhesion to the plate coated with silicone (right). Cell, L; medium, MEM + 5% calf serum; 2-day cultivation.

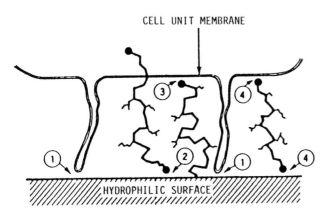

Figure 24 Bonding mode between cell and bioceramics. 1, microvilli; 2, glycoprotein and main protein chain; 3, grafting molecule; 4 molecule in sandwich layer.

and porous surface of the implant. There are three methods for surface modification, as follows (Table 7):

1. Coating by plasma spraying, dipping, sintering, and cementing
2. Chemical treatment with acidic solution, anodic oxidation, and ion plating
3. Mechanical treatment by sand blasting and machining

A. Coatings

Four benefits are considered in the coatings on metallic implant: (1) increase of bonding strength at implant–bone interface, (2) prevention of metallic ion release, (3) masking of surface contaminations caused by fabrication process, and (4) making a rough and porous surface.

1. Hydroxyapatite and Bioactive Glass/Glass Ceramics

In recent years, many papers have reported osteoconductivity of new bone growth [108–112] and increase of bone bonding strength [113–117] in coated metallic implants with hydroxyapatite (HA) and bioactive glass/glass ceramics (BG). Coated HA [118–122] and BG [123–128] exhibit direct bone bonding without any soft tissue intervention at the implant–bone interface and have higher bone bonding strength, compared with that of uncoated metallic implants. *In vitro* examinations have reconfirmed that HA or BG is a favorable coating material for the growth and differentiation of osteoblast-like cells [129,130].

HA coatings have a dominant effect on the initial fixation of the coated implants. Bonding strength of coated HA to bone is higher than that of the Ti–HA interface, with failed first under mechanical stress. It was suggested that the bonding at HA–bone interface is chemical in nature (active), in contrast to the osseointegration of titanium (passive) [118]. On the bonding mechanism between bone and HA, Mouroe et al. [95] and Bagambisa et al. [122] reported that bone tissue components bonded directly to HA crystals via a recrystallization layer on coated HA.

Many reports have indicated that significant increase of bone growth speed and bone bonding strength was realized by coating HA and BG on metallic implants. However, these new bone conductivities are limited to the early stage, up to 26 weeks. Signs of partial dissolution of HA layer are evident at the HA–bone interface 12 weeks postimplantation [118,119,131]. Gottlander et al. [132,133] reported a greater degree of bone contact area to HA-coated Ti implant 6 weeks postimplantation, compared to uncoated

Table 7 Surface Modification

Method	Process	Material
Casting	Plasma spray, sintering, dipping	HA, BG, Ti, Zr, Al_2O_3
Chemical treatment	Acidic solution, gas, ion plating, anodic oxidation, heating	HF, H_2O_2, H_2SO_4, HCl, HNo_3, H_3PO_4, N, O
Mechanical treatment	Sand blasting, machining	Alumina, garnet, corundum, silicon, carbide, TiO_2, Fe_2O_3

Ti implant, but the bone contact area decreased 12 months postimplantation compared with the increase of bone contact in uncoated Ti implant. Burr et al. [134] confirmed that HA-coated implants demonstrated significantly more rapid bone growth than uncoated implants through 16 weeks postimplantation but not by 26 weeks. They concluded that the Ti implant surface catches up to and may surpass the degree of osseointegration seen with HA-coated implant. These conclusions are in agreement with statistical investigations on long-term animal tests and clinical observations, which indicate no significant difference in bone contact area and bonding strength between the coated and uncoated Ti implants at 3 years postimplantation [135]. These phenomena may be explained by a bonding mechanism at the HA–bone interface involving dynamic equilibrium [136], gel–sol equilibrium [137], and constant ion exchange [138]. Thus HA is not stable but has a bioactive surface that may produce bone conductivity and chemical bonding. On the other hand, bioactive materials deteriorate and weaken within the living body.

The same findings have been reported for bioactive glass/glass ceramics, whose leaching and corrosion in living bone have been pointed out [83] and observed with backscattered electron images [139]. Quality control to minimize product failure caused by the coating process is difficult. Also, micromotion within the living body may lead to continuous disruption at the glass–tissue and/or glass–substratum interfaces with continuous leaching of the coated layer of bioactive glass/glass ceramics [140,141].

2. Is HA Coating Beneficial?

Natural HA in a tooth structure of homo/hetero graft is resorbable in the host alveolar bone. Of course, almost all synthetic HAs undergo slow dissolution and decomposition after the implantation, except for ultradense HA. As mentioned in Section III.B, "Tissue Response and Molecular Structure," the more the implant material resembles host tissue in molecular/atomic structure, the more severe are the reaction and the material's decomposition in the living body. Synthetic HAs that are closely similar (not equal) to natural HA will be easily dissolved and digested in the host bone. This is a natural phenomenon, considering the resorption of heterogeneous/homogeneous transplanted tooth in the host bone [142,143]. Ducheyne et al. reported that

> Calcium phosphate ceramics can bond to bone tissue, can enhance bone formation and can be employed as synthetic bone grafts. However, one is hard pressed to formulate mechanisms of action, or indicate that the sequence of reactions is the same in each case. In short one does not know the fundamental reason why, biologically these materials perform the way they have been observed to do in the numerous animal and clinical studies reported in the literature [144].

On the other hand, Kawahara et al. [12,59] and Steflik et al. [60] presented electron micrographs from which one could estimate the chemical bond formation at the Ti–bone interface. And Kasemo [145] described bone bondings to titanium with weak, stable van der Waals forces as well as strong, unstable bonds, and concluded that there was a dynamic equilibrium occurring between the titanium oxide layer and living bone. As described above, titanium/alloys are not *bioactive but bioreactive.**

Zablotsky et al. [146] concluded that HA-coated implants have a significant place in implant dentistry, especially for those sites that are considered initially compromised.

Bioactive means (bone)tissue conductivity. *Bioreactive* does not mean (bone)tissue conductivity but bioreactivity at the material–tissue interface, which can produce chemical bonding.

However, no implant should ever be used at the compromised site. If it is to be used in a compromised case, prior repair of poor alveolar bone by bone augmentation and guided tissue regeneration (GTR) must be undertaken.

In summary, HA/BG coatings have the following advantages at the early stage of implantation: (1) direct contact at the bone–HA interface without any soft tissue intervention, (2) higher bonding strength in HA/BG-coated implants than in uncoated implants, (3) wider bone contact area in the coated than the uncoated implants. However, it is not clear that the HA/BG coatings are of benefit for long-term clinical use, because of their detachment, leaching, deterioration, and dissolution in living bone. Many failures have been experienced in our clinical evaluation because of fracture at the HA–Ti interface or within the coated HA layer, which may be caused by deterioration and leaching over the 3 years following implantation.

Long-term clinical data have indicated problems associated with the use of HA/BG-coated dental implants at compromised sites, uncoated Ti/alloy implants would be more effective in dental reconstructions.

3. Surface Roughness and Porous Structure

The lack of sufficient bone bonding to the implant surface has created a demand for new and improved materials that can be firmly and permanently attached to bone tissue. There are two possible solutions. One is osseointegration created with bioactive materials. The other is mechanical anchor with bone ingrowth into a rough and porous surface. Osseointegration may be produced by chemical and/or physical bonding at the bone–implant interface. Many papers have been published on osseointegration in bioactive ceramics of bioglass/glass ceramics, hydroxyapatite, and TCP, and even in metallic materials of titanium, zirconium, and their alloys. However, these osseointegrations have been confirmed only under non-load-bearing conditions. On the contrary, evidence of bone resorption was observed constantly under load-bearing conditions [83,147]. The resorption area with soft fibrous tissue may have a buffering effect upon the functional impact. It is, however, apt to induce extensive bone resorption and loosening of the implant, due to uncontrollable phenomena of bone physiology under the load-bearing condition.

A number of clinical reports indicate that reliable fixation of dental implants depends upon macroscopic mechanical interlocking at the implant–bone interface with rough surface, porous structure, vents, holes, polycapillaries, and screws rather than osseointegration produced by changes in physicochemical characteristics of the implants surface modified with HA, BG, and other bioactive materials. In short, HA and BG coatings are of less merit when compared with macroscopic designs having an implant–bone mechanical anchor [148]. Surface roughness, porous, or polycapillary structures are essential factors for long-term stability of dental implants that depend upon mechanical interlocking at the implant–bone interface.

An implant–bone-mechanical anchor produced with surface roughness ($Rz = 38$ μm or more) formed by Ti plasma spray coating and a porous structure (pore size 320 μm or more) by Ti formed bead coating, created reliable fixation of the Ti implant in bone tissue. The bone bonding shear strengths were 10–20 times higher in the push-out test than that of polished and acid-pickled surfaces ($Rz = 0.6$–5.3 μm). The acid-pickled surfaces produced higher shear strength in proportion to the increase of surface roughness, but the difference is not significant due to large variations that may be caused by soft tissue capsulation (Figs. 25–28). These shear strengths in the push-out test may be 2 or 3 times higher than those of implants inserted into human maxillomandibular bone. In

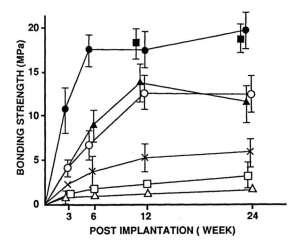

Figure 25 Bone bonding strength and surface roughness of titanium implant. Test method, push out; Bone, cortical bone of dog tibia; test piece: ϕ5 mm × 15 mm, cpTi. $\triangle$: barrel polish, Rz 0.6 μm, Rmax 3.5 μm; $\square$: 4% HF 60 sec, Rz 2.0 μm, Rmax 4.1 μm; ×: 8% HF 180 sec, Rz 5.3 μm, Rmax 10.2 μm, $\bigcirc$: Ti plasma spray, Rz 38 μm, Rmax 52 μm, $\bullet$: porous, pore size 460 μm, $\blacktriangle$: HA rod, Rz 0.7 μm, Rmax 3.5 μm, $\blacksquare$: polycapillary Ti plate, capillary size, ϕ500, 1000, 2000 μ.

fact, clinical usage tests demonstrated lower shear strengths in the implants inserted into cancellous bone in monkey's mandibles than that in the cortical bone of dog's tibia [149–157].

B. Chemical Treatment

The aim of chemical treatment is to remove any contamination occurring during the fabrication process and clinical preparation for implant surgery, and also to ensure a clean, aseptic, and biocompatible condition of the implant surface. In general, the following three methods are used: (1) pickling with acidic solution, hydrofluoric (HF), sulfuric (H_2SO_4), nitric (HNO_3), phosphoric (H_3PO_4), and organic acids; (2) anodic oxidation with electric current in acidic solutions; (3) nitrification with nitrogen gas and heating at 750°–900°C and TiN ion plating.

In this chapter, as representatives of chemical treatments, three effective and economic methods of HF pickling, anodic oxidation, and nitrification technique are described.

1. Pickling with Hydrofluoric Acid Solution

Ti plate are treated for up to 120 sec in 4% or 8% HF solution (4% HF, 8% HF), and then with 4% HF + 8% H_2O_2 solution to obtain suitable surface roughness and biocompatibility. The H_2O_2 may produce a more stable surface by oxidizing the reactive site exposed on the HF-treated Ti surface [56].

Microgeometric structures of the treated surface resemble a crowd of craters with 2–10 μm diameter and 2–4.8 μm height of somma, which changed with treating time and HF concentration. Higher HF concentration and longer treating time increased the diameter and height of the crater's somma, and were accompanied by corrosion micropits at the bottom of the enlarged craters. Treatment with 4% HF, for 60 sec, followed by 4%

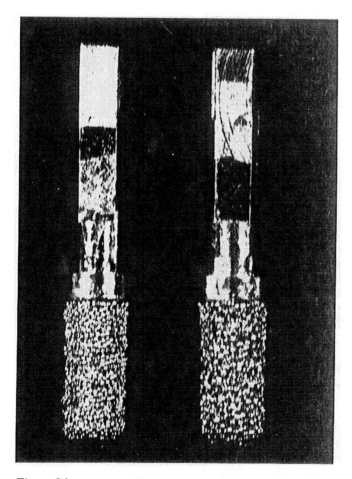

Figure 26 Porous implants with various pore sizes made by sintering with titanium beads. Test piece, ϕ5 mm × 15 mm; pore size, 320 μm (left) and 460 μm (right).

HF + 8% H_2O_2 for 15 sec produced the most suitable microgeometric structures. This treatment may make the implant–bone microanchor less likely to loosen due to tissue irritation caused by the sharp crest of somma under micromovement and also may impede progressive corrosion (Figs. 29 and 30) [158].

Corrosion resistance of a titanium surface with the double treatment of 4% HF, 60 sec → 4% HF + 8% H_2O_2, 15 sec, designated treatment (a), was measured by electrochemical corrosion analysis in Ringer's solution and tissue culture medium of MEM, and compared with that of four other treatments: (b) a sound specimen, (c) barrel-polished specimen, (d) nitrified specimen, and (e) a specimen treated with 4% HF, 60 sec. Ranking of their corrosion resistances is as follows: d > e > b > c > a in Ringer's solution and d > a > b > c > e in MEM (Fig. 31) [158]. An inversion of the ranking of (a) in MEM and Ringer's solution may be caused by the more active adsorption of macromolecular substances in MEM due to the H_2O_2-posttreated surface, which has Ti-superoxide

Figure 27   Bone ingrowth into Ti porous implant, 24 weeks postimplantation.

and Ti-peroxide. Animal experiments demonstrated that bone bonding strength of dog's tibia to the Ti surface was increased by HF pickling [159].

In conclusion, the double treatment of 4% HF, 60 sec pickling and 8% H_2O_2 post-treatment is an effective method for removing contaminations occurring during the fabrication process and to ensure cleanliness and sterility of metallic implants, Ti alloys, and Co–Cr–Mo in chair side work for metallic implant surgery.

2.   Anodic Oxidation

Anodic oxidation can produce a stable, biocompatible, and colorful oxidation layer on Ti alloy surfaces by an electrochemical reaction between an anode of Ti alloy and a cathode of Pt, Ag, Ti, stainless steel, etc., in an electrolytic medium, such as H_2SO_4, H_3PO_4, citric acids, etc., and in complex solutions containing organic substances.

For example, when Ti plate was treated by an electric current in the voltage range 10–100 V and 10 mA/cm^2 in a medium of 10% dextrine aqua dest. solution (10 parts) and H_3PO_4 solution (1 part), 10 interference colors were demonstrated from light brown to pinky-blue, according to the applied voltage (Table 8). When Ti plate was treated at 50–70 V and 25°C in the medium, the surface changed to a golden color by anodic oxidation. The golden-colored layer had a highly stable and biocompatible surface that demonstrated suitable cell attachment and cytoplasmic adhesion of epithelial cell of HeLa S3 and epithelial-like cells derived from human gingiva [160–162].

Compared with untreated Ti implant, a healthier appearance and a natural pink tone of the gingival tissue around the golden Ti postneck were demonstrated in animal experiments using monkeys as well as in clinical observations [156].

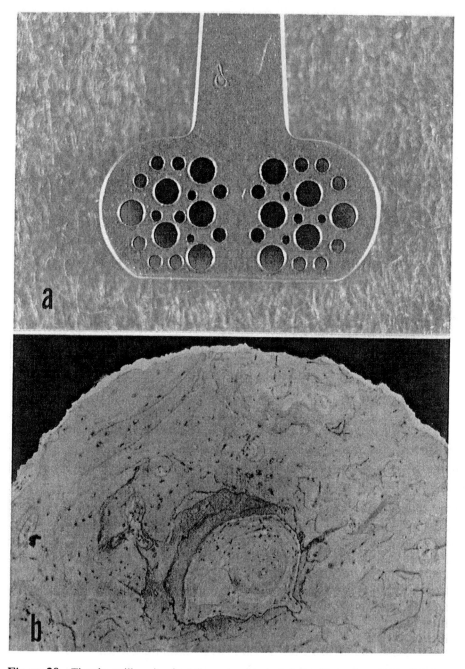

Figure 28   Ti polycapillary implant for animal test. (a) Capillary size: ϕ300, 600, 1200 μm. (b) New bone growth into the ϕ1200 μm capillary.

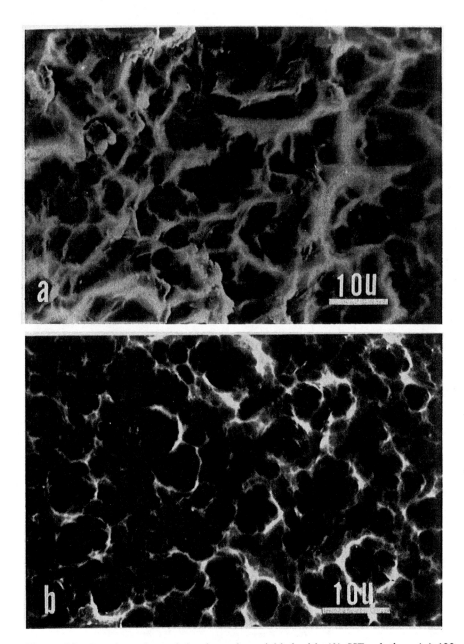

Figure 29   Rough surface of titanium plate pickled with 4% HF solution. (a) 120-sec pickling produced large diameter and depth of craters. (b) Double treatment with 4% HF 60 sec and 4% HF + 8% H₂O₂ 15 sec produced round form of suitable craters.

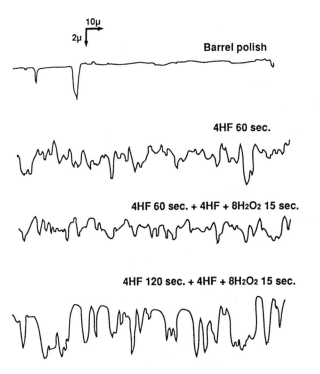

Figure 30 Surface roughness of pure titanium treated with 4% hydrofluoric acid solution (4HF) and 8% hydroperoxide solution ($8H_2O_2$).

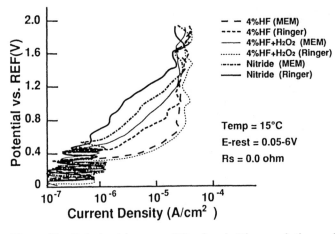

Figure 31 Polarization curve of titanium in Ringer solution and tissue culture medium of MEM.

Table 8 Interference Color of Titanium Surface Treated by Anodic Oxidation

Voltage	Interference color
10	Light brown
20	Dark violet
30	Cobalt blue
40	Light blue
50	Golden-green
60	Golden-yellow
70	Golden-red
80	Light violet
90	Violet
100	Pinky-blue

Note: Test piece, $\phi300$ mm $\times$ 1.0 mm of cp Ti. Medium, 10 parts of dextrine 1.0% water solution and 1 part of H_3PO_4. Temperature, 25°C. Electric current, 10 mA/cm^2

3. Nitrogen-Ion Implantation and Biological Seal

There are two submethods in the Ti alloys nitrogen-ion method. One is heating at 750°–900°C in an atmosphere of nitrogen gas, and the other is the ion-plating method in which evaporated metal of Ti and nitrogen gas are used in a vacuum chamber to form a TiN intermetallic compound on the Ti alloy surface. Nitrogen-ion implantation can produce a golden color and high physicochemical stability.

For dental implants it is effective to have nitrogen-ion treatment at 830°C for 3 h in nitrogen gas atmosphere, giving a nitrified layer 300–400 Å thick that has a suitable golden color and high wear resistance. The nitrified surface of Ti implant demonstrated a favorable biological seal with close adhesion (attachment) of epithelial cells in the same manner as pure titanium (Figs. 32 and 33).

In clinical cases, the nitrified golden surface demonstrated a healthy, natural color tone of gingival tissue, and high chemical stability and high wear resistance to brushing and scaling, which may maintain the suitable biological seal for long-term implantation [163].

C. Scientific Considerations

1. Porous Structure

In vitro tests on the cell ingrowth mechanism into porous disks (average pore size, 20–250 μm; porosity, 27.5–36.5 Hg%) made of alumina [164] and titanium alloys [52] were carried out by using L292 cells and bone marrow cells. The disks were cut into four pieces and the cut surface was divided into five areas to calculate numbers of ingrowth cells in the pore after 24- and 48-h cultivation.

Ingrowth cells had normal morphological appearance and close adhesion to the pore wall. And the cell number in the pore increased as the cultivation time increased from 24 to 48 h. After 48-h cultivation, no cell ingrowth was observed deeper than 1.2 mm in the small pore size of 20 or 50 μm. However, in the large pore size of 135 and 150 μm, 9–32

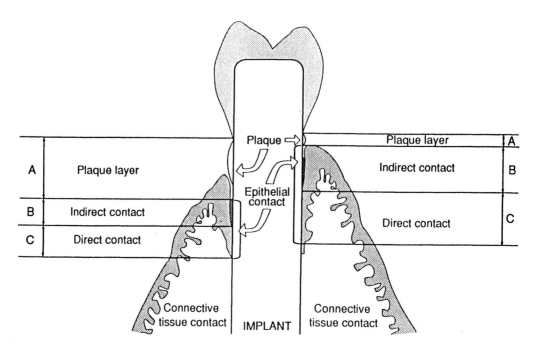

Figure 32 Three layers in biological seal of titanium implant. Plaque layer was located at the shallow area of implant crevice. Indirect contact layer with viscous substance was observed at the epithelial cell–implant interface. This layer may function as protection against bacterial attack from the plaque layer. Direct contact layer with chemical bond at epithelial cell–implant interface was located at the deepest area beyond the bottom of crevice [161].

cells could be counted in the deep area (Fig. 34). Summarizing the results, the increase of cell number in the 0.4 mm depth of the 150 μm pore size was larger than that of the cell growth curve in the normal cell culture commonly used in an open system of petri dishes. From these findings, it is seen that the increase of cell number in the pore depends upon both cytoplastmic locomotion and cell mitosis; the former factor may have a dominant role in the deep area of the porous structure [164].

In vivo tests reported that a porous structure of metals, plastics, and ceramics could create reliable fixation for dental implant by bone ingrowth into the pores. The first studies on the speed of bone ingrowth into various kinds of porous ceramics were carried out by Klawitter and Hulbert [165] and by Hulbert et al. [166]. Selting and Bhaskar [167] measured the structural strength of the interface between bone and porous alumina implant, and concluded that normally healed cortical defects 10 weeks postoperatively had about 80% of the uninjured cortical strength. Predecki et al. [168] investigated the dependency of pore channel diameter on bone ingrowth kinetics using alumina and titanium. They reported that the most rapid bone ingrowth was obtained with 500, 525, and 1000 μm diameter channels in all the titanium and alumina samples. However, the fraction of any given channel cross section filled with bone was observed to decrease as the channel diameter increased. The presence of interconnecting fine porosity was found not to be essential for bone ingrowth, nor did it have a major effect on ingrowth kinetics

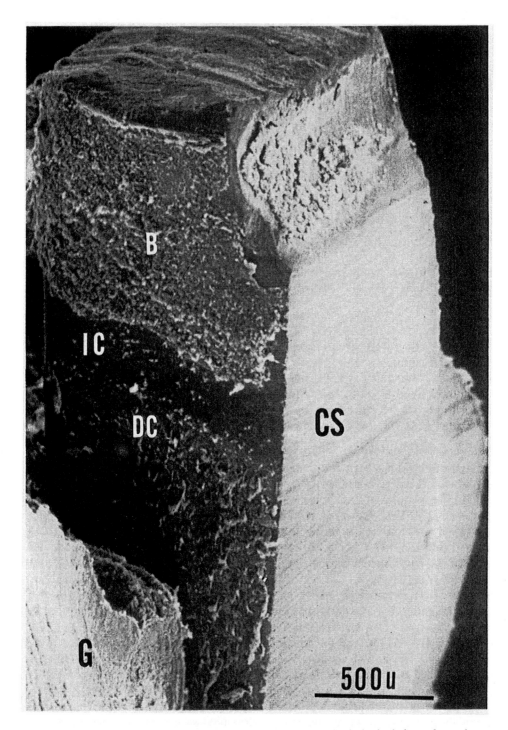

Figure 33 SEM of titanium implant surface after removal of gingival tissue from the post. B, bacterial plaque layer; IC, indirect contact layer; DC, direct contact layer of epithelial cells; CS, cut surface; G, removed gingival tissue [161].

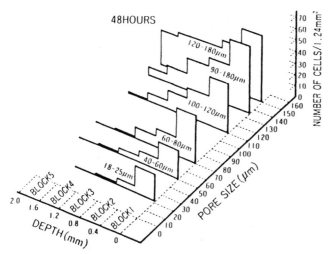

Figure 34 Cell ingrowth into porous alumina block. Cell, L; inoculated cell number: 5×10^4 cell/ml; medium, YLH + 5% calf serum; 48-h cultivation.

[168]. Clinical radiographic observations and histological examinations of porous titanium alloys and alumina implants in the mandibles of dogs and monkeys revealed ingrowth of bone tissue and fibrous tissue into the interconnecting channels. No adverse tissue reaction was found in the pore, and excellent data on reliable fixation of the dental implants with porous structure were presented by Pedersen et al. [169], Deporter et al. [170], Pilliar et al. [171], and Kawahara [151,172,173].

2. Surface Modification by Chemical Treatment

Anodic Oxidation and Nitrogen-Ion Implantation.   Light and electron microscopic studies on epithelial tissue response to the Ti/alloy surface have shown that the crevicular epithelium have a structure similar to that of natural tooth surface in all aspects of the biological seal [174, 175]. However, the Ti/alloy implants are apt to have a grayish color at the gingival tissue, which is unfavorable esthetically. Ti/alloys treated with anodic oxidation (Table 8, 50–70 V) or nitrogen-ion implantation developed a golden color on their surface and established a healthy natural color tone in the gingival tissue around the posts of Ti/alloy implants.

In *in vitro* tests, cytotoxicity, and cell attachment and adhesive strength to Ti treated with anodic oxidation and nitrogen-ion implantation were investigated and measured by photopattern analysis [176] and the viscometric method [177], and compared with that of Pyrex glass, HA, Al_2O_3, and polyvinylchloride. No cytotoxicity was observed in all materials, except PVC, and no significant difference in the cell attachments and adhesive strengths of treated and nontreated Ti was recognized (Figs. 35 and 36).

In vivo tests reconfirmed that the golden surface color of Ti implants treated by anodic oxidation or nitrification is effective in promoting a healthy color tone of the gingival tissue around the implant post while maintaining a suitable biological seal [178]. On the bone tissue response, Johansson et al. [179] reported that light microscopic morphometry did not reveal any significant difference in bone–metal contact area between nitrogen-ion-treated and non-treated Ti implants.

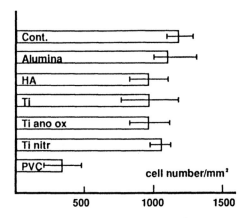

Figure 35   Cell multiplication of HeLa S3 on the test plate, 4-day cultivation.

Hydrofluoric Pickling and H_2O_2 Treatment.   *In vitro* tests to investigate better methods of surface treatment of cleaning and sterilization for Ti implants were made by measurement of contact angles to water and the critical surface tension in relation to the initial cell attachment to the treated surfaces. The results showed that cleaning using acetone or acetone with 4% hydrofluoric acid yielded smaller contact angles and high critical surface tension values. On the other hand, detergent cleaning gave a large contact angle and a low critical surface tension. It became apparent that the surfaces cleaned with detergent were not appropriate even when treated with water rinse and sterilization by autoclave, dry air, or UV irradiation, from the viewpoints of contact angle and critical surface tension. Only acetone cleaning may be significant as a potential candidate. It is revealed that the detergent cleaning commonly recommended in chemical and biological experiments may not be favorable for a titanium implant [180]. However, the results on initial cell attachment and morphological observation did not differ among any of the surface cleanings or sterilization methods (Fig. 37). The reason for the lack of difference in the initial cell attachment may be the regulating effect of the sandwich layer, that is, the extracellular

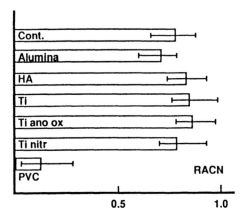

Figure 36   Adhesive strength (RACN) of HeLa S3 on the test plate, 4-day cultivation.

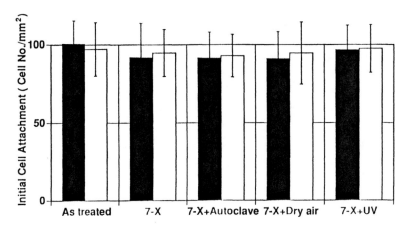

Figure 37   Initial cell attachment to titanium treated by different methods.

matrix and conformation layer of adsorbed albumin, amino lipids, and other macromolecular substances in the serum (Figs. 4 and 11) [12,181].

In vivo examination using dog tibia demonstrated a slightly greater bone bonding strength to HF-treated surface (2.5–5.0 MPa) compared with that of a nontreated surface (0.8 MPa). However, these differences may not be significant due to the wide variation in the dental implants inserted into alveolar trabecular bone (Fig. 25).

VII.   BIODYNAMIC INTERFACE

In vitro and *in vivo* studies on the relationship between interface mechanics and biology from the standpoint of wound healing, bone remodeling, and resorption after implantation have been carried out by many researchers. In many of these studies a multitude of shapes, sizes, materials, and animal models has precluded any generally accepted rules for "favorable/unfavorable" interfacial stress transfer conditions, as reported in the review papers of Brunski [182,183].

As one approach, Morita [184] investigated the influence of mechanical stimulation on self-healing behavior of bone in relation to fatigue life of living bone. The histometric analysis was carried out by measuring the accumulation rate of bone damage caused by cyclic loadings and its remodeling activities at the rest time after the loading. From the results, the following formula was established.

$$T = 1/\{f/N\text{-}(V_f/V_0)/t_0\}$$

where

T = fatigue life of living bone
f = frequency of cyclic load
N = repeating number of cyclic load
V_f = volume of new bone formation/cm^3
V_0 = volume of original bone/cm^3
t_0 = testing time

Clear-cut evidence of bone resorption is observed constantly, even under functional biting stress. The resorption area, with soft fibrous tissue, may have a buffering effect on the impact stress during mastication. Long-run histological observations have stated that implants are supported with a biodynamic interface representing both the status of bone resorption and the simultaneous remodeling occurring at different sites of the implant-bone interface; and at the same site, a turnover phenomenon between bone and soft tissue was observed within 48- to 72-h intervals [147]. To explain functional fixation of the implant, a hypothesis of fibro-osseous integration has been suggested by Weiss [185]. However, the attempt to create a peri-implantium is apt to introduce loosening of the implant, because it is difficult to maintain a favorable thickness of the peri-implantium in conditions for a successful implant. A porous or polycapillary structure [186] may be recommended for more reliable fixation for long-term implantation, which is produced through use of a combined fibrous tissue soft anchor and bone tissue hard anchor at the implant-bone interface [186].

Recent studies using finite element analysis [187,188], orthodontic force in tooth displacement, and also biochemical analysis [189] on function and differentiation of osteoblast-like cells cultured on a collagen plate with 6–24% stretching and shrinking, suggested that new bone formation around the implant materials may be accelerated under conditions of suitable micromovement. These suggestions bring up an uncertain problem regarding the usefulness of submergible-type implants.

It is difficult to appreciate long-term interactions at the implant-tissue interface under functional loads based on data from pinpoint observations by light and electron microscopy, and even by the histometric method. The implant-tissue interface is a *biodynamic interface* having turnover phenomena of bone resorption/remodeling with 24- to 72-h intervals and controlled by the fatigue life of living bone. Further long-term studies on the biodynamic interface are needed to increase the reliability of dental implants.

ACKNOWLEDGMENT

The author is grateful to the staff members at the Institute of Clinical Materials, and is indebted to Dr. D. Kawahara and Ms. Y. Ohta for their efforts in the preparation of this manuscript.

REFERENCES

1. Kawahara, H., S. Ochi, and A. Yamagami, Biological testing of dental and surgical materials by means of tissue culture, in *Second Proceedings of the International Academy of Oral Pathology*, 1965, pp. 79–92.
2. Kawahara, H., Biological problem of implant materials, in *First Proceedings of the Japanese Society of Implant Dentistry*, 1975, pp. 205–246.
3. Kawahara, H., Biomedical materials for dental implants, in *Implantology for Dental Clinics* (H. Kawahara, ed.), Ishiyaku Pub., Tokyo, 1975, pp. 122–148.
4. McKinney, R. V., D. E. Steflik, and D. L. Koth, Evidence for a biological seal at the implant-tissue interface, in *Dental Implant* (R. V. McKinney and J. E. Lemons, eds.), PSG Pub., Massachusetts, 1983, pp. 25–55.
5. Brånemark, P-I., B-O. Hansson, R. Adell, V. Beine, J. Lindstrom, O. Hallen, and A. Ohman, Osseointegrated implants in the treatment of edentulous jaw; Experience from a 10-year period, *Scand. J. Reconstr. Surg., Suppl. 16*, 1–7 (1977).
6. Kawahara, H., Today and tomorrow of biomedical polymeric materials, *Gakujutsu Geppho, 31*, 805–811 (1979).
7. Park, J. B., and R. S. Lakes, *Biomaterials: An Introduction*, Plenum Press, New York, 1992, pp. 146–157.

8. Skinner, E. W., and R. W. Phillips, *The Science of Dental Materials*, W. B. Sanders, Philadelphia, 1968, pp. 157–158.

9. Smith, R., C. Oliver, and D. F. Williams, The enzyme degradation of polymers *in vitro*, *J. Biomed. Mater. Res., 21*, 991–1003 (1987).

10. Smith, R., D. F. Williams, and C. Oliver, Bio-degradation of polyether urethanes, *J. Biomed. Mater. Res., 21*, 1194–1966 (1987).

11. Kawahara, H., Future vision of implantology, in *Proceedings of the First Congress of Implantology and Biomaterials in Stomatology* (H. Kawahara, ed.), Ishiyaku Pub., Tokyo, 1980, pp. 1–17.

12. Kawahara, H., Cellular responses to implant materials: Biological, physical and chemical factors, *Inter. Dent. J., 33*, 350–375 (1983).

13. Kawahara, H., Materials for hard tissue replacement, *Inter. J. Oral Implantol., 3*, 17–27 (1985).

14. Kawahara, H., A. Yamagami, and M. Nakamura, Biological testing of dental materials by means of tissue culture, *Inter. Dent. J., 18*, 443–467 (1986).

15. Kawahara, H., Biological evaluation of implant materials — Cell adhesion to material, *Trans. 10th Annual International Biomaterials Symposium*, San Antonio, 1978, pp. 11–15.

16. Kawahara, H., T. Maeda, and M. Nakamura, Adhesive strength of cells to biomaterials by using viscometric method, *in vitro, Trans. 7th Annual Meeting of the Society for Biomaterials*, Troy, NY, 1981, p. 50.

17. Nakajima, A., T. Hayashi, and Y. Hata, Surface structure and properties of block copolypeptide membranes having microheterophase structure, and evaluation as biomedical materials, in *Design of Multiphase Biomedical Materials* (T. Tsuruta, ed.), 1985, pp. 3–6.

18. Okano, T., Control of cell function by block and graft copolymers having hydrophilic and hydrophobic microdomains, in *Design of Multiphase Biomedical Materials*, (T. Tsuruta, ed.), 1983, pp. 9–12.

19. Shinohara, I., Synthesis of block and graft copolymers having hydrophilic–hydrophobic chains and their blood compatibility, in *Design of Multiphase Biomedical Materials* (T. Tsuruta, ed.), 1983, pp. 11–13; 1985, pp. 15–19.

20. Ikada, Y., Studies on hydrophilic multiphase biomaterials having biocompatibility, in *Design of Multiphase Biomedical Materials* (T. Tsuruta, ed.), 1985, pp. 53–58.

21. Hodosh, M., W. Montans, and M. Pover, Implants of acrylic teeth in human beings and experimental animals, *Oral Surg., 18*, 569–579 (1964).

22. Hodosh, M., M. Pover, and G. Shklar, Periodontal tissue acceptance of plastic tooth implant in primates, *J. Am. Dent. Assoc., 70*, 362–371 (1965).

23. Hodosh, M., G. Shklar, and M. Pover, Syntactic, porous, polymethacrylate–vitrous carbon tooth-replica implants as abutments for fixed partial dentures, *J. Prosth. Dent., 36*, 676–684 (1976).

24. Gettleman, L., D. Nathason, and R. L. Myerson, Effect of rapid curing procedures on polymer implant materials, *J. Prosth. Dent., 37*, 74–82 (1977).

25. Ehrlich, J., and B. Azaz, Immediate implantation of acrylic resin teeth into human tooth sockets, *J. Prosth. Dent., 33*, 205–210 (1975).

26. Otis, D., and O. D. Rackley, Cold-curing resin tooth implant: Report of case, *J. Am. Dent. Assoc., 86*, 1355–1357 (1973).

27. Herschfus, L., The use of self-curing resins for direct implant impressions, *J. Implant Dent., 2*, 22–25 (1956).

28. Pennisi, V. R., The use of Marlex 50 in plastic and reconstructive surgery, *Plast. Reconstr. Surg., 30*, 247–251 (1962).

29. Pennisi, V. R., Marlex 50 as a replacement for mandibular condyle, *Plast. Reconstr. Surg., 35*, 212–217 (1965).

30. Kawahara, H., Biomaterials for dental implant, in *Dental Implantology* (H. Kawahara, ed.), Ishiyaku Pub., Tokyo, 1987, pp. 15–42.

31. Kawahara, H., Polymeric materials for dental implant, in *Oral Implantology* (H. Kawahara, ed.), Ishiyaku Pub., Tokyo, 1992, pp. 124–130.

32. Boss, J. H., I. Shajrawi, and D. G. Merdes, Histological patterns of the tissue reaction to polymer in orthopaedic surgery, *Clin. Mater., 13*, 11–17 (1993).

33. Bowen, R. L., Dental filling material comprising vinyl-silane treated silica and binder consisting of the reaction product of bisphenol and glycidyl-acrylate, U.S. patent 33,66,112 (1962).

34. Kawahara, H., Y. Imanishi, and S. Takeda, A new composite resin of TMM–Si₃N₄ for de-amalgam in dentistry, in *Proceedings of the First Congress of Implantology and Biomaterials in Stomatology* (H. Kawahara, ed.), Ishiyaku Pub., Tokyo, 1980, pp. 74–79.

35. Homsy, C. A., J. N. Kent, and E. C. Hinds, Materials for oral implantation—Biological and functional criteria, *J. Am. Dent. Assoc., 86*, 817–822 (1973).

36. Homsy, C. A., Implant stabilization—Chemical and biomechanical consideration, *Orthop. Clin. North Am., 4*, 295–305 (1973).

37. Kent, J. N., P. D. Long, C. A. Homsy, and E. C. Hinds, Intraosseous dental implants: Nature of the interface, *J. Am. Dent. Assoc., 89*, 1142–1151 (1974).

38. Kawahara, H., S. Ochi, and K. Tanetani, Biological test of dental materials. Effect of pure metals upon the mouse subcutaneous fibroblast, strain L cell in tissue culture, *J. Japn. Soc. Dent. Apparat. Mater., 4*, 65–85 (1963).

39. Kawahara, H., On the cytotoxicity of mercury and cadmium for dental use, *Dent. Outlook, 37*, 43–48 (1971).

40. Kawahara, H., and T. Nishida, Recent study of tissue culture in dentistry, metal uptake by strain cell, by electron probe analysis, *J. Electron Microscopy, 23*, 244–245 (1974).

41. Ishizaki, N., and H. Kawahara, Corrosion of dental alloys in tissue culture medium by paper chromatography and circular dichroism, in *Report of the 25th Annual Meeting of the Japanese Society for Dental Materials and Equipment*, Kokura, Japan, 1973, p. 50.

42. Kawahara, H., H. Okuda, and T. Hosohama, Dissolution mechanism of mercury from dental amalgam into tissue culture medium by atomic absorption spectrometry, *Dent. Outlook, 48*, 692–698 (1976).

43. Kawahara, H., H. Okuda, and S. Isomura, Uptake mechanisms of mercury vapour in cells, *in vitro*, by atomic absorption spectrometry, *Dent. Outlook, 48*, 871–876 (1976).

44. Kawahara, H., and M. Nakamura, Cytotoxicity of mercury and cadmium by electron probe analysis, *Mater. Sci., 11*, 261–262 (1974).

45. Takeda, S., and H. Kawahara, Study on chemical reaction between amino-acids and metal by circular dichroism method (Group VIII metal and L-cysteine), in *Report of the 39th Annual Meeting of Japanese Society for Dental Materials and Equipment*, Kokura, Japan, 1980, p. 7.

46. Takeda, S., and H. Kawahara, Study on the chemical reaction between L-cysteine and metals of zinc, cadmium and mercury by circular dichroism method, in *Report of the 40th Annual Meeting of the Japanese Society for Dental Materials and Equipment*, Tokushima, 1981, p. 20.

47. Kawahara, H., M. Nakamura, N. Ishizaki, and S. Takeda, Cytotoxicity and solubility of cobalt from Co-Cr binary alloys in tissue culture, in *12th Meeting of Society for Biomaterials*, Minneapolis, 1986, p. 181 (abstract).

48. Kawahara, H., and S. Tomii, Cellular response to Ag-Sn amalgam, *J. Japn. Soc. Prev. Dent., 2*, 328–334 (1959).

49. Kawahara, H., M. Nakamura, A. Yamagami, and T. Nakanishi, Cellular responses to dental amalgam *in vitro*, *J. Dent. Res., 54*, 394–401 (1975).

50. McBridge, E. D., Magnesium screw and nail transfixion in fracture, *South Med. J., 1*, 508–510 (1938).

51. Aragon, P. J., and S. F. Hulbert, Corrosion of Ti-6Al-4V in simulated body fluids and bovine plasma, *J. Biomed. Mater. Res., 6*, 155–164 (1972).

52. Kawahara, H., and I. Miura, Titanium alloys and zirconium alloys for biomaterials—Tissue

adhesion to the alloys, in *Report of the First Annual Meeting of The Japanese Society for Biomaterials*, Sendai, Japan, 1979, p. 157.

53. Kawahara, H., and H. Takeuchi, Jr., Ti-Fe-O-N high strength titanium alloy for hard tissue replacements, *Inter. J. Oral Implantol.*, *7*, 82 (1990).

54. Okuno, O., I. Miura, H. Kawahara, M. Nakamura, and K. Imai, New porous zirconium titanium alloys for implant, in *Transactions of the 2nd World Congress on Biomaterials*, Washington, DC, 1984, p. 121.

55. Civijan, S., E. F. Hught, and L. B. Desimon, Potential application of certain nickel-titanium alloys (Nitinol), *J. Dent. Res.*, *54*, 89-96 (1975).

56. Tengvall, P., and I. Lundström, Physico-chemical considerations of titanium as a biomaterials, *Clin. Mater.*, *9*, 115-134 (1992).

57. Kawahara, H., K. Imai, I. Miura, and O. Okuno, Cellular responses to new bio-alloys of titanium for porous structure, *in vitro*, in *Transactions of the 15th International Symposium for Biomaterials*, Birmingham, USA, 1983, p. 68.

58. Kawahara, H., Bioceramics for dental implants, in *Oral Implantology*, Ishiyaku Pub., Tokyo, 1992, pp. 162-176.

59. Kawahara, H., Y. Takashima, A. Tominaga, and K. Tan, Electron microscopic observations on the tissue ingrowth and osteogenesis in polycapillary implant of titanium, in *Oral Implantology and Biomaterials* (H. Kawahara, ed.), Elsevier, Amsterdam, 1989, pp. 177-183.

60. Steflik, D. E., A. L. Sisk, G. R. Parr, L. K. Gardner, P. J. Hanes, F. T. Lake, D. J. Berkery, and P. Brewor, Osteogenesis at the dental implant surface: High-voltage electron microscopic and conventional transmission electron microscopic observations, *J. Biomed. Mater. Res.*, *27*, 791-800 (1993).

61. Steflik, D. E., A. L. Sisk, G. R. Parr, P. J. Hanes, F. T. Lake, P. Brewor, J. Horner, and R. V. McKinney, Correlative transmission electron microscopic and scanning electron microscopic observations of the tissues supporting endosteal blade implants, *Inter. J. Oral Implantol.*, *13*, 110-120 (1992).

62. Listergarten, M. A., D. Buser, S. G. Steinemann, K. Donath, M. P. Lang, and H. P. Weber, Light and transmission electron microscopy of the intact interfaces between nonsubmerged titanium-coated epoxy resin implants and bone or gingiva, *J. Dent. Res.*, *71*, 364-371 (1992).

63. Budd, T. W., K. Nagahara, K. L. Bielat, M. A. Meenaghan, and N. G. Schaaf, Visualization and initial characterization of the titanium boundary of the bone-implant interface of osseointegrated implants, *Inter. Oral Maxillofac. Implants, 7*, 151-160 (1992).

64. Ferguson, A. B., P. G. Laing, and E. S. Hodge, The ionization of metal implants in living tissues, *J. Bone Joint Surg.*, *42-A*, 77 (1960).

65. Kawahara, H., M. Nakamura, and S. Takeda, Toxicity and solubility of nickel from Ni-Cr alloys in tissue culture, in *11th Meeting of Society for Biomaterials*, San Diego, 1985, p. 175 (abstract).

66. Giorgio, R., Allergic reaction to steel prosthesis, *Minerva Stomat.*, *9*, 53 (1960).

67. McDougall, A., Malignant tumor rat, steel of bone planting, *J. Bone Joint Surg.*, *38*, 709-715 (1956).

68. Magnusson, B., M. Bergman, B. Bergman, and R. Söremark, Nickel allergy and nickel-containing dental alloys, *Scand. J. Dent. Res.*, *90*, 163-167 (1982).

69. Tamaki, T., H. Kawahara, and M. Miyake, Unilateral subperiosteal implant-Bionium casting system, in *First Proceedings of Japanese Society of Oral Implantology* (H. Kawahara, ed.), Ishiyaku Pub., Tokyo, 1975, pp. 261-278.

70. Kawahara, H., Y. Gonda, and M. Nakamura, *Co-Cr-Mo Alloys and Casting*, Ishiyaku Pub., Tokyo, 1979.

71. Meyer, A., R. E. Baier, J. R. Natiella, and M. A. Meenaghan, Investigation of tissue/implant interactions during the first two hours of implantation, *J. Oral Implantol.*, *14*, 363-367 (1988).

72. Baier, R. E., Surface preparation, *J. Oral Implantol.*, *14*, 208-210 (1988).

73. Miyakawa, O., K. Watanabe, S. Osaka, S. Nakano, M. Kobayashi, and N. Shiokawa, Layered structure of cast titanium surface, *Dent. Mater. J., 8*, 175–185 (1989).

74. Woodman, J. L., J. J. Jacobs, R. M. Urban, and J. O. Galante, Vanadium and aluminum release from fiber metal composites in baboons—A long term study, *Trans. ORS, 8*, 238 (1983).

75. Ohnishi, H., Shapememory alloy of Ti-Ni in orthopaedic surgery, *Artificial Organ (Japan), 12*, 862–871 (1983).

76. Sunderman, F. W., A Review of the metabolism and toxicity of nickel, *Ann. Clin. Lab. Sci., 7*, 377–381 (1977).

77. Pegum, J. S., Nickel allergy, *Lancet, 1*, 674–677 (1974).

78. Hueper, W. C., Nickel can produce rat tumors *in vitro*, *Texas Rep. Biol. Med., 10*, 167 (1952).

79. Kawahara, H., Cytotoxicity of metallic implants, *Kinzoku (Metals), 50*, 45–50 (1980).

80. Taylor, A. R., Scialom needle implants, *Dent. News, 15*, 6 (1968).

81. Plenk, H., The advantages of tantalum as implant material, *Implandent (Austria), 1*, 1–4 (1989).

82. Hulbert, S. F., J. J. Klawitter, and R. B. Leonard, Compatibility of bioceramics with the physiological environment, in *Ceramics in Severe Environments, Mater. Sci. Res.* (W. W. Kriegel, ed.), Plenum, New York, 1970, pp. 5–7.

83. Kawahara, H., Bioceramics for hard tissue replacements, *Clin. Mater., 2*, 181–206 (1987).

84. Kawahara, H., Hard tissue replacements: Biomedical materials research and development, in *Advanced Biomedical Materials* (M. Senoo, ed.), R&D Pub., Tokyo, 1986, pp. 419–457.

85. Nagai, N., N. Takeshita, and J. Hayashi, Biological reaction of zirconia ceramic as a new implant material in the dental field, *Japn. J. Oral Biol., 24*, 759–762 (1982).

86. Tateishi, T., and H. Yunoki, Research and development of alumina and zirconia artificial hip joint, *Clin. Mater., 12*, 219–225 (1993).

87. Kawahara, H., and M. Hirabayashi, Fabrication of bioceramic material using single crystal alumina, Japanese patent 1,073,745 (1975–1976).

88. Hirabayashi, M., H. Kawahara, and Y. Taniguchi, Somatic element of single crystal sapphire ceramics, U.S. patent 4,122,605 (1978).

89. Kawahara, H., M. Hirabayashi, and T. Shikita, Single crystal alumina for dental implant and bone screw, *J. Biomed. Mater. Res., 14*, 597–605 (1980).

90. Hulbert, S. F., J. N. Kent, and J. C. Bokros, Design and evaluation of LTI-Si carbon endosteal implants, in *Transactions of the 7th Annual International Biomaterials Symposium*, Clemson, 1975, p. 18.

91. Cranin, A. N., H. Silverbrand, and J. Sher, Vapor deposited carbon dental endosteal implants—A study of 80 human cases, *1st World Biomaterials Congress*, Baden, 1980, Poster 1.32 (abstract).

92. Shim, H., W. Ellis, and A. Haubold, The mechanical behavior of LTI pyrolytic carbon post type endosseous dental implants, in *Implantology and Biomaterials in Stomatology* (H. Kawahara, ed.), Ishiyaku Pub., Tokyo, 1980, p. 36.

93. Hench, L. L., and H. S. Paschall, Histochemical responses at a biomaterials interface, prosthesis and tissue, *The Interface Problem*, Clemson, 1973.

94. Brömer, H., K. Deutshcer, and B. Blencke, Ceravitar®-bioactive glass–ceramics, *Sci. Ceramics, 9*, 219–223 (1977).

95. Mouroe, E. A., W. Votava, and D. B. Bass, New calcium phosphate ceramic material for bone and tooth implants, *Os Om Op, 50*, 860–865 (1970).

96. Fischer-Brandies, E., and E. Dielert, The response of cortical bone to hydroxyapatite ceramic, *Clin. Mater., 1*, 23–28 (1986).

97. Kokubo, T., M. Shigematsu, and Y. Nagashima, Apatite and woolastonite containing glass-ceramics for prosthetic application, *Bull. Inst. Chem. Res., 60*, 3–4 (1982).

98. Gross, U., and V. Strunz, Interface of various glasses and glass ceramics in a bony implantation bed, private communication (1984).

99. Boyde, A., E. Maconnachie, C. Muller-Mai, and U. Gross, SEM study of surface alterations of bioactive glass and glass-ceramics in a long implantation bed, *Clin. Mater., 5*, 73–88 (1990).

100. Rawlings, R. D., Bioactive glasses and glass ceramics, *Clin. Mater., 14*, 155–179 (1993).

101. Kawahara, H., Designing criteria of bioceramics for bone and tooth replacement, in *Ceramics in Surgery* (P. Vincenzini, ed.), Elsevier, Amsterdam, 1983, pp. 49–60.

102. Klawitter, J. J., and S. F. Hulbert, Application of porous ceramics for the attachment of load bearing internal orthopaedic applications, *J. Biomed. Mater. Res. Symp., 2*, 161 (1971).

103. Kawahara, H., Biocompatibility of bioceramics, *Med. Philosophica, 4*, 523–529 (1985).

104. Dearnaley, G., Diamond-like carbon: A potential means of reducing wear in total joint replacements, *Clin. Mater., 12*, 237–244 (1993).

105. Kawahara, H., T. Maeda, and M. Nakamura, Studies on cell adhesion to biomaterials by viscometric method, *in vitro*, Rep.1: On the visconic properties of tissue culture medium, *J. Japn. Soc. Biomater., 2*, 43–52 (1984).

106. Kawahara, H., and A. J. Sánchez, Studies on cell adhesion to biomaterials by viscometric method, *in vitro*, *J. Japn. Soc. Biomater., 2*, 53–58 (1984).

107. Nakamura, T., and H. Kawahara, Biodynamic response at implant–bone interface, Part 4: Radiographic measurements on implant–bone interspace in long time implantation, in *Proceedings of JSOI Annual Meeting*, Sapporo, Japan, 1978, B34, p. 99.

108. Meffert, R. M., M. S. Block, and J. N. Kent, What is osseointegration? *Int. J. Periodont. Rest. Dent., 7*, 9–21 (1987).

109. Weinlaender, M., E. B. Kenney, and J. Beumer, Comparison of implant–bone interfaces in three different implant systems, *J. Dent. Res., 68*(special issue), 963 (1989).

110. Zablotsky, M. H., R. M. Meffert, and R. F. Caudill, Histological and clinical comparisons of guided tissue regeneration of dehisced HA-coated and titanium endosseous implant surfaces. A pilot study, *Int. J. Oral Maxillofac. Implants, 6*, 294–303 (1991).

111. Knox, R. J., R. F. Caudill, and R. M. Meffert, Histological evaluation of dental endosseous implants placed into surgically created extraction defects, *Int. J. Periodont. Rest. Dent., 11*, 365–375 (1991).

112. Sevor, J., and R. M. Meffert, Regeneration of bone on dental implants with a resorbable membrane, *J. Dent. Res., 70*(special issue), 347 (1991).

113. Cook, S. D., J. F. Kay, and K. A. Thomas, Interface mechanics and histology of titanium and hydroxylapatite-coated titanium for dental implant applications, *Int. J. Oral Maxillofac. Implants, 2*, 15–22 (1987).

114. Cook, S. D., G. C. Baffes, and K. A. Thomas, Comparison of models for evaluating interface characteristics of HA-coated implants, *J. Dent. Res., 70*(special issue), 530 (1991).

115. Steinemann, S., J. Eulenberger, and P. Maeusli, Adhesion of bone to titanium, in *Biological and Biomechanical Behavior of Biomaterials* (P. Christel, A. Meunier, and A. Lee, eds.), Elsevier, Amsterdam, 1986, pp. 409–414.

116. Gross, U., W. Roggendorf, and H. Schmits, Biomechanical and morphometric testing methods for porous and surface reactive biomaterials, in *Quantitative Characterization and Performance of Porous Implants for Hard Tissue Application* (J. E. Lemons, ed.), American Society for Testing and Materials, Philadelphia, PA, 1987, pp. 330–346.

117. Taylor, J., J. Brunski, and S. Hoshaw, Interfacial bond strengths of Ti–6Al–4V and hydroxylapatite-coated Ti–6Al–4V implants in cortical bone, in *Proceedings of the 2nd International Congress on Tissue Integration in Oral, Orthopedic, and Maxillofacial Reconstruction*, Quintessence Pub., Lombard, IL, 1992, pp. 125–132.

118. Spivak, J. M., J. L. Ricci, N. C. Blumenthal, and H. Alexander, A new canine model to evaluate the biological response of intra-medullasy bone to implant materials and surfaces, *J. Biomed. Mater. Res., 24*, 1121–1149 (1990).

119. Wang, B. C., T. M. Lee, E. Chang, and C. Y. Yang, The shear strength and failure mode of plasma-sprayed hydroxyapatite coating to bone: The effect of coating thickness, *J. Biomed. Mater. Res., 27*, 1315–1327 (1993).

120. Hayashi, K., T. Inadome, T. Mashima, and Y. Sugioka, Comparison of bone–implant interface shear strength of solid hydroxyapatite and hydroxyapatite-coated titanium implants, *J. Biomed. Mater. Res., 27,* 557–563 (1993).

121. Pilliar, R. M., D. A. Deporter, P. A. Watson, M. Pharoah, M. Chipman, N. Valiquette, S. Carter, and K. de Groot, The effect of partial coating with hydoxyapatite on bone remodeling in relation to porous-coated titanium-alloy dental implants in the dog, *J. Dent. Res., 70,* 1338–1345 (1991).

122. Bagambisa, F. B., U. Toos, and W. Shilli, Mechanisms and structure of bone between bone and hydroxyapatite ceramics, *J. Biomed. Mater. Res., 27,* 1047–1055 (1993).

123. Hench, L. L., and H. A. Paschall, Direct chemical bond of bioactive glass-ceramic materials to bone and muscle, *J. Biomed. Mater. Res., 7,* 25–42 (1973).

124. Kokubo, T., T. Hayashi, S. Sakka, T. Ketsugi, T. Yamamuro, M. Takagi, and T. Shibuya, Surface structure of a load-bearable bioactive glass-ceramic A-W, in *Ceramics in Clinical Applications,* Elsevier, Amsterdam, 1987, (P. Vincenzini, ed.), pp. 175–184.

125. Muller-Mai, C., H. J. Schmitz, V. Strunz, G. Fuhrmann, T. Fritz, and U. Gross, Tissues at the surface of the new composite material titanium/glass-ceramic for replacement of bone and teeth, *J. Biomed. Mater. Res., 23,* 1149–1168 (1989).

126. Kamegai, T., F. Ishikawa, H. Nakao, and Y. Seino, Clinical applications and short term results of dental root implantation using materials coated with bioactive glass, in *Oral Implantology and Biomaterials* (H. Kawahara, ed.), Elsevier, Amsterdam, 1989, pp. 121–126.

127. Gross, U., R. Kinne, H-J. Schmits, and V. Strunz, The response of bone to surface-active glass/glass-ceramics, *CRC Crit. Rev. Biocompatibility, 4,* 155–179 (1988).

128. Maruno, S., S. Ban, Y-F. Wang, H. Iwata, and H. Itoh, Properties of functionally gradient composite consisting of hydroxyapatite containing glass coated titanium and characters for bioactive implant, *J. Ceramic Soc. Jpn., 100,* 362–367 (1992).

129. Davies, J. E., N. M. Price, and T. Matsuda, *In vitro* biocompatibility assays which employ bone derived cells, in *Oral Implantology and Biomaterials* (H. Kawahara, ed.), Elsevier, Amsterdam, 1989, pp. 197–204.

130. Massas, R., S. Pitaru, and M. Weinreb, The effects of titanium and hydroxyapatite on osteoblastic expression and proliferation in rat parietal bone cultures, *J. Dent. Res., 72,* 1005–1008 (1993).

131. Buser, D., R. K. Schend, S. Steinemann, J. P. Fiorellini, C. H. Fox, and H. Stich, Influence of surface characteristics on bone integration of titanium implant. A histomorphometric study in miniature pigs, *J. Biomed. Mater. Res., 25,* 889–902 (1991).

132. Gottlander, M., and T. Albrektsson, Histomorphometric analyses of hydroxyapatite-coated and uncoated titanium implants, *Clin. Oral Implant Res., 3,* 71–76 (1992).

133. Gottlander, M., and T. Albrektsson, Histomorphometric studies of hydroxyapatite-coated and uncoated CP titanium threaded implants in bone, *Int. J. Oral Maxillofac. Implants, 6,* 399–404 (1991).

134. Burr, D. B., S. Mori, R. D. Boyd, T. C. Sun, J. D. Blaha, L. Lane, and J. Parr, Histomorphometric assessment of the mechanisms for rapid ingrowth of bone to HA/TCP coated implant, *J. Biomed. Mater. Res., 27,* 645–653 (1993).

135. Kawahara, H., Y. Mimura, and K. Hashimoto, *Long term animal experiments on bone bonding to HA coated and non-coated titanium implants,* paper presented at the Institute of Clinical Materials Meeting, Osaka, Feb. 1993.

136. Jarcho, M., Retrospective analysis of hydroxyapatite development for oral implant applications, *Dent. Clin. North Am., 36,* 19–26 (1992).

137. Jarcho, M., Calcium phosphate ceramics as hard tissue prosthetics, *Clin. Orthop., 157,* 259–278 (1981).

138. Weinlaender, M., Bone growth around dental implants, *Dent. Clin. North Am., 35,* 585–601 (1991).

139. Boyde, A., E. Maconnachie, C. Muller-Mai, and U. Gross, SEM study of surface alterations

of bioactive glasses and glass-ceramics in a bony implantation bed, *Clin. Mater., 5*, 73–88 (1990).

140. Rawling, R. D., Bioactive glasses and glass-ceramics, *Clin. Mater., 14*, 155–179 (1993).

141. Klein, C. P., A. A. Drissen, and K. de Groot, Biodegradation behavior calcium phosphates, *J. Biomed. Mater. Res., 17*, 769–805 (1983).

142. Schwartz, O., Atlas of replantation and transplantation of teeth, in *MEDI* (J. O. Andreasen, ed.), Globe, London, 1992, pp. 224–239.

143. Nakamura, T., and H. Kawahara, Clinical observations on autogeneous tooth transplantation, *Dent. Outlook, 79*, 609–619, 869–881, 1469–1473 (1992).

144. Koeneman, J., J. Lemons, P. Ducheyne, W. Lacefield, F. Magee, T. Calahan, and J. Kay, Workshop on characterization of calcium phosphate materials, *J. Appl. Biomater., 1*, 79–90 (1990).

145. Kasemo, B., Biocompatibility of titanium implants: Surface science aspects, *J. Prosthet. Dent., 49*, 832–837 (1983).

146. Zablotsky, M. H., Hydroxyapatite coatings in implant dentistry, *Implant Dent., 1*, 253–257 (1992).

147. Kawahara, H., Biodynamic interface, in *Oral Implantology* (H. Kawahara, ed.), Ishiyaku Pub., Tokyo, 1992, pp. 197–200.

148. Saadoun, A. P., and M. Le Gall, Clinical results and guidelines on Steri-Oss endosseous implant, *Int. J. Periodont. Rest. Dent., 12*, 487–499 (1992).

149. Kawahara, H., Porous structure of biometals and bioceramics, in *Transactions of the 3rd World Biomaterials Congress*, Kyoto, 1988, p. 3.

150. Okuno, O., N. Shibata, T. Nakano, I. Miura, and H. Kawahara, Biomechanical and morphological evaluations of the porous zirconium-titanium implants, in *Transactions of the 3rd World Biomaterials Congress*, Kyoto, 1988, p. 396.

151. Kawahara, H., Porous implants, in *Oral Implantology* (H. Kawahara, ed.), Ishiyaku Pub., Tokyo, 1992, pp. 773–803.

152. Pilliar, R. M., Porous-surfaced metallic implants for orthopedic applications, *J. Biomed. Mater. Res., Appl. Biomat., 21*A1, 1–33 (1987).

153. Maniatopoulos, C., R. M. Pilliar, and D. C. Smith, Threaded versus porous surfaced designs for implant stabilization in bone-endodontic implant model, *J. Biomed. Mater. Res., 20*, 1309–1333 (1986).

154. Selting, W. J., and S. N. Bhaskar, Structural strength of the interface between bone and nondegradable porous ceramic implants, *J. Dent. Res., 52*, 91–95 (1973).

155. Kawahara, H., Porous alumina reinforced with sapphire core, in *Bioceram Porous Implant* (H. Kawahara, A. Yamagami, and Y. Asai, eds.), Ishiyaku Pub., Tokyo, 1989, pp. 20–42.

156. Kawahara, H., Y. Mimura, and S. Moriwaki, Polycapillary implants made of pure titanium, in *Dental Implant* (H. Kawahara, ed.), Ishiyaku Pub., Tokyo, 1987, pp. 99–106.

157. Clemow, A. J. T., A. M. Weinstein, and J. J. Klawitter, Interface mechanics of porous titanium implants, *J. Biomed. Mater. Res., 15*, 73–82 (1981).

158. Kawahara, H., T. Iwao, K. Hashimoto, and Y. Takashima, Corrosion resistance of titanium treated with hydrofluoric acid and hydrogen peroxide, in *Transactions of the 19th Society for Biomaterials Meeting*, Birmingham, 1933, p. 84.

159. Mühlebach, J., K. Müller, and S. Schwargenbach, The peroxo complexes of titanium, *Inorg. Chem., 9*, 2381–2390 (1970).

160. Kawahara, H., Y. Mimura, M. OKi, K. Kubo, Y. Soeda, and Y. Nomura, *In vitro* study on cell adhesive strength to titanium with anodic oxidation and nitridation, in *Oral Implantology and Biomaterials* (H. Kawahara, ed.), Elsevier, Amsterdam, 1989, pp. 169–175.

161. Kawahara, H., Y. Mimura, and K. Hashimoto, *Biological seal and the epithelial cell attachment to titanium implant*, ACOI Meeting, Philadelphia, Sept. 1992 (abstract).

162. Hayakawa, N., N. Wakumoto, K. Kawahara, and H. Kawahara, Experimental studies on the biological seal. Rep.4: Effect of human plaque extract upon the cell adhesion of epithelial

cell and fibroblastic cell derived from human gingiva, in *Transactions of the 3rd World Congress for Oral Implantology*, Yokohama, 1994 (in press).

163. Kawahara, H., K. Tanaka, and Y. Ashiura, Coloured titanium dental implant, Japanese patent, application 61-304829 (1986).

164. Kawahara, H., K. Imai, and D. Kawahara, Porous polycrystal alumina implant reinforced with single crystal alumina, in *High Tech Ceramics* (P. Vincenzini, ed.), Elsevier, Amsterdam, 1987, pp. 253-264.

165. Klawitter, J. J., and S. F. Hulbert, Application of porous ceramics for the attachment of load bearing internal orthopaedic applications, *J. Biomed. Mater. Res. Symp., 2*, 161 (1971).

166. Hulbert, S. F., J. J. Cooke, and R. B. Klawitter, Attachment of prostheses to the musculoskeletal system by tissue ingrowth and mechanical interlocking, *J. Biomed. Mater. Res., 7*, 1-23 (1973).

167. Selting, W. J., and S. N. Bhaskar, Structural strength of the interface between bone and nondegradable porous ceramic implants, *J. Dent. Res., 52*, 91-94 (1973).

168. Predecki, P., J. E. Stephan, and B. A. Auslaender, Kinetics of bone growth into cylindrical channels in aluminium oxide and titanium, *J. Biomed. Mater. Res., 6*, 375-400 (1972).

169. Pedersen, K. N., H. R. Haanaes, and S. Lyng, Tissue ingrowth into mandibular intrabony porous ceramic implants, *Int. J. Oral Sur., 3*, 158-165 (1974).

170. Deporter, D. A., P. A. Watson, R. M. Pilliar, A. H. Melcher, J. Winslow, T. P. Howley, P. Hansel, C. Maniatopoulos, A. Rodriguez, D. A. Abdulla, K. Parisien, and D. C. Smith, A histological assessment of the initial healing response adjacent to porous-surfaced titanium alloy dental implants in dogs, *Int. J. Oral Implantol., 4*, 39-45 (1987).

171. Pilliar, R. M., Porous-surfaced metallic implants for orthopedic applications, *J. Biomed. Mater. Res., 21*A1, 1-33 (1987).

172. Kawahara, H., *In vitro* study on the mechanism of cell ingrowth into porous alumina, in *Bioceram Porous Implants* (H. Kawahara, A. Yamagami and Y. Asai, eds.), Ishiyaku Pub., Tokyo, 1989, pp. 2-7.

173. Kawahara, H., and K. Hashimoto, Porous structure of Biometals and bioceramics, in *Transactions of the 8th International Research Committee of Oral Implantology*, Modena, 1990, p. 31.

174. Gould, T. R. L., D. M. Brunette, and L. Westbury, The attachment mechanism of epithelial cells to titanium in vitro, *J. Period. Res., 16*, 611 (1981).

175. Gould, T. R. L., L. Westbury, and D. M. Brunette, Ultrastructural study of the attachment of human gingiva to titanium *in vivo*, *J. Prosth. Dent., 52*, 418 (1984).

176. Kawahara, H., and K. Imai, Photopattern analysis and computation on cell density under the viable condition *in vitro*, *Tissue Culture Dent. Res.*, No.20: 1983, pp. 1-5.

177. Kawahara, H., T. Maeda, T. Iseki, and A. Sánchez, Studies on cell adhesion to biomaterials by viscometric method *in vitro*. Report 1: Viscoelastic behavior of tissue culture medium; Report 2. Adhesive strength of cells to glass surface, *J. Japn. Soc. Biomater., 2*, 43-52 (Report 1); 53-58 (Report 2) (1984).

178. Kawahara, H., T. Iwao, and Y. Nomura, *In vitro* study on cell adhesive strength to titanium with anodic oxidation and nitrification, in *Transactions of the 20th Japanese Society for Oral Implantology*, Ube, 1990, p. 54.

179. Johansson, C. B., J. Lausmaa, T. Röstlund, and P. Thomsen, Commercially pure titanium and Ti 6Al 4V implants with and without nitrogen-ion implantation: Surface characterization and quantitative studies in rabbit cortical bone, *J. Mater. Sci. Mater. Med., 4*, 132-141 (1993).

180. Kawahara, D., Y. Kimura, M. Nakamura, and H. Kawahara, Studies on the tissue adhesive capability to titanium by dynamic wettability test and cell attachment, *in vitro*, *Clin. Mater., 14*, 229-233 (1993).

181. Kawahara, D., K. Kokita, Y. Kimura, M. Nakamura, and T. Iwanoto, *In vitro* effect of proteins on cell attachment to titanium plate, *J. Dent. Mater. Devices, 9*, 137 (1990).

182. Brunski, J. B., A. F. Moccia, Jr., S. R. Pollack, E. Korostoff, and D. Trachtenberg, The influence of functional use of endosseous dental implants on the tissue–implant interface. I. Histological aspects, *J. Dent. Res., 58,* 1953–1969 (1979).

183. Brunski, J. B., Biomechanical factors affecting the bone–dental implant interface, *Clin. Mater., 10,* 153–201 (1992).

184. Morita, M., Fatigue life of live bone, in *Oral Implantology* (H. Kawahara, ed.), Ishiyaku Pub., Tokyo, 1991, pp. 75–90.

185. Weiss, C. M., Tissue integration of dental endosseous implants — Description and comparative analysis of the fibroosseous integration system, *J. Oral Implantol., 12,* 169–214 (1986).

186. Kawahara, H., K. Tanaka, and Y. Ashiura, Endosseous implant having polycapillary structure, U.S. Patent 4,964,801 (1990).

187. Kawahara, H., T. Yokobori, and S. Sasaki, Finite element stress analysis on supporting tissue around polycapillary implants, in *Transactions of the 16th Annual Meeting, Society for Biomaterials,* Charleston, 1990, p. 55.

188. Kawahara, H., Y. Mimura, and T. Yokobori, Micro-geometric study on the impact stress at implant–bone interface by finite element analysis, in *Transactions of the 4th World Biomaterials Congress,* Berlin, 1992, p. 630.

189. Saitoh, S., Biochemical studies on the effect of external force upon the cells derived from human periodontal ligament and alveolar bone, in *Transactions of the 1st Meeting, Oral and Maxillofacial Biomechanics Association,* Tokyo, 1993, p. 31–39.

51
Dental Porcelains and Ceramics

Waldemar G. de Rijk
University of Illinois at Chicago
Chicago, Illinois

I. INTRODUCTION

Many different products made of porcelain and ceramics are used in the dental profession. The applications range from filler particles as used in casting investments and composite restorative resins, to porcelain teeth, veneering laminates, and cosmetic veneers for metallic restorations. In this discussion, only the use of porcelains and ceramics as restorative biomaterials is addressed. Since the casting investments are industrial tools in the production of dental prostheses and the clinical characteristics of the composite resins are almost entirely determined by the physical properties of the resin, these materials have been excluded from the present discussion.

A. Nomenclature

When reference is made in the dental literature on dental ceramics, often the materials described are in fact porcelains, modified glasses, or glass ceramics. This designation is in agreement with the definition of ceramics by Kingery et al. [1]; however, the materials have properties that are more related to glasses such as the aluminosilicates than to sapphire or mullite.

In general, the term *porcelain* refers to a modified glass containing small crystalline particles. These particles act as reinforcing agents and as a large group of optical scattering centers that diffuse and scatter light. The diffuse scattering of light from deeper layers of the material is the same mechanism that enhances the beauty of fine china. The term *ceramic* refers to a highly crystalline material that is far more optically opaque even though it is quite reflective, often with a particular color associated with the material. The reader is cautioned that the dental literature does not make this distinction consistently and may refer to ceramics as all materials derived from blends of metal oxides as opposed to metal alloys.

Restorative dental biomaterials are used to replace lost tooth structure, thereby restoring the physiological function of the dentition. This encompasses all aspects of the oral physiology, facial appearance, speech, and mastication. To date, no dental biomaterial has succeeded in fully meeting all the physiological demands; therefore a group of materials have been developed to meet certain clinical needs more or less successfully. Some hybrid materials that use metal alloys as well as porcelains have shown good promise in clinical trials.

B. Considerations in Favor of Using Ceramics

The reasons for favoring the use of porcelains in dentistry can be identified as:

- Porcelains can be made to truly look like healthy natural teeth, having a lifelike appearance.
- Porcelains are highly biocompatible with oral tissues.

A good cosmetic appearance is a highly desirable property for a biomaterial that is to become a part of a person's physiognomy. The public's pursuit of a youthful and healthy appearance has caused a strong demand for cosmetic dental materials. The dental porcelains have proven to be a mainstay of the profession in meeting this demand. Many dental technicians are masters at blending material science with porcelain artistry to produce highly aesthetic results. The need for aesthetic materials is many centuries old and has led to attempts to use ivory and porcelains in dental restorations [2]. It is the similarity in appearance between whitewares and human teeth that made porcelain the material of choice when aesthetic demands have to be met. The many centuries of experience with whitewares as food containers and cooking vessels have also shown that the interaction with foodstuffs is minimal, making the material highly desirable and a logical choice for intraoral use. Most glazed porcelains and glasses are chemically inert at ambient temperatures and thus highly biocompatible as a restorative material, with little or no likelihood of causing inflammatory reactions within the soft tissues of the oral cavity.

C. Considerations Against the Use of Ceramics in Dentistry

The disadvantages of using ceramics as restorative materials can be summarized in three specific phenomena:

1. Adherence

Ceramics do not readily adhere to a supporting structure, whether it is a natural tooth or a denture base.

For many years the bonding of denture teeth to the denture base was a real problem. Often protruding, gold-plated pins were fired into the porcelain to act as retentive devices. To this date the retention of porcelain teeth in dentures, especially thin dentures, remains a problem for dental laboratories. A similar situation is true for cementing porcelain restorations to natural teeth. Most dental cements are luting materials, which rely on micromechanical retention to lock a restoration in place. Recent developments in resin-based cements have improved the bonding to the porcelain, however, as of now, there is no chemical bonding between these resins and human dentin. The problems arising from absence of good adhesion are magnified by the next point of discussion, the absence of close fitting porcelain restorations.

2. Difficulty of Manufacture

The manufacturing process for high-precision (< 50 μm tolerance), all-ceramic prosthodontic devices of custom design is inherently very difficult.

The firing shrinkage and warping of ceramics and porcelain is well known from the whiteware industry [1]. For mass-produced items a compensation for such changes can be incorporated into the mold used to produce the green forms. For the design of individual dental restorations this is not possible. For metallic restorations, it is generally thought that a margin discrepancy of less than 25 μm is clinically well tolerated. It is for this reason that in the revised ANSI/ADA (American National Standards Institute/American Dental Association) specification No. 8, the maximum cement film thickness is specified at 25 μm for zinc phosphate cements [3]. As newer products evolve the precision of the fit of the crowns seems to improve.

3. Fracture Strength

Ceramic materials have an intrinsically low fracture strength (in tension).

When compared to industrial ceramics, the dental porcelains and ceramics have low flexural strengths. This is due in part to the low-temperature processing of the ceramics and to material choice based on optical properties rather than mechanical considerations.

The variety of the products available to the clinical dentist reflect the various methods that have been tried to overcome these problems, with varying degrees of success. Not all solutions to the aforementioned problems have necessarily been achieved by improving the ceramic materials; for example, the precision of the fit of the restoration becomes less critical when the erosion or dissolution of the cementing material is no longer a problem.

II. DENTAL APPLICATIONS OF CERAMICS

Dental restorations may replace (1) a part of a tooth, (2) one or more teeth, or (3) an entire dentition. Each of these three tasks imposes requirements on the materials that are sufficiently different for each of the tasks to warrant a brief description of each of the types of restoration.

The types of restorations for replacing part of a tooth are:

- Inlays: replacing less than one surface of a tooth; an example of a Class V inlay is shown in Fig. 1.
- Onlays: replacing the occlusal surface and one or two of the proximal surfaces.
- Laminate veneers: replacing the facial surface of anterior teeth.
- Partial and complete crowns replacing the most or all of the coronal part of a tooth.

In each of these restorations, the missing part of the tooth is reconstructed in a dental biomaterial and then bonded to the tooth. This discussion is limited to all-ceramic and ceramometal restorations. Inlays and onlays are intracoronal structures in which the tooth supports the restoration. Because these restorations are rather small, the load-bearing area is usually small and fracture of the porcelain is not a problem. This is in contrast to a crown, which must add mechanical strength and support to a tooth. In the case of the laminate veneers, a very thin porcelain layer is bonded to the facial surface of the tooth for the purpose of masking discolorations and enamel imperfections. The laminate does not contribute to the mechanical strength of the tooth.

When one tooth or a number of teeth are to be replaced, with the replacement tooth/

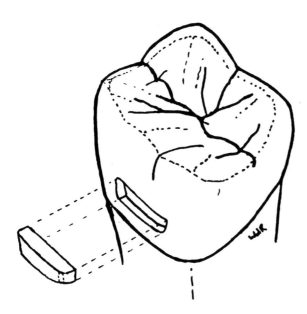

Figure 1 A schematic representation of a Class V ceramic inlay in a non-load-bearing area of the molar.

teeth entirely supported in a fixed manner by adjacent teeth or implants, the restorations are referred to as *fixed dental prostheses* (commonly called fixed bridges).

When most or all teeth are to be replaced by a device that is partly or totally supported by the oral soft tissues, the prosthetic device is labeled as a *partial or full denture*.

The full and the partial denture consist of acrylic and/or a metal framework that supports a set of artificial teeth. These denture teeth can be made of acrylate or porcelain. The use of porcelain teeth in dentures is declining in current clinical practice because of poor retention of the teeth by the acrylic resin of the denture, requiring more bulk of the resin. It is also believed that the high elastic modulus and the high indentation hardness of the porcelain teeth is a major factor in observed bone loss in patients with full dentures. Denture teeth are produced from standard molds using medium to high fusing porcelains. The production process is entirely under manufacturer control, making for a highly reproducible and high-quality product. A large catalog of colors, shapes, and sizes exists to accommodate the needs of patients.

For each of the listed applications, special materials have been developed to meet a specific need. For the inlays, onlays, and laminate veneers the tensile strength requirements are sufficiently low that all-porcelain or glass ceramic restorations can be made. In that case, the tooth supports the restoration instead of the other way around. The problem of precision fit for these restorations is in part overcome by recent developments in computer-driven machining procedures and by the use of resin-based cements.

For use in full crowns and bridges, different solutions have been developed. The problem of the low rupture strength of the dental porcelains and the difficulty in manufacturing precision custom devices has been overcome by making a hybrid restoration. The lost wax casting technique is used to produce precision castings of high-fusing metals to produce prosthodontic devices that have precise dimensions and the strength to be used in load-bearing areas. These devices are then veneered with a layer of porcelain to obtain the desired aesthetic qualities. This solution has been very successful and the porcelain-fused-to-metal (PFM) restoration is by far the most frequently used method for crowns and bridges. The difficulties seen with the PFM restorations arise from the morphology of teeth, which dictates that near the margin (the tooth–restoration interface), the restoration must be very thin. This in turn compromises the masking effect of the porcelain, often leading to exposed metal near the margin. A cross section of an anterior tooth with a PFM crown is shown in Fig. 2.

In order to avoid the problem of the metal edge showing on the PFM crowns, all-ceramic restorations have been developed. There are several systems now on the market that are entirely based on ceramics. These systems are:

Castable glass ceramics. Products in this group are:
    DICOR, Dentsply International, York, PA
    CERAPEARL, Kyocera Bioceram, Kyoto, Japan
    Olympus Castable Ceramic, Olympus Optical Company, Tokyo, Japan
All-ceramic copings that are subsequently covered with a porcelain veneer. These products are:
    IN-CERAM, Vident, Baldwin Park, CA
    Alceram (formerly Cerestore)
    IPS Empress, Ivoclar, Schaan, Lichtenstein
    Ducera, Degussa Company, South Plainfield, NJ.
    Optec-HSP, Jeneric/Pentron, Norwalk, CT

Each of these products achieves a better cosmetic result than most of the PFM restorations. A drawback common to all these restorations is their brittle nature, leading to clinical fractures and the need to have a larger amount of tooth structure removed to create the space to accommodate the bulk of these restorations.

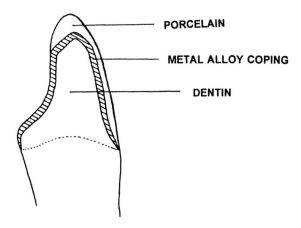

PORCELAIN

METAL ALLOY COPING

DENTIN

Figure 2   A schematic cross section of an incisor, showing the layers in a porcelain-fused-to-metal crown.

III. CLINICAL REQUIREMENTS FOR DENTAL PORCELAINS

The clinical requirements for a dental porcelain are described in the following sections.

A. Optical Properties

All porcelains must have optical properties that resemble natural dental enamel. The optical properties include the index of refraction, translucency, transparency, and color. The dental porcelains are categorized by one of three shade guides. The BIOBOND (Dentsply International, York, PA), the CERAMCO (Ceramco/Dentsply, Burlington, NJ), and VITA (Vita Zahnfarik, Bad Saeckingen, Germany). In Table 1 values for these optical properties are given for dental enamel and two commercially available porcelains, showing the relationship between the proprietary shade guides and the CIE indices (International Commission on Illumination, Paris, France) and the position in the Munsell catalog (Munsell Color, MacBeth Division of Kollmorgen Co., Baltimore, MD).

B. Chemical Durability

The durability of the porcelains must be such that the material is inert with respect to foodstuffs, saliva, and dentifrices. Restorative materials are exposed to the oral environment, which includes saliva, foodstuffs, toothpastes, and a variety of medications. When the restorations are placed below the gingivae, tissue reactions have to be considered. Tissue reactions are evaluated via animal models, and to date the dental porcelains have not shown any problems with host tissue responses.

The discussion of chemical durability includes the solubility of porcelains in fluids. Experiences with enamel coatings of water tanks have shown that light-alkali aluminosilicates may be subject to aqueous dissolution [4]. Table 2 reports the solubility of several experimental and commercial dental porcelains [5]. The solubility is measured by means of the weight loss of porcelain tablets exposed to various solutions for the duration of 2 months. There exists at the moment no standard solubility test. A test was proposed for dental porcelains that used exposure of a glazed porcelain to 5% acetic acid at 80°C for 60 h, and limited the total weight loss to < 10 mg/cm^2. Even though it appeared to be a practical screening test [6], this procedure was never adopted for porcelain evaluation.

Table 1   Comparison of Optical Characteristics

	Human enamel	Vita VMK	Ceramco
Index of refraction[a]	1.65		
Translucency (%) at 550 nm	45	19[b]	23
Hue[c]	2Y	2Y	2.1Y
Value[c]	7.0	7.2	7.1
Chroma[c]	/3	/2.8	/3.2

[a]From Ref. 2
[b]From D. L. Menis, M.S. thesis, Northwestern University, 1987
[c]From L. L. Miller, A scientific approach to shade matching, in *Perspectives in Dental Ceramics*, Quintessence Publishing, Chicago, IL, 1988

Table 2   Solubility of Dental Porcelains Exposed to Six Different Solutions at Room Temperature for 12 Months, Weight Loss (mg/cm^2)

Type of solution	Porcelain				
	PC23 (exp)	PC29 (exp)	PC310 (exp)	Ceramco B	Neydium B
Instant tea	0.140	10.699	0.135	0.017	0.064
Artificial saliva	0.161	0.254	0.050	0.003	0.414
Ringer's solution	0.001	0.129	0.233	0.003	0.084
Dist. H$_2$O	0.001	0.003	0.120	0.001	0.169
Acetic acid (4%) at 20°C	0.023	19.870	0.385	<0.001	11.923
Acetic acid (4%) at 80°C for 162 h	0.161	13.914	1.173	0.039	6.182

Source: Ref. 5

In the past there have been observations of interactions between the metallic substrate and the veneering porcelain, resulting in a gradual discoloration of the porcelain veneer. The alloy components have been identified and the problems are no longer seen [7].

A concern for the dental porcelains is the application of topical fluorides as a caries prevention mechanism, in particular the acidulated NaF gels, which have shown clinical evidence of etching the porcelain after a 4-min application. Most clinicians are aware of this problem and avoid the use of the acidulated NaF for patients with extensive porcelain restorations by using SnF (stannous fluoride)-based gels instead.

C. Decay Considerations

The material must not enhance dental decay and perhaps can inhibit carious lesions. The prevention of recurrent decay is usually not an action of the material used in restorations but a function of the physiological contour of the restoration and the integrity of its margins.

The precision fit of dental restorations is dictated by the fact that any discrepancy in size between the restoration and the tooth had to be filled with a luting medium that over time would wash away in the saliva. This in turn would open up an area that by its nature would be prone to secondary decay, which would jeopardize the survival of the tooth. The advent of resin-based cements and the availability of silanizing agents have made it possible to bond ceramics to dental structures with greatly reduced degradation of the margins. This development in turn has prompted the use of all-ceramic inlays made with computer-controlled equipment (CAD–CAM) that can be machined to tolerances of about 80 μm. The use of ready-made glass inserts is also considered as a form of ceramic inlays [8]. In this case the resin cement is entirely relied on to provide the margin integrity, while reducing the bulk of the resin, thus eliminating the problems seen with bulk resins (i.e., polymerization shrinkage, high coefficient of thermal expansion).

For the porcelain laminates, the contour is very critical for aesthetic as well as periodontal reasons. The need for higher flexural strength of the porcelains becomes very evident in these restorations. Often the thickness of the all-ceramic restorations is dictated by the strength of the material. Thus the all-ceramic restorations, by virtue of their thicker walls, are more difficult to produce with the proper contour. Current research efforts are aimed at reducing the required thickness for these systems. There is some evidence that one glass ceramic (DICOR) inhibits the adherence of microorganisms to its surface, thereby enhancing a favorable periodontal response. In general the host response to porcelain materials has been very favorable.

D. Mechanical Strength

Restoration materials must have the mechanical strength to survive the clinical masticatory demands. There are many things that can cause a dental restoration to fail, such as loss of periodontal bone support or the recurrence of dental decay. The failures discussed are only those that are directly related to material properties.

The rate of failure seen with the PFM restorations due to the fracture of the porcelain from the metal substrate is very low (less than 5% after 7 years according to Glantz [9]) and has steadily improved as the technology matures. The improper matching of the coefficients of thermal expansion of the porcelain and the metal was believed to be the cause of the failures [10].

The material failures seen with the all-ceramic crowns are catastrophic fractures [11,12] and occur at a rate of up to 22% after 3 years for nonbonded restorations. This area is of continuing interest, since these restorations offer the advantage of the best cosmetic results. In addition, these restorations are radiolucent (to dental x rays), permitting early detection and remediation of recurrent carious lesions.

E. Machinability

The material must be to some degree machinable. The human masticatory system is highly sensitive to discrepancies in the relation between the temporomandibular joint and the dentition. Occlusal interferences of as little as 50 μm in the functional and parafunctional excursions of the mandible may have severe consequences in mandibular function during speech and mastication. It is very difficult for the dental laboratory to produce this kind of precision for the occlusal contour with the current bite registrations and articulators (simulators of human mandibular motion). For this reason, the dentist is often required to make adjustments to the anatomical contour of the restorations at the chairside. The adjustments are made by abrasive machining and it is necessary that the porcelain material can withstand the grinding and the reglazing without cracking and crazing.

With the glass ceramic restorations, the machinablitiy of the material has become so feasible that it is now a major feature of the material, to the point that ceramic inlays can be made by computer-controlled machining of the entire restoration.

IV. STRUCTURE AND COMPOSITION OF DENTAL PORCELAINS

A. Porcelains for Porcelain-to-Metal Restorations

The porcelains used for the PFM restorations are modified glasses, that is, a glass matrix filled with crystalline particles. The particles act as optical scattering centers and to some extent as inhibitors of crack growth. Some of the filler crystals (e.g., Leucite) have a high coefficient of thermal expansion, thereby making it feasible for the porcelain to match the expansion of the metal. The glass matrix is primarily based on a blend of the oxides of SiO_2, Al_2O_3 and K_2O. The crystalline component of the porcelains is $K_2O \cdot$

$Al_2O_3 \cdot (SiO_2)_4$ or leucite. Other oxides include B_2O_3 and both Na_2O and Li_2O as possible light alkali oxides. Other oxides such as TiO_2 and SnO_2 are included as part of the pigments and opacifiers needed to create a tooth color. Table 3 lists the actual compositions of some PFM porcelains. Only one dental porcelain is based on nepheline syenite (Ney Porcelain).

The porcelains are produced in three separate forms: an opaquing agent, a body porcelain, and an incisal porcelain. Each of these components comes in a variety of colors or shades adhering to one of three recognized shade guides. The opaquing agent is a porcelain, highly loaded with crystalline particles, with the primary function of bonding with the metal substrate and masking the metal color. The opaque porcelain has a sintering temperature that is at least 40°C higher than the subsequent porcelain layers. The firing temperatures for a few commercial products are listed in Table 4. The body porcelains, sometimes called the *dentine porcelains*, provide the bulk of the enameling layer, often with more than one shade used to produce a certain aesthetic appearance. This is especially true when a central incisor is to be restored adjacent to a natural incisor. The blending of porcelain colors to match existing teeth is still an art and relies on the craftsmanship of the dental technician and dentist.

Incisal porcelain is used in the incisal portion of the restoration where no metal backing is present and the more transparent part of the tooth is to be mimicked, including the transparency that makes an incisal edge appear darker. The effect of the pigments in the porcelain is such that when the shades are too drastically different, the materials will not sinter properly during the firing cycle.

The raw materials used in the production of the PFM porcelains are feldspar, clays, kaolin, and quartz, hence the occasional reference to *feldspathic porcelains*. The actual ratios of the ingredients and the fritting process are proprietary for the manufacturers and not disclosed. (The results presented in Table 3 were obtained by independent investigators.) The final products are sold in powder form with a particle size distribution ranging from <1 μm to 100 μm. The wide distribution of particle sizes is designed to produce dense packing during the condensation process of the green product.

Table 3   Chemical Composition of Dental Porcelains

Compound	Ceramco[a]		Vita[a]		Ceramco: dentin	Vita: dentin	Biobond: dentin
	Opaque	Dentin	Opaque	Dentin			
SiO_2	51.3	61.2	53.1	62.7	62.2	56.8	56.9
Al_2O_3	11.2	13.0	14.9	13.0	13.4	16.3	11.8
K_2O	7.0	9.4	8.7	7.6	11.3	10.3	10.0
Na_2O	3.2	6.8	5.0	5.2	5.4	8.6	5.4
BaO	–	2.6	5.8	3.5	–	–	3.52
CaO	0.06	–	0.06	0.08	1.0	2.0	0.6
TiO_2	0.25	–	3.3	1.65	–	0.27	0.61
ZrO_2	5.77	–	–	–	0.34	1.22	1.46
SnO_2	10.75	0.51	0.51	–	0.5	–	–
Other	10.5	6.5	14.5	6.3	5.9	2.7	10.4

Source: J. N. Nally, and J.-M. Meyer, Recherche expérimentale sur la nature de la liason céramo-métallique, *Schweiz. Monatschrift Zahnheilkunde,* 80, 250 (1970)
[a]From *Dent. Mater. Japan,* 83(1), 1331 (1983)

Table 4 Typical Firing Procedures for Dental Porcelains

Porcelain	Drying time	Starting temp.	Vacuum (mm Hg)	Heating rate	Temperature during vacuum and air phases		Glaze
					Vacuum	Fire under air	
Ceramco							
Opaque	3–5 min at 650°C	650°C	25–100	35°C/min	650°–925°C	925°–1000°C	N/A
Dentin	3–5 min at 650°C	650°C	25–100	35°C/min	650°–925°C	925°–985°C	No vacuum 985°C
Microbond							
Opaque	5 min, room temp.	750°C	No vacuum	35°C/min	N/A	750°–925°C	N/A
Dentin	Until dry	750°C	75	30°C/min	750°–950°C	N/A	980°C, 0.5 min hold
Vita							
Opaque	Not needed	800°C	10–100	15°C/min	800°–960°C	N/A	N/A
Body	10 min, room temp.	750°C	10–100	15°C/min	750°–960°C	N/A	960°C, 2–4 min hold

Note: Values are for horizontal furnaces; for vertical furnaces, subtract 25°C from stated values.

The production of a PFM restoration begins with the casting of a metal substrate or coping. The casting alloy must be suitable for intraoral use and have a melting temperature well above the sintering temperature of the porcelain. The sag resistance of the alloy also must be adequate for the task. The casting process uses the lost wax technique to produce a thimble that fits the tooth quite accurately but does not have the bulk of the final product. The precision of the fit of the restoration is entirely determined by the cast metal part. Often the coping is tried for fit in the mouth of the patient prior to porcelain application. The metal coping is then subjected to an oxidizing procedure (also known as the *degassing cycle*) to produce a stable adherent metal oxide layer that will bond with the oxides in the porcelain through oxygen bridging. The opaque powder is mixed with distilled water (sometimes containing some lubricants) to form a paste that is applied to the metal surface. Vibrating and blotting of the coping are used to obtain a dense layer of the opaque material, which is made as thin as possible while still masking the metal surface. The opaque layer is dried at about 80°C (making sure that no steam develops, which would cause blistering) and then fired at the prescribed temperature, following the recommended pressure, time, and temperature profile. Many of the dental porcelain furnaces are capable of doing a porcelain fire under vacuum ($P < 10$ mm Hg) at 1020°C. Whether a vacuum is needed for the procedure is specified by the manufacturer. The opaque application may be repeated if it is found that the density of the layer is insufficient. The opaque layer has a rather dead appearance and is therefore kept to a minimum thickness. For physiological reasons, as stated earlier, the total porcelain thickness usually cannot exceed 1.5 mm and in reality is often about 1 mm. The opaque layer then, by necessity, must be less than 0.5 mm thick. The firing process initiates a sintering process with surface diffusion creating the bonds between the individual particles [1] and the development of oxygen bridges linking the porcelain to the metal [13].

The body porcelain is applied in a similar manner and fired at a lower temperature. The porcelain has a firing or sintering shrinkage of nearly 40% by volume, which makes repeated applications a necessity to obtain an acceptable final product. Dental porcelains will contract toward the bulk of the material; hence only small amounts can be processed at a given time.

Laboratory experience shows that two to three applications of the body porcelain will produce the desired results of a restored contour and a proper occlusal contact. The firing temperature is decreased by about 10°C for each subsequent firing. This step is important for those porcelains that use the air inclusions in the condensed powder as optical scattering centers. Overfiring of the porcelain leads to a much more glass-like appearance of the material. This last fact is used in the last step of the laboratory procedure, the glazing procedure, where the porcelain is rapidly heated to a temperature 20° to 40°C above the sintering temperature to produce a glaze, a thin glassy layer on the surface. The restoration is then air quenched to produce a surface layer in compression [14], which, to some degree, increases the fracture resistance due to tensile stresses.

If the cosmetic result obtained by blending various shades of body porcelain (shades of dentin, gingival, and incisal porcelain) is not satisfactory, the clinician can still add some coloring to the surface by using a staining kit. These stains are lowfusing porcelains that can be fired onto the glazed surface. The corrections that can be made by this method are limited because the human eye expects the pigmentation in the deeper layers of the tooth. The durability of the surface stains has been questioned; however, clinical acceptance seems to indicate that this is not a major problem.

B. Porcelains Used in All-Ceramic Restorations

For aesthetic and physiologic reasons there exists an impetus to develop restorations that are entirely produced from ceramic materials. This is especially true because there is still a reluctance to use nickel-containing alloys in the oral environment and possible systemic reactions are still being investigated [16]. This has been achieved with varying degrees of success depending on the material and the application. The load-bearing aspect of the anterior teeth is much less a concern than it would be with the molars. Also the requirements of strength are less for an individual crown than they are for a bridge replacing multiple teeth. Manufacturers usually indicate the scope of applications of their products.

With the exception of the porcelain jacket crown discussed in the next section, all the materials have been introduced into the dental profession since 1980. At that time the gradual shift in dental practice philosophy from on-site laboratory procedures to total reliance on a commercial laboratory had apparently proceeded sufficiently to make the purchase of expensive specialized equipment economically justifiable for dental laboratories. (Presently dental students are still being trained to be fully proficient in casting technique for noble metals.)

C. Porcelain Jacket Crowns

The porcelains used in porcelain jacket crowns—crowns that are entirely made of feldspathic porcelains without the metal coping—are the same as the porcelains discussed above for the PFM restorations. These porcelains do not require an opaque layer. The initial support for the green product is provided by a platinum foil swagged and formed over the replica of the tooth. Because of the uneven firing shrinkage, the sintering shrinkage and the flow toward the bulk of the restoration, these crowns have a very poor fit to the tooth. The gap between the tooth and the crown could range from $<50~\mu m$ to over 0.5 mm, leading to high concentrations of stress within the restoration. This, in turn, produced high failure rates and declining clinical use of these restorations.

The other all-ceramic restorations have specific mechanisms designed to overcome the specific problem of obtaining a precision fit to the tooth. Recent developments in refractory materials have produced casting investments and die materials that are thermally and dimensionally stable at the melting and sintering temperatures of the porcelains and ceramics. This development has made a higher precision possible for the conventional feldspathic porcelains. The thin porcelain laminates could not be made without the refractory materials.

D. Castable Glass Ceramic Crowns

The castable glass ceramics overcome the problem of the precision of fit by using the same technique used for the all-metal cast restorations: the lost wax casting technique. The desired shape of the restoration is produced in wax, on a tooth replica mounted on a dental articulator. All the features of the final product are incorporated, as accurately as possible, in the wax form. The wax pattern is placed in a casting investment and then burned out, leaving a mold cavity shaped like the crown. This mold cavity is then filled with a molten glass in a centrifugal casting machine. The glass crown is recovered from the investing material and then subjected to a heat treatment called the *ceramming phase*. During the heat treatment, crystals precipitate within the glass body, producing a glass ceramic [1,17]. The products currently available produce mica crystals ($KMgAlSi_3O_{10}F_2$)

as seen in DICOR®, mica and spodumene ($LiAlSi_2O_6$) crystals in Olympus OCC, or hydroxyapatite crystals [$Ca_5(PO_4)_3 \cdot OH$] in CERAPEARL. The DICOR system is the only system that has gained acceptance in the U.S. market. The reported constituents of CERAPEARL and other products are listed in Table 5.

The glass ceramics are available in one shade only. Both systems can be modified by surface stains. In addition, the DICOR system, being moderately translucent, has a series of colored cements that can be used to alter the outward cosmetic appearance. The cementation kit includes nonsetting cements for the intraoral determination of the proper shade. These resin-based cements have the additional advantage that by means of etching the interior part of the crown with HF and coating with a silanizing agent, strong bonds between the resin and the tooth can be obtained. The effect of bonding has had a remarkable effect on the clinical failure rate [12] when compared to a luting agent. These findings are substantiated by *in vitro* data [18]. The castable ceramics perform equally well for dental veneering laminates. Many clinicians do not recommend the use of glass ceramics in long-span bridges [19].

E. The All-Ceramic Coping

For the all-ceramic coping, the cosmetic success of the porcelains used in PFM restorations is combined with the advantages gained by eliminating the metal substrate. The metal is replaced by an all-ceramic material, which can be produced by different methods. In general, the applications of these materials are restricted due to the limitations imposed by the bulk requirements of the material. The design of the tooth preparation needs to be altered, and where a metal coping can be as thin as 0.5 mm, these materials require a thickness of 1 to 1.5 mm prior to veneering with a feldspathic porcelain. The materials in this category are: Alceram (previously Cerestore), IPS Empress, and In-Ceram.

The Alceram system is based on injection molding of a high alumina and magnesia content material with a resin binder [20]. A model of the prepared tooth is made of an expanding epoxy resin upon which a wax model is made of the coping. The wax coping, together with the epoxy, are then invested. The wax is melted from the mold and the aluminous material is injected under high pressure and at elevated temperature (300°C).

Table 5 Major Components of Some Commercial All-Ceramic Systems (wt%)

Component	In-Ceram[a]		CeraPearl[b]	Cerstore[c] (after firing)
	Powder	Glass		
Al_2O_3	98	16	—	60
SiO_2	—	14	34	8
La_2O_3	—	60	—	—
CaO	—	—	45	—
$MgAl_2O_4$ (spinel)	—	—	—	22
MgO	—	—	5	—
P_2O_5	—	—	15	—
$BaMg_2Al_3(Si_9Al_2O_3O_{30})$	—	—	—	10

[a]From Ref. 21
[b]From Ref. 15
[c]From Ref. 20

The material is taken from the mold and, at this time, has a high green strength. The coping is then sintered at a high temperature (1100°C) for 8 h to achieve complete sintering. The alumina undergoes a spinel transformation during this firing sequence, which, together with the expansion observed with the epoxy die, compensates for the sintering shrinkage. The coping is then veneered with a feldspathic porcelain. The use of this material has been restricted to single-unit crowns. The material was originally introduced by Coors Biomedical, marketed by Johnson and Johnson under the name Cerestore. The new owner of the process now markets under the name Alceram. The contents of the material are tabulated in Table 5.

The production method for crowns and laminates with the IPS system Empress is very similar; however, the sintering shrinkage is in this case reduced by dense packing of the green product, with additional compensation from the thermal expansion of the mold material. The material itself is a feldspathic porcelain with a moderate leucite content [19]. The material is actually a glass-ceramic that can be veneered with a thin layer of feldspathic porcelain, if so desired.

The latest development in the all-ceramic coping is based on a glass-infiltrated ceramic. The material is marketed by Vita GMBH under the name In-Ceram. The system has its own die material for making a duplicate of the prepared tooth. Alumina powder is applied to the die as an aqueous slurry and then dried. The powder is then fired at 1100°C, at which temperature the alumina undergoes a limited amount of sintering with little or no shrinkage (0.24%) [1], while the die material disintegrates. After cooling, the material can be trimmed and finished on the original die, having chalk-like mechanical properties. The alumina coping is then coated with a lanthanum-based glass powder and heated again to the melting temperature of the glass, which is soaked into the porous alumina coping. The alumina structure provides the dimensional stability through the process. The entire structure is then cooled and excess glass is trimmed. The content of the glass is La (60%), Al_2O_3 (16%), and SiO_2 (14%) as reported by Pober et al. [21]. A subsequent heat cycle is used to prepare the coping for porcelain application. The final product has a dull yellowish appearance and needs an aesthetic porcelain veneer. Not all feldspathic porcelains will adhere to this coping, even if they are obtained from the same manufacturer. The Vitadur-N porcelain adheres nicely; the Vitadur-VMK porcelain does not. Short-span bridges, as well as single-unit crowns, are made with this material. Clinical experience seems to indicate that it appears to have sufficient strength.

F. Machinable Glass Ceramics

The development of the glass ceramics finds its origin, in part, in the development of machinable glass ceramics, such as the Corning product MACOR® (Corning Glassworks, Corning, NY). This product also relies on the precipitation of mica plate-like crystals within the glass phase. The crystals inhibit crack propagation within the glassy matrix, thereby strengthening the material. The resultant material is machinable with ordinary machine tools. The same has been done for the DICOR porcelain, resulting in DICOR MGC®, a machinable glass ceramic from Dentsply International. A competing product is marketed by Vita under the name BlockCeram.

The machinable glass-ceramics were developed as a response to the need for machinable ceramics to be used in computer-assisted design and computer assisted manufacture (CAD–CAM) of dental porcelain inlays. The first commercially available machine to produce machined inlays from ceramic blocks is the CEREC unit, developed and built by

Siemens, based on the design by Möhrmann and Brandestini [22]. This unit scans the prepared cavity with a video camera, coupled to a computer; after design modifications by the dentist, the unit proceeds to carve an inlay from the ceramic block. As described before, the inlay is then etched, silanized, and bonded in place with a resin-based cement. The production time of an inlay is typically in the order of 20 min. The dentist still needs to contour the occlusal surface, which again involves abrasive machining, using a diamond particle coated rotating instrument. The apparatus is appearing in more and more dental practices. A review of dental CAD–CAM is presented by Rekow [22].

V. PHYSICAL PROPERTIES OF DENTAL PORCELAINS

The physical properties of dental porcelains include the optical properties and the chemical durability, which have been discussed in a previous section. The feldspathic porcelains are quite capable of simulating natural tooth substance, especially when reflections and scatter from deeper layers can be achieved. This is one of the reasons that the all-ceramic restorations—whether castable glass ceramic or all-ceramic coping enameled with a feldspathic porcelain—are so satisfactory from a cosmetic viewpoint. The light scattered and reflected from the deeper layers of the ceramic resembles the light reflected by dentin under the dental enamel. An opaque masking layer over a metal surface does not have the same effect because the optical path of the reflected/scattered rays is inherently shorter. The highly personalized nature of the restorations and the labor intensity required to produce a cosmetic prosthesis pose economic requirements that make these restorations costly. Future research may very well address means to reduce production costs.

The thermal properties of the dental porcelains are of interest for the manufacturing process. The porcelains used in PFM restorations must have a coefficient of thermal expansion that is nearly the same as that of the metal substrate. If this is not the case, high residual stresses will occur during the cooling from the sintering temperature (950–1000°C) to oral temperature (36°C). For a difference in coefficients of 1×10^{-6}, a discrepancy of about 0.1% will exist in size between the porcelain and the metal.

For the all-ceramic restorations, the only requirement for the coefficient of thermal expansion is that it is equal to that of the casting investment or internal transformations (e.g., spinel transformations) to compensate for the thermal shrinkage due to the cooling from the sintering temperature. This shrinkage is in addition to any firing or sintering shrinkage that may occur.

All porcelains and ceramics have a very low thermal conductance K that is comparable to dentin. Actual values of K for several materials are listed in Table 6. The low thermal conductance is a beneficial property because it protects the tooth, and particularly the nerve tissue in the pulp, from thermal shock caused by various foods (ice cream, hot coffee, etc.). Metallic restorations are known to cause painful reactions to temperature extremes.

The mechanical property that is of utmost importance for the porcelains used in PFM restorations is the adhesion to the metallic substrate. In the past, a debonding of the porcelain from the metal was a clinical problem. Either the oxide on the alloy surface was instable or the bond between the opaque layer and the alloy was inadequate; nevertheless the restoration needed to be redone. For the PFM systems in current use, the failure rates due to porcelain failure are very low. In a current clinical study by de Rijk, Wood, and Thompson [23], one porcelain failure has been seen after 2 years in a population of 96 patients. Laboratory data show that in a modified four-point bend test for eight PFM

Table 6 Thermal Conductance of Dental
Ceramics and Natural Dental Structures
[cal/sec/cm^2/°C/cm ($\times 10^3$)]

Enamel	2.2
Dentin	1.5
Dicor	4.0
Porcelain (feldspathic)	2.4
Glass[a]	2.0
PFM alloy	700.0
Siver amalgam	54.0

Source: Ref. 17
[a]From R. Resnick and D. Halliday, *Physics*, Part
1, Wiley and Sons, New York, p. 553

systems, the bond strength of the porcelain to the metal exceeds the rupture strength of the porcelain itself [24]. The modified beam consisted of a cast metal strip with porcelain inserts. The bending moment is applied to the metal part only (see Fig. 3). In the uniform stress field through the beam, the failures occurred in the body of the porcelain only, not at the PFM interface. The uniform stress field finds the weakest link in the system, which was the porcelain rupture strength and not the porcelain–metal interface. The observed characteristic strength (using the Weibull distribution) was about 60 MPa for conventional PFM systems. The Weibull distribution was used because it is an extreme value distribution, appropriate for the weakest link determination of the rupture stress.

The mechanical properties for the all-ceramic restorations are related to the resistance to fracture. As indicated before, after correction for other hazards not associated with the material, such as recurrent decay and loss of periodontal support, catastrophic fracture is the most common cause for replacing all-ceramic restorations. The three-point flexure strength or rupture strength is considered an indicator of clinical performance. Table 7 gives the rupture strengths (modulus of rupture) of dental porcelains. It is clear that for the In-Ceram material, the three-point flexural or rupture strength is significantly larger than that of the other materials. In part, this explains why fewer fractures are seen during clinical trials.

The ratio of the elastic modulus of the dental ceramic and the supporting tooth structure plays a role in the fracture resistance of all-ceramic crowns. In a recent study

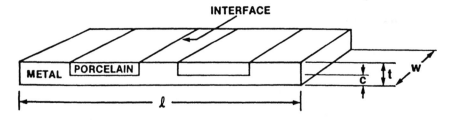

Figure 3 Diagram of a four-point bending specimen for the determination of the porcelain-to-metal bond strength. One of the four interfaces that are under a pure tensile load is indicated by the arrow.

Table 7 Transverse Strength T of Dental Porcelains,
Glass Ceramics and Ceramic Core Materials

Product	T (MPa)	
Biobond	50[a]	
Ceramco	68[b]	51[a]
CeraPearl	150[c]	
Cerestore	142[b]	125[d]
Dicor	145[d]	152[e]
In-Ceram	446	354
Optec	105	
Vitadur-N	107[d]	

[a]From Ref. 24
[b]From Ref. 2
[c]From Ref. 15
[d]J. W. McLean, and M. I. Kedge, High strength ceramics, in
Perspectives in Dental Ceramics (J. D. Preston, ed.), Quintessence Publishing, Chicago, IL, 1988
[e]From Ref. 17

[25], the fracture resistance of all-ceramic crowns was measured for three different materials: DICOR, Ceramco, and In-Ceram. The fracture resistance was measured as a function of the crown length. The crown lengths and the spherical steel indenter, used to simulate routine occlusal contacts, are shown schematically in Fig. 4. It was found that the load at fracture increased remarkably when the crown length was increased. In a subsequent study [26] the elastic modulus of the supporting material (the part that simulates the actual tooth) was increased. In this case the fracture strength became independent of the crown length. This provides indirect evidence that perhaps just improving the tolerance of flexure may be more beneficial to clinical results than increasing flexural strength. The results of the study on the effect of crown length are tabulated and shown in Fig. 5. It should be realized that the values in Fig. 5 represent the characteristic loads as calculated by the maximum likelihood estimate of the Weibull distribution for the fracture data. The range of the elastic modulus varied from 2.9 to 4 GPa. The results

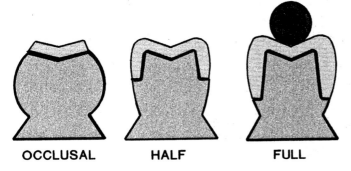

OCCLUSAL   HALF   FULL

Figure 4 Schematic representation of three crown configurations for the determination of the effect of crown length on fracture resistance. The full crown diagram shows the position and proportions of the steel indenter.

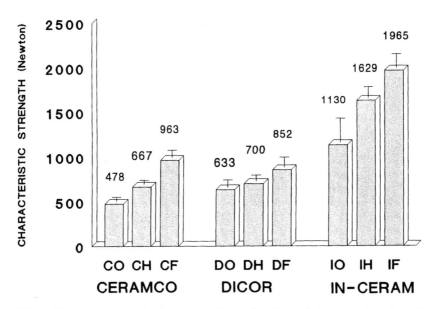

Figure 5 Results from the fracture study on the effect of the crown length on fracture resistance for three different all-ceramic crown systems.

of that study, combined with the results in Fig. 5, are shown in Fig. 6. The literature value for the elastic modulus of dentin ranges from 5.5 to 23, with our own evidence showing a value of 6.0 GPa. It is clear from Fig. 6 that in that range of elastic modulus, the porcelain carries most of the occlusal load. The effect of crown length would be diminished if most of the load was transferred to the dentin. Dental restorations have been designed

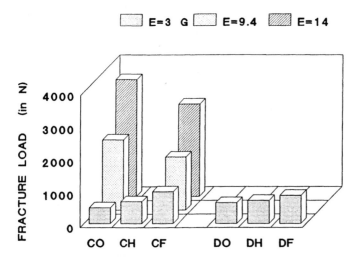

Figure 6 Fracture loads for two all-ceramic systems for three different moduli of elasticity E (values for E were 3.0, 9.4, and 4 GPa) and two different crown lengths.

with an assumed elastic modulus of 6 GPa for the dentin. This assumption may have to be reconsidered.

In order to improve the fracture resistance of porcelains, attempts to put the surface under compression by means of ion exchange are advocated [27]. The effect of exchanging the small Na ion with a K ion (using, e.g., Tufcoat, GC America, Scottsdale, AZ) on the clinical durability of a porcelain is not clear. Stronger ceramic materials are still under development to counteract the high clinical fracture rate.

REFERENCES

1. Kingery, W. D., H. K. Bowen, and D. R. Uhlman, *Introduction to Ceramics*, 2nd ed., Wiley and Sons, New York, 1976.
2. O'Brien, W. J., *Dental Materials: Properties and Selection*, Quintessence Publishing, Chicago, IL, 1989.
3. Council on Dental Materials and Devices, Revised ANSI/ADA Specification No. 8 for zinc phosphate cement, *J. Am. Dent. Assoc., 96*, 121 (1978).
4. Eppler, R. A., *Bull. Am. Ceram. Soc., 61*(9), 989–995 (1982).
5. De Rijk, W. G., K. A. Janning, and D. L. Menis, A comparison of chemical durability test solutions for dental porcelains, in *Biomedical Engineering. IV: Recent Developments* (B. W. Sauer, ed.), Pergamon Press, New York, 1985, pp. 152–155.
6. Mabie, C. P., D. L. Menis, E. P. Whittenton, R. L. Trout, and C. H. Ferry, Gel route preparation of low fusing dental porcelain frit, *J. Biomed. Mater. Res., 17*, 691–713 (1983).
7. Bertolotti, R. L., Rational selection of casting alloys, in *Perspectives in Dental Ceramics* (J. D. Preston, ed.), Quintessence Publishing, Chicago, IL, 1988.
8. Eichmiller, F. C., Clinical use of β-quartz glass-ceramic inserts, *Compendium Cont. Educ. Dentistry, 13*(7), 568–576 (1992).
9. Glantz, P.-O. J., The clinical longevity of crown and bridge prostheses, in *Quality Evaluation of Dental Restorations* (K. J. Anusavice, ed.), Quintessence Publishing, Chicago, IL, 1989.
10. Prasad, A., G. P. Day, and R. G. Tobey, A new dimension for evaluation of porcelain-alloy compatibility, in *Perspectives in Dental Ceramics* (J. D. Preston, ed.), Quintessence Publishing, Chicago, IL, 1988.
11. Scherrer, S. S., P. Mojon, U. Belser, and J.-M. Meyer, *J. Dent. Res., 67*, 214, IADR abstracts, Abstract No. 811 (1988).
12. Grossman, D. G., V. F. Beckwith, D. K. Smith, and K.A. Malament, *J. Dent. Res., 71*, AADR abstracts, Abstract No. 1721 (1992) .
13. Fairhurst, C. W., Metal surface preparation sand bonding agents in porcelain–metal systems, in *Alternatives to Gold Alloys in Dentistry* (T. M. Valega, ed.), U.S. Department of Health, Education, and Welfare (now HHS), U.S. Public Health Service, National Institute of Health, DHEW Publ. No. (NIH)77-1227, 1977.
14. Van Vlack, L. H., *Elements of Materials Science and Engineering*, 5th ed., Addison-Wesley, Reading, MA, 1985.
15. Hobo, S., Castable hydroxyapatite ceramic restorations, in *Perspectives in Dental Ceramics*, Quintessence Publishing, Chicago, IL, 1988.
16. Hensten-Pettersen, A., Casting alloys: Side-effects, *Adv. Dent. Res., 6*, 38–43 (1992).
17. Grossman, D. G., Processing a dental ceramic by casting methods, in *Ceramic Engineering and Science Proceedings* (W. J. O'Brien and R. C. Craig, eds.), American Ceramic Society, 1985.
18. Scherrer, S. S., and W. G. de Rijk, *J. Dent. Res., 70*, 434, IADR abstracts, Abstract No. 1342 (1991).
19. Lehner, C. R., and P. Schärer, All-ceramic crowns, *Curr. Opinion Dentistry, 2*, 45–52 (1992).
20. Starling, L. B., Transfer molded "all-ceramic crowns": The Cerestore system, in *Ceramic*

Engineering and Science Proceedings, Vol. 6, No. 1-2 (W. J. O'Brien and R. G. Craig, eds.), American Ceramic Society, 1985.

21. Pober, R. L., R. Giordano, S. Campbell, and L. Pelletier, *J. Dent. Res. 71*, 253, AADR abstracts, Abstract No. 1179 (1992).

22. Stachniss, V., and R. Stoll, Computer technologist in dentistry, computerized restorations, CEREC and other methods, *Proceedings, International Symposium on Computer Restorations* (W. H. Möhrmann, ed.), Quintessence Publishing, Chicago, IL, 1991 pp. 33-49.

23. De Rijk, W. G., M. Wood, and V. P. Thompson, *J. Dent. Res., 72*, 295, IADR abstracts, Abstract No. 1532 (1993).

24. De Rijk, W. G., J. A. Tesk, M. Conner, K. A. Jennings, and G. E. O. Widera, *J. Dent. Res., 62*(SI), IADR abstracts, Abstract No. 777 (1983).

25. Scherrer, S. S., and W. G. de Rijk, The effect of crown length on the fracture resistance of posterior porcelain and glass-ceramic crowns, *Int. J. Prosthodont., 5*, 550-557 (1992).

26. Scherrer, S. S., and W. G. de Rijk, The fracture resistance of glass ceramic crowns on supporting structures with different elastic moduli, *Int. J. Prosthodont., 6*, 462-467 (1993).

27. Southan, D. E., Cosmetic strengthening for dental porcelains, in *Perspectives in Dental Ceramics* (J. D. Preston, ed.), Quintessence Publishing, Chicago, IL, 1988.

52

Glassy Polymers as Dental Materials

S. Kalachandra*
Virginia Polytechnic Institute and State University
Blacksburg, Virginia

I. INTRODUCTION

Glassy polymers are used as dental materials because their rigidity enables them to support loads and to resist forces imposed on them in service in the oral cavity. Crosslinked polymers such as polydimethacrylates are the most commonly used materials in dentistry [1–3]. Perhaps the most challenging application is in the restoration of teeth [4]. The monomers for this purpose must be nontoxic and capable of rapid polymerization in the presence of oxygen and aqueous and biological environment because restorations are polymerized *in situ*. The products should have a service life of 10 years or more in a complex environment characterized by cyclic forces, aqueous fluids, and alternating temperature conditions. They should have properties comparable to tooth enamel and dentin. In addition, the products are expected to have optical and mechanical properties comparable to tooth structure. In current restorative materials such properties are sought using so-called dental composites that consist of three essential components: a crosslinked polymeric matrix, a high volume fraction of a particulate silicate filler, and a bonding agent to promote matrix–filler adhesion [5–9]. The most commonly used matrix component is 2,2-bis-[4-(2-hydroxy-3-methacryloxyprop-l-oxy)phenyl] propane (BIS-GMA); see Fig. 1. The filler is usually a combination of a finely divided crushed silicate glass and colloidal silica while the bonding agent is a modified vinyl silane.

Glassy polymers may fail in service in aqueous environments because of microcracking. A contributory cause is a stress gradient caused by progression of swollen networks made by ethylene glycol dimethacrylate in methylene chloride. This presents an extreme case because equilibrium swelling corresponds to a relatively large uptake of solvent

---

Current affiliation: University of North Carolina, Chapel Hill, North Carolina

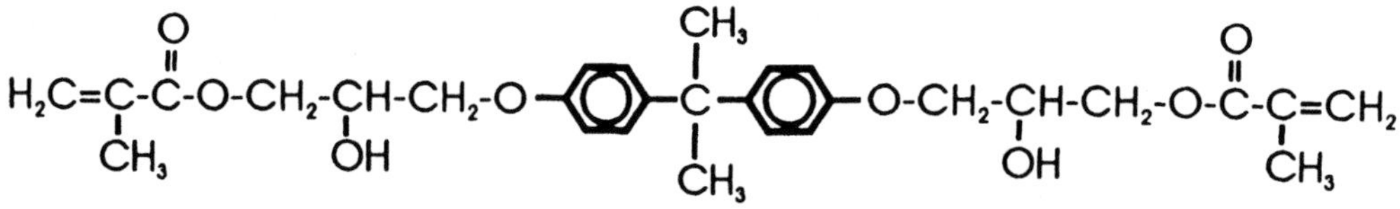

Figure 1 BIS-GMA [2,2-bis-(4-2-hydroxy-3-methacryloxyprop-1-oxy) phenyl], the most commonly used matrix component.

(25%) as compared to 7% uptake in the case of water. The rate of desorption is much greater than the rate of sorption and is seen to be accompanied by extensive microcracking.

As long as bonding between the matrix and filler is adequate and the quantity of the matrix is sufficient to fill the spaces between the filler particles, increased filler content tends to improve mechanical properties, and to reduce curing shrinkage and thermal expansion coefficient. Since the amount of filler that can be incorporated is limited by the viscosity of the resultant pastes, it is customary to dilute the highly viscous (~ 1200 Pa) BIS-GMA with more fluid difunctional monomers [10,11]. The added diluent monomers such as triethylene glycol dimethacrylate (TEGDM) tend to adversely affect the properties of the matrix material, increasing water sorption and curing shrinkage. Up to the present time, polymerization is initiated by free-radical mechanisms to yield crosslinked products [12] by redox systems such as benzoyl peroxide/aromatic amine and by photopolymerization with either ultraviolet or longer-wavelength light. Due to steric isolation or vitrification, many of the double bonds remain unreacted owing to immobilization [13–18]. The shrinkage involving conversion of double to single bonds is believed to be the main cause of failure in service performance [19,20].

One approach to this problem is to reduce the concentration of double bonds in the reactants by using dimethacrylates of higher molecular weight such as urethane dimethacrylates [21,22] and BIS-GMA. A second approach is to find less viscous substituents for BIS-GMA that would reduce the need for diluents. A more ambitious approach is to eliminate shrinkage by resorting to ring-opening polymerization reactions, but this recourse requires a long-term effort to achieve the application stage [23,24].

The main deficiency of these polymeric materials for the restoration of teeth is their poor resistance to abrasion [5–9]. This can be improved by inclusion of particulate fillers that are harder than the polymeric matrix. An ambitious goal would be to match the remarkable properties of dental enamel, which contains more than 75 vol% of hydroxyapatite crystallites tightly packed into an intricate microstructure. In comparison, the current composite restorative materials have a crude microstructure with no more than 65 vol% inorganic filler. A wide range of fillers of varying shapes and sizes, ranging from collodial dimensions to tens of microns, are being used in varying combinations. Silane coupling agents are being used, in an extension of Bowen's pioneering work [25], to bond these particulates to the polymeric matrix. Yet, the current composite materials have much lower wear resistance than the silver amalgams that they are designed to replace. Despite this deficiency, there are diverse cogent reasons, such as aesthetics and avoidance of mercury pollution of the environment that spur on their further development.

Current composite filling materials have also been used for the restoration of posterior teeth, but they have not been as successful in this application as for the anterior teeth. They have certain deficiencies in properties and technique that limit their adoption as posterior restorative materials and they are in direct competition with dental amalgam for this use. Amalgam has its own combination of advantages and disadvantages, but in comparison to composites, amalgam has the advantage of being a well-established and simple technique. Amalgam is used three to four times as frequently as composites for posterior fillings in spite of being nonadherant and less aesthetic than the composites.

There has been considerable concern for a number of years about the use of mercury in dental amalgams constituting a health hazard for the patient and for personnel in the dental office. Gradually a consensus seems to have developed that with the exercise of appropriate precautions, the risk from this source can be controlled, and that it is small in comparison to other potential sources of exposure. Thus to date, amalgam has remained a major portion of the treatment armamentarium.

This situation appears likely to change rapidly in the near future. A new concern has arisen in regard to gradual increases in the amount of mercury released to the environment from all sources including dentistry. No fully successful way has been found to prevent mercury from dental offices entering the waste stream and eventually the environment. As a result, serious consideration is being given worldwide to eliminating the use of amalgam. Sweden has already started a 3-year phaseout of its use, and a number of other European countries are expected to soon take similar action. It seems reasonable to expect that amalgam may be banned in the United States as well as elsewhere within 10 years. Thus there is an urgent need for the development of alternative materials suitable for all of the applications currently utilizing dental amalgam as the material of choice.

Bifunctional monomers of new formulations have been synthesized and tested for potential use in dental composite filling materials. The monomers were designed to have lower viscosity than the standard basic monomer, thus requiring less dilution with low molecular weight monomer to achieve useful viscosities [26]. It was hypothesized that the presence of the diluent is a main source of adverse effects on curing shrinkage and water sorption. The newly synthesized monomers (3F BIS-GMA and CH_3 BIS-GMA) did exhibit reduced viscosity (3.713 and 0.321 Pa, respectively) relative to BIS-GMA (1200 Pa) and required less dilution by triethylene glycol dimethacrylate (TEGDM). They also exhibited less curing shrinkage (5.35% and 7.53%, respectively) and water sorption ($\sim 0.9\%$ for both 3F BIS-GMA and CH_3 BIS-GMA).

II. WATER SORPTION

Water sorption of dental materials can result in undesirable changes in dimensions and in a deterioration in mechanical properties. Studies were made of BIS-GMA copolymers of the kind mentioned above [27,28] and also of polymers of potential use in crown and bridge work [29–31]. Here attention is confined to more systematic results obtained with networks made by copolymerization of methyl methacrylate (MMA) and triethylene glycol dimethacrylate (TEGDM) using a redox initiator [32].

The diffusion coefficient for water was found to be rather insensitive to the proportion of the dimethacrylate crosslinker, especially in sorption (Fig. 2). This insensitivity might be rationalized in a number of ways: One of these is consistent with the view that these systems have a lightly crosslinked matrix that does little to impede the diffusion of water molecules.

The significant increase in water uptake, as much as twofold (Fig. 3) caused by replacing MMA with TEGDM is rather surprising. It does not seem to be due to incomplete polymerization because similar results were calculated from data previously reported

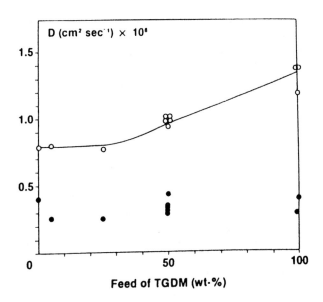

Figure 2   Dependence of diffusion coefficient, D on proportion of TGDM in a monomer mixture polymerized by a redox initiator. Sorption (●); desorption (○). The other monomer is MMA.

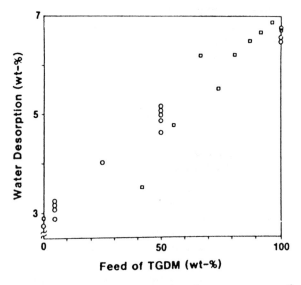

Figure 3   Dependence of uptake of water on proportion of TGDM in a monomer mixture with MMA.

[29] for specimens made at much higher temperature, 120°C. Nor can it be attributed to the more hydrophilic nature of TEGDM because similar results were obtained for copolymers of MMA and triethylene glycol dimethacrylate. An alternative hypothesis was proposed that crosslinking results in less efficient macromolecular packing and hence in increased accommodation of water in microvoids. However, this hypothesis was tested by monitoring changes in density [33] and judged to be inadequate [34]. Thus influence of crosslinking in increasing water uptake remains unexplained.

Water sorption characteristics have been studied in terms of diffusion coefficient, equilibrium uptake, and solubility as a function of the volume loading of pyrolytic silica filler for two different resins. Equilibrium water uptake decreased with filler loading but if calculated on the basis of original volume of monomer present gave consistent values for each resin. TEGDM-based materials had much higher water uptake and diffusion coefficients than the urethane dimethacrylate resins. Furthermore, the former exhibited higher concentration dependence of the diffusion coefficient. Diffusion coefficients were sensibly independent of filler loading. Hence the filler appears to take little part in the absorption process [35].

A study of diffusion of water in composite filling materials was made. The uptake of water by composite filling materials seems to be a diffusion-controlled process; also, diffusion coefficients are generally lower than in the composites based on difunctional methacrylates compared with MMA, presumably because of the highly crosslinked nature of the former [36].

Water sorption of unfilled resins was studied [37]. The water sorption behavior of polymers used as the basis of composite filling materials indicates that water-soluble impurities substantially increase the water sorption of composite filling materials above the level due to the polymer phase itself, on the basis of the work of Muniandy and Thomas [38]. It appears that this enhanced uptake is governed by an equilibrium between osmotic and elastic pressures. Hence water absorption of composites probably induces internal strains.

In other work related to crosslinked networks, dimethacrylate monomers were polymerized by free-radical chain reactions to yield crosslinked networks that have dental applications. These networks may resemble those formed by step-growth polymerization reactions in having a microstructure in which crosslinked particles are embedded in a much more lightly crosslinked matrix. Consistently, polydimethacrylates were found to have very low values of glass transition temperature T_g by reference to changes in modulus of elasticity determined by dynamic mechanical analysis (DMA). Also, the mechanical data on the influence of low volume fractions (0.03–0.05) of rigid filler particles provide evidence of a localized plastic deformation that would not seem understandable by reference to a uniformly crosslinked network. A nonuniformly crosslinked matrix might also be invoked to account for the insensitivity of rate of diffusion of water to the apparent degree of crosslinking. However, an observed increase in the uptake of water with apparent degree of crosslinking remains unexplained [4].

In another recent study on water sorption by triethylene glycol dimethacrylate and MMA monomers (determined by Karl Fischer analysis) and their corresponding polymers, water sorption of the polymers conformed approximately to Fick's law of diffusion. The weight percent oxygen content (WPO), which is indicative of hydrophilic character of these materials, was calculated on the basis of affinity to constituent groups of atoms and correlated with the total oxygen present in each monomer molecule. A linear relationship between the WPO and % water uptake was observed in a series of four glycol dimethacrylate and eight linear MMA monomers. Their corresponding polymers conformed to a linear correlation between the WPO and contact angle (ϕ). One of the dimethacrylate monomers, BIS-GMA, sorbed less water than expected on the basis of its WPO. This was interpreted as due to the existence of intramolecular hydrogen bonds between the hydroxyl of the secondary alcohol group ($-$CHOH) and the oxygen of the ester carbonyl. Nuclear magnetic resonance (NMR) spectral evidence was obtained for the presence of the intramolecular hydrogen bond in BIS-GMA monomer. The D values were generally lower for the networks based on difunctional methacrylates than for methyl methacrylates, presumably because of the highly crosslinked nature of the former [39].

Recently in our laboratory a study (unpublished) has been made to determine how the amount of water taken up by poly(methyl methacrylate) (PMMA) affects mechanical properties such as indentation hardness number, diametral tensile strength, elastic modulus (E), ultimate tensile strength (UTS), and tensile strain to fracture. The amount of water taken up is controlled by equilibration of PMMA specimens over various saturated salt solutions. Specimens were tested after annealing, drying, and conditioning at each RH–temperature combination using 3-point bending [40]. The moduli were higher than those calculated for unannealed tensile samples but lower than the annealed tensile samples. The results obtained with reference to Knoop hardness, E, UTS, and strain measurements provide no evidence that marked changes occur in the physical properties between 1% and 2% water uptake due to cluster formation, while they confirm some of the previous observations.

Another study involved the characterization of the water sorption of dental composites in terms of water uptake, diffusion coefficients (D), and polymer content; it also studied how these parameters are influenced by the nature of the filler and the presence of 4-META [41]. Four anhydrous composites—(1) tribasic calcium phosphate (TCP, 50%) and TEGDM, (2) silanated lithium aluminum silicate (SS, 75%) and TEGDM, (3) barium sulfate [$BaSO_4$, (70%)] and PMMA, and (4) silane-coated barium glass (SBG, 75%) and PMMA—were employed in water sorption studies at 37°C. The effect of 5% 4-META on the diffusion and uptake of water was studied at 37°C. The data conformed approximately to Fick's laws of diffusion. The values of D were found to be significantly smaller with SS, suggesting an effective coupling. The uptake of water by the filled specimens was about twofold higher than would be expected on the basis of the PMMA content. As an explanation, it was suggested that the additional amount of water is perhaps accommodated at the interface between the filler and the PMMA matrix. The D values for water in filled specimens were considerably larger than those in the unfilled specimens. It appears that the filler–matrix interface provides paths of facile diffusion similar to grain boundaries [41,42] (Fig. 4).

A preliminary investigation has been conducted [26] to confirm or disprove the critical role of matrix viscosity in determining the properties of dental composites. Three basic matrix monomers were used: BIS-GMA; a BIS-GMA analogue with the pendant hydroxyls replaced by CH_3 (designated $-CH_3$BIS-GMA); and a monomer identical to $-CH_3$BIS-GMA except that the central $-CH_2-$ group was replaced with $-C-$ (designated 3F BIS-GMA). They were diluted as needed with TEGDM to produce viscosities of 2000 cP and 1000 cP, values bracketing the viscosity of commercial composite matrices. The monomer mixtures were evaluated for contact angle against glass, PMMA, dentin, and enamel. Polymer samples were prepared, and curing shrinkage and water sorption were determined. The experimental monomers had lower initial viscosity, required

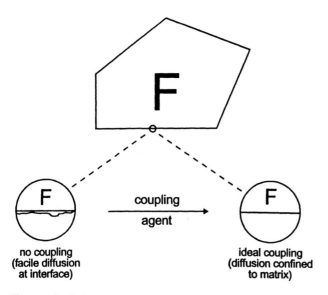

Figure 4 Schematic representation of a coupling agent in the diffusion process around a filler particle (F) involving facile diffusion at the interface and diffusion confined to the matrix.

less dilution, and had lower curing shrinkage and lower water sorption than BIS-GMA (Table 1).

III. DYNAMIC MECHANICAL ANALYSIS

The Rheovibron DDV-II is a dynamic viscoelastometer that gives direct readings of complex modulus, elastic modulus, loss tangent, and related compliances. Dynamic testing of proprietary dental composites for both dry and wet applications ($-120°$ to $+120°C$; Table 2), has been used to determine Young's modulus and damping [43], and to assign a value of T_g from a maximal value of δ [44] (Fig. 5). However, for present purposes, attention is concentrated on unfilled networks made from mixtures of known components [44]. The following mixture was polymerized by exposure to light of wavelength >400 nm: BIS-GMA (75 wt%) + triethylene glycol dimethacrylate (25%) containing *dl*-camphoroquinone (0.2%) and *N,N*-dimethylaminoethyl methacrylate (0.1%).

Table 1 Properties of Diluted and Undiluted Monomers

Monomers	Undiluted			Diluted to 1000 cP	
	Viscosity, poise	% water sorption	% shrinkage	% TEGDM	% shrinkage
Bis-GMA	2000 poise	3.05	6.40	33.5	8.3
3F BIS-GMA	3713 cP	0.92	5.40	14.0	6.3
CH₃ BIS-GMA	321 cP	0.91	7.50	No dilution required	
TEGDM	11 cP	6.00	11.9	No dilution required	

Table 2   Dynamic Mechanical Properties of Proprietary Materials ($-120°$ to $+200°$C)

System	E', dry, $\times 10^{-10}$ (dyne/cm^2)	E', wet, $\times 10^{-10}$ (dyne/cm^2)	ΔE (% change)	ΔT_g (change in T_g, °C)
Silux	5.37	3.48	-35	-7
Silux Plus	3.95	1.36	-66	-13

Only about 50% of the double bonds reacted, as determined by calorimetry [45]. For seven light-cured proprietary composite restorative materials, [P-50 (3M), P-30 (3M), Fulfil (Caulk), Herculite (Kerr), Silux Plus (3M), and Silux (3M)] and several of the model systems, both dry and water saturated, elastic modulus and glass transition temperature were evaluated with a dynamic mechanical analyzer (Autovibron DDV-II-C). BIS-GMA copolymer model systems were also studied with reference to water sorption, for example: BIS-GMA (75%) + triethylene glycol dimethacrylate (25%), BIS-GMA (30%) + triethylene glycol dimethacrylate (10%) + lithium aluminum silicate (60%), BIS-GMA + triethylene glycol dimethacrylate (14%) + barium glass (silanated) (86%), and BIS-GMA + triethylene glycol dimethacrylate (14%) + zinc glass (silanated) (86%). It was concluded that the changes in elastic modulus tend to confirm the hypothesis that the matrix–filler interface contains water. Silux and Silux Plus accommodated the greatest amount of water at the interface. Silux Plus displayed a dramatic reduction in elastic modulus at 0°C, indicating possible melting of water clusters (Fig. 6). It appeared that the only available hole for water clusters was at the matrix–filler interface [6]. The percentage change in length for Silux Plus is shown in Fig. 7. The change in slope on the length–temperature plot is taken as glass transition temperature, T_g [44].

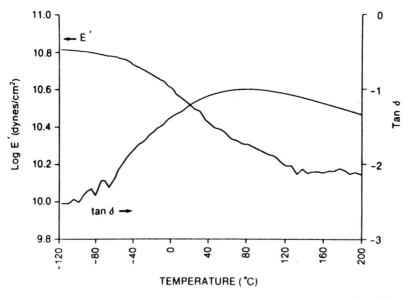

Figure 5   Elastic modulus (E') and loss tangent (tan δ) as a function of temperature for Silux Plus, dry.

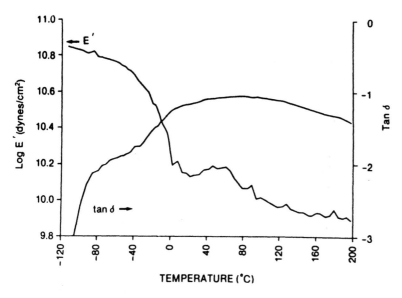

Figure 6   Elastic modulus and loss tangent (tan δ) as a function of temperature for Silux Plus, water saturated.

IV. FILLERS

The effect of fillers has been studied at high volume fractions [46–48]. However, in addition, it is instructive to study low volume fractions in order to test the conformity to theoretical prediction that certain mechanical properties should increase monotonically as

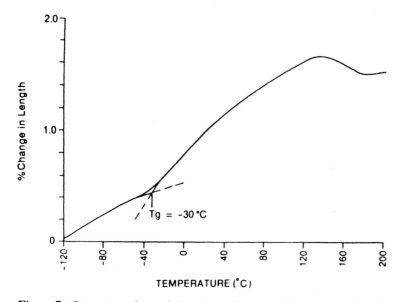

Figure 7   Percentage change in length as a function of temperature for Silux Plus, dry.

the volume fraction of filler is increased [49]. For example, Einstein's treatment of fluids predicts a linear increase in viscosity with an increasing volume fraction of rigid spheres. For glassy materials, related comparisons can be made by reference to properties that depend mainly on plastic deformation such as yield stress or, more conveniently, indentation hardness. Measurements of Vickers hardness number were made after photopolymerization of the BIS-GMA recipe, with various amounts of a silanated silicate filler with particle size in the range of tens of microns. Contrary to expectation, a minimum value was obtained [50,51], for a volume fraction of 0.03–0.05 (Fig. 8). Similar results were obtained with all other fillers examined [4]. The particles varied in size from collodial dimensions up to tens of microns and differed in surface characteristics, including treatment with silane coupling agents. A general explanation was based on the knowledge that rigid inclusions result in a highly localized stress concentration on application of an external force [52]. It was suggested that this may result in a localized plastic deformation and hence in a reduction of macroscopic properties that depend on yielding, such as yield stress and indentation hardness. With higher volume fractions of filler, effects due to isolated particles become unimportant and there is an eventual increase in property values as predicted theoretically.

A practical consequence of this work is that an isolated filler particle can act as a site of mechanical weakness. In lightly crosslinked materials this can result in plastic deformation. Presumably, in highly crosslinked materials this might result in brittle fracture, especially in fatigue. Isolated filler particle situations can be envisaged in a variety of service applications. In the field of dental materials, this might occur at an interface between a dentin bonding agent and a composite filling, and thereby constitute a zone of mechanical weakness.

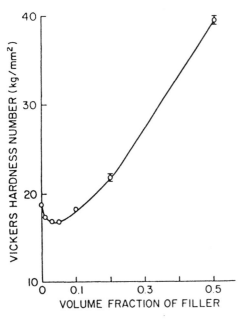

Figure 8   Influence of volume fraction of rigid finer particles on Vickers hardness number of BIS-GMA copolymer.

Work on the adhesion of resins to teeth has been reviewed in previous symposia [53,54]. In the case of tooth restorations, it has been recognized that it is important to effect good bonding both to enamel and to dentin. Considerable success has been achieved in bonding to enamel while bonding to dentin has been more of a problem. Yet in the case of deep cavity preparations for posterior composites, it has been pointed out that it is especially important to get good adhesion to dentin, inasmuch as this would favor more conservative cavity preparations as well as the prevention of marginal leakage in areas most vulnerable to recurrent caries [55]. In the usual approach to this problem adhesion to both enamel and dentin is effected via a resin cavity liner that, subsequently, can be bonded easily to a composite filling material. Therefore, achieving a strong bond between a resin cavity liner and dentin is of practical importance.

Most of the experience on bonding resins to teeth has been with the use of self-curing systems. Excellent results have been obtained using tri-*n*-butylborane with methyl methacrylate, where it has been demonstrated that still further improvement can be made by including 4-methacryloxyethyl trimellitate anhydride (4-META) [56]. Excellent results have also been obtained in bonding a composite resin to dentin that had been treated with various reagents, including 4-META [57]. Much less information is available concerning systems in which polymerization is initiated by exposure to light. In the case of visible ("blue") light there is only the claim that an impressively high adhesive strength (13.8 MPa) was achieved in bonding a proprietary composite material (Fotofil) to enamel. However, details of the testing procedure were lacking and no data were reported for dentin [58].

The use of visible light is becoming increasingly important in the restoration of posterior teeth. It has been observed that Bowen's resin and a composite made from it can be polymerized on exposure to "blue" light. Simply by inclusion of 1% camphoroquinone, 4-META can be used to increase the adhesive strength of Bowen's resin to bovine enamel by approximately 55% (to 9.6 ± 3.8 MPa) and to bovine dentin by approximately 20% (to 3.9 ± 0.8 MPa). It was seen that a stronger adhesive bond was obtained using dried dentin (8.3 ± 3.4 MPa). The fracture surface showed fracture of resin tags. This is in contrast to the observation that with moist dentin, the resin tags were pulled out of the dentinal tubules without fracture. The efficacy of 4-META in increasing diametral tensile strength by 12% (to 62 MPa) is believed to be due to its ability to wet filler particles. Consequently it acts as a coupling agent, although it is not known whether the bonds to filler particles are primary or secondary (van der Waals). It should be mentioned that the increase in optimum compressive strength was only 3% to (315 MPa).

It is pertinent to mention with reference to copolymers in models that in water sorption studies it was seen that the system (BIS-GMA–TEGDM) 14% + zinc glass (silanated) (86%) exhibits the least water uptake (0.61%), suggesting that matrix and the filler particles were most effectively coupled. Variations in the water uptake observed among the copolymer model systems may be attributed to the nature of the filler particle and the coupling agents, that is, differences in filler–matrix bond strengths [6,41].

V. CONCLUDING REMARKS

It seems that dimethacrylate networks are not uniformly crosslinked, as was often assumed in pioneering studies [59]. Instead, they have some resemblance to the "porridge" microstructure first attributed by Houwink to Bakelite (60) and subsequently adopted to account for microstructural observations on other networks prepared by stepwise polymerization reactions [61]. Changes in modulus (E') for the composites based on dimethacrylate copolymer systems, confirm the hypothesis that the filler–matrix interface accommodates water. Variations in the uptake of additional water in the spaces at the interface between filler particle and the matrix may be attributed to use of coupling agents, method of polymerization, and nature of filler particles.

ACKNOWLEDGMENTS

The author wishes to express his grateful thanks for the inspiration, constant support, and encouragement of the late Professor Derek T. Turner; and to thank Professor Duane F. Taylor of the University of North Carolina Dental School, Chapel Hill, NC., for his valuable suggestions. Finally he wishes to thank the NSF Science and Technology Center for the secretarial support.

REFERENCES

1. Phillips, R. W., *Skinner's Science of Dental Materials*, 9th ed., Saunders, Philadelphia, 1982.
2. Craig, R. G., *Restorative Dental Materials*, 6th ed., C.V. Mosby, St. Louis, 1980.
3. Greener, E. H., Harcourt, J. K., and Laurenschlager, E. P., *Materials Science in Dentistry*, Williams and Wilkins, Baltimore, 1972.
4. Turner, D. I., Haque, Z. U., Kalachandra, S., and Wilson, T. W., in *Cross-Linked Polymers: Chemistry, Properties, and Applications*, (R. A. Dickie, S. S. Labana, and R. S. Bauer, eds.), American Chemical Society Reprints Series 367, Washington, DC, 1988, pp. 427–438.
5. Taylor, D. F., and Leinfelder, K. F., eds., *Posterior Composites: Proceedings of an International Symposium*, University of North Carolina, Chapel Hill, 1982.
6. Kalachandra, S., and Wilson, T. W., *Biomaterials, 13*, 105 (1992).
7. Braem, N., Doctoral Thesis, University of Leuven, Belgium, 1985.
8. American Dental Association (CDMIE), *J. Am. Dental Assoc., 112*, 707 (1986).
9. Roulet, J. F., *Degradation of Dental Polymers*, Karger, Basel, 1987.
10. Braden, M., *Oper. Dent., 3*, 97 (1978).
11. Davy, K. W. M., and Braden, M., *Biomaterials, 12*, 406 (1991).
12. Ruyter, I. E., and Oysaed, H. J., *Biomed. Mater. Res., 21*, 11 (1987).
13. Loshaek, S., and Fox, T. G., *J. Am. Chem. Soc., 75*, 3544 (1953).
14. Hwa, J. C. H., *J. Polym. Sci., 58*, 715 (1962).
15. Horie, K., Otawaga, A., Muraoka, M., and Mita, I., *J. Polym. Sci. (Chem. Ed.), 13*, 445 (1975).
16. Ruyter, I. E., and Svendsen, S. A., *Acta Odont. Scand., 36*, 75 (1978).
17. Asmussen, E., *Acta Odont. Scand., 33*, 337 (1975).
18. Ferracane, J. L., and Greener, E. H., *J. Dental Res., 63*, 1093 (1984).
19. Going, R. E., *JADA, 84*, 1349 (1972).
20. Bausch, J. R., deLange, K., Davidson, C. L., Peters, A., and deGee, A. J., *J. Prosthetic Dent., 48*, 59 (1982).
21. Asmussen, E., *Acta Odont. Scand., 33*, 129 (1975).
22. Ruyter, I. E., and Sjovik, I. J., *Acta Odont. Scand., 39*, 133 (1981).
23. Thompson, V. P., Williams, E. F., and Bailey, W. J., *J. Dental Res., 58*, 1522 (1979).
24. Bailey, W. J., and Amone, M. J., *Polym. Preprints, 28*(1), 45 (1987).
25. Bowen, R. L., Dental filling material comprising vinyl silane treated fused silica and a binder consisting of the reaction product of bisphenol A and glycidyl acrylate, *U.S. Pharmacopeia*, 30661/2 et seq., 1962.
26. Kalachandra, S., Taylor, D. F., DePorter, C. D., Grubbs, H. J., and McGrath J. E., *Polymer, 34*, 778 (1993).

27. Soderholm, K. J., *J. Biomed. Mater. Res., 18,* 271 (1984).
28. Kalachandra, S., and Turner, D. J., *J. Biomed. Mater. Res., 21,* 329 (1987).
29. Atsuta, M., Hirasawa, T., and Masuhara, E., *J. Japan Soc. Dent. Appar. Mater.* (in Japanese), *10,* 521 (1969).
30. Suzuki, S., Nakabayasji, N., and Masuhara, E., *J. Biomed. Mater. Res., 16,* 2175 (1982).
31. Cowperthwaite, G. F., Foy, J. J., and Malloy, M. A., in *Biomedical and Dental Applications of Polymers* (C. G. Gebglein and F. K. Koblitz, eds.), Plenum Press, New York, 1981, p. 379.
32. Turner, D. T., and Abell, A. K., *Polymer, 28,* 297 (1987).
33. Turner, D. T., *Polymer, 23,* 197 (1982).
34. Haque, Z. U., and Turner, D. T., unpublished work.
35. Braden, M., *Biomaterials, 5,* 373 (1984).
36. Braden, M., Causton, E. E., and Clarke, R. I., *J. Dent. Res., 55,* 730 (1976).
37. Braden, M., and Davy, K. W. M., *Biomaterials, 7,* 474 (1986).
38. Muniandy, K., and Thomas, A. G., in *Conference Proceedings, Institute of Physics and Institute of Marine Engineering,* 1984.
39. Kalachandra, S., and Kusy, R. P., *Polymer, 32,* 2428 (1991).
40. Kusy, R. P., Whitley, J. W., and Kalachandra, S., *J. Dent. Res., 71,* 313 (AADR Abstract 1657) (1992).
41. Kalachandra, S., *Dent. Mater., 5,* 283 (1989).
42. Braem, M., *An in vitro investigation into the physical durability of dental composites,* Doctoral Thesis, Leuven, Belgium, 1985.
43. Greener, E. H., and Bakir, N., *J. Dent. Res., 65,* 219 (1986).
44. Wilson, T. W., and Turner, D. T., *J. Dent. Res., 62,* 121 (1983).
45. Antonucci, J. M., and Toth, E. E., *J. Dent. Res., 62,* 121 (1983).
46. St. Germain, H., Swartz, M. L., Phillips, R. W., Moore, B. K., and Roberts, T. A., *J. Dent. Res., 64,* 155 (1985).
47. Atsuta, M., and Turner, D. T., *Polym. Compos., 3,* 83 (1982).
48. Atsuta, M., Nagata, K., and Turner, D. T., *J. Biomed. Mater. Res., 17,* 679 (1983).
49. Nielsen, L. E., *J. Compos. Mater., 1,* 100 (1967).
50. Kalnin, M., and Turner, D. T., *J. Mater. Sci. Lett., 4,* 1479 (1985).
51. Kalnin, M., and Turner, D. T., *Polym. Compos., 7,* 9 (1986).
52. Goodier, J. N., *J. Appl. Mech., 1,* 39 (1993).
53. Manly, R. S. (ed.), *Adhesion in Biological Systems: Adhesives for Hard Tissues,* Academic Press, New York, 1970, pp. 225–89.
54. Moskowitz, H. D., Ward, C. T., and Woolridge, E. D. (eds.), *Dental Adhesive Materials: Consideration of Adhesion to Tooth Structure,* U.S. Department of Health, Education, and Welfare, New York, 1973, pp. 95–205.
55. Going, R. E., *J. Am. Acad. Gold Foil Operators, 18,* 16–21 (1975).
56. Nakabayashi, N., Kojima, K., and Masuhara, E., The promotion of adhesion by the infiltration of monomers into tooth substrates, *J. Biomed. Mater. Res., 16,* 265–273 (1982).
57. Bowen, R. L., Cobb, E. N., and Rapson, J. E., Adhesive bonding of various materials to hard tooth tissues: Improvement in bond strength to dentin, *J. Dent. Res., 61,* 1070–1076 (1982).
58. Denyer, R., The physical and mechanical properties of Fotofil, in *Proceedings of the International Symposium on Fotofil Dental Restorative,* Franklin Scientific Projects, London, 1978, pp. 7–10.
59. Loshaek, S., and Fox, T. G., *J. Am. Chem. Soc., 75,* 3544 (1953).
60. Houwink, R., *Elasticity, Plasticity and Structure of Matter,* 2nd ed., Dover, New York, 1958.
61. Morgan, R. J., and O'Neal, J. E., *J. Mater. Sci., 12,* 1966 (1977).

53
Alveolar Ridge Maintenance Using Endosseously Placed Bioglass® (45S5) Cones

Harold R. Stanley
University of Florida
Gainesville, Florida

I. THE HISTORY OF ALVEOLAR RIDGE RESORPTION

Continuous alveolar ridge resorption (ARR) after tooth extraction affects millions of patients whose loss of denture stability and retention produces impaired masticatory efficiency, oral and systemic health problems, and compromised aesthetics, especially in elderly patients who have been edentulous for many years (Bell, 1986; Quinn and Kent, 1984; Stanley 1987). A significant number of older patients will also have neuromuscular deficits, making it difficult to learn to adapt to a mandibular complete denture (Ettinger, 1993). Figure 1 (Keipper, 1988) demonstrates an example of extreme ARR in an 80-year-old female with chronic dementia due to Alzheimer's disease and a history of stroke with a residual left hemiparesis. When she occluded her edentulous resorbed alveolar ridges, her nares would collapse and severely impede her air exchange, making her hypoxic and cyanotic. These episodes would occur when the patient would subconsciously purse her lips while her mouth remained closed (a snout-like reflex). Merely placing a bridge of tape from the tip of her nose to the forehead would prevent this crisis.

Many dentists readily accept the fact that a significant number of their patients will never wear a removable mandibular complete denture with any degree of satisfaction (Rothstein, 1984a). A survey in 1984 estimated that of 20 million totally edentulous persons in this country, 70% were dissatisfied with their mandibular dentures (Misch, 1984). However, due to the recent decline in the number of teeth being extracted in the 55- to 64-year age group, some researchers (Ettinger, 1993; Weintraub and Burt, 1985) have projected a continuous reduction in the prevalence of edentulous adults, to a low of about 15% of the adult population by the year 2020. Nevertheless, Meskin, Brown, Brunelle, and Warren (1988) estimated that a constant subpopulation of about 9 million older adults will need complete dentures in the next 30 years. Also, in developed countries, people are living longer and this increased longevity means that the condition of

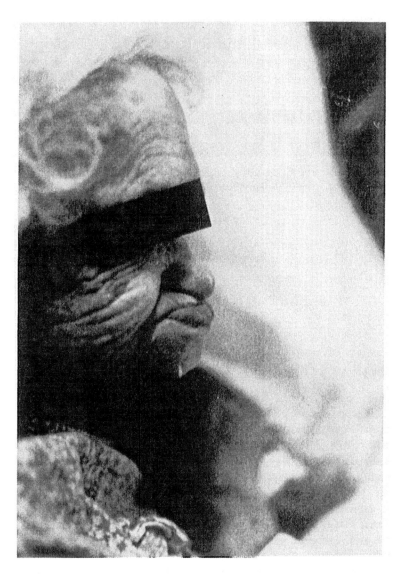

Figure 1   An 80-year-old woman with "snout suffocation" syndrome. The patient was supine when photographed; the photograph has been rotated 180°. (From Keipper, 1988. Reprinted, by permission of the *New England Journal of Medicine, 319*:1097, 1988.)

extreme alveolar ridge resorption will become more common in this edentulous population and may prolong or increase such dissatisfaction (Brook, Sattayasanskul, and Lamb, 1988; Zarb, 1988). One should appreciate that there was estimated to be about 25,000 centenarians in the United States in 1988, and that their number will climb to approximately 100,000 by the year 2000 (Feinberg, 1987).

Clinically, ARR can progress so rapidly that dentures cannot be worn for more than a short period of time before a reline or rebase is necessary; or in later years, there may be so little alveolar bone remaining that no denture can be retained (Tallgren, 1967,

1969). In many individuals, the process continues seemingly until death, resulting in the removal of massive amounts of bone (Atwood, 1979; Quinn et al., 1985). Nevertheless, dentists and their edentulous patients are inclined to believe that their situation will be different and their well-made complete dentures will be worn satisfactorily and perhaps even indefinitely, not like the experience of many of the other patients (Zarb, 1988). Because of its adverse effect on the quality of life for the elderly, extreme ARR has been termed a "major disease entity" (Atwood, 1971). The expression "dental cripples" says it all (Brewer and Morrow, 1980).

Atwood and Coy (1971) attributed the progressive, irreversible ARR process to a loss of anatomic, biologic, and mechanical factors. Maintenance of viable alveolar bone and ridges is dependent on stimulation of the periodontal membrane about teeth or roots (Atwood, 1962, 1963, 1971; Veldhuis, Driessen, Denissen, and de Groot, 1984).

A person can tolerate dentures only when the repair process of the oral tissues outpaces the trauma induced by functioning dentures. As every complete denture moves in function, the health and resistance of the denture-supporting tissues are critical to successful denture wearing. The consequence of loss of teeth and long-term denture wearing is ARR (Ettinger, 1993).

Following the extraction of teeth, the sockets fill with blood, which subsequently clots and is replaced with new bone, which undergoes continuous remodeling (bone resorption and subsequent structural rearrangement) according to the degree of stress applied (Atwood, 1963; Bell, 1986; Kangvonkit, Matukas, and Castelberry, 1986; Lam, 1960; Pietrokovski and Massler, 1967). The absence of stimulation to the softer alveolar cancellous bone or the overloading from denture pressure and masticatory trauma, particularly after the denture becomes ill-fitting, produces varying degrees of resorption and atrophy of the soft cancellous portion of the alveolar bone until basilar cortical plates are continuous and the less dense cancellous alveolar bone is completely resorbed (Atwood, 1979; Hench and Ethridge, 1982; Neufeld, 1958; Quinn and Kent, 1984; Quinn, Kent, Hunter, and Schaffer, 1985). The factors involved in this complex mechanism are not clearly understood, but the initiating factor seems to be the loss of teeth (Simon and Kamura, 1974).

Atwood (1963, 1971) and Atwood and Coy (1971) described in detail the sequence for the reduction of the residual ridge after tooth extraction:

1. The high, well-rounded residual alveolar ridge becomes narrower labiolingually. The labial and lingual cortical plates do not disappear, but the ridge narrows, mostly at the expense of the cancellous bone. The height of the edentulous alveolar ridge is almost unchanged until the stage of the so-called knife edge is reached, at which time the labial and lingual cortical plates merge into one.
2. As the reduction in size continues, the knife-edge alveolar ridge becomes shorter and eventually disappears, leaving a flat ridge without any residual alveolar bone (Denissen and de Groot, 1979).

The total amount of bone resorbed and the rate of resorption not only are different for each individual, but also vary greatly in the same individual at different times (Atwood, 1962; Sobolik, 1980; Veldhuis et al., 1984). The reduction of residual alveolar ridges occurs most rapidly in the first 6 to 24 months after extraction (Quinn and Kent, 1984). Height loss in the first 5 years is more than twice the height loss in the succeeding 20 years (7.6 mm/3.1 mm) (Veldhuis et al., 1984). However, there is not much difference after 10 years. Ettinger (1993) states that the rate of resorption is about four times faster in the mandible than the maxilla. Atwood (1971) pointed out that the ARR rate in 20% of his subjects was 1 mm/year or more. Although the highest rate of ARR, which occurred over 3 years, was 2.2 mm/year, the mean ARR for the maxilla was 0.1 mm per year and the mean for the mandibular anterior ridge was 0.4 mm per year. The mean rate of ARR was 0.5 mm per year for both jaws when combined. Consequently, the average rate of anterior vertical ARR of 0.5 mm per year could represent an average loss of 5.0 mm in anterior ridge height in 10 years. According to Tallgren (1967, 1969), even when wearing full dentures, the resorption rate of the mandibular ridge is still three times greater than the maxilla.

Atwood, in 1971, listed the stages in his anatomic classification for ARR in a horizontal direction as follows:

Stage I. Before extraction.
Stage II. Directly after extraction.
Stage III. There is a high and well-rounded ridge.
Stage IV. There is a knife-edge ridge.
Stage V. There is a low and rounded ridge.
Stage VI. There is a negative ridge.

Cawood and Howell (1988) classified edentulous mandibles according to a three-dimensional analysis of the mandibular anatomy. Three different cross sections were used: in the symphysis region, through the mental foramen, and in the molar region. They concluded that the alveolar ridge resorbs in the anterior and premolar regions horizontally as well as in a vertical direction. A remarkable finding of their study was that the resorption process was almost entirely confined to the alveolar ridge of the mandible, and that the basal part, which is the part caudal to the mental spina, does not significantly change after extraction. Cawood and Howell (1988) proposed the following progressive stages of resorption:

Class I. Dentate ridge
Class II. Ridge directly after extraction.
Class III. Broad and rounded ridge with adequate height and width.
Class IV. Knife-edge ridge with sufficient height but insufficient width.
Class V. Flat ridge with insufficient height and width.
Class VI. Depressed ridge with a cup-shaped surface.

They suggested that two more classes of resorption could be added to the above classification: Class VII, the labial part at the site of the symphysis is resorbed in a vertical direction to a height of 10 mm; and Class VIII, a further reduction of the height occurs, to a value of 5 mm (Cawood and Howell, 1988; Denissen, Kalk, Veldhuis, and van Waas, 1993).

II. THE BENEFITS OF ROOT RETENTION

Since the beginning of the 19th century, clinicians have observed the persistence of alveolar bone around retained roots and impacted teeth in otherwise edentulous mouths. On September 22, 1826, Dr. Fay, an American dentist, received a medal of honor from the English Society of Arts for stating "What is of the greatest moment, the stump or stumps left in the jaw afford a firm support to the adjoining teeth, and without which [i.e., such] support, the alveolar process corresponding to the part of the interstice structure of the jaws becomes absorbed." (Veldhuis, 1988) (Present author's addition in brackets).

III. OVERDENTURES

During the last decade, overdenture treatment has become one of the main strategies of preventive prosthodontics because the retained teeth are assumed to improve denture function by reducing bone resorption (Brewer and Morrow, 1980; Kalk, Van Rossum, and Van Waas, 1990).

Overdentures can be considered a preferred alternative to complete dentures, especially in patients with insufficient alveolar bone support (Brewer & Morrow, 1980; Langer and Langer, 1993; Prieskel, 1985). Experienced prosthodontists have long been wary of radical tooth extraction in the mandible, but in view of the present developing knowledge, the prevention of complete maxillary edentulousness is equally advisable, even when anatomic conditions for making a complete denture appear satisfactory.

The overdenture is not a new concept. In 1861 at the American Dental Convention in New Haven, Connecticut, a symposium was held entitled "Surgical Preparation of the Mouth for Artificial Dentures: Should the Roots of Broken and Decayed Teeth Always Be Removed?" The consensus was that in many situations retention of the roots, or "fangs," would enable the dentist to give the patients needing complete dentures treatment superior to that obtained after extraction of all roots (Brewer and Morrow, 1980). However, it was Miller (1958) whose foresight was of prime importance in convincing the profession that the overdenture was a superior treatment modality (Brewer and Morrow, 1980; Denissen and de Groot, 1979). Miller pointed out that:

1. The architecture of the maxillae and mandible was designed to house the roots of teeth, not to act as support for artificial dentures.
2. It does not need to be debated that roots of teeth offer a better medium of support for an artificial replacement than the mucoperiosteum. The fixed partial denture, when properly formulated, has been dentistry's best means of replacing missing teeth.
3. In nearly every instance of a full-mouth extraction, two, three, or four teeth could be salvaged and restored to a degree of health to act as support for complete dentures. Even if a short life expectancy of the teeth is anticipated and their value in preventing resorption is not considered, the assistance they would lend in helping patients become accustomed and habituated to dentures is worth the effort.
4. The use of teeth as support for dentures is aimed at reducing the load on the osseous portions of the denture-bearing area and at minimizing the process of resorption.
5. The biologic maintenance of a neuromuscular mechanism, the temporomandibular articulation, and the supporting structures of a denture can be accomplished better by teeth than by the mucoperiosteum.

Miller's technique (1958) required full crown preparation of isolated natural teeth covered with copings to prevent the exposed tooth structure from decaying while supporting the complete dentures with crowns (thimbles) attached to the dentures fabricated to fit over the jacketed abutment teeth. The abutment teeth played no part in the retention of the dentures but acted only as "stabilizers." Frictional retention was not required or desired. Miller (1958) reported less resorption of alveolar ridge tissue over a period of 6 years in 46 overdenture patients. No mention was made as to whether the abutment teeth were left vital or non-vital, but it is assumed that they were maintained as vital teeth.

Kabcenell (1971) described how remaining teeth could be reduced to the level of the gingiva after endodontic treatment and fitted with posts, copings, and male–female retentive connectors. This technique also provided horizontal stabilization and improved retention of the denture.

However, in situations where the conventional overdenture approach (thimbles) was not advisable (poor oral hygiene, periodontitis, caries, economics, etc.), the concept of root submersion developed (Casey and Lauciello, 1980). Subsequently, three methods of utilizing retained but submerged roots in animals and humans were reported:

1. Endodontically treated roots (Goska and Vandrak, 1972 – 3 roots in a single human subject; O'Neal, Gound, Levin, and del Rio, 1978)
2. Intentionally extracted roots (after endodontic treatment) reimplanted below the alveolar crest (Gound, O'Neal, del Rio, and Levin, 1978; Simon and Kumura, 1975)
3. Vital roots reduced 1–2 mm below the level of alveolar bone and covered by a mucoperiosteal flap (Delivanis, Esposito, and Brickley, 1980; Garver and Fenster, 1980; Garver, Fenster, Baker, and Johnson, 1978; Guyer, 1975: Herd, 1973; Johnson, Kelly, Flinton, and Cornell, 1974; Murray and Adkins, 1979; Welker, Jividen, and Kramer, 1978; Whitaker and Shakle, 1974)

Although Garver et al. (1978) at first advocated retention of vital roots covered with mucoperiosteum, later Garver and Fenster (1980) recommended dealing only with healthy pulps treated with calcium hydroxide capping agents.

Guyer (1975) presented the results on a single human subject, a 52-year-old male where the crowns of two mandibular cuspids were severed and the gingival tissue sutured so as to completely cover the retained roots. Both roots were radiographically normal after a little over 2 years.

After reviewing the world's relevant literature, Casey and Lanciello (1980) (reporting no work of their own) concluded that of the three techniques, vital submerged roots provided the most acceptable results.

It was unfortunate that such enthusiasm developed for these different techniques when the data to support them were sometimes based on just a single case (Goska and Vandrak, 1972; Guyer 1975). As mentioned earlier, Guyer only severed the crowns of two roots in one patient.

In 1978 Crum and Rooney compared 8 patients with mandibular overdentures with 8 patients with conventional mandibular dentures (Van Waas, Jonkman, Kalk, Van'T Hof, Plooij, and Van Os, 1993). When mandibular canines were left for retention of overdentures, they found an average vertical loss of alveolar bone of only 0.6 mm in the anterior portion of the mandible, as compared to an average loss of 5.2 mm of vertical bone in patients with conventional dentures after 5 years.

Steen (1984) compared 22 patients receiving immediate overdentures on two lower canines with patients receiving immediate complete dentures. After the first year, he found 50% less bone resorption in all parts of the mandible in the overdenture group as compared to the immediate complete denture group.

Van Waas et al. in 1993 reported on 74 patients with severely decayed and/or periodontally involved teeth. An immediate maxillary denture was indicated or a complete denture was already present. For the mandible, the patients were treated with either an immediate overdenture on the roots of the lower canines or with an immediate complete denture. All abutment teeth were still present 2 years later. The amount of bone resorption was always smaller in the immediate overdenture group compared with that in the immediate complete denture group, except in cluster A (mandibular posterior ridge) in the second year. Analysis of the data showed that the average bone reductions in the lower canine regions in the first years was 0.9 mm in the immediate overdenture group and 1.8 mm in the immediate complete denture group. In the posterior parts of the mandible, the bone reductions were, respectively, 0.7 mm and 1.9 mm.

The amount of bone reduction in the frontal region of the mandible in the immediate denture group was lower in this study than that reported in previous studies. Atwood and Coy (1971), Tallgren (1972), Carlsson and Persson (1967), and Steen (1984) had found after the first year about 4.0 mm. of bone reduction in the frontal region of the mandible after extraction of teeth as compared to the 1.9 mm lost with retained canine roots. There was also significantly less bone reduction in the overdenture group, compared with the complete denture group, after 2 years. This suggests that, in patients with severely decayed and periodontally poor teeth, when the indications for overdentures are not obvious, retaining roots of canines has a positive effect on the reduction of the alveolar bone loss in the period during which the bone resorption is usually the largest.

The preservation of just two canines—although often in poor condition—influenced the remodeling process in a positive way (less bone resorption) not only in the frontal region, near the remaining canines, but also at some distance from the remaining teeth because of reduced forces on the adjacent bone. In many cases, the mandibular anterior teeth, which are usually the last to survive, may have traumatic effects on an opposing maxillary edentulous jaw bearing a complete denture and may give rise to a specific clinical condition known as the "combination syndrome." In advanced stages, the sequelae of this condition include pathological changes in both jaws, such as extrusion of the anterior teeth and bone resorption under the free extension denture saddles in the mandible. It is also accompanied by overgrowth of tuberosities, papillary palatial hyperplasia, and bone loss in the anterior ridge area of the maxillae. With an edentulous maxillae, development of the combination syndrome may be prevented or checked by turning the offending mandibular teeth into overdenture abutments, thereby removing them from direct occlusion with the opposing complete denture (Langer and Langer, 1993). Reducing the clinical crown close to ridge level also reverses the unfavorable crown-to-root ratio, decreasing the leverage on the periodontal and osseous structures.

The proprioception benefit provided through remnants of the periodontal ligament about the retained roots also serves as an important factor in maintaining the discrimination and transmission of sensory information to the central nervous system (Langer and Langer, 1993). Mushimoto (1981), Loiselle, Crum, Rooney, and Stuever (1972), Rissen, House, Manley, and Kapur (1978), and Nagasawa, Okane, and Tsuru (1979) all showed that retained roots improved neuromuscular activity and coordination, control of mandibular movement, chewing power and efficiency, and sensory discrimination in differentiating between objects of various sizes placed between teeth. Retaining, the terminal roots also has important psychological implications because overdenture patients do not feel completely edentulous. The improved performance of their dentures boosts their self-confidence and relieves the frustration stemming from physical invalidity. Although it is theoretically possible to keep bare roots caries-free and in good periodontal health, epidemiologic studies suggest that many patients are negligent in oral hygiene and consequently develop root pathoses. Most overdenture patients are advanced in age. Attempting to retain vital teeth under denture bases significantly increases the risk of subsequent periapical pathosis (Langer and Langer, 1993). Denture bases inhibit self-cleansing of the underlying tissues. In the absence of painstaking oral hygiene, the accumulation of food and plaque severely threatens bare abutments through exposure to caries and periodontal pathosis. Although meticulous maintenance regimen can maintain roots for many years in healthy condition (Davis, Renner, Antos, Schlissel, and Baer, 1981; Toolson, Smith, and Phillips, 1982), clinical findings show that neglect of oral hygiene in overdenture patients because of lax recall regimens is widespread and can lead to abutment failure (Toolson and Taylor, 1989). Also, the diminished self-cleansing properties of saliva, often associated with advanced age, can contribute to retained root failures in overdenture wearers (Langer and Langer, 1993).

In a 2-year longitudinal study, Toolson and Smith (1978) found that 34.8% of the unprotected roots were affected by caries after a year, and another 19% a year later. Ettinger, Taylor, and Scandrett (1984) found a progressive increase in the caries rate of up to 20.6% in bare root abutments during a 5-year observation period.

In 254 overdenture patients recalled at 6-month intervals for evaluation and maintenance treatment during a period of 12 years, Ettinger (1988) found abutment loss of only 4.2%. Almost all of the roots were bare (97%). Root loss was equally attributable to periodontal disease and caries. The recall rate of more than 90% and the periodic follow-up regimen were crucial for maintaining abutment health.

However, the treatment of the bare roots with applications of stannous fluoride, stannous fluoride gels, and fluoride gels provides significant caries preventive effects (Derkson and MacEntee, 1982; Fenton and Hahn, 1978; Langer and Langer, 1993; Toolson and Smith, 1978).

Modern endodontics and periodontics have made it possible to preserve the roots of caries-destroyed or periodontally compromised teeth to be used as overdenture abutments.

In preparation for overdentures, the devitalized teeth are reduced to root level or below; obturated with silver amalgam, glass ionomer cement, or composite resin; and then rounded, contoured, smoothed, and polished. Sealing endodontically treated root canals is essential to prevent periapical microleakage of microorganisms. Swanson and Madison (1987) have shown that with incomplete obturation it takes dye only 3–7 days to reach the apex, implying that microorganisms could rapidly penetrate the periapical areas (Langer and Langer, 1993).

However, replanted submerged tooth roots eventually tend to fail because of ankylosis and resorption (Lam, 1972; Simon and Kimura, 1974). Consequently, when very little ridge height remains, various surgical procedures have been attempted to increase the basal area for support of prostheses (Bell, 1986). The dental literature is replete with articles that describe various techniques used in an attempt to preserve or rebuild the edentulous alveolar ridge, mostly with limited success (Kwon, El Deeb, Morstad, and Waite, 1986). For many years autogenous bone grafting has been an established technique and considered a satisfactory treatment. Its main advantage is the maintenance of a stable relationship between mucoperiosteum, cortex, and underlying bone to provide a sound denture base. Although useful, these operations are not without morbidity, occasional total failure, and a significant relapse rate even in successful patients. Other disadvantages are associated with the need for a donor site (the iliac crest), including increased blood loss, increased operation time, and potential local pain and morbidity at the donor site. Although preservation techniques have sharply reduced the antigenicity of freeze-dried bone, there are still the theoretical disadvantages of disease transmission, graft infections, and rejection when it is used for ridge augmentation (Stanley, Hall, Colaizzi, and Clark, 1987).

IV. IMPLANTS

A. The Need for Replacements

Since the loss of tooth roots sometimes cannot be avoided because of the extent of caries and periodontal disease, the concept of using alloplastic implants as a substitute for natural roots developed (Denissen, Rejda, and de Groot, 1978). In a number of publications, the earliest one dating from 1826, it was suggested that the filling of extraction sockets with space-maintaining implants would slow down mandibular ARR (*Nottingham Review*, 1826).

If the loss of natural tooth roots is the basic cause of various complaints, the most straightforward approach to solving this problem would be to replace natural tooth roots with artificial implants in the sockets of extracted teeth (Denissen et al., 1978) and hope that the presence of an implant might simulate a tooth root and preserve the alveolar bone (Lam, 1972).

Based on the past and current literature and the anatomic classifications of mandibular resorption (Atwood, 1971; Cawood and Howell, 1988; Tallgren, 1972), Denissen, Kalk, Veldhuis, and van Waas (1993) developed the following four-stage classification of ARR (Fig. 2) for "preventive implant therapy" (to forestall mandibular resorption):

Preventive stage I: Anatomic situation directly after tooth extraction.
Preventive stage II: After the initial resorption of the edentulous ridge – the ridge still has
    width and height. No bone removal is necessary for placing cylindrical implants 3 or
    3.75 mm in diameter.
Preventive stage III: Ridge has atrophied to a knife-edge shape. Bone removal is necessary
    for the placement of implants.
Preventive stage IV: Only basal bone is present in a vertical direction.

Based on their study it was recommended that implantation be performed much earlier than is now usually the case. Preventive implantation should be started before the edentulous alveolar ridge has been resorbed to such a degree that only a knife-edge ridge is

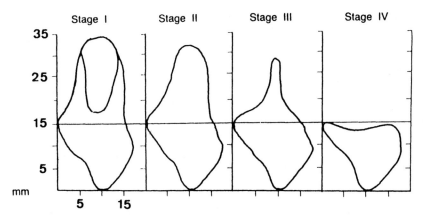

Figure 2 Cross sections of edentulous mandibles at the sites of the central incisors, indicating preventive stages I to IV. (From Denisson et al., Anatomic Considerations for Preventive Implantation, *Int. J. Oral Maxillofac. Implants, 18*:191–196, 1993. Reprinted by permission of Quintessense Publishing Co., Inc.)

present (Preventive stage III). When resorption has progressed to this stage, the remaining parts of the joined alveolar cortical plates must be removed, leaving no ridge whatsoever. Therefore, implantation should be done in Preventive stage II, when the ridge is still bulky and high so as to prevent the development of the knife-edge ridge of Stage III.

B.   Submerged Implant Materials

The ideal socket implant material should have the following criteria: (1) no evidence of early resorption of the implant material, (2) acceptable strength, (3) strong attachment to the soft and hard tissue, and (4) no adverse host reactions. Implants can provide mechanical support as a scaffolding and prevent the collapse of both the labial and lingual plates of bone (Denissen and de Groot, 1979; Stanley, 1987).

The following synthetic products have been tried: acrylic resin (Lam and Poon, 1968, 1969); hydroxyapatite (Quinn et al., 1985); coraline hydroxyapatite (Frame, 1984), carbon, calcium phosphate ceramics (James, 1984; Misiek, Kent, and Carr, 1984); tricalcium phosphate and Bioglass® (Hall, Stanley, King, Colaizzi, Spilman, and Hench, 1985; Stanley, Hall, Colaizzi, and Clark, 1987; Stanley, Hench, Bennett, Chellemi, King, Going, Ingersoll, Ethridge, Kreutziger, Loeb, and Clark, 1981; Stanley, Hench, Going, Bennett, Chellemi, King, Ingersoll, Ethridge, and Krautizer, 1976).

In recent years bioactive materials have received increasingly wide attention. Bioactive glasses and calcium phosphate ceramics show excellent bone-bonding properties. Calcium phosphate ceramics resemble hydroxyapatite, the most important mineral in living bone (Vrouwenvelder, Groot, and de Groot, 1993).

V.   REVIEW OF LITERATURE REGARDING CONE IMPLANTS

Waerhaug and Zander (1956) showed in dogs that acrylic resin root implants created little or no irritation in hard and soft tissues. Waerhaug and Loe (1958) also found in dogs no basic difference in tissue reaction to cold-curing acrylic resin root implants and heat-cured acrylic resin root implants.

Lam and Poon (1968, 1969) appear to be the first to use alloplastic cones in a clinical study. They utilized a cold-curing acrylic resin (a self-polymerizing methylmethacrylate resin). Six patients needing the extraction of maxillary anterior teeth were selected. Impressions were taken of the roots of the extracted teeth and replicas made. One patient had solid-cone implants, two had thimble-type implants, and three had perforated thimble type implants. After the cones were totally embedded into the sites of extraction, sutures were placed only in the last four patients. Acrylic resin partial dentures were then placed using closed face techniques. The two unsutured patients had their implants removed the first week because of exfoliation. Two other patients had their implants removed after 3 months due to gradual dehiscence. Even in those patients who were sutured initially there was incomplete primary union of soft tissue over the socket area. The fifth patient had the implants removed the fourth month because of gradual dehiscence and incomplete primary union. After 20 months, the sixth patient still maintained the implants with healed covering mucosa. Cortical bone had formed over one of the implants. The cones induced the formation of "bony bumps" clinically. The authors did not specify which type of cone implant was successful. The amount of bone loss in the area of the implants was markedly reduced as compared to the control areas. The implants did not stop bone resorption but certainly minimized its rate and amount. Lam and Poon (1968, 1969) recommended that the cones be placed 3 mm or more below the crest.

Mills, Chan, Voss, and Grenoble appear to have been the first, in 1974, to study the cellular response of alveolar bone to vitreous carbon cones placed in fresh mandibular extraction sites of rabbits, dogs, and man. In dogs there was bone ingrowth into the implant surfaces with normal bone remodeling. After 18 months the dogs demonstrated well-fitting vitreous carbon implants in well-maintained alveolar bone. The number of specimens was not given. In man and rabbits, however, the implants became encapsulated. The fibrous membrane consisted of an acellular, fibrous material that interdigitated with the collagen fibers approximating bone.

A. Hydroxyapatite

In 1978 Denissen, Rejda, and de Groot presented clinical data on cones fabricated from a composite of calcium hydroxyapatite (HA) and *p*-hydroxyethyl methacrylate (pHEMA): HA is known to stimulate bone growth, while pHEMA is a well-studied biocompatible biopolymer. The HA was sintered to produce a porous material (porosity 40%, diameter of pores 100 μm). The pores were then impregnated with a polymerizing mixture of 67% HEMA (hydroxyethyl methacrylate), 3% of a crosslinker (ethylene glycol dimethacrylate–EDMA) and 30% of an aqueous solution of initiator (0.5% ammonium persulfate). The sintering or hot-pressing process produced a structure with low porosity and with little or no potential for biodegradation (Denissen and de Groot, 1979). In this study Denissen and de Groot placed cones in 10 fresh sockets of four patients, sutured the wounds, and found new bone forming as soon as 2 months. Healing was observed radiographically for 5 months postextraction.

Since then, hydroxyapatite (implant materials based on calcium phosphate salts) has received the most attention as cone implants (Brook and Lamb, 1987). Although these materials are not osteogenic (implants of HA in the soft tissues of experimental animals show no evidence of new bone formation), if placed in subperiosteal sites, they act in an osteoconductive way by stimulating the laying down of bone about the implant.

In 1979 Denissen and de Groot placed 50 nonbiodegradable dense calcium hydroxyapatite cones into empty mandibular premolar sockets of dogs. The wounds were permitted to heal by primary closure of the mucosa. Radiographic studies showed the surrounding bone to be closely adapted to the implant after 3 months, even over the implants. All implants were retained up to 18 months.

Also, in 1979–1980, Denissen and de Groot (1979) and Denissen, Veldhuis, Makkes, van den Hoof, and de Groot (1980) placed 100 dense HA ceramic cone implants into 20 patients with severe periodontal disease or teeth with persisting draining fistulae following endodontic treatment and apical curettage. The cones were placed in fresh sockets, which were closed before the seating of full mandibular dentures or fixed and removable partial prostheses. Nine of the 20 patients received immediate lower dentures. Seventy-one implants were placed in nine patients under mandibular full dentures. Implants were inserted in such a way that their cervical plane was situated just below the most apical part of the socket crest (mostly the vestibular buccal part) to permit primary closure of the extraction wounds. After 1 year, all the implants were retained. No radiolucencies developed. The residual alveolar bone closely adhered to the implants even in those patients that received immediate dentures. The physical presence of the implants maintained a bulky ridge; the ridges collapsed only at sites where no implants were present.

Sixty-nine implants were still retained at 18 months. One was removed at 18 months due to a long-term dehiscence. A total of six dehiscences occurred under mandibular full dentures through 24 months. This process could be arrested by shortening the implant. No dehiscences recurred during the 12 months following subsequent shortening. Even those implants requiring shortening appeared to be very strongly attached to the bone at the time of surgery. It appeared that the cones became ankylosed as would natural roots. The remaining 29 implants were placed in 11 patients under fixed or removable partial prostheses. All were retained in a submerged position without dehiscences. In order to obtain primary closure, an implant must be submerged in a single socket surrounded by healthy tissue, and this presented a problem. Dehiscences in the mandible due to the periodic mechanical trauma inflicted by the mandibular denture occurred with a frequency of 3-4%/year. Denissen, Veldhuis, Makkes, van den Hoof, and de Groot (1980) continued to follow these patients up to 30 months.

In 1984 Veldhuis, Driessen, Denissen, and de Groot reported on the continued evaluation of the patients of Denissen and de Groot (1979), and added several more patients to make a grand total of 212 implants in 24 patients over a 5-year period. Initially the implants were individually fabricated to match the roots of the extracted teeth but later on the patients received the most appropriately sized prefabricated cones, placed at least 2 mm below the alveolar crest and primarily closed with approximating tissues.

Although Denissen and de Groot (1979) and Denissen et al. (1980) demonstrated that cone implants reduced the bulk of mandibular ridge loss, they observed that dehiscences began to occur after approximately 10 months and increased in number for approximately 2.5 years, after which new dehiscences occurred generally at a much lower rate.

By 1984 Veldhuis et al., from their experience, expected that all implants placed a few millimeters below the socket crest would eventually develop dehiscences. However, since this occurred in only 30% of the patients after 5 years, they concluded that 70% of the implants succeeded in reducing or preventing significant height loss, and that the height loss seemed to stop when the receding alveolar bone reached the occlusal or incisal surface of the implant. It took approximately 2 years on the average for the receding bone to reach the implants.

It seemed to Denissen and associates that beyond doubt the physical presence of such implants, simulating the principle of an ankylotic root, prevented collapse of the cortical plates and guaranteed a residual bulk or volume of the denture bearing region of the mandibular alveolar ridge whether the implants were submucosal or permucosal. But they again recommended that the implants be tightly wedged at least 2 mm below the occlusal aspect of the alveolar crest in order for occlusal bone formation to occur (Denissen and Kalk, 1991; Denissen et al., 1980).

After an 11-year follow-up, Denissen, Kalk, Veldhuis, and van den Hoof (1989) reported that during the first observation period, of 5 years, 16 of 81 bulk HA implants placed in 11 patients had become permucosal (can be seen through thinned mucosa) but were ankylosed to the bone and therefore could not be removed. In the second observation period (5-11 years), 4 of these same 16 implants that had become permucosal were lost and 16 of the remaining 77 implants were lost (a total of 20 lost implants—24.7%). Ten implants still present were ankylosed but became permucosal. All of these implants had been placed into fresh empty sockets of the mandible, the gingivae sutured, and immediate dentures placed.

The authors pointed out that some patients stated that the loss of implants left a temporary feeling similar to that of a sore spot caused by the pressure of the denture.

Von Wowern, Harder, Hjorting-Hansen, and Gotfredsen (1990), studying the mineral content of bone, concluded that there was less loss of bone mass at sites with implants and a slowing of the alveolar resorption pattern generally when loaded by an implant-supported prosthesis rather than a conventional mucosa-supported denture.

Denissen, van Beek, Papapoulos, Lowik, Kalk, and van den Hooff (1993) displayed their frustrations with HA cones by initiating research related to incorporating bone resorption inhibitors into the cone implants. No resorption-inhibiting influence on bone can be attributed to the HA implants. This could be overcome theoretically by adding a resorption-inhibiting agent to the HA implants. Bisphosphonates are known inhibitors of bone resorption, and a new bisphosphonate [(3-dimethylamino-1-hydroxypropylidene)-1,1-bisphosphonate; dimethyl-ADP] has been shown to effectively suppress bone resorption *in vitro* and *in vivo* and to be devoid of toxicity (Papapoulos, 1989).

The combination of a HA carrier with the bisphosphonate dimethyl-ADP combines the favorable biological properties of an alveolar bone-bonded ceramic delivery system with a proven systemic resorption-inhibiting factor.

The HA implant carrier was designed in the form of a cylindrical tube. The tube was impregnated with dimethyl-ADP and placed in fetal long bones of mice. Dimethyl-ADP did not change the bulk properties of the HA ceramic but the surface properties were altered. The fetal mouse bones were not adversely affected but Ca release was suppressed, as expected. The release rate of the incorporated dimethyl-ADP was high in the beginning but slowed down after 1 week. The average amount released in 2 weeks was about half of the incorporated amount. The HA–dimethyl-ADP carriers were compatible, stable, and bonded strongly to alveolar bone. It appeared that an incorporated dose of 0.78 ± 0.20 μg of dimethyl-ADP, which inhibits bone resorption effectively, can be delivered to the alveolar bone without any adverse effects on either the implant or the bone.

In an initial study by Boyne, Rothstein, Cook, Stutz, and Gumaer (1982), root replica implants of solid hydroxyapatite (durapatite[1]) were placed in both fresh and healed extraction socket sites in beagle dogs for 6 to 24 months. They were very careful to approximate the mucoperiosteal flaps (Boyne, 1990). New bone formation occurred at the crest of the ridge and the alveolar bone height was maintained. The new alveolar bone was firmly adherent to the ceramic surface of the implant after 24 months.

In another study Boyne, Rothstein, Gumaer, and Drobeck (1984) placed ceramic (durapatite)[1] implants in fresh extraction sites of beagle dogs at the level of the lingual crestal bone and approximately 1 mm above the level of the labial crestal bone. The mucoperiosteal tissues were then approximated. In another group the surgical extraction site was permitted to heal for 5 weeks, after which a socket was trephined and an implant placed. No wound dehiscence or exfoliation of the implants developed. After 36 months, attempts to elevate or separate the implants at the bone–implant interface were unsuccessful and resulted in flaking and fracturing of the implants. Thirty percent of the specimens showed new bone completely covering the implants; 60% of these implants had been

---

[1]Durapatite—a nonproprietary term designated for a particular form of hydroxyapatite manufactured by Sterling Drug, Inc. (Rothstein, 1984b). Durapatite is a synthetic hydroxyapatite [$Ca_{10}(PO_4)_6(OH)_2$] which has physical and chemical characteristics similar to vertebrate tooth and bone mineral. The material is a nonresorbable, dense, pure polycrystalline form of hydroxyapatite that has a higher compressive strength (133,000 psi) and transverse strength (28,400 psi) than previously reported for sintered hydroxyapatites. Durapatite has been proven to be biocompatible and capable of stimulating new bone formation without itself being resorbed, and it does not exhibit a change in strength after treatment in different solutions or in various implant sites. The material is easily shaped and is available in particulate and solid form. It is capable of sterilization by routine gas or autoclave technique without alteration of its physical or chemical properties (Kangvonkit et al., 1986).

placed in the fourth premolar and first molar areas. Microscopic examination of the ridge demonstrated increased thickness in the superior portion after both immediate and delayed implantation, compared with the controls. In the second and third premolar regions, apposition of bone in the crestal region favored the lingual crest over the labial crest.

Quinn and Kent (1984) placed 36 ultrasonically cleaned, solid nonporous durapatite hydroxyapatite root implants 1 to 2 mm below the alveolar crest into fresh extraction sockets of baboons without soft tissue closure. The results of custom-made (tapering root) implants, which were contoured to fit the entire length of the sockets, were compared with untapered (cylindrical) root implants that filled only the occlusal half of the sockets. Gingival healing across the occlusal surfaces of the implants was observed at 4 weeks. At 3 months nearly all tapered and untapered (cylindrical) root implants were covered with bone on all surfaces. At 6 months, the bone prevailed, with no intervening soft connective tissue membrane. The new bone demonstrated a normal alveolar trabecular pattern. The assumption that a custom-shaped root implant replicating the natural tooth would be necessary to preserve the alveolus was not substantiated. No significant differences were apparent after 6 months.

In another phase of the study in the same publication, beagle dogs received 128 Calcitite® RM[2] root implants. The implants were contoured to fit the sockets snugly 2 mm below the crestal bone level. No attempt was made to cover the implants with soft tissue. The animals were observed for a period of 18 months. The placement of Calcitite® root implants in fresh sockets preserved an average of 2 mm more alveolar bone height as compared to control sites. However, if the implants were placed into extraction sites 3 months after healing, there was no alveolar bone height advantage over the control sites. Rohrer, Swift, and Penwick (1988) demonstrated in dogs small amounts of osteoid around solid HA implants as early as 2 weeks.

In a study by Quinn, Kent, Hunter, and Schaffer (1985), Calcitite® RM root forms were fitted tightly into fresh extraction sites of 49 patients 2–3 mm below the alveolar crest without gingival tissue closure. After a period of 3 to 31 months (apparently only 2 patients were in the 24 to 31 month range), 42 patients with 81 implants revealed a loss of 9 (9.7%) implants in 9 patients. Twenty-four implants (26.0%) in 17 patients became prominent submucosally, necessitating elevation of a small mucoperiosteal flap and recontouring of the implant. Approximately twice as much alveolar bone was maintained according to height and width measurements as compared to the control sites. After 23 months, ridge height loss was 2.7 mm in the implant group and 5.5 mm in the control group; ridge width loss was 2.6 mm in the implant group and 4.5 mm in the control group.

In 1984 Cranin and Shpuntoff, working with 10 dental residents, placed 100 dense, nonresorbable HA cones at least 1 mm below the alveolar crest into fresh sockets of 10 patients without mucosal closure. Denture impressions were taken no sooner than 8 weeks postsurgery. Sixty-two cones were placed in the maxilla and 38 in the mandible. At the end of 1 year 55% of the cones were lost: 32 (51.6%) of 62 maxillary cones and 23 (60.5%) of 38 mandibular cones.

---

[2]Calcitite RM (ridge management): Calcitite, a synthetic ceramic material, is pure calcium phosphate (Calcitek, Inc., 1984 advertisement). During the process of its manufacture, HA powder is sintered to a dense, microporous form that, although hard and brittle, is suitable for implantation (Jarcho, 1986; quoted by Brook and Lamb, 1987).

Block and Zieman (1985) studied bone preservation with particulate and solid HA in canine extraction sites and concluded that bone resorption was not prevented by these HA implants, but that both the root and particulate forms did maintain alveolar bulk and form because of their nonresorptive presence.

In 1986 Bell implanted 51 cones of a solid form of HA into fresh sockets of eight patients (23–62 years) immediately after the extraction of teeth. The tooth sockets were required to have a depth of at least 6 mm in bone. Forty-eight cones were implanted in the mandible between the second premolar regions and three cones were placed at other sites. The cones were modified to fit snugly and seated at least 1 mm below the most superior aspect of the alveolar crest. After suturing, immediate dentures were placed. The patients were followed for 42 months. Two of the cones were lost within the first 30 days. Eighteen of the cones became prominent submucosally between 3 and 21 months. Cones that became submucosally prominent were not treated unless they caused patient discomfort. Cones that eroded through the mucosa were removed only if mobile. If immobile, the implant was resubmerged by shortening the cone below bone level or removed. After 42 months, 12 (23.5%) of the 51 cones had been removed. The task of removing an immobile cone was similar to that of removing an ankylosed tooth root. From these results Bell felt it was obvious that the cones were not completely effective in preserving the alveolar ridge and preventing the resorption that occurs immediately after tooth extraction.

In 1986 Kangvonkit et al. submerged 96 durapatite implants in the symphysis region of the mandible of 15 patients. Most of the implants were placed in the sockets of three teeth on each side of the midline. The teeth before removal had to have at least 8 mm of alveolar bone in a vertical dimension or periodontal bone loss less than 2/3 of the root length. The length and labiolingual width of the natural tooth were measured immediately after extraction and a corresponding appropriately sized cone was selected. The cone was then shaped and placed snugly into the socket wound at least 1 mm below the most superior aspect of the alveolar crest. All extraction sites were sutured. Complete maxillary and mandibular dentures were placed at the time of surgery.

Of 15 patients receiving 96 implants, Kangvonkit et al. (1986) experienced an implant loss of 6 (6.2%) and 17 (13.6%) dehiscences after 2 years. They found a ridge height loss of only 1.4 mm after 2 years in the experimental group as compared to 4.2 mm in the control group. Despite dehiscence, the implants were nevertheless strongly attached to the surrounding bone. Dehiscence most frequently occurred on the labial or lingual surface where the implants had sharp angles at the corners of the occlusal and proximal surfaces. Because of this finding, the cones were subsequently rounded on the buccocervical and linguocervical surfaces to the level of the surrounding bone to prevent erosion of the alveolar mucosa. Dehiscence occurred less frequently after using this technique. The wound then healed either by primary closure or secondary intention.

Resorption occurred very rapidly during the first 3 months. After 1 year, the mean resorption rate of the anterior part of the mandible in the implant group was 1.4 mm, compared with 4.2 mm in the control group. This mean resorption value of 4.2 mm compared with the values of 4.4 mm reported by Carlsson and Persson (1967) and 4.7 mm reported by Atwood (1963) for the same postoperative period (quoted by Kangvonkit et al., 1986).

In 1986 Kwon et al. placed 70 HA ceramic cones (durapatite) into fresh extraction sockets of 10 patients at least 1 mm below the alveolar crest. All the cones were placed in the mandible between the second premolars. All patients needed maxillary and mandibular complete dentures. Like Kangvonkit et al. in 1986, these authors required that the root of the tooth to be extracted have a minimum of 8 mm of alveolar bone support in a vertical dimension, or a loss of periodontal bone less than two thirds of the length of the root. Immediately after tooth extraction an appropriately sized cone was selected to fit snugly and usually without contouring. The socket was not touched. Following the final positioning of the cones, any interrupted interproximal papillae were sutured, wherever possible. Dentures were inserted immediately. The follow-up period ranged from 12 to 24 months (mean, 20.6 months). Thirty-seven (53%) of the implant cones became exposed and 19 (27%) had to be removed. Of 10 patients, 2 patients lost no cones, and 1 patient had 8 of 9 cones removed.

At 21 months, the mean decrease in the height of the alveolar bone on the anterior mandible was 0.33 ± 0.28 cm in the experimental group and 0.57 ± 0.08 cm in the control group, with no statistically significant difference. During the follow-up period, dentures were lined an average of five times in the experimental group and three times in the control group. Obviously Kwon et al. (1986) felt that the HA cones did not significantly preserve the alveolar bone.

In 1987 Filler and Kentros placed 86 (56 maxillary/30 mandibular; 63 anterior and 23 posterior) dense, nonporous HA cones into fresh sockets of 15 patients 1.0 mm below the crest while attempting a snug fit. Mucoperiosteal flaps were not elevated and primary closure was not obtained. Some patients received immediate dentures, some after 6–10 weeks, and some received no dentures. It did not appear that the placement of the dentures, immediate or otherwise, had much effect. After 1 year, 12% of the cones were lost and many cones required reduction due to dehiscences. A higher percentage of mandibular cones (22%) was lost than maxillary cones (8%). Thirteen percent of the anterior and 90% of the posterior cones were lost. No ridge preservation was demonstrated. At 6 months 25% of the patients had developed dehiscences; between 6 and 12 months approximately another 25% developed dehiscences. During this time period multiple appointments were required for reductions or removal of cones (Filler, 1989).

Brook et al. (1988) reported on the placement of 81 dense sintered polycrystalline HA root-replicated implants (Calcitite®) in 21 healthy patients following extraction of mandibular teeth and prior to the insertion of immediate complete dentures. No soft tissue flaps were raised. To obtain an accurate fit, the corresponding HA root replica was trimmed and the coronal end domed and inserted 2–3 mm below the crest of the alveolar bone. After suturing the interdental papillae to control bleeding, the dentures were inserted. Eight (10.0%) implants were lost during the first month. Of the eight implants lost, seven were lost from cuspid sockets. Lower cuspid roots are spindle shaped, and poor correspondence between the extracted canine roots and the simplified root shapes of the replicas is perhaps inevitable, which leads to failure. Despite careful adjustment, it was difficult to achieve a tight fit and the high rate of cuspid loss during the first month (17%) could be explained as a consequence of a lack of mechanical wedging into the socket. The loss of the cuspid implant was invariably sudden, without the implant first becoming prominent beneath the mucosa. After 1 year, approximately 20.0% of the cones had been lost. The proportion of cuspid cones lost (30.0%), however, was significantly greater than that of noncuspids (8.8%).

Other than those lost within the first month, the mode of loss was that the implant first became prominent and palpable beneath the mucoperiosteum, with exposure following later. On palpation, the exposed implant often felt fixed to the underlying bone, and

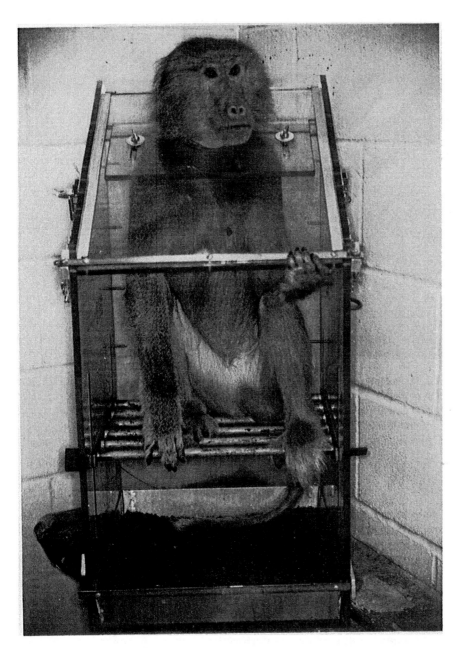

Figure 3 Photograph of a baboon in a customized restraining chair, a modification of the model first described by Glassman, Negrao, and Doty 1969. (Reprinted, by permission Mosby-Year Book, Inc., Stanley et al., The Implantation of Natural Tooth From Bioglass in Baboons, *Oral Surg., 42*: 339–356, 1976.)

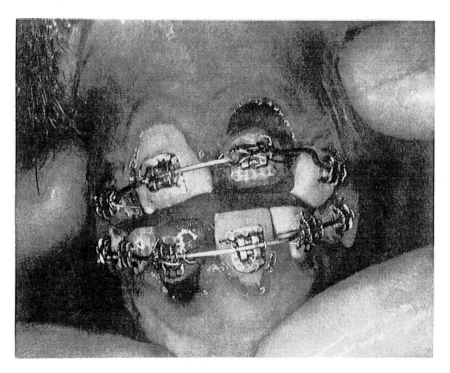

Figure 4 Anatomic implants seated in baboon socket sites of maxillary left central and right lateral incisor and mandibular right central and left lateral incisor; implants splinted to remaining teeth for 3-month healing period by brackets luted with composite resins and secured as individual units by arch wires and ligature ties. (Reprinted, by permission Mosby-Year Book, Inc., Stanley et al., The Implantation of Natural Tooth From Bioglass in Baboons, *Oral Surg., 42*:339–356, 1976.)

loss occurred only over a period of weeks. Earlier workers (Veldhuis et al., 1984) had stressed the need to retain these implants by surgical intervention. Consequently, in three cases a buccal flap was raised and the prominence of the implant reduced. However, the results were disappointing. Patients tolerated the procedure well, and all implants lost were symptomless.

In 1988 Sattayasanskul, Brook, and Lamb reported on 32 patients (20 experimental and 12 control) who required extraction of their remaining mandibular teeth and immediate complete dentures. To simulate clinical conditions, a minimum amount of alveolar bone support (4 to 5 mm) was required. Only patients with incisors, canines, and premolars were included. Sharp socket margins were smoothed, but no soft-tissue flaps were raised. Dense sintered polycrystalline HA (Calcitite®) root replica implants (cones) were fitted as accurately as possible 2 to 3 mm beneath the alveolar crest. After suturing the interdental papillae, the dentures were inserted. Seventy-four root replica implants were placed in the mandibles of 20 test patients. The patients were followed for 1 year. Twenty percent of the cones were lost. During the first year significantly less ARR took place in the implant group (2.5 mm) than the control group 4.1 mm). The mandibular alveolar width loss in the implant group was 2.9 mm, versus 3.4 mm in the control group.

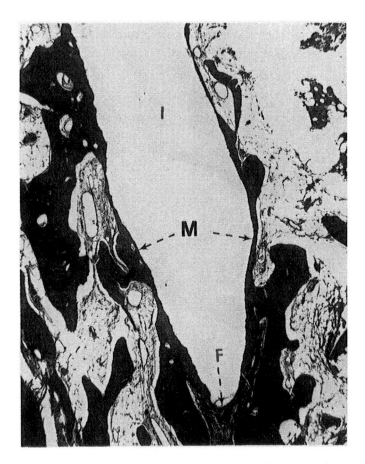

Figure 5   Implant formula F. Low-power magnification outlines baboon socket lined with mineralized tissue (M) all along the implant (I) surface 6 months after implantation. Only at the apex does some fibrous tissue persist (F) (×25). (Reprinted, by permission Mosby–Year Book, Inc., Stanley et al., The Implantation of Natural Tooth From Bioglass in Baboons, *Oral Surg.*, *42*:339–356, 1976.)

B.   Bioglass

Another promising biomaterial used for cone implants, offering controlled or bioactive features, is Bioglass®. In 1986 Stanley, Clark, Hall, King, Colaizzi, Spilman, and Hench presented the initial results of a Bioglass cone implant study that began as a clinical trial in 1983.[3] The project was the outgrowth of a 2-year baboon study where Bioglass® replicas of incisal teeth were implanted in fresh sockets (Figs. 3–7) (Stanley et al., 1976, 1981). Two hundred and sixteen Bioglass® cones of the 45S5 formula of Bioglass® had been placed in 26 patients and the follow-up period averaged 6.7 months (range 0–18 months),

---

[3]Bioglass: registered trademark, University of Florida, Gainesville, Florida 32601.

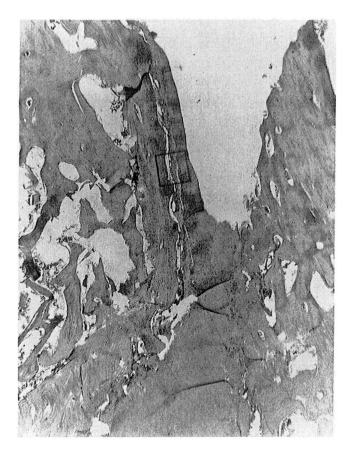

(a)

Figure 6   (a) A 45S5 formula implant in mandibular left lateral incisor site of a baboon 2 years after implantation. Most of socket lined by bone (×9.6). (b) A higher power magnification of (a) showing bony surface approximating implant (×100). (From Stanley, H.R., Hench, L.L., Bennett, C.G., Jr., et al. The implantation of natural tooth from Bioglass in baboons—long term results. *Int. J. Oral Implantology*, 1981; 2:26–38 with permission.)

at which time 1.8% of the implants had been lost and 2.8% had developed dehiscences. Functioning dentures had been placed in 13 patients after an average postimplantation time of 6.5 months.

In 1987 Stanley, Hall, Colaizzi, and Clark gave a follow-up report on 242 Bioglass cone implants in 29 patients. After an average postimplant period of 19.9 months (range 12–32 months), 7 implants (2.9%) had been lost and 9 implants (3.7%) had developed dehiscences. Twenty-seven patients had been fitted with dentures after an average of 14 months (range 2–27 months). Seven patients had required a reline after an average of 14 months of denture function.

(b)

VI. HOW BIOGLASS WORKS

Most commercial ceramics contain some glassy phase that makes them susceptible to attack by aqueous solutions, simulated body fluids (SBFs—designed to emulate human plasma in an inorganic composition with similar ion concentrations and pH values) or body (tissue) fluids (Li, Ohtsuki, Kokubo, Nakanishi, Soga, Nakamura, and Yamamuro, 1993). The hypothesis underlying the use of bioglass is that it is possible to take advantage of time-dependent, kinetic[4] modifications of the surface corrosive chemical reactions upon implantation, and that this results in the development of an outer surface silica-rich gel layer that subsequently mineralizes (Andersson and Kangasniemi, 1991; Hench, 1991; Hench and Paschall, 1974).

In the 1980s, an explanation of how bioactive glasses and glass ceramics worked was a rather simple thing. However, the learning curve has developed, and the numerous

---

[4]Kinetic: putting in motion; the branch of dynamics that pertains to turnover, or rate of change; the study of the rates and mechanisms of chemical reactions (*Dorland's Illustrated Medical Dictionary*, 27th ed.).

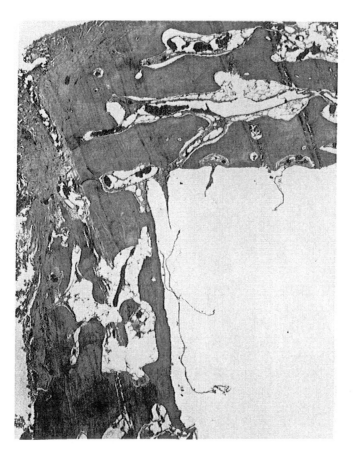

Figure 7   An F formula implant in maxillary left central site of a baboon 2 years after implantation. Except for a few tags of fibrous tissue extending into irregularities of the implant surface, the
bulk of the socket is bone-lined (×25). (From Stanley, H.R., Hench, L.L., Bennett, C.G., Jr., et
al. The implantation of natural tooth from Bioglass in baboons—long term results. *Int. J. Oral
Implantology*, 1981; 2:26–38 with permission.)

intricacies that have become known, detailing descriptive phases or stages, have increased
the explanation from 2 to 6 to 11 different stages (Andersson and Kangasniemi, 1991;
Filgueiras, La Torre, and Hench, 1993; West and Hench, 1993).

One of the main components of bioactive glasses and glass ceramics is silica (Li et
al., 1993). A bioactive glass can be regarded as a three-dimensional silica (SiO_2) network
that can be modified by incorporating other oxides such as sodium oxide (Na_2O), calcium
oxide (CaO), phosphorus pentoxide (P_2O_5), aluminum oxide (Al_2O_3), and barium oxide
(B_2O_3); or halides such as calcium fluoride (CaF_2) (Andersson, Rosenqvist, and Karlsson, 1993). Ogino, Ohuchi, and Hench (1980) have shown that the formation of apatite
on the surface of glasses and ceramics depends strongly upon the composition of their
components in the range of 42–60% silica, 0–29% calcium oxide, 17.7–26.3% sodium
oxide, and 0–2.6% phosphorus pentoxide. Bioglass® itself is composed of 45.0% SiO_2,
24.5% CaO, 24.5% Na_2O and 6.0% P_2O_5; signifying the 45.0% SiO_2 as the network
former and a 5-to-1 molar ratio of Ca to P (Hench, 1991; Stanley et al., 1987).

Originally it was believed that in order for an implanted artificial material to be able to bond to living bone, the material must have the ability to form an apatite layer on its surface unless this surface was already apatite. This is the only common condition (characteristic) that had been conclusively shown to be necessary for bone bonding (the growth of new bone onto the apatite) to occur on all known bioactive implant materials (Hench, 1991; Kangasniemi, Vedel, de Blick-Hogerworst, Yli-Urpo, and de Groot, 1993). The formation of a carbonate-containing hydroxyapatite (HCA) layer on the surface of the implanted material bridged it to the host tissue. Some researchers refer to this layer as a hydroxy carbonate apatite (HCA) reaction layer on the surface of the material (Filgueiras et al., 1993; West and Hench, 1993) while others refer to it simply as a calcium phosphate (CaP) layer. The HCA phase is equivalent chemically and structurally to the mineral phase in bone and is responsible for interfacial bonding (Hench, 1991). It is the rate of the HCA formation and the time for onset of crystallization that vary so greatly. The time for apatite formation is dependent upon the SiO_2 content of the glass, but the growth rate of apatite crystal formation is enhanced by the amount of calcium and phosphate incorporated into the glass (Kangasniemi et al., 1993). When the rate becomes excessively slow, no bond forms and the material is no longer considered bioactive (Hench, 1991).

When bioactive glasses are contacted by solutions, reaction occurs; it has been proposed that three main reactions occur within a controlled sequence: dissolution, leaching, and precipitation (Andersson et al., 1993). When the silica on the surface of an implant

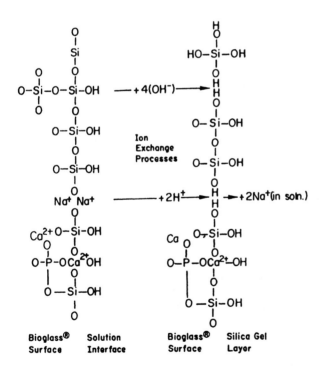

Figure 8   Reaction of (OH^-) and H^+ ions (i.e., pH effect) on a schematic Bioglass® implant surface. Sodium is depleted from the surface, leaving behind a hydrated silica gel upon which a hydroxyapatite layer forms. (From Stanley, 1987.)

comes in contact with fluids there is a *dissolution*, or breakdown of the silicon–oxygen bonds (Si—O—Si), with release of other elements (Ca, Na, P) within it (*leaching*) (Fig. 8). This leaves behind a silica-rich gel of approximately 200 μm, which generates the formation of the HCA layer (Andersson, Li, Karlsson, Niemi, Miettinen, and Juhanoja, 1990; Hench, 1981; Hench and Clark, 1982; Hench and Ethridge, 1982; Ogino and Hench, 1980; Stanley, Hall, Colaizzi, and Clark, 1987). It has been shown that hydrogen ions from the body fluids replace Na^+ (alkaline ions) in the glassy network, which are released into solution (Hench, 1981; Hench and Paschall, 1974; Ogino et al., 1980). This reaction produces hydroxyl ions, which destroy Si—O—Si groups on the Bioglass® surface, and permits calcium and phosphorus within the implant to migrate to its surface, thereby forming an amorphous calcium phosphate-rich gel layer that attracts additional Ca, P, and CO_3 (carbonate) from the body fluids (Fig. 8) (Hench, 1981; Hench and Ethridge, 1982). Incorporation of carbonate ions within the amorphous calcium phosphate gel results in the crystallization (*precipitation*) of HCA on the implant surface. The HCA layer then acts as an osseoconductive agent, guiding the location for new bone formation.

Kim, Kim, and Chang (1993), using four different reaction solutions – pure water; SBF (containing Na^+, K^+, Mg^{2+}, Ca^{2+}, Cl^-, HCO_3^-, SO_4^{2-}, HPO_4^{2-}); TBS (tris-hydroxymethyl aminomethane that contains $(CH_2OH)_3CNH_2$, HCl, and H_2O); and TBSP (TPS containing 100 ppm of HPO_4^{2-} ion) – studied the reactions of glass disks for 100 h at 37°C. During the first few hours, extensive dissolution of the glass network occurred.

The silanol condensation reaction provides an energetically favored pathway for incorporating Ca and P from body fluids or SBFs onto the hydrated silica-rich surface (Andersson et al., 1993; Filgueiras et al., 1993; West and Hench, 1993). The silanol groups (SiOH) on the surface of the silica provide the base for apatite nucleation. The silanol groups, being negatively charged, attract the Ca ions to the negatively charged surface while the phosphate groups are kept in the vicinity of the surface through hydrogen bonds. These reactions create a local accumulation of both Ca and P about the surface of the silica gel, which establishes a supersaturated solution of Ca and P adjacent to a bioactive glass implant (Li et al., 1993). If one places Bioglass into a SBF at an initial pH of 7.4, due to the presence of Ca^{2+} and HPO_4^{2-} in the SBF, the growth of the precipitated layer is rapid (Kokubo, Ito, Huang, Hayaski, Sakki, Kitsugi, and Yamamuro, 1990). West and Hench (1992) recently proposed that pentacoordinated silicon ions formed by condensation of silanol groups are responsible for the nucleation of HCA.

Andersson and Kangasniemi (1991) suggested that phosphate ions bind first to the silicate network because of a condensation reaction with the silicate network, resulting in an initial Ca/P ratio near 1 for the Ca,P precipitate. As the carbonate hydroxyapatite forms in bone or near bone, new bone firmly attaches to it (Andersson et al., 1990, 1993; Hench, 1980). Clark, Kim, West, Wilson, and Hench (1990) showed, using Auger electron spectroscopy and Ar-ion beam milling, that even by 2 min, calcium and phosphate enrichment occurred on the surface of 45S5 Bioglass® to a depth of approximately 20 nm. Filgueiras et al. (1993) described the reaction of 45S5 glass when placed in SBFs for 10 min as a rapid ion exchange of Na^+ on the surface of the glass, with H^+ or H_3O^+ from the solutions. The breaking of bonds in the Si—O—Si groups led to the formation of silanols (SiOH) at the glass–solution interface. The condensation and repolymerization of the surface silanols and the repolymerization of a hydrated SiO_2-rich layer, depleted of alkalis and alkaline earth cations, formed the hydrated silica gel surface.

Kim, Clark, and Hench (1989) showed that as the 45S5 glass surface was leached of Na, Ca, and P ions, a silica-rich layer developed with SiO_4 units containing a nonbridging oxygen. As the Ca_2^+ and PO_4^{3-} groups migrated up to the surface of the Si-rich layer an amorphous film of Ca_2 and PO_4^3 (a CaP) began to form on top of or in the Si-rich layer, within a minimum of 15 min. Growth of this film also occurred by incorporating soluble Ca and phosphates from the surrounding solution (Filgueiras et al., 1993; Kangasniemi et al., 1993).

Ogino, Ohuchi, and Hench (1980) showed that by 1 h, the calcium phosphate layer grew to the thickness of 200 nm. The bilayer films of calcium phosphate on top of the polymerized silica gel are shown in Fig. 9. The calcium phosphate rich layer extends to a depth of nearly 0.8 μm after only 1 h in rat bone. The silica-rich film was already several micrometers thick. Organic constituents containing C and N were also incorporated into the growing calcium phosphate rich film.

Kangasniemi et al. (1993), using various reaction solutions, found that during the first few hours, extensive dissolution of a disk of bioactive glass occurred and became totally dissolved in pure water after 100 h at 37°C. According to the composition of the glass and the type of reaction solution, the dissolution rate varied in the first 8 h between 1 and 4 μm. But as the surrounding solution became saturated with silica after about 15 h, the dissolution rate slowed considerably (Andersson et al., 1993).

This has a favorable aspect, however, because as Hench (1988) has stated, the failure

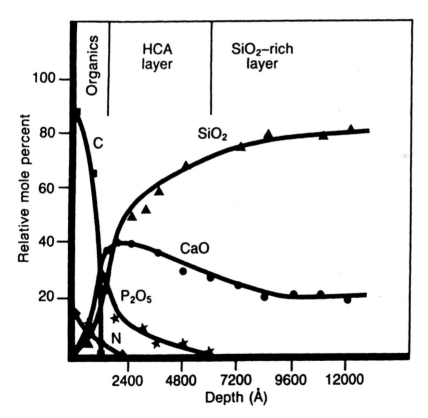

Figure 9 Bilayer films formed on 45S5 Bioglass® after 1 h in rat bone, *in vivo* (1Å = 10^{-1}nm). (From L. L. Hench. Bioceramics: From Concept to Clinic. *J. Am. Ceram. Soc.*, 74:1487–1510, 1991. Reprinted by permission of the American Ceramic Society.)

strength of the bioactively fixed bond is inversely proportional to the thickness of the bonding zone. Thus, prolonged leaching of the glass results in a poorer bonding fixation (Andersson et al., 1988).

By 10 h, the HCA layer had grown to 4 μm in thickness (Hench, 1991). By 2 weeks the crystalline HCA layer was equivalent to biological apatites grown *in vivo* (Le Geros, Bone, and Le Geros, 1978).

By 10 days some foci of newly formed immature bone were visualized on Bioglass implant surfaces in the femurs and tibias of rats. By 4 weeks, foci of mature bone appeared at numerous points, and after 12 weeks a completely mature laminated bone interface was established (Beckham, Greenlee, and Crebo, 1971; Greenlee, Beckham, Crebo et al., 1972; Hench and Clark, 1982). Apparently bone formation depends entirely on the sequence of events described above, and any interruption of the sequence—for instance, as related to excessive mobility of the implant—may inhibit the bone formation. Also, as soon as the gel layer is mineralized, the Bioglass®-serum interaction ceases and the deeper part of the Bioglass® implant is protected from further action (see Fig. 10: the layers after 24 months). The composition of the Si-rich layer appeared to be relatively constant while the region referred to as the Ca,P-rich layer resembled an interfacial diffusion profile with gradual changes in the ionic levels that increase in value to match that of mineralized bone (Fig. 10).

Wahl (1986) showed that the remodeling activity of bone around implants as detected by technetium 99 scintigraphy increased during the first 4 weeks after implant insertion and again 4–6 weeks after bridge placement (Newman and Flemming, 1988).

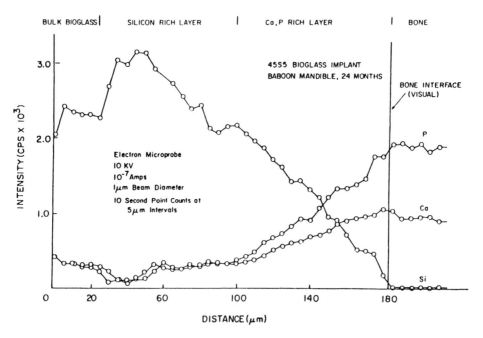

Figure 10   An electron microprobe profile of 45S5 formula implant after 24 months. (From Stanley, H.R., Hench, L.L., Bennett, C.G., Jr., et al. The implantation of natural tooth from Bioglass in baboons—long term results. *Int. J. Oral Implantology*, 1981; 2:26–38 with permission.)

Kangasniemi et al. (1993), working with Ca,P ceramics with different dissolution rates, prepared composites of them using the same glass composition. They expected to find different dissolution behaviors than with the mother glasses and a faster Ca,P precipitation with the composites because of the increased Ca and P concentration on the surface. HA $Ca_{10}(PO_4)_6(OH)_2$[HA] and rhenanite Ca $NaPO_4$ [Rh] ceramic particles were used because of their different dissolution rates and because they both exhibited the ability to bond to bone.

Square pieces were sawed out of the glass block with a diamond saw. The remains of the same glass batch were then ground and sieved for particles below 45 μm in size to make a powder. The Ca,P ceramics (HA and Rh) were prepared the same way.

The powders were mixed in a 7 : 3 weight ratio. The materials were placed in polystyrene cups and liquid was added, 0.1 ml/mm^2 (SBF and Tris-buffer solutions). The cups were closed with a lid and placed in a static water bath at 37°C for 3 h to 6 days. The Ca, P, and Si concentrations in the liquids were measured before and after treatment.

The results showed that the rhenanite-containing glass AOBO composite had the highest potential for forming a Ca,P layer on its surface, followed by various compositions of glass such as AOBO/HA, AOBO, A2B3.3/Rh and A2B3.3/HA, in decreasing magnitude; and finally glass A2B3.3 without any Ca,P precipitation demonstrated after 72 h.

The first change at 6 h was caused by accumulation of Ca, P, and Si ions from the liquid onto and into the Si-rich layer created on the surface of both glass and composite materials. At 24 h, a thin Si,Ca,Na-rich layer was observed on the surfaces of both materials. Only small amounts of phosphorus were present in this layer at this time and thus a Ca/P ratio as high as 6 was measured. It seemed that the Si,Ca,Na film soon attracted the phosphates, resulting in an intermediate phase with a Ca/P ratio near 1.

After periods of time longer than 90 h, the two materials developed similar surfaces. The precipitates grew from small needles into spheres of several microns. Ca and P seemed to migrate through the Si-rich layer into the bulk material following the front of the reaction layer.

The following conclusions were drawn from this study:

1. Inherently, hydrolyzed silicates enabled the beginning of Ca,P precipitation. The number of nonbridging oxygens resulting from the alkali ion exchanges influenced the time of onset for the precipitation. Alternatively, onset time depended upon the silicon concentration in close vicinity to the material surface as a result of silicate dissolution from the material.

2. Once precipitation started, silicates no longer had a function and only Ca and P ion concentrations influenced the precipitation.

3. A bioactive glass phase with a Ca,P phase that dissolves in SBF should have a higher rate of Ca,P precipitation than the bioactive glass itself. The Ca,P phase increased the local availability of Ca^{2+} and H_2PO_4 ions.

In an apparently continuous process, the following stages were observed: (1) a Si-rich layer; (2) a Si,Ca,Na-rich film; (3) P ions binding with the Si,Ca,Na film; (4) Ca,P spheres precipitating; and (5) the remaining Si,Ca,Na layer is redissolved. In the beginning, it seemed that, topographically, two mechanisms for the onset of the precipitation were possible. One was that of direct precipitation in and on the Si-rich layer (gel) created by leaching of the glass surface. The other required absorption of Si ions from the liquid onto the surface of the glass (or other material) and binding of Ca to this film, which then acted as a substrate for the precipitation.

In 1992 Li, Ohtsuki, Kokubo, Nakanishi, and Soga showed, however, that a pure silica gel containing no Ca or P within it was also capable of inducing apatite formation on its surface. They (Li, Ohtsuki, Kokubo, Nakanishi, Soga, Nakamura, and Yamamuro, 1993) nevertheless supported the concept that a silica hydrogel must first form to generate the formation of the HCA layer. They produced a pure silica gel through a sol-gel process from tetraethoxy-silane and polyethylene glycol. Specimens of the silica gel were then immersed in various SBFs for varying time periods. These SBFs did not precipitate their own mineral content spontaneously during the first 5 weeks.

Li et al. pointed out that the silica component must be readily hydrolyzed to produce sufficient silanol groups for the induction of apatite nucleation. The apatite nucleation began as a small crystallite structure with spherulite features similar to bone apatite. The apatite nuclei grew spontaneously at the expense of Ca and P in the surrounding SBFs, once they were stimulated by the gel-derived silica.

No apatite formation was initiated at any time in SBFs at a pH of 7.0, but these systems did begin to form apatite on the surfaces of silica gel at 1 week at pH 7.4, at 2 weeks at pH 7.3, and at 5 weeks at pH 7.2. The induction period for apatite formation was also shortened with increases of Ca or P, or both in the SBFs. The results showed that the gel-derived silica is a potent apatite inducer.

Although the reactions were controlled by pH of the solution, McGrail, Pederson, and Petersen (1986) found the ion exchange to be independent of pH in the interval 6–9. The silica dissolution was constant but small below pH 9. It increased rapidly above this pH, and above pH 9.5 the silica dissolution was dominant. Thus, at physiological pH the ion exchange dominated over the silica dissolution, although both reactions occurred simultaneously. Bioglass 45S5 in the system $Na_2O-CaO-SiO-P_2O_5-$, when implanted into living tissue releases sodium, calcium, and phosphate ions into the surrounding body fluid, leaving a silica gel layer on its surface. The pH value of the fluid around the silica gel ascends due to the liberation of sodium oxide (Na_2O).

A. The Importance of Collagen Fibers

Since the area becomes slightly more basic (alkaline) due to the release of OH ions, the environment of the gel layer favors the replication and differentiation of osteoblasts rather than just fibroblasts (Hench and Ethridge, 1982; Raisz, 1985). As proteoglycans and glucosaminoglycans (formerly called *mucopolysaccharides*) accumulate and interact with the gel layer, collagen fibers, those already present and developing ones, are coated, entangled, incorporated, and bonded to the gel surface (Hench and Paschall, 1974). These collagen fibers now incorporated into the gel and developing matrix, become part of the implant surface. The minerals released from the implant surface and those deposited by the body fluids collect about the original and newly formed collagen fibers to create the equivalent of bundle bone with Sharpey's fibers. In rodent femurs and tibias, it takes about 12 weeks for mature bone to form.

Transmission electron micrographs have demonstrated within the amorphous gel layer of Bioglass® highly elongated hydroxyapatite crystals (Fig. 11) bridging the gap between the implant and the mature bone that had formed around the implant after 3 weeks in a rat femur. When the interface was demineralized for 30 min, removing the hydroxyapatite crystals, an amorphous layer approximately 800–1000 Å in width ap-

Figure 11 Transmission electron micrograph of the mineralized surface gel layer between the Bioglass® and ceramic contiguous with mineralizing bone (B) after 3 weeks in a rat femur interface area (I). (Courtesy of L. L. Hench, from Stanley, 1987.)

peared to cement collagen fibers of the matrix to the implant (Beckham et al., 1971; Hench and Ethridge, 1982; Hench, Splinter, Allen, and Greenlee, 1972; Stanley, 1987).

Albrektsson (1985) demonstrated that as the implant surface was approached, the collagen bundles gave way to randomly arranged filaments at 0.1 to 0.5 μm from the titanium implant. Collagen filaments were observed at a distance of about 200 Å from the titanium surface. There was a partly calcified amorphous ground substance consisting of proteoglycans and glucosaminoglycans covering the last 200 to 300 Å of the tissue interface toward the titanium implant. Cell processes from osteoblasts or osteocytes approached the titanium surface but were always separated from it by a 200 to 300 Å thick proteoglycan layer. Thicker proteoglycan layers have been demonstrated around Ti alloy as compared to pure titanium. Other metals such as gold or zirconium induce proteoglycan layers without any calcification to a thickness of 500–5000 Å while polymerized bone cement induced a proteoglycan layer of 20,000 Å . Glass ceramics generate proteoglycan layers from 600 to 1000 Å (Gross, Brandes, Struntz, Bab, and Sela, 1981) and from 2000 to 3000 Å (Clark, Hench, and Paschall, 1976).

Hench and Clark (1982) and Clark, Stanley, Acree, and Kreutziger (1979) confirmed that the bone is not deposited directly on the as-implanted Bioglass® surface, but that the gel layer can be divided into two zones based upon their chemical composition. The inner zone, closer to the bulk Bioglass, is a silica-rich layer approximately 120 μm in width and an outer, Ca,P-rich layer, approximately 70 μm in width, lying between the Si-rich layer and the bone (Stanley, 1987). In 1980 Denissen, Veldhuis, Makkes, van den Hooff, and de Groot, with transmission electron microscopy, described a similar amorphous layer of about 60 μm of interface tissue between hydroxyapatite implants and the newly formed bone. Hench and Clark (1982) and Clark and associates (1979) also showed that the mechanical properties have a continuous gradient across the gel layer. Hench and Ethridge (1982) proposed that the gel bonding layer results in a gradient in elastic modulus and bioelectric potential that mimics a natural junctional interface between hard and soft tissue. The elastic modulus gradient between an HCA implant and bone is nearly 1000 times higher than the gradient between Bioglass® and bone.

An ideal implant should have the same modulus of elasticity (springiness) as bone, much like a wooden peg holding two pieces of wood together. Unfortunately, we have at present no such material (Hench and Ethridge, 1982). Most implant materials are more like a stiff nail introduced into a plank subjected to continuous stress, and one can picture the nail loosening up (Bokros, 1979). The greater the mismatch between the implant material and the adjacent bone, preventing stress distribution, the greater the amount of bone resorption, with fibrous tissue replacement and gradual loosening ("wobbling") of the implant. Mismatch of elastic moduli causes bone resorption. An appreciation of this problem related to mismatched elastic moduli in the absence of a suitable implant material led to design changes in implants from spirals and screws to blades and anchors intended to diminish high stress concentration and distribute the load over larger areas and reduce bone resorption (Hench and Ethridge, 1982). Nevertheless, some such implants have still demonstrated only limited success in achieving lasting stability because of mechanical interface failure between the bone and the implant, for the reasons given above.

A gradation of mineralization within the Bioglass® gel layer, decreasing from the outermost surface inward, provides an elastic compliance (a springiness) that permits a favorable stress (load) transfer from the implant to the newly formed bone much like a natural tooth situated within a normal periodontal membrane (Clark et al., 1979; Hench and Clark, 1982; Stanley, 1987; Weinstein, Klawitter, and Cook, 1980).

The creation of a bond built into the implant surface explains why fractures occur either within the implant itself or in the newly formed surrounding bone but not at the interface (Piotrowski, Hench, Allen, and Miller, 1975; Stanley et al., 1976). The interface region induced by Bioglass®, instead of being an actual line of bonding, evidently acts as a shock absorber minimizing the transmission to bone of tension and compressive forces when a mechanical load is applied to the implant (James, 1984; Weinstein et al., 1980). Consequently, stress-induced ARR should be minimized (Stanley et al., 1987).

B. Bioactivity Index

Bioglass® has a unique property that appears to provide an advantage over other substrates (Wilson and Low, 1992). Hench (1988b), having established a "bioactivity index" (I_b) for bioactive ceramics, gave Bioglass® 45S5 a very high grade because its reaction layers begin to develop within minutes following implantation. As osteogenic cells are freed in the implantation site by surgery, they can colonize about the surfaces of the Bioglass® implant, now coated with a HCA layer, and produce bone. This supplements the bone that grows by osteoconduction from the alveolus.

At present Bioglass® appears to have the highest rate of bioactivity of the bioactive materials hitherto known and consequently induces rapid formation of apatite on the surfaces of the buried implant (Li et al., 1993). Important differences between bioactive substrates can affect their behavior (Hench, Stanley, Clark, Hall, and Wilson, 1991).

A comparison of the relative chemical activity of different types of bioceramics is given in Fig. 12. The original bioactive glass 45S5 has been used as a baseline for most surface studies, largely because it is single phase and has only four components (Na_2O, CaO, P_2O_5, SiO_2) (Hench, 1991). The relative reactivity shown in Fig. 12a correlates very closely with the rate of formation of an interfacial bond of implants with bone (Fig. 12b).

By changing the compositionally controlled reaction kinetics, the rates of formation of hard tissue at a bioactive implant interface can be altered, as shown in Fig. 12. Thus, the relative bioreactivity of the material is composition dependent. The level of bioactivity of a specific material can be related to the time for more than 50% of the interface to be bonded. Gross and his colleagues (Gross, Brandes, Strunz, Bab, and Sela, 1981; Gross, Kinne, Schmitz, and Strunz, 1988; Gross, Schmitz, Strunz, Schuppan, and Termine, 1986b; Gross and Strunz, 1980) have shown that the initial concentration of cells present at the interface (stem cells, osteoblasts, chondroblasts, and fibroblasts) varies as a function of the fit of the implant and the condition of the bony defect. Consequently, all bioactive implants require an incubation period before bone develops and bonds, which is evident in Fig. 12. The length of the incubation period varies over a wide range depending on implant composition, which controls the kinetics of the surface reactions.

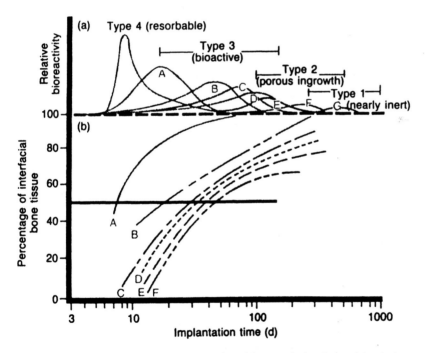

Figure 12 Bioactivity spectrum for various bioceramic implants: (a) relative rate of bioreactivity and (b) time dependence of formation of bone bonding at an implant interface [(A) 45S5 Bioglass®, (B) KGS Ceravital®, (C) 55S4.3 Bioglass®, (D) A/W glass-ceramic, (E) HA, (F) KGX Ceravital®, and (G) Al_2O_3–Si_3N_4]. (From L. L. Hench. Bioceramics: From Concept to Clinic. *J. Am. Ceram. Soc.*, *74*:1487–1510, 1991. Reprinted by permission of the American Ceramic Society.)

In contrast, HA implants bond much more slowly and form a bonding zone directly to bone, resulting in a much less natural stress transfer gradient into the alveolar ridge. Interfacial strength appears to be inversely dependent on the thickness of the bonding zone. For example, 45S5 Bioglass® with a very high I_b value develops a gel-bonding layer of 100 μm (Fig. 13) that has a relatively low shear strength.

Figure 13 shows an optical micrograph of the interface of 45S5 Bioglass® bonded to bone (rat tibia) after 1 year and an electron microprobe analysis of the interface. The bulk bioactive glass implant (BG), silica-rich layer (S), HCA layer (Ca,P), and bone (B) are indicated. Living bone cells (osteocytes, labeled O) are at the interface. There is no

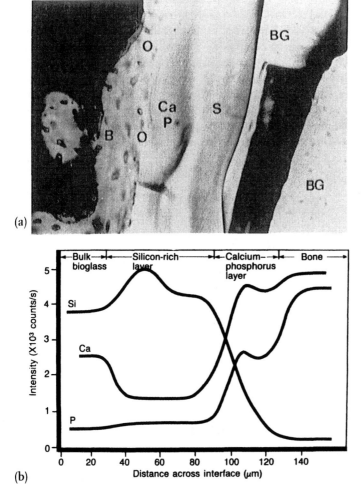

Figure 13 (a) Optical micrograph of a (BG) 45S5 Bioglass® implant bonded to (B) rat bone after 1 year showing (O) osteocytes or bone cells in conjunction with the (Ca-P) HCA layer formed on top of the (S) silica gel. (b) Electron microprobe analysis across the implant–bone interface shown in (a) [electron microprobe at 20 kV, 100 nA (specimen current), ~1-μm beam diameter, and 20 μm/(min in.) scan rate]. (From L. L. Hench. Bioceramics: From Concept to Clinic. *J. Am. Ceram. Soc.*, *74*:1487–1510, 1991. Reprinted by permission of the American Ceramic Society.)

seam of interfacial fibrous tissue. Even at the electron microscopic level there is almost no unmineralized tissue at the bonding interface.

Also, when Vrouwenvelder et al. (1993) cultured osteoblasts for different time periods on four different bone-replacing materials (Bioglass® 45S5, HA, a titanium alloy, and stainless steel), they found that the process of corrosion with the Bioglass® 45S5 starts almost immediately and that after 6–8 days of culturing, a higher proliferation rate of osteoblasts is apparent, with a significantly higher alkaline phosphatase activity, as compared to the other materials.

Previously, they had shown that osteoblasts cultured upon bioactive glass demonstrated superior histological and biochemical parameters, and exhibited a more compact structure with a so-called "stand-off" morphology, having many dorsal ruffles and fibropodia, indicative of a better osteoblastic condition (a better expression of the osteoblast phenotype in comparison to the other substrates).

As described earlier, the silanol condensation reaction provides an energetically favored pathway for incorporating Ca and P from the body fluids onto the hydrated silica-rich surface (Andersson et al., 1993; Filgueiras et al., 1993; West and Hench, 1993). Since the surface of the silica, because of the silanol groups, is negatively charged, the osteoblasts are probably able to approach, attach, and spread themselves more effectively on the Bioglass® surfaces in comparison to the other materials due to the high free surface energy resulting from the characteristic corrosion process.

In 1970 Carlisle showed that both calcium and silicon values are invariably low in the periosteum of young mouse and rat bone, but that the adjacent osteoid layer contains up to 25 times as much silicon and 9 times as much calcium as the periosteum. The amount of silicon present appears to be related to the "maturity" of the bone. In the advanced stages of bone formation, the percentage of silicon falls markedly, as the proportions of Ca and P increase and approach the values of hydroxyapatite. Silicon is then present only at minimal detection limits.

It is crucial that the HCA layer forms, without being disturbed, at the implant–tissue interface at the right time to match the repair rate (sequence of healing) of the implant site (Li et al., 1993). The amount of future bone development is dependent on the function of the implant unit itself (Hench and Ethridge, 1982).

VII. CLINICAL PROTOCOL

A. Procedures

Stanley, Hall, Hench, King, Colaizzi, and Wilson in 1983 began a clinical trial using cone-shaped devices (Fig. 14) made of the 45S5 formula of Bioglass® placed in fresh sockets to prevent ARR.

Normal, healthy individuals scheduled for the extraction of teeth for removable partial or complete dentures in either or both jaws were selected. Patients were excluded if they were under treatment for cancer, had uncontrolled diabetes mellitus, or were taking corticosteroids, immunosuppressive medications, or warfarin (Coumadin).

The teeth to be extracted received a prophylaxis approximately 1 week in advance of extraction to improve the tissue tone for healing and to remove any calculus or debris that might enter a socket during tooth extraction. A week later the patient returned and the teeth were extracted under local anesthesia with either 2% lidocaine with 1 : 100,000 epinephrine or 3% mepivacine (Carbocaine). Preoperatively, most patients received 2 g of Pen V-K [the potassium (K) salt of penicillin V].

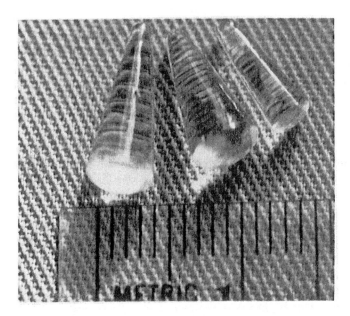

Figure 14   Typical Bioglass® cones ready for implantation. (From Stanley et al. Remedial Alveolar Ridge Maintenance with New Endosseous Implant Material, *Journal of Prosthetic Dentistry, 58:* 607–613, 1987. Reprinted by permission Mosby-Year Book, Inc.)

Buccal/labial-lingual mucoperiosteal flaps were reflected with appropriate releasing incisions to allow for primary closure over the extraction sites. The teeth were removed as atraumatically as possible to preserve the bony walls of the alveolar sockets. During the extraction procedure, every attempt was made to avoid having fractured bone chips enter the sockets, as these would become nonvital and create the activation of inflammation and resorption, interfering with the desired chemical–implant–tissue reaction.

For placement of the implants, a graduated series of matched bone burrs (four) and implants was developed (Fig. 15). The appropriate-sized burr to accommodate the appropriate-sized implant was selected for each socket in which an implant was to be placed. If necessary, the socket was widened and lengthened by removing a small amount of buccolabial crestal bone and interdental bone by using a slow-speed contraangle handpiece with copious irrigation.

The implant system consisted of a set of 12 conical implants, with rounded tops and four burs. The implants ranged in size and shape from 4 to 12 mm in length, 4° to 6° of taper, 1.5 to 2.6 mm tip (apical) diameter, and 1.96 to 4.57 mm crestal diameter. The larger implants were used more frequently (to derive a tight fit).

Low speed (less than 2000 rpm) with controlled torque, a pumping motion, and thorough irrigation was used. It was recommended that the bone paste in the drill flutes be removed at least once during each preparation to improve cutting efficiency and avoid overheating (more than 43°C), crushing, and smearing the bone (Schnitman, Rubenstein, Woehrle, DaSilva, and Koch, 1988). Just enough bone was removed to allow placement of the appropriate-sized implant 2 mm below the level of the crestal bone, producing maximum contact with the more intact bone (Stanley, Hall, Hench, King, Colaizzi, and Wilson, 1983).

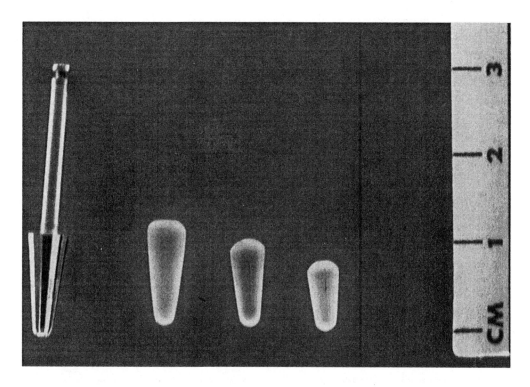

Figure 15 Drill and matching Bioglass® cone implants. (From Stanley et al. Remedial Alveolar Ridge Maintenance with New Endosseous Implant Material, *Journal of Prosthetic Dentistry*, *58:* 607–613, 1987. Reprinted by permission Mosby-Year Book, Inc.)

The implant cones made of Bioglass® formula 45S5 (U.S. Biomaterials Inc., Alachus, FL) were provided in sterile packs (ethylene oxide sterilized) and were handled and pressed into the prepared sockets by using rubber-tipped forceps (Figs. 16 and 17). Because the cone implants function endosseously to maintain the alveolar ridge, the acronym ERMI (endosseous ridge maintenance implant) was coined. The FDA approved the product in November 1988 on the basis of a 510-K application.

The eventual goal of placement was (1) to have the crestal portion of the implants approximately 2 mm below the crestal bone, (2) to maximize bone–implant contact (a tight fit), and (3) to avoid any recontouring of the implant itself. Implants were best placed below the crest to protect them from permucosal dehiscence due to bone resorption and pressure mobilization induced by removable prostheses (Shulman, 1988). In sockets where the implant did not occupy the entire buccal/labial-lingual dimension, the implant was placed in the center or lingual (palatal) aspect of the socket.

A primary mucosal closure was obtained by making buccal and labial periosteal releasing incisions and suturing with 4-0 vicryl sutures. According to clinical judgment, many patients received 500 mg Pen V-K g.i.d. (4 times a day) for 7 days postoperatively. Panoramic and periapical radiographs were taken at 6-month intervals for 2 years and annually for 5 years for documentation.

Dentures were not placed until at least 6 weeks after healing to allow time for the implant–tissue reaction to develop without disturbance. Only micromovement of the

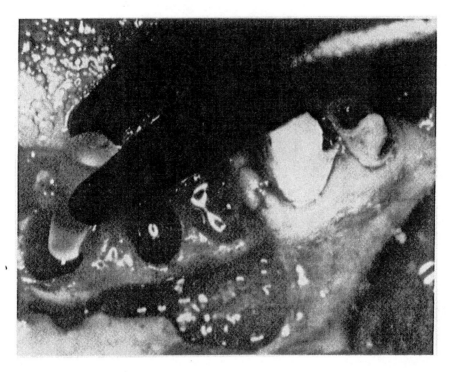

Figure 16 The placement of Bioglass® cone implants in prepared sockets. (From Stanley et al. Remedial Alveolar Ridge Maintenance with New Endosseous Implant Material, *Journal of Prosthetic Dentistry, 58:*607–613, 1987. Reprinted by permission Mosby-Year Book, Inc.)

implants is permitted during the first 6 weeks; otherwise the gel layer–tissue interface reaction is aborted.

At periodic recall, observations were made as to whether the implants had become lost, developed dehiscence, tilted in position, or resorbed; and whether alveolar bone resorption affected the fit of the dentures (Fig. 18). A control group of patients matched to the implant group in terms of age, sex, and race were monitored over similar time periods. Clinical data such as needs for denture adjustments and/or relines were compared between the two groups.

B. Results of Recent Patient Recall

The findings of a recent recall of 20 of the original 29 patients produced the following data. The first patient received implants on 3/26/84. The postimplantation period averaged 54.1 months (ranging from 36 to 71 months). One patient refused to wear the dentures as originally agreed upon. The 20 patients consisted of 12 males and 8 females; 17 Caucasians and 3 Blacks with an average age of 43.2 years (range 27–67 years).

1. Lost Implants

Of the 168 implants placed in the 20 recalled patients, 21 (12.5%) had been lost between 6 and 69 months (average 30.9 months).

Of the 168 implants, 83 were placed in the maxilla, 57 in anterior sockets, and 17 in

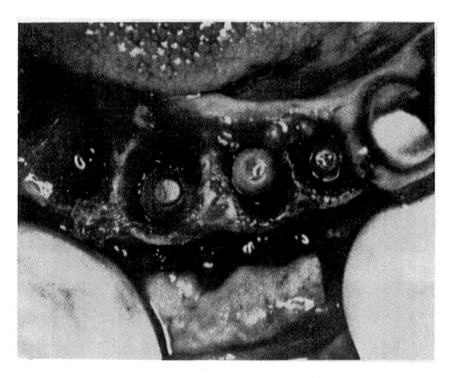

Figure 17   Cone implants in place. (From Stanley et al. Remedial Alveolar Ridge Maintenance with New Endosseous Implant Material, *Journal of Prosthetic Dentistry*, *58:*607–613, 1987. Reprinted by permission Mosby-Year Book, Inc.)

posterior sockets. A total of 3 (5.3%) implants were lost from the anterior sockets and 3 (17.6%) from posterior sockets.

Of 85 implants placed in the mandible, 52 were placed in anterior sockets and 33 in posterior sockets. Nine (17.3%) were lost from anterior sockets and 4 (12.1%) from posterior sockets. In the cases successfully recalled, the maxillary anterior implants did the best (94.7% retained), and the mandibular posterior implants second best (87.9% retained). The retention rates of the mandibular anterior implants (82.7%) and the maxillary posterior implants (82.4%) were similar.

Of the 20 recalled patients, 12 received complete maxillary and mandibular dentures and 8 received the combination of a complete maxillary denture and a removable mandibular partial denture. Of the 21 lost implants, 8 were lost from the maxilla under complete dentures and 13 from the mandible under complete dentures. No implants were lost when covered by removable mandibular partial dentures.

2.   Recontoured Implants

Thirteen (7.7%) implants in both jaws were recontoured for dehiscence after initial implantation between 4 and 46 months (average 17.0 months the first time). Two implants were recontoured twice in the same patient; the first time at 4 months and again at 39 months (the average time interval then increased to 23.0 months).

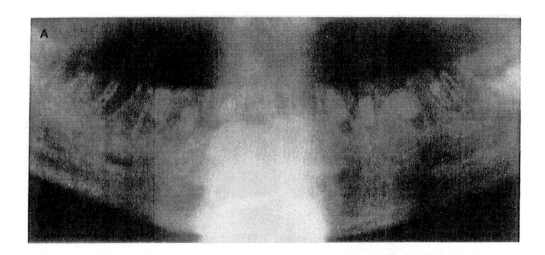

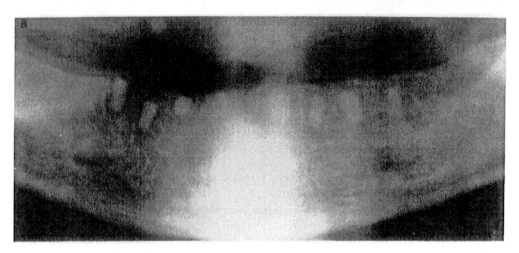

Figure 18 Portions of panoramic radiographs of mandible of a 49-year-old woman who received seven Bioglass® cone implants. (A) Immediately after placement and wedging of cones into fresh sockets as deeply as possible. Radiolucent areas can be seen above and below cone implants. (B) Ten months after implantation, five cone implants are completely surrounded by new bone and the original socket topography has been obliterated. The cone implant on extreme left has developed a peripheral radiolucency and the cone implant third from left appears to have no new bone on its occlusal surface. (From Stanley et al. Remedial Alveolar Ridge Maintenance with New Endosseous Implant Material, *Journal of Prosthetic Dentistry, 58:*607–613, 1987. Reprinted by permission Mosby-Year Book, Inc.)

Of 83 implants placed in the maxilla, 57 in anterior sockets and 17 in posterior sockets, 2 (3.5%) anterior implants and only 1 (5.9%) posterior implant required recontouring. None of the recontoured implants in the maxilla was lost.

Of 86 implants placed in the mandible, 52 in anterior sockets and 34 in posterior sockets, 10 (11.6%) required contouring; 2 (3.8%) in anterior sockets and 8 (23.5%) in posterior sockets. One recontoured mandibular posterior implant was recontoured at 7 months but lost at 14 months.

About a dozen implants in the first 6 patients were contoured on their coronal aspects at the time of placement, which evidently created sharp edges. After 7 months, 9 of 14 implants in the first 6 patients had to be recontoured because sharp edges were palpable under the mucosa and caused irritation. In anticipation of a continuing future problem because of sharp edges on the implants caused by recontouring, the practice of recontouring at the time of implantation ceased after the first 6 patients. In the treatment of subsequent patients, the need for recontouring did not occur again until 30 months.

Evaluation of the few implants reexposed surgically revealed new bone in direct contact with the implants circumferentially, with no mobility (Clark, Stanley, Hall, King, Colaizzi, Spilman, and Hench, 1986; Hall et al., 1985).

3. Relining Dentures

The need for a reline was not unexpected inasmuch as the Bioglass® cones were placed approximately 2.0 mm below the superficial, ragged, possibly diseased alveolar bone.

Twenty-one of the original 29 patients were fitted with two complete dentures and 8 with one complete maxillary denture and a removable partial mandibular denture. The time period for receiving complete dentures averaged 14.0 months (range 2–27 months). Seven patients required a reline of both dentures postdelivery after an average of 12.8 months (range 10–15.5 months). Four patients required a reline of one denture after an average postdelivery interval of 14.0 months (range 8–19 months). Of 18 control patients with both complete dentures (average age 69 years; range 52–89 years), 4 required relines within the first 5 months; 10 required relines within 1 year; and 2 required relines after 2 years. Two patients did not require relining.

Of the 20 recalled patients, indication for the need for reline was recorded in 17 (1 patient died, 1 was not wearing his maxillary denture, and 1 did not need a reline). Of the 17 patients, the time interval between insertion of dentures and the recognized need for reline averaged 14.5 months (range 3 months, 25 days to 22 months, 21 days). At the present time no patient has received a second reline. There does not appear to be a great rush initially or in delivering the dentures once the ERMI cones are placed, in terms of slowing down the rate of ARR. The initial minimal early rate of ARR occurred regardless of whether the dentures were placed sooner or later.

VIII. DISCUSSION

One should remember that animal studies concerning cone implants without the stress of overlying functioning dentures may give extremely favorable results that may be very misleading. Dogs, contrary to humans and monkeys, do not demonstrate as much resorption of the residual alveolar ridge after extractions. Also, bony healing occurs 2.5 to 3 times more rapidly in dogs than in man (Denissen and de Groot, 1979). Also, one should appreciate that in animal studies one is usually dealing with younger, healthier animals, quite different from the healing capacity of aging humans with advancing periodontal disease.

As each investigator or team of investigators goes through the experience of cone implantation studies, each seems to come up with similar recommendations. However, some investigators point out that there are too few long-term completed clinical trials for definitive claims to be made (Sattayasanskul et al., 1988).

All investigators have commented on the number of implants that emerge through the mucosa and have emphasized the need to preserve them by surgical means, but few have established correlations between the prevalence of emergence and the site of implantation (Brook et al., 1988).

The problem of implant dehiscence that occurred in the Denissen et al. projects was felt to be due to sharp occlusal facial line angles and failure to place the implant sufficiently deep below the crestal bone level. At first these researchers felt that the depth (level) of the implant and the correction of the crest (contouring and flattening the bony peaks and removing the sharp edges) were less importance than achieving primary closure. But from their total experience, they eventually recommended that the cones be placed approximately 2.0 mm below the lowest part of the circumference of the socket to prevent a protrusion above the bony crest level after the initial vertical resorption of the crest, and they concluded that primary closure of soft tissue over the implant was a necessity (mandatory).

Denissen et al. also found that dehiscences in the mandible were enhanced by the continuous mechanical trauma inflicted by the complete mandibular denture with an increasing frequency of 3–4% per year. Fixed and removable (clasped) prostheses sufficiently prevented pressure on the mucosal tissues overlying the implants and avoided dehiscences, whether the implants had been placed above or below the bone crest level (Denissen et al., 1980; Veldhuis et al., 1984).

After their experience of many years, Denissen et al. in 1989 made the following recommendations:

1. Immediate implantation after extraction is complicated because of the difficulty in closing the extraction wounds over the implants. Delay of implantation 3–4 months after extraction, when the mucoperiosteum has covered the extraction wounds in the course of wound healing, solves this problem and avoids mobilization of mucosal tissue with consequent loss of vestibular depth. By that time the initial vertical resorption has already occurred, thus avoiding permucosal exposure and the need for early rebasing of the denture.
2. The design of the bulk HA implant cone needs to be changed. The implants have been bullet shaped, with a rounded top. Denissen et al. now prefer a smooth cylindrical (nontapering) design consisting of a coating of HA ceramic cemented to a core of titanium. The titanium core lends strength to the implant and can serve as an anchoring site for supporting future superstructures. However, because of a failure rate of approximately 25.0% over a period of 8 years due to loosening of the cement layer between the titanium and the HA coating, they are now working with plasma-sprayed coatings of HA on the titanium cores.

Quinn, Kent, Hunter, and Schaffer (1985) lost 9.7% (9 of 92) of their implants and had to recontour 26% (24 of 92) of their implants between 3 and 31 months as they became submucosally prominent. They suggested that their total 36% complication rate could have been reduced by following three basic procedures:

1. Not placing HA roots in sockets with inadequate bony depth, especially 3-rooted maxillary molar sockets
2. Contouring and resubmerging HA cones as soon as possible after their emergence from the ridge
3. Using HA roots with rounded occlusal–facial angles

Although Kwon, El Deeb, Morstad, and Waite (1986) found their results [cone loss 27% (19 of 70); dehiscence 53% (37 of 70)] higher than those of Quinn et al. (1985) (9.7% implant loss; 26% dehiscence), they felt their results were essentially similar to those of six other institutions (studies) with an average cone loss of 23% and a range of dehiscence of 26–30% (Kwon et al., 1986; Quinn et al., 1985; Veldhuis et al., 1984).

The crucial question asked by Kwon et al. (1986) was: Does dehiscence result from migration of the implants or from resorption of the bone? If it is true that HA bonds at its interface with bone (Denissen and de Groot, 1979; Jarcho, Bolen, Thomas et al., 1986; Jarcho, Kay, and Gumaer, 1977), then migration should not be the reason for dehiscence. This leaves bone resorption as the explanation (Kwon et al., 1986).

According to Hench (1991), Bioglass has a much more rapid rate of bonding to bone than HA. Consequently, there is a lower loss of Bioglass® cones and less loss of alveolar ridge bone since stress begins to be transferred to the Bioglass® implants within 5 to 7 days after implantation.

Kangvonkit et al. (1986) lost 6.0% of the implants over a 2-year period. They noticed that dehiscences occurred most frequently on the labial or lingual surface where the cones had sharp angles at the corners of the occlusal and proximal surfaces created by horizontal one-plane sectioning of the implant. They recommended that the cones be rounded on the buccocervical and linguocervical surfaces to prevent abrasion of the alveolar mucosa. Although they recontoured the implants with a high-speed diamond that could produce microcracks and surface irregularities, they believed these factors were not clinically significant. They felt that removal of the sharp edges, coupled with sealing (seating) the implant at least 2 mm below the alveolar crest to accommodate vertical resorption of the alveolar bone during the initial stage of healing, reduced the trauma to the mucosa from the functioning denture. Although they recognized that vital submerged natural root techniques required mucoperiosteal flapping, they did not feel that it was required with durapatite submerged cones. However, they did suture interrupted interproximal papillae whenever possible.

Bell (1986) lost 23.5% of the implants over a 20- to 42-month period and felt his failure rate was due to the fact that he only had four sizes of cones to select from. This appears to confirm the hypothesis of our study (Stanley et al., 1987) that an increased loss occurs if the shape of the cone does not correspond closely to the shape of the socket.

Bell (1986) identified four factors as major causes of submucosal prominence:

1. Depth of sealed (buried) implant not far enough below the bony crest of the tooth socket
2. Inadequate thickness of the labial or lingual alveolar plate, or both
3. The rate of alveolar resorption
4. Denture trauma to the mucosa above the implant site

Bell also felt that many of the clinical complications encountered could have been minimized or delayed if his clinicians had intervened more often once submucosal prominence was identified. Overall, he felt that residual alveolar resorption continued, but at a slower pace than in those regions that did not receive implants.

Filler and Kentros (1987) lost 12% of their cones after 1 year, with many additional implants requiring recontouring due to dehiscence.

Brook, Sattayasanskul, and Lamb (1988)—who lost 20% by 1 year but did not flap over their cones—felt their high loss rate (30%) was due to the cuspid site where the shape of the human cuspid socket was so unusual that a good fit was impossible. They recommended redesigning the cones for cuspid sockets. Brook et al. (1988) pointed out that Quinn, Kent, Hunter, and Schaffer (1985) had a cone implant loss of only 9.7% because they included maxillary sockets. They felt this created a bias because the smaller degree of resorption in the maxilla would prevent cones from becoming prominent and would make them less likely to develop dehiscences.

In our own study, initially, there were too few implant sizes and no custom-made burrs. This required contouring of the implants as well as the sockets. As the proper delivery system was developed, the retention rate improved and the number of implants requiring recontouring decreased (8 by 7 months, but only 4 from 7 through 64 months). Two implants that had been contoured at 4 months were recontoured at 39 months.

It has been reported that the anterior mandible, with its greater amount of compact bone, provides an ideal site for implant success (Schnitman et al., 1988). Clark, Hall, Turner, West, and Wilson (1989) have shown in dogs that the amount of bone formed about Bioglass® was greater the closer the implant was to the peripheral compact bone of femurs as compared to the implants located in the medullary cavity. There is much more cortical bone in the anterior mandible. Smiler (1990) also stressed that the success rate is highest in the anterior mandible and that it declines when implants are placed in the posterior mandible. In the maxilla, implant success is less because of the porous quality of the bone. In addition, the labial/buccal cortical plates are thinner.

Our results at present are contrary to the literature. The anterior maxillary area lost the least number of implants (5.3%); the posterior mandibular area was second (12.1%), with the anterior mandibular area (17.3%) and the posterior maxillary area (17.6%) losing about the same number of implants. Apparently Bioglass® cone implants can succeed in any anatomic location if enough bone is present at that particular individual site and the implant is placed deep enough. Evidently in this study (Stanley, Brandt, Hall, Clark, and King, 1994) the trauma of the dentures to the crest of the anterior mandible caused excessive bone resorption only until the implant was approached.

According to Bahat (1993), in dealing with Noebelpharma implants, the role of posterior maxillary implants remains to be clarified because of the poor quality of the bone and low position of the maxillary sinus. As a result, posterior maxillary implants have had a lower success rate than those placed in the mandible or the anterior maxilla.

When teeth are lost, the maxillary bone is resorbed posteriosuperiorly, reducing the radius and circumference of the anterior arc. Bone concavities may become accentuated and irregular, and the anterior ridge often becomes knife-edged. These changes create unfavorable relationships with the posterior mandible, in which bony resorption proceeds inferolaterally.

Anterior implants may be placed with unfavorable emergence profiles and at exaggerated angles, necessitating the use of angle abutments and encouraging nonaxial loading when fixed restorations are used, often predisposing the implants to failure. The principal rule is that implants must not be used unless it is possible to place a suitable number with sufficient length at the appropriate location and with the proper angulation to support the desired final restoration. The surgeon must consider the three-dimensional features of a potential site for both anatomic and functional adequacy.

The guidelines for tooth preservation in patients receiving periodontal prostheses are appropriate here, albeit with greater strictness in view of the future presence of implants, which do not respond to loading and splinting in the same way as teeth. Bone is anisotropic[5]; the strength and elastic properties depend on the orientation of its microstructure in relation to the direction of loading. An implant placed so that it is subjected to nonaxial (i.e., not directly compressive) loading no matter what the bone's structure has unavoidably been placed in weakened bone and is at greater risk of osseointegrative and mechanical failure (Bahat, 1993).

During the initial phase of osseointegration, transmucosal loading must be kept to a minimum or be prevented. In the maxilla, it appears that much of the bone remodeling subsequent to implant placement takes place during this time rather than after occlusal loading, which seems to occur in the mandible. Sequential radiographs show greater loss of bone in the maxilla than in the mandible during osseointegration, which Adell, Lekhoml, Rockler, and Branemark (1981) attribute to the cancellous character and rich vascular supply of the maxilla.

Although one tries to obtain the tightest fit of the implant as possible, the number of entries must be minimized. Any implant that remains or becomes mobile should be removed immediately.

The implant should be surrounded by compact or trabecular bone without radiolucency. The latter indicates formation of a fibrous capsule and is the harbinger of implant failure even if the implant is at first stable to manipulation. Clinically, one finds these implants are surrounded by soft tissue that can be removed from the bone cavity like the capsule of a cyst. As mentioned above, the results of implant placement anywhere in the maxilla generally have been poorer than the results in the mandible. Jaffin and Berman (1991) reported a loss of 8.3%, and Adell's group (1981), in their 5-year experience, had a failure rate of almost 20% for maxillary implants. Results in the posterior maxilla have been even less impressive. In one series, 1 of 6 maxilla molar implants was lost as compared with 2 of 45 in the corresponding area of the mandible. Da Silva, Schnitman, Wohrle, Wang, and Koch (1992) had a 6-year survival rate of posterior maxillary implants of 74%, compared to 94% for posterior mandibular implants. Kopp (1989) wrote that the replacement of bilateral posterior maxillary edentulous areas with osseointegrated implants is difficult if not impossible.

In the posterior maxilla, cancellous bone is said to be of Type III or IV, although in practice, intermediate types (II-III, III-IV) are not unusual. The bone quality may be even poorer in older patients. The cortex is thin, and the trabecular bone is often of low density. Such highly porous bone (Type IV) is less tolerant of the repetitive compression coincident with occlusal loading of the implants because there can be fewer points of contact between the bone and the implant. Implants placed in Type IV bone therefore have had a higher failure rate. Jaffin and Berman (1991) reported a 35% failure rate in Type IV bone in either jaw, compared with 3% of those placed in Type I, II, and III bone.

Nevins and Langer (1993) recently pointed out that posterior areas of the jaw were initially avoided because of anatomic structures—the inferior alveolar nerve in the mandible and the maxillary sinuses. Also, there is a questionable volume of bone available in

---

[5]Anisotropic: Greek; unequal, uneven, dissimilar, and turning toward; changing—having unlike properties in different directions, as in any unit lacking spherical symmetry (*Dorland's Illustrated Medical Dictionary*, 27th ed.).

these posterior areas, and the nature of cancellous bone itself places the success rate at a disadvantage. They found, however, that if they waited 3 months in the mandible and 6 months in the maxilla before exposing and loading the buried bone embedded implants with superstructures, their success rate greatly improved, with a large number of free-standing implants and a variety of other osseointegrated procedures.

Also, Meffert (1993) has stressed that you should not go from uncovering an implant to full function immediately but must "baby it"—take at least 18 weeks to subject the implant to maximum loading. This approach is especially important when dealing with the posterior maxillary area.

We started out in our original baboon studies hoping to find a typical periodontal membrane about the implants. Although periodontal membranes formed initially in some instances, with increasing time such tissue was replaced by ankylotic bone. Now we settle for a combination of ankylotic bone and fibrous tissue provided there is sufficient anky-lotic bone.

No matter how hard we tried surgically, even in dogs, to place the implant in or near cortical bone, to maximize contact with approximating or existing alveolar bone, we did not obtain total true ankylosis all around the implant. We strove for total osseointegra-tion but we missed in enough areas, evidently fortunately, that the fibro-osseous relation-ship was sufficient to save the day. Nature seems to know when to help us out with the combination of ankylotic bone and fibrous tissue. Obviously, we do not need solid bone completely around an ERMI for long-term retention. Since bone is a dynamic tissue, the amount of ankylotic bone is also subject to change. With Bioglass® implants, the increas-ing gradation of mineralization of the 200 μm outer layer resulting from the formation of the gel layer also may help with the problems of stress transfer (Clark et al., 1979; Stanley, 1987; Weinstein et al., 1980).

We also appreciate the findings of Harrell, Keane, Acree, Bates, Clark, and Hench (1978) that, for chemical bonding to occur, the space between the implant surface and the approximating socket alveolar bone cannot be greater than 0.5 mm; otherwise, the so-called dead space occurs, which fills in with persisting fibrous tissues. Recognition of this requirement for Bioglass® implants and the need for tight wedging of the cones may explain the low rate of dehiscence in our study. Some of our patients have been wearing dentures over ERMIs for more than 6 years.

Carlsson, Rostlund, Albrektsson, and Albrektsson (1988) evaluated osseointegration of implants with different gaps between the bone and the surface of commercially pure titanium implants in rabbit tibias. Sites 3.7 mm in diameter were prepared and smooth titanium cylinders with diameters of 3.7, 3.0, and 2.0 mm were placed. After 6 weeks, those implants with a diameter of 3.7 mm (an exact fit initially) exhibited a direct bone-to-implant contact without any intervening soft tissue. However, only 1 of 4 im-plants with a diameter of 3.0 mm displayed direct bony apposition to the implants; the others exhibited mainly fibrous tissue with a few scattered spicules of trabecular bone. None of 4 implants with a diameter of 2.0 mm displayed bony apposition to the implants, and again patchy trabeculated bone was found without cortical bone connection. The margins of the original proximal cortical bone were rounded and exhibited no evidence of osteoblastic activity.

As suggested by Brunski (1988a), it would be valuable to know what combination of bone and fibrous tissue is adequate; what percentage of the implant's surface needs support by bone versus soft tissues.

We have developed an appreciation for the fact that the Bioglass® cones must be placed symmetrically, as much as possible. Since the usual alveolar ridge resorption continues to occur in areas too far removed from the stimulatory effects of the cone implants, we must take this factor into consideration when spacing the implants for long-term successful results. When the bone recedes in some areas and not others, the dentures begin to rock (become wobbly).

It appears that an implant with a good, but not exact, surgical fit in bone may initiate enough partial or patchy direct contact between the implant and cortical bone that it will suffice to stabilize the implant. The more exact the surgical fit, the more appositional bone lamellae will form. Our study demonstrated the importance of a close fit between the bone and implant at the time of initial implant placement.

Von Steenberghe (1988) stressed those factors that were most detrimental to bone deposition: micromovement, lack of biocompatibiltiy or osseoconductivity, and bacterial contamination. Poor stability at the time of implantation favors micromovement at the interface and leads to interfacial fibrous tissue formation.

Brunski (1988) has also stressed that fibrous tissue is a by-product of mechanical interference (overloading) with bone formation around an implant. Therefore, intimate contact is especially important to provide immobility during the initial healing period. Denture wearing early on causes deformation of the jaw bone, which can induce micro-movements at the interface. Some disappointing results in the maxilla might be due to the poor corticalization of the jaw bone, especially at the side of the floor of the nasal or sinusal cavities. The initial stability is thus poor. By allowing intraosseous implants to heal and become stabilized in a bony matrix for a few weeks before the wearing of a denture, the heavy occlusal loads are avoided during the critical healing time.

Mohammed, Atmaran, and Schoen (1977), using the photoelastic technique *in vitro* to model three types of root configuration, showed that the cylindrical root form in ankylotic conditions yielded the most favorable stress conditions when compared with a natural-tooth root form and tapering conical shapes (Hench and Ethridge, 1982). The recent favorable results of Denissen, Kalk, Veldhuis, and van Waas (1993a) seem to support the recommendations of Mohammed's group.

In regard to Bioglass®, bone formation depends apparently entirely on the sequence of events described earlier, and any interruption of the sequence—for instance, as related to excessive mobility of the implant—may inhibit bone formation (Stanley, 1987). If the sequence is interrupted, the action is aborted and fibrous tissue forms. Experience has shown that no matter how long the Bioglass® implant remains in place after the initial chemical reaction between the tissue and implant is aborted, no boney attachment will occur.

Bioglass® has one tremendous advantage over other similar products. When the fracture line of an implant of any of the retained Bioglass® implants occurred below the gingival surface of the baboons, new bone was deposited on the fresh incisal surface of the fractured root implant irrespective of the postoperative implant interval, just as when first implanted. Such surfaces became completely covered with new bone. In other words, when a fresh surface of Bioglass® appeared, in this case due to a fracture, the osteoconductive activity was reinitiated just as it was when first implanted. This is the opposite to fatigue fracture with metals and vitreous carbon implants. Distorted fragments of metal implants work their way out of the tissues. It is possible the Bioglass® is the only implant material that can heal itself (see Figs. 19 and 20) (Stanley et al., 1981).

The question is asked: Why not implant a single free-standing implant initially, instead of a cone? Then you will not need another implant subsequently. Not every patient is a candidate for implant reconstruction at this time. Implants generally add

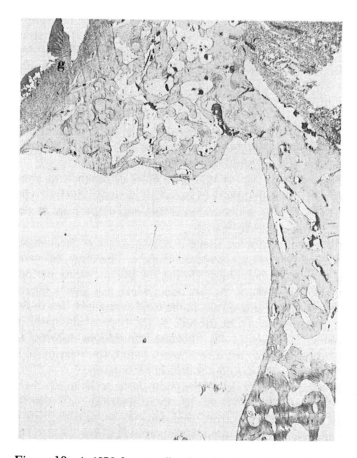

Figure 19 A 45S5 free-standing individual tooth-form, implant previously fractured off below the gingival mucosa and now covered with new bone formation (2 years after implantation) in a baboon mandibular left lateral site. Note the bone-lined socket, gingival mucosa in upper left hand corner (g) and new bone formation above and approximating fractured root surface ($\times$8.1). (From Stanley, H.R., Hench, L.L., Bennett, C.G., Jr., et al. The implantation of natural tooth from Bioglass in baboons—long term results. *Int. J. Oral Implantology*, 981; 2:26–38 with permission.)

another level of complexity and cost, placing them out of reach of many patients. Also, some patients who are comfortable with their removable prosthesis may not want to undergo or be able to undergo the surgical procedures of implant placement (Schnitman et al., 1988).

Some patients may not be financially stable at the time of extraction, some may have too many sockets, and some may not even be able to afford fixed prostheses. An effort should be made to encourage the placement of Bioglass® cones into every fresh extraction socket to maintain the alveolar ridge, whether or not an intraoral attachment is to be placed immediately or added subsequently with more extensive restoration. Bone resorption will be retarded until the patient completes appropriate treatment. Where natural teeth are to be used for abutments, it will help prevent approximating pontics ending up

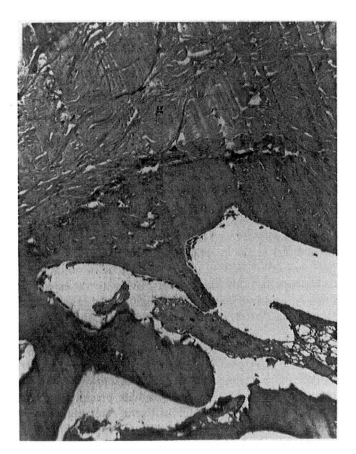

Figure 20 New solid bone formation covering the fractured root surface of a once free-standing individual baboon incisor tooth form Bioglass® implant. Healed gingiva (g) is seen above. At 6 months the implant was desplinted. Sometime after that the implant fractured and new bone formed on the freshly exposed surface (× 100). (From Stanley, H.R., Hench, L.L., Bennett, C.G., Jr., et al. The implantation of natural tooth from Bioglass in baboons – long term results. *Int. J. Oral Implantology*, 1981; 2:26–38 with permission.)

in midair as the alveolar bone resorbs. As total edentulism disappears, so will the need for cone implants. But in the meantime, there is a large market.

As simple as the clinical procedures appear to be, clinicians must be trained in the proper use of cone implants. Cranin and Shpuntoff (1984) felt that the disparate skills of the clinicians, the precision of the implant fit, and the lack of tissue coverage played a part in the poor results of their study.

Despite their poor biomechanical properties (mainly brittleness), bioactive glasses may be regarded as potential alternatives to replace HA as a coating for bone implants (Ducheyne, 1985; Hench, Pantano, Buscemi, and Greenspan, 1977; Hench, Paschall, Allen, and Piotrowski, 1977; Vrouwenvelder et al., 1993). Because of the development of the silica-rich gel layer, there is no need for gross or microscopic surface configurations – the smoother the surface the better.

IX. NEW CONCEPTS

Although it has been assumed that the formation of a Si-rich layer on the surface of the glass is a necessary prerequisite for the formation of a Ca,P layer, Kokubo (1990, 1991) stated that this is not necessarily so. In apatite–wallastonite glass ceramics, no Si-rich layer was observed after treatment in aqueous solutions. Ca and silicate ions dissolved into the solution and then a Ca,P layer formed on the surface of the material. This led to the conclusion that apatite precipitation may occur after absorption of silicate from an appropriate solution onto the surface of a material without a Si-rich layer.

In 1990, Abe, Kokubo, and Yamamuro also showed that when a bioactive glass ceramic piece was held close to the surface of other nonbioactive materials in SBF, an apatite layer would form on those surfaces. This indicated strongly that silicate absorption would also facilitate Ca,P precipitation.

The recent work of van Blitterswijik, Leendero, Brink, Bovell, Gallard, Heintze, and Bakker (1992) is very important because they showed that an increased fluid uptake by copolymers resulted in an intimate contact at the interface with bone and triggered the formation of new bone. Since there was no calcium in the copolymer, none could have been released, so all the calcium came from the body fluids of rats. Evidently the extensive movement of fluids into the polymer delivered sufficiently high quantities of calcium into the polymer for the sorption of Ca by the PEO (polyethylene oxides) fragment to take place through a chelating effect.

Bone bonding was found only on those PEO/PBT copolymer formulations that showed substantial calcifications at their interface. Bone bonding occurred only after calcification on the polymer. Although the copolymer did not contain Ca prior to implantation, it still bonded to bone provided sufficient calcium phosphate precipitation occurred near the interface with bone. This seems to be in line with the emerging opinion that the postoperative calcium phosphate precipitation found with the known bioactive ceramics and glasses is more important than the initial presence of calcium. Here we have an elastomeric bone-bonding biomaterial, in contrast to the stiff and fragile ceramics, that could have great promise in bone replacement surgery.

X. SUMMARY AND CONCLUSIONS

Eventually, most investigators with their increasing experience with cone implants develop the following recommendations to avoid or minimize dehiscences and exfoliation:

1. Implant cones must fit tightly (snugly) and be wedged. In order to do so, there must be a selection of multiple implant sizes to choose from. One must recognize that three-rooted sockets and cuspid sockets are difficult to deal with. Use burrs that match the implants in size to provide a tight fit.
2. Do not make any adjustments on the cone implant itself initially. All cutting, grinding, or shaping of cones leaves sharp angles, irregularities, or grinding debris that irritate the mucosal tissues. Make minimal adjustments to sockets.
3. Cone implants must be placed at least 2.0 mm below the lowest level of circumference of bone crest. There is a spontaneous loss of alveolar bone initially, due not only to fracture of alveolar bone from extraction procedures, but to resorption of irregularities of bone contour due to periodontal disease.
4. Every effort must be made to obtain primary closure of soft tissue over the implanted cone.

5. Delay placing removable full dentures at least 6 weeks to minimize mechanical trauma. Fixed prostheses and clasp-supported partial dentures can be placed earlier.
6. Minimally recontour an exposed implant surface, provided the implant is ankylosed, as soon as it begins to be exposed. Leave an exposed surface as smooth as possible. It is not necessary to provide primary closure the second time.
7. Reline removable dentures as soon as the need is apparent (ill-fitting) to minimize increasing mechanical trauma, which enhances dehiscence, bone resorption, and exfoliation.

Our long-term results are evidently superior because Bioglass® cones were placed deep and dentures were not inserted into the original 29 patients for an average of 14 months. The lower ARR rate in the maxilla may explain part of our success rate as compared to those studies that only used mandibles and inserted dentures immediately.

The low rates of dehiscence and implant loss over a period of 71 months (average 54.1 months) with Bioglass® continue to exhibit very favorable clinical results. While specialists in biomaterials continue to cope with the problem of coating Bioglass onto stronger supporting cores (vitallium, titanium, aluminum oxide, and stainless steel), solid pieces of Bioglass® can be used successfully as ERMIs.

The success rate of the ERMI project offers long-term advantages over other commercial products of a similar nature in preserving the alveolar ridges under functioning dentures. Bioglass® cones placed into fresh extraction sockets under functioning dentures have been followed in 20 patients for periods over 54 months, on average, and have been able to maintain edentulous alveolar ridges with only a 12.5% loss of implants. Other products, made of either solid or porous hydroxyapatite, have not been able to approach this feat clinically.

We accepted most of the patients presented for evaluation, mostly indigent. We did not have a highly selective group as required by some implant systems. Also, the dentures were fabricated by dental students, many performing this service for the first time. Possibly our results would have been even better with more experienced clinicians.

REFERENCES

Abe, Y., Kokubo, T., and Yamamuro, Y. (1990). Apatite coating on ceramics, metals and polymers utilizing a biological process. *Mater. Med., 1*, 233–238 (quoted by Kangasniemi et al., 1993).

Adell, R., Lekholm, U., Rockler, B., and Branemark, P-I. (1981). A 15-year study of osseointegrated implants in the treatment of the edentulous jaw. *Int. J. Oral Surg., 10*, 387–416 (quoted by Bahat, 1993).

Albrektsson, T. (1985). The response of bone to titanium implants. *CRC Crit. Rev. Biocompatibility, 1*(1), 53–84.

Albrektsson, T., Zarb, G., Worthington, P., and Eriksson, A. (1986). The long-term efficacy of currently used dental implants: A review and proposed criteria of success. *Int. J. Oral Maxillofac. Implants, 1*, 11–25 (quoted by Hartman et al., 1989).

Andersson, O. H., and Kangasniemi, L. (1991). Calcium phosphate formation at the surface of bioactive glass *in vitro. J. Biomed. Mater. Res., 25*, 1019–1030.

Andersson, O. H., Karlsson, K. H., Kangasniemi, K., and Yli-Urpo, A. (1988). Models for physical properties and bioactivity of phosphate opal glasses. *Glastech. Ber., 61*, 300–305 (quoted by Andersson et al., 1993).

Andersson, O. H., Liu, G., Karlsson, K. H., Niemi, L., Miettinen, J., and Juhanoja, J. (1990). *In vivo* behavior of glasses in the $SiO_2-Na_2O_2-CaO-P_2O_5-Al_2O_3-B_2O_3$ system. *J. Mater. Sci.: Mater. Med., 1,* 219–227.

Andersson, O. H., Rosenqvist, J., and Karlsson, K. H. (1993). Dissolution, leaching, and Al_2O_3 enrichment at the surface of bioactive glasses studied by solution analysis. *J. Biomed. Mater. Res., 27,* 941–948.

Atwood, D. A. (1962). Some clinical factors related to rate of resorption of residual ridges. *J. Prosthet. Dent., 12,* 441–450.

Atwood, D. A. (1963). Postextraction changes in the adult mandible as illustrated by microradiographs of midsagittal sections and serial cephalometric roentgenograms. *J. Prosthet. Dent., 13,* 810–824.

Atwood, D. A. (1971). Reduction of residual ridges: A major oral disease entity. *J. Prosthet. Dent., 26,* 266.

Atwood, D. A. (1979). Bone loss of edentulous alveolar ridges. In *Oral Perspective on Bone Biology.* The 8th James English Symposium (E. Hausman, Chairman). *J. Periodontol., 50,* 11–21.

Atwood, D. A., and Coy, W. A. (1971). Clinical cephalometric, and densitometric study of reduction of residual ridges. *J. Prosthet. Dent., 26,* 280–299.

Bahat, O. (1993). Treatment planning and placement of implants in the posterior maxilla: Report of 732 consecutive Nobelpharma implants. *Int. J. Oral Maxillofac. Implants, 8,* 151–161.

Beckham, C. A., Greenlee, T. K., Jr., and Crebo, A. R. (1971). Bone formation at a ceramic implant interface. *Calcif. Tissue Res., 8,* 165.

Bell, D. H. (1986). Particles versus solid forms of hydroxyapatite as a treatment modality to preserve residual alveolar ridges. *J. Prosthet. Dent., 56,* 323–326.

Block, M. S., and Zieman, G. A. (1985). A comparison of particulate and solid hydroxylapatite in canine extraction sites (abstract). *Proc. 67th AAOMS Annual Meeting,* p. 50.

Bokros, J. C. (1979). *Carbon in dentistry. I. Material characteristics and biocompatibility of carbon.* Unpublished paper, CarboMedics, Inc., San Diego, pp. 14–16 (cited by King, K. O. Overview of currently used major dental implant modalities).

Boyne, P. J., Rothstein, S., Cook, V., Stutz, T. I., and Gumaer, K. I. (1982). Fluorescence microscopy of durapatite implants. *Proc. 34th Annual Session American Institute of Oral Biology,* pp. 53–57.

Boyne, P. J., Rothstein, S. S., Gumaer, K. I., and Drobeck, H. P. (1984). Long-term study of hydroxylapatite implants in canine alveolar bone. *J. Oral Maxillofac. Surg., 42,* 589–594.

Boyne, P. J. (1990). Personal communication.

Brewer, A. A., and Morrow, R. M. (1980). *Overdentures.* C. V. Mosby, St. Louis, MO, pp. xi–xii, 3–14.

Brook, I. M., and Lamb, D. J. (1987). The use of particulate and block forms of hydroxylapatite for local alveolar augmentation. *Int. J. Oral Maxillofac. Implants, 2,* 85–89.

Brook, I. M., Sattayasanskul, W., and Lamb, D. J. (1988). Dense hydroxyapatite root replica implantation: Tooth site and success rate. *Br. Dent. J., 164,* 212 (abstracted in *Int. J. Oral Maxillofac. Implants, 3,* 295, 1988).

Brunski, J. B. (1988a). Biomaterials and biomechanics in dental implant design. *Int. J. Oral Maxillofac. Implants, 3,* 85–97.

Brunski, J. B. (1988b). Biomechanical considerations in dental implant design. *NIH Consensus Development Conference on Dental Implants,* June 13–15, pp. 81–83.

Carlisle, E. M. (1970). Silicon: A possible factor in bone calcification. *Science, 167,* 279.

Carlsson, G. E., and Persson, G. (1967). Morphologic changes of the mandible after extraction and wearing of dentures: A longitudinal, clinical and x-ray cephalometric study covering 5 years. *Odontol. Rev., 18,* 27–34.

Carlsson, L., Rostlund, T., Albrektsson, B., and Albrektsson, T. (1988). Implant fixation improved by close fit. *Acta Orthop. Scand., 59,* 272–275 (abstracted in *Int. J. Oral Maxillofac. Implants, 3,* 295, 1988).

Casey, D. M., and Lauciello, F. R. (1980). A review of submerged-root concept. *J. Prosthet. Dent., 43*, 128–131.

Cawood, J. J., and Howell, R. A. (1988). A classification of the edentulous jaws. *Int. J. Oral Maxillofac. Surg., 17*, 232–236 (quoted by Denissen et al., 1993).

Clark A. E., Hall, M., Turner, G., West, J., and Wilson, J. (1989). Bioglass® coated and HA coated titanium alloy implants in dogs. *Transactions of Society for Biomaterials*, Lake Buena Vista, FL, April 28–May 2, p. 175.

Clark, A. E., Hench, L. L., and Paschall, J. (1976). The influence of surface chemistry on implant interface histology: A theoretical basis for implant materials selection. *J. Biomed. Mater. Res., 10*, 161–174.

Clark, A. E., Kim, C. Y., West, J. K., Wilson, J., and Hench, L. L. (1990). Reactions of fluoride- and non-fluoride-containing bioactive glasses. In *Handbook of Bioactive Ceramics*, Vol. 1, *Bioactive Glasses and Glass-Ceramics* (T. Yamamuro, L. L. Hench, and J. Wilson, eds.) CRC Press, Boca Raton, FL.

Clark, A. E., Jr., Stanley, H. R., Acree, W. A., and Kreutziger, K. (1979). Thickness of bonding layers on Bioglass® dental implants. *J. Dent. Res., 58*, Special Issue A, abstract 824.

Clark, A. E., Stanley, H. R., Hall, M. B., King, C., Colaizzi, F., Spilman, D., and Hench, L. L. (1986). Clinical trials of Bioglass implants for alveolar ridge maintenance (abstract). *J. Dent. Res., 65*, Special Issue, 304.

Cook, S. D., Kay, J. F., Thomas, K. A., and Jarcho, M. (1987). Interface mechanics and histology of titanium and hydroxylapatite-coated titanium for dental implant applications. *Int. J. Oral Maxillofac. Implants, 2*, 15 (quoted by Brunski, 1988).

Cranin, A. N., and Shpuntoff, R. (1984). Hydroxylapatite (HA) cone implants for alveolar ridge maintenance – One year follow-up. *J. Dent. Res., 63*, Special Issue, Abstract 269.

Crum, R. J., and Rooney, G. E. (1978). Alveolar bone loss in overdentures: Five-year study. *J. Prosthet. Dent., 40*, 610–613 (quoted by Kwon et al., 1986).

Da Silva, J. D., Schnitman, P. A., Wohrle, P. S., Wang, H. N., and Koch, G. G. (1992). Influence of site on implant survival: 6-year results (abstract). *J. Dent. Res., 71*, 256 (quoted by Bahat, 1993).

Davis, R. K., Renner, R. P., Antos, E. W., Schlissel, E. R., and Baer, P. N. (1981). A two-year longitiudinal study of the periodontal health status of overdenture patients. *J. Prosthet. Dent., 45*, 358–363 (quoted by Langer and Langer, 1993).

Delivanis, P. O., Esposito, C., and Brickley, R. (1980). Clinical considerations for root-submergence procedures. *J. Prosthet. Dent., 43*, 487–490.

Denissen, H. W., and de Groot, K. (1979). Immediate dental root implants from synthetic dense calcium hydroxylapatite. *J. Prosthet. Dent., 42*, 551–556.

Denissen, H. W., and Kalk, W. (1991). Preventive implantations. *Int. Dent. J., 41*, 17–24 (quoted by Denissen et al., 1993).

Denissen, H. W., Kalk, W., Veldhuis, A. A. H., and Van Den Hoof, A. (1989). Eleven-year study of hydroxyapatite implants. *J. Prosthet. Dent., 61*, 706–712.

Denissen, H. W., Kalk, W., Veldhuis, H. A. H., and van Waas, M. A. J. (1993). Anatomic considerations for preventive implantation. *Int. J. Oral Maxillofac. Implants, 8*, 191–196.

Denissen, H. W., Rejda, B. V., and de Groot, K. (1978). *Calcium hydroxyapatite/p-hydroxy–ethyl methacrylate (HA/p-HEMA) composites as natural tooth root substitutes*. Presented at the 4th Annual Meeting, Society for Biomaterials, San Antonio, TX.

Denissen, H., van Beek, E., Papapoulos, S., Lowik, C., Kalk, W., and van den Hooff, A. (1993). Hydroxyapatite implants as carriers for the bone resorption inhibitor (3-dimethylamino-1-hydroxypropylidene)-1, 1-bis-phosphonate. *Trans. Soc. Biomaterials, 19th Annual Meeting*, Birmingham, AL, April 28–May 2, p. 252.

Denissen, H. W., Veldhuis, A. A. H., Makkes, P. C., van den Hooff, A., and de Groot, K. (1980). Dense apatite implants in preventive prosthetic dentistry. *Clin. Prev. Dent., 2*, 23–28.

Derkson, G. D., and Mac Entee, M. M. (1982). Effect of 0.4% stannous fluoride gel on the gingival health of overdenture abutments. *J. Prosthet. Dent., 48*, 23–26 (quoted by Langer and Langer, 1993).

Dorland's Illustrated Medical Dictionary, 27th Edition, W.B. Saunders, Philadelphia, 1985.

Ducheyne, P. (1985). Bioglass coatings and bioglass composites as implant materials. *J. Biomed. Mater. Res., 19*, 273–291.

Ettinger, R. L. (1988). Tooth loss in overdenture population. *J. Prosthet. Dent., 60*, 459–462.

Ettinger, R. L. (1993). Managing and treating the atrophic mandible. *JADA, 124*, 234–241.

Ettinger, R. L., and Krell, K. (1988). Endodontic problems in an overdenture population. *J. Prosthet. Dent., 59*, 459–462 (quoted by Langer and Langer, 1993).

Ettinger, R. L., Taylor, T. D., and Scandrett, F. R. (1984). Treatment needs of overdenture patients in a longitudinal study. *J. Prosthet. Dent., 52*, 532–537 (quoted by Langer and Langer, 1993).

Feinberg, L. (1987). Report: More Americans living longer. *The News Journal*, Daytona Beach, FL, July 6, p. 4B.

Fenton, A. H., and Hahn, N. (1978). Tissue response to overdenture therapy. *J. Prosthet. Dent., 40*, 492–498 (quoted by Langer and Langer, 1993).

Filler, S. J. (1989). Personal communication.

Filler, S. J., and Kentros, G. A. (1987). Hydroxylapatite cone implants for alveolar ridge preservation—One year follow-up. *J. Dent. Res., 66*, Special Issue, Abstract #1140, p. 249.

Filler, S. J., Kentros, G. A., and Heberer, D. D. (1987). Hydroxylapatite particulate implants for alveolar ridge preservation—One year follow-up. *J. Dent. Res., 66*, Special Issue, Abstract #1139, p. 249.

Filgueiras, R. M., La Torre, G., and Hench, L. L. (1993). Solution effects on the surface reactions of a bioactive glass. *J. Biomed. Mater. Res., 27*, 445–453.

Frame, J. W. (1984). Augmentation of an atrophic edentulous mandible by interpositional grafting with hydroxylapatite. *J. Oral Maxillofac. Surg., 42*, 88–92.

Garver, D. G., and Fenster, R. K. (1980). Vital root retention in humans: A final report. *J. Prosthet. Dent., 43*, 368–373.

Garver, D. G., Fenster, R. K., Baker, R. D., and Johnson, D. L. (1978). Vital root retention in humans; a preliminary report. *J. Prosthet. Dent., 40*, 23–28.

Glassman, R. B., Negras, M., and Doty, R. W. (1969). A safe and reliable method for temporary restraint of monkeys. *Physiol. Behav., 4*, 431–432.

Goska, F. A., and Vandrak, R. F. (1972). Root submerged to preserve alveolar bone: A case report. *Mil. Med., 137*, 446–447 (quoted by Casey and Lanciello, 1980).

Gound, T., O'Neal, R. B., del Rio, C. E., and Levin, M. P. (1978). Submergency of roots for alveolar bone preservation. II. Reimplanted endodontically treated roots. *Oral Surg., 46*, 114–122.

Greenlee, T. K., Jr., Beckham, C. A., Crebo, A. R., and Malmborg, J. C. (1972). Tissue responses at the interface of a ceramic. *J. Biomed. Mater. Res., 6*, 244.

Gross, U. M. (1988). Biocompatibility: The interaction of biomaterials and host response. *NIH Consensus Development Conference on Dental Implants*, June 13–15, pp. 97–99.

Gross, U., Brandes, J., Strunz, V., Bab, I., and Sela, J. (1981). The ultrastructure of the interface between a glass ceramic and bone. *J. Biomed. Mater. Res., 15*, 291 (quoted by Albrektsson, 1985, p. 67).

Gross, U., Kinne, R., Schmitz, H. J., and Strunz, V. (1988). The response of bone to surface active glass/glass-ceramics. *CRC Crit. Rev. Biocompatibility, 4*, 2.

Gross, U., Roggendorf, W., Schmitz, H. J., and Strunz, V. (1986a). Testing procedures for surface reactive biomaterial. In *Biological and Biomechanical Performance of Biomaterials*. (P. Christel, A. Meunier, and A. J. C. Lee, eds.), Elsevier, Amsterdam, p. 367.

Gross, U., Schmitz, H. J., Strunz, U., Schuppan, D., and Termine, J. (1986b). Proteins at the interface of bone-bonding and non-bonding glass-ceramics. *Trans. 12th Annual Meeting Soc. for Biomaterials*, Algonquin, IL, p. 98.

Gross, U. M., and Strunz, U. (1980). The anchoring of glass ceramics of different solubility in the femur of the rat. *J. Biomed. Mater. Res., 14*, 607.

Guyer, S. E. (1975). Selectively retained vital root for partial support of overdentures: A patient report. *J. Prosthet. Dent., 33*, 258–263.

Hall, M. B., Stanley, H. R., King, C., Colaizzi, F., Spilman, D., and Hench, L. L. (1985). Early clinical trials of 45S5 Bioglass for endosseous alveolar ridge maintenance implants. *Excerpta Medica Proceedings International Congress on Tissue Integration and Maxillofacial Reconstruction*, Elsevier Science Publishers, Amsterdam, pp. 248–252.

Harrell, M. S., Keane, M. A., Acree, W. A., Bates, S. R., Clark, A. E., and Hench, L. L. (1978). Thickness of Bioglass bonding layers. *Trans 4th Annual Meeting Society for Biomaterials*, San Antonio, TX, p. 111.

Hench, L. L. (1980). The interfacial behavior of biomaterials. *J. Biomed. Mater. Res., 14*, 805–811.

Hench, L. L. (1981). Stability of ceramics in the physiological environment. In *Fundamental Aspects of Biocompatibility* (D. F. Williams, ed.), Vol. 1, CRC Press, Boca Raton, FL, pp. 67–85.

Hench, L. L. (1988a). Bioactive fixation. In *Trans. 3rd World Biomaterials Congress*, Abstract, p. 24.

Hench, L. L. (1988b). In *Bioceramics: Materials Characteristics Versus in vivo Behavior* (P. Ducheyne and J. E. Lemmons, eds.), Annual of the New York Academy of Sciences Vol. 523, pp. 54–71 (quoted by Wilson and Low, 1992).

Hench, L. L. (1991). Bioceramics: From concept to clinic. *J. Am. Ceram. Soc., 74*, 1487–1510.

Hench, L. L., and Clark, A. (1982). Adhesion to bone. In *Biocompatibility of Orthopedic Implants* (D. F. Williams, ed.), Vol. 2. CRC Press, Boca Raton, FL, Chap. 6.

Hench, L. L., and Ethridge, E. C. (1982). *Biomaterials: An Interfacial Approach*. Academic Press, New York, Chaps. 5, 7, 14.

Hench, L. L., Pantano, C. G., Jr., Buscemi, P. J., and Greenspan, D. C. (1977). Analysis of bioglass fixation of hip prostheses. *J. Biomed. Mater. Res., 11*, 267–282.

Hench, L. L., and Paschall, H. A. (1974). Histochemical responses at a biomaterial's interface. *J. Biomed. Mater. Res., 3*, 49.

Hench, L. L., Paschall, H. A., Allen, W. C., and Piotrowski, G. (1977). *An investigation of bonding mechanisms at the interface of a prosthetic material*. Report No. 8. Document published by the Department of Materials Science and Engineering, University of Florida, Gainesville.

Hench, L. L., Splinter, R. J., Allen, W. C., and Greenlee, T. K. (1971). Bonding mechanisms at the interface of ceramic prosthetic materials. *J. Biomed. Mater. Res., 5*, 117–141.

Hench, L. L., Stanley, H. R., Clark, A. E., Hall, M., and Wilson, J. (1991). Dental applications of Bioglass® implants. In *Bioceramics*, Vol. 4 (W. Bonfield, G. W. Hastings, and K. E. Tanner, eds.). Proc. 4th Int. Symp. Ceram. Med., Butterworth–Heinemann, London.

Herd, J. R. (1973). The retained root. *Aust. Dent. J., 18*, 125 (quoted by Casey and Lauciello, 1980).

Jaffin, R. A., and Berman, C. L. (1991). The excessive loss of Branemark fixtures in type IV bone: A 5-year analysis. *J. Periodontol., 62*, 2–4 (quoted by Bahat, 1993).

James, R. (1984). Tissue response to dental implant devices. In *Clinical Dentistry*, Vol. 5 (J. W. Clark, ed.). Harper & Row, Philadelphia, Chap. 48, pp. 8–9.

Jarcho, M. (1986). Biomaterial aspects of calcium phosphates. *Dent. Clin. North Am., 30*, 25–47 (quoted by Brook and Lamb, 1987).

Jarcho, M., Bolen, C. H., Thomas, M. B., et al. (1986). Hydroxylapatite synthesis and characterizations in dense polycrystalline form. *J. Mater. Sci., 11*, 2029 (quoted by Kwon et al., 1986).

Jarcho, M., Kay, J. F., Gumaer, K. I., et al. (1977). Tissue: Cellular and subcellular events at a bone–ceramic hydroxylapatite interface. *J. Biol. Eng., 1*, 79 (quoted by Kwon et al., 1986).

Johnson, D. L., Kelly, J. F., Flinton, R. J., and Cornell, M. T. (1974). Histologic evaluation of vital root retention. *J. Oral Surg., 32*, 829–833.

Kabcenell, J. L. (1971). Tooth-supported complete dentures. *J. Prosthet. Dent., 26,* 251.

Kalk, W., Van Rossum, G. M. J. M., and Van Waas, M. A. J. (1990). Edentulism and preventive goals in the treatment of multilated dentition. *Int. Dent. J., 40,* 267–274 (quoted by Van Waas et al., 1993).

Kangasniemi, I. M. O., Vedel, E., de Blick-Hogerworst, J., Yli-Urpo, A. U., and de Groot, K. (1993). Dissolution and scanning electron microscopic studies of Ca,P particle-containing bioactive glasses. *J. Biomed. Mater. Res., 27,* 1225–1233.

Kangvonkit, P., Matukas, V. J., and Castleberry, D. J. (1986). Clinical evaluation of Durapatite submerged-root implants for alveolar bone preservation. *Int. J. Oral Maxillofac. Surg., 15,* 62–71.

Keipper, V. L. (1988). Snout suffocation syndrome. *N. Engl. J. Med., 319,* 1097–1098.

Kim, C. Y., Clark, A. E., and Hench, L. L. (1989). Early stages of calcium-phosphate layer formation in bioglasses. *J. Non-Cryst. Solids, 113,* 195–202.

Kim, C. Y., Kim, J., and Chang, S. Y. (1993). Hydroxyapatite formation on bioactive glass in a solution with different pH and P-ion concentration. *Trans. 19th Annual Meeting Soc. Biomaterials,* Birmingham, Abstract, p. 4.

Kokubo, T. (1990). Surface chemistry of bioactive glass-ceramics. *J. Non-Cryst. Solids, 120,* 138–151 (quoted by Hench, 1991).

Kokubo, T. (1991). Bioactive glass-ceramico-properties and applications. *Biomaterials, 12,* 156–163 (quoted by Kangasniemi et al., 1993).

Kokubo, T., and Itou, S. (1985). A new glass-ceramic for bone replacement; evaluation of its bonding to bone tissue. *J. Biomed. Mater. Res., 19,* 685–698 (quoted by Kang et al., 1993).

Kokubo, T., Ito, S., Huang, Z. T., Hayaski, T., Sakka, S., Kitsugi, T., and Yamamuro, T. (1990). Ca,P-rich layer formed on high strength bioactive glass-ceramic A-W. *J. Biomed. Mater. Res., 24,* 331–343 (quoted by Li et al., 1993).

Kokubo, T., Shigematsu, M., Nagashima, Y., et al. (1992). Apatite- and wallastonite-containing glass ceramics for prosthetic application. *Bull. Inst. Chem. Res., Kyoto Univ., 60,* 260–268.

Kopp, C. D. (1989). Branemark osseointegration: Prognosis and treatment rationale. *Dent. Clin. N. Am., 33,* 701–731 (quoted by Bahat, 1993).

Kwon, H. J., El Deeb, M., Morstad, T., and Waite, D. (1986). Alveolar ridge maintenance with hydroxylapatite ceramic cones in humans. *J. Oral Maxillofac. Surg., 44,* 503–508.

Lam, R. V. (1960). Contour changes of the alveolar processes following extractions. *J. Prosthet. Dent., 10,* 25–32.

Lam, R. V. (1972). Effect of root implants on resorption of residual ridges. *J. Prosthet. Dent., 27,* 311–323 (quoted by Kangvonkit et al., 1986).

Lam, R. V., and Poon, K. Y. (1968). Acrylic resin root implants: a preliminary report. *J. Prosthet. Dent., 19,* 506–513.

Lam, R. V., and Poon, K. Y. (1969). Acrylic resin root implants: continuing report. *J. Prosthet. Dent., 22,* 657–662.

Langer, Y., and Langer, A. (1993). Treatment of terminal dentitions: Epidemiologic and clinical aspects of overdenture application. *Compend. Contin. Educ. Dent., 14,* 876–889.

Lee, D. R., Lemons, J. E., and LeGeros, R. Z. (1989). Dissolution characterization of commercially available hydroxylapatite particulate. *Trans. Soc. Biomaterials,* Lake Buena Vista, FL, p. 161.

LeGeros, R. F., Bone, G., and LeGeros, R. (1978). Type of H_2O in human enamel and in precipitated apatites. *Calcif. Tissue Res., 26,* 111 (quoted by Hench, 1991).

Li, P., Ohtsuki, C., Kokubo, T., Nakanishi, K., and Soga, N. (1992). Apatite formation induced by silica gel in a simulated body fluid. *J. Am. Ceram. Soc., 75,* 2094–2097 (quoted by Kangasniemei et al., 1993).

Li, P., Ohtsuki, C., Kokubo, T., Nakanishi, K., Soga, N., Nakamura, T., and Yamamuro, T. (1993). Effects of ions in aqueous media on hydroxyapatite induction by silica gel and its relevance to bioactivity of bioactive glass and glass-ceramics. *J. Appl. Biomat., 4,* 221–229.

Loiselle, R. J., Crum, R. J., Rooney, G. E., and Stuever, C. H., Jr. (1972). The physiologic basis for overlay denture. *J. Prosthet. Dent., 28*, 4–11 (quoted by Langer and Langer, 1993).

Mc Grail, B. P., Pederson, L. R., and Petersen, D. A. (1986). The influence of surface potential and pH on the release of sodium from $Na_2O_3SiO_2$ glass. *Phys. Chem. Glasses, 27*, 59–64.

McQueen, D., Sundgren, J. E., Ivarsson, B., Lundstrom, B., af Ekhsatam, B., Svensson, A., Branemark, P-I., and Albrektsson, T. (1982). Auger electron spectroscopic studies of titanium implants. In *Clinical Applications of Biomaterials* (A. J. C. Lee, T. Albrektsson, and P-I. Branemark, eds.). Wiley and Sons, New York, p. 179 (quoted by Brunski, 1988).

Meffert, R. M. (1993). *How to maintain the dental implant (Part I) and How to save the ailing, failing implant (Part II)*. Oral presentation, Florida National Dental Congress, June 4, Orlando, FL.

Meskin, L. H., Brown, L. J., Brunelle, J. A., and Warren, G. B. (1988). Patterns of tooth loss and accumulated prosthetic treatment potential in U.S. employed adults and seniors 1985–86. *Gerodontics, 4*, 126–135 (quoted by Ettinger, 1993).

Miller, P. A. (1958). Complete dentures supported by natural teeth. *J. Prosthet. Dent., 8*, 924–928 (quoted by Brewer and Morrow, 1980).

Mills, B. G., Chan, C. C., Voss, R., and Grenoble, D. E. (1974). Effects of vitreous carbon implants on alveolar bone morphology. IADR abstracts, No. 298.

Misch, C. E. (1984). Core-Vent implant: An alternative treatment for the edentulous mandible. *J. Mich. Dent. Assoc., 66*, 219–223.

Misiek, D. J., Kent, J. N., and Carr, R. F. (1984). Soft tissue responses to hydroxylapatite particles of different shapes. *J. Oral Maxillofac. Surg., 42*, 150–160.

Mohammed, H., Atmaram, G. H., and Schoen, F. J. (1977). Photoelastic stress analysis of single tooth implants with different root configurations. *Biomed. Mater. Res. Symp. Trans., 1*, 71 (quoted by Hench and Ethridge, 1982).

Murray, C. G., and Adkins, K. F. (1979). The elective retention of vital root for alveolar bone preservation: A pilot study. *J. Oral Surg., 37*, 650–656.

Mushimoto, E. (1981). The role in masseter muscle activities of functionally elicited periodontal afferents from abutment teeth under overdentures. *J. Oral Rehabil., 44*, 441–455 (quoted by Langer and Langer, 1993).

Nagasawa, T., Okane, H., and Tsuru, H. (1979). The role of the peridontal ligament in overdenture treatment. *J. Prosthet. Dent., 42*, 12–16 (quoted by Langer and Langer, 1993).

Neufeld, J. O. (1958). Changes in trabecular pattern of the mandible following loss of teeth. *J. Prosthet. Dent., 8*, 865 (quoted by Quinn and Kent, 1984).

Nevins, M., and Langer, B. (1993). The successful application of osseointegrated implants to the posterior jaw: A long term retrospective study. *Int. J. Oral Maxillofac. Implants, 8*, 428–432.

Newman, M. G., and Flemmig, T. F. (1988). Periodontal considerations of implants and implant-associated microbiota. *NIH Consensus Development Conference on Dental Implants*, June 13–15, Bethesda, MD, pp. 57–64.

Nottingham Review, 1826 (cited by Veldhuis et al., 1984).

Ogino, M., and Hench, L. L. (1980). Formation of calcium phosphate films on silicate glasses. *J. Non-Cryst. Solids, 38–39*, 673–678.

Ogino, N., Ohuchi, F., and Hench, L. L. (1980). Compositional dependence of the formation of calcium phosphate fibers on Bioglass®. *J. Biomed. Mater. Res., 14*, 55–64 (quoted by Hench, 1991).

O'Neal, R. B., Gound, T., Levin, M. P., and del Rio, C. E. (1978). Submergence of roots for alveolar bone preservation. I. Endodontically treated roots. *Oral Surg., 45*, 803–810.

Ortman, H. R. (1982). Factor of bone resorption of the residual ridge. *J. Prosthet. Dent., 12*, 440.

Pantano, C. G., Jr., Clark, A. E., Jr., and Hench, L. L. (1974). Multilayer corrosion films on glass surfaces. *J. Am. Ceram. Soc., 57*, 412–413.

Papapoulos, S. E. (1989). *J. Bone Min. Res., 4*, 775–781 (quoted by Denisson et al., 1993).

Pierce, W. S. (1989). Cardiothoracic surgery. *J. Am. Med. Assoc., 261*, 2827–2829.

Piotrokovski, J., and Massler, M. (1967). Alveolar ridge resorption following tooth extraction. *J. Prosthet. Dent., 12,* 21–27.

Piotrowski, G., Hench, L. L., Allen, W. C., and Miller, G. J. (1975). Mechanical studies of the bone bioglass interfacial bond. *J. Biomed. Mater. Res., 6,* 47–61 (quoted by Hench and Ethridge, 1982).

Prieskel, H. W. (1985). Precision attachments. In *Prosthodontics: Overdentures and Telescopic Prostheses,* Vol. 2. Quintessence Publishing Co., Chicago, (quoted by Langer and Langer, 1993).

Quinn, J. H., and Kent, J. N. (1984). Alveolar ridge maintenance with solid nonporous hydroxylapatite root implants. *Oral Surg., 85,* 511–521.

Quinn, J. H., Kent, J. N., Hunter, R. G., and Schaffer, C. M. (1985). Preservation of the alveolar ridge with hydroxylapatite tooth root substitutes. *J. Am. Dent. Assoc., 110,* 189–193.

Raisz, L. G. (1985). Personal communication.

Rissin, L., House, J. E., Manly, R. S., and Kapur, K. K. (1978). Clinical comparison of masticatory performance and electromyographic activity of patients with complete dentures, overdentures and natural teeth. *J. Prosthet. Dent., 39,* 508–511 (quoted by Langer and Langer, 1993).

Rohrer, M., Swift, J., and Penwick, R. (1988). The Donath technique used for comparing alloplastic and allogenic implants and autogenous grafts in dogs. *Oral Surg., 66,* 572 (abstract).

Rothstein, S. S. (1984a). Use of durapatite for the rehabilitation of resorbed alveolar ridges. *J. Am. Dent. Assoc., 109,* 571–574.

Rothstein, S. S. (1984b). Appropriate use of the term durapatite. *J. Oral Maxillofac. Surg., 42,* 141.

Sattayasanskul, W., Brook, I. M., and Lamb, D. J. (1988). Dense hydroxyapatite root replica implantation: Measurement of mandibular ridge preservation. *Int. J. Oral Maxillofac. Implants, 3,* 203–207.

Schnitman, P. A., Rubenstein, J. E., Woehrle, P. S., DaSilva, J. D., and Koch, G. G. (1988). Implants for partial edentulism. *NIH Consensus Development Conference on Dental Implants,* pp. 53–56.

Shulman, L. B. (1988). Surgical considerations in implant dentistry. *NIH Consensus Development Conference on Dental Implants,* pp. 43–48.

Simon, J. H., and Kimura, J. T. (1974). Maintenance of alveolar bone by the intentional replantation of roots. *Oral Surg., 37,* 936–945 (quoted by Kangvonkit et al., 1986).

Simon, J. H., and Kimura, J. T. (1975). Histologic observation of endodontically treated replanted roots. *J. Endocrinol., 1,* 178–181 (quoted by Kangvonkit et al., 1986).

Smiler, D. G. (1990). Socket-lift procedure with immediate placement of an implant in an extraction socket. *Implant Digest,* Summer, pp. 2–3.

Sobolik, D. F. (1980). Alveolar bone resorption. *J. Prosthet. Dent., 10,* 612–619.

Stanley, H. R. (1987). Implantology and the development of Bioglass. In *The 1987 Dental Annual* (D. D. Derrick, ed.). John Wright & Sons, Bristol, pp. 209–230.

Stanley, H. R., Brandt, R. L., Hall, M., Clark, A., and King, C. (1994). Alveolar ridge maintenance using endosseously placed Bioglass® cones (4.5 yr. evaluation). *J. Dent. Res., 73,* Special Issue, Abstract 2376, p. 399.

Stanley, H. R., Clark, A. E., Hall, M. B., King, C., Colaizzi, F., Spilman, D., and Hench, L. L. (1986). *Clinical trials of bioglass implants for alveolar ridge maintenance.* Presented at 12th Annual Meeting of the Society for Biomaterials, Minneapolis.

Stanley, H. R., Hall, M. B., Colaizzi, F., and Clark, A. E. (1987). Residual alveolar ridge maintenance with a new endosseous implant material. *J. Pros. Dent., 58,* 607–613.

Stanley, H. R., Hall, M., Hench, L. L., King, C., Colaizzi, F., and Wilson, J. (1983). Research protocol and consent form for project entitled: Preservation of alveolar ridge with the intraosseous implantation of root-shaped cones made of Bioglass. University of Florida, J. H. Miller Health Center, Gainesville, FL.

Stanley, H. R., Hench, L. L., Bennett, C. G., Jr., Chellemi, S. J., King, C. J., III, Going, R. E.,

Ingersoll, N. J., Ethridge, E. C., Kreutziger, K. L., Loeb, L., and Clark, A. E. (1981). The implantation of natural form Bioglass in baboons – Long term results. *Int. J. Oral Implant, 2,* 26–38.

Stanley, H. R., Hench, L. L., Going, R., Bennett, C., Chellemi, S. J., King, C., Ingersoll, N., Ethridge, E., and Kreutziger, K. (1976). The implantation of natural tooth form bioglasses in baboons. *Oral Surg., 42,* 339–356.

Steen, W. H. A. (1984). *Measuring mandibular ridge reduction.* Ph.D. thesis, University of Ultrecht, The Netherlands (quoted by Van Waas et al., 1993).

Swanson, K., and Madison S. (1987). An evaluation of coronal microleakage in endodontically treated teeth. Part I: Time periods. *J. Endod., 13,* 56–59 (quoted by Langer and Langer, 1993).

Tallgren, A. (1967). The effect of denture wearing on facial morphology; a 7-year longitudinal study. *Acta Odontol. Scand., 25,* 563–592.

Tallgren, A. (1969). Positional changes of complete dentures; a 7-year longitudinal study. *Acta Odontol. Scand., 27,* 539–561.

Tallgren, A. (1972). The continuing reduction of the residual alveolar ridges in complete denture wearers: A mixed-longitudinal study covering 25 years. *J. Prosthet. Dent., 27,* 120–132.

Toolson, L. B., and Smith, D. E. (1978). A 2-year longitudinal study of overdenture patients: Part I: Incidence and control of caries on overdenture abutments. *J. Prosthet. Dent., 40,* 486–491 (quoted by Langer and Langer, 1993).

Toolson, L. B., Smith, D. E., and Phillips, C. (1982). A 2-year longitudinal study of overdenture patients. Part II: assessment of the periodontal health of overdenture abutments. *J. Prosthet. Dent., 4,* 4–11 (quoted by Langer and Langer, 1993).

Toolson, L. B., and Taylor, T. D. (1989). A 10-year report of a longitudinal recall of overdenture patients. *J. Prosthet. Dent., 62,* 179–181 (quoted by Lander and Lander, 1993).

van Blitterswijik, C. A., Leenders, H., Brink, J. v.D., Bovell, Y. P., Gallard, M., Heintze, P., and Bakker, D. (1992). Calcium phosphate precipitation causes the bone-bonding of PEO/PBT copolymers (Polyactive™). *Trans. 19th Annual Meeting, Soc. Biomaterials,* Birmingham, p. 297 (abstract).

Van Steenberghe, D. (1988). Periodontal aspects of osseointegrated oral implants modum Branemark. *Dent. Clin. N. Am., 32,* 355–370.

Van Waas, M. A. J., Jonkman, R. E. G., Kalk, W., Van'T Hof, M. A., Plooij, J., and Van Os, J. H. (1993). Differences two years after tooth extraction in mandibular bone reduction in patients treated with immediate overdentures or with immediate complete dentures. *J. Dent. Res., 72,* 1001–1004.

Veldhuis, H. Personal communication, 1988.

Veldhuis, H., Driessen, T., Denissen, H., and de Groot, K. (1984). A 5-year evaluation of apatite tooth roots as means to reduce residual ridge resorption. *Clin. Prev. Dent., 6,* 5–8.

von Wowern, N., Harder, F., Hjorting-Hansen, E., and Gotfredsen, K. (1990). ITI-implants with overdentures: A prevention of bone loss in edentulous mandibles. *Int. J. Oral Maxillofac. Implants, 5,* 135–139 (quoted by Denissen et al., 1993).

Vrouwenvelder, W. C. A., Groot, C. G., and de Groot, K. (1993). Histological and biochemical evaluation of osteoblasts cultured on bioactive glass, hydroxylapatite, titanium alloy, and stainless steel. *J. Biomed. Mater. Res., 27,* 465–475.

Waerhaug, J., and Loe, H. (1958). Tissue reaction to self-curing acrylic resin implants. *D. Pract. D. Rec., 8,* 234–240 (quoted by Lam and Poon, 1968).

Waerhaug, J., and Zander, H. A. (1956). Implantation of acrylic roots in tooth sockets. *Oral Surg., 9,* 46–54 (quoted by Lam and Poon, 1968).

Wahl, G. (1986). Postoperative Knochenstoffwechselaktivitaten und Ihre Bedeutung bei der Belastung von Implantaten. *Z. Zahnarzt. Implantol., 2,* 140–144 (quoted by Newman and Flemmig, 1988).

Weinstein, A. M., Klawitter, J. J., and Cook, S. D. (1980). Implant–bone characteristics of Bioglass dental implants. *J. Biomed. Mater. Res., 14,* 23–29.

Weintraub, J. A., and Burt, B. A. (1985). Oral health status in the United States: Tooth loss and edentulism. *J. Dent. Educ., 49*, 368–376 (quoted by Ettinger, 1993).

Weiss, C. M. (1988). Fibro-osteal and osteal integration. *NIH Consensus Development Conference on Dental Implants*, pp. 37–41.

Welker, W. A., Jividen, G. J., and Kramer, D. C. (1978). Preventive prosthodontics — Mucosal coverage of roots. *J. Prosthet. Dent., 40*, 619–621.

Welsh, R. P., Pilliar, R. M., and MacNab, J. (1971). Surgical implants — role of surface porosity in fixation to bone and acrylic. *J. Bone Jt. Surg.*, Am. Vol. 53A, 963–977 (quoted by Hench and Ethridge, 1982).

West, J. K., and Hench, L. L. (1992). Reaction kinetics of bioactive ceramics. Part V: Molecular orbital modelling of bioactive glass surface reaction: In *Bioceramics — 5* (T. Yamamuro, T. Kokubo, and T. Nakamura, eds.), Kyoto Kokunshi Kankakai, Kyoto, pp. 75–86 (quoted by Li et al., 1993).

West, J. K., and Hench, L. L. (1993). Molecular orbital modeling of bioactive glass reactions of stages 3 and 4. *Trans. 19th Annual Meeting Soc. Biomaterials*, Birmingham, p. 2 (abstract).

Whitaker, D. D., and Shakle, R. J. (1974). A study of the histologic reaction of submerged root segments. *Oral Surg., 37*, 919–935.

Wilson, J., and Low, S. B. (1992). Bioactive ceramics for periodontal treatment: Comparative studies in the Patus monkey. *J. Appl. Biomater., 3*, 123–129.

Zarb, G. A. (1988). Implants for edentulous patients. *NIH Consensus Development Conference on Dental Implants*, pp. 49–52.

54
Single-Crystal Alumina (Sapphire) for Dental Implants

Akiyoshi Yamagami
Kyoto Institute of Implantology, Kyoto, Japan
Haruyuki Kawahara
Institute of Clinical Materials, Osaka, Japan

I. INTRODUCTION

Dental implants have been made of metallic materials—316 stainless steel, Co–Cr–Mo alloy, pure titanium, titanium alloys, tantalum; ceramic materials—alumina (Al_2O_3), zirconia, hydroxyapatite, bioactive glass/ceramics; and composite materials and other materials. Ceramic implants, especially porous alumina implants reinforced with single-crystal alumina (Al_2O_3), have been studied for clinical applications by many researchers. Alumina ceramics are highly stable as biomaterials in the human body. They cause neither tissue irritation nor toxicity, and are superior in adhesion to cells and surrounding tissues. Polycrystalline alumina has some disadvantage in mechanical strength; it can be fractured when inserted as an artificial dental root with a small diameter.

Various kinds of polycrystalline alumina implants have been developed: in Switzerland (1965) by Sandhaus [1], in the United States (1973) by Driskell [2], in Germany (1976) by Schulte and Heimke [3], and (1977) by Mutschelknaus and Dörre [4]. These materials have been widely used in clinical applications. In Japan, Kawahara, Yamagami, and Hirabayashi developed a single-crystal alumina that has bending strength three times higher than that of polycrystalline alumina. Data on this material were presented at an ICOI meeting [6].

Since their introduction, various types of single-crystal alumina sapphire implants (BIOCERAM®) have been manufactured in cooperation with many researchers, and they are now widely applied in orthopedic and dental applications.

Animal studies on and clinical applications of these single-crystal alumina and alumina porous implants are discussed in this chapter.

II. PHYSICAL AND BIOLOGICAL PROPERTIES OF ALUMINA CERAMICS

There are many types of ceramics for dental use, and their features vary according to their applications. Kawahara [7] has classified alumina ceramics as bioinert materials, and Hulbert [8] as nearly bioinert. In any case, there is no doubt that alumina ceramics are among the most stable biomaterials. Being used as dental implants, the biomaterials are required to be both biomechanically strong and biologically stable. The characteristics of alumina ceramics are as follows:

1. Physicochemical stability within the body
2. Hydrophilic surface
3. Low thermoconductivity
4. Low electroconductivity
5. High mechanical strength and hardness
6. Wear resistant
7. Aesthetically pleasing

However, alumina has some disadvantages in designing dental implants. It cannot be custom made easily and it cannot be bent like metal. It has high modulus of elasticity compared with bone. Consequently, if an excess stress is loaded, a wide radiolucent zone develops around the implant because of bone resorption and an increase in connective tissue at the implant–bone interface. However, if the implant has a suitable loading of functional stress, it can maintain functional fixation with biodynamic responses of bone resorption and remodeling.

The composition of single-crystal alumina is as follows:

Composition: Al_2O_3
Crystal system: Hexagonal closest-packed structure
Density: 4.0 g/cc
Melting point: 2053 °C

Table 1 summarizes the mechanical strength of bioceramics. As for bending strength, single-crystal alumina is about three times as strong as polycrystalline alumina, so it can be used in small-diameter designs. Yamagami, Kawahara, and Hirabayashi [6]; Kawahara, Yamagami et al. [9–12]; Yamane [13]; Nishijima et al. [14]; McKinney and Koth [15]; Sugimoto [16]; Kawahara [17]; and Steflik, McKinney, and Koth [18] have reported its suitable compatibility to alveolar bone and gingival epithelium.

Table 1   Mechanical Strength of Bioceramics

Materials	Bending strength (kg/cm²)	Compressive strength (kg/cm²)	Elastic modulus (×10⁴ kg/cm²)
Polycrystalline alumina	3,500	10,000	371
Single-crystal alumina (sapphire)	13,000	30,000	385
Polycrystalline zirconia	8,000	35,000	240
Hydroxyapatite (HAP)	2,200	6,000	35
Human bone	300–1,900	900–2,300	17

Figure 1 shows a rhesus monkey mandible. A BIOCERAM W-type implant was inserted in the posterior area of the mandible and the implant was sectioned buccolingually over the post portion 6 months postimplantation. It is observed that the pocket depth of the mandible is very shallow and that the adhesion of gingival tissue is excellent. From the SEM image in Fig. 2, it is clear that the bone tissue is well adhered to the surface of the sapphire implant. Kawahara [17] reported that when water adsorbs on the surface of highly polished bioinert ceramics, the water changes into a frozen structure and creates an amorphous water layer. This amorphous water layer builds hydrogen bonds between the ceramics and surrounding tissues.

We have confirmed that alumina ceramics are superior in epithelial tissue adhesion, using tissue cultures and animal experiments, compared with metallic implant materials.

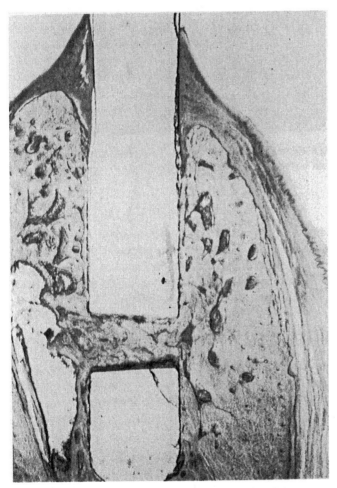

Figure 1 Tissue surrounding a BIOCERAM® plate-type implant. BIOCERAM® W type was implanted in mandible posterior area of a rhesus monkey and metal crown was set. The mandible was buccolingually sliced. Occlusal force was applied at 6 months postimplantation. Gingival sulcus is shallow; no epithelial tissue downgrowth can be observed; biological seal is excellent (HE stain). New bone growth is observed around the implant surface; some connective tissue is also observed.

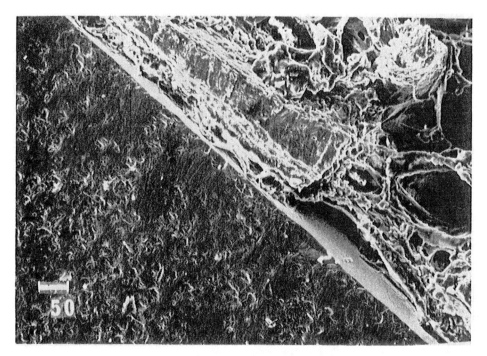

Figure 2 Tissue surrounding a BIOCERAM® plate-type implant. This SEM image shows that the bone tissue is attaching to the smooth surface of the BIOCERAM® implant.

Regarding this adhesion mechanism, Steflik et al. [18] have reported a transmission electron micrograph (TEM) image showing hemidesmosome attachment. Also, this study indicated that the measurement values at the implant cervix, evaluated by clinical probing and PERIOTRON® analysis, were smaller than those of natural teeth. Because of this observation, it seems that the epithelial tissue attachment is excellent.

III. VARIOUS TYPES OF SINGLE-CRYSTAL ALUMINA IMPLANTS

Designs of single-crystal alumina (sapphire) implants have been improved since 1975, and their clinical applications have been widened. They have been manufactured as endodontic endosteal implants, endosteal implants, and intramucosal inserts.

A. Endodontic Endosteal Implants

Endodontic endosteal implants are used to maintain and stabilize mobile teeth by inserting a pin through the tooth to the bone for the purpose of improving the crown-and-root ratio. The following types of these implants have been developed by Yamagami and Kawahara [6] (Osaka Dental University, Department of Biomaterials), IRG (Implant Research Group) of Osaka Dental University (Department of Biomaterials), and by Sekiya, Nakamura, Kuroda, Asai, Shimodaira, and others: (1) tapered pin, (2) anchor pin, (3) post pin, (4) lock pin.

B. Endosteal Implants

Various designs of endosteal implants have been developed by Kawahara, Yamagami, and IRG of Osaka Dental University, the Yamane study group, Kyoto Institute of Implantology, and by members of the BIOCERAM oral implantology study group. These designs have been used for patients by many dentists.

BIOCERAM endosteal implants are classified as screw type and plate type. Applications of endosteal implants are based on anatomical shapes and edentulous areas; and, of course, the patient's age, sex, occupation, physical and psychological state, and local conditions. When anatomical diagnosis by x-ray is done, it must be recognized that the x-rays are not real images but shadows. It is important that they should always be taken with a 5-mm precision stainless steel ball bearing at the intended implant site. The true vertical dimension of bone is then calculated. It is also necessary to determine the condition of the gingiva and the shape of the bone by optical observation and palpation.

1. Screw-Type Implants

The screw-type has three variations: S, E, and A (see Fig. 3). They are used as abutments for small edentulous cases or intermediary cases. A variety of screw diameters are available, from 3 mm to 5 mm; a size is selected based on the buccolingual width of alveolar bone. It is ideal to keep a ratio of 2 : 1 for the implant length below the bone crest and the crown portion (length from bone crest to opposing teeth), and the procedure is considered indicated when the implant has at least a ratio of 3 : 2.

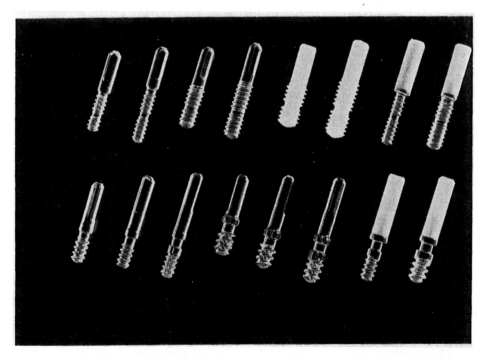

Figure 3 Screw-type BIOCERAM® implants. From upper left: 3S1S, 3S1L, 4S1S, 4S1L, 5S1S, 5S1L, 3A0L, 4A0L; and from lower left: 3E1S, 3E1M, 3E1L, 4E1S, 4E1M, 4E1L, 3A1L, 4A1L.

2. *Plate Type Implants*

The surgical procedure for plate-type implants is similar to that for metal blade and anchor-type implants. These plate implants are effectively used as abutments for free-end bridges or intermediate cases. The buccolingual width of a plate-type implant is 1.8 mm; therefore, they are applicable even when the buccolingual width of alveolar ridge is small. This type has three variations—T, U, and W (see Fig. 4), which differ in mesiodistal length. The criterion for its use is the mesiodistal space of applicable alveolar bone.

3. *Porous Alumina Implants*

Porous alumina implants were developed by Kawahara, Yamagami, and Hirabayashi [30]. They are made of single-crystal alumina cylindrical core around which porous poly-crystalline alumina is fused at high temperature. The core of the prototype was made of polycrystalline alumina but it did not have enough strength. By applying single-crystal alumina for the core, this implant was made strong enough to bear occlusal force. Details on this implant will be reported later.

C. Intramucosal Inserts

BIOCERAM buttons are made of polycrystalline alumina. Compared to conventional metal inserts, these are light, and they present neither any corrosion possibility in the oral cavity nor any gustatory problem caused by metal ionization. They are also excellent in oral hygiene. (See Fig. 5.)

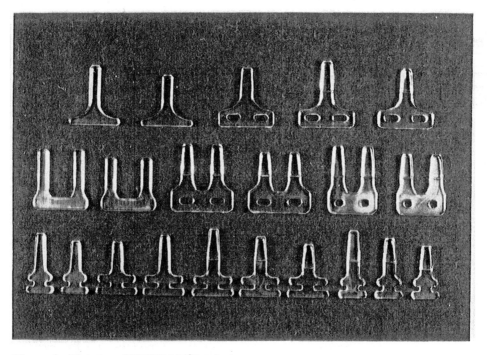

Figure 4   Plate-type BIOCERAM® implants.

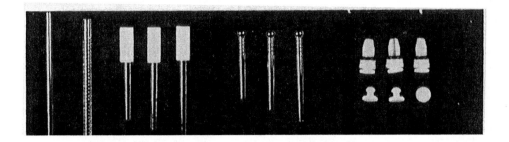

Figure 5   Endodontic endosteal implants and intramucosal inserts of BIOCERAM®.

As mentioned above, alumina ceramic implants exist in a great variety of types. That is because one design alone is unable to cover all edentulous sites for oral rehabilitation. For this reason, the diagnosis of implant indication and the criteria for selecting a particular implant design become very significant. Most implant failures can be attributed to maldiagnosis of the indications. For example, if a large implant is placed where the quality of bone is not sufficient, it will probably fail. For diagnosis of implant indication, it is recommended that reference be made to the revised edition of *BIOCERAM Sapphire Implant Manual* [19], as space in this chapter is too limited to cover the details.

IV. STATISTICAL ANALYSIS OF CLINICAL INVESTIGATIONS

BIOCERAM sapphire implants have been developed for the treatment of partial edentulous cases. They have been used in connection with natural teeth except porous-type and single-tooth cases. We made a retrospective clinical study of patients whose implants and prostheses were placed from April 1975 to December 1991, extending over 16 years for the longest case. In this report, we discuss the indications and contraindications for implants applying to partial edentulous cases.

A. Evaluation Method

Patients were chosen basically from those who requested implants for their partial edentulous condition. Evaluation and diagnosis of patients — such as systemic, intraoral, and psychological condition; financial situation; and so on — were performed carefully before a determination of the optimum treatment procedure. As a general rule, implant operations were carried out after decay, periodontal, and other treatments had been performed. Surgery was carried out following the standard procedure for BIOCERAM implants.

After the surgery, a temporary crown or bridge was placed. After confirming the stability of implants, a few weeks to 3 months after the surgery, a final prosthesis was cemented. It is generally agreed that patients with dental implants should be provided maintenance care at 3-month intervals: EPP, tissue condition by probing, gingival crevicular fluid volume check by PERIOTRON, and mobility check (PERIOTEST®) were performed. Dental x-rays around implants was taken every 6 months or 1 year. Also, panoramic x-rays and oral photographs were taken every other year.

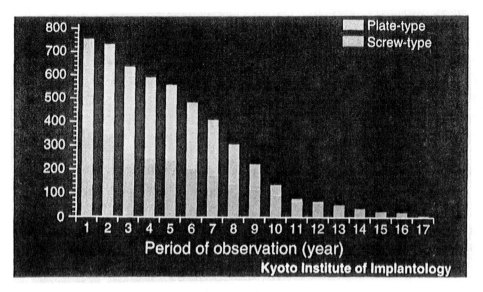

Figure 6   Clinical application of BIOCERAM® implants from 1975 to 1991.

B.   The Criteria for Removal

Conditions that require implant removal are:

1. Chronic occlusal or percussive pain related to the implant devices, which continued after a temporal mitigation of occlusal force or could not be alleviated with adequate treatment.
2. Excess buccolingual, and mesiodistal and vertical mobility of the device: more than 1.0 mm.
3. Inflammatory gingivitis, tissue markedly red, edematous, bleeding on finger pressure or spontaneously.
4. Progressive bone resorption observed radiographically around the implant.
5. Aesthetic problems.
6. Continuous use is impossible because of damage to the implant.

X-rays should be taken periodically. If bone resorption is observed around an implant, these should be compared with the x-rays taken at the surgery. We have also computed a quantitative analysis, but we do not refer to this in this chapter.

Some implants were removed as failure cases according to the criteria listed above. If the mobility of a device is in the range 0.5–1.0 mm, it may be saved by removal of old granulation tissue or occlusal equilibration. The implants were not classified as failures at the last observation period. Those impossible to follow up because of patients' relocation, death, or other reasons were removed from the statistics. The criteria of success used in the study were based on the NIH-Harvard Consensus Development Conference.

Sex, age, and anatomical location were analyzed. Also, prosthesis shape, number of abutments, and the condition of opposing teeth were categorized. The relationship between survival term and implant survival rate was examined.

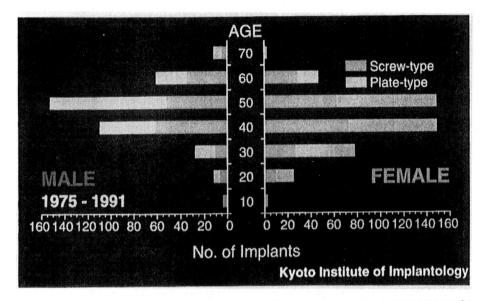

Figure 7   Distribution of patient age, sex, and implant type (screw, plate). BIOCERAM® implants were placed from 1975 to 1991. The age at which the greatest number of patients were indicated for implants was: male, the 50s; female, the 40s.

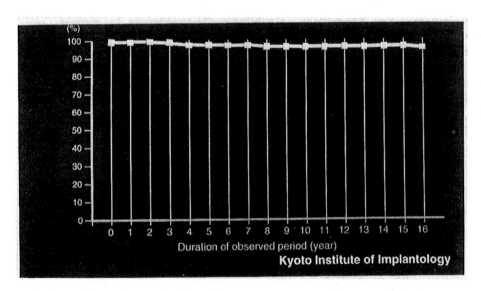

Figure 8   Long-term cumulative survival rate of BIOCERAM® screw-type implants.

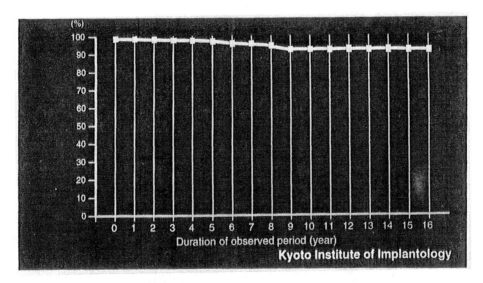

Figure 9   Long-term cumulative survival rate of BIOCERAM® plate-type implants.

C.   Survival Rates

In the period 1975–1991, 337 pieces of screw-type BIOCERAM made of single-crystal alumina or polycrystalline alumina and 452 pieces of the plate type were used, mainly for reconstructing partial edentulous cases.

This study evaluated 536 patients (screw type: 377 implants in 237 patients, plate type: 452 implants in 299 patients). (See Figs. 6–9 and Tables 2 and 3.) At least two postoperative examinations were performed after implant and prosthesis placement. Regarding the 452 implants/299 patients of plate-type BIOCERAM that were placed during this period: seven patients (8 pieces) were not sufficiently followed or dropped out throughout a maximum of 15 years. Thirteen pieces out of 444 implants in the rest (8 male patients, 3 female patients) were removed since they exceeded the criteria for removal and were judged as failures. Since 5 male patients died after the final evaluation, the follow-up was discontinued on 8 pieces (5 patients) of plate type and 2 pieces (2 patients) of screw type.

The accumulative survival rate for all plate-type pieces was 98.4% ($n = 383$) at 1 year, 96.6% ($n = 217$) at 5 years, and 94.3% ($n = 55$) at 10 years. Of BIOCERAM screw type, 314 pieces were placed and follow-up on 1 piece (1 patient) was discontinued

Table 2   BIOCERAM® Screw-Type Implants (1975–1991)

All cases	Number of patients	Average age	Number of implants
Male	116	50.5	174
Female	121	47.6	203
Total	237	49.0	377

Table 3 BIOCERAM® Plate-Type Implants (1975–1991)

All cases	Number of patients	Average age	Number of implants
Male	136	52.7	204
Female	163	47.5	248
Total	299	49.7	452

during the period. Six pieces (4 patients) were removed. An accumulative survival rate of more than 94% was obtained at 15 years for both screw and plate types, and no significant difference was observed between screw and plate types.

The accumulative survival rate was 99.2% ($n = 252$) at 1 year, 97.8% ($n = 137$) at 5 years, and 95.1% ($n = 60$) at 10 years. As shown in Fig. 2 indicating sex, age, and applied implant type, the implants were applied to healthy patients from teenage to more than 70 years of age. The age at which the greatest number of patients were indicated for implants, was 40–50. For male, the 50s had the greatest number of patients and for females, it was the 40s. Regarding implant site, the applied number of screw type was large for anterior sites in maxilla and posterior sites in mandible, and was small for posterior sites in maxilla and anterior sites in mandible. On the other hand, the applied number of plate type was large for posterior sites of both maxilla and mandible. That is attributed to anatomical configuration of bone and its quantity. It leads us to the conclu-

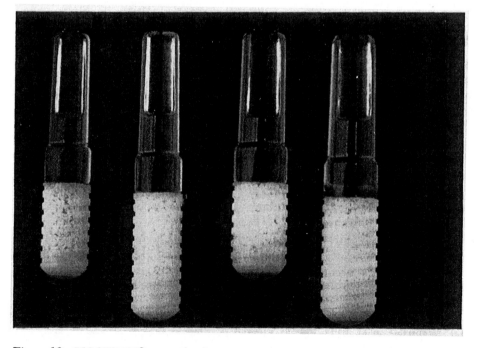

Figure 10 BIOCERAM® porous implants.

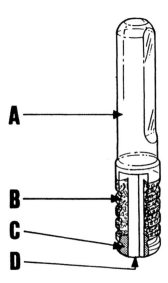

A

B

C

D

Figure 11   Basic design of BIOCERAM® porous implants. Design for animal studies is illustrated diagrammatically. Porous alumina layer and apex portion illustrated with fine dots are made of polycrystalline alumina (B, C). Screw is designed for self-tapping. Root core is made of single-crystal alumina (D). (A) Cervical portion.

sion that a single implant design can never be applied to all edentulous cases. Many researchers [20–22] have reported good results in clinical applications.

V.   POROUS ALUMINA CERAMIC IMPLANTS

Various kinds of studies on porous implants have been reported by many researchers. According to the report of Klawitter and Hulbert [23] on studies of cases using calcium aluminate, ingrowth of fibrous tissue was observed when the pore size was 5–15 μm, slight ingrowth of cortical bone tissue was observed in the cases of 44–100 μm, and when the pore size was more than 100 μm, ingrowth of new bone was observed. Hammer et al. [24] reported that when the pore size was more than 150 μm, formation of haversian canal structures was possible. Predecki, Stephan, Auslaender [25] studied the dependence of ingrowth kinetics of bone tissue on the pore diameter. They reported that in the cases of 500 μm, the invasion speed of the bone ingrowth was quickened if the diameter was widened, and delayed if it was contracted; but if the pore size was under 500 μm, diameter dependency was scarcely observed.

Pedersen, Haanaes, and Lyng [26] have implanted porous alumina implants in rhesus monkey mandible and made observations of tissue samples and images by microradiography. According to their evaluation of porous alumina implants, both fibrous connective tissue and bone tissue invaded the pores, and there was no adverse influence on surrounding tissues.

Attempts to apply these porous alumina implants to clinical use have been made by Klawitter, Weinstein et al. [27], and by Schneider, Lutz et al. [28]. The failures in the previous studies of the porous implant (of similar pore-size distribution [27–29]) can be

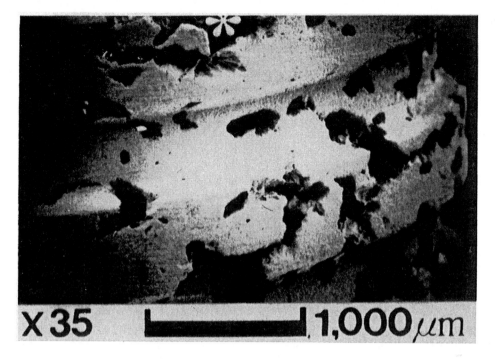

Figure 12 Surface structure of porous layer (SEM image). Pore with various kinds of holes can be seen on the surface of pore. (Original magnification ×35.)

(b)

100μ

(a)

Figure 13 Longitudinal section of BIOCERAM® porous implant. Porous portion representing pore structure that interconnects three dimensionally. (a) single-crystal alumina core; (b) polycrystalline porous alumina.

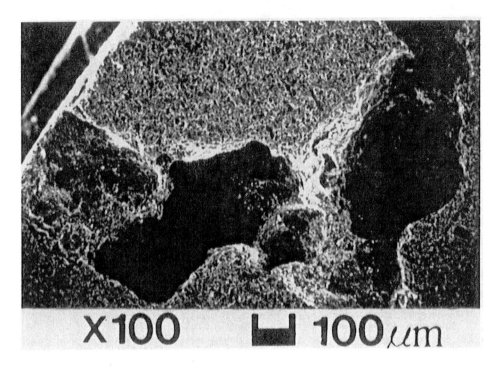

Figure 14 Horizontal section of porous alumina implant (SEM image). Pore running three dimensionally through the inside of porous implant (×100). This implant contains both macro- and micropores.

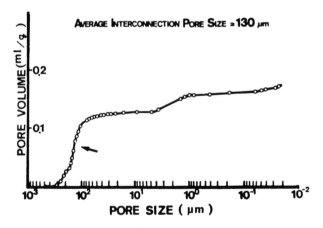

Figure 15 Porosity of porous alumina implant.

explained by the presence of micropores on the cervical portion. The porous portion was contaminated in the gingiva by invasion of oral microorganisms.

The one failure in this animal experiment was attributed to technical error at surgery, resulting in a part of the porous layer being exposed to the oral cavity [26]. This suggests that the anatomic suitability of the design as well as the selection of the implant materials is indispensable in establishing an adequate biological seal by deep insertion into alveolar bone.

A.   BIOCERAM Porous Implants

As shown in Figs. 10–15, a porous alumina implant was designed by Kawahara, Yamagami, and Hirabayashi [30–37]. The implant is composed of a core of cylindrical single-

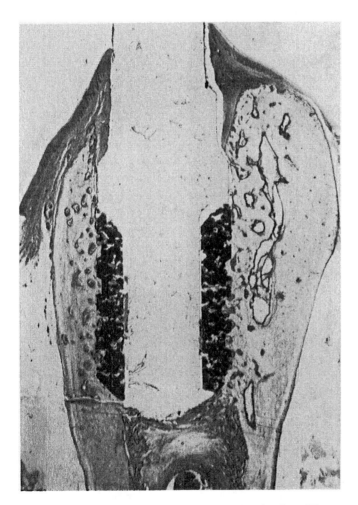

Figure 16   Tissue surrounding porous alumina implant. Three months after the implantation of a rhesus monkey: bone tissue, osteoid, and fibrous tissue are growing into various sizes of pores and the surface of the porous implant.

crystal alumina, an outer porous layer, and a smooth apex made of polycrystalline alumina. The porosity, interconnecting pore size, and range of pore sizes were evaluated by water-replacement and mercury intrusion porosity analysis, which showed the following results: volume porosity = 35%; average interconnecting pore size = 130 μm; and the range of pore sizes = 10–300 μm. Most of the pores are not closed but are open pores.

For the animal studies, specially designed porous alumina implants were used. These implants consist of a total length of 20 mm, a 7 mm long porous portion, and a 4 mm diameter. The porous layer is 1 mm wide.

For human clinical studies, four types of porous implants are available on the market. There are two diameters of porous implants, 4.2 mm and 4.8 mm; and two lengths, short (20.5 mm) and long (24.0 mm). Four variations, therefore, are available. The length of the root portion in the short type is 12.5 mm; and in the long type, 16.0 mm. If they are

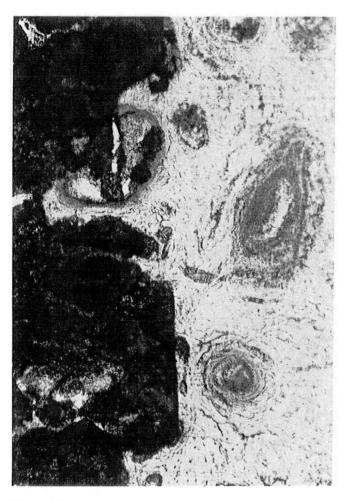

Figure 17 Bone interfacing an alumina porous ceramic implant. New bone ingrowth deep into pore. (Enlargement of Fig. 16.)

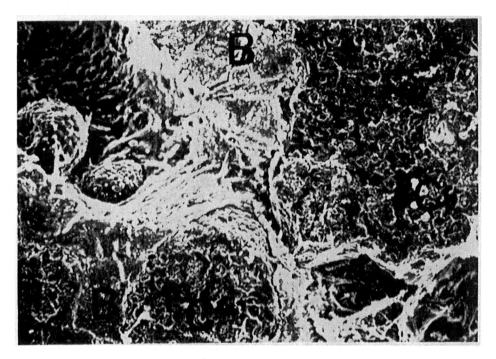

Figure 18   Bone tissue inside porous pores. Direct contact of bone tissue to ceramic surface is observed (SEM image). (B) bone tissue; (C) alumina ceramics.

placed properly according to standardized surgical procedure, the length of the post above the alveolus is 8.0 mm in all cases.

B.   Animal Studies

Implants were placed in extracted premolar and molar sites in 15 male rhesus monkeys (average age 5.1 years). The implants were placed 3 months after tooth extraction. All the opposing teeth were natural and the implants were free-standing at all times. A gold alloy crown was mounted on the implant.

The implantations were followed up by clinical examination, x-ray examination, assessment of the pocket depth, and a diagnosis of mobility and gingival color. After 4, 6, and 8 months the monkeys were sacrificed randomly. Tissue reaction was evaluated by light microscopy (LM), scanning electron microscopy (SEM), and electron probe microanalysis (EPM). (See Figs. 16–20.)

C.   Clinical Applications

Clinical applications of the BIOCERAM porous implant have been made in 90 patients for a total of 148 implants (male 53/99, female 37/49), from 1977 to 1991. (Figures 21 and 22, and Table 4 provide a summary of results.) Average age at implantation was 46.7 for males and 41.3 for females. Fourteen implants (in 12 patients) were removed. The

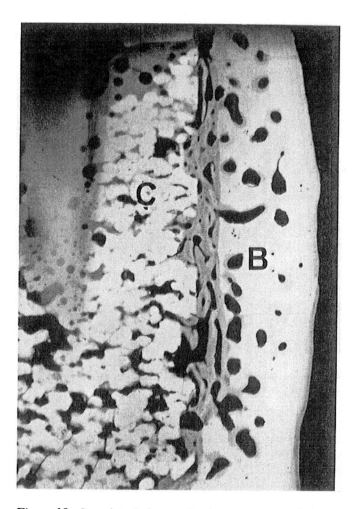

Figure 19 Bone interfacing an alumina porous ceramic implant (CMR image). Porous implant was implanted in mandibular posterior area of a rhesus monkey 3 months postoperation, and occlusal force was applied for 1 month. This is a photo of CMR image after this implant functioned for 1 month. We can observe that the surrounding bone tissue is newly developed and is growing into the porous surface or inside the pores. (C) alumina ceramics; (B) bone.

main causes of these removals were: 5 implants lost their initial fixation; 5 were damaged; in others, movement of connecting natural teeth created implant overloads.

Takahashi, Umehara et al. [38] have reported that 4 out of 56 cases (61 implants) were removed from 1984 to March 1993. These results indicate the high probability of successful clinical application of BIOCERAM porous implants. We are now trying to apply a two-stage BIOCERAM implant, but more examinations must be made regarding mechanical strength.

Figure 20   Line analysis by electron probe analyzer: boundary of the surface of alumina porous and bone tissue by EPA.

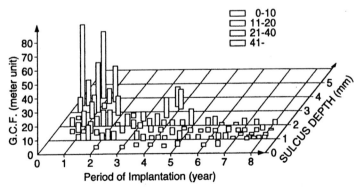

Figure 21   Longitudinal comparison between sulcus depth and periotron value.

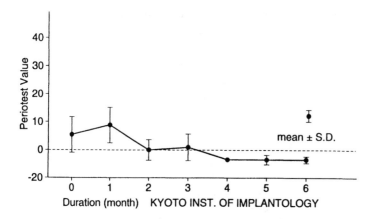

Figure 22   Longitudinal change of periotest value.

Table 4   BIOCERAM® Porous Implants: Clinical Applications

All cases	Number of patients	Average age	Number of implants	Failure cases	
				Patients	Implants
Male	53	46.7	99	11	13
Female	37	41.3	49	1	1
Total	90	44.0	148	12	14

ACKNOWLEDGMENT

The authors thank Y. Isobe (staff, Kyoto Institute of Implantology) for his significant suggestions for translation of this paper; N. Matsumura (staff, Kyoto Institute of Implantology) and K. Ohta (staff, Institute of Clinical Materials) for their valuable assistance in preparation of this manuscript; other staff members of Kyoto Institute of Implantology; and Kyocera Co., Kyoto, for its donations of materials to this study.

REFERENCES

1. Sandhaus, S., Nouveaux aspects de l'implantologie, Compte rendu des journées implantaires de Lausanne, *L'implantologie, Lausanne,* 1969, pp. 17–21.
2. Driskell, T. D., The present status of ceramic and viterous carbon implants, in *Proc. Symp. Dent. Biomater. Res. Priorities,* 1973, p. 199.
3. Schulte, W., and G. Heimke, *Die Quintessenz. Sounder Druck Report No. 5456,* June 1976, p. 27.
4. Mutschelknauss, E., and E. Dörre, Extensions-implantate aus aluminumoxid-keramik (1), 5623:1–6 (1977).
5. Nagai, N., A basic study on the new dental implant material of zirconia – Histopathological aspects on the endosseous implant in monkey, *J. Jpn. Prost. Soc., 28,* 498–514 (1984).
6. Yamagami, A., H. Kawahara, and M. Hirabayashi, *BIOCERAM® – A New Type of Ceramic,* ICOI, Kyoto, August 18, 1975.

7. Kawahara, H., Biological problem of implant material, *Proc. Jpn. Soc. Implant Dentistry*, 1975, pp. 205–246.

8. Hulbert, S. F., L. L. Hench, D. Forbes, and L. S. Bowman, *History of Bioceramics, Ceramics in Surgery*, Elsevier, Amsterdam, 1983, pp. 3–29.

9. Kawahara, H., A. Yamagami, and M. Hirabayashi, BIOCERAM®—A new type of ceramic, *Proc. Jpn. Soc. Implant Dentistry*, 1975, pp. 187–196.

10. Kawahara, H., A. Yamagami, and K. Shibata, New bone adhesion to polycrystal alumina implant, *Trans. 3rd Annual Meeting Soc. Biomater.*, New Orleans, LA, April 15–19, 1977, p. 133.

11. Yamagami, A., and H. Kawahara, Electron microscopic studies on the tissue-implant interface of single crystal and poly-crystalline of alumina *in vitro* and *in vivo*, *Trans. 5th Annual Meeting Soc. Biomater.*, 1979, p. 42.

12. Yamagami, A., and H. Kawahara, Electron microscopic studies on the tissue-implant interface of single crystal and polycrystal of alumina—*in vitro* and *in vivo*, *Trans. 5th Annual Meeting Soc. Biomater.*, Clemson, SC, April 28–May 1, 1979.

13. Yamane, T., Clinical studies of BIOCERAM® sapphire implant with animal experiment, *Shikai Tenbo, 50*, 1179–1184 (1977).

14. Nishijima, K., T. Ishida, K. Maeda et al., Experimental studies on vascularization of blood vessels after endosseous BIOCERAM® implant, *J. Oral Implant Res.* (in Japanese), *1*, 18–26 (1978).

15. McKinney, R. V., Jr., and D. L. Koth, The single-crystal sapphire endosteal dental implant: Material characteristics and 18 months experimental animal trials, *J. Prosthet. Dent., 47*, 6984 (1982).

16. Sugimoto, T., Experimental studies of BIOCERAM® implant, *Kyushu Dent. J., 38*(1), 1–25 (1984).

17. Kawahara, H., Today and tomorrow of bioceramics, *Oral Implant, 8*, 411–423 (1979).

18. Steflik, D. E., R. V. McKinney, Jr., and D. L. Koth, Correlative approaches to examine the interface between a bio-material and biological tissues, *Trans. 11th Annual Meeting Soc. Biomater.*, San Diego, CA, April 24–28, 1985.

19. Kawahara, H., A. Yamagami, and T. Yamane, *BIOCERAM® Sapphire Dental Implants*, Ishiyaku Pub. Inc., Tokyo, 1986.

20. Yamagami, A., Single-crystal alumina dental implant—13 year long follow-ups and statistical examination, *Bioceramics* (Ishiyaku Euro America, Inc., Tokyo, St. Louis), 1, 332–337 (1989).

21. Koth, D. L., D. E. Steflik, R. V. McKinney, and Q. B. Davis, A clinical and statistical analysis of human clinical trials with the single crystal alumina oxide endosteal dental implants: Five year results, *J. Prosthet. Dent., 60*, 222–234 (1988).

22. Söremark, R., and B. Fartash, Long term clinical study of single crystal Al_2O_3 implants (BIOCERAM®) in the lower jaw, *Abstr. World Congress Implantol. Biomater.*, Paris, March 8–11, 1989.

23. Klawitter, J. J., and S. F. Hulbert, Application of porous ceramics for the attachment of load bearing internal orthopaedic applications, *J. Biomed. Mater. Res. Symp., 2*, 161 (1971).

24. Hammer, W. B., R. G. Topazian, R. V. Kinney, and S. F. Hulbert, Alveolar ridge augmentation with ceramics, *J. Dent. Res., 52*(2), 356–361 (1973).

25. Predecki, P., J. E. Stephan, B. A. Auslaender et al., Kinetics of bone growth into cylindrical channels in aluminium oxide and titanium, *J. Biomed. Mater. Res., 6*, 375–400 (1972).

26. Pedersen, K. N., H. R. Haanaes, and S. Lyng, Tissue ingrowth into mandibular intrabony porous ceramic implants, *Int. J. Oral Sur., 3*, 158–165 (1974).

27. Klawitter, J. J., A. M. Weinstein, F. W. Cooke, L. J. Peterson, B. M. Penneland, and R. V. McKinney, An evaluation of porous alumina ceramic dental implants, *J. Dent. Res., 56*, 768–776 (1977).

28. Schneider, P. H., F. Lutz, and H. R. Mühlemann, Clinical evaluation of porous alumina ceramic dental implants, *Oral Implantol., 8*, 371–379 (1979).

29. Karadianes, M. T., R. E. Westerman, J. J. Rasmaseen, and A. M. Lodmell, Development and evaluation of porous dental implants in miniature swine, *J. Dent. Res., 55*, 85–93 (1976).

30. Kawahara, H., A. Yamagami, K. Imai, and M. Nakamura, Porous alumina implant with single crystal core, *J. Dent. Res., 58* (Special Issue C), 1289 (1979).

31. Yamagami, A., H. Kawahara, and M. Hirabayashi, Studies on the tissue response to Al_2O_3 porous ceramic implant – Animal experiment and clinical application, in *Ceramics in Surgery*, Elsevier, Amsterdam, 1983, pp. 73–78.

32. Kawahara, H., K. Imai, and D. Kawahara, Porous polycrystal alumina implant reinforced with single crystal alumina in *Ceramics in Clinical Applications* (P. Vincenzini, ed.), Elsevier, Amsterdam, 1987, pp. 253–264.

33. Yamagami, A., S. Kotera, Y. Nishio, and H. Kawahara, Porous alumina for free-standing implant – Long term clinical observation, *Proc. Book 13th Annual Meeting Soc. Biomater.*, New York, June 23, 1987.

34. Yamagami, A., Porous alumina ceramics implant – Research, development and its clinical application, *J. Dent. Med.* (Japan), *25*(5), 635–646 (1987).

35. Yamagami, A., S. Kotera, and H. Kawahara, Studies on a porous alumina dental implant reinforced with single-crystal alumina: Animal experiments and human clinical applications, in *ASTM STP 953* (J. E. Lemon, ed.), American Society for Testing and Materials, Philadelphia, 1988, pp. 399–408.

36. Yamagami, A., S. Kotera, and Y. Ehara, Porous alumina for free-standing implants. Part I: Implant design and *in vivo* animal studies, *J. Prosthet. Dent., 59*(6), 689–695 (1988).

37. Kawahara, H., A. Yamagami, Y. Asai et al., *BIOCERAM® Porous Implant*, Ishiyaku Pub. Inc., Tokyo, 1989.

38. Takahashi, T., K. Umehara et al., Clinical process of BIOCERAM® porous implant, *J. Jpn. Soc. Oral Implantol., 6*(2), 167–173 (1993).

55
Bonding to Dental Substrates

John M. Powers
University of Texas–Houston Health Science Center
Houston, Texas

I. INTRODUCTION

This chapter reviews various types of *in vitro* and *in vivo* bond tests for the study of bonding to dental substrates and reports data on substrates such as enamel, dentin, dental amalgam, composites, glass ionomers, alloys, ceramics, and acrylic resins and plastic orthodontic brackets. The chapter is organized as follows: first-level headings (e.g., III. ENAMEL SUBSTRATE) indicate the substrate, and the second-level headings (e.g., A. Composite–Enamel Bonding Agents) indicate adherands or bonding agents. Chemistry, product names, and companies that produce common adhesives used to bond to various dental substrates are listed in Table 1. Acronyms and the chemistry of various dental materials are described in the Glossary. Other topics of interest, not covered here, are microleakage, biocompatibility of bonded systems, and clinical durability of bonded restorations.

II. BOND TESTS

A. *In Vitro* Bond Tests

Various types of bond tests are summarized in Table 2. Most *in vitro* bond tests are made in tension or shear using a flat substrate. A tensile bond test reported by Barakat and Powers (1986) uses an inverted, truncated cone bonded to a flat substrate or a second inverted, truncated cone. The advantage of this specimen design is that the lowest cross-sectional area of the specimen occurs at the bonded interface. Recently, use of a three-dimensional cavity as a substrate has been reported (Prati, Simpson, Mitchem, Tao, and Pashley, 1992).

An alternative to the measurement of bond strength is the measurement of the plane-strain fracture toughness of the joint. A miniature short-rod fracture toughness specimen

Table 1 Chemistry, Products, and Producers of Common Adhesives for Various Dental Substrates

Chemistry	Product	Producer
Adhesive restorative materials		
Glass ionomers (water	Fuji II	GC America
based)	Ketac-Fil	ESPE
Hydbrid ionomers	Fuji II LC	GC America
	Photac-Fil	ESPE
	Variglass	Caulk/Dentsply
	Vitremer	3M Dental Products
Adhesive cements		
Bis-GMA, NTG–GMA,	Geristore	Den-Mat
PMDM		
Organic phosphonate	Panavia 21	Kuraray
	CR Inlay Cement	Kuraray
4-META	Super-Bond C&B	Sun Medical
	C&B Meta-Bond	Parkell
All-purpose bonding agents		
Bis-GMA/HEMA	Optibond	Sybron/Kerr
	Scotchbond Multi-Purpose	3M Dental Products
Bis-GMA/HEMA/PENTA	Probond (formerly Prisma	Caulk/Dentsply
	Universal Bond 3)	
NTG–GMA, BPDM	All-Bond 2	Bisco
NTG–GMA, PMDM	Tenure	Den-Mat
Organic phosphonate	Clearfil New Bond	Kuraray
	Clearfil Photo-Bond	Kuraray
Bonding composite to por-		
celain		
Organic phosphonate and	Clearfil Porcelain Bond	Kuraray
silane	Scotchprime	3M Dental Products
4-META and silane	Porcelain Liner M	Sun Medical
Denture acrylic for bonding		
to alloy		
4-META	Meta-Dent	Sun Medical
	Meta-Fast	Sun Medical
	Acrylic Solder	Parkell
Bonding amalgam to teeth		
NTG–GMA, BPDM	All-Bond 2	Bisco
4-META	Amalgambond Plus	Parkell
Silica coatings for pretreat-		
ment of alloys		
Blasted silica coating	Rocatector	ESPE
Pyrogenic silica coating	Silicoater	Heraeus Kulzer
Thermal silica coating	Silicoater MD	Heraeus Kulzer

Source: Farah and Powers (1991, 1993a); Powers (1991)

Table 2   Types of *In Vitro* and *In Vivo* Tensile Bond Strength Tests

Type of test	Ref.
In vitro tensile bond test	
Cylinder on flat substrate	Fowler et al. (1992); Stanford, et al. (1985)
Fracture toughness specimen	Tam and Pilliar (1993)
Three-dimensional cavity	Prati et al. (1992)
Truncated cone on flat substrate	Barakat and Powers (1986)
In vitro shear bond test	
Cylinder on flat substrate (shear ring knife)	Stanford et al. (1985)
Cylinder on flat substrate (cable loop)	McGuckin et al. (1994)
Truncated cone on flat substrate (cable loop)	Berry (1991)
In vivo tensile bond test	
Orthodontic button cemented on monkey incisor dentin	Tyler et al. (1987)
In vivo shear bond test	
Cylinder on dog canine and molar enamel and dentin (cable loop)	McGuckin et al. (1991)
Cylinder on goat incisor dentin	Gray and Burgess (1989)

has been developed by Tam and Pilliar (1993). Values of K_{IC} and tensile bond strength do not appear to correlate.

B.   *In Vivo* Bond Tests

Bond strengths have been measured *in vivo* using various animal models including dogs, goats, and monkeys as shown in Table 2.

III.   ENAMEL SUBSTRATE

Human enamel is 96–97% by weight hydroxyapatite, with the remainder being water and about 1% organic matter. In the mouth, enamel is covered by an organic pellicle (biofilm) with a critical surface tension of about 28 dynes/cm (Jendersen and Glantz, 1981).

A.   Composite–Enamel Bonding Agents

Acid etching of enamel to create a durable bond with resin was introduced by Buonocore (1955). Etching results in increased surface energy of the enamel (Jendersen and Glantz, 1981), increased surface roughness, increased bonding area, and decreased contamination. Factors that affect the acid etching of enamel include: chemical composition of the enamel, type and concentration of acid, application time of acid, consistency of the etchant, active versus passive etching, and the instrumentation of enamel before etching (Van Meerbeek, 1993). Some of these factors are described in Table 3. Bond strengths of composites with an enamel bonding resin to etched enamel generally exceed 20 MPa.

B.   Composite–Dentin Bonding Agents

Dentin bonding agents were not used routinely on enamel surfaces before placement of composite restorations until 1992. The primers of some dentin bonding agents improved bond strength to enamel, whereas others decreased bond strength (McGuckin, Powers,

Table 3   Factors That Influence Bonding of Composites to Human Enamel

Factor	Observation	Ref.
Consistency of etchant[a]	Acid gel preferred over liquid acid	Asmussen et al. (1989)
Etching time[a]	20 to 30 sec effective	Asmussen et al. (1989); Gwinnet (1990)
Washing times[a]	10 to 20 sec effective	Asmussen et al. (1989); Gwinnet (1990)
Type and concentration of acid	30 to 40% phosphoric acid	Asmussen et al. (1989); Gwinnet (1990)
	10% phosphoric acid effective	Kanca (1991); Suh (1991)
	2.5% nitric acid for 30 sec effective	Blosser (1990)
Use of ethanol to dry	Enhances penetration of monomers	Qvist and Qvist (1985)

[a]Conditions for 37% phosphoric acid

and Li, 1994), as shown in Table 4. Many composites now in use, when used with their bonding agents, bond equally well to both enamel and dentin with bond strengths greater than 20 MPa, as shown in Table 5 (Farah and Powers, 1993a). An adhesive composite (Geristore) bonds adequately to enamel (Berry and Powers, 1994b), as shown in Table 6. The bond strength is improved slightly when the adhesive composite is used in conjunction with an NTG–GMA/PMDM dentin bonding agent (Tenure).

Most manufacturers recommend acid etching of enamel with 37% phosphoric acid for times varying from 5 to 120 sec before application of a resin bonding agent. Acid etching of enamel with 10% maleic acid during the simultaneous preparation of enamel and dentin for bonding is recommended by the manufacturer of one bonding agent (Scotchbond Multi-Purpose). Bond strengths of Scotchbond MultiPurpose with several types of phosphoric acid etchants range from 13.6 to 17.2 MPa, whereas the bond strength with the maleic acid etchant is 17.3 MPa (Swift, Denehy, and Beck, 1993).

Table 4   24-h Shear Bond Strengths of Dentin Bonding Agent/Composite Systems to Human Enamel and Dentin

Bonding system	Bond strength, MPa			
	Enamel		Dentin	
	Primer	No primer	Etched	Unetched
Denthesive/Charisma	12.3	14.0	10.0	5.9
Gluma/Pekalux	6.7	9.2	8.1	7.3
Prisma Universal Bond 3/Prisma APH	13.7	9.7	12.3	8.2
Scotchbond 2/Silux Plus	8.5	8.7	5.3	5.9
Tenure/Visar Seal/Perfection	6.0	6.1	9.9	8.0
XR Bond/Herculite XRV	16.1	14.0	12.7	10.7

Source: McGuckin et al. (1994)

Table 5   *In Vitro* 24-h Bond Strengths of Commercial Composite/
Bonding Agent Systems to Human Enamel and Dentin

	Bond strength, MPa	
Bonding agent/composite system	Dentin	Enamel
All-Bond 2/Bisfil-P	24	23
Denthesive II/Charisma	25	25
Gluma 2000/Pekafill	12	20
KB-200/Clearfil Photo Anterior	24	30
Optibond LC/Herculite XRV	27	26
Probond/Prisma TPH	19	26
Scotchbond Multi-Purpose/Z100	18	20
Syntac/Tetric	27	27

Source: Farah and Powers (1993a)

Contamination of enamel after etching but before application of the bonding agent reduces the *in vitro* bond strength of bonding agents tested (Xie, Powers, and McGuckin, 1993). Contaminants including artificial saliva, human plasma, zinc oxide–eugenol cement, and non-eugenol–zinc oxide cement interfere with bonding of All-Bond 2, Gluma 2000, and Scotchbond Multi-Purpose, as shown in Table 7. All three bonding agents bond well to enamel exposed to air or slight moisture (damp); however, water reduces the bond strength of Gluma 2000 and Scotchbond Multi-Purpose but does not affect the bond strength of All-Bond 2 as much.

Fracture toughness of composite–bonding agent bonded to bovine enamel using a miniature short-rod specimen is 1.1 $MNm^{-3/2}$ (Tam and Pilliar, 1993). Bond failures are adhesive and cohesive in both composite and enamel.

C.   Glass Ionomers and Hybrid Ionomers

Hybrid ionomers have a low bond strength (5 MPa) to abraded enamel but bond much better (21 MPa) to dentin conditioned with a solution of 10% polyacrylic acid, as shown in Table 8. Air abrasion improves the bond to untreated enamel but decreases the bond to conditioned enamel (Berry, Rainey, and Powers, 1993).

Table 6   *In Vitro* Tensile Bond Strengths of Adhesive Composites to Dentin, Enamel, Amalgam, Porcelain, and Ni–Cr–Be Substrates

	Bond strength, MPa		
Primer	Adhesive composite (Geristore)	Adhesive composite with bonding agent (Geristore/Tenure)	Composite with bonding agent (Marathon/Tenure)
Enamel	13.9	15.6	21.3
Dentin	9.0	14.2	9.6
Amalgam	12.8	21.8	9.7
Ni–Cr–Be alloy	22.9	25.1	27.1
Porcelain	10.4	13.9	10.4

Source: Berry and Powers (1994b)

Table 7   Tensile Bond Strengths of Dentin Bonding Agent/
Composite Systems to Contaminated and Decontaminated
Human Enamel

Contaminant	Bond strength, MPa		
	AB2[a]	G2000[a]	SBMP[a]
Air	17.9	19.9	19.9
Water	20.6	8.3	12.0
Damp	23.1	15.9	22.0
Saliva	10.5	8.2	12.5
Saliva reetched	20.5	21.4	20.3
Plasma	9.9	10.8	13.3
Plasma reetched	18.8	11.7	18.6
Lubricant	22.9	19.1	15.7
Lubricant reetched	17.2	25.2	7.2
ZOE	10.0	3.3	6.6
ZOE reetched	19.1	16.8	20.1
Non-ZOE	5.1	0.0	2.4
Non-ZOE reetched	23.1	15.9	15.4

Source: Xie et al. (1993)
[a]AB2, All-Bond 2/Bisfil-P; G2000, Gluma 2000/Pekalux; SBMP,
Scotchbond Multi-Purpose/Z100

D.   Amalgam with Adhesive Cements

In the restoration of a posterior tooth by an amalgam, a small portion of enamel is
available for bonding. Without the use of an adhesive, however, amalgam will not bond
to enamel. Several adhesive systems have been tested for bonding amalgam to bovine or
human enamel (Shimizu, Takashi, and Kawakami, 1986; Staninec and Holt, 1988; Varga,
Matsumura, and Masuhara, 1986). Bond strengths range from 8.8 to 13.4 MPa, as shown
in Table 9.

Table 8   *In Vitro* 24-h Bond Strengths of Hybrid Ionomer
(Fuji II LC) to Human Enamel and Dentin

Treatment	Bond strength, MPa	
	Dentin	Enamel
No conditioner	8	5
10% polyacrylic acid	19	22
Air abraded	—	10
Air abraded/10% polyacrylic acid	—	17
Etched/bonding agent	22	—

Source: Berry et al. (1993)

Table 9   *In Vitro* Bond Strength of Amalgam Bonded to Bovine and Human Enamel and Dentin Treated with an Adhesive System

Tooth structure and adhesive system	Bond strength, MPa	Ref.
Bovine enamel		
Phosphonate resin (Panavia)	8.8	Shimizu et al. (1986)
4-META/MMA–TBB resin (Super-Bond C&B)	10.0	Shimizu et al. (1986)
Human enamel		
Phosphonate resin (Panavia)	9.7	Staninec and Holt (1988)
4-META/MMA–TBB resin (Super-Bond C&B)	13.4	Varga et al. (1986)
Bovine dentin		
Phosphonate resin (Panavia)	1.3–1.7	Shimizu et al. (1986)
4-META/MMA–TBB resin (Super-Bond C&B)	2.2–6.4	Shimizu et al. (1986)
Human dentin		
Glass ionomer (Ketac-Cem)	3.3	Covey and Moon (1991)
NTG–GMA, DMA (All-Bond)	11.0	DeSchepper et al. (1991)
Phosphonate bonding agent (Scotchbond 2)	4.9	Covey and Moon (1991)
Phosphonate resin (Panavia)	3.2	Staninec and Holt (1988)
Phosphonate resin (Panavia)	3.5	Covey and Moon (1991)
4-META (Amalgambond)	3.6	Covey and Moon (1991)

E.   Zinc Phosphate and Zinc Polyacrylate Cements

Zinc phosphate and zinc polyacrylate cements are commonly used for retaining metallic crowns and bridges and orthodontic bands. Zinc phosphate cement bonds to enamel primarily through mechanical undercuts with very low bond strengths. Zinc polyacrylate cement bonds to untreated enamel as shown in Table 10, but the bond strength can be improved by treatments with polyacrylic acid liquid, ferric chloride, or ferric oxalate (Barakat and Powers, 1986). A polyacrylate cement containing sulfate ions as an impu-

Table 10   *In Vitro* Tensile Bond Strengths of a Polyacrylate Cement to Treated Human Enamel and Dentin

Bonding technique	Bond strength, MPa	
	Enamel	Dentin
Untreated	3.7	2.0
Polyacrylic acid liquid	7.3	5.7
Ferric chloride (15%)	4.2	2.1
Ferric oxalate (5%)	4.5	4.1

Source: Barakat and Powers (1986)

rity bonds chemically to enamel by formation and growth of calcium sulfate crystals on the enamel. Although bond strengths of polyacrylate cements can be measured in the laboratory and appear to have a chemical component, the clinical importance of this chemical bonding for retention of crowns and bridges has not been established. Direct-bonding orthodontic brackets can be retained by the zinc polyacrylate cements, but the bond strengths are considered too low to be practical.

F. Resin and Adhesive Resin Cements

Resin-bonded bridges can be bonded to enamel using resin and adhesive resin cements. Bond strengths of filled Bis-GMA resin cements to enamel range from 7.3 to 12.0 MPa (Aksu, Powers, Lorey, and Kolling, 1987), whereas those for adhesive resin cements range from 10.5 MPa for a phosphonate cement to 15.2 MPa for a 4-META/MMA–TBB cement (Powers, Watanabe, and Lorey, 1986). Bond strength is improved by rough surfaces. Increasing the film thickness of a paste-primer cement from complete seating to 0.51 mm decreases bond strength by more than 70% (Aksu et al., 1987). New adhesive and aesthetic resin cements have been reviewed by Farah and Powers (1993b).

G. Resin Cements for Orthodontic Brackets

The resin cements for bonding orthodontic brackets to enamel consist of three types (Craig, 1993). Unfilled acrylics are monomer–polymer formulations based on methyl methacrylate and comonomers with an amine–peroxide accelerator/initiator system. Highly filled diacrylate cements contain diacrylate oligomers, diluent monomers, and more than 60% by weight of silanated inorganic fillers. The slightly filled diacrylates contain about 28% colloidal silica. These diacrylate cements are either accelerated by amines or are light activated. Bond strength of these cements to enamel is adequate clinically if proper techniques of isolation and etching are followed. Penetration of the resin into the etched enamel is the bonding mechanism. Bond failures typically occur at the cement–bracket interface rather than the cement–enamel interface.

Paste-primer resin cements require no mixing and thus are convenient. A priming liquid is applied to the etched enamel, and the paste is applied to the bracket base. Polymerization initiates when the bracket is placed against the primed tooth. As the thickness of the paste between the bracket and tooth increases, the bond strength decreases (Evans and Powers, 1985). Eventually, failures are characterized by incomplete polymerization of the resin. If the primer is exposed to a simulated oral environment for a minute or more before the bracket with the paste is placed, bond strengths decrease. Refer to subsequent discussions of metal, ceramic, and plastic substrates for further information on the bonding of resin cements to orthodontic brackets.

IV. DENTIN SUBSTRATE

Human dentin is about 70% by weight hydroxyapatite, 18% organic material (primarily collagen), and 12% water. By volume, dentin is 50% mineral, 30% organic matter, and 20% fluid (Driessens and Verbeeck, 1990) and is estimated to be a mixture of 40% polyproline (collagen) and 60% hydroxyapatite (Ruse and Smith, 1991). The surface energy of dentin is about 45 dyn/cm (Douglas, 1989). The structure of dentin at three different levels (near enamel, central, and deep) has been examined by scanning electron microscopy (Olsson, Oilo, and Adamczak, 1993). Dentin shows an increasing number of tubules with diameters increasing from 0.8 μm at the dentinoenamel junction to 2.5 μm at the pulp. The superficial dentin surface is 96% intertubular dentin, 3% peritubular dentin, and 1% water (Garberoglio and Brannstrom, 1976). Deep dentin is 66% peritubular dentin, 12% intertubular dentin, and 22% water by weight. Microstructure and characterization of dentin are reviewed by Marshall (1993). Pashley, Ciucchi, Sano, and Horner (1993) have reviewed permeability characteristics of dentin.

After mechanical preparation, dentin is covered by a smear layer that varies from 1 to 5 μm thick (Eick, Wilko, Anderson, and Sorensen, 1970). Its composition is dentin mixed with saliva, bacteria, and other debris (Pashley, 1990). EDTA is the best conditioner for removing the smear layer, followed by phosphoric acid, lactic acid, polyacrylic acid, and citric acid (Meryon, Tobias, and Jakeman, 1987). Rueggeberg (1991) has reviewed a number of other variables that affect bonding to dentin.

A. Composite–Bonding Agents

Anterior and posterior composites used with a phosphonate bonding agent (2-methacryloxyethyl phenyl phosphoric acid) have bond strengths after 24 h of 10 to 14 MPa to unetched dentin, 7 to 10 MPa to dentin conditioned with 10% polyacrylic acid, and 6 to 8 MPa to dentin etched with 37% phosphoric acid (Yamaguchi, Powers, and Dennison, 1989). Use of phosphoric acid can cause dissolution of the mineral component of dentin, and this can reduce the effectiveness of chemical bonding, which is dependent on the calcium in the hydroxyapatite.

Until 1992, bonding of composites to dentin was accomplished without the recommended etching of dentin with the exception of bonding agents such as the phosphonates (Clearfil New Bond, Clearfil Photo-Bond) and a bonding agent based on NTG–GMA and DMA (All-Bond 2). For some dentin bonding agents, etching of dentin increased bond strength, whereas for others etching decreased bond strength (McGuckin et al., 1994), as shown in Table 4. Currently, many bonding agents (Table 5) simultaneously etch enamel and dentin or use an acidic primer for pretreatment of dentin. Fluoride iontophoresis of dentin does not adversely affect bonding of composite to dentin (Mangum, Jeske, Chan, and Powers, 1993).

Contamination of dentin before application of the bonding agent reduces the *in vitro* bond strength of several bonding agents tested (Xie et al., 1993). Contaminants including artificial saliva, human plasma, zinc oxide–eugenol cement, and non-eugenol–zinc oxide cement interfere with bonding of All-Bond 2, Gluma 2000, and Scotchbond Multi-Purpose, as shown in Table 11. Both All-Bond 2 and Scotchbond Multi-Purpose bond best to damp dentin, although Scotchbond Multi-Purpose is less affected by dentin exposed to air and water than All-Bond 2. Water and slight moisture (damp) interfere with the bonding of Gluma 2000.

An adhesive composite (Geristore) bonds to dentin (Berry and Powers, 1994b), as shown in Table 6. The bond strength is improved by about 60% when the adhesive composite is used in conjunction with an NTG–GMA/PMDM dentin bonding agent (Tenure).

Fracture toughness of composite/bonding agent bonded to bovine dentin using a miniature short-rod specimen varies from 0.20 to 0.69 $MNm^{-3/2}$ for the bonding agents tested (Tam and Pilliar, 1993), as shown in Table 12. Bond failures are adhesive and cohesive in both composite and enamel.

Postulated mechanisms of bonding of dentin adhesives to dentin include microme-

Table 11   Tensile Bond Strength of Dentin Bonding Agent/
Composite Systems to Contaminated and Decontaminated
Human Dentin

Contaminant	Bond strength, MPa		
	AB2[a]	G2000[a]	SBMP[a]
Air	11.9	11.6	18.5
Water	11.3	0.0	15.9
Damp	24.5	5.4	20.4
Saliva	8.8	0.0	12.6
Saliva reetched	20.1	12.2	18.8
Plasma	3.9	0.0	5.5
Plasma reetched	19.7	10.6	18.4
Lubricant	8.9	5.9	23.2
Lubricant reetched	16.7	11.1	22.2
ZOE	13.8	9.5	6.9
ZOE reetched	21.9	12.2	17.2
Non-ZOE	9.8	6.7	10.4
Non-ZOE reetched	20.8	13.2	19.1

Source: Xie et al. (1993)
[a]AB2, All-Bond 2/Bisfil-P; G2000, Gluma 2000/Pekalux; SBMP,
Scotchbond Multi-Purpose/Z100

chanical interlocking to acid-etched dentin, chemical adhesion, and micromechanical attachment through formation of a hybrid layer. Micromechanical interlocking or tag formation in acid-etched dentin results in only a slight improvement in bond strength (Torney, 1978). Chemical adhesion as studied by spectroscopy has not been demonstrated for recent bonding agents (Edler, Krikorian, and Thompson, 1991; Eliades, Palaghias, and Vougiouklakis, 1990).

The formation of a hybrid (resin-impregnated) layer at intertubular dentin is considered to be the primary bonding mechanism of current bonding agents (Erickson, 1989; Nakabayashi, Kojjima, and Masuhara, 1982). Critical to formation of this hybrid layer is the pretreatment of the dentin with a hydrophilic conditioner or primer (Soderholm, 1991). Other factors that affect dentin bonding include: presence of smear layer; dentinal tubule density, size, and length; dentin sclerosis; lesion shape and size; structure of

Table 12   Fracture Toughness (K_{IC}) of Composite/Dentin Bonding Agent Systems to Bovine Dentin

Storage condition	Fracture toughness (K_{IC}, MNm$^{-3/2}$)		
	Scotchbond 2	Scotchbond Multi-Purpose	All-Bond 2
P-50	0.20	0.34	0.69
Bis-Fil	0.31	0.51	0.67

Source: Tam and Pilliar (1993)

Table 13 *In Vitro* Bond Strengths of a Water-Based
Glass Ionomer to Dentin with Different Conditioning
Techniques and Acid Concentrations

Treatment	Polyacrylic acid concentration, %	Range of bond strengths, MPa
Untreated	—	1.9–2.1
Passive	10	1.9–3.3
	25	1.8–3.4
Active	10	2.3–4.3
	25	2.1–3.7

Source: Barakat et al. (1988)

enamel and dentin; flexure of the tooth and the restorative material; and tooth location (Heymann and Bayne, 1993). A review of current dentin bonding agents is given by Eichmiller (1993).

B.   Glass Ionomers and Hybrid Ionomers

The bond strength of a water-based glass ionomer to dentin depends somewhat on passive versus active conditioning of the dentin and on the concentration of the polyacrylic acid, as shown in Table 13 (Barakat, Powers, and Yamaguchi, 1988). The highest bond strengths result from active etching with 10% polyacrylic acid. Dentin tubules are opened to a greater extent by active etching with 25% polyacrylic acid than by passive conditioning with 10% acid. Overetching of bovine dentin can deplete surface calcium and affect bonding (Smith and Ruse, 1987).

The bond strengths of a water-based glass ionomer base and a luting cement are dependent on the type of dentin, on the smoothness or roughness of the surface, and on conditioning of the dentin (Berry, 1991; Berry and Powers, 1992, 1994a). As shown in Table 14, the following conclusions may be drawn about the bonding of water-based glass ionomers to dentin:

Table 14   Shear Bond Strengths of Water-Based Glass Ionomer Cements to Smooth and Rough Coronal, and to Radicular Dentin With and Without Conditioning

Bond enhancer	Bond strength, MPa			
	Coronal dentin		Radicular dentin	
	Smooth	Rough	Smooth	Rough
Base cement				
No conditioning	1.5	2.5	2.3	3.2
Conditioned (25% PAA)	3.4	5.0	3.1	5.3
Luting cement				
No conditioning	1.7	1.9	3.1	2.7
Conditioned (25% PAA)	2.0	3.4	2.4	4.9

Source: Berry (1991); Berry and Powers (1992, 1994a)

1. The bond strength is better to radicular than to coronal dentin.
2. The bond strength is better to a rough than a smooth surface.
3. The bond strength is better with a base than with a luting consistency.
4. The bond strength is better to a surface conditioned with 25% polyacrylic acid than to a nonconditioned surface.

The bond strength of hybrid ionomers to dentin depends on the surface treatment of the dentin, as shown in Table 8. Although the bond strength of one hybrid ionomer (Fuji II LC) to unconditioned dentin is more than twice that of water-based glass ionomers, its bond strength is improved by conditioning the dentin with polyacrylic acid or by etching with phosphoric acid and applying a dentin bonding agent (Berry et al., 1993).

C.   Amalgam with Adhesive Cements

Amalgam is used as a core buildup material beneath crowns because of its strength, dimensional stability, and insolubility in oral fluids. Unfortunately, amalgam does not adhere to dentin but relies on mechanical undercuts to achieve retention. The use of various adhesive systems produces bond strengths of amalgam to bovine or human dentin ranging from 1.3 to 11.0 MPa (Covey and Moon, 1991; DeSchepper, Cailleteau, Roeder, and Powers, 1991; Shimizu et al., 1986; Staninec and Holt, 1988), as shown in Table 9. Pretreatment of bovine dentin by 8% SnF_2 followed by etching with 10% citric acid/3% ferric chloride results in increased bond strength of amalgam used with a 4-META/ MMA–TBB resin, whereas the use of 38% $Ag(NH_3)_2F$, etching, and lining with a glass ionomer liner works best with a phosphonate resin (Shimizu et al., 1986). The mechanisms of bonding are not well understood.

D.   Zinc Phosphate and Zinc Polyacrylate Cements

Zinc phosphate cements only bond mechanically to dentin. Zinc polyacrylate cements show measurable bond strengths to dentin, as shown in Table 10, but the values are lower than those to enamel (Barakat and Powers, 1986).

E.   Resin and Adhesive Resin Cements

The retention of metal post-and-core restorations on dentin surfaces of endodontically treated teeth can be improved over results with the traditional zinc phosphate cement by the use of adhesive cements and silica coating of the post, as shown in Table 15 (O'Keefe, Powers, McGuckin, and Pierpont, 1992).

V.   DENTAL AMALGAM SUBSTRATE

Dental amalgam is a metallic restorative material used for the restoration of posterior teeth subjected to high biting forces. It is an alloy that results from the combination of mercury with an alloy containing silver, tin, and copper (Craig, O'Brien, and Powers, 1993).

Although amalgam restorations are usually replaced when fractured, the repair of an amalgam restoration could be advantageous in certain clinical situations. Amalgam can be repaired with fresh amalgam with or without the application of an adhesive liner. Recently, composites have been bonded to amalgam to provide improved aesthetics.

Table 15 *In Vitro* Bond Strengths of Abrasive-Sprayed and Silica-Coated Posts Bonded in Roots of Extracted Teeth with the Application of Resin and Adhesive Cements

| | Bond strength, MPa | |
Cement	Abrasive sprayed	Silica coated
Phosphonate resin (Panavia)	9.8[a]	7.5[a]
Resin cement/Bonding Agent (Dicor/Prisma Universal Bond 2)	8.8[a]	10.9[a]
4-META/MMA–TBB resin (Super-Bond C&B)	10.8[a]	14.5[a]
Zinc phosphate cement	4.4[b]	5.4[b]

Source: O'Keefe et al. (1992)
[a]Bonds fail primarily at the dentin–cement interface.
[b]Bonds fail cohesively in the cement.

A. Amalgam

Fresh amalgam is often used to repair an amalgam restoration. The strength of a repaired specimen is about 49% of the unrepaired specimen (Hibler, Foor, Miranda, and Duncanson, 1988). The tensile bond strength of amalgam repaired with fresh amalgam using an intermediate NTG–GMA/resin bonding agent (All-Bond with Liner-FX) ranges from 3.4 to 8.8 MPa, as shown in Table 16. Bond failures are primarily adhesive. Preparation of the amalgam by sandblasting results in a higher bond strength than abrasion by 120-grit silicon carbide (SiC) paper. Repair by a spherical amalgam is more effective than repair by an admixed amalgam, possibly because the spherical amalgam allows easier condensation into small mechanical undercuts. Thermocycling appears to have no effect on bond strength (DeSchepper et al., 1991).

Table 16 *In Vitro* Tensile Bond Strengths of Admixed Amalgam Repaired by Admixed and Spherical Amalgams after Sandblasting or Abrasion by SiC and the Application of an Adhesive Resin (All-Bond with Liner-FX)

| Core amalgam/repair amalgam | Bond strength, MPa | |
	Sandblasted	Abraded by SiC
24-h storage, 100% RH, 37°C		
Admixed/admixed	8.8	3.4
Admixed/spherical	8.5	4.8
Thermocycled		
Admixed/admixed	5.5	3.5
Admixed/spherical	8.4	6.1

Source: DeSchepper et al. (1991)

B. Composites

Successful bonding of composite resin to an amalgam restoration requires the use of an adhesive. The shear bond strength of composite resin bonded to amalgam with a 4-META bonding agent (Cover-Up II) is 4.3 MPa, about five times that of the bond without the adhesive (Hadavi, Hey, and Ambrose, 1991). The tensile bond strength of an adhesive composite (Geristore) to amalgam is about 13 MPa (Berry and Powers, 1994b), as shown in Table 6. It is improved by 70% when the adhesive composite is used in conjunction with an NTG–GMA/PMDM dentin bonding agent (Tenure).

VI. COMPOSITE SUBSTRATE

Composites are particle-reinforced polymeric restorative materials used for the restoration of anterior teeth and for posterior teeth subjected to low to medium biting forces where aesthetics is important. The polymer matrix is usually Bis-GMA or UDMA oligomers filled with irregularly shaped filler particles from 0.04 to 5 μm in diameter. The fillers are mostly radiopaque glasses or colloidal silica (Craig et al., 1993). Willems, Lambrechts, Braem, and Vanherle (1993) have reviewed the properties of composites.

Fractured, discolored, and worn composite restorations are often repaired by the addition of new composite. When composite restorations such as inlays and onlays are fabricated indirectly on a die, they must be bonded to the tooth using a resin cement.

A. Composites

Composite restorations can be repaired by mechanically roughening the surface of the old composite, cleansing it with 30–50% phosphoric acid, treating with an unfilled bonding agent, and adding new composite (Boyer, Chan, and Reinhardt, 1984). *In vitro* tensile bond strengths of various composite bonding agents to old composites range from 6.9 to 14.0 MPa (Pounder, Gregory, and Powers, 1987).

B. Resin Cements

One measure of the clinical success of indirect composite inlays is the reliability of bonding of the resin cement to both the inlay and tooth structure. The bond between the resin cement and the composite inlay may be the weakest interface, particularly with postcured microfilled composite inlays. Surface pretreatment of the inlay is necessary, and a methyl methacrylate pretreatment (Special Bond II) results in a higher bond strength than pretreatments with either a dimethacrylate resin (Heliobond) or silane, as shown in Table 17 (DeSchepper, Tate, and Powers, 1993; Tate, DeSchepper, and Powers, 1993). A hybrid composite inlay should be cleaned with phosphoric acid rather than etched with hydrofluoric acid gel (Tate et al., 1993). The highest bond strengths to light-cured, hybrid composites are obtained with resin cements (see Table 17).

VII. GLASS IONOMER SUBSTRATE

Water-based glass ionomer restorative materials are formulated from fluoroaluminosilicate glass and a water solution of polymers and copolymers of acrylic acid, and set by an acid–base reaction (Craig et al., 1993). Recently, hybrid ionomers have become available. They are composed of polycarboxylic acid with pendant methacrylate groups, a crosslink-

Table 17 Bond Strengths of Resin Cements to Light-Cured and Postcured Microfilled and Hybrid Composites Treated with Phosphoric Acid and Various Bond Enhancers

	Bond strength, MPa			
	Microfilled composite		Hybrid composite	
Bond enhancer	Light-cured	Postcured	Light-cured	Postcured
Adhesive resin cement				
Special Bond II	13.5	6.2	13.0	13.5
Heliobond	10.0	5.1	12.4	10.5
Silane (Kerr)	7.5	5.1	11.6	9.8
Resin cement				
Special Bond II	11.2	4.5	22.6	12.2
Heliobond	7.3	5.0	19.9	15.1
Silane (Kerr)	6.9	4.5	12.1	13.0

Source: DeSchepper et al. (1993); Tate et al. (1993)

ing component (HEMA), silicate glass, and photoinitiators. They set by a acid–base reaction and a light- and chemical-activated polymerization of the methacrylate groups and HEMA.

When glass ionomers are used as a liner or base, bonding by the restorative material applied subsequently may be desirable. Fractured cores of reinforced glass ionomer are sometimes repaired by the addition of new glass ionomer.

Core buildups of reinforced glass ionomer occasionally fracture. Repair is an alternative to replacement. *In vitro* tensile bond strengths of reinforced glass ionomer bonded to a substrate of set reinforced glass ionomer range from 1.7 to 5.2 MPa, with primarily interfacial failures (Roeder, Fulton, and Powers, 1991).

VIII. ALLOY SUBSTRATE

Dental alloys are typically made of noble metals such as gold, platinum, or palladium; or of base metals such as chromium, cobalt or nickel. They are used for crowns, bridges, and partial denture framework (Craig et al., 1993). Most dental implants are made from pure titanium or titanium alloys. Certain orthodontic appliances are fabricated from stainless steel wires or meshes.

A. Resin and Adhesive Resin Cements

Resin-bonded bridges can be bonded to tooth structure using resin and adhesive resin cements. Resin cements can be also used to lute fixed, removable bridges to implant abutments.

Bond strengths of adhesive resin cements are reported to be higher to Ni–Cr–Be alloy and Type IV gold alloy (Watanabe, Powers, and Lorey, 1988) than to human enamel or dentin (Powers, Watanabe, and Lorey, 1986), as shown on Table 18. These data suggest that the use of sandblasted alloys may be clinically acceptable for resin-bonded bridges. Manufacturers of the cements recommend tin plating of gold alloys and electrolytic etching of base metal alloys for optimum bonding. The bond strength of the adhesive resin cements is less susceptible to deterioration by thermocycling than a resin cement.

Table 18   24-h *In Vitro* Bond Strength of Resin Cements to Tooth Structure,
Sandblasted and Tin-Plated Type IV Gold Alloy, and Sandblasted and Electroetched
Ni–Cr–Be Alloy

Storage condition	Tensile bond strength, MPa		
	Phosphonate[a]	4-META[a]	Resin[a]
Dentin, unetched	4.4 af	4.2 af	0 af
Enamel, etched	10.5 cf	15.2 cf	10.2 cf
Ni–Cr–Be alloy			
Sandblasted	19.8 cf	24.0 cf	14.1 af
Electroetched	21.8 cf	27.4 cf	25.3 cf
Type IV gold alloy			
Sandblasted	11.0 cf	22.0 cf	9.4 af
Tin plated	18.4 cf	25.5 cf	12.8 af

Source: Powers et al. (1986); Watanabe et al. (1988)
[a]af, adhesive failure; cf, cohesive failure of cement

    Bond strengths of a resin cement to a Ni–Cr–Be alloy appear to be higher when the
alloy is treated with a silica coating or a phosphonate resin cement than with a 4-META
resin cement or all-purpose bonding agents (Chang, Powers, and Hart, 1993), as shown
in Table 19.

    The bond strength of an adhesive resin cement (Geristore) to Ni–Cr–Be alloy is about
23 MPa (Berry and Powers, 1994b), as shown in Table 6. The bond strength is improved
by about 10% when the adhesive composite is used in conjunction with an NTG–GMA/
PMDM dentin bonding agent (Tenure).

Table 19   24-h Tensile Bond Strengths of Composite Cement to
Ni–Cr–Be Alloy Treated with Various Bonding Systems

Bonding system	Bond strength, MPa
Adhesive resin cements	
Phosphonate resin (Panavia)	17.0
Phosphonate resin (Panavia TPN)	22.1
4-META/MMA–TBB resin (Super-Bond C&B)	14.2
All-purpose bonding agents	
HEMA, PENTA (Prisma Universal Bond 3)	13.2
NTG–GMA, DMA (All-Bond 2)	11.4
NTG–GMA, PMDM (Tenure)	14.6
Silica coatings	
Blasted silica coating (Rocatector)	20.2
Pyrogenic silica coating (Silicoater)	18.6

Source: Chang et al. (1993)

B. Composites

In the past, bonding of composite to metal has required mechanical retention devices such as bars, loops, and beads. A reliable chemical bond between composite and metal in prostheses designed for the aesthetic restoration of anterior teeth is desirable, because it reduces the amount of tooth to be removed and reduces or eliminates interfacial staining. Progress has been made toward achieving adhesive bonding, although actual bond mechanisms have not been verified.

Bonding of composite veneering material to nickel–chromium–beryllium (Ni–Cr–Be) and gold–palladium (Au–Pd) porcelain fusing alloys can be accomplished by the use of silica-coating techniques or etching, although etching is much less effective with Au–Pd alloys (Schneider, Powers, and Pierpont, 1992). The thermal silica coating (Silicoater MD) and the pyrogenic silica coating (Silicoater) result in higher *in vitro* bond strengths than the blasted silica coating (Rocatector), as shown in Table 20. The application of silane and an opaquer is critical to these techniques, and with current materials, bond failures occur mostly in the opaque layer. Improvements in bonding composites to alloy will require stronger opaquers with improved polymerization. An adhesive opaque resin based on 4-META-MMA-TBB has been reported to bond well to various alloys (Matsumura, Kawahara, Tanaka, and Atsuta, 1991).

C. Acrylic Denture Resins

Bonding denture acrylic to Ni–Cr–Be partial denture alloy requires the use of adhesive denture resins (Meta-Dent, Meta-Fast) or traditional denture acrylic with adhesive primers (CR Inlay Cement, Super-Bond C&B) and alloy pretreatment (sandblasted, acid etched, blasted silica coating). Bond strengths range from 7 to 23 MPa with the adhesive acrylics (Nabadalung, Powers, and Connelly, 1991a) and between 14 and 19 MPa with acrylic bonded to primed and etched or silica-coated alloy (Nabadalung, Powers, and Connelly, 1991b).

D. Ceramics

The porcelain fused to metal restoration is the most aesthetic, durable restoration in dentistry. Its success depends on the adequacy of the ceramic–metal bond, which is greatly influenced by differences in the coefficients of thermal expansion of the alloy and porcelain during cooling after sintering of the porcelain (Craig, 1993). Other factors include the composition and morphology of the alloy oxide layer.

Table 20   *In Vitro* Tensile Bond Strengths of Composites Bonded to Silica-Coated Ni–Cr–Be and Au–Pd Alloys

Bonding technique	Bond strength, MPa	
	Au–Pd alloy	Ni–Cr–Be alloy
Etched (hydrofluoric acid – gel)	6.5	25.4
Blasted silica coating (Rocatector)	16.4	14.6
Pyrogenic sillica coating (Silicoater)	17.9	21.9
Thermal silica coating (Silicoater MD)	22.0	22.5

Source: Schneider et al. (1992)

Table 21   *In Vitro* Bond Strengths of Direct-Bonding Resin
Orthodontic Cements to Metal, Ceramic, and Plastic Bracket
Bases

Resin cement	Bond strength, MPa		
	Metal	Ceramic	Plastic
Unfilled acrylic	7.8	11.0	10.8
Slightly filled diacrylate	8.8	4.6	8.3
Highly filled diacrylate	13.0	5.1	8.1

Source: Buzzitta, Hallgren, and Powers (1982); de Pulido and Powers (1983)

E.   Resin Cements for Orthodontic Brackets

The bond strength of resin cements to mesh metal orthodontic bracket bases ranges from 8 to 13 MPa and depends on the type of cement, as shown in Table 21. Metal bracket bases require a highly filled resin cement, because of stress concentrations resulting from imperfections in the mesh bases (Dickinson and Powers, 1980). Bonding is mechanical in nature, and *in vitro* and *in vivo* failures typically occur at the resin–bracket interface. Reconditioning of mesh metal bases by thermal treatment, chemical treatment, and by grinding with a green stone results in a 20–50% decrease in bond strength (Wright and Powers, 1985).

Bonding to metal bases has been improved by the use of photoetched and grooved metal bases. Bond strengths of resin cements to these bases have been tested with various surface treatments (Siomka and Powers, 1985), as shown in Table 22. Resin cement bonds better to the etched, grooved metal base than to the mesh or photoetched bases.

IX.   CERAMIC SUBSTRATE

Porcelain is an aesthetic, biocompatible restorative material based on silica, feldspar, and alumina, and is used for inlays, onlays, and veneers (Craig et al., 1993). Recently, ceramic brackets have been used in adult orthodontics where aesthetics is a prime concern.

Table 22   *In Vitro* Bond Strengths of Direct-Bonding Resin Orthodontic
Cements to Different Types of Treated Metal Bracket Bases

Treatment	Bond strength, MPa		
	Mesh	Photoetched	Grooved
No treatment	8.3	8.7	9.4
Silanation	10.7	8.9	9.2
Etching	8.4	8.5	14.7
Activation	9.4	9.4	10.5
Etching plus silanation	9.7	7.6	11.2
Etching plus activation	8.1	8.4	12.2

Source: Siomka and Powers (1985)

Modern porcelain restorations are usually bonded to tooth structure with aesthetic resin or adhesive resin cements. The brittleness of porcelain leads to failures clinically. Composites are a reliable method of repair intraorally. In both situations the ceramic restoration requires pretreatment by etching and/or priming.

A. Composite Repair Materials

Composites are an ideal repair material for porcelain because of their shade-matching ability, strength, and durability; but bonding of the composite to the porcelain has always been a problem.

Typically, *in vitro* bond strengths are reported between 1 and 15 MPa (Council on Dental Materials, Instruments, and Equipment, 1991). Recently, several primers based on silane and phosphonate or 4-META monomers have become available with bond strengths as much as 24 MPa (see Table 23).

Bonding of composites to porcelain is affected by various factors, including surface pretreatment and storage conditions (Wolf, Powers, and O'Keefe, 1992). Blasting with alumina results in a better bond strength than pretreatment with diamond abrasion or etching with hydrofluoric acid gel. The particle size and shape of the alumina is important, as is the time of etching with hydrofluoric acid gel (Wolf, Powers, and O'Keefe, 1993). Cohesive failures occur in porcelain during debonding when the porcelain is over-etched by hydrofluoric acid gel. Blasting with 48-μm alumina results in higher bond strengths than use of 10-, 34-, or 78-μm particles. Sharp alumina particles appear more effective for bonding pretreatment than rounded particles.

The bond strength of an adhesive composite (Geristore) to porcelain is about 10 MPa (Berry and Powers, 1994b), as shown in Table 6. The bond strength is improved by about 33% when the adhesive composite is used in conjunction with an NTG–GMA/PMDM dentin bonding agent (Tenure).

B. Resin Cements for Orthodontic Brackets

The bond strength of resin cements to ceramic orthodontic bracket bases ranges from 5 to 11 MPa and depends on the type of cement, as shown in Table 21. Bonding may be mechanical or chemical, or both. The failure of ceramic brackets depends on the retention

Table 23 *In Vitro* Bond Strengths of Composite to Porcelain with Various Surface Pretreatments and Primers

Primer	Bond strength, MPa		
	Sandblasted	HF etch	Diamond abrasion
Silane/phosphonate (Clearfil Porce-lain Bond)	23.4[a]	20.0[a]	23.5[a]
Silane/phosphonate (Scotchprime)	23.7[b]	17.2[a]	15.0[b]
Silane/4-META (Porcelain Liner M/ Super-Bond C&B)	20.4[b]	16.9[a]	19.1[b]

Source: Wolf et al. (1992)
[a]Cohesive failure in porcelain
[b]Adhesive failure

Table 24   *In Vitro* Tensile Bond Strengths of Resin and Hybrid Ionomer Cements to Ceramic, Metal, and Plastic Orthodontic Brackets

	Bond strength, MPa	
Bonding technique	Resin cement	Hybrid ionomer cement
Ceramic bracket (silanated)	6.9	7.4
Ceramic bracket (silanated) with polycarbonate base	7.1	4.5
Ceramic bracket (mechanical)	7.4	5.8
Metal bracket (silanated)	6.0	3.7
Plastic bracket (metal reinforced)	10.3	1.4

Source: Blalock (1993)

features of the base and on the presence of silanation. Recently, Blalock (1993) has measured the bond strength of resin and hybrid ionomer cements to polycrystalline alumina brackets as shown in Table 24. Bond failures of these ceramic brackets occur mostly at the bracket–cement interface.

X.   SUBSTRATES OF ACRYLIC DENTURE RESIN AND PLASTIC ORTHODONTIC BRACKETS

Acrylic resins based on poly(methyl methacrylate) are used extensively in prosthetic dentistry in the construction of full and partial dentures (Craig et al., 1993). Plastic orthodontic brackets are usually polycarbonate, although brackets fabricated from composite resin are available.

A.   Acrylic Repair Resins

Repair materials for acrylic dentures are usually powder–liquid acrylic plastics that are heat accelerated or chemically accelerated (Craig, 1993). The transverse strength of a heat-accelerated repair is about 80% of that of the original acrylic, whereas the strength of a chemically accelerated repair is only about 60% of the original acrylic. Recently, a light-activated acrylic has been reported to be fast and effective.

B.   Resin Cements for Orthodontic Brackets

The bond strength of resin cements to plastic orthodontic bracket bases ranges from 8 to 11 MPa and depends on the type of cement, as shown in Table 21. Bonding to plastic bases by resin cements appears to be chemical and is improved dramatically by the use of monomeric bracket primers (de Pulido and Powers, 1983). Plastic brackets tend to fail at the wings or within the bracket rather than debonding.

ACKNOWLEDGMENTS

The author thanks Lisa Jacob for assisting with the review of the literature and Upasana Vaid for proofreading the manuscript.

GLOSSARY

Acronyms

Bis-GMA: 2,2-bis[4(2-hydroxy-3-methacryloyloxy-propyloxy)-phenyl] propane or bis-phenol A–glycidyl methacrylate
BPDM: biphenyl dimethacrylate
EDTA: ethylene diamine tetra-acetic acid
HEMA: hydroxyethyl methacrylate
MMA: methyl methacrylate
NTG–GMA: *N*-tolylglycine-glycidyl methacrylate
PAA: polyacrylic acid
PENTA: phosphonated penta-acrylate ester
PMDM: pyromellitic diethylmethacrylate
4-META: 4-methacryloxyethyl-trimellitic anhydride
TBB: tri-*n*-butylborane
UDMA: urethane dimethacrylate

Terminology

Amalgam: alloy that results from the combination of mercury with an alloy containing silver, tin, and copper
Composites: restorative material or cement, usually based on Bis-GMA resin filled with silane-treated, irregularly shaped glass particles
Dentin (*human*): 70% by weight hydroxyapatite, 18% organic material (primarily collagen), and 12% water
Enamel (*human*): 96–97% by weight hydroxyapatite; remainder is water and about 1% organic matter
Glass ionomer (water-based): restorative material or cement composed of ionic polymer (copolymeric carboxylic acid), calcium aluminum fluorosilicate glass, water, tartaric acid
Hybrid ionomer: restorative material or cement composed of polycarboxylic acid with pendant methacrylate groups, a crosslinking component (HEMA), silicate glass, and photoinitiators
Ni–Cr–Be alloy: nickel–chromium–beryllium alloy
Oligomer: high molecular weight monomer
Phosphonate: methacryloxyethylphenyl phosphate

REFERENCES

Aksu, M. N., Powers, J. M., Lorey, R. E., and Kolling, J. N. (1987). Variables affecting bond strength of resin-bonded bridge cements. *Dent. Mater., 3*, 26–28.

Asmussen, E., De Araujo, P. A., and Peutzfeldt, A. (1989). *In vitro* bonding of resins to enamel and dentin: An update. *Trans. Acad. Dent. Mater., 2*, 36–63.

Barakat, M. M., and Powers, J. M. (1986). *In vitro* bond strength of cements to treated teeth. *Aust. Dent. J., 31*, 415–419.

Barakat, M. M., Powers, J. M., and Yamaguchi, R. (1988). Parameters that affect *in vitro* bonding of glass-ionomer liners to dentin. *J. Dent. Res., 67*, 1161–1163.

Berry, E. A., III. (1991). *Factors affecting the bond strength of glass ionomer cement to dentin*. Master's Thesis, University of Texas Health Science Center at Houston, Dental Branch.

Berry, E. A., III, and Powers, J. M. (1992). Bond strength of glass ionomers to smooth and rough dentin (abstract). *J. Dent. Res., 71* (Special Issue), p. 632.

Berry, E. A., III, Rainey, J. T., and Powers, J. M. (1993). Bond strength of hybrid ionomer to enamel treated by KCP-2000. Trans. Second International Congress on Dental Materials, p. 302.

Berry, E. A., III, and Powers, J. M. (1994a). Bond strength of glass ionomers to coronal and radicular dentin. *Oper. Dent., 19*, 122–126

Berry, E. A., III, and Powers, J. M. (1994b). Bond strength of adhesive composites to dental substrates. *J. Prosthodont., 3*, 126–129.

Blalock, K. A. (1993). *In vitro bond strength of new esthetic orthodontic brackets and cements*. Master's Thesis, University of Texas Health Science Center at Houston, Dental Branch.

Blosser, R. L. (1990). Time dependence of 2.5% nitric acid solution as an etchant on human dentin and enamel. *Dent. Mater., 6*, 83–87.

Boyer, D. B., Chan, K. C., and Reinhardt, J. W. (1984). Build-up and repair of light-cured composites: Bond strength. *J. Dent. Res., 63*, 1241–1244.

Buonocore, D. H. (1955). A simple method of increasing the adhesion of acryl filling materials to enamel surfaces. *J. Dent. Res., 34*, 849–853.

Buzzitta, V. A. J., Hallgren, S. E., and Powers, J. M. (1982). Bond strength of orthodontic direct-bonding cement-bracket system as studied *in vitro. Am. J. Orthod., 81*, 87–92.

Chang, J. C., Powers, J. M., and Hart, D. (1993). Bond strength of composite to alloy treated with bonding systems. *J. Prosthodont., 2*, 110–114.

Council on Dental Materials, Instruments, and Equipment. (1991). Porcelain repair materials. *J. Am. Dent. Assoc., 122*, 124–130.

Covey, D. A., and Moon, P. C. (1991). Shear bond strength of dental amalgam bonded to dentin. *Am. J. Dent., 4*, 19–22.

Craig, R. G. (ed.) (1993). *Restorative Dental Materials*, 9th ed. Mosby-Year Book, St. Louis.

Craig, R. G., O'Brien, W. J., and Powers, J. M. (1993). *Dental Materials Properties and Manipulation*, 5th ed. Mosby-Year Book, St. Louis.

de Pulido, L. G., and Powers, J. M. (1983). Bond strength of orthodontic direct-bonding cement-plastic bracket systems *in vitro. Am. J. Orthod., 83*, 124–130.

DeSchepper, E. J., Cailleteau, J. G., Roeder, L. B., and Powers, J. M. (1991). *In vitro* tensile bond strengths of amalgam to treated dentin. *J. Esthet. Dent., 3*, 117–120.

DeSchepper, E. J., Tate, W. H., and Powers, J. M. (1993). Bond strength of resin cements to microfilled composites. *Am. J. Dent., 6*, 235–238.

Dickinson, P. T. , and Powers, J. M. (1980). Evaluation of fourteen direct-bonding orthodontic bases. *Am. J. Orthod., 78*, 630–639.

Douglas, W. H. (1989). Clinical status of dentine bonding agents. *J. Dent., 17*, 209–215.

Driessens, F. C. M., and Verbeeck, R. M. H. (1990). *Biominerals*. CRC Press, Boca Raton, FL.

Edler, T. L., Krikorian, E., and Thompson, V. P. (1991). FTIR surface analysis of dentin and dentin bonding agents (abstract). *J. Dent. Res., 70*, (Special Issue), p. 458.

Eichmiller, F. C. (1993, Sept.–Oct.). A clinical perspective on dentin adhesives. *Ind. Dent. Assoc. J.*, pp. 22–24.

Eick, J. D., Wilko, R. A., Anderson, C. H., and Sorensen, S. E. (1970). Scanning electron microscopy of cut tooth surfaces and identification of debris by use of the electron microprobe. *J. Dent. Res., 49*, 1359–1368.

Eliades, G. C., Palaghias, G., and Vougiouklakis, G. J. (1990). Surface reactions of adhesives on dentin. *Dent. Mater., 6*, 208–216.

Erickson, R. L. (1989). Mechanism and clinical implications of bond formation for two dentin bonding agents. *Am. J. Dent., 2*, 117–123.

Evans, L. B., and Powers, J. M. (1985). Factors affecting *in vitro* bond strength of no-mix orthodontic cements. *Am. J. Orthod., 87*, 508–512.

Farah, J. W., and Powers, J. M. (eds.) (1991). Dentin bonding agents and adhesive cements. *The Dental Advisor, 8*(3), 2–8.

Farah, J. W., and Powers, J. M. (eds.) (1993a). Dental ceramics. *The Dental Advisor, 10*(3), 8.

Farah, J. W., and Powers, J. M. (eds.) (1993b). Dental cements. *The Dental Advisor, 10*(4), 7–8.

Fowler, C. S., Swartz, M. L., Moore, B. K., and Rhodes, B. F. (1992). Influence of selected variables on adhesion testing. *Dent. Mater., 8*, 265–269.

Garberoglio, R., and Brannstrom, M. (1976). Scanning electron microscopic investigation of human dentinal tubules. *Arch. Oral Biol., 21*, 355–362.

Gray, S., and Burgess, J. (1989). *In vivo* and *in vitro* comparison of dentin bonding agents (abstract). *J. Dent. Res., 68* (Special Issue), p. 375.

Gwinnet, A. J. (1990). Interactions of dental materials with enamel. *Trans. Acad. Dent. Mater., 3*, 30–54.

Hadavi, F., Hey, J. H., and Ambrose, E. R. (1991). Shear bond strength of composite resin to amalgam: An experiment *in vitro* using different bonding systems. *Oper. Dent., 16*, 2–5.

Heymann, H. O., and Bayne, S. C. (1993). Current concepts in dentin bonding. *J. Am. Dent. Assoc., 124*, 27–36.

Hibler, J. A., Foor, J. L., Miranda, F. J., and Duncanson, M. G. (1988). Bond strength comparisons of repaired dental amalgams. *Quintessence Int., 19*, 411–415.

Jendersen, M. D., and Glantz, P.-O. (1981). Clinical adhesiveness of selected dental materials — An *in vivo* study. *Acta Odontol. Scand., 39*, 39–45.

Kanca, J., III. (1991). Dentin adhesion and the All-Bond system. *J. Esthet. Dent., 3*, 129–132.

Mangum, F. I., Jeske, A. H., Chan, J. T., and Powers, J. M. (1993). Effect of fluoride iontophoresis on dentinal bonding. *Gen. Dent., 41*, 139–142.

Marshall, G. W., Jr. (1993). Dentin: Microstructure and characterization. *Quintessence Int., 24*, 606–617.

Matsumura, H., Kawahara, M., Tanaka, T., and Atsuta, M. (1991). Surface preparations for metal frameworks of composite resin veneered prostheses made with an adhesive opaque resin. *J. Prosthet. Dent., 66*, 10–15.

McGuckin, R. S., Powers, J. M., and Li, L. (1994). Bond strengths of dentinal bonding systems to enamel and dentin. *Quintessence Int., 25*, 791–796.

McGuckin, R. S., Tao, L., Thompson, W. O., and Pashley, D. H. (1991). Shear bond strength of Scotchbond *in vivo*. *Dent. Mater., 7*, 50–53.

Meryon, S. D., Tobias, R. S., and Jakeman, K. J. (1987). Smear removal agents: A quantitative study *in vivo* and *in vitro*. *J. Prosthet. Dent., 57*, 174–179.

Nabadalung, D. P., Powers, J. M., and Connelly, M. E. (1991a). Bond strength of traditional and adhesive denture resins to Ticonium. *Trans. Acad. Dent. Mater., 4*, 116–117.

Nabadalung, D. P., Powers, J. M., and Connelly, M. E. (1991b). Bond strength of denture resin to treated partial denture alloy (abstract). *J. Dent. Res., 70* (Special Issue), p. 433.

Nakabayashi, N., Kojjima, K., and Masuhara, E. (1982). The promotion of adhesion by infiltration of monomers into tooth substrates. *J. Biomed. Mater. Res., 16*, 265–273.

O'Keefe, K. L., Powers, J. M., McGuckin, R. S., and Pierpont, H. P. (1992). *In vitro* bond strength of silica-coated metal posts in roots of teeth. *Int. J. Prosthodont., 5*, 373–376.

Olsson, S., Oilo, G., and Adamczak, E. (1993). The structure of dentin surfaces exposed for bond strength measurements. *Scand. J. Dent. Res., 101*, 180–184.

Pashley, D. H. (1990). Interactions of dental materials with dentin. *Trans. Acad. Dent. Mater., 3*, 55–73.

Pashley, D. H., Ciucchi, B., Sano, H., and Horner, J. A. (1993). Permeability of dentin to adhesive agents. *Quintessence Int., 24*, 618–631.

Pounder, B., Gregory, W. A., and Powers, J. M. (1987). Bond strengths of repaired composite resins. *Oper. Dent., 12*, 127–131.

Powers, J. M. (1991). Adhesive resin cements. *Shigaku, 79*, 1140–1143.

Powers, J. M., Watanabe, F., and Lorey, R. E. (1986). *In vitro* evaluation of prosthodontic adhesives. In *Adhesive Prosthodontics—Adhesive Cements and Prosthodontics* (L. Gettleman, M. M. A. Vrijhoef, and Y. Uchiyama, eds.), Academy of Dental Materials, Nijmegen.

Pratt, C., Simpson, M., Mitchem, J., Tao, L., and Pashley, D. H. (1992). Relationship between bond strength and microleakage measured in the same class I restorations. *Dent. Mater., 8*, 37–41.

Qvist, V., and Qvist, J. (1985). Effect of ethanol and NPG–GMA on replica patterns on composite restorations performed *in vivo* in acid-etched cavities. *Scand. J. Dent. Res., 93*, 371–376.

Roeder, L. B., DeSchepper, E. J., and Powers, J. M. (1991). *In vitro* bond strength of repaired amalgam with adhesive bonding systems. *J. Esthet. Dent., 3*, 126–128.

Roeder, L. B., Fulton, R. S., and Powers, J. M. (1991). Bond strength of repaired glass ionomer core materials. *Am. J. Dent., 4*, 15–18.

Rueggeberg, F. A. (1991). Substrate for adhesion testing to tooth structure—Review of the literature. Report of ASC MD 156 Task Group. *Dent. Mater., 7*, 2–10.

Ruse, N. D., and Smith, D. C. (1991). Adhesion to bovine dentin—Surface characterization. *J. Dent. Res., 70*, 1002–1008.

Schneider, W., Powers, J. M., and Pierpont, H. P. (1992). Bond strength of composites to etched and silica-coated porcelain fusing alloys. *Dent. Mater., 8*, 211–215.

Shimizu, A., Takashi, U., and Kawakami, M. (1986). Bond strength between amalgam and tooth hard tissues with application of fluoride, glass ionomer cement and adhesive resin cement in various combinations. *Dent. Mater. J., 5*, 225–232.

Siomka, L. V., and Powers, J. M. (1985). *In vitro* bond strength of treated direct-bonding metal bases. *Am. J. Orthod., 88*, 133–136.

Smith, D. C., and Ruse, N. D. (1987). Adhesion to dentin—characterization of the substrate. *Trans. Soc. Biomater., 10*, 153.

Soderhohn, K.-J. M. (1991). Correlation of *in vivo* and *in vitro* performance of adhesive restorative materials: A report of the ASC MD 156 Task Group on Test Methods for the Adhesion of Restorative Materials. *Dent. Mater., 7*, 74–83.

Stanford, J. W., Sabri, Z., and Jose, S. (1985). A comparison of the effectiveness of dentine bonding agents. *Int. Dent. J., 35*, 139–141.

Staninec, M., and Holt, M. (1988). Bonding of amalgam to tooth structure: Tensile adhesion and microleakage tests. *J. Prosthet. Dent., 59*, 397–402.

Suh, B. I. (1991). All-Bond: Fourth generation dentin bonding system. *J. Esthet. Dent., 3*, 139–146.

Swift, E. J., Denehy, G. E., and Beck, M. D. (1993). Use of phosphoric acid etchants with Scotchbond Multi-Purpose. *Am. J. Dent., 6*, 88–90.

Tam, L. E., and Pilliar, R. M. (1993). Fracture toughness of dentin/resin–composite adhesive interfaces. *J. Dent. Res., 72*, 953–959.

Tate, W. H., DeSchepper, E. J., and Powers, J. M. (1993). Bond strength of resin cements to a hybrid composite. *Am. J. Dent., 6*, 195–198.

Torney, D. L. (1978). The retentive ability of acid-etched dentin. *J. Prosthet. Dent., 39*, 169–172.

Tyler, M., Charbeneau, G., Dennison, J., Heys, D., and Fitzgerald, M. (1987). *In vivo* and *in vitro* tensile bond strengths of a glass ionomer cement (abstract). *J. Dent. Res., 66* (Special Issue), p. 112.

Van Meerbeek, B. (1993). *Dentine adhesion: Morphological, physico-chemical and clinical aspects.* Thesis, Katholieke Universiteit te Leuven, Leuven, Belgium.

Varga, J., Matsumura, H., and Masuhara, E. (1986). Bonding of amalgam filling to tooth cavity with adhesive resin. *Dent. Mater. J., 5*, 158–164.

Watanabe, F., Powers, J. M., and Lorey, R. E. (1988). *In vitro* bonding of prosthodontic adhesives to dental alloys. *J. Dent. Res., 67*, 479–483.

Willems, G., Lambrechts, P., Braem, M., and Vanherle, G. (1993). Composite resins in the 21st century. *Quintessence Int., 24,* 641–658.

Wolf, D. M., Powers, J. M., and O'Keefe, K. L. (1992). Bond strength of composite to porcelain treated with new porcelain repair agents. *Dent. Mater., 8,* 158–161.

Wolf, D. M., Powers, J. M., and O'Keefe, K. L. (1993). Bond strength of composite to etched and sandblasted porcelain. *Am. J. Dent., 6,* 155–158.

Wright, W. L., and Powers, J. M. (1985). *In vitro* tensile bond strength of reconditioned brackets. *Am. J. Orthod., 87,* 247–252.

Xie, J., Powers, J. M., and McGuckin, R. S. (1993). *In vitro* bond strength of two adhesives to enamel and dentin under normal and contaminated condition. *Dent. Mater., 9,* 295–299.

Yamaguchi, R., Powers, J. M., and Dennison, J. B. (1989). Parameters affecting *in vitro* bond strength of composites to enamel and dentin. *Dent. Mater., 5,* 153–156.

56
Fracture Mechanics of Dental Composites

J. L. Drummond, D. Zhao, and J. Botsis
University of Illinois at Chicago
Chicago, Illinois

I. INTRODUCTION AND REVIEW

The prediction of lifetime service of dental restorations and the characterization of material properties have assumed positions of prominence in dentistry. Consequently, the understanding and modeling of restorative systems are issues of central importance to dental practitioners and material scientists alike.

In general, dental composites fail when their load-bearing capacity is compromised by damage due to materials degradation. Damage in dental composites may take the form of matrix and/or filler deterioration due to mechanical and/or environmental loads, interfacial debonding, microcracking, etc. Progressive degradation eventually leads to crack initiation and growth. Under the continuous application of mechanical and/or environmental loads, catastrophic failure of a dental restoration may occur.

The use of fracture mechanics to explain and understand the fracture and clinical properties of dental composite restorative materials has increased in the past 10 years. Most of the published work is concerned with mode I (Fig. 1) straight-line crack growth and with toughness characterization of various composites that have been exposed to air, water, ethanol, and other environments [1].

Söderholm and Roberts found that specimens stored dry were significantly stronger then those stored wet for 6 months or those stored wet and then dried [2]. The aging in water appears to increase filer particle pullout on the fractured surface, possibly due to breakdown of the silane bond between the resin and the filler particle. Storage in a lactate buffer (pH = 4) caused a decrease in surface hardness of dental composites over a 12-month span [3]. Depending on the solution and the commercial composite, the onset of change in surface hardness varied. Ferracane and Marker used fracture toughness specimens 17 mm × 3 mm × 1.5 mm with an a/w ratio of 0.5 and utilized a razor blade insert to form the starter crack [4]. Following the ASTM standard E399, they found no

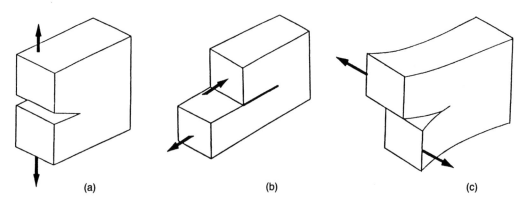

Figure 1 Three different possible modes of fracture: (a) normal mode, (b) shear or sliding mode, and (c) tearing mode.

difference in wet versus dry testing. Aging in water for 14 months had no statistically significant effect on K_{IC} for the filled composites or the unfilled resin. Observations on the fracture surface showed no difference due to aging conditions. A significant reduction in K_{IC} was observed in ethanol after 14 months of aging, but the value at the end of 14 months was the same as after 2 months of aging. Ferracane and Condon used single edge notched specimens 25 mm × 5 mm × 2.5 mm with an a/w ratio of 0.5 and a loading rate of 1.27 mm/min for heat-treated specimens at 120°C and found increases in K_{IC} (16–47%) and E (12–60%) [5]. The heat treatment of the small particle hybrid enhanced crack propagation through the matrix, resulting in less filler–matrix debonding. For the microfill, the heat treatment resulted in less filler–matrix delamination and enhanced fracture of the prepolymerized resin filler. The increases in K_{IC} and E were attributed to an increase in the degree of conversion. Kovarik et al. found no difference from the ASTM standard for small (6.3 mm × 6.6 mm × 1.7 mm) versus large (14 mm × 15 mm × 4 mm) fracture toughness specimens [6]. Lloyd found that microfine fillers lowered fracture toughness more than small or large filler particles, with fracture occurring through the resin matrix [7]. As the filler content increased there was an increase in K_{IC} and then a decrease. Ferracane et al. used two sets of single edge notched (SEN) specimens, precracked and not precracked, 30 mm × 4.5 mm × 2.1 mm, with no surface polish [8]. Specimens were stored in water 24 h before testing. Precracks were formed immediately after curing because the specimens were too brittle to form a precrack after 24 h. There was no correlation between degree of cure and fracture toughness. Correlation ($r = 0.79$) was higher for percent volume filler versus fracture toughness. The fracture toughness was greater for the filled composites than the unfilled resin. This increase in fracture toughness was attributed to increased fracture energy due to crack pinning and bowing between particles rather than increased fracture surface energy, as compared to unfilled resin. Pilliar et al. found a decrease in fracture toughness with aging over 1 month using a short rod fracture specimen, 9 mm with a 6-mm diameter, but no change under 1 month [9]. Specimens were kept moist but not immersed. Microfills had a lower fracture toughness than small-particle composites. Aging in ethanol caused a significant decrease in the fracture toughness [10]. Goldman used a double torsion specimen 30 × 10 × 2 mm with a groove along the length of the specimen, with no surface polish [11]. He attributed fracture toughness to a crack-pinning mechanism of hard particles in a soft matrix. The fracture toughness reached a maximum, then fell off due to overlapping of particle strain energy fields. The dispersed phase increased E and K_{IC} and lowered tensile strength, with the particles acting as stress concentrators and crack precursors. An increase in the filler particle size resulted in an increase in the inherent flaw size. Mazer et al. attributed the marginal degradation of composites to tensile fatigue failure [12]. Water absorption caused a 1.0% increase in volume; and polymerization shrinkage, a 2.5% decrease. The hydroscopic expansion may generate a positive pressure up to 15 MPa [13]. Cook and Johannson attributed wear to thermomechanical fatigue, that is, the differences in the thermal coefficient of expansion of the resin matrix, the inorganic filler particles, and the supporting tooth structure [14]. Tyas found a significant correlation between fracture toughness, tensile strength, and elastic modulus [15]. Kovaric and Ergle evaluated long-term aging in water and air of dental composites [16]. The fracture toughness for dry specimens at 7 days was 1.69 ± 0.13 MPa·m$^{1/2}$ and at 4 years 1.68 ± 0.19 MPa·m$^{1/2}$, and for wet specimens at 7 days 1.94 ± 0.13 MPa·m$^{1/2}$ and at 4 years 1.55 ± 0.18 MPa·m$^{1/2}$. Jones et al. stored composite specimens in water up to 29 months and found no effect on the elastic modulus [17]. A more recent study on apparent fracture toughness, K_Q, found that specimens with a starter crack had a $K_Q = 2.20$ MPa·m$^{1/2}$; whereas specimens with a crack generated by compressive fatigue had a value of $K_Q = 1.69$ MPa·m$^{1/2}$, leading the authors to state that using a compressive fatigue crack would give a more conservative value of the fracture toughness [18].

Although mode I fracture toughness characterization (Fig. 1) of dental composites has been well investigated, research on mechanical and/or environmental fatigue, fracture toughness, and wear under complex loading conditions has been very limited.

The existence of real mechanical fatigue in brittle composite materials has been evident for some time. However, the mechanism(s) for crack growth is still in a stage of speculation and the contribution of mechanical fatigue to the overall reliability and lifetime in a dental restoration has not been resolved. It has been suggested that plastic deformation and fracture surface aspiration under compressive loading may cause crack growth. Friction heating at the crack tip has also been proposed as a possible source of mechanical fatigue [19]. Wright and Burton found that fracture mechanics play a large part in determining the wear of the harder and more brittle materials (i.e., the filler) [10]. Fan and Powers observed that the surface morphology became more severe with an increase in accelerated aging [20]. Small-particle composites were characterized by formation of microcracks and exposure and loss of filler particles. Microfills initially formed isolated microcracks that became more extensive with aging.

It has been suggested that fatigue resistance of dental composites may be directly related to clinical wear. Factors that influence wear are hardness, adhesion efficiency at the interface between the reinforcing particles and the matrix, level and speed of applied load, surface roughness, and temperature. Commercial products have also been classified according to their morphological and mechanical characteristics, with Young's modulus, surface roughness, and Vickers hardness determined to be the most important parameters [21].

Several key aspects of wear have been characterized in the last two decades. A quantitative description is given, on a phenomenological level, in terms of the volume of material worn away during wear. It is obtained from measurements before and after wear, and subsequently correlated with the applied force, a wear coefficient, sliding distance, and hardness. Another type of wear that has been considered recently in composite research is that of fretting wear, that is, materials response at joints with other parts. In fretting wear tests, the material is cut into a small rectangular specimen, which is fixed on a specimen holder. Fretting pins slide in an oscillating manner with their front sides against the specimen surface. A measure of fretting wear is taken as the mass loss divided by the material's density, applied force, and number of loading cycles [22].

Wu et al. found subsurface damage beneath all surfaces exposed to intraoral environments for all *in vivo* composite restorations studied [23,24]. The *in vitro* observations of wear were: (1) low wear with no subsurface damage, (2) high wear with considerable damage attributed to fatigue, and (3) regions of catastrophic wear resulting from extensive fatigue damage and the existence of fatigue cracks. The absorption and desorption of water also has an effect on the mechanical properties of composite dental materials [25,26].

In addition, Troung and Tyas's theoretical predictions for good wear resistance of composites required: (1) a high fracture toughness, (2) a small inherent flaw size, and (3) a high crazing stress [27]. They used double torsion specimens ($20 \times 30 \times 2.3$ mm) with a precrack of $a = 0.5$ mm. No difference was found between specimens tested wet or dry, but for those tested in ethanol–water, the ethanol acted as a solvent on the composite. The fatigue properties increased with filler content, degree of cure, and degree of silanization.

Although there is no direct quantitative link between the mechanics of wear and fracture, wear loading introduces surface and/or subsurface damage, which leads to cracks that are at an angle with the surface [28]. Accordingly, these cracks are under modes I and II fracture (Fig. 1). That is, in addition to a normal stress component, shear stresses drive the fracture process. At this point mixed-mode fracture mechanics can be applied to characterize and predict the behavior of these surface or subsurface cracks. Therefore, failure of a composite restoration due to wear may be characterized to a certain extent with the use of fracture mechanics parameters if the stress state is properly defined. The fracture toughness of ceramics in combined mode I and mode II loading has been able to differentiate an increased fracture resistance in mode II loading due to grain interlocking and abrasion [29]. Fractography verified an increased percentage of transgranular fracture of the grains in mode II loading.

Indeed, the theoretical considerations of Hills and Ashelby indicate that two crack growth processes proceed simultaneously under wear conditions [30]. The movement and application of a force over a subsurface flaw causes it to experience two distinct growth periods. The first growth period occurs just before the contact point reaches the flaw and consists of mode II stress intensification. The other growth period occurs following the contact point and consists of both mode I and mode II openings (Fig. 1).

Sakaguchi et al. stated that low wear rates predispose to fracture those composites that have higher crack propagation rates than porcelain or natural teeth [31]. The calculated crack propagation rates were based on the assumptions of complete bonding and a homogeneous material. They determined that a defect of 1 μm initiates fatigue fracture within a dental restoration under physiological loads. Since the composite wear rate is less than the crack propagation rate, the composite is inclined to fatigue fracture. Truong et al. stated that fatigue wear is the main mechanism of clinical wear of composites in posterior teeth [32]. Crack growth was studied using a double cantilever beam specimen with a 0.55 mm deep machined groove (1/4 of specimen thickness) down the center of the specimen to avoid a curved crack path [32]. They assumed that the groove's effect was not significant. This experimental setup, however, forces the crack to follow the direction of the groove rather than the crack direction being determined by the microstructure of the composite.

Curved crack paths have been observed in our studies and in brittle composite materials in general. This is shown in Fig. 2 where crack trajectories of different fatigue experiments have been superimposed [33]. These results clearly indicate that the groove may be appropriate for determining fracture toughness, but not for crack growth studies.

The results of numerous investigations have indicated that linear elastic fracture mechanics (LEFM) could be employed to characterize the fracture toughness of dental composites. However, only mode I loading conditions have been investigated thoroughly. Fatigue crack growth studies as well as fracture toughness characterization under complex loading conditions have been very limited. This is perhaps due to difficulties associated with crack growth along a straight line in dental composites, the simplicity of mode I fracture toughness testing, and the fact that the dissipative mechanisms, such as irreversible deformation and/or microcracking, needed to stabilize the crack are very limited [34]. Accordingly, when the crack tip approaches a void, segregated particles, or a fiber, it easily deviates from the linear path without returning to its initially straight crack growth mode.

Thus, failure due to fatigue crack growth in dental composites could fall into one of the following two categories depending on the loading conditions, specimen configuration, and material.

1. Crack initiation from a notch of predetermined size is followed by fast crack growth and specimen fracture.
2. Crack initiation is followed by a limited crack growth along a straight line and then by a curved crack growth before rapid specimen fracture.

To avoid the possibility of a curved crack path, researchers have introduced grooves on both sides of the specimen along lines that originate from the notch tip [32]. The presence of the groove forces that crack to grow along the groove and, thus, on a straight path. Such an experimental design is not realistic for crack growth studies since the crack path may not result from the material morphology but from the groove itself. Thus, such studies may not reflect the actual fatigue fracture behavior of these materials.

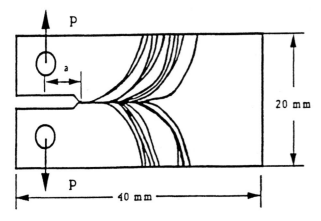

Figure 2   Superimposed crack trajectories for 18 specimens fractured under cyclic fatigue loading.

In many instances, fracture and failure of dental composites occur from a surface or subsurface crack, or a flaw that is oriented at an angle with respect to that applied load (mode I and mode II loading) [28]. These flaws are usually introduced during material processing. Accordingly, normal (mode I) and shear (mode II) loads drive the crack. Inclined flaws or cracks are also observed in wear loading in many composite materials. In fact, it is believed that fracture characteristics of any composite material could be realistically investigated under combined fracture modes because the highly heterogeneous materials' microstructures give rise to curved crack paths (Fig. 2). In addition, it offers a more realistic approach to fatigue and fracture of dental restorations since it is more likely that a flaw is at an angle with the force of mastication.

An important factor in the overall chemical, physical, and mechanical response of any composite system, is the adhesion efficiency at the interface of the matrix and reinforcing particles. It is well recognized in the mechanics of composites community that a complex situation develops around a reinforcing particle. Namely, there are areas that consist of imperfect bonding, areas of adsorption interaction in polymer surface layers onto the filler particle, residual stresses due to thermal mismatch, and high stress concentrations due to the imperfections around the reinforcing particle and diffusion of water in the filler and resin.

Although theoretical treatments addressing the problem of interphase have been proposed, these are for ideal systems and incorporate some unrealistic assumptions regarding the distribution of reinforcing particles and the material at the interphase [35]. This is because, in a real composite system, it is very difficult or even impossible to measure the local properties of the interphase as well as the spatial distribution of the reinforcing particles. Namely, the particles are not uniformly distributed and of a spherical shape, and in many instances clusters of particles are formed during processing. This situation makes application of the existing theoretical models difficult.

During exposure to various environments, dental composites are subjected to material property changes due to degradation and aging. The changes of the composite properties are due to the diffusion of the environment through the matrix and/or applied load. The diffusion of moisture through the resin may lead to nucleation and growth of microcracks at the interphase and the resin, which assist further the path for the environment. These cracks, however, are not always normal to the applied load. In addition, diffusion could be affected by the reinforcing particles in a number of different ways. Firstly, the diffusion path is dependent upon the arrangement of the reinforcement particles as well as the volume fraction. Secondly the interface has different diffusion characteristics than the matrix material. Accordingly, wicking, or capillary action, may occur along the interface, causing rapid diffusion. Moreover, the diffusion process and hence the rate of property degradation are increased by the applied stress. Therefore, irreversible processes, such as the following, contribute to irreversible material degradation.

1. Chemical breakdown by hydrolysis [24,25,36–42]
2. Chemical breakdown by stress-induced effects associated with swelling and applied stress [36–45]
3. Chemical composition changes by leaching [36–45]
4. Precipitation and swelling phenomena to produce voids and cracks [24,25]
5. Loss of strength due to corrosion [23,24]

The environment could also affect the strength of the reinforcing glass particles. That is, glass particles are susceptible to water leaching of the soluble oxides such as K_2O, SrO_2, and Na_2O and hydrolysis of the $Si-O$ bonds.

In an attempt to expedite aging in various environments, researchers performed accelerated aging. Reported studies, however, showed that accelerated aging of dental composites causes microcracks, possibly due to residual stresses and exposure of filler particles, which results in changes in the mechanical properties of the surface layer. The glass–water reaction is accelerated by stress that allows water to enter glass during deformation [46]. Subsurface damage during fatigue is accentuated by oral fluids and water. Polymerization shrinkage or mastication would serve as the source of stress allowing water to enter silica glass during slow crack growth. The degraded surface would be removed more easily by abrasion, erosion, or other mechanisms of wear. In normal function or abnormal bruxism, a freshly worn surface would continue the cycle, providing foci for microcrack initiation on the surface.

Although progress has been reported towards a better understanding of the mechanical and fracture behaviors of dental composite materials, a number of issues still remain unresolved. In particular, the responses of dental restorative materials under complex loading conditions and various environments need further experimental and analytical work.

The specific aims of our research were to investigate fracture pathways, fracture characteristics, and chemical compositional changes of an admixed hybrid dental composite and unfilled resin. Flexure strength and modulus, mode I and combined modes I and II fracture toughness characteristics, and fatigue crack propagation were addressed on specifically made dental composites. In addition, the effects of aging in air and water on strength modulus and mode I fracture toughness were studied.

II. EXPERIMENTAL PROCEDURES

A. Material Compositions and Specimen Preparation

The composite materials were 75% or 79% by weight strontium glass filler (small particle 1-8 mm) without or with colloidal silica (0.04 mm) using a 60% Bis-GMA and 40% TEGDMA resin. The colloidal silica when used was 10% by weight of the glass filler in a ratio of 9 (small particle) to 1 (microfill). The starting composition of the glass filler as determined by energy dispersive x-ray analysis was 59% SiO_2, 10% Al_2O_3, 30% SrO_2, and 1% BaO_2. The specimens were aged (6 or 12 months) in either air or distilled water at 37°C and were tested in their respective aging media at a loading rate of 1.22 mm/sec. Details of specimen preparation are given by Drummond et al. [47].

Four different composites as well as the unfilled resin were used in this investigation. The compositions for each material are shown in Table 1. The degree of polymerization was analyzed with the use of a Fourier transform infrared spectrometer (FX-6200 FTIR, Analect Instruments, Irvine, CA). It was found that the degrees of polymerization in the 75Sr and 75Sr10 materials were 52% and 56%, respectively. The results for these specially prepared composites were very close to reported ranges for the degree of conversion of most commercial composites [48].

Prismatic bars with dimensions 2 × 4 × 70 mm were prepared for flexure strength tests (Figs. 3a and 3b). Mode I fracture toughness tests were performed on single edge notched (SEN) specimens with the same dimensions. A 60° V notch of 0.5 mm in depth

Table 1   Composite Filler Compositions

| | Strontium | Colloidal | Total (% by weight) ||
Composite	glass (%)	silica (%)	Filler	Resin
Resin	0	0	0	100
75Sr	100	0	75	25
75Sr10	90	10	75	25
79Sr	100	0	79	21
79Sr10	90	10	79	21

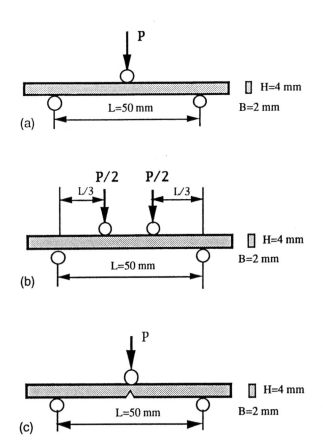

Figure 3   Specimen geometries and loading configurations: (a) three-point loading, (b) four-point loading, (c) single edge notched specimen, (d) precracked specimen in diametral compression, and (e) double cantilever beam.

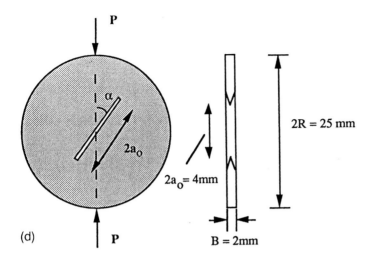

(d)

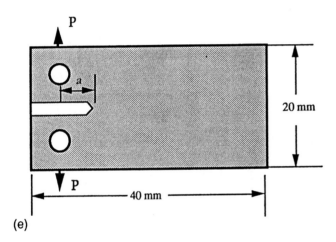

(e)

was machined at the midspan of the bars (Fig. 3c). Studies on the effects of notch tip geometry on fracture toughness were carried out on specimens with straight and 60° V notches using the 75Sr10 composite. In each case, three notch depths were examined; a = 0.5, 1.0, and 2.0 mm.

Modes I and II as well as mixed-mode fracture toughness tests were carried out on precracked disk specimens in diametral compression using the 75Sr and 75Sr10 composites (Fig. 3d). The dimensions of the disks were 25 mm diameter and 2 mm thickness. A chevron notch with an initial crack length of $a_O = 2$ mm was machined in the middle of the specimen using steel blades 20 mm in diameter and 0.75 mm in thickness.

Fatigue crack growth studies were conducted on the 75Sr10 composite material. A double cantilever beam (DCB)-type specimen was used with dimensions 40 × 20 × 2 mm. The specimens were prepared following the procedures for the preparation of the prismatic bars. The initial crack was introduced as follows: First, a straight notch was cut using a straight-head disk cutter 0.2 mm in thickness, then followed by a disk cutter of the same thickness and a V-shaped edge to give a symmetric and relatively sharp tip. A schematic of the specimen geometry is shown in Fig. 3e.

B. Experimental Methods

All experiments were conducted on an MTS material testing system (MTS, Minneapolis, MN), controlled by Instron digital electronics (Instron Corporation, Canton, MA), at room temperature in laboratory air. A 100-N Instron load cell was used to maximize the sensitivity of the outputs. Flexure experiments were performed under displacement control mode with a constant cross-head speed of 1.22 mm/sec. Considering the scatter of mechanical properties in brittle materials, more than 10 specimens per batch (25 in most cases) were tested under the same conditions to provide sufficient statistical information.

Three-point bending (3-P-B) and four-point bending (4-P-B), both with a support span of 50 mm, were used for strength and moduli measurements. In the 3-P-B configuration, the load was applied at midspan of the specimen (Fig. 3a). The 4-P-B configuration was loaded symmetrically at two locations, with the distance between the loading points being one third of the support span (Fig. 3b). Flexure strength s_f and flexure modulus E were determined from simple beam theory [49]:

$$S_f = \frac{MH}{2I} \tag{1}$$

$$E = \frac{PL^3}{4BH^3\delta} \qquad \text{(3-P-B testing)} \tag{2}$$

$$E = \frac{5PL^3}{27BH^3\delta} \qquad \text{(4-P-B testing)} \tag{3}$$

here H is the specimen height, I is the moment of inertia of the cross section, M is the moment due to the applied load P, and B is the thickness. The load P, and the corresponding deflection δ, were determined experimentally.

For 3-P-B, $M = PL/4$, provided the specimen fails at the midspan; and for 4-P-B, $M = PL/6$, as long as the specimen fractures within the pure bending portion of the beam (Figs. 3a and 3b).

The statistical analysis consisted of a one-way analysis of variance followed by a Tukey multiple means analysis [50]. The statistical results listed in the tables are composed of horizontal comparisons between the five different composites (Table 2) with the F value and probability listed in the right-hand column; and the vertical comparisons between the aging media and aging times with the F value and probability listed at the bottom of the column. The higher the F value, the greater the difference between the means of the variables. The letter codes indicate which means are significantly different from each other. For example, three means coded A, B, C are all significantly different; whereas three coded A, A, A are not statistically different. For clarity of presentation the letters V, W, X, Y, and Z are used to represent the horizontal comparisons and the letters A, B, C, D, and E are used to represent the vertical comparisons.

The data on fracture strength were analyzed using Weibull statistics. This model, sometimes known as "weakest link theory," is appropriate for a brittle isotropic and homogeneous material under uniaxial tension that is governed by volumetric flaw distribution. According to this model, the cumulative probability distribution of failure at a stress level, is s_f, given by the following empirical formula [51]:

$$F(s_f) = 1 - \exp\left[- \int_v \left(\frac{s_f - s_u}{s_0}\right)^m dV \right] = 1 - e^{-R} \qquad (4)$$

Here, s_u is the stress level at which $F(s_f)$ approaches zero; s_0 is the characteristic strength, which corresponds to 63.2% failure probability of a unit volume or area subjected to a uniform tensile load; m is a flaw density exponent or the Weibull modulus that reflects the extent of variability in strength; V is the specimen volume; and R is the so-called risk of rupture, which is a function of geometry and loading conditions. Weibull analysis has been also utilized to describe strength data from flexure tests. For rectangular bars where $V = bhL$ and assuming $s_u = 0$, R is expressed as

$$R = \frac{V}{2(m + 1)^2} (s_{f,3p}/s_0)^m$$

for three-point bending, and

$$R = \frac{V(m + 3)}{6(m + 1)^2} (s_{f,4p}/s_0)^m$$

for four-point bending [52]. These expressions can be written as $R = \alpha s_f^m$ where s_f stands for $s_{f,3p}$ or $s_{f,4p}$, and a is a constant.

The function $F(s_f)$ was determined by performing n tests and numbering the resulting fracture stresses s_f, in ascending order. Assuming that $F(s_{f,i}) = i/(n + 1)$, $i = 1, 2, \ldots n$, parameters m and $\ln(\alpha)$ are determined as slope and intercept in a linear regression analysis:

$$\ln\left[- \ln\left(\frac{n + 1 - i}{n + 1}\right)\right] = m\ln(s_{f,i}) + \ln(\alpha); i = 1, 2, \ldots n \qquad (5)$$

At the failure probability of 63.2%, the left-hand side of Eq. 5 is zero. This allows for the evaluation of a characteristic strength, $s_f = \exp(-\ln(\alpha)/m)$, at that probability which corresponds to a specimen of a given volume and applied load. Equation 5 can also be utilized to predict the fracture strengths at certain failure probabilities when m and $\ln(\alpha)$ are known. Discussion on the Weibull distribution and its application to brittle materials is reported elsewhere [53].

To obtain a better insight into the trend of the statistical data, the experimental values on strength, modulus, and mode I fracture toughness were presented with box-and-whisker plots. The box in these displays encloses the interquartile range with the lower line identifying the 25th percentile and the upper line the 75th percentile. A line sectioning the box displays the 50th percentile and its relative position within the inter quartile range. If the line sectioning the box is not in the geometric middle, the data distribution is not normal. The whiskers at either end extend to the extreme values of the experimental data [54].

Mode I fracture toughness K_{IC} was evaluated with the use of fracture mechanics formulas:

Table 2 Mechanical and Fracture Properties of Five Composites Aged in Air and Distilled Water

Aging	Resin	N	75Sr	N	75Sr10	N	79Sr	N	79Sr10	N	F value/prob.
					Flexure Strength (MPa ± SD)						
Air											
Control	AB[a] 86.2 ± 15.4 Y[a]	22	AB 110.0 ± 15.4 X	25	A 101.8 ± 8.6 X	24	A 95.9 ± 10.5 XY	14	B 101.5 ± 16.7 X	25	8.4/0.00
6 months	A 97.0 ± 13.7 X	22	AB 101.8 ± 13.7 X	25	B 88.4 ± 18.6 Y	25	AB 92.5 ± 13.5 XY	25	B 101.3 ± 19.1 X	25	3.2/0.02
12 months	B 80.5 ± 14.5 Y	22	AB 103.0 ± 17.8 X	24	AB 96.7 ± 19.5 X	24	AB 88.4 ± 12.2 XY	25	B 92.3 ± 17.3 XY	25	6.1/0.00
Water											
Control	AB 88.4 ± 16.5 Y	22	B 96.1 ± 14.9 Y	25	AB 94.8 ± 9.9 Y	25	A 99.6 ± 12.5 Y	24	A 118.0 ± 15.6 X	24	14.9/0.00
6 months	C 68.8 ± 15.4 X	22	C 78.6 ± 9.0 X	25	C 76.5 ± 16.0 X	25	BC 80.7 ± 11.1 X	25	C 77.3 ± 17.9 X	25	2.3/0.06
12 months	C 58.6 ± 12.9 X	22	C 70.7 ± 12.9 X	25	C 69.8 ± 16.1 X	25	C 65.1 ± 11.5 X	25	X 66.7 ± 17.1 X	25	2.6/0.04
F value/prob.	18.4/0.00		29.3/0.00		16.3/0.00		25.7/0.00		27.9/0.00		
					Flexure Modulus (GPa ± SD)						
Air											
Control	B[a] 3.5 ± 0.2 Z[a]	22	C 11.1 ± 0.9 Y	22	B 12.0 ± 0.7 X	24	C 12.0 ± 0.7 X	14	C 11.8 ± 0.8 X	25	53.2/0.00
6 months	A 3.8 ± 0.2 Z	22	B 12.9 ± 0.6 X	22	B 11.9 ± 1.1 Y	25	B 13.9 ± 0.9 W	25	A 16.3 ± 2.0 V	25	397.0/0.00
12 months	A 3.9 ± 0.3 Z	22	A 13.7 ± 1.2 Y	22	A 13.7 ± 1.2 Y	24	A 17.1 ± 1.5 W	24	B 15.1 ± 1.1 X	25	438.0/0.00

Water						
Control	C 3.0 ± 0.3 Z — 22	D 9.2 ± 0.8 Y — 25	C 9.9 ± 0.9 X — 25	C 11.0 ± 0.9 W — 24	C 11.7 ± 0.8 V — 24	462.4/0.00
6 months	D 2.3 ± 0.2 Z — 22	E 6.2 ± 0.7 Y — 25	D 6.5 ± 1.2 Y — 25	C 10.2 ± 0.9 W — 25	D 8.8 ± 0.8 X — 25	299.3/0.00
12 months	D 2.3 ± 0.2 Z — 22	E 6.7 ± 0.8 Y — 25	D 7.2 ± 0.9 Y — 25	D 8.7 ± 0.8 X — 25	D 8.2 ± 1.0 X — 25	221.8/0.00
F value/prob.	223.6/0.00	312.0/0.00	195.2/0.00	229.4/0.00	191.9/0.00	

Fracture Toughness (MPa · $m^{0.5}$ ± SD)

Air						
Control	BC[a] 1.0 ± 0.2 Z[a] — 25	A 1.4 ± 0.1 XY — 25	B 1.4 ± 0.1 XY — 25	C 1.3 ± 0.2 Y — 11	B 1.5 ± 0.1 X — 18	53.2/0.00
6 months	BC 1.0 ± 0.1 Y — 25	B 1.4 ± 0.1 X — 25	B 1.6 ± 0.1 W — 24	A 1.6 ± 0.1 W — 24	AB 1.6 ± 0.2 W — 25	106.0/0.00
12 months	C 0.9 ± 0.1 Z — 25	A 1.5 ± 0.1 W — 24	C 1.4 ± 0.1 X — 24	B 1.4 ± 0.1 X — 24	B 1.4 ± 0.2 XY — 24	124.0/0.00
Water						
Control	A 1.2 ± 0.2 Z — 25	A 1.5 ± 0.1 WX — 25	B 1.5 ± 0.1 WX — 25	B 1.4 ± 0.2 XY — 25	A 1.7 ± 0.1 W — 25	63.6/0.00
6 months	D 0.7 ± 0.1 Z — 25	C 1.3 ± 0.1 X — 25	A 1.3 ± 0.1 X — 25	C 1.2 ± 0.1 X — 24	C 1.3 ± 0.1 X — 25	125.7/0.00
12 months	D 0.7 ± 0.1 Z — 25	C 1.2 ± 0.2 X — 25	A 1.2 ± 0.2 X — 25	C 1.2 ± 0.1 X — 24	C 1.2 ± 0.1 X — 25	73.3/0.00
F value/prob.	53.4/0.00	58.1/0.00	28.4/0.00	35.6/0.00	50.7/0.00	

[a] ANOVA statistical analysis followed by Tukey analysis was conducted between the different compositions (V, W, X Y, Z) with the F value/prob. in the right-hand column (horizontal comparisons) and between aging times (A, B, C, D, E) with the F value/prob. at the bottom of the column (vertical comparisons). Mean values connected by the same letter were not statistically significant at the 0.05 level.

$$K_{IC} = \frac{PLf_1(a/H)}{BH^{1.5}} \qquad \text{(3-P-B testing)} \qquad (6)$$

$$K_{IC} = \frac{PLa^{0.5}f_2(a/H)}{BH^2} \qquad \text{(4-P-B testing)} \qquad (7)$$

Here P is the load at specimen fracture, L is the support span, B is the specimen thickness, H is the specimen height, and a is the notch depth. Functions $f_1(a/H)$ and $f_2(a/H)$ are correction factors appropriate to the specimen geometry [55,56].

Three-point-bending configuration has been recommended by the American Society for Testing and Materials (ASTM Standards, 1990 [57]) for fracture toughness testing, but still some concern exists about the effects of shear stress. Several check tests were performed and only a 1–3% difference was observed between the toughness values obtained from 3-P-B and 4-P-B. These differences were considered small and the 3-P-B loading configuration was employed in all mode I fracture toughness testing.

Mixed-mode fracture toughness was measured on two dental composites (75Sr and 75Sr10) using Chevron-notched disk specimens in diametral compression. Different stress states at the notch tip were obtained by aligning the center notch of the disk at an angle α relative to the loading direction (Fig. 3d). The crack inclination angle α varied from $\alpha_I = 0$ for pure mode I loading to approximately $\alpha_{II} = 30$ for pure mode II loading (note that α_{II} depended on the initial crack length and the disk radius). The stress intensity factors under combined modes I and II were calculated using the expressions proposed by Atkinson et al. [58]:

$$K_I = \frac{P\sqrt{a}}{\sqrt{\pi r B}} N_I \qquad (8)$$

$$K_{II} = \frac{P\sqrt{a}}{\sqrt{\pi r B}} N_{II} \qquad (9)$$

In these equations, r and B are the disk radius and thickness, respectively. N_I and N_{II} are dimensionless coefficients that depend upon the crack length and loading direction [58]. For each material, five specimens were fractured under displacement control mode with a constant cross-head speed of 0.25 mm/sec.

Fatigue crack propagation tests were performed in laboratory air at room temperature. Double cantilever beam (DCB) specimens were used in these experiments (Fig. 3e). For this particular specimen geometry, the stress intensity factor can be expressed as [59]:

$$K_1 = \frac{\sqrt{12}P}{BH^{0.5}}\left(\frac{a}{H} + 0.7\right) \qquad (10)$$

where P is the applied load, B is the thickness, and a is the crack length.

To avoid extraneous stresses, which could result from clamping misalignment and/or unloading, the specimens were mounted using pins. In fracture mechanics, it is assumed that a crack tip should be "atomically" sharp. The crack-like notch tip obtained by the notching technique mentioned above is definitely not sharp enough. To avoid a "pop-in" effect that could lead to catastrophic failure after crack initiation, a crack was initiated by fatigue under tension displacement controlled mode with a range of 0.08 mm applied at the grips of the specimen. After crack initiation, the specimens were fatigued under load controlled mode at a frequency of 1 Hz with a sinusoidal loading wave form. A constant load range ($P_{max} - P_{min} = 35\ N$) was maintained during crack propagation. The crack lengths were measured on the polished specimen surface to an accuracy of $\pm 100\ \mu$m using an optical microscope and a video display unit. In all experiments, accurate specimen dimensions were measured just before testing using a micrometer. Load deflection curves were recorded on an Instron plotter. All data were analyzed on the basis of linear elastic behavior.

C. Chemical Analysis

The amount of water absorption in the five different materials was measured from sets of 10–11 prismatic bars that were immersed in 200 ml of distilled water. The bars were aged under the same conditions as the mechanical property bars. The percentage of water absorbed was determined, at appropriate time periods, by the difference between the weight after being in water (w_w) minus the original dry weight (w_d) divided by the dry weight (w_d); that is, [percentage of water absorbed = $(w_w - w_d)/w_d$]. The bars were removed from the distilled water and dried to remove any excess water on the surface and then weighed on a balance (Mettler H30 Balance, Mettler Instrument Corporation, Princeton, NJ).

A second procedure was the chemical analysis of the aging solutions (distilled water) in which the specimens were aged. These liquids were stored in the original polystyrene containers along with representative samples of distilled water with no specimens for the same aging periods. The inorganic analysis for the ions Si, Ba, Al, and Sr used atomic absorption spectroscopy with background correction by the Smith–Hieftje technique. Acetylene/nitrous oxide was used for flame determinations.

III. RESULTS AND DISCUSSION

A. Flexure Strength

In the 4-P-B testing, the specimens always failed between the load application points, with only a few failures at either point of load application. Moreover, the load deflection curves were linear up to the maximum load [60]. Analysis of the strength data with the use of a two-parameter Weibull model for all composites is shown in Figs. 4a and 4b. (Note that for the sake of visual clarity, the data are presented in two plots.) The number of specimens, material type, average values, and standard deviation as well as the coefficients of the two-parameter Weibull distribution and the corresponding correlation coefficients are shown in Table 3. For comparison, the strength data for the resin are also presented in Table 3. It is interesting to note that the Weibull modulus varies from 6.43 for the 79Sr10 to 12.82 for the 75Sr10 material. These differences demonstrated certain variations in strength of these composites since a low value of m is indicative of a wide distribution with a long tail at low stress levels, and a high value of m shows a close grouping of failure stresses. This is especially true for the 75Sr10 material, which showed the largest value of m. The characteristic strength s_f, which is also the stress at 63.2% failure probability [53] did not change significantly from composition to composition (largest difference from the mean was about 9%). This was also true for the stress levels at 50% and 90% failure probabilities. However, the stress levels at 1% failure probability were substantially greater in all composites in comparison to the resin, with the 75Sr10 material giving the highest failure stress. On the basis of the average values, the 75Sr gave the highest failure stress.

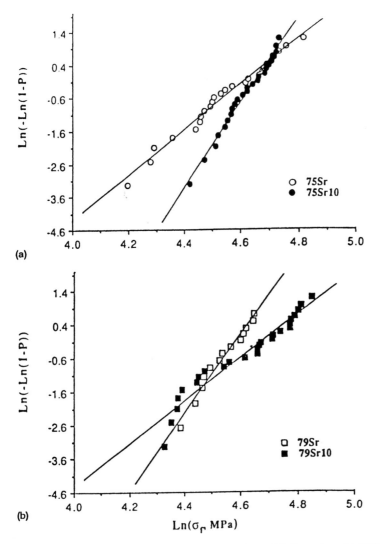

Figure 4   Weibull analysis of strength data from 4-P-B testing: (a) 75Sr and 75Sr10, and (b) 79Sr and 79Sr10.

Three point bending testing is very often employed by different investigators to evaluate strength characteristics of composite materials. It has been argued, however, that this test configuration may not be appropriate due to the fact that fracture does not occur at the specimen midspan where the moment attains its maximum value, and that the center section is not under pure bending. To examine the suitability of the 3-P-B testing and to compare the results with those from the 4-P-B testing, strength studies were also carried out on the resin and the 75Sr and 75Sr10 materials using the 3-P-B flexure testing. While the load deflection curves were linear up to fracture [60], the specimens did not always fail at the midspan or, for that matter, at the section of the maximum

Table 3   Strength Data Obtained from Four-Point Flexure Tests [a]

Material	Sample size	m	Correlation coefficient	s_f (MPa)	$\sigma_{0.01}$ (MPa)	$\sigma_{0.50}$ (MPa)	$\sigma_{0.90}$ (MPa)	$\sigma_f^a \pm$ SD (MPa)
Resin	22	4.44	0.96	95	34	88	114	86 ± 18
75Sr	25	7.33	0.99	117	62	111	131	110 ± 15
75Sr10	24	12.82	0.99	106	74	103	113	102 ± 8
79Sr	14	11.73	0.97	98	66	96	106	96 ± 10
79Sr10	25	6.43	0.97	109	53	103	125	102 ± 16

[a] s_f, characteristic stress, which corresponds to the stress level for a 63.2% probability of failure; $\sigma_{0.01}$, stress level for a 1% probability of failure; $\sigma_{0.50}$, stress level for a 50% probability of failure; $\sigma_{0.90}$, stress level for a 90% probability of failure; σ_f^a, average fracture stress

moment. Instead, 95% of the specimens fractured at a distance of 2.64 ± 1.60 mm from the midsection. Thus, an error is introduced when the maximum moment is used to evaluate the stress at failure. To evaluate that error, the bending moment was calculated with the use of the following expression.

$$M_f = \frac{P}{2}\left(\frac{L}{2} - l\right)$$

where l is the distance from the midspan to the fracture plane. Taking a measure of the differences as $(M_{max} - M_f)/M_{max} = 2l/L$, an average error in the stresses caused by assuming that fractured occurred at the middle of the beam was less than 10%.

Analysis of the strength data obtained from the 3-P-B testing for the 75Sr and 75Sr10 composites with a two-parameter Weibull distribution is shown in Table 4. Note that the Weibull modulus m is substantially lower in comparison to the respective values obtained in the 4-P-B, except for the resin. These results indicated a wider distribution at low stress levels obtained in the 3-P-B testing. It has been recognized, however, that Weibull statistics treats only the flaw variability of the material. Therefore, the errors caused by the experimental fixture and by assuming that the specimen failed at midspan are not taken into account in the Weibull analysis. A comprehensive analysis of experimental error can be found elsewhere [61]. The data in Table 4 indicate that the 75Sr material was best for 90% failure probability while the 75Sr10 was best for 1% probability of failure. These

Table 4   Strength Data Obtained from Three-Point Flexure Tests[a]

Material	Sample size	m	Correlation coefficient	s_f (MPa)	$\sigma_{0.01}$ (MPa)	$\sigma_{0.50}$ (MPa)	$\sigma_{0.90}$ (MPa)	$\sigma_f^a \pm$ SD (MPa)
Resin	10	7.62	0.97	108	59	103	120	102 ± 13
75Sr	25	6.31	0.97	121	58	114	138	113 ± 18
75Sr10	25	7.67	0.97	113	62	108	126	106 ± 15

[a] s_f, characteristic stress, which corresponds to the stress level for a 63.2% probability of failure; $\sigma_{0.01}$, stress level for a 1% probability of failure; $\sigma_{0.50}$, stress level for a 50% probability of failure; $\sigma_{0.90}$, stress level for a 90% probability of failure; σ_f^a, average fracture stress

trends are similar to those shown in Table 2 for the 4-P-B testing, although the stress levels corresponding to the failure probabilities were different.

In Weibull analysis, strength data obtained from different test configurations are often related. This is achieved by considering the same risk of rupture R, modulus m, and scale parameter σ_O. Furthermore, it is tacitly assumed that surface and volume flaws have the same effects on strength. Accordingly, for the cases investigated herein [52]:

$$\frac{\sigma_{f,3p}}{\sigma_{f,4p}} = \left(\frac{m + 3}{3}\right)^{1/m}$$

Data analysis, however, for two materials tested under 4-P-B and 3-P-B did not result in similar values of m (Tables 3 and 4). These variations may be due to different contributions of the surface and volume flaws to the overall strength response, and to errors introduced by the experimental fixtures. This was supported by the experimental observations in the 3-P-B testing, where most specimens did not fracture at the midsection. Thus in the particular case investigated, the Weibull parameters obtained from one type of testing, that is, 4-P-B, cannot be utilized to predict the probability of failure in the 3-P-B testing.

B. Flexure Modulus

The experimental data on flexure modulus obtained from the 4-P-B test configuration are provided in Table 5 and a graphical presentation using the box-and-whisker plots is shown in Fig. 5. The experimental data in Table 5 demonstrate that the addition of fillers resulted in a threefold increase of the flexure modulus. 75Sr shows the greatest variability and the lowest mean, with the other three compositions having essentially the same mean and variability. The addition of microfill on the 79Sr composite slightly reduced the mean value and increased the variability.

It is interesting to note that the variations of the strength data (Table 3) are larger than those in the flexure modulus values (Table 5). These findings may be explained in terms of the greater sensitivity of strength in comparison to the flexure modulus on local fluctuations of the material's microstructure [62].

C. Mode I Fracture Toughness

According to the ASTM standards for fracture toughness [57], certain requirements should be satisfied for a valid fracture toughness measurement. In addition to the linearity of the load displacement curves and the plane strain conditions, the crack should be "atomically" sharp. Although the first two requirements can be easily satisfied, the third

Table 5   Experimental Values of Flexure Modulus

Material	Sample size	Range (GPa)	Mean ± SD (GPa)
Resin	22	3.00–3.98	3.52 ± 0.18
75Sr	25	8.95–12.64	11.16 ± 0.94
75Sr10	24	10.49–13.17	11.92 ± 0.71
79Sr	14	10.71–13.53	11.99 ± 0.74
79Sr10	25	10.17–13.51	11.85 ± 0.82

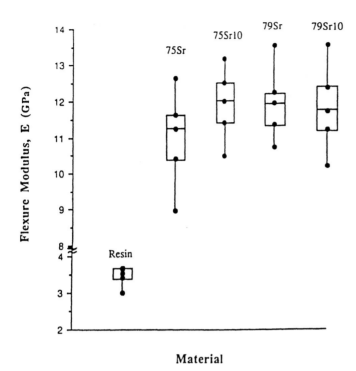

Figure 5   Box-and-whisker displays for flexure modulus.

one is very difficult to meet in brittle materials. This is partially due to problems associated with growing a sharp crack by fatigue normal to the applied load. To circumvent this problem, various techniques are usually implemented to introduce a sharp notch in specimens of brittle materials [63]. In the present studies, specimens of the 75Sr10 composite with two types of machined notches and three depths were first fractured to investigate any effects of notch tip and its depth on fracture toughness. Specifically, a relatively sharp 60° V notch tip and a straight notch with depth-to-beam height ratios $a/H = 0.125$, 0.250, and 0.500 were considered.

Table 6 presents fracture toughness measurements for two notch geometries and three depths for each notch type, together with the size requirement for plane strain obtained from the expression $2.5(K_Q/s_f)^2$, (where K_Q is an apparent fracture toughness). Note that

Table 6   Effects of Notch Geometry Data on Mode I Fracture Toughness

a/H	Straight notch			V-type notch		
	Sample size	Mean ± SD (MPa√m)	$2.5K_Q^2/\sigma_y^2$ (mm)	Sample size	Mean ± SD (MPa√m)	$2.5K_Q^2/\sigma_y^2$ (mm)
0.125	10	2.01 ± 0.11	0.96–1.29	10	1.19 ± 0.14	0.26–0.57
0.250	10	1.74 ± 0.17	0.60–1.13	10	1.12 ± 0.12	0.22–0.45
0.500	10	1.57 ± 0.17	0.43–0.88	10	1.11 ± 0.13	0.20–0.47

due to the lack of a yield point in any of these dental restorative composite (DRC) materials, the average fracture strengths s_f^a for each material were used instead. The values of toughness for each notch depth are also shown in Figure 6. These data demonstrate that toughness values obtained from specimens with the straight notch were more than 41–69% greater in comparison to the corresponding values obtained from the V notch. Similar results have been reported for other brittle materials [63]. The toughness values (Table 6) observed in the case of straight notches were presumably due to the bluntness of the notch, which results in higher fracture stresses, as well as to the variations of local notch tip geometry of a straight notch from specimen to specimen. Furthermore, the size requirements for plane strain are met in all cases except in the straight notch with $a/H = 0.125$.

In addition to the notch type, notch depth has an effect on fracture toughness (Table 6). These effects were noticeable in the case of the straight notch (differences ~28%) in comparison to the differences in the V notch (differences ~7%) (Fig. 6). These results, and toughness values on similar materials reported in the literature [64], indicated that the V notch could be appropriate for mode I fracture toughness measurements. Thus, assuming that the same trend would hold in all DRC material investigated herein, toughness measurements were carried out on specimens with a notch depth between $a = 0.5$ and 2.0 mm. According to the ASTM standards for fracture toughness testing [57], an initial crack size should be in the range $a = 0.45$–$0.55 H$ (i.e., $a = 1.8$–2.2 mm). Taking into consideration the results of notch depth effects and the difficulties associated with machining relatively deep notches in a consistent manner in brittle materials, a V notch

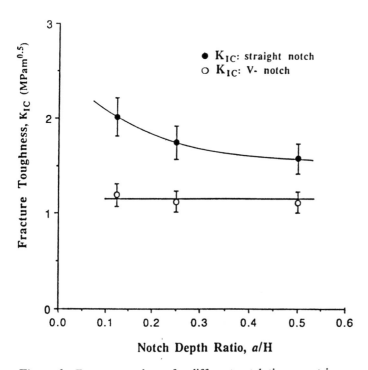

Figure 6   Fracture toughness for different notch tip geometries.

Table 7 Experimental Values of Mode I Fracture Toughness

Material	Sample size	Range (MPa√m)	Mean ± SD (MPa√m)	$2.5\,K_Q^2/\sigma_{ys}^2$ (mm)
Resin	25	0.64–1.27	1.00 ± 0.18	0.2–0.9
75Sr	25	1.14–1.44	1.30 ± 0.08	0.4–0.6
75Sr10	25	1.11–1.64	1.43 ± 0.14	0.3–0.8
79Sr	11	1.04–1.52	1.34 ± 0.16	0.4–0.8
79Sr10	18	1.34–1.67	1.55 ± 0.09	0.6–0.9

with depth $a = 0.5$ mm was used in all mode I toughness measurements of DRC materials employed in these studies.

Experimental data on apparent fracture toughness K_Q, and the size requirements for plane strain conditions, that is, a, $B > 2.5\,(K_Q/s_f)^2$ are provided in Table 6. The data presented in Table 7 show that the experimental data of K_Q are valid values of K_{IC}. Box-and-whisker displays of toughness are shown in Fig. 7. The data in Figure 6 show that, while the mode I fracture toughness in all composites was higher than that of the resin material, the 79Sr10 material displayed the highest mean value for toughness, with relatively small variability. Moreover, the addition of colloidal silica in the 75Sr and 79Sr composites improved their toughness by 10% and 15.7%, respectively (Table 7). It should be noted that the toughness values of the 75Sr10 material were 17% larger in comparison to the toughness values of the same material measured in the notching effects

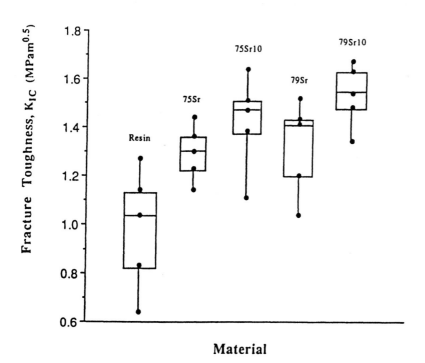

Figure 7 Box-and-whisker displays for fracture toughness.

studies (Table 6). These discrepancies were attributed to variations in specimen preparation.

D. Aging and Chemical Analysis

The effects of aging on flexure strength, s_f, flexure modulus, E, and fracture toughness, K_{IC}, on the resin and four specially made dental restorative materials are shown in Table 2. The data indicate that an increase of about 10–28% in s_f of the resin occurred with the addition of the filler. However, the amount and type of filler had a relatively small effect on s_f. For all composites and resin, testing in air and water did not affect significantly the value of s_f. A substantial decrease in the values of s_f were observed when all materials were aged in water for 6 or 12 months. The reduction was larger for 12 months of aging. The differences were as large as 35% for the resin and all composites. While the addition of the fillers increased the value of E threefold, aging in water for 6 and 12 months decreased the value of E for the resin by about 35%. A similar trend was observed in all composite materials. Although the addition of the filler increased the value of K_{IC} by 30–50%, the different fillers and amounts did not alter K_{IC} significantly except in some cases for the 75Sr material. A decrease in K_{IC} occurred in the resin when aged in water for 6 or 12 months. Similar trends were recorded for the 75Sr, but those for the rest of the composites were insignificant. Since small or no differences were seen in the values of s_f, E, and K_{IC} between the 6 and 12 months in water, it may be assumed that water absorption reached saturation before this time. More detailed analysis of the aging can be found in Ref. 33.

This initial study indicated that aging of the composites is not significant after 6 months, but that treatment and aging is significant between 0 and 6 months. The addition of the microfill resulted in improved fracture properties. Testing in water lowered s_f and E but increased K_{IC}, and aging in water lowered all properties investigated in this work. The flexure modulus was the parameter most significantly affected by the testing variables.

Atomic absorption analysis of the aging solutions found only SiO_2 and SrO_2 (Table 8). The concentrations of SrO_2 and SiO_2 indicate that differences occur according to the weight percent of filler available. This is logical since the resin appears to take up most of the water absorption and by limiting the amount of resin, the exposure of the filler to the water was limited. The data on water absorption are presented in Table 9. The bars were weighed at 0, 14, 35, 50, and 63 days. The water absorption results for 14 and 35 days were statistically different from each other. For 50 and 63 days the water absorption results were not statistically different from each other, although they were statistically different from the 14- and 35-day results. This trend of the data suggests that water absorption reached a saturation level before 50 days. Only the 50 and 63 days of water

Table 8 Atomic Absorption Results ($\mu g/ml$)

	Resin		75Sr		75Sr10		79Sr		79Sr10	
	Si	Sr	Si	Sr	Si	Sr	Si	Sr	Si	Sr
6 months	0.00	0.00	5.21	26.96	7.02	22.29	3.40	12.71	5.66	15.88
12 months	0.00	0.00	7.02	37.54	11.99	37.32	5.66	21.42	9.95	24.37

Table 9   Water Absorption Results

Composite	Percent water increase			
	14 days	35 days	50 days	63 days
79Sr10	2.2 ± 0.1	3.0 ± 0.1	3.8 ± 0.1	3.8 ± 0.2
79Sr	2.2 ± 0.3	3.1 ± 0.1	3.9 ± 0.1	3.8 ± 0.1
75Sr	2.3 ± 0.1	3.2 ± 0.2	4.1 ± 0.1	4.1 ± 0.1
75Sr10	2.4 ± 0.1	3.4 ± 0.1	4.2 ± 0.1	4.2 ± 0.1
Resin	6.3 ± 0.2	8.1 ± 0.2	9.5 ± 0.2	9.4 ± 0.2

absorption data were clearly delineated to show a significant difference between the 75 wt% filler, the 79 wt% filler, and the resin composites. If the assumption is made that water absorption is proportional to the material present, the total water absorption can be separated into a resin and filler portion. This is done by taking 0.25 or 0.21 of the resin value (to correspond to the 75 or 79 wt% filler) and subtracting this from the total value to obtain the filler portion. Then it is clear that the water absorption occurs in all constituents of the composite system. Note that the interfaces may absorb water. However, such measurements were not carried out in these studies. Thus, in the preceding argument, it is implied that the water absorbed by the interface is relatively insignificant or that it is distributed between the amounts absorbed by the filler and matrix.

The data presented in the foregoing section showed that water absorption is completed between 35 and 50 days, and mostly occurs in the resin and less in the filler. The results of the present studies clearly demonstrate that the uptake of water is probably the main cause for material property degradation. The underlying mechanisms for degradation may, thus, be due to chemical breakdown of the silane bond between the glass filler and the resin; chemical breakdown induced by stresses due to swelling of the polymer matrix, leading to voids and cracks; changes in chemical composition; and so forth. It is difficult, however, to assess their relative importance because of their interaction and the interaction with the microstructure; that is, diffusion of water may be affected by the amount of the reinforcing particles, preexisting defects, etc. Moreover, the results show very small differences in properties between 6 and 12 months in water. This indicates a time where the material reaches equilibrium with regards to transport of water and the stabilization of degradation of its properties and the extent of damage. Additional research is needed to elucidate this type of behavior.

E.   Mixed-Mode Fracture Toughness

In this preliminary study, the mixed-mode fracture toughness of two dental composites (75Sr and 75Sr10) was examined using chevron-notched disk specimens in diametral compression (DC). Different stress states were obtained by aligning the center notch of the disk at an angle α relative to the loading direction (Fig. 3d). The stress intensity factors under combined modes I and II were calculated using expressions 8 and 9 proposed by Atkinson et al. [58]. The values of fracture toughness for modes I and II are presented in Table 10.

The results of these studies indicated that the addition of the microfiller increased the values of K_{IC}, while the effects on K_{IIC} were minimal. Note that values of K_{IC} obtained form this work were very close to the results obtained from three-point fracture testing of

Table 10   Mixed-Mode Results

| Material | Diametral compression | | 3 point: |
	K_{IC} (MPa√m)	K_{IIC} (MPa√m)	K_{IC} (MPa√m)
75Sr	1.18 ± 0.03	1.83 ± 0.11	1.30 ± 0.08
75Sr10	1.35 ± 0.08	1.80 ± 0.18	1.43 ± 0.14

specimens with a V notch. Moreover, the K_{IIC} values were larger that those of K_{IC} by 30–50% (Table 10).

Fracture envelopes for these two composite materials are shown in Fig. 8. It is worth noting that the data fit the following empirical equation:

$$\frac{K_I}{K_{IC}} + \left[\frac{K_{II}}{K_{IIC}}\right]^2 = 1 \tag{11}$$

Here K_{IC} and K_{IIC} are the pure mode I and mode II fracture toughness. Relations similar to Eq. 11 have been found for different materials [29] and can be looked upon as fracture envelopes for mixed-mode fracture.

Figure 9 shows typical fractured test pieces for the 75Sr composite under mode I (Fig. 9a), combined mode (9b), and pure mode II (9c) loading. Similar fracture paths were

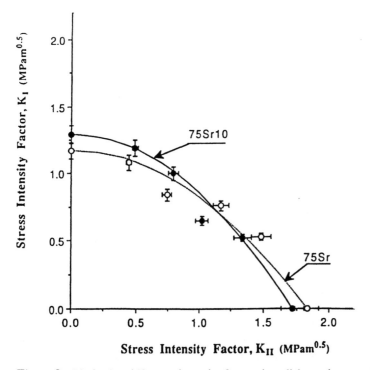

Figure 8   Modes I and II stress intensity factors in a disk specimen under diametral compression.

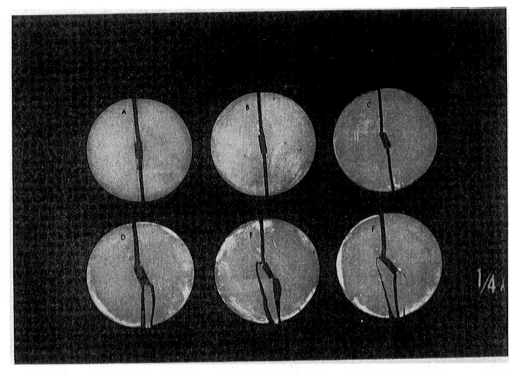

Figure 9   Typical fracture paths under (A) mode I, 0°; (B) modes I and II, 5°; (C) modes I and II, 10°; (D) modes I and II, 15°; (E) modes I and II, 20°; and (F) mode II, 30° (disk diameter 25 mm).

observed in all specimens. A photograph of the chevron notch geometry of a specimen fractured under mode I is displayed in Fig. 10. Note here the sharpness of the notch, which was typical in all specimens and may account for the small standard deviation of the fracture toughness recorded in these preliminary experimental results of mixed-mode fracture.

F.   Fatigue Crack Growth Behavior

The main objectives of these preliminary studies were to investigate the possibilities and conditions for crack initiation and stable crack growth in DRC. Since the 75Sr10 composite material displayed the best mechanical behavior, it was chosen for the fatigue crack growth studies. The parameters for this material obtained in the monotonic tests were: fracture toughness, 1.43 ± 0.14 MPa·m$^{0.5}$ and flexure modulus, 11.92 ± 0.71 GPa. The confidence interval for the mean value of K_{IC} was 1.36–1.50 MPa·m$^{0.5}$ at a probability of 99%. The observed experimental extreme values were 1.11–1.64 MPa·m$^{0.5}$ (Table 7).

The time to crack initiation in this material was a function of the applied maximum stress intensity factor K_{1max} and frequency. Under a load controlled mode with $K_{1max} \sim 0.6\, K_{IC}$ and at a frequency of 1 Hz, a crack did not initiate in more than 15 h. In some specimens, crack initiation did not occur even after 45 h. On the other hand, when $K_{1max} \sim 0.9\, K_{IC}$, crack initiation was followed by unstable crack growth in about 1800 cycles. To circumvent this problem and grow a fatigue crack, displacement controlled loading

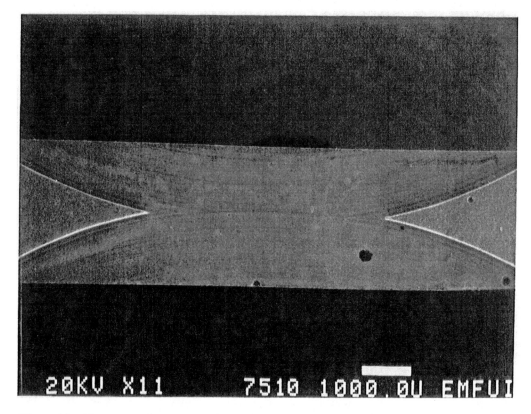

Figure 10   Chevron notch geometry of a specimen fractured under mode I.

conditions with $0.6\ K_{IC} \leq K_{1max} \leq 0.9\ K_{IC}$ were applied on the specimens. Under these conditions, crack initiation occurred in about 3 h. Accordingly, displacement controlled conditions were employed to initiate the crack in all fatigue tests. After crack initiation, the displacement controlled mode was changed to a loading controlled mode with a frequency of 1 Hz and applied loadings of $P_{max} = 55$ and $P_{min} = 20$ N. In all test pieces, stable crack growth was observed under these conditions. The crack paths were divided into a stable portion and a unstable portion. The stable portion was dominated by crack growth along a straight path. Towards the end of the stable propagation phase, crack growth departed from the straight growth mode. Subsequently, rapid fracture occurred along a curved path. Eighteen specimens with slightly different thickness were investigated using the same loading conditions during slow crack propagation. The crack trajectories obtained after specimen fracture are shown in Fig. 2. Micrographs of fracture surface of a typical specimen are exhibited in Fig. 11a. Higher-magnification observations, ahead of the notch tip, revealed relatively rough surfaces in the middle of the specimen (Fig. 11b) and relatively smooth surfaces near the specimen edges (Fig. 11c). This morphology suggests that the crack first initiated slowly near the surface of the specimens and relatively quickly in the middle of the specimen. In the stable crack growth phase, defects in the form of voids along the crack path are seen (Fig. 11a). Local variations in notch tip geometry from specimen to specimen, a distortion of the stress field in the vicinity of the crack tip due to voids, and lack of dissipative mechanisms in this material needed to stabilize the crack may have resulted in the nonreproducible curved crack paths. The results of these preliminary studies suggest that it may be more appropriate to investigate crack growth in these materials under simultaneous I and II fracture modes.

Crack growth kinetics are shown in Fig. 12 for two specimens fractured under the same loading conditions, plotted against the stress intensity factor range $\Delta K = K_{Imax} - K_{Imin}$. The data in Fig. 12 indicate large fluctuations in crack speed and the appearance of large differences in crack speed at the same level of ΔK. This behavior is attributed to the effects of voids induced during materials preparation as well as the lack of dissipative mechanisms (i.e., damage). Relatively large scatter on crack speed has also been observed in other dental composites [32]. The large fluctuations and nonreproducibility of fatigue crack growth in DRC materials requires that a large number of specimens be tested and statistically analyzed and modeled for a reliable design approach to fatigue life prediction.

In summary, dental composites investigated in these studies exhibited brittle behavior. The nature of brittle behavior requires the employment of statistical theories. The box-and-whisker displays of both flexure modulus and fracture toughness were not symmetric, indicating that the distributions of these values were not normal.

Preliminary experimental results have demonstrated that both notch tip geometry and depth have important effects on mode I fracture toughness. That is, fracture toughness values obtained from a straight notch were more than 41–69% greater than those obtained from a sharp V notch tip (Table 6). In the case of a V notch, toughness values were practically independent of depth.

The average values of toughness obtained from specimens with a V notch were 30–55% higher in all composites as compared to the values of the resin. The largest variability in these data was observed in the 75Sr10 and 79Sr composites, and the smallest in the 75Sr and 79Sr10 materials. In addition, the highest average value was seen in the 79Sr10 material (Fig. 7).

Mixed-mode fracture toughness testing indicated that for the two composites tested herein, K_{IIC} were 55% (75Sr10) and 33% (75Sr) higher in comparison to K_{IC} values. Moreover, the K_{IC} values obtained from the disk specimens were close to the values recorded in the SEN bar specimens. These data suggested a notch insensitivity in the K_{IC} values (i.e., chevron and V notches).

The 75Sr10 composite material exhibits true cyclic stress induced fatigue crack growth behavior. The slow crack growth was observed in the stress intensity factor range 60–90% of K_{IC}. The results of the investigations indicated that the cyclic crack growth can be divided into two parts: stable and unstable crack growth. During the stable portion, the crack propagated along a straight line. Towards the end of the stable portion, the crack path deviated to a curved crack path. The fluctuation of crack growth rate is attributed to the voids or agglomeration of particles around the crack path, and to a lack of dissipative mechanisms in the materials investigated herein.

ACKNOWLEDGMENTS

The authors wish to acknowledge the financial support from NIDR under grant RO1-DEO7979. Thanks are also due to Diana Racean, John Van Scoyoc, Jeff Samyn, John Wozny, and John Gramsas for assistance in the specimen preparation; to Bisco Inc. for supplying the dental composite; and to the Electron Microscopy staff of the Research

(a)

(b)

Figure 11 (a) Micrograph of a fractured specimen, (b) higher magnification of area (A), (c) higher magnification of area (B).

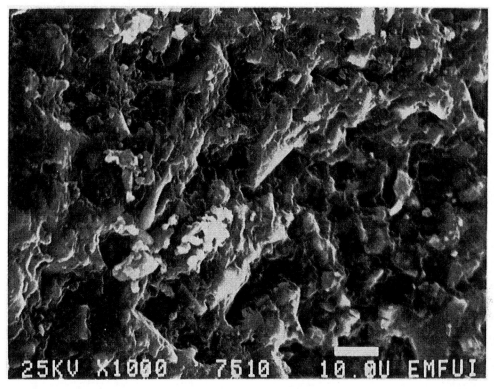

(c)

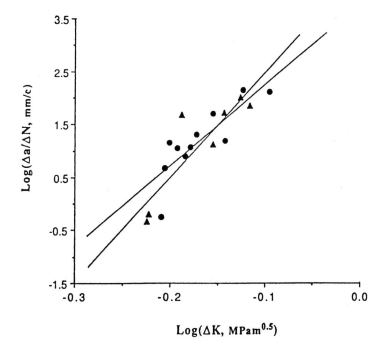

Figure 12 Crack growth rates plotted against the stress intensity range.

Resources Center of the University of Illinois, Chicago, for use of the electron microscope.

REFERENCES

1. K. Hellan, *Introduction to Fracture Mechanics*, McGraw-Hill, New York, 1984, pp. 154–160.
2. K. J. M. Söderholm and M. J. Roberts, Influence of water exposure on the tensile strength of composites, *J. Dent. Res., 69*(12), 1812–1816 (1990).
3. R. G. Chadwick, J. F. McCabe, A. W. G. Walls, and R. Storer, The effect of storage media upon the surface microhardness and abrasion resistance of three composites, *Dent. Mater., 6*, 123–128 (1990).
4. J. L. Ferracane and V. A. Marker, Solvent degradation and reduced fracture toughness in aged composites, *J. Dent. Res., 71*(1), 13–19 (1992).
5. J. L. Ferracane and J. R. Condon, Post-cure heat treatments of composites: Properties and fractography (abstract 1661), *J. Dent. Res., 71*, 313, (1992).
6. R. E. Kovarik, J. W. Ergle, and C. W. Fairhurst, Effects of specimen geometry on the measurement of fracture toughness, *Dent. Mater., 7*, 166–169 (1991).
7. C. H. Lloyd, The fracture toughness of dental composites. II, *J. Oral Rehab., 9*, 133–138 (1982).
8. J. L. Ferracane, R. C. Antonio, and H. Matsumoto, Variables affecting the fracture toughness of dental composites, *J. Dent. Res., 66*(6), 1140–1145 (1987).
9. R. M. Pilliar, D. C. Smith, and B. Maric, Fracture toughness of dental composite determined using the short-rod fracture toughness test, *J. Dent. Res., 65*(11), 1308–1314 (1986).
10. K. H. R. Wright and A. W. Burton, Wear of dental tissues and restorative materials, in *The Wear of Non-Metallic Materials* (D. Dowson, M. Godet, and C. M. Taylor, eds.), Mechanical Engineering Publication, London, 1976, pp. 116–126.
11. M. Goldman, Fracture properties of composite and glass ionomer dental restorative materials, *J. Biomed. Mater. Res., 19*, 771–783 (1985).
12. R. B. Mazer, K. F. Leinfelder, and C. M. Russel, Degradation of microfilled posterior composite, *Dent. Mater., 8*, 185–189 (1992).
13. Y. Momoi and J. F. McCabe, Positive pressure developed in composite resins by a hydroscopic expansion (abstract 664), *J. Dent. Res., 71*, 598 (1992).
14. W. D. Cook and M. Johannson, The influence of postcuring on the fracture properties of photo-cured dimethacrylate based dental composite resin, *Biomed. Mater. Res., 21*, 979–989 (1987).
15. M. J. Tyas, Correlation between fracture properties and clinical performance of composite resins in class IV cavities, *Aust. Dent. J., 35*(1), 46–49 (1990).
16. R. E. Kovarik and J. W. Ergle, Effects of long term aging on fracture toughness of composites (abstract 1662), *J. Dent. Res., 71*, 313 (1992).
17. D. W. Jones, A. S. Rizkalla, E. J. Sutow, and G. C. Hall, Elastic moduli of wet and dry experimental composite materials (abstract 666), *J. Dent. Res., 71*, 599 (1992).
18. K. J. Jones, M. Sherriff, T. F. Watson, A. Greasley, and J. Moffatt, Compressive fatigue cracking and the fracture toughness of posterior composites (abstract 958), *J. Dent. Res., 71*, 635 (1992).
19. M. J. Reece, Cyclic fatigue crack propagation in alumina under direct tension–compression loading, *J. Am. Ceram. Soc., 72*(2), 348–352 (1989).
20. P. L. Fan and J. M. Powers, Wear of aged dental composites, *Wear, 68*, 241–248 (1981).
21. G. Willems, P. Lambrechts, M. Braem, J. P. Celis, and G. Vanherle, A classification of dental composites according to morphological and mechanical characteristics, *Dent. Mater., 8*, 310–319 (1992).

22. N. P. Suh, An overview of the delamination theory of wear, *Wear, 44*, 1–16 (1977).
23. W. Wu, E. E. Toth, J. F. Moffa, and J. A. Ellison, Subsurface damage layer of *in vivo* worn dental composite restorations, *J. Dent. Res., 63*(5), 675–680 (1984).
24. J. E. McKinney and W. Wu, Relationship between subsurface damage and wear of dental restorative composites, *J. Dent. Res., 61*(9), 1083–1088 (1982).
25. R. A. Draughn, Compressive fatigue limits of composite restorative materials, *J. Dent. Res., 58*(3), 1093–1096 (1979).
26. G. Carfagna, G. Guerra, L. Nicholais, and S. Tartars, Effects of postcuring and water sorption on the mechanical properties of composite dental restorative materials, *Biomaterials, 4*, 228–229 (1983).
27. V. T. Truong and M. J. Tyas, Prediction of *in vivo* wear in posterior composites: A fracture mechanics approach, *Dent. Mater., 4*, 318–327 (1988).
28. J. L. Ferracane, J. R. Condon, and J. C. Mitchem, Evaluation of subsurface defects created during the finishing of composites, *J. Dent. Res., 71*(9), 1628–1632 (1992).
29. D. Singh and D. K. Sheety, Fracture toughness of polycrystalline ceramics in combined mode I and mode II loading, *J. Am. Ceram. Soc., 72*(1), 78–84 (1989).
30. D. A. Hills and D. W. Ashelby, On the application of fracture mechanics to wear, *Wear, 54*, 321–330 (1979).
31. R. L. Sakaguchi, M. Cross, and W. H. Douglas, A simple model of crack propagation in dental restorations, *Dent. Mater., 8*, 131–136 (1992).
32. V. T. Truong, D. J. Cook, and N. Padmanathan, Fatigue crack propagation in posterior dental composites and prediction of clinical wear, *J. Appl. Biomater., 1*, 21–30 (1990).
33. D. Zhao, J. Botsis, and J. L. Drummond, Fracture studies of selected dental restorative composites, *Dent. Mater.* (in press).
34. J. G. C. Arridge, Particulate composites, in *Concise Encyclopedia of Composite Material* (Anthony Kelly, ed.), Pergamon Press, 1988, pp. 227–231.
35. G. C. Papanicolaou and P. S. Theocaris, Thermal properties and volume fraction of the boundary interphase in metal-filled epoxies, *Coll. Polym. Sci., 257*, 239–246 (1979).
36. J. M. Powers, P. L. Fan, and M. Marcotte, *In vitro* accelerated aging of composites and a sealant, *J. Dent. Res., 60*(9), 1672–1677 (1981).
37. W. Wu and J. E. McKinney, Influence of chemicals on wear of dental composites, *J. Dent. Res., 61*(10), 1180–1183 (1982).
38. J. M. Powers, M. D. Ryan, D. J. Hosking, and A. J. Goldberg, Comparison of *in vitro* and *in vivo* wear of composites, *J. Dent. Res., 62*(10), 1089–1091 (1983).
39. J. L. Drummond, M. A. Khalaf, and G. Randolph, *In vitro* aging of composite restorative materials, *Clin. Mater., 3*(3), 209–221 (1988).
40. J. L. Drummond, Cyclic fatigue of composite restorative materials, *J. Oral Rehab., 5*, 509–520 (1989).
41. J. F. Roulet, *Degradation of Dental Polymers*, Karger, Basel, 1987, pp. 1–228.
42. C. H. Lloyd, The fracture toughness of dental composites, *J. Oral Rehab., 11*, 393–398 (1984).
43. D. C. Evans and J. K. Lancaster, The wear of polymers, in *Treatise on Materials Science and Technology* (D. Scott, ed.), 1979, Vol. 13, pp. 86–140.
44. A. R. Lansdown and A. L. Price, *Materials to Resist Wear*, Pergamon Press, Oxford, 1986, pp. 1–127.
45. M. Braden and G. J. Pearson, Analysis of aqueous extract from filled resins, *J. Dent., 2*(9), 141–143 (1981).
46. M. Tomozawa, W. T. Han, and W. A. Lanford, Water entry into silica glass during slow crack growth, *J. Am. Ceram. Soc., 74*(10), 2573–2576 (1991).
47. J. L. Drummond, D. Zhao, and J. Botsis, Fracture properties of aged and postprocessed dental composites, *Dent. Mater.* (in press).
48. J. L. Ferracane and E. H. Greener, The effect of resin formulation on the degree of conver-

sion and mechanical properties of dental restorative resins, *Biomed. Mater. Res., 20*, 121–131 (1986).

49. J. M. Gere and S. P. Timoshenko, *Mechanics of Materials*, 2nd ed., Brooks/Cole Engineering Division, 1984.

50. L. Wilkinson, *SYSTAT: The System for Statistics*, Systat, Inc., Evanston, IL, 1989, pp. 446–471.

51. W. Weibull, A statistical distribution function of wide applicability, *J. Appl. Mech., 18*, 293–297 (1951).

52. N. A. Weil and I. M. Daniel, Analysis of fracture probabilities in non-uniformly stressed brittle materials, *J. Am. Ceram. Soc., 47*(6), 268–274 (1964).

53. G. D. Quinn and R. Morrell, Design data for engineering ceramics: A review of the flexure test, *J. Am. Ceram. Soc., 74*(9), 2037–2066 (1991).

54. R. V. Hogg and J. Ledolter, *Applied Statistics for Engineers and Physical Scientists*, 2nd ed., Macmillan, New York, 1992, pp. 1–45.

55. D. Broek, *Elementary Engineering Fracture Mechanics*, 4th ed., Kluwer Academic Publishers, Dordrecht, 1986, pp. 179–186.

56. B. Gross and J. E. Strawley, *Stress intensity factors for single-edge notch specimens in bending and tension by boundary collocation of a stress function*, NASA Technical Note, D2603, 1965.

57. *1990 Annual Book of ASTM Standards*, Part 10, E-399-90, 1990.

58. C. Atkinson, R. E. Smelser, and J. Sanchez, Combined mode fracture in biaxial stress state: Application of the diametral-compression (Brazilian Disk) test, *Int. J. Fract., 18*, 279–291 (1982).

59. R. W. Davidge, *Mechanical Behavior of Ceramics*, Cambridge University Press, 1979.

60. D. Zhao, Master's Thesis, Department of Civil Engineering, Mechanics, and Metallurgy, University of Illinois at Chicago, Chicago, IL, 1992.

61. F. I. Baratta, G. D. Quinn, and W. T. Matthews, *Errors associated with flexure testing of brittle materials*, U.S. Army Report No. 87-35, U.S. Army Material Lab., Watertown, MA, 1987.

62. P. M. Duxbury, Breakdown of diluted and hierarchical systems, in *Statistical Models for the Fracture of Disordered Media* (H. J. Herrmann and S. Roux, eds.), Elsevier, Amsterdam, 1990, pp. 189–228.

63. A. Ghosh, M. G. Jenkins, K. W. White, A. S. Kobayashi, and R. C. Bradt, Elevated temperature fracture resistance of a sintered a-silicon carbide, *J. Am. Ceram. Soc., 72*, 242–247 (1989).

64. D. C. Smith, Posterior composite dental restorative materials: Materials development, in *Posterior Composite Resin Dental Restorative Materials* (G. V. Vanherle and D. C. Smith, eds.), 1985, pp. 47–60.

57

Evaluation of the Titanium–Bone Interface of Dental Alveolar Implants

Philip J. Boyne
Loma Linda University Medical Center
Loma Linda, California

I. INTRODUCTION

The status of titanium metal implant to bone interfaces depends upon many factors including those involving the composition of the titanium or the titanium alloy implant, the characteristics of the surface of the implanted material, and the histologic response of the surrounding bone.

The surgical bone implantation site may be composed of cortical lamellated bone or cancellous or trabecular bone. The site may contain a large number of marrow vascular spaces, which in turn may be highly cellular or hemopoietic marrow that usually is responsive to surgical and functional changes brought about by the placement of implants. The surrounding marrow vascular spaces in an implant site may also contain largely fatty tissue, fibrous tissue, or even a cellular exudate resulting from chronic inflammation. The cellular contents of such spaces do not respond well in the formation of bone matrix when stimulated by the placement of metal implant surfaces. Thus, appropriate characterization of host bone sites becomes essential in determining the eventual implant surface contact interaction.

In addition, many patients presenting for reconstruction with dental alveolar root-form implants have insufficient bone to accommodate the implantation fixtures and, therefore, require bone grafts. The grafts may be: (1) entirely autogenous grafts taken from the iliac crest in case of restoration of large defect areas or taken from another area of the oral cavity in restoring smaller defects for the reception of the metal implant, (2) allogeneic grafts, (3) a combination of autogenous bone and various bone graft substitutes (alloplastic, allogeneic, or xenogeneic) [1], or (4) the bone graft substitute materials used alone.

Allogeneic grafts or implants are banked bone materials taken at autopsy from another individual of the same species and usually preserved by freeze-drying. These in turn may be divided into nondecalcified banked allogeneic osseous materials and decalcified banked freeze-dried bone materials [1].

Alloplastic bone substitutes are for the most part synthetic hydroxyapatites or tricalcium phosphates or carbonate containing hydroxyapatite materials, which can be used alone or in combination with autogenous grafts [1].

Xenogeneic grafts or xenografts are bone substitutes of animal origin placed in a recipient site of another species. Usually these bone minerals have all the organic materials removed to produce essentially a hydroxyl-deficient carbonate apatite. This has proved to be an excellent material for use in combination with an autograft [2,3].

These various materials used to reconstruct the host bone site for the reception of the implant affect the interface and the type of bone formation around the implant initially and also the reaction of the bone and the general status of the host bone over the long-term service of the implant in function.

The use of bone inductor materials and bone growth factors offers possible future surgical applications in enhancing the initial and long-term character of the metal-bone interface [2-4].

In this chapter we present examples of these various types of interfaces and the prognosis in terms of these interfaces responding to changes in functional load on a long-term basis.

II. IMPLANT SITES

A. Titanium Implants in the Mandible

The bone of the mandible is composed of an outer cortical layer that tends to be thicker on the lateral and inferior surfaces than that of the maxilla, and of an inner trabecular pattern that tends to be quite thin (Fig. 1). This thinning of the trabecular pattern in the edentulous area militates against rapid osteointegration of the implant in such areas. However, this thin trabecular pattern can be thickened and increased in density by the use of appropriate bone graft materials. The use of substitute bone graft substance has markedly affected the character of metal implant to bone interfaces.

When implants are placed in the mandibular alveolar ridge, the surgeon relies upon the dense outer cortical bone surface to "fix" the implant initially, pending the formation of new bony matrix against the implant surface in deeper more distal areas of the bony mass. Thus the initial fixation is brought about more easily by a more concentrated lamellar bone at the crest of the mandible (Fig. 2). When this is not present or insufficient, the implant tends to be slightly mobile upon insertion and the resultant metal implant micromovement produces a lack of bony interface integration. Thus when studying implant interfaces in the mandible, one is presented with a host test site of great variability. For this reason one should utilize a more standardized model to test implant-bone interfaces. Such a model is available in the femoral shaft of experimental animals, particularly the subhuman primate. Examples of the study of implant interfaces in the femur are given.

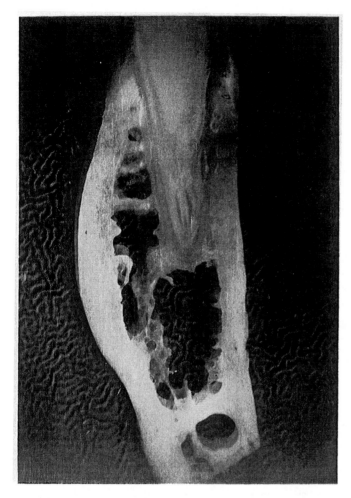

Figure 1   Cross-section of a mandible of a Rhesus monkey showing the thick lamellated outer cortex and the thin trabecular pattern.

B.   Maxilla

The maxilla presents the same problem as that existing in the mandible. That is, the maxilla, particularly the posterior portion, possesses an outer thin cortical layer and an inner trabecular pattern with large marrow vascular spaces (Fig. 3). The outer cortical surface of the maxilla is much thinner than that of the mandible, therefore the possibility of early mechanical retention of the implant when placed by the surgeon is diminished. As stated previously, such mechanical early fixation is important in keeping the implant stable until the normal host bone physiological remodeling process can take place to effect a stable bone-to-implant interface. This lack of sufficiently thick cortical bone surface is one of the reasons for the higher incidence of failure of osteointegration of implants when placed in the maxilla.

Additionally, the thin trabecular pattern is more marked in the posterior aspect of

Figure 2   Cross-section of the mandible of a Rhesus monkey with a root-form implant in place. The initial fixation of the implant is being brought about by the cortical layer, with relatively little cancellous bone in the deeper portions available at this time to immobilize the implant and to produce the necessary osteointegration.

the maxilla than in the anterior. This, plus the lack of sufficient surface cortical bone to give adequate fixation, produces a situation in which the implant is not stabilized sufficiently initially to obtain a proper bone interface (Fig. 3). As mentioned earlier, an increased bone density can be brought about by the appropriate use of graft materials.

C.   Femoral Long Bone Sites

Weight-bearing long bones such as the femurs of subhuman primates (e.g., *Macaca fascicularis*) offer ideal experimental sites for evaluation of the interface of titanium root-form implants with surrounding osseous tissue [2,3]. The histologic anatomy of such areas present both cortical and cancellous bone, and large areas of sparsely mineralized medullary canal for study. In such sites, one has the opportunity to evaluate different types of bone responses to the surgical insult of placement of titanium implant and the long-term effect of function [4]. The femoral metaphysis of the Rhesus monkey presents

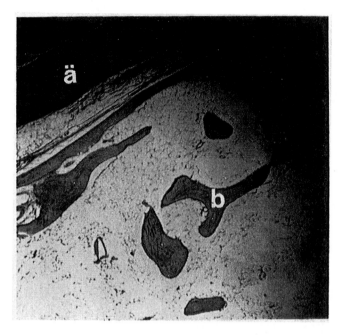

Figure 3   Cross-section of a posterior portion of an edentulous *Macaca fascicularis* maxilla show-ing the thinning of the cortical bone at the lateral aspect and crest of the ridge (a). In addition, the trabecular bone is extremely thin (b), offering a poor site for integration of root-form implants. This is one of the reasons for the inadequate integration of titanium implants in the posterior portion of the maxilla.

a cross section of approximately 3 mm thickness of circumferential lamellated bone and approximately 2 mm of cancellous bone in the area immediately surrounding the medul-lary canal (Fig. 4). The evaluation of implants in such areas offers the unique opportunity to use histomorphometric analysis to determine the characteristics of cortical bone to implant interface and cancellous bone to implant interface, as well as the nature of an empty medullary canal with largely marrow vascular spaces in interacting with the tita-nium implant surfaces (Fig. 4).

III.   MODIFICATION OF THE INTRAOSSEOUS IMPLANT RECIPIENT SITE BY USE OF BONE GRAFT MATERIALS

A.   Changes in the Healing Pattern of the Root-Form Intraosseous Implant to Bone Interface in the Mandible and Maxilla

The tissue reaction at the surface of root-form metal implants resulting from the surgical procedure of implant placement in the alveolus of the mandible and maxilla that have been bone grafted depends to a large extent on the type of osseous graft material used in rebuilding the recipient site.

An investigation was made in Rhesus monkeys to study the repair patterns of the bone of the mandible and maxilla when various types of bone graft substitutes were used

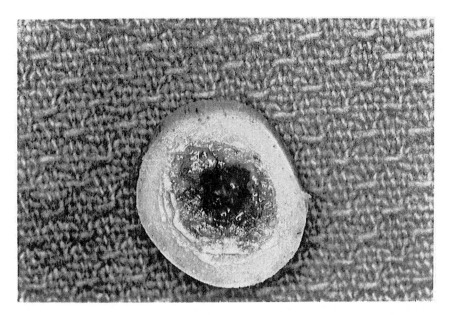

Figure 4   Cross section of a femur showing a relatively stabilized outer lamellated cortical bone layer, cancellous bone next to the cortical layer, and a large empty femoral shaft in the center portion. When this area is used as a test site, these different zones of bone tissue would be interacting with any implant placed for evaluation.

to prepare the area, either at the time of the placement of the root-form implants or in staged procedures with the titanium implants being placed at a later date after healing of the bone-grafted site. The results indicate that the nature of bone graft substitute materially affects for a long period of postoperative time the osseous response at the surface of the implant to physiologic changes manifested by changes in occlusal function brought about by the titanium implant fixtures.

To understand the histologic nature of these changes it is necessary to first review the types of bone graft substitutes commonly used in the mandible and maxilla.

B.   Types of Bone Graft Substitutes Used in Implant Sites

Hydroxyapatite (HA), as commonly used, may be a solid nonporous, nonresorbable type, or a partially resorbable composite type, or a third type that is porous and slowly resorbable.

1.   *Nonporous HA*

Nonresorbable solid nonporous hydroxyapatite particles placed in a surgical area tend to adhere well to the newly forming repair bone, producing a bony interface that is titanium implant compatible, but this material has the disadvantage of not remodeling and not being appropriately responsive to changes in function that may affect the titanium surface when the implant is placed in full functional load (Fig. 5). This is particularly true in long-term functional situations [6]. Such particles of nonresorbable HA may be seen next to the titanium or next to the HA-coated implant adhering to the root-end fixture but not contributing to the remodeling process [6]. The technique of placing large masses of

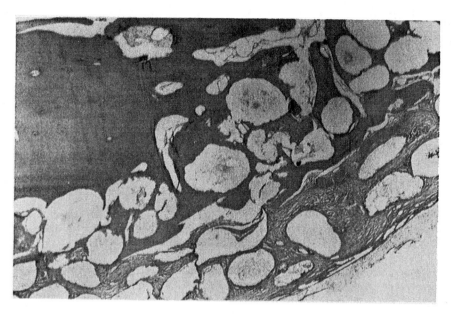

Figure 5 A view of nonresorbable dense nonporous particulate hydroxyapatite placed in a healing tooth extraction socket. The response of connective tissue around most of the particles can be seen, with only those particles actually touching the bone presenting an osseous interface. This type of graft material does not remodel and is not capable of being resorbed and replaced by normal bone with passage of time and with the changes necessitated by subsequent functional alterations.

non-porous HA on the surface of edentulous ridges to support conventional prostheses has usually led to a decrease in strength of the ridge due to the fibrous tissue, which interacts with these nonporous particles. In addition, if an intraosseous root-form implant is placed in such an area, there is a decrease in strength in the area and a tendency toward fracture of the mandible (Fig. 6).

2. *Partially Resorbable HA*

The partially resorbable type of hydroxyapatite particles usually contain tricalcium phosphate. The tricalcium phosphate (TCP) portion of such a product is resorbable whereas the hydroxyapatite tends to remain behind as a slowly resorbable or completely nonresorbable fraction. This condition also affects the nature of the result of the response of the bone to changes in function. Host response to this HA–TCP material is different from that of normal repair, because the residual particles of nonporous HA are not remodeled as the cancellous bone remodels and adapts in responding to the influence of function.

   An additional significance of this difference in peri-implant bone response occurs in the case of TCP containing HA "grafts." Here resorption occurs by nonphysiologic lysis (i.e., nonosteoclastic resorption) of the TCP, usually resulting in a tissue "space" being left after the resorptive phase. These tissue spaces occur because the osteoblastic bone-forming phase is not synchronized with the degree of bone loss. Such incompatible reaction between resorption and bone formation classically is seen in the use of other types of resorbable materials such as plaster of paris (i.e., $CaSO_4$ mixed with HA). The

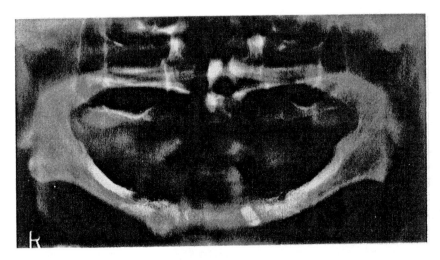

Figure 6 Pantomographic view of an atrophic mandible that had been implanted with a large amount of nonporous particles of HA, with subsequent placement of two root-form implants. The implant on the right side of the mandible has been removed, after the patient sustained a fracture at the site. The particulate HA used in this manner does not strengthen the mandible since most of the mass of HA is merely surrounded by connective tissue. Thus root-form implants placed in this mass of HA will not integrate with osseous tissue, and failure of function of the implant with a probability of bone fracture is the result.

$CaSO_4$ is lysed rather quickly (2–3 weeks) without cellular resorption, leaving spaces around the HA particles, which the osteoblasts are not able to fill with bone matrix because of their slower maturation rates as compared with fibroblasts. This results in spaces that rapidly fill with connective tissue, leading to an inappropriate graft result. It is true that at times, later remodeling may eventually form some new bone matrix in these areas, but there always remains significant connective tissue resulting from a graft material that contains $CaSO_4$, TCP, and other rapidly resorbable elements.

3. Porous Hydroxyapatite

Porous hydroxyapatite products usually have the characteristic of being very slowly resorbable over a long period of time. Porous hydroxyapatite having no tricalcium phosphate will at times have polynucleated cells that appear on the ceramic implant surface [7]. These cells are also found in the soft tissue and appear to be foreign-body megakaryocytes. They may be involved in the partial destruction of the so-called nonresorbable hydroxyapatite [7].

However, this form of resorption is not similar to the normal resorption–remodeling process of bone seen in the autogenous grafts, in which osseous matrix is resorbed by osteoclasts and remodeled or rebuilt with osteoblasts that are acting in concert by a "coupling" type of cell interaction. This is a mutually complementary system. Thus, this method of resorption or dissolution of hydroxyapatite is very unlike normal bone remodeling when one is assessing function over a long term.

The immediate response of the host bone to this type of hydroxyapatite is generally favorable in that a bone interface will form next to the HA particles if they have been placed adjacent to a host bone surface. However, this does not really markedly enhance the formation of new bone or the rate of bone formation in the area.

4. Porous Bone Mineral — Xenogeneic Bone

This material is derived from bovine bone by a chemical process that removes all organic material, leaving the inorganic matrix. The residual matrix (BioOss®) is more than simple hydroxyapatite, being composed instead of all the elements and structures of bone mineral (carbonates and phosphates as well as hydroxyapatite). Secondly the porosity of this product is exactly that of natural bone, with cancellous spaces and the appropriate diameters of the Haversian and interconnecting channels. Thus perfusion by neoangiogenesis is facilitated by this product. Further, the carbonate fraction invites osteoclastic resorption, producing the histologic cell synchronization that is necessary for normal repair, and that is lacking in the type of resorption commonly seen in porous HA (as described above).

Such repair process is also quite different from the noncellular "lysis" seen in certain types of materials such as $CaSO_4$ or Ca_3PO_4, which tend to provide residual areas of connective tissue rather than new bone matrix and thus result in lack of appropriate bone remodeling. The final goal of osseous reconstruction is to obtain a remodeling bone matrix surface next to the implant rather than a mixture of bone and connective tissue and a noncalcified matrix.

As mentioned, synthetic calcium phosphate material in particular does not undergo physiological remodeling, characterized by osteoclast-mediated resorption followed by bone formation. Thus bone graft substitutes of this nature may lyse, leaving the opportunity for fibrous soft tissue to intrude on the bone regenerating area. The evidence of osteoclast-like cells on the surface of BioOss suggests that this natural bone mineral implant remodels in a more physiological manner [6] than the synthetic bone graft substitutes.

The mechanical properties of BioOss (including stiffness, porosity, and surface area) are also important. Synthetic nonporous, nonresorbable hydroxyapatite implants are very dense and stiff, and can therefore adversely influence the remodeling of surrounding bone. Moreover, their density and hardness prevent subsequent surgical preparation of the implant site. Porous bone mineral (PBM — BioOss) has properties that more closely resemble those of natural bone and, therefore, present less surgical problem of this nature [8]. Moreover, the large surface area of porous bone mineral commends its use as a delivery system or "carrier" for antimicrobials and other agents such as growth factors [8].

The characteristics and type of bone formation on and around an implant surface may be related to the period required for the deposition of biological calcified bone matrix. In light of this observation, a substance such as natural bone mineral (BioOss), which already is composed of biological apatite with the naturally occupying carbonate fraction, has appropriate porosity, the optimal surface area, and the presenting chemical and physical properties necessary to establish an environment for bone repair (Fig. 7). The unique chemistry and structure of BioOss supports its use as a bone substitute material.

IV. HA COATINGS ON THE SURFACE OF TITANIUM IMPLANTS

The interface of HA coatings on titanium intraosseous implants is the subject of much discussion. Advocates of HA coating believe that the HA-to-bone interface is far superior to a plasma-sprayed titanium surface [9]. It is stated that repairing bone forms more rapidly on HA-coated implants.

Of equal or increased importance, however, is the final or the long-term response of the bone matrix bonding to the hydroxyapatite and the participation of the hydroxyapa-

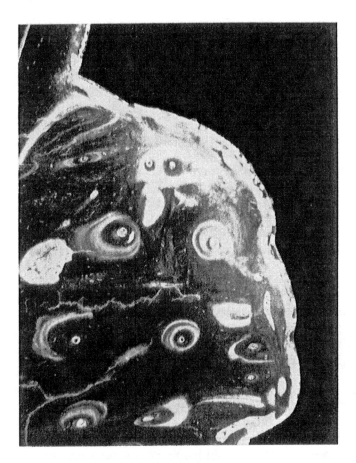

Figure 7 Porous bone mineral (BioOss®) particles (at the top of the section) are undergoing resorption and remodeling 2 years after placement next to a titanium intraosseous implant. The porous bone mineral is contributing to the underlying process by enhancing the trabecular bone density in the area of the titanium implant. Thus the long-term interface of titanium to bone is changed by (1) an increased vital bone matrix density and (2) the presence of still unremodeled particles of porous bone mineral. The net effect is supportive of the longevity of the titanium implant system.

tite surface in the osseous remodeling process as the implant is subjected to changes in function and variations in pressure–tension vectors of forces during months and years of subsequent use. Thus the final outcome of the use of the HA-to-metal bond would depend upon the role of the ceramic material in influencing overall bone remodeling and in affecting the responsiveness of surrounding bone to functional changes.

Additionally, it is important to study the relationship of the HA to the bone within the context of the response of bone to the HA surface coating itself and the use of bone graft substitutes in conjunction with the placement of root-form implants at two possible times: at the time of bone grafting and as a staged, delayed procedure, with the titanium implant being inserted in the alveolar ridge after a suitable period of healing and bone remodeling of the surgical defect that had received bone substitute materials at the time of tooth extraction.

A. Effect of HA Coating on Bone Repair of Extraction Socket

Reports vary as to the effect of the HA surface on the healing of the extraction socket itself. It has been reported that if more than 0.5 mm of space remains between the titanium implant and the healing socket wall, that distance will fill with soft tissue rather than with bone [9]. In our work with the Rhesus monkey in the placement of implants in healing sockets that have not received a bone graft or a bone graft substitute [10], we have found that bone would appropriately osteointegrate the implant whether HA coating or titanium was used. Additionally, increased rates of bone formation from the socket wall to the implant surface have been reported when HA-coated implants were used. Our work does not show this, and repair rates are approximately equal between HA- and non-HA-coated implants. Thus there is a difference of opinion as to the enhancement of the initial bone-to-HA bonding resulting from these cases.

B. Effect of HA and PBM Particles Between Implant Surface and Socket Wall

The use of particulate bone graft substitutes that include porous hydroxyapatite as well as porous bone mineral in the healing extraction socket does enhance the rate of formation of bone from the socket wall to the HA-coated or titanium plasma sprayed implant, and thus does improve the predictability of the formation of bone in the space between implant and the socket wall. This phenomenon of enhanced repair occurs in the case of both titanium-surfaced implants and HA-coated implants. There appears, from our work, to be no statistical difference in the rate of bone formation between the two forms of surfaces.

Thus the use of bone substitute particulate material in the surrounding surgical site also materially affects the rate at which bone forms on the HA-coated surface. Resorbable HA, or resorbable bone mineral of the porous variety, tends to incorporate in the clot and be revascularized through the optimal porosity of the particles involved. The revascularized porous particulate material then offers increased surface area for new bone formation (conduction). This tends to lead to an optimal healing of the defect. In the same manner as in the case of immediate implant placement in tooth extraction sockets — when healing extraction sockets are implanted with porous bone mineral and allowed to heal in the mature Rhesus monkey for approximately 2–3 months — implants can be placed in the healed socket area with excellent bony integration.

C. Nature of the Implant Recipient Site in Determining the Bone-to-Implant Interface

It is important to note in establishing a consensus regarding formation of bone on HA-coated surfaces that it is recognized that bone response varies markedly with the recipient site. When one places titanium implants in the femoral shaft as previously mentioned, the root-form implant actually is interacting with three zones of bone in the recipient sites: The first zone is the outer cortical layer of the femur. The second is cancellous bone surrounding the medullary canal; and the third zone is the almost empty space of the medullary canal, which contains very little if any calcified bone matrix but which does contain marrow and is highly cellular, containing many pleuripotential cells. Thus in this case, the implant is separated from osseous matrix wall by considerable distance at the apical end of the implant surface (an area to be compared with the placement of implant extraction sockets at some distance from the alveolar bone wall), yet these implants do produce cancellous bone formation that tends to engulf the entire apical implant, resulting in good osteointegration and excellent integration of the bone to the implant surface. Again, formation of bone against the implant surface is without noticeable difference in rate of bone formation on the implant surface area regarding HA versus non-HA coatings. Thus even in this highly structured anatomic model, there is no histologic difference between HA- and non-HA-coated titanium implant surface to bone interface.

D.   Clinical Extrapolation of Animal Studies

Our clinical protocols call for the placement of implants as an immediate procedure after the extraction of the tooth provided that there is approximately 4 mm of bone available beyond the fundus of the socket so that the implant can be perfectly stable in residual bone. We then place porous bone mineral or porous hydroxyapatite in the space between the implant surface and the socket wall, and close the mucoperiosteum over the surgical site of the healing socket. We have found no difference between osteointegration of the implant surfaces in the placement of the implants in this manner and osteointegration in the delayed placement of implant in the sockets that have received the BioOss particles immediately after tooth extraction and have been subjected to a 3-month delay prior to preparation of the socket for the placement of the root-form implant. Again the limiting factor in these cases does not appear to be the HA coating but the immobility of the implant within the socket area. Provided the implant is sufficiently immobile, equal success can be obtained with HA coating or non-HA-coated implants in using this procedure. While many investigators and clinicians still advocate waiting 2 to 3 months after extraction of teeth prior to placement of an implant, we feel that immediate implant placement can be highly successful provided the porous mineral particles are appropriately used between the implant and host bone wall to produce an osteoconductive effect from the socket wall to the root-form implant. It is interesting to note that the original recommendation for the healing of the extraction socket prior to the placement of root-form implants was between 9 and 12 months. This has now been reduced to approximately 3 months. As noted previously, we feel that the immediate placement of root-form implants can be equally successful. Again in this technique, it is interesting to note that the emphasis on the importance of interface characterization of hydroxyapatite and bone has shifted from the HA-coated implant to bone interface to the effect of surrounding porous bone particles on bone formation and the conductive properties of this particulate material placed in the clot between the implant and the host bone wall of the socket.

Thus we are dealing with the HA-to-bone interface discussion, but within the context of particles rather than the metal ceramic surface. This would seem to be appropriate since bone conduction is really the problem being discussed. One would expect the particles of hydroxyapatite would conduct the bone from the socket wall to the implant more effectively than a simple surface coating of HA with a distance of several hundred microns to the nearest bone wall. That is, the space between the HA surfaces and host-calcifying and noncalcifying bone matrix is markedly reduced if one uses the particulate form of porous mineral. Discussion of the effect of HA particles on bone in this context opens new possibilities for clinical care.

Even more important is the final result of the long-term effect of HA and the HA-to-bone interface, in terms of HA particulate surfaces in sustaining bone remodeling in the area of the root-form implant, and in maintaining an area of increased bone density surrounding the osteointegrated implant.

V. CHANGES OF THE METAL-TO-BONE INTERFACE WITH ALTERNATIVES IN FUNCTION

The titanium- or HA-coated interface is constantly changing in response to the demands of occlusal forces and functional loads. In general, areas of compression bring about increase in bone density (Figs. 8 and 9), and the surfaces of the implant facing the tension load respond with decrease in bone matrix and a thinning of the trabecular pattern. If porous HA or porous bone mineral or fragments of HA, having separated on the titanium surface, are present in the area of increased occlusal load, one tends to see an enhancement of bone formation, with the HA particles or the porous bone mineral particles serving as a nidus for lamellated bone. Thus the long-term interface effect of implant particles appears to be to improve, at least from the histologic standpoint, the calcified matrix and to bring about an optimal interface longevity.

VI. SUMMARY

The dynamics of the intraosseous implant-to-bone interface offers extensive possibilities for developing improved implant systems. Changes in matrix density, trabecular patterns, and ratios of compact to cancellous bone at the interface are evidencing adaption to functional loads and physiologic success of the implant system.

Figure 8   Increased bone density next to a titanium implant surface after a long-standing increase in compression along this surface. The compressive forces have resulted in compact bone at the implant interface, and the particles of BioOss® are being replaced with osteonal bone top arrows.

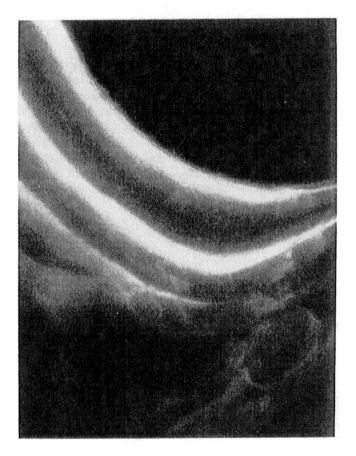

Figure 9   Another view of the implant shown in Fig. 8, showing sequential osteonal apposition of bone.

REFERENCES

1. Boyne, P. J., Maxillofacial surgery, in *Bone Grafts and Bone Substitutes* (Mutaz Habal, ed.), W. B. Saunders, 1992, pp. 291–298.
2. Boyne, P. J., The study of interface bone formation resulting from the use of intraosseous titanium implants, *Mater. Res. Soc. Symp. Proc., 110*, 561–569 (1989).
3. Boyne, P. J., and Sheer, P. M., Maintenance of alveolar bone by the implantation of hydroxylapatite and anorganic bone in extraction sockets, *JADA, 114*(5), 594–597 (May 1987).
4. Boyne, P. J., Host response to intraosseous implants placed in HA grafted mandibles, *Mater. Res. Soc. Symp. Proc., 110*, 219–227 (1989).
5. Boyne, P. J., Comparison of porous and non-porous hydroxylapatite and xenografts in the restoration of alveolar ridges, in *Proceedings of ASTM Symposium on Porous Implants*, Nashville, TN, 1987, pp. 359–369.
6. DeLange, G. L., Deputter, C., and DeWifs, F. L., Histological and ultrastructural appearance of the hydroxylapatite bone interface, *J. Biomed. Mater. Res., 24*, 829–845 (1990).
7. Donath, K., *The Implanting of Bone Replacement Materials in the Jaws* (Oral Surg.), Carl Hanser, Munich, 1988.

8. Spector, M., Characterization of calcium phosphate bioceramic implants, in *Hefte zur Unfall- heilkunde*, Vol. 216 (A. H. Huggler and E. K. Kuner, eds.), Springer-Verlag, 1991, pp. 11–22.
9. Knox, R., Caudill, R., and Meffert, R., Histologic evaluation of dental endosseous implants placed in surgically created extraction defects, *Int. J. Periodont. Rest. Dent., 11*(5), 365–375 (1991).
10. Boyne, P. J., Analysis of performance of root-form endosseous implants placed in the maxillary sinus, *J. Long-Term Effects Med. Implants, 3*(2), 143–159 (1993).

58
Surface Characterization
of Dentin and Enamel

Amy Lin, N. S. McIntyre, and R. D. Davidson
The University of Western Ontario
London, Ontario, Canada

I. INTRODUCTION

One of the most important aspects of restorative dentistry is the removal of decaying tooth substrate and the attachment of aesthetic replicas to the remaining part of the dentition. Marginal apposition between the filling materials and walls of the prepared cavity is essential if leaking of oral fluids, cells, and bacteria is not to result in recurrent caries. The attachment of the materials to the tissue surface is therefore of the greatest significance in many dental procedures and is often the most important factor in controlling clinical success.

The tooth is a composite material. The main structural component of the tooth is the dentin. Dentin is composed of a mineralized collagen matrix penetrated by long, narrow parallel channels called *dentinal tubules* that are oriented perpendicular to the dentino-enamel junction. This makes the dentin a very porous structure. The composition of dentin consists of about 18% collagen, 69% calcium hydroxyapatite, and 13% water. The dentin is covered peripherally by enamel on the crown and by cementum on the root surfaces. The tough enamel surface has a much higher mineral content. There is about 95% calcium hydroxyapatite, 4% water, and 1% organic content in enamel. Dentin is the only innervated hard tissue of teeth, but it remains relatively insensitive as long as it is covered. However, it should be noted that any substance applied to the dentin has direct access to the pulp cavity and thus to nerve tissues, which can cause irritation.

When the dentin is cut or shaped in preparation for a restoration, a layer of cutting debris called the *smear layer* is created. This smear layer is composed of the underlying matrix. As the composition of the underlying matrix may change as the dentin becomes thinner (i.e., near the pulp), the smear layer composition may also change as deeper dentin is cut. The thickness of smear layers is only about 1 μm. This layer also adheres well to the underlying matrix. Smear layers cannot be rinsed off or scrubbed off the dentin surface, and thus become a consideration when applying materials to this newly prepared surface.

Adhesion can be defined as the attraction exhibited between the molecules of different materials at their interface. The criteria for achieving adhesion include: (1) a clean substrate for intimate access of adhesive to its surface, (2) complete wetting of the substrate surface, and (3) liquid-to-solid transformation of the adhesive. Toxicity, moisture, temperature, and the chemical nature of the substrate present further restrictions that need to be dealt with when designing an adhesive component. Chemical variation of the substrate, differences in the coefficient of thermal expansion and volume, and wetting changes on solidification will all affect the final performance of the adhesive.

Thus, adhesion to the tooth substrate requires intimate contact to be established between the tooth and the adhesive material, good wettability of the adhesive on both the dentin and the enamel in the fluid state, and minimal shrinkage of the adhesive upon polymerization. The adhesives are usually designed to bond chemically to a component of the dentin and to the resin of the restorative filling, which would be layered on top of the adhesive. They can either bond to the calcium in the inorganic component of the tooth or to the organic collagen. The constraints imposed on these adhesives can greatly affect their function in an intraoral environment. A great deal of research and development has been carried out to date on understanding and creating strong bonds between the tooth and restorative material.

Pretreatment producing a modified tooth surface often improves adhesion between the tooth structure and a dental restorative. Various treatments have been used routinely to prepare the surface to improve receptivity for adhesion. Acids, alkalis, enzymes, and sequestrant have been used to "clean" the cavity surface by removal of the loose or partially attached debris, as well as to remove selectively some of the organic or inorganic portion of the dentin and enamel. Other techniques have modified tooth surfaces by chemical means to improve adhesion.

Bond strength tests of adhesives and cements to dentin and enamel surfaces have been the predominant method of evaluating the effectiveness of the material outside of a clinical study. A variety of test methods are available and they are performed under all types of conditions. Unfortunately, because of this, it is difficult to compare the results of these studies and to make sound conclusions. It is the purpose of this chapter to introduce a more comprehensive approach to studying adhesion of dental materials to tooth structures. The results of surface modification of the tooth structure and the interfacial interactions between the material and tooth surfaces can be analyzed using several surface characterization techniques. The following techniques are discussed in detail: scanning electron microscopy, X-ray photoelectron spectroscopy, laser scanning confocal microscopy, and secondary ion mass spectrometry. Specific applications of these techniques have been carried out in the authors' laboratory in relation to modified surfaces of dentin and enamel, as well as a study on the adhesion of glass ionomer cements to dentin surfaces. These examples are described in the following sections.

II. SCANNING ELECTRON MICROSCOPY

Scanning electron microscopy (SEM) is the best known and most widely used of the surface analytical techniques. High-resolution images of surface topography, with excellent depth of field, are produced using a focused, scanning electron beam. The electrons entering a surface with an energy of 1–40 keV generate many low-energy secondary electrons. The intensity of these secondary electrons is governed largely by the surface topography of the sample. An image of the sample surface can be constructed by measuring secondary electron intensity as a function of the position of the scanning primary electron beam. An important feature of the SEM is the three-dimensional appearance of the specimen image, a direct result of the large depth of field, as well as of the shadow-relief effect of the secondary and backscattered electron contrast. High spatial resolution is possible because the primary electron beam can be focused to a very small spot (less than 10 nm). In addition to low-energy secondary electrons, backscattered electrons and X-rays are also generated by electron bombardment. The intensity of backscattered electrons can be correlated to the atomic number of the element within the sampling volume, and some qualitative elemental information can be obtained.

An ISI Model DS-130 scanning electron microscope was used to examine surface dentin and enamel structures before bonding and after different pretreatments, and also for examining the interfacial structure between the adhesive and the dentin. The failed surfaces from bond strength tests for glass ionomer cements on dentin were examined. Prior to analysis, all sample surfaces were air dried and sputter gold coated.

Figure 1 shows four SEM micrographs of the dentin surface. The fractured specimen show the densely populated dentinal tubules within the bulk of the dentin structure. The polished appearance of the dentin surface (low and high magnification) is typical of SEM images when the dentin is covered with a smear layer. There is no longer any indication of underlying dentinal tubules. The final image is of the same polished dentin surface after ultrasonication for 15 min. Most of the smear layer debris and smear plugs have been removed. The smear layer trapped on Nucleopore filters (0.2 μm) can be seen in

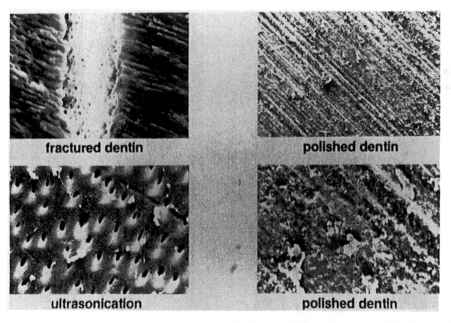

Figure 1 SEM micrographs showing freshly fractured dentin, a low and high magnification of polished dentin surface with smear layer attached, and the dentin surface after ultrasonication to remove the smear layer.

Fig. 2. The particles appear to be fibrous in nature and many of them are in the form of long narrow plugs that may have come from the tubules. The smear layer particles should represent mineralized collagen fibers, which fracture when the mineralized dentin is cut. Globular-like structures of smear layers have been previously observed by Pashley from dentin smear layer debris created by a carbide-steel burr on a high-speed handpiece [1]. Excellent SEM images were presented in this work before and after sonication and the collection of filtered particles. In Fig. 3, closer examination of the smear layer shows penetration of a smear plug (see arrow) into the dentinal tubule for about 10 μm in length.

Different pretreatments on the surface of dentin and enamel have been performed and subsequent SEM images produced. Phosphoric acid (10%) was used to etch the surfaces of enamel (Fig. 4) and dentin (Figs. 5 and 6). Etching of the enamel surface creates an increased surface area and opens up pores into which adhesive resin can flow. This greater porosity allows an easier ingress of uncured resin and therefore better mechanical interlocking of the hardened resin with the enamel phase. The SEM images show preferential dissolution of portions of the enamel surface and exposure of the enamel rods. Acid-etched dentin surfaces show the smear layer removed from the surface and the exposure of the dentinal tubules. The dentin surface appears rough and irregular.

An aqueous solution of maleic acid (2.5%) and hydroxyethylmethacrylate (HEMA, 58.5%) was used to pretreat the dentin surface in Fig. 7. This effectively removes the smear layer. The dentinal tubules are occluded and the walls appear fibrous in nature. A solution of 16% EDTA (ethylenediaminetetra-acetate) in a water buffer was used to treat the surface of the dentin in Fig. 8. This shows removal of the smear layer and only some removal of smear plugs from the tubules. Scratch marks from the polishing procedure can be seen on this surface. The last micrograph in this series depicts the surface of dentin

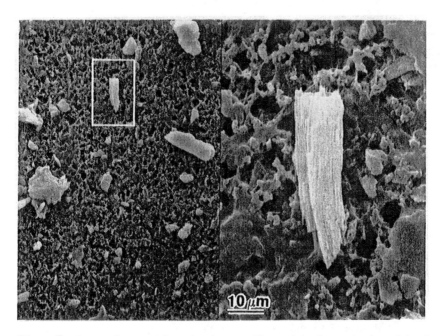

Figure 2   Smear plugs on 0.2-μm Nucleopore filters after ultrasonication of dentin.

Figure 3   SEM micrograph of smear plug showing penetration into the dentinal tubule.

Figure 4   SEM micrograph of an enamel surface after phosphoric acid etching.

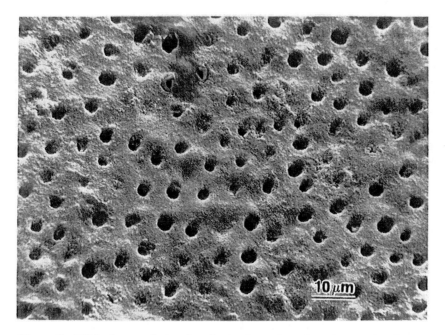

Figure 5   SEM micrograph showing the dentin surface after phosphoric acid etching.

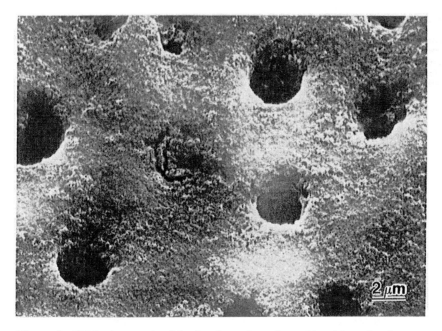

Figure 6   SEM micrograph of the dentin surface after acid etching, showing a roughened surface and exposure of the dentinal tubules.

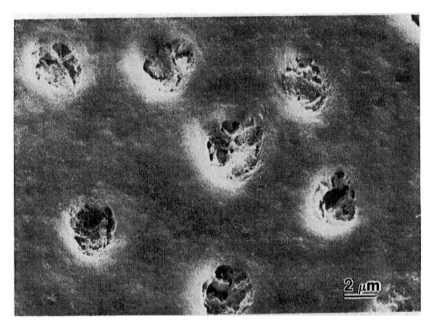

Figure 7   SEM micrograph of dentin after a pretreatment of maleic acid and hydroxyethyl methacrylate.

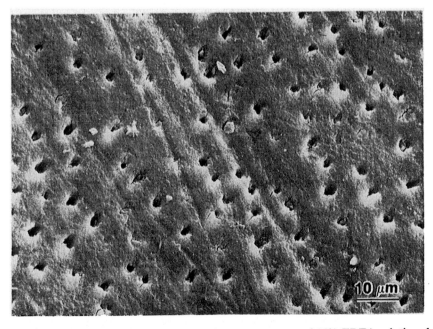

Figure 8   SEM micrograph of dentin after pretreatment of 16% EDTA solution showing removal of the smear layer.

treated with a solution of phosphonated penta-acrylate ester (PENTA) and hydroxyethyl methacrylate (HEMA) in ethanol, Fig. 9. This surface is similar to the maleic acid/ HEMA treatment, showing a clean surface with no smear layer and smear plugs still visible.

With the aid of the SEM, the prepared surfaces of dentin can be examined in great detail. The degree of etching and cleaning is variable with the treatment, and very different appearances of the surface result from such treatments. This is an important consideration when determining the permeability of dentin and the effectiveness of a resin to adhere to the surface from a mechanical point of view.

The SEM can be used to examine the interfacial structure between an adhesive material and the dentin surface. SEM was used to examine failed bond strength test samples for both dentin and glass ionomer cement surfaces. This type of examination of the fracture site can aid in the understanding of bonding mechanisms and the nature of the bond failure. For the light-curable glass ionomer cement, the bonded area showed the presence of a thin layer of glass ionomer on the surface of the dentin, as seen in Fig. 10. The chemically cured glass ionomer cement resulted in a bonded area with very little residual glass ionomer, as seen in Fig. 11. However, a smear layer of dentin debris was observed on the bonded area of the glass ionomer side, showing that this system failed cohesively within the smear layer. Mechanical interlocking is evident from SEM observations of the bond fracture surfaces between the ionomer and dentin. Figure 12 is a micrograph of the failed bonded surface of dentin after fracture. Many tubules are occluded with the light-cured glass ionomer cement, and ionomer resin tags are protruding from this surface.

Figure 9   SEM micrograph of a dentin surface after pretreatment with a solution of PENTA and HEMA in ethanol.

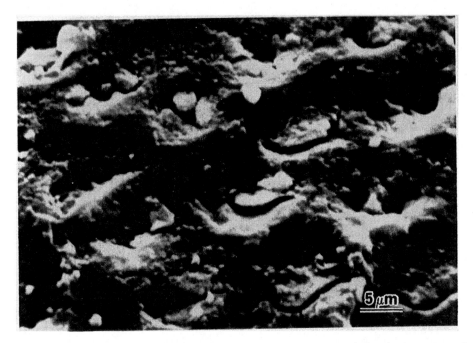

Figure 10   Bond site on dentin after fracture of a chemically-cured glass ionomer cement.

Figure 11   Bond site on dentin after fracture of a chemically-cured glass ionomer cement.

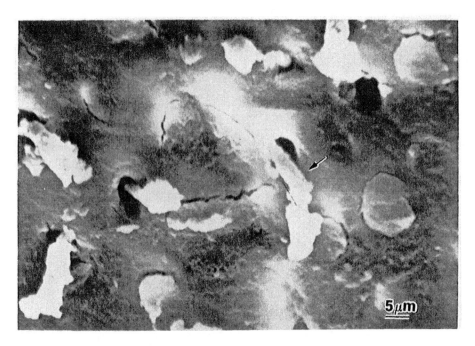

Figure 12 SEM micrograph showing occluded dentinal tubules and resin tags at the bond site of glass ionomer cement on dentin.

III. X-RAY PHOTOELECTRON SPECTROSCOPY

In electron spectroscopic techniques, the samples are exposed to a beam of either electrons or electromagnetic radiation (X-rays or UV light), which results in ionization of sample atoms and the ejection of electrons from levels corresponding to binding energies that are less than the energy of the incident radiation. The basis of X-ray photoelectron spectroscopy (XPS—the exciting source is an X-ray beam) is the detection and energy analysis of these photoelectrons. The hole state that is produced by the ejection of a photoelectron is then filled by an electron from a higher-lying state and the energy associated with this transition is liberated in the form of either an emitted Auger electron or an X-ray photon.

The probability or cross section of obtaining photoelectrons and Auger electrons depends upon the type of incident particle, its energy, and the particular target atom in question. By measuring the kinetic energy of ejected electrons, one can deduce information concerning the nature of the atom from which the electron was omitted and, in certain cases, the local chemical environment of that atom in the solid. The escaping photoelectrons originate from very near the surface (typically 0.5 to 10 nm).

An XPS spectrum contains peaks from core-level excitations, Auger transitions, and from excitation of valence electrons. Compositional analysis is generally carried out using the core-level peaks. The intensity of a specified core-level peak is determined by the number of atoms in the analyzed volume, the ionization cross section, and matrix effects such as the escape depths of the photoelectrons. One of the main advantages of the XPS is that the position of the peaks can provide information concerning local chemical environment. This shift is caused by changes in the screening of core-level electrons as the valence electron density changes during chemical bond formation.

XPS analyses were carried out to detect chemical changes on the surface of the dentin and enamel after surface pretreatment. Using a slow-speed saw, samples were cut to a size of 2 mm × 2 mm. Immediately prior to analysis by XPS, the samples were pumped in a vacuum chamber to reduce the water vapor pressure to allow sample introduction into the UHV (ultra high vacuum) spectrometer analysis chamber. The spectrometer used was a customized SSL SSX-100 X-ray photoelectron spectrometer that uses a monochromatized AlK$_\alpha$ X-ray gun to excite photoelectrons from areas as small as 150 μm. It is possible to adjust the position of the spot where analysis is to take place to a precision of 10 μm. Control specimens of dentin were always analyzed to monitor the effect of dehydration of samples in the vacuum chamber of the spectrometer. For each sample analyzed, broadscans were accumulated to detect all the elements. During these analyses, a 600-μm spot size and a pass energy of 148 eV were used. High-resolution spectra of peaks of interest (carbon, oxygen, and nitrogen), were obtained so that chemical effects on the photoelectron peak positions could be assessed. For these analyses of the core-level spectra, a 150-μm spot size and a pass energy of 50.0 eV were used. The charging effect observed during photoelectron analysis of nonconducting surfaces was controlled by utilizing an electron flood gun and a metal grid placed in close proximity to the surface.

The first set of XPS data is used to describe the chemical composition of dentin surfaces. XPS spectra from freshly fractured dentin are shown in Fig. 13. Calcium, phosphorous, and oxygen represent the inorganic portion of the dentin. The Ca : P ratio is 1.32. The Ca : P ratio of pure synthetic apatite is 2.15 whereas the ratio in bone and teeth is lower. The lower ratio is due to substitution of calcium by other ions such as sodium and magnesium, or substitution of hydronium ions for two adjacent calcium ions in the crystal lattice, or the absence of some calcium ions. The surface of the tooth has a greater latitude for substitutions than in the interior, and the surface has a net electrical charge that must be balanced by ions of the opposite charge.

Dentin is known to contain about 20% organic matter. The fractured dentin analyzed here contains 7.4% nitrogen, of which about 90% is present in collagen. Collagen works out to be about 18% of the dentin. Figure 14 shows the dentin surface after polishing. This surface was covered with the smear layer. The smear layer is made up of some Ca and P from hydroxyapatite (from both dentin and enamel), nitrogen from the organic matter in dentin, and a large amount of carbon from contaminants in the debris. When this smear layer is removed by ultrasonication, the surface has the chemical composition seen in Fig. 15. This dentin surface has increased amounts of Ca, P, and N, and reduced amount of C, and is essentially the same as the fractured dentin surface.

As mentioned earlier, the advantage of using XPS is the additional information obtained from binding energy shifts of core-level electrons. The high-resolution spectra of carbon 1s is shown in Fig. 16 for both the fractured dentin and the polished dentin surface. The widths of the peaks are slightly large but this is expected for these types of samples, which exhibit charging. The peaks are labeled to show the shifts associated with the various carbon groups that exist in the organic part of dentin—mainly the amino acids in collagen and other proteins. The large amount of hydrocarbons present as contaminants on the surface of the polished sample makes it more difficult in this case to distinguish the other carbon groups.

XPS analysis has been used to monitor chemical changes on the surface of prepared dentin after different pretreatments. This type of study would aid in understanding how

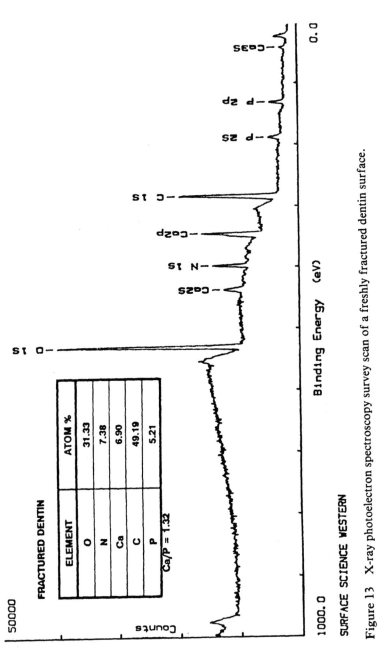

SURFACE SCIENCE WESTERN

Figure 13 X-ray photoelectron spectroscopy survey scan of a freshly fractured dentin surface.

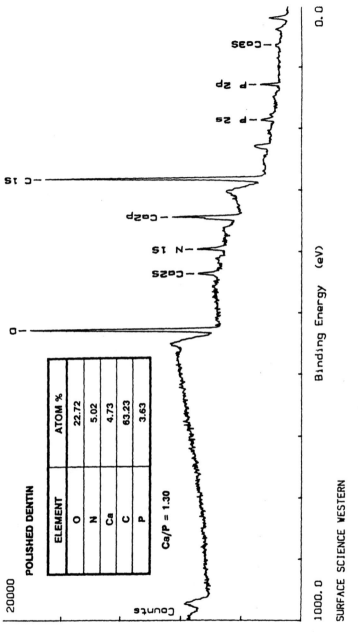

POLISHED DENTIN

ELEMENT	ATOM %
O	22.72
N	5.02
Ca	4.73
C	63.23
P	3.63

Ca/P = 1.30

SURFACE SCIENCE WESTERN

Figure 14 X-ray photoelectron spectroscopy survey scan of a polished dentin surface covered with the smear layer.

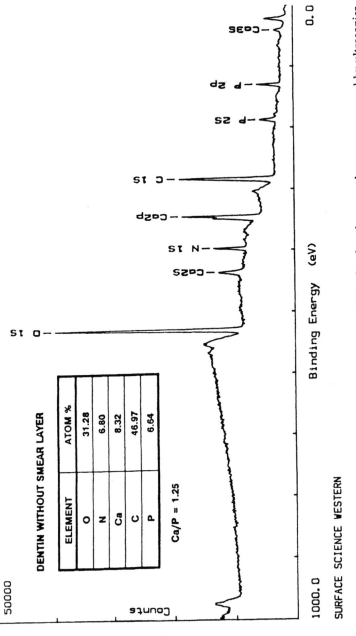

SURFACE SCIENCE WESTERN

Figure 15   X-ray photoelectron spectroscopy survey scan of dentin after the smear layer was removed by ultrasonication.

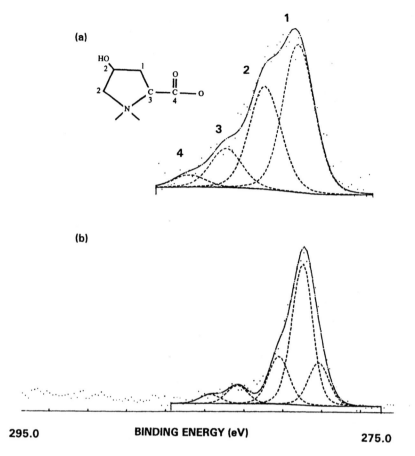

Figure 16 XPS high-resolution carbon 1s spectra for (a) fractured dentin and (b) polished dentin with smear layer.

surface alterations of dentin affect the overall bond strength of the adhesive. Table 1 shows the XPS surface composition results for various pretreatments of polished dentin. The phosphoric acid gel preferentially dissolves the inorganic portion of the dentin, leaving only organic constituents on the surface for further bonding. It is also interesting to note that a significant amount of Si has been left on the surface even after several washings with water. Nitrogen is also reduced, suggesting that some of the protein component of the dentin has been removed by the acid.

Maleic acid–HEMA solution on the dentin surface has also been shown to remove the smear layer and the large debris on the surface. The acid has also dissolved almost all of the inorganic portion of the dentin. However, the nitrogen content remains the same, leaving the proteins untouched by this treatment. Methacrylate groups on the surface are now present and represented by the large increase in the carbon content.

PENTA–HEMA–glutaraldehyde in ethanol solution has a different effect on the surface. This primer is designed to alter the smear layer on dentin to facilitate interaction with the restorative by chemical processes. The large debris has been removed, which is seen with the reduced calcium and phosphorous. However, a thin layer, along with the

Table 1   XPS Surface Composition (at. %) for Various Pretreatments of Polished Dentin

Dentin pretreatment	C	O	N	Ca	P	Other
Untreated dentin	49.2	31.3	7.4	6.9	5.2	—
10% phosphoric acid	62.7	22.0	4.4	0	0	Si
Maleic acid–HEMA	69.4	21.7	7.2	0.74	0.98	—
PENTA-HEMA-glutaraldehye	67.9	22.9	1.9	2.7	2.4	Si
16% EDTA in water buffer	68.2	17.7	10.7	0.75	0.69	Si
16% EDTA + HEMA-glutaraldehyde	69.7	26.4	3.2	0.28	0.4	—

smear plugs in the dentinal tubules, remains present on the surface. This surface is further altered by the glutaraldehyde, which denatures the protein and partially dissolves the collagenous matrix of the dentin. The nitrogen content is reduced on the surface. A large amount of carbon is seen on the surface from the application of HEMA.

A 16% EDTA solution was first used to treat the dentin surface and was followed by treatment with a HEMA–glutaraldehyde solution. The EDTA treatment removes the smear layer and also preferentially dissolves the inorganic phase of the dentin surface. A decrease in the amount of calcium and phosphorous on the surface to less than 1 at.% was observed. The nitrogen content has increased, suggesting that the chemical composition of this surface has a significantly reduced inorganic component relative to the organic component. The slightly larger amount of nitrogen than untreated dentin may be from residual EDTA on the surface. After further treatment with HEMA–glutaraldehyde solution, the surface exhibited a decrease in nitrogen content, as seen previously with glutaraldehyde treatments, and an increase in the carbon content, showing coverage of the surface with HEMA.

XPS analyses were carried out to detect chemical changes on the surface of the dentin after bond strength tests [2]. These results are presented in Figs. 17 and 18. The dentin side of the failed bond area for light-cured glass ionomer cement to dentin appears to attract additional silicon, aluminum, strontium, fluoride, and zinc ions, presumably from the glass powder component. Calcium and phosphorous concentrations are also reduced in this area. The dentin side of the failed bond area for chemically cured glass ionomer cement, which was chemically cured, showed mostly carbon and oxygen, and small amounts of calcium and phosphorous. However, there was no indication of an increased amount of glass particles since only a small amount of silicon was detected at this surface.

High-resolution spectra for carbon 1s are shown in Fig. 19 and compared for surfaces of ionomer, dentin, and the failed bond areas of dentin for light-cured and chemically cured glass ionomer cement. In comparing the chemical state of carbon from the C(1s) spectra for the dentin surface, light-cured glass ionomer bond fracture site, and chemically cured glass ionomer bond fracture site, carbon contribution from different functional groups can be detected for each surface. Singly bonded C—OH or C—O—C groups were observed on the surface. Polar groups bonded to carbon were expected, due to the nature of the wet dentin surface. Carbonyl groups are present on the surface of dentin and are due to the collagen content in the dentin smear layer. A higher amount of

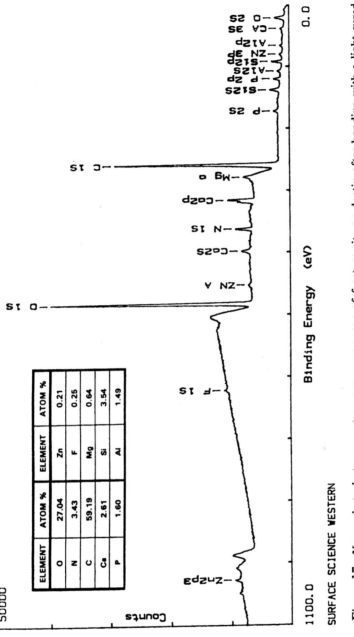

ELEMENT	ATOM %	ELEMENT	ATOM %
O	27.04	Zn	0.21
N	3.43	F	0.25
C	59.19	Mg	0.64
Ca	2.61	Si	3.54
P	1.60	Al	1.49

SURFACE SCIENCE WESTERN

Figure 17 X-ray photoelectron spectroscopy survey scan of fracture site on dentin after bonding with a light-cured glass ionomer cement.

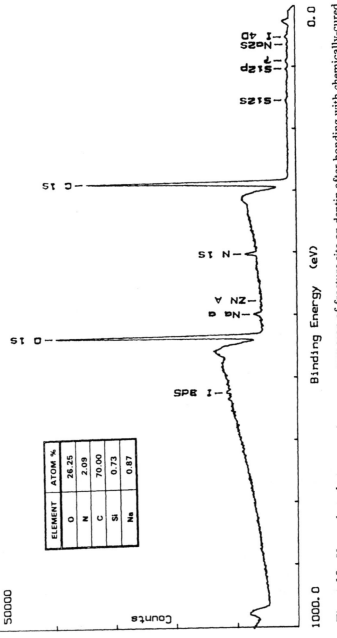

ELEMENT	ATOM %
O	26.25
N	2.09
C	70.00
Si	0.73
Na	0.87

Figure 18 X-ray photoelectron spectroscopy survey scan of fracture site on dentin after bonding with chemically-cured glass ionomer cement.

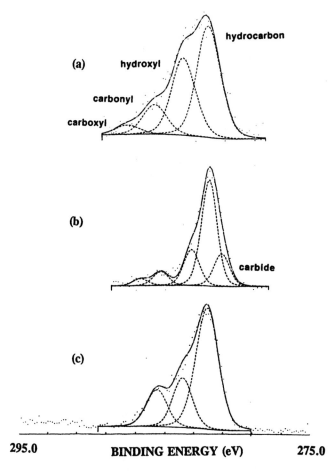

BINDING ENERGY (eV)

295.0   275.0

Figure 19 High-resolution carbon 1s spectra for (a) dentin, (b) light-cured glass ionomer bond area, and (c) chemically-cured glass ionomer cement bond area. The moieties responsible for the various peaks are indicated in (a) and (b), and are shown as dotted lines.

carboxyl groups was detected on the chemically cured ionomer failed-bond site compared with that on the light-cured ionomer. This indicates a low amount of ionic bonds present where the bond failed for the light-cured material. Chemical analysis of interfaces showed that different bonding mechanisms may have been responsible for the bonding of the light-cured and chemically cured systems.

IV. LASER SCANNING CONFOCAL MICROSCOPY

The essential feature of a confocal microscope is that the illumination and detection are confined to the same spot in the specimen at any one time. If the spot is so small that its limits are set by diffraction, the resolution in a confocal microscope is greater than in a conventional one. The lateral resolution (*xy* direction) may approach closely the theoretical optimum of 0.7 times the conventional resolution. The key feature in confocal imaging is that only what is "in focus" is detected. Out-of-focus regions of the sample appear black.

There are many possible forms of confocal microscopes, but the epi-illumination design as shown in Fig. 20, is particularly effective, since the same lens functions as both condenser and objective, obviating the need for exact matching and coorientation of two lenses. Light from an aperture is reflected into the rear of the objective lens and is focused on the specimen. Light returning from the specimen, as a result of either reflection or fluorescence, passes back through the lens and is focused on a second aperture, which allows a portion of the beam to pass to a detector such as a photomultiplier.

A confocal microscope does not contain a complete optical image of the specimen: To build up an image it is necessary to scan the point probe over the field of view. This can be done either by scanning the specimen across the beam or scanning the beam over a fixed specimen. Instruments use a laser as a light source in order to improve the resolution and sensitivity, in particular with the application to fluorescent specimens. Lasers provide more adequate intensity for imaging. The use of a single scanning beam allows the formation of two or more simultaneous images, using different optical modes. Since the beam passing over a specimen feature affects all the detectors simultaneously, the reconstructed images show perfect registration with each other.

A Bio-Rad MRC-600 (Bio-Rad, Inc., Cambridge, MA) was used in the studies at Surface Science Western. This system allows simultaneous imaging of two different fluo-

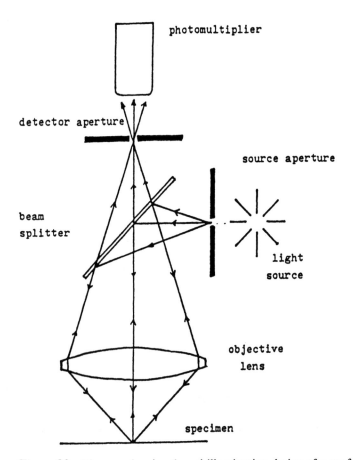

Figure 20   Diagram showing the epi-illumination design of a confocal microscope.

rescent stains. Nonfocal transmission imaging in phase contrast, DIC (differential inter-ference contrast), polarizing, and all other conventional light microscope modes can also be performed.

Confocal microscopy was used to obtain thin optical sections below the surfaces of tooth restoration specimens. This technique provides a unique opportunity of imaging the penetration and thickness of the cement and adhesive in the dental restoration. Freshly extracted third molar teeth were used for the restorations. Class V cavities were prepared to reproduce those in normal clinical practice. All cavities were 2 mm wide and 2 mm high, and cut with a type 012 diamond burr running wet in an ultrahigh-speed handpiece. Enamel surfaces were etched with 37% phosphoric acid gel for 15 sec and then rinsed with distilled water for 30 sec. A light-cure glass ionomer cement (Vitrebond, 3M) and conventional glass ionomer cement (Glass Ionomer Cement Liner, 3M) were both pre-pared according to the manufacturer's instructions. Over the thin layer of ionomer, dentin primer (Scotchprep, 3M) was applied to any exposed dentin surfaces with a sponge for 60 sec. The adhesive resin (Scotchbond 2, 3M) was then applied to the glass ionomer cement, primed dentin, and etched enamel surfaces. The adhesive was light cured for 20 sec. The cavities were then filled with a composite material (P-50, 3M) and light cured.

These restored teeth samples were stored in distilled water at 37°C for 24 h. Using a slow speed saw, samples were sectioned longitudinally through the center of the restora-tions. Thus, the interface was exposed for examination with the confocal microscope.

To aid in the visualization of the penetration of the glass ionomer cement and the adhesive, fluorescent labels were incorporated into the materials used. Double labeling by fluorescent isothiocyanate (FITC) and Texas Red (Molecular Probes, Inc., Eugene, OR) was used. The FITC is maximally excited by the 488-nm (blue) laser line and emits at wavelengths to which the photomultipliers are highly sensitive. Texas Red is maximally excited at the 514-nm (green) laser line, thus producing perfect registration for two-channel confocal imaging. FITC was used to label the Scotchbond 2 adhesive resin, and Texas Red was used to label the polymer portion of the glass ionomer cement. This produced the best double-labeled images. The scanned samples were examined for gas formations, penetrations of materials into the tooth surfaces, and distributions of the two materials relative to each other. Images were recorded with a 35-mm camera with T-MAX 100 black-and-white high-contrast film (Kodak).

The scanned images of the sections taken with the confocal microscope showed close adaptation of the light-cured glass ionomer cement to the dentin surface. Penetration of the polymer into the dentinal tubules is clearly shown in Fig. 21. The chemically cured glass ionomer cement exhibits less effective penetration into the dentin surface, as seen in Fig. 22. The labeled polymer has not moved down along the dentinal tubules to the same extent as the light-cured material. By double labeling, the distribution of both the adhe-sive and the ionomer components of the restoration system can be observed. No adhesive penetrated the ionomer layer and smear layer to enter the tubules, except where no ionomer was present. In Fig. 23, a gap is seen within the light-cured glass ionomer cement layer. Although there is still penetration into the tubules and close apposition of the ionomer to the dentin surface, a failure is seen within the ionomer itself.

V. SECONDARY ION MASS SPECTROMETRY

Energetic ion bombardment of the surface of a material causes atoms to be ejected by the sputtering process. A fraction of sputtered species are ejected as either positive or negative ions. This fraction can be mass analyzed in a mass spectrometer to determine the elemen-

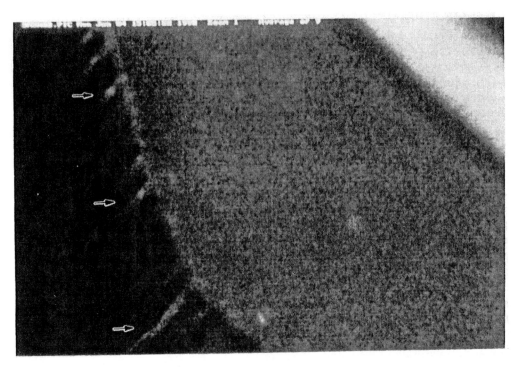

Figure 21 Confocal scanning microscope image showing the interface between the light-curved glass ionomer cement and dentin.

tal composition of the surface. Because of the very high signal/background ratios that are peculiar to mass spectrometry, this technique is extremely sensitive and can thus detect trace quantities of elements on a surface. Secondary ion mass spectrometry (SIMS) extends the surface sensitivity range provided by XPS. The elemental detection limit for SIMS often extends to 0.1 ppm (atomic) and below. Areas as small as 50 μm in diameter can be analyzed. All elements, even hydrogen, can be detected with SIMS. Because of the sputtering nature of the technique, measurements of the ion currents as a function of time provide a depth-composition profile, from the original surface into the bulk of the material. Such profiles generate much information concerning trace and major elemental concentration gradients in the outer surface. This technique has been used to measure changes in the concentration of essential elements at and near the interface between the dental material and the dentin.

In order to analyze the interface between light-cured glass ionomer cement and the dentin surface, SIMS depth profiles were carried out through these layers. Depth profiles were carried out on light-cured glass ionomer cement, polished dentin, and the interfaces between the two substrates. A Cameca IMS-3f secondary ion mass spectrometer was used. An O^- primary beam provided a stable charge condition on the nonconductive samples. Positive secondary ions were monitored, which provided good ion yields for the elements of interest. The elements monitored were ^1H, ^{12}C, ^{19}F, ^{27}Al, ^{28}Si, ^{31}P, ^{40}Ca, and ^{64}Zn.

The resulting SIMS depth profiles for ionomer and dentin interfacial profiles are

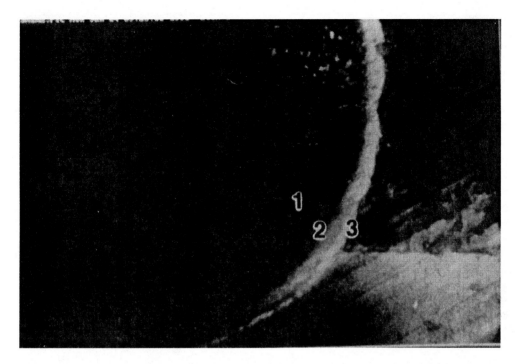

Figure 22 Confocal scanning microscope image showing the interface between chemically-cured glass ionomer cement and dentin. Area 1 is dentin, area 2 is the fluorescence-labeled light-cured glass ionomer cement, and area 3 is the fluorescence-labeled adhesive.

shown in Fig. 24. On a control dentin surface, calcium, phosphorous, and hydrogen were of intensities similar to those observed for dentin near the bond interface. However, carbon, aluminum, silicon, and fluorine were sharply higher at the surface of the dentin near the interface. Zinc intensities were moderately higher at the interface. Aluminum, silicon, fluorine, and zinc were of similar intensity in the control ionomer when compared with the ionomer close to the interface. Near the interface, the glass ionomer had increased amount of carbon, hydrogen, fluorine, phosphorous, and zinc. SIMS depth profiling confirmed the ionic exchange process between the light-cured glass ionomer cement and the dentin surface. There was evidence for the movement of ions from the glass ionomer cement 1.5 μm into the dentin surface and for the movement of calcium and phosphorous ions from dentin 1 μm into the ionomer.

VI. SUMMARY

Adhesives and cements used in the dental profession have gone through many changes since their introduction. The understanding of the mechanisms of bonding agents and cements to enamel and dentin surfaces has become a popular topic of research. With the existence of sophisticated surface techniques such as those described in this chapter, studies involving dentin and enamel surfaces and interfaces between these surfaces and

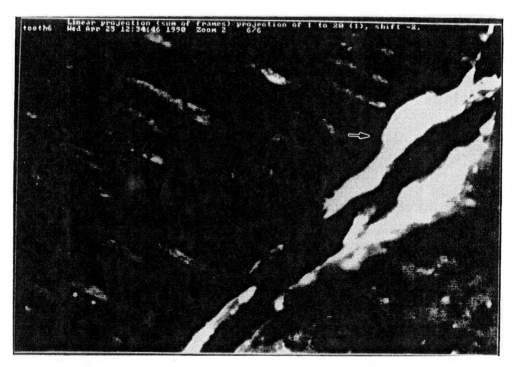

Figure 23 Confocal scanning microscope image showing failure within the fluorescence-labeled glass ionomer cement.

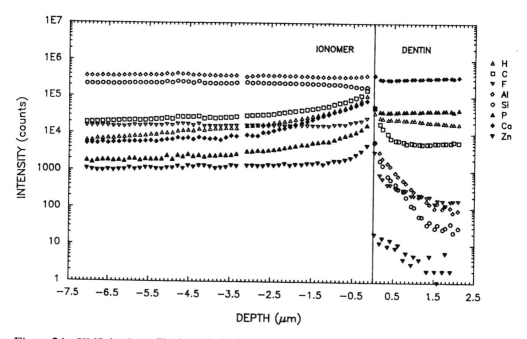

Figure 24 SIMS depth profile through the light-cured glass ionomer cement and into the dentin.

the dental materials will become more informative and comprehensive. It is obvious that studies based only on bond strength data would not provide enough information regarding the mechanism of failure. The analyses of light-cured and chemically cured glass ionomer cements provide a more complete picture of how these materials bond differently to the dentin surface, both chemically and structurally. As well, surface techniques have been successfully used here to aid in the understanding of how the surface of the tooth is modified by polishing and chemical pretreatments.

In order to design functional adhesives with long-term success, there is a need to improve our understanding on a basic level of the adsorption of macromolecules and of living cells at solid and semisolid surfaces. Lack of the required knowledge of these fundamental processes limits our attempts at any practical control of adhesion in the biological environment. Understanding of the affects of different surface properties — including texture, charge, chemistry, and surface energy — on adsorption of dissolved macromolecules is imperative. As well, understanding how particular components or subfractions of materials selectively adsorb on the surface of the tooth and assessing the orientation of the adsorbed molecular entities can aid in determining the final outcome. Another important aspect of these studies is being able to define the modifications to the original surface properties resulting from various treatments. Further investigations on the nature of chemical bonding at the interface, and on the role of mechanical interlocking, would be useful for the purpose of optimizing adhesion to the components of tooth structure in an aqueous environment. These type of studies can take place with the use of surface characterization techniques in combination with bond strength tests and clinical evaluations.

ACKNOWLEDGMENTS

The authors would like to gratefully thank Dr. M. Hubert, Ontario Laser and Lightwave Centre, University of Toronto, for his skilled technical assistance and use of the scanning laser confocal microscope; and Dr. Kent Nielsen, 3M Canada, Inc., for his assistance with the fluorescence labeling. As well, the authors would like to thank Dr. Ben Balevi for sharing his knowledge and expertise in the area of dentin bonding agents.

REFERENCES

1. Pashley, D. H., Dentin: A dynamic substrate — A review, *Scan. Micros., 3*, 161–176 (1989).
2. Lin, A. L., N. S. McIntyre, and R. D. Davidson, Studies on the adhesion of glass-ionomer cements to dentin, *J. Dent. Res., 71*, 1836–1841 (1992).

59

In Vitro and *In Vivo* Evaluations of Base Metal Dental Casting Alloys

Joel D. Bumgardner and Linda C. Lucas
University of Alabama at Birmingham, Birmingham, Alabama

Berit I. Johansson
University of Umeå, Umeå, Sweden

I. INTRODUCTION

Metals and alloys currently used for dental restorative procedures may be divided into two broad categories: the precious and the base metals and alloys (Phillips, 1991). The designation *precious* refers to the metal's intrinsic monetary value. The precious metals used in dentistry are gold, palladium, platinum, and silver. These metals and their alloys are also referred to as *noble* (i.e., nonreactive), due to their position in the standard EMF series. Silver, which is reactive in the oral environment, is not considered noble in the dental community. The designation *base* refers to alloys composed of nonprecious elements such as iron, cobalt, nickel, and copper. These elements are reactive with their environments. Through alloying, base metals can be made to demonstrate behavior more like the noble metals.

Base metal alloys were introduced to the dental profession over 50 years ago, primarily as a substitute for precious alloys for partial dentures, and have expanded since then to be used for crowns, bridges, dental implants, and orthodontic appliances, and for porcelain veneering (Leinfelder and Lemons, 1988). In fact, Eichner (1983) reported that due to the increase in costs of gold and other precious metals in 1978–1979, processing of precious metals for dental restorations substantially decreased while the use of base metal alloys increased. Other advantages of the use of base metal alloys over the gold alloys include better mechanical properties and increased availability of the base metal constituents (Leinfelder and Lemons, 1988). This discussion is concerned with the nickel- and copper-based metal alloys.

II. NICKEL-BASED DENTAL CASTING ALLOYS

A. Physical and Mechanical Properties

The nickel-based dental alloys have a silver–platinum color and are low in density (Moffa and Jenkins, 1974). The neutral coloring is desirable for the aesthetics of porcelain veneering, and the low density increases the volume of material per dollar to decrease overall costs (Eichner, 1983; Leinfelder and Lemons, 1988; Smith, 1983). As shown in Table 1, the alloys contain approximately 60–80% nickel and 10–25% chromium (Leinfelder and Lemons, 1988; Phillips, 1991). The addition of various minor alloying constituents such as iron, molybdenum, manganese, aluminum, silicon, gallium, niobium, and beryllium as well as others makes possible a broad spectrum of castable alloys. Thus, a correspondingly wide range of microstructures and physical properties exist (Moffa and Jenkins, 1974; Smith, 1983). Due to the relatively high chromium content of the nickel-based dental alloys, as compared to other secondary alloying elements, these alloys are often referred to in the literature as *nickel–chromium* or *nickel–chrome* dental alloys.

Depending on the alloy's composition, the microstructures of these alloys may be described as solid solution matrixes with precipitated intermetallic compounds or as a typical dendritic arrangement with solidified eutectic between the dendrites (Fig. 1). The nickel-based alloys have higher melting and fusion temperatures than the gold-based alloys and thus have greater dimensional stability and resistance to sag at porcelain firing temperatures. Problems with castability and decreased marginal accuracy of these alloys have been reported (Eichner, 1983; Gregory, 1982; Moffa and Jenkins, 1974). Beryllium and gallium additions to the alloy chemistries have improved castability and marginal accuracy by reducing alloy melting temperature ranges and increasing fluidity.

Table 2 gives representative ranges of the physical and mechanical properties exhibited by nickel-based alloys (Phillips, 1991). The wide range of physical properties is reflective of the widely varying compositions of these alloys. In general, the nickel-based alloys demonstrate increased values for hardness, modulus of elasticity, yield strength,

Table 1 Composition of Ni-Based Dental Casting Alloys (wt%)

Nickel	63.36–80.86
Chromium	11.93–20.95
Molybdenum	0–8.40
Gallium	0–7.04
Iron	0.10–5.18
Manganese	0–4.28
Aluminum	0.16–4.15
Niobium	0–4.10
Silicon	0.18–2.72
Beryllium	0–1.67
Copper	0–1.54
Tin	0–1.25
Cobalt	0–0.42

Source: Adapted from Philips, *Science of Dental Materials*, 9th ed., 1991, p. 373; and Bumgardner and Lucas, *Dent. Mater., 9,* 252–259, 1993.

sag resistance, and bond strength to porcelain as compared to the gold-based alloys (Gregory, 1982; Moffa and Jenkins, 1974; Moffa et al., 1973b). The increase in rigidity and resistance to deformation makes these alloys suitable for use in dental bridge constructions, especially when thinner castings are required (Gregory, 1982). Percent elongation is used as an indication of the burnishability, or the ability to finish the edges and margins of a crown. There is a wide range of percent elongations for the nickel-based alloys due to their wide composition ranges. However, with their greater strength and modulus, a larger force is required to cause flow in these alloys, and thus burnishability may be more difficult (Brockhurst and Cannon, 1981; Gregory, 1982; Johanson, 1983).

Nickel–chromium alloys provide a suitable oxide substrate during firing of porcelain-to-metal restorations. Additionally, their expansion coefficients are similar to those of the porcelains (Eichner, 1983; Gregory, 1982). However, the bonding of porcelain veneers to these alloys may be very technique sensitive (Carter et al., 1979; Gregory, 1982; Johanson, 1983; Moffa and Jenkins, 1974).

B. Corrosion and Surface Properties

Chromium possesses the ability to form a protective oxide layer on its surface and thereby greatly reduce its corrosion rate (Pourbaix, 1984). This ability of chromium is utilized to provide corrosion resistance for the nickel-based alloy system in the oral environment (Pourbaix, 1984). It has been suggested that a chromium content from 16% to 27% will provide corrosion resistance to nickel-based alloys, and that the addition of manganese and molybdenum will also further increase corrosion resistance (Brune, 1986; Meyer, 1988; Morris et al., 1992; Muller et al., 1990a, 1990b; Pourbaix, 1984).

Electrochemical and dissolution corrosion studies have shown that the nickel–chrome alloys do corrode in physiological solutions such as balanced salt, proteineous, artificial saliva, human saliva, and artificial sweat solutions (Bumgardner and Lucas, 1993b; Covington et al., 1985b; Geis-Gerstorfer and Weber, 1987; Herø et al., 1987; Johansson et al., 1989a, 1989b; Lee et al., 1985; Lucas et al., 1991; Meyer, 1988; Pfeiffer and Schwickerath, 1991; Randin, 1988; Sarkar and Greener, 1973; Tai et al., 1992). These studies have shown that alloys, with chromium in the ranges suggested by Pourbaix (1984) and Brune (1986), and with molybdenum additions greater than 6 wt%, demonstrated improved corrosion properties and resisted accelerated corrosion better than alloys with lower chromium and molybdenum content. However, the addition of beryllium to the alloys composition was shown to reduce their corrosion properties and to increase their susceptibility to pitting and crevice corrosion regardless of the chromium and molybdenum alloying content (Bumgardner and Lucas, 1993b; Covington et al., 1985b; Geis-Gerstorfer and Pässler, 1993; Johansson et al., 1989a; Lee et al., 1985; Meyer, 1988). Cyclic polarization curves representative of different nickel-based dental alloy compositions are shown in Fig. 2. The curves of the Ni–20Cr–8Mo alloy show lower corrosion rates and higher breakdown potentials than either the Ni–13Cr–7Mo or the Ni–12Cr–5Mo–1.7Be alloys. Furthermore, based on electrochemical corrosion theory, resistance to pitting and crevice corrosion of the Ni–20Cr–8Mo alloy was shown due to the lack of hysteresis behavior of the corrosion curves. The curves for the other two alloys demonstrated large hysteresis behavior indicative of susceptibility to pitting and crevice corrosion processes.

Previous studies have suggested that the increased susceptibility to accelerated corrosion of nickel-based alloys with lower chromium content than that recommended, was

(a)

Figure 1   Nickel–chromium alloy microstructures electrolytically etched. (a) Alloy with solid solution matrix with precipitated intermetallics. (b) Alloy with solidified eutectic in a dendritic arrangement. (From Bumgardner and Lucas, *Dent. Mater., 9*, 252–259, 1993.)

due to the inability to develop adequate surface oxides for corrosion resistance (Sarkar and Greener, 1973). In a recent study, surfaces of a series of nickel-based alloys polished to simulate clinical conditions were evaluated using Auger electron spectroscopy (Bumgardner and Lucas, 1993b). The study showed that non-beryllium-containing alloys had homogeneous distributions of nickel, chromium, molybdenum, and oxygen on the alloys' surfaces. The observed increase in corrosion resistance of the higher chromium and molybdenum alloy (Fig. 2a) as compared to the lower chromium and molybdenum alloy (Fig. 2b), was attributable to an increase of the chromium and molybdenum content in the alloys' surface oxides. In contrast, beryllium-containing alloys demonstrated nonhomogeneous nickel, chromium, molybdenum, and beryllium surface oxides. As shown in Fig. 3, these nonhomogeneous surface oxides revealed areas that were depleted in chromium, molybdenum, and oxygen, and elevated in nickel and beryllium (Bumgardner and

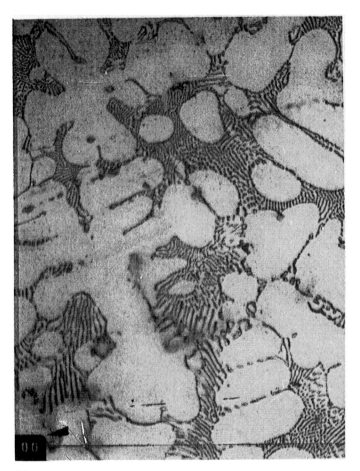

(b)

Table 2   Physical and Mechanical Properties of Ni-Based Alloys

Tensile strength (MPa)	1139–703
Yield strength (MPa)	260–838
Modulus of elasticity [MPA ($\times 10^3$)]	154–210
Percent elongation (%)	2.3–27.9
Vicker's hardness (DPH)	175–357
Density (g/cm^3)	7.9–8.7
Porcelain bond (MPa)	51.0–106.2

Source: Adapted from Philips, *Science of Dental Materials*, 9th ed., 1991, p. 374.

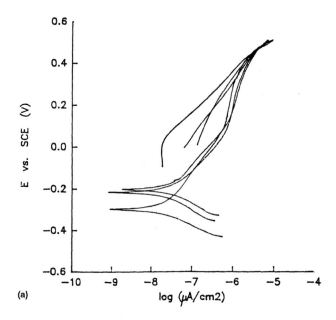

(a)

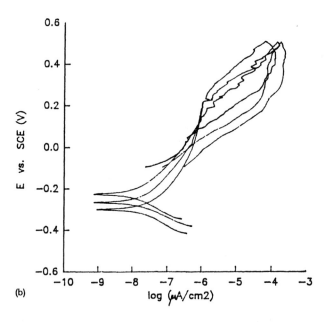

(b)

Figure 2 Cyclic polarization curves of nickel–chromium dental alloys conducted in cell culture media at 37 °C in a humidified 5% CO_2 atmosphere. (a) 63Ni–21Cr–8.4Mo–4.1Nb alloy. (b) 67 Ni–13Cr–7Mo–7Ga alloy. (c) 77Ni–12Cr–4.8Mo–1.7Be alloy.

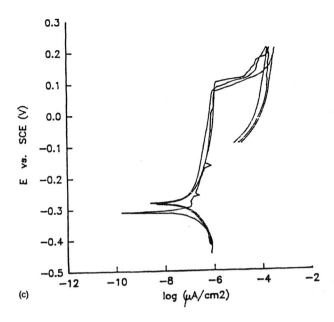

(c)

Lucas, 1993b; Covington et al., 1985a). The addition of beryllium to the alloy chemistries caused the formation of a nickel–beryllium eutectic phase in the alloy's microstructure (Fig. 1), which resulted in an easily disrupted, nonhomogeneous surface oxide (Fig. 3) that provided little resistance to accelerated corrosion, regardless of the alloy's chromium and molybdenum content (Fig. 2c). Indeed, the areas on the surfaces that were decreased in chromium, molybdenum, and oxygen were shown to be initiation sites for pitting corrosion and corresponded to observations of preferential corrosion of the alloys' eutectic nickel–beryllium phase (Fig. 4).

The release of increased levels of metal ions is of concern to the health of local and systemic tissues. Thus, from these investigations it may be concluded that nickel-based alloys require a minimum chromium and molybdenum content as previously suggested, as well as homogeneous surface oxides for improved corrosion resistance and reduced metal ion release.

As long as the surface oxides remain intact, the nickel-based alloys have demonstrated relatively low corrosion rates. However, they still have the ability to release nickel, chromium, and other corrosion products into the body, especially if their surface oxides become damaged. Over the course of time, these corrosion products may become clinically significant. It is already well known that there is a risk of developing an allergic reaction to nickel from nickel-containing devices, and there are also concerns over the ability of chromium to elicit an allergic reaction (Brockhurst and Cannon, 1981; Gregory, 1982; Mitchell, 1984; Moffa, 1984). Furthermore, a greater dissolution of beryllium from the bulk alloy composition is expected due to the corrosion-prone eutectic. Possible long-term adverse effects of the nickel and beryllium ions are of concern. (Bumgardner and Lucas, 1993b; Covington et al., 1985b; Tai et al., 1992). Thus, due to the potential toxicity of corrosion products from these alloys in both adjacent and systemic tissues, the question regarding the biocompatibility of these alloys remains open.

Figure 3   Auger electron surface map of a beryllium-containing nickel–chromium dental alloy. Light areas show the increased level of nickel and beryllium on the surfaces of the alloy's eutectic structure. The dark areas correspond to dendritic surfaces enriched in chromium and molybdenum. (From Bumgardner and Lucas, *Dent. Mater., 9,* 252–259, 1993.)

C.   *In Vitro* Cell Culture Studies

In vitro cell culture techniques have been used to evaluate the cellular response to nickel–chromium dental alloys, their constituents, and leachable components. Various investigators have used metal salt solutions to evaluate morphology, proliferation, protein synthesis, sister chromatid exchanges, colony-forming ability of cells *in vitro*, hemolysis of human RBCs, and effects on bacterial growth (Bearden and Cooke, 1980; Joshi and Eley, 1988; Merritt et al., 1984a; Rae, 1978; Sen and Costa, 1986; Wataha et al., 1991b). These investigations found that nickel salt solutions are generally more toxic to cells than the chromium solutions. The valence state of chromium must be considered as a factor in the results of these studies. It has also been shown that nickel ions may be taken up by fibroblasts *in vitro* and bound to the nuclei, cytosol, mitochondria, and microsomes (Hensten-Petersen, 1984; Wataha et al., 1993). Using particulates, disks, or thin sheets of pure nickel and chromium in cell cultures, nickel was generally found to interfere with various enzyme systems, disrupt intracellular organelles, alter morphology, decrease cell viability and numbers, and increase hemolysis to a significantly greater extent than chro-

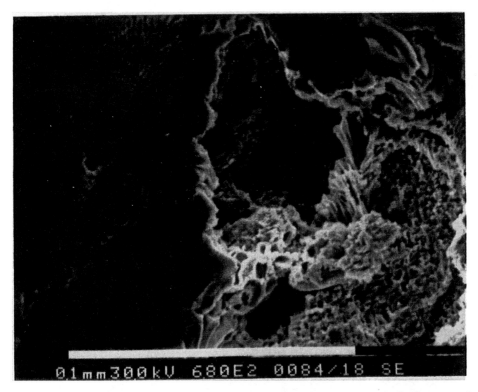

Figure 4 A beryllium-containing nickel–chromium alloy after cyclic polarization corrosion testing, showing severe pitting of the nickel–beryllium phase in the eutectic structure.

mium (Craig and Hanks, 1990; Evans and Thomas, 1986; Kawahara, 1983; Rae, 1975, 1978).

Evaluations of particulate or solid samples of nickel-based alloys revealed a low cytotoxic response of the cultured cells to these materials as evidenced through morphological and ultrastructural evaluations, synthesis of various proteins, analysis of enzyme activity, viability, and proliferation (Bumgardner and Lucas, 1993a; Craig and Hanks, 1988, 1990; Exbrayat et al., 1987; Kawahara et al., 1968; Woody et al., 1977;). The authors attribute the reduction in cytotoxicity of the nickel in the nickel-based alloys, as well as the relative "inertness" of the pure chromium samples, to the protective chromium oxide layer. Indeed, atomic absorption analyses have shown that metal ions released *in vitro* from test alloys were lower than levels required to elicit adverse responses to cultured cells using metal salt solutions.

Atomic absorption studies showed that metal ions were not released at levels proportional to bulk alloy compositions, but rather correlated well with the alloys' chromium and molybdenum content, surface characteristics, microstructures, and corrosion properties (Bumgardner and Lucas, 1993a; Covington et al., 1985b; Geis-Gerstorfer and Weber, 1987; Meyer, 1988; Pfeiffer and Schwickerath, 1991; Tai et al., 1992). Bumgardner and Lucas (1993a) measured the metallic ions released from a series of nickel-based alloys in cell culture experiments (Fig. 5). Cell culture media were analyzed by graphite atomic

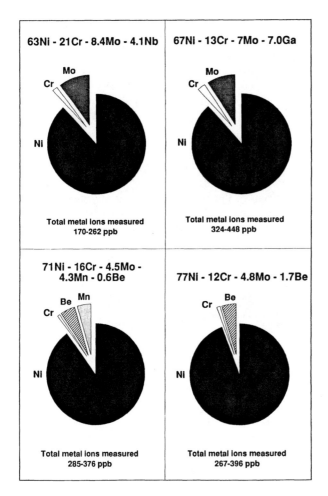

Figure 5   Atomic absorption analyses of metal ions released from nickel–chromium dental casting alloys after 3 days in cell culture tests.

absorption for release of nickel, chromium, and molybdenum from all alloys as well as niobium, gallium, manganese, and beryllium if present in the alloy chemistry. In this study, the Ni–20Cr–8Mo alloy, which demonstrated improved corrosion resistance (Fig. 2a), released lower levels of metal ions as compared to the other alloys. The lower levels of metal ions released were associated with smaller reductions in the uptake of radiolabeled thymidine by cultured cells. Additionally, these *in vitro* atomic absorption analyses showed that beryllium-containing alloys released not only increased levels of metal ions but highly elevated levels of beryllium ions as compared to bulk alloying content as well.

Comparison of the released metal ion profiles of the alloys in Fig. 5 showed that the nonberyllium alloys released mostly nickel, followed by molybdenum and then chromium; while the beryllium-containing alloys released mostly nickel, followed by beryllium, manganese, if present in the alloy, and then chromium. This and other studies have clearly shown that metal ion release from the nickel-based alloys is not proportional to bulk alloy composition (Bumgardner and Lucas, 1993a; Covington et al., 1985b; Geis-Gerstorfer and Weber, 1987; Meyer, 1988). The preferential release of beryllium ions from these alloys was due to the eutectic, nickel–beryllium phase and the concomitant nonhomogeneous distribution of chromium and molybdenum in their surface oxides. Under simulated wear conditions, beryllium ion release was shown to be further enhanced (Tai et al., 1992). The increased release of nickel and beryllium ions from the two beryllium-containing nickel-based alloys evaluated by Bumgardner and Lucas (1993a) was associated with larger reductions in ATP levels of cultured cells than the non-beryllium-containing alloys (Bumgardner et al., 1994).

Corrosion evaluations have shown that nickel-based alloys have low breakdown potentials, and that if their surface oxide layer is damaged, increased corrosion may occur. Metal ions released from these alloys, especially the beryllium-containing alloys, have been shown to inhibit cellular proliferation and reduce intracellular ATP levels at subcytotoxic levels. Considering, then, the increased lifetime cancer risk from the released beryllium ions, the long latency period for adverse reactions due to either nickel or beryllium ions, and the possible adverse synergistic effects of released metal ions, it is imperative that only nickel-based alloys that demonstrate reduced metal ion release and resistance to accelerated corrosion be used for dental restorative purposes.

D. *In Vivo* Evaluations

In vivo animal studies have investigated the tissue reaction of oxide powders, individual constituents, and various alloy compositions of the nickel-based alloys. Nickel, nickel oxides, nickel sulfides, and a nickel–gallium alloy were found to be carcinogenic in mice and rats (Gilman, 1962; Mitchell et al., 1960). When Soremark et al. (1979) injected radiolabeled chromium salts into mice, they found that the chromium was rapidly distributed by the blood, to be accumulated in the gastric and intestinal mucosa, and in the renal cortex of the kidneys. These sites of accumulation indicated modes of elimination but may also have pointed to more specific tissue–organ interactions. Additionally, chromium was taken up in both the reticuloendothelial and central nervous systems, and was shown to be able to cross the placental membrane (Soremark et al., 1979).

Bergman et al. (1980) studied the dissolution of nickel from nickel-based alloys implanted in mice for 5 months. They found significant nickel accumulation in the fibrous capsules around the alloys and, to a limited extent, nickel accumulation in the blood, kidney and pancreas. Hensten-Pettersen (1984) reported that neutron-activated nickel and nickel alloys implanted in rats released small but significant amounts of nickel to the adjacent tissues, and that the nickel was distributed via the bloodstream to the liver, kidneys, and lungs. Merritt et al. (1984b), by injecting metal salts and corrosion products from alloys containing nickel and chromium intramuscularly into hamsters, also showed that nickel and chromium were rapidly distributed by the blood. The metallic ions, Cr^{6+} and to a lesser extent Ni^{2+}, were associated with blood cells, while Cr^{3+} was found mostly in serum.

Moffa et al. (1973a) and Adrian and Huget (1977) evaluated the histological response to various nickel-based alloys, with and without beryllium additions for up to 1-year implantation in rodents. The results of their investigations indicated that the alloys were associated with mild tissue reactions, surrounded by fibrous capsules, and in general did not behave significantly different from surgical-grade base-metal- or gold-based alloys. But, when Sandrik et al. (1974) compared the nickel-based alloys to 316L stainless steel, they found chronic inflammatory reactions associated with the nickel-based alloys, leading them to conclude the nickel-based alloys were more reactive than the stainless steels.

In an *in vivo* usage study, full nickel–chromium and nickel–chromium–beryllium crowns were cast, placed in canines for 1 year, and evaluated for corrosion and tarnish (Johansson et al., 1989b; O'Neal et al., 1988). The nickel–chrome alloy showed no, or minor, signs of corrosion and tarnish, while the nickel–chromium–beryllium alloy showed significant tarnish and preferential phase dissolution (Fig. 6). These results were in agreement with earlier corrosion studies in which nickel–chromium alloys containing beryllium were not as corrosion resistant as the beryllium-free nickel-based alloys (Bumgardner and Lucas, 1993b; Covington et al., 1985b; Geis-Gerstorfer and Weber, 1987; Herø et al., 1987; Lee et al., 1985).

Spiechowicz et al. (1983) investigated the sensitivity reactions to nickel in sensitized guinea pigs. Interestingly, they found no mucosal reactions at the site of contact with nickel-containing plates during the 3-week evaluation period.

It is apparent that the nickel-based alloys may release corrosion products *in vivo* and that these products may be distributed and accumulated throughout the body. The consequences of these accumulations long term are unknown. Further research is needed to elucidate this information.

E. Clinical Considerations

The clinical use of nickel-based alloys has certain distinct advantages. Some of the advantages associated with these alloys include: lower thermal conductivities, increased aesthetics of porcelain veneers due to neutral coloring, thinner interproximals due to increased strength; and improved porcelain-to-metal bonding (Kelly and Rose, 1983; Weber, 1983). However, there are potential problems in that: (1) there is a risk of metal sensitivity; (2) components of these alloys have been shown to be carcinogenic in animals; and (3) exposure to nickel, chromium, and other alloy components in industrial environments has been implicated in the development of various cancers in humans (Bullough et al., 1988; Jacobsen et al., 1981; Merritt and Brown, 1981; Mitchell, 1984; Moffa, 1984; Morris, 1987). In addition, problems may occur due to casting and polishing methods, which may affect the fit of the restoration (Gregory, 1982; Moffa and Jenkins, 1974). There is a wide range of nickel-based alloy compositions with a wide range of physical properties, and generalizations on usage of different nickel-based alloys are consequently difficult to make (Kelly and Rose, 1983; Morris, 1987; Weber, 1983).

Morris (1987) reported that nickel-based alloy usage for restorations increased from 32% in 1970 to 80% in 1987, in his study of 696 patients with nickel-based dental restorations. In that study, only one patient converted to a positive patch test for nickel sensitivity at 4-year follow-up. Spiechowicz et al. (1984) evaluated 10 nickel-sensitive patients treated with nickel-based prosthetic appliances for hypersensitivity reactions. No adverse general, oral, or histological reactions were observed in the patients after 12–40 months. Weber (1983) reported on the successful use of nickel-based alloys in his practice for 5 years without localized or generalized tissue reactions. Indeed, van Loon et al. (1988) reported that only pure nickel resulted in clinically as well as histologically defined contact stomatitis in nickel-sensitive patients, while no significant activity could be shown using nickel-based alloys. The authors did indicate, though, that selected nickel-based alloys might cause allergic reactions under more severe electrochemical conditions. They

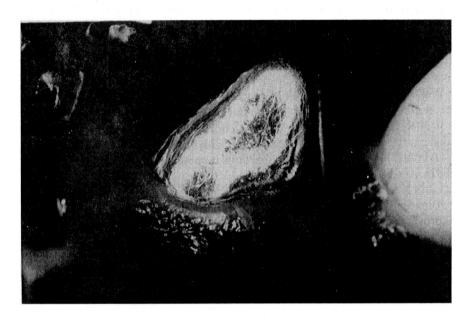

(a)

(b)

Figure 6   (a) A non-cleaned nickel–chromium crown, 1 year after placement in a canine model. (b) A non-cleaned beryllium-containing nickel–chromium crown showing adherent deposits 1 year after placement in a canine model.

concluded that some of the nickel-based alloys were stable and thus very likely to function clinically without complications.

However, Magnusson et al. (1982) reported positive skin patch tests to nickel-containing dental alloys in 9 out of 10 women with known nickel sensitivity. Environmental differences between skin and oral exposure to nickel and other metal ions have been discussed as possible reasons for the lack of mucosal reactions in metal-sensitive patients (Spiechowicz et al., 1984). Nielsen and Klaschka (1971) have reported that higher concentrations of allergen were required to elicit reactions in the oral mucosa than on the skin. Nevertheless, the release of nickel from dental alloys may increase the oral intake of nickel and thus contribute to adverse tissue reactions in metal-sensitive patients (Magnusson et al., 1982).

Garuet et al. (1988) reported severe tissue reactions, the presence of metal fragments and increased levels of nickel and chromium ions in tissues surrounding failed nickel-based dental prostheses. Questions were raised concerning the contribution to the adverse tissue reactions by mechanical forces exerted by the implant, corrosion and metal ion release, and damage to the prostheses during implantation.

In evaluations of orthopedic devices containing nickel, Evans et al. (1974) implicated metal hypersensitivity, in metal-sensitive patients, as playing a role in irreversible blood vessel damage leading to bone necrosis and implant loosening. Further, patients with failed orthopedic protheses who were also metal sensitive showed positive *in vitro* lymphocyte transformation tests to nickel, chromium, or cobalt (Christiansen et al., 1979; Carando et al., 1984). Thus, while nickel-based alloys appear to be clinically successful, there remain concerns over the potential for these alloys to elicit adverse effects.

III. COPPER-BASED DENTAL CASTING ALLOYS

Copper-based alloys, used for a number of industrial and marine applications, are also used as dental casting alloys. Brasses, which are mainly copper and zinc, and bronzes, primarily alloys of copper and aluminum, are two types of copper-based alloys used for dental crowns and prostheses. Other elements added to these alloys include iron, nickel, and manganese. Iron acts as a grain refiner and produces small precipitates in the microstructure of the aluminum bronzes. Nickel improves the corrosion resistance, and the mechanical properties are controlled by the addition of aluminum and manganese (American Foundrymen's Society, 1984).

A. Physical and Mechanical Properties

Lacefield et al. (1988) evaluated and compared the physical and mechanical properties of a series of commercially available and experimental copper-based dental casting alloys, two gold-based alloys, and two nickel-based dental alloys. The compositions of the alloys evaluated are provided in Table 3. The results of the mechanical tests are provided in Table 4. Duncan's multiple range test was used to determine if significant differences existed between the means of the various groups, and vertical lines were used to connect alloys in which no such differences existed. The two nickel-based alloys had yield strengths significantly higher than any of the gold-based or copper-based alloys evaluated. Three of the copper–aluminum alloys had higher yield strengths than the type II gold alloy.

The ultimate tensile strength results showed a similar pattern. The nickel-based alloy

Table 3   Compositions of Proprietary and Experimental Alloys (wt%)

Alloy (code)	Copper Alloys							
	Cu	Al	Fe	Zn	Ni	Co	Mn	Other
Duracast MS[a] (MS)	81.6	8.3	3.9		4.1		1.2	
Trindium[b] (TR)	87.0	11.0			1.0		1.0	Ga, In
Goldent[c] (GD)	76.0	6.5		12.0	5.0		0.5	
Experimental 1 (E1)	88.0	11.63	0.16					
Experimental 2 (E2)	82.0	9.54	0.81			4.6	3.1	
Experimental 3 (E3)	58.0			42.0				

	Gold Alloys				
	Au	Ag	Cu	Pd	Zn
Modulay[d]	77	14	8	1	
Midas[d]	46	39.5	7.5	6	1

	Nickel Alloys			
	Ni	Cr	Mo	Be
Litecast[e]	68.5	15.5	14.0	
Litecast-B[e]	77.5	12.5	4.0	1.7

[a]Duracast Inc., Brasilia, São Paulo, Brazil
[b]Trindium Corp. of America, Los Angeles, CA
[c]Goldent Inc., Brasilia, São Paulo, Brazil
[d]J. F. Jelenko, Armenak, NY
[e]Williams Gold, Buffalo, NY

that contained beryllium had a significantly higher tensile strength than all other alloys tested. The copper-based alloy that contained cobalt was the strongest of the copper-based alloys and was significantly stronger than either of the gold-based alloys tested. In general, the copper-based alloys were 20–30% stronger than the gold-based alloys, although the differences were not significant in every case.

As might be expected, the elongation values were higher for the alloys that exhibited the lowest yield and tensile strength values, with the exception of Litecast, the nickel-chromium alloy without beryllium. The copper–zinc alloy, E3, the high-gold alloy, Modulay, and the nickel–chromium alloy, Litecast, had elongation values higher than 30%.

B.   Corrosion Properties

Previously, German (1985) reported that two different aluminum–bronze alloys were significantly less resistant to corrosion and tarnish than a control gold-based alloy. Stoffers et al. (1987) evaluated the corrosion and tarnish resistance of a bronze alloy; they concluded that the alloy exhibited poor tarnish and corrosion resistance, and that the use of this alloy in prosthetic dentistry was questionable. Mueller (1987) investigated the corrosion characteristics of a copper–aluminum alloy using a series of solutions with varying concentrations of human salivary dialysate. Results in this study showed that the copper alloy displayed different corrosion rates depending on the dialysate concentration.

Table 4 Comparison Tests of Mechanical Properties of Copper-, Nickel-, and Gold-Based Dental Casting Alloys

Alloy[a]	Ultimate Tensile Strength	
	UTS (MN/m^2)	Duncan group[b]
Litecast B	1089	
E2	797	
Litecast	689	
Trindium	491	
Goldent	460	
Duracast MS	438	
Modulay	385	
E3	362	
Midas	359	
E1	333	

Alloy[a]	0.2% Yield Strength	
	YS (MN/m^2)	Duncan group[b]
Litecast B	655	
Litecast	286	
E2	224	
Duracast MS	214	
Goldent	195	
Midas	180	
TR	166	
Modulay	160	
E1	159	
E3	148	

Alloy[a]	Elongation	
	Elongation (%)	Duncan group[b]
E3	50.8	
Litecast	40.3	
Modulay	36.9	
Trindium	18.7	
Litecast-B	16.3	
Duracast MS	15.8	
Goldent	12.6	
E2	11.9	
E1	10.0	
Midas	9.2	

[a]Alloy codes are shown in Table 3.
[b]Lines indicate groups of alloys for which no significant difference was found.

Although the corrosion rates of the copper-based alloys varied, their rates were still significantly higher than values observed for other dental casting alloys.

Johansson et al. (1989a) investigated the corrosion behavior of three commercially available copper-based alloys under both *in vitro* and *in vivo* conditions. The corrosion data obtained for the copper-based alloys was compared to the corrosion data obtained for a high-gold-based alloy. Table 3 provides the compositions of these alloys. The *in vitro* testing involved conducting anodic and cathodic polarization curves for each of the alloys as described in the American Society for Testing and Materials specification (ASTM G5) for corrosion testing. The electrolyte was an artificial saliva solution consisting of chlorides, phosphates, sulfides, and urea. The electrolytic solution was purged with oxygen during the test and the temperature was maintained at 37 ± 1°C. The average curves for each of the alloys are provided in Fig. 7. The corrosion potentials for the three copper-based alloys were more active and the corrosion currents for the copper-based

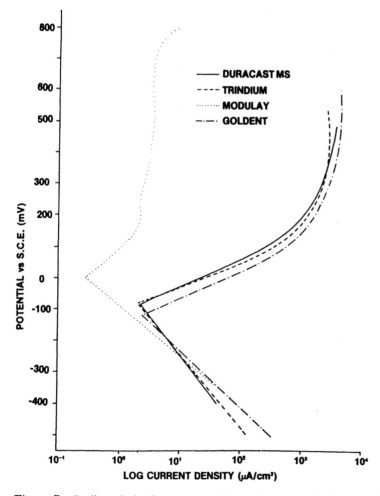

Figure 7 Cyclic polarization curves of three copper-based alloys and a gold alloy in artificial saliva solution.

alloys were significantly higher than for the gold-based alloy. At potentials greater than the corrosion potential, the current magnitudes exhibited by the copper-based alloys were as much as 1000 times greater than for the gold alloy. Johansson et al. (1989b) evaluated the *in vivo* corrosion behavior of several copper-based alloys as compared to a gold-based alloy in a canine crown study. Crowns were placed in each of the four canine quadrants. After 12 months, the crowns were retrieved and evaluated using scanning electron microscopy. The gold crowns showed no, or minor, signs of corrosion; however, after 12 months the copper crowns exhibited significant surface alterations. Corrosion products were observed on all of the copper alloys. SEM micrographs of the copper crowns revealed pitting corrosion. A SEM micrograph of a copper crown after 12 months *in vivo* is provided in Fig. 8. Clearly the *in vitro* and *in vivo* studies referenced in this chapter indicate that the copper-based dental casting alloys can release significant quantities of metallic ions to the oral and systemic tissues.

C.   *In Vitro* Cell Culture Evaluations

The potential for elevated metallic ion release from copper-based dental casting alloys raises the question of biocompatibility. *In vitro* cell culture studies of epithelial and gingival cells have demonstrated reductions in viability and proliferation when exposed to copper salt solutions, high-copper amalgams, and copper-based dental alloys (Bumgardner, 1989; Bumgardner et al., 1989; Kawahara et al., 1968; Leirskar, 1974; Lucas et al., 1984).

Bumgardner et al. (1989) evaluated the cellular response to a series of copper-based alloys. They determined that only the copper–zinc experimental alloy resulted in altered cellular morphologies or decreased cell viabilities of human gingival fibroblasts exposed to the alloys for 24 to 72 h. However, after only 24 h, all the copper-based alloys evaluated caused decreases in ^{3}H-thymidine uptake. The investigators determined that

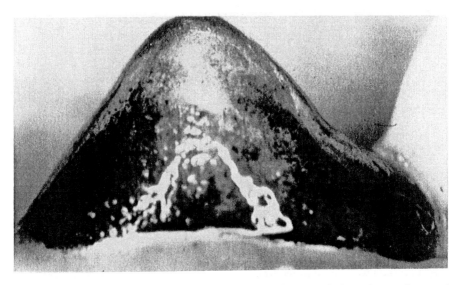

Figure 8   Copper-based crown showing deposits and accumulation of corrosion products after 1 year in a canine model.

changes in DNA synthesis occurred at much lower released copper and zinc concentrations than changes in cellular morphology and viability.

In vitro studies involving human T- and B-cell lines showed that copper-based dental alloys affected their cellular viability and proliferation (Bumgardner et al., 1990a, 1990b, 1991). Inhibition of fresh canine peripheral blood lymphocyte–monocyte division by copper salt solutions and decrease of ^{3}H-thymidine uptake in human mononuclear cells by copper-based dental alloys have also been reported (Bumgardner et al., 1990a; Shifrine et al., 1984). Additionally, *in vitro* investigations have shown metal cations, including copper, zinc, and nickel, can modulate normal immune function (Smith and Lawrence, 1988; Warner and Lawrence, 1988). The susceptibility of copper-based alloys to high corrosion rates, and the release of elevated levels of metal ions, raises concerns over the maintenance of the normal cell-mediated immune response in the oral cavity. Consequently, Bumgardner et al. (1993) hypothesized that oral immunohomeostasis may be altered by metal ions released from the copper dental alloys.

D. *In Vivo* Tissue Response

The *in vivo* response of copper-based alloys has been reported in several publications. Niemi and Hensten-Petersen (1985) reported that the implantation of a copper and palladium alloy in guinea pigs elicited an acute inflammatory response. Frykholm et al. (1969) reported a lichen planus lesion in the oral cavity of a 45-year-old woman in response to dental restorations containing copper.

Lemons et al. (1988) evaluated the biocompatibility of a series of copper-based dental alloys using a variety of test methods. In LD_{50} rodent tests, no significant adverse symptoms or deaths related to the ingestion of copper particulate were observed. The particulate passed with the feces and collection showed recovery of the samples. Lemons et al. (1988) also evaluated skin contact reactions to the copper-based alloys in rabbits. These studies revealed only a passive response, with no significant irritation. They did observe some tissue discoloration after exposure to the copper-based alloys but not at a level of significance.

Lemons et al. (1988) implanted cylindrical shaped copper alloy rods in the back muscles of rats. After 1- and 3-week time periods, tissue around the implants showed discoloration and a serous fluid interface within an evolving fibrous tissue capsule. In another study, tissues adjacent to the copper-based alloys were evaluated histologically after 7 and 21 days implantation in rabbit tibia and femur. Hematoxylin and eosin-stained sections showed foreign-body responses, corrosion–degradation products, and regions where bone was being altered to fibrous granuloma.

Hao (1989) evaluated the histological responses to copper- and gold-based crowns in a canine model. After 12 months the adjacent bone and gingival tissues were retrieved and decalcified for evaluation. The dentin, cementum, bone, and periodontal ligament tissues showed no significant tissue alterations as a result of being exposed to the copper-based dental alloys. In contrast, in the gingival tissues, the copper-based alloys induced a more severe inflammatory reaction than did the gold-based alloys. Figure 9 shows the histological response to a gold-based alloy, Modulay, as well as the response to a copper-based alloy, Goldent. The thickness of the pocket epithelium in control teeth (teeth not exposed to any material) was in the range of 0.02–0.06 mm. This is the same thickness as shown in Fig. 9a for the pocket epithelium adjacent to the gold crown. In contrast, the epithelium response to the copper alloy (Fig. 9b) was significantly thicker and registered

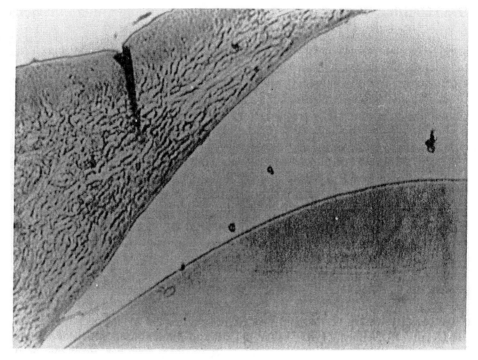

(a)

(b)

Figure 9   (a) Gingival tissue response to a gold-based alloy. The thickness of the junctional epithelium is in the range 0.02–0.06 mm. (b) Gingival tissue response to a copper-based alloy. The thickness of the junctional epithelium reached a maximum value of 0.4 mm.

a maximum thickness of 0.4 mm. The pocket epithelium around the copper-based alloys was almost always associated with a severe and chronic inflamed submucosa.

Soileau et al. (1990) evaluated the release of copper ions from the crowns placed by Hao. Atomic absorption analyses of both gingival and systemic tissues were conducted. They measured elevated concentrations of copper in buccal and lingual gingival tissues adjacent to the copper crowns. Tissues adjacent to the copper crowns contained copper concentrations as high as 6.5 ppm. These elevated metal ion concentrations were associated with thicker inflammatory tissue responses. For systemic tissues, Soileau et al. reported no elevated copper concentrations in whole blood, liver, lymph nodes, or spleen; however, elevated copper ion concentrations were measured in kidney tissues as compared to the control dogs. These data may suggest a mode of elimination of corrosion products from these alloys and/or a site of organ-specific toxicity. Further investigations are required to evaluate these possibilities.

The clinical use of copper-based alloys is limited in the United States; however, the use of the alloys is extensive in South America. Currently, controlled clinical studies are needed to fully ascertain the clinical response to these base metal alloys.

IV. SUMMARY

In summary, the copper-based alloys and selected nickel-based alloys have been shown to be susceptible to accelerated corrosion processes. Thus, these alloys may release elevated quantities of potentially toxic metallic ions to adjacent and systemic tissues. While some clinical success has been reported with the use of these materials, there are still concerns over their long-term performance based on present *in vitro* and *in vivo* studies.

REFERENCES

Adrian, J. C., and Huget, E. F. (1977). Tissue response to base-metal dental alloys, *Mil. Med., 142*, 784–786.

American Foundrymen's Society (1984). *Casting Copper-Base Alloys*, Brass and Bronze Institute, Des Plaines, IL, pp. 132, 159.

Bearden, L. J., and Cooke, F. W. (1980). Growth inhibition of cultured fibroblasts by cobalt and nickel, *J. Biomed. Mater. Res., 14*, 289–309.

Bergman, M., Bergman, B., and Soremark, R. (1980). Tissue accumulation of nickel released due to electrochemical corrosion of non-precious dental casting alloys, *J. Oral Rehab., 7*, 325–330.

Brockhurst, P. J., and Cannon, R. W. S. (1981). Alloys for crown and bridgework, *Austral. Dent. J., 26*, 287–291.

Brown, S. A., Farnsworth, L. J., Merritt, K., and Crowe, T. D. (1988). *In vitro* and *in vivo* metal ion release, *J. Biomed. Mater. Res., 22*, 321–338.

Brown, S. A., and Merritt, K. (1980). Electrochemical corrosion in saline and serum, *J. Biomed. Mater. Res., 14*, 173–175.

Brune, D. (1986). Metal release from dental biomaterials, *Biomaterials, 7*, 163–175.

Bullough, P. G., DiCarlo, E. F., Hansraj, K. K., and Neves, M. C. (1988). Pathologic studies of total joint replacement, *Orthop. Clin. N. Am., 19*, 611–625.

Bumgardner, J. D. (1989). *Biocompatibility evaluations of copper-based dental alloys in cell culture*, Thesis, University of Alabama at Birmingham.

Bumgardner, J. D., Doeller, J., Messer, R., and Lucas, L. C. (1994). Effects of Ni–Cr dental casting alloys on cellular metabolism, Proceedings of the 64th General Session of International Association for Dental Research, Seattle, WA, *J. Dent. Res.*, Abstr. 1696.

Bumgardner, J. D., and Lucas, L. C. (1993a). Cell culture evaluation of nickel-based dental casting alloys, Proceedings of the 63rd General Session of International Association for Dental Research, Chicago, IL, *J. Dent. Res., 72* (DMG 2116), 1–22.

Bumgardner, J. D., and Lucas, L. C. (1993b). Surface analysis of nickel chromium dental alloys, *Dent. Mater., 9,* 252–259.

Bumgardner, J. D., Lucas, L. C., Alverson, M. W., and Tilden, A. B. (1993). Effects of copper-based dental casting alloys on two lymphocyte cell lines and the secretion of interleukin 2 and IgG, *Dent. Mater., 9,* 85–90.

Bumgardner, J. D., Lucas, L. C., and Tilden, A. B. (1989). Toxicity of copper-based dental alloys in cell culture, *J. Biomed. Mater. Res., 23,* 1103–1114.

Bumgardner, J. D., Lucas, L. C., and Tilden, A. B. (1990a). Effects of copper-based dental alloys on lymphocyte proliferation, *J. Dent. Res., 69,* 265, Abstr. 1249.

Bumgardner, J. D., Lucas, L. C., and Tilden, A. B. (1990b). The effects of corrosion products from copper dental alloys on lymphocyte proliferation, *Trans. 74th Ann. Mtg. FASEB, 4,* 2480, Abstr. 2480.

Bumgardner, J. D., Lucas, L. C., and Tilden, A. B. (1991). Variable effects of copper dental alloys on lymphoid cell lines, *Trans. 17th Ann. Mtg. Soc. Biomater., 17,* 233.

Burns, J. K., Lucas, L. C., and Johansson, B. I. (1989). Biocompatibility evaluations of copper-based dental casting alloys, *Trans. 15th Ann. Mtg. Soc. Biomater., 15,* 57.

Carando, S., Cannas, M., and Rossi, P., et al. (1984). The lymphocyte transformation test in the evaluation of intolerance on prosthetic implants, *Ital. J. Orthop. Traumatol., 10,* 33–41.

Carter, J. M., Al-Mudafar, J., and Sorensen, S. (1979). Adherence of a nickel–chromium alloy and porcelain, *J. Prosthet. Dent., 41,* 167–172.

Christiansen, K., Holmes, K., and Zilko, P. J. (1979). Metal sensitivity causing loosened joint prostheses, *Ann. Rheum. Dis., 38,* 476–488.

Covington, J. S., McBride, M. A., Slagle, W. F., and Disney, A. L. (1985a). Beryllium localization in base metal dental casting alloys, *J. Biomed. Mater. Res., 19,* 747–750.

Covington, J. S., McBride, M. A., Slagle, W. F., and Disney, A. L. (1985b). Quantization of nickel and beryllium leakage from base metal casting alloys, *J. Prosthet. Dent., 54,* 127–136.

Craig, R. G., and Hanks, C. T. (1988). Reaction of fibroblasts to various dental casting alloys, *J. Oral Pathol., 17,* 341–347.

Craig, R. G., and Hanks, C. T. (1990). Cytotoxicity of experimental casting alloys evaluated by cell culture tests, *J. Dent. Res., 69,* 1539–1542.

Eichner, K. (1983). Applications of metal alloys in dentistry — A review, *Int. Dent. J., 33,* 1–10.

Evans, E. J., and Thomas, I. T. (1986). The *in vitro* toxicity of cobalt-chrome-molybdenum alloy and its constituent metals, *Biomaterials, 7,* 25–29.

Evans, E. M., Freeman, M. A. R., Miller, A. J., and Vernon-Roberts, B. (1974). Metal sensitivity as a cause of bone necrosis and loosening of the prothesis in total joint replacement, *J. Bone Joint Surg., 56B,* 626–642.

Exbrayat, P., Couble, M. L., Magloire, H., and Hartmen, D. J. (1987). Evaluation of the biocompatibility of a Ni-Cr-Mo dental alloy with human gingival explant culture *in vitro*: Morphological study, immunodetection of fibronectin, and collegen production, *Biomaterials, 8,* 385–392.

Frykolm, K. O., Frithiof, L., Fernstrom, A. I., Moberger, G., Blohm, S. G., and Bjorn, E. (1969). Allergy to copper derived from dental alloys as possible cause of oral lesions of lichen planus, *Acta Derm. Venereol., 49,* 268–281.

Garuet, A., Simonoff, M., Berdeu, B., Llabador, Y., Michelet, F. X., and Caitucoli, P. F. (1988). Measurement of nickel and chromium at the site of metallic dental implants, in *Biocompatibility of Co-Cr-Ni alloys* (H. F. Hildebrand and M. Champy, eds.), Plenum Press, New York, pp. 161–174.

Geis-Gerstorfer, J., and Pässler, K. (1993). Studies on the influence of Be content on the corrosion behavior and mechanical properties of Ni-25Cr-10Mo alloys, *Dent. Mater., 9,* 177–181.

Geis-Gerstorfer, J., and Weber, H. (1987). *In vitro* corrosion behavior of four Ni–Cr dental alloys in tactic acid and sodium chloride solutions, *Dent. Mater., 3*, 289–295.

German, R. M. (1985). *Evaluation of two aluminum–bronze dental casting alloys*, Report to the Gold Institute.

Gilman, J. P. W. (1962). Metal carcinogenesis. II. A study on the carcinogenic activity of cobalt, copper, iron, and nickel compounds, *Cancer Res., 22*, 158–165.

Gregory, D. O. (1982). Nickel–chromium alloys in casting, *Miss. Dent. Assoc. J., 38*, 18–20.

Hao, S. Q. (1989). *Reaction of dental tissues to copper, nickel, and gold based alloy crowns in dogs*, Thesis, University of Alabama at Birmingham.

Hensten-Pettersen, A. (1984). Metabolism of degradation/corrosion products from tissue–material interactions, *Biomaterials, 5*, 42–46.

Hensten-Pettersen, A. (1992). Casting alloys: Side effects, *Adv. Dent. Res., 6*, 38–43.

Herø, H., Valderhaug, J., and Jorgensen, R. B. (1987). Corrosion *in vivo* and *in vitro* of a commercial NiCrBe alloy, *Dent. Mater., 3*, 125–130.

Jacobsen, N., Hensten-Pettersen, A., and Hofsoy, H. (1981). Some biological aspects of nickel, in *Systemic Aspects of Biocompatibility* (D. F. Williams, ed.), CRC Press, Boca Raton, FL, pp. 115–133.

Johanson, L. N. (1983). The physical properties of some alternative alloys, *Int. Dent. J., 33*, 41–47.

Johansson, B. I., Lemons, J. E., and Hao, S. Q. (1989a). Corrosion of dental copper, nickel, and gold alloys in artificial saliva and saline solutions, *Dent. Mater., 5*, 324–328.

Johansson, B. I., Lucas, L. C., and Lemons, J. E. (1989b). Corrosion of copper, nickel, and gold dental casting alloys: An *in vitro* and *in vivo* study, *J. Biomed. Mater. Res.: Appl. Biomater., 23*, 349–361.

Joshi, R. I., and Eley, A. (1988). The *in-vitro* effect of a titanium implant on oral microflora: Comparison with other metallic compounds, *J. Med. Microbiol., 27*, 105–107.

Kawahara, H. (1983). Cellular response to implant materials: biological, physical, and chemical factors, *Int. Endodon. J., 33*, 350–375.

Kawahara, H., Yamagami, A., and Nakamura, M. (1968). Biological testing of dental materials by means of tissue culture, *Int. Dent. J., 18*, 443–467.

Kelly, J. R., and Rose, T. C. (1983). Nonprecious alloys for use in fixed prosthodontics: A literature review, *J. Prosthet. Dent., 49*, 363–370.

Lacefield, W. R., Lucas, L. C., Wendt, S. L., and Gray, S. A. (1988). Microstructure and mechanical characteristics of copper–aluminum alloys, *J. Dent. Res., 68*, 303, Abstr. 977.

Lee, J., Lucas, L., O'Neal, J., Lacefield, W., and Lemons, J. (1985). *In vitro* corrosion analyses of nickel-base alloys, *J. Dent. Res., 63* (DMG 1285), 1–21.

Leinfelder, K. F., and Lemons, J. E. (1988). *Clinical Restorative Materials and Techniques*, Lea and Febiger, Philadelphia, pp. 139–157.

Leirskar, J. (1974). On the mechanism of cytotoxicity of silver and copper amalgams in a cell culture system, *Scand. J. Dent. Res., 82*, 74–81.

Lemons, J. E., Lucas, L. C., Henson, P. G., and Hill, C. J. (1988). Biocompatibilies of copper base alloys, *J. Dent. Res., 67*, Abstr. 1197.

Lucas, L., Hensten-Pettersen, A., and Niemi, L. (1984). Biocompatibility testing of low-gold alloys for dental applications, *Programme and Abstract Book, First International Biointeractions Conference*, London, Abstr. 7.

Lucas, L. C., Dale, P., Buchanan, R., Gill, Y., Griffin, D., and Lemons, J. E. (1991). *In vitro* vs. *in vivo* corrosion analyses of two alloys, *J. Invest. Surg., 4*, 13–21.

Lucas, L. C., and Lemons, J. E. (1992). Biodegradation of restorative metallic systems, *Adv. Dent. Res., 6*, 32–37.

Magnusson, B., Bergman, M., Bergman, B., and Söremark, R. (1982). Nickel allergy and nickel-containing dental alloys, *Scand. J. Dent. Res., 90*, 163–167.

Merritt, K., and Brown, S. A. (1981). Hypersensitivity to metallic biomaterials, in *Systemic Aspects of Biocompatibility* (D. F. Williams, ed.), CRC Press, Boca Raton, FL, pp. 33–45.

Merritt, K., and Brown, S. A. (1988). Effect of proteins and pH on fretting corrosion and metal ion release, *J. Biomed. Mater. Res., 22,* 111–120.

Merritt, K., Brown, S. A., and Sharkey, N. A. (1984a). The binding of metal salts and corrosion products to cells and proteins *in vitro, J. Biomed. Mater. Res., 18,* 1005–1015.

Merritt, K., Brown, S. A., and Sharkey, N. A. (1984b). Blood distribution of nickel, cobalt, chromium following intramuscular injection into hamsters, *J. Biomed. Mater. Res., 18,* 991–1004.

Meyer, J. M. (1988). The corrosion of dental Ni–Cr alloys: An *in vitro* evaluation, in *Biocompatibility of Co–Cr–Ni alloys* (H. F. Hildebrand and M. Champy, eds.), Plenum Press, New York, pp. 305–320.

Mitchell, D. F., Shankwalker, G. B., and Shazer, S. (1960). Determining the tumorigenicity of dental materials, *J. Dent. Res., 39,* 1023–1028.

Mitchell, E. W. (1984). The biocompatibility of metals in dentistry, *CA. Dent. Assoc. J., 12,* 17–19.

Moffa, J. P. (1984). Biocompatibility of nickel based dental alloys, *CA. Dent. Assoc. J., 12,* 45–51.

Moffa, J. P., Guckes, A. D., Okawa, M. T., and Lilly, G. E. (1973a). An evaluation of nonprecious alloys for use with porcelain veneers. Part II. Industrial safety and biocompatibility, *J. Prosthet. Dent., 30,* 432–441.

Moffa, J. P., Lugassy, A. A., Guckes, A. D., and Gettleman, L. (1973b). An evaluation of nonprecious alloys for use with porcelain veneers. Part I. Physical properties, *J. Prosthet. Dent., 30,* 424–431.

Moffa, J. P., and Jenkins, W. A. (1974). Status report on base-metal crown and bridge alloys, *J. Am. Dent. Assoc., 89,* 652–655.

Morris, H. F. (1987). Veterans Administration cooperative studies project no. 147. Part IV: Biocompatibility of the base metal alloys, *J. Prosthet. Dent., 58,* 1–5.

Morris, H. F., Manz, M., Staffer, W., and Weir, D. (1992). Casting alloys: The materials and "the clinical effects," *Adv. Dent. Res., 6,* 28–31.

Mueller, H. J. (1987). The effects of human salivary dialysate upon ionic and electrochemical corrosion of a copper-aluminum alloy, *J. Electrochem. Soc., 134,* 575–580.

Mulford, S. J., and Tromans, D. (1988). Crevice corrosion of nickel-based alloys in neutral chloride and thiosulfate solutions, *Corrosion, 44,* 891–900.

Muller, A. W. J., Maessen, F. J. M. J., and Davidson, C. L. (1990a). The corrosion rates of five dental Ni–Cr–Mo alloys determined by chemical analysis of the medium using ICP-AES, and by the potentiostatic de-aeration method, *Corrosion Sci., 30,* 583–601.

Muller, A. W. J., Maessen, F. J. M. J., and Davidson, C. L. (1990b). Determination of the corrosion rates of six dental NiCrMo alloys in an artificial saliva by chemical analysis of the medium using ICP-AES, *Dent. Mater., 6,* 63–68.

Nielsen, C., and Klaschka, F. (1971). Teststudien an der Mundschleimhaut bei Ekzemallergikern, *Dtsch. Zahn-, Mund-Kieferheilkd., 57,* 201–218.

Niemi, L., and Hensten-Pettersen, A. (1985). The biocompatibility of a dental AgPd–Cu–Au-based casting alloy and its structural components, *J. Biomed. Mater. Res., 5,* 535–548.

O'Neal, S. J., Leinfelder, K. F., Lemons, J. E., Ratanapridakul, K., Isenberg, B., and Henson, P. (1988). Clinical studies of copper alloys in dogs, *J. Dent. Res., 67,* Abstr. 174.

Pfeiffer, P., and Schwickerath, H. (1991). Nickelabgabe von Dentallegierungen in Abhänggigkeit vom pH-Wert der Korrosionslösung, *Dtsch. Zahnärztl. Z., 46,* 753–756.

Phillips, R. W. (1991). *Science of Dental Materials,* 9th ed., W. B. Saunders, Philadelphia.

Pourbaix, M. (1984). Electrochemical corrosion of metallic biomaterials, *Biomaterials, 3,* 122–134.

Rae, T. (1975). A study on the effects of particulate metals of orthopedic interest on murine macrophages *in vitro, J. Bone Joint Surg., 57B,* 444–450.

Rae, T. (1978). The hemolytic action of particulate metals (Cd, Cr, Co, Fe, Mo, Ni, Ta, Ti, Zn, Co-Cr alloy), *J. Pathol., 125,* 81–89.

Randin, J. P. (1988). Corrosion behavior of nickel-containing alloys in artificial sweat, *J. Biomed. Mater. Res., 22*, 649–666.

Sandrik, J. L., Kaminski, E. J., and Greener, E. H. (1974). Biocompatibility of nickel-base dental alloys, *Biomat. Med. Dev. Art. Org., 2*, 31–39.

Sarkar, N. K., and Greener, E. H. (1973). *In vitro* corrosion resistance of new dental alloys, *Biomat. Med. Dev. Art. Org., 1*, 121–129.

Sen, P., and Costa, M. (1986). Incidence and localization of sister chromatid exchanges induced by nickel and chromium compounds, *Carcinogenesis, 7*, 1527–1533.

Shifrine, M., Fisher, G. I., and Taylor, N. G. (1984). Effect of trace elements found in coal fly ash on lymphocyte blastogenesis, *J. Environ. Pathol. Toxicol. Oncol., 5*, 15–24.

Smith, D. L. (1983). Dental casting alloys: Technical and economic considerations in the USA, *Int. Dent. J., 33*, 25–34.

Smith, K. L., and Lawrence, D. A. (1988). Immunomodulation of *in vitro* antigen presentation by cations, *Toxicol. Appl. Pharmacol., 96*, 476–481.

Soileau, R. L., Gantenburg, J. B., and Lucas, L. C. (1990). Metallic ion release and distribution from copper-based dental alloys, *J. Dent. Res., 69*, 26, Abstr. 1244.

Soremark, R., Diab, M., and Arvidson, K. (1979). Autoradiographic study of the distribution patterns of metals in which occur as corrosion products from dental restorations, *Scand. J. Dent. Res., 87*, 450–458.

Spiechowicz, E., Glantz, P.-O., Axel, T., and Chmielewski, W. (1984). Oral exposure to a nickel-containing dental alloy of persons with hypersensitive skin reactions to nickel, *Contact Dermatitis, 10*, 206–211.

Spiechowicz, E., Nyquist, G., Goliszewska, E., and Chmielewski, W. (1983). Experimental investigations on sensitivity to nickel present in alloys used in dentistry carried out on guinea pigs previously sensitized to this metal, *Swed. Dent. J., 7*, 39–43.

Stoffers, K., Strawn, S., and Asgar, K. (1987). Evaluation of properties of MS dental casting alloy, *J. Dent. Res., 66*, 205, Abstr. 785.

Tai, Y., De Long, R., Goodkind, R. J., and Douglas, W. H. (1992). Leaching of nickel, chromium, and beryllium ions from base metal alloys in an artificial oral environment, *J. Prosthet. Dent., 68*, 692–697.

van Loon, L. A. J., van Elsas, P. W., Duysters, P. P. E., Bos, J. D., and Davidson, C. L. (1988). Allergic contact stomatitis from Ni-alloys: A histological, immunolohistological and electrochemical relation, in *Biocompatibility of Co-Cr-Ni alloys* (H. F. Hildebrand and M. Champy, eds.), Plenum Press, New York, pp. 213–224.

Warner, G. L., and Lawrence, D. A. (1988). The effect of metals on IL-2-related lymphocyte proliferation, *Int. J. Immunopharmac., 10*, 629–637.

Wataha, J. C., Craig, R. G., and Hanks, C. T. (1991a). The release of elements of dental casting alloys into cell-culture medium, *J. Dent. Res., 70*, 1014–1018.

Wataha, J. C., Hanks, C. T., and Craig, R. G. (1991b). The *in vitro* effects of metal cations on eukarotic cell metabolism, *J. Biomed. Mater. Res., 25*, 1133–1149.

Wataha, J. C., Hanks, C. T., and Craig, R. G. (1993). Uptake of metal cations by fibroblasts *in vitro, J. Biomed. Mater. Res., 27*, 227–232.

Waters, M. D., Gardner, D. E., and Aranyl, C., et al. (1975). Metal toxicity for rabbit alveolar macrophages *in vitro, Environ. Res., 9*, 32–47.

Weber, H. (1983). The clinical acceptance of dental nickel–chrome alloys, *Int. Dent. J., 33*, 49–54.

Woody, R. D., Huget, E. F., and Horton, J. E. (1977). Apparent cytotoxicity of base metal casting alloys, *J. Dent. Res., 56*, 739–743.

60
Metal Release from Dental Casting Prosthodontic Alloys

Berit I. Johansson and Joel D. Bumgardner
University of Umeà, Umeà, Sweden
Linda C. Lucas
University of Alabama at Birmingham, Birmingham, Alabama

I. INTRODUCTION

A definition that might be used to describe the biocompatibility of a material is "the ability of a material to perform with an appropriate host response, in a specific application" (Williams, 1990). Many processes of interaction between the material and the tissue are involved in the description: The material should have an ability to perform a function in the body. This function depends on the mechanical and physical properties of the material and on its interaction with the tissues. The host response should be appropriate or acceptable when the material is performing its function. The situation where the material is functioning is also of importance.

One aspect to consider when describing the biocompatibility of a metal or alloy is the need to clarify the release of substances from the material to its environment. The rates and quantities of released substances and their chemical forms are of importance since, as Paracelcus wrote in 1538, "All substances are poisons. There is none that is not a poison. The right dose differentiates a poison and a remedy."

II. COMPOSITION OF CLASSIFIED ALLOYS

Metals have a long history as dental materials. Gold was used as early as 700 BC (Jones, 1988). The introduction of the "lost wax casting method" in dentistry radically improved the precision of dental prostheses. Today, a variety of metals suitable for casting are used. Casting alloys for inlays, crowns, and bridges can be classified according to their composition and function. Based on their composition, alloys may be divided into precious and base metal alloys (Phillips, 1991). Precious alloys include: (A) the high-gold alloys with at least 75% by mass of gold and metals from the platinum group, Pt, Pd, Ir, Rh, Ru, and Os; (B) low-gold alloys, with a gold content less than 60 but at least 40 wt%, with gold as the major component; and (C) the silver–palladium alloys. Precious alloys used for ceramic veneering include: (D) gold–platinum–palladium alloys, (E) gold–palladium alloys, (F) high-palladium alloys with up to 88% palladium; and (G) palladium–silver alloys. Figure 1 shows typical concentrations of gold, palladium, silver, and copper for alloys divided into groups A–G. The gold content decreases and the silver content increases from group A to group B to group C. The alloys in group C have a higher content of palladium than the alloys in groups A and B. For the metal ceramic alloys, the gold content decreases and the palladium content increases from group D to group E to group F. The alloys in group G have high contents of palladium and silver. Platinum is a constituent in gold–platinum–palladium alloys and is often included in high-gold alloys. In addition, cobalt, iridium, ruthenium, rhenium, indium, zinc, tin, iron, and gallium can be included in precious alloys. Iron, tin, and indium are used to produce an oxide film on the alloy surface to achieve a chemical bond between the ceramic and the alloy. Thus, these elements are essential for the aesthetics of metal ceramic alloys. There are commercial precious alloys available that do not match fully with the above-classified groups.

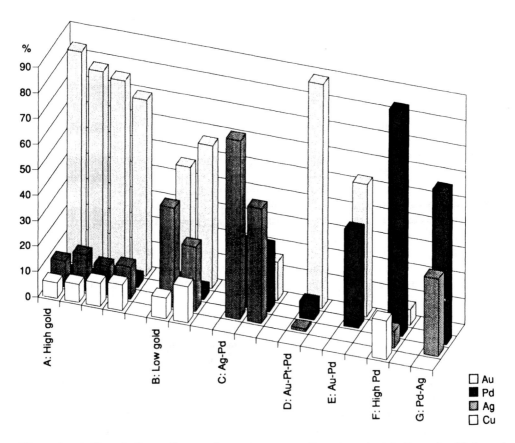

Figure 1   Gold, palladium, silver, and copper content of precious dental alloys classified as: A, high gold; B, low gold; C, silver–palladium; D, gold–platinum–palladium; E, gold–palladium; F, high-palladium; and G, palladium–silver alloys. The maximum content of the components is shown for group F. (Adapted from Phillips, 1991.)

The reduced cost, lighter weight, and, in general, improved mechanical properties have made base metal alloys an alternative to the precious alloys. Base metal alloys with nickel as the major component can be used for fixed prostheses with and without resin cover or ceramic veneers. The nickel content of these alloys may range from 70% to 80% and the chromium content from 13% to 22%. Small amounts of iron, aluminum, molybdenum, silicon, beryllium, copper, manganese, cobalt, and tin may be added (Phillips, 1991). Cobalt–chromium alloys have been introduced for fixed prosthodontics but they are mainly used today for removable partial dentures.

Other metals/alloys are also available for dental prosthodontics. Copper-based alloys are frequently used in some countries but not at all in others. Chemical compositions for the copper alloys are 76–87 wt% copper, 6.5–11% aluminum, 1.0–5.0% nickel, and 0.5–1.2% manganese. They may also contain zinc, iron, gallium, and indium (Johansson, Lucas, and Lemons, 1989). The compositions of the aluminum bronzes are similar to the alloys used for marine components (Upton, 1963).

Titanium has been introduced as a prosthodontic material. Both unalloyed titanium and titanium alloys have successfully been used for implants for a long time. Titanium crowns and bridges can be made using different techniques. Two methods used are casting and fabricating with machine duplication and spark erosion. Difficulties with casting titanium are its high melting temperature, low thermal conductivity, high reactivity, and gas absorption at elevated temperatures. Recent developments have made it possible to cast unalloyed titanium with the precision and quality required for fixed prosthodontics (M. Bergman, 1990). Fabrication of titanium crowns with machine duplication and spark erosion was developed by Matts Andersson. Two principles are involved in the fabrication: machine duplication of models and electric discharge machining. With these combined techniques, shortcuts associated with casting have been eliminated (Andersson, Bergman, Bessing, Ericson, Lundquist, and Nilson, 1989).

III. THE ENVIRONMENT

In function, dental alloys are exposed to a complex milieu that is influenced by external factors such as how often and what you eat and drink, and your tooth-brushing habits. The saliva contributes a great deal to the complexity. The composition of saliva is different from individual to individual and varies also during the day. The saliva composition is location dependent since saliva is excreted from several glands at different positions. Problems with acid regurgitation and vomiting also influence the composition and pH of the solution found in the mouth.

To study the influence of the surrounding solution on the electrochemical behavior of dental alloys *in vitro*, several electrolytes have been used. Meyer and Nally (1975) studied the corrosion of three dental alloys in natural saliva, Ringer's solution and five artificial saliva solutions. A modified Fusayama artificial saliva solution was found to give the best correlation with the results obtained with natural saliva. The compositions of natural saliva, two artificial saliva solutions, and Ringer's solution are shown in Table 1. The chloride content in Ringer's solution is much higher than in natural saliva. Both Darvell's and Fusayama's solutions include chlorides and phosphates in the same range as natural saliva. No bicarbonate is found in the Fusayama's solution.

Marek and Topfl (1986) compared various electrolytes for corrosion testing of dental alloys. They concluded that standard and diluted Ringer's solution and 1% NaCl can be used for screening tests. These electrolytes are more aggressive than natural saliva. Artifi-

Table 1   Composition (mM) of Natural Saliva, Artificial Saliva[a], and Ringer's Solution[b]

	Natural saliva[c]	Darvell's solution	Fusayama's solution	Ringer's solution
Cl^-	5–40.9	29.8	23.0	164.5
PO_4^{3-}	1.9–7.1	4.7	5.0	
Bicarbonate	1.0^d–13.5^d	7.1		
Na^+	0.4–40.3^d	40.4	11.9	155.0
NH^{4+}	0.6–14.7	4.1		
K^+	7.7–36		5.4	5.7
Ca^{2+}	0.8–2.7		5.4	1.9
S^{2-}	NA^e		0.02	
SCN^-	0.0–5.5	2.5		
Citrate	0.0–0.1	0.07		
Urea	2.3–12.5	6.7	16.7	
Uric acid	0.03–0.2	0.2		
Lactate	NA			

[a]Adapted from Holland (1992)
[b]Butler and Dawson (1992)
[c]The values for natural saliva are for mixed saliva (Haeckel, Walker, and Colic, 1989; Shafer, Hine, and Levy, 1974).
[d]Values for parotid saliva (Shannon, 1983; Chauncey, Feller, and Kapur, 1987)
[e]NA: data not available

cial saliva solutions containing chlorides, phosphates, and bicarbonate in the same range as natural saliva best approximate the corrosion-related solution parameters of saliva. Furthermore, they found that the use of complex solutions with minor ingredients was not justified.

Holland (1992) studied nine dental alloys in various electrolyte solutions by generating potentiodynamic polarization curves. He used two artificial saliva solutions and an aqueous solution of 1% sodium chloride buffered to pH 4, 7, or 10 as electrolytes. Darvell's artificial saliva solution was slightly less corrosive to the casting alloys than Fusayama's solution. For the high-gold alloys, the corrosion sensitivity was low and was not influenced by changes in the electrolyte. The low-gold and nonnoble alloys exhibited corrosion sensitivity with greater variation depending on the electrolyte solution. The study showed that the corrosion activity of the alloys is dependent on the pH of the solution.

The influences of solutions on the corrosion of alloys are illustrated in Fig. 2. After 4 weeks in a 0.9% sodium chloride solution, the copper–aluminum alloy was discolored and pitting was observed, Fig. 2a. Exposure of the same alloy to an artificial saliva solution made the grains visible, Fig. 2b. The sodium chloride solution was more aggressive than the artificial saliva solution (Johansson et al., 1989).

Components and the temperature of food and drinks may also influence the corrosion of dental alloys, as well as smoking and the use of tobacco products. Herø and Niemi (1986) found that specimens made of a multiphase silver–palladium–copper–zinc alloy placed in acrylic dentures tarnished in the oral cavity in 4 out of 10 people. *In vitro*, the alloy had a low corrosion and tarnish resistance when immersed in 2% Na_2S. The higher sulfide ion concentration *in vitro*, compared with an individual variation and low sulfide

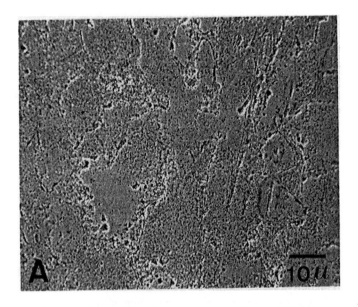

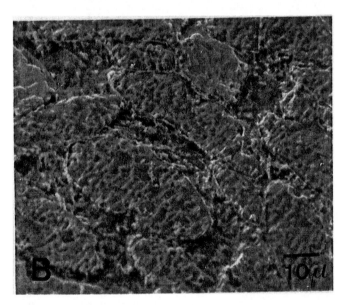

Figure 2 Surface of copper–aluminum crowns after 4 weeks in (A) 0.9% NaCl solution; (B) artificial saliva solution.

ion concentration in natural saliva as well as the absence of a protein film may be factors explaining the difference between the *in vitro* and the clinical findings. Sulfide-containing amino acids in food may contribute to an aggressive environment that promotes tarnish and corrosion (Statens Livsmedelsverk, 1986).

A protein film is formed on all solid surfaces in the oral cavity (Jendresen and Glantz, 1981). The film, *pellicle*, may act as a diffusion barrier for species involved in the corrosion process (Holland, 1984). Pellicle on a gold alloy in contact with conventional amalgam has been found to decrease the currents generated by the specimens. No effect was found when a conventional amalgam was covered or when high-copper amalgams were used. The pellicle was formed after an 1-h exposure to saliva in the oral cavity.

Koivumaa and Mäkilä (1970) showed that, after 3 days, the accumulation of plaque was more frequent on the side of full dentures where specimens of amalgam and gold alloy were placed. A study on plaque accumulation on metal ceramic restorations cast from noble and nickel-based alloys showed no observable differences between the investigated alloys after 36 months. However, all patients in the study had good oral hygiene and no unfavorable periodontal conditions (Morris, 1989). Under deposits such as plaque, dissolved oxygen is depleted, causing a more aggressive local environment that favors corrosion; that is, a concentration cell is formed. Concentration cell corrosion is often part of crevice corrosion, which can occur in spaces such as interproximal areas where the transport of solution is restricted and differences in solution chemistry such as acidification occur (Marek, 1985).

Bacteria are always present in the oral cavity. *Streptococcus mutans* from plaque have been found to have a higher adherence to corroded specimens compared to uncorroded ones (Miyajima, 1984). The corrosion of dental alloys and amalgams increased in cultures of *Bacteroides corrodens* compared with a sterile growth medium (Palaghias, Söremark, and Nord, 1982). Corrosion was less pronounced for the gold and cobalt-chromium alloys as compared with the amalgams.

Teeth and dental restorations wear in the mouth. Wear associated with these materials is influenced by such factors as contact load, duration, and the velocity of the movement. External factors such as particles in a dusty environment that mix with saliva might also affect the wear of teeth and restorations. Some people grind and clench their teeth extensively. Bruxism results in an increased loss of substance due to contact wear of prosthodontic materials (Ekfeldt, 1989). Nickel, chromium, and beryllium ions were found to be released from nickel-based alloys by both dissolution and occlusal wear in an artificial environment capable of reproducing three-dimensional force–movement cycles of human mastication. The release was most pronounced for metal versus porcelain contacts (Tai, Long, Goodkind, and Douglas, 1992). Abrasion of alloys and their protective films may accelerate corrosion tremendously.

To clean teeth and oral tissues, people brush more or less frequently. The surfaces of dental restorative materials can be affected by brushing and the abrasive toothpaste that is often used (Johannsen, Redmalm, and Rydén, 1989). The use of abrasive prophylactic devices also causes surface alterations on restorative materials (Eliades, Tzoutzas, and Vougiouklakis, 1991). Products released due to corrosion and wear, and a combination thereof, should be considered when estimating the total amount of released substances from dental alloys.

IV. TYPES OF CORROSION

The creation of concentration cells and crevice corrosion are dependent on the situation in the oral cavity. A crevice must be created; the gaps between dental amalgam restorations and the teeth are frequent sites for crevice corrosion (Sutow, Jones, and Hall, 1989). The area between veneered porcelain, alloy, and gingiva might also favor the formation of crevice corrosion.

Types of corrosion that are more dependent on the alloy per se are pitting, stress, and galvanic corrosion. Alloys forming a protective surface film are susceptible to localized corrosion in the form of pitting, crevice, or phase corrosion. If the protective film breaks down or the rate of dissolution of the film is increased, the unprotected area corrodes. Base alloys are susceptible to localized corrosion and their protective oxide surface layers are susceptible to breakdown (Johansson et al., 1989).

Corrosion of alloys may be affected by stress. Application of stress on amalgams was found to increase their corrosion (Gjerdet and Espevik, 1978). The increase was higher for amalgams with high creep than for amalgams with low creep. Rupture of the protective oxide on the amalgams due to the increased strain was discussed as a possible mechanism for the phenomenon.

Galvanic corrosion occurs when dissimilar metals are in contact and is a common type of corrosion in the oral cavity. Different metallic restorations in functional or antagonistic contact, or soldered constructions may suffer from galvanic corrosion. The corrosion of the less noble alloy is enhanced by the increased area for the cathodic reaction (Marek, 1985). Galvanic corrosion has been investigated in several studies. The quantity and the quality of released substances are influenced by the type of alloys in contact. Moberg (1985) found that the release of copper from a high-gold alloy increased when placed in contact with a gold–platinum–palladium alloy. Nickel base alloys were found to be susceptible to crevice corrosion when in contact with a gold alloy. Furthermore, the release of nickel, chromium, and molybdenum increased when the alloys were in contact compared with being separate. The release was significant for one of the studied nickel alloys.

A great number of combinations of alloys occur when units are brazed together. Certain combinations were shown to be unsuitable when brazed joints were examined under a light microscope, and the junctions between the solder and the gold alloys were classified (M. Bergman, 1977). Angelini, Pezzoli, Rosalbino, and Zucchi (1991) found that a cobalt–chromium alloy brazed with a gold alloy had a reduced tensile strength after 60 days immersion in Ringer's solution. The authors attributed the reduction in tensile strength to galvanic corrosion. The cobalt–chromium alloy was largely unaffected when brazed with a nonprecious alloy. Thus, the galvanic corrosion might in some cases be severe enough to adversely affect the mechanical properties of the alloys.

V. CHARACTERIZATION WITH POLARIZATION CURVES

Polarization curves may be used to obtain information on the corrosion characteristics of a metal/alloy in a selected electrolyte. Polarization is the difference between the potentials of an electrode with and without current. Polarization curves show the relationship between electrode potential and current density (Wranglén, 1972). Various instruments and an electrochemical cell containing a reference electrode, the specimen or the working electrode, and a counterelectrode are used to polarize the sample and to register the currents and the potentials. The potential of the specimen is measured versus the reference electrode, which is an electrode with a constant potential. The counterelectrode allows currents to flow in the system, and its surface is unaffected by changes of potentials and currents (Meyer, 1988). The potential of a metal/alloy may reflect its affinity for electrochemical reactions to occur, and the current is a measure of its velocity. Potential current curves provide information on the corrosion resistance of the metal, the ability to form protective films, and the susceptibility to accelerated corrosion processes such as pitting and galvanic corrosion.

Four different types of anodic polarization curves are shown in Fig. 3, which includes both anodic and cathodic polarization curves. Figure 3a denotes an alloy that undergoes general corrosion (Pourbaix, 1984). The magnitude of the anodic current density increases with more positive or noble electrode potentials. The corrosion potential E_{cr} is defined to be the potential when the added currents due to oxidation and reduction processes are zero. The corrosion potential is close to the open circuit potential, that is, the potential observed in the mouth without any influence by other metallic restorations (Meyer, 1988). More active or negative corrosion potentials indicate that these alloys are more susceptible to corrosion and might experience accelerated corrosion if placed in contact with metallic materials with a more noble or positive potential (Lemons, Lucas, and Johansson, 1992). The corrosion current is the current due to oxidation processes at the corrosion potential. A high corrosion current indicates that the alloy may release significant quantities of metal ions in the oral environment.

Figure 3b denotes an alloy that exhibits minimal corrosion. The anodic current density is low and invariable over a broad potential range. Figure 3c denotes an alloy that corrodes with the formation of a protective film, making the alloy passive. The passive current is the current necessary to maintain the protective layer. Figure 3d denotes an

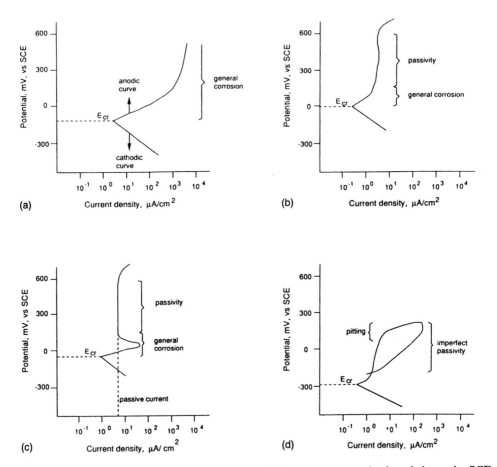

Figure 3   Schematic polarization curves. Potential is versus saturated calomel electrode, SCE.

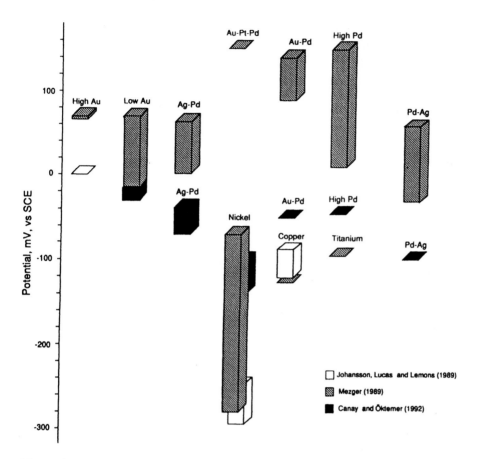

Figure 4 Corrosion or open circuit potentials versus saturated calomel electrode, SCE, for alloys classified in groups.

alloy that forms a protective film. The reverse scan of the anodic polarization curve indicates susceptibility to localized corrosion in the form of pitting, crevice, or phase corrosion.

Electrochemical potentials and currents may be used to rank dental casting alloys. The corrosion or the open circuit potentials for the alloys classified by groups in artificial saliva are shown in Fig. 4, which is based on corrosion potentials reported by Johansson et al. (1989) and by Canay and Öktemer (1992); and on open circuit potentials reported by Mezger (1989). Even though the same test parameters were not used in these studies — for example, the temperature of the solutions, the gases used for bubbling, and the artificial saliva solutions were different — similar trends could be observed. In general, the precious alloys had more positive potentials than the base metal alloys and unalloyed titanium. Mezger (1989) registered both open circuit potentials, OPC, and zero current potentials, ZCP. The OPC values for the alloys were higher than the corresponding ZCP values. Formation of thin films after polishing or upon immersion may be the reason for this discrepancy. The ZCP values tended to be lower for the base metal alloys than for the precious alloys, with some overlap existing.

The less noble or more active alloy might suffer enhanced corrosion when in metallic contact with a more noble alloy. Galvanic corrosion is dependent on the relative magnitude of the potential difference, the environment, and the relative surface area ratios. For example, if a more active nickel-based alloy is in contact with a more noble high-gold alloy, it would be expected that the nickel alloy would experience enhanced galvanic corrosion. The oral fluids would be the electrolyte and the nickel alloy could corrode at a higher rate than if it were not in metallic contact with the gold alloy (Lemons et al., 1992).

Figure 5 shows the highest corrosion current registered within each alloy group by Johansson et al. (1989) and Mezger (1989).

A.   High-Gold Alloys

The high-gold alloys exhibited low corrosion currents, Fig. 5. The polarization curves for high-gold alloys were similar to the curves in Fig. 3b. The recorded currents were low at all potentials.

B.   Low-Gold Alloys

The corrosion currents were about the same or somewhat higher than for the high-gold alloys. The polarization curves for low-gold alloys were similar to the schematic curves in

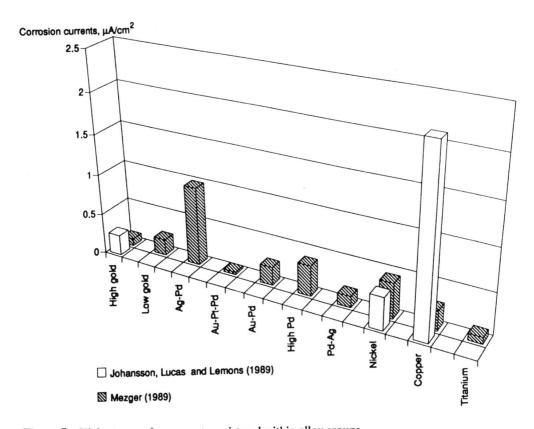

Figure 5   Highest corrosion currents registered within alloy groups.

Figs. 3b and 3c. The low-gold alloys exhibited passivation even though the passivation was not distinct for all alloys. The passive current densities were relatively low (Johnson, Rinne, and Bleich, 1983).

C. Silver-Palladium Alloys

The corrosion currents for the silver-palladium alloys were about the same or higher than the corrosion currents for the high-gold alloys. Mezger (1989) studied six silver-palladium alloys, of which two contained approximately 20 wt% gold. One of the silver-palladium-gold alloys had three times the corrosion current density than the other. The lower nobility of the second phase occurring in the microstructure of this alloy is thought to be the cause for the higher corrosion current density.

Of the 13 prosthodontic alloys studied by Canay and Öktemer (1992), one of the silver-palladium alloys had the highest passive current density registered. The anodic polarization curve for another silver-palladium alloy in the same study exhibited a hysteresis effect when the potential scan was reversed. Johnson et al. (1983) found that the silver-palladium alloys exhibited passive behavior to some degree depending on the minor constituents. For example, copper has been shown to enhance passivity. However, a silver-palladium alloy containing 20% indium demonstrated deleterious corrosion resistance at potentials 100 mV more noble to the corrosion potential. The corrosion resistance seems also to vary depending on whether the alloys are flame or electric cast.

D. Gold-Platinum-Palladium Alloys

The corrosion current for the studied gold-platinum-palladium alloy was low, Fig. 5. Low currents at all potentials were also registered for a gold-platinum alloy studied by Canay and Öktemer (1992).

E. Gold-Palladium Alloys

The gold-palladium alloys had about the same or somewhat higher corrosion currents than the studied gold-platinum-palladium alloy, Fig. 5. The anodic polarization curve for the gold-palladium alloy studied by Canay and Öktemer (1992) is similar to the one in Fig. 3c. The alloy exhibited active/passive behavior.

F. High-Palladium Alloys

The high-palladium alloys exhibited low currents or were passive in the anodic potential range of oral interest, that is, +300 mV versus SCE (saturated calomel electrode) (Corso, German, and Simmons, 1985; Ewers and Greener, 1985). The corrosion behavior of high-palladium alloys was similar for all tested alloys (Mezger, 1989). Composition and constitutional variations did not have any substantial effect. The electrochemical behavior indicated resistance to pitting or the buildup of an effective protective layer. Canay and Öktemer (1992) found that the high-palladium alloy had the same magnitude of the passive current as the 22-carat gold alloy studied.

G. Palladium-Silver Alloys

The palladium-silver alloys had low currents or were passive up to 300 mV (SCE). The microstructural and compositional differences between the investigated alloys did not have any significant effect on the electrochemical behavior (Mezger, 1989). However, Canay and Öktemer (1992) observed a selective dissolution of the silver-rich phases of the palladium–silver alloys.

H. Nickel Base Alloys

One of the two nickel alloys studied by Johansson et al. (1989) had approximately the same corrosion current as a high-gold alloy. The nickel alloy containing beryllium had a higher corrosion current than the alloy without. A typical curve for a Ni–Cr–(Be) alloy is seen in Fig. 3d. The hysteresis behavior observed was more significant for the beryllium-containing nickel base alloy than for the non-beryllium-containing alloy. The microstructure of a beryllium-containing nickel–chromium alloy after 1 month in an artificial saliva solution revealed corroded nickel–beryllium rods in eutectic areas, Fig. 6. The corrosion resistance of nickel base alloys is dependent on the formation of a passive oxide film. High chromium and molybdenum contents favor the formation of stable passive films within the potential range of oral interest (Canay and Öktemer, 1992).

I. Copper–Aluminum Alloys

The copper–aluminum alloys had much higher corrosion currents than the nickel base and high-gold alloys in the study by Johansson et al. (1989). Mezger (1989) found that the corrosion current for the tested copper alloy was within the general alloy spectrum range. However, in saline solution the corrosion current was the highest for all alloys studied. The polarization curves for copper-based alloys were similar to the curves in Fig. 3a, indicating susceptibility to general corrosion.

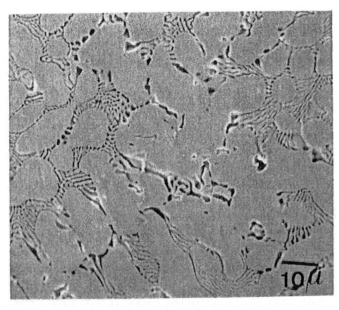

Figure 6   Surface of crown made of Ni–Cr–(Be) after 4 weeks in artificial saliva solution.

J. Titanium

Unalloyed titanium had a low corrosion current, Fig. 5. Titanium was passive over the whole potential range and no tendency for pitting was observed. Titanium develops an oxide surface for corrosion resistance. However, the film can be damaged and has been found to not withstand acidic fluoride agents (Pröbster and Hutteman, 1992).

VI. CORROSION RATES – CURRENT DENSITIES – ION RELEASE

Corrosion rates determined by electrochemical techniques are expressed by current density. These expressions can be converted into amount of ions released according to Faraday's law, $\Delta m = MQ/zF$, where Δm is the mass released, M is the molecular weight of the element, z is its valence, Q is the quantity of charge obtained by integrating the current–time curve, and F is Faraday's constant.

Johansson (1986) studied galvanic corrosion and found in a laboratory study that the quantity of tin ions released from a conventional amalgam was related to the charge transferred in ampere-seconds (Coulombs) between the amalgam and gold alloy. The same tendency was found in a clinical study using a saliva sample that confined and collected the released tin ions into a small volume when conventional amalgam was in contact with a gold alloy, Fig. 7. The gold alloy and amalgam specimen were fixed in an acrylic bite splint that made it possible to collect the released ions and register the current for up to 1 h. The copper content was below or close to the detection limit and could not be related to the charge transfer.

Brown, Farnsworth, Merritt, and Crowe (1988) conducted an *in vitro* and *in vivo* study to measure the weight loss and to calculate the net charge transfer using an accelerated corrosion test. The weight loss and the ion release from stainless steel in saline, 10% serum, and in subcutaneous space in hamsters could be calculated using Faraday's law assuming release in proportion to alloy composition. The results using the cobalt–chromium–molybdenum alloy indicated that the release rates *in vitro* could be used to determine the proportionality of release *in vivo*.

A galvanic corrosion study included measurements of released substances from various nickel-based alloys in contact with titanium. The study also calculated the release of metallic ions from the alloys using current density–time curves. A good correlation was established for the two methods and a differentiation of the corrosion behavior of the nickel alloys could be made (Geis-Gerstorfer, Weber, and Sauer, 1989). Thus, electrochemical studies seem to be useful in making comparisons and in extracting knowledge about corrosion rates and amounts of released ions from dental alloys.

VII. IMMERSION STUDIES

Immersion studies on the quantity and quality of released substances from dental alloys are few. Long exposure times, very aggressive solutions, or very sensitive analysis techniques are needed since the substances from the alloys sometimes dissolve very slowly. Long exposure times make it difficult to keep all test parameters constant. Very aggressive solutions may give different results from those obtained using more adequate solutions. Sensitive analysis techniques demand proper preparation and care to avoid contamination problems. Nevertheless, trace amounts of released substances *in vitro* from an alloy might be of importance when describing the biocompatibility of the alloy in function in the oral cavity.

Tin, μg

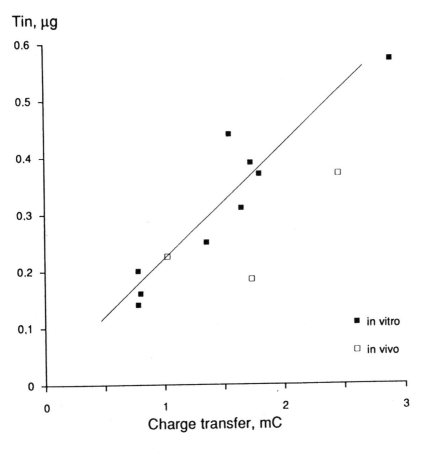

Charge transfer, mC

Figure 7 Released amount of tin when charge was transferred between conventional amalgam and high-gold alloy. *In vitro* regression line: slope 0.19 (0.17–0.21) μg/mC, intercept 0.03 (−0.01–0.06) μg with 95% confidence levels. (Adapted from Johansson, 1986.)

In an immersion test, the total loss of substance from four nickel–chromium alloys varied between approximately 8 and 4700 $\mu g/cm^2$ after 7 days in a lactic acid–sodium chloride solution (Geis-Gerstorfer and Weber, 1987). The molybdenum-free alloy and the beryllium-containing alloy released the highest amount of substance. Figure 8 shows the composition of two alloys, A and B, and the loss of substance from the alloys. Alloy A is a beryllium-free alloy with high chromium and molybdenum contents, and alloy B is a beryllium-containing alloy with lower chromium and molybdenum contents than alloy A. The loss of substance from alloy A was much lower than the loss from alloy B after 7 days in the used solution. Approximately 6% of the released substance from alloy B was beryllium even though the alloy content of beryllium was only 1.7%. The amounts of components released were not always in proportion to the bulk composition of the alloy.

Polarization and potential time curves were also used to characterize the nickel base alloys. The alloys that released high amounts of substance had high corrosion currents in the potential area of interest. The alloys that released low amounts of substance had corrosion characteristics similar to precious alloys (Geis-Gerstorfer and Weber, 1987).

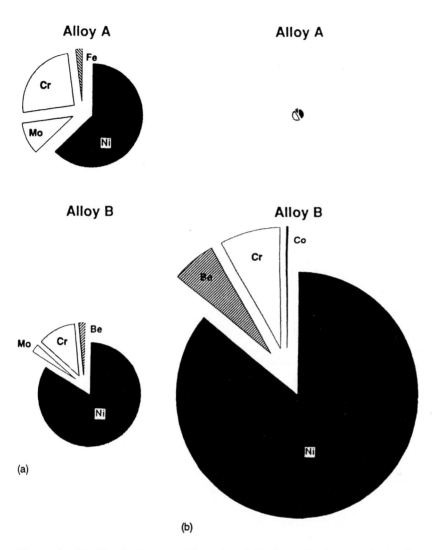

Figure 8   (a) Chemical composition of and (b) loss of substance from alloy A and B. Total amount, released after 7 days in lactic acid–sodium chloride solution from A and B are indicated by circle areas. The substances released from alloy A were Mo, Ni, Cr, and Fe. (Adapted from Geis-Gerstorfer and Weber, 1987.)

VIII.   CLINICAL STUDIES

Clinical problems and complications with the use of cast dental alloys for fixed prosthodontic works are few. The majority of inlays, crowns, and bridges are in function and continue to do so without problems after years of clinical service (Glantz, Ryge, Jendresen, and Nilner, 1984; Morris, Manz, Stoffer, and Weir, 1992). Unrestored periodontal teeth are slightly more affected by plaque than cast restorations. However, effects on the gingiva, the pocket depth, and the loss of gingival attachment are somewhat higher for cast restorations than for periodontal controls (Morris et al., 1992). Factors such as overcontouring and secondary caries are associated with unacceptable crowns. Marginal periodontitis is observed with both acceptable and unsatisfactory fixed restorations but mainly with the latter (Glantz et al., 1984). Individuals who had been treated with extensive metallic reconstructions and who reported symptoms that could be alleged to have been caused by oral galvanic actions had undergone more restorative procedures that were conducted at a lower level of technical quality as compared with a control group not suffering from any oral discomforts (Nilner, Glantz, Ryge, and Sundberg, 1982). Complaints from patients as well as from dental practices have focused an interest on corrosion of cast dental alloys. A study was initiated after personnel in dental practices had noted frequent corrosion of a nickel–chromium alloy (Oehmichen and Klötzer, 1984). Signs of corrosion were present in approximately 20% of the 429 prosthodontics made of the nickel-based alloy. The corroded crowns and bridges were produced in three different commercial laboratories and seated in three different dental practices. The authors concluded that the alloys were susceptible to general corrosion and possibly sensitive to processing errors.

Different types of corrosion such as crevice, pitting, stress, and galvanic corrosion could be observed on studied prosthodontic restorations. Metal ions released from cast conventional high-gold alloy restorations have been detected in dentine and enamel after short periods. The use of pure cohesive gold or nearly pure cast gold, or homogenization of the gold castings for sufficiently long periods, reduced the corrosion and the penetration of corrosion products into the dental tissues. A cavity liner was found to prevent corrosion products from migrating into the teeth (Söremark, Wing, Olsson, and Goldin, 1968). Sections of soft tissues in contact or next to dental restorations made of high-gold, low-gold, palladium, and base metal alloys have been analyzed (Kratzenstein, Sauer, and Weber, 1988; Kratzenstein, Sauer, Weber, and Geis-Gerstorfer, 1986). Components released from the alloys could be detected in the tissues. The low-gold alloys were less resistant to corrosion processes than the high-gold alloys. However, the nonnoble components (Fe, Sn, In) were detected in the affected gingiva close to the high-gold alloys made for porcelain veneering. The majority of intolerance reactions such as burning sensation and metallic taste occurred where low-gold and base metal alloys and solders were combined.

Titanium for crowns and bridges has been used for a short time. Titanium copings covered with resin or veneered with low-fusing ceramic have been examined. The clinical findings were promising, with some shortcuts associated with the resin or the ceramic veneer (B. Bergman, Bessing, Ericson, Lundquist, Nilson, and Andersson, 1990; Nilson, Bergman, Bessing, Lundquist, and Andersson, 1994). Cast titanium crowns were tested clinically. Only one case each of discoloration and occlusal wear was observed of more than 100 titanium crowns seated in the mouth during a 2-year period (Ida, Tani, Tsutsumi, Togaya, Nambu, Suese, Kawazoe, Nakamura, and Wada, 1985).

To determine the quantity of released metals from fixed prosthodontic restorations in the oral cavity is a delicate technical problem since each individual is unique and has his/her own habits. Furthermore, detection methods should not destroy the metallic restorations or be extremely inconvenient for the individuals. Attempts to register the currents or the charge transfers between different metallic restorations, inlays, crowns, and bridges have been made (Johansson, 1986; Lukas, 1981). According to Faraday's law, the charge transfer could be related to amount ions released. Johansson (1986) found no significant differences between a patient group with complaints related to corrosion and a control group when each individual was represented by the highest charge transfer or by the total added charge transfer registered in the oral cavity. However, the highest charge transfers recorded were within the patient group. This was similar to the results obtained by Lukas (1981), who compared a patient group and a control group concerning mean values and distribution of currents.

IX. CONCLUSION

Knowledge of corrosion characteristics of dental casting prosthodontics alloys is necessary to be able to estimate the release of metal from the alloys in the mouth. Furthermore, knowledge about different types of corrosion, the complex environment, interaction with other restorations, and the wear situation is also essential in efforts to optimize the biocompatibility of the prosthodontic alloys.

ACKNOWLEDGMENT

The authors wish to express their gratitude to Ms. Helén Agdahl for her most patient work with the figures.

REFERENCES

Andersson, M., Bergman, B., Bessing, C., Ericson, G., Lundquist, P., and Nilson, H. (1989). Clinical results with titanium crowns fabricated with machine duplication and spark erosion, *Acta Odontol. Scand., 47*, 279–286.

Angelini, E., Pezzoli, M., Rosalbino, F., and Zucchi, F. (1991). Influence of corrosion on brazed joints' strength, *J. Dent., 19*, 56–61.

Bergman, B., Bessing, C., Ericson, G., Lundquist, P., Nilson, H., and Andersson, M. (1990). A 2-year follow-up study of titanium crowns, *Acta Odontol. Scand., 48*, 113–117.

Bergman, M. (1977). Combinations of gold alloys in soldered joints, *Swed. Dent. J., 1*, 99–106.

Bergman, M. (1990). *Gjutning av titan*, DP Nova AB, Malmö, pp. 6–18.

Brown, S. A., Farnsworth, L. J., Merritt, K., and Crowe, T. D. (1988). *In vitro* and *in vivo* metal ion release, *J. Biomed. Mater. Res., 22*, 321–338.

Butler, M., and Dawson, M. (1992). Balanced salt solutions and buffers, in *Cell Culture Labfax* (M. Butler and M. Dawson, eds.), BIO's Scientific Publishers, Oxford, pp. 135–140.

Canay, S., and Öktemer, M. (1992). *In vitro* corrosion behavior of 13 prosthodontic alloys, *Quintessence Int., 23*, 279–287.

Chauncey, H. H., Feller, R. P., and Kapur, K. K. (1987). Longitudinal age related changes in human parotid saliva composition, *J. Dent. Res., 66*, 599–602.

Corso, P. P., German, R. M., and Simmons, H. D. (1985). Corrosion evaluation of gold-based dental alloys, *J. Dent. Res., 64*, 854–859.

Ekfeldt, A. (1989). Incisal and occlusal tooth wear and wear of some prosthodontic materials. An epidemiological and clinical study, *Swed. Dent. J.*, Suppl. 65.

Eliades, G. C., Tzoutzas, J. G., and Vougiouklakis, G. J. (1991). Surface alterations on dental restorative materials subjected to an air-powder abrasive instrument, *J. Prosthet. Dent., 65*, 27–33.

Ewers, G. J., and Greener, E. H. (1985). The electrochemical activity of the oral cavity—a new approach, *J. Oral Rehabil., 12*, 469–476.

Geis-Gerstorfer, J., and Weber, H. (1987). *In vitro* corrosion behavior of four Ni–Cr dental alloys in lactic acid and sodium-chloride solutions, *Dent. Mater., 3*, 289–295.

Geis-Gerstorfer, J., Weber, H., and Sauer, K.-H. (1989). *In vitro* substance loss due to galvanic corrosion in Ti implant/Ni–Cr supraconstruction systems, *Int. J. Oral Maxillofac. Implants, 4*, 119–123.

Gjerdet, N. R., and Espevik, S. (1978). Corrosion and creep of dental amalgam, *J. Dent. Res., 57,* 21–26.

Glantz, P.-O., Ryge, G., Jendresen, M. D., and Nilner, K. (1984). Quality of extensive fixed prosthodontics after five years, *J. Prosthet. Dent., 52,* 475–479.

Haeckel, R., Walker, R. F., and Colic, D. (1989). Reference ranges for mixed saliva collected from the literature, *J. Clin. Chem. Clin. Biochem., 27,* 249–252.

Herø, H., and Niemi, L. (1986). Tarnishing *in vivo* of Ag–Pd–Cu–Zn alloys, *J. Dent. Res., 65,* 1303–1307.

Holland, R. I. (1992). Corrosion testing by potentiodynamic polarization in electrolytes, *Dent. Mater., 8,* 241–245.

Holland, R. I. (1984). Effect of pellicle on galvanic corrosion of amalgam, *Scand. J. Dent. Res., 92,* 93–96.

Ida, K., Tani, Y., Tsutsumi, S., Togaya, T., Nambu, T., Suese, K., Kawazoe, F., Nakamura, M., and Wada, H. (1985). Clinical application of pure titanium crowns, *Dent. Mater. J., 4,* 191–195.

Jendresen, M. D., and Glantz, P.-O. (1981). Clinical adhesiveness of selected dental materials. An *in-vivo* study, *Acta Odontol. Scand., 39,* 39–45.

Johannsen, G., Redmalm, G., and Rydén, H. (1989). Surface changes on dental materials, *Swed. Dent. J., 13,* 267–276.

Johansson, B. I. (1986). Tin and copper release related to charge transfer between short-circuited amalgam and gold alloy electrodes, *Scand. Dent. Res., 94,* 259–266.

Johansson, B. I., Stenman, E., and Bergman, M. (1986). Clinical registration of charge transfer between dental metallic materials in patients with disorders and/or discomfort allegedly caused by corrosion, *Scand. J. Dent. Res., 94,* 357–363.

Johansson, B. I., Lucas, L. C., and Lemons, J. E. (1989). Corrosion of copper, nickel, and gold dental casting alloys: An *in vitro* and *in vivo* study, *J. Biomed. Mater. Res.: Appl. Biomater., 23,* 349–361.

Johnson, D. L., Rinne, V. W., and Bleich, L. L. (1983). Polarization-corrosion behavior of commercial gold- and silver-base casting alloys in Fusayama solution, *J. Dent. Res., 62,* 1221–1225.

Jones, D. W. (1988). The future of biomaterials, *Can. Dent. Assoc. J., 54,* 163–173.

Koivumaa, K. K., and Mäkilä, E. (1970). The effect of galvanism on accumulation of bacterial plaque *in vivo, Suom. Hammaslaak. Toim., 66,* 367–371.

Kratzenstein, B., Sauer, K.-H., and Weber, H. (1988). *In-vivo* Korrosionserscheinungen von gegossenen Restaurationen und deren Wechselwirkungen mit der Mundhöhle, *Dtsch. Zahnärztl. Z., 43,* 343–348.

Kratzenstein, B., Sauer, K.-H., Weber, H., and Geis-Gerstorfer, J. (1986). *In-vivo* Korrosionsuntersuchungen goldhaltiger Legierungen, *Dtsch. Zahnärztl. Z., 41,* 1272–1276.

Lemons, J. E., Lucas, L. C., and Johansson, B. I. (1992). Intraoral corrosion resulting from coupling dental implants and restorative metallic systems, *Implant Dent., 1,* 107–112.

Lukas, D. (1981). Elektrische Strommessungen und Erkrankungen der menschlichen Mundschleimhaut, *Dtsch. Zahnärztl. Z., 36,* 144–147.

Marek, M. (1985). *Corrosion in a biological environment. Literature review,* International Workshop on Biocompatibility, Toxicity, and Hypersensitivity to Alloy Systems Used in Dentistry, Ann Arbor, MI.

Marek, M., and Topfl, E. (1986). Electrolytes for corrosion testing of dental alloys, *J. Dent. Res., 65,* 301, Abstr. 1192.

Meyer, J.-M. (1988). The corrosion of dental Ni–Cr alloys. An *in vitro* evaluation, in *Biocompatibility of Co–Cr–Ni Alloys* (H. F. Hildebrand and M. Champy, eds.), Plenum Press, New York, pp. 305–320.

Meyer, J.-M., and Nally, J.-N. (1975). Influence of artificial salivas on the corrosion of dental alloys, *J. Dent. Res., 54,* 678, Abstr. 76.

Mezger, P. R. (1989). *Corrosion behaviour of dental casting alloys*, Dissertation, University of Nijmegen, The Netherlands.

Miyajima, T. (1984). A bacteriological study of plaque formed on the amalgam restorations—The relationship between the electrical potentials of amalgam restorations and the prevalences of *Streptococcus mutans* from plaques formed on them, *J. Nihon Univ. Sch. Dent., 26*, 165.

Moberg, L.-E. (1985). Long-term corrosion studies *in vitro* of gold, cobalt–chromium, and nickel–chromium alloys in contact, *Acta Odontol. Scand., 43*, 215–222.

Morris, H. F. (1989). Veterans Administration Cooperative Studies Project No. 147. Part VIII: Plaque accumulation on metal ceramic restorations cast from noble and nickel-based alloys. A five-year report, *J. Prosthet. Dent., 61*, 543–549.

Morris, H. F., Manz, M., Stoffer, W., and Weir, D. (1992). Casting alloys: The materials and "the clinical effects," *Adv. Dent. Res., 6*, 28–31.

Nilner, K., Glantz, P.-O., Ryge, G., and Sundberg, H. (1982). Oral galvanic action after treatment with extensive metallic restorations, *Acta Odontol. Scand., 40*, 381–388.

Nilson, H., Bergman, B., Bessing, C., Lundquist, P., and Andersson, M. (1994). Titanium copings veneered with Procera®-ceramics. A longitudinal clinical study, *Int. J. Prosthodont., 7*, 115–119.

Oehmichen, A., and Klötzer, W. T. (1984). Klinische Nachuntersuchung von Korrosionserscheinungen einer NEM-Legierung, *Dtsch. Zahnärztl. Z., 39*, 828–831.

Palaghias, G., Söremark, R., and Nord, C. E. (1982). The effect of *Bacteroides corrodens* on some dental alloys, *J. Dent. Res., 61*, 576, Abstr. 104.

Phillips, R. W. (1991). *Skinner's Science of Dental Materials*, 9th ed., W. B. Saunders, Philadelphia, pp. 359–384.

Pourbaix, M. (1984). Electrochemical corrosion of metallic biomaterials, *Biomaterials, 5*, 122–134.

Pröbster, L., and Hutteman, H. (1992). Effect of fluoride prophylactic agents on titanium surfaces, *Int. J. Oral Maxillofac. Implants, 7*, 390–394.

Shafer, W. G., Hine, M. K., and Levy, B. M. (1974). *A Textbook of Oral Pathology*, 3rd ed., W. B. Saunders, Philadelphia, p. 379.

Shannon, I. L. (1983). Inorganic components of human saliva, in *Handbook of Experimental Aspects of Oral Biochemistry* (E. P. Lazzari, ed.), CRC Press, Boca Raton, FL.

Söremark, R., Wing, K., Olsson, K., and Goldin, J. (1968). Penetration of metallic ions from restorations into teeth, *J. Prosthet. Dent., 20*, 531–540.

Statens Livsmedelsverk. (1986). *Food Composition Tables*, Liber Tryck AB, Stockholm, pp. 221–254.

Sutow, E. J., Jones, D. W., and Hall, G. C. (1989). Correlation of dental amalgam crevice corrosion with clinical ratings, *J. Dent. Res., 68*, 82–88.

Tai, Y., Long, R. D., Goodkind, R. J., and Douglas, W. H. (1992). Leaching of nickel, chromium, and beryllium ions from base metal alloy in an artificial oral environment, *J. Prosthet. Dent., 68*, 692–697.

Upton, B. (1963). Corrosion resistance in sea water of medium strength aluminum bronzes, *Corrosion* (NACE), *19*, 204–209.

Williams, D. F. (1990). Biocompatibility: An overview, in *Concise Encyclopedia of Medical and Dental Implants* (D. F. Williams, ed.), Pergamon Press, Oxford, pp. 51–52.

Wranglén, G. (1972). *An Introduction to Corrosion and Protection of Metals*, Butter & Tanner, London.

Index